T0181395

CAMBRIDGE LIBRARY COLLECTION

Books of enduring scholarly value

Technology

The focus of this series is engineering, broadly construed. It covers technological innovation
from a range of periods and cultures, but centres on the technological achievements of the
industrial era in the West, particularly in the nineteenth century, as understood by their
contemporaries. Infrastructure is one major focus, covering the building of railways and canals,
bridges and tunnels, land drainage, the laying of submarine cables, and the construction of
docks and lighthouses. Other key topics include developments in industrial and manufacturing
fields such as mining technology, the production of iron and steel, the use of steam power,
and chemical processes such as photography and textile dyes.

The International Exhibition of 1862

Replete with detailed engravings, this four-volume catalogue was published to accompany
the International Exhibition of 1862. Held in South Kensington from May to November,
the exhibition showcased the progress made in a diverse range of crafts, trades and industries
since the Great Exhibition of 1851. Over 6 million visitors came to view the wares of more
than 28,000 exhibitors from Britain, her empire and beyond. Featuring explanatory notes and
covering such fields as mining, engineering, textiles, printing and photography, this remains
an instructive resource for social and economic historians. The exhibition's *Illustrated Record*,
its *Popular Guide* and the industrial department's one-volume *Official Catalogue* have all been
reissued in this series. Volume 4 continues to catalogue the Foreign Division. Notable is the
appearance of early exhibits from Steinway & Sons in the brief section for the United States.
In contrast, Austrian and German exhibits occupy more than 400 pages.

Cambridge University Press has long been a pioneer in the reissuing of out-of-print titles from its own backlist, producing digital reprints of books that are still sought after by scholars and students but could not be reprinted economically using traditional technology. The Cambridge Library Collection extends this activity to a wider range of books which are still of importance to researchers and professionals, either for the source material they contain, or as landmarks in the history of their academic discipline.

Drawing from the world-renowned collections in the Cambridge University Library and other partner libraries, and guided by the advice of experts in each subject area, Cambridge University Press is using state-of-the-art scanning machines in its own Printing House to capture the content of each book selected for inclusion. The files are processed to give a consistently clear, crisp image, and the books finished to the high quality standard for which the Press is recognised around the world. The latest print-on-demand technology ensures that the books will remain available indefinitely, and that orders for single or multiple copies can quickly be supplied.

The Cambridge Library Collection brings back to life books of enduring scholarly value (including out-of-copyright works originally issued by other publishers) across a wide range of disciplines in the humanities and social sciences and in science and technology.

The International Exhibition of 1862

The Illustrated Catalogue
of the Industrial Department

VOLUME 4: FOREIGN DIVISION

ANONYMOUS

CAMBRIDGE
UNIVERSITY PRESS

CAMBRIDGE
UNIVERSITY PRESS

University Printing House, Cambridge, CB2 8BS, United Kingdom

Published in the United States of America by Cambridge University Press, New York

Cambridge University Press is part of the University of Cambridge.
It furthers the University's mission by disseminating knowledge in the pursuit of
education, learning and research at the highest international levels of excellence.

www.cambridge.org
Information on this title: www.cambridge.org/9781108067317

© in this compilation Cambridge University Press 2014

This edition first published 1862
This digitally printed version 2014

ISBN 978-1-108-06731-7 Paperback

THE

ILLUSTRATED CATALOGUE

OF THE

INTERNATIONAL EXHIBITION.

FOREIGN DIVISION.

THE INTERNATIONAL EXHIBITION *of* 1862.

THE ILLUSTRATED CATALOGUE

OF THE

INDUSTRIAL DEPARTMENT.

VOL. IV.

FOREIGN DIVISION.

PRINTED FOR HER MAJESTY'S COMMISSIONERS.

Printed for Her Majesty's Commissioners by

CLAY, SON, & TAYLOR, Bread Street Hill.
CLOWES & SON, Stamford Street

PETTER & GALPIN, La Belle Sauvage Yard.
SPOTTISWOODE & Co. New Street Square.

CONTENTS.

VOLUME IV.

GERMANY.

CONTENTS.

FOREIGN DIVISION.

AUSTRIA

AT

THE INTERNATIONAL EXHIBITION

OF

1862.

UPON ORDERS FROM THE I. R. MINISTRY FOR COMMERCE AND NATIONAL ECONOMY

BY

Prof. Dr. JOS. ARENSTEIN.

VIRIBVS VNITIS.

Indian-corn paper.

VIENNA.

IMPERIAL ROYAL COURT AND STATE PRINTING-OFFICE.

1862.

Space allotted to Austria in the northern part of the western Transept.

Groundfloor.

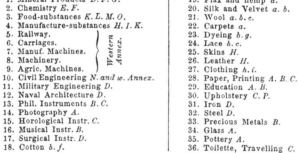

DISTRIBUTION OF SPACE.

Great Italics refer to the Ground-floor, little ones to the Gallery.

1. Mineral Products *D. F. G.*
2. Chemistry *E. F.*
3. Food-substances *K. L. M. O.*
4. Manufacture-substances *H. I. K.*
5. Railway.
6. Carriages.
7. Manuf. Machines.
8. Machinery.
9. Agric. Machines.
10. Civil Engineering *N. and w. Annex.*
11. Military Engineering *D.*
12. Naval Architecture *D.*
13. Phil. Instruments *B. C.*
14. Photography *A.*
15. Horological Instr. *C.*
16. Musical Instr. *B.*
17. Surgical Instr. *D.*
18. Cotton *b. f.*

19. Flax and hemp *d.*
20. Silk and Velvet *a. b.*
21. Wool *a. b. c.*
22. Carpets *a.*
23. Dyeing *b. g.*
24. Lace *b. c.*
25. Skins *H.*
26. Leather *H.*
27. Clothing *h. i.*
28. Paper, Printing *A. B. C.*
29. Education *A. B.*
30. Upholstery *C. P.*
31. Iron *D.*
32. Steel *D.*
33. Precious Metals *B.*
34. Glass *A.*
35. Pottery *A.*
36. Toilette, Travelling *C.*

Western Annex.

Some extensive Exhibitions in the Dome.

Gallery 1. Floor.

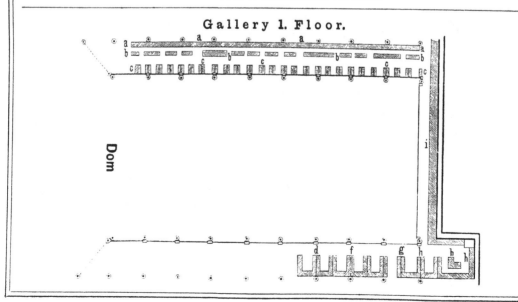

The I. R. Central Committee for the Agricultural, Industrial and Art Exhibition at London, in 1862, presided over by His Excellency, the Minister for Commerce and National Economy, intended to publish not a dry catalogue, but rather a supplement to the Austrian department of the International Exhibition, the sum and substance of all those circumstances that are not apparent at the mere sight of the exhibited objects.

The aim pointed at has been, to give a general account of the state of Agriculture, Mining, Manufacturing Industry, Commerce and Intellectual Progress in the provinces of Austria, besides a special illustration of the exhibited productions of Austrian Industry.

The subjects to be treated of have been arranged in a General and in a Special Part.

The General Part presents outlines intended to elucidate the condition of Austria with respect to national economy, commercial legislature, tariff of customs, institutions for commerce and credit, trade and agriculture, assurance companies and means of communication, etc., in addition to which there is an insight offered into the intellectual culture of Austria, such as it is represented by public educational institutes and literary intercourse, nor were charitable institutions forgotten to be duly noticed.

The Special Part has been classified in accordance with the sections of the Programme of the Exhibition.

The General Part reviewing the condition of the whole Monarchy, leaves to the introductory chapters of the several classes the task of giving a concise description concerning important trades in especial and with respect to their extent or other particulars.

It depends upon too many individual circumstances, whether any one class be represented adequately to its national or industrial importance at an Exhibition so remote from the country that ought to share in it, particularly so, if the exhibitors are themselves to bear the charges of the transport. One not initiated in such circumstances might be apt to draw conclusions from the small number of exhibitors to the little extent of industry in the class in question. It may fairly be supposed that the introductory chapters of each class will avert such a disadvantageous judgment. Thus, for instance, in class 6 (Carriages not connected with Railroads), Austria numbers but two exhibitors, probably owing to the high charges of transport; but the chapter introductory to this class shows that 25,000 workmen are occupied with the manufacture of carriages, bringing a capital of 20,000,000 florins into circulation, a statement which will save this branch of industry from being underrated, though it is represented by only two exhibitors.

Nearly the same remarks may be made concerning the exhibitors individually. Many circumstances relating to the extent of a trade in question, or having particular reference to the saleableness of certain productions, do not directly appear on a superficial view of the exhibited articles, and are discerned only by him who is well acquainted with the special circumstances connected with them. Therefore, since it is desirable to bring circumstances of this nature to bear at the emporium of London, such particulars about many exhibitors and their productions have been inserted as could be gathered from authentic sources or directly from the exhibitors themselves. Publicity seems also to be the best control for the veracity of such statements. If, in a national point of view, acknowledgment is due to all those who readily assisted the editor in this first attempt to collect materials for such statements, it is, on the other hand, evident that the interests of the parties concerned are essentially promoted by their own contributions.

In compiling these particulars, affording a clear insight into the real state of industry in that empire, the editor principally aimed at making foreigners acquainted with all they may wish to know, in order to enable them either to apply their capitals in, or extend their mercantile transactions to, some agricultural, mining or manufacturing branch of industry in Austria.

The sketch of the national economical condition of Austria (General Part, I—XXIII) has been composed by Mr. Ficker, secretary to the Ministry of Finances, and the introductory chapters to the several classes were contributed by Mr. Schmitt, clerk to the Ministry of Finances, both gentlemen having worked under the immediate guidance of his Excellency, Baron Czoernig, Chief of the Office for Administrative Statistics. The data for the epitome of public instruction were furnished by the Educational Section of the State Ministry. The introduction to the catalogue for the section of Art was composed by Professor Dr. Eitelberger.

The editor begs leave to claim the generous indulgence of the English Nation for the present translation of the German original into English. To the short time of a few weeks during which this very difficult task was to be performed, and to the circumstance of there not having been even time enough for making the necessary corrections, must be attributed the numerous errors that are encountered, but which will be corrected in the next edition.

But there is particular admiration due to the excellent organization and superior artistic capacity of the I. R. Court and State Printing Office, which alone was able to print three volumes of this extension in four weeks and in three languages.

The alphabetic index contains all names; but of other references the heading words only. In order to retain the same number for each exhibitor in all the three catalogues, which they have in the English Official Exhibition Catalogue, it was sometimes necessary to deviate from the alphabetic order to which some exhibitors belong, but there is always reference to be found at the proper place.

Vienna, April, 1862.

Arenstein.

CONTENTS.

—∞∞∞—

General Part.

Special Part.

GENERAL PART.

SKETCH

OF THE NATIONAL-ECONOMICAL CONDITION

OF THE

AUSTRIAN EMPIRE,

IN THE BEGINNING OF THE YEAR 1862.

><·

Extent.

The Austrian Empire embraces an area of 11,252·9 Austrian square miles, distributed as follows:

	sq. miles
Archduchy of Austria, Lower	344·5
„ „ „ Upper	208·5
Duchy of Salzburg	124·5
„ „ Styria	390·2
„ „ Carinthia	180·3
„ „ Carniola	173·6
Littorale (Principality of Goritz and Gradisca, Margraviate of Istria and City of Trieste)	138·8
Principality of Tyrol and Vorarlberg	509·0
Kingdom of Bohemia	902·8
Margraviate of Moravia	386·3
Duchy of Silesia	89·4
Kingdom of Galicia	1364·1
Duchy of Bukowina	181·7
Kingdom of Dalmatia	222·3
„ „ Lombardy and Venice	436·9
„ „ Hungary	3727·7
„ „ Croatia and Sclavonia	335·0
Grand Principality of Transylvania	954·3
The Military Borderland	583·0

Mountains and Plains.

The Austrian Monarchy is the most mountainous State of Europe, Switzerland excepted, full three quarters of its area comprising mountain-regions. The extensive Alpine system (Tyrol, Salzburg, southern border of Austria, Styria, Carinthia and northern Carniola) is, in the south, steeply sloping towards the Venetian plain, advancing south-eastward, terrace-like, into the Karst land and the Istrian-Liburnian-Dalmatian coast-borders which continue the former, crossing to the north-east, without any such gradation, into the mountain-region of southern Austria-proper and western Hungary, and reaching its northern termination in the Danubian basin which is, in the archduchy, divided into the Upper Danubian basin and the Vienna basin. On the other side of the Danube the Bohemian-Moravian-Silesian highland expands itself, encompassed by border-mountains, and is, within, interlinked in an undulating form. From the Mark eastward, the Carpathian mountains commence and stretch thence in a semi-circular course between Hungary and Galicia, on both sides of which the High-Carpathians are connected with a mountain-region passing northward over the Galician gradient levels into the Polish-Russian plain, southward down to the Danube and Theiss belonging to Hungary, and in the south-east hanging together with the mountain fastnesses of the Transylvanian highland. The Alps, the Carpathians and the Transylvanian highland enclose the extensive Hungarian lowland. The Galician gradient levels occupy an area of 1,000, the Hungarian lowland an area of 2,000 sq. miles; the Venetian plain is confined to a little more than 100 sq. miles.

Rivers.

The system of the soil is also fairly balanced by the range of the Austrian river-system. Tributary to the Adriatic Sea are 1,215 sq. miles, to the North Sea 1,050 sq. miles, to the Baltic 825 sq. miles of river area; the remainder is tributary to the Black Sea, principally through the Danube system the area of which comprises 7,600 sq miles within the empire.

Sea.

The Adriatic Sea washes 250 miles of the Austrian coast, and above 300 miles of Islands belonging to Austria. The eastern coast is mostly high and rocky, offering however natural road-steads to vessels by means of numerous creeks, inlets and channels. The western coast, on the other hand, is flat; the Lagoons before the Bacchiglione and the Brenta being separated from the open sea by small dikes. A strong northerly current passes along the Dalmatian coast

a *

another, taking a southern direction, passes along the Venetian.

Geognostic division.

With respect to geognostic relations the Austrian Monarchy is divided into four systems:

   a) the system of Plains;
   b)   „   „   „ Alps;
   c)   „   „   „ Carpathians;
   d)   „ Bohemian-Moravian-Silesian system.

System of Plains.

With the exception of those horizontal deposits of diluvial and alluvial formations, which usually accompany rivers both in the plain and in other ranges, and of which gault clay and silt prevail in the plain, — the hilly parts consist almost only of neogenite tertiary strata of clunch, sand, sandstone, silt-conglomerate and Leitha-lime.

Alpine System.

The central chain of the Alps is composed of links of the crystalline slate mountains, particularly of gneiss, mica-slate, clay-slate, granular lime-stone, with here and there some serpentines. In the north and south of this central chain the „Lime-alps“ appear as sedimentary formations. Between the lime-alps and the central chain, slate of the gray-wacke and coal formation is found stratified, whereas in an opposite direction there are added, in the south and north of the above-mentioned lime-alps, cretaceous formations where slate and sandstone prevail. Eocene and neogenite tertiary formations fill up more or less extended basins, the former consisting mostly of numulitic lime and sandstone, the latter of clunch, Leitha-lime and sandstone. Of block-stone there occur insulated groups of granite, porphyr and basalt.

Carpathian System.

Geognostically, this system is subdivided into the northern (Galician) and southern (Transylvanian) range. The geognostical components of the northern Carpathian range are in general similar to those of the Alpine range, with the difference that the Vienna sandstone is here denominated Carpathian sandstone, and that near the borders of the lower Danube basin, besides components of crystalline slate formation, great masses of granite, porphyr and diorite are spreading and, of neoteric eruptive stone, basalt and trachyte are found to a great extent. The mountains enclosing Transylvania consist of rocks of crystalline slate, gneiss, mica, clay-slate, etc., partly of block-stone such as granite, porphyr, trachyte, and basalt; the less frequent trias, lias, and chalk formations resemble those of the Alpine range. In the interior, there is a great extent of eocene, but especially of neogenite, tertiary deposits where also basalts and trachytes occur.

Bohemian-Moravian-Silesian System.

The ranges of this system (the Bohemian forest, Pine-, Ore- and Giant-mountains, Sudetes, and Bohemian-Moravian border mountains) likewise consist, in their higher parts, mostly of crystalline slate strata (gneiss, mica-slate, clay-slate, granulated lime-stone, and serpentine), but there is also block-stone (granite, granilite, diorite, porphyr, and syenite). Towards the inland of Bohemia and Moravia are joined the crystalline slate strata, gray-wacke formation, and in larger basins coal formation and adamic earth are to be met with. Trias and Jura formation is entirely wanting in this range; chalk formation, however, is represented by the strata of upper chalk (plener and square stone). The middle mountains of north-western Bohemia belonging to the tertiary time consist of numerous neogenite tertiary basins, grouping about basalts and phonolithes; moreover, neogenite tertiary formation appears also in some smaller basins as e. g near Budweis.

Climate.

The greatest part of the Austrian monarchy lies in the temperate zone, the northmost parts of Bohemia, Silesia, Galicia, and Bukowina alone stretching out into the subarctic zone.

Considered in general, the isotherm of 11 degrees R. takes its course through the north of Dalmatia; that of 10 degrees enters the empire at Lodrone, embraces a considerable portion of southern Tyrol and crosses over Padua to Rovigno and Fiume; the curve of 9 degrees of mean annual temperature touches Verona and Temesvár; that of 8 degrees joins Bregenz with Marchfeld and central Transylvania; that of 7 degrees ascends in north-western Bohemia up to Saaz, approaches, past Sternberg, the Carpathian mountains on the southern declivity of which it stops; the isotherm of 6 degrees traverses the gradient country of the Sudetes and turns to the Pruth through the midst of Galicia.

Upon the whole it is ascertained that, in Austria, with every degree of increasing latitude, a decrease of the mean annual range of temperature by 0·44 degree takes place, and the difference of heat between the farthest west and east averages 1 degree. Southern Tyrol, Venice, Dalmatia, Hungary and the lowland plains of Bohemia and Moravia are, in proportion to their latitude, distinguished by a higher mean annual range of temperature, whereas northern Tyrol, Salzburg, Carinthia, Upper Styria, and Austria-proper are much less favoured.

In a vertical direction, the annual range of heat within Austria decreases 1 degree, at an altitude of 110 fathoms above the level of the sea. The height of the line of perpetual snow is found in the Alps at about 1,360 fathoms, in the Carpathian mountains at about 1,330 fathoms.

The greatest deviations from the mean annual range of temperature are exhibited in those provinces that share the character of the continental climate in a higher degree.

Meteorological observations.

a) Winds.

The Austrian monarchy is situated almost entirely in the region of the westerly winds. The mean direction of the winds in north-western Bohemia, is from west to south; in the rest of Bohemia, in Moravia, Galicia, Hungary and Austria-proper, from west to north; within the Alpine regions the general laws are, at nearly every point, variously modified by local circumstances. In Venice northerly or north-easterly winds prevail, and on the eastern coast of the Adriatic, Bora and Sirocco (winds of equalization between land and sea temperature) struggle for ascendency nearly the whole year.

The amount of atmospheric pressure is on an average in:

Vienna . . . . 27″ 6·61‴
Prague . . . . 27″ 5·76‴
Buda . . . . 27″ 9·85‴
Venice . . . . 28″ 0·69‴

b) Rains.

The Austrian monarchy is, nearly throughout all parts, favoured with a sufficient quantity of atmospheric descents of moisture, in which respect the Alpine regions may be considered among the richest in Europe.

The isohyetose (line of equal quantity of rain) of from 40″ to 45″ ascends from the Garda lake up to Roveredo, passes over Vicenza to Treviso, Trieste, Fiume, Ragusa, Cattaro, and returns over Carlstadt, Cilli and Gailthal, including the ranges of Belluno, Goritz, Laibach, Tarvis with from 45″ to 50″, those of Udine and Adelsberg with from 55″ to 60″, and that of Tolmezzo with 70″ annual quantity of rain. The line of from 35″ to 40″ turns closely around the first; the curve of from 30″ to 35″ accompanies the valley of the river Po, extending itself northwardly to Verona, Padua and Chioggia, whilst, on the other hand, it takes its course from Trient to Klagenfurt, Windischgratz and Petrinia. The isohyetose of from 25″ to 30″ moves along the northern boundary of the Alps, bends from the Kahlengebirge down to the Semmering and south-westwardly to Murau, and finally turns over Marburg to Agram and Brod. This latter alone makes its appearance as a recurvating line at two points of the Hercynian mountain system, in southern and western Bohemia, from Hohenfurt to Rumburg, and in the Sudetes, the lowland and the Moravian mountains, from Reichenberg to Zwettl and Göpfritz. Both within the Hercynian and the Carpathian mountain system the isohyetose of from 30″ to 35″ occurs frequently, and within the Alpine range the isohyetoses succeed each other rapidly. The Ortles with 92″, Aussee with 62″, Starkenbach with 81″ constitute uncommonly pluvious regions. The isohyetose of from 20″ to 25″ takes its direction through northern and eastern Bohemia, past Vienna to the Platten lake, finally to Esseg and Semlin, but is, with a narrow strip, also traversing central Hungary and accompanying the Carpathian mountains in a semi-circular course down to Weisskirchen. The interior of Bohemia, southern half of Moravia, Vienna valley, and Hungarian plains are situated beyond the curve of 20″, the environs of Prague, the country about Raab, Komorn, and Tokai forming insulated spots with from 12″ to 15″ quantity of rain.

c) Electrical Phenomena.

The number of thunder-storms diminishes from the northern coast of the Adriatic sea (where the number of 50 is reached at Udine) both towards the south and the interior, so that Ragusa stands with 12·5, and Czernowitz with 7·5 at the opposite end of the scale.

d) Hail-showers.

Hail-showers are generally least in number in the Alpine lands; but there are countries, as Upper-Austria, where the Insurance Companies find insurers against hail risks paying 4%, whereas in other parts, as in the Marchfeld, they vainly offer their service at ¾%.

Population.

The last census of 1857 has shown a population of 34,439,069, for the actual extent of the monarchy, the army not included.

Based upon a strict registration of births and deaths, it is to be computed that Austria numbers in the beginning of 1862 (the army not included) 35,795,000 inhabitants, who are distributed in the several Kingdoms and dependencies of the Crown, as follows:

	Population	Number of inhab. to sq. mile
Austria, Lower	1,719,000	4,990
„ Upper	714,000	3,424
Salzburg	147,000	1,186
Styria	1,077,000	2,760
Carinthia	338,000	1,875
Carniola	459,000	2,644
Littorale (Goritz, Gradisca, Istria, Trieste)	545,000	3,927
Tyrol and Vorarlberg . .	862,000	1,694
Bohemia	4,952,000	5,485
Moravia	1,955,000	5,061
Silesia	471,000	5,268
Galicia	4,900,000	3,592
Bukowina	487,000	2,680

Dalmatia	437,000	1,966
Lombardy and Venice .	2,523,000	5,735
Hungary	10,172,000	2,729
Croatia and Sclavonia . .	920,000	2,746
Transylvania	2,027,000	2,124
The Military Borderland	1,090,000	1,870

Nationalities.

This population divided with respect to race and language: —

Germans	8,200,000
Bohemians, Moravians and Slovacks . .	6,300,000
Poles	2,200,000
Russians	2,800,000
Slovenians	1,210,000
Croats	1,360,000
Servians	1,470,000
Bulgarians	25,000
Magyars	5,050,000
Italians (inclusive Ladins and Friauls) .	3,050,000
Eastern-Romans	2,700,000
Members of other races	1,430,000

Religions.

Regarding religion there are the following distinctions:

Roman Catholics		24,874,000
Greek	„	3,600,000
Armenian	„	10,000
Schismatic Greeks		3,000,000
„	„ Armenians	4,000
Protestants of the Augsburg Persuasion		1,250,000
„	„ „ Helvet. Persuasion .	2,000,000
Jews		1,052,000
Members of other religions		5,000

Habitations.

The inhabitants of Austria live in 878 towns, 2,264 market-towns, and 66,376 villages; among the towns, Vienna numbers above 500,000 inhabitants, Pesth-Buda near 200,000, Prague 150,000 and Venice 120,000. There are furthermore 10 towns containing above 50,000, and again 15 containing above 25,000 inhabitants.

Government.

The form of government in the Austrian empire is a constitutional monarchy.

Council of the Empire.

The rights which, in consequence of the Diploma of Oct. 20th 1860 and the Fundamental Law (Constitution) of Feb. 26th 1861, are conferred upon the Joint Council of the Empire are as follows:

a) Consent to all laws relating to military duty.

b) Co-operation in the legislature on coinage, matters of money and credit, on duties and commercial transactions, on the principles of banking, posting, telegraph, and railway concerns.

c) Examination of the estimates for the public expenditure, and the balance of public accounts; granting of new taxes or heightening existing taxes; approving of new loans, converting previous stocks, the sale, commutation or mortgaging of public estates.

The public debt is put under the control of the Council of the Empire.

Diets.

As for the right of giving, amending or abrogating the rest of the laws, the Emperor participates in this with the Partial Council of the Empire (to which the members of the provinces belonging to the Hungarian Crown are not joined) for public concerns common to all the German, Slavonian and Italian provinces, and relatively also with the Hungarian, Croatian-Slavonian, or Transylvanian diets for the public concerns of these respective provinces.

For those Kingdoms and dependencies, which are represented in the Partial Council of the Empire, there are moreover instituted 16 Separate Diets (besides the Common Council of Trieste performing also the office of a Separate Diet). The public business submitted to the co-operation of such a diet is: —

a) All regulations concerning culture in the respective province, public buildings or charity institutions at the charge of the province, and its economical administration in general.

b) Special regulations within the range of general laws concerning communities, churches and schools, relays, provisions and quarters for the army.

c) Dispositions about other business and propositions, specially directed to the diet, regarding general laws and institutions particularly bearing upon the common welfare of that province.

Autonomy of Municipal Corporations.

In the same provinces the autonomy of municipal corporations is of a very liberal extent. The Municipal Corporation Act of March 17th 1849, issued for all German and Slavonian provinces, declares that all concerns and interests of a municipal corporation and all that is to be enforced within their limits, belongs to the natural sphere of activity of a municipal corporation; and the same Act entitles to vote for the representation of a municipal corporation all burgesses that have, the year previous, actually paid at least 1 florin 5 kreutzers direct taxes for a house or land, situated within the municipality, or for a trade, for the exercise of which a stationary abode within the municipality is stipulated; furthermore all ecclesiastical ministers, teachers, doctors, government and military officers are entitled to vote. Only in such places as possess statutes of their own, there is a higher scale of valuation.

Lombardy and Venice also enjoy great liberty of autonomy in their municipal corporations. All landowners (ecclesiastics and military gentlemen excepted) have equal votes in the representation of a municipal corporation (convocato generale), only the election of the chief delegate of the corporation (executive organ) must be made from among the three landowners that pay the highest taxes. The law of March 5ᵗʰ 1862 again secures the autonomy of municipal corporations and leaves the special regulations to the several diets.

Statutes of Diets.

Dalmatia possesses a separate diet, but its position relatively to political law and its connection with Croatia and Slavonia is subject to further negociations.

Till the time when a constitution for the kingdom of Lombardy and Venice shall be published upon principles equal to those of the other provinces, the sphere of action of the Central Assembly existing since 1815 and the new Provincial Assembly has been liberally enlarged.

The diet of each German or Sclavonian province is composed of only one assembly, the members of which are:

1. The archbishops and bishops of the Catholic and Oriental-greek Church and the rectores magnifici (chancellors) of the universities.

2. The representatives of great estates convened through the direct elective right of such owners of lands, registered at the board of States, as are, within the respective province, assessed at a certain sum of property tax (generally 100 florins, in Tyrol 50 florins, in Lower Austria 200 florins, in Bohemia, Moravia and Silesia 250 florins); in Tyrol, Silesia and Bukowina also through the direct elective right of certain landowners without scale of valuation, and in Dalmatia through the direct elective right of those that pay the highest taxes (50—100 florins).

3. The representatives of towns and other places of comparatively large extent, the direct election of whom devolves upon those members of a municipality who are qualified to be electors for the representation of the municipal corporation and who, in municipalities with three bodies of electors, belong to the first or the second body, or who, when belonging to the third, pay at least 10 florins direct taxes; in the remaining municipalities, those that constitute the first two thirds of all municipal electors ranked according to the amount of their direct taxes.

4. The representatives of Boards of Commerce and Trades, who are elected by a body of electors composed of the members of each Board and their substitutes.

5. The representatives of country municipalities, the election of whom devolves upon chosen electors, which latter are appointed by such members of a municipality as are qualified electors for the representation of a municipal corporation, and either belonging to the first or the second body in municipalities with three bodies of electors, or ranking among the first two thirds of those that pay direct taxes.

Regarding the composition of the Hungarian and the Croatian-Slavonian diet, the Act of 1848 has been revived; regarding the Transylvanian diet, the election-franchise has been enlarged to the valuation of eight florins, direct taxes.

Members of the Council of the Empire.

a) House of Commons.

The diets of the German and Sclavonian provinces, and the central assembly of the kingdom of Lombardy and Venice elect the deputies for the Council of the Empire from among their members, the former by groups, the latter by a terniary of the provincial assemblies. The deputies thus elected constitute the House of Commons of the Partial Council, and, together with the deputies of the Hungarian, Croatian-Slavonian, and Transylvanian diets, form the House of Commons of the Total Council of the Empire.

b) House of Lords.

The House of Lords of the Council of the Empire is composed of the adult Princes of the Imperial House; the chiefs of those families of nobility that are appointed to the hereditary dignity of Councillors of the Empire; the catholic archbishops and lord-bishops; and those gentlemen that are appointed members for life by the Emperor.

Rights of the representative bodies.

Both the Houses of the Total as well as Partial Council of the Empire, and the separate diets, within their sphere of action, have the right of proposing first. A law of the empire requires the united consent of the two Houses of the Total Council of the Empire; a law for the German, Sclavonian and Italian provinces requires that of the two Houses of the Partial Council of the Empire; but both kinds of laws require the sanction of the Emperor; a provincial law requires the Monarch's consent after the decision of the diet.

Diets and the Council of the Empire are convened annually, prorogation and dissolution are imperial prerogatives. When the Council of the Empire or the Diets are not assembled, and when, in the meanwhile, urgent measures must be taken in matters belonging to their sphere of action, the ministry has to state the reasons and success of the enactment to the next convention. Sessions are public; a decision is only accomplished by absolute majority of the members present.

Equality before the Law.

All subjects are equal before the law. The state maintains inviolate the personal security, private property and rights of every subject.

The attainment to offices, dignities and honours is not connected with any prerogative of birth; military duty is equally incumbent upon every male person capable to bear arms, on his entering the age of 21 till he is 26 years of age, so that a claim to exemption from military service can only be founded upon his being indispensable for the support of a family, his taking orders, entering the service of the state as civil officer, or upon eminent distinction in higher studies.

An alien is excluded only from the political rights of a liege subject; concerning personal rights, the acquisition of property, exercise of agricultural, commercial or trade industry, he is on a level with a natural born subject of Austria.

Religion.

Lawfully recognized congregations of faith (the Catholic, Oriental-Greek, Armenian-Gregorian Churches, Protestants of the Augsburg, Helvetic and Anglican Persuasion, Unitarians and Jews) are allowed the public observances of their religious doctrines, unrestricted autonomy and disposition of their funds and establishments. A share in political rights is conditional to no religious creed.

Particular position of the inhabitants of the Military Borderland.

A peculiar position within the empire is held by the inhabitants of the Military Borderland.

The borderers, indeed, enjoy the full right of property of their land and premises, but with the general obligation to do military service for a considerable length of time. They also stand under military government and have a kind of municipal constitution only for their towns and market-towns (Border-Municipalities).

Central Offices.

The highest central offices of the administration are:
1. The Ministry for Foreign Affairs
2. „ „ „ Finances
3. „ „ „ Commerce
4. „ „ „ War
5. „ „ „ the Navy
} For the whole Empire

6. „ „ „ the State
7. „ „ „ Police
8. „ „ „ Justice
} For the provinces represented in the Partial Council of the Empire.

9. „ Hungarian Court-Office
10. „ Croatian-Slavonian Court-Office
11. „ Transylvanian Court-Office
} For their respective provinces.

On a level with the above mentioned offices is still the Accountant General's Office for the direction of the extensive business of public accounts.

Provincial offices.

The political administration in the German, Slavonian and Italian provinces is managed, under the Ministry of the State (Home Office), by 10 vicegerencies and 6 provincial-governments, subordinate to which are 967 district-offices. In Bohemia, Galicia and Dalmatia there besides exist sheriffs, in the Lombardo-Venetian kingdom delegations, serving as middle organs between the vicegerencies and district-offices. The management of police is, in general, assigned to the organs of Political administration; in larger cities only there are distinct police-offices. The Municipality Act hints at the institution of District-municipalities.

Administration of Justice.

a) In the German and Slavonian Provinces.

The judicial administration of civil right is, in the first instance, assigned partly to special courts (many of which are at present still connected with the political district-offices), partly, for cases of higher importance, to 71 tribunals entitled Provincial or District Courts. For cases in commercial and maritime law, there exist several distinct Commercial Courts; in other respects the Courts decide after consulting men of experience, a usage also followed in questions of miners' statutes. In penal cases the examination and decision about minor offences is within the province of special courts, when they are not assigned to political or police jurisdiction.

Examination in cases of crime and misdemeanor is conducted by distinct courts of inquisition, whereas public trials and the passing of sentences is incumbent on tribunals (or provincial courts). There is a prospect of the aid of a Jury for decision in questions of fact in all important cases, when a new penal code will be instituted.

In the event of an appeal, the case is referred to a Higher Court of Justice of which there are nine in activity; there is a third appeal to the Chief Court of Chancery at Vienna.

The Army and Navy, as well as the inhabitants of the Military Borderland, are, in all instances, subject to military jurisdiction.

b) In Hungary, Croatia, Slavonia, and Transylvania.

For political administration and jurisdiction there are in these provinces the chosen officers of the comitates (counties), districts and cities, who, regarding political administration, are respectively subordinate to the Hungarian vicegerency, the Croatian, Slavonian council of vicegerency, and the Transylvanian gubernation, — and for jurisdiction, there are the Royal Hungarian Board, the Compulsory Board (penal board) and the two courts of appeal of the three Transylvanian nations (from which appeals are referred to the Septemvir Board and the Hungarian Court Office), the Chief Croatian-Slavonian Court of Justice and the Transylvanian Gubernation.

Financial Offices.

The administration of direct taxes is managed by 12 provincial directories for financial matters, the financial prefecture at Venice, and 7 offices of taxes in the minor crown-dependencies, subordinate to which are the district-offices and tax-commissions, and subordinate to these again, there are offices of assessment and administration of taxes. The administration of i n d i r e c t t a x e s is managed by the provincial offices for financial matters and the financial prefecture, subordinate duty is done by district offices for financial matters and by financial intendencies, and subordinate to these again, there are various tax-gathering offices.

The offices for distinct R e v e n u e s (lottery, tobacco, stamps, excise, duties), for the exchequer mining and smelting works, salines and montanic manufactures, forests, coining and assaying and stamping business, etc. are very numerous.

Navigation Offices.

For N a v a l m a t t e r s and C o m m e r c i a l N a v i g a t i o n there exists, as Chief Office, the Central Navy Office at Trieste, subordinate to which are 13 port-offices and boards of health, besides numerous port-commissions and commissions of health, agencies and reference-offices.

Agricultural condition.

More than two thirds of the population of the empire are engaged in husbandry. In Bohemia only, the number of individuals (and their families) thus occupied sinks below a half of the total population; and in Lower Austria and Moravia one half is but little exceeded.

The component parts of the soil, the abundance of water and the quick succession of isothermal lines, are circumstances eminently favourable to the growth of those products which Austria is daily studying to improve.

The peaks and ridges of the various systems of high mountains in Austria, the elevated portions of the Karst and the Dalmatian ranges of heights, some marshes and moors (particularly near the Hungarian lakes and within the river system of the Theiss), and some small tracts covered with quicksand are, the only portions of land that may be said to be entirely without vegetation.

Repartition of the soil.

Numerical details respecting the repartition of the soil cannot be exhibited here, but the fact is established that many e x t e n s i v e landed estates can be pointed out in Austria, and that the division of the soil into little plots of property is, in general, far from that excess to which it has been carried in western Germany as well as in France. However, the number of great landed estates is comparatively very little in the Littorale, Carniola, part of Tyrol and Vorarlberg,

where the ancient laws for the preservation of farms in sufficient consistency were not enforced during the French and Bavarian domination and these regulations never having had any efficiency in Venice and Dalmatia; the foregoing remarks are also to be referred to the last mentioned provinces. In Lower Austria, the metropolis, and in Bohemia and Moravia the settlements of a numerous manufacturing population, have, though only partially, accelerated division into small lots. In general, it may be computed that to one individual paying land-tax (to one taxed plot) in Salzburg, there falls an average share of between 40 and 50 acres (1,600 square fathoms per acre); in Carinthia, Bukowina and Galicia between 20 and 30; in Hungary 20; in Croatia, Slavonia and Transylvania 18; in Styria and Upper Austria 17; in Silesia 14; in Carniola and Dalmatia 13; in Bohemia 12; in Lower Austria and the Littorale 11; in Moravia 10; in Venice 7½ acres of land. But there is to be taken into consideration that many a landowner must be repeatedly en umerated (even ten times perhaps if not oftener) as possessor several of taxed plots, and, therefore, as many times as a payer of land-taxes, so that the above-quoted cyphers might be heightened by nearly a third.

In the Hungarian provinces, the segregation of compound plots, and the aggregation of scattered ones was, during the years 1854 to 1860, regulated energetically.

Exoneration of tenures from statute-labour.

The exoneration of tenures having brought in its train the exemption from all burdens upon landed property, occasioned by the feudal and patrimonial system, commissions have been appointed in all provinces in order to achieve the redemption of the service of tenant and other encumbrances of relative rights upon landed property.

Value of the soil.

It is extremely difficult to ascertain an average value of the soil throughout the whole monarchy. But in a gross estimate, the following average value may be quoted: an acre of arable land 150 florins; an acre of garden ground 400 florins; an acre of vineyard 300 florins; an acre of meadow land 150 florins; an acre of pasture ground 50 florins; an acre of wood land 40 florins.

But these mean numbers stand, of course, between two extremes, which are fluctuating respecting arable land between 30 and 1,500 florins, meadow land rising up to 3,000 florins, and wood land fetching from 10 to 800 florins. Farm rent naturally tallies with the value of the soil.

Woodland.

There are 3,186·5 Austrian square miles of wood-land, which are distributed among the several pro-vinces as follows:

	Absolute scale of sq. miles	Relative scale of p. C. of the whole area.
Austria, Lower	111·0	32·2
„ Upper	70·3	33·7
Salzburg	67·1	53·9
Styria	183·2	47·0
Carinthia	96·8	53·7
Carniola	74·6	43·0
Littorale	35·3	25·4
Tyrol and Vorarlberg	142·7	28·0
Bohemia	261·5	29·0
Moravia	100·0	25 9
Silesia	28·4	31·8
Galicia	346·9	25·4
Bukowina	83·3	45·8
Dalmatia	47·6	21·4
Lombardo-Venetian Kingdom .	52·2	11·9
Hungary	841·8	22·6
Croatia and Slavonia	125·3	37·4
Transylvania	356·3	37·3
Military Borderland	462·2	27·8

Thus the actual Alpine regions, besides Buko-wina, Croatia-Slavonia and Transylvania are espe-cially rich in forests. The prevailing forest tree in the high mountain-ranges is the Pine, in the middle heights intermingled with Firs and Pitch-trees, the latter everywhere accompanying the calcareous moun-tain-ranges and frequently covering sandy plains The principal indigenous tree of the lower lying tracts in the Carpathian mountains and Alps is the Beech, in the north-west of the empire partly still the Oak, in the south the Elm, the Chestnut and Walnut tree, and farther southward thrive the Laurel, the Olive, and lastly the Fig tree. Even after the loss of Lombardy, Austria is still in possession of 4 sq. miles of Olive groves, and 6 sq. miles are covered with Laurel and Chestnut forests. Scarcity of wood is apparent only in the south of Venice, in Dalmatia and Istria (where from the 15th to the 18th century, forests were utterly devastated), likewise also in the Hungarian moorlands and in the north-east of Galicia. The reverse, however, is the case in Upper Austria and Salzburg, in Bohemia, Moravia and Silesia, where the cultiva-tion of forests is for the greater part exemplary.

Forest Products.

Austria produces, on an average, per annum, 30,000,000 Vienna fathoms of wood, mostly of excellent quality.

Notwithstanding the extensive use of oak-forests for the masts there are 500,000 Cwts of gallnuts gathered per annum. Besides, the forests of Austria yield for commerce 100,000 Cwts of potash, 250,000 Cwts of turpentine and resin, and 4 millions Cwts of tanning-bark.

Grass-land.

Within the forests there are considerable tracts used as pasture; besides, the monarchy is endowed with 2,820·3 sq. miles of real grass-land, distributed among the several provinces as follows:

	Absolute scale of sq. miles	Relative scale of p. C. of the whole area.
Austria, Lower	69·7	20·2
„ Upper	45·3	21·7
Salzburg	21·0	16·8
Styria	83·0	21·3
Carinthia	38·1	21·2
Carniola	64·8	37·4
Littorale	61·6	44·4
Tyrol and Vorarlberg	128·4	25·2
Bohemia	175·0	19·4
Moravia	72·1	18·7
Silesia	16·1	18·0
Galicia	291·0	21·3
Bukowina	48·8	26·9
Dalmatia	127·6	57·4
Lombardo-Venetian Kingdom .	128·5	29·4
Hungary	972·2	26·1
Croatia and Slavonia	70·1	20·9
Transylvania	248·9	26·1
Military Borderland	158·1	27·1

The extent of grass-land is, therefore, relatively very considerable in Tyrol, Venice, Istria and Dalma-tia, in Bukowina, a large part of Hungary, Tran-sylvania and the Military Borderland. That natural pastures occupy nearly half of that area, is to be understood from the former state of farming, which has but begun to give way to a more scientific management, since the exoneration of tenures. In Bohemia alone, the culture of meadows has, partly already in previous times, attained a high degree of perfection. In the Venetian province, where restrictions upon the im-provement of the soil have, in the main, already disappeared long ago, there is a partially cultivated irrigation system which is of great advantage for the culture of meadows. Altogether Dalmatia possesse as much grass-land as the Lombardo-Venetian King-three-fourths of it are used only as pasture.

Upon the whole, there are made about 363,000,000 Cwts of hay, and 200,000,000 Cwts of various herbage for fodder are cropped, forming the basis of the rearing dom, but of cattle in Austria.

Arable land.

The area of soil continuously or intermittingly used as tillage land comprises 3,582·4 sq. miles, distributed among the several provinces as follows:

	Absolute scale of sq. miles.	Relative scale of p. C. of the whole area,
Austria, Lower	141·4	41·1
„ Upper	73·6	35·3
Salzburg	11·7	9·4
Styria	86·3	22·1
Carinthia	23·9	13·3
Carniola	23·7	13·6
Littorale	24·1	17·3
Tyrol and Vorarlberg	26·0	5·1
Bohemia	433·1	48·0
Moravia	196·1	50·8
Silesia	41·6	46·6
Galicia	555·1	40·7
Bukowina	44·6	24·5
Dalmatia	21·4	11·0
Lombardo-Venetian Kingdom .	169·8	38·9
Hungary	1265·3	33·9
Croatia and Slavonia	89·0	26·6
Transylvania	216·1	22·6
Military Borderland	136·6	23·4

There is excellent cornland to be found in the alluvial soil of the Danube valley, the flat country about Salzburg, the Windian hills in Styria, the environs of Laibach and Wippach in Carniola, the lowland portions on both sides of the middle Elbe and the lower Eger in Bohemia; but egregiously rich cornlands are in the Moravian Hanna, the north-east of Galicia, the level part of Bukowina, a large portion of the Hungarian lowland inclusive the Banat, eastern Slavonia and the contiguous Military Borderland, finally the south-east of Transylvania.

The central Alps and Carpathian mountains generally form the polar boundary of extended growth of maize and buck-wheat as after-crop.

The Lombardo-Venetian Kingdom contains 5 sq. miles of rice-fields, and also in the Littorale and in southern Hungary rice is cultivated.

Extraordinarily great was the success in agriculture in the course of the last ten years. Extended application of drainage, manufacture and use of artificial manures, frequent substitution of manual labour by agricultural machines, thorough improvement of implements, rendered it possible to make the reforms that took place in the circumstances of the soil and its cultivators particularly available for the furtherance of the productiveness of tillage land in especial.

Agricultural Produce.

a) Production of corn.

All European kinds of grain are cultivated in the Austrian monarchy. There are grown: —

Wheat	50,000,000	Metzen [1]
Mixed corn	15,000,000	„
Rye	65,000,000	„
Barley	50,000,000	„
Oats	100,000,000	„
Maize	44,000,000	„
Millet and Buckwheat	10,000,000	„
Rice	500,000	Cwts.

With the surplus of their produce, Upper-Austria, Bohemia, Moravia, Galicia, but chiefly the Hungarian provinces not only meet the wants of less productive regions of the empire, but furnish also a considerable quantity for export.

In the culture of wheat, Hungary (with the Banat) and Bohemia stand foremost; but the latter even surpasses fertile Hungary in the production of rye; Galicia follows next in both kinds of crops, succeeded by Moravia and Lower-Austria. Four-fifths of all the barley and oats are supplied by Hungary, Galicia, Bohemia and Moravia; a large quantity of oats being produced in Lower-Austria. Hungary, Croatia, Slavonia, Transylvania, the Military Borderland and Lombardo-Venetian Kingdom contribute extensive crops of maize the growth of which is also considerable in Bukowina.

The mean market-price in Austria, during the year 1861, was: for wheat 4·5 florins, rye 2·7 florins, maize 2·6 florins, barley 2·5 florins, oats 1·8 florins Austrian Currency.

The straw of cerealia, exclusive of that which serves for stubble-pasture, is estimated at 40,000,000 Cwts.

b) Turnips, Potatoes, Pulse, etc.

Generally speaking, the line cutting off the copious produce of maize towards the north may be considered the limit where the extensive cultivation of potatoes begins; pulse, cabbages and turnips form a very general and abundant crop in all districts of the monarchy down to the Adriatic Sea.

The amount of that produce is:

Potatoes	120,000,000	Metzen
Pulse	5,000,000	„
Cabbages	60,000,000	Cwts.
Beet roots	20,000,000	„
Turnips etc.	30,000,000	Metzen

c) Trade-plants.

The soil of a large extent of the monarchy is well adapted for the culture of various trade-plants.

The whole north-west is excellently qualified for the growth of flax, the south and east for that of Hemp.

Hops constitute a staple crop throughout Bohemia and have till now defied all competition.

Saffron of Lower-Austria, Moravian Mustard, Hungarian Safflower etc. are superior products, but of inconsiderable quantity.·

The culture of Rape-seed has become of very great importance to Bohemia and the Hungarian provinces, and has, moreover, become very extensive in Austria-proper, Moravia and Galicia.

Since, besides the actual olive groves, there are still 9 or 10 sq. miles of arable and grass-land on the shores of the Adriatic Sea planted with those trees, the produce of olive oil exceeds the wants of these parts.

Although Tobacco is a monopoly, the cultivation of it has, during the last ten years, in point of quantity as well as quality, risen to an extraordinary height in the Hungarian provinces, in Transylvania and the Military Borderland, in Galicia and Bukowina, in Tyrol and Venice.

The quantities produced are:

Hemp and flax . . . . 3,000,000 Cwts
Hops . . . .   40,000 „
Tobacco . . . . 1,000,000 „ [1]
Other trade-plants . . . .   230,000 „

The special amount of crop in Linseed and Hemp-seed is 2,500,000 Metzen; rapeseed 1,200,000 Metzen; the produce of olive oil is 100,000 Cwts.

d) Vegetables.

Besides pulse, Turnips and Cabbages, there are also grown other pot-herbs, especially on tillage ground, so that special kitchen-gardens are to be met with only in the north-west and in the environs of large towns. In the south, cucurbitaceous fruits are also raised in the fields.

The gross amount of vegetable crop is 16,000,000 Cwts.

Culture of Fruit trees.

All sorts of Middle-European fruit-trees thrive in Austria particularly well: Austria-proper, Styria, Carinthia, Northern Tyrol, but especially Bohemia and Moravia cultivate fruit-trees with assiduous care, and the west and south of the Hungarian provinces is also herein very productive. On the shores of the Adriatic Sea fruits of the south grow abundantly. The total amount of crop fluctuates between 13,000,000 and 14,000,000

Growth of the Vine.

Besides France, Austria is, among the European states, most distinguished for the culture of the vine to which 110 sq. miles of real vineyards and 140 sq. miles of fields planted with vine are allotted. The variety of the elements of the soil influence the diversity of the products.

The superior development of the vine is promoted by the favourable soil of the Hegyallya hills near Tokai with their continuation to Erlau; the tract of the Elbe valley from Melnik and Cernosek downward; the country round about the Neusiedl lake, Rust and Oedenburg included; the heights near the Middle-Danube, together with St. Georgen, Neszmil, Buda, Szekszard, Villány, up to Schomlau and Menesch; the Fruska-Gora and the extent to Karlowitz; the Kahlen-gebirge, besides Vöslau, Gumpoldskirchen, Grinzing, and Weidling; the Bisamberg; the Styrian Drave and lower Mur vallies, inclusive of Kerschbach and Luttenberg; the southern slopes of the Karst and the Istrian and Dalmatian coasts.

Hungary, Lower-Austria and Stiria are the principal wine-growing provinces of the monarchy. The abolition of the custom-laws which formerly placed a bar between the Hungarian provinces and the rest of the monarchy, has on both sides of the empire awakened a lively emulation in the improvements of wine-growing.

The produce of wine fluctuates between 30,000,000 and 40,000,000 Eimers (12·449 Engl. gallons per Eimer), the husks offering besides an important lateral product.

Total estimate of the productions of the soil.

The gross amount of value of all hitherto enumerated products of the soil in Austria averages per annum 1,600,000,000 florins Austrian Currency.

Breeding of Cattle.

Austria is possessed of all the fundamental conditions necessary for the successful breeding of cattle, in which pursuit much exertion is being displayed, the more so as the improvement of husbandry is depending upon it.

A numeration of the cattle was also included in the census of October 31st 1857.

a) Horses.

The number of Horses in Austria reaches three millions and a half. The distinctive race is in the east of the monarchy, the Hungarian breed being excellent for riding-horses and the Transylvanian for coach-horses. The Noric breed of the Alpine regions is considered very good for draught-horses. Besides these two regions, there is particular care bestowed on the breeding of horses in the Bohemian, Moravian and Silesian provinces, whereas in the Littorale they pay less attention to it. In addition to numerous private studs,

---

[1] In 1858 1,707,671 Cwts, in 1859 755,080 Cwts, in 1860 692,012 Cwts.

there are two studs belonging to the Imperial Court, and six military studs with an average number of 7,200 horses among which there are 4,800 foals. Dalmatia and the Lombardo-Venetian Kingdom excepted, there are in the rest of the provinces nearly 600 horsing stations where about 2,200 stallions are kept.

For further encouragement of the breeding of horses there are numerous races held, and considerable prizes settled by the state as well as by private gentlemen.

b) Mules and Asses.

Mules and hinnies (24,000) are almost limited to Dalmatia, the Lombardo-Venetian Kingdom and Southern Tyrol; besides there are 100,000 common asses in Hungary.

c) Oxen.

The number of oxen amounts to 15,000,000 of which nearly 30 p. C. are bred in the Hungarian provinces, 16 p. C. in Galicia, 14 p. C. in Bohemia, 7 p. C. in Transylvania.

The majority of oxen indigenous in the empire belong to three kinds: the lowland race, the mountain race, and the domestic breeds.

The Hungarian cattle and the Podolic race are classified to the lowland race. A transition to the mountain race is formed by the Mürzthal cattle, annexed to which is that of Mariahof or Lavantthal.

The real mountain races are represented by those of Pinzgau and Pongau, Zillerthal, Duxerthal and Oberinnthal, next in succession to which follows the breed of Montafone as transition to the Switzer-race.

As belonging to the domestic breeds, may be mentioned: the ancient German breed, the ancient Bohemian breed, the Carpathian mountain breed and the cattle of Pusterthal.

For fattening purposes, the Hungarian and Podolic races are best, in Galicia and Bukowina the fattening of oxen imported from the east is considerable. Productiveness of milk is greatest in the richly fed Alpine cattle, with the breeding of which Alpine cow-husbandry is in close connection. All kinds of Middle-European cheeses are successfully imitated in Austria, the Emmenthal, Bohemian and Liptau cheeses deserving special mention.

Buffaloes are chiefly bred in South-western Hungary and in the Transylvanian Saxon district, because they are contented with inferior pasture.

d) Sheep.

A prominent feature in the husbandry of Austria is the breeding of sheep. To judge by the quantity of wool produced in Austria, the number of sheep undoubtedly amounts to 30,000,000

Lower-Austria, Bohemia, Moravia and Silesia number already 2,000,000 sheep of improved breeds; in Hungary and Galicia too, the breed of fine-woolled sheep has made considerable progress, although the restrictions upon the right of pasture, on ground not his own, constrain the wool-stapler (since 1848) to direct his attention rather to a good middling sort of dense fleece-staple. In Transylvania where sheep are commonly driven to pasture in the Danubian Principalities and Bulgaria, they also bestow a high degree of care to sheep-breeding; thus we may consider the region along the shores of the Adriatic Sea, and in part also Carniola and the Military Borderland, the only districts where sheep are only bred for butcher's meat.

Besides their wool and skins (the latter frequently used for peasant's clothing) their milk is of great importance for the production of Brimza-cheese.

e) Goats.

The breeding of goats, economically speaking, is of little importance, yet for certain mountainous districts it is quite indispensable, one million and a half are bred yearly, most of which in Dalmatia.

f) Swine.

Much importance is to be attached to the breeding of 8—9 millions of hogs which is managed in flocks, principally in the Hungarian provinces, with periodical migrations of those beasts, and in connection not only with the fattening of 500,000 hogs imported from Servia and Wallachia, but also with Alpine cow husbandry and Bohemian, Moravian and Silesian industry which offers much waste for food, and lastly connected with the husbandry of every landowner for the realization of his waste or for his own consumption.

The freight for hogs forms a considerable quota of receipts in the business of railways and the steam-navigation on the Danube. Though the German, Yorkshire and Essex breeds are spread in Austria, the Podolic (Galician), Servian (Mangalizan) and Szalontan breeds are prevalent.

Total estimate of cattle and its production.

The value of all this cattle amounts to 1,000,000,000 florins; and the annual produce from them embraces: 100,000,000 Eimer of milk (partly made into butter and cheese), 20,000,000 head of young cattle, 18,000,000 Cwts of meat and fat from grown cattle, 12,000,000 skins and furs, 700,000 Cwts of wool, with a total value of 450,000,000 florins Austrian Currency.

Fowls.

The number of fowls is very difficult to be ascertained; however, there may safely be quoted an average number of 60,000,000 estimated at 10,000,000 florins; those annually consumed are replaced, and 2,400,000,000 eggs are supplied, representing a value of 40,000,000 florins.

Bees.

Breeding of bees, a favourite culture in nearly all provinces, but particularly in Austria-proper, Carinthia, Bohemia, Moravia and Silesia, yields a considerable annual produce of honey and wax, the returns of which cannot be determined numerically.

Silk-worms.

Notwithstanding the disjunction of Lombardy, the rearing of silkworms is cultivated to a great extent in the south of the Austrian monarchy, the Lombardo-Venetian Kingdom and Southern Tyrol still supplying annually about 270,000 Cwts of Cocoons, estimated at 22,000,000 florins.

Hunting and Fishing.

In order, finally, not to omit mentioning the products of hunting and fishing, we must remark that the results of hunting are by far more lucrative in the regions of the Hercynian and Carpathian mountains than in the Alpine countries; fishing, on the other hand, is of great account in consequence of the attention paid to pond-storing in Austria-proper, Bohemia and Moravia, furthermore for reason of the great abundance of fish in almost all rivers and lakes of the empire, and extensive fishing in the sea, which is very important for the coasters of Istria and Dalmatia. Hunting and fishing together yield on an average product estimated at 50,000,000 fl. annually.

Mineral Products.

In the Austrian monarchy, there is a great diversity of mineral productions deserving attention both on account of the nature of the minerals and the geognostic conditions of their beds. The following summary exhibits the most important of these mineral productions and their most abundant beds:

a) Gold and Silver.

Auriferous and argentiferous ores, as well as copper and lead ores, partially also ores containing both the last-named metals, are found in trachyte in Hungary; in diorite in Hungary; and in porphyry in Hungary and Transylvania; furthermore in crystalline schistous rocks in Hungary, the Banat, Transylvania, Bukowina, Bohemia, Styria, Salzburg, and Tyrol; in gray-wacke in Bohemia and Styria; and in other sedimentary strata of a later period in Hungary, Transylvania and Tyrol, mostly in trias, lias and jura formations.

Gold by itself is found in porphyry in Transylvania; in crystalline schistous rocks in Bohemia and Tyrol; and in sedimentary strata of a later period in Transylvania. Numerous rivers and brooks depose auriferous sand, which especially in the border-mountains between Hungary, Transylvania and the Banat increases to real wash-gold.

b) Copper.

Copper-ores, without contents of silver, are extracted from the crystalline schistous rocks in Transylvania, Bukowina and Salzburg; from gray-wacke in Tyrol, Styria and Venice; and from red-rocks in Bohemia.

c) Lead.

Lead-ores, without contents of silver, occur in the crystalline schistous rocks of Bohemia; in the gray-wacke of Carniola; and in the later sedimentary formations in Tyrol and Carinthia.

d) Quicksilver.

The occurrence of Quicksilver in greater quantity is restricted to the coal-formation in Carniola; in lesser quantity it is also found in the crystalline schistous rocks of Hungary, and in gray-wacke in Carinthia.

e) Zinc.

Ores of Zinc are extracted only from later sedimentary formations in Cracow, Tyrol, Carinthia and Venice, and from crystalline schistous rocks in Bohemia; Tin-ores, however, are found in crystalline schistous rocks of Bohemia.

f) Iron.

A great variety of Iron-ores, especially sparry iron-ore, clay iron-ore, magnetic iron-ore, brown iron-ore, red iron-ore, argillaceous iron-stone, bog iron-ore, are stratified partially in diorite and trachyte in Hungary and Transylvania; more extensively in crystalline schistous rocks in Bohemia, Moravia, Silesia, Hungary, the Banat, Bukowina, Lower-Austria, Styria, Tyrol, and Carinthia; in gray-wacke in Styria, Lower-Austria, Salzburg, Tyrol, Bohemia, Carniola, and Croatia; in coal-formations in Styria, Carinthia, Salzburg, and the Banat; in later sedimentary beds in Salzburg, Tyrol, Carinthia, Carniola, Hungary, Bukowina, Galicia, and Moravia; likewise also in tertiary rocks of Bohemia, Hungary, and Carinthia. In Styria, Carinthia, Carniola, and Tyrol, sparry iron-stones prevail; in Silesia, clay iron-ore; in the rest of the provinces, the other ores, among which again the brown and red iron-ores prevail.

g) Nickel.

Ores of Nickel are obtained from diorite in Hungary; from crystalline schistous rocks in Styria and Bohemia; and from gray-wacke in Salzburg.

h) Antimony, Arsenic.

In crystalline schistous rocks in Hungary and Bohemia, Antimony is worked, and Arsenic in Salzburg and Bohemia.

i) Sulphur.

Besides iron-pyrites in Hungary and Bohemia, native Sulphur is drawn in volcanic formations of Hungary and Transylvania, and in tertiary rocks of Croatia and Cracow.

k) Graphite.

Graphite is contained in the crystalline schistous rocks of Bohemia, Moravia, Styria, Lower-Austria and Carinthia.

l) Alum.

For the production of Alum, alum-rock is drawn in Hungary; and alum-slate is produced partly from the gray-wacke and pit-coal formations of Bohemia, partly from the tertiary formations in Bohemia, Styria, and Lower-Austria, and from the chalk formations of Moravia and Istria.

m) Salt.

The rock-salt of the Alps is stratified in trias formations, that of the Carpathian mountains in tertiary formations..

n) Nitre.

In the north-eastern portion of the Hungarian lowland, between the Theiss and the Berettyö, there is a saltpetre district of 130 square miles of extent.

o) Soda.

The same region decomposes Soda; and a strip of land, traversing the midst of the comitates of Pesth, Pilis, Solt, and Bacska, also forms a bed of Sodium.

p) Coal.

Of Coals, there are worked Anthracites as well as Pit-coal and Brown-coal. Anthracites are produced, in small quantity, in one locality only (at Turrach, in Styria); but the other sorts of pit-coal formations are mostly worked in Bohemia, Moravia, Silesia, Cracow, and the Banat. Of the secondary fossil coal, fletzes are worked in the lias formations of Hungary, the Banat and Lower-Austria; and in chalk formations of Moravia and Istria. Brown-coal, for the most part belonging to neogenite tertiary time (in part lignites), is deposited in large quantities in Bohemia and Styria, and besides also in Hungary, Transylvania, Moravia, Lower-Austria, Galicia, Carinthia, Carniola, Venice, Dalmatia, and Croatia.

q) Turf.

Turf is found not only in the alluvial formations on the banks of rivers (in Hungary, Bohemia, Austria-proper, Salzburg, Tyrol, Styria, Carinthia, Carniola, Galicia, and Transylvania), but also in the upland plains of Bohemia and Styria.

r) Asphaltum.

Asphaltum is produced partly from cretaceous formations, but mostly from tertiary strata in Tyrol, Dalmatia, Hungary, and Galicia.

s) Argillaceous earths.

Besides Loam and Potter's Clay to be met with nearly in all parts of the monarchy, Bohemia, Moravia, Hungary, and Venice produce Porcelain-clay; Bohemia, Galicia, Hungary, and Styria, Fuller's earth; Bohemia, Tyrol, and Venice various Colour-earths.

t) Rocks.

Free-stone and Quarry-stone are to be me with over the whole of the empire, and Austria is rich in Limestone of various kinds. The Alps contain beautiful varieties of Marble; the Sudetes and the southern Alps, Roof-slate.

u) Gypsum and Chalk.

Gypseous beds abound in the north and west, Cretaceous beds in the south and east of the empire.

v) Gems.

Among the various precious stones, the Hungarian Opals and the Bohemian Garnets, Agates and varieties of Jasper rank foremost.

Quantities of Mining Products.

a) Gold.

Gold-mining is successfully managed in Transylvania and Hungary. The production of gold by washing, along of the Maros, Szamos, Aranyos, Körös etc., has decreased in consequence of defective management. That of Transylvania amounts to 5,340 marcs, and that of Hungary to 2,220 marcs. 80 marcs may be quoted for the gold production in the valleys of Rauris and Gastein in Salzburg, and 27 marcs for that of the Ziller valley in Tyrol.

b) Silver.

Hungary furnishes 66,500 and Bohemia 48,000 marcs of silver. Next in succession come Transylvania with 7,500, Tyrol with 660, and Salzburg with 340 marcs.

The law ordaining the sale of gold and silver mining products to the mint, has been repealed, everybody's products being at his free disposal now.

c) Copper.

Not alone the ores hid in the depths of the earth are assiduously sought after, and cement-waters zealously made use of, for the production of Copper, but also those ores are converted into products, that formerly were left quite disregarded. Nearly four-fifths of the production of coarse copper (about 40,000 Cwts) are furnished by Hungary; further 4,500 Cwts are the produce of Agordo in Venice, nearly 4,000 Cwts that of Tyrol; 2,000 Cwts are returned from Salzburg, 1,700 Cwts from Transylvania, 1,300 Cwts from Božorita in Bukowina.

d) Lead.

The production of Lead and Litharge is partly connected with the production of silver; there are, however, lead-ores, containing no silver, drawn in Carinthia, Bohemia, Tyrol, and Carniola. About one-half of these products are derived from Carinthia (above 80,000 Cwts); another fifth (32,000 Cwts) from Hungary; one-sixth (25,000 Cwts) from Bohemia; the rest comes from Carniola, Tyrol, the Lombardo-Venetian Kingdom, and Transylvania.

e) Quicksilver.

Idria furnishes from 4,000 to 5,000 Cwts (and sometimes more) of Quicksilver; Hungary and the Lombardo-Venetian Kingdom add a fourth of this production.

f) Calamine and Sulphuret of Zinc.

Calamine and Sulphuret of zinc mining is of little extent, and even masses of the former, already drawn up, are waiting to be smelted. Three-fifths of the zinc produced are derived from the district of Cracow (15,000 Cwts). About half of this quantity is taken from the rubbish of some lead-mines in Carinthia; both Tyrol and the Lombardo-Venetian Kingdom return 1,500 Cwts each.

g) Tin.

The production of tin-ores is restricted to Schlaggenwald, Zinnwald, and Graupen in Bohemia, which furnish 1,300 Cwts. of this metal.

h) Iron.

With the exception of Upper-Austria, the Littorale, Dalmatia and Venice, all the provinces of the empire yield quantities of Iron. Styria and Carinthia furnish more than two-fifths of the ores; Hungary above three millions, Bohemia more than two millions and a half, Moravia above one million and a half Cwts of ores. From the different nature of the iron ores proceeds the fact that the northern provinces produce many cast iron-wares, whereas the southern provinces are principally engaged in the production of pig-iron for being worked into refined iron and steel. At present nearly all the Austrian smelting-works use charcoal, some few however coke; pitcoal and turf are nowhere exclusively used. Of the 6,300,000 Cwts of pig-iron and cast-iron, the average produce in the empire, Hungary, the Banat and Transylvania furnish 1,800,000 Cwts, Styria about the same quantity, Bohemia 1,250,000 Cwts, Carinthia 640,000 Cwts, Moravia 300,000 Cwts, Carniola and Silesia 100,000 Cwts each, Galicia 92,000 Cwts, Salzburg, Lower-Austria, Tyrol and Bukowina furnish the rest.

i) Other kinds of metals.

Such metals as are produced only in little quantities are: 8,000 Cwts of Antimonium crudum and Antimonium regulus mostly derived from Hungary; 1,200 Cwts of Arsenic drawn from Salzburg; 200 Cwts of Nickel amalgam worked in Hungary, Salzburg and Styria; 1,300 Cwts of Manganese furnished by Salzburg, Tyrol, Carniola, and Hungary. Tungsten and Uranium of Joachimsthal in Bohemia, Chromium of Styria, and others are rising into importance.

k) Sulphur.

The produce of Sulphur from earths or pyrites is rapidly increasing, and capable of still greater development; the present production being 30,000 Cwts principally derived from Salzburg, Galicia, Venetia and Croatia.

l) Graphite.

In the production of 100,000 Cwts of Graphite, Bohemia is concerned with three-fifths, Moravia with one-fourth, Styria and Lower-Austria in equal shares with the rest.

m) Alum.

Of Alum and Vitriolic schist there are produced in Bohemia, Lower-Austria, and partly also in Styria and Moravia: about 40,000 Cwts of Alum, 44,000 Cwts of fuming Sulphuric acid, and 72,000 Cwts of Sulfate of iron (green copperas).

n) Salt.

The production of Salt is a monopoly of the state; Sea-salt alone may be produced by private persons, who are, however, bound to sell their product to the State. 4,000,000 Cwts of rock-salt are drawn up annually; of which 1,500,000 Cwts are procured from Wieliczka and Bochnia in Western Galicia, about 1,100,000 Cwts from the salterns at Marmaros and 1,360,000 Cwts from Transylvania. From the artificial brine at Hallstadt, Ischl, Aussee, Hallein, and Hall, and from the native brine of the East-Galician saltworks, the Bukowina salt-pan at Kaczyka and the Hungarian at Soóvár, — 2,650,000 Cwts of Culinary salt are produced, of which three-sevenths are manufactured in Upper-Austria, one-fifth in Galicia, and the remainder in Styria, Salzburg, Tyrol, and Hungary, in nearly equal parts.

The production of sea-salt on the Istrian and Dalmatian coast has, during the last ten years, risen to double its former quantity exceeding now 1,300,000 Cwts. The production of Cattle-salt and Dung-salt, as well as the sale of salt for chemical-purposes, at moderate prices, has reached 1,000,000 Cwts.

o) Nitre and Soda.

The production of Saltpetre is limited to 40,000 Cwts, entirely consumed in the manufacture of gunpowder; but the production of Hungary is increasing. The reverse is, for the moment, the case with efflorescing Soda, its quantity having diminished to 6,000 Cwts.

p) Coal.

The production of Fossil coals (pit-coal, brown-coal, lignite and some anthracite) has, within 30 years, increased to more than twelve times its former extent, and already exceeds 72,000,000 Cwts. Almost one-half of that quantity is supplied by Bohemia; Hungary yields one-eighth, and Silesia a trifle more one-tenth is provided by Styria, one-fifteenth by Moravia. The remainder is jointly produced in Lower-

Austria, Galicia, Carniola, Upper-Austria, Carinthia, the Littorale, the Lombardo-Venetian Kingdom, Tyrol, and Dalmatia; Transylvania and Croatia-Sclavonia furnishing but an inconsiderable quantity, Salzburg and Bukowina none at all. To the secondary period of coal-formation belong the coalfields of Silesia and Western Galicia with their whole proportion, those of Bohemia and Moravia with two-thirds; those of Hungary with one-half, those of Lower-Austria with one-third of their proportions. Upper-Austria is nearly limited to the production of lignite, the Lombardo-Venetian Kingdom is so entirely, lignites being, moreover, of great importance and extent in Lower-Austria, Styria, Moravia, and Hungary. The remainder belongs wholly to the secondary period of brown-coal-formation. Of Anthracites there are obtained only 10,000 Cwts in Styria.

q) Turf.

Turf-moors are only availed of to any extent in Upper-Austria, Salzburg, Styria, Carinthia, Carniola, Tyrol, Bohemia, Hungary, and Galicia.

r) Asphaltum.

Asphalt-stones are obtained but in inconsiderable quantity (15,000 Cwts) at Seefeld in Tyrol and Dalmatia.

s) Porcelain-earth

of superior quality is obtained at Zettlitz in Bohemia and also in Moravia, Hungary, and Venice.

t) Freestone.

The working of Quarries, Loam-pits etc. only takes place in order to supply the local demands of landowners. It is but in the vicinity of great towns that Building stones are quarried in the regular way of industrial pursuits. 344 quarries supply better examples of sand-stone and limestone, granite, serpentine, marble etc. fit for Free-stone. Large quarries of Roof-slate are found at Flietsch in Tyrol, at Rabenstein in Bohemia, and at Durstenhof in Silesia.

u) Lime.

The most important sorts of Cement in the empire are those produced of the lime-marl of Kufstein in Tyrol, that produced of the aptychose lime of the vicinity of Vienna, and that of Sagor and Prague. The total production extends to 400,000 Cwts. Lime-kilns of some extent are found in the neighbourhood of great towns.

v) Gypsum.

The Gypsum layers of the empire that are most worked, are those of Austria-proper, Salzburg, Silesia, and Tyrol.

w) Gems.

Precious Opal is only found at Vörösragas in Hungary. Bohemian semi-precious stones, Tyrolese and Bohemian Garnets are not so much sought after at present.

Total value of the minerals.

The value of all the minerals produced (salt not taken into account as being an object of monopoly) reaches the sum of 60,000,000 florins Austrian Currency.

Trade Industry.

(Detailed statements respecting each branch of Industry are contained in the Catalogue at the head of the respective class.)

The Industry of Austria is in our period, with few exceptions, extending to all varieties of manufactured products. If we leave out of consideration the mining industry of the Alpine and Carpathian districts, the glass manufacture of Bohemia, the linen manufacture in the valleys of the Giant mountains, Sudetes and Carpathian mountains, carried on as an additional occupation in connection with agricultural pursuits, and cloth weaving transplanted from Lausitz to Bohemia at the end of the 17th century: we may in Austria date the development of an organic factory system of unceasing growth from the beginning of the present century. But a particularly rapid progress did not manifest itself until the year 1830, when, through a judicious reorganization of the frontier-guardship, the prohibitory system accepted since 1784 was fully brought to bear; and since the settlement of the Protective Tariff System of 1852, the scale of factory production as well as organization of labour has extended largely. If we direct our attention to manufactures destined for immediate use, such as, webs and textures, paper and productions of the same, leather and leather-wares, wood-, glass-, and earthen-ware, metal manufactures, vehicles and vessels, machines, implements, musical instruments, fancy-goods, chemical products, colours, oils, fats and chemical lights, literary and art productions, we shall find that the export of them has risen from 78 millions florins, in 1852, to 140 millions florins in 1861, whereas the import of such manufactures has risen from 24 to only 36 millions florins. Up to the year 1830, the two most important branches of productions, the chemical and mechanical, were almost wholly wanting in the organism of Austrian industry. The latter did not enter the circle of home industry, with anything like efficiency, till after the beginning of railway-building in 1838. With such a short existence of these two branches, it is obvious why several kinds of chemicals and machines are to this very period still imported.

The best and surest scale for a ratio of the development of factory system and production by machinery is the consumption of mineral coal, this fuel having been comparatively little used in the households of Austria.

Now the consumption of coals has, since 1839, increased from 10 to nearly 70 millions Cwts. Moreover, there is in the industrial districts of the Giant and Ore mountains scarcely any water power left that is not availed of, so that at present water works make their way out of the valleys into the plains; and in the immediate vicinity of coal-fields industrial districts are rising into existence where the works are carried on with steam-power.

An account of the condition of the several branches of Austrian industry and the estimation of the value of the resulting productions being reserved for the introductions to the several classes of the exhibition, there are only some strictures to be made on the distribution of industrial activity over the several provinces of the monarchy.

Bohemia.

Endowed by nature not alone with a favourable condition of the soil regarding hydrographical circumstances and raw products of the most various kinds, but also raised to eminence by an unremittingly assiduous population, Bohemia has, of old, cultivated all branches of industrial activity. Although industry is spread over the whole province, the more industrious districts are generally the slopes of mountains occupied by Germans. Here the several branches of industry have gradually centred in distinct districts, and the density of population increased to a degree that is only equalled in the most industrious countries of Europe. Linen, woollen, cotton, and mixed fabrics; glass, metal, porcelain, and earthen ware; chemicals; beet-root sugar; paper; beer and brandy, constitute the most prominent branches of Bohemian industry.

Moravia and Silesia.

In these provinces features similar to those of Bohemia present themselves; but, in consequence of the orographical condition of the country, water power is used in a lesser proportion, and woollen fabrics predominate over linen fabrics. Besides these two industrial pursuits, the manufacture of iron-ware and inland sugar, beer-brewing and brandy-distilling stand foremost.

Lower-Austria.

Quite different circumstances are to be met with in Lower-Austria, which owes its flourishing industrial condition principally to the Imperial Metropolis, and to the profusion of its river-system the tributaries of which emanate from the Alps. A favoured site on the most important stream of the country, and extensive direct communications with all parts of the empire, have elevated Vienna to the rank of the first city for the staple commerce of the monarchy. Immense consumption of the most various products, and a brisk sale of its own manufactures, especially fashionable fabrics and fancy articles, have likewise made this city the first industrial place of the empire, more than two-thirds of the value of Lower-Austrian manufactures falling to the share of Vienna. — The industrial activity of this province supplies a manifold variety of productions, the most prominent of which are: in Vienna, silk manufactures, shawls, fancy articles (of gold, silver, alloys, leather), fashionable fabrics, leather, machines, implements and musical instruments, as well as chemical products; in the environs of the capital and the rest of the lowland plain (partially also in Vienna), cotton spinning manufactures, printed cotton and woollen fabrics, paper and manufactures of paper, refined colonial-sugar and iron-ware.

The Lombardo-Venetian kingdom.

In the Venetian province, which nature has provided with all that is required for richly productive agricultural pursuits, there are also several branches of manufacturing industry in a flourishing condition. Profusion of pecuniary means and excellent roads of communication, liberty of trade and a dense population, cannot but exercise a favourable influence upon the liveliness of industry. Yet, the silk and glass production of Venice, and paper, leather and metal manufactures excepted, there is no particular extent of manufacturing industry, notwithstanding the variety it is exhibiting; even its flax and hemp fabrics yield but inconsiderable results.

The other Provinces of the Empire.

Though more or less extensive industrial pursuits are carried on in the other Crown-dependencies, only the following deserve particular mention: — Manufactures of iron, linen and cotton in Upper-Austria; iron industry in Styria, Carinthia and Carniola; paper manufacture in Styria; silk and cotton fabrics and metal wares of Tyrol and Vorarlberg; ship-building in the Littorale and in Dalmatia; weaving of linen, hemp and woollen stuffs, besides brandy distilling in Galicia and Bukowina; liqueur production in Dalmatia. — In Hungary and its dependencies, in Transylvania and the Military Borderland (where the pursuits of husbandry are capable of immense development, provided sufficient manual labour can be procured) there were till now wanting the fundamental requisites for eminent industrial undertakings. Want of capital, operative hands and means of communication have occasioned these provinces as well as Dalmatia (where similar drawbacks exist) to remain behind the rest of the empire with respect to industry.

Nevertheless, the domestic weaving of linen, hemp and woollen stuffs, production of chemicals (soda, potash, alum, saltpetre), distilling, leather, paper, wooden and glass ware in Hungary, the woollen and linen fabrics, wooden and leather wares, and brandy distilling in Transylvania, spirits, earthen-ware, hempen fabrics and wooden articles in the Military Borderland — are by no means irrelevant.

Commerce.

The geographical position of the whole monarchy in the heart of Europe, and the interchanging diversity of the several provinces together with their sub-divisions, warrants a lively commercial intercourse bringing the most affluent amounts of capital into uninterrupted circulation. Whilst even the mountain-ous tracts put no insurmountable barriers to modern means of communication, there are to be taken into so much the greater account all those manifold con-trivances for transport that are offered by the features of the surface, the river-system and the coast-exten-sion of the empire.

Inland Trade.

The home trade of the Austrian monarchy is, in general, subject to no restriction or surveyance, there-fore no numerical estimate of its extent or its details can be given.

The most important trading staples for the inland commerce of Austria, in the alternate succession of her provinces, are: Vienna, Linz, Gratz, Botzen, Prague, Reichenberg, Pilsen, Brünn, Olmütz, Bielitz, Lemberg, Mantua, Verona, Padua, Pesth, Oedenburg, Pressburg, Kaschau, Debreczin, Agram, and Hermannstadt

The communication between the Austrian ports is, in general, restricted to the Austrian flag, and annually averages 66,000 arrivals and departures of vessels of 32 tuns each, mean-estimate. The most important ports are: Trieste, Venice, Fiume, Grado, Capo d'Istria, Rovigno, Pirano, Pola, Zengg, Porto Rè, Zara, Spalato, Ragusa, and Cattaro.

Foreign Trade.

The outward trade of Austria is partly Land trade, partly Maritime trade.

Of the trade by land, three-fifths belong to transactions with the German States; one-seventh to the commerce with Turkey and its Protectorate-States; one-tenth to the commercial intercourse with Italy, and nearly the same ratio to that with Switzer-land; lastly one-twentieth to the trade with Russia and Poland. The places most distinguished for land-trade with foreign countries are: Vienna, Salzburg, Prague, Reichenberg, Pilsen, Troppau, Cracow, Brody, Pesth, Semlin, and Kronstadt.

The Austrian trade by sea, carried on with coun-tries abroad, averages 4,500 arrivals and departures of vessels, annually, of 130 tuns each, mean esti-mate. Of this trade, one-fifth is carried on with Great-Britain, the same severally with Turkey and Italy; one-fifteenth of it belongs to France. Besides, the maritime trade with the Russian coasts of the Black Sea, and the Atlantic range of America, is of some relevancy. In the trade with all of these states, foreign flags compete with the Austrian, so that only one-fourth of the number of vessels and little more than two-fifths of the tonnage fall to the share of the Austrian flag.

The result of the import and export trade in the course of the years 1851 to 1860, within the Austrian tariff-range, appears, in general, from the following table:

	Import	Export
	Florins Austrian Currency	
1851	158,074,663	136,524,944
1852	209,329,849	195,814,828
1853	207,262,160	228,440,293
1854	219,165,017	228,924,871
1855	248,288,157	244,134,142
1856	301,194,829	263,928,641
1857	292,995,251	242,363,721
1858	308,285,925	275,599,871
1859	268,227,783	292,651,240
1860	231,226,702	305,197,493

In 1861 the value of the whole import amounted to 232,732,554 florins, that of the export 310,687,250 florins distributed among the 22 classes of the custom-tariff as follows:

	Import	Export
	Florins Austrian Currency	
Colonial wares and fruits from the South	15,849,737	5,780
Tobacco and Manufactures of tobacco	2,723,647	588,806
Garden and Field fruits .	12,299,963	45,538,698
Cattle	17,507,927	9,309,386
Animal products	7,189,941	4,355,125
Fats and Oils	13,357,885	3,842,076
Drinks and Food	2,203,119	3,940,300
Fuel, Building, and Manu-facturing Substances . .	5,6705,34	24,897,510
Medicine, Perfumery, Co-louring, Tanning and Chemical Substances . .	17,4343,67	4,303,386
Metals, raw, and half manu-factures	33,505,135	35,541,670
Weaving and Textile Stuffs	47,523,596	35,646,949
Yarns	21,169,668	2,582,215
Woven Fabrics	9,968,830	48,721,535

	Import	Export
	Florins Austrian Currency	
Manufactures of Straw, Bast etc. Paper and Manufactures of paper	1,222,295	5,126,321
Leather and Manufactures of leather etc.	5,333,000	3,505,300
Wooden, Glass and Earthen Ware	3,270,866	22,872,411
Metal Manufactures . . .	2,670,058	16,362,674
Vehicles and Vessels . .	324,000	3,953,000
Instruments, Machines and Fancy goods	6,238,074	22,493,852
Chemical Products, Colour, Fat and Chemical Light Manufactures . .	2,018,442	4,499,938
Works of Literature and Art	5,251,470	2,493,560
Waste	. . . .	106,758

In order to illustrate the economical circumstances of the empire more nearly, the most important objects of import and export trade shall be submitted to a separate investigation.

Trade in Agricultural Products.

a) Trade in Wood.

Trade in wood forms a very essential part of the Austrian commerce which is principally carried on by means of navigation.

In 1861, 5,272,400 cubic feet of wood for fuel and 6,088,900 cubic feet of common tmber, estimated together at 3,600,000 florins, were brought, mostly on the Danube, from southern Germany to Austria, whereas 4,813,500 cubic feet of the former, and 43,079,400 cubic feet of the latter, valued at 23,000,000 florins, went abroad, partly on the Adriatic Sea, partly on the Po, the Elbe, the Oder, the Vistula and the Dniester.

b) Trade in Corn.

An important rank is held by the Corn-trade. In years of less favourable harvests in Central Europe contiguous to the western boundary of Austria, the latter country exports upwards of 6 and 7 millions Cwts of her various kinds of corn; in the year 1861 this number increased to more than 8 millions.

Wheat is foremost in this export. There were indeed 478,450 Cwts of it imported from Southern Russia and the Danubian Principalities, but, on the other hand, 5,863,847 Cwts were exported, mostly to Northern Germany, France, and England. The trade in Rye is of little importance only, the import of 626,067 Cwts (partly from the Tariff Union States for the consumption of Austrian boundary districts, partly from Southern Russia) being opposed to an export of 333,306 Cwts (mostly to the countries of the Tariff Union). The trade in M a i z e i limited to the south of the empire: 106,362 Cwts came partly from Italy, partly from Turkey; 1,416,815 Cwts went most of it to Italy. The import of B a r l e y amounted to 106,253, the export 473,158 Cwts, though the sale to Southern Germany decreased much in comparison with former years. Of O a t s, 246,529 Cwts were imported, 765,342 Cwts exported.

The total value of the imports of all kinds of corn (besides the irrelevant trade with pulse) amounted to 6,700,000 florins; tht of export to 28,600,000 florins.

Trade in Rice is not included, it being carried on with the south alone and having, in 1861, comprised 63,576 Cwts of imports and 99,004 Cwts of exports.

c) Trade in manufacturing plants.

Among plants, in 1861, export articles of considerable value were supplied. Hops for 2,000,000 florins, Linseed, Hempseed, and Rape for 3,400,000 florins; whereas of Tobacco, the manufactories of the exchequer imported for the value of 3,000,000 florins; and of Hempseed oil, Linseed oil and Rapeseed oil, the value of 2,000,000 florins, of Olive-oil the value of 8,000,000 florins was brought to Austria. Other articles of import were : — medicinal herbs and perfume-substances, valued at 900,000 florins; dyeing woods valued at 1,500,000 florins, Madder valued at 2,200,000 florins, Cocheneal and Indigo valued at 7,000,000 florins Gums and Resins valued at 11,400,000 florins.

d) Trade in fruit.

Fruit, fresh, and in still greater quantities dried, is exported from Bohemia and Moravia to the countries of the Tariff Union. Even in the less favourable harvest of 1861, this export amounted to 320,724 Cwts, valued at one million and a half florins.

e) Wine trade.

A considerable article of trade is supplied by the production of Wine. The import, as far as quantity is concerned, comes from countries not belonging to the Tariff Union, viz. common Italian, Moldavian and Wallachian wines, besides fine Southern and Western European sorts. Opposed to the total import, valued at 800,000 florins, is an export, valued at more than 2,000,000 florins, which has become nearly a regular one, in spite of the great restrictions put upon it by former treaties of customs; and superior sorts are even sent across the Ocean.

B. Results of Cattle-breeding, Fishing, etc.

a) Trade in Horses.

Trade in horses is subject to remarkable fluctuations brought about partly by the unsettledness of the want of them or their new breed, partly by temporary prohibitions of export. In 1861 the import of 7502 head, valued at 560,000 florins (most of them from the contiguous countries), was counter-balanced by an export of 13,045 head, estimated at about one million, directed mostly to the states of the Tariff Union.

b) Trade in Bovine Cattle.

Trade in cattle is of great consequence. The import of oxen and bulls from the eastern contiguous countries is continually increasing, and reached in 1861 (a great want of fodder occurring in Servia) the number of 86,220 head, valued at 5,200,000 florins. There is, however, a striking decrease in the import of cows from the same direction, and of calves (especially) from the states of the Tariff Union, for the purposes of rearing, in one year embracing only 43,164 head of the former and 17,266 of the latter, estimated at not quite 2,000,000 florins. The export is nearly exclusively directed to the countries of the Tariff Union, and comprised, in 1861, 47,662 oxen and bulls, 31,679 cows, and 42,382 calves, with a total value of nearly 4 millions and a half florins.

c) Trade in Sheep.

Import and export of sheep are, in general, on a balance. In 1861, the import preponderated, being opposed with the value of 600,000 florins to the export valued at 400,000 florins.

d) Trade in Pigs.

The trade in pigs forms a very important item. The import of 549,379, in 1861, came mostly from Servia and Wallachia, the export of 263,223 having gone as far as the Rhine, Hamburg, and Stettin. But the value of the droves that come and those that go is very different, the former entering Austria for fattening, the latter being exported for the butcher. Constant increase of the export is obvious, for, at other times, it amounted to not quite one-fourth of the total trade, whereas now it increased to nearly a half of the value.

e) Trade in Fish.

Fish are an article of considerable import, having amounted, in 1861, to 125,934 Cwts, valued at 1,200,000 fl. (two-thirds of them codfish and herrings).

f) Trade in Animal Products.

As articles of trade there are to be particularized:

Tallow, of which 86,347 Cwts were imported from the eastern contiguous countries, representing a value of 1,640,000 florins.

Bacon, of which 65,246 Cwts, estimated at 1,600,000 florins were exported to the Tariff Union States.

Butter, amounting to an export quantity of 53,320 Cwts, value 1,600,000 florins.

Cheese, to the imports of which, 29,562 Cwts coming from Switzerland and Italy, valued at 1,000,000 florins, can be opposed scarcely half that quantity of exports from Tyrol to Southern Germany.

Skins and Hides which, entering Austria mostly from Russia and Turkey, amounted, in 1861, to 187,948 Cwts, and represented a value of 5,800,000 florins; whilst even of the exported 17,260 Cwts, not quite 600,000 florins in value, a part returned as Leather to Austria from the States of the Tariff Union.

Sheep's Wool, the imports and exports of which are pretty equal in quantity (220.000 Cwts against 224,000 Cwts), but the value of which is widely different, the imports comprising mostly coarse commodity from the eastern neighbouring lands, estimated only at 11,000,000 florins whereas the exports supply Prussia, Saxony, and Southern Germany with fine sorts of more than 25,000,000 florins value.

Feathers, an article of constantly increasing export, 42,329 Cwts, value 3,000,000 florins, having gone to the Tariff Union States.

Raw Silk, not thrown, was much more exported (for the value of 7,000,000 fl.) than imported (for the value of 750,000 fl.).

Train-oil was imported in quantities amounting to 48,206 Cwts, valued at 800,000 fl.

C. Mineral Products.

a) Trade in Precious Metals.

The import of precious metals in Ingots is limited principally to that which the National Bank receives, having amounted, in 1861, to 900 Cwts of four millions and a half in value.

b) Trade in Iron, Copper and Tin.

The traffic in Pig-iron was almost limited to the import of 223,748 Cwts of 600,000 florins value, destined particularly for Bohemian and Silesian Smelting Works.

Respecting other metals, Copper and Tin alone are articles of considerable import, of the former having entered 35.407 Cwts, of the latter 8.443 Cwts, from the States of the Tariff Union, and representing a value of 2,700,000 florins.

c) Trade in other products of the mineral kingdom.

The import of Salt for the exchequer stores or for chemical purposes does not reach by far the value of the exports which comprised 1,289,639 Cwts.

The import of Soda amounted to 128,013 Cwts. valued at more than one million florins; that of Chili-saltpetre 59,390 Cwts (value 600,000 florins), and of Sulphur 114,063 Cwts (value 570,000 florins).

Finally, Fossil Coals are an article of very brisk trade. In 1861, imports and exports fluctuated in quantity between 5,360,000 and 5 900,000 Cwts; the exchange in value amounting pretty equally to 1,400,000 florins.

D. Products of Manufacturing Industry.

What has been said before in the remarks on manufacturing industry, must be repeated also here, viz. that the statistical summary preceding each class of exhibited articles in the Catalogue renders any detailed explanation superfluous.

For this reason, only those tariff articles will be mentioned that, either in the imports or exports of 1861, brought into commercial intercourse values of more than half a million florins, as under:

In the Tariff Class:	Import	Export
	Florins Austrian Currency	
Tobacco and Manufactures of —	560,229	8,057
Drinks and Eatables:		
Beer	76,274	568,264
Brandy and Spirit of Wine	27,180	1,178,085
Fine Eatables	802,890	. . . .
Textile Materials:		
Spun Raw-Silk	4,143,100	1,198,900
Yarns:		
Cotton Yarns	13,346,238	339,930
Linen Yarns	1,529,850	1,568,305
Woollen Yarns	6,293,580	673,980
Woven Manufactures:		
Ropes and Pack-Cloth .	66,840	500,590
Linen Fabrics, coarsest .	25,585	2,848,720
„ „ coarse . .	14,240	3,492,300
„ „ middle-fine	10,880	1,478,800
„ „ fine . . .	96,000	4,120,700
Cotton Fabrics, middle fine	750,000	5,483,025
„ „ fine . . .	348,490	1,647,610
Woollen Fabrics, coarse .	69,680	3.346,920
„ „ middle-fine . .	2,186,000	7,276,320
„ „ fine . .	256,650	2,525,100
„ „ superfine	33,600	2,655,000
Silk Fabrics, coarse . .	532,800	3,023,540
„ „ fine	4,563,200	1,161,300
Clothing and Millinery, coarse	48,420	857,310
„ „ „ fine .	90,000	4,465,800
„ „ „ super-fine	339,000	2,805,300
Manufactures of Straw etc.:		
Paper, coarse	86,125	701,822
„ fine	325,470	3,193,365
Leather, Manufactures of:		
Leather, coarse	3,669,680	1,314,600
Manufactures of Leather and Gum, coarse . . .	387,480	2,410,500
Gum, fine	784,340	8,975,400
Gloves	35,200	580,800
Wooden, Glass and Earthen Wares:		

	Import	Export
	Florins Austrian Currency	
Wooden Ware, coarsest .	488,540	777,550
„ „ fine . .	251,365	2,896,240
„ „ finest . .	727,310	1,544,970
Glass, coarse	276,953	1,303,700
„ middle-fine . . .	107,270	2,910,145
„ fine	61,350	10,861,290
Mirrors	255,780	546,840
Porcelain Ware, finest .	508,800	792,680
Metal Ware:		
Ironware, coarsest . . .	325,600	1,202,520
„ coarse . . .	438,768	3,689,064
„ fine	1,899,400	10,495,890
Manufactures of Copper, Zinc, Tin, etc. etc. . . .	. . . .	975,200
Vehicles and Vessels:		
Vessels	. . . .	2,918,400
Carriages and Sledges . .	. . . .	998,600
Instruments, Machines, Fancy Goods:		
Instruments, astronomical, surgical, optical, philosophical	. . . .	671,700
Pianofortes	. . . .	484,632
Machines and Parts of —	2,898,845	1,060,460
Short Merchandise, coarsest	415,400	659,160
„ „ coarse .	313,600	10,337,400
„ „ fine . .	446,629	4,294,500
„ „ super-fine . .	2,163,600	4,986,000
Chemical Products:		
Colours, fine	350,740	532,560
Chemicals and Colours mixed	1,067,700	957,125
Stearine Ware etc. . . .	. . . .	551,265
Chemical Lights, coarse .	. . . .	2,027,375
Literary and Art Works:		
Books and Maps	4,615,870	2,273,400

E. Colonial Products and Fruits from the South.

The articles, which Austria will ever be obliged to have imported from abroad, principally comprise: Coffee, Tea, Sugar, Spices, Fruits from the South and Cotton.

The import of Coffee is regularly increasing and has augmented, during the last 8 years, by 80 per cent, comprising, in 1861, 408,830 Cwts, estimated at 10,629,580 florins.

Also the import of Tea rises continually; in comparison with the year 1851 it rose to more than double the former quantity, amounting, in 1861, to 4,130 Cwts, value 702,100 florins.

The reverse is the case with the import of Colonial sugar, which has become almost superfluous through the exertions made in home produce. In 1831, there were imported 395,800 Cwts of sugar powder, and, in 1853, 779,500 Cwts. In 1861, on the contrary, this import amounted to only 31,700 Cwts of sugar-powder, 34,578 Cwts of molasses, and 10,068 Cwts of refined Colonial sugar, total value of 830,619 florins.

The import of spices keeps pretty firm, comprising, in 1861, 27,455 Cwts of coarse, and 8,038 Cwts of fine and superfine spices, the former representing a value of 549,000 florins, the latter a value of 643,040 florins.

Also the import of fruits from the South (especially Oranges and Citrons) is pretty much the same as usual, comprising 117,300 Cwts valued at about 600,000 florins, but the import of fine fruits of the South (raisins, cubebs, almonds and dates) rises, amounting, in 1861, to 59,386 Cwts estimated at 119,809 florins.

Finally, the import of Raw Cotton increases regularly, heightening during 20 years about 216 p. C. and amounting, in 1861, to 879,535 Cwts, valued at 28,155,629 florins.

Commerce in Dalmatia.

The commercial intercourse of the separate Dalmatian range of customs comprised, in 1861, in values of imports 7,997,684 florins, in values of exports 4,490,731 florins, distributed among the following eight Tariff Classes:

	Import	Export
	florins	
Colonial and pharmaceutical articles and spices	276,536	55,210
Field and Garden fruits, Products of the vegetable and mineral kingdoms	2,795,177	115,186
Cattle and Animal Products not mentioned elsewhere	778,392	1,620,261
Drinks, Eatables, Oils, Fats and Manufactures of them . .	400,052	2,352,514
Textile Materials, Yarns, Woven Fabrics, Clothing, etc. . . .	1,966,505	163,950
Metal, Earthen, Glass Ware . .	568,378	9,050
Manufactures not contained in other Tariff Classes	1,212,644	118,286
Waste		56,274

In the number of items, of more than a quarter of a million value, there are to be mentioned:

	Florins
a) For Import.	
Tobacco Manufactures, valued at	359,415
Corn and Pulse, „ „ . . .	916,676
Rice „ „	284,184
Flour and Mill Products, „ „	921,760
Beeves „ „	404,440
Flax and Hemp Manufactures valued at, .	338,590
Cotton Manufactures, „ „ .	706,385
Woollen Manufactures, „ „ .	344,410
Leather and Manufactures of-, „ „ .	263,780
Wooden and Stone Ware, „ „ .	256,488
Short Merchandise, „ „ .	451,250
b) For Export.	
Fish and other aquatic animals, valued at .	355,659
Meat, „ „ .	261,540
Skins and Hides, „ „ .	770,560
Wine, „ „ .	760,424
Fats, „ „ .	250,380
Olive Oil, „ „ .	1,307,880

In 1860, the export of Olive Oil reached 2,481,930 florins, the unfavourable harvest of 1861 having occasioned the above shown diminution.

Commercial and Trade Laws.

Liberty of Trade.

There is Liberty of commerce and trade in Austria. Foreigners, too, are admitted to the absolute exercise of manufacturing and commercia industry, upon obtaining consent of the Ministry. There are now but very few licensed trades, for the exercise of which a special permission is necessary mostly dependent on certain references or proofs of personal capacity. To such trades belong: — the Press; circulating libraries and reading-rooms; undertakings of periodical conveyance of persons; building trades; manufacture and sale of arms, munition and fire-work materials; inns and taverns; commercial travellers; pedlers; finally butchers in the Military Border-land.

The right of exercising a trade or craft is, in general, personal. But there still exist a considerable number of Trades that, by the modes of acquisition of civil right, may be transmitted to other persons, being either connected with immoveable goods or alienable trades.

Personal licensed trades are not transmissible. Only in the interest of a widow or of heirs under age (till the time of their becoming of age) the licensed trade of the husband or testator may be carried on.

Trades, exercised by means of furnaces, steam-engines, or water-works, or otherwise apt to endanger or inconvenience the neighbourhood by rendering the same unhealthy or unsafe, spreading offensive smell, or causing uncommon noise, require a permission concerning the locality of that trade, which, for certain trades specified by the law (such as a flayer's occupation, manufacturing of chemicals, etc.) is granted only after judicial summons of the parties in question, when the municipal representatives and neighbours have an opportunity to offer reasons for any objection to the intended exercise of such trade.

Uniting several trades.

Every tradesman has the right jointly to execute all works requisite for the complete performance of his manufactures, and to employ the workmen of other trades necessary for his own. This right, formerly granted only to large factories, has now become general; but the privilege of using the Imperial Eagle upon the sign-board and in the seal, and the denomination of Imp. Roy. Patented Manufactory, or Wholesale-Business is bestowed only upon those business undertakings that are of eminent importance for the development of national industry and apt to enliven commerce. The right of manufacturing an article includes also the right of trafficking with the like productions of others. Multiplication of the premises of a trade in separate places within the community where the business is established, as well as the erection of branch-establishments and warehouses without the community, is, generally, considered as a new establishment of some trade.

Removal of business establishments.

The removal of a business establishment to another place is only allowed, if the trade in question is not, concerning its exercise, assigned to a fixed locality, there being required no renewed proof of technical qualification, necessary to be given in other cases.

Commission.

Any tradesman may give in commission, or deliver upon orders, the articles of his trade, beyond the range of the community where his business is located, to any trader who is allowed to deal in such commodities; and he may anywhere execute the work for which he has received orders. Artisans domiciled in a foreign country may, in case of reciprocity, upon orders given, execute within the empire such works of industry as require no license. The importation of works produced abroad and the delivery of them to the committers is allowed on observance of the custom-regulations. The proprietor of an industrial establishment may also exercise his trade through an agent or let it on lease. The agent or lessee, however, must always be possessed of the qualifications necessary for the absolute exercise f that trade, and is to be notified to the authorities in the case of a licensed trade. The new Austrian Trade Legislation does not particularize the factory-system as a defined notion. There exist only a few regulations significative of a factory-system for industrial undertakings on an extensive scale, where usually more than 20 people, without distinction of sex and age, are working in common in the same work-shops. Such establishments are to keep a register of all the work-people, to have in the work-shop a regulation of service containing the rights and duties of the labourers posted up, and if, considering their great number or the nature of their occupation, there should be an apparent necessity of specially providing for them in case of accident or disease, the factory people ought either to lay up insurance funds for their relief through periodical contributions of their own, or they ought to join some other insurance company.

Employment of Children.

Children under ten years of age are not allowed to be employed at all; and, upon a license made out by the parish magistrate being produced, children over ten years but under twelve, are permitted to be employed in large industrial establishments for such work only as does not stint their physical development. Another consideration to be taken in making out such a license is, that a child's attending an elementary school be compatible with his or her employment in the industrial establishment, or that the proprietor of the factory have, consistently with the regulations of the authority for public instruction, sufficiently provided for the instruction of the children by having established a special school.

Working-time of the Children.

For individuals under 14 years of age, the time of work must not exceed 10 hours a day; for such as are 14 past and under 16, it must not exceed 12 hours a day; sufficient time of repose to intervene with adequate distribution.

Night-work.

Individuals under 16 years of age must not be employed for night-work after 9 o'clock p. m. nor before 5 o'clock a. m. But for factories where work is continued night and day, or where the management of the business would otherwise be obstructed, the magistrate may allow of hands under 16, but not under 14, years of age being employed for night-work; the magistrate may likewise, in cases of extraordinary exigency, consent to a temporary prolongation of working-time by 2 hours a day for hands under 16, for the space, however, of 4 weeks at the utmost.

Fellowships.

Among those that exercise the same trade, or trades of a similar nature, in one and the same municipality, or in several communities jointly, there is to be kept up a common union, a Fellowship; and wherever it does not yet exist, such a one ought to be organized as complete as possible. This is, however, not brought to bear upon Lombardo-Venetia, nor are such fellowships in existence in several other provinces.

Fairs.

Everybody is entitled to visit the Austrian fairs with commodities licensed for trade and assigned to the special nature of such fair. With regard to the visiting of fairs, foreigners are, upon practice of reciprocity, treated like inland traders. All wares admitted in open commercial transactions, are also admitted at fairs inasmuch as the respective market-privileges are not expressly restrictive to a special kind of articles. Market-dues are only to be paid in compensation for the granted space, the use of stalls and furniture, and other expenditures connected with the holding of fairs.

Patents.

In order to animate progress in industry, discoveries, inventions and improvements, if of recognized novelty, are distinguished by grants of Patents for a term not exceeding 15 years, upon payment of a tax of from 21 to 735 florins, according to the duration of the patent. For patent articles to be introduced from abroad into the Austrian State, an exclusive patent can be granted only, if the same are still patented abroad, and such a grant can be bestowed only upon the proprietor, or lawful claimant, of the foreign patent.

Protection of patterns and models.

In like manner, the exclusive right of a person availing himself of some Pattern or Model may be acquired, but only for the term of three years upon payment of a tax of 10 florins. The warrantee of this protection must, however, within a twelvemonth apply the pattern or model to some industrial production at home, and bring the latter into trade; and in the event of his importing into the Austrian dominions any articles manufactured abroad after his pattern or model, he shall forfeit his right.

A tradesman, in order to distinguish his productions and wares intended for trade from those of other tradesmen, may mark his own articles with his name, firm,

ensigns armorial, or denomination of his establishment, being unconditionally protected against any abuse of the same. Moreover, if he wishes to make use of symbols, ciphers or other devices, he may do so upon payment of a fee of 5 fl. and, for the term of his exercising his trade, acquire the right of exclusively using such devices. This privilege of signs is adherent to the undertaker of a trade, and, being serviceable to his successor, the latter may avail himself of it, if he gets the mark in question transferred to his name within three months.

Boards of Commerce and Trade.

After the example of Lombardo-Venetia, there are, since 1850, in all parts of the empire Boards of commerce and trade. They emanate from the commercial and trading classes by election, and are official business organs, through which the aforesaid classes present their desires to the Government, and are enabled to second the endeavours of the latter in promoting industrial intercourse.

Exhibitions.

To this department belong also Industrial Exhibitions which are to take place every five years. The last exhibition was in 1845, the three, that ought to have taken place since, not having been attempted for reason of Austrian industries having had so extensive a share in the London, Paris, and Munich exhibitions.

Registration of Firms.

In the interest of the credit indispensable to merchants and tradesmen, there exist legal regulations concerning the right and duty of keeping books, the registration of firms, procurations and partnerships for joint trading. In some provinces of the empire, wives must also have those claims registred which according to their marriage-articles, they wish to make upon the funds of their husbands.

Establishments for Credit.

Amongst the number of establishments, where merchants and tradesmen may apply for necessary supplies and relief, are reckoned: — the Austrian National Bank Patented, the Credit Institution for Commerce and Trade Patented, the Lower Austrian Discount Institution, the Commercial Banks of Trieste and Pesth, and a considerable number of Savings-Banks.

Exchange.

For the furtherance of commercial intercourse, there exist the general Exchanges at Trieste and Venice, the Stock-Exchange at Vienna, and the Merchandise Exchanges at Vienna and Prague. For

E

transactions on Change and elsewhere there are exclusively authorized brokers and agents.

Means of communication.

Great attention has been paid to the facilitating of transmission of goods by means of high-ways, by preserving rivers navigable, and by a system of railways extending over the whole empire. Railways are at present, almost without exception, private undertakings having a certain rate of interest warranted. The construction of a rail-road destined for public transport depends on a special concession usually given only for a term of 90 years, at the expiration of which the property of the railway, grounds and buildings, devolve to the state immediately and without compensation. The latter revises the tariffs every three years, and, if the clear profits of the railway exceed 15%, the rate of transport is lowered. Patents for steam-navigation are no longer granted.

Navigation.

Austrian Sea-navigation is divided into a short and a long line of coasting-navigation (cabotage), and into an open-sea-navigation. The short line of cabotage extends to the whole Adriatic gulf. The long line of cabotage extends to the promontory of Otranto, and on the eastern side to the Ionian islands and the port and canal of Zante. Any farther course is reckoned open-sea-navigation. These courses require a special permission, called a pass-port for shipping. Shipowners or freighters desirous of obtaining a warrant of navigation, as well as Masters or Captains must be Austrian subjects. The vessel must have been built or rebuilt in Austrian docks or yards, or, if built elsewhere, she must be proved, by deeds and vouchers, to be lawful and exclusive property of an Austrian subject, in all cases, however, at least two-thirds of her crew must be of Austrian nationality.

Austrian Mercantile Marine in the year 1860.

	Number	Tonnage	Manning
Fishing-smacks, numbered craft and lighters . . .	6,479	23,091	16,908
Small coasting vessels . .	2,345	44,304	7,656
Large „	349	34,487	2,093
Ships for long voyages . .	571	218,752	6,359
Steam-boats	59	21,338	1,701
Total .	9,803	341,972	34,717

In all the sea-ports of Austria (114), there have in 1860:

	Ships	Tonnage	Value of import
			florins
sailed into harbour	89,512	3,410,937	185,500,000
of these fall to the share of the Austrian flag	84,601	2,912,283	112,500,000
			Value of export
sailed out of harbour	91,623	3,436,965	139,500,000
of these fall to the share of the Austrian flag	86,528	2,913,545	105,500,000

Assurance on Ships.

In order to heighten the credit of Austrian assurance on ships, there were prescribed for Assurance Companies engaged in this business funds of at least 157,500 fl., and it is settled that, in case the funds of the company should be diminished by a third and the reserved funds also exhausted, affairs are to be liquidated and wound up, unless the share-holders pay in new capital.

Consuls.

Numerous navigation treaties, and consuls appointed to the most important places abroad whither Austrian vessels sail, provide for the interests of Austrians in their commercial intercourse and for the maintenance of their respective relations with foreign countries.

Banks.

The following five Banking and Credit institutions exist in Austria:

	Joint-Stock	Stock issued to the amount of	Rest-Capital
	Florins Austrian Currency		
Austrian National Bank	109,384,590	109,384,590	10,871,961
Lower Austrian Discount Company . . .	21,000,000	7,000,000	95,862
Credit-Institution for Commerce and Trade .	105,000,000	60,000,000	822,513
Commercial Bank at Triest . .	5,000,000	2,500,000	. . . . . .
Commercial Bank of Hungary .	2,100,000	2,100,000	. . . . .
Sum total .	242,484,590	180,984,590	11,790,335

The most influential establishment for the monetary transactions of the whole empire is the Austrian National Bank, uniting within itself the branches of banks of issue, discount, loan, and circulation. The National Bank has, indeed, the character of a patented private establishment, but its foundation emanated from the government, who had in view, by the mediation of the National Bank, to call in the then dishonoured notes issued by the State („Payable and Promissory Notes", or the then current „Vienna Value"), and to regulate monetary circulation by creating a new paper currency which, based upon ready funds of bullion, was to be at any time exchanged for specie and thus rendered safe from pernicious fluctuations of value. Also after the establishment of the Bank, the paper money of the „Vienna Value" was taken up by the National Bank, Bank-notes being issued instead, and although this exchange was suspended in 1817, it was again resumed in 1820 and continued till now, by which means the whole of the paper money of the „Vienna Value" was called in. The National Bank received in exchange Bonds of the National Debt, which are gradually sunk by the aid of cash raised by loans, and yearly instalments successively increasing through the addition of interests for the Bonds thus secured. In this manner was contracted the first debt of the State to the National Bank, which is as yet not quite sunk, but, since the plan for sinking it is strictly acted upon, it will be so soon.

In many other ways also did the National Bank aid the financial operations of the government, partly by intervention in public loans, and partly by discounting Central Cash Assignations which were issued payable three months after date.

But a closer connection of the state with the National Bank commenced with the year 1848. Although forced to suspend payment in specie, the National Bank, during the years 1848 to 1859, yielded the State any advances required. Already in 1849 endeavours were begun to unravel the perplexities caused by this compliance, conducing to reiterated agreements the completion of which was, however, again and again delayed by political circumstances, so that there is still a debt owing by the State to the National Bank amounting to 249,847.213 fl., of which 40,955,255 fl. are the remainder of the „Vienna Value" debt in course of sinking, 89,891,958 fl. secured upon part of the public estates, and 99,000,000 fl. covered with Tickets of the State Lottery Loan of 1860.

On the other hand, the activity of the National Bank, during the same period, enlarged and extended itself in all directions. What principally contributed to enliven its transactions was, that the Joint-Stock was doubled, a special division for advances on securities established, and numerous branches for discount and loan business established.

In 1861 there were:

	Austrian Currency
Checks issued for	85,160,118 fl.
Bills discounted for	321,759,678 „
Advances granted to the amount of	218,747,600 „
Clearings transacted for	47,805,757 „
Loans on securities granted up to	4,413,350 „

At the end of 1861 there were:

	Austrian Currency
Bank-notes circulating to the amount of	475,182,853 fl.
Stock of Bullion and Specie	99,148,381 „
Deposits amounting to	99,646,692 „
Mortgages circulating to the amount of	33,286,980 „
Net profits of stock-holders for 1861	10,982,471 „

The Lower-Austrian Discount Company, founded upon Most High Permission Nov. 16th 1853, first of all pursues the purpose of allowing the benefit of credit to those merchants and tradesmen of Lower-Austria who cannot procure credit through bills accepted by the Bank, but yet can give sufficient security. Their joint-stock amounts to 7 million florins, their transactions being, however, greatly enlarged by the National Bank granting the Company a credit, reaching double the amount of their joint-stock, at the rate of interest regulated by the Bank.

The results of 1861 are as under:

	Austrian Currency
Total of bills of exchange	40,240,734 fl.
Other outstanding debts	1,520,003 „
Funds for security } of credited traders {	2,261,078 „
Rest capital } {	107,254 „
Net profits for 1861	389,040 „

The business of the last-named Company is restricted to minor bill transactions for the metropolis and the province of Lower-Austria. But to extend the same over the whole empire, and especially to industrial undertakings on a grand scale, that is the task of an establishment like the Credit Institution for Commerce and Trade. The finishing of the Austrian railway system, in favour of which it has since 1858 carried on an operation of a great lottery-loan, has been one of its chief transactions.

On December 31st 1861, the Credit Institution possessed:

	Florins Aust. Curr.
Stocks and other obligations	6,978.297
Railway-Shares	15,800.228
Priority Obligations	177,579
„ Tickets	8,690,323

Net profits of the Shareholders amounted in 1861 to 2,302,209 florins.

d *

Austrian Tariff of Duties.

Already in the year 1851, Austria abandoned the prohibitory system and took to the Protective System of Duties. According to the latter there are imposed low duties on products, especially agricultural products, imported from abroad; but higher import customs levied for manufactures and works of art.

The export duties are insignificant, only some raw products have a high duty laid on. Transit duties are but control-duties. After the Hungarian and Croatian intermediate-custom-line having been abolished in 1850, Austria has been divided into two ranges of customs, one of which embraces Dalmatia, the other (the general Austrian range of customs) embraces the rest of the Crown-dependencies, except some custom-exemptions such as the free ports: Triest, Venice, Fiume, Zengg, Carlopago, Portorè, Buccari, part of the town of Brody in Galicia, and the community of Jungholz in Tyrol.

After the Tariff and Commercial Treaty with Prussia and the Tariff Union States of the German Confederacy had, on the 19th of February 1853, been concluded for the term of 12 years, a new general Austrian Tariff of Duties was issued for the general Austrian range of customs.

This Tariff started into life on January 1st 1854, and contains under 22 classes the articles subject to duties.

According to the diversity of the merchandise, the Ratio for the Tariff is adapted to the weight, measurement or numbers. There is no ad valorem duty in Austria.

By weight (Zollcentner-Tariff Hundred-weight) is to be understood, for import, partly the gross weight, partly the neat weight; but for export and transit always the gross weight.

It is necessitated by special political and national economical circumstances, that free import and transit are interdicted, being permitted only on particular conditions for: culinary salt, gun-powder, tobacco and its manufactures; the last being only imported duty-free for the exchequer.

The following are wholly exempt from duties: means of transport, packing-cases or casks, merchandise weighing under $5/10.000$ Cwt (less than $5/100$ pound), or such articles as are not rated higher than $1\frac{3}{4}$ creuzer Austr. Currency; travelling effects, emigrants' property, newspapers, etc. etc.

Free export is the rule. Exceptionally only, some articles are subject to an export duty, viz. leather-waste, horn and bones (75 creuzers per Cwt, neat), gall-nuts, wood for fuel (42 cr. per 100 cubic feet), some sorts of timber, sulphur, tartar, hair, rags (4 fl. 20 cr.), cocoons (13 fl. 12 cr. per Cwt, gross).

Through special decisions concerning articles passing the line of customs, conditional exemptions from duty were introduced, viz. for corn sent or received for grinding, for articles sent or received to be dressed, refined, reformed, etc., or imported for uncertain sale.

In transit there exist for some distances exemptions from transit duty.

Besides the tariff duty, there are still some office fees to be paid, for weighing, cocket, and seal.

The rates of duties contained in the tariff are not the same for all the entrance stations; for many articles, when imported from the Tariff Union States, are quite duty-free or enjoy some abatement, and others are favoured with privileged duties in several ranges of the custom-line, such as in stations towards Bosnia, Italy, Switzerland and the sea-coast.

Some instances taken from the tariff will best show these differences and the rate of duties:

			Import-duty	
Wheat per Cwt, gross	. . . .	35	cr.	
„ from Modena per Cwt, gross	. . . .	21	„	
„ „ Switzerland per Cwt, gross	. . .	17½	„	
„ „ Bosnia „ „ „	. .	11⅔	„	
„ „ the Tariff Union States exempt from duty.				

		Import-duty		
Wine: in bottles per Cwt, neat	. . .	13 fl.	15	cr.
„ from Modena „ „ „	. . .	7 „	35	„
„ in pipes or leather-bags, per Cwt,				
gross	. . . .	10 „	50	„
„ coming from free towns	. . . —	„	95	„
Piedmontese Wines	. . . .	1 „	22½	„
Servian Wines	. . . . . .	6 „	30	„

		Export-duty	
Staves (deal) per 100 cubic feet:			
exported by water	. . . .	40 cr.	
„ „ land	. . . .	8	„

		Import-duty		
Oxen and Bulls, per head:				
in general	. . . . .	4 fl.	20	cr.
from the Tariff Union States	3	„	75	„
from Modena	. . . .	2	„	20	„
from Bosnia	. . . .	1	„	40	„
Iron, per Cwt, gross:				
pig-iron	. . . .		42	cr.
by sea, and from Italy	. . .		63	„
from the Tariff Union States	.		37½	„
refined iron, per Cwt, neat	.	2 fl.	10	„
by sea, and from Italy	. . .	2 „	63	„
from the Tariff Union States	1	„	—	„
Cotton Manufactures, per Cwt, neat:				
coarse	. . . . .	42 fl.	—	cr.
middle fine	. . . .	78 „	75	„
from the Tariff Union				
States	. . . .	45	„	—	„

Transit-duty, 26 cr., or 10½ cr. per custom-unit.

Among the many modifications of tariff-rates introduced within the last few years, the manifold concessions for facilitating commercial transactions with regard to duties, there is specially to be mentioned that in 1856 there was introduced by way of experiment (and since January 15th 1862 definitively settled) a 3 months' or 6 months' Credit, without Interest, of due amounts of import-dues, granted to great commercial houses on certain conditions, to a maximum amount of 60,000 fl., against a security of the amount of duty, by giving as security either stocks or other bond-debts of the state, or bonds of joint and separate obligation of at least two well-accredited firms.

Another grant, for encouraging the inland sugar manufacture, consists in there being given a drawback for debentured sugar, without distinction, viz. 3½ fl. for raw sugar, and 4 fl. 30 cr. for refined sugar, per Cwt neat, and also the extraordinary (war) tax of 20% in consideration of sugar manufactured from home-products. — Powdered sugar for trade pays 9 fl. 45 cr. import duty; powdered sugar for refiners pays 6 fl. 30 cr.; and refined sugar 13 fl. 15 cr. per Cwt neat.

For the Dalmatian range of customs, the Dalmatian Tariff has been introduced since May 1st 1857; suitable to the geographical position and conducive to the internal prosperity of this province, there have been fixed much lower duties, and the import of many articles subject to duties in the general range of customs as well as exports are here declared exempt.

For the intercourse of the Dalmatian range of customs with the general Austrian range, there are instituted special reliefs.

The amounts of duties are fixed in the imperial standard and are now to be paid in silver.

The official management attending duties is referred to 515 custom-house offices (chief and secondary offices) at the frontiers as well as in the interior of the empire.

In 1861, the total revenue of duties (additional fees included) amounted, in both ranges of customs, to the sum of about 13 millions and a half florins.

The total value of commodities imported into the general Austrian range of customs in 1861 amounted to about 232 millions florins, but that of exported goods 310 millions florins; consequently the value of the export, in comparison to that of the import, shows a plus of 78 millions.

In the Dalmatian range of customs the value of imports amounted to about 8 millions florins; that of the exports to about 4 millions and a half; whereby a much more unfavourable balance is resulting than that of the general Austrian range of customs.

Freightage and Fares.
Railways.

The Austrian railway plants have, since the year 1827, from year to year increased in extent. From 0·50 of an Austr. mile in 1827, 34·19 Austr. miles in 1836, 141·14 m. in 1845, they have increased, in 1861, to 756·00 m., distributed in 44 lines and carried on by 13 Companies. 700 miles more are partly building, partly in projection, so that the Austrian railway plant will in a few years embrace 1400 miles.

The Emperor Ferdinand Northern Railway and the Charles Lewis Railway connect Poland, Prussia, Silesia, and Russia with the centre of the empire; the State Railway brings about the communication between Northern Germany and Bosnia and Turkey; the Southern Lombardo-Venetian and Central Italian Railway joins Vienna with the Adriatic Sea and Italy; finally, the Empress Elizabeth Railway links Vienna with Southern Germany and France; whilst the rest of the railways accomplish the intercommunication within and without the country. With the exception of Dalmatia and Transylvania, nearly all provinces are in connection with the metropolis by means of railroads.

The fares for persons are on an average, per Austrian mile, in the 1st class 45 creuzers, in the 2d class 34 cr., in the 3d class 22 cr. Luggage 25 pounds free, for every 20 pounds more 2 cr. per mile. For express trains and military transports (the latter more than one-third lower in price), or certain distances and places, the prices increase or decrease.

In 1860, there were 12,223,968 persons conveyed on the whole of the Austrian railways.

The freightage for goods is on an average, per Austrian mile and Tariff Cwt, for the 1st class of goods (raw products) $2^6/_{10}$ creuzers, for the 2d class of goods (manufactures) $3^3/_{10}$ cr., for the 3d class of goods (voluminous objects) 5 cr.

Assurance and other fees are extra and vary in rate, being mostly determined either per Cwt, without distinction of distance, or according to the declared value.

For speedy conveyance of some goods, for separate trains, or for more extensive transports, there are various modifications.

For instance there is an average of freightage both per head and per Austr. mile. as under:

Live cattle, common freight	15 cr.
„ „ speedy conveyance	24 „
Horses	75 „
Dogs	6 „
Hogs	4 „
When the transport is extensive:	
Wood for fuel per Cwt and mile	1 cr.
Coals	$^9/_{10}$ „
Corn	2 „

(In 1860, the whole of the Austrian railroads carried 142,147,128 Tariff Cwts.)

Those 13 railway Companies mentioned above had in 1860:

Income about 66,000.000 fl.
Expenses   „   33,000.000 „

Postage.

Where there are no railroads, the communication is mediated by the „diligence“ and mail-post, extra-post, omnibus, and freight conveyances which are established by the government. In the diligence, a person pays 56 cr. per Austr. mile, 32 pounds luggage free. Conveyance fees for each 100 fl. and each pound of luggage amount to 2 cr. per 5 miles. The terms for extra-post are on an average, fees excepted, 1 fl. 60 cr. per stage (= 2 German miles) and one horse.

Letter-postage.

The postage for a simple letter (weighing half an ounce) is settled at 3 cr. within the distance of the post-office district; at 5 cr. to the distance of 10 miles; 10 cr. to the distance of 20 miles; and 15 cr. beyond this distance. Papers sent open in paperbands need a stamp of 2 cr. per half an ounce; periodicals a 1 cr. stamp.

The following table exhibits the increase of Post office transactions:

Posting Transactions.

a) Official Intercourse.	Correspondences	Packages. Pounds	Money. Florins
Year 1830 .	3,058,375	454,038	60,244,041
„ 1845 .	6,527,689	2,235,422	114,942,895
„ 1860 .	26,314,314	6,261,016	1,607,283,011
b) Private Intercourse.			
Year 1830 .	15,536,801	2,881,468	120,829,424
„ 1845 .	23,698,750	2,950,682	187,131,744
„ 1860 .	79,267,550	8,117,051	1,845,613,835

A comparison of the following numbers will show, that, in spite of the rapid growth of railway-plants and increase of steam-boat lines, travellers with the post, as well as intercourse in general, are continually augmenting:

Passengers conveyed by post.

1830 53,615
1845 216,796
1860 206,592

Telegraph Returns.

	Miles of communication	State Despatches	Private Despatches
Year 1850 .	487·2	10,004	3,045
„ 1860 .	1,661·9	115,249	471,216

River Steam-Navigation.

The Danube Steam-Navigation Conpany run their steamers the Danube and its tributaries; besides there are some smaller and more recent undertakings of this kind, such as, the Bavarian Danube Steam-Navigation Company, and that of the river Inn, added to which are several private boating concerns. Next follow steam boats plying on the inland lakes.

The fares per person and per Austrian mile, are rated by the Danube Steam-Navigation Company, as under:

1st class down the stream 22   cr., up the stream 18 cr.
2d   „   „   „   „   14½ „ „   „   „   11 „
3d   „   „   „   „   11   „ „   „   „   8 „

Freightage per Cwt and per mile:
General tariff up the stream . . 2   cr.
„   „   down the stream . 1⅛ „
Moderate „   up the stream . . 1¹₂ „
„   „   down the stream . 1¹⁄₁₀ „

In 1860, the number of travellers amounted to 1,152,531, that of shipped goods to 37,106,641 Cwt.

Steam Navigation on the Sea.

The communication of Austria with the Levant and the Danubian principalities, and of Trieste with Venice, Croatia and Dalmatia is mediated by the Austrian Lloyds at Trieste.

Passengers prices average per person and league:
1st class . .   1 fl. 35 cr.
2d   „   . . — „ 95 „
3d   „   . . — „ 59 „
Freightage per Cwt and league:
1st class of goods . . . . 1⁹⁄₂₀ cr.
2d   „   „   „   . . 2⁵⁄₁₀ „
3d   „   „   „   . . 3²⁄₁₀ „
4th   „   „   „   . . 3⁸⁄₁₀ „

(In 1860, the number of passengers amounted to 345,362, that of shipped goods to 1,615,387 Ctw.)

Assurance Offices.

The Empire numbers 120 Assurance Companies, based partly upon the mutual, partly upon the proprietary principles, assuring all sorts of risks.

For sea risks of all descriptions, there exist in the Littorale, in Dalmatia and Venice alone 28 Unions. 27 Unions undertake assurance for damage in transports within certain local districts. 8 Unions undertake to assure the lives of cattle within certain local districts.

Particularly worthy of note are the 42 private fire-assurance unions existing in Austria-proper, as well as many peasants' fire assurance associations based upon the mutual principle, which, not included in the afore-quoted number, exist in Upper-Austria and Tyrol.

The most important corporations of this nature, extending their benevolent activity, with the help of general agencies and agents, to all parts of the monarchy are:

a) the proprietary „Azienda assicuratrice" in Trieste, taking fire and storm risks. Assured capital exceeding 1,100 millions florins, yearly damages indemnified amounting to more than 1 million florins.

b) The proprietary „Riunione adriatica di sicurtà" in Trieste. Assured capital about 520 millions florins, yearly damages indemnified amounting to more than 1½ millions florins.

c) The mutual Fire Assurance Institution in Vienna. Assured capital exceeding 70 millions florins; 60,000 assurers; 120,000 assured buildings.

d) The (proprietary) First Austrian Assurance Company in Vienna. Assured capital 335 millions florins; the yearly damages indemnified exceeding the sum of 1 million florins.

Besides these general assurance establishments among which the newly founded „Austrian Phoenix" must also be mentioned, there exists almost in every province a provincial assurance institution, the most prominent of which are: — the recently established Hungarian Assurance Company, and those of elder standing in Linz, Salzburg, Prague, etc.

Sum of the Assurance Capital (1861):

	Florins Austr. Curr.
a) Of the first-rate mutual assurance companies (17)	297,500,000
(Accumulation 1,600,000 fl., Assurances paid 1,400.000 fl.)	
b) Of the first-rate proprietary assurance companies (27)	3,064,000,000
Total of the assurance capital . . .	3,361,500,000

(Receipt of premiums 19,000,000 fl. Assurances paid 11,000.000 fl. Capital in shares [nominal] 27,000,000 fl.)

General Benefit and Annuity Societies.

Apart from the many institutions and unions for the benefit and provision of their members as well as their widows and orphans, there are also several general benefit and annuity societies.

But, notwithstanding some institutions of this kind existing since some time, the development of assurance transactions, annuities, life policies, reversions, etc. is owing to such first-rate institutions as the Vindobona, the Anchor, the Phoenix, and the Austria etc.

Savings-Banks.

Savings-banks are, with respect to erection and statutes, dependent on a concession from, and on regulations settled by, the Government. In the 143 savings-banks that are now established, there are at present lodged capitals amounting to 150 millions florins, of which by far the greater balance (about 90 millions florins) is in favour of credit on real property and is employed on loans against mortgages. The lowest deposition taken in these savings-banks is, with few exceptions, the sum of 26 creuzers, the maximum deposition is not everywhere fixed. The rate of interest is on an average 5%. The regulations for return-payments are settled so that especially persons in less easy circumstances are, in cases of need, accommodated, if not with their whole deposits, at least with the greater part of them, without previous notice.

Regarding the participation of the population in these savings-banks, there is to be ascertained that in 94 inhabitants of the whole empire there is one depositor, but in Vienna there is in 3 inhabitants one depositor.

The most extensive institution of this kind is the „First Austrian Savings-Bank" at Vienna (founded in 1819). It had, at the end of 1861, deposits of 39½ millions florins, and the property of its own amounted to 4 millions florins.

This institution pays 4½% interest for deposits.

Besides this, there are of note the Bohemian and Hungarian (5% rate of interest), and the Venetian Savings-Banks. The latter stand in immediate connection with the loan-institutes (Monte di Pietà or pawn-broking establishments) and cede their funds to them.

But the greatest influence on the material accommodation of the population at large may be said to be exercised by the smaller savings-banks established in provincial communities.

Total of capital invested in all savings-banks of Austria, Hungary included (1857) . . 133,500.000 fl.
Hungary excepted . . . . 108,000.000 „
Number of parties 627,000.

Intellectual-Culture.

Direct and indirect products of the soil constitute the physical, the moral disposition and mental parts of the inhabitants, the intellectual power of a State.

The development of the last mentioned power has been advancing in Austria with surer and quicker steps during the last 14 years than in many a period previous to this. Before all, due attention has been awarded to the culture of practical sciences and, in accordance with the spirit of this progress, both the primary and middle stages of instruction have been liberally enlarged.

The educational system of a State is of no less importance than its national-economical system, and as the latter is exhibited in its industrial progress, so is the former in the state of its Public Instruction.

Public Instruction.

Infant Nurseries.

Infant Nurseries are charitable institutions for poor children of the working classes, standing under the superintendence of school-authorities, without the state contributing to their support. They are established in all the large towns of the empire.

Primary Schools.

In 1859, there were in the empire of Austria (without Lombardy) 29,972 public lower Primary Schools, among which are included 824 higher Elementary-Schools and Civic-Schools. The number of salaried teachers and assistant-teachers appointed to these public schools amounted to 39,825, teachers of religion not included.

Classified with respect to religion, there were 23,217 Catholic Primary Schools, 6,367 belonged to other Christian Churches, and 388 were Jews' Schools. Of the teachers, 31,966 were catholics.

There were, 3,900,900 boys and girls obliged by the law to attend daily-schools; of these, 2,723,400 had frequented the public primary schools; the total number of the population having been 34,439,069.

Primary Schools are either public or private. Conformably to the extension and purport of instruction, there are three kinds of public primary schools: the Lower Elementary-School with 3 classes; the Higher Elementary-School with 4 classes; and the Civic School with 6 and 7 classes. The Lower Elementary School teaches Religion, Vernacular Language (i. e. Reading, Grammar, Orthography, Composition, etc.), Writing, Arithmetic, Singing; for girls also Needle-work etc., Natural History, Geography etc., is imparted to children in the course of their readings in books adapted to that purpose. Consistent with local circumstances, particular branches of practical husbandry (such as: Horticulture, Culture of Fruit-trees, Management of Bees, Rearing of Silk-worms, etc.) are taught in many schools as additional objects. The laws of the state have made school-attendance in this elementary stage obligatory on every one, and public Lower Elementary Schools must be erected at every place where there is a want them.

Higher Elementary Schools.

The Higher Elementary School includes the Lower, enlarges the circle of information, affords instruction in a second, occasionally in a third, language of the empire, and teaches drawing. It is, by statute, a preparatory school for the Middle-Schools

(i. e. Gymnasium or Realistic School), and usually exists in towns of more than 3,000 inhabitants.

Civic Schools.

The Civic School comprises the higher elementary school besides two and frequently three classes for such realistic knowledge as is indispensable for minor town or rural industry. They commonly exist in manufacturing and commercial towns of more than 6,000—7,000 inhabitants, without there being any coercion for the erection of such schools.

Elementary schools of a special character are: Trade Schools established in many factories; Institutes and Schools for the Deaf and Dumb, and for the Blind; and numbers of industrial schools for females.

Language used in Instruction.

The law prescribes that the language used as medium of tuition must, in all elementary schools, be the mother-language of the children; where the language of the population is mixed, one and the same elementary school must suffice, in case there should be no possibility of establishing separate schools for each language. Hence follows that agreeably to the diversities of languages and idioms in Austria there are: German, Bohemian, Hungarian, Italian, Polish, Slovene, Illyrian, Servian, Croatian, Ruthene, East-Roman, Bohemian-German, Italian-German, Polish-German, Illyrian-German, Slavonic-German, Illyrian-Italian, Hungarian-Roman, Hungarian-Ruthene — Schools, besides others with different combinations.

The sexes of the children offer reasons for another division of the elementary schools; according to the sex these schools are: Boys' Schools, Girls' Schools, or Schools for both Boys and Girls. — Civic Schools are throughout Boys' Schools.

Every public school is either a Daily-school, or a Sunday-school in connection with the former. The time of instruction in daily-schools (lawful vacations excepted) is 4 or 5 hours a day, in sunday-schools from 2 to 4 hours. Conformably to the law that renders school-attendance compulsive, all children that do not receive necessary instruction at home or in private institutes, must attend a public daily-school, from their 6th to their 12th year of age; and if, on leaving the Elementary School, they do not frequent a higher school, they must furthermore attend a Sunday-school, till full 15 years of age.

School Books.

School-books (text-books and reading-books) are prescribed by the government; and it is only in the Civic Schools, that other than the prescribed books are admitted, when requested by the staff of the school. The sale of the prescribed schoolbooks is administered by the government: no gain being sought. Poor children receive the necessary books gratis.

For training teachers in the principles and practice of teaching in lower and higher elementary schools, there are Courses of Pedagogy in immediate connection with Model or Normal-Schools.

There are special, richly endowed state-institutes for the accomplished education of governesses destined for the higher ranks.

For training, instructing, and practising teachers for Civic Schools, there exist, according to necessity, biennial Courses in connection with Higher Realistic Schools.

For the accomplishment and improvement of teachers already employed, numerous teachers' libraries and school-libraries have been provided, and conferences of the several educational staffs are to be held regularly by order of law.

The whole of the teachers of elementary schools are appointed by the government either directly or indirectly, or under its supervision. They can be dismissed only with the consent of the government; a retiring salary and pension for their widows or orphans is allowed by law.

The charges for maintaining elementary schools are borne jointly by the public exchequer and the municipalities. Of the latter, many receive aids by school-patrons and by means of fees which such children as are able to pay must contribute, amounting from one to six (occasionally eight) florins per annum, rated suitably to the circumstances of the school and its local situation; poor children, however, receive instruction gratis.

Inspection.

Direct superintendence and guidance of every lower elementary school is confided to the ecclesiastical minister and to an inspector selected from the members of the municipality; the presbytery surveys and directs Protestant Schools. For the purpose of higher inspection, all elementary schools are divided into districts, where an ecclesiastical dean surveys the instruction, and the imperial district-office, the external order and economical administration, of these schools; next in succession come diocesan school-authorities and vicegerenqies, having school-counsellors as professional officials adjoined; the Ministry and the Court-Offices of Hungary, Croatia-Slavonia and Transylvania constitute the Central-Direction of the complete organization of Elementary Schools.

Next in order follow the **Middle Schools.**

Gymnasia.

The plan of organization of 1849 united the former gymnasia of 6 classes with the then existing philosophical colleges into Gymnasia of 8 classes, extended the range of studies in Greek and German, distributed to each of the 8 annual classes the disciplines of History, Mathematics, Natural Philosophy, etc., assigned vital importance to instruction in Vernacular Languages, referred the study of Philosophy to universities, and called substantial improvements of method into existence.

In 1859, there were in the whole empire 240 gymnasia (164 with 8 classes, and 73 with 4 classes), with 2,454 teachers in ordinary, and 493 extraordinary teachers; these schools were frequented by 51,121 pupils.

A Complete Gymnasium requires 12, a Lower Gymnasium 6 teachers in ordinary; one of them being at the same time Director. For an appointment to the place of a teacher the candidate is to undergo the examination prescribed for the competency of a public teacher at a Gymnasium. The immediate superintendence of Gymnasia is entrusted to a provincial School-counsellor.

The choice of the language used as medium of instruction depends on those who are to maintain the gymnasium and who, consequently, have the right to appoint the teachers.

Regular, prescribed objects of instruction are: Religion, Latin, German, a second Language that may happen to be prevalent in the province in question, History and Geography, Greek, Natural History, Physics, Philosophical Propaedeutics. Instruction in other living languages, in calligraphy, drawing, singing, stenography, and gymnastics is optional.

The number of hours for lessons in prescribed objects averages from 22 to 27, weekly, for each class. Vacations last 6 or 8 weeks (August and September).

The number of pupils in one class must not exceed 50.

Regular annual examinations do not take place; ascending to a higher class being allowed or refused, according to the annual maximum credits obtained by the pupil. However, an examination of maturity is required of such scholars as are going to leave the gymnasium for their matriculation at a university.

Text-Books are not prescribed, all good ones deemed admissible being approved of by the Ministry.

The amount of annual fees alternates between 8·4 fl. and 12·6 fl. The sum total of fees is conveyed to that cash whence the gymnasium derives its maintenance. In 1859, the fees of all the Austrian gymnasia amounted to 274,372 fl. Special reliefs to scholars in gymnasia are afforded by numerous Exhibitions founded by the State, by Corporations and private persons, besides also also by exemptions from payment of fees. In 1859, 2,834 pupils enjoyed Exhibitions amounting to the total of 207,098 fl.

Educational and Scientific Collections, almost entirely wanting up to the year 1849, have since been founded and enriched to such a degree, that the Libraries of Gymnasia, their Cabinets of Philosophical Instruments, Apparatus, and Collections for Natural History, are mostly answering their purpose at present.

The average expenditure for a complete state gymnasium amounts to 12,000 fl., for a lower gymnasium of 4 classes, to 5,000 fl. The pay of teachers varies between 735 fl. and 1,050 fl.; that of directors, between 1,155 and 1,365 fl.; besides, after every 10 years' service, every director and teacher is allowed an increasing salary of 105 fl., and after 30 years' service he is, if disabled, entitled to a retiring salary of the full amount of his previous pay.

Realistic Schools.

A reform of the Realistic Schools was undertaken, based upon an Imperial Decree of March 2d, 1851, in consequence of which they were organized as Special Institutes under their own direction, and subdivided into:

Lower Realistic Schools with triennial courses, and Higher Realistic Schools with sexennial courses. The destination of these Schools is to afford their pupils, besides a liberal education of general nature, without the aid of the Classics, both a middle degree of preparation for trading and manufacturing industry, and a sufficient degree of preparation for studies at Technical Institutes or at a Polytechnic School.

Regular, prescribed objects of instruction in Lower Realistic Schools are: Religion, Vernacular Language, History, Geography, Arithmetic, Natural History, Physics, Chemistry, Drawing, Book-keeping, Rudiments of Architecture and Calligraphy. A second Austrian language is prescribed to those only, whose mother-tongue it may happen to be. Modern languages, Singing, Stenography, Gymnastics, are optional. — The prevailing method is a popular one and based upon Demonstration and Illustration (practical Intuition).

In the three higher classes (of Higher Realistic Schools) the knowledge acquired in the three lower classes is enlarged, instruction continued in rather a more scientific method, and, the general polite education within the scope of these schools attained, special preparation or technical studies is afforded. Here, Elementary Mathematics take the place of Arithmetic, Engineering is added as a prescribed object; Calligraphy and modern languages are optional.

The number of complete organized Realistic Schools, with the right of being public institutes, amounted in 1859 to 32; there were 520 teachers instructing 9,939 pupils.

The weekly number of lessons amounts to 30 or 34 for the prescribed objects. There are 12 teachers in ordinary for a Higher, 6 or 7 for a Lower Realistic School.

The number of pupils must not exceed 80 in one class. For Drawing, the teacher is allowed an Assistant, when there are more than 50 pupils.

Every candidate must prove his scientific competency for the vocation of a teacher before a Board of Examination; and his methodical capacity, by teaching, during a year of trial, at a public Realistic School. Text-books for the several objects are not prescribed, the teachers being at liberty to choose among such as are approved of by the Ministry.

The fees to be paid by the pupils are, agreeably to local circumstances, fixed from 8 to 20 fl. annually. The pay of teachers varies from 630 to 1,260 fl. with an increase of 210 fl. after 10 years' service.

The yearly expenditure for a Lower Realistic School averages from 8,000 to 11,000 fl., and that for a Higher Realistic School, from 15,000 to 20,000 fl.

The establishment of any Elementary or Middle School is dependent on the capacities of the applicant.

Universities.

The Austrian empire has 8 Universities: those of Vienna, Prague, Pesth, Cracow, Padua, Lemberg, Gratz, and Innsbruck.

There are four faculties in the first four of them, viz. the theological, jurisprudential-political, medical, and philosophical. The University of Padua has a fifth, viz. the mathematical faculty. At the three last-named, the medical faculty is wanting.

In 1848 and the following years, the Austrian Universities underwent a complete reform consistent with the principle of Liberty for Teaching and Learning. The Staff of Teachers at a University consist of the ordinary and extraordinary professors, habilitated private lecturers, and teachers of modern languages, besides teachers of gymnastics etc.

From the Staff of Teachers of each faculty emanates the College of Professors, as the immediate authority directing the respective division of studies. This College consists of all the ordinary professors, the senior extraordinary professors and two private lecturers. Their president is the Dean, annually elected from the number of ordinary professors.

From the College of Professors emanates annually the Academic Senate, as the chief authority of the university. The Senate consists of the Rector, prorector, dean, and pro-dean of the several Colleges of Professors. The rector is elected annually according to a settled alternation of faculty.

In Vienna nd in Prague each faculty is divided into two colleges, one of professors and one of incorporate doctors. At the head of the colleges of doctors is an elected dean having seat and vote in the respective college of professors and in the academic senate.

Professors have the rights of civil officers and are nominated by the Emperor. Their income consists in fixed stipends (partly increasing after 10 years' service, partly not), in fees from their students, and in fees for examinations. The minimum stipends, upon which a free competition for a vacant professorship is founded, differ at various universities, amounting to 1,050 fl., 1.260 fl., 1,365 fl. and 1,680 fl. with allowances of 210 fl. and 315 fl. after 10 years' service. The extraordinary professors have fixed stipends, from 600 to 1,200 fl.

He who wishes to deliver lectures at a university, in the quality of a private lecturer, is to undergo an examination of his scientific qualifications, in the respective college of professors. The institution of private lecturers is properly a kind of nursery of university professors.

A scholastic year begins on the first of October and ends on the last day of July, being divided into two terms.

A student is for each term separately registered for the several courses of lectures he frequents.

The students are at liberty to choose those Courses of lectures, and among the professors or lecturers those, that suit them best. They are also allowed to frequent a foreign university, where there is liberty of teaching and learning and the terms they pass there are computed as valid; moreover foreigners may be matriculated at Austrian universities.

The university time of studies settled by law is: for the theological, and jurisprudential-political faculties 4 years; for the medical faculty 5 years.

Students at universities have to pay no fees to the State, but only fees for the lectures they attend, the lawful minimum being, per term, as many florins as there are delivered lectures per week by the professor or teacher who is, however, when salaried by the state, bound to deliver one course of lectures gratis.

In regard to their civil relations to society at large, students are subject to the general laws, but in regard to their academic behaviour they are to comply with the special academic regulations and disciplinary rules. Penal justice is exercised by the academic senate.

The total number of students having attended the 8 Austrian universities in 1858/59, was 8,030.

The expenditure connected with the maintenance of the universities, inclusive of charges for university libraries and grants for scientific collections and institutions, but exclusive of the costs of Exhibitions for poor students, buildings, taxes, passive interests, — averages in: Vienna 355,000 fl., Prague 201,000 fl., Padua 140,000 fl., Pesth 150,000 fl., Cracow 102,000 fl., Lemberg 115,000 fl., Gratz 66,000 fl., Innsbruck 63,000 fl.

Theological Faculties.

The whole of the Austrian universities have Theological faculties. Besides these, there are computed 120 theological colleges in the empire.

In a considerable number of dioceses, there are Seminaries for boys.

For the degree of Doctor of Divinity, candidates must undergo four rigorous examinations.

The total expenditure for Catholic theological colleges and educational institutes amounts to 600,000 fl. annually.

The Protestant theological faculty at Vienna obtained, in 1861, the right of promoting to academic degrees.

The total number of the teaching staff at the theological institutes of all creeds amounts to above 600, that of students to 5,000 and odd.

Jurisprudential-Political Faculties.

For students at Law, devoting themselves to the public service or desirous of undergoing the examinations for the degree of doctor, the scientific branches and their order are prescribed.

Such as wish to enter the service of the state as clerks must undergo three theoretical state-examinations.

In order to attain the degree of doctor, a candidate must undergo 4 (in Pesth only 3) rigorous examinations before the professors of the faculty in question.

The degree of doctor dispenses with the theoretical state-examinations for obtaining an employment in the service of the State.

Medical Faculties.

In order to attain the degree of doctor, which is now still limited to medical science, and never extended to surgery separately, candidates must show proofs of 5 years' attendance at a university, 4 years of which are required for exclusive studies at the medical faculty, and further two years of clinical studies.

Two rigorous examinations, valid as state-examinations and entitling to the exercise of practice, are requisite for candidates to the degree of Doctor of Medicine. There exists besides a Doctor's degree for chirurgery, a degree of Master of obstetric art, of ophthalmic therapeutics, and of dentist-surgery.

Philosophical Faculties.

The task and tendency of philosophical faculties is to educate teachers for universities, and for such branches of teaching in schools as include classical knowledge; also to afford to students devoted to other professional pursuits the necessary basis of philosophical knowledge For the degree of doctor, three years' continual studies are required.

In this faculty those sciences are taught that are within the province of philosophical knowledge.

Lectures are partly theoretical, partly practical (interlocutory, experimental, exercising), for the behalf of which there exist seminaries, museums, laboratories, etc. The most deserving of notice are in Vienna: the philological-historical seminary; the seminary for Austrian historiology; the institute for physics; the cabinet of physics; the central institute for meteorology and terrestrial magnetism; the observatory; the chemical laboratory of the philosophical faculty; the natural-historical museum for zoology and mineralogy; the botanical garden.

Similar institutions exist also for the philosophical faculties at the other Austrian universities.

At the universities of Vienna and Prague, where the number of lecturers is greater and more students attend such lectures than at any other university, there are, in proportion, also more lectures delivered.

For both half-yearly terms of the philosophical faculty in Vienna there may be computed 200 separate courses of lectures, in Prague at least 150, and in like proportion at the other universities.

The Government spends, annually, about 45,000 fl. in extraordinary Exhibitions for students of the philosophical faculties, averaging from 300 to 500 fl. each, suitable to the wants and merits of the grantees. Besides, there are still systematized government exhibitions in various numbers and amounts.

Care has been taken that the most important branches of philosophical, philological, historical, literary and natural-philosophical sciences may be represented by salaried professors. But many sciences are taught by several professors simultaneously, each of whom is making some part or a special direction of the science the object of his instruction.

The total number of teachers in the philosophical faculty of the university at Vienna amounts to 60.

The Mathematical Faculty at Padua.

The fifth faculty of that university, has, in a triennial course, the task of educating E n g i n e e r s and other technologists for their vocation. There is also an annual course for land-surveyors.

Higher Special Institutes.
Law-Academies.

For the education of a proportionate reinforcement for the service of the State in Hungary, Croatia, and Transylvania, there exist, besides the jurisprudential-political faculty at the university of Pesth, five law-academies for these provinces, those of Pressburg, Kaschau, Grosswardein, Agram and Hermannstadt, where the course of studies lasts three years.

Surgical Institutes.

There are only 6 special institutes for surgery, those of Gratz, Innsbruck, Klausenburg, Lemberg, Olmütz and Salzburg, where the courses are triennial.

The total expenditure for these 6 surgical institutes is computed at 47,363 fl. inclusive of grants for collections and instruments

A right of practice is to be attained by undergoing two examinations.

Technical Institutes.

Such institutes exist in Vienna, Prague, Gratz, Lemberg, Cracow, Brünn, and Buda. They receive as regular pupils such persons as have finished their studies with good success in a higher gymnasium, or higher realistic school. The choice of objects taught in these institutes is free to any one that is sufficiently prepared.

The number of teachers at technical Institutes amounts to 170, that of students, to 2,800. The latter may submit themselves to an examination in those sciences they have improved themselves in, and receive certificates of their success.

There are annual fees to be paid at these institutes, varying from 10 to 24 fl. The expenditure for the Polytechnic Institute of Vienna, to which is joined a work-shop for mechanical-astronomical instruments, amounts to 132,000 fl. annually. The remaining technical institutes require, in proportion to the degree of completeness, a budget of 20,000 to 50,000 fl. per annum.

Institutes for special branches of science, or industrial pursuits.
Astronomy.

Theoretical instruction in astronomy is given at universities, practical one at the observatories in connection with universities. Observatories maintained by the State are in Vienna, Prague, Cracow, and Padua. Of the number of private observatories, that of Kremsmünster holds a high rank.

Expenditure inclusive of stipends: 17,000 fl. annually.

Central Institute for Meteorology and Terrestrial Magnetism.

This institute is in Vienna and forms the central point of all stations of observation, which are spread over the different parts of the empire and are a few hundred in number — yearly expenditure 10,000 fl.

Commercial and Nautical Academy at Trieste.

This institute is divided in two sections, one for Commerce, another for Navigation. The commercial section consists of a preparatory course, and two annual courses for commercial knowledge. — The nautical section has two annual courses, destined for the accomplishment of officers of the mercantile marine.

Besides, there is for officers a half yearly popular course of navigation; and another on cabotage (coasting trade) for mariners in general. The total expenditure for this Academy is 30,000 fl.

For instruction in Ship-building there exists in this section a special annual course for such as wish to improve themselves in the theory of ship-building.

Nautical Schools.

Besides the above-named nautical section, there are 5 Nautical Schools more, viz. in Venice, Fiume, Lussin-piccolo, Spalato, and Cattaro; the expenditure for these averaging 8,000 fl. At the nautical schools of Fiume and Venice exist also courses for ship-building.

Commercial Academies.

There are Commercial Academies in Vienna, Prague, and Pesth, which are much frequented. Moreover, the Commercial Unions in all populous trading towns of the empire maintain numerous lower commercial schools for apprentices and shopmen.

Trade-Schools.

Trade-Schools for apprentices and journeymen were instituted in many towns at the expense of Trade Unions. Particularly distinguished among these are the Trade-schools in Vienna and Brünn.

In Brünn, private persons founded also a higher Weaving-School.

Agricultural Schools.

1. The Higher Institute for Husbandry at Ung. Altenburg for the improvement of landowners, farmers, stewards, or administrators of large estates, as well as teachers of husbandry. There are two annual courses, for the reception into which the necessary preparation required is: — finished studies at a higher gymnasium or realistic school and practical elementary knowledge of husbandry. — There are 9 teachers and 147 students, paying 63 fl. fees each; the annual costs of maintenance being 19,400 fl. supplied by the State.

2. Agricultural Middle Schools for the improvement of functionaries to estates, of owners or farmers of estates. There are 4 such schools, viz. at Grossau in Lower-Austria, Tetschen-Liebwerd in Bohemia, Kreuz in Croatia, and Dublány in Galicia, with courses of 2 and 3 years, requiring as preparation studies at a lower gymnasium or realistic school; yearly fees from 30 to 52 fl.; number of pupils 164; annual expenditure averaging 9,250 fl. for one, supplied by the provincial funds, by agricultural societies, and partly by grants of the State.

3. Lower Agricultural Schools for the improvement of peasants' sons in managing small farms, and qualifying them for under-stewards of larger estates. There are 7 schools of this kind (some of them connected with middle-schools), viz. at Grossau in Lower-Austria, Libějič-Rabin and Tetschen-

Liebwerd in Bohemia, Gratz in Styria, Kreuz in Croatia, Czernichow in Galicia, and Laibach in Carniola. Courses last from 2 to 4 years, elementary school required as preparation. Average number of scholars 230; fees from 30 to 40 fl., partially to be paid with work; free scholarships 72; annual expenditure 6,385 fl., where not in connection with middle-schools, from 4,000 to 5,000 fl.

4. Higher Schools for the science of Forest culture are instituted for the purpose of educating scientifical managers of forests and officials. There are two, viz. at Maria-Brunn in Lower-Austria, and at Schemnitz in Hungary; the courses last 2 and 3 years, and the preparation required is: higher studies of a gymnasium or realistic school and drawing, or the mathematical and natural-historical objects of a polytechnic institute. 160 students pay 10 fl. fees each, the state bearing costs amounting to 14,000 fl. for the school at Maria-Brunn.

5. Middle-Schools for Forest-culture, destined for the instruction of forest keepers and other subalterns employed in the management of forests, are 3, viz. at Weisswasser in Bohemia, Aussee in Moravia and Kreuz in Croatia, with courses from 2 to 3 years, for which the schooling of a lower gymnasium or realistic school, besides the practice of a year or two, is required as preparation. Total number of pupils 160 (Kreuz not yet in activity); fees from 20 to 62 fl.; annual costs of each about 8,000 fl. contributed by foresters-unions, private persons and provincial funds.

6. Schools for the culture of meadows, fruit-trees, vines, and rearing of silk-worms, five in number (at Gratz and its environs, at Klosterneuburg and at Buda), have courses of one year or two, and expect the pupils to have finished the elementary school. An average number of 100 pupils compensate for fees, board and lodging either wholly or partially with labour. 32 free places cost 2,726 fl.; the annual expenditure varies, averaging 6,000 fl. for one school, board included, and is at the charge of provincial funds, agricultural societies, and founders of free places.

Public lectures on husbandry and agriculture are delivered at most of the universities and polytechnic schools.

The Veterinary Institutes at Vienna and Pesth, several Horse-shoeing schools (at Gratz and Laibach) may also be considered as auxiliary to schools for husbandry.

7. Higher Schools for Mining are at Schemnitz in Hungary, Leoben in Styria and Přibram in Bohemia. Courses from 2 to 4 years; preparation: higher gymnasium or realistic school; number of pupils: 255; fees 10 fl.; exhibitions 70; expenditure 14,700 fl., borne by the state.

8. Lower Schools for Mining, purposed for instructing assistants; 3 in number (Přibram in Bohemia, Windschacht in Hungary, Nagyag in Transylvania); courses from 2 to 3 years; preparation: elementary school, and practice in mining; number of pupils: 87.

Schools of Art.

The Academy of Fine Arts at Vienna was founded by the Emperor Leopold I. in the year 1704. The object of this institute is, through instruction in the principal branches of the Fine Arts, to train proficient artists, and exert an influence upon the improvement of art.

The instruction is in part preparatory and theoretical, in part practical For practical instruction in the Arts, there are the schools of historical painting and sculpture, and the schools of landscape painting, copper-engraving and architecture.

The Collection of Plastic Works consists of objects of ancient and mediaeval art. Nearly all the realistic schools of the empire have, during the last ten years, been provided with casts from these mouldings. This Academy is also in possession of a Library and Collection of Prints. The former numbers above 8,000 volumes, most of them folios, sets of copperplates among which are copies from the greater number of picture galleries in Europe. The impressions, prints and illustrations in all exceed 80,000.

The Picture Gallery consists of a collection of 840 pictures.

The course of instruction for architecture usually lasts three years, but there is no prescribed term for any branch of the Arts.

Number of pupils 180—240. (See: Introduction to the IV. Section.)

I. R. Military Educational Institutions.

The Military Schools comprise Institutes for Education and Institutes for Instruction.

The Military Institutes for Education are divided into Institutes for non-commissioned officers, and Institutes for commissioned officers.

The Institutes for non-commissioned officers embrace: the Lower houses of education; the Upper houses of education; the School-companies.

The Institutes for commissioned officers embrace: the Cadet Institutions, and the Military Academies.

Of Senior and Special Military Schools there are: the Military Teachers' Training-School; the Higher Artillery Course; the Higher Engineers Course; the Central Cavalry School; the War School; the Military School for Administration; the Medico-Surgical Academy, and the Veterinary Institute.

There is also an educational institute for daughters of Officers at Hernals, near Vienna.

The object in view by the military institutions for education is to afford, to sons of deserving and indigent parents being or having been connected with the army, the necessary education for commissioned or non-commissioned officers.

Concerning the places or foundations in these military educational institutions, there are places granted from military foundations, others from provincial foundations, and others again from private foundations. Lastly, there are places for pupils whose parents pay for them.

For admission into each several institute, the normal age, physical qualification, and proper preparation is required.

In the month of September, the pupils of the last annual course in School Companies enter the army in the quality of non-commissioned officers; the pupils of the Academies, in that of commissioned officers. The pupils of the last annual course in the Lower houses of education proceed to the Upper houses of education, from the latter to the School Companies; and from the Cadet Institutions to the Academies.

The register of educational institutions shows:

5 Lower Houses of Education, each for 100 pupils, in 4 annual courses.

5 Upper Houses of Education, each for 200 pupils, in 4 annual courses.

2 Infantry School Companies, each for 120 pupils, in 2 annual courses.

4 Artillery School Companies, each for 120 pupils, in 3 annual courses.

1 Pioneer School Company for 120 pupils, in 3 annual courses.

1 Engineer School Company for 120 pupils, in 3 annual courses.

4 Cadet Institutions, each for 200 pupils, in 4 annual courses.

3 Military Academies, that of Neustadt for 400 pupils, in 4 annual courses.

The Artillery and Engineers Academy, each for 160 pupils, in 4 annual courses.

In the Military educational institutions, the Bohemian, Hungarian, Italian, and East-Roman languages are treated as regular objects of instruction.

The Military Institute for Training Teachers has to train teachers, serving as non-commissioned officers and inspection-sergeants in the educational institutions for officers, besides fencing-masters and teachers of gymnastics for Military Schools and the Troops.

There are 50 non-commissioned officers of the troops attending this Institute, where the course of instruction lasts one year.

The Higher Artillery and Higher Engineers Courses, and the War School are destined for training skilled officers, and affording to them a superior degree of education for the higher branches of service in the army. Those who attend these Institutes are second or first lieutenants, whose admission depends on a previous examination before a commission. The course to last 2 years in each institute.

The Central Cavalry School purposes to instruct talented officers of cavalry not only in equitation, in the art of breaking, treating, and improving horses, but also in the science of leading on, and making use of cavalry. The course of instruction lasts one year, one officer of every regiment of cavalry being admitted.

The School for Military Administration consists of two divisions, viz. one for officers of military courts of justice, and another for the administrative business of the Military Border-land.

The Medico-Surgical Academy is founded for the education of field-physicians and surgeons serving in the army. It is divided into a higher course for the education of doctors of medicine and chirurgery, and a lower course for training surgeons intended for the lower branches of the profession.

This Academy stands in equal rank with the medico-surgical faculty of the Vienna university.

The higher course extends to 5, the lower to 3 years.

From the higher course the students enter the army as Physicians with the degree of Doctors, bound to do military service for 10 years; from the lower course, as Surgeons with the degree of Chirurgeons, bound to a military service of 8 years.

The Military Veterinary Institute instructs individuals for the calling of veterinary surgeons, or farriers, both for the military and civil; and is moreover the chief scientific authority in that profession in judicial and police cases on veterinary regulations.

This Institute comprises the Veterinary School for educating veterinary surgeons and farriers, and the School for Shoeing of Horses for the improvement of ordinary horse-shoe-smiths.

Courses for veterinary surgeons last 3 years, for farriers, 2 years.

The pupils of the Veterinary Surgeons School enter the army as patented under-veterinary-surgeons, bound to 6 years' military service.

The military pupils of the biennial course join their regiment and are, according to necessity, promoted to army-farriers.

The Educational Institute for daughters of Officers at Hernals, near Vienna, has been founded for the purpose of facilitating to deserving officers or their widows the education of their daughters The pupils, amounting to 70, are maintained at the public cost, or enjoy foundation places.

I. R. Navy Educational Institutions.

There exists in the I. R. Navy:
A School for Ship's Boys.
A School for Marine Infantry.
A School for Marine Artillery.
A School for Marine Pupils of the 1st Class.
A School for Marine Pupils of the 2d Class.
A Theoretical Course for Navy Cadets (Midshipmen).
A Higher Course for Navy Officers.

The School for ship's boys has the task of educating young men for warrant-officers. The highest promotion to be attained is that of an upper boatswain.

In the School for marine infantry and artillery, men that are in active service are educated for non-commissioned officers. Those who are apt for commissioned officers may be promoted to such on leaving the school.

The School for marine pupils of the 1st class is on board of a man-of-war designated for that purpose. Young men who, with a proficient technical education, enter at the age of 16 or 18 years as marine pupils of the 1st class, receive here the education necessary for Navy Officers. After a course of one year's length, they enter the service as Navy Cadets; and after these cadets having been 2 or 3 years on board a man-of-war, they enter the theoretical Course for Navy Cadets.

The School for marine pupils of the 2d class has the exclusive purpose of educating for Navy Officers those who are admitted as pupils of this class. There are free places as well as such as must be paid for, and instruction is given on a school-ship fitted out for that purpose.

After the 3d annual course, the pupils enter service as navy cadets to be practically employed on board; and after 2 or 3 years time more, they are admitted into the theoretical course for navy cadets. The number of these pupils varies from 40 to 50.

The navy officers of the School-ship are appointed for directing practical exercises and manoeuvres.

The theoretical Course for Navy cadets is on land, lasting one year. At the end of that course every cadet is to undergo the prescribed examination for officers.

The higher Course for Navy officers is instituted for the purpose of accomplishing in the higher branches of science those young officers who are particularly distinguished for scientific parts and inclined to further mathematical and hydrographical studies. The course extending to one and two years' duration.

Musical Institutes and Unions.

The Austrian empire is profusely rich in musical proficients, the races of Sclavonian and Italian tongues being prominent above others in natural gifts for music. Bohemian instrumental musicians are to be met with in most of the European orchestras, and even in those of the principal towns of North-America. Austrian Italy sends numerous superior voices to all Italian operas. The German provinces of the empire abound most in educational institutes and unions for the cultivation of music. Every central provincial town has a Musical Union which, maintained by private means, forms the centre of social and pedagogical endeavours for the promotion of music. Complete Academies of Music in Vienna and Prague afford well-grounded instruction, given by first-rate teachers, in playing all instruments in every theoretical and practical branch of the art.

The Academy of Music in Vienna is the foundation and property of „the Union of Austrian friends of Music", that started into existence in 1816. The unmber of pupils in 1861 was 179. Besides this, the „Union of Austrian friends of Music" has founded, and brings forward, periodic orchestral concerts; joined with this Union is a special Orchestral Union and a Vocal Union, the former with 55, the latter with 207 members.

Of small private music-schools, Lower-Austria alone possesses 76, which may serve as average scale for most of the Austrian provinces. No fewer than 54 Unions exist in the small province of Lower-Austria, applied to the cultivation of Church choral-music and lay music. In the metropolis the „Union for the promotion of classic Church-music" exerts itself specially for the musical education of school-teachers and directors of choirs. There are besides 7 other Unions for Church-music and 10 Vocal Unions.

For the scientific treatment of the History and Aesthetics of Music, there is an extraordinary professorship at the university of Vienna Valuable means for the study of music are found in the rich Collections of music, books, and instruments belonging to the „Union of friends of Music", and in the highly valuable musical collection in the Library of the I. R. Court.

For instruction in Singing there are teachers at the greater gymnasia and realistic schools of the empire.

Those of the central provincial towns that have eminently distinguished Unions for Music, are: Prague, Gratz, Pesth, Brünn, and Innsbruck. In Vienna appear two journals exclusively devoted to Music; in Prague, a musical journal in the Bohemian language. Besides, every news-paper of some extent has regularly columns devoted to music.

A far spread and well deserved renown is enjoyed by the Austrian Military Music Bands, distinguished for strict harmonious play and for the excellence of their instruments. The Austrian Military Music consists of one Chief Conductor of the Bands, and 178 regimental music bands with one conductor for each, together 10,000 men.

Tabular Survey of Public Instruction in the Austrian Monarchy.

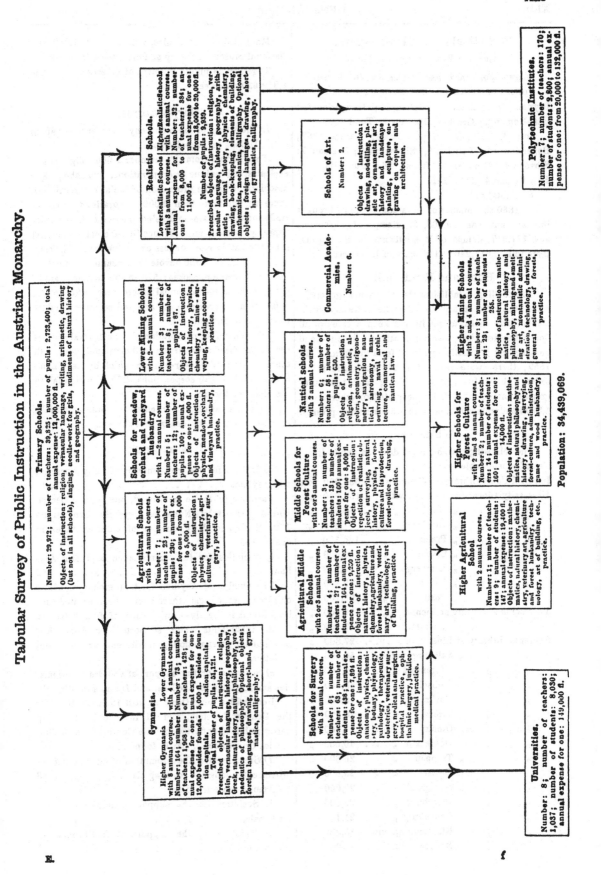

Primary Schools.

Number: 29,972; number of teachers: 39,825; number of pupils: 2,723,400; total annual expense: 13,000,000 fl.

Objects of instruction: religion, vernacular language, writing, arithmetic, drawing (but not in all schools), singing, needle-work for girls, rudiments of natural history and geography.

Gymnasia.

Higher Gymnasia with 8 annual courses. Number: 164; number of teachers: 1,968; annual expense for one: 12,000 fl. besides foundation capital.

Lower Gymnasia with 4 annual courses. Number: 75; number of teachers: 428; annual expense for one: 5,000 fl. besides foundation capital.

Total number of pupils: 54,121.

Prescribed objects of instruction: religion, latin, vernacular language, history, geography, Greek, natural history, natural philosophy, propaedeutics of philosophy. Optional objects: foreign languages, short-hand, gymnastics, calligraphy.

Agricultural Schools with 2–3 annual courses.

Number: 7; number of teachers: 73; number of pupils: 230; annual expense for one: from 4,000 to 5,000 fl.

Objects of instruction: physics, chemistry, agriculture, veterinary surgery, practice.

Schools for meadow, orchard and vineyard husbandry with 1–2 annual courses.

Number: 5; number of teachers: 12; number of pupils: 100; annual expense for one: 6,000 fl.

Objects of instruction: physics, meadow, orchard and vineyard husbandry, practice.

Lower Mining Schools with 2–3 annual courses.

Number: 3; number of teachers: 8; number of pupils: 87.

Objects of instruction: natural history, physics, chemistry, , mine - surveying, keeping accounts, practice.

Realistic Schools.

Lower Realistic Schools with 3 annual courses. Annual expense for one: from 8,000 to 11,000 fl.

Higher Realistic Schools with 6 annual courses. Number: 32; number of teachers: 834; annual expense for one: from 12,000 to 20,000 fl.

Number of pupils: 9,959.

Prescribed objects of instruction: religion, vernacular language, history, geography, arithmetic, natural history, physics, chemistry, drawing, book-keeping, elements of building, mathematics, mechanics, calligraphy. Optional objects: foreign languages, drawing, short-hand, gymnastics, calligraphy.

Schools of Art.

Number: 2.

Objects of instruction: drawing, modelling, plastic art, ornamental art, history and landscape painting, sculpture, engraving on copper and architecture.

Agricultural Middle Schools with 2 or 3 annual courses.

Number: 4; number of teachers: 27; number of students: 164; annual expense for one: 9,250 fl.

Objects of instruction: natural history, physics, chemistry, agriculture and forest husbandry, veterinary art, technology, art of building, practice.

Middle Schools for Forest Culture with 2 or 3 annual courses.

Number: 3; number of teachers: 13; number of students: 100; annual expense for one: 8,000 fl.

Objects of instruction: repetition of realistic objects, surveying, natural history, physics, forest-culture and its protection, forest-police, drawing, practice.

Nautical schools with 2 annual courses.

Number: 6; number of teachers: 58; number of pupils: 650.

Objects of instruction: religion, arithmetic, algebra, geometry, trigonometry, navigation, nautical astronomy, manœuvring, naval architecture, commercial and nautical law.

Commercial Academies.

Number: 6.

Schools for Surgery with 3 annual courses.

Number: 6; number of teachers: 63; number of students: 438; annual expense for one: 7,893 fl.

Objects of instruction: anatomy, physics, chemistry, botany, physiology, pathology, therapeutics, obstetrics, veterinary surgery, medical and surgical hospital practice, ophthalmic surgery, juridico-medical practice.

Higher Agricultural School with 2 annual courses.

Number: 1; number of teachers: 9; number of students: 147; annual expense: 19,400 fl. Objects of instruction: mathematics, natural history, chemistry, veterinary art, agriculture and forest husbandry, technology, art of building, etc., practice.

Higher Schools for Forest Culture with 2 and 3 annual courses.

Number: 2; number of teachers: 14; number of students: 160; annual expense for one: 14,000 fl.

Objects of instruction: mathematics, natural philosophy and history, drawing, surveying, forest-culture, administration, game and wood husbandry, practice.

Higher Mining Schools with 2 and 4 annual courses.

Number: 3; number of teachers: 23; number of students: 255.

Objects of instruction: mathematics, natural history and philosophy, mining and smelting art, montanistic administration, technology, drawing, general science of forests, practice.

Polytechnic Institutes.

Number: 7; number of teachers: 170; number of students: 2,800; annual expense for one: from 20,000 to 132,000 fl.

Universities.

Number: 8; number of teachers: 1,037; number of students: 8,030; annual expense for one: 149,000 fl.

Population: 34,439,069.

Periodicals and Newspapers.

From the Post-catalogue of 1862.

Inland, political journals: German 76; Bohemian 8; Polish 5; Slavonian 2; Servian 2; Croatian 3; Illyrian 1; Ruthene 2; Italian 11; Hungarian 15; East-Roman 2; Greek 1; Hebrew 2. Total 130.

Inland, not political periodicals: German 162; Bohemian 16; Polish 11; Slavonian 2; Servian 5; Slovene 4; Croatian 4; Italian 26; Hungarian 47; East-Roman 4. Total 281.

Sum total 411.

In 1860 the number of news-paper stamps brought into account was: 49,808,238 stamps of 1 kreutzer, and 177,347 stamps of 2 kreutzers. Of the former, there fell 30,817,240 to the share of Lower-Austria.

Book and Print Selling.

Book selling establishments in Austria generally contribute to the propagation of intellectual accomplishments. The first-rate publishers keep a stock of all books besides their own publications. Second-hand book-selling is usually separated from the former. In 1859, there were in Austria 472 book and print-sellers; about 400 printing-offices; and 180 lithographic institutions. They computed together about 1,800 handpresses and more than 300 machine-presses.

Libraries.

The following may be noted down as public libraries of considerable extent:

Imperial Court Library at Vienna, with . 358,000 vols
University Library at Vienna, with . . 159,700 „
„   „   „ Gratz   „ . . 49,900 „
„   „   „ Innsbruck, with . 57,800 „
„   „   „ Prague   „ . 126,100 „
„   „   „ Cracow   „ . 102,300 „
„   „   „ Lemberg   „ . 39,200 „
„   „   „ Padua   „ . 89,500 „
„   „   „ Pesth   „ . 98,100 „
Polytechnic Institute at Vienna „ . 26,800 „
„   „ Gratz   „ . 47,500 „
Students' Library at Linz, with . . . . 27,800 „
„   „   „ Salzburg, with . . 50,100 „
„   „   „ Klagenfurt „ . . 31,700 „
„   „   „ Laibach   „ . . 33,600 „
„   „   „ Goritz   „ . . 10,000 „
„   „   „ Triest   „ . . 28,000 „
„   „   „ Olmütz   „ . . 54,400 „
„   „   „ Debreczin „ . . 40,000 „
„   „   „ Troppau   „ . . 26,800 „
„   „   „ Eperies   „ . . 25,600 „
„   „   „ Sarospatak „ . . 25,000 „
„   „   „ Pressburg „ . . 23,000 „
„   „   „ Oedenburg „ . . 21,100 „
„   „   „ Klausenburg, with 13,400 „
„   „   „ Mantua, with . . 41,000 „

St. Marco Library at Venice, with . . . 114,230 vols
National Museum Library at Pesth, with 100,000 „

Together 28 public libraries of greater extent with 1,819,630 volumes.

Corporations.

In the whole of the monarchy there are now about 4,450 Unions or Associations.

If we deduct from this number 1.600 religious and ecclesiastical unions or corporations, and about 850 associations for the promotion of social entertainment, there are still more than 2,000 corporations which, as trade societies, share-holders' companies, or in any other capacity, protect and advance intellectual or material interests within more or less extensive local districts.

Of the latter, 1350 devote their zeal to the further-ance of humane purposes; 518 to the advancement of production and commercial intercourse; 120 to the cultivation or promotion of arts and sciences; and 27 to miscellaneous purposes.

The number of corporations in the several provinces is in direct proportion to their population, with the exception of Lower-Austria which by itself numbers more than 450 corporations, 260 of which fall to the share of Vienna.

An enumeration of particular associations may be conducive to give an idea of that subject.

I. Human Societies.

a) Unions for the maintenance of Baby Nurseries and Asylums 10

Particularly to be mentioned, the Vienna Central Union for Baby Nurseries, yearly expenditure 132,000 fl. and daily number of children taken in charge 153.

b) Societies for the maintenance of Infant Asylums 68

The number of Asylums for taking charge of little children is very great, there having been erected such institutions in nearly every place of from 5,000 to 6,000 inhabitants, and supported by benevolent individuals, corporations, institutes, or industrial societies.

c) Societies for the support of Children's Infirmaries and Asylums for the Deaf and Dumb and Blind, 16

But besides these there are still 19 Institutions for the Deaf and Dumb, and 11 for the Blind, being partly public (supported by the Central or some Provincial Government), partly foundations.

d) Societies for the promotion of educational improvement among the working classes 58

Among these there are:

1. Societies for the support of industrial schools for females 5

2. Societies for the support of institutions for employing boys 2

3. Societies for the education of poor children, training of good servant-maids, bestowing gratuitous instruction 33

4. Unions for the promotion of trade and agriculture among Jews 9

5. Unions for the protection of animals 4

6. Societies for the propagation of good popular literature 5

The St. Stephan's Union in Pesth, yearly expenditure 36.000 fl.

e) Protectorial Unions for reclaiming neglected children, fallen females, and discharged prisoners . 9

The Vienna protectorial union for reclaiming neglected children, endowed with 2 asylums; yearly expenditure 13,000 fl.

f) Catholic Journeymen's Unions 29

The foremost amongst which is that in Vienna, membering 4,000 members.

g) The Army Institutions, amounting to 7,251, with an annual expenditure of 3 million florins and 500,000 individuals receiving donations.

h) Asylums for the maintenance of the indigent, 1,289 in number, with an annual expenditure of 1½ million florins and 30,000 beneficiaries.

i) There are 336 public hospitals, 40 lunatic asylums, 38 lying-in hospitals, 30 foundling, and 60 orphan asylums.

k) Miscellaneous Benefit Societies 157

l) Unions for mutual support of sick or temporarily unprovided for members 359

m) Unions for contributing towards funerals . 349

n) Societies for maintaining their members, their widows and orphans 114

II. Savings-Banks and Savings-Unions.

1. Savings-banks 143

2. Unions for saving and supplies 12

(On savings-banks there is a separate section.)

III. Pawn-broking, Loan and Advance Unions . . 29

(There are 6 public pawn-broking offices, and 43 monti di Pietà in Lombardo-Venice.)

IV. Societies for the Promotion of Agricultural Industry.

a) Unions for Agriculture and Forest Management 72

In this number are not included the Unions of South Tyrol and 102 Unions in Lombardo-Venice; there are also in every province, Agricultural Societies, with Branch Unions and Agricultural Schools supported by them, Horticultural Societies and Unions for Forest Culture.

b) Unions for the Promotion of Mining . . . 13

They are partly purely scientific societies, partly unions for industry.

c) Unions for the advancement of industry and commerce in general, and for the exercise of industrial undertakings in especial 160

1. Industrial Unions 20

2. Companies for the exercise of particular undertakings 120

d) Societies for the traffic and preservation of railways, bridges, highways, and waterways . . 33

Of these, there are railway societies 26

Suspension-bridge joint-stock companies . . 6

Tunnel company 1

e) Unions for regular conveyances on water and on land 36

Steam-navigation concerns 6

Miscellaneous conveyance unions 24

f) Credit Institutions 5

g) Assurance Companies 120

h) Institutions for annuities and life interests 15

V. Societies for the promotion of Arts and Sciences 120

These are divided into scientific, art and musical societies, securing Austria's intellectual position, in conjunction with public educational institutions, academics and schools.

VI. Miscellaneous Unions 27

Table

of

Austrian, English and French Money, Weights and Measures of Length, Superficies, Capacity, and Liquids.

I. Money.

In Silver: 1 £ = 100 Francs = 40·25 florins Austrian Currency.

Austr. Curr. fl. kr.	English s. d.	French F. C.	Austr. Curr. fl. kr.	English s. d.	French F. C.	Austr. Curr. fl. kr.	English s. d.	French F. C.
·5	·1·	·12	1·	·2·	2·48	20·	2·	49·69
·10	·2·1	·24	2·	·4·	4·97	30·	3·	74·53
·20	·4·3	·48	3·	·6·	7·45	40·	4·	99·38
·30	·7·	·75	4·	·8·	9·94	50·	5·	124·22
·40	·9·2	·99	5·	·10·	12·42	60·	6·	149·6
·50	1··	1·24	6·	·12·	14·91	70·	7·	173·91
·60	1·2·1	1·49	7·	·14·	17·39	80·	8·	198·75
·70	1·4·3	1·74	8·	·16·	19·88	90·	9·	223·60
·80	1·7·	1·99	9·	·18·	22·36	100·	10·	248·44
·90	1·9·2	2·24	10·	1··	24·84	1000·	100·	2484·40

In Banknotes at the rate of exchange: 10 £ = 135 fl. 100 Francs = 53 fl.

Austr. Curr. fl. kr.	English L. s. d.	French F. C.	Austr. Curr. fl. kr.	English L. s. d.	French F. C.	Austr. Curr. fl. kr.	English L. s. d. f.	French F. C.
·5	·3·	·9	1·	1 5 3	1·89	20·	1 9 7 2	37·73
·10	1 3·	·19	2·	2 11 2	3·77	30·	2 4 5 1	56·60
·20	3 2·	·38	3·	4 5 1	5·66	40·	2 19 3 ·	75·47
·30	5 1	·57	4·	5 11 ·	7·55	50·	3 14 · 3	94·34
·40	7 ·	·75	5·	7 4 3	9·43	60·	4 8 10 2	113·21
·50	8 3	·94	6·	8 10 2	11·32	70·	5 3 8 1	132·8
·60	10 2	1·13	7·	10 4 1	13·21	80·	5 18 6 1	150·94
·70	1 · ·	1·32	8·	11 10 ·	15·9	90·	6 13 3 3	169·81
·80	1 2 ·	1·51	9·	13 3 3	16·98	100·	7 8 1 2	188·68
·90	1 3 3	1·70	10·	14 9 3	18·87	1000·	74 1 5 2	1886·79

Austrian. — English. — French.

II. Measures of Length.

- Ell (Vienna) à 29·578 Inches = 2 Feet and 6·67 Inches = 0·779 Mètre.
- Ell (Bohemian) à 22·548 Inches = 1 Foot and 11·35 Inches = 0·592 „
- Foot = 12 Inches, Inch = 12 Lines = 1·037 Feet = 0·316 „
- Fathom = 6 Feet = 6 Feet and 2·67 Inches = 2 Yards and 2·67 Inches = 1·896 „
- Mile (Post) = 8297 Yards = 7586·0 „
- Inch = 12 Lines = 1·037 Inches = 0·026 „

III. Measures of Superficies.

- Foot (Square) = 144 Square inches = 1·07 Square foot = 0·001 Are.
- Yoke = 1600 Square fathoms = 5·68 Roods = 1·42 Acre = 57·6 „
- Yoke (of Hung. Woods) = 1200 Square fathoms = 4·26 „ = 1·06 „ = 43·2 „
- Fathom (Square) = 36 Square feet = 4·28 Square yards = 0·036 „
- Mile (Square) = 14200 Square acres = 5755·6 Hectare.

IV. Measures of Capacity.

- Foot (Cubic) = 1728 Cubic inches = 7·0 Gallons = 31·6 Litre.
- Fathom (Cubic) = 216 Cubic feet = 23·6 Quarters = 6822·4 „
- Metzen = 13·6 Gallons = 61·5 „

V. Measures of Liquids.

- Eimer (Beer) = 42½ Mass = 13·2 Gallons = 60·0 Litre.
- Eimer (Wine) = 41 Mass = 12·76 „ = 56·6 „
- Mass (Vienna) = 1 Quart and ½ Pint = 1·41 „

VI. Weights.

- Loth = 4 Quentchen = 9·84 Drachmes = 0·017 Kilogramme.
- Pound (Vienna) = 32 Loth = 1·23 Pound = 1 Pound 3 Ounces 11 Drachms = 0·559 „
- Pound (Tariff) = 1·10 „ = 1 „ 1 Ounce 10 „ = 0·50 „
- Hundredweight (Vienna) = 100 Pounds = 123·0 Pounds = 7 Ounces 5 Drachms = 55·9 „
- Hundredweight (Tariff) = 110·0 „ = 3 „ 10 „ = 50·0 „

English-Austrian.

I. Money.

1 Pound Sterl. = 20 Shillings à 12 Pence = 10 florins Austrian Currency in Silver.
1 Guinea = 21 Shillings.
1 Shilling = 50 kr. Austr. Curr. Silver.
1 Penny = 4½ „ „ „ „

II. Measures of Length.

Fathom = 5·78 Vienna Feet.
Foot = 0·96 Vienna Feet = 11·57 Vienna Inches.
Furlong = 106·06 Vienna Fathoms.
League 979·2 Vienna Fathoms = 73/80 Miles.
Mile (English) = 804·86 Vienna Fathoms = 1666 Yard.
Mile Statute = 847·7 Vienna Fathoms.
Pole Perch Rod = 2·65 Vienna Fathoms.
Yard = 2·89 Vienna Feet = 1·174 Vienna Ell.

III. Measures of Superficies.

Acre = 1124·93 Vienna Fathoms square.
Foot square = 133·9 Vienna Inches „
Rod square Pole = 7·02 Vienna Fath.
¼ Acre = Rod Acre = 281·288 Vienna Fathoms square.
Yard square = 8·35 Feet square.

IV. Measures of Capacity.

Bushel = 0·59 Austrian Metzen.
Ghaldron = 21·27 „ „
Callon = 3·21 „ Mass.
Last = 47·27 „ Metzen.
Pint = 1·60 „ Seitel.

Quart = 3·20 Austrian Seitel.
Quarter = 4·73 „ Metzen.
Sack = 1·77 „ „
Wey = 23·63 „ „

V. Measures of Liquids.

Gallon = 3·21 Austrian Mass.
Pint = 1·60 „ Seitel.
Quart = 3·21 „ „

	Wine, Spirits, Cider, Vinegar	Ale	Beer
	Austrian Eimer		
Barrel	·	2·57	2·89
Butt	10·11	7·70	8·66
Firkin	··	··	0·72
Hogshead	5·05	3·85	4·33
Kilderkin	··	1·28	1·44
Puncheon	6·74	5·14	5·78
Rundlet	1·44	··	··
Tun	20·23	15·42	17·34

VI. Weights.

Drachm = 0·10 Vienna Loth.
Hundredweight = 90·72 Vienna Pounds.
Pound = 0·81 Vienna Pounds = 25·92 Vienna Loth.
Quarter = 22·68 Vienna Pounds.
Stone = 11·34 „ „
Ton = 1814·4 „ „
Ounce = 1·62 „ Loth.

French-Austrian.

I. Money.

1 Napoléon = 20 Francs = 8 fl. 6 kr.
1 Franc = 20 sous = 40 3/10 kr.
1 Sous = 2 kr. Austrian Currency Silver.
1 Centime = 0·4 kr. „ „ „

II. Measures of Length.

Kilomètre = 1000 Mètre = 527·25 Vienna Fathoms.
Mètre = 10 Decimètre = 100 Centimètre = 1000 Millimètre = 3·1635 Vienna Feet = 1·28 Vienna Ell.
Myriamètre = 10000 Mètre = 5272·4 Vienna Fathoms = 1·3181 Mile Post.

III. Measures of superficies.

Are = 27·80 Fathoms (square).
Hectare = 2780 Fathoms square = 1·7375 Yoke.

IV. Measures of Capacity.

Hectolitre = 10 Decalitre = 100 Litre = 1·6259 Austrian Metzen.
Litre = 0·130 Vienna Achtel = 0·26 Austrian Massl.

V. Measures of Liquids.

Litre = 10 Decilitre = 100 Centilitre = 0·71 Vienna Mass = 2·83 Austrian Seitel.

VI. Weights.

Kilogramme = 10 Hectogrammes = 100 Decagrammes = 1000 Grammes = 1·7857 Vienna Pound = 1 Pound 25·14 Vienna Loth.

SPECIAL PART.

I. R. Austrian Central-Committee for the International Exhibition at London. 1862.

Vienna, City 666.

President.

Count Constantine Capello **WICKENBURG**, Minister for Trade and National Economy, Director of the Navy Office, I. R. Privy-Counsellor and Chamberlain; Grand Cross of the Leopold Order, Knight of the Iron Crown I. cl., Gr. Cr. of the Bavarian House Order and Bav. Crown Order, Knight of the Russian Anne Order and the Prussian Red Eagle I. cl., First Treasurer of the Duchy of Styria, etc. etc.

Proxy.

Adam Chevalier de **BURG**, I. R. Counsellor, Professor of Mechanics at the Polytechnic Institute of Vienna; Knight of the Leopold Order. Civ. Mer. O. of the Bavarian Cr., Portuguese Chr. O., the Belgian Leop. and Russian Anne O., Officier de la Légion d' Honneur, Parma Lew. O.,. Saxon Ern. H. O.; Member of the Academy of Sciences; President of the Lower-Austrian Trade Union. Member of Com. and Jury, Vienna 1845, Pres. and Jury London 1851, Munich 1854, Paris 1855.

Members.

Dr. Joseph **ARENSTEIN**, I. R. Professor, Chevalier de la Légion d' Honneur, Member of many Learned Societies. Editor of the „Allgemeine Land- und forstwirthschaftliche Zeitung" (General Journal of Agriculture and Forest Economy), Jury and Reporter, Paris 1855, Commiss. and Jury at the International Agricultural Exhibitions. Paris 1856 and Vienna 1857. *Vienna.*

Ralph **EITELBERGER** von Edelberg, I. R. Professor of the History and Archaeology of Arts at the University of Vienna, and Lecturer of the same at the Academy of Arts; Reporter on Arts, Paris 1855 and Munich. *Vienna.*

Robert **HAAS**, Manufacturer. *Ebergassing near Vienna.*

Anthony **HARPKE**, Manufacturer, Member of the Council of the Exchange and the L. A. Board of Com. and Trade. *Vienna.*

Baron Otto **HINGENAU**, I. R. Chamberlain, Counsellor of Mines and Professor at the University of Vienna; Founder and Editor of „Austrian Journal for Mining and Smelting." Com. at the Agric. Exhib. Vienna 1857. *Vienna.*

Adolphus **PARMENTIER**, I. R. Counsellor to the Ministry of Trade and National Economy; Knight of the Bav. Michael Ord.. the Würtemb. Cr. O., and the Sax. Alb. O.

Nicholas **RABE**, I. R. Counsellor to the Ministry of Finances. *Vienna.*

Joseph **RECKENSCHUSS**, Manufacturer, Member of the Lower-Austrian Diet and Board of Trade; Jury and Reporter, Munich 1854. *Vienna.*

Ignatius von **ROHONCZY**, Aulic Counsellor to the R. Hung. Court-Chancery, Knight of the I. R. Leop. Ord. *Vienna.*

Augustus **ROTH** v. Telegd, Aulic Counsellor to the Transylvanian Court-Chancery. *Vienna.*

Dr. Anthony **SCHRÖTTER**, I. R. Professor of Chemistry at the Polytechnic Institute, and General Secretary to the I. R. Academy of Sciences at Vienna; Knight of the I. R. Francis Joseph Order, and Chevalier de la Légion d'Honneur. Jury, Vienna 1845 and London 1851. *Vienna.*

Dr. Ferdinand **STAMM**, Member .of the house of Commons. Editor of the Journal: „Neueste Erfindungen" and „Der Nährstand". *Vienna.*

John **STREICHER**, Vice-president of the L. A. Union of Trade, Pianoforte Maker to the I. R. Court, Mb. Com. Vienna Exhib. 1845, Rep. Lond. 1851, Jury, Munich 1854. *Vienna.*

Ognieslaw **UTIEŠENOVIĆ**, Aulic Counsellor to the Croatian Slavonian Court-Chancery; Knight of the I. R. Francis Joseph Order. *Vienna.*

Ernest **WERTHEIM**, Exporter, Member of the L. A. Board of Com. and Trade, Rep. Lond. 1851, Jury Munich 1854 and Paris 1855. *Vienna.*

Francis **WERTHEIM**, Vice-president of the L. A. Board of Com. and Trade; Knight of the I. R. Francis Joseph Order, the R. Han. Guelph O.. G. G. Med., I. Russian St. Anne O., Officer of the J. Ottom. Medjidié O. etc. etc. Jury, Lond. 1851, Munich 1854, Rep. Paris 1855. *Vienna.*

Maurice Chevalier de **WODIANER**, Banker and Proprietor, Knight of the I. R. Iron Crown Order III. class and of the I. R. Francis-Joseph Order, President of the I. R. Exchange, Director of the Austrian National-Bank etc. etc.

Charles **ZIMMERMANN**, Merchant, Member of the L. A. Trade Union. *Vienna.*

Secretary:

Dr. Edward **FALB**, I. R. Secretary to the Ministry of Trade and National Economy.

Adjoint.

Charles **BOLESLAWSKI**, I. R. Clerk to the Ministry of Trade and National Economy.

I. R. Austrian Commissioners at the International Exhibition in London.

(6, Onslow Crescent, Brompton, S. W.)

President.

Baron Anthony ROTHSCHILD, I. R. Austrian General Consul etc. etc. *London.*

William Chevalier de SCHWARZ, Dr., I. R. Counsellor, Dir. of the I. R. Austr. Gen. Consular Office at Paris, and I. R. First Austr. Commissioner at the International Exhibition in London, Knight of the I. R. Iron Crown O. III. cl., Officier de la Légion d'Honneur, K. Bav. Mich. O. Com. Würtemb. Fred. O. *Paris.*

Members.

Dr. Joseph ARENSTEIN, I. R. Professor (see p. 3).

Chevalier Adam BURG (see p. 3).

Anthony HARPKE (see p. 3).

Chevalier Ignatius SCHÄFFER, Director of the I. R. Austrian Gen. Consular Office in *London.*

Count Eugen SZÉCHÉNY. *Gross-Zinkendorf, Hungary.*

Count HARRACH.

Count John WALDSTEIN, I. R. Chamberlain, Aulic Counsellor, Knight of the Orders of St. Stephan and sov. Johannite. *Vienna.*

Francis WERTHEIM (see p. 3).

Charles ZIMMERMANN (see p. 3).

Austrian Members of the International Jury.

1. Class:

Peter TUNNER, I. R. Counsellor, Director of the Mining Academy at Leoben in Styria, Knight of the I. R. Iron Crown O. III cl. and the R. Bav. Civ. Mer. O. of Mich. Jury, Vienna 1845, London 1851, Munich 1854, and at Paris 1855. *Leoben.*

2. Class,

Sub-Class a):

Frederick ANTHON, Chemist. *Prague.*

Sub-Class b):

Anthony SCHRÖTTER (see p. 3).

3. Class:

Sub-Class a):

Count Henry ZICHY, *Oedenburg, Hungary.*

Sub-Class b):

Baron RIESE-STALLBURG Frederick Werner, Chamberlain, Knight of the Prussian Johanniter-Order and of the Légion d'Honneur. Jury, Paris 1855, Vienna 1857. *Prague.*

Sub-Class c):

Robert SCHLUMBERGER, Knight of the R. Würtemb. Fred. O. *Vöslau, L. A.*

4. Class:

Sub-Class a):

Emil SEIBEL, Chemist, Miner and Manufacturer, Chevalier de la Légion d'Honneur, Member of the Board of Com. and Trade Vienna. Jury, Munich 1854 and Paris 1855. *Vienna.*

Sub-Class b):

Dr. Robert CZILCHERT. *Gutor, Com. of Pressburg, Hungary.*

Sub-Class c):

Chevalier William SCHWARZ (as above). Chairman of the Jury for this class.

8. Class:

Chevalier Adam BURG (see p. 3). Chairman of the Austrian Members of the Jury.

9. Class:

Edward EGAN, *Bernstein, near Güns, Hungary.*

10. Class:

Maurice LOEHR, I. R. Counsellor. G. Cross of Merit with the Crown. *Vienna.*

13. Class:

C. E. KRAFT sen., Machinist. *Vienna.*

16. Class:

Ernest PAUER, Professor. *London.*

18. Class:

Maximilian DORMITZER Cotton-Printer. President of the Board of Com. and Trade at *Prague.*

19. Class:

Charles OBERLEITHNER jun., Manufacturer. Jury, Paris 1855. *Schönberg, Moravia.*

20. Class:

Anthony HARPKE (see p. 3).

21. Class:

Charles OFFERMANN, Cloth Manufacturer; Chevalier de la Légion d'Honneur. Proxy-Jury London 1851, Jury 1854 and Paris 1855. Chairman of the Jury for this class. *Brünn.*

23. Class:

Frederick LEITENBERGER, Cotton Printer; Member of the Board of Com. & Trade at Reichenberg. *Josephsthal near Cosmanos, Bohemia.*

24. Class:

Ralph LAPORTA, Merchant and Purveyor to the Court. *Vienna.*

26. Class:
Maurice POLLAK, Wholesale Dealer. *Vienna.*

27. Class:
Sub-class c):
Joseph GUNKEL, *Vienna.* Chairman of the Jury for this class.
Sub-class d):
Fred. SUESS, Leather Manufacturer. *Vienna.*

28. Class:
Charles GIRARDET. *Vienna.*

29. Class:
John Geoffrey MÜLLER, Professor at the Law-Academy at *Hermannstadt, Transylvania*

30. Class:
Michael MARKERT, Door and Window-frame Manufacturer. *Vienna.*

31. Class:
Sub-class a):
Chevalier Francis FRIEDAU. *Vienna.*
Sub-class b):
Dr. Ferd. STAMM. (see p. 3).

32. Class:
Francis WERTHEIM (see p. 3).

33. Class:
Ferdinand FRIEDLAND, Gas Manufacturer, K. Brazil R. O. Rep. Paris 1855. *Prague.*

34. Class:
Frederick SCHMITT. J. R. Clerk to the Direction of Administrative Statistics. Rep. Paris 1855. *Vienna.*

Proxy Jurymen.

Professor ARENSTEIN (see p. 3).

Charles CESSNER, Dr. Med. and Chir., I. R. Professor at the University of *Vienna.*

Dr. JENNY, Counsellor of Mines and Professor at *Schemnitz.*

J. NAGEL. *Vienna.*

Sigmund POPPER, Manufacturer. *Brünn.*

Joseph RECKENSCHUSS (as above).

J. B. STREICHER (as above).

Official Reporters.

Professor Arenstein (as above), Chief-Reporter.

Charles Cessner (as above).

Baron Ebner, I. R. Lieutenant-Colonel, Knight of several Orders.

Ralph Eitelberger von Edelberg (as above p. 3).

Dr. Charles Güntner, Professor at the Realistic School Wieden, *Vienna.*

Dr. Edward Hanslik, I. R. Professor of History and Aesthetics of Music. *Vienna.*

Dr. Robert Heym, Secretary to the Board of Commerce and Trade at Brünn.

Dr. Jenny (as above).

Dr. E. Jonak, I. R. Professor at the University of *Prague.*

Dr. Peter Mischler, I. R. Professor of Political Economy at the University of *Prague.*

Frederick Müller, Captain in the I. R. Artillery.

J. Nagel (as above).

Joseph de Partenau, Manufacturer. Vienna.

S. Popper (as above).

J. Reckenschuss (see p. 3).

Dr. Edm. Schebek, Secretary to the Board of Commerce and Trade at Prague.

F. Schmitt (see p. 5).

J. B. Streicher (see p. 3).

Ernst de Teschenberg, I. R. Min. Clerk, Vienna.

Exhibition Office.
(6, Onslow Crescent, Brompton, London, S. W.)

Director: Chevalier William Schwarz (see p. 4).

Secretary: Aloysius Heinrich, Secretary to the L. A. Trade Union.

Adjoints.

Charles Boleslawski, I. R. Min. Clerk; Baronet Arthur Hohenbruck; Dr. A. Thaa; Dr. J. Woerz, I. R. Min. Clerk.

For Decoration:

Commissioner: Frederick Stache, Architect.
Assistant: Frederick Setz.

Augustus Weber, Architect.
Lewis Neustätter, Painter (for Sect. IV).

For the Hungarian Articles.

Jankò Vincent, Manufacturer and Secretary of the Hungarian Exhibition-Committee. *Pesth.*

Szabó Joseph, Dr., Professor and member of the Hungarian Exhibition-Committee. *Pesth.*

Commercial-Agents:
Peter Bender.
Pierce.
Ignatius Steinberger.

Inspectors of Divisions:
Theodore Arnemann (for the Machine Room).
Grieselitsch (for the Ground-floor)
Werlicsek (for the Galleries).

Exhibitors' Productions are placed in the following forty Classes.

Section I.

Class 1.
Mining, Quarrying, Metallurgy and Mineral Products.

Class 2.
Chemical Substances and Products, and Pharmaceutical Processes.

Sub-Class a) Exhibitors of Chemical Products.
" " b) Exhibitors of Medical and Pharmaceutical Processes.

Class 3.
Substances, used for Food.

Sub-Class a) Exhibitors of Agricultural Produce.
" " b) Exhibitors of Drysaltery, Grocery etc.
" " c) Exhibitors of Wines, Spirits, Beer and other Drinks and Tobacco.

Class 4.
Animal and Vegetable Substances, used in Manufactures.

Sub-Class a) Exhibitors of Oils, Fats, and Wax, and their Products.
" " b) Exhibitors of other Animal Substances, used in Manufactures.
" " c) Exhibitors of Vegetable Substances, used in Manufactures etc.
" " d). Exhibitors of Perfumery.

Section II.

Class 5.
Railway Plant, including Locomotive Engines and Carriages.

Class 6.
Carriages not connected with Rail or Tram Roads.

Class 7.
Manufacturing Machines and Tools.

Sub-Class a) Exhibitors of Machinery, employed in Spinning and Weaving.
" " b) Exhibitors of Machines and Tools employed in the Manufacture of Wood, Metal etc.

Class 8.
Machinery in general.

Class 9.
Agricultural and Horticultural Machines and Implements.

Class 10.
Civil-Engineering, Architectural, and Building Contrivances.

Sub-Class a) Exhibitors of Civil Engineering and Building Contrivances.
" " b) Exhibitors of Sanitary Improvements and Constructions.
" " c) Exhibitors of Objects, shown for Architectural Beauty.

Class 11.
Military Engineering, Armour and Accoutrements, Ordnance and Small Arms.

Sub-Class a) Exhibitors of Clothing and Accoutrements.
" " b) Exhibitors of Tents and Camp Equipage.
" " c) Exhibitors of Arms, Ordnance, etc.

Class 12.
Naval Architecture — Ships' Tackle.

Sub-Class a) Exhibitors of Ship Building for purposes of War and Commerce.
" " b) Exhibitors of Boat and Barge Building, and Vessels for amusement, etc.
" " c) Exhibitors of Ships, Tackle, and Rigging.

Class 13.
Philosophical Instruments, and Processes depending upon their use.

Class 14.
Photographic Apparatus, and Photography.

Class 15.
Horological Instruments.

Class 16.
Musical Instruments.

Class 17.
Surgical Instruments and appliances.

Section III.

Class 18.
Cotton.

Class 19.
Flax and Hemp.

Class 20.
Silk and Velvet.

Class 21.
Woollen and Worsted, including Mixed Fabrics generally.

Class 22.

Carpets.

Class 23.

Woven, Spun, Felted, and Laid Fabrics, when shown as specimens of printing or dyeing.

Class 24.

Tapestry, Lace, and Embroidery.

Sub-Class a) Exhibitors of Skins and Furs.
     " " b) Exhibitors of Feathers.
     " " c) Exhibitors of Manufactures from Hair.

Class 25.

Skins, Fur, Feathers, and Hair.

Class 26.

Leather, including Saddlery and Harness.

Sub-Class a) Exhibitors of Leather.
     " " b) Exhibitors of Saddlery, Harness, etc.
     " " c) Exhibitors of Manufactures, generally made of Leather.

Class 27.

Articles of Clothing.

Sub-Class a) Exhibitors of Hats and Caps.
     " " b) Exhibitors of Bonnets and General Millinery.
     " " c) Exhibitors of Hosiery, Gloves and Clothing in general.
     " " d) Exhibitors of Boots and Shoes.

Class 28.

Paper, Stationery, Printing, and Bookbinding.

Sub-Class a) Exhibitors of Paper, Card, and Millboard.
     " " b) Exhibitors of Stationery.
     " " c) Exhibitors of Plate, Letterpress, and other modes of Printing.
     " " d) Exhibitors of Bookbinding.

Class 29.

Educational Works and Appliances.

Sub-Class a) Publishers.
     " " b) Apparatus Makers.

Sub-Class c) Toy and Games Manufacturers.
     " " d) Exhibitors of Specimens and Illustrations of Natural History.

Class 30.

Furniture and Upholstery, including Paper Hangings and Papier Mâché.

Sub-Class a) Exhibitors of Furniture and Upholstery.
     " " b) Exhibitors of Paper Hangings and General Decoration.

Class 31.

Iron and General Hardware.

Sub-Class a) Exhibitors of Iron Manufactures.
     " " b) Exhibitors of Manufactures in Brass and Copper.
     " " c) Exhibitors of Manufactures in Tin, Lead, Zinc Pewter, and General Braziery.

Class 32.

Steel, Cutlery, and Edge Tools.

Sub-Class a) Exhibitors of Steel Manufactures.
     " " b) Exhibitors of Cutlery and Edge Tools.

Class 33.

Works in Precious Metals, and their Imitations, and Jewellery.

Class 34.

Glass.

Sub-Class a) Exhibitors of Stained Glass, and Glass used in buildings and decorations.
     " " b) Exhibitors of Glass for household use and fancy purposes.

Class 35.

Pottery.

Class 36.

Manufactures not included in previous Classes.

Sub-Class a) Exhibitors of Dressing Cases and Toilette articles.
     " " b) Exhibitors of Trunks, and Travelling apparatus.

Section IV.

Class 37.

Architecture. Architectural Models and Designs.

Class 38.

Paintings in Oil and Water Colours, and Drawings. Art Desings and Manufactures.

Class 39.

Sculpture by various processes, Models, Die Sinking, and Intaglios. Fine Art in Repoussée, Chasing etc.

Class 40.

Etchings and Engravings.

Table of the Austrian exhibitors in London 1862 (without the IV. Section).

Section	Class	Boards of Trade exhibiting in this class	Exhibitors	Area horis.	Area vertle.	Agram	Botzen	Brody	Brünn	Budweis	Czernowitz	Eger	Knsegg	Feldkirch	Fiume	Goritz	Gratz	Innsbruck	Klagenfurt	Cracov	Kronstadt	Laibach	Lemberg	Leoben	Lins	Olmütz	Pesth, Provincial-Comitee	Plisen	Prague	Reichenberg	Roveredo	Rovigno	Salzburg	Trieste	Troppau	Udine	Venice	Verona	Vicenza	Vienna	Zara
I.	1.	22	81	633	578	2	·	·	1	1	·	1	·	·	·	·	5	3	4	1	2	1	4	2	3	2	12	·	9	·	·	2	2	1	1	·	1	·	·	21	·
	2.	21	90	401	332	1	·	·	2	·	2	·	2	·	1	1	·	1	4	·	4	·	4	·	·	4	5	2	10	4	1	·	2	2	3	·	·	·	·	34	1
	3.	21	234	1.353	289	2	3	·	10	·	·	·	5	·	2	2	11	·	·	1	14	1	5	·	3	6	91	·	22	4	1	·	·	2	4	·	·	·	·	42	3
	4.	23	139	941	1.183	1	1	·	6	·	2	·	5	·	·	3	3	3	2	·	3	1	3	1	4	2	14	1	8	6	·	·	·	3	8	·	·	·	·	58	1
Total of Sect.			**544**	**3.328**	**2.382**	6	4	·	19	1	4	1	12	·	3	6	19	7	10	2	23	3	16	3	10	14	122	3	49	14	2	2	4	8	16	·	1	·	·	155	5
II.	5.	3	7	881	160	·	·	·	·	·	·	·	·	·	·	·	·	·	·	·	·	·	·	·	·	1	1	·	·	·	·	·	·	·	·	·	·	·	·	5	·
	6.	1	2	72	8	·	·	·	·	·	·	·	·	·	·	·	·	·	·	·	·	·	·	·	·	·	·	·	·	·	·	·	·	·	·	·	·	·	2	·
	7.	6	13	191	8	·	·	·	1	·	·	·	·	·	·	·	1	·	·	·	·	1	·	·	·	1	·	2	·	·	·	·	·	·	·	·	·	·	·	7	·
	8.	6	25	727	32	·	·	·	·	·	·	·	·	·	·	·	2	3	·	1	·	·	·	·	·	·	·	3	·	2	·	·	·	·	·	·	·	·	·	14	·
	9.	8	21	786	25	·	·	·	·	1	·	1	·	·	·	·	1	2	1	·	·	·	·	·	·	6	·	3	·	·	·	·	·	·	·	·	·	·	·	6	·
	10.	7	21	530	223	·	·	·	·	·	·	·	·	·	·	·	2	2	·	·	·	·	·	2	·	2	·	·	·	·	·	·	1	1	·	·	·	·	·	11	·
	11.	7	14	231	32	·	·	·	·	·	·	·	·	·	·	·	1	1	·	1	·	·	·	·	1	2	·	·	1	·	·	·	·	·	·	·	·	·	·	7	·
	12.	3	3	40	··	·	·	·	·	·	·	·	·	·	·	·	·	·	·	·	·	·	·	1	·	1	·	·	·	·	·	·	·	·	·	·	·	·	·	1	·
	13.	6	21	162	170	·	·	·	·	·	·	·	·	·	·	·	1	·	·	·	·	·	·	·	1	2	·	1	·	·	1	·	·	·	·	·	·	·	·	15	·
	14.	4	12	30	329	·	·	·	·	·	·	·	·	·	·	·	·	·	·	·	·	·	·	1	1	·	·	·	·	·	·	·	1	·	·	1	·	·	9	·
	15.	4	10	15	348	·	·	·	·	·	·	·	·	·	1	1	·	·	·	·	·	·	·	·	1	·	·	·	·	·	·	·	·	·	·	·	·	·	7	·
	16.	7	42	744	325	·	·	·	·	·	·	·	·	·	·	·	1	·	·	·	·	·	·	2	·	3	·	8	1	·	·	·	·	·	·	·	·	1	·	26	·
	17.	4	8	49	15	·	·	·	·	·	·	·	·	·	·	·	1	·	·	·	·	·	·	·	·	3	·	1	·	·	·	·	·	·	·	·	·	·	·	3	·
Total of Sect.			**199**	**4.458**	**1.675**	·	·	·	1	·	1	·	1	·	·	·	6	10	3	3	·	1	·	6	2	24	·	20	2	1	·	1	2	·	·	1	1	·	113	·
III.	18.	5	24	210	423	·	·	·	·	·	·	·	1	·	·	·	·	·	·	·	·	·	·	·	·	1	·	2	11	·	·	·	·	·	·	·	·	·	9	·
	19.	4	22	210	581	·	·	·	·	·	·	·	·	·	·	·	·	·	·	·	·	1	·	·	·	·	·	9	·	·	·	6	·	·	·	·	·	·	6	·
	20.	13	29	490	1.254	1	1	·	·	·	·	·	1	·	4	1	·	·	·	3	·	·	·	·	·	·	·	2	·	1	2	·	·	1	·	·	1	·	10	1
	21.	10	135	806	2.715	·	·	·	35	·	·	·	·	·	2	·	1	·	1	·	1	·	·	2	2	·	67	·	·	9	·	·	·	·	·	15	·
	22.	3	4	135	603	·	·	·	·	·	·	·	·	·	·	·	·	·	·	·	·	·	·	·	·	1	·	·	1	·	·	·	·	·	2	·
	23.	7	21	811	886	·	·	·	1	·	·	2	·	1	·	·	·	·	·	·	·	·	·	·	1	·	4	6	·	·	·	·	·	·	6	·
	24.	6	20	118	121	·	·	·	·	·	·	1	·	·	·	·	·	·	·	·	·	·	3	·	5	·	3	·	·	1	·	·	·	·	7	·
	25.	7	17	70	109	·	·	·	·	·	·	·	·	·	·	·	4	1	1	·	·	4	1	4	·	·	·	·	·	·	·	2	·
	26.	9	23	160	297	·	·	·	·	·	·	1	3	1	·	1	·	·	·	·	4	·	4	1	1	·	·	·	·	·	·	7	·
	27.	13	72	620	1.476	·	·	1	2	·	2	·	1	1	·	2	·	·	·	·	19	1	9	3	2	·	·	·	·	·	1	28	·
	28.	9	30	226	661	1	·	·	1	·	·	1	·	·	1	1	·	·	·	3	·	1	1	·	·	·	·	·	20	·
	29.	39	64	403	1.589	1	2	1	1	1	1	1	1	·	1	1	2	1	·	1	1	1	1	1	1	2	1	5	1	2	·	1	1	1	1	1	·	27	·
	30.	5	23	1.132	2.052	·	·	·	·	·	·	·	·	1	1	2	·	·	2	·	·	·	·	17	·
	31.	10	56	742	1.541	·	·	1	·	·	·	5	1	2	·	1	·	1	6	2	6	7	·	·	·	·	29	·
	32.	7	59	93	536	·	1	·	·	·	3	2	·	·	1	38	·	1	·	·	·	·	13	·
	33.	5	22	67	56	·	·	·	·	·	1	·	2	9	1	·	1	·	·	9	·
	34.	9	23	499	577	·	·	·	1	·	1	1	·	2	1	2	12	·	·	·	·	2	·
	35.	6	10	363	135	·	·	·	·	·	1	3	1	·	1	1	·	·	3	·
	36.	2	16	676	464	·	·	·	·	·	1	·	·	·	·	15	·
Total of Sect.			**670**	**7.831**	**16.076**	3	4	1	39	4	1	5	1	3	2	8	13	6	5	1	14	3	3	3	46	9	53	4	57	114	13	·	2	1	20	1	1	1	1	227	1
Sum total			**1.413**	**15.617**	**20.133**	9	8	1	59	5	6	6	14	3	5	14	38	23	18	6	37	7	19	6	62	25	199	7	126	130	16	2	7	11	36	1	3	2	1	494	6
Numb. of Cl. in which the Board District exhibits						7	5	1	10	4	4	5	5	2	4	8	14	16	8	6	12	7	7	5	11	12	30	6	26	17	10	1	5	7	11	1	3	2	1	36	4

SECTION I.

Class 1.

Mining, Quarrying, Metallurgy and Mineral Products.

Stones and Earths. The inland consumption of building and working Stones is completely provided for by numerous Quarries of Limestone, Marble, Sandstone, Granite, Gneiss and other Rocks, spread over the whole surface of the Austrian Empire. Gypsum, hydraulic Lime, refractory Clay, Sulphate of Baryta, Kaolin, coloured Ea ths and other raw mineral products for Industry are likewise occurring in more or less abundant deposits within the boundaries of this Empire. In general, the commerce of both these materials between Austria and foreign parts is comparatively of little importance; in 1859, its value did not exceed 201,600 florins Austrian money for imported, and 360,000 florins for exported articles. Raw Alabaster and Marbles, Emerly, „Meerschaum" and Millstones were the chief objects of import.

Graphite. Graphite holds the first place among the objects of export, its productions having more than doubled since 1851 and amounted to nearly 200,000 Cwts in 1860. Considerable quantities of whetstones are likewise exported from Galicia to Russia. Among precious Stones, the Garnets of Bohemia and the noble or iridescent Opals of Hungary are objects of export.

The breaking and working of Stones and mineral products (Mining in the strict sense of the term excepted) gives occupation to a total of 25,000 persons and these productions represent yearly a value of 38 Millions. In this total, the value of the produced raw materials is represented by 28 Millions and the value of worked products by 10 Millions.

Peat. Nearly one Million Cwts of Peat are yearly produced in the Austrian Empire, which are almost entirely consumed for domestic use, except a small portion, used as fuel in Carinthia, Styria and Salzburg for the refining of Iron, or in Bohemia for the high-furnaces. In general, the consumption of this sort of fuel has hitherto remained far behind the rich layers accumulated in the extensive peat-bogs of Carinthia, Bohemia and Styria etc.

Mineral Coal. Since the year 1831, the mining of the rich layers of Black and Brown-Coal in the Austrian Empire and the consumption of this fuel for industrial purposes has increased in enormous proportions. The quantities dug up are as under: in 1831 . . . . 4,000,000 Cwts.

        „ 1851 . . . . 24,000,000 „
        „ 1860 . . . . 72,000,000 „

The most extensive working of mines has taken place in the Coal-fields of Bohemia where the products amounted in 1860 to: Black Coal 18,000.000 Cwt, Brown 14,000,000 Ctw.

The increase of the commercial movement in Coal is represented by the following numbers in Cwt.

    Import: 1851 . . . 1,516.700   1,064.100
    Export: 1860 . . . 4,687.000   5,576.000

Salt. The production of Salt is a monopoly of the Imp. Government of Austria. The Alpine Provinces produce, nearly exclusively, Culinary Salt; the Carpathian Salt-works furnish chiefly Rock-Salt, and Sea-Salt is produced in establishments along the coast of the Adriatic. The whole amount of Salt produced by these three methods, is shown in the following numbers:

    in 1841 . . . . 6,470,000 Cwts.
    „ 1851 . . . . 6,731,100 „
    „ 1859 . . . . 7,754,000 „

Beside these quantities, an average amount of 300,000 Cwts of Sicilian Sea-salt is imported yearly for Italian consumption, on the other hand, rock and culinary salt is sold to neighbouring Governments at prices fixed by special agreement. The amount of Salt thus exported nearly reaches 1,000.000 Cwts.

Iron. In quantity as in value, Iron holds by far the first place amongst all Metals, produced by the mining and metallurgical Establishments of the Austrian Empire. In 1860, nearly 20,000.000 Cwts of Carbonate of Iron, Sphaerosiderites, brown Hydrate, red oxyd and magnetic oxydule of Iron have been dug up to be worked in 279 smelting furnaces, fed, for the most part, with charcoal. In 1860, these furnaces gave 6,200.000 Cwts of smelted Iron, among which were 700.000 Cwts of cast articles.

This production not being sufficient for fully providing the Austrian Refining-establishments with raw materials, an amount of

        smelted Iron . . . . . 182.300 Cwts.
        and raw cast articles . 21.800 „

were imported in the course of 1860.

Leaving aside the consumption of smelted Pig Iron for railroads and machinery (which will be spoken of in another place), the Austrian establishments for refining Iron have produced, in 1860, an amount of 2,000.000 Cwts of all sorts of hammered or rolled Iron including:

    Puddled Iron . . . . 832.000 Cwts.
    Square Iron („Winkel-Eisen") and plates
        for ships and other purposes . . . . 200.000 „
    Plate Iron . . . . 240.000 „

Steel. The Steel-works of the Austrian Empire produced in 1860

        Raw Steel (for sale) . . . . . . 110.000 Cwts.
        Hammered Steel and refined Steel 112.000 „
        Cemented Steel . . . . 18.000 „
        Cast Steel . . . . 20.000 „

The quantities of refined Iron and Steel produced in the Austrian Empire exceeding by far the demands of home Industry, they take a conspicuous place among the articles of export, as shown by the following numbers, relating to the year 1860:

	Import	Export	Value
Refined Iron, Steel	29.500 Cwts.	—	400.000 flor.
Plate Iron and wire .. — —		211.500 Cwt.	3,440.000 „

Metals. The amount of o t h e r M e t a l s and m e t a l l i c O r e s, produced in 1860, stands, as follows:

Antimony (metallic and Ores) .	8.767 Cwts
Cadmia	131.043 „
Chrome-ores	18.974 „
Copper	52.665 „
Gold	32 „
Lead	140.020 „
Litharge	25.342 „
Mercury.	4.697 „
Nickel (reguline)	35 „
Nickel, unrefined („Nickelspeise")	67 „
Nickel and Cobalt-Ores	5.711 „
Silver	6.797 „
Tin	1.305 „
Wolfram (Tungsten) Ores . .	112 „
Zinc	26.028 „

With the exception of Mercury and Lead, being, in their unrefined metallic state, notable articles of export, considerable quantities of the above-named raw Metals, and especially Gold, Silver, Copper, Tin and Zinc, are imported in large quantities, partly to supply the wants of the Mint offices, partly to provide inland Industry with raw materials.

The import and export movement during the year 1861 is represented by the following numbers in Cwts.:

	Import	Export
Gold in bars . . .	11·68	0·29
Silver „ „ . . .	2.812·861	2·94
Mercury . . . . ,	—	2.515
Lead-unrefined . .	1.433	2.461
Litharge	266	3.603
Tin, unrefined . .	9.086	382
Zinc	18.589	4.061
Copper	26.504	5.996
Nickel	62	12

The total number of w o r k m e n, employed in Mining establishments, has been proved to amount to 100,000. Fifty-thousand other workmen find employment in Smelting and Refining-establishments.

---

1. BAADER'S Son, Commerce of Minerals. *Vienna, Wieden 797.* Minerals, Rocks and Organic remains from the Austrian Empire.

Specimens of Collections for Students.

The Exhibitor keeps for selection a most extensive store of Minerals, Rocks and Organic remains, especially from the Austrian Empire and sells them in detail, and in whole collections, in any size and at most moderate prices. Specially to be commended for School establishments and Students, to whom are sold every year some hundreds of collections, such as are exhibited here.

2. BOCHDANOVICS Alexander, Possessor of Mining and Metallurgical establishments. *Zimbró, Comitate of Arad, Hungary.*

• Specimens of Mining productions.

3. BRUNICKI Maurice, Baron. Village of *Pisarzowa, Circle of New-Sandec, Galicia.*

Petroleum.

(See Class 2.)

4. BRUNN-TAUBITZ MINING-COMPANY. *Brunn am Wald,* next to *Gfoehl, Lower Austria.* Agent at Vienna: John Preindelsberger and Son, City 1138.

Graphite.

5. CURTI, Dr. Alexander. Manufacturer of Portland-Cement. *Muthmannsdorf next to Wiener Neustadt, Lower Austria.*

Portland-Cement under varied forms, as: a lion's head, a paper-presser in form of a dog, bricks (with and without admixture of Sand).

This manufactory is calculated for an annual production of 40,000—60,000 Cwt of Portland-cement of the same quality as the exhibited objects. The bricks (either of pure cement, or with an admixture of coarse Sand in the proportion of 1 to 3, 4, or 6), are attested by the stamp they bear, to have been cast (not „pressed") in presence of a Commission of the Industrial Society of Lower Austria. Those exhibited are intended to be submitted to a trial of their absolute cohesion.

6. D'ELIA Joseph. *Alt-Orsova, Military Frontier.* Specimens of Chromium-ore, No. 1—121.

These ores occur in the mountain-ranges near Alt-Orsova, either in veins or in irregular accumulations, varying in extent and in thickness from one to five feet. Some of them include more than 1,000 Cwt of these ores.

7. DOMOKOS, I. R. and Societary Copper Mining-, Smelting- and Hammering-establishment. *Balánbánya, Transylvania.*

Ores and Slags.
(See Class 31.)

8. DOPPLER John, Stone-cutter and owner of Marble quarries. *Salzburg.* (Medal II. Class 1855, Paris).

Specimens of polished Marble from the principal quarries in the Dukedom of Salzburg and Fancy-wares of Marble.

(See Class 10.)

In 1859, the exhibitor has laid open several new Marble-quarries at Adneth near Salzburg), from which two and three coloured (white, red and blue) Marbles are extracted. These varieties are particularly fit for objects bearing a sumptuous character and have already been used for fountain and stair-balusters in the new Exchange of Vienna, and for an altar in St. Sepulchre's Chapel at Jerusalem. The Exhibitor sells this Marble in rawly hewn pieces (up to 150 Cubic feet) for 4—7 Austrian florins per Cubic foot (loco Salzburg) and executes drawings of every sort in finished stone-cutter's work at most moderate prices.

9. EGGER S. Ware-house of Coins, Antiquities and objects of Natural History. *Pesth, Dorotheer-Gasse 11.* Depot at Vienna: City 1134. Agent at London: M. Davidsohn, 32, Wilson Street, Finsbury.

Minerals from Hungary and other countries: Euchroïte, Coelestine, Aragonite, Libethenite, Lettsomite etc. A Clasp for a Hungarian-Mantle („Mente") in antique style, beset with genuine Pearls, Emeralds and Rubies. A Letter-presser in shape of a table, with Fruits composed of Oriental Pearls on a plate of Lapis lazuli.

(See Class 33.)

This ware-house possesses an extensive selection of the rarest Antiquities, of Greek, Roman, Celtic and Mediæval (especially Hungarian) Coins and Medals, of Armours, Instruments, Goblets of gold, silver and bronze Ornaments in antique and „Rococo" style. The Natural-History division is constantly provided with a choice store of Minerals, Rocks and Petrifications, of Lepidoptera and Coleoptera, of marine, fresh-water and land Shells, of Plants and Algae, skins and eggs of Birds etc. Catalogues are delivered gratis.

10. EGGERT A. and Comp. *Mugrau, District of Krumau, Bohemia.*

Graphite.

11. EIBISWALD (I. R. Steel- and Ironwork-Administration). *Eibiswald, Stiria.*

Cast-Puddling and Cemented Steel; Iron.

12. ESCHER Henry. Stabilimento industriale. *S. Andrea di Rovigno, Trieste.*

Specimens of Cement and Cement-mortar from S. Andrea.

These specimens of Cement-mortar are partly pure Cement, partly a mixture of $1/3$ of Cement with $2/3$ of Sand, having lain in water between 24 hours and 8 days, for trying their relative power of resistance. Price of Cwt sporco loco Trieste: 3 florins 50 Kreutzers Austrian money (at the rate of 13 florins = 1 £ = 4 s. 4 d. per English Cwt.)

13. FALKENHAIN Count Theodore, Possessor of Landed estates. *Kyowitz, Silesia.*

Slate-plates for paving and roofs.

(See Class 4.)

The Slate-quarry of Kyowitz has been opened two years ago. The numerous remains and prints of Sagenariae, Calomites and Trichomanites, occurring in these Slates, show them to belong to the „Culm" formation. Besides local demands, the slate-plates find their way to Vienna, Pressburg, Pesth, Cracow and Warsaw.

14. FINANCES (I. R. Ministry of). *Vienna.* For the I. R. Administration of Saltworks at Hallein, *Upper Austria.*

„ Mining-Forest- and Saltwork-Direction at Klausenburg, *Transylvania.*

„ Direction of Salt-works at Maros-Ujvár, *Hungary,*
„ „ „ „ „ „ Marmaros, *Hungary,*
„ „ „ „ „ „ Wieliczka, *Galicia,*
„ „ „ „ „ and Forests at Gmunden, *Upper Austria.*

Rock- and boiled Salt from the Government Saltworks. Iron, Marble, Sandstone, Chalk and Gypsum.

(See Class 4.)

The Rocksalt-formation of Marmaros occupies an area of many square miles, in which the Rock-salt occurs in shape of a continuous mass with comparatively insignificant interruptions. Its thickness varies between 40 and 100 Vienna fathoms; it must be, however, remarked, that complete working takes only place, where the salt exists with the least possible admixture of extraneous substances. Wherever this rocksalt-mass has been opened, the nethermost strata have proved to be the purest and those most worth being worked, the upper portions being generally impure and will not fit for consumption, although the clay mixed with them does not exceed the average proportion of one-fifth. The exploration is conducted in ascending order; next to the soil of the mine, masses („benches") of rock-salt, 1—2 fathoms in length, 16 inches in breadth and 9 inches in depth, are detached and then subdivided into regular parallelopipeda of 75—80 ℔ each. The salt comes into commerce under four different forms. 1) in regular fragments: *a)* in pieces of 75—80 ℔ each; *b)* in pieces between 74 and 50 ℔; 2) in irregular fragments: *a)* from 49 down to 25 ℔; *b)* under 25 ℔. Presently three saltworks (Rhónaszék, Szlatina and Sugatagh) with 12 Salt-mines, of a subterraneous area of 18.073 square fathoms, are in activity within the district of Mamaros. The annual production of Salt, as regulated by the probable demands, amounting to 1,000,000 Cwt; and the specific weight of this substance being = 2·16; the yearly working, as described above (supposing it to go on at an equal rate over the whole area), would take away a layer of 14 inches in thickness and, consequently, lower the soil of the Salt-mines at the rate of 14 inches every year. The breaking, however, is conducted alternately, as it appears more profitable with respect to the over-ground strata. Only Rock-salt is produced in the Marmaros; the evaporation of the subterraneous waters, although sometimes supersaturated with salt, being too expensive. These waters, containing in solution above 200,000 Cwt of Salt, are therefore conducted into over-ground water-currents and, only for a small proportion, used for curative baths. According to a calculation comprising a series of years, the average proportion between pure and impure salt is as 100 to 20 for Rhónasség, 100 to 25 for Szlatina, and 100 to 5 for Sugatagh. The Marmaros Salt (especially of quality 2·6) is converted into economical salt for stabled cattle by the admixture of 2 p. C. of Oxyd of Iron and $1\frac{1}{2}$ p. C of powdered Charcoal; about 80,000 Cwt of this mixture come yearly into commerce. The workmen, employed in the breaking of Salt, number 1,121 persons.

The Iron casting establishment of Fejérpatak has one high-furnace of 46 feet height and a mechanical work-shop. The yearly production is: raw Iron: 17,000 Cwt; cast Iron: 2,000 Cwt. The raw Iron is converted into soft Iron (13,000 Cwt every year) at Kobolopojana. The ores feeding the high-furnace, are: 1) Red Oxyd of Iron with 18—38 p. C. of metal; 2) Carbonate of Iron (Iron-spar) with the maximum of 68 p. C. of metal 3) Sphaero-

siderites with 15—26 p. C. of metal; 4) Diorites with 13 to 18 p. C. of metal. A mixture of 75—84 p. C. of these ores with 25—16 p. C. of Limestone gives 22 p. C. of raw Iron. The consumption of Charcoal for one Cwt of Iron, is 10 cubic feet for the high-furnace and 24 cubic feet for the soft Iron. The raw Iron, generally of the grey sort, gives soft Iron of very good quality. The Iron-works give employment to 163 persons.

15. GEOLOGICAL INSTITUTE, Direction of the I. R. *Vienna.* Silver Medal I. Class at the Paris Exhibition, 1855. (See **646.**)

Objects brought to Exhibition:

Special Geological Map of the Archdukedom of Austria beneath and above the Enns. Special Geological Map of the Dukedom of Salzburg. Special Geological Map of the Kingdom of Illyria and of the Dukedom of Styria. Special Geological Map of the Kingdom of Bohemia. General Geological Map of the Kingdom of Hungary. General Geological Map of the Lombardo-Venetian Kingdom. General Geological Map of the Principality of Tyrol and Vorarlberg. General Geological Map of the Serbian Wojwodina, Temes Banat and Banatian Military Frontier. General Geological Map of the Kingdom of Galicia and the Dukedom of Bukowina. General Geological Map of the Grand-Principality of Transylvania. Annals of the Imp. R. Geological Institute, 1850—1860, 11 volumes octavo. Transactions of the I. R. Geological Institute. Volumes I, II and III quarto. A collection of specimens of Fossil Fuel occurring within the Austrian Empire: Coal, Brown coal, Lignite, and Peat. A collection of artificial Crystals, produced in the Laboratory of the Imp. R. Geological Institute by Chevalier Charles de Hauer, Superintendent of this establishment.

(See Class 29.)

The above-mentioned collection of specimens of Austrian coals contains also those from the coal-pits at Fünfkirchen in Hungary belonging to the Danube Steam Navigation Society. These coal-pits, which were first begun to be worked in 1853, lie half a mile north-east from Fünfkirchen. The entire area of this coal-field, granted to the society conformably to the laws of mining, amounts to 1,109.339 square fathoms, or 884/5 Austrian square rods of mines. The quantity of coals, that may be dug out from the present coal-field, is to be estimated at a minimum of 500 millions Cwts, if only calculated for a depth of 150 fathoms. Most of the layers afford small and grit coal among which there are scarcely more than 5% lump-coal. The coal of Fünfkirchen is very good caking-coal, excellent for coke-burning and much liked as smithy-coal. On account of its great proportion (from 80 to 85 %) of carbon and surplus of carburet of hydrogen, this coal emits great heating power, is easily inflammable and burns with a brisk flame. The results of assaying the different kinds of coals in these layers are: contents of ashes from 5 to 15 %, sulphur from 1 to 3 %, and carbonized coal (coke) from 78 to 84 %. Coking in the furnaces of those works amounts to an average weight of 75 %. In order to remove the inconveniences in using grit-coal for grate firing, there were, in 1861, made experiments to transform grit-coal, by means of pressing, into artificial lump-coal, and to mix coal-grit with materials which on the one hand prevent close caking in firing, but on the other hand serve as cement to bring about the requisite firmness. The steam-engines employed in the working of coals have nominally a total of 75 horse-power, with which the swallet is also lifted. The production of coal consisted in:

1859	2,043,248 Cwts.
1860	2,323,570 „
1861	2,580,200 „

In 1861, about 800 workmen were employed herein. Besides there were, in 1861, produced about 10,000 Cwts of cokes in 2 furnaces, and 1,600 Cwts of pressed coal (briquets). At the end of 1861, 4 newly built coke-furnaces, that were building in the course of the same year, were put to use. so that now 6 furnaces for the production of coke are working for providing the locomotive-engines of the Mohács-Fünfkirchen railway with cokes.

The coke-furnaces are to be filled from above, the uncharging being done by means of a horizontal machine for squeezing them out. The side-walls and soles of the furnaces are provided with flues through which the combustible gases pass into a common tube and through this into the chimney.

For lodging a multitude of workmen necessary for the mining-works, persons that were to be procured from remote places, since there was want of them in these parts, the society found reasons for building proper dwelling-houses for them on the society's own grounds. The present mining colony numbers already 63 houses, with a church of their own, a cemetery, a school and an hospital. In order to provide the colony and the cauldrons with fresh water, a costly aqueduct was laid. For the support of sick and disabled miners, their widows and orphans, a companion-fund (mutual funds for sustenance) was instituted in 1855, that rose to the sum of 20.000 fl. at the end of 1861. Thus the material and moral benefit and weal of the miners has been taken care of, and it is to be hoped that these coal-mines which, under the direction of the manager, Mr. Schroll, during the short period of 8 years, rose already now to a first-rate mining concern in Austria, with a developed opening, considerable production and, besides the requisite working contrivances, with a colony exceeding 1.000 souls; it is to be hoped that with such progressive extension conformable to the area and productiveness of the coalfield, coupled with other favourable circumstances, such as the immediate vicinity of the railway, these works will prosper so far as to become a great industrial undertaking useful to the Danube Navigation Company, by constantly providing their ships with superior fuel at comparatively cheap prices, and by the sale of their products to other consumers to exercise a beneficial influence on the animation of commerce and industry in Hungary and even in other provinces.

16. GESSNER Edward, **POHL** William, **ULLBRICH** Otto. *Müglitz, Moravia.*

Graphite.

(See Class 2.)

17. GOLDSCHMIDT S., Jeweller and trader in Jewels. *Vienna, City 260,* Possessor of the Emerald-mine in the Dukedom of Salzburg.

A series of Emeralds in their matrix, of Smoke-Topazes (yellowish-brown Rock-Crystal) and of Rock-Crystal.

These Emeralds are found in the dukedom of Salzburg in an altitude of 8,700 feet, on a steep cliff, hitherto only accessible by aid of ropes. Fragments of Emeralds found on several occasions in avalanches having rolled down into the neighbouring valley, had led to the discovery of their native locality. This place having never been made before an object of technical investigation, the Exhibitor, in the last days of August 1861, ventured an excursion to this inaccessible cliff and, after a search of a few days, was fortunate enough to discover the Emerald specimens brought to exhibition. He has now purchased from Government the whole of the surrounding surface in an extent of 175 Austrian acres („Joch") and has opened a regular breaking of Emeralds and other Minerals, as white and yellowish-brown Rock-crystals. Orders for the articles, of which specimens are exhibited, will meet with prompt execution. The Exhibitor has never yet brought to exhibition any other articles of his trade, existing since the year 1839.

18. GROHMANN Adolphus. *St. Wolfgang near Ischl, Upper Austria.*

Coal and Peat.

19. HOFMANN Ernest de. *Alt-Orsova, Austrian Military Frontier.* Ware-houses at Alt-Orsova and Galatz. Agent-general for foreign parts: Guillaume Brand, at Paris.

Four Specimens of Chromate of Iron, containing 60·51, 56·43 and 52·19 p. C. of Oxyd of Chromium.

20. IDRIA, I. R. Mining-office. *Idria, Carniola.*

Cinaber, raw and evaporated Aethiops mineralis, Cinaber in pieces, Vermillion.

21. JACOMINI-HOLZAPFEL-WAASEN, Francis, Knight of the Empire, Possessor of Manufactories at *Nötsch, Villach, Carinthia.* Bronze M. Vienna 1845. Honourable mentions: Munich 1854 and Paris 1855.

Lead and Sulphuret of Zinc.

The softness and ductility of the lead produced at Bleiberg makes it particularly appropriate for tubes, plates and leaftin, as also for the manufacture of Cerussa. Its superiority is a generally known and never contested fact and the oxyds, prepared out of it, find their way even to England. The exploration of this metal is in the hands of a number of owners of mines, formed into distinct Mining Societies. Government has an extensive share in this branch of Mining Industry comprising a total

of 470 mining fields of the first class („Haupt-Grubenmassen"), 80 workshops and smelting-houses, 14 mining forges and 44 overground-fields („Tagmassen") called: „Bachstall-Lehen." Of these, 378 mining-fields of the first class, 13 workshops and smelting-houses, 7 mining forges and 12 overground-fields are the Exhibitor's property. The total production of the Bleiberg works, with little changes since nearly 50 years, when Carinthia had been restored to the Austrian Empire, comes to a yearly amount of about 40,000 Vienna Cwt, subordinate circumstances being left out of consideration. The Exhibitor's establishment, according to a decennal average, produces about $1/6$ (6,500 Vienna Cwt), of refined Lead, besides a yearly average production of about 3,000 Vienna Cwt of Sulphuret of Zinc. The annually produced 6,500 Cwt of refined Lead are obtained from about 9,700 Cwt of Lead - „Schlichs", the residuum of 50,000 to 60,000 Cwt of raw ores, more or less rich in Lead, after they had passed through the crushing-cylinders, the stamps and the washing-tables. For a period of ten years, the average yearly amount of Government-taxes of every sort, contributions for the repair of roads and for districtual funds etc. is 5,343 Austrian florins. For the same period the average of the annual contributions for the pensionaries-fund („Bruderlade") amounts to 2,393 Austr. florins. The expenses for salaries, wages and materials are equivalent to a yearly average sum of 68,603 Austr. florins and, with the expenses for postage, stamps, transports and lawyers, to a total of 76,534 florins, this total within the decenial period from 1852 to inclus. 1861 having reached the sum of 765.346 florins. These numbers, taken from regular accounts, together with the quality of the objects of production, may give an idea of the relative importance of the Exhibitor's mining-establishments. These give employment to 13 officials and 359 workmen (among them 93 females) besides a number of wood-cleavers and colliers. The productions of the mines and smelting-houses find their way into the manufactories at Villach and Klagenfurt or serve to supply the Exhibitor's own industrial establishments at Noetsch in the Gail valley. The Sulphuret of Zinc either supplies the wants of smelting-houses in Carniola and Stiria or is worked out into a pigment of the Exhibitor's own invention in his establishment at Noetsch. This pigment, examined and publicly commended by the Industry-Society of Lower Austria for its good quality, resistance to atmospheric agents and cheapness, is an object of extensive demands in Carinthia and in the neighbouring Provinces. In the principal mines of the Exhibitor's and his share-holders' mining territory, three water-column engines are in permanent activity, one of them bringing up the subterraneous waters from a vertical depth of about 1,300 Vienna feet. Four other engines, moved by water-power bring to day-light, from a depth of nearly 1,300 Vienna feet, the ores cut out in the principal mines. An adit opened in 1789 in a deeper slope of the valley, has already penetrated into the interior within a length of more than 2,800 Vienna fathoms (about 5,600 mètres) from W. to E. and in a vertical depth of 288 Vienna feet beneath the soil of the valley. This adit, receiving the waters of 8 chief mines, providing them with respirable air and making possible the exploration of their deeper horizons, is intended to be driven still about 1.000 Vienna fathoms (2,000 mètres) farther, so as to reach the deeper subterraneous regions in the Eastern masses of the Bleiberg metalliferous mountain-group. This mine is in the Exhibitor's possession since a period of 32 years, and he has constantly endeavoured, not only to concentrate it and enlarge it by purchases, but also to adopt any improvements, issuing from scientific or practical progress, as far as they proved compatible with economical advantage and local circumstances. He has first introduced the use of railways in his workshops, superseded the old ore- and „schlich"-troughs by those of improved construction, as considerably diminishing the loss of valuable substances. He has first used crushing-engines instead of the old imperfect ore-mills, introduced the use of cast-iron stamps with a changeable nucleus of hard iron, besides other highly profitable improvements, whose enumeration would take too much room here, and which have been generally adopted by the owners of the neighbouring mining establishments. Here may be also mentioned the construction of an extensive establishment, under the name of „Jacomini-Hütte", in the valley of „Windischen Graben", comprising a crushing- and a stamp-work; „Schlich"-works, hree oscillating washing-tables and five smelting-furnaces; the construction of a manufacture of Litharge, Minium- and Zinc-pigment, and of a forge with three fires and one anvil at Noetsch in the Gail valley and of a manufactory of wire-ropes at Bleiberg-Kreuth, supplying the wants of the whole of private and Government mining establishments in the Province. Besides the distinctions enumerated above, the „Société nationale d'Encouragement" of Paris has, in 1856, presented the Exhibitor with a silver Medal.

22. JENBACH, Iron-work-Direction. *Jenbach, Tyrol.*

Refined Steel, refined Steel-rails.

23. JOACHIMSTHAL, I. R. Superior Mining-office. *Joachimsthal, Bohemia.* M. of honour, Munich 1854. M. I. Cl. Paris 1855.

Oxyd of Uranium (Uraniate of Ammonia), Oxyd of Uranium combined with Soda (light and orange yellow); Vanadate of Soda.

24. KAISERSTEIN Baron Francis. Graphite Mine. *Raabs, Lower Austria.*

Graphite, raw and washed in larger and smaller regular and irregular fragments.

25. KERTSCHKA Francis. *Brunn am Walde, District of Gföhl, Lower Austria.*

Graphite.

26. KOETTIG A. H., Mining-master and Possessor of a Mine. *Teplitz, Bohemia.*

Brown coal, Cokes.

27. KRATZER Leopold, Miner. *Rosenberg, Comitate of Liptau, Hungary.*

Regulus Antimonii.

28. KRIEG Charles, Possessor of Graphite-works. *St. Marein, Inferior Austria.*

Graphite, raw and washed.

29. KRONSTADT Mining and metallurgical Association of Shareholders. *Zillthal, Transylvania.*

Coal and Iron-ores.

30. LACZAY SZABÓ Charles, Possessor of Landed estates. *Sárospatak, Comitate of Zemplin, Hungary.* Agencies: at Vienna, Joseph Schwarz, City 771, at Bude: John Blum, possessor of a Steam-mill.

Quarz-millstones first quality.

The Millstones exhibited here may enter into competition with the best at present known, and may even be preferable to those of France, as they surpass them in hardness and porosity and want less sharpening. They are far cheaper in price to those from France, which they have partly superseded in water-mills as in steam-mills. The occurrence of this material in Hungary is therefore of some importance for the neighbouring countries.

Prices in Austrian florins with an earnest of 50 florins for the pair, to be delivered in 6 weeks. Loco Sáros patak.

per inch Vienna measure the pair:

30 inches	. . . 180 florins.	40 inches	. . . 230 florins.
31 "	. . . 184 "	42 "	. . . 252 "
32 "	. . . 190 "	45 "	. . . 290 "
33 "	. . . 195 "	48 "	. . . 325 "
34 "	. . . 200 "	50 "	. . . 350 "
35 "	. . . 205 "	52 "	. . . 370 "
36 "	. . . 210 "	54 "	. . . 390 "
38 "	. . . 220 "	56 "	. . . 400 "

31. LENGYEL And., I. R. Docimast. *Nagybánya, Comitat of Száthmar, Hungary.*

Crystallized Pyrites and Antimony.

32. LEOBEN Chamber of Commerce and Industry. Collective Exhibition of Mining productions, Iron- and Steel-Manufactures.

(See Classes 31 and 32.)

The individual partners are:

Drasche Henry, Coal mines. *Leoben, Stiria.* Domicile: Vienna, Wollzeile.

Specimens of Coal; Drawings and Plans.

Fürst Ignace and Mary, possessors of Iron-works. *Büchsengut and Thörl near Aflenz, Stiria.*

Nail- and Cement-steel-Iron, worked Iron for gun-barrels, Iron wire, bars for nails, rolled Gun-barrel.

Eisenerz I. R. Chief Association ("Hauptge-werkschaft") *Eisenerz, Stiria.*

Iron and Steel.

Henkel von Donnersmark Count, Iron-work. *Zeltweg, near Judenburg, Stiria.*

Iron.

Neuberg I. R. Administration. *Neuberg, Stiria.* M. of honour: Munich 1854. Silver M. I. Cl. Paris 1855.

Specimens of Ores and raw Iron, Plate Iron for cauldrons, angular Iron, Iron for parts of Machinery, Iron axle-trees for Locomotives, fracture-specimens of Puddling-steel and Iron-tyres, Puddling-steel Tyres, Black-smith articles.

Pfeifer John. *Spitzenbach St. Gallen, Stiria.*

Iron and Steel.

Radmeister (Work-owners) Community. *Vor-dernberg, Stiria.*

Raw Iron.

Schwarzenberg Prince Adolphus, Possessor of Iron-works. *Murau, Stiria.*

Iron articles.

This Iron-work is in existence since nearly 200 years. Steel-like raw Iron is converted into superior Steel (renowned under the name of „Paalstahl") for cutting instruments, sabre blades etc., as also into Brescian steel for scythes and sickles. Hammered Iron of every sort, Black-smith articles and Cement-steel Iron of superior quality are likewise produced there. The annual production amounts to 70,000—80,000 Vienna Cwt of raw and cast Iron, 25,000—30,000 Vienna Ctw of Steel of different sorts and 20,000—25,000 Vienna Cwt of hammered Iron of every description.

Sessler Victor Felix, Iron-work. Krieglach.

Iron.

33. LIECHTENSTEIN Prince John. Iron-work. *Alois-thal, Moravia.*

A thin plate, cast out of the high-furnace.

(See Class 4.)

34. MANGER Rodolphus, Possessor of Mines. Domicile: *Prague 853-2.*

Iron Pyrites.

35. MAYR Francis. Possessor of Iron-works. *Leoben, Stiria.* Puddling and Rolling-works at Dona-witz, Rolling-work at Gemeingrube, Manufactory of Cast-steel at Waasen, Hammer-works at Waasen, Töllerl, Goss, St. Michael, Höll, Bruck on the Mur; Coal mines in the Tollinggraben.

Assortment of hammered Iron, common hammered Iron tested for its quality and fracture, refined hammered Iron, Cement and Puddling-steel, black sheet Iron, assortment of Cast-steel, Cast-steel plate for Steam-boilers, Cast-steel Rails, quality- and fracture-samples of Cast-steel; drawings of patented Furnaces and Machinery.

The Iron and Steel sorts are axclusively produced out of Vordernberg and Innerberg raw Iron; the superior quality of these articles has become notorious. The annual production of Iron, Steel, melted, cemented and puddling Steel amounts to 180,000—200,000 Vienna Cwt, of Cast-steel to 12,000 Vienna Cwt.

36. MAYR Max, I. R. Mining-officer. *Vienna, Land-strasse 703.*

Specimens of common and hardened Steel.

37. MERAN Count. Iron-works. *Krems, Stiria.*

Iron, Steel articles and Coal.

Objects brought to exhibition: 1. Raw materials: raw Iron and Coal; 2. Materials for farther work: plate and coarse Iron, raw Rails; 3. stretched Iron (current commercial Iron, made exclusively out of Puddling Iron); 4. rolled Iron of Puddling Iron, plates made with one heating (particularly soft and smooth), 5. Puddling steel, raw and refined (Scythe-Steel and „Zeugmacher"-Steel), prepared with raw Lignite containing 20—30 p.C. of water; 6. Cast-Steel in 8 assortments, among which No. 7 is eminent for its hardness and tenacity.

38. MITTROWSKY Count Wladimir, *Pernstein, Moravia.*

A cube of Lepidolite.

39. MOSLAVINA Mining Association. *Borik, Croatia.*

Petroleum, Mineral Tar, raw Asphalt.

The establishments for working of Naphta, Bitumen, Asphalt, and Brown coal existing in Croatia near Moslavina and in the Military Frontier near Petravoselo are to be sold. Farther information to be obtained from F. S. Steiner at Agram (Croatia).

40. NAGY Charles de, Possessor of Mines. *Vienna 276.*

Grey Manganese-ore, with 80—90 p. C. of Peroxyd of Manganese from the mine of Ponor, Comitate of Bihar, Hungary.

41. NICKEL George. *Vienna, City 1088.*

Mining and metallurgical Productions.

42. NOWICKI Constantine and **HAUSOTTER** Franois, possessors of Mines. *Grasslitz, Bohemia.*

Plan illustrating the operations intended for re-opening the Copper-mines of Eibenberg, Grünsberg and Schwaderbach. Copper-ores from these localities.

The Copper-mines of Grasslitz, after a period of high prosperity between 1570 and 1632, gradually came into decay in consequence of civil disturbances and, under the pressure of unintelligent management, were completely abandoned in the course of the 18 Century. In the last month of 1858, the Exhibitors began the operations for their re-opening, with special regard to, deeper horizons, where, as they had hoped, they met with still unexplored regions. In the course of these operations, the old works were converted into use as much as possible and recent improvements of more powerful economical effect were adapted to them. The number, extent, level and shape of the re-opening operations are made conspicuous in the Plan. The Copper-ores are resting on a greenish-grey Argillaceous Slate — frequently of chloritic nature — accessorily impregnated with these ores. The metalliferous bed, together with the layers of barren Slate, has a total thickness of 1—2½ fathoms. The ten metalliferous beds, presently known, are prepared for mining in descending order on six more or less extensive areae. The maximum for separated rich ores has proved to vary between 8 and 14 p. C. of metallic Copper; the average percentage of the whole of raw ores hitherto opened for mining, was stated by mining counsellor Knell, at Saalfeld, and Professor Fritzsche, at Freiberg, to be two per cent. According to Mr. Knell's estimate, 738,500 Vienna Cwt of Copper may be extracted between the surface and the soil of the deepest gallery, and about 2,900,000 Cwt of this metal out of the whole metalliferous circum ference. About 15,000 Cwt of Copper are ready for further working in the ores obtained in the course of the re-opening operations. The construction of a stamp-work and of a smelting-house is intended to be begun in the present year. The concessioned area of 301,056 Vienna square fathoms is to be doubled by means of further concession and purchase.

43. PILLERSEE I. R. Mining-smelting- and hammering-establishments. *Tyrol.* M. of honour: London 1851; Bronze M. Munich 1854; Silver M. I. Cl. Paris 1855.

16 specimens of refined steel in 4 qualities.

44. PRAGUE Association for Iron-industry. Chief store-houses at Prague, Vienna, Pilsen, Brünn and in 22 towns of Bohemia.

Coal, Cokes, Iron-ores, Limestones, Raw Iron, articles of cast Iron, hammered, and rolled Iron.

(See also Class 31.)

Coal from the Association's own mines at Kladno, Rappitz, Rakonitz, Wilkischen, Blattnitz, Dobraken and Steinaugezd, with Cokes produced out of it. Iron-ores from Nutschitz, Zbusan, Swarow, Chrbina, Chimiava and Libicow (own mines near Kladno), from the Circle of Pilsen near Rokitzan, Mies, Marienbad, Tachau and other localities (not in the Association's own possession).

Lime stones from Tachlowitz, Tachau, Amberg, Zditz. Raw iron: 1. grey; 2. white, a) made with raw Coal from Kladno; b) with Cokes from Kladno; c) refined raw Iron out of the varieties a) and b); d) made with Charcoal in the Association's own establishment at Carolinengrund and in the leased high-furnaces at Bras, Mitrowitz, Ferdinandsthal, Sorghof and Schwarzhammer.

Cast articles of different sorts, from the Association's own casting-works at Kladno and Carolinengrund. Hammered and rolled Iron in every stadium of working and in every dimen-

sion required for trade, according to the profile extant in the Association's own Rolling-works at Nürschau and Hermannshütte near Pilsen.

Plate Iron (black, white and for Boilers) in every variety of dimension and thickness, from the Association's own Rolling mills at Nürschau, near Pilsen, and Josephshütte, near Plan.

The two tables, with their top-sets, serving for the exhibition of the above named objects are, at the same time, specimens of workmanship in the establishments of the Prague Association for Iron-industry.

Coal	6,954,664	Cwt.
Cokes	686,731	"
Iron-ores	1,403,280	"
Raw Iron made with Coke	379,294	"
" " " Char- coal	103,427	"
Cast articles	46,839	"
Rails for Rail-roads	206,750	"
Tyres and Axle-trees	1,110	"
Rolled Iron		
" common and fine for trade	44,481	"
Hammered parts of Machinery	524	"
Plate Iron f. Boilers	10,330	"
" " black	12,326	"
" " "	6,355	" in boxes of 150 ℔ each.
" " white	5,687	"

At the end of December 1861, the Association was in possession of the following establishments:

Pits for the extraction of Coal: in activity 16; in way of being sunk 3; Pits for extraction of Iron-ores 43; a locomotive-road for communication between Coal- and Iron-pits and the metallurgical establishments 20,851 fathoms in length, with 7 Locomotives and 92 Waggons; roads for mining-carfs: overground 5,358 fathoms; underground 20,907 fathoms; 212 Coke-furnaces with 5 washing-works for Coal and 6 for Iron-Pyrites; 8 furnaces for roasting Iron-ores, 6 high-furnaces heated with Coke and 6 heated with Charcoal; 7 Cupolo-furnaces; 2 Flame-furnaces; 49 Puddling-furnaces; 23 Soldering-furnaces; 9 furnaces for heating Plate Iron; 2 engines for crushing lumps of soft Iron; 6 Hammers moved by steam; 1 large Hammer; 3 Forges for stretching lumps of soft Iron, with 9 hustings; 3 Forges for stretching coarse Iron and Rails, with 15 hustings; 2 Forges for stretching refined Iron with 14 hustings.

The workshops comprise:

Forging-fires 73; Turning-lathes 34; Boring-engines 21; Lever-engines 5; besides a number of other auxiliary engines.

The moving powers are: Steam-engines 68, of 2,164 horse-power and 113 boilers; Water-wheels 9, of 212 horse-power. Total of persons: 74 officials and 5,019 workmen.

45. PRASCHNIKER A. *Stein, Carniola.* Depot of manufactures: H. Waschnitius, Vienna, Wieden 901.

Cement-lime.

(See Class 2.)

Annual sale: 60,000 Cwt. Prices per Cwt. loco Stein 80 kreuzers (package included); loco Vienna 1 florin 80 kreuzers (package included). This Cement-lime, equally to be used in and out of water, has succeeded in coming into good credit. Mixed with sand and gravel, it is particularly fit for „Béton" construction. A portion of this lime, kept in a separate subdivision of the Exhibition-buildings, is intended for being submitted to trials.

46. PŘIBRAM I. R. Mining-office of the I. R. and private Associates; Silver- and Lead-Mines. *Birkenberg, Bohemia.*

Mining productions; specimens taken from the Přibram metalliferous veins. General subterraneous Plan of the Přibram metalliferous region. Model of a continuous Washing-table in use for the concentration of Přibram ores.

47. PUTZER Paul de, Direction of Mining establishments, Coal-mines. *Gonze, near Tüffer, Stiria.*

Brown coal.

48. PUTZER Paul de, Direction of Iron-works. *Store, Cilli, Stiria.*

Cast-Steel, hammered and raw Iron.

49. PUTZER Paul de, Direction of Mining-establishments, Coal-mines. *Peconye, Cilli, Stiria.*

Brown coal.

50. QUAGLIO Julius, Civil Engineer. *Vienna, Weiss-gärber 140.*

Objects made of Hauyne-rock (a new rock, discovered in Transylvania).

(See Class 9.)

51. RAUSCHER & Comp. Hammer-works. *Heft and Mosing, Carinthia.*

Raw Iron and Ores. Sulphate of Baryta.

52. RIEDMAYER Eliza, first Tyrolian manufactory of Asphalt and Mineral productions. *Giessenbach near Seefeld, Tyrol.* Factory: Innsbruck 582.

Asphalt-stone and Petroleum.

(See Class 2.)

53. RIEGEL Anthony. Possessor of Mines. *Fünf-kirchen, Hungary.* Manufactory at Mohacs, Hungary.

Patent compressed Coal, prepared of grit and minute fragments out of the Exhibitor's own mines.

This compressed Coal may be obtained in any required form and size with presses constructed according to the Systems of Middleton, Exter, Corard, Falgière etc. and has been found, by practical proof, to be highly appropriate for domestic uses, for the heating of Locomotive, Steam-boats and fixed Boilers. The proceeding for obtaining this compressed Coal is a new one as to the choice of the connecting substance, procurable in any quantity whatever and at the lowest prices, and has been patented for the Austrian Empire. The connecting substance, composed of organic and inorganic ingredients, is combustible without leaving a residuum of ashes.

54. ROTT V. J. Merchant. *Prague.*

Stones for Polishing.

(See Class 4.)

55. (Omitted.)

56. SALGÓ-TARJANY, St. Stephan Association for Coal-mining. *Pesth, obere Donauzeile.*

Coal and Mining productions.

57. SALT-WORKS (Possessors of). *Pirano, Istria.*

Sea-salt.

58. SALT-WORKS (Direction of). *Venice.* M. I. Cl. Paris 1855.

Specimens of Sea-salt, made 1861. Plans of the Salt-works and of the buildings belonging of them.

The Salt-works of Venetia were the only establishment of this kind, decorated with the I. Class Medal at the Paris Exhibition of 1855, when their production scarcely came to the amount of 50,000 metrical Cwt (about 100,000 Vienna Cwt). In 1861, the production fell but little short of 150,000 metrical Cwt (about 300,000 Vienna Cwt). The climate of Venice, comparatively cold and rainy, is by far less favourable to the production of Sea-salt, than the warmer and drier climate of Sicily.

59. SAULLICH Angelo, Firm: Saullich and Kraft, Patented Portland-Cement-Manufactory. *Perlmoos, near Kufstein, Tyrol.* Ware-house: Vienna, Wieden 791. Prize-M. Linz 1861.

Perlmoos Portland Cement powdered; Perlmoos Portland Cement burnt. not powdered. Bricks for trial of relative strength, Prisms for trial of absolute strength. Testimonials and reports on its usefulness and strength.

(See class 10.)

This manufactory has increased its productiveness, during two years, to 30,000 tuns of cement (a tun = 400 or 450 pounds) and is continually aggrandizing. The manufacture of tiles, pavings and conduit pipes of Perlmoos Cement has already taken a considerable extent and earned full acknowledgment. Prize-Medal of the Agricultural Society Vienna 1861

60. SCHEMNITZ, I. R. Direction of Mines, Forests and Estates. *Schemnitz. Windschacht, Comitat of Hont. Rhonitz, Comitat of Zolyom, Hungary.*

Productions of Mines.

61. SCHWARZENBERG, Prince John Adolphus, from his dominions.

Mining and mineral productions.

(See Classes 3 and 4.)

62. SEYBEL Aemilius, possessor of the first privileged Manufactory of Chemical preparations, of the Mining and Smelting Establishment at Bösing near Pressburg in Hungary and of the Chromium-mines next Kraubath in Stiria. *Vienna, Wieden 26.*

Pyrites from Bösing. Chromium-ores and Magnesite from Kraubath.

(See Class 2.)

Very extensive deposits of Pyrites, although of little percentage in ores, are being worked at Bösing. These Pyrites are used for the manufacture of Sulphuric Acid, Iron-vitriol and Sulphate of Alumina. The Chromium-ores, occurring in the Serpentine of Kraubath, give the best percentage among all the ores of this kind now known; two thousand Cwt of them, holding 25% of Oxyd of Chromium, are extracted every year, to serve for the preparation of Chromate of Potash in the establishment of Liesing next Vienna.

63. SIEGL Adolphus. *Lemberg, Galicia.*

Naphtha (Petroleum), unrefined.

64. SILBERNAGL Baron Julius, Hammer-work. *Waidisch, Carinthia.*

Slags, raw Iron and soft Iron, prepared in open fire.

65. STATE-RAILROAD-COMPANY, I. R. privileged Austrian. *Vienna.*

Páraffine, Mineral oils for burning, Slate-oil containing Benzine, Minerals of different sorts, Sulphate of Copper, Sulphuric Acid, Copper, Iron-ores, raw and cast Iron, half-ready materials, ready Iron articles, impregnated Railroad-sleepers and transversal sections of trunks.

(See Classes 2, 4, 5, 8, 10, 13.)

All these objects come from the territory which the Company possesses in Banat. The Paraffine and the Oils are extracted from carboniferous Slates. The Paraffine enters into fusion at a temperature of 49° C. The Mineral oils for burning are between 0·82 and 0·84°; the Slate-oil, containing Benzine, is 0·755°. The impregnated specimens are beechwood, treated with a solution of Sulphate of Copper according to Dr. Boucherie's method.

66. STEELWORK-ASSOCIATION, I. R. privileged Stirio-Austrian. *Vienna.* For the Steel-works of the I. R. Innerberg principal Association at Weyer, Klein-reifling, Hollenstein, and Reichraming.

Carbonate of Iron, raw Iron, Iron in bars, raw, refined and cast-Steel, Brescian or Milanese Steel.

(See Class 32.)

67. SWOSZOWICE, I. R. Mining and metallurgical Office. *Swozowice, Galicia.*

Sulphur, raw and refined.

68. SZUMRÁK Julia. Chief-shareholder of the Cobalt-Nickel- and Copper-Mines and Stamp-works at Libethen, Kolba and Podlipa. Lives at *Tapio Szele, Comitat of Pesth, Hungary.* Store-house at Libethen. Agent: Williams & Norgate, London, 14, Henriette-street, Covent-Garden.

Cobalt-Nickel-ores, containing: Nickel: 10.4, Cobalt: 12.2, Copper: 5.1, Iron: 9.2, Bismuth: 11.4, Arsenic: 12.1, Sulphur: 8.6, earthy particles: 31.0 per cent.

69. THROR & Comp. First Hungarian I. R. privileged manufactory of Millstones. Specimens to be seen at Mr. Fr. X. Spannraft, *Vienna, Schottenfeld 341.*

Millstones.

In the course of these last years, the Exhibitors succeeded in discovering in Hungary Quarz beds, similar to those, used in France for mill-stones. They entrusted Mr. Frederick Haszlinszky,

Professor of Natural Sciences at the College of Eperies, with researches for such Quartz beds, and, after considerable expense of time, trouble and money, such a bed was discovered, at last, in the last days of September 1857. The bed of freshwater Quarz here in question, is situated at the upper extremity of the village of Fony, in the Comitat of Abaujvár, 3 miles from the railroad-station of Forro-Encs and 6 miles from Tokay. Specimens had been sent into Germany and France, for trying their qualities. The results having proved favourable, whole composed Mill-stones were brought to trial and these experiments having likewise succeeded, the present Company resolved to establish a Mill-stone-manufactory after French models and principles. In spring 1858, this establishment was first called into activity. Since this time, above 500 pair of Mill-stones found their way into several mechanical steam- and water-mills of the Austrian Empire, where they superseded the use of French mill-stones. The first of these pairs had been sent to the Vienna I. R. privileged Steam-mill. These Hungarian mill-stones, known under the name of „Fony mill-stones" offer the following advantages: They are pure freshwater Quarz, without any extraneous substance. Their texture, instead of being spongeous, as are the French mill-stones, is compact; consequently they act more by c u t t i n g than by friction, and the substances to be ground are not subject to be over-heated. The Fony Quarz is harder than the French variety in the proportion of nearly 25%, is less worn out by grinding, leaves the substance it acts upon without admixture of dust, and gives more and better flour. The flour, thus obtained, is f i n e, white, of far better lustre, may be kept without alteration, and possesses a greater proportion of nutritive substance; when kneaded into dough, it takes up more water, to the great advantage of consumers. The Fony stones do far more work in the same time and keep their edge for a long period. They want sharpening but twice in about 3 weeks and, if duely managed, may work incessantly for 12, 15 and even 20 years, before they become unfit for farther use. The average thickness, together with the plaister cover, is 12 to 13 inches for the upper moveable and 11—11½ inches for the lower fixed stone. Altering these proportions would not be advisable, the Quarz-mill-stones being intended for a c u t t i n g rather than. for a crushing action and, on this account, not to be paralleled with the above-mentioned Sandstones. Besides mill-stones for grinding corn, the Exhibitors commend their mill-stones for grinding mineral substances, ores, pigments, bones, gypsum, cements, barks for tanning, etc.

Those purchasers, who are not in the state to make their choice by personal inspection or through persons entrusted by them, may be pleased to indicate, for what kind of corn and for what mode of grinding the stones in demand are intended. If sharpening or fitting out the stones is required, precise written dates or drawings would become necessary and the direction of rotation (to the right or to the left) to be given to the moveable stones is to be indicated. Prices for a pair of stones, first quality: Diameter in inches: 36, 38, 40, 42, 45, 48, 50, 52, 54: 180, 190, 200, 210, 225, 250, 275, 300, 325 Austrian florins. For second quality stones, the prices are 25% lower; for superior quality articles, a 25% augmentation takes place. The above prices are fixed for payments in ready money loco Fony or loco Tokay. Sharpening, according to drawings, charged from 8 to 15 Austrian florins; charges for package as moderate as possible. On orders given, half the price is expected to be paid on account, the other half on articles received. Costs of transport at charge of the purchasers. Delivery, if possible, between 1 and 3 months.

70. THURN-VALLE-SASSINA Count George. *Streieben, Carinthia.*

Puddled Steel.

71. TÖPPER Andrew, I. R. privileged I. Austrian Manufactory of Steel, rolled Iron, Nails and Gas-pipes. *Neubruck near Scheibbs.* Gas-pipe manufactory at *Gaming.* Hammer- and Puddling-works at *Lunz.* Coalmines at *Gresten, Lower Austria.* Great Gold M. *Vienna* 1839 and 1845, Gratz and Linz 1841 and 1846. Prize-M. London 1851. M. of honour: Munich 1854. M. I. Cl. Paris 1855.

Plate Iron articles, Gas-pipes, rolled Iron and Nails.

This manufactory has been built in 1818. Water is the generally adopted moving power and, therefore, is carefully stored up in tanks and reservoirs, from which the special motor, attached to every mechanical apparatus, is provided with an adequate quantity. The Iron undergoing the softening operations is, exclusive of all other sorts, Raw Iron from Eisenerz and Vorderrberg, made with Charcoal. After having passed through the preparatory operations of hammering, it is worked out into commercial goods of every kind, from the largest steam-boilers down to the most subtle plates, pipes and other rolled articles. The softening is done on an artificial soil of smelted slags („Schwall-

boden-Arbeit") according to the Austrian method; good, soft Charcoal is exclusively used for fuel. The establishment at Neubruck, near Scheibbs, comprises: 5 softening fires with hammers; 8 flame-furnaces, heated by the superfluous heat, emanating from the softening-fires and still sufficient for heating plate Iron; 6 rolling-mills with a stretching-table, („Streckstrasse") for grosser and another for finer sorts, this last with three cylinders, moved by a Jonval turbine, besides 4 small and 1 large pair of mechanical scissors; 2 pressing-engines; a turner's workshop with 3 turning-engines and a blacksmith's establishment; 1 forge for tools; dry stamp-mill, a manufactory of nails with 6 engines and a Gasometer for lighting the establishment. The great Hammering-establishment and Gas-pipe manufactory at Gaming (the first and still the only industrial enterprize of this description in Austria), comprises: 2 softening-fires with hammers for preparing the Iron to be worked out in the establishment at Neubruck; and (for the special purposes of the Gas-pipe manufacture): 2 soldering-furnaces; 1 stretching-apparatus; 1 rotatory saw; 1 boring- and lathe-engine; 4 engines for cutting screws and 1 engine for giving the last finish. The great Hammering- and Puddling-establishment at Lunz is in possession of 4 softening-fires with their hammers, preparing material for the Neubruck establishment, 2 flame-furnaces fed with the superfluous heat from the softening-fires, 1 stretching-table, 1 large and 2 small pairs of mechanical scissors.

The Coal-mines give excellent Liassic Coal fit for soldering Gas-pipes, as also for the production of Lighting-gas. Specimens of this Coal may be seen among the collection exhibited by the I. R. Geological Institute. The manufactories only working for satisfying express demands, keep no stores of their articles. These have got such a credit for their acknowledged superiority, that demands from great Railroad-Associations, Mechanical establishments and private commercial enterprises follow each other without interruption. The present annual production, providing with employment a standard personal of 220 workmen, amounts to 12,000—15,000 Cwt. of soft Iron articles of every sort. By a rational organization and distribution of labour, the same number of persons, in 1856, proved sufficient for a production double the just-quoted amount.

The superior quality of the articles manufactured in the Exhibitor's establishments may be considered to have been sufficiently attested by the distinctions he received at the Vienna, Gratz, Linz, London, Munich and Paris Exhibitions.

72. (Omitted.)

73. WALDBÜRGERSCHAFT of Upper Hungary, Mining and Metallurgical Works of Copper, Silver and Mercury. *Iglo, Comitate of Zips, Hungary.* Director: John Terientsik. Stores at the Copper smelting-houses, Vienna: at J. G. Sina. Agent: Szumrath Ernest, for Messrs. Williams and Norgate, London. Honourable mention: London 1851. M. I. Cl. Paris.

Ores, metallurgical productions and Metals.

From the smelting of Pyritic ores („Gelferze"): Pyritic ores, raw („Gelf") Copper, purified („Rosette") Copper, Copper in plates, Laminated Copper, bars for tests. From the smelting of Grey Copper-ores („Fahlerze"): Grey Copper-ore, metallic Mercury, raw smelted Sulphuret („Lech"), raw „Speise", raw Antimony („Speise" with a large proportion of Antimony), roasted artificial Sulphurets („Rohleche and Oberleche"), Black Copper argentiferous); smelted Sulphurets („Oberleche"), Amalgam of Silver from the Amalgamating operations, residuous Copper (raw Copper free of Silber), purified („Rosette") Copper, Copper-plates, Copper in laminae and in bars for tests, Cement-Copper from the Copper-Extraction operations. — The „Waldbürger-schaft" of Upper Hungary is a Mining-Association, comprising the totality of private miners of Copper, Silver and Mercury within the Mining-district of Schmöllnitz in Upper Hungary. Each mine stands under the personal and independant administration of its private possessor, who has to bear alone the profits and losses of it. The purpose of the Association is: to facilitate the working of ores by means of communal smelting-establishments, to distribute the profits, thus obtained among the mining establishments, proportionally to the quantity and quality of the metalliferous raw materials delivered by each of them to the smelting-works and to keep up certain establishment for the general benefit of the mining-district. The merely cupriferous ores („Golferze") are treated separately from the „Fahlerze" containing Silber, Copper and Mercury. Wood and Charcoal are managed in common; the purchases of Wood and and its Charring being under the superintendence of the Director of the Metallurgical establishments. Water is exclusively used as mechanical motor, as also Wood and Charcoal are the only sorts of fuel, subservient to the extraction of Metals. This Association is the most extensive of Mining-Companies working noble Metals within the limits of the Austrian Empire. The annual production amounts to: Copper: 18,000—19,000 Cwt. Silver 3.000 Mint ₰, Mercury: 600 Cwt. A half-ready article, the raw Antimony („Hüttenspeise"), resulting from the metallurgical treatment of the „Fahlerze" deserves special mention; it contains per Cwt Antimony: about 90 ℥, Copper: 4—7 ℥, Iron: 2—4 ℥ and small quantities of some other Metals. The quantity of Copper, produced by the „Waldbürgerschaft" of Upper Hungary, together with the 4,000—5,000 Cwt of this metal, annually issuing from the Government's smelting-works in the Mining district of Schmöllnitz, amounts to no less than one half of the total of Copper, produced in the Austrian Empire.

The method for extracting Mercury is different from any other presently known; it is based on a roasting operation, and is so simple and cheap, that even a quantity of ¼ ℥ of mercurial „Fahlerz" may be submitted to it with profit. The treatment of Grey Copper („Fahlerz"), a rather complicated operation, as it intends the simultaneous and profitable extraction of Silver, Copper, Mercury and raw Antimony („Hüttenspeise"), deserves some attention. Besides the metallurgical operations viâ siccâ (smelting in pit- and flame-furnaces), a new method, hitherto quite unknown in Austria called „Extraction" had been set into practice by the present Director of Metallurgical establishments. By treating the ores viâ humidâ, Cement-Silver and Cement-Copper are obtained, each separated from the other, in their metallic state. The Association possesses a rent-free fund of above 1,200,000 Austrian florins, a Pension-fund for their officials, a charity-fund („Bruderlade") for pensioning subaltern masters and workmen and a special fund for mineral searchings. The sum yearly brought into circulation, is no less than 5—6 millions of Austrian florins.

The establishments in possession of the Association are: 3 smelting-houses with 12 pit- and 5 flame-furnaces; 2 double roasting-furnaces for preparing argentiferous Black-Copper for Amalgamation to obtain the Silver contained in it and 1 large double roasting-furnace for Extraction-operations. The personal is composed of 1 Director of Forests and Metallurgical establishments, 3 Forest- and 9 acting Metallurgical officials and 260 workmen. The leading Committee („Assessorium"), under the presidence of His Excellency Count George Andrassy, highly deserving for having founded this Association amidst extremely difficult circumstances and for unceasingly promoting their interests, is composed of 1 Representative (President of the „Waldbürgerschaft" of Upper Hungary), 1 Vice-representative, 4 Assessors, 4 Trustees, 2 Notaries and 1 Superior Attorney („Ober-Fiscal"), assisted by Boards of Accounts and Cash. The Association has been constituted in 1835 and, till 1850, has exclusively operated on pyritic ores („Gelferze"). The treatment of Grey-Copper ores („Fahlerze"), decided in 1844, came not into practice before the year 1850.

74. VOLDERAUER Frederick, *Rothgulden, Lungau, Salzburg.*
Arsenic-ores.

75. WAIDHOFEN Town, Iron and Steel-works. *Klein-Hollenstein, Lower Austria.*
Iron in bars, raw Iron, refined Steel.

76. WALDSTEIN Count Ernest, Possessor of Mines and Metallurgical establishments. *Boros-Sebes, Comitat of Arad, Hungary.*
Mining and metallurgical productions.

77. WALLAND Ignace, Possessor of Mines. Domicile: *Vienna, City 303.*
Lead and Lead-ores.

78. WANG Joseph, Possessor of Mines and Metallurgical establishments. *Schönberg, Bohemia and Kniowitz pr. post Seltschau, Bohemia.* Store-house: Prague 821-2, Wenceslas Cejka.
Antimony.

a) 2 specimens of Grey Antimony-ore;
b) 3 „ „ Antimonium crudum;
c) 2 „ „ metallic Antimony (Regulus)-
Prices loco Prague, each constant for two consecutive months:
Ad a) Antimony-ore: 8 Austrian florins per Vienna Cwt;
Ad b) Antimonium crudum: 15 Austr. flor. per Vienna Cwt;
Ad c) Regulus Antimonii: 44 „ „ „ „ „
The establishments produces between 800 and 1,000 Cwt of Antimonium crudum every year.

79. (Omitted).

80. WISSIAK Charles. *Vienna, Burg-Glacis 134.*
Quartz, refractory Clay, Gold-ocre, Satinober, Benzine-earth.

81. ZEH John, *Lemberg.*
Mineral tar, Montina and Naphtha (unrefined).
(See Class 2.)

Class 2.

Chemical Substances and Products and Pharmaceutic Processes.

Sub-Class a) Exhibitors of Chemical Products.
„ „ b) Exhibitors of Medical and Pharmaceutical Processes.

Since the prevalence, in Austria, of the immediate production of white („English") Sulphuric Acid out of Pyrites, over the ancient complicated method of first obtaining Sulphur by distillation, then converting it by combustion into Sulphuric Acid and provoking the oxydation of this Acid by the presence of water, this Acid the most important among auxiliary chemical substances has of late entirely distanced the import of Sicilian Sulphur, by being now produced in such quantity (300,000 Cwt), that, together with the home production of 50,000 Cwt of brown („Nordhausen") Sulphuric Acid, it not only suffices for inland wants, but also for the demands of export.

The same is the case with Muriatic Acid (150,000 Cwt), the use of which, is now rapidly increasing in Austria, owing to the preparation of Ores viâ humidâ having become more extensive, and to the chemical Establishments being provided with salt at considerably lower prices.

In 1860, scarcely 400 Cwts of these two Acids were imported, whilst their export exceeded the amount of 17,800 Cwts.

The most important chemical substances and pigments, produced in the industrial Establishments of the Austrian Empire are: Nitric Acid (exclusively out of Chili Saltpetre): 30,000 Cwt, refined Saltpetre (Nitrate of Potash): 25,000 Cwt,

refined Tartrate of Potash: 15,000 Cwt; Taratic Acid: 1,200 Cwt; Carbonate of Potash: 100,000 Cwt; Carbonate of Soda (natural and artificial): 150,000 Cwt; Sulphate of Soda (Glauber's Salt); 110,000 Cwt; Alum (Sulphate of Alumia with Potash or Ammonia): 40,000 Cwt, Iron-Vitriol. 75,000 Cwt, Carbonate of Lead (Ceruse): 36,000 Cwt, Cinaber: 2,000 Cwt, artificial Ultramarine: 1,000 Cwt.

Besides these home-produced materials, nearly 100,000 Cwt of Carbonate of Soda, together with some Ammoniacal Salts, Chloruret of Lime and Verdegris, are the only chemical productions imported in any considerable quantities, to supply the wants of Austrian Industry.

The Inland consumption of Candles and Soaps of every description is abundantly supplied, by the combined activity of industrial Establishments and individuals working in small trade-concerns, without the import and export of these articles being significant, with the only exception of Stearine Candles, of which one tenth of the quantity produced (about 100,000 Cwt) is going abroad.

The production and export of chemical lights is of by far greater importance. The chief seat of this branch of Industry is Bohemia; its Establishments, however, are spread over all the Provinces of the Empire. The enormous quantity of 15,000 „Klafter" of wood, is worked out into sticks for chemical matches. The whole yearly production of these matches is about 150,000 Cwt; their export amounts to about one half of this total.

Pharmaceutical preparations are produced, partly in chemical Manufactories, partly in Laboratories.

But very recently, some Apothecaries, especially at Vienna, have made arrangements for producing such preparations on a larger scale and have succeeded in procuring a good number of Customers among their country-colleagues.

The branches of Industry here in question give employment to more than 50,000 w o r k m e n; the pecuniary value of their annual production amounts nearly to the sum of 50 Millions Austrian florins.

The results of foreign trade during 1860, stand thus:

	Value of	
	Import	Export
Auxiliary chemical substances . .	2,997,300 flor.	2,725,600 flor.
Chemical Productions and Pigments	1,819,000 „	1,426,000 „
Candles and Soap	319,600 „	812,900 „
Pyrogene articles („Zündwaaren")	— „	2,647,000 „
Total . .	5,136,400 flor.	7,612,100 flor.

---

82. ACHLEITNER Louis. *Salzburg.*
Wooden Lighting-matches.

83. AUSTRIAN ASSOCIATION FOR CHEMICAL & METALLURGICAL PRODUCTION. *Vienna.* Manufactory Aussig on the Elbe, Bohemia.

The Association has been founded in 1857 on the principle of shares, of which presently 2,000 (at 500 Austrian florins) are emitted and fully paid in. The establishment at Aussig is the most extensive Soda-manufacture presently existing in the Austrian Empire..Present annual production: Sulphuric Acid 96,000 Custom-Cwt, Sulphate 117,500 Custom-Cwt, Crystallized Glauber-salt 2,200 Custom-Cwt, Calcinated Soda 70,000 Custom-Cwt, Crystallized Soda 13,000 Custom-Cwt, Chloruret of Calcium 20,000 Custom-Cwt, Nitric Acid 3,300 Custom-Cwt, Hydrochloric Acid 145,000 Custom-Cwt, Super-phosphate 11,000 Custom-Cwt, Hyposulphite of Soda 1,200 Custom-Cwt, representing an annual circulation of 1,191,209 Austrian florins. The light-gas establishment of the manufacture supplies the wants of the town of Aussig.

84. BANYAY Charles, Merchant. *Klausenburg, Transylvania.*

Pomatums and Perfumery, Victoria-Gold-Cream, Transylvanian blossoms, Venus-Pomatum, Bear's-grease Pomatum, shaving Soap, washing-water for Hair, fluid for Fumigations, Mouth-water, Mustachio-wax, Transylvanian Medicated Sources effervescent Powders.

(See Classes 3 and 25.)

Authenticated testimonials from the Transylvanian Provincial Protomedicus, from the City-Physician of Klausenburg and from Professor Dr. John Meizner, of the same place, attest these compositions to have been scrupulously examined and to have been found perfectly free of noxious ingredients.

85. BATKA Wenceslas. *Prague.*
Chemical preparations.

86. BAUER James, Dentist. *Brünn.*
Odontic paste.

87. BLUMBERG & RINDKOPF. *Zuckmantel, Silesia.*
Chemical preparations.

88. BRAZZANOWITZ Nicholas. *Zara.*
Insectifuge powder.

89. BREITENLOHNER Dr., for the Archducal Manufactory of Peat-productions at *Chlumetz, Bohemia.*

Photogen, Pyrogen (Solar Oil), with I. R. exclusive patent, a block of Paraffine.

This establishment, the only one in Austria preparing Paraffine, Photogen, Pyrogen, Creosote, Asphalt etc. out of Peat-tar, has succeeded, by overcoming considerable difficulties, in extracting lighting Oils of superior quality out of this raw material. It is patented for a proceeding and peculiar apparatus, by which 60 — 65 p. c. of Solar oil or Pyrogen may be profitably extracted out of considerable and nearly unprofitable accumulations of residua of raw oils. This proceeding, besides destroying or removing impurities, lowers the specific density of the produced oils from 0.895 — 0.945 down to 0.825 — 0.845, gives them a light wine-yellow tint, unalterable by effect of time and makes them fit to endure cold without thickening. Used in Lamps for mineral Oils (especially those of Ditmar, Vienna) they burn without any disturbance, giving an intense white light. This patented proceeding deserves the more attention,

as the residua of this kind, notwithstanding repeated purification and distillation, give only oils of reddish tints and difficultly accessible to absorption. The Paraffine coming into commerce is generally of greasy touch, not perfectly inodorous and mixed with about 20 p. c. of Stearine. The establishment uses a simple and practical method for preparing out of Peat-tar a sort of Paraffine, of perfect whiteness, of dry touch, free from any smell or heterogeneous substances. Since the short time of existence of this manufacture, the demands of Paraffine have taken such an extent, that it has become impossible to satisfy the increasing number of demanders.

90. BRUNICKI Baron Maurice. *Village of Pizarzowa, District of Neu-Sandec, Galicia.*
Preparations of Petroleum.
(See Class 1.)

91. BURKHART G., Manufacture of Ultramarine. *Aussig, Bohemia.*
Ultramarine and green Pigment free of metallic substances.

92. COLLECTIV-EXHIBITION of mineral waters, through Well William, Noble de, Dr., General store of Mineral Waters. *Vienna, City, Wildpretmarkt 557.*

A. Alkaline Mineral waters: 1. Alkaline Acidules. 2. Alkaline muriatic Acidules. 3. Alkaline-saline Mineral waters. 4. Alkaline ferrugineous Acidules. *B.* Bitter Mineral waters. *C.* Muriatic Mineral waters. *D.* Ferrugineous Waters.

This establishment, existing since nearly a Century, brings into commerce every year some hundred thousands of bottles of Mineral waters. Demands of larger quantities are satisfied at prices some percents below those, marked on the list regularly emitted in spring.

93. DIEZ Ernest. *St. John near Villach, Carinthia.*
Cerussa and small Shot.
(See Class 11.)

94. ENGELMANN Samuel. *Carolinenthal 168, Bohemia.*
Albumine, Dextrine, Laiogome, Amidon and succedanea for Gum.

95. FICHTNER J. & Sons, I. R. privileged first Manufactory of Bone-powder and Glue. *Atzgersdorf near Vienna.* Bronze M. Munich 1854. I. Cl. M. Paris 1855. Great silver M. Vienna 1857.

Bone-powder and preparations for manuring. Several sorts of inodorous Joiner's Glue.
(See Class 9.)

96. FINANCES (I. R. Provincial Direction of) in Bukowina. Establishment at *Gurahumora.*
Potash.

97. FISCHER F., I. R. privil. Manufactory of Candles and Soap. *Wien and Simmering.* Store-houses at Vienna, Simmering and Prague. The commercial relations with foreign parts are entrusted to travelling commissioners paid by the establishment.

Vienna Tallow; different sorts of Soap for industrial, common washing and cosmetical purposes; Nitrate of

Potash, crystallized and in powder; Soda, raw and calcinated.

(See Class 4.)

The Vienna establishment, founded more than 20 years ago, and the Simmering manufacture, having more than 10 years of existence, have not yet appeared at any Exhibition. Both, calculated for extensive activity, act by steam-power and possess extensive tallow-smeltings, in which about 18,000 Vienna Cwt of tallow are smelted every year; about 5,000 Vienna Cwt to be worked out into candles and the rest to be brought into Commerce under the denomination of „Vienna Tallow" („Wiener Kernscheiben-Unschlitt"). The production of Soap extends to any sort whatever, from common washing up to the most exquisite toilet soap. Special attention, however, is paid to sorts used for industrial purposes. The annual production amounts to: Washing-soaps about 17,000 Vienna Cwt, Toilet-soaps about 30,000 dozens, Nitrate of Potash about 40,000 Vienna Cwt. The Soda, resulting from the manufacture of this last substance, is, for the most part, consumed in the manufactory itself for the production of Soap.

98. FRAUENDORFER Ferdinand. *Gaudenzdorf near Vienna.*

Chemical preparations.

99. FÜRTH Bernard, I. R. privil. Manufactory of Pyrogenous preparations. *Schüttenhofen and Goldenkron in Bohemia.* Store-houses: Vienna (City 898), Prague (Fleischmarkt). Own depots and agencies: Alexandria, Bukarest, Beyrut, Constantinople, Cairo, Hamburg, Hong-kong, London, Melbourn, New-York, Odessa, Piraeus, Patras, Rio-Janeiro, San Francisco, Shanghai, Sydney, Trapezunt, Valparaiso, Warsaw. Silver M. Vienna 1843. Bronze M. London 1851 and Munich 1854. Great silver M. Paris 1855.

Pyrogenous articles.

The establishment numbers at present about 20 years of existence; it entered into activity at a time, when the production of wooden matches was still an exclusive attribute of petty industry. By the extensive activity of the establishment at Schüttenhofen and the moderate prices effected by it, these matches have taken a conspicuous place among Austrian industry and among its export articles; a result, acknowledged soon after, the foundation of this manufactory, by the Silver Medal (the only one distributed at the Vienna Exhibition of 1845) imparted to the Exhibitor. Notwithstanding the grand organization and the spacious extent of the manufactory at Schüttenhofen, this establishment proved inadequate for satisfying the continually increasing outward demands. A second establishment of this kind became necessary at Goldenkron next to Krumau, offering as favourable local circumstances as Schüttenhofen. Both establishments employ about 2,000 persons. It may be considered as a fact, creditable to these establishments, that they have opened a new source of profitable labour to the destitute inhabitants of a region of the Böhmerwald, whose soil is scarcely accessible to any sort of culture. The turned wooden boxes, used for the packing of matches, and the manufacture of other tools required for their production, have become regular and extensive branches of industry in this district. The constantly improved quality of articles, the immense increase of wooden matches (by millions every day), Cigar lighters, small wax-candles etc. and the unparalleled cheapness of these articles, resulting from this unbounded production, have greatly increased the credit of these establishments; so that there may scarcely exist any commercial place on the whole Globe, where they are not in demand. To quote but one instance: the detail-sellers at Sidney and Melbourn, to commend a certain sort of wooden matches, announce them under the denomination „real Fürth's". A particular advantage of the Schüttenhofen and Goldenkron articles is the absence, or at least a minimum, of offensive smell of matches, when lighted or kept. Even unvarnished articles, when transported over sea, resist the influence of climate and moist air, and this may partly account for their being preferred (as above said), in tropical regions to any other articles of the same kind. The establishments are careful to keep up connexions with the chief European and transmarine commercial places, either by immediate correspondence or by the intervention of special Agents, so, as to be directly acquainted with the demands and habits of the consumers and ready to satisfy them at any time.

A particular advantage of the establishments here in question is the really surprising cheapness of their articles, for the most part a consequence of their uncommonly extensive production, of the adoption of every mechanical and chemical improvements presently known, and especially of their favourable situation and of the practice of home-labour for any secondary operations whatever ex. gr. for the manufacture of packing-chests,

for which purpose the establishments are provided with board saws of their own.

100. GESSNER Edward, POHL William, UPLBRICH Otto, *Müglitz, Moravia.*

Starch.

(See Class 1.)

101. GÖDL C. & Comp., Manufactory of Pyrogenous and Wood-articles. *Bärn, Moravia.*

This establishment produces wooden friction-matches of every sort, packed at the demanders choice, wooden sticks for matches, wooden Boxes, steeping-frames and any other implements required for the production of pyrogenous articles. Direct questions readily answered, price lists sent gratis.

102. GOSLETH Francis, Chevalier de. Manufactory of Chemical preparations. *Trieste — Hrastnigg.*

Chemical preparations.

103. GRAF M. *Vienna, Schaumburger-Grund, Linien-Gasse 62.*

Shoe-black.

104. GUST Adolphus. Patentee. *Kronstadt, Transylvania.*

Economic Matches.

105. HEINDEL John Bapt. Manufactory of Chemical preparations. *Ottakring near Vienna.*

Photogen and Solar-oil.

106. HERBERT Baron F. P. Cerussa-manufactory. *Klagenfurt and Wolfsberg, Carinthia.* Gold M. Vienna 1845; Prize-M. 1851. Medal Munich 1854; I. Cl. M. Paris 1855.

Cerussa of different sorts.

This establishment, the oldest in the Austrian Empire, dating from 1759, produces Cerussa of every sort, acknowledged to be of superior quality.

107. HERRMANN & GABRIEL, I. R. privil. Manufactory of Pyrogenous requisites. *Vienna, City 426.* Agent at London Mascha & Comp., 38, Bow-lane.

Pyrogenous articles.

The Exhibitors produce water-proof pyrogenous articles of every description, particularly adapted for export. They export every year to the Levant, Egypt, England and America considerable quantities of „Saloon-matches" in round elegant boxes of different size, containing each 100, 150 or 200 matches. These articles have been acknowledged, in all transmarine market-places, to stand in the first rank and are consequently very much in demand. Price-lists to be had in the Exhibition-buildings.

108. HIRSCH Francis, Wool- and Yarn-trade. *Vienna, City 689.* Agent in London J. L. Mayer Sons & Comp. 21, 22, Ironmonger-lane, Cheapside.

Preparations for washing and scouring Wool.

The Exhibitor has invented a preparation, varying in composition, for washing wool on the body of Sheep, shorn wool for manufacturing purposes (combed wool and fleeces excepted) and sheep-skins. These preparations may serve likewise for undyed wool-yarns and tissues intended to be bleached by sulphur, for cleansing dyed wool-yarns and tissues, Printer's types and cylinders, kitchen and table-plate of any kind, as also for half-fulling and washing of body- and toilet-linen. Informations, directions for use and testimonials to be obtained at the Agent's.

109. JÄCKLE George. *Gratz, Styria.*

Tartrate of Potash.

110. KAISER J. E., Manufactory of chemical Pigments and Starch. *Pesth, Hungary.*

An assortment of blue, yellow and green Pigments, red Laque-pigments, Varnishes and Wheat-starch.

The Exhibitor's establishment, still in continual increase consumes every day 40 metzens of Wheat for the production of Starch. In the present year, about 2,000 Cwt of linseed-oil has been used for the manufacture of Varnishes, among which a sort of Oil-varnish of eximious transparence, particularly adapted for Zinc-pigments, may scarcely be produced in better quality by any existing establishment. The Essence of Vinegar (saturating 50—55 p. C. of Potash), is exclusively prepared for the home-manufacture of green and yellow Pigments (Acetates of Copper and Lead, Saccharum Saturni), requiring daily 80—100 eimers of Spirits, purchased from productors at Pesth. Indigo excepted, the whole of the raw materials, used for the preparation of Pigments, are inland-productions.

111. KEIL Alois, Manufacturer of Alcoholic Lacquers and Polishing-preparations. *Vienna, Wieden 2.* Storehouses in most of the towns of the Austrian Empire.

An assortment of Lacquers and Polishes.

The Exhibitor has been patented for Lacquers for the floors of rooms, for engines and models in a variety of colours and equally adapted to wood and iron. These Lacque-varnishes dry up readily and are in odorous and permanent. Besides, the Exhibitor produces all sorts of Lacque-varnishes, used by joiners, ebonists, turners, pipe-carvers, modellers, hat-makers and in the ateliers of sculptors, oil- and water-colour-painters, photographers etc. He commends farther his new-invented waterproof and elastic Lacquer for Leather, adapted to every colour for fancy leather objects, and for the restoration of leather-covered furniture and his black Lacquer for horses' harnesses, leather belts and shoeings. Particular attention may be paid to some refined Polishes,. preparations lately brought into commerce by the Exhibitor and already acknowledged to be of profitable use. These preparations, fit for all sorts of wood, may be obtained in ruby, brown, fine orange and transparent tints. A newly invented black preparation of this kind, darkening the natural tint of wood, may not be unwelcome to Ebonists and Wood-carvers. The Exhibitor's grinding Oil-varnish for grinding the surface of wood intended to receive a polish, may be commended to any branch of industry connected with such proceedings; as proved by inspection of the best-provided stores of furniture, in which objects, artistically finished out of most valuable materials, but with polish deeply altered after some few months by the superficial efflorescence of the grinding-oil, are of no rare occurrence and, consequently, have lost a good deal of their elegant aspect and of their commercial value. An excedent of 1% of expenses is, in the present case, surely compensated for by a 10% increase in the value of manufactured objects. Price-lists, if applied for, gratis.

112. KOPPERL Simon, Confectioner, *Jungbunzlau. Bohemia.*

Albumine of Eggs.

113. KÜHN Edward & Charles. Chemists and Patentees. *Sechshaus near Vienna.*

Prussian Blue, Ammonia.

(See Class 31.)

114. KUNERLE Francis, Merchant. Chief store-house of Mineral waters. *Prague 550-1.*

Mineral waters of Marienbad, Franzensbad, Karlsbad, Bilin, Giesshübel and Püllna; Substances extracted and prepared from the sources of Marienbad, Franzensbad and Bilin; specimens from the Mineral Moor-deposit near Franzensbad.

115. KURZWEIL Frederick, I. R. privil. Manufacturer of Chemical preparations and Pigments. *Freudenthal, Silesia.*

Cobalt-blue Paper for washing; assortment of Cobalt-blue for washing and of New-blue; Paris and Prussian Blue; prepared Indigo and Paris Blue.

116. KUTJEVO, I. R. Administration of Forests and Dominions. *Kutjevo, Slavonia.*

Potash.

(See Class 4.)

117. KUTZER Joseph. Manufacturer of Ultramarine, Pigments and Chemical preparations. *Prague.* Bronze M. Leipzig 1850, honourable mentions: London 1851, Paris 1855.

42 different sorts of Ultramarine; mineral and vegetable Pigments.

This establishment produces every sort of Ultramarine for printing cotton-tissues, bleaching, paper-manufactures, together with Pigments of every kind for painting and daubing, Varnishes and Lacquers of every description, genuine Alizarine-Copying-and Gall-nut Ink and the best quality of Pearl-Sago of perfect whiteness and homogeneïty of grain. Specimens and price-list, if applied for, gratis.

118. KWIZDA Francis John, Veterinery and Apothecary. *Korneuburg.* Medals: Munich 1854, Paris 1855, Vienna 1857.

Korneuburg Curative Cattle-powder, powder for diseases of Hoofs, powders for Sheep and Swine, Blossom-resin, Restorative Fluid.

The medicament, invented by the Exhibitor and brought into commerce as „Korneuburg Cattle-powder" has proved equally effectual on horses, horned cattle and sheep. Fourteen-hundred depots of this medicament exist within the Austrian Empire and 800 others in the Prussian and Saxon states. In the years between 1857 and 1861 no less than 3,560,000 packets, of $^3/_4$ Vienna ß each, have been brought into commerce. The varieties of this medicament, particularly to be commended to persons dealing with domesticated Animals of every kind, are : the curative hoof-powder for putrid and cancrous affections of horses' hoofs, for chronical hoof-ache of horned cattle, for malignant hoof-epidemies among Sheep; the powder for Dysenteria of Lambs, for frequently occuring diseases of Swine. The „Blossom-resin" is a proof remedy against barenness of domesticated Animals in general. The „Restorative Fluid" keeps Horses fiery and capable of the hardest fatigues, even to their utmost old age, and heals also their inveterated diseases, as lameness, rheumatic affections, and diseases which have resisted the use of hot iron, setum or corrosive frictions. Besides the objects enumerated here, the Exhibitor keeps a complete store of all the articles used in Veterinary practice.

119. LAMATSCH Dr. John, Manufacturer of Chemical preparations. *Vienna, Wieden 8.*

Chemical preparations.

This establishment produces Chemical substances of every description, together with all those admitted into the „Pharmacopoea Austriaca" and all sorts of chemical, technical and photographical preparations. The objects brought to Exhibition are :

Chemico-technical division.

Acetamide. Caustic Potash, Soda, and Baryta, Allantoïne, Allorantine, Formiates of Lime, Copper, Manganese, Strontia and Zinc. Arseniate of Copper. Benzoïc, Succinic and Boric Acids. Borate of Soda. Butyrates of Lime and Zinc. Bromo-Ammonia of Cadmium and Copper. Chromium-Alum. Chloride and Chlorure of Chromium. Chromic Acid. Chromates of Silver, Magnesia and Strontia, Bichromate of Ammonia. Bichromate of Ammonia and Mercury. Camphoric Acid. Chloruret of Calcium. Chlorate of Strontia. Chloroform. Cholesterine. Citrates of Iron and of Magnesia. Cyanurets of Copper and Potassium. Coumarine. Iron-Alum. Acetate of Copper and Potash. Acetate of Cobalt. Oxyd of Methyle. Oxyds of Manganese, Sodium and Zinc. Fluoruret of Ammonia. Fluorurets of Calcium and Sodium. Guanine. Joduret of Lead. Jodoform. Cobalt-laques (rose-coloured, yellow and purple). Chloruret of Cobalt. Carbonates of Ammonia (condensated in glass tubes) and of Lithion. Mannite. Manganese-Alum. Lactates of Iron and of Zinc. Molybdenic Acid. Molybdate of Ammonia. Myricine. Naphthaline. Oxalic Acid. Oxalates of Ammonia and Cererium. Chromic oxyd-Potash. Oxydule of Cobalt (red and blue). Oxamyde. Phosphoric Acid (free of water). Phosphate of Soda. Picric Acid. Picrate of oxydulated Nickel. Platino-Cyenurets of Baryum, Magnium and Copper-Ammonia. Chloride and Chlorure of Mercury. Jodide and Jodure of Mercury. Cyanuret, Oxydule and Nitrate of Mercury. Rubidium (Chloruret and Sulphate). Nitrates of Lead, Cobalt, Uranium-Bismuth. Nitrite of Lead and Potash. Seleniates of Baryta and Uranium. Cupro-ammoniacal Sulphate. Sulphites of oxydulated Iron, of Manganese and of oxydulated Nickel with Ammonia. Sulphate of Copper. Sulphureted Antimony with 5 equivalents of Sulphur. Mustard-oil (artificial). Tannate of Iron. Titanic Acid. Racemate of Soda. Grape-sugar. Uranic Acid. Urea pura. Uranium-yellow (in 3 varieties). Oxyd of Uranium combined with Ammonia and the same Oxyd combined with Soda. Hypermanganate of Potash. Hypophosphite of Lime. Valerianate of Zinc. Scheelium-Lime. Scheelate of Ammonia and of Soda.

Photographic division.

Bromurets of Ammonia, of Cadmium and of Calcium. Chloruret of Calcium. Jodurets of Potassium, of Lithium and of Zinc. Clorurets of Jode and of Iron. Collodion duplex. Collodioned Alcohol. Jodurated Collodion. Jodo-bromed Collodion. Jodo-lithioned Collodion. Collodion-wool. Bichromate of Ammonia. Fluorurets of Ammonia, of Potassium and of Sodium. Gallic and Tannic Acids. Traumaticine. Glycerine. Chloruret of Gold. Chloruret of Gold with Sodium and Potassium. Jodurets of Ammonia, of Calcium, of Cadmium, of Potassium, of Lithium, of Sodium. Uranio-Sodium Joduret. Chloruret of Platin. Pyrogallic Acid. Nitrates of Lead, of Baryta, of Copper, and of Silver (white smelted, grey smelted and crystallized), of Uranium. Sulphate of Zinc Hyposulphate of Soda. Phosphate of Soda. Aluminia.

Pharmaceutic division.

The totality of Extracts, comprised in the new „Pharmacopoëa Austriaca". Aqua Lauro-cerasi. Pepsine.

120. LARISCH-MÖNNICH Count. Soda - manufacture. *Petrovitz, Silesia.*

Calcinated Soda and calcinated Glauber-salt.

121. LEHNER Edward. *Vienna, Gumpendorf 71.*

Aniline and products of Aniline.

122. LEHRER Adolphus. Manufacturer of chemical Pigments. *Türnitz, Bohemia.*

Store-house: Prague. Silver M. Leipzig 1850, honourable mention. London 1851.

Ultramarine.

This pigment is the chief object of the manufacture here in question. Pigments for painting and daubing, Varnishes and Luque-Varnishes are also produced. In no other Austrian manufactory, the production of Varnish-pigments, adapted for lining the inside of sugarmoulds and chests, has hitherto taken place. The Soda-Ultramarine-manufacture converts the mixed raw materials, with a loss of 5 to 6 p. c. for every operation, into a blueish-red substance. The sort (marked S) prepared out of it, withstands the action of Alum; the first-rate sorts are eximiously adapted for bleaching and manufactoring of paper. The manufactory possesses ten furnaces, each making ready for further preparation 6 Cwt at once, and disposes of a steam force of 20 horse-power for moving mills, sieves and other mechanical apparatus. The raw materials come from the mineral works and chemical manufactories of Bohemia, the fuel comes from the neighbouring coal-mines. The Ultramarine and the other Pigments find their way into all the Provinces of the Austrian Empire.

123. LEWINSKY Brothers. Manufactory of Vinegarsprit and Saccharum Saturni. *Dobrzisch, Bohemia.*

Store-house of Saccharum Saturni at Joseph Geitler's, Prague.

Acetic Acid and Acetates.

This establishment, existing since 1856, produces the following articles: Vinegar-sprit, common table Vinegar, composed Vinegars (with estragon, rasp-berries etc.), concentrated Acids of chemical purity and fine, white crystallized Saccharum Saturni. The Vinegar, produced by the Exhibitors, is proved by testimonials of accredited persons and authorities, to be free of any noxious substance and to be adapted, not only to domestic and technical, but also to medical purposes. The same has been proved by official analysis as to the Exhibitor's Saccharum Saturni. Present prices (very variable): Saccharum Saturni 29—30 florins pr. Cwt, Vinegar-sprit 5 florins pr. Eimer. Conditions of sale 4 months or 4 p. c. Cash-sconto. The travelling Agents of the manufactory are regularly provided with samples.

124. MADACS Edward & Comp. *Plojeschti.* Through Charles Stiehler, Kronstadt, Transylvania.

Photogen-gas.

125. MANUFACTORY of Chemical products (Stabilimento prodotti chimici). *Fiume.*

Chemical preparations.

126. MARASPIN Brothers, I. R. priv. Manufacture of chemically pure Alum, free of Iron. *Trieste.*

Alum chemically pure and free of Iron, purified Saltpetre in powder and in crystalline pieces(58 degrees), calcinated Soda.

This Alum is preferred to the best commercial sorts at present known and, for delicate operations, even to Roman and Toscan Alum. Besides this article, the manufactory produces calcinated Soda of 52—58° and refined Saltpeter in powder or crystallized, as required by demanders.

127. MILLER & HOCHSTÄDTER, first Austrian Soda-manufacture. *Hruschau, Silesia.*

Chemical preparations.
(See Class 35.)

128. MOLL Augustus, Apothecary and possessor of a Chemical manufactory. *Vienna, Inner City 562.*

Chemical and pharmaceutic preparations.

129. MÜLLER Ant. John, Manufacturer of Pigments for high and deep printing. *Schwechat near Vienna 90.* II. Cl. M. Paris 1855.

Printing-Pigments and specimens of prints.

130. NACKH Joseph & Son, I. R. priv. Chemical manufactory. *Vienna, Wieden, Heugasse 118.*

Chemical preparations.

131. NEUBURG & ECKSTEIN. *Pilsen, Bohemia.*

Chemical preparations.

132. NICOLAUS Constantin, Merchant and Gas-manufacturer. *Kronstadt, Transylvania.*

Photogen-gas.

133. NOWAK Joseph. Manufacturer of extracted Pigments and Succedanea for Gum. *Carolinenthal near Prague.* Agency: Wesser and Müller, London.

Extracts of Pigments (Pigment-pastes), Gummata or condensing substances.

These Extracts or pastes are characterized by a perfect purity of the pigments contained in them and, as to their chemical properties, by imparting to tissues, under cooperation of vapours, very solid and bright colours, at low prices, if mixed with a convenient condensing substance and with a very small portion of accessory matters. The Gummata dissolve easily and completely; for this reason, Patent-Gum and Dextrine is, in some cases, preferred to the costly Arabic Gum and, whenever Ultramarine is to be condensed, they frequently do better service than Albumine of Eggs, as they are not liable to emit foam and facilitate printing operation. First-rate manufactures in the Austrian Empires, as in the neighbouring regions of the Tariff-Union, are satisfied with the use of these preparations and, indeed, the splendid tints of the tissues printed in them, plead in favour of the Extracts and Gummata used for their preparation. A greater diffusion of these substances would prove highly profitable to manufacturing-establishments. Samples and detailed directions for use to be transmitted, if required.

134. OTTENREITER Leopold, manufacturer of fancy Leather-articles, *Pesth.* Chief store-house Vienna: Wittmann & Freyler.

Leather Trunks.
(See Class 26.)

135. PANESCH Ant., Manufacturer of Shoeing-articles and Patentee *Vienna, City 1100.*

Shoeblack and Lacquer.
(See Class 27.)

136. PARGER John, Shoeblack and Lacquer manufacturer. *Vienna, City 851.*

Lacquer- and Indigo-lacquer-Shoeblack.

137. PEUSENS Francis Charles & **WERSCHOWETZKY,** *Vienna, Alservorstadt 73.*

Benzine.

138. PFEFFERMANN Dr., Dentist. *Vienna. Lugeck.*

Perfumery-articles.

139. PIERING C. F. *Carolinenthal near Prague 157.*

Saccharum Saturni.
(See Class 3.)

140. POLLAK A. M., I. R. priv. manufactory of Pyrogenous articles. *Vienna, City 768.* Manufactories: Vienna, Prague and Budweis. Store-house: London, Albion Place, London Wall. Great M. Munich 1854, I. Cl. M. Paris 1855.

Pyrogenous articles of different sorts.

The Exhibitor's establishments (among which the Vienna manufactory numbers 24 years of existence), produce every sort of coloured Saloon-matches in paste-board and wooden boxes, ten sorts of Pipe- and Cigar-lighters of the Exhibitor's own invention, varnished Wax-candles, as also the new-invented waterproof-varnished matches, their keeping being warranted for a 10 years' period. Since about 30 years, the credit of the Exhibitor's articles — as well as all the pyrogenous articles produced at Vienna — is established everywhere. His articles have found their way over the whole surface of the Globe, even into regions still inaccessible to civilization. The circumnavigating expedition on board H. M's. Frigate „Novara" found everywhere in trade and use the matches manufactured by A. M. Pollak. Besides the above-mentioned Exhibition-distinctions, His Maj. the Emperor of Austria has been pleased to confer on the Exhibitor His Gold Merit Cross with the Crown, in acknowledgement for his exertions in giving extent to this branch of industry.

141. POLLAK Bernard jun., Manufacturer of Wood for Gun-carriages, Builders, Cartwrights and Coopers. *Vienna, Praterstern 396.* Laudative mention Vienna 1857.

Potash.
(See Class 4.)

A quantity of about yearly 2,000—3,000 Vienna Cwt of Potash, all of superior quality, is prepared out of the best appropriate trunks of Elm, Beach and Ash, in 4 establishments, situated within a complex of 30 square miles of forests, belonging to the State-dominions in the I. R. Croato-Slavonian Military Frontier and in Civil-Slavonia. On account of its superior quality, this Potash is constantly demanded at the highest market-prices at present 24 Austrian florins in Banknotes per Cwt, loco Vienna). It is packed in 10 Eimer tuns and goes generally into foreign parts, either for ready payment or for bills on one of the first-rank Vienna banking-houses.

142. POLLEY Charles. Manufactory for distillation of Tar. *Simmering near Vienna.* Store-house Vienna 924.

Preparations from Coal-tar, Mineral-tar and Petroleum, raw materials for manufacturing, drawings of Apparatus.

This establishment prepares out of Coal and Mineral-tar, and Asphaltic earth and stones, the following preparations, exhibited here: Benzine (for cleansing tissues), Benzole (for production of Aniline), Naphtha (for burning in self-regulating gas-lamps and fabrication of volatile gaseous substances), Naphthaline, Photogen, Solar-oil, raw Paraffine, Antifriction- and Cohesion-Engine-oils and greases, Naphthaline, Gome for greasing vehicles, Greases for wire-ropes, wheel-combs and Loury-grease, Mineral Linseed-oil-varnish and Lacquer (a succedaneum of genuine Linseed-oil-varnish) to be used for walls, wood and iron, together with pigments prepared with it, Creosote and Creosote-soda (for antiseptic preparation of wood), Asphalt-tar, genuine Asphalt-goudron, ready-Asphalt, endless Roofing-pasteboard, Daubing for roofs, water-proof Packing-paper, water-proof elastic sheets for covering objects against moisture, Benzine-Caoutchouc-varnish (for the production of water-proof tissues), raw and calcinated Soot, Briquette-pitch, Liquid Coke-briquettes. This establishment works with Tar produced in the Gas-works of the Imperial-Continental-Gas-Association, with native Naphtha of Galicia and Wallachia, Asphaltic earths etc., according to Civil-Engineer Paul Wagenmann's method (patented for England 1853, for Austria 1860 and 1861). In 1861, the establishment has consumed 40,000 Cwt of Tar. The value of annual production (now 120,000 fl.), is expected to increase to 400,000 fl. by the full activity of Briquette-manufacturing.

143. POPP J. G., practical Dentist and Patentee, *Vienna, City 357.* Chief store-house at the Producent's. Stores in all inland and foreign pharmaceutic establishments; for London: Marschall & Comp., Chemist, 21, Prince-street, Hannover-square.

Anatherine Mouth-water. Odontic Paste, Filling and Powder.

The „I. R. patented and first American patented Anatherine Mouth-water" invented by the Exhibitor 14 years ago, has been acknowledged as a superior prophylactive and curative of Tooth- and Mouth-affections by common consent of the first medical and chemical authorities. The Exhibitor refers to the testimonials of Dr. Oppolzer, Professor at the I. R. Clinic of Vienna, Dr. V. Kletzinsky, I. R. Chemist to the Provincial Tribunal of Vienna, Dr. John Flor. Heller, Superintendent of the I. R. Institute for teaching Pathological chemistry, and I. R. Chemist to the Provincial Tribunal of Vienna and Dr. Schillbach, Assistant to the Surgical Polyclinic of Jena. The specific prophylactic action of the Anatherine Mouth-water is manifest, wherever offensive smell from hollow or artificial teeth or from tobacco-smoke comes into question; it makes a fragrant breath and refreshes the mouth. It acts as a medicament for teeth and gums by appeasing tooth-ache, healing loose, spungy and easily bleeding gums, fastening loose teeth by increased contraction of gums and preventing their putrid affection. For this reason, it is specially commendable on voyages against scurvy, for blisters and pustules in the mouth and in general, for any morbid affection of the mucous membranes of this organ. The Exhibitor exports every year above 150,000 flagons of his Anatherine Mouth-water to Germany, Switzerland, Sweden, Russia, Valachia, Syria, Denmark, Orient (especially Egypt) and, since a year, also to America, where he has obtained an exclusive patent. The Exhibitor's I. R. patented Anatherine Odontic paste, his Odontic Filling and Powder deserve particular commendation.

144. POSELT L., *Vienna Landstrasse 772.*
White Shell-lac.

145. PRASCHNIKER A. *Stein, Carniolia.* Store-house at H. Waschnitzius, Vienna, Wieden 901.
Powder for cleansing Metals.

Yearly sale 3,000 Cwt, exported to all parts of Europe, price, loco Vienna, 12 Austrian florins pr. Custom Cwt. This cleansing-powder is acknowledged to stand first among all similar preparations, being a natural production, free from any noxious substance. Used on metals, glass and china, it operates quickly and imparts to them specular brightness. Under the microscope, this powder appears as an aggregation of spungy substance, in which even 300 fold magnifying cannot discover crystalline angles or sharp edges noxious to perfect polish. A quantity of this powder, deposited in a separate room of the Exhibition-buildings, may be used for trials.

146. PRIMAVESI Paul, I. R. privil. Sugar-manufacture. *Gross-Wisternitz and Bedihoscht, Moravia.*
Potash.
(See Class 3.)

147. PUNSCHHART Francis & **RAUSCHER** *St. Veit, Carinthia.*
Cerussa.

148. QUAPILL Ralph, possessor of a Manufactory, *Znaim, Moravia.*
Assortment of Honey-wood juice.

149. RADAUTZ, I. R. Military Stud. *Bukowina.*
Simple calcinated Potash.
(See Class 4.)

150. RICHTER & KLAR Brothers. *Hernskretschen near Tetschen, Bohemia.*
Orseille (herb and extract), Indigo, Carmine.

151. RIEDMAYER Eliza, First Tyrolian manufactory of Asphalt and Mineral preparations. *Giessenbach near Seefeld, Tyrol.* Factory: Innsbruck 582.
Photogen and Asphalt-tar.
(See Class 1.)

152. SAPIEHA Prince Adam. *Krasiczyn, Circle of Przemisl, Galicia.*
Turpentine.

153. SCHAUMBURG-LIPPE Prince. *Veröcze, Croatia.*
Potash.
(See Classes 3 and 4.)

154. SETZER John. *Weitenegg on the Danube. Inferior Austria.*
Specimens of Ultramarine.

155. SEYBEL Emilius, possessor of a privil. Chemical manufactory and of a Mining- and Metallurgical establishment at *Bösing near Pressburg, Hungary.* Vienna, Wieden 26.
Chemical preparations made of Pyrites.
(See Class 1.)

This manufactory, existing since 1847, produces every year: Sulphuric Acid: 6,000 Cwt., Vitriol: 10,000 Cwt. etc.

156. (Omitted.)

157. STATE RAILROAD-ASSOCIATION, I. R. privil. Austrian.

Substances obtained by distillation of Bituminous Slate. Paraffine in blocks and in candles. Sulphate of Copper and Sulphuric Acid.

(See Classes 1, 4, 5, 8, 10 and 13.)

STABILIMENTO PRODOTTI CHIMICI See No. 125.

158. STELZL Francis, manufacturer of Pyrogenous preparations. *Bärn, Moravia.*

Coloured Saloon-matches in mosaics, common Pyrogenous articles and wooden sticks for matches.

159. STOÉ Ante. *Samobor.* Manufacture at Delnice, Croatia.

Potash.

160. STROBENTZ Brothers, Manufacturers of Starch and Chemical Pigments. *Pesth, Zrinyi-Gasse.*

Chemical preparations.

161. TOBISCH Anthony jun., Apothecary. *Königgrätz. Bohemia.* Agent for London A. Neumann & Comp., 75, Cannon street, West E. C.

Pomatum for the growth of hair. Odontic balm.

Both objects are prepared out of exclusively mineral substances. The Exhibitor, scorning mountbank-like encomia, has merely relied on experience for stating the good qualities of his preparation. The Pomatum promotes the growth of hair with striking quickness, especially after exhausting maladies, cases of complete obliteration of capillar bulbs of course excepted. As for the Odontic balm, a single application is frequently sufficient to calm in a moment the severest pains. Pomatum for growth of hair: 2 shill. per pot; Odontic Balm: 1 shill. per flagon and case.

162. TSCHELIGI R., I. R. priv. manufacture of Lead-preparations. *Villach, Carinthia.* I. Cl. M. Munich 1854, Paris 1855.

Oxyds of Lead in glasses.

This establishment, founded in 1787, is the most ancient in date of this kind in Carinthia; it produces oxyds and other preparations of Lead of superior quality. Manufacture-stamp: a Double Eagle within a circle. The manufacture consumes every year 8,000 Vienna Cwt of Lead. Average number of workmen: 18—22.

163. UNTERWEGER G. B. *Trient. Tyrol.*

Varnish.

164. (omitted.)

165. WAGENMANN Gustavus, I. R. priv. manufactories of substances extracted by distillation of Petroleum, Mineral Tar and Bitumen, of water-proof tissues and incombustible Roofings and of Asphalte. *Vienna, Inner City 1047.*

Petroleum (raw Naphtha) and solid and liquid Greasings for steam- and auxiliary engines, waggons, etc. prepared out of it; Mineral oil (Photogen, Solaroil, inodorous Paraffine-oil) „for lighting Goudron" (inodorous Asphalt-pitch), hard and soft for preparation of Asphalt and Lacquers, common Grease for vehicles.

This establishment, although in existence only since the end of 1860, has to satisfy a multitude of demands of the first establishments in all parts of the Austrian Empire, numbering among those in foreign parts many consumers of oils, greases and lighting preparations, produced in it. These results plead the more for the good quality of its articles, as the establishment had to overcome considerable difficulties before it succeded in introducing in Austria the extensive use of Mineral oils and greases, previously unknown in this Empire. Many prejudices, for the most part originating from unpleasant experience, made on account of Resin-oils, were to be contended with. The good quality of articles, the really enormous saving compared with the expenses for Olive-oil, succeeded at last in refuting any objection. The greasing-oils, prepared in this establishment, keep the sockets constantly clean, as they secure them against being worn out or over-heated, and especially in spinning-establishment they have achieved amazing results by the diminution of friction. By using these oils, the same number of spindles may be moved with less expense of power (ex. gr. at low water), than it would be possible, where olive-oil is in use. The oils, prepared in this establishment, have proved excellent for the greasing of waggon axle-trees. Rolling-works of importance use the solid greasings for their special purposes. The Lighting-oils, prepared in the manufacture, have nothing of the penetrant smell peculiar to analogous combinations of Carbon with Hydrogen extracted from Brown Coal; some of them are quite inodorous and burn with a splendid white flame in the lamps adapted for them. The saving, in comparison to colza-oil, amounts to more than one half, if the greater photogenic power of mineral oils is taken into account.

Asphalt-„goudron" (inodorous) for the preparation of Asphalt and Asphalt-Lacquers. Grease for vehicles out of the results of distillation of coal-tar, Naphtal and Resin, equal in quality to the so-called „Patent Paraffine-grease" and about 30% cheaper.

Waterproof Tissues of first-rate sail-linen, impregnated with Caoutchouc-linseed-oil-varnish after a new method, have soon been demanded by Railroad- and Steamboat enterprizes on account of their excellent quality and moderate prices.

166. WAGENMANN, SEYBEL & Comp., I. R. privil. manufactory of Chemical preparations. *Liesing near Vienna.* Silver M. Vienna 1845, Prize-M. London 1851.

Chemical preparations.

This establishment, founded 1828 by Dr. Charles Wagenmann and enlarged on a grande scale since 1841 under Aemilius Seibel's superintendence, produces about 50,000 Cwt of Sulphuric Acid, both out of Pyrites and by combustion of Sulphur. The establishment itself consumes a large portion of this quantity for the decomposition of 10,000 Cwt of common Salt and of 7,000 Cwt of Saltpetre, as also for the preparation of 3,000 Cwt of Sulphate of Ammonia out of the waters used by the Vienna Gas-establishments. One of the chief objects is Tartric Acid, about 2,000 Cwt every year. Another important object is the production of Saccharum Saturni, Acetic Acid and various Acetates by means of the Vinegar-essence prepared out of spirits. Sulphates of Iron, Copper and Zinc are prepared in notable quantities, as also Chlorure and Chloride of Zinc, Stannate of Soda and other Tin-preparation. In these last years, Ae. Seybel introduced the fabrication of crystallized and sublimated Sal-Ammoniac, of Chromate of Potash and of other Chromium-preparations. But quite lately, the preparation of Aniline and of the blue and red pigments produced by operating on this substance, have been tried with complete success. The total value of Chemical preparations, produced in the course of 1861, represents a sum of about one million of Austrian florins. Besides the honorific distinction mentioned above, the establishment received in 1844, the Gold Medal of the Industrial Society of Lower Austria, as an acknowledgement for having introduced in Austria the fabrication of Tartric Acid. The establishment was excluded from competition at the Munich and Paris Exhibitions, Ae. Seybel, at this time, having acted as a member of the Jury.

167. WAGNER Daniel, Dr., Pharmaceutic and Technico-chemical Institute. *Pesth.* Agent Hermann Rubeck, 12, little Tower-street, London, E. C.

Essences and Oils of different sorts, subservient for the production and improvement of Wines and Liquors, or for pharmaceutic, photographic and perfumery purposes, as: Oils of Cognac Valeriana, Chamomile, Carraway, Genever and Anise, Rum-, English-bitter-, Muscat-Lunel-, Rosoglio-Cream-, Curaçao-Cream-Oils, Maraschino-Cream-essence, several chemical and medical preparations for photographic and pharmaceutic uses, as: Melissa-oil, Carmine, Santonine, Collodium, Cyanuret of Potassium, Glycerine, Cedar-resin for string-instruments, Hungarian medical plants, raw and prepared, as: Alcanna-root, white Mustard-seed, Cantharids, white Helleborus, white Saponaria-root.

168. WILHELM Francis & Comp. *Vienna, City 1100.*

Indigenous productions and chemical preparations of different kinds.

169. ZACHERL John, Produces of the genuine Persian Insecticide Powder from Tiflis in Asia. *Vienna, 624.* Store-house: Tiflis, Erivan-place.

Blossoms, whole and pulverized, of Pyrethrum roseum Caucasicum.

This is the only genuine Persian Powder, destroying Insects of every kind, its unfailible effects having met with universal acknowledgement and having been confirmed by official and private testimonials. It comes directly from the Exhibitor's estate in Asia, where it is collected and prepared from the above-named plant by himself or by his local agent, so as to come into commerce from its genuine source. Extensive stores of it are constantly kept; one for wholesale, in originary ballots; the other for detail, in flagons and packets. Prices always most moderate.

Price-lists delivered gratis, whenever demanders may be pleased to apply for them.

170. ZARZETSKY Joseph, manufacturer of chemical Pyrogenous articles. *Pesth, zwei Mohren-Gasse.*
Pyrogenous articles.

171. ZEH John. *Lemberg.*
Benzine-leather, Grease for vehicles and purified Montina.
(See Class 1.)

Class 3.

Substances used for Food.

Sub-Class a) *Exhibitors of Agricultural Produces.*
 „ „ b) *Exhibitors of Drysaltery, Grocery etc.*
 „ „ c) *Exhibitors of Wines, Spirits, Beer and other Drinks and Tobacco.*

More than one third (36 millions of Austrian acres — the Austrian acre being to the English acre nearly as 16 is to 11) of the whole area of the Empire is under cultivation. The extensive Danubian plains of central Hungary in the East, those along the Vistula in the North and the Lombardo-Venetian plains along the Po in the South; the first for wheat, the second for rye, the third for rice and maize, may be considered as the chief granaries of the Austrian Empire. The numerous valleys, fertilized by rivers, in the NW. Provinces of the Empire, are likewise more or less fit for agricultural purposes; so that their amount of production is nearly every where adequate to the wants of their own population, the more so, as in these Provinces, Agriculture has come to a higher degree of improvement than in those of the East.

Only the Alpine regions in the strict sense of the term (Upper-Styria, Tyrol, Salzburg, Carinthia and North-Carniola) depend on importation of Cereals from Hungary and Southern Germany for the sustenance of their population.

The yearly average production of Cereals in the whole Austrian Empire is expressed, in Vienna Metzens (the Vienna Metzen to the British Imp. Standart Bushel nearly as 80 to 47), by the following numbers:

Wheat . . . . 50,000,000 of Vienna Metzens
Half-corn[1] („Halbfrucht") and Rye 80,000,000 „ „ „
Barley . . . . 50,000,000 „ „ „
Oats . . . . 100,000,000 „ „ „
Maize . . . . 44,000,000 „ „ „
Rice . . . . 550,000 Cwts

To these sums must be added a yearly average amount of:
Buck-wheat and Millet . . nearly 10,000,000 of Metzens
Pease, Beans, Lentils etc. „ 5,000,000 „ „
Potatoes . . . . „ 120,000,000 „ „

Potatoes constitute the chief food of the population in mountainous regions of the northern Provinces, and are also extensively used for the extraction of alcoholic liquors.

With such abundance of Productions, the corn-export from Austria — even in less favorable years — exceeds the importation (for the most part from S. Germany into the W. Alpine regions and, under the influence of abundant crops, rises to very considerable dimensions, as shown by the following numbers.

		Import	Export
Wheat {	1860	448,110	2,796,100 Cwts
	1861	478,300	5,863,800 „
Rye {	1860	327,700	1,796,200 „
	1861	626,000	333,300 „
Barley {	1860	180,330	1,139,200 „
	1861	203,800	473,300 „
Oats {	1860	155,300	685,300 „
	1861	148,000	765,300 „

The grinding of corn is generally performed by common grinding mills moved by water-power. Latterly, the construction of mechanical mills (moved either by water or by steam) has made progress in Austria, and the Wheat-flour, produced by them, has become an export-article of increasing importance, as shown by the following numbers.

	Import	Export
Flour:		
1860	237,500	584,000
1861	258,700	757,800

The Manufacture of Beet-root-Sugar has advanced so far in Austria, that the establishments generally perfectly well organized for this purposes — provided they have sufficient raw materials at their disposal — are adequate for satisfying the total home-consumption of Sugar. In consequence of this progress, the import of Colonial Raw Sugar, which, in 1851 amounted to 678,400 Cwt has diminished in 1860 down to the comparatively small quantity of 35,200 Cwt. In 1860 the number of Beet-root-Sugar manufactories—most of them in Bohemia, Moravia and Hungary, was 124, consuming together a total of 15,900,000 Cwt (in 1859, nearly 19·3 millions of Custom Cwt) of Beet-root.

The Molasses from the Beet-root-Sugar-manufactures, Corn, Potatoes and Fruits (these last especially in the SE. Provinces of the Empire) are the raw materials consumed in the extensive production of Brandy and Alcohol, not only quite sufficient for the considerable demands of home-consumption, but preparing also great quantities (in 1860 nearly 65,000 Cwt for export.

In 1860 the number of active Distilleries was 103,000; among which, however, scarcely 6,000 deserved the name of manufactories, the rest being mere distilleries with a very simple appa-

---

[1] A mixture of Wheat and Rye, cultivated and reaped simultaneously.

ratus, only intended to provide for the home-consumption of their owners.

The total yearly consumption of Brandy and Spirits, reduced to 20° Beaumé, nearly reaches the sum of 4 millions of Eimers (the Vienna Eimer to the English Gallon as 400 to 32 or 25 to 2).

The Austrian Empire produces yearly between 30 and 40 Millions Eimers of Wine. The whole Empire (even its Alpine Provinces) is more or less cultivating wine; with the only exception of the two northern most Provinces: Galicia and Silesia. The greatest quantity and partly the best qualities of Wine come from the terminations of the Alpine and Carpathian ranges into the central Hungarian plain. Austria, Styria and Bohemia likewise produce good qualities, and, in these latter times, wine-culture and the management of wine are most conspicuously progressing in these parts.

The production of Beer is chiefly concentrated within the NW. Provinces of the Empire, gradually progressing from these into the E. and S. regions.

In 1840, the production of Beer remained under 8 millions of Eimers; in 1859, it rose to 12·6 millions of Eimers; a quantity, not only sufficient for the increased demands of home consumption, but also for extensive export. (Import in 1860: 12,000 Cwts;

export in the same year: 57,800 Cwts) Several of the 3,300 Breweries now in activity, especially in the environs of Vienna, are very extensive; some of them producing yearly to the amount of 200,000 to 300,000 Eimers.

Hydromel, in the E. parts of Galicia, and Cider, in Upper Austria and Styria, are objects of only local and little production.

The production of Snuff, Tabacco and Cigars, and the trade of these articles in Austria being a State monopoly, the Tobacco-culture cannot take place without the consent of the authorities. In 1860 this culture extended over 74,138 Austrian acres (66,769 acres in Hungary). In the same year the production was 778,000 Cwts, increasing in favourable years (as in 1858) to nearly 2 millions of Cwts.

Twenty-four State-manufactories with 21,800 persons employed in them, worked out, in 1860, a quantity of 733,600 Cwts of inland and foreign tobacco-leaves. The sale, in the same year, amounted to:

Snuff	60,700	Cwt
Tobacco	581,700	„
Cigars, inland . .	823,400,000	pieces
„ foreign . .	14,500,000	„

---

172. ALTHANN Count Michael Charles. *Swoischitz, District of Czaslau, Bohemia.*
Productions of the soil.

173. ANDRASSY Count George, Possessor of Landed estates. *Hosszurét, Comitate of. Gömör, Hungary.*
Wines.

174. AUGUSZ Baron Ant., Possessor of Landed estates. *Pesth.*
Szegszard Wines, from his own vineyards.

175. BÖCK William, Manufacturer of Liqueurs and Rosoglio. *Gross-Meseritsch, Moravia.*
Several flagons of Liqueurs, called: „Austrian Ladies-Saloon-Liqueurs".

176. BALÁS George, Curate. *Ipoly-Nick, Com. Hont, Hungary.*
Wines and Slivovitz.

177. BALTZ Philip, Wine grower. *Pesth, Donauzeile.*
Wines.

178. BANYAY Charles L., Merchant. *Klausenburg, Transylvania.*
Wines.
(See Classes 2 and 25.)

179. BARBER'S Sons, Possessors of a Steam-mill. *Buda.*
Assortment of Flour.

180. BARTHA Ladislav, Possessor of Landed estates. *Fehér-Gyarmath, Comitate of Száthmár, Hungary.*
Tobacco.

181. BARZDORF, Manufactory of Sugar. *Barzdorf near Jauernig, Silesia.*
Sugar, raw and in loaves.

182. BATHYANYI Count Gustavus, Possessor of Landed estates. *Hungary and Croatia.* London, 130, Park-steeet, Grosvenor-square.
Corn and Seeds of different sorts. — Wine, 6 bottles.

183. BAUER Casimir, Distiller of Liqueurs. *Vienna Leopoldstadt 316.* Prize-M. Munich 1854 and Paris 1855. Gr. Silver M. Vienna 1857.
Liqueurs.

The Exhibitor's establishment is in existence since 1841. This period of 21 years has been employed in constantly improving the articles produced in it. These exertions have not been ineffectual, as proved by the distinctions granted to the Exhibitor (Prize-medals at Munich and Paris, great Silver Medal at the Vienna Agricultural Exhibition, and honourable testimonials for objects sent to the Pesth Exhibition), and by the credit, his articles have obtained, being demanded not only in every Province of the Austrian Empire, but also exported into Moldavia, Wallachia, Serbia, Turkish Empire, and other foreign countries. The Exhibitor has taken special care to produce in equal superiority, but at a cheaper price, those Liqueurs, as Curaçao, Marasquine, Anisette and Absinth, which had been previously imported from Holland, France and Switzerland, and he thinks himself entitled to rank his own productions with any foreign articles of the same description. Prices per flagon in Austrian money: Caraways: 1 fl.; Vanille: 1 fl. 10 creutzers; Marasquine: 1 fl. 15 cr.; Curaçao: 1 fl. 15 cr.; Anisette: 1 fl. 15 cr.; Ananas: 1 fl. 15 cr.; Raspberry: 1 fl. 15 cr.; Allasch: 1 fl.; Rostopschin: 1 fl.; Absinth: 1 fl. 10 cr.; Peach: 1 fl. 15 cr.; Spirit of agriots: 1 fl. 10 cr.; Orange: 1 fl. 15 cr.; Essence of Punch: 1 fl. 10 cr.; Tea-liqueur: 1 fl. 50 cr.; Coffee: 1 fl. 15 cr.

184. BAUER Theodore, Sugar-manufactory. *Königsfeld and Tischnowitz, Moravia.*
Sugar.

185. BELGIAN-TURBINE-GRINDING-MILL. *Carlstadt, Croatia.* Property of V. Pleiweiss, Vienna, and G. Pongratz, Agram.
Articles obtained by grinding.

This establishment produces out of best Banat wheat in perfectly dry condition, 100,000 Cwts of first-rate flour, much in demand for inland consumption, and for sea-export into every part of the Globe.

186. BERCHTOLD JUN. Count Sigismond, Possessor of Landed estates and Glass-manufactory. *Buchlau, district of Hradisch, Moravia.*
Wine (from his own vine-yards) in bottles from the Exhibitor's own Glass-manufactory.

187. BIZEK Francis. Producer. *Alt-Becse, Comitate of Bacs-Bodrogh, Hungary.*
Agricultural productions.
(See Class 4.)

188. BLAESS Julius, Vinegar-manufacturer. *Waitzen, Comitate of Pesth, Hungary.*
Vinegar.

189. BLASCHÜTZ'S Brothers, Distillers. *Bogsan, Comitate of Krassó, Hungary.*
Spirituous Liquors.

190. BLÜCHER VON WAHLSTADT Countess Mary, born Countess Larisch-Mönnich. *Radun near Troppau, Silesia.*
Sugar.
(See Class 4.)

191. BLUM John, Possessor of a Steam-mill. *Buda, Haupt-Gasse.*
Mill-productions of various kinds.

192. BOHEMIAN STEAM-MILL-ASSOCIATION, Fürst I. R. privil. *Smichow near Prague, Bohemia.*
Flour.

The Association has constructed, in 1846, two Steam-mills, the one at Smichow near Prague, the other at Lobositz in Bohemia, working with a total of 22 pair of mill-stones moved by 4 steam-engines of 120 horse-power and employing 128 workmen. These mills consume every year 350,000 metzens of Cereals from Bohemia or Banat. They produce six sorts of fine flour. two middle and two inferior qualities, together with 3 sorts of grits. The six first sorts have been alone sent to the exhibition, as giving the best idea of the quality and cheapness of prices. A baking-house with 5 ovens at Smichow, and another with one oven at Lobositz, both according to Ignace Korde's patent-method, are connected with these mills. These establishments convert every year into bread 60,000 Cwt of flour, supplying the daily wants of 50,000 of the inhabitants of Prague. Flour is exported to Germany, England, Scotland, Russia and even to Siberia.

193. BOGESDORF, local Office *near Mediasch, Transylvania.*
Wines.

194. BOGYAI Lewis, Possessor of Landed estates. *Village of Haláp, Comitate of Zala, Hungary.*
Wines.

195. BRAUN Brothers. Liqueur- and Champaign-manufacturers. *Pesth.*
Wines, Liqueur and Slivovitza.

196. BREUNNER Count Augustus. *Zelisz, Comitate of Bars, Hungary.*
Wines.

197. BUDACKER Theophile, Director of Gymnasium. *Bistritz, Transylvania.*
Heidendorf Wines of 1859.

198. BURCHARD & Comp. Stephen, Wine growers. *Tokay, Comitate of Zemplin, Hungary.*
Genuine Tokay wines.

199. CALIGARICH Gasparo. *Zara.*
Marasquine.

200. CATTICH brothers. *Zara.*
Wine and Olive-oil.
(See Class 20.)

201. CHWALIBOG John. *Lipovice, Circle of Zloczow, Galicia.*
Brandy and Liqueurs.

202. COMER Fr. *Trieste.*
Marasquine and Rosoglio.

203. CSERNOVICS Paul, Possessor of Landed estates. *Szatárd, Comitate of Bihar, Hungary.*
Wheat.

204. CSALOGOVIC Balthasar. *Essegg.*
Maize and Wheat.

205. DEGENFELD Count Emerick, Possessor of Landed estates. *Nyir-Bakta, Comitate of Szabolcz, Hungary.*
„Ausbruch" of Tokay wines.

206. DESENSY Maurice, possessor of Distillery. *B. Komlos near Kikinda, Hungary.*
A bottle of 37° Spirit.

207. DEVRIENT Augustus, Wine grower. *Pesth.*
Wines.

208. DIETZL & PONINGER, Wine-merchants. *Prague.*
Central store-house for Bohemia. *Carolinenthal 145.*
Commandite at Hamburg, John Dotzauer.
Wines of different sorts.

209. DREHER Anthony, possessor of Landed estates. Brewery, *Klein-Schwechat near Vienna.* Gold. M. Vienna 1857.

This establishment, one of the eldest, was first authentically mentioned as a brewery in an official document in 1634, but it is only since it came into possession of the present owner, who spent his youth in earnest professional studies in the chief places of Europe that it has risen to the eminent position it now holds as the largest Brewery carrying on the most extended trade on the continent of our part of the world. In reference to its size it stands not even second to the greatest London Brewery, since its area, covered with buildings, extends to more than 8 acres 85 poles, comprising therein 8 acres and 45 poles of vaulted room, mostly subterranean; a circumstance not occurring in England, where the preparation of malt forms a separate trade and where the largest quantities of Beer are stored above ground. Dreher's Brewery contains 31 malt floors, most of them underground, of such dimensions that even more than 15,717 bushels of raw barley may be spread therein, 10 double floored malt-kilns of 52 square-poles and granaries for 219,700 bushels. A farther portion of the malt required for the Brewery is prepared on the Exhibitor's neighbouring estate „Freyenthurm at Mannswörth where 13 malt floors and 2 kilns are in use. The removal of these quantities from the floors to the kilns, from these to the dust-brushes, to the malt beans, to the crushing-rollers and to the colle ting boxes, for bruised malt over the mash-tuns, is effected by mechanical power and steam. Three steam-boilers of 50, 36 and 30 horse-power, a water-power of 20 horse, brought to by a steel-wire-rope; and 300 workmen, work together to effect the object of this establishment, which activity however is confined to the winter-months as the time most appropriate for the brewing of keeping beer in the way of under-fermentation, here solely in practice. One copper of 5,971 gallons, a second of 4,578 gallons and four smaller ones of 3,483 gallons capacity each together with 8 mash-tuns and mashing machines are mostly situated in the high room for preparing the extract and boiling the worts, the roof of which is 5 poles 10½ feet wide. Twenty four copper coolers are placed partly above this vaulted room, mostly however in separated cooling houses skilfully adapted to the purpose The fermenting-localities contain 1,056 fermenting-tuns with a total capacity of 558,000 gallons. Such extensive means make it alone possible to produce 42,500 gallons of Beer within 24 hours. Eleven store-cellars, containing besides adequate quantities of ice, space enough for 4,067,000 gallons of beer, in vats varying in capacity from 750 to 1,900 gallons (one of 434,000 gallons another 682,000, a third of 992,000, a fourth of 1,148,000 and the seven others taken together of 511,000 gallons capacity), stand among the most valuable establishments connected with this brewery, covering alone an area of 5 acres 110 poles. The present state of the business requires 21,000 casks of the size of 12½ or 25 gallons each, and 3,500 store-vats, containing 3,720,000 gallons. The colossal ware-houses, required for keeping part of these implements, after their having been emptied, during the months of greatest heat, stand next to these store-cellars, where also a pitching-house spacious enough for pitching at once two of the largest vats, has been erected. The transport of raw materials to, and manufactured goods from, the Brewery is effected by aid of 124 teams whose stables, for the most part of very appropriate and wholesome construction are placed on the South-Westend of the buildings. During the first year (from April 1 1836 to March 31. 1837) in which the Brewery was going on under direction of its present owner, the production amounted to 329,314 gallons, paying a tax of 33,953 florins, whilst in the year gone by, it rose to 5,015,000 gallons and the taxes paid for the production including the octroi for the quantity brought to Vienna (about ⅗ of the whole) were no lesser a sum than 833,997 florins 34 creutzers. A small part of the whole production went into Turkey, Egypt, Greece and some islands of the Archipelago. The agricultural usefulnes of this establishment, the raw materials of which are all products of agriculture, retributing the agricultor's exertions not only by raised consumption, but also by notable differences in price for superior qualities, has been acknowledged by the I. R. Agricultural Society at Vienna by affording their highest distinction, the Gold Medal, to the owner of this Brewery, as being the undoubtedly most eminent promoter of this branch of industry. The Brewery at Klein-Schwechat led the way for introducing and propagating the improved method of brewing Keeping Beer by under-fermentation in Austria instead of the hitherto used method by bung-or upper-fermentation. In consequence of this improvement, Beer, previously confined to the nutritive wants of the poorer classes of the population, has become in Austria a favorite beverage of the wealthier classes. To this circumstance alone it is attributable that this article could bear a eiterated raising of its former duty without a sinking consumption. Previous to the agricultural exhibition in Vienna 1857 the Brewery at Klein-Schwechat has not yet appeared at any Exhibition whatever.

210. DRESSLER Francis. *Obora, Bohemia.*
Sugar.

211. DUDA Joseph Francis, *Prague 707-2.*
Chocolate.

212. EDER F. M., Merchant and Champaign - manufacturer. *Pesth, Waitzner-Gasse.*
Hungarian Champaign.

213. EGAN Edward, Lease-holder of Landed estates. *Bernstein, Hungary.*
Cereals of various sorts, Seeds, varieties of Potatoes, Photographs of races of cattle.

214. ELTZ Count Charles. *Vukovar, Slavonia.*
Wine, Maize, productions of spirit-distillery with materials produced by a new method.

215. (Omitted.)

216. ERDÖDY Count Francis, Possessor of Landed estates. *Somlovár, Comitate of Veszprim. Hungary.*
Wine.

217. FABER C. M., Possessor of Wineyards. *Marburg, Styria.*
Frauheim-Buchberg-Moselle Wines of own growth. 12 bottles of 1857, 12 bottles of 1861.

Wine-culture in Styria occupies 54.600 Joch, of 1,600 fathoms each, and produces an annual average of 1,360.000 Austrian eimers. The best Vine-yards are those in the Pachermountains (Frauheim, Radisell, Pikern etc.) with an area of 1,300 Joch and an annual production of 35,000 eimers. The sorts, cultivated in these vine-yards, are all of superior quality, chiefly Moselle (denominated: „Joannina princeps" by v. West, a Styrian Botanist), Tramine, Cleve, Italian Rissling etc. The vernal and autumnal temperatures of this region (between 6 and 14° R.) is highly favourable to wine culture. The soil of the Pacher range, prevalently composed of Gneiss and Micashist, generally ferrugineous, as manifested by the reddish tints of the Pikern mountains, are an excellent soil for vine-yards. The late Archduke John, of imperishable memory in Styria, had transformed his extensive dominion of Johannesberg at Pikern into a model-vine-yard, at the same time bringing his superior wine into high credit by means of a rational management of cellars, and by founding prizes for the encouragement of industry and activity among the wine growers. The crop of grapes does not take place before the last days of October or the beginning of November; the juice of grapes is tapped 24—36 hours after pressing (consequently, before the beginning of fermentation) and this operation takes again place in the last days of March. The wine, thus improved, may be commended for strength, sweetness and flavour. Pacher wines sell immediately from the press for 8—13 Austrian florins, and, if old and tapped, for 14—20 and more florins per eimer. With regard to particulars of Styrian wine-trade, detailed information will be given on inquiry.

218. FARKAS Lewis, Wine grower. *Grosswardein, Comitate of Bihar, Hungary.*
Wines.

219. FEHÉR Brothers, Agricultural producers. *Török-Becse, Comitate of Torontal, Hungary.*
Agricultural productions and Wines.

220. FEKETE & Comp., Wine-merchants. *Erlau, Comitate of Heves, Hungary.*
Wines.

221. FESTETICS Count George, Possessor of Landed estates in Hungary. *Oedenburg.*
Cereals of various sorts, Colza-seed, Tobacco in leaves, Wines, Beet-sugar.

222. FESTETICS Count Tassilo, Possessor of Landed estates. *Kesthely, Comitate of Zala, Hungary.*
Wines.

223. FISCHER C. F. *Pesth, Serviten-Platz.*
Stewed and preserved Fruits.

224. FISCHER Edward de Röslerstamm. *Vienne, City, Current-Gasse 407.*
Flour meats (maccaroni, vermicelli etc.)

225. FIUME, FABBRICA DI PASTA A VAPORE. *Fiume, Croatia.* Store-houses: Trieste, Pesth and Laibach, Agent for London: Drake, Kleinwort and Cohen.
Flour meats.

This establishment, founded in 1845, has a steam-engine of 24 horse-power, a mill with 5 pair of stones, 4 kneading-engines and 5 presses. It works with wheat, rich in gluten („Grain dur") from the Azowian Sea and Asia minor, according to the Neapolitan method, converting it into two qualities of flour meats various in form. In these last years, the production, with regard to the increase of inland and transmarine demands, has been raised to the double, and, in 1861, amounted to 9.000 Cwt.

226. FIUME, STABILIMENTO COMMERCIALE DI FARINA. *Fiume, Croatia.* Agencies: Odessa, Pesth and Singapore. Store-houses: Trieste and sea-ports of Istria and Dalmatia. Agent for London: Drake, Kleinwort and Cohen. Honourable Mention: London 1851. Great Bronze Medal Vienna 1857.
Articles produced by the mills and Ship-biscuit.

This establishment, founded in 1839 at half an hour's distance from the sea-shore, possesses 18 pair of French millustones of 4½ to 5½ feet in diameter, together with all the newest auxiliary engines for fine grinding, moved by three water-wheels of 95 horse-power, several workshops for repairs and a workshop for coopers. It works with the best wheat from Banat, Russia and Romagna, producing every year 198,000 Cwts of flour. Rational commixtion of these different sorts of wheat and appropriate treatment make this flour capable to bear long sea-transports and even to remain unaltered after having lain stored up for years in tropical regions. These valuable properties and the scrupulously kept up equality of this article have imparted a high credit to the mark „Fontana Hungarian" in the sea-ports of South-Africa, the Levant, East-India, and England. In 1860 the export of flour into the above-named countries amounted to 30.000 barrels A baking-establishment for Ship-biscuit, organized according to the English system and working with English machinery, is connected with the mill; it produces daily 96 Cwts of biscuit in round and square form for the approvisionment of military and naval forces (as in the Crimea war and in the Italian campaign of 1859), as also for transmarine export.

227. FLANDORFER Ignace, Wine-merchant. *Oedenburg, Hungary.* Agent: Frederick Class, London, 36, Crutched Friars, E. C. Medal of honour: Munich 1854. Honourable mention Paris 1855.
Hungarian Wines.

The Exhibitor's Colonial Goods Commission establishment, existing since the year 1814, exports Wines to Germany, England, Sweden, Norway, Russia, etc. The Wines brought to Exhibition are sweet „Ausbruch", Dessert wines, genuine Tokay of first quality, as also fine, white and red unprepared Wines.

228. FORGACH Count Coloman, Possessor of Landed estates. *Nagy-Szalant, Comitate of Abauj, Hungary.*
Specimens of Wheat and of Soils.

229. FRANK S. Joseph, priv. Manufactory of Rolled barley and mechanical Mill. *Kronstadt, Transylvania.* Store-houses: Hermannstadt and Schässburg.
Mill Products of various sorts.

230. FRÖHLICH William, Wax-manufacturer. *Klausenburg, Transylvania.*
Honey.

231. FUCHS & Comp. Gustavus, Merchants. *Pesth.*
Wines.

232. FÜNK Edward. *Gratz.*
Styrian Liqueurs.

233. FÜNK Henry, Manufactory of Liqueurs, Rum, Spirits and Vinegar. *Gratz.*
Spirits, Liqueurs etc.

This establishment, existing since 1845, the productions of which are in high demand, intending to open transmarine markets for them, has brought to the Exhibition a series of articles, the qualities, fittings out and prices of which will procure farther demands. These articles are kept in bottles fit for export (see: Advertisement).

234. FÜNK Max, Francis. *Gratz.*
Liqueurs.

235. GALLINY Francis, Apothecary and Wine grower *Lugos, Comitate of Krassó, Hungary.*
Wines.

236. GANZ J. & Comp., Wine-merchants. *Pesth.* Commission store-house: Hamburg: M. L. Würzburg & Comp. Agent: Philip Bonfort, London, 31, Bushlane, Cannon-street.
Wine, „Ausbruch“, Spirit of wine, Slivovitz.

237. GAVELLA N., Commerce of Drugs, Spices, Pigments, Wines, Spirits and Comestibles. *Agram, Croatia, and Lipnik, Slavonia.*
Wine from Sveli-duh near Agram.

The Exhibitor keeps extensive stores of his articles at Agram and at Lipnik, a watering-place in high credit for its good effects in chronical affections.

238. GEMPERLE John, I. R. Patented manufacturer of Succedanea for Coffee. *Lettowitz, Moravia.* Vienna, Neubau 137.
Succedanea for Coffee.

The articles issuing from this manufactory are well known for their good quality, and moderate prices. They are, to the present day, the object of frequent adulterations; a circumstance, which could not have taken place, had not the genuine articles come into high and general credit. Since its first beginning, in 1836, the establishment has never yet appeared at any Exhibition, neither national nor international, and if, on the present occasion, it submits its articles to public judgement, it is only with the intention, to contribute its share to the Austrian section of the Exhibition. The yearly production is about 10,000 Cwts of Coffee-succedanea in more than 60 different sorts.

239. GÖGL Zeno. *Krems, Austria.*
Mustard.

240. GOLDNER & ZINNER, Expedition- and Commission-trade. *Pesth.*
Paprika (Cayenne-pepper).

241. (Omitted.)

242 a. GREIS John. *Vienna, Jesuiterhof on the Glacis, Laimgrube.*
Fine durable Flour and Grits.

242 b. GREGER Max, Wine-Merchant. *Vienna and London, 17, Colchester Street.*
Wines.

243. GYÖNGYÖS, Agricultural Society of the Comitate of. *Heves, Hungary.*
Collective exhibition of Agricultural products.

244. GYULAY Gál Dionysius, Possessor of Landed estates. *Csombóra, Comitate of Somogy, Hungary.*
Wheat and Wines.

245. HAACK Anthony, through Hyppolite **MLEKUS**, Manager of Haack's Liqueur-manufactory. *Gratz.*
Liqueurs.

246. HAAS Godfrey, Wine grower and merchant. *Grosswardein, Comitate of Bihar, Hungary.*
Wines.

247. HAASE Alois sons Wine-merchant and Winegrower. *Znaim, Moravia.*
Wines.

248. HAASE Aloysius son, Wine-merchants. *Znaim 162, Moravia.* Prize-medals: Vienna 1845 and Paris 1855. Honourable mention: Munich 1854.
Moravian Wines in various sorts.

The whole of the exhibited Wines are of Moravian growth (partly of the Exhibitor's own growth), of the years 1811, 1822, 1834, 1841, 1846, 1848, 1850, 1852, 1854, 1856 to 1859, and 1861. The Exhibitor is successor to the house „Alois Haase“.

249. HANNAK G. *Brandeis, Bohemia.*
Flour.

250. HAVAS Joseph de, Royal Court-counsellor and Vine-cultivator. *Bude.*
Wines.

251. HEGYALLJA VINE-CULTURE-ASSOCIATION. *Mád, Comitate of Zemplin, Hungary.*
Tokay Wines of various sorts.

252. HITZGERN J. G., Commerce of Cremor tartari and Mustard. *Krems, Lower Austria.*
Cremor tartari in large pieces and Mustard in pots.

Cremor tartari, most in use for dyeing and medicinal purposes, numbers among the most important commercial articles in Austria. Numerous and extensive connexions are kept up by the Exhibitor with traders in Germany, especially with those of Hamburg, Stuttgart, Francfort, etc. Austrian pure and unadulterated Cremor tartari has become so much demanded for in these last two years, that extensive orders for it could only be promptly effected by means of the Exhibitor's numerous connexions. Krems Mustard is a well known article, of which the Exhibitor sends away 500 eimers every year in parcels of whole, half and quarter eimers.

253. HOFGRÄFF John, Provincial Advocate. *Bistritz, Transylvania.*
Heidendorf-Steiniger Wine.

254. HOLZER Ignace. *Vienna, Alt-Lerchenfeld 52.*
Vienna Tea-and Wine-rolls, 1000 pieces.

255. HOPS-PRODUCERS. *Auscha, Bohemia.* Honourable Mention: Vienna 1857.
Auscha Hops.

The culture of Hops, being the chief branch of production in the district of Auscha, has for centuries been attended to with the utmost care and, since these last ten years, has got highly honourable credit for having improved its production. The so called: „Red Hops Lands“ of Auscha, comprising portions of the Leitmeritz, Wegstadl and Böhmisch-Leippa districts, yield a crop of about 30,000 Cwts of Hops, of which nearly 10 pCt. may be counted for the culture of Green Hops. The district of Auscha, with 82 inhabited places, contributes, to this crop with an amount of 20,000 Cwts for the most part Red Hops. At the Exhibitions of Vienna and Teplitz, the Red Hops of the district of Auscha met with full acknowledgement for their favour and superior quality and found purchasers at the rate of 10 pCt. above the normal price. Extensive quantities of Auscha Hops are sent every year into foreign parts under the denomination of „Saatz Hops“; a proof of their being by no means inferior to these. In 1860, the production of the Saatz district, as stated in official reports, did not exceed $1/_8$ crop; under such circumstances, this district may be inferred to have acquired large quantities from the Auscha district in order to complete the annual consumption of $1/_2$ crop of Saatz Hops. The town of Auscha is itself the chief seat of commerce in Hops. The persons, taking a share in it, are: Behr Francis, Diehl Joseph, Dobiasch Joseph and Ignace, Friedrich Francis, Guth Joseph, Gäube Joseph, Grundmann Jos., Hor Francis, Jäger Joseph, Klimpel Wenceslas, Knesel Anthony, Jarich Jos., Kriesche Wenceslas, the Common of Linka, Maier Anna, Müller Wenceslas, Pawlik Francis, Schmid Jos., Sigmund Jos., Schrödlein Charles, Schulz Jos., and Tischler Jos.

256. HUMMER John, Wine-merchant. *Brünn.*
Austrian and Moravian Wines.

257. JALICS & Comp. Francis, Wine-merchants
Pesth.
Wines.

258. JANKO Vincent de, at the head of the Engine-manufactory „Vidáts István Gépgyara". *Pesth, Zwei Hasen-Gasse 9.*
Collective Exhibition, systematically arranged, of RawProductions of theKingdom ofHungary, with Statistical notices and illustrated with photographic representations of the places most in renown for Wine culture, of national Costumes, races of Cattle and Agricultural implements of Hungary.

259. JANKOVITS Vincent, Possessor of Landed estates. *Galabócs, Comitate of Neograd, Hungary.*
Wines.

260. JASZAY Daniel, Captain on halfpay. *Abauj-Szánthó, Comitate of Abauj, Hungary.*
Szt. Andras Muscat-wine and Tobacco.

261. JECKEL John Paul, I. R. Steam-mill, Brewery and Distillery of Spirits. *Hoszufalu, Transylvania.* Store-houses: Kronstadt, Hermannstadt, Thorda, Mediasch and Schässburg.
Flour and Pearl-barley of various sorts.

262. JOÓ János, Wine grower. *Erlau, Hungary.* Bronze M. Vienna 1857.
Erlau Wines.

263. JORDAN & sons. *Tetschen, Bohemia.*
Flour.

264. JORDAN & TIMAEUS, Patented Chocolate-manufacturer. *Bodenbach, Bohemia.* Store-houses: Vienna, Prague, Pesth. Agencies: Brünn, Olmütz, Lemberg, Linz, Gratz, Agram, Pressburg.
Chocolate and succedanea for Coffee.

This establishment, working in Austria, since 1855, found till now no occasion yet to avail itself of Exhibitions. The Chocolate - manufactory works with a steam-engine of 14 horse-power and with an auxiliary apparatus of recent invention, offering security for the purity and good quality of the materials submitted to its action. Every sort of Cacao, brought into commerce, is made use of; so that the assortment of cacao-preparations, issued by the establishment, is adequate to any demand whatever. The assortment of Chocolates for drinking, with or without vanille or spices, comprise about 50 numbers of different qualities; the number of Medicinal Chocolates, requiring additions of Chinine, Iron, Arrow-root, Salep, Semen Zedoariae, Osmazome, Carageen and Island-moss, amounts nearly to ten. Cacao in cakes or pulverized is offered in various sorts, as: Portocabello, Carracas, Angostura, Maracaibo, Martinique, Trinidad, Guayaquil, Maranham, Pará, Bahia, and Domingo. Cacao free of oil, Cacoigna, Racahout de l'Orient, Racahout des Arabes and the Chocolate-powder, quoted in the price-list under Nr. 93, may be commended for their easy digestiveness to persons, to whom the use of tea and coffee is prohibited. An assortment of about 100 numbers of Chocolates for immediate consumption (Chocolate-bonbons, Chocolat-praliné, Chocolate for dessert etc.) of every form and quality are offered for sale in elegantly labelled packets, boxes and cartoons. An assortment of about 1500 fancy articles (brown and coloured), may serve to testify the efforts made by the establishment for attaining perfection in its speciality. The spare room allowed to Austria in the Exhibition-buildings is an obstacle to the full exhibition of specimens illustrating the establishment's activity to its full extent; all that could possibly be done, was to bring into sight a limited assortment of its articles. The store-house at Vienna („am Peter" 577) offers in itself a permanent exhibition. Professor Oppolzer and Dr. Heller, Chemist to the Provincial Tribunal of Vienna, have delivered certificates doing high honour to the establishment. The manufactory of succedanea for coffee, provided with a steam-engine of 12 horse-power, is working up on the best raw materials, as succory root and turnips, with addition of genuine coffee, cacao or figs (according to the price of the fabricated sort) and is continually endeavouring to satisfy any equitable claims respecting the quality of articles brought into commerce. The establishment begs leave to recommend to special attention the good quality of the flour used for its articles. Price-lists in German and in English are added to the exhibited objects.

265. JURENAK Alex., Possessor of Estates. *Pressburg.*
Wheat.

266. (Omitted.)

267. KAMMERGRUBER John, Wine-grower. *Lugos, Comitate af Krassó, Hungary.*
Wines and Slivovitza.

268. KAMPLMILLNER Henry, Wine-merchant. *Gratz, Styria.* Store-houses: Bruck and Judenburg.
Five sorts of first-rate Styrian Wines.

The Exhibitor, whose establishment has existed these ten years, keeps an extensive store of Styrian table- and dessert-wines of every sort, especially bottled Wines of first-rate quality, made out of Rhenish-Muscat- and „Kleinriesling" grapes.

269. KAPPEL Frederick de, Possessor of Landed estates. *Kis-Tur, Comitate of Hont, Hungary.*
Productions.

270. KASSOWITZ J. H., Wine-merchant. *Pesth, Felber-Gasse 5.*
Wines.

271. KATTUS John, Merchant. *Vienna, City 330.* — Tokay Wines, Essences and Vinegars. Wines from Mediasch for the town Mediasch in Transylvania.

Prices loco Vienna: Tokay Wines 1 florin 50 creutzers to 2 florins: 50 creutzers per original bottle; Tokay Essences: 2 flor, 50 creutz. to 6 flor. pr. bottle; Tokay Vinegars: 1 to 3 florins Austrian money per bottle holding about 1 Lire = $^7/_{10}$ Austrian mass.

The Analysis of the Wines brought to Exhibition, has given the following results:

Tokay	Water	Alcohol	Acetic acid	Organic extractive substances	Ashes	Fermentable Sugar (Glycose)	Tartric acid (with traces of Succinic acid)	Extractive substance free of Azote	Extractive substance holding Azote	Phosphoric acid in the Ashes	Specific weight of original Wine or Vinegar	of Wine or Vinegar after boiling	Remarks
				in percental proportion									
Sack No. I	83.780	12.485	. .	3.433	0.302	1.519	0.675	1.723	0.516	0.105	0.9968	1.0138	Particularly flavoured.
„ No. II	86.696	11.154	. .	1.855	0.295	0.201	0.554	0.650	0.352	0.098	0.9919	1.0074	Completely fermented.
„ No. III	82.052	12.525	. .	5.115	0.308	1.723	0.586	2.324	0.482	0.112	1.0035	1.0205	Heaviest and most spirited sort.
Wine No. I	77.869	10.285	. .	11.334	0.512	1.911	0.872	7.828	0.723	0.127	1.0315	1.0460	Highly flavoured.
„ No. II	85.183	9.377	. .	5.125	0.315	0.486	0.766	3.488	0.385	0.095	1.0070	1.0205	Gone through complete fermentation.
„ No. III	76.774	9.000	. .	13.415	0.811	4.384	0.792	7.252	0.987	0.152	1.0425	1.0552	Heaviest sort.
Wermuth No. I	70.956	0.505	. .	27.615	0.924	9.913	0.957	4.528	2.217	0.322	1.1347	1.1355	$^1/_2$ percent of Alcohol.
„ No. II	74.288	0.145	. .	24.752	0.815	8.590	0.962	4.125	1.075	0.301	1.1113	1.1115	Traces of Alcohol.
Essence No. I	51.934	7.000	. .	39.112	1.954	25.512	0.602	9.883	3.115	0.544	1.1669	1.1770	Middle sort, richer in Extracts.
„ No. II	69.800	8.114	. .	20.955	1.131	12.432	0.595	6.785	1.213	0.311	1.0761	1.0875	„ „ richer in Alcohol.
„ No. III	52.706	9.297	. .	36.212	1.785	23.930	0.605	8.952	2.725	0.478	1.1474	1.1604	„ „ most spirited.
„ No. IV	46.934	8.000	. .	42.851	2.215	23.786	0.708	13.185	5.172	0.525	1.1815	1.1930	„ „ heaviest.
Vinegar No. I	86.142	6.298	1.730	5.499	0.331	3.308	0.830	1.065	0.296	0.132	1.0210	1.0235	Poorer in Extract; Acetic acid 8 p. c.
„ No. II	79.464	4.293	1.304	13.714	0.725	4.509	1.125	1.001	0.354	0.118	1.0500	1.0560	Abundant in Extract and Tartaric acid; Acetic acid 6 p. c.

272. KIS DE NEMESKED Paul, Possessor of Landed estates. *Miszlán, Comitate of Tolna, Hungary.*
Wines.

273. KLEIN D. W. *Vukovar, Slavonia.*
Camelina Sativa (gross-grained), Maltese wheat.

274. KLEINOSCHEGG Broth., Wine-merchants. *Gratz 100, Styria.* M. II. Cl. Paris 1855. Silver M. Vienna 1857.
Styrian Champaign- and Dessert-Wines.

Vine-culture and improvement of vines have notably progressed in some vineyards of Styria in the course of these last 20 years, and it has been observed, that the sorts prevailing in Champaign, as „Rissling“, „Traminer“, Portuguese and Cleves, not only thrive in Styria, but also, if submitted to rational technical treatment, give a sparkling wine, quite equal in respect to power to real Champaign and merely different in peculiar taste. The Exhibitors, after having taken up the practical side of the question and studied the treatment of wines in Champaign itself, have succeeded in extracting from the better sort of grapes, grown in Styria, an imitation of Champaign of best quality, and which has met with approval everywhere. Encouraged by this good success and general acknowledgement, the Exhibitors resolved to take their share in the Paris Exhibition of 1855, where, the Examining Jury having pronounced their wines „not inferior in quality to genuine Champaign“, they obtained a II. Class Medal. The Styrian Champaign remaining constantly in favour, the Exhibitors were enabled gradually to extend their new enterprise and thought convenient to send wines to the Vienna Industrial Exhibition of 1857, where they succeeded in obtaining the „first Silver Medal“ in acknowledgement of the excellent result of their untiring exertions. Since this time, the Exhibitors taking incessant care to remove any defect whatever, their Styrian Wines have constantly remained in favour. The following list shows the prices, in Austrian florins, of the different sorts of these Wines:

Champaign Wines:
„Crême de Styrie blanche“: 1 fl. 50 cr. up to 1 fl. 80 cr. per bottle;
    „   „   „ rose“   1 „ 50 „

Dessert-Wines:

		Per bottle
Luttenberg, best quality:	. .	60—70 cr. per bottle;
Kerschbach,	„	. . 60—70 „ „ „
Stadtberg,	„	. . 50—60 „ „ „
Murberg,	„	. . 50—60 „ „ „
Türkenberg,	„	. . 50—60 „ „ „

Table-Wines:
Any sort of Styrian growth: per Austrian Eimer 10, 12, 14, 16, 18, 20, 40 etc. Austrian florins, packing and tons free, loco Gratz, railroad-terminus.

Proportion of Measures:
1 Austrian Eimer is equal to about 48 Prussian Quarts, 13 1/3 Engl. Gallons, about 52 Munich Mass, 14 1/2 Polish Garnetz and nearly 56 1/2 Litres; Four Austr. Eimers = 1 Oxhoft; Weight of 1 Austr. Eimer: nearly 120 Vienna or 134 1/2 Custom Pounds.

275. KLOSTERNEUBURG ABBEY, Possessor of extensive Landed estates. *Klosterneuburg, Lower Austria.*
Assortment of Wines (own growth).

This Abbey stands in the first rank among Austrian Vine cultivators, no less for the excellent management of its cellars (quantity being less regarded than quality) than for the extension of its Vine-yards around the Kahlenberg near Vienna and at Neszmély in Hungary. The enormous Cellars in the Abbey of Klosterneuburg occupying three stories above each other, have a temperature between 6 and 8° R. The procrastinated fermentation and ripening of the Wines under such circumstances is compensated by their keeping for a long time their Carbonic acid and by their obtaining exquisite flavour, not to be obtained in any other way. The Abbey has highly promoted culture in Lower Austria by admitting into its buildings the School for Culture of Wine and Fruit-trees, founded by the Agricultural Society of Lower Austria and by bestowing all possible attention to this school. The results of the analysis of the Wines brought to the Exhibition are:

	Year	Alcohol	Acids
Kahlenberg	. . . 1797	10%	0.75%
Weidling	. . . 1808	10 „	0.85 „
Kahlenberg	. . . 1811	10 „	0.75 „
„	. . . 1834	10 „	0.75 „
„	. . . 1811	10 „	0.85 „
„	. . . 1846	10 „	0.71 „
Grinzing	. . . 1848	10 „	0.70 „
Kahlenberg	. . . 1852	10 „	1.06 „
Weidling	. . . 1857	10 „	0.75 „
Nezmély	. . . 1859	10 „	0.76 „
„	. . . 1856	12 „	0.62 „
„	. . . 1857	12 „	0.65 „

276. KONKOLY Francis. *Bekes, Comitate of Bekes, Hungary.*
Wines.

277. KORNIS Joseph, Curate. *Gyorok-Ménes, Comitate of Arad, Hungary.*
Menes Wines and „Ausbruch“ at prices of 18, 25, 30, 40, 50, 60, 180 and 350 florins.

278. KOSZTKA John, Wine-merchant. *Pesth.*
Ermelek Wines.

279. KRAETSCHMAR C. A., Merchant. *Rima-Szombath, Hungary.* M. London 1851 and New-York 1853. M. II. Cl. Paris 1855. Agent: Ernest Szumrák at Williams & Norgate, London, 17, Henriette-street, Covent-Garden.
Tokay „Ausbruch“ (dreibuttig), Szamarodna and Hegyallya „Wermuth“ (zweibuttig), the same from Tallya in the Hegyallya.
(See Classes 4, 26 and 27.)

280. KRATOCHWILL Wenceslas, Cultivator and trader in Hops. *Louky, Bohemia.*
Hops.

281. KRAUS Brothers, Possessors of a mechanical Mill. *Carolinenthal near Prague.* Store-houses: Prague, Carolinenthal, Smichow, Brandeis and Radnitz.
Articles produced by grinding.

282. KREMS, Town, *Lower Austria.*
Collective exhibition of Wines.

283. LAIBACH, I. R. Agricultural Society of Carniola, *Laibach.*
Cereals.

284. LARISCH-MÖNNICH Count, Sugar-manufactory. *Ober-Suchau.*
Molasses and refined Beet-sugar.
(See Class 4.)

285. LAY'S Michael heirs, Possessor of agricultural establishments. *Essegg, Slavonia.* M. I. Cl. Paris 1855. Great Silver M. Vienna 1857.
Cereals (7 sorts) and Madder.
(See Class 4.)

The Maize, silver-medalled in 1855 and 1857, is the result of rational cross grafting between other varieties; its culture is systematically conducted to a definite purpose. Seed-wheat has been introduced from Taganrog by the importers. Madder has been experimentally sown out at the instigation of the Industrial Society of Lower Austria, with seeds from Avignon; results, as to quality, not yet known. The Oleagineous seeds have been produced in the way generally used in Croatia

286. LEIBENFROST Francis & Comp., Wine-purveyors to the I. R. Court and Wine-merchants. *Vienna, 1107.* Great M. Munich 1853. M. II. Cl. Paris 1855. Silver M. Vienna 1857.
Wines.

Since the establishment of this trade in 1780, its commercial sphere has been continuously enlarged, so that presently its cellars at Döbling near Vienna contain a standard store of 20,000 eimers, completely assorted with every variety of Austrian and Hungarian Wines, from those of first-rate superiority down to those of common use on account of their moderate prices. The sales of the Exhibitor's Wines extend to the whole Austrian Empire, the Orient, Russia, Northern Germany, and, of late also to America. The fundamental principle for keeping up the credit obtained, is to bring the Wines into commerce in their state of original purity and in conditions securing them against adultering agents. The establishment has gradually succeeded (not without great sacrifices) to get in possession of a continuous area of 18,000 square fathoms (called „Nussberg“) on the best-situated slope towards the Kahlenberg near Vienna, where the best sorts of Rhenish and other Vines are cultivated. Every new practicable experience in Vine culture and management of Cellars has been availed of for ensuring to the Austrian Wines the share in universal commerce, which they are entitled to claim. The Wines, exhibited under the denomination of „Oesterreich's Donauperlen" (Danubian pearls of Austria) are made with „Traminer" grapes of 1859 and ready for transport. The totality of stores of red and white Austrian and Hungarian Wines, „Ausbruch," Tokay and Essences is completely organized for export business, an adequate amount of it is drawn into bottles, and every measure taken for the most extensive commercial intercourse.

287. LEITHNER John, I. R. patented manufactory of succedanea for Coffee. *Gratz, Stiria*. Store-houses: Vienna, Pesth, Temesvár, Linz, Salzburg, Pressburg, Oedenburg, Wiener-Neustadt, Klagenfurt, Laibach, Gorizia.

44 sorts of succedanea for coffee.

The establishment produces for the most part succedanea of refined quality, especially Fig-coffee. It is one of the most extensive of this kind in the German provinces of the Empire, producing every day 50 Cwts of Raw materials and finished articles are transported from one floor to an other by a machinery moved by steam, a power used for the most part of the operations performed in the establishment.

288. LENK Samuel. *Oedenburg, Hungary*. Honourable mention: Paris 1855. Agent: Charles Schäffer, G. B., New Broad Street, London.

Hungarian Wines, „Ausbruch" and Essences.

289. LENKEY Achaz de & Comp. *Vienna*.

Bottled Wines.

290. LIEBL Vincent & Son, Wine-merchants. *Rötz, Lower Austria*. Store-house: Vienna 899. Honourable mention: Paris 1855, Gr. silver M. Vienna 1857.

Various sorts of white and red Wines from the crops of 1834—1861 (own growth).

This establishment, existing since 1800, keeps up a very active commercial intercourse. The Retz Wines may undoubtedly rank with the best among the Austrian sorts, being even preferred to some of them on account of their mildness and freshness, because they act less on the circulation of the blood. These Wines ripen and are fit for use at an earlier time than other sorts, and, if rationally treated, remain unaltered for years in bottles and barrels or tons. Retz is one of the chief places for Austrian wine-trade. Liebel's establishment produces also „Ausbruch" Wine in the greatest variety, acknowledged for their quality and good preservation by competent authorities, and keeps a store of several thousand eimers, at all prices, from the most moderate up to the highest. Liebl's Vinegar-manufactory produces genuine Wine-vinegar with exclusion of all other sorts; a manufactory of Mustard is connected with it. The establishment here in question, one of the first and most ancient in Austria, stands in solid and far-spread renown.

291. LITTKE John and Joseph, Wine-merchants. *Fünfkirchen, Comitate of Baranya, Hungary*.

Wines.

292. LÖFFLER John Paul, possessor of Landed estates, *Langhalsen near Neufelden, Upper Austria*.

Hops.

(See Class 4.)

293. LONYAI Melchior de, Possessor of Landed estates. *Pesth Zweiadler-Gasse 3*.

Wines.

294. LOVASSY Albert, Possessor of Landed estates. *Nagy-Szalonta, Comitate of Bihar, Hungary*.

Wheat.

295. LÖWENFELD Brothers **& HOFMANN**, I. R. priv. Mechanical mill. *Kleinmünchen near Linz, Upper Austria*. Agent: C. W. Groos, London, 65, Leadenhall-street. Gr. Silver M. Vienna 1855 and Linz 1861.

Productions of the Mill (23 sorts) and different kinds of Cereals.

This establishment, dating from 1854, is at present in possession of two Turbines of Jonval's system, acting with 150 horse-power by a water-fall of 10 feet, obtained by an express conduit from the river Traun, 14 pair of 4 feet French and 3 pair o inland millstones, together with an auxiliary apparatus of flour-cylinders, engines for cleansing flour and grits, etc. The cereals

are measured on a patented apparatus. Grinding is performed on the principle of absolute dryness and continual movement by mechanical means. The articles, produced in the establishment have got a good credit on all continental commercial staples for their nutritiveness and easy preservation.

Table of production:

Years	Metzens of Cereals	Number of Turbines	Number of Pair of Millstones	Production in Vienna Cwt
1855	57,284	1	6	45,784
1856	73,614	1	6	55,178
1857	80,844	2	10	67,066
1858	119,313	2	10	84,421
1859	174,619	2	10	126,779
1860	166,000	2	12	132,832
1861	188,000	2	14	151,000

296. LUKÁCS Sigismond, Possessor of Mills. *Balosa, Comitate of Stuhlweissenburg, Hungary*. Domicile: Stuhlweissenburg.

Assortment of Flour.

297. LUXARDO Girolamo, I. R. priv. Maraschino-manufactory. *Zara, Dalmatia*.

Maraschino.

This liqueur, justly taking the first rank among all others on account of its superior flavour, is distilled out of a species of Cherries, called „Marasca", of frequent occurrence in Upper-Dalmatia. Among the Maraschino-distilleries of Zara, where this branch of industry is practised since centuries, the Exhibitor's establishment enjoys particular credit, sending its articles into every part of Europe, to America and even to the Eastern Peninsula of India.

298. MAGYAR Emeric, Possessor of Landed estates. *Maria-Theresiopel, Comitate of Bacs-Bodrogh, Hungary*.

Wheat.

299. MAKAY Alexander. *Lugos, Hungary*.

Plum-Slivovitza from the years 1817, 1820, 1845, 1846 and 1847.

300. MÁLNAY Dr. Ignace, Vine cultivator. *Tóthfalu, Hungary* (lives at Pesth, Museumgasse 1).

Wines.

301. MARBURG, first Association of Styrian Vine Cultivators. *Marburg, Styria*. Silver M. Vienna 1857.

12 sorts of bottled and Table-Wines from the most eminent wine growing regions of Styria.

The Association has been constituted in 1855, with the purpose to make Styrian Wines, hitherto scarcely known beyond the provincial boundaries, although equal to the best sorts, an object of commerce by means of rational treatment and to open ways for their export. The Association began their activity with forming a solid store and with imparting commercial regularity to the intercourse with the surrounding vine culture districts. In consequence of successive extension of connexions, the demands from Bohemia, Moravia, Silesia, Galicia, the Venetian and the Adriatic Littorale are progressively increasing. Connexions with the states of the German Tariff-union have been tried more than once; in this direction, however, the high customs, charged on Austrian Wines, have proved an absolute obstacle. The North of Europe opens a good market for Styrian Wines, which, if once known, may certainly find favour for their strength and fire, as they already did in the Orient. The Association is composed of 12 members, possessing a vine culture area of about 1,120,000 Austrian square fathoms, so that their own production may prove constantly adequate to the most extensive demands-

302. MATKOWSKA Zofia. *Jezierzani, Circle of Stanislav, Galicia*.

Barley.

303. MAUTHNER A. J. & son., Brewery and Malt-manufactory. *Vienna and Göding*. Distillery of Spirits and Pressed Yeast. Vienna and Simmering. Chief-store-house and Agent: Leopold Wimmer, Vienna, Inner City 838. Gr. Silver M. Vienna 1857.

Vienna St. Mark's Pressed Yeast, St. Mark's Saloon-beer (bottled).

The exhibitors have, in 1847, first introduced into Austria the manufacture of artificial yeast, the first impulse for it having emanated from the Industrial Society of Lower Austria, who had proposed for a prize their Great Gold Medal of 50 ducats in value, and from the Corporation of Bakers at Vienna, who had subscribed a sum of 1,000 fl. for the same purpose. Messrs. Mautner and Son obtained both prizes in 1849, as also, in 1857, the Great Silver Medal of the I. R. Agricultural Society of Lower Austria. Their produce is purest corn-yeast, extracted from the materials prepared for the distillation of spirit nor is it to be identified with pressed beer-yeast, brought into commerce under the designation of „Natural pressed yeast“. The St. Mark's Pressed Yeast offers the following advantages: a proportionally small quantity of it provokes spirituous fermentation, wherever it may be required, with full security and in briefest delay; added to paste, it neither darkens its tint nor imparts to it any heterogeneous taste and, lastly, it remains unaltered for several weeks, even in summer, and, for this reason, is particularly fit for export. Nearly all the bakers at Vienna use this yeast, and to this circumstance the acknowledged superior quality of their products may be chiefly ascribed. The establishments of the Exhibitors are continually increasing; at present 50 Cwts of yeast and 150 eimers of spirits are extracted every day from 3,000 eimers of materials prepared for distillation; in 1861, the consumption of Cereals were 216,000 metzens, and the taxes, paid for it, amounted to 290,000 fl. The Exhibitors may be said to have completely solved the difficult problem of preparing Pressed yeast, tried without success by toher industrial establishments. Besides these articles, the extensive Brewery of the Exhibitors produces several Beers of superior quality, among which the sort, brought into commerce under the denomination of „St. Mark's Saloon Beer,“ deserves special attention, as being the only sort of Vienna Beer brought into sale in bottles, according to the use prevailing in England. This Beer is brewed by a new and practically acknowledged method out of select malt and hops, and, if duly reposed, is eminent for p u r i t y, p l e a s a n t t a s t e, n u t r i t i v e n e s s and q u a n t i t y of C a r b o n i c acid. Conveniently bottled, as it is offered for sale, it may safely be recommended as the sort most appropriate for appearing on elegant tables. Since March 1861, when this sort was first brought into commerce, a great quantity has been consumed at Vienna and a still more considerable amount has found its way into the Lower Danubian regions, Wallachia and Asia. In 1861, the Breweries of Messrs. Mautner and son at Vienna and Göding have produced 200,000 eimers.

304. MERAN Count, Cellar-office. *Marburg, Styria.*
Wine.

305. MOLNÁR George, Possessor of Landed estates. *Debreczin, Comitate of Bihar, Hungary.*
Wheat, Cereals, Tobacco and Wine.

306. MOLNÁR & TÖRÖK, Possessors of Landed estates and Vine Cultivators. *Pesth.*
Tokay Wine („Ausbruch“)

307. MÜNCH-BELLINGHAUSEN Count. *Vienna, Inner City 543.*
Vöslau red Wines.
(See Class 4.)
The red Wines, brought to the Exhibition, are grown on the Exhibitor's own vine yards, situated on the south side of the hill-range of Vöslau. Pure natural Wine from the crops of 1857 and 1858, under the label: „Vöslauer Wein, Merkensteiner Gewächs“ This red Wine has kept up its credit for good quality and long preservation even in Trans-atlantic regions, having been repeatedly sent to Egypt without any loss whatever in its qualities, while other Wines from northern parts generally suffer under the action of tropical heat. Prices: 64 kreutzers per bottle; up to 20 florins per eimer.

308. NAGY Michael, Manufacturer. *Raab, Hungary.*
Several sorts of Flour Meats.

309. NAWRATIL Peter. *Romanowka, Circle of Zloczow, Galicia.*
Hops.

310. NITSCH Joseph, Superintendent of Landed estates. *Strassnitz, Moravia.*
Wine, Seeds and commercial Plants.
(See Class 4.)

311. OROSZY Nicholas de, Possessor of Landed estates. *Szentes, Comitate of Csongrád, Hungary.*
Wheat.

312. PANETH Jos. Siegfried. *Vienna, Inner City 154.*
Wines.

313. PARAGH Gabriel de, Wine-merchant. *Pesth, Sebastian-Platz.*
Hungarian Wines from various localities and years

314. PASTNER Joseph, I. R. priv. Spirit-Liqueur and Vinegar manufacturer. *Gratz.*
Liqueurs.

315. PERGHAMMER Felix de, Possessor of Landed estates. *Eppan, Tyrol.*
Champaign, manufactured by the Exhibitor.

316. PESTH, Cylinder-mill Association. *Pesth.*
Assortment of Flour.

317. PIERING C. F. *Carolinenthal near Prague 157.*
Vinegar-spirit and Wine-vinegar.
(See Class 2.)

318. PIERING & DONAT. *Carolinenthal near Prague 157.*
Mustard.

319. PODMANITZKY Baron, Armin, Possessor of Landed estates. *Aszod, Comitate of Pesth, Hungary.*
Wines.

320. POHRLITZ, Archducal Superintendency of Mills. I. R. priv. mechanical Mill. *Pohrlitz, Moravia.*
Articles produced in the mill.

321. POTOCKI Count Adam, Steam-mill. *Tenczynek, Galicia.*
Flour.

322. PRASCH Leander, Possessor of Landed estates. *Vienna.*
White Wines from Mitterretzbach, Lower Austria, crops of 1857, 1858, 1859, and 1860.
Extent of his own vine yards: 30,400 square fathoms. Treatment with „Kahlschnitt“ (bald cut) and partly according to Hooibrenk's method. Cellar-room for 12,000 eimers. Commerce exclusively with own growth, improved from year to year. Prices: 15 and up to 30 fl. per eimer; sale in bottles, if demanded.

323. PRIMAVESI Paul, for his I. R. priv. Sugar-manufactories. *Gross-Wisternitz and Bedihoscht, Moravia.*
Sugar-loaves and Spirits.

324. RAMASETTER Vincent. *Sümegh, Comitate of Zala, Hungary.*
Various sorts of wine,
The exhibitor's vineyards are:
at Sümegh . . . . 20,000 square fathoms,
at Badacsony on the Platten lake 36,000   „   „
at Somlau 12,000   „   „
situated on the south side, producing annually from 1,000 to 2.000 Eimer of wine from the best sorts of vine. Sale in barrels and bottles in the empire and to Saxony and Prussia. Price, in barrels, from 10 to 50 fl. per Eimer. Orders by letter, for sales of larger quantities to known firms 4 months' credit.

325. RANOLDER Bishop Dr. John. *Veszprim, Hungary.* Agent for London Exhibition: Vincent Jankó in the Austrian board.
Wines and „Wermuths“ of own growth: Schomlau Wines from the crops 1853, 1856, 1858, 1861. Badacsony Wines from the crops 1858, 1861, Badacsony „Wermuth“ from the crop 1859.

326. RAUSSNITZ E., Liqueur-manufactory. *Hernals near Vienna.*
Assortment of Liqueurs.

327. REHBERG I. R. priv. mechanical Mill (possessed by Chevalier F. de Kleyle). *Krems, Lower Austria.*

Store-houses: Krems, St. Pölten, Steyr, Linz, Salzburg.
Honour-prize: Vienna 1857.

Articles produced in the Mill.

A thousand metzens of Cereals are consumed every week. The articles, brought to Exhibition, are: Wheat-flour and Grits in 9 sorts, Corn, in 3 sorts, all kept in glasses.

328. REISENLEITNER Jos., Wine-merchant. *Vienna 599 and 600.*

Austrian Wines of various sorts.

The Exhibitor's establishment, existing nearly 60 years without change of locality, ranks among the most ancient in Vienna. Mr. Reisenleitner has it made a point of honour to keep up the well-founded credit of his house and continually to enlarge its commercial sphere. His cellars contain a store of 6,000 to 8,000 eimers of superior Wines of Lower Austria. These Wines, coming from every wine region of this Province and, submitted to rational management without any artificial admixture, may be warranted for purity and long preservation. The sale chiefly made in barrels and tons; also in bottles, if expressly demanded. Conditions of sale as usual. The house has never yet brought its articles to Exhibitions.

329. REISSENBERGER William. *Hermannstadt.*
Wines.

330. RICHTER Anthony, I. R. priv. Sugar-refinery and manufactory of Earthen ware. *Königsaal, Bohemia.*
Sugar.
(See Class 35.)

331. RIEMERSCHMIDT Anthony, Refiner of Spirits, manufacturer of Liqueurs and Vinegar. *Vienna, Wieden 237.*

Potato-spirit, perfectly free from Amyl-Alcohol.

Since the time, that this article, serving for all purposes requiring Alcohol perfectly free of empyreumatic oil, has been first produced in Austria, the import from French spirit of wine, formerly very considerable, has sunk to a minimum; so, that the manufacture came to the necessity to enlarge its productiveness from year to year.

332. RIESE-STALLBURG Baron. *Vienna, City, Mölker-Bastei.*
Photographs of Cattle.

333. RIGYITZKY Paul de, Possessor of Landed estates. *Skribestye, Banat. Hungary.*
Slivovitz, Maize in bunches, and Potatoes.

334. RIMANOCZY Paul, Possessor of Landed estates. *Village of Töhöl, Comitate of Bars, Hungary.*
Agricultural products.

335. RINGLER Jos. Sons, manufactory of preserved and candied Fruits, Drugs and Sugar-ware. *Botzen, Tyrol.*
Preserved, dried and candied Fruits in elegant boxes.

336. RIPKA Jos. Max & Comp. Mill-industry. *Brünn.*
Articles produced in Mills.

337. RITTER Hector de **ZAHONY**, possessor of a Mechanical mill. *Strazig, Istria.*
Flour.
(See also Class 20.)

338. ROBERT & COMP., Sugar-manufactory and distillery of Spirits. *Gross-Seelowitz near Brünn, Moravia*, Store-house: Vienna, City 797. Great M. Munich 1854, M. 2. Cl. Paris 1855.
Sugar and Spirits.

The manufactory, constructed in 1837, is in possession of 18 steam-boilers with 1,800 horse-power, 15 steam engines with 120 horse-power, 16 hydraulic presses, 16 reservoirs for macerating green and dried Beets; so, as to being about, in the course of the six winter months, to work up a quantity of 60 millions pounds of Beets. Every year about 10 millions pounds of raw Sugar are refined and about 10,000 eimers (6,000 hectolitres) of 95 p O. Spirits, perfectly free from empyreumatic oil, are distilled out of molasses. The establishment uses, since more than 6 years the apparatus for measuring

Spirits. The evaporation-apparatus, known and generally used under the denomination of „Robert's Steam-apparatus" works without interruption since the expedition of 1851/52. It has been constructed in the establishments's own mechanical workshop under the superintendence of Mr. Jacquier, attached to it as Mechanician. The use of this apparatus has been unrestrictedly opened to Industry, nor has the inventor taken a protective Patent for it,; so that Austria is the only beet sugar manufacturing country, in which its free and unrestrained use is set at public disposal. In winter, the number of persons employed is 600—800, in summer, about 300, those employed for beet-culture left out of account.

339. ROHONCZY Ignace de, Possessor of Landed estates. *Nagy-Bogdány, Comitate of Veszprim, Hungary.*
Spirits and Liqueurs.

340. ROSCONI Nicholas, Wine grower. *Bude, Wasserstadt 538.*
Hungarian Wines.

341. RULIKOWSKY, Cajetan. *Zwitazow, Circle of Zolkiew, Galicia.*
Wheat.

342. RUNGG Albert. *Triente, Tyrol.*
Wine.

343. RUPP Francis X. Hops-merchant. *Neufelden near Linz. Upper-Austria.*
Neufeld Hops for the sort of beer called „Lagerbier" (own growth).

344. SAAZ Agricultural (branch-) Society *Saaz, Bohemia.*
Various sorts of Cereals.
(See Class 4.)

345. SAUER Dr. Ignace, Proto-physician. *Pesth.*
Wines.

346. SCHARY J. M. Brewer. *Prague.*
Beers „Lager" and „Bock", Barley-malt.

347. SCHÄSSBURG Common. *Schässburg, Transylvania.*
Wines.

348. SCHAUMBURG-LIPPE, Prince. *Veröcze, Slavonia.*
Wine, Flour and Grits.
(See Classes 2 and 4).

Mr. Garthe, manufactural Chemist, has obtained the following results by analysing the Wines from the dominion of Veröcse:

Wines	Specific gravity	Alcohol, pCt. of volume	Sugar	Acids
Red	0·995	10·7	1·0	0·598
White	0·990	9·9	0·9	0·575

349. SCHLUMBERGER Robert, Manufacturer of foaming and select Table-wines. *Vöslau near Vienna.* Counting- and store-house: Vienna 238. Bronze M. Vienna 145, Gr. M. Munich 1854, II. Cl. M. Paris 1855, Gr. Silver M. Vienna 1857.
Assortment of Wines of own growth.

The establishment, founded 1842 by Mr. Schlumberger, is situated at Vöslau in the centre of its own extensive Vineyards, comprising the best grounds for Vine-culture (as Steinberg, Goldeck etc.). It has been founded for the purpose to improve, by own Vineculture purchase of products of the best renowned regions of Austria, and to convert Austrian Wines into foaming sorts (according to the method used in Champaign) or into table-wines. Above 100 persons find constant employment in its different branches. Its articles find their way to all Provinces of the Austrian Empire, especially to Hungary and Italy, to the Danubian Principalities, the Orient, Poland, Russia, German Tariff-union States etc., as far as excessive custom-duties are not an obstacle to export. These Wines have proved capable to be kept for a long time and to endure transport by sea. H. M. Frigate **N o v a r a**, when setting out for her Circumnavigation-expedition, had taken on board nearly 4,000 bottles of these Wines. After a voyage of $2\frac{1}{2}$ years and having repeatedly passed the Aequator, the samples brought home had highly improved in quality and flavour. Brought to Exhibition:

	per dozen bottles, frank Docks at London
Foaming Wines at	1 L. 6 s.
„ „ „	1 „ 11 „
„ „ „	1 „ 16 „

	per dozen bottles frank Docks at London	Percent of Alcohol
Fine red Vöslau Table-wines	— L. 12 s.	11·05
„ „ „ „ „ . . .	— „ 14 „	11·70
„ „ „ „ „ . . .	1 „ 6 „	12·10
„ „ „ „ „ . . .	1 „ 8 „	11·80
„ „ „ „ „ . . .	1 „ — „	11·75

Alcohol de termined at + 15° C. in percents of Volume by Gay-Lussac's method.

350. SCHMITT Adolphus, *Deutschbrod, Bohemia.*
White and red Clover-seed.
(See Class 33.)

351. SCHNEIDER Augustus. *Vienna , City 727.*
Wines.

352. SCHÖFFEL Jos. *Saaz, Bohemia.* Bronze M. London 1851, Gr. Gold M. Vienna 1857.
Hops (own growth).

353. SCHÖLLER Alexander, possessor of the Steam-mill and Rolled-barley-manufacture at Ebenfurt. *Ebenfurt, near Vienna.* Counting-house: Vienna, Leopoldstadt 4. Store-houses and Agencies at Vienna and most of the ·Provincial Capitals. Stuttgart: Julius Kraemer, Hamburg: G. Löning and Kaufmann, London: Agent W. Adolph and Comp.
Flour in every stadium of production.

This establishment, built in 1854, disposing of 2 Turbines with a summary power of 150 horse and of an auxiliary Steam-engine of 60 horse-power in case of deficiency of water, works every year 300,000 Cwts of Wheat and 40,000 Cwts of Barley. It is at present the most extensive establishment of this kind within the Austrian Empire. The objects exhibited are samples of the results obtained by grinding 5,000 metzens of wheat and 1,000 metzens of barley. Samples, if applied for, gratis.

354. SCHORKOPF Charles Henry, Superintendent of Forests and Estates in Count Michael Bethlén's service. *Alsó-Bakos, Transylvania.*
Wheat, giving flour of superior quality and being three times and above more productive than any other sort of wheat.

355. SCOTCH ABBEY, *Vienna.*
Various sorts of Wines from Lower-Austria (own Growth).

The analysis of the Wines exhibited here, has led to the following results:

Year	Sorts	Alcohol in percents of weight	Sugar in percents
1858	Grinzing	13	2·5
1852	Nussberg	13	2·58
1857	Grinzing	13·5	2·56
1854	„	12	2
1855	Nussberg	12	2·5
1851	„	13	2·5
1852	Grinzing	13	2
1859	„	11	2·58
1854	Pulkau	9	2·5
1856	Zellendorf	8	2·5
1852	Enzersdorf	12	2·5
1846	Grinzing	12	2·58
1855	„	11	2
1848	Enzersdorf	13	2
1846	Ottakring	9	2·5
1848	Grinzing	13	2

Year	Sorts	Alcohol in percents of weight	Sugar in percents
1857	Ottakring	9	2
1857	Enzersdorf	11	2·5
1857	Perchtoldsdorf	10	2
1859	Parmesthal	10	2
1859	Nussberg	11	1·8
1858	„	10	1·8
1846	„	10	2·5
1834	„	13	1·5
1834	Grinzing	13	1·58
1827	„	13	1·5
1834	Enzersdorf	12	1·58

356. SCHÜRER Francis Paul. *Stein on the Danube, Lower Austria.*
Wine.

357. SCHWARZENBERG Prince Joseph Adolphus. From his dominions.
Hops.
(See also Classes 1 and 4.)

This sort is held equal in value with the „Stadtgut" Hops of Saaz; about 1,000 Cwts of it are produced on the Prince's extensive grounds not far from the town of Saaz. It is very much in demand, as standing in high credit for its homogeneity and special care bestowed on its culture, desiccation and selection. For farther information apply to Prince Schwarzenberg's Superintendent of Estates at Postelberg and Zittolieb near Saaz.

358. SCHWARTZER'S A., successor. *Vienna, Spiegelgasse 1102.*
Wine.

359. SEEGER Frederic, Vine-cultivator. *Pesth, Marokkaner-Haus.*
Wines.

360. SELBSTHERR Brothers. Wine-commerce for the Hegyallya in Upper Hungary. *Mád, near Tokay.*
Tokay Wines of different sorts (2-, 3- and 4-„buttig") Szamarodna and Maszlás Wines.

361. SIMON Anthony. Cake-baker. *Vienna, Neubau; 145.*
Biscuit in 14 sorts.

362. SINA Baron John, I. R. priv. Sugar and Spirit-manufactories ·*St. Miklos, Hungary, and Rossitz, Moravia.* Store-houses at Vienna and Pesth. I. Cl. M. Paris 1855.
Loaf-sugar and Spirits.

The Sugar-manufactory at St. Miklos (Comitate of Wieselburg) in Hungary, constructed in 1848 and provided with machinery and apparatus of newest invention, is capable of operating on 500,000 to 600,000 Cwts of beet-root, for the most part of own growth. Steam is used in every way possible. Movement is imparted to every agent and in every direction by 8 steam-boilers of 70 horse-power each, and by 13 acting steam-engines. Peat of excellent quality used for fuel, comes from the extensive peat-diggings on the dominion of St. Miklos; its conveyance immediately to the manufacture is effected by nearly 100 boats, of 200 Cwts bearing each, on a navigable channel of 2 German miles length, expressly constructed for this purpose. The digging and preparation of Peat employs nearly 3,000 workmen, besides a number of subaltern superintendents and officials; the annual production amounts to about 600,000 Cwts. The manufactory works with hydraulic presses and their auxiliary machinery and apparatus of improved construction. The greatest part of Sugar is converted into Melis and Raffinade up to the highest refined sorts; another is converted into Juice-melis. The refining-establishment, connected with the manufactory, possesses means to make ready for sale 6,000 Cwts per month. In another subdivision, the Molasses are worked by 2 large distilling and 2 rectifying-apparatus, assisted by a steam-mill with 4 pair of stones, for extracting Spirits of the best sort out of them. This part of the establishement has been fitted out so as to enable it to produce every year about 15,000 eimers of Spirits ready for export. Another subdivision is intended for the burning of Spodium, the grinding of bones and the revivification of Spodium; the quantity of the revivified Spodium, issuing from this last operation, is sufficient to cover the wants of the Sugar-manufactory. Workshops for machinery, copper-smiths, casting, joiners, coopers, varnishing of plate iron and other auxiliary operations for keeping everything in good repair, are joined to the manufactory. Security against fire is procured by a uniformed corps of Fire-men, organized according to the Badish system, and recruited among the carpenters, masons, lock-smiths, and other workmen, attached to the establishment. The persons employed in the manufactory, amount to 600; a School, an Infirmary anApothecary, and several Physicians are exclusively devoted to their use. Six large bridge-balances, placed in special houses around the manufactory-buildings, serve to control the raw materials entered and issued. The Sugar-manufactory at Rossitz near Brünn (Moravia), constructed in 1849 and organized according to the newest system, is capable for working every year 300,000 Cwts of beet-roots of own growth, by means of hydraulic presses. Steam is the only moving power, 8 boilers of 30 horse-power each, feed 5 steam-engines, acting as motors for mechanical apparatus of every description. The Refinery produces 5,000 Cwts per month. The Molasses are worked in a Distillery, producing yearly 8.000 eimers of first-rate Spirits, fit for export. A well-organized set of apparatus for burning and revivifying Spodium acts as an auxiliary to this establishment, assisted by workshops for engine-repairing, copper smiths, joiners and coopers. The number of persons employed amounts to 350. Rossitz Coal, from the collieries in the next vicinity of the manufactories, serve for fuel. Two large bridge-balances, for the control of raw materials, stand at the entrance and at the issue of the buildings.

363. SMETANA John, Superintendent of Countess Rosa Lazansky's Estates. *Dominion of Mileschau, Bohemia.*

Productions of the Mittelgebirg in Bohemia.

Winter-Wheat: 3 sorts; Varying wheat (for summer and winter-culture); Summer-Wheat: 2 sorts; Winter-Corn: 2 sorts; biennial Shrub-Corn; Corn for seed; Barley: 2 sorts; Oats: 3 sorts; Pease: 2 sorts; Vesces; Lentils; Millet: 2 sorts; Summer- and Winter-Colza; Mustard; Turnips: 4 sorts; Clover: 4 sorts; Dried fruits: 6 sorts. All these objects have been produced within the highest horizons of the Bohemian Mittelgebirg.

364. a) SPRINGER Max, I. R. priv. manufactory of Spirits and Pressed yeast. *Reindorf near Vienna.* Agencies in all provinces of the Austrian Empire.

Pressed yeast; Spirits; raw materials required for the production of both these articles.

This establishment, founded in 1851, exhibits now for the first time. It employs an average number of 70 workmen and produces yearly 24,000 eimers of raw Spirits with 32—36% of Alcohol per Eimer and 7,000 Cwts of Pressed yeast, by working yearly 73,000 Cwts of corn, maize and malt, for the most part produced on Hungarian grounds, held in lease by the Exhibitor. The Spirits are sold to the refining establishments of Vienna, the Pressed yeast is sent into all directions in wholesale and detail trade, at the rate of 42 Austrian kreutzers per pound. This pressed yeast holds an eminent rank among other articles of this kind on account of its fermenting power, its fitness for every purpose, and its easy preservation. It may be preserved for 4 to 6 weeks, as temperature permits, coming into trade in such a state of dryness, that the manufacture-mark may be impressed on every single parcel, and this may be sent away simply wrapped up in paper. The packets are sent away from the weight of $1/4$ pound up to any weight demanded. The manufactory has crushing-mills of its own and works with two boilers, of 40 horse-power each, one of them being constantly kept in reserve. The wastes of Spirit-distillation amount every year to 660,000 eimers of diluted liquid matters, sold to persons keeping cattle for milk at Vienna and environs. The establishment is charged with public taxes to the amount of about 160,000 florins every year.

364. b) STAMM Martin, *Wartmann near Kronstadt, Transylvania.*
Wheat.

365. STEAMMILLS, Company, patented Shareholdes. *Vienna.* Agencies: in every larger provincial town of the Austrian Empire. Silver M. Vienna 1845. Prize-M. London 1851. Great M. Munich 1854. Honourable Mention Paris 1855.

Articles produced by the Mills.

This Association, together with its establishment at Vienna (Schüttel)has been founded in 1842,by means of a ready paid sum of 1,260,000 fl., represented by the emission of 3,150 shares of 500 fl. each, constituting a total of 1,575,000 fl. The establishment is provided with 22 pair of mill-stones, 4 feet in diameter, with their accessory apparatus, moved by means of two Wolf's steam-engines of 80 horse-power each and one of 40 horse-power, representing a total of 200 horse-power. The

minimum of wheat, made ready in a 24 hour work, is 1,100 Austrian metzens = 104,720 pounds. The quantities yearly produced since 1846, together with their qualities, are made evident in totals and in percental proportions by the objects brought to Exhibition. Grinding is generally performed dry; raw materials procured from several regions of Hungary; sale for the most part at Vienna and in the Austrian Provinces. Agencies exist in most of the provincial Capitals; export is made in every direction, especially into North- and South-Germany, Switzerland, and, in 1861, also into France and England.

366. STEAM-MILL-ASSOCIATION „STEPHAN", *Debreczin, Hungary.* Store-houses at Tokay, Grosswardein, Kaschau, Eperies, and Szathmár.

Wheat, and Grits: 5 sorts, Wheat-flour: 7 sorts; Corn-flour: 2 sorts; Barley-grits: 6 sorts.

367. STEPNICZKA Francis. *Prague, Neustadt, Spitalgasse 1269.*
Rock-drops.

368. STIASNY Hermann. *Vienna, Jägerzeile 503.*
Liqueurs.

369. STIRIAN I. R. AGRICULTURAL SOCIETY. *Gratz.*
Collective Exhibition of Stirian Agricultural productions. Contributors: Abbey of Admont, Brandstetter Frederick, Falk Joseph, Fischer Anthony, Heinrich Alexander, Kamplmillner Henry, Klement Francis, Kodolitsch Richard Noble de, Neuhold Anth., Ottenig George, Pauer John Paul, Perko Francis, Pistor John Chevalier, Rebenburg Louis Noble de, Rochel Cajetan, Sartory Francis, Schilder John, Schmiedbauer Joseph, Sixt Ferdinand, Stöger George, Tschebull Vincent, Walterskirchen Baron William, Washington Baron Maximilian, Wokaun Joseph.

370. SUCHANEK Alexander, Liqueur-manufacturer. *Brünn.*
Assortment of Radix Rhei (Rhabarb) and Moravian productions.

371. SZALAY Paul, Possessor of Landed estates. *Tisza-Varkony, Comitat of Szolnok, Hungary.*
Wheat from the Theiss-regions.

372. SZEGSZARD, Shareholders' Association for Wine-commerce. *Szegszard, Comitat of Tolna, Hungary.*
Wines.

373. SZIRANYI Andr. & Gomp., Confectioner. *Gyöngyös, Comitat of Heves, Hungary.*
Confectioner's articles.

374. TAUBER Max & **BETTELHEIM** Charles, Spirit-manufacturers. *Vienna, Jägerzeile 522.* Manufactory: Czenyi, Banat.
Slivovitz and Cherry-spirit.

375. TAUSSIG J. A. and Brothers, Possessors of Spirit-refinery and of the I. R. priv. Vinegar-, Rum- and Liqueur-manufactory. *Sechshaus 16, near Vienna.*

The establishment refines every year 55,000 eimers of Spirits to 90—96 p C. of Alcohol, employs 40 workmen and 2 steamboilers, brings its articles to inland and foreign markets and has Agencies in all Provinces of the Austrian Empire and in the most considerable commercial places of Italy. The yearly production of Vinegar-essence amounts to 10,000 eimers. Extensive stores of Liqueurs, Rum, Slivovitz, from the most common up to the most exquisite sorts, are constantly kept.

376. TELEKY Count Dominic, Possessor of Landed estates. *Klausenburg, Transylvania.*
Wines.

377. THUN-HOHENSTEIN Count Francis, possessor of a Mechanical Mill. *Tetschen, Bohemia.* Store-house of the Sugar-manufactory: Prague, Jerome Albert. M. I. Cl. Paris 1855. Great Silver M. Vienna 1857.

Articles, produced by the Mill, and Sugar.

The Mechanical Mill of Teschen, constructed on the newest principles, is employed with the production of fine Flour by dry grinding. It is provided with 12 pair of stones and the necessary apparatus for grinding 3,000 metzens of Cereals per day. Movement is imparted to the machinery by Turbines constructed according to Jonval's system. About 30,000 metzens of Corn and 7,0000 metzens of Wheat are worked every year, giving articles in a total average value of 800,000 fl.

Hundred Vienna pounds of Wheat are converted into:

fine flour	41·64 pounds
middle „	18·80 „
common flour	20·08 „
bran	16·20 „
waste	3·28 „
Total .	100 pounds.

In the Sugar-manufacture at Peruc, near Libochowitz, the juice is obtained by means of Centrifugal engines. Beets may be worked to the amount of 150,000 Cwt every year.

378. THUN-HOHENSTEIN Count Francis, priv. Sugar-manufactory. *Peruc, Bohemia.*

Assortment of Sugar.

(See Class 4.)

379. TOBACCO-MANUFACTORY, Central Superintendence and offices of. *Vienna.*

Inland Tobacco-leaves and articles produced by the I. R. Tobacco-manufactories.

The culture of Tobacco in Austria is subject to restrictions drawn by the existence of monopoly. This culture is confined to only a few provinces: Hungary, Galicia, Tyrol, Venice; and even there it is limited to certain districts, every planter being, moreover, dependent on a licence from the proper authorities. In 1860, the area of tobacco-planting amounted to 74,138 yokes, each yoke 1,600 sq. fathoms, of which 66,769 yokes fell to the share of Hungary alone; the produce of the crop having been 692,012 Cwts. In fertile years the crop rises much higher, for instance, in 1858, it was 1,707.671 Cwts from an area of tobacco-planting of 137,414 yokes, in 1859, it was 755.080 Cwts from an area of tobacco-planting of 74363 yokes. The manu-

facture is exclusively reserved for the manufactories of the state, of which there were, in 1860, 24 in the several provinces; a staff of 268 officials, 329 servants, and 21,751 hands having been employed; 979 men and 17,284 females of the latter number worked in the manufacture of cigars. The production of these manufactories amounted, in 1860, to 55,175 Cwts of snuff, 520,456 Cwts of tobacco for pipes, and 905,488,000 cigars, weighing 76,688 Cwts, for which were used 557,889 Cwts of inland and 96,422 Cwts of imported tobacco-leaves. The sale, too, is depending on licences from the administration. For the transactions of sale, there were, in 1858, 69 storehouses and 70,162 retailers in the whole empire. The consumption of tobacco amounted, in 1860, to 54 192 Cwts of snuff with a receipt of 6,166,115 fl., 519,427 Cwts of tobacco for pipes with a receipt of 24,996,536 fl., 837,864,680 cigars with a receipt of 21,511,383 fl. In the latter are included 14,511,900 cigars imported from the Havanna and sold for 1,345,148 fl. The net revenue of the tobacco-monopoly amounted, in 1860, to 34,695,419 fl., in 1861, to 36,436,344 fl.

380. TOMASI Alois, Vinegar-manufacturer. *Gaya, Moravia.*

Assortment of Vinegar and Spirits.

381. TÖRÖK Gabriel de, possessor of Landed estates, together with a number of Co-exhibitors.

Wines.

382. TRAYTLER L. A., possessor of a Steam-Sawing and Mechanical Mill. *Arad, Hungary.*

Flour.

383. TROPPAU Sugar-refining Association. *Troppau.*

Sugar in loaves.

384. TSCHINKEL'S Augustus Sons. *Schönfeld and Lobositz.*

Succedanea for Coffee and Chocolate.

385. URBANEK Brothers, Mechanical Millers. *Prerau, Moravia.*

Cereals and Flour.

386. VALERIO Angelo. *Trieste.* Chocolate.

387. VARGA Lewis, Possessor of Landed estates. *B. Füred, Comitate of Veszprim, Hungary.*

Wines.

388. VASICS, married **SZILAGYI** Eleonora. *Klausenburg, Transylvania, Königsgasse 51.*

Rosoglio (Brandy).

389. VERSCHETZ, Royal Free-town. *Hungary.* Prize-M. Vienna 1857.

Wines: red 1861, changeant (light red) 1857 and 1861, white 1857 and 1861, „Wermuth" 1861, Mustard 1861.

The art of improving Wines, so important in most wine regions, has but begun to come into practice at Verschetz, everything being left there to natural agencies, although this place might be expected to rank among the first wine growing places in the world, the production in middle good years amounting to 350,000 eimers. Of this enormous quantity, generally an amount of 10,000 eimers reaches the next following crop; a proof of the Verschetz wine being highly in demand. The Verschetz Wines are quite fit for consumption, 3 months after the crop, and 5 months after this epoch, their constituent parts, without any artificial aid, have achieved their full developement. Speculators being frequently in the necessity to keep other sorts stored up for four, five and more years, before they become fit for consumption and, by this delay, losing nearly the whole amount of their capital, may take advantage of this quality peculiar to Verschetz Wines. In favourable years, naturally sweet red and palered Wines, are produced merely through the excellent quality of grapes, without any artificial aid.

390. WALLIS Count, Superintendence of Estates. *Kolleschowitz, Bohemia.*

Hops.

(See Class 4.)

Count Fred. Wallis' estate of Koleschowitz is situated in one of the better hop-districts of Bohemia, yielding a yearly crop of from 100 to 150 Cwts of hops excellent for heavy store-beers (Lagerbier).

391. WANKA Francis, I. R. first Prague Steam-mill and Brewery. *Prague.* M. II. Cl. Paris 1855. Silver M. Vienna 1857.

Bottled Beer („Bock" and „Lager"), Flour, samples of Malt.

392. WARHANEK Charles. *Vienna, Leopoldstadt 4.* Manufactory: Fiume.

Conserves of every description in vinegar and oil, in boxes.

393. WAWRA Vincent jun., possessor of Mills. *Prague 207-2.*

Compressed Flour („Slacenka"), known under the denomination of „Stone-flour".

This invention, patented for Austria and France, enables Flour to be preserved for years and to bear the transport to hot countries, generally deleterious to every sort of this article. The compressed Flour, offering a 50—60 percent saving of volume, is especially good for naval transport and for the approvisionment of fortified places and magazines, being easily stored up in the same way as blocks of sugar or salt. It may be no less appropriate for troops on marches. It has the farther advantage to be ready for transport without being packed up in tons or otherwise. Compressed Flour may be produced in whatever degree of hardness demanded, without adding any heterogeneous substance or impairing its quality. Bran and coarsely ground Oats may likewise be solidified with a 15—20 percent saving of volume. This invention is to be sold to private persons or to Governments. Information to be obtained from the I. R. Austrian Commissioner for Exhibition, Chevalier de Schwarz.

394. WEINMAYER Leopold. *Grinzing near Vienna.* Wine.

395. WEISS John, Provincial Advocate. *Bistritz, Transylvania.*

Heidendorf-Steinig Wine of 1859.

396. WERTHER Frederick, Possessor of manu-factories. *Buda.*

Articles produced in the Exhibitor's Steam-mill and Gruel-manufactory; Genever-liqueur prepared according to the Dutch method; Sugar-cane and Cotton, grown in open air in the neighbourhood of Buda.

The establishment belonging to the Exhibitors, whose industrial merits H. M. the Emperor of Austria has been most graciously pleased to acknowledge by granting him His Gold Cross of Merit decorated with the Crown, comprises an I. R. priv. Engine manufactory, a considerable Distillery of Spirits with production of Pressed yeast, an extensive mechanical Steam-mill' organized by a method not yet set into practice by any private enterprise in Hungary and competing, in a manner most advantageous for the consuming public, with the Joseph Cylinder-mill at Pesth, founded on shares. The Exhibitor is further in possession of a grit manufactory, the only one in Hungary, organized in imitation of similar foreign establishments and producing 12 sorts of grit of best quality, of an extensive Baking-establishment acting by steam and providing above 200 detail-sellers at Pesth and Buda with bread of best quality and agreeable taste, and of Malt-baths of most salutary curative effects, as attested by the first medical authorities. Besides these extensive and multifarious technical establishments, the Exhibitor is employed with rational Agriculture, paying special attention to the acclimatizing of foreign Vegetables of technical use. Some of these experiments, among others those with Sugar-cane and Cotton, have been crowned with success.

397. WIDTER Joseph, Miller. *Perchtoldsdorf, Lower Austria.*

Flour.

398. WINNICKI Sylvan. *Boryszkovice, District of Zloczow, Galicia.*

Tobacco-leaves and Maize.

399. WÖHRL J. G. *Vienna.* Counting - house: City 511.

Bottled Wines.

400. WOLF Sigismond & Comp., I. R. priv. Liqueur-manufactory. *Weisskirchen, Moravia.*

Spirituous liquors.

401. ZALLINGER Charles de, Possessor of Landed estates. *Botzen, Tyrol.*

Wines of own growth.

The exhibitor produces wine, from grapes of Rhinegau Riesling, at the steep mountains-slopes, deserving to be recommended as table wine form its peculiar generosity and special mildness.

402. ZICHY-FERRARIS Count Emmanuel, possessor of Landed estates. *Nagy-Szöllös, Comitate of Veszprim, Hungary.*

Wines.

403. ZWICKL Joseph, Veterinarian in the service of Prince Eszterházy. *Eszterház, Hungary.*

Rust „Wermuth"-essence.

Class 4.

Animal and Vegetable Substances, used in Manufactures.

Sub-Class a) Exhibitors of Oils, Fats, and Wax, and their Products.

    „   „ *b) Exhibitors of other Animal Substances, used in Manufactures.*

    „   „ *c) Exhibitors of Vegetable Substances, used in Manufactures etc.*

    „   „ *d) Exhibitors of Perfumery.*

The production of Olive-oil in the Southern Provinces of the Austrian Empire (Istria, Lombardo-Venetia and Dalmatia) amounts to a yearly average - quantity of about 100,000 Cwts Besides this home-supply, the demands of consumers require a yearly importation of about 200,000 Cwts, among which 80,000 Ctws, intended for technical purposes, are favoured with a deminution of Import-duty.

The culture of an other oleaginous plant, the Colza (Rape), is far more extensive, especially in Hungary and Bohemia; its annual crop giving, at least, 1,200,000 Vienna Metzens. A great portion of the crop of Lin and Hemp-seed (an average of 2,300,000 Metzens yearly) is either exported into foreign parts or expressed for obtaining Oil. In 1860, the export of all these three oleaginous seeds amounted to 670,000 Cwts In the same year, the home production rose to 500,000 Ctws of rape-seed oil and 200,000 Cwts of Linseed oil. These oils are expressed either in common stamp - mills or in improved mechanical mills.

Besides the 1 million Cwts of Tallow, produced within the Austrian Empire, large quantities of this material are imported from the Danubian Principalities, to supply the wants of Austrian candle — and soap manufactures. In 1860, this importation nearly reached the amount of 48,000 Cwts. The home production of Skins and hides (in 1860 twelve millions of pieces of ox, sheep, goat — and other skins and hides) is still inadequate to the wants of Austrian industry; so that, every year, above 100.000 Cwts of this material must be imported from foreign parts.

Austria produces every year nearly 700,000 Cwts of Wool of any quality. One half of this quantity is of Hungarian origin; Moravia, Silesia and Bohemia chiefly producing fine and superfine sorts, the most part for export. The ordinary qualities are the object of inland industry; inferior sorts are being chiefly imported from foreign parts.

Taken as a total and under normal circumstances, the export is greater than the import, as shown by the following numbers:

	Cwts	
	imported	exported
In 1859	188.200	279.700
„ 1860	219.000	244.300

The culture of Flax in the mountainous regions of the Giant Mountains, the Sudetes and Carpathian mountains and in the Alpine ranges, existed for a long time in its primitive state, until quite lately when the establishment of extensive flax waterings and the increased demands of mechanical yarn-spinning-factories gave a powerful impulse to the culture of flax both in quantity and quality. In the Eastern and Southern parts of the Empire the culture of Flax gives place to that of Hemp. Southeastern Hungary and the Venetian Provinces must be named here as producing this material in a superfine quality. The average yearly production, amounting to 2 millions Cwts. for Flax and to 1 Million for Hemp, remains behind the demands of Austria which still requires a supplementary import of these materials. In 1860, the commercial movement in these articles is represented by the following numbers in Cwts.:

	Import	Export
Flax	67.000	8.300
Hemp	100.000	51.700

Silk culture is confined to the Southern and Adriatic Provinces of the Empire, having reached its highest development in the Venetian Provinces and Southern Tyrol. Trials made in the other Provinces have hitherto remained without results worthy of note.

The yearly average production of Cocoons amounts to 270.000 Cwts, of which, in 1860, above 6.000 Cwts have been exported to foreign parts. The Austrian Empire is in possession of nearly 32 millions Austrian acres of Forests, producing yearly about 30 millions klafters of Wood of every description, consumed in extensive proportion as fuel for domestic purposes or in industrial establishments. Pines, Firs, Beaches or Oaks are partly used for building purposes, partly worked out into boards, posts etc. Timber and worked wood is exported for the most part from the ports of Trieste and Fiume and the rest on the navigable rivers (Elbe, Vistula, Pruth, Dniester and Danube); in 1860 to an amount of nearly 32 millions of cubic feet. In the same year only a comparative small quantity of this article ($6\frac{1}{2}$ millions of cubic feet, coming down the Danube from Bavaria) has been imported. The import of wood for fuel in 1860 (5,700.000 cubic feet) was nearly equal to the export (6,400.000 cubic feet) in the same year.

---

404. AFH Frederick, Manufactory of Basket and Cane Works to the I. R. Court. *Vienna, Neubau 19.* Bronze M. Vienna 1839, 1844 and 1852. Ware-house: Vienna, City 1111.

Chinese Bird-cage serving at the same time as fish- and flower-stand; Paper-basket; Work-table; diverse little Fancy articles.

That establishment is capable of enlarging wholesale business to that extent which may satisfy any orders of customers abroad. There is continually a store of various articles belonging to this line, from the finest to the coarsest, manufactured of various materials and trimmings.

405. ALTHANN Michael Charles, Count. *Swoischitz, District of Czaslau, Bohemia.*
Fleeces.

406. BARATTA-DRAGONO Charles, Knight; Landed proprietor. *Budischau, Moravia.* Silver M. Paris 1855.
Fleeces.

The Budischau parent stock of 3,000 head consist of offsprings of Padua silk-woolled sheep and improved Spanish from the parent sheep-breeds at Holitsch, the improvement of which was promoted by considerable purchases of ewes and rams of superior Saxon flocks and inland breeds at Fulnek, Hennersdorf and Gross-Herlitz, and was since that time brought to the present pitch of perfection by constant crossing. Since 1825 fostered after sound principles by the same hand, it is particularly distinguished by hereditary silk-gloss and superior fineness of the wool, perfect equality in any consideration as well as profusion of wool relatively to weight of body. In 1827 was founded the register of the flock enjoying a remunerative sale of sires and grandams to Moravian, Bohemian and Hungarian wool-husbandries.

407. BARTENSTEIN Joseph, Baron; Landed proprietor. *Hennersdorf, Silesia.*
Fleeces.

408. BAUER Matthew. *Warasdin, Croatia.*
2 cross cuts of Oak-trunks; 10 staves.

409. BAUMGARTEN-FÜRST Michael, Hammer-works Owner. *Unterweissenbach, Upper-Austria.* Honourable Mention: Vienna 1857.
Flax.
(See Classes 31 and 32.)

410. BEISIEGEL Philip, Turner. *Vienna, Wieden 925.*
Meerschaum Manufactures.

411. BELLEGARDE Rodolphine, Countess, born Countess Kinsky. *Gross-Herrlitz, Silesia.*
Fleeces.

412. BEZEREDY Nicholas. *Cseb, Comitate of Bacs-Bodrogh, Hungary.*
Flax.

413. BIACH Emanuel & Comp. Patented Manufactory of Oil and Paraffine Fat. *Theresienfeld, near Wiener-Neustadt, Lower-Austria.* Ware-house: Vienna, City 464.

Colophony, white pitch, turpentine, turpentine-oil, resin-oil, lubricator's oil, cart-grease, turpentine-varnish, iron-lacquer, cobbler's wax, smith's pitch, brewer's pitch.

That manufactory was rebuilt two years ago, and has turned its chief attention to introducing all the improvement made in the industrial branch in question, since a series of years, both at home and abroad. There are worked up about 12,000 Cwts of colophony annually, one third of which is derived from Lower-

Austria, chiefly from the environs of Mödling (Hinterbrühl), Pottenstein and Guttenstein, where the black fir (pinus austriaca) is indigenous. The remainder of the requisite raw-products comes from the southern states of America, Trieste being the main staple for them. Moreover, there are yearly worked up about 1,000 Cwts of wood-tar and 1,000 Cwts of coal-tar. 10 large distillation apparatus, besides other contrivances serve for producing the above-quoted articles. There are consumed from 7,000 to 8,000 Cwts of coal, got from the environs of Wiener-Neustadt, per annum. The productions of this manufactory are sent to all parts of the empire, the patented grease for wheels and axes is generally used as being of superior quality.

414. BIENERT D. & Son. Patented Manufactory of Sounding Boards. *Maderhäuser, Bohemia.* Honourable Mention, London 1851. M. I. Cl. Paris 1853.

Wood for musical instruments; Sounding-boards.

Francis Bienert of Hamburg established this manufactory in 1826 under many difficulties, the Bohemian Wood being at that time little accessible and as little explored. In 1827, the first consignment of sounding-board wood produced in this establishment was sent to Mr. H. Bradwood in London, to his entire satisfaction. In consequence of this and on account of practical improvements of raw wood, this manufactory soon got orders from the most distinguished pianoforte makers in London, Vienna and other towns of Germany, Paris, as well as in Russia, Naples and America. At present the following sorts of wood are produced in that establishment. Cut sounding-board wood of from 2 to 6 feet length; cleft sounding-board wood of from 2 to 6 feet length; top-wood of from 4 to 6 feet length; key-wood of from 18 to 32 inches length; rib-wood of 6 or 6½ feet length; frame-wood of 4½ feet length; violoncello and double-bass tops, violin and bass-viol tops, guitar tops. The manufactory yearly sends off about 150 to 200 boxes of cut sounding-boards or about 35,000 to 40,000 pieces, 300 to 400 bundles of cleft sounding-boards or about 6,000 to 8,000 pieces, 200 to 300 bundles of tops, 36 sq. feet each, or about 2,500 to 3,000 pieces in all, 12,000 to 15,000 bundles of key wood, of from 18 to 32 inches length, 9 up to 15½ sq. ft. with about 18,000 to 20,000 pieces, 300 pieces of rips, 100 bundles of frame wood with 2,400 pieces, 50 pieces of violoncello and double-bass tops, 500 pieces of violin and bass-viol tops, 500 pieces of guitar tops, all together representing an average value of 17,700 fl. mostly sold to countries abroad. The raw material is derived from the forests at the estates of Stubenbach and Krumau belonging to his Serenity the sovereign Prince of Schwarzenberg. The works are done with three sawing and two circular-saw mills driven by water-power; 100 to 150 persons being employed.

415. BIONDEK Michael, Agriot-sticks Maker. *Baden, Lower-Austria.*

Agriot-sticks.

416. BIRNBAUM James, Merchant. *Pesth.*

Assortment of Flax.

417. BISTRITZ, Veneer-cutting Company at; through Fluger Ferdinand & Comp. *Bistritz, Transylvania.* Agents: Dietrich & Comp. at Klausenburg, Hügel & Comp. at Gallatz.

Nut-wood and Ash-wood Veneers, 2 gun-stocks of nut-wood.

418. BIZEK Francis, Producer. *Alt-Recse, Com. of Bacs-Bodrogh, Hungary.*

Wool.
(See Class 3.)

419. BLÜCHER v. **WAHLSTADT** Maria, Countess, born Countess Larisch-Mönnich. *Radun, Silesia.*

Fleeces.
(See Class 3.)

420. BÖHM John & **GENSCHEL** Joseph, Guilloshers. *Vienna, Neubau 59.*

Guilloshed fancy articles of all sorts of metal, as also of mother-of-pearl and ivory.

421. BOSCHAN'S Joseph Sons. Oil Manufactory. *Angern, Lower-Austria.*

Oils.

422. BRICHTA Adolphus, Perfumer. *Prague 36-2 Breitegasse.*

Patented newly invented Glycerine Borax Toilet-soap.

Being the result of various experiments, this soap is unique on account of its superior qualities. It purifies the skin completely, bestows upon it a velvet-like softness and gloss, and, at the same time, it makes freckles vanish entirely.

423. BUKOWINA, I. R. Provincial Finance Direction at, for the Upper Forest-office *Kimpolung.*

Samples of Timber.

424. BURKHART Caspar, Manufactory of Oil. *Osterberg, Carniola.*

Oils.

425. BURSCHIK Andrew, Turner. *Vienna, Wieden 852.*

Pipe-sticks.

426. BÜTTNER C. F. *Steieregg, Upper-Austria.*

Teasels.

The growing of fuller's thistles, introduced since 1827, furnishes a yearly produce of about 40 to 60 millions of teasels, representing a value of about 100,000 fl. of which 200 up to 300 fl. gross-income is fetched per yoke of land. In commerce these teasels, rivalling the Styrian and Bavarian in quality, are packed in boxes. Price between 1 and 3 fl. per mille.

427. CHIOZZA Charles, Aloysius & Son. Patented Manufactory of Soap. *Trieste.* Honourable Mention: Vienna 1835. Gold M. Vienna 1839. Silver M. Vienna 1845. Prize M. London 1851.

Various sorts of Olive-oil, Tallow, Cocoa-nut, Palm-oil Soaps and others, serving for purposes of industry, washing and toilet.

That manufactory exists since 1781, producing 10,000 Cwts of soap per annum. But it could manufacture greater quantities, there being 20 boilers, 34 ash-pits for preparing lie, 14 coolers and considerable oil-cisterns and other vats. To the circumstance of the establishment being situated within the district of the free-port and entirely exposed to foreign competition may be attributed that its production is not more speedily increasing than it actually does, since the favours allowed to imports into the empire are not on a level with the duties regulated abroad.

428. CZILCHERT Robert, Dr., Landowner. *Gútor, Com. of Pressburg, Hungary.* Premium of Money and Prize-M. Paris 1855.

Merino Fleeces.

The exhibitor's Electoral-Negretti parent-stock of Gútor yields about 16 Cwts of superfine wool and 100 head of fine-fleeced rams per annum. It received a premium of 450 francs and a prize-medal at Paris in 1855 and also in Pesth 1857, 1858, 1859. Rams are sent off, per railway, in all directions even so far as to sea-ports. The Gútor stock has been purely bred from the most improved Silesian blood, and preserves softness and all superior qualities of superfine cloth-wool, appropriating from the Negretti-breed propensity only fine shape of body and great weight of shearings. Perfectly free from the tetter and other distempers, their rams are much sought after, because their natural treatment makes them apt to be acclimatized everywhere under more severe circumstances of climate and economy. Average-price 25 ducats per head; wool, in 1861, 250 fl. per Austrian Cwt.

429. DAPSY William, Landowner. *Rima-Szombáth, Com. of Gömör, Hungary.*

Gall-nuts.

430. DAUN Henry, Count. *Vöttau, Moravia.*

Fleeces.

431. DIEDEK'S A. C. Son. *Vienna Altlerchenfeld 117.*

Washing and Toilet Soap.

432. DOBLINGER Francis, Waxdrawer. Manufactory at *Vienna, Matzleinsdorf.*

Wax.

433. ENGL Henry, Professor and Curate. *Traiskirchen near Riedau, Upper-Austria.* Agent in London: Baron A. Hohenbruck, of the Austrian Commission.

Collection of Wood-samples-of all kinds of wood in Upper-Austria.

The exhibitor has, in the course of many years, collected pieces of speckled wood of all kinds of wood in Upper-Austria, and caused the instrument maker Stratzinger at Linz to compose an inlaid table-top of 29 different speckled pieces of wood representing the scutcheon of Austria. A list of German and of botanical denominations of all the speckled wood-sorts is lying besides the table-top.

434. FALKENHAIN Theodore, Count. *Kiowitz, Silesia.* Medal I. Cl. Paris 1855.

Fleeces of a flock of 1500 head improved by crossing Hennersdorf ewes with Lichnowsky rams.

(See Class 1.)

435. FERENCZY John, Turner. *Pesth, Herrengasse.*

Billiard Dakos.

436. FINANCES, I. R. Ministry of the. For the I. R. Direction of Forests at *Vienna.*

Forest Objects.

437. FINANCES, I. R. Ministry of the. For the I. R. Provincial Direction of Finances at *Gratz.*

Forest Objects.

438. (Omitted.)

439. FINANCES, I. R. Ministry of the. For the I. R. Direction of Forests at *Montana in Istria.*

Forest Objects.

440. FINANCES, I. R. Ministry of the. For the I. R. Direction of Salterns and Forests at *Gmunden in Upper-Austria,* and for the I. R. Forest Office at *Aussee in Styria.*

Gmunden: Forest Objects.

Aussee: Pine for sculptor's works, Larch for ship-building, Pine and Fir for instruments.

441. FINANCES, I. R. Ministry of the. For the I. R. Direction of Salterns and Forests at *Salzburg (Hallein.)*

Forest Objects.

442. FINANCES, I. R. Ministry of the. For the I. R. Provincial Direction of Finances at *Lemberg.*

Forest Objects.

443. FINANCES, I. R., Ministry of the. For the I. R. Provincial Direction of Finances at *Pressburg.* For the I. R. Forest Office at *Znyo-Várallya, Com. of Turocz, Hungary;* Ware-house: Znyo-Várallya. For the I. R. Forest Office at *Rosenberg, Com. of Liptau;* Ware-house: Hradek.

Pressburg: Forest Objects.

Znyo-Várallya: Cuts and samples of white firs, white pines and larches.

Rosenberg: Cuts and samples of firs, white pines, white firs, larches, stone-pine, fir tan-bark in rolls, samples of boards, window-posts, stamped tan-bark.

444. FINANCES, I. R. Ministry of the. For the Financial Direction at *Venice.*

Forest Objects.

445. FISCHA Peregrinus, Wax-drawer and Gingerbread Baker. *Bistritz, District of Iglau, Moravia.*

Bleached Wax.

446. FISCHER F., Patented Manufactories of Candles and Soap. *Vienna, Landstrasse 280 and Simmering 248.*

Tallow and Soap.

(See Class 2.)

447. (Omitted.)

448. FRIEDRICH John. *Vienna, City 1037.*

Meerschaum Wares.

449. FRÖHLICH William, Wax Manufacturer. *Klausenburg, Siebenbürgen.*

Wax.

(See Class 3.)

450. GÄRTNER jun. J. F., Patented Manufactory of Oil, Starch, and Gum-succedaneum. *Rannersdorf near Vienna.* Bronze M. Leipzig 1850. M. II. Class Paris 1855.

Oil; Starch; Gum-succedanea.

This manufactory works with 40 horse-power of water and 36 horse-power of steam, producing during 24 hours 200 Cwts of oil, 330 Cwts of rape-cakes, 100 Cwts of starch, and 30 Cwts of gum-succedanea. The exhibitor possesses also a branch manufactory at Prague, where likewise 200 Cwts of oil and 320 Cwts of rape-cake are produced per diem. There are exhibited: Samples of raw and refined rape-seed oil, refined rape-seed oil without any acid whatsoever for lubricating locomotive-engines and other machines, linseed oil, tobacco oil, maize oil; samples of rape and lin-seed cakes; samples of starches of wheat, maize and potatoes, dextrine-gum, leyogom, fécule, amidon grillé, starch-paste, and albumine végétale. In 1847, the exhibitor received the great silver medal of the Lower-Austrian Trade Union; but he did not send to the London and Munich exhibitions in 1851 and 1854.

451. GERMER William & Co., Owner of Agriot Plantations at *Baden near Vienna 62.*

Agriot Sticks.

452. GOLDMANN Maurice. *Vienna, Gumpendorf 621.*

Pipes of Meerschaum-paste.

453. GORITZ, Agricultural Society at. *Goritz.*

Agricultural Products (sorts of Wood).

454. HANNIG M. L., Commercial Agent. *Debreczin. Com. of Bihar, Hungary.*

Mats of rush and cane.

455. HARTL George & Son. Patented, Manufactory of Candles and Soap. *Vienna.* Honourable Mention: Vienna, 1845. Laudatory Mention, Munich 1854. M. II. Cl. Paris 1855. Ware-houses in Vienna: Rossau, Schmidtgasse 98; City, Weihburggasse 895; Wieden, Adlergasse 1028; Lichtenthal, Kirchengasse 75; Windmühle, Mariahilfer Hauptstrasse 19.

Cocoa-nut-oil soap, manufacture soap, several sorts of washing and toilet soaps.

The soaps No. 1 and 2 are prepared of cocoa-nut-oil by means of concentrated caustic lye of soda. The manufacture soap No. 4 is produced of elaine-acid and palm-oil by means of caustic lye of soda. The washing soap No. 5 is boiled of tallow by means of caustic lye; No. 7 of elaine-acid and waste of tallow-meltings; No. 11 of tallow and resin by means of caustic lye of soda. The various kinds of toilet-soap are produced of tallow, cocoa-nut-oil, and some with addition of palm-oil. The cocoa-nut-oil soap and also the toilet-soaps are prepared by means of concentrated etching-lye of soda and preserve the skin soft and smooth, being mixed with the glycerine contained in the fat and only perfectly pure materials being used for the production. The manufacture soap is especially employed in dying, its virtue of preserving stuffs soft and smooth making it serviceable in silk-dying for the process of boiling out as well as of dying, since particularly the silk does not lose its lustre and is also preserved soft and glossy when treated with this soap. The washing soaps are perfectly neutral, possessing, since they have no admixture of foreign ingredients, cleansing power without injuring linen. The exhibitors produce yearly more than 4,000 Cwts of various washing soaps, 1,200 up to 1,500 Cwts of manufacture soap for silk-dyers and more than 20,000 dozen of toilet-soaps. There are about 10,000 Cwts of raw tallow melted with steam, 3,000 Cwts of which being used for the production of various sorts of tallow-candles, another part for the manufacture of soap, the rest coming into commerce as cake-tallow.

456. HARTMANN Lewis. Manufactory of Turner's Works. *Vienna, Laimgrube 25.* Prize-Medal London 1851. Medal I. Cl. Paris 1855.

Diverse Smoking Articles of meerschaum and amber, in forms fit for exports.

This establishment exists since 1829, occupying at present 25 journeymen and 4 apprentices at home, and 6 masters with 10 apprentices out of the house. The principal productions are smoking articles of meerschaum and amber, both for inland sale and particularly also in forms adapted for export trade, the latter being sent, by means of direct connexions, to Germany, France, Belgium, England, Switzerland, and America. This manufactory, at present known as the most extensive in this line, keeps rich stores of pipes and cigar-pipes of meerschaum, amber and bruyère-wood, as well as of all sorts of pipe-sticks, walking-sticks and other turner's works, and is capable of executing the most extensive orders at the shortest notice and at the cheapest prices.

457. (Omitted.)

458. HERTRUM Joseph. *Trieste.*
Fats and Volatile Oils.

459. HILLINGER John, Turner. *Vienna, Alservorstadt 113.*
Turner's Manufactures of bone and wood, adapted for export. Artistically finished Chess-board.

460. HIMMELBAUER Anthony & Co. Patented Manufactory of Candles, Soap and Perfumery. *Stockerau, near Vienna.* Medal I. Class Paris 1855. Ware-house: Vienna, City 850.
Articles of Perfumery, Stearine-candles and Soap.

Among the exhibited articles, there are: stearine-candles, soaps, glycerine-productions, and the newest compositions of toilet-soaps and other perfumery, such as, glycerine-soap Austrian-soap, Queen Victoria soap, Eau de Vienne, Bouquet de l'Impératrice d'Autriche, Odeur Parisienne, London perfume, Toilet-glycerine, etc. This establishment offers, by the variety of its productions, the greatest choice in this branch, works with 300 people and propagates its articles in all directions at home and abroad.

461. HOFMANN J., Patented Manufactory of Stearine-Candles, Soap and Glycerine. *Gratz, Stiria.*
Stearine-candles, Glycerine, Toilet-articles.

462. HOFRICHTER Francis. *Reichenau, Bohemia.*
Boxes.

463. HOSCHEK Andrew, Manufactory of Meerschaum-pipes. *Vienna, Neubau 263.*
Pipe-sticks, Meerschaum-pipes, cigar-pipes, and amber-works.

464. HOSSNER Jos. John. *Schluckenau 118, Bohemia.*
Spanning-plaits, viz. mattings and coverings.

465. HOYOS-SPRINZENSTEIN Ernest, Count, Proprietor of *Stixenstein and Gutenstein, Lower-Austria.*
Great Silver M. Vienna 1857.
Cuts of trees.

1. Black fir, pinus laricio, poir, from the forests near the castle of Stixenstein in Lower-Austria, standing-place: alpine chalk, northern site, 1,600 feet above the level of the sea. Price per cubic foot, loco Payerbach, 1 fl. Austr. Curr. 2. Red fir, abies excelsa, decand. — 3. Silver fir, abies pectinata, decand, from the forest Neuwald at the source of the Mürz in Lower-Austria; standing-place: grey-wacke; bottom of the valley: 3,000 feet above the level of the sea. Price per cubic foot 30 kreutzors. — 4. White fir, pinus silvestris, L., from the forests near the castle at Gutenstein in Lower-Austria; standing place: dolomite; northern site, 1.600 feet above the level of the sea. Price per cubic foot 30 kreutzers. — 5. Red beech, fagus silvatica, L., from the forests near the castle Gutenstein in Lower-Austria; standing-place: alpine chalk; southern site, 2,000 feet above the level of the sea. Price per cubic foot 18 kreutzers.

466. JABUREK Francis, Manufactory of Meerschaum Articles. *Vienna, Mariahilf 92.*
Meerschaum Articles.

467. JOSEPHSTHAL, Paper Oil and Dying Wood Manufactory. *Josephsthal, near Laibach, Carinthia.* Honorary M. Munich 1854. M. 2 Cl. Paris 1855.

Ware-houses in Vienna, City 579, in Trieste and Agram.
Dying Wood and Oil.
(See Class 28.)

This manufactory, the present owners of which are Fidelis Terpinz, owner of Kaltenbrunn, Valentine Zeschko, owner of Freudenthal, Charles Galle, landowner, and Valentine Krisper, was erected in 1841 by the three first named gentlemen in the centre of ten villages of the former estate of Kaltenbrunn on the Laibach-river. The whole of the works are put in motion, through the Laibach-river that never freezes, by means of 8 large iron turbines and 3 large iron English water-wheels having together more than 500 horse-power. The oil manufactory consists of 7 English and Swiss double presses and other machines, with double pressing daily producing 80 Cwts of rape-seed oil which enjoys a brisk sale, because of its double, pure machine refining. The rape-seed used for it is derived from Hungary, the oil is mostly sold in the inland, and the cakes are sent to Italy and England. — In connexion with the paper and oil manufactory are the dying-wood sawing, planing, rasping and grinding works, on Hamburg and French stones, and fitted up with the best known foreign machines. There are worked more than 100 Cwts of all sorts of dying wood, per diem, imported via Trieste, and sent to all Austrian provinces where they are highly prized; madder-wort, too, is solidly and finely milled to the full satisfaction of distinguished dying establishments, being worked with proper stamping engines set up for this particular use.

468. JUNG John George, Comb-maker. *Braunhirschen, near Vienna 161.*
Combs.

469. KEIL Giles & Edward **RUDZINSKY-RUDNO.** *Endersdorf, Silesia.*
Fleeces.

470. KLEIN Caroline. *Wedtzierz, Galicia.*
Indestructible incrustations on wood, metal and stone.

471. KORIZMICS Ladislaus, R. Hungarian Counsellor and Vice president to the Hungarian provincial-commission of exhibition. *Pesth, Uelloer Strasse 12.*
Fleeces.

472. KOUFF Francis. Patented Manufactory of Pitch and Turpentine-Oil. *Hinterbrühl 114, near Mödling, environs of Vienna.*
Black-fir resin, Turpentine-oil, Pitch. A resined black-pine of 3 years.

The manufactory, working with steam-engines, possesses a stock of 60,000 trunks of black-firs, adjacent to the establishment, for which 50 pitch-makers are employed. The prices were, during a series of years, for turpentine-oil, about 28 fl. per Cwt for colophony and red pitch from 5 to 6 fl. On account of the present high rate of silver agio and the political events in America as well as the last dry summer, the raw product was sought at high prices, and turpentine-oil rose to the price of 42 fl. colophony and red pitch to the price of 9 fl., per Cwt In like manner all branch products underwent a comparative rise in price. Although that manufactory received orders from many parts abroad, it was not capable to accept them, various inland consumers having completely engrossed its activity, because, having previously used American colophony and turpentine-oil for their productions, they were now referred to Austrian products for reason of the above-mentioned circumstances. The space for the exhibition being scanty, the manufactory cannot send its pitch loaves in the usual form of 100 pounds each, but only in lumps of 25 pounds. Joined to the exhibited articles is a resined black-pine trunk of 3 years, and the hand-tools of a pitchmaker. The exhibitor received the Silver Medal of the Lower Austrian agricultural Society, at the exhibition of agricultural and forest industry which took place at Mödling in 1861.

473. KRAETSCHMAR C. A. *Rima-Szombath, Hungary.* M. London 1851; M. New-York 1853, M. II. Cl. and Diploma Paris 1855. Agent: Mr. Ernest Szumrák at Messrs. Williams & Norgate, London, 17, Henriette Street, Covent Garden.
Gall-nuts, Hungarian red clover-seed, yellow wax, wool (first shearing, second shearing, summer and lambs-wool).
(See Classes 3, 26 and 27.)

E.

474. KREUTER Francis, Civil Engineer, Cessionary of Boucherie's patent. *Vienna, Landstrasse 744.*

Impregnated fir and pine telegraph-poles.

These poles are exclusively employed for conducts by the I. R. Telegraph Direction, and are impregnated with blue vitriol after Boucherie's patent. There are 5 stations for impregnation in Austria, furnishing about 20,000 poles per annum; one pole being paid with 2 fl. 60 kr. loco railway-station.

475. KUMPF Pius. *Schluckenau 150, Bohemia.*

Matting and Spanning Plaits.

476. KUTJEVO, I. R. Forest and Demesne Office af. *Kutjevo, Slavonia.*

Products of Forest Husbandry, such as: oaks, beeches, birches, chestnut-trees, etc., oaken staves, roof-shingles, rafters and boards, vine yard-poles, hedge-poles, squared and cut timber.
(See Class 2.)

477. KUZEL John, Turner. *Vienna, Schottenfeld 23.*

Turner's Manufactures of wood mounted with bronze.

478. LANGE Wenceslas. *Vienna, Schottenfeld 3.*

Mother-of-pearl Buttons.

479. LARISCH-MÖNNICH John, Count. *Freistadt, Silesia.*

Fleeces.
(See Class 3.)

The exhibitor's Merino Sheep, 30,000 head, are derived from a purchase, made about 65 years ago, from among the I. R. Austrian Merino stock at Holitsch, and from the purchase of a flock of Malmaison in 1812; they have, since this time, received no admixture of other blood, but were, up to this period, improved by themselves. The weight of shearing of a ewe amounts from 1¾ to 3¼ pounds, and that of a ram from 3²/₄ to 5³/₄ pounds with clean natural washing, as shown by the exhibited fleeces. The sale of the yearly quantity of wool is mostly transacted with wool-dealers of Breslau; there having been attained, in 1861, an average price of 264 fl. per Cwt. of wool.

480. LAY'S Michael, Heirs, Manufactory of Oil. *Essegg, Slavonia.* Honourable Mention: Vienna 1857.

3 Sorts of Oil.
(See Class 3.)

This establishment exists since 1826 and sells most of its oils and fodder-cakes to Trieste. At the Agricultural Exhibition in Vienna 1857, its refined gold-of-pleasure and rape-seed oils were distinguished with an Honourable Mention. Chief of that firm: J. C. Pruckner, member of the Board of Trade and Commerce at Essegg.

481. LEOPOLD J., Trader in Products. *Pesth, Hochstrasse 7.*

Gall-nuts.

482. LIECHTENSTEIN John, Prince. *Vienna.*

Fleeces of the Electoral Merino-stocks at his several demesnes.
Cuts of white beech.
(See Class 1.)

The exhibitor's total number of sheep, on his Hungarian, Austrian, Moravian, Silesian, and Bohemian estates, amounts to 65,000 head, the wool of which has reached an average price of 215 fl. per Cwt. Weight of shearing 2 pounds and more per head.

483. LÖFFLER John Paul, *Langhalsen, near Neufelden, Upper Austria.*

Flax.
(See Class 3.)

484. LUKASCH Joseph, Manufactory of Mother-of-pearl Fancy Articles. *Vienna, Spittelberg 129-143.*

Fancy Articles of mother-of-pearl, and clocks of the same. Agents: Messrs. J. L. Mayer Sons & Comp. London, 21, 22, Ironmonger lane, Cheapside.

485. MACHÁŽEK Joseph, Farmer and Owner of a Sugar Manufactory. *Königshof, Bohemia.*

Fleeces and Samples of Wool.

486. (Omitted.)

487. MEINERT'S Hugh, Heirs, Owners of the estates *Partschendorf and Erb-Sedlitz, Moravia.* M. I. Cl. Paris 1855. Little Silver M. Vienna 1857.

Fleeces of the shearing, 1860/61, of the parent-stock at Partschendorf.

The parent-stock at Partschendorf consists of about 2,500 head of ewes and rams, the origin of which, like that of all Moravian pure blood breeds, may be traced directly or indirectly to the I. R. original Spanish breeds. That stock is, therefore, to be declared of the Negretti breed. The weight of shearing, the numerous summer and winter lambs included, averages per head full 2¾ Vienna pounds, or 3 Tariff Union pounds, when very clean washed. With the usual Silesian trade intercourse, the wool fetches, according to junctures, a price between 230 up to 245 fl. On account of the eminent quality of wool and body of the sheep of this parent-stock, numerous sales of rams and ewes are contracted at home and abroad, particularly to Prussia and Prussian Silesia, to Poland and Russia.

488. MERGENTHALER Charles. *Essegg.*

Wax.

489. MITTAK John, Producer. *Szkalitz, Com. of Pressburg, Hungary.*

Hungarian Woad (isatis tinctoria.)

490. MÜNCH-BELLINGHAUSEN, Count. *Vienna, City 543.*

Cut of a trunk of a black fir (pinus austriaca.)
(See Class 3.)

Illustration of the process of procuring resin from the black fir, which is native only in the ancient large forests of Lower-Austria, and furnishes resin in great quantity and of superior quality.

Cut of a hazel-nut tree of 4 feet 6 inches diameter.

Unique in Europe. Age of the tree 280 years, breadth of the top 76 feet, height of the tree 70 feet. The like of this kind, approximating an age of 100 years, are planted in great number at the estate of Merkenstein and are an ornament of their standing places. The wood, of leather-brown colour, is employed for cabinet-work.

491. MUNDY John, Baron, *Ratschitz, Moravia.* Gold M. Paris 1855. Gold, Silver and Bronze M. Paris 1856.

Ram and Ewe Fleeces.

The parent-ewes of these sheep were acquired from Saxony in 1823, and, during some 20 years, the parent-rams were taken from the parent-stocks of Prince Lichnowsky at Kuchelna and Borutin. During 25 years breeding is conducted by pure home breeding. Of this parent-stock-breeding, rams are sold for breeding as well as ewes. The sheep-husbandry of Ratschitz is four hours' way distant from Brünn, the provincial central town of Moravia.

492. MURALT Daniel. *Vienna, City 1075.*

Cockle Fan y Articles.

493. NITSCH Joseph, Administrator of estates. *Strassnitz, Moravia.*

Fleeces; dried Herbs and Seeds.
(See Class 3.)

494. NOCKER Peter, Sculptor. *Botzen, Tyrol.*

Haut-relief in wood.

495. PAGET F. Manufactory of Water-proof Stuffs. *Vienna, City 487.*

Water-proof Stuffs for Awnings.
(See Class 27.)

496. PARTEL & ZESCHKO. Manufactory of Soap. *Laibach, Polana-Vorstadt 67.*

Various sorts of soaps.

The manufactory produces: New National Soap, white and yellow; Oriental Soap, rose-coloured, white, yellow, brown; Canea, Kernel, American, Cocoa, Palm-oil, and Resin Soaps, fine and common; — besides also the current sorts of toilet-soaps. Stores of choice tallow kept.

497. PETRI C. A. Original Spanish Parent-sheep Husbandry in the I. R. Colony of *Theresienfeld, Lower-Austria.* Silver M. Petersburg 1845. Great Golden M. Berlin. Great Gold M. Munich 1854. Silver M. Paris 1855. Gold M. Paris 1856. Agent: Professor Dr. Joseph Arenstein, of the Austrian Commission.

Fleeces of Wool-silk Merinos in their natural state, and washed, carded and spun wool-silk gained from them.

The parent-stocks were, in 1803, selected by Mr. C. A. Petri himself from the Cavagna of Spain: Negretti, St. Paular and Guadeloupe. These wool-silk Merinos are a constant breed now, an artificial breeding from the fore-named parent-stocks, and furnish in the cheapest way a great deal of wool similar to silk and Angora-down, applicable for the most precious shawls and refined weavings. In spite of very unfavourable circumstances of climate and locality, these healthy animals seek their food always on pastures, as may be seen by their dusty thick fleeces. Many of the most renowned Prussian, Russian and Austrian wool husbandries purchase parent rams and ewes from this stock.

498. PETRICIOLI-SALGHETTI, Brothers. Manufactory of Salghetti Wax Candles. *Zara, Dalmatia.* Silver M. Vienna 1857.

Several sorts of table-candles and church-tapers of pure wax without any admixture.

This manufactory has existed these 150 years, but only taken part in the Vienna exhibition of 1857. There is no stearine mixed with their products, as is now usually the case to the prejudice of the article. The bleaching of wax is performed in a natural method without application of acids, which are apt to give the production a fine appearance of white for the moment, but which will soon change on coming in contact with air. There is no need of snuffing the wicks which are prepared in a peculiar way.

499. PFEIFFER Henry. *Trieste.* Manufactory of Sheep-horn Combs.

Sampler of Combs.

500. PICHELMAYER Tobias. *Leoben, Stiria.*

Oily Liquids.

501. PITTNER John. *Algersdorf, near Gratz, Stiria.* Flower-resin (flores resinarium). Remedy against the infecondity of domestic animals.

The flower-resin, principally produced from resin and wild honey, exercises a very favourable influence on the pituitous tunics of animals, especially heightens the activity of the kidneys and all organs in connexion with them, augments and improves the milk, and is particularly to be recommended as after-cure in such distempers where cattle-powder was used with strong ingredients. Genuine flower-resin dyes water and other beverages yellow.

502. POLLAK Bernard jun., Preparer of Timber for carriages of cannons, for buildings, cooper's and wheelwright's works. *Vienna, Praterstern 396.*

Timber for carriages of cannons, for builder's, cooper's and wheelwright's purposes.

This timber is prepared of wood derived from the colossal forests in Slavonia, of the finest and best trunks without the heart. The exhibitor is yearly producing and selling hundreds of thousands of cubic feet of timber for carriages of cannons and other purposes of artillery furnished to the I. R. Arsenal of Vienna; then immense quantities for staves and sleepers for railroads, for building and wheelwright's uses; 400 to 500 people being employed for their preparation.

503. PRANDAU Gustavus, Baron Hilleprand. *Valpo, Slavonia.*

Wood Products.

The exhibitor's group of woods, consisting of nearly 60,000 yoke, contain the stalk-oak (quercus pedunculata) which is their main growth. It is there, for the most part, of an age fit for felling, of considerable dimensions of strength and height, and possessed of such a degree of qualities peculiar to this kind of wood as to make it especially applicable for technical purposes of any nature. The situation of these woods is level throughout and 1/2 up to 3 miles distant from the navigable Drau-river. Previous to the exhibitor's production of the so-called French staves having attained the present extent, the production consisted mostly of German cooper's wood and all sorts of timber for building on water and land, ship, railway and artillery constructions. Besides these last-named productions, the most prominent place is taken in by the French staves, the trade of which increased materially since several years, going mostly over the sea via Trieste.

504. RADAUZ, Military Studs at. *Radauz, Bukowina.*

Specimens of Timber.
(See Class 2.)

505. RÖMER Charles, Wholesale Dealer and Owner of a Glue Manufactory. *Pressburg 129, Hungary.* Silver M. Pesth 1843. Honourable Mention: Vienna 1845.

Various sorts of white, yellow and black glue, in 13 packets, and 3 parcels of hart's and ox bones.

The exhibitor produces about 1,000 Cwts of glue per annum in his manufactory existing since 1829. The hart's horn and ox-bones prepared there serve for the uses of turners and pianoforte-key cutters.

506. ROSCHÉ M. & PRAUSE A. (formerly J. Schoffer, M. Rosché), Manufactory of Dorsh's Patented Cod-oil. *Vienna, Landstrasse 478.* Ware-houses in the most towns of Austria and Russia and at Alexandria.

Dorsh's Cod-oil.

This manufactory is engaged in purifying cod-oil. The leading principles of the production are unimpaired preservation of the oil's remedial virtues and removal of only the rancid and nauseus contents, in order to render it easy and agreeable for the patient to take it. The oil itself is of a bright yellow colour, clear and transparent like the finest olivo-oil, and has a fine taste not unlike that of sardines. The universal remedial virtues of this oil are verified by many testimonials, especially by chemical analyses and opinions of Professor Dr. Heller, Chief of the pathologico-chemical institute at Vienna, and Prof. Kletzinsky, sworn chemist to the I. R. Provincial Court of Justice. This manufactory alone is in possession of a patent on this invention.

507. RÖSNER Henry, Wax-drawer and Ginger-bread Baker. *Olbersdorf, Silesia.*

Basket with flowers of wax; Wax-taper ornamented with flowers.

508. ROTSCH Francis. *Gratz.*

Teasels.

509. ROTT V. J. Merchant., *Prague.*

Turner's Manufactures in Wood.
(See Class 1.)

510. SAAZ, Agricultural Branch Society at. *Bohemia.*

Clover, Wood and Commercial Seeds.
(See Class 3.)

511. SAPIEHA Adam, Prince. *Krasiczin, District of Przemisl, Galicia.*

Electoral Fleeces of the shearing in 186 1/2.
(See Class 2.)

512. SARG F. A. Patented Manufactory of Milly-Candles and Glycerine. *Liesing near Vienna.* Warehouses: Vienna City 1047; Pesth at Mr. Kochmeister's and Mr. Oszwald's; Lemberg at Mr. Schellenberg's; Cracow at Mr. Bartl's; Prague at Mr. Jeitcle's; Trieste at Mr. Seravallo's; as well as in all larger provincial towns. Gold M. Vienna 1845. Prize-M. London 1851. Great M. Munich 1854. M. I. Cl. Paris 1855.

Candles, Soap, Glycerine.

Founded by Mr. Gustavus de Milly of Paris, this manufactory was the first and is the oldest stearine manufactory that was established in the empire. In 1839, it went into the possession of a joint-stock company, that built, in 1854, a new manufactory on a great scale at Liesing (South-railway station, half an hour's distance from Vienna). In 1858, this establishment was bought by Mr. F. A. Sarg, the present owner, at a public sale. The productions of this manufactory have always enjoyed the best reputation and have been honoured with 7 medals at the various exhibitions at home and abroad. 4 warm and 6 cold presses are working and the yearly consumption of tallow amounts to 25,000 and even 30,000 Cwts. Besides the various sorts of candles for churches, lustres, tables and carriages, night-candles, stearic acid in cakes and blocks, the manufactory also produces eleïne-soaps. The present owner was the first that brought about the production of glycerine, on a great scale, formerly imported from abroad at high prices, and now introduced into trade, at considerably cheaper prices, by home industry, in connexion with the before-named establishment, for manufacturing glycerine in all stages, from the raw articles up to the purest, clear and scentless, to the concentrated quality of 30°. Testimonials of the first authorities in chemistry and medicine at home and abroad have acknowledged and certified the faultless quality of this important article, and the beneficent virtues, so healing for the affected skin, of the patented glycerine toilet articles prepared from it, such as: toilet-glycerine, glycerine-crême, liquid glycerine-soap (40% glycerine contents), solid glycerine-soap (containing 24% glycerine).

513. SARTYNI Anthony. *Smorce, District of Stry, Galicia.*

Woodsticks for Lucifer Matches.

514. SCHÄDELBAUER Vincent, Turner. *Ottakring, near Vienna, 105.*

Mother-of-pearl Buttons.

515. SCHAUMBURG-LIPPE, Prince. *Veröcze, Slavonia.*

Products of wood; Staves.

(See Classes 2 and 3.)

The stock of trees in the exhibitor's woods is such that, at any time, there may be sold about 150,000.000 Austrian cubic feet of holm-oak (quercus robur) of from 200 to 250 years of age, about 10,000.000 cubic feet of cluster-oak (quercus sessiflora) of from 130 to 150 years of age, 300,000 cubic feet of ash and as many of elm.

516. SCHINDLER Emil Frederick von. *Kunewald, Moravia.* Silver M. Vienna 1857.

Fleeces and Specimens of the Negretti parent-stock breeding at Kunewald.

The parent-stock of the Kunewald breed is on the estate of Zauchtl, lying contiguous to Kaiser Ferdinand Northern Railway station Zauchtl. This parent-stock husbandry was founded in 1834 by the present owner, Fred. E. Schindler von Kunewald, and selected from the richest woolled, most improved and renowned stocks of Moravia and Silesia; it was distinguished with the Silver Medal at the exhibition of 1857 in Vienna.

From that period the stock was continually cultivated and improved with solicitude and skill, so that now it may rival the richest woolled and best flocks.

Parent rams and ewes are warranted excellent health. Extent of the flock 1,000 head; product of wool 20 Cwts; price of sale 220 fl. per Cwt. Price of rams 100 rising up to 500 fl., price of ewes from 20 to 50 fl.

517. SCHMIDT Edward, Turner. *Vienna, Leopoldstadt 484.* Agents: Messrs. J. L. Mayer Sons & Comp., London, 21, 22, Ironmonger lane, Cheapside.

Meerschaum cigar-pipes with carved work, Smoke-cooling-cylinder pipes, tobacco-pipes, and cigar-pipes.

518. SCHREIBER Emanuel. *Vienna, Margarethen 201.*

Pipes of meerschaum-paste, and cigar-pipes.

519. SCHÜPLER Joseph, Turner. *Vienna, Schottenfeld 17.* Honourable Mention: London 1851.

Diverse wood fancy articles.

520. SCHWARZENBERG John, Adolphus, Prince.
Fleeces.

(See Class 1 and 3.)

521. SEILLER Anthony & Comp. Patented Manufactory of Cream-Soap and Candles. *Goritz.* Warehouses: Udine, Treviso, Padua, Verona, Trient, Bergamo, Brescia, Trieste.

Various sorts of fine toilet cocoa-nut-oil and common washing soaps, singly and doubly refined cream.

522. SOAP-BOILERS' COMPANY, First Austrian. *Vienna, Schottenfeld 343, and Penzing 92, near Vienna.* Warehouses: Vienna, Brünn, Pesth. Gold M. Vienna 1845. Prize-M. London 1851. M. Munich 1854. M. Paris 1855.

Apollo Candles and Soaps.

The First Austrian Soap Boilers' Company was constituted in 1839, was united with the Vienna Stearine Candle Manufactory in 1840, and purchased, in 1842, the then existing Margarin Candle Manufactory, and in 1844 the Crystal Candle Manufactory. The productions of the fore-named firm of „Austrian Apollo Candles" are only manufactured, in superfine quality, of completely pure stearine acid, and table or candelaber candles, church candles according to the Catholic and Greek. rites, coach lantern candles are furnished in Vienna, Tariff-Union, English and Polish weight, but night-candles only in Vienna weight. The Vienna Apollo Candles are renowned in all parts of the world and sought for their superior quality and comparatively cheap prices; wherefore attempts were made by others to take advantage of the good reputation, enjoyed by these candles, by appropriating the orange coloured packing paper, originally employed in the Apollo Candle Manufactory, previously never used for similar purposes in any manufactory at home or abroad, and by imitating even stamped marks and devices, so that purchasers not knowing to read, or such as did not more nearly examine the marks, were deceived. — Moreover, this Company produces the well-known Apollo Soap manufactured of the elaïne gained in the production of stearine. This soap is chemically pure and perfectly neutral, therefore of superior quality. On account of its cheapness, this soap is much used for washing linen, for dyeing wool and silk, as well as for toilet uses. As was the case with the above mentioned candles, the renown of this soap was also availed of, in order to sell substitute sorts of soap composed of extraneous ingredients under similar names. In order to obviate these practises and to protect their great reputation, the Company has, of late inserted a warning, styled in 12 languages, on the inside of the packing-paper, characterizing the adulterations and imitations. That establishment has in its manufactory 20 hydraulic presses, 2 steam apparatus for the manufacture of candles, 3 steam apparatus for melting tallow, 4 large soap boilers etc., and works up annually about 6 millions pounds Vienna weight (that is 7,407.400 pounds English weight, or 3,364.300 Kilogrammes) of tallow for Apollo Candles, and part of its gained elaïne, viz. 1,600.000 pounds, is used for soap, the rest being sold. Considerable quantities of glycerine, gained in manufacturing candles; are sold by the Company at moderate prices, partly in Austria, partly abroad.

523. SPRINGER & BECHER. *Vienna, Mariahilf 71.*
Perfumes.

524. STATE-RAILWAY COMPANY, I. R. Patented Austrian.

Forest Products, and Impregnated Wood.

(See Classes 1, 2, 5, 8, 10, and 13.)

525. STABILE Anthony. *Goritz.*

Sumach.

526. THUN-HOHENSTEIN Francis, Count. *Bohemia.* M. I. Cl. Paris 1855.

Fleeces and Specimens of Wool from the parent-stock husbandries at Peruc (estate of Peruc) and at Krögliz (estate of Tetschen).

(See Class 3.)

The parent-stock husbandry Peruc (adjacent to the railway-station Lobositz), consisting of 450 head of ewes and the corresponding number of rams, full blood Merino breed, has attained in its breeding results, superfine wool products, connected with uncommon richness of wool and size of body; breeding yearly 200 rams for sale, whose value for breeding was favourably acknowledged at home and abroad both by repeated premiums for the parent and descendant animals at the exhibitions in London, Paris, Berlin, Vienna, Prague, and Brünn, and by many demands for rams. The sale of rams begins every year in December, off hand. The present state of the flock (the above-named parent-stock included) is about 1600 head of full grown animals; the weight of shearing is from 6 to 8 pounds Austr., clean washed, of rams of from 1 to 2 years of age; from 4 to 5 pounds of ewes. The total quantity of shearing amounts to 50 Cwts per annum. The last price, in 1861, was 205 fl. per Cwt without accessory conditions. The wool ranks in quality with the best in the country, and is brought to sale by the authorized counsellor of economy and central director of estates, Anthony E. Komers, Prague, 633-2 Here is also settled the sale of the shearing results of Count Francis Thun's estate of Tetschen, and of Count Henry Chotek's estates of Weltrus, Neuhof and Bloschitz, amounting to a total of

200 Cwts. Besides the production of fine wool and valuable parent stocks, there is separately managed mutton grazing of 2,000 up to 3,000 head, with the use of wastes from sugar manufactories. The parent-stock of Krögliz belongs to the Merino Electoral breed. The present state of it is: 1300 head of full grown animals, besides 350 lambs. The weight of shearing is 2 pounds 6 loth (¹/₂ ounce = 1 loth) clean washed wool per head of full grown animal, and the price of the wool is 210 fl. per Vienna Cwt.

527. THURN and **TAXIS** Hugh, Prince. *Laucin, Bohemia.*

Fleeces.

528. TREBITSCH Arnold, Manufactory of Meerschaum-paste pipes. *Vienna, Wieden 723.* Prize-M. Munich 1854. Honourable Mention: Paris 1855.

Pipes and Cigar-pipes of meerschaum-paste.

The exhibitor produces all sorts of tobacco and cigar-pipes, according to the taste and fashion of all parts of the world. He employs for carving of pipes, turning and sculpturing about 60 persons in his establishment where the paste used for his manufactures is produced from pure meerschaum waste.

529. TRENNER Joseph sen. & jun., Turners and Makers of agriot pipe-sticks. *Baden, near Vienna.*

Agriot pipe-sticks.

530. TYROL and **VORARLBERG**, Central Union of Agriculture at. *Innsbruck.*

Flax of Oetzthal and Axam.

531. VITORELLI Peter. *Bòrgo di Valsugana, Tyrol.* 100 cuts of Nut-wood.

(See Classes 8, 9 and 30.)

532. WALDSTEIN-WARTENBERG Ernest, Count of the Empire. *Münchengrätz, Bohemia.*

Unwashed Fleeces of a ram and a ewe of the Electoral breed.

533. WALLIS Count, Administration of estates. *Koleschowitz, Bohemia.* Honourable Mention: London 1851. M. I. Cl. Paris 1855. Silver M. and Prize of Honour: Paris 1856. Golden M. Vienna 1857.

Wool.

(See Class 3.)

The Koleschowitz Merino flock was founded by the present owner Count Wallis in 1835, more than 25 years ago, by the purchase of a considerable stock of improved parent animals from the then much renowned original Merino flock at Gross-Herrlitz in Austrian Silesia, followed by yearly additional purchases of smaller parties of parent rams and ewes from the same flock, up to the year 1845, when the last ram was bought for 1,000 fl. Since that time the Koleschowitz flock, secluded within itself, fostered and improved by inter-breeding, has not only its uncommon constant transmission secured, but also trained and imprinted the highest improvement and richness of wool among all the animals of this flock. It must, moreover, not be omitted that the Merino flock at Gross-Herrlitz in Austrian Silesia derives its origin indirectly from the original Negretti Merino stock, that was directly procured from Spain and last century colonized at Jarmoritz in Moravia by the Austrian premier minister Prince Kaunitz. The total stock of Count Wallis, the parent flock included, amounts now to 4,600 head, affording a yearly result of shearing of from 105 to 110 Cwts of superfine, perfectly clean-washed wool. The parent-stock, besides, parts with a considerable number of improved rams and ewes annually, not only for supplying the Count's own estates, but also for sale to other wool-husbandries at home and abroad. The rams and ewes are particularly distinguished for beautiful wool and staple - improvements, highly ennobled wool, fineness and strength of its fibres, as well as profusion of wool and overgrowth of the whole body. The parent animals and wools, exhibited at the greater agricultural and industrial exhibitions (viz. London 1851, Paris 1855 and 1856, Vienna 1856 and 1857), were already advantageously distinguished by having awarded 3 gold, 2 silver and 4 bronze medals, besides considerable prizes of money and honour. The average weight of shearing amounts with a ewe to 3 up to 4 pounds, with a ram to 4 up to 7 pounds of quite clean-washed wool.

534. WEISS John, Provincial Attorney for Hügel & Comp. *Bistritz, Transylvania.*

Specimens of Wood.

(See Class 3.)

535. WERTHEIM Ernest. *Vienna, City 1009.*

Meerschaum Articles for smoking produced by Ph. Rothenstein, Turner, and an Ivory Goblet with artistic carvings.

Ph. Rothenstein, turner, established since 1859 at Vienna, neue Wieden 749, works with 25 persons, selling his productions to all parts of the world and attaining a monthly exchange of 10,000 fl. The honorary goblet, dedicated to the exhibitor by the turners of Vienna in acknowledgment for his endeavours at the international exhibition of Paris in 1855, is carved of one piece, bearing the inscription carved in relief: „Thanks to your perseverance that procured us honour and distinction", in masterly execution according to all rules of calligraphy. This goblet was made by G. Flöge, turner in Vienna.

536. WIDMANN Leopoldina, Baroness, born Countess Sedlnitzky, Owner of the estates of *Lodnitz and Stremplovitz, Silesia.*

Fleeces unwashed.

537. WIROVATZ Brothers. Hemp-Hatchelry. *Apathin, Hungary.*

Shoemaker's Hemp, half rubbed hemp, point-hemp.

538. WOLF Joseph. *Vienna, Mariahilf 164.* Agent in London: Messrs. Wolf & Baker, 1, Sambrook-Court, Basinghall Street.

Turner's fancy articles of meerschaum and amber.

539. WOLLNER Adolphus. Patented Hemp Hatchelry. *Vienna, Leopoldstadt 564.*

Very finely combed Hungarian hemp (done in Vienna).

540. WSETIN, Administration of estates at; through Raikem Emil. Steward. *Wsetin, Moravia.*

Cuts of silver-leaved firs.

541. WÜNSCHE Anthony's Widow & Son. *Ehrenberg 39, Bohemia.*

Wood bottoms, wood tops, wood plaits and caps.

542. ZICHY-FERRARIS Felix, Count. *Oroszvár Com. of Wieselburg, Hungary.*

Fleeces.

SECTION II.

Class 5.

Railway Plant, including Locomotive Engines and Carriages.

The following, in Austrian miles (= 24.000 Austr. feet), expresses the increase of rail-roads within the Austrian Empire:

In 1831 . . . . 11·5 miles (by horse power);
„ 1841 . . . . 85·0 „
„ 1851 . . . . 309·6 „
end of 1861 . . . . 755·0 „ (33 of these miles by horse power.)

The constructions, achieved between 1841 and 1851, have been, nearly entirely defrayed by the State. Subsequently (between 1855 and 1857), they have come by purchase into the hands of private Societies, so that the State is now only in possession of the Vienna connecting-road (0.7 mile). The share, taken by private individuals in the construction of Railroads began with the publication of the Railroad-concession-law of 1854; from 1856 to the end of 1861 no less than 383·2 Austrian miles of Railroads having been achieved and set into activity.

The rise and developement of Rail Manufacture progressed in proportion with the extent and importance of Railroad-construction. In 1857 the production of these manufactures had nearly reached the amount of 2 millions Ctw. of Rails. In the same year, the production of Rails came to its maximum of 1,300.000 Ctw.; and, besides these, 62.000 Ctw. of Tyres and 56.000 Ctw. of Iron Axle-trees were issued by inland establishments.

The further demands of Austrian Railroads were supplied by the importation:

in 1857 of . . . . 811·000 Ctw.
„ 1858 „ . . . . 1,562.000 „
„ 1859 „ . . . . 528.000 „
„ 1860 „ . . . . 162.000 „
foreign Rails.

The other Iron constituent parts of Railroad constructions (among which Iron Bridges have but lately reached any degree of importance) are almost entirely supplied by inland Ironworks and Engine manufactories. Since 1859, the production of unfinished articles (rod and sheet iron etc.) has become an inland industry.

Waggons for Passenger traffic have been but little manufactured in this country, since the import of them from the Tariff Union States (Zollverein), in 1857 & 1858, has by far exceeded the inland demand. Waggons for the conveyance of Goods are nearly all of Austrian manufacture.

Only 3 establishments-two at Vienna, one at Wiener-Neustadt are employed in the construction of Locomotives, producing yearly about 150 of them. In 1859, only 20 Locomotives were imported from foreign parts. Latterly Traction-engines of Austrian workmanship have been exported to Russia.

The construction of Railroads and the manufacture of articles required for their management in Austria, together with the repairs going on in the local mechanical establishments of the Railroad-Societies, give employment to about 25.000 workmen. The average annual value of their production represents a sum of nearly 20 millions of Austrian florins.

---

543. GANZ Abraham, Iron-Foundery and Machine Manufactory. *Buda.*

Foundery Manufactures for Railways.

544. HERZ Dr., General Secretary to the Carl Ludwig Railway. *Vienna, Haidenschuss.*

Plans and Drawings.

545. KÖSTLIN Augustus and **BATTIG** Anthony. Railway Engineers. *Vienna, Rennweg 541 and Wieden 89.*

Model of a rail-system without wood, besides a drawing and description.

The invention of this rail-system is quite new and therefore not yet put to practical use. The inventors have acquired patents in all great states. Particulars may be derived from the French and German description joined to the model. The costs of plant of this rail-system without wood are not greater than those of the present systems upon sleepers of wood. The economy in maintaining it is evident since there is no need of wood, but particularly also on account of the rail-head which, subject to wearing off and forming a separate part of it, can be exchanged by itself, and thus requires only half of the present material in case of new fittings. As to power of bearing, the solidity of this system is equal to that in present use. The basis of pression has remained the same. The junction of the rails' ends has been improved in such a way that the most important influence on the conservation of the means of management may be expected from it. The separate rail-head may be made of one piece of iron without any welding and, therefore, much more solid and durable, or also of steel because of its small dimensions.

546. ROTHSCHILD Baron. Ironworks. *Wittkowitz, Moravia.* Silver Medal Vienna 1845, London 1851, Paris 1855. Raw and rounded off axes and tyres; locomotive engine and waggon wheels. Links of stiff suspension bridges after Schnirch's patent, produced by pressure without welding.

Model of a steam hammer.

(See Class 31.)

These works comprise the coal-pits at Ostrau in Moravia, at Jaklowetz, Hruschau, Orlau and Dombrau in Austrian Silesia, and at Petřzkowitz in Prussia, their joint annual produce being 3,500,000 Cwts of coals, of which 500,000 Cwts of cokes are produced. The smelting-house consists of 3 high furnaces, a foundery, with 2 cupola and 1 puddling furnace, then the puddler's and rolling mill with 30 puddler's and 24 rechafing furnaces, with a yearly production of 150,000 Cwts of raw iron, 30,000 Cwts of cast-ware, 330,000 Cwts of various rod-iron, especially rails, tyres and other requisites for railways. The machine workshop, connected with the works, consumes annually about 50,000 Cwts of cast and rod iron for boilers and steam-engines, bridges and other iron constructions for halls and roofs, water-stations, slide turn-rail and switch for railways; wheels, axes and tyres for locomotive engines and waggons, and all other machines and requisites for railways, mining, iron works and other manufactures. These works are the eldest in Austria adapted for mineral fuel, both for high furnace and puddling and rolling.

547. (Omitted.)

548. STATE-RAILWAY SOCIETY, I. R. Patented Austrian. *Vienna.*

Fast-train locomotive-engine (duplex engine) according to Haswell. Mountain locomotive engine according to Engerth. Photographs, detail drawings and model of a lattice-crossing of two lattice-bridges executed in Hungary. New intercalary system of telegraph stations. Signal-disk with illumination. Screw-brake. Large Snow-plough.

(See Classes 1, 2, 4, 8, 10 and 13.)

The lattice-bridges are at Szolb and Gran in Hungary and executed after the system of Charles von Ruppert, director of buildings to the State Railway Society. The new intercalary system of telegraph stations (by Ferdinand Teirich) dispenses with the batteries on intermedial stations. The screw-brake (after Alex. Lindner) renders a speedy effect practicable. The signal-disk with illumination for siding-standers is contrived by Mr. Wolf Bender, Upper Inspector of the State Railway at Vienna.

549. STATE-RAILWAY SOCIETY I. R. Patented Austrian. Vienna.

Fast-train locomotive-engine with 4 members together with tender, besides a drawing of such a one with two members; then, models of brake improvement and discharging slide with drawing and explanation. Model of a signal disk for switches with illumination and a snow-plough.

(See Classes 1, 2, 4, 8, 10 and 13.)

Class 6.

Carriages not connected with Rail or Tram Roads.

Carts, Waggons, Sledges and other means of conveyance used in agricultural districts are either manufactured by the Peasants themselves or by Smiths and Cartwrights. The use of Iron in their construction is greatly diminishing in the Eastern parts, so much so that in Bukowina, even at the present day, small carts without any iron whatever being used in their manufacture are being constantly made.

In the W. provinces, however, Iron has become a prevalent constituent of every sort of carriage and agricultural implement; so, that a carriage without iron axle-trees is rarely met with in these parts.

The local demands for the common sort of Carriages are supplied by Cartwrights and Smiths. In some few towns, a certain form (called „Pritschka") of such carriages, are made in greater number and at cheap prices, for foreign export. In 1860, about 1.300 such carriages, representing a value of 250.000 florins, found their way into foreign parts.

The manufacture of Coaches and Carriages of elegant and refined making is confined to Vienna and to some few of the larger provincial towns. This branch of industry, however, has only risen to importance in Vienna, Prague and Pesth; and even in these places, it cannot strictly claim the denomination of a manufacture as they are not made in special establishments, but rather by the cooperation of a number of independent tradesmen.

The Travelling-carriages of Vienna, owing to their construction, combining lightness with solidity, and to their moderate prices, have succeeded in keeping up their old reputation, forming still a considerable article of export to the Danubian Principalities, Turkey and Russia. In 1861, their export amounted to 1.300, equivalent to a sum of 500.000 Austrian florins.

Some complete Carriage-manufactories, established within the last few years, owe their origin and existence to the uncommon increase of Omnibuses and Cabs required for the traffic within and around Vienna.

The construction of Carriages of every description in the Austrian Empire gives employment to nearly 25.000 workmen, whose yearly productions represent a sum of 20 millions of Austrian florins.

550. LOHNER James (formerly Laurenzi & Lohner). Patented Manufactory of Coaches. *Vienna, Rossau 237.* Bronze Medal London 1851. Honorary M. Munich 1854. M. 2. cl. Paris 1855.

This manufactory has existed these 40 years and turns its attention to the production of all sorts of coaches and carriages for luxury and transport both in raw and finished condition, and likewise produces all component parts of carriages. His Majesty the king of Sweden and Norway appointed this manufactory to serve His court.

551. WAENZEL Francis & Son. *Vienna.* Chaise and Waggon Axle-trees. (See Class 11.)

Class 7.

Manufacturing Machines, Tools and Implements.

Sub-Class a) Machinery employed in Spinning and Weaving.

„ „ *b) Machines and Tools employed in the Manufacture of Wood, Metal etc.*

Tools of the coarsest kind such as: hatchets, hammers, hoes etc., are in Austria for the most part produced in iron-works, called tool-smithies. Tools of the smaller kind such as: chisels, gimlets, awls and the like, are mostly manufactured by toolsmiths, who constitute a particular trade or handicraft. But not being able to compete with several newly established manu-factories fitted up with every kind of working machines, they have of late turned their industry into new channels, the more so, as the above-mentioned manufactories provide the said iron-tools with the necessary wood-appurtenances. A material rise took place, of late, in the manufacture of apparatus for beet-sugar-manufactories, spirit distilleries and breweries. The production of the Austrian establishments in this line do not only cover the wants of the home market, but even export apparatus for spirit-distilleries with all the necessary mechanical appurtenances to Russia, the Danubian Principalities and Turkey.

Engines for mining, mill-works, presses and pumps of all kinds, as well as working machines for machine workshops are for inland wants almost exclusively furnished by the Austrian machine manufactories. In like manner, the manufacture of sewing-machines has already extended to such a degree, that only inconsiderable imports of these machines are taking place, although the want of them has extraordinarily increased. On a somewhat smaller scale is conducted the manufacture of spin-ning and weaving machines, most of them being imported from abroad, though several establishments have undertaken to pro-duce this particular article, and several spinning establishments begun in their factories to construct spinning mills and produce preparatory machines. Articles of import are power-looms and paper-machines.

The Austrian tariff of duties making no difference between manufacturing machines, motors and agricultural machines, it is hardly possible to point out the precise quantity and value of the imported machines. Only in respect to round-looms and powerlooms there is a separate account of the (free from duty) import, which in the year 1860 amounted to 7.842 Cwt.

The total production of tools and manufacturing machines in the Austrian Empire represents an average value of 20 millions of florins, occupying 13,000 workmen.

---

552. BEARZI J. B. Manufactory of Reeds. *Vienna, Wieden 114.*

Reeds.

553. BLOBEL Augustus. *Vienna, Schottenfeld 238.*

Velvet-rods.

554. BRAND C. & **L'HUILLIER** F. Manufactory of Apparatus and Machines. *Brünn.*

Sugar Moulds.

The Moulds produced in this manufactory are stamped and embossed out of one piece. There is nothing left to be wished for in the system after which these moulds are worked: solidity is warranted in all respects, since they rank with the best that ever were produced in this branch. The enamelled moulds present the advantage that the loaves coming out of them are of a beauty not to be attained with lacquered moulds, and, if carefully handled, are of almost illimited duration. This estab-lishment has provided the refinery of Messrs. Robert & Comp. with enamelled moulds, which have been incessanly used these 8 years and were not yet in need of repairs. The sheet iron employed by this manufactory, rivals the English sheet-iron of the best quality, and is manufactured in the „Joseph" works of the Prague Industrial Society.

555. FICHTNER Anthony, Member of the Lower-Austrian Trade Union. *Vienna, Leopoldstadt 551.*

Lithographic and Autographic Presses.

556. GASTEIGER Anthony von. *Wilten, near Inns-bruck, Tyrol.*

Pantograph (a very simple and cheap wheel indentation tool).
(See Classes 11 and 27.)

557. KRONIG Charles. *Vienna.*

Patented Sugar Moulds.
(See Classes 30 and 31.)

558. PORTHEIM Joseph von, **KÜNDIG** Henry and **BERTSCHY** George, Patentee. *Prague, Bohemia.*

Electro-coppered iron rollers for engraving and cylinder printing.

This proceeding is patented and employed in the calico manufactory of Brothers Porges at Prague. There are presented great advantages, both because of the much simpler and con-siderably cheaper production of cylinders in comparison with the usual massive copper rollers. The copper produced in this way proves more uniform in engraving and more durable in printing; moreover this proceeding includes also a constant periphery of the cylinders, which of course facilitates the production of variegated patterns. A $^6/_4$ roller of 91 centi-meter breadth and 14 centim. diameter weight 110 pounds (of which only 10 lb. copper, the rest cast iron); price about 25 fl.

559. REZNICEK A. M. Tin Ware Manufactory. *Smichow, near Prague.*

Sugar Moulds of Tin-plate.
(See Classes 8 and 31.)

560. RIEFFEL Charles, Machinist. *Pesth, Toleranz-gasse.*

Accommodating Nut Key.

561. SCHALLER Joseph. Patented Manufactory of Bellows and Army Forges. *Vienna, Leopoldstadt 426.*

Bellows; Army Forges.

562. SCHRAMM Willibald. Patented Manufactory of Jacquard Machines. *Vienna, Gumpendorf 486.*

Jacquard Machines. Bored „Galir" Board with 20.000 holes.

One of the exhibited machines is a double Jacquard machine in connection with a treddle machine and double cylinder for pattern double stuff weaving, offering the advantage that, instead of 4.800 cards, only 600 are necessary, and that the front equipage of a loom is to be dispensed with. Price 70 fl. The second is a simple machine with 1.200 fine divisions, with which the economy, in comparison with a machine of coarse division, amounts to 20% in the pattern cards. The manufactory is worked only by help of engines. The intercourse of their trade extends to Germany, Italy, Russia and Turkey.

563. THUMB V. Manufactory of Machines. *Unterhimmel, near Steyr, Upper-Austria.*
Mechanical littering rods for weaving.

564. WALLERSTEIN Salomon. *Vienna, Landstrasse 128.*
Machine for cutting out linen, 600 pieces an hour.

Class 8.

Machinery in general.

Manufacturing machines (working-engines) having already been spoken of in the preceding class and the statement of the produce of agricultural machines being referred to the subsequent class, there remains nothing for this division but a report on the manufactures of motors in Austria.

The rapid rise in the production of machines in Austria, dating from the year 1838, has not yet advanced so far as to bring to bear a division of labour among the different manufacturing-establishments. Hence it happens that even now-a-days each establishment receives and effectuates orders for all kinds of machines. It is only the construction of steam-engines, which is confined to the larger establishments, there being only 40 manufactories, out of the 112 in the whole of the Austrian Empire, that are now actively engaged in the manufacture of these motors, and enabled to produce about 400 engines a year, averaging 8,000 horse-power. Most of these establishments are to be found at Vienna and Prague; their erection in the manufacturing and mining districts falls within a recent period. Water-motors, turbines and iron water-wheels are supplied by machine manufactories; their application is now progressively increasing in preference to those till now furnished by carpenters and joiners and constructed of wood only, according to empirical principles.

Wooden treddle whims and windmills are in inconsiderable numbers constructed by carpenters, whereas the production of horse whims is for the most part connected with the manufacture of agricultural machines.

The Austrian manufactories of machines, engaged in the construction of steam-engines, range among the greatest establishments of this kind, most of them possessing their own cupola furnaces for machine founding and all appropriate apparatus.

The manufacture of boilers has of late been ceded to special establishments, the industry of which is chiefly directed to all sorts of plate work and plate rolling etc.

The production of motors engages upon the whole 8,000 workmen, the value averaging 10 millions florins

---

565. BEER Lawrence, Fire-Engine Maker. *Vienna, Wieden 51.*
Models.

566. DINGLER Henry, Manufactory of Machines. *Vienna, Wieden 120.* Bronze M. Vienna 1839. Silver M. Vienna 1845. 2 small Silver M. Vienna 1857.
Hydraulic Press Pump, and Wine Press.

The construction of the press-pump is new and obviates two great disadvantages of the construction in use till now, the ratch-contrivance of the valves being open to inspection, wherefore any untightness is instantly perceivable, which was not the case heretofore and caused many breakings of cylinders. The second improvement consists in the position of the water reservoirs, by means of which it is possible entirely to do away with the till now inevitable defilement of the water through the grease of the moving parts, thereby dispensing with the frequent cleaning of valves and consequent interruption of work. Similar pumps, finished since some time, have been working up to the present time without any stoppage. The improvement in question is also materially promoted by the circumstance, that all constituent parts are manufactured of the stero-metal, which is as yet little known, and when warm, to be hammered like copper and possesses great compactness and durability. There is also to be observed that, on account of the water-reservoir allowing of a higher position, there never takes place a failure in scooping, because of the sole pressure of the water, a circumstance of great consideration with so little a diameter. The manufacturer is convinced that, by means of these contrivances, the greatest perfection is attained in the construction of those pumps. Price, loco London, £ 75.

The wine-press, constructed by Lerol, presents great advantages in the production of juice, it occupies less space than the common beam presses, is working quicker and exercising greater pressure, wherefore, as practice has shown, a greater result of juice is gained. Pressure: 50 pounds per square inch. This press has been introduced in a very short time in many places of Austria, which afforded the exhibitor opportunities for improvements that are useful and to the purpose in practice. Price, loco London, £ 40. The exhibitor's manufactory of machines is one of the eldest in Vienna, was the first that introduced steam-power for its works, had the first planing-machine in Austria (constructed in his establishment, for 13 feet table length), was the first that successfully managed cupola furnace founding, also the first that employed the ventilator in Austria. The chief productions of the manufactory are: — hydraulic presses for all branches of manufacturing; fittings and machines for the manufacture of stearine candles, sugar and oil; mills; agricultural machines (among which there were sold 1,400 Seidl's thrashing machines), steam engines, transmissions, adjutory machines etc. Out of this establishment were first furnished a great number of machines for sugar manufactories, the first of which is at Rusin in Bohemia (property of the exhibitor), also the mechanical mill in Moravian Ostrau was rationally outfitted by him. The exhibitor's arrangements and contrivances for stearine manufactories have been much sought these 22 years and enjoyed many exports.

567. DOBBS William. *Vienna, Landstrasse 286.*
Steam-boiler Fire-box.

568. HORAK Anthony, Establishment for Construction of Machines and Manufacturing of Mathematical Instruments. *Vienna, Gumpendorf 324.*
Punching Machine, together with centre-cut, for the simultaneous self-acting productions of 4 plate coffee-sieves.

The holes of those sieves run in a spiral line, the sieves having a diameter of 2—12 inches. The complete machine is sold for 650 fl. That manufactory keeps a store of all kinds of geometrical and mathematical instruments, and furnishes surveying apparatus of the best construction, mechanical lathes of the newest contrivance, slide-rests, blank-cutting presses, etc. at the cheapest prices.

E

569. HUBAZY George, Manufacturer of Agricultural Machines. *Vienna, Erdberg 152.*

Improved Locomobile of 4 horse-power, furnished with a new apparatus for heating with straw, without obstructing flues and grate, also to be exchanged for contrivance to heat with coals, wood, saw-dust, etc.

The patented improvement of the locomobile with the new apparatus consists in the arrangement that any straw may be burned in it, and that as much steam is produced as with any other fuel, whereby the machine is peculiarly adapted for Hungarian localities where no other fuel but straw is at hand.

The straw is regularly put into the locomobile through the plate tube fitted to the stoke-hole. A boy may do the feeding.

Inside the fire-box, close to the boiler flues, is a grate of quadrangular hammered iron, the openings of which are $1/4$ inch, so that half-burnt flying straw does not immediately rush into the flues, but rebounds at the beforementioned red-hot iron grate, takes fire, and allows only clear flame to pass through the flues. But the straw ashes, constantly rebounding in greater masses from the grate, fall upon the grating, and can at any time be removed by means of the cleaning swivel apparatus fitted to the ashes box, on turning a little at a crank fitted to the grate, in consequence of which the straw ashes lying on the grate fall into the ashes box beneath filled with water.

The exhibitor has received first prizes at all the great exhibitions at home.

570. JENBACH, Iron Works administration at. *Jenbach, Tyrol.*

Horizontal Turbine; Stamp for Coinplates.
(See Class 1.)

571. KLOTZ Joseph, Professor of Agriculture. *Gratz, Styria.*

Patented Safety Valves for Steam-boilers.

572. KOHN Hermann & Son, Patented Manufactory of Machines. *Vienna, Weissgärber 18.* Agent in London: Mr. A. Heinrich, of the Austrian Commission.

Transportable Spirit and Wine Pump.

This manufactory, directed by Mr. Joseph Roy, principally turns its attention to the production of newly improved wine, spirit and water pumps, English scentless house and room closets, watering-niches, kitchen kennels, etc. — The exhibited pump fetches 2 Eimer (kilderkins) of liquid per minute, when worked by one man only, may be used for warm as well as cold liquids, and is easily to be transported, two men being able tonremove it. By the construction of an elastic piston, the power of this pump is greatly heightened. Moreover it is not subject to so many repairs as those constructed till now. Of especial use for breweries, spirit distilleries, and wine trading concerns. Sale mostly to Hungary and Transylvania.

573. MICHALEK Francis. *Vienna, Wieden 428.*

Boring Machine for iron with treadle, straight and round Shears. Middle kind of Plate Shears for iron.

574. NEMELKA Lawrence. Mill Building and Machine Constructing Establishment. *Fischamend near Vienna.* From August 1862 beginning, the Manufactory will be at Simmering near Vienna.

Model of a complete mechanical mill with 5 sets of Stones, together with all requisite machines for cleaning corn and assorting grit, flour, and sorting cylinders, transport-contrivances, and necessary transmissions, according to the Vienna milling system.

This manufactory has, during a space of 20 years, completely and most successfully built steam and water mills, in all provinces of Austria as well as abroad, having employed the newest construction of the Vienna milling system warranting the most extensive power of flour production, and introduced all experiences and improvements that occurred in practice.

This model of a machine-mill contains, besides the sets of stones, all adjutory machines in use, in order to clean the corn before grinding, to gain the various sorts of grit and to sort them, and to produce the various sorts of flour up to the finest quality by means of cylinders, also the requisite contrivances for transport in order to bring the mill product to its place of destination without special labouring powers.

575. NETREBSKY John, Knight. *Cracow.*

Model of a Steam Engine, the distributing-regulator of which is in the piston.

576. PLUMER Melchior. *Borgo di Val Sugana, Tyrol.*
Component Parts of a Hydraulic Pump.

577. POSZDECK Joseph, Bellows Maker. *Pest, Gezagasse 3.*

Model of a Bell with Belfry.

578. REACH Frederick, Decimal Weight and Metal Ware Manufactory. *Carolinenthal near Prague.* Warehouses: Pilsen, Reichenberg, Bukarest.

Model of a Scale Beam 16 feet long for loaded waggons (centesimal system), price 100 fl. or £ 10. Iron blued Decimal Kitchen Weight per 25 ℔, price 25 fl. 2 £ 10 s. Common Decimal Weight per 500 ℔, price 25 fl. or 2 ℔ 10 s. Model of a new Decimal Cattle Weight with iron rails per 2,000 ℔, price 25 fl. or 2 £ 10 s.

(See Class 31.)

This manufactory existing since April 1861 has during this short period furnished 800 weights to all the Austrian provinces, to Moldavia and Wallachia, and produces also turned axles for waggons and coaches, and vices of the very best Stirian iron. Prices franco Prague:

Superior well-constructed decimal weight with 8 brass weights, warranted one year:
Power of bearing:

$1/2$,	1,	2,	3,		5,	6—8,	10,	15,	20,	25,	30,	40 Cwts.
$11^1/_2$,	$13^1/_2$,	15,	$16^1/_2$,	20,	22,	27,	32,	48,	60,	70	85 fl.

New decimal Cattle Weight with iron rails, per 20 Cwts power of bearing; 80 fl. Scale Beam for weighing loaded 4 wheel waggons for:

50, 60, 80, 100, 120, 140, 160 Cwts
380, 400, 450, 500, 600, 800, 850 fl.

The costs of wall-work and erecting, viz. journey of the machinist and 3 fl.per day allowance, at the charge of the customer. The plans for the wall-work are furnished by the manufactory

579. REZNICEK A. M. Smichow near Prague.
Candy-sugar Vats, Fire-engines.

(See Classes 7 and 31.)

580. SASSE Frederick. Gratz.
Jacks for heaving great loads.

581. SCHEMBER C., Manufactury of Scale Beams. Vienna, Leopoldstadt 779.
Centimal and Decimal Weighing Machines.

The patented Centimal Scale Beam serves for weighing locomotive engines, having a power of bearing up to 2.000 Cwts at the same time stating the pressure that each wheel of the engine exercises separately. The exhibited Centimal Scale Beam forms only one wing for weighing a couple of wheels. The patented transportable Centimal Scale Beams have a power of bearing from 20 to 80 Cwts, affording considerable saving of weights and facilitating the manipulation. The Decimal Weight has 1 Cwt power of bearing with a proportion of 1 : 1.

The Manometer attached to these beam-weights, manufactured by Schember L. &. Posner in Vienna, serves for determining the pressure and the force of tension, exercised by the steam of a steam-engine, a locomobile, a dyer's boiler, etc. the protective plate of aluminium being especially worthy of notice. The Patented Manometer is provided with a contrivance for prevénting over-heating, thus serving to save fuel and to preserve the boilers.

582. SCHEMBER Lewis. Vienna, Leopoldstadt 779.
Autographic Multiplication Press and Satinizing Machine.

This autographic press multiplies handwriting and is serviceable for producing circulars, lists of prices, labels, bills of lading, invoices, market-reports, drawings, music, etc., in the simplest, surest and speediest manner. By means of this machine, 1,000 copies, 80 and 100 an hour, may be got. The Satinizing Machine is for satinizing paper, photographs, stereoscopes etc., all improvements made in this line having been adapted.

583. SCHMID H. D., Successor to Rollé & Schwilgué. Patented Manufactory of Machines. Simmering near Vienna. Silver M. Vienna 1835. Gold M. Vienna 1839. Silver and Bronze M. Munich 1854, London 1851, Paris 1855 and Order of the Legion d'honneur. Gold M. Vienna 1857.
Horizontal Steam Engine, 20 horse-power, price 2,400 fl. in silver, Scale Beam for locomotive engines, price 800 fl. silver, Screw-cutting machine, price 500 fl. silver, Model of a fire-engine, Models of scale-beams.

This establishment is active in manufacturing of railway machinery and fittings, such as: railway carriages and waggons, 1,800 waggons for goods and 200 carriages for persons being annually built, more than 8,000 having been till now furnished by that establishment; there are also made sand-waggons, snow-ploughs, and station-fittings. This firm was the first in Austria, that engaged in this branch of industry, and it has in general materially contributed to the advancement of machinery in Austria. Moreover there are yearly manufactured an average number of 60 various steam-engines and boilers, diverse adjutory

tools, 1,000 scale-beams and many fire-engines. This establishment is provided with means to answer all demands to their full extent. The mechanism of most beet-root sugar manufactories in the Austrian states, steam-engines and machine-whims for mining-works have issued from this establishment. It has the merit of having introduced in Austria the high-pressure steam-engine with variable expansion, hundreds of which in all parts of the empire bear testimony to their superior construction. Likewise the sugar-moulds of iron-plate, Schützenbach's cases, of the former 500,000 and of the latter 100,000, have been first produced in this manufactory. Also in the agricultural line, there has been much done in this establishment, mainly by producing the first locomobiles in Austria, wherefore it has been distinguished with the Gold Medal at the exhibition in 1857. As to the portable and conveyable scale-beams, after an exclusively patented system, 10,000 of which were produced, they too were introduced by this firm into Austria; moreover, the manufacture of fire-engines, having before still been in its primitive state, was first attempted after rational principles in this establishment, where also the first steam fire-engine was constructed. An important industrial branch of this manufactory, is the production of ammunition and carriages for it and for cannons etc. furnished to the army; finally cast wares of which the foundery connected with it produces 5,000 Cwts a month. The following machines new in Austria are at present going to be introduced by this establishment: — the caloric machine, system Bellon; the gas-machine, system Lenoir; the Lemercier shoe-machine and the Vouillon machine for felting of yarns. After what has been mentioned may be surmised that this establishment is enabled, on a grand scale, to produce all objects within the scope of machinery. Extended on an area of 8,000 square fathoms, it comprises all the various departments of forging with 60 fires worked with ventilators, foundery with 3 cupola furnaces, turning and locksmith works, copper-smith, grinding, lackering, wood-seasoning, wheelwright, carpenter and joiner's work-shops, the latter with very well adapted wood working machines, carriage building sheds, etc. The whole is worked by means of 3 large steam-engines of 120 horse-power, and from 1,200 to 1,500 people being employed. The total value of the production in this establishment averages 2,000,000 fl. per annum.

584. STATE-RAILWAY-SOCIETY, I. R. Patented Austrian.
Drawing of a new steam forge press.
(See Classes 1, 2, 4, 5, 10 and 13.)

585. SZABO Francis, Maschinist and Mill Builder. Pesth, Schlangengasse 5.
Designs of Mills.

586. TIBÉLY Francis, Director of the Iron Ware Manufactory. Bries, Comitat of Zomlyo, Hungary.
Joint-lock of earth-borers without using screws.

587. VITTORELLI Peter. Borgo di Val Sugana, Tyrol.
Meat-cutting machine for sausage Makers.
(See Classes 4, 9 and 30.)

588. WARCHALOWSKI James. Vienna.
New Sewing Machine.

589. WINIWARTER George, Knight, Civil-Engineer. Ware-house: Vienna 816. Agent: Mr. William Andrews (Messrs. Royle & Comp.) London, 32. Kingstreet, Cheapside.
Punching Machine, of the exhibitor's own construction, for punching five holes at once.

This Machine was constructed and manufactured in the Iron and Smelting Work Petrovogora near Topusko in the Croatian Military Borderland. This establishment produces pig-iron, raw and finished cast wares, besides soft and hard rollers.

Class 9.

Agricultural and Horticultural Machines and Implements.

The iron component parts for agricultural and horticultural implements are in Austria generally manufactured in hammerworks or tool forges, and furnished ready for use, except the wooden parts, which are provided either by the purchaser himself or made by carpenters and cart-wrights.

Thus also plate iron for ploughs is worked in tool forges in flat forms only, it being left to smiths or trade-concerns of smaller extent to adapt it to the different kinds of coulters, such as they are, conformably to the use of the country. In the eastern parts of the empire, plough implements are made of wood cased with thin sheet-iron.

Scythes, chopping-blades and sickles, manufactured in particular scythe-forges in Austria-proper, Styria, Carinthia and Carniola, enjoy great renown respecting their superior quality; the quantity produced amounts to 6 and even 8 millions per annum, great numbers of which are exported to Russia and the Danubian Principalities.

The establishment of manufactories for the production of agricultural machines belongs to our present period, and was encouraged by landed proprietors of Bohemia and Silesia, who called such manufactories into existence, in order to supply the wants of their own extensive estates. But the rise of independent extensive manufactories of this kind dates no farther back than 14 years, when, in consequence of the abolition of soccage, the wages for day-work rose considerably, and the use of machines for extensive husbandries has become an urgent necessity. Number and industrial activity of these manufactories, especially of those in Hungary, is increasing from year to year, so that the value of their annual productions, a staff of 5,000 workmen being employed therein, is to be estimated at 6 millions florins at the very least.

The manufacture of agricultural and horticultural implements of all kinds, as well as agricultural machines, occupies, in Austria, a minimum of 15,000 persons, their joint production representing a value of 12 millions florins.

---

590. BOKOR Ferdinand, Machine Smith. *Gross-Zinkendorf, Com. of Oedenburg, Hungary.* Great M. Vienna 1857.

Hoeing and Clumpering Plough.

That establishment undertakes the manufacturing of all sorts of agricultural machines and implements, employing an average of 10 people. The exhibited plough is a new invention and even unknown in this construction in England. In clumpering the tray plate can, according to want, be set more narrow or broad, and so can the shares in hoeing. There is a contrivance at the wheel for setting the plough to go higher or deeper. Both contrivances can be taken off, but it is not necessary the one not impeding the other in its working.

591. BORROSCH & EICHMANN, Manufactory of agricultural machines and implements. *Prague.* Bronze M. Munich 1854. Great Silver M. and Diploma Paris 1855. Gold M. Prague 1856. Great Silver and Bronze M. Vienna 1837.

Agricultural Machines and Implements.

The proprietors of this manufactory, erected in 1852, emanated in the selection of their exhibited objects from the principle that an International Exhibition, where, of course, numbers are represented, must not be abused as an emporial market, wherefore every exhibitor should restrict himself to send only few objects, but those to be such as characterize his country and its mode of manufacturing. In accordance with this the exhibitors have made their choice, and on purpose avoided to be ostentatious with respect to number, variety and object of show only; on the contrary they have selected from their current articles only such as are either of their own invention or at least essentially improved by them and at the same time the most spread in Bohemia.

1. The Kleyle plough (1 ℒ 18 s.) This plough, going a depth of 14 Vienna inches, is with us particularly used for culture of beets, and was awarded a premium at the international exhibition of Paris in 1855.

2. The Zugmaier plough (1 ℒ 2 s.). In spite of its many defects, this plough is still, especially with small farmers, one of the most used in Austria and the Eastern countries adjacent to it.

3. The Bohemian Ruchadlo (15 s.). A Bohemian invention scarcely 40 years old, spread with wonderful rapidity over all Germany and the Western agricultural states of North-American Union, many times modified and baptized upon the names of supposed inventors, through which circumstance the true Ruchadlo lost its peculiar qualities without acquiring those of a perfect plough. In Bohemia, it was preserved in its simple primitiveness, and is in fact a highly valuable ploughing implement for soft soil, though it does not draw deeper than 4 to 5 Vienna inches, not exactly raising up, but in exchange completely crushing the heaved clods and at the same time turning them, it requires with very easy management the least power of draught.

4. Transferable plough-cart (19 s.). Entirely our construction, the usefulness of which has proved general by its far spread sale, yearly increasing, and extending even to the inland parts of Russia. The elongated axles serve for setting the wheels into the requisite distance of the furrows.

5. Turning Ruchadlo for hill-sides, 1 ℒ 4 s.

6. Horsky plough with subsoil diggers (2 ℒ 18 s.). This plough is also very popular in Northern Germany, the inventor having succeeded to unite the advantages of the Ruchadlo with the qualities of a good deep-drawing plough by means of suitable exchanging shares. Thus is may be used without the coulter, as Ruchadlo, with the common Ruchadlo share only a little larger, with the turning share and coulter as deep-going plough, and in both cases, according to one's pleasure, as subsoil digger. The digging shares do not raise up the subsoil, which is often disadvantageous, but crush in a depth of 16 inches, in consequence of which the atmospheric constituents can penetrate. Simple and ingenious is the contrivance for setting the plough higher or lower, to the right or left, by means of a vertical and a horizontal screw. Finally the cheek-pin is shaped both as hammer and screw-wrench, to be near at hand for any arrangement.

7. Horsky double potato-moulding plough, 2 ℒ 12 s.

8. Five-share marker to this implement, 1 l. 6 s. This very practical implement has been warmly recommended by the R. Prussian College of Economy. With two workmen and two horses the moulds are formed over the laid potatoes on a square area of 4 or 5 Vienna yokes (according to the distance of rows of 18, 21 or 24 inches).

9. Horsky potato-cultivator 2 ℒ 18 s. It is applied a straddle over each mould. The hoeing-shares serve for weeding, the settable digging coults for loosening the mould beneath the roots of the plants, and the moveable mould-boards for throwing up the heaps.

10. Fritsch potato-cultivator, 1 ℒ 9 s. does likewise, but only between the rows, the service of a weeder, loosener, and heaper, to be separated from it or used at pleasure.

11. Horsky three-share drill-rake 1 ℒ 9 s. On account of its perfectly equal seed-scattering in any desirable depth and its little want of draught-power, this implement has gained extensive use in Bohemia and Moravia.

12. Beet-weeder with protecting plates 1 ℒ 11 s. An implement very popular in Bohemia and Moravia, the rotatory protecting transferrable plates of from 12 to 20 inches distance of rows not only protecting the plants from being overwhelmed with earth, but also having the consequent important use of cutting the crust of the soil and preventing the uncovering of the roots through cracks which will arise.

13. Horsky clover drill-plough 1 ℒ 13 s. Serves unfailingly for equally dense dissemination. The wheels inward turned prevent the occurrence of empty strips, and the filler to be closed underneath has exactly the capacity of the room of a seed-drum The spilling of seed is exactly to be regulated by the sliding of the drum-hoops.

14. Alban broad drill-plough (large) 9 ℒ 12 s. A comparison of this machine with one manufactured elsewhere will show the many improvements applied by the exhibitors. For instance, the regulator is not to be displaced by jolts during the drive, and a perfectly equal distribution of the seeds even upon sloping plains is secured by the conducting ledges upon the strewing-board. This drill-plough by a third smaller (8 instead of 12 feet broad) is in growing use with small farmers.

15. Regulation thrashing machine with whim, 24 ℒ. This machine, much used by small farmers, which together with the whim belonging to it, is entirely of our own construction, with a contrivance not yet brought into use anywhere before. The aim of it was, not to make the larger or smaller position of the thrashing apparatus dependent on the husbandman setting it at random but to fit an unfailing regulation apparatus visible on the outside, in connection with a setting mechanism at once convenient and exact. Any body may manage this apparatus with sure success, having only to take care that the numbers on the scale coincide.

16. Green-malt bruising-mill, 11 ℒ 8 s. It frequently occurring in green-malt, that by too long germination barley is clotted and bruising rendered difficult, we contrived an apparatus separating the clotted malt, before it goes to the bruising rollers, which apparatus can be taken off when barley of short germination is bruised, in which case the same machine serves also as dry malt bruising mill. Its most extensive sale is to Galicia.

17. Chaff cutter, 5 ♙, likewise cutting grass, clover, esparsette etc. to a length of 4 inches.

18. Decimal weight for 15 Vienna Cwts of load 4 ♙ 10 s. Besides other minor improvements, it is peculiarly distinguished by a contrivance of the exhibitors' for instant fixing of the scalebeam during the loading and unloading, the disturbances, of the weighing mechanism, otherwise frequently occurring, being thus avoided.

19. Centesimal cattle-weight, 25 Cwts power of bearing, 15 ♙. Our own construction, calculated for convenient use and long duration, joined with precision and sensibility of the weighing mechanism as a matter of course.

20. Fischer's mangle, 4 ♙ 16 s., is our property of invention relatively to the practical execution of the principle of a concave face with self-acting load to be regulated by spring power conformable to wishes. The little space taken up by this mangle, its appearance suitable to any room, and its easy management not even fatiguing the weakest female, have procured for it a sale spread far beyond the boundaries of the empire.

21. Hand thrashing machine, 12 ♙. Whereas in Great Britain and in countries, where proprietors of great estates and farms are not intermingled with the peasantry and farmers of small husbandries, hand thrashing machines have been entirely superseded by steam and whim thrashing machines, the manufacture of the former is in Austria still a branch of industry of brisk activity, we alone having, during 8 years, already constructed near a thousand of them and successively brought them to the present perfection.

22. Meadow moss-harrow, 4 ♙. Whereas of all implements formerly contrived for weeding moss from meadows, none fully answered this purpose, the present harrow invented by Mr. Semsch, counsellor of economy at Swoischitz, completely fulfills its purpose.

592. (Omitted.)

593. CHRISTALNIGG Charles Count, Iron Foundery and Manufactory of Machines. *St. Johann am Brückl, Carinthia.* M. II. Cl. and Honourable Mention Paris 1855.
Maize-disgraining Machine.
(See Class 31.)

594. FARKAS Stephen, Manufactory of Machines *Pesth.*
Ploughs.

595. FICHTNER J. & sons. *Atzgersdorf, near Vienna.*
Drill-plough, Rotatory harrow.
(See also Class 2.)

596. GOLARZEWSKY Ladislaus, Knight. Manufactory of agricultural machines and implements. *Targowiska, District of Sanok, Galicia.* Prize M. of the agricultural societies at Warsaw, Lemberg, Cracow.
Thrashing Machine with round flails, together with whim, the exhibitor's own system.

This thrashing machine is going much more easily than any other, thrashing perfectly clean and crushing no corn (the usual disadvantage of all thrashing machines); wherefore it is peculiarly adapted for thrashing out grains destined for sowing, and barley destined for the preparation of malt. A great number of such machines working in this country to the full satisfaction of the owners, bear testimony to the virtues of this system. 4 Vienna Metzen per hour done by it. Price, the iron transmission belonging to the whim included, 450 fl.

597. GUBITZ Andrew, Manufactory of Machines. *Pesth.*
Fore-saddles of ploughs.

598. HOFMANN Francis. William, Counsellor of Economy. *Vienna, Josephstadt 52.*
Model and Description of his dung-bed for a farm on a small scale; Model and Description of his new method of training hops on treillis.
(See Class 29.)

599. KOLB Francis Joseph. *Muria-Enzersdorf.*
Bees' products; implements, machines etc.

600. LEICHT Melchior. *Essegg.*
Plough.

601. MELICHER Lewis, Dr. *Vienna, Alservorstadt 166.*
Bee-hives and products.

602. MĒSZÁROS John, Machine Smith. *Gross-Zinkendorf, Hungary.*
Ploughs.

603. NACHTMANN James, Apothecary. *Hermagor Carinthia.*
Carinthian Bee Hives.

604. ÖTTL John Nep. Curate. *Puschwitz.*
Mechanical Bee-hive of straw; strawprince with tubes of the newest kind; wax-filtration pot of tin.

605. QUAGLIO Giles, Civil-Engineer, *Vienna, Weissgärber 140.*
Corn Measuring Machine.
(See Class 1.)

606. REACH Frederick. *Prague 1049.*
Cattle Weighing-Machine.
(See also Class 8 and 31.)

607. REISS A., Tin-man to the I. R. Court. *Vienna, Laimgrube 87.*
Garden Implements.
(See Class 31.)

608. SZABÓ Francis, Machinist. *Pesth.*
Corn Cleaning Machine.
(See Class 8.)

609. VIDATS Brothers, Machine Manufacturers. *Pesth, Zweihasengasse 9.*
Field Implements.

610. VITTORELLI Peter. *Borgo di Val Sugana 80, Tyrol.*
Butter-churn. Model of a thrashing machine.
(See Class 4, 8 and 30.)

Class 10.

Civil Engineering, Architectural and Building Contrivances.

Sub-Classe a)   *Exhibitors of Civil Engineering and Building Contrivances.*
        „   „   b)   *Exhibitors of Sanitary Improvements and Constructions.*
        „   „   c)   *Exhibitors of Objects, shown for Architectural Beauty.*

The Austrian Empire offers specimens of every degree of architectural progress, from the wooden cottages in the Riesen- and Erzgebirg and the „Sennhütten“ (Châlets) of the Alpine regions to the monumental stone buildings for sacred and other purposes, from the lofty brick-houses of Vienna and the provincial Capitals down to the Loamhuts („Pisé“) of the East Romans.

The materials for covering roofs are as diversified as those employed for the buildings themselves. In the open country, straw- and rush-roofs are prevalent, in smaller towns boards or shingles serve the same purpose, and tiles, slate, sheet iron or copper in cities of first and second class. In the Western Provinces, the laws regulating the construction of new buildings greatly favour and, even enforce, the extensive use of bricks and other materials offering security against devastations by fire.

The covering of roofs with fire-proof paste-board or incombustible felt, lately invented, has hitherto gained very little, if any, ground. In Austria, brick, or stone-work is generally joined with Lime; the use of Hydraulic Lime and Cements for overground constructions has but lately begun, to be used on a somewhat extensive scale.

About 500.000 Cwts of Hydraulic Lime of superior quality are produced every year in Lower Austria, and Tyrol. The inland production of artificial Cement has not yet made progress sufficient to supersede the annual import of about 100.000 Cwt of Portland Cement.

Architecture and the trades connected with it, have been satisfactorily advancing in Austria. In one of these trades,

Austria ranks first on the continent. The Carpenters of Vienna have acquired such renown, especially in timber structure, combining elegance with solidity, that their fellow-workmen from all the parts of Germany resort to Vienna, as their best practical school.

The outsides of buildings are decorated with hewn stone-work or with Terra-cotta ornaments. The inside ornamentation lies chiefly in the hands of decorators; and it is only of late that the taste for paper-hangings and the decoration of walls with artificial marbling has become more general and is increasing from year to year.

Building, and the trades connected with it, employ nearly 750,000 persons during the greatest part of the year. The total annual earnings of this class of the working population may be approximately esteemed at 200 millions florins. The value of the materials, employed for building purposes, has been already mentioned under other classes of raw material production (Working of stone- and Lime-quarries, Ceramic industry etc.).

---

611. BENCZUR Joseph & Co., Cement Manufactory. *Mogyoróska, Com. of Zemplin, Hungary.* Bronze Medal Paris 1855. Ware-houses: Eperies, Pesth, Tokay, Szegedin, Grosswardein. Bukarest.

Cement Products.

This manufactory was founded in 1854 by Joseph Benczúr. Price per Cwt of cement 1 fl. (1 cubic foot pressed 80 pound). The Jury of Paris said that „it was one of the most noteworthy foreign cements". The Paris analysis showed: lime 58·03, traces of magnesia, silica 27·44, alumium and peroxyde of iron 14·53.

612. CURTI Alexander, Dr., Portland Cement Manufactory. *Muthmannsdorf, near Wiener Neustadt, Lower-Austria.*

Ornaments of Portland Cement: Balcony Socle (cement to sand like 3:1); Cornice.

(See Class 1., where particulars about the manufactory.)

613. DOPPLER John, Stone Mason and Owner of Marble-quarries. *Salzburg.* Medal II. Cl. Paris 1855.

Marble Fountain.

(See Class 1.)

614. EGGSPÜLER Anthony, Filtering Apparatus Maker. *Vienna, Wieden 920.* Ware-houses: Agram, Laibach, Linz, Pesth, Wieselburg.

Three Apparatus for filtering diverse liquids; filtering substance for them.

These apparatus can be made in all sizes for large and small quantities, losing neither in quality nor in quantity. Such an apparatus was made, at the expense of the city of Vienna, for the Albertin aqueduct adapted for a daily penetration of 28,000 Eimer. Price from 9 fl. upwards. Explanation for use and list of prices gratis.

615. FILLUNGER John, Civil Engineer. *Vienna, Landstrasse.*

Model of an arched bridge with cast iron ribs, patented in Austria and Great Britain.

616. HALL, I. R. Administration of Saltern at. *Tyrol.* Models of Hearths for burning coal-grit.

617. HALLER Christopher. *Gratz 217.* Roof-felt, and Artificial slate.

618. LANGER Joseph, I. R. Engineer. *Vienna, Landstrasse 472.* Agent: Dr. Ferd. Stamm, Vienna, Josephstadt 50.

Iron Bridge Model of 32 feet length.

The model is destined for making trials of the power for bearing loads. It is 32 feet long, built in $^1/_{20}$ of the natural size, therefore, representing a large bridge of 640 feet length in three panes, the middle of which having to occupy a distance of support of 360 feet joined is a pamphlet: „Iron Constructions for Bridges and Roofs" by the exhibitor, Vienna 1862, edited by Förster, and „Report on the results of Trials of Load-bearing made by an I. R. Commission at Vienna".

619. LÖSSEL Frederick, Knight, Engineer. *Linz, Bahnhof.* Honorary Medal Munich 1854.

Isopedical Ground Relievos.

This new way of representing the ground supplies one of the most important departments of engineering art. Such a relievo, grounded upon levelling surveys, divides all elevations into separate strata of determinate height, and affords hereby a reliable and abstract insight into the presented formation of surface, of proportions of heights and slopes, an easy and exact ascertainment of their horizontal and vertical projections or profiles, so that its thorough usefulness for professional scientific and technological purposes is self-evident. With the incomparable advantage of both an exact mathematical and plastically ocular objectivity, it may be laid down as a basis for all plans of building railways, roads, canals, river-regulations, water-powers, well-conduits, cultures, drainages, mining-works, fortifications, etc., and herein it is essentially different from the usual geoplastic or only typographic representations. Moreover its application is independent of time and in general always alike. The exhibited relievos show some of the most interesting parts of ground of the now finished railway from Munich to Salzburg. The photographic appendix represents the exhibitor's pressing and cutting apparatus constructed for his artistic works.

MILITARY HOSPITALS, I. R. Commission for ventilation, heating and furnishing of.

(See Nr. 630.)

620. NAPHOLTZ Matthew, Manufacturer. *Nagy-Becskerek, Comitate of Torontal, Hungary.*

Bricks.

621. PALESE. *Trieste.*
Stone Mason's Work.

622. PLEISCHL Adolphus M. *Vienna, Alservorstadt 109.*

Drawing of a new system for suspension bridges, spanning distance : 150 Vienna fathoms; computation of the strength of chains and of the weight.

623. RAMSAUER, I. M., Wear Master. *Ischl.*
Model for ramming in piles by means of water power.

624. SAULLICH Angelo, for the firm Kraft & Saulich *Kufstein, Tyrol.* Warehouse: Vienna, Wieden 791.
Building Ornaments.
(See Class 1.)

625. SCHAUMBURG Charles, I. R. Chief Engineer and Architect. *Gratz.*
Architectural Drawing.

626. SCHNIRCH Frederiek Chief Engineer, and **FILLUNGER** John, Civil Engineer, Inspector of Railway Buildings. *Vienna, Landstrasse 355.*
Model of an arched bridge with cast-iron ribs patented in Austria and Great Britain. Plan and photographic representations of the first suspension bridge over the Vienna Danube canal.

627. STATE MINISTRY (Section for Building). *Vienna.*
Charts of the Theiss and Danube. Explanation.
(See Class 29.)

628. STATE RAILWAY SOCIETY, I. R. Patented Austrian.
Plans of Bridges, Drawings and Photographs. Crossing-piece of grating-bars in nature.
(See Classes 1, 2, 4, 5, 8, and 13.)

629. STREFFLEUR Valentine, I. R. War Commissary General, and Editor of the Austrian Military Gazette. *Vienna.*
Hypsometrico-plastic Maps and Charts, and Town-plans, viz. City of Vienna; Vienna with its Suburbs; Railway Vienna-Trieste; Lower-Austria orographical; Lower-Austria geological; Lyons; Canal la Manche; Bottom of the Sea about Malta; Corfu.

630. VIENNA COMMISSION of the I. R. **MILITARY HOSPITAL** for ventilation, heating and furnishing of them. *Vienna.*
Plans and drawings of the ventilation and heating contrivances in the I. R. Garrison Hospital No. 1 at Vienna. Model of a new Ventilator by Dr. Heger.

The contrivance shown in the plans solves the problem of completely ventilating and heating an hospital. The defects under which similar arrangements executed before are labouring, are removed by this system. A well planted pulsion-system, appropriately connected with steam and vapour heating, are means, known in their nature, with which, as appearing from a close and comparative study of that plant, the purpose in question has been realized after a new principle. Heating and Ventilation of the localities are executed quite independent from each other. The Ventilator of Dr. Heger propels great volumes of air with comparatively little force of work, since every irregular current of air is avoided. The construction of this superior ventilator is quite new.

631. WINIWARTER George, Chevalier, Civil Engineer, and **PFANNKUCHE** Gustavus, Machinist. *Vienna, Althan 2.* Agent: Mr. William Andrews (Messrs. Royle & Co.), London, 32, King Street, Cheapside.
Fire and burglary-safe office completely furnished, as a little house by itself, of zinked iron-plate.

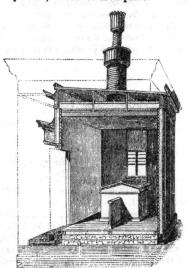

The accompanying drawing shows this business office in profile and in perspective view. Quite a new system is represented here, according to which dwelling houses are constructed of zinked iron-plate in such a way as to be easily transported and constructed again, at the same time affording as efficient protection against heat and cold as against attacks of burglars, like any strongly walled house. The walls of this house are double, outside zinked chamfered iron-plate, inside a wall of zinked oval plate-tubes closely approaching each other, enwrapped with straw ropes and loam. This inner wall is sized over with common lime mortar, whitewashed or papered. This office is destined for a merchant in the colonies and completely fitted up for that purpose- The wood works are by Mr. J. Spale, joiner in Vienna, Landstrasse 733, and the pendulum clock by Mr. J. Vorauer, clock-maker in Vienna, City 324, who was distinguished with medlas at the exhibitions of Vienna and Munich for his chronometers. Particular attention was paid to the proper ventilation in this office, so that this room in also to be inhabited with door and shutters closed. Particulars respecting the various productions of J. G. Winiwarter are contained in a special treatise illustrated with many wood-cuts, sold at the Exhibition and at the booksellers for 2 s. 6 d.

Class 11.

Military Engineering, Armour and Accoutrements, Ordnance and Small Arms.

Sub-Classe a) *Exhibitors of Clothing and Accoutrements.*
     ,,   ,,   b) *Exhibitors of Tents and Camp Equipage.*
     ,,   ,,   c) *Exhibitors of Arms, Ordnance, etc.*

The representatives of Military Engineering n Austria are the Imp. Corps of Engineers and Pioneers, together with the technical divisions of the Artillery and Navy, whose members receive their professional education in special establishments.

Military Accoutrement and Armour are entrusted to Commissions residing at Stockerau (Lower Austria), Gratz (Styria), Prague (Bohemia), Brünn (Moravia), Jaroslau (Galicia), Venice (Lombardo-Venetian Kingdom), Altofen (Hungary) and Carlsburg (Transylvania). The raw materials, purchased from private contractors, are worked out by tailors, shoe-makers, saddlers, taken out of the different corps of the Army and put at the disposal of these Commissions in a number adequate to the momentary wants. The produced arms, furniture and linen are distributed among the regiments and corps. If required, these Commissions may furnish, fit for immediate use, a monthly amount of 91.000 complete uniforms, full equipement for 86.000 men and 3.000 horses and above this 250.000 complete suits of linen. And it is only in cases of pressing necessity that the trade corporations of large towns are invited to take their share in military arming and equipping operations.

Gun-barrels, Bayonets and Sabres are manufactured by private industry and delivered to military authorities. The completion of military guns by providing them with fire-locks and stocks is exclusively in the hands of the military workmen of the Imp. Arsenal at Vienna, a new establishment and one of the grandest amongst technical establishments in general. It is here also that the whole brass Artillery of the Army and Navy is cast, bored and provided with carriages. The casting and fitting out of iron Artillery is entrusted to the Imp. Iron-works near Mariazell. Projectiles are cast, partly in these Iron-works, partly in private establishments of the same kind.

The production of Gun-powder is a State-monopoly. Private powder-mills, are provided with their raw materials by the proper military authorities and are bound to deliver, at a fixed price, their whole production to the same authorities. From these, other local authorities receive the quantity of powder intended for civil purposes.

The manufacture of sporting arms is entirely in the hands of private industry. A great number of Gun-makers, established at Vienna, Prague and in the other provincial Capitals provide the public with articles of superior quality and great elegance. Similar articles for common use are manufactured at extremely cheap prices by the numerous Gun-makers of Ferlach in Carinthia.

About 11.000 persons (besides those, employed for the same purpose in military and government establishments) are employed in the manufacture of military and civil arms. The value of the articles produced by them, may be very moderately esteemed at 2 millions florins for every year.

---

632. DIEZ Ernest. *St. Johann, near Villach Carinthia.*
Small Shot.
(See Class 2.)

633. FENDT Anthony. *Steyr, Upper-Austria.*
Rifled Gun and Cross Bow.

634. GASTEIGER Anthony von. *Wilten, near Innsbruck, Tyrol.*
Gasteiger's and Lang's, in Austria exclusive-patented arms, viz: *a)* Revolver of 54 shots with automatic loading on mere action of the forefinger; *b)* Pistol, with which 200 shots can be discharged in one hour; *c)* Mountain Cannon, model, for speedy loading behind; *d)* the bullet mould, belonging to them, which throw out the ball finished, when, after the cast, a slight pressure of the hand is made.
(See Classes 7 and 27.)

635. (Omitted.)

636. JUNG Francis, Armourer, Possessor of the Golden M. for Art. *Vienna, City 1049.*
Armour Steelworks inlaid with gold and silver and engraved.

637. JURMANN Charles. Patented Manufactory of Arms. *Neunkirchen and Mürzzuschlag, Lower-Austria.*
Arms and Components of arms.

638. KIRNER Joseph, Gun-maker. *Pesth, Servitenplatz.*
Guns.

639. LENGYEL Stephan, Mechanical Locksmith. *Pesth, Spitalgasse.*
Newly invented guns and arms.

640. MASCHEK Wenceslas. *Vienna, St. Ulrich 75.*
Guns; Pistols in cases.

641. MAURER, Brothers. *Vienna, City 1146.*
Arms; Sportman's Equipage.
(See Class 30.)

642. OHLIGS B. W., Armourer to the I. R. Court. *Vienna.*
Fire-arms and Swords for luxury, in steel, silver and gold.

643. STRIBERNY Anthony. *Vienna, City 1134.*
Arms.

644. UMFAHRER Francis. *Klagenfurt, Carinthia.*
Guns and Pistols.

645. WAENZEL Francis and son. *Vienna, Margarethen 176.*
Guns.
(See Class 6.)

Class 12.

Ship-building, Rigging and Tackle.

Sub-Classe a)   *Exhibitors of Ship Building for purposes of War and Commerce.*
    "   "   *b)*   *Exhibitors of Boat and Barge Bulding, and Vessels for amusement etc.*
    "   "   *c)*   *Exhibitors of Ships, Tackle, and Rigging.*

There are 37 dock-yards in active operation along the Austrian coast of the Adriatic Sea; their productions not only suffice to supply the wants annually arising in the Austrian trading-navigation by ship-wreck or breaking up, but are also sold to the neighbouring states. The military operations of 1859 had materially impaired and delayed the construction of larger vessels, and even in 1860 the consequences were still apparent.

From the following comparison it may be seen that in the Austrian dock-yards the productions amounted

in 1858 to 471 ships of 14,618 tonnage
   " 1859 " 542   "   " 6.143   "
   " 1860 " 601   "   " 9.402   "

Among the productions of the last-named year, there are 15 vessels for the high seas, of 4.733 tonnage; the remainder were large coasting-vessels (partly fit for navigation in the Adriatic Sea, partly for courses to the Black Sea and to the straits of Gibraltar) and small coasting vessels destined only for intercommunication between the Austrian ports.

Trade in ships with foreign countries shows a balance in favour of Austria; there were

	bought from abroad		sold to foreign parts	
	Ships	of Tonnage	Ships	of Tonnage
in 1858	13	of 2.331	28	of 3.423
„ 1859	9	„ 1.805	42	„ 9.169
„ 1860	9	„ 1.604	19	„ 5.151

The ship-timber made use of, is, without exception, of home growth, and only inland rigging and sail-cloth is used for fitting out.

Sea-going Steam-vessels are likewise of late built in the dock-yards of the Austrian Lloyds and fitted out with home-made machines.

For the construction of men-of-war, the arsenals at Pola and Venice, and the ship-building establishment at Muggia, are actively employed.

Besides the construction of vessels destined for the sea, the production of ships for the navigable rivers of the Austrian empire is a special branch of industry, so much the more as the vessels, going down the Danube, Vistula and Elbe and their tributaries, are, for the most part, not taken back against the stream, but either sold abroad for further service as ships, or broken up and converted into money by being sold as timber.

An exception to the last-mentioned particulars are the iron trading vessels, of late more extensively used, and built in special larger docks for the navigation on the Danube and Elbe, component parts of steam-boats being likewise made in these dock-yards. The arsenal of the Austrian Lloyds at Trieste is for naval-architecture, what the dock-yard of the Danube Steam-Navigation Company at Buda is for the production of iron ships and parts of steam-boats for river navigation, both these establishments being the most important of their kind in the land.

There is an average number of 15.000 workmen employed in the building and rigging of sea and river ships, the value of which productions being estimated at 15 millions florins at the least.

---

646. DANUBE STEAM-NAVIGATION SOCIETY, I. R. Patent. *Vienna* (Manufactory at *Old Buda, Hungary*). Ship Models. (See No. 15.)

647. JÄGER Francis John, Rope-maker. *Prague 349-1.* Rope Manufactures.

648. REICH S. & *Comp.* Manufactory of Rope Wares. *Zaschau, near Wallach. Meseritsch, Moravia.* Ware-house: Brünn. Agents: Vienna, Prague, Dresden, Leipzig, Berlin, and Hamburg.

Rope-maker's Works, Hoses, Pack-thread, Running-Mats, Chain-pump Girths, Transmission Girths, Fire Buckets.

Class 13.

Philosophical Instruments, and Processes depending upon their use.

The production of Scientific Instruments of every description has established its chief seat at Vienna, but there are yet some Mechanicians and Opticians residing in the principal provincial towns, especially at Venice, Pesth, Lemberg, Prague and Gratz.

Instruments for Measuring are made in high perfection in a large establishment at Vienna and by other Mechanicians. Balances and Cases of mathematical instruments („Reisszeuge") are in the hands of special trade-corporations.

Chemical Apparatus are either directly purchased from Glass or Porcelain Manufactories, or fitted up in the establishments of Mechanicians, and so are Barometers, Thermometers, Arecometers and Saccharometers.

The production of Physical Apparatus has been lately extensively developed in consequence of the establishment of a great number of Realistic Schools („Realschulen") each of them endowed with a collection of philosophical instruments.

For some time these articles were in great demand, so that a large quantity of them were to be imported from foreign parts especially from Berlin.

The Superiority of Optical Instruments, manufactured by many Vienna Opticians, is sufficiently confirmed by the unanimous verdict of men of Science in all parts of the world; there is scarcely a country, where Observatories are not in possession of a Telescope of Vienna make. Second-rate long-sights and opera-glasses generally of French make, are the only objects of import worth noticing. Spectacle-glasses manufactured at Gratz and Spectacles are exported ot the countries nearest to the eastern frontiers of the Austrian Empire.

The production of the articles here in question gives employment to more than 2.000 persons and reaches a yearly value of about 3 millions florins.

The average amount of import with 400.000 florins is more than counterbalanced by an export of 500.000 florins.

---

649. (Omitted.)

650. HAAS C. & *Comp.*, manufactory of Electrometallurgical objects and precipitates. *Gratz.* Storehouse: Vienna 1062.

Objects plated with Tin, Zinc and Copper and Galvanoplastic objects.

651. (Omitted.)

652. (Omitted.)

E.

653. KAVALIER Joseph, I. R. privil. Glass-manufactory. *Sazawa, near Kohljanowitz, Bohemia.*

Honourable mention: Munich 1854; M. II. Cl. Paris 1855.

Implements and apparatus for Chemical and Pharmaceutic operations, made of Potash-glass free from lead and inaccessible to the effect of Acids. Combustion-tubes with above 80 p. c. of Silicia, for the analysis of Organic substances.

The manufactory herein question has to satisfy demands from Europe and America, at a rate increasing from year to year. The objects, produced, have no reason to fear competition with objects of French manufacture, as, in the Exhibitor's establishment materials are treated with the utmost care, and the glass articles, issued by it, being commendable for their purity, lightness and especially for their moderate prices. On occasion of the Exhibition in 1855, the Superintendence of the manufactures of Sèvres and Javelle, in a letter to the Exhibitor, transmitted to him by the I. R. Consulate-general, required his authorization for taking copies of the glass objects, then brought to exhibition. This fact may prove a sufficient acknowledgment of the superior quality of his articles.

654. KRAFT E. & son, Mathematical and Physical Instruments. *Vienna and Trieste.* Store-house: Vienna Bürgerspital 1043. Gold M. Vienna 1845. Bronze M. Munich 1854. Bronze M. Paris 1855.

3 Levelling-instruments, 1 Geodaetic apparatus; 2 Instruments for Mining-survey; 1 apparatus for calculation .of Area, 1 Pantograph, 3 Miner's compasses, 1 Sextant, 1 apparatus for measuring the celerity of Streams; Mathematical Instrument-cases and Compasses of various kinds, Inch-rules, 4 Ophthalmoscopes and other Instruments for Engineers, Mining-Surveyors and Lithographers.

This establishment has 36 years of existence and provides with the diversified objects, produced there, a great number of Military Institutes and authorities, the I. R. Navy, the I. R. Board of Survey (with exclusion of any other, the I. R. Mining-department, a number of Railroad and other Building establishments, private Engineers, Mining-surveyors and medical men through the whole Austrian Empire. It exports also articles to Russia, Italy, France, and Turkey. The two large Levelling-instruments of the Exhibitor's own construction have splendidly gone through practical trial and 480 of the same kind have already been construed. The smaller Instrument, being easily manageable, may be preferred to the Dioptra-instruments formerly in use. Above 500 numbers in several other constructions. The Geodaetic table is construed according to E. Kraft's improved principles, which, by nearly 1830 of these instruments, constantly in use during a space of 36 years, have proved to be highly commendable for the solid, durable and appropriate connexion of all the component parts.

The Telescope-rule belonging to it, which, since about 20 years, has superseded the use of the simple Dioptra in the operations of the I. R. Board of Survey, has been acknowledged by the demands for 1,480 of these instruments. Provided with a suspensive Lead and a graduated Sector, it serves for levelling and hypsometrical operations. Among the instruments of Mining survey, those with fixed suspensive apparatus are most in use, and 410 of them, in two different sizes, have been produced. One hundred and forty-two of the Pantographs, simplified and improved in construction by E. Kraft (as the one exhibited here) have been constructed. A hundred of the Mariner's Compasses, Sextants and specular Circles, as exhibited here, have been delivered to the I. R. Navy. Measuring-rules, exactly sub-divided down to $^1/_{1000}$ of an inch, and Inch-rules of any form and for any metrical system known, are standard articles of fabrication. As for Mathematical Instrument-cases, E. Kraft and Son hope not to remain behind the best objects of this kind, issued by other establishments. Among the Ophthalmic Specula, so much in esteem since a few years, those of Professor J. de Jäger, whose technical execution had been exclusively entrusted to the Exhibitor, have most of all met with acknowledgment on account of their superiority and easy practical use; 6.000 of them have been sold into all countries. Lithographic Presses of the Exhibitor's own construction and other Lithographic instruments are much in demand on account of the work they can do. Balances and weights for the I. R. Mint-offices, as for other physical and chemical operations, Balances to be charged up to 6 Cwt, Telegraphic apparatus, a variety of tool-engines and tools of delicate construction have procured to the establishment an abundance of employement and acknowledgment. It employs 70 first-rate workmen in executing Instruments of precision of any description whatever.

655. LEITER Joseph, manufacturer of Surgical and Physical Instruments and Apparatus; Purveyor to the University of Vienna. *Vienna, Alservorstadt 150.*

Surgical and physical Instruments.
(See Class 17.)

656. LENOIR G. A. *Vienna, City 1019.*
Scientific Instruments.

657. LEOPOLDER John, Mechanician. *Vienna.*
Pneumatic pump. Electrical bell-work for Railroad-keepers. Electrical bell-work for Railroad-termini. Morsian Writing-apparatus. Galvanic battery. Three Gas-burners for chemical use.

This establishment produces physical, mathematical and chemical Instruments of every description. Its speciality are Electrical Telegraphs and Bell-works for Railroads, and, having got credit for them with all the leading authorities of Austrian Railroads, has provided them, since a number of years, with any apparatus, requisites and materials in use for the construction of Telegraphs. An active connexion is constantly kept up with foreign parts, especially with Moldavia and Wallachia. Scientific and practical progress is never left out of sight. The Pneumatic pump, exhibited here, has two glass cylinders. The exclusion of air is effected by screws and leather instead of cocks. The two emboli are composed of leather, equally pressed against the inner surface of the glass cylinders by means of springs. The valves are adapted for being taken out of the cylinder from above, whenever they want cleansing, so, as to avoid a frequent unscrewing of the engine. Price 140 Austr. florins in Bank-notes. The Electrical Bell-work for Railroad-keepers has 4 essential component parts: an Electro-magnet, an apparatus for interrupting communication, a moving-apparatus and a roof with a bell. Prices, with weight, rolls of pullies and hooks for fixing on a wall: 105 Austrian florins Bank-notes a-piece. Bell-work for Railroad-termini; price: 96 Austr. florins in Bank-notes. But lately, the superintendence of the I. R. privil. Southern State-Railroad Association for Lombardy, Venetia and Central Italy had circulated among all the Mechanicians of Vienna, an invitation to send in specimens of such Bell-works. On this occasion, the apparatus exhibited here obtained the preference before all others, on account of its simple construction and its unerring mode of action. The principle, on which Morse's Writing-apparatus is construed, is the winding-up of the wheels by the movement of the writing lever. Price: 130 Austr. flor. in Bank-notes. The Galvanic battery, composed of 12 Smee's elements is easily decomposed into its parts. Price: 42 Austr. flor. in Bank-notes. The Gas-burners are construed according to the principle suggested by Dr. Böhm, I. R. Regiment-surgeon at Vienna; a description of them may be seen in the „Proceedings" of the Imp. Academy of Sciences, Mathematico-physical class, Vol. XIX, p. 374. The Telegraphic apparatus, exhibited by the I. R. priv. State-Railroad Association at Vienna (Class 13) and the two Air-Manometers, exhibited by the I. R. Military Infirmary at Vienna (Class 10), have also been fabricated in the Exhibitor's establishments. Further information to be obtained at the Exhibitor's Agent, Ferdinand Teirich, superintendent of Telegraphic service to the I. R. privil. Austrian State-Railroad Association at Vienna.

658. MARCUS Siegfried. Engineer, Mechanician and Patentee. *Vienna, Neubau 313.*

Telegraphs and other Electrical and Physical apparatus.

659. MARITIME BOARD CENTRAL I. R. *Trieste.*
Nautical Instruments, invented by Prof. Zescevich, at Trieste.

a) Triëdrometer for resolution of Spherical Triangles. *b)* Nautical Rule for taking, without a Windrose, courses and positions from a Nautical map. *c)* Dromoscope, for transferring Compass-courses into real courses and for the inverse operation. All these Instruments are invented by Prof. John Zescevich, at Trieste. Moveable Wind-roses for solving current problems by means of Nautical maps, on which rumbs or points of the Compass are not made evident, invented by Prof. Rob. Zamarra, at Fiume.

660. NEUHÖFER & FRIEDBERGER, fabricator of Optical Instruments. *Vienna, City 1149.*

Optical articles.

661. NUSS Anthony, Mechanician. *Pesth, Blaue Hahn-Gasse 12*, and Gallitzer James, Merchant, *Pesth, Elisabeth-place 4*.

Improved Morse's Telegraphic apparatus.

662. PERFLER Joseph, Mechanician. *Vienna, Landstrasse 311*.

Levelling-instruments, Pantographs, Planometers, Instruments for Mining-survey, Calculating apparatus, Rule for drawing, Measuring rules, and Mathematical Instrument-cases.

663. SCHABLASS Joseph, Mechanician. *Vienna, St. Ulrich 136*.

Mathematical Instruments.

664. SEYSS Lewis, Mechanician and Patentee, *Atzgersdorf near Vienna*.

Weight - Manometer, Double - tubed Manometer, two Spring - balances, two Counters for heavings, Indicator for Locomotives, Apparatus for signalling by bells.

The Weight-Manometer, equal in precision to any open Mercury-Manometer, is of the Exhibitor's original invention (price: 25 Austrian florins). The double-tubed Manometer (price: 18—25 florins, to be moderated in proportion to the extent of demands), may be better depended on than any other Spring-manometer for its simple construction, in which any friction and soldering is avoided and boring has been practiced for keeping the condensation-water within the tubes. The patent, granted to the Apparatus for bell-signals, relates to the heaving of hammers and to the way, in which the interrupting apparatus is secured against commotion.

665. SIEMENS & HALSKE, *Vienna, Erdberg 46*.

Apparatus for setting fire to charged Mines, according to L. C. Baron Ebner's system. Telegraphic apparatus of Austrian construction.

666. STATE RAILROAD SOCIETY, I. R. privil. Austrian.

Model of the Intercalation-system of Telegraphic stations, without using Batteries at the intermediary stations, with drawings and description.

(See Classes 1, 2, 4, 5, 8, and 10.)

667. TELEGRAPHS, I. R. Direction. *Vienna*.

Morse's apparatus with Electromotor. Telegraphic Map of the Austrian Empire.

668. VOIGTLÄNDER & son, *Vienna, City 949*. Optical articles.

(See Class 14.)

669. WALDSTEIN J., Optician. *Vienna, City 5*.

Raw Flint- and Crown-glass in disks up to 16 inches diameter, Optical tubes, Theatrical tubes, Field-tubes, Spectacles and Lorgnettes with lenses of Bohemian Rock-Crystal, Spectacle: frames of various matters; in sum about 400 numbers.

First and only manufactory of Crown- and Flint-glass in Austria at Ottakring next Vienna. I. R. priv. manufactory of Optical Instruments. Oculistic Spectacle-Institute at Vienna. Great Gold Medal of the Industrial Society of Lower Austria. Produces Optical Instruments of every description, especially Pocket-field-tubes, patented in Austria and France, serving at pleasure as double tubes or as simple field-tubes. Spectacle-frames of gold, silver, tortoise-shell, buffalo-horn, and steel. Thread-counters, as they are in demand for the use of the I. R. Custom-offices.

Class 14.

Photographic Apparatus and Photographs.

Besides a great number of „Dilettanti" Vienna numbers no less than 400 Photographers ex professo, paying taxes as such, with a nearly equal number of auxiliaries. This branch of Industry is amongst those of earliest origin in Austria, and its rapid developement dates only from the last three years. It concentrates the greatest portion of its activity in the production of photographic visit-cards; nevertheless eminent copies of landscape-scenery and of celebrated pictures have been produced.

In the provincial towns of Austria, Photographers are increasing in number. Vienna however is still the centre of Austrian Photography, represented there by a Photographic Society and a special monthly Periodical („Photographisches Journal").

Photographic Apparatus, are mostly produced by Vienna Opticians, but the papers and chemical preparations are generally of foreign import. Latterly some Vienna chemical manufacturers have successfully attempted to provide Photographers with materials prepared in their own establishments.

The number of persons employed in photographic Industry may be esteemed at nearly 2,000; the value of the objects produced by them may amount to more than three millions florins every year.

---

670. ANGERER Lewis, Photographer. *Vienna, Wieden, Feldgasse 1061*.

Photographs in frames.

671. DIETZLER Charles, Mechanician and Optician. *Vienna, Wieden 102*.

Photographic apparatus and Astronomical Double-objects, Camera with case, Statives. Photographic specimens and pictures of various sorts.

672. GIESSENDORF Charles de, Lithographer. *Vienna, Wieden 508*.

Paniconographs and Photographs.

(See Class 28.)

673. LEMANN Charles. *Vienna, Gumpendorf 24*.

Photographic reproductions of archaeological and artistical objects.

674. MAYER George, Photographer. *Pesth*.

Photographs of various kind.

675. MELINGO Achilles. *Vienna, Prater-Strasse 512*.

Photographs.

676. OBERHAUSEN Edward de, Artistical and Industrial etablishment. *Vienna, City 613*.

Works of Photographic Art.

677. PERINI Antonio, Photographer. *Venice*.

Facsimile of the celebrated „Breviario Grimani preserved in the Library of St. Marco.

(See Class 29.)

678. RUPP William, Photographer. *Prague*.

Photographs.

8*

679. VOIGTLÄNDER & Son, *Vienna, City 949.*
Photographic apparatus with specimens of Photographs executed with it.
(See Class 13.)

680. WIDTER Anthony, *Vienna, Landstrasse 136.*
Monuments of History and Architecture in Austria reproduced by Photography. 2 parts.

The first part comprises Armours and Weapons, especially of Austrian Princes and Emperors from the XIV. to the XVIII. Century.
The second part contains Architectural monuments from various parts of the Austrian Empire, from the Roman period down to the XVII. Century. (These objects are not intended for sale.)

681. WÜNSCH Gustavus. *Vienna, City 684.*
Photographs in frames.

Class 15.

Horological Instruments.

The manufacture of Watches is visibly declining in Austria, through the combined and simultaneous effect of both the enormous development of watch manufacture in Switzerland and the abolition of the Prohibition Custom-system in Austria.

The Watch-makers in this Empire are chiefly employed in adjusting the watches imported from Switzerland and in repairs.

On the other hand the manufacture of Clocks, especially at Vienna, Prague and Gratz, is advancing. The cheapness of second and third rate Clocks, the component parts of which are imported from Switzerland, and then adjusted and put together by Austrian workmen, have nearly put an end to the once extensive trade in wooden („Schwarzwald") clocks (Dutch clocks).

The Chronometers of several Vienna masters deserve particular mention.

Clocks for Steeples etc. are manufactured at Vienna, Prague etc. by special Clock-makers, and are exclusively of Austrian materials and workmanship.

Watches and Clocks going through the Austrian Customs under the general denomination of fancy goods („kurze Waare"), the amount and value of the importation is not to be expressed with any degree of precision. The number of Watch and Clock makers is about 3,000, with as many assistant workmen; the value of the articles yearly produced by them, may amount to nearly 750,000 florins.

---

682. EFFENBERGER F. *Vienna, City 1148.* Watches.

683. FRÖHLICH Fr. C., Watch-maker. *Hartberg in Stiria.*
Oil for watches of superior quality.

684. KRALIK Samuel, Watch-maker. *Pesth, Schlangen Gasse 5, Hungary.*
Honourable mention: Vienna 1845. M. II. Cl. Medal. Munich 1854.
I. Regulator-clock-work, Echappement in precious stones: Price. 40 L. St. II. Clock-work. Price 25 L. St.

The Exhibitor, who has, in 1840, taken in his hands this establishment, founded in 1822, not contented with procuring watches from the best-famed manufacturers, adjusting them and bringing them thus into trade, has it made a point of honour to raise his profession to the rank of an art, by paying special attention to Regulators of highest precision, constructed according to the newest improvements and to his own principles. A special establishment, provided with the best tools and mechanical apparatus, makes it possible to him, to execute with the utmost precision any demands for Astronomical watches and Regulators, and in general, orders of any nature. Besides those above-quoted, the Exhibitor obtained the following distinctions: Honourable mention Pesth 1841, Bronze M. Pesth 1843; Great Silver M. Pesth 1816. In consequence of the objects brought by him to the London Exhibition 1851, the Exhibitor, in 1852, has been elected a member of the „Académie nationale. etc." of Paris.

685. KRESPACH Anthony, *Vienna, Schottenfeld 293.*
Watch-cases, partly provided with clock-wock, Pendula.

686. LUZ Ignace. *Vienna, Alser-Vorstadt 276.*
Small Control-watch.

687. MARENZELLER Ignace, Watch-maker. *Vienna. 641.* Great Medal. Munich 1854.
Pendulum-watches of his own construction.

688. MAYER. A. W., Watch-maker. *Vienna, City 647.*
Pendulum-watch for being placed on a table.

689. MÖSSLINGER Francis, Manufacturer of Dials for watches. *Vienna, Laimgrube 195.*
Dials for watches and Inscription-plates of various kind.

690. OBERLEITNER J., Possessor Assmann, teacher for Drawing in the establishement for eduction of deaf and dumb persons. *Hall, Tyrol.*
Watches.

691. SCHÖNBERGER Manufactory of Regulators, *Vienna, City 648.*
Regulators going 8 days, 1, 3 and 6 months, up to a whole year; Regulator going 8 days or 1 month with striking-work for hours and half-hours; Regulators repeating hours and quarters of hour.

These Regulators are mounted in elegant wooden boxes diversified in form and colour.

The establishment manufactures also precise Pendulum-clocks going 8 days; price: 15—26 fl.; ditto with striking mechanism („Schlagwerk"): price: 32—63 florins) and at higher prices, when intended for adorning elegant appartments. Price-lists, if applied for, gratis.

Class 16.

Musical Instruments.

The simplest amongst these Instruments — Jewsharps and Harmonicas at remarkably cheap prices — are exported from Austria to every corner of the Globe. The Jewsharps are produced by the Iron workmen of Stadt Steyr; the seat also of the Harmonica-manufacture. In 1860 no less than 6 millions of the first and nearly 100,000 of the latter instruments were turned out.

The manufacture of Stringed and Wind Instruments is practised on an extensive scale at Vienna, Prague and Königgratz (Bohemia). The production of such articles at low prices is an important industrial resource, maintaining the population in the barren regions of the Western Bohemian Erzgebirg. The wind-instruments for military bands, produced in Austria, have come into great renown in foreign parts and are frequently exported to supply the wants of other Armies.

Pianofortes (generally horizontal ones) have become an important branch of Vienna industry. The simplicity of their mechanism (known under the special denomination of Vienna mechanism), the superior quality and the low prices of the inland materials used for their manufacture (resounding-boards from the Bohemian forests, nut-wood and Hungarian ash-wood for inlaid work, and steel strings of superior quality), make it possible to bring them into commerce at prices, which — with equal perfection of workmanship and tune — would be impossible under different circumstances. Vienna produces every year an average amount of 2,600 Pianofortes, besides those made by some Pianoforte-makers residing at Salzburg, Pest, Prague etc. In 1860, 1000 of these instruments have been exported.

The production of musical Instruments in Austria gives employment to about 10,000 persons; its average annual value falls but little short of 5 millions florins.

---

692. AST John, Pianoforte-maker. *Vienna, Wieden 412.*
Two Pianofortes.

693. BAUER Edward John. *Prague 780-2.*
Musical Instruments.

694. BEREGSZASZY Lewis, Pianoforte Manufactory. *Pesth.* Bronze Medal: Vienna 1845. Silver Medal: Pesth 1846. Laudatory Mention: Munich 1854. M. I. Cl. Paris 1855.
Two Pianofortes of the finest Hungarian speckled oak and nut wood, with sounding-board constructed after his own system.

The exhibitor produces from 80 to 90 instruments a year and exports them to Germany, Russia and Italy.

695. BITTNER David. Violin-maker. *Vienna, City 1038.*
Bow Instruments.

696. BLÜMEL Francis, Pianoforte-maker. *Vienna, Wieden 300.*
Pianoforte.

697. BOCK Francis, Metal Wind Instrument Maker. *Neu-Lerchenfeld near Vienna 171.* M. II. Cl. Paris 1855.
Machine Instruments of all kinds and newest construction.

698. BOHLAND Gustavus. Manufactory of Musical Instruments. *Graslitz, Bohemia.*
Brass Wind Instruments, partially inventions of the exhibitor's.

699. BÖSENDORFER Lewis, Pianoforte Manufacturer to the Imp. Court at the Brazils. *Vienna, Neu Wien 377.* Gold Medals: Vienna 1839 and 1845.
Pianoforte, Vienna and Patent Mechanism of the exhibitor's invention.

This manufactory was established in 1828, is at present in point of numerical production the most extensive, incontestably holding the first rank in Austria. There are from 60 to 70 workmen employed in the manufactory and as many out of it who furnish the metal parts.

700. BRANDL Charles, Violin-maker. *Pesth.*
Two Violins after forms of Straduarius and Guarnerius.

701. CERVENY V. F. Patented Manufactory of Metal Musical Instruments. *Königgrätz, Bohemia.* Bronze Medal and Praise: New-York 1853. Great Medal: Munich 1854. M. I. Cl. Paris 1855.
Metal Musical Instruments.

Octavin, Cornets, Winghorns of all kinds, Alto and Tenor Horns, Euphonium, Trumpets, Buglehorns, Cornon, Phonicon, Baroxytom, Bombardons, Doublebass with newly improved cylinder machine, Tritonicon, Bassoon and Drums. Of the exhibitor's invention are: 1. Cornons, instead of buglehorns, patented 1844; 2. doublebass, in the year 1845; 3. tone exchanging machine, instead of eking pieces, patented 1846; 4. phonicon, baryton and solo-instrument for concert and orchestra 1848; 5. baroxyton, instead of the large bombardons, with small mouthpieces as baryton solo-instrument, patented 1853; 6. obligato-alto-horns for solos, 1859. Improvements introduced by him are: 1. Buglehorns received easy address to a sure and powerful tone 1853; 2. the tronbone had its tone strengthened with preservation of its character 1854; 3. tritonicons, instead of double bassoon, 20″ shorter, more convenient and practical for every music, at the same time easy management for the right hand 1856; 4. cylinder machine, pat. 1861, which improvement joins simplicity with elegance and durability, and, what is still mere to the purpose, by having the valve fitted tight into the hollow cylinder, a perpetually easy address is effected for every metal instrument, besides easy movement of the valve. According to the laws of physics this result must take place, if the cone of the valve gravitates in a more narrow space and the propelling power is applied to its heavier end. The reverse was the case with these machines before, because the cone was directed from below to upward, in consequence of which it was sinking into wider dimensions when worn off, and on account of the airtight shutting was to lose the more as the side parts of the valve and the hollow cylinder were rubbed off; moreover the valve lost considerably in agility by applying the propelling power to its lighter end.

702. CRAMER Geoffrey, Pianoforte-maker. *Vienna, Landstrasse 37.*
Pianoforte.

703. EBERLE Charles. *Verona.*
Musical Instrument Strings.

704. EHRBAR Fred., Pianoforte-maker to the I. R. Court. *Vienna, Wieden 753.*
Pianoforte.

705. HOLZSCHUCH Matthew, Mechanical Locksmith, Patentee. *Vienna, Laimgrube 100.* Honourable Mention: Vienna 1845.

Pedal treddles.

706. HÖRBIGER Aloysius, Organ Builder. *Atzgersdorf near Vienna.*

Vox humana (musical instrument).

707. KANDLER Daniel. *Vienna, Strozzengrund 20.*

Metal Wind Instruments.

708. KIENDL Anthony, Manufactory of Instruments and Strings. *Vienna, Josephstadt 18.* Medal: Munich 1854. M. II. Cl. Paris 1855.

Two Citherns, with usual mechanism. A travelling and a song cithern (elegy-cithern).

This manufactory is now restricted to the production of pinching and bowing citherns, and the strings etc. belonging to it. The exhibitor's newest invention deserves particular mention, viz. his elegy cithern, having become very popular, on account of its strong durable tone and harmonious sound, and which was sold to many customers. A very advantageous reputation is also attached to the exhibitor's cithern strings.

709. KLUIBENSCHÄDL Joseph, Instrument-maker. *Innsbruck 114.*

Violins.

710. KOHLERT Vincent. *Graslitz 309, Bohemia.*

Wind Instruments of wood.

711. LAUSMANN J. W. *Linz.*

Wind Instruments.

712. LEMBÖCK Gabriel, Instrument-maker. *Vienna, City 838.*

Bow Instruments.

713. MARTIN Joseph, Machinist for Brass Wind Instruments. *Graslitz 198, Bohemia.*

Machines for Doublebass (Helicon), Euphonion, Winghorn.

714. MEINL Daniel, Patented Manufactory of Instruments. *Vienna, Landstrasse 94.*

Metal Wind Instruments.

715. MILLER Martin's son. *Vienna, Gumpendorf 351.*

Pianoforte Strings.

(See Class 32.)

716. PATZELT John Ferd., Violin-maker. *Vienna, City 31.*

Violins, Viol, Violoncello, Cithern with case.

717. PICK Ralph's Widow, Manufactory of Musical Metal-tongue Instruments (accordions). *Vienna, Alservorstadt 27.*

Musical Instruments.

Particular attention is requested to the instruments of one, two, and three rows, tuned with settled accords, because to each of these instruments is annexed a great number of self-teaching schools and music books, improve-

ment in playing on these instruments being attained in a very short time, also by those that are not trained in music. The chromatic tuned three rowed and pianoforte hand instruments are distinguished above all others, especially the French instruments (flutins or harmonie flûtes), by their having a complete bass and accord accompaniment, whereas on the French only treble tones can be played.

718. POHL Emanuel. *Kreibitz, Bohemia.*

Glass Harmonica.

719. POTTJE John, Pianoforte-maker. *Vienna, Wieden 88.*

Pianoforte.

720. RIEDEL John, Musical Wind Instrument-Maker. *Pesth.*

Diverse Wind Instruments.

721. RÖDER Charles. *Vienna, Wieden 1.*

Case with Bassoon Tubes.

722. ROTT Aug. Henry, Manufactory of Musical Instruments. *Prague 799-2.*

Brass Wind Instruments.

723. SCHAMAL Wenceslas, Musical Instrument-Maker. *Prague 1025-2.*

Brass Wind Instruments.

724. SCHNEIDER Joseph, Pianoforte-maker. *Vienna, Wieden 447.*

Pianoforte.

725. SRPEK Joseph. *Vienna, Mariahilf 2.*

Wind Instruments.

726. STEHLE John, *Vienna, Landstrasse 379.*

Various Wind Instruments, and Photographs of such.

727. STÖHR Joseph. *Prague 391-1.*

Brass Wind Instruments.

728. STOWASSER Ignatius, Musical Instrument-Maker. *Vienna, Josephstadt 66.*

Musical Instruments.

729. STRATZINGER Ralph, Instrument-maker. *Linz 129.*

Cithern with table.

730. STREICHER J. B. & son. Pianoforte-maker to the I. R. Court. *Vienna, Landstrasse 375.*

Pianofortes; String Balance for ascertaining the power of bearing of strings, as the power of a complete pianoforte stringing is called.

731. THIE William, Harmonica-maker. *Vienna, Mariahilf 95.*

Harmonicas.

732. UHLMANN L. Manufacturer of Instruments. *Vienna, Mariahilf 25.*

Wind Instruments.

733. ZIEGLER J. *Vienna, Leopoldstadt 693.*

Wind Instruments.

Class 17.

Surgical Instruments and Appliances.

As is the case with scientific Instruments in general, the manufacture of Instruments and apparatus for Surgical purposes is nearly exclusively concentrated in Vienna and Prague, more or less powerful impulse being given by the professional heads of establishments of Surgical Clinics. From these two places, the Instruments either invented by foreign professional celebrities, or imported from foreign parts and improved in Austria, find their way over the whole extent of the Empire and even, to a great amount, into foreign parts.

Surgical Instrument-cases and obstetrician Instruments are the chief articles of manufacture. Among this last category, Professor Braun's Decapitation-hook scalari-form Repository and Colperintor, then Professor Chiari's Excerebration Pincette, may be quoted as specific Austrian inventions.

Orthopædical Apparatus and artificial Limbs are manufactured at Vienna. Craoutchouc objects for surgical purposes are, for the most part, imported. This branch of industry, employs about 500 workmen, producing, in yearly average, of $^1/_2$ million of florins.

---

734. CZERMAK John, Dr. Med. and Professor. *Prague*. Honourable Mention, and Monthyon Prize of the Academy of Sciences at Paris 1861.

Three Tables with plan drawings in illustration of his method of examining the larynx (laryngoscopy) and the nostrils (rhinoscopy) of patients and of one's self (autoscopy). Laryngoscopic mirrors in all sizes. Concave Illustration-mirrors with mouth-handle, forehead-tie, palate-spring, for rhinoscopy; Injection-tubes, caustics holder, for local treatment with the aid of the mirror. Apparatus for self-observation and demonstration; case for practical physicians.

Photographs of larynx and nostril after life, the former taken stereoscopical. Printed works: „Physiological investigations with Garcias Laryngoscopic mirrors." Three Tables, Report of session of the Imp. Academy of Sciences in Vienna Vol. XXIX, 1858. „The Laringoscopic Mirror and its value for physiology and medicine", Leipzig, Engelmann 1860. The same, French Edition, J. B. Baillière, Paris 1860; English Edition, The New Sydenham Society, London 1861.

735. DREHER Ignatius, Surgical Instrument-maker. *Pesth*.
Surgical Instruments.

736. FISCHER Peter, Surgical Instrument-maker. *Pesth, Franciskanerplatz*.
Surgical Instruments.

737. HEBRA Ferd. Dr. and I. R. Professor to the General Hospital at *Vienna*.
Apparatus for the use of continuous baths for medical purposes.

738. KRANZELBAUER John. *Pesth*.
Patented Surgical Band.

739. LEITER Joseph, Manufacturer of surgical and physical instruments and apparatus, Purveyor to the University of Vienna. *Vienna, Alservorstadt 150*.
Surgical and Physical Instruments.
(See Class 13.)

740. TEICHMANN Lewis, Dr. I. R. Professor. *Cracow*.
39 anatomical preparations for examining the osseous nasal apparatus of various mammalia; 10 microscopic injection preparations for demonstrating the lymphiduct and chyliduct capillaries of various organs; injection preparations exhibiting the ductus thoracicus of the following species of mammalia: Talpa, mus ratus, cavia, and erinaceus europ. These preparations are from the collection of the exhibitor and performed by himself.

741. TÜRCK Lewis, Dr. I. R. Head Physician to the General Hospital of *Vienna*.
Laryngoscopic and rhinoscopic apparatus.

6 laryngoscopic mirrors in different sizes and forms. 2 illustration apparatus for laryngoscopical and rhinoscopical examinations. Tongue-holder with three braces of leaves for laryngoscopic examination, besides 3 tongue-spatula for rhinoscopic examination. Uvula-stringer for rhinoscopic examination.

Publications relating to the above. Magnifying apparatus for laryngoscopic examinations. Fixator of the laryngoscopic mirror. The whole of the apparatus invented by the exhibitor, according to his plan executed by the machinists Mr. Hauck, Vienna, Wieden 820, and Mr. Thurrigl, Vienna, Alsergrund 205

SECTION III.

Class 18.

Cotton.

The increase of the manufactures of cotton in Austria is best exhibited by the returns concerning foreign trade in raw cotton and cotton yarns. There were imported:

Years, on an average,	Cwts of	
	Cotton	Cotton yarn
1831—1840	213,715	45,826
1841—1850	430,282	46,233
1831	492,868	46,768
1860	896,651	112,950
1861	879,500	182,700

The quantity of imported raw cotton quoted above is almost entirely worked (without further export) to yarns (mostly to No 40) in the Austrian cotton-manufactures. There are now 170 establishments of this kind working with 1·8 millions of spindles [1]), the greater number being worked by water power and a reserve steam-engine being availed of only in the event of scarcity of water. Many of these spinning houses join twisting to the former manufacture, one factory alone, that of Haratitz in northern Bohemia, the largest on the continent, is doing nothing else and using 8,000 spindles. Besides there are great quantities of cotton thread produced with hand twisting machines, an industry which likewise has its seat in northern Bohemia.

The exports of cotton yarn and thread to foreign countries are inconsiderable (in 1860, 8,555 Cwts), nearly all the home production of yarns and threads besides the afore-named imports being worked up by Austrian weavers.

Cotton-weaving, in Austria, originated in linen-weaving carried on to a great extent long before as domestic industry.

Both branches exist jointly to this very day in the valleys of the Ore and Giant Mountains and Sudetes; according to the alternating circumstances of sale, the same loom is being successively employed for either of those branches.

Carried on as domestic industry, cotton-weaving suffered from the evil that improvements in the loom could be but slowly and partially introduced: In spite of the endeavours made by Boards of Commerce and Trade, and notwithstanding the assistance of employers, the majority of domestic looms still belong to the primitive kind, the number of dandy-looms used in domestic industry scarcely amounting to 15,000, whilst that of common looms extends to more than 200,000.

Since 1830, extraordinary progress has been made in the production of cotton fabrics of all sorts, and in linen, woollen and silk fabrics, mixed with cotton yarn or thread. More than 25,000 dandy-looms and 15,000 power-looms have been set up since 1851, most of them also managing the finishing, dyeing and printing of their own manufactures or of the fabrics furnished by domestic weavers.

Cotton industry covers not only the demands at home, but exports considerable quantities of cotton manufactures, mostly of middling quality, to foreign countries, the latter amounting together, in 1860, to 34,000 Cwts, valued at 8 millions florins.

There are in Austria about 350.000 hands employed in the spinning and twisting of cotton, and in manufacturing raw cotton and mixed fabrics, the value of the annual productions being estimated at 100, rising to 120 millions florins.

---

742. BREVILLIER and Comp. Cotton Mill. *Schwadorf, Lower-Austria.* Ware-house: *Vienna 803.*

Cotton Ware. Warpcops and Pincops.
(See Class 31.)

This cotton spinning mill is worked only with water-power, has 21.366 jenny spindles (self-actors) occupying in all about 300 hands. It produces only No. 10/26 warp, No. 36 warpcops and 42 pincops. Average number in 1861 = 24.

743. BURCKHART Charles, *Vienna, Schottenfeld 219.*

Cotton Chenille Manufactures.

744. CORDELLA Brothers. *Weissbach, Bohemia.*
Cotton Yarns.

745. EHINGER Adalbert, Bleaching and Finishing Establishment. *Oberlangenau, near Hohenelbe, Bohemia.* Honorary M. Munich 1854.

Bleached Cotton Wares: Cotton Damask, Cambridge, Chiffon, Shirting and Yarns.
(See Class 19.)

746. FECHNER Philip, White-ware Manufactory. *Vienna, City 541.*
Cotton Manufactures.

747. FIAL John. *Vienna, Gumpendorf 538.*
Chenille Wares.

748. FÖRSTER John, jun. *Rumburg, Bohemia.*
Cotton Manufactures.
(See Class 21.)

749. GARBER John and Son. *Vienna, Gumpendorf 206.*
Cotton Manufactures.
(See Class 20.)

---

[1]) The local distribution of Cotton Spinning and Cotton Industry in Austria has been illustrated on a special Map, exhibited Nr. 1196, together with other Maps of Industry, by the J. R. Direction of Administrative Statistics.

750. GOLDBERG C. R., Manufactory of Linen, Woollen and Cotton Wares. *Warnsdorf, Bohemia.*
Cotton Stuffs for coats and trousers.

751. GOLDBERGER and Sons, Samuel F. Cotton Printers. *Old Buda, Hungary.*
Cotton Manufactures.

752. HELLMANN N. *Prague, Altstadt 710.*
Raw Weaving Stuffs.

753. KLAUS J. E. *Niedergrund, Bohemia.*
Cotton Stuffs for coats and trousers.
(See Classes 19 and 21.)

754. LANG John Nep. *Vienna, Mariahilf 45.*
Cotton Lama, Fustian, Cotton Stuffs for gowns.
(See Class 21.)

755. MICHEL Philip, Patented Manufactory of Woollen, Linen and Cotton Weavings. *Gärten near Schönlinde, Bohemia.*
Cotton weavings of different fineness and size.
(See Classes 19 and 21.)

756. MILLER Francis. *Georgswald, Bohemia.*
Muslins and Cambrics, raw, 9/8 broad.

757. MITSCHERLICH Adolphus and Robert. *Teplitz, Bohemia.*
Assortment of Webs and Cotton Manufactures.

758. MÜLLER'S Joseph Son. *Rumburg, Bohemia.*
Cotton Manufactures.
(See Class 19.)

759. POTTENDORF, Patented Cotton Spinning and Weaving Mills at. *Lower Austria.* Ware-house: Vienna. Great Silver M. Vienna 1839 and 1845. M. I. Cl. Paris 1855.

Cotton Yarn and Weavings.

The patented Cotton Spinning and Weaving Mills at Pottendorf owe its existence in the first instance to their Graces the Princes Jos. Schwarzenberg and Colloredo-Mannsfeld, who, as Directors of the then existing I. R. licensed Loan and Exchange Bank, resolved, in 1801, to establish in Austria a cotton spinning mill, not yet existing at that time, and engaged Mr. John Thornton of Lancashire in England to execute that plan. In 1803, a society was formed to which the above mentioned bank transferred the letters patent, and after the I. R. Court Counsellor Jos. Hartl von Luxenstein having been put at the head, there was bought an extensive area of land at Pottendorf, a market place 3 miles distant from Vienna, where the requisite buildings were erected and put in connexion with the rivers Leitha and Fischa by means of a canal of 2.400 Vienna fathoms length, from which the manufactory received a water-power equal to about 400 horse power. Under the direction and according to the design of Mr. Thornton, the necessary machines were constructed in Pottendorf, and in 1804 there were already

 18.000 mules and
  430 throstle spindles a-going, furnishing about
 12.000 knots of yarn (5 lb. Engl. each) annually. Progressively enlarging, the establishment numbered, in 1832,
 40.321 mules and
  2.944 throstles, with a yearly production of
 186.000 knots.

In 1841 Z. C. Chevalier Popp de Böhmstetten, was entrusted with the direction of the establishment, which underwent a complete reorganisation on the motion of the then representative, Baron George Sina. New machines were procured from England and Switzerland, 4 new iron water-wheels and 3 turbines set up, a yarn bleaching and spacious workshops for repairing machines erected. These improvements were, since that period, unceasingly continued; in 1857 after the flax-spinning mill existing in Pottendorf having been acquired, power looms were procured, and, scarcity of water having been sadly felt since years, the efficaciousness of the management of these united mills was promoted by two coupled steam-engines of 120, and a little one of 20 horse power. For lighting the manufactory, gasworks were planted for 1.500 burners. The extent of the whole establishment, in 1862, may be seen from the following statements: the area of the premises amounts to 73 yokes (1.600 sq. fathoms

each). The buildings belonging to the manufactory cover 5.855 sq. fathoms. There are going the following number of spindles:
  11.800 self-actors,
  4.200 half self-actors,
  32.000 hand mules,
  14.000 throstles,
  2.000 twisting throstles,
 total: 64.000 spindles. Besides there are
  332 looms in the weaving department.
The yearly production of yarn amounts to
  600.000 knots (5 lb. Engl. each), Nr. 6 to 40, in various qualities, then
  180.000 knots (1 lb. Engl. each), of threefold knitting yarn, and
  2,000.000 ells of heavy calicos. The sale of these articles extends to all parts of the Austrian empire, and that of calico also to the Tariff Union States, mediated by a ware-house office in Vienna. Yearly consumption of raw material:
  6.000 bales of America,
  1.000 bales of Surate,
 total 7.000 bales.
That establishment is now employing
  497 males,
  565 females,
  218 children,
total number of hands 1.280, who are well provided for: 4 large dwelling-houses for working people contain 220 healthy and fine lodgings, 2 physicians and 1 surgeon are appointed to attend the institution for the sick, the working people receiving adequate support during their illness from the funds. There is an hospital with 24 beds and a bath. For the children of the working people there is an endowed school with a teacher for religion and one for the elementary objects, in connection with which there is a drawing-school and for girls a needle-work school; little children and babies being guarded, nursed and fed in an asylum during the working time of their parents. An area of land, measuring 5 yokes, is divided into 230 garden plots, the use of which is bestowed upon the most active workmen gratuitously. The central direction of the whole establishment (being the property of a dormant partnership of 25 shareholders among whom are his Grace Prince John Lobkowitz, his excellency Baron John Sina, Royal Ambassador of Greece, and the Counts Spangen) resides in Vienna, represented by the shareholders Baron John Sina and two directors, Z. C. Chevalier Popp de Böhmstetten, and Mr. Themistokles Metaxa; the technical direction being confided to two local directors: Mr. Bruno Henneberg and Baron Anthony Seiller.

760. RICHTER F. and Comp. Cotton Spinning Mill. *Smichow near Prague.* Silver M. Leipzig 1850. Honorary M. Munich 1854, M. II. Cl. Paris 1855.

Cotton Yarn.

This manufactory was established in 1845—46, and is worked with a steam-engine of 110 horse-power, numbering 16.000 spindle self-actors and 1.100 half self-actors, besides the requisite preparation works of new construction. It occupies 250 hands and produces 240 Cwt Engl. weekly, average number 24. The exhibited 1 knot No. 20, 1 knot 24, 1 knot 30 are spun of American cotton, ordinary, good ordinary to low middle.

761. RICHTER Ignatius and Sons. *Niedergrund near Rumburg, Bohemia.*
Cotton-Velvets and Trousers Stuffs.

762. SPINNING Mill and Dyeing in Adrianople red, Patented, *Aidussina.*
Raw Cotton Yarns.
(See Class 23.)

763. SPITZER M. A., Patented Manufactories of Silk and Chenille Wares. *Vienna and Neuhaus in Bohemia.* Ware-house Vienna, Gumpendorf 327. Agents at Leipzig, Hamburg, London, New-York.

Chenille Wares.

This manufactory, exhibiting for the first time, chiefly produces cotton chenille stuffs finding an extensive sale in all parts of the world, and victoriously entering the lists of competition on account of their beauty and cheapness in particular. It employs about 500 hands and has the especial merit of producing the most tasteful patterns and elegant stuffs, by means of a new patented invention, out of chenilles that were formerly only used in flowered patterns and in gay stuffs. Attention is begged to be given to the exhibited „Patented Scarfs".

764. WITSCHEL and **REINISCH**. Patented Manufactory of Cotton, Linen and Woollen Fabrics. Wares-

E.   9

dorf, Bohemia. Honorary Medal Vienna 1845. London 1851. Honourable Mention Munich 1854. Ware-houses : Vienna, City 350, Brünn and Prague.

Cotton Manufactures.

(See Class 19.)

This manufactory produces cotton stuffs for trousers, nankeens, shoe-stuffs, all sorts of white-wares, linen stuffs for trousers ¹/₂ and ¹/₁ ell broad, raw ticken and tela russa. This establishment enjoys a far-spread reputation, and may be said to be the first in Austria with respect to manufacture of fine articles, especially nankeens. The fabrics are worked on dandy-looms, 600 hands being employed.

765. WOLF C. *Bielitz, Silesia.*

Machine Carding.

Class 19.

Flax and Hemp.

Nearly 2 millions Cwts of flax and 1 million Cwts of hemp, home-grown and imported raw products (see Cl. 4), are annually worked up by the Austrian flax and hemp industry.

In the eastern portions of the empire, where flax is grown in great quantities on the slopes of the Carpathian mountains, the working of this material devolves entirely on domestic industry.

The same landowner that reaps the plant, also rets and brakes the fibre, spins the yarn and, during the winter-months, works it up into coarse linen for his own use. It is but in the higher regions of the Carpathian mountains (in Upper Hungary and Galicia) where flax is the principal product of the soil and where nearly every house has its own loom, that the production of linen exceeds the wants of the household, hence considerable quantities of coarse webs are coming from those parts. Commercial weavers in particular are but to be met with in the western provinces of the empire, they usually receiving the hand-yarn bought up by merchants in the markets, or such machine-yarn as they procure for themselves, and furnishing the finished webs to merchants.

Hand-spinning of flax-yarn is to this day the occupation of a very great number of individuals (particularly women and children) during the winter months. Machine flax-yarn spinning is now being exercised in 33 establishments possessing all together 200.000 spindles, and are for the most part settled in Bohemia, Moravia and Silesia, where weaving of finer sorts of linen and damask may be said to have its home.

Austrian linen industry, which before had gradually lost all exports of better sorts of webs, has in recent times, especially through improvements in machine spinning, gained ground again. In 1860, there were woven 220.000 Cwts of inland, and 12.000 Cwts of imported, machine-yarn. The exports of inferior and fine linen-wares having in 1851 scarcely amounted to 1,000.000 florins, rose in 1860 to above 4,000.000 florins. Of linen-wares in general, there were in 1860 imported for less than 300.000 florins, but, on the contrary, exported for 9,600.000 florins.

Flax growing, in the east and south of the empire, giving way to the culture of hemp, the working up of hemp consequently prevails in these regions. Whilst in the remaining provinces some inconsiderable local trade only is transacted, hempen wares are, in the eastern provinces, to a great extent produced for domestic wants; and, on the other hand, the demands of sea-navigation have called into existence an extensive production of ropes, sail-cloth etc. in the maritime provinces. Home consumption deducted, there are still from 10.000 to 15.000 Cwts of ropemaker's wares exported annually to foreign parts.

The number of hands employed in Austrian linen and flax industry is difficult to be ascertained, because of the fact that it mostly forms an agricultural branch occupation carried on during part of the year only. There may be nearly 4,000.000 of persons who have a share in the production of flax and hemp yarn its webbings and twistings. The value of these productions is estimated at 150 millions florins about half of which circulates in commerce, the other half being for home use.

766. EHINGER Adalbert, Bleaching and Finishing Establishment. *Oberlangenau, near Hohenelbe, Bohemia.* Honorary M. Munich 1854.

Bleached Linen Wares and Yarns.

(See Class 18.)

767. GROHMANN J. and Son, Manufactory of Canvass for sails and Ticking. *Sternberg, Moravia.* Honorary M. Munich 1854.

Linen Wares.

The exhibitors produce all sorts of sail-cloth, of cotton and linen yarn for steam and sailing vessels, water-proof and air-tight covers for goods in railway-waggons, measuring from 4 to 5 Vienna ells in breadth, moreover various sorts of tickings and linens for the army, and many descriptions of corn-sacks (woven in one piece on common hand looms without steam or water power). The sail-cloths are produced by hand and by one weaver and the waggon-cloths of 5 ells breadth by two weavers on the power-loom.

768. HEIDENPILTSCH, Administrative Counsellor of Flax-spinning at. *Heidenpiltsch near Freudenthal, Silesia.*

Machine Yarn.

769. HEINZ F. and A., Patented Manufactory of Linen, Ticking and Damask Wares. *Freudenthal,* *Silesia.* Honourable Mention Munich 1854 and Paris 1855.

Linen Wares.

In 1833, Francis Heinz began manufacturing linen stuffs on few looms, adding, a few years after, the production of tickings, always sticking to the principle of manufacturing only genuine wares. In 1852, Anthony Heinz entered the business as public partner, and then damask-wares were introduced besides the above-named manufactures. In 1858, that firm obtained the patent of a free manufactory. The wares produced here are as follows : ⁴/₄ and ⁵/₄ broad, linen stuffs of all kinds of length and quality, per 30 ells, varying in price from fl. 10 to 85 ⁸/₄, ⁹/₄, ¹⁰/₄ broad, linen stuffs, per 30 ells from fl. 33 to 64.

Ticking and damask towels, 30 ells and in dozens from fl. 4·35 to 15.

"	"	"	huckaback, 30 ells from fl. 15 to 23.
"	"	"	napkins, per doz. from fl. 4·40 to 24.
"	"	"	table-cloths in all sizes ⁵/₄ long from fl. 2 to 8, longer than ⁸/₄ dearer in proportion.
"	"	"	sets for 6, 12, 18, 24 persons from fl. 4·35 to 30.
"	"	"	dessert napkins, per doz. from fl. 2·15 to 8.

Tea-table linen ⁸/₄ long, from fl. 3·50 to 10, longer than ⁸/₄ dearer in proportion; handkerchiefs, white-linen, per dozen from fl. 2·30 to 15.

770. JERIE. W. *Hohenelbe, Bohemia.*

Linen Yarns.

771. JERIE Wenceslas. *Marklow, Bohemia.*
Linen Weavings.

772. KLAUS J. E. *Niedergrund, Bohemia.*
Linen Stuffs for coats and trousers.
(See Classes 18 and 21).

773. KRATKY James. *Freudenthal, Silesia.*
Tea-table Cloths.

774. KÜFFERLE Augustus and Comp., Patented Manufactory of Linen, Ticking and Damask Wares. *Freywaldau, Silesia.* Ware-house Vienna 360. Partners: Küfferle Augustus and Wiesner Joseph, under the firm of Küfferle and Comp. Honorary M. Munich 1854, M. I. Cl. Paris 1855.
Ticking, Damask and Plain Manufactures.

Founded in 1818, this establishment applied, up to the year 1850, to the production of Silesian white-yarn linen, about which time the manufacture of ticking and damask was added. From this period dates the rapid rise of this manufactory, with which few establishments of this kind of Austrian industry can rival to this extent. In a few years the owners succeeded to carry the extensive field of their activity to that pitch which gives a new impulse to a branch of industry so important for Austria. Their success at the Industrial Exhibitions in Munich and Paris as well as the rising sale of their manufactures at home prove their victorious competition. The imports of Saxon table-linen, so popular in former times, have now ceased, on the other hand, new markets for sale opened in South-Italy, Russia and America, which are likely to continue so even after the rate of agio shall have settled down to nothing. This manufactory does not exclude the production of articles of luxury, but prefers cultivating those of current middle quality. By concentrating their powers, they succeeded in uniting superior quality with cheapness. In order to give an idea of the extent of their manufactures, we mention that ticking and damask wares, furnished mostly by domestic system of labour, amount to about 500,000 ells of ^{5}/$_8$ to 5 ells breadth, without making particular mention of plain linen wares, the manufacture of which numbers among the most extensive at home. Bleaching and Finishing contrivances according to the latest experiences (for 3,500 Cwt of yarn), as well as the employment of all improvements in manufacturing make their sales rapidly rise, and, as shown by the above-quoted numbers, are surpassed by no other manufactory of Austria.

775. KÜHNEL Joseph. *Engelsberg, Silesia.*
Linen and Thread.

776. MAY'S J. Son. *Schönlinde, Bohemia.*
Linen-thread dyed and bleached.

777. MICHEL Philip, Patented Manufactory of Woolen, Linen and Cotton Webs. *Gärten, near Schönlinde, Bohemia.*
Linen Webs of various fineness and size.
(See Classes 18 and 21.)

778. MÜLLER Joseph, Patented Manufactory of Linen-thread and Linen-webs. *Schönlinde, Bohemia.*
Linen-thread of all colours and descriptions. Linen-thread socks and stockings.

779. MÜLLER Joseph, Son. *Rumburg, Bohemia.*
Linen Manufactures.
(See Class 18.)

780. PICK Isaak D. *Nachod, Bohemia.*
Linen Manufactures.

781. PLISCHKE John. *Freudenthal, Silesia.*
Tea-table Cloths.

782. RAYMANN & REGENHART, Purveyors to the I. R. Court, Patented Manufacturers of Linen, Ticking and Damask Wares, and Owners of a Power Flax-yarn Spinning-mill at *Freywaldau in Austr. Silesia.* Warehouses: Vienna, Prague, Lemberg, Gratz, Arad. Gold M. Vienna 1839 and 1845. Gold M. Munich 1854.
Rumburg Linen, Ticking, Damask and Table-cloth Manufactures.

One damask table-cloth, 3^1/$_2$ V. ells (3 yards), round with imperial eagle and coat of arms. 2 damask table-cloths 3^1/$_2$ ells square. The middle one of the three tableaux is framed in by 3 linen tea-table cloths interwoven with blue silk; the two lateral tableaux each by 3 white and natural linen tea-table cloths. Assortment of ticking, brillantine and damask table-cloths, the latter to a breadth of 5 V. ells (4^1/$_2$ yards). Assortment of Rumburg linen 53 ells (45 ^{1}/$_2$ yards) long, and ^{17}/$_{16}$ ells (33 inches) or ^{7}/$_8$ ells (36 inches) broad. Assortment of 2^1/$_4$ ells (67 inches) and 2^1/$_2$ ells (76 inches) broad bed-cloth. Assortment of Irish linen, 50 ells (43 yards) long, and ^{9}/$_8$ ells (36 inches) broad. I. sort (strong family). Assortment of Irish linen, 49 ells (42 yards) long, and ^{6}/$_4$ and ^{9}/$_8$ ells (31 and 36 inches) broad. III. sort (medium) of from middling to superfine quality. Assortment of Irish creas (heavy linen), 38 ells (32^1/$_2$ yards) long, and ^{9}/$_8$ ells (36 inches) broad. 12 damask napkins natural with white, 12 scutcheons and devices.

This manufactory, existing since 1820 and having since 1827 introduced damask weaving into Austria, on account of which imports of this branch of manufactures from Saxony are dispensed with, is weaving up about 6,500 three-scores of yarns annually, of which 4,000 three-scores are yarns of No. 14 to 65 mostly spun in Austria in their own machine spinning mill; 2,500 three-scores English and Belgian yarns of No. 70 to 200. Out of these 6.500 thrss-scores machine-yarns there are yearly produced: 75,000 sq. ells (55,100 sq. yards) of damask of from 1 to 6 ells (31 to 198 inches) breadth; 175,000 sq. ells (128,570 sq. yards) of Jacquard and ticking of from 1 to 3 ells (31 to 108 inches) breadth; 600,000 sq. ells (440,820 sq. yards) of plain linen of from 1 to 2^1/$_2$ ells (31 to 78 inches) breadth. Total productions: 850,000 sq. ells (624.490 sq. yards).

This total production of 850.000 sq. ells, besides the quantity of from 6.000 to 7.000 three-scores yarn is bleached completely on a bleachery of their own, contrived according to the English system, and finished according to the English way of finishing, with the help of water-power equal to 40 horse-power, 1 steam-engine of 20 horse-power and 3 boilers, 5 atmospheres each. The establishment inclusive of the spinning mill employs for weaving and bleaching about 400 families of weavers, besides a staff of hands amounting to 1.000 persons, who, mostly by fine weaving, get good earnings, compared with other jobbing weavers, in consequence of which, the wealth of that district has materially increased, especially since 3 years when Rumburg and Irish linen manufactures (22 to 140 fl. per piece) were introduced. The sale of that establishment, only accommodated for heavy good productions, extends, besides the empire of Austria which is the chief consumer, to Russia, Turkey, Greece and America. The coats of arms and devices on those 12 napkins that are exhibited, represent the houses of Count Wenkheim, Count Wimpfen, Archduke Albrecht, Count Schönborn, Count Olam-Gallas, Baron Orczy, Prince Fürstenberg, Archduke Lewis Victor, Prince Lichnowsky, Prince Ypsilanti, Count Nako, Prince Schwarzenberg Duke of Krumau.

That firm was till now represented only in the exhibitions of 1839, 1845 and 1851.

783. SCHÖNBAUMSFELD Dominik, Thread-button Manufactory. *Vienna, Altlerchenfeld.*
Thread-buttons.

784. TÄUBER Ferdinand. *Untermeidling near Vienna.*
3 hemp-yarn-hoses (pipes) weaved water-proof, 200 feet long, diameter 1^1/$_8$, 1^7/$_8$, 2^1/$_2$ inches.

785. VONWILLER and Comp. Linen and Cotton Manufactory. Ware-house *Vienna, City 727.*
Thorough-linen, linen and cotton mixed stuffs for trousers and coats.
(See Class 21.)

786. WITSCHEL & REINISCH, Patented Manufactory of Cotton, Linen and Woollen Fabrics. *Warnsdorf, Bohemia.* Honorary M. Vienna 1845 and London 1851. Honourable Mention at Munich 1854. Ware-houses: Vienna, City 350, Brünn and Prague.
Linen Manufactures.
(See Class 18 where particulars about that establishment.)

787. WURST John Nep. and Sons. Patented Manufactory. *Freudenthal, Silesia.*
Linen; Tea-table cloths of cotton and linen-yarn, wool and silk.

Class 20.

Silk and Velvet.

The whole of 250.000—300.000 Cwts of Cocoons, produced every year in the whole Austrian Empire, with exception of an insignificant portion (6.000 Cwt in 1860 (exported into foreign parts) is worked up into Raw-silk by the inland spinning-establishments („Filande") generally united with the establishments of Silk-worm breeding and, in this respect, constituting a secondary branch of Rural Economy.

Eighty-three establishments of this kind, with nearly 4.000 boilers (among which 37, with 2.021 boilers, in Southern Tyrol) are working on an extensive scale, using steam power for heating their boilers and moving their mechanical appliances. Including the smaller „Filande" the reeling of raw-silk and offals („Struse") requires, for an average yearly period of 50 days, the operation of nearly 80.000 boilers, as many yarn-spindles and about 60.000 individuals, producing a total of 20.000—23.000 Cwts of raw silk and offals; partly for export (9.000 Ctw. in 1860) and the rest (11.000—14.000 Cwts) for the use of inland establishments („Filatoji"), working up these materials into Spun Silk („Organzin" and „Trama"). In general, these establishments are moved by water power. The Venetian Silk, on account of its peculiar quality, is chiefly converted into „Trama" for weaving, sewing and knitting purposes, the Tyrolese establishments twist up into „Organzin" a considerable portion of their home-produced raw materials. A small amount of „Frawa"(in 1860 about 1.200 Cwts) finds its way into foreign parts; the production of „Organzin" or twisted Silk, remaining behind the demands of inland Industry, this want

being regularly filled up by an import from foreign parts (about 3.000 Cwts in 1860).

Taken in the general average, (spinning-offals („strazze") set apart) the inland-establishments are provided every year with 9.000—10.000 Cwts of spun Silk, to be worked out into Silk-tissues. Some few establishments in S. Tyrol, in the Venetian Provinces, in Bohemia and Moravia excepted, the manufacture of Silk-tissues, velvets, and ribbons is nearly exclusively limited to Vienna and its environs, where (in 1860) the looms at work exceeded the number of 7.500. Among them were, 2.283 plain looms, 3.720 Jacquard looms, 1.466 mill and 40 mechanical looms.

The number of persons, permanently employed in Austrian Silk-industry, or acting as auxiliaries for the space of 1—2 months immediately after the gathering of Cocoons, is no less than 100.000, and the value of their yearly production may be fairly estimated at 25,000.000 of Austrian florins.

The following numbers exhibit the trade done in Silk during 1860:

	Import:	Export:
	Cwts	
Silk articles of choice quality	2.088	632
ditto of ordinary quality (mixed with threads ot another nature)	997	6.620

The Turkish Caps („Fesz") of wool tissue with silk tassels have the greatest share in the export of ordinary Silk articles. They are manufactured in Bohemia and Lower Austria.

---

788. BACKHAUSEN John. *Vienna, Schottenfeld 206.*
Patent Grenadin Stuffs.

789. BALOGH, born **POPOVITS** Maria. Silk Producer, *Klausenburg, Transylvania.*
Raw Silk Hanks.

790. BRAUN Aloysius. *Vienna, Schottenfeld 280.*
Silk Ribbon.

791. BURGSTALLER John Lewis, *Hruskovatz, Com. of Agram, Croatia.*
Silk.

792. CATTICH, Brothers. *Zara.*
Raw Silk.
(See also Class 3.)

793. COLLECTIVE EXHIBITION of Silk Wares of the Viennese Silk Stuff Manufacturers: —
Bujatti Francis, Manufactory of Silk Stuffs. *Vienna and Moravian Schönberg.* Bronze M. Vienna 1845. Gold. M. Leipzig 1850. Honourable Mention and M. London 1851. Great M. Munich 1854. Gold. M. Paris 1855.
Assortment of Stuffs for furniture and damasks for churches, Lampas; bed and table covers with cut out borders; printed foulard shawls, plain satins, gaze diaphane for windows etc. etc.
The manufactory, established in the beginning of this century by the father of the exhibitor, and since a few years enlarged with the branch establishment at Schönberg, enjoys

great renown for its solid and cheap productions, completely finished after original patterns and in accordance with the taste of our times. The exhibitor's productions are not only far spread in Austria, but are also much in request abroad. He was especially often honoured with orders from the Most High Court and the Nobility.

Flemmich A.'s Widow. Manufactory of Silk Wares. *Vienna.* Honourable Mention London 1851, Paris 1855.
Plain and patterned stuffs for gowns, umbrellas and parasols. A picture with flowers and figures produced on the Jacquard loom.
Active since 1837, this manufactory occupies on an average from 60 to 80 hands, producing a total of 40,000 ells of stuff, valued at about 80,000 fl. The yearly consumption of raw silk amounts to about 2,000 pounds. The stuffs produced are plain umbrella stuffs, plain and patterned parasol stuffs, besides patterned stuffs for gowns.

Frischling, Arbesser & Comp. *Vienna, City 1105.*
Silk-Velvet.
Giesauf John & Sons. Manufactory of Half-Silk Stuffs. *Vienna, Margarethen 38.*
Half-Silk and Turkish Stuffs.
Hornbostel C. G. & Comp. Manufactory of Silk Stuffs. *Vienna.* Gold M. Vienna 1835, 1839, 1845. Honourable Mention London 1851. Prize-M. Munich 1854.
Plain and fashionable Stuffs of thorough and mixed silk, partially produced on self-acting loams, which, continually improved, are since 1819 working in their branch establishment at Leobersdorf.

Lehman Joseph & Son. *Vienna, Gumpendorf 24.*
Church stuffs and Fashionable Wares.

Siebert Fred. & Son. *Vienna, Wieden 656.*
Chenille Shawls.

Woitech Francis, Patented Manufactory of Banners and Silk Wares. *Vienna, Gumpendorf 411.* Prize-M. Munich 1854, Paris 1855.

Waistcoats of black patterned and plain grenadin, patterned silk-piqué and fashionable waistcoats. Plain and patterned gentlemen's neckcloths, patterned ladies broach neckerchiefs, colours for the I. R. Austrian Army, silk-weaver's work. with the I. R. Eagle on one side and the image of Maria on the other, cavalry standard (silk, weaver's work), I. R. Eagle on both sides.

The colours for the I. R. Austrian army are made double-sided on the loom. On one side of the stuff is the Imp. Eagle with all the escutcheons of the crown-dependencies and orders, exactly made according to heraldry; on the other side of the same stuff is executed in liturgic colours the image of Maria, standing in a gold-star above the clouds on the globe. The gold and silver used for it is genuine, the weight of the colours with borders is 1½ pounds; price 54 £. In most countries, colours for armies are painted on silk stuff with oil-colours, and thus it was also in the Austrian army, till the exhibitor succeeded in making colours on the loom, answering all expectations. After many trials and two years' use, they were declared perfect and introduced in the whole army upon orders of His Majesty the Emperor of Austria. The cavalry standard is produced in the same way, but the same double eagle is on each side, price 13 £.

Wottitz William D. Manufactory of Silk Wares. *Vienna, Schottenfeld 154.*

Patterned gown on black ground, with white squares laid on; plain wood-brown gown; glacé gown stuff.

This manufactory produces the following articles: patterned and plain stuffs for gowns; all sorts of stuffs for lining, such as: croisé, renforce, marcelline ³/₄ and ⁴/₈ broad, Florence bast ³/₄ and ⁴/₈ broad, satins and linings for gentlemen's hats. For the exhibited patterned gown on black ground worked in 4 colours, a complete shading of the bouquet was attained by the help of the colour of the ground stuff.

794. DUSSINI Joseph, Trader and Silk Spinner. *Lana, Tyrol.*
Ten skeins of fine spun silk.

795. FIGAROLLI Lewis. *Riva, Tyrol.*
Silk.

796. FRANZOT Anthony, Spinner. *Farra near Gradisca.*
Raw Silk.

797. GANAHL Charles. *Feldkirch, Vorarlberg.*
Raw Silk produced at Feldkirch.

798. GARBER John & Son. *Vienna, Gumpendorf 206.*
Silk Stuffs and Dyed Silk.
(See Class 18.)

799. GENTILLI Jeromy J., Spinner. *Goritz.*
Raw Silk.

800. GIAVONI Francis. *Verona.*
Raw Silk.

801. GROHMANN, Brothers. *Schönlinde, Bohemia.*
Silk petinet Manufactures.

802. HAAS Philipp & Sons, Owners of the Furniture Stuff and Carpet Manufactory at *Vienna, Gumpendorf 201 and 202,* the Furniture Damask Manufactory at Mitterndorf, near Vienna, the Cotton Spinning Mill and Powerloom at Ebergassing near Vienna, the Branch Manufactory of Damask and Utrecht Velvet at Hlinsko in Bohemia, and the Furniture Stuff Manufactory at Bradford in England. Ware-houses: Vienna 718 Pesth, Prague, Milan; Commission stores in all large commercial towns of Europe, London at Messrs. Jordan & Parker, 26, Noble Street, E. C. Gold M. Vienna 1839 and 1845. Silver M. Berlin 1844. Prize-M. London 1850. Large M. Munich 1854. M. 1. Cl. Paris 1855.
(See Classes 21 and 22.)

The above mentioned manufactories of the exhibitors produce all sorts of furniture stuffs of silk, mixed silk, wool, mixed wool, and cotton, Utrecht velvet, goblins, table-covers, curtains, carpets, calicos; their sale extending to all parts of the world. The beginning of this business dates from the year 1810 at which time Philip Haas commenced the manufacture of white calico with but a trifling capital. To this was subsequently joined the production of organtins, mousselins and linons. Improvements in bleaching, dyeing and finishing soon raised this little establishment to such a height as to distance foreign manufactures of their kind from the Austrian market. Philip Haas was the first that introduced the long printing-tables into Austria. He also invented an improvement in the contrivance for „broche" work, after which about 1000 similar machines were constructed in Austria. At the general exhibition in Vienna 1839 his cotton weavings were awarded the gold medal. In the same year he established with his two sons the present firm, now principally attending to the manufacture of furniture stuffs and carpets, considerably enlarging their business and earning acknowledgment of their productions in all circles. At the exhibition of Berlin in 1844 the exhibitors received the silver medal for damask weavings. In 1845, they established at Mitterndorf, near Vienna, a manufactory of furniture damasks, with 30 mechanical wooden looms after a peculiar system, besides a dyeing concern for yarns and stuffs. At the general exhibition of 1845 in Vienna the exhibitors received the gold medal. In 1850 they erected a branch manufactory at Hlinsko in Bohemia, with 50 looms for damask and 50 for Utrecht velvet. At the German industrial exhibition of Leipzig in 1850, the chief of the firm was decorated with the knight's cross of the Royal Saxon Order of Merit. In the same year the exhibitors acquired the cotton spinning mill at Ebergassing, with 12,000 spindles and a considerable water-power not yet properly availed of till then, but subsequently rendered useful for a manufactory with mechanical looms. The jury of the international exhibition at London in 1851 awarded the prize medal to the exhibitors. At this time there were purchased in England a considerable number of powerlooms for calico, damask and carpets and put in activity, likewise the Vienna manufactory at Gumpendorf was materially enlarged and destined for the production of silk damask and carpets, lastly a new manufactory was established at Bradford in England. At the Munich Exhibition in 1854, the exhibitors received the large medal and at the Paris exhibition in 1855 the 1ˢᵗ class prize.medal. At present there are working, in the manufactories at Vienna, Ebergassing, Hlinsko and Bradford, about 12,000 spindles and more than 500 various looms, about 1,500 persons being occupied. At Mitterndorf a new spinning and weaving manufactory is building. The exhibitors annually work up about 1,000 Cwt of wool, 5.000 Cwt of cotton, 600 Cwt of linen and 50 Cwt. of silk etc., out of which an average number of more than a million ells of various stuffs are produced. Besides the various descents of water amounting to more than 300 horse-power, there are used two large boilers for dyeing, dressing, and finishing. That firm is enabled to match any other at home and abroad in respect of quick performance of the most extensive orders. Lists of prices and samples gratis on demand; for purchasers of great quantities the usual allowances.

803. HETZER Charles & Sons, Ribbon Manufactory. *Vienna, Gumpendorf 534.*
Silk Ribbons.

804. HOFMANNSTHAL Ignatius von, Dr. *Vienna, City 733.*
Case containing Silk of his own produce.

805. INSTITUTE OF DEAF AND DUMB, gen. Austr. Jews. *Vienna.*
Plans of the Institute. Specimens of Silk.

There are two printed treatises joined to the former objects. One containing a report of the foundation, development and method of teaching of this establishment, composed by Dr. J. v. Hofmannsthal, Chief of the institute, in German, French and English; the other, being the textbooks for the instruction of the deaf and dumb, is composed by H. Deutsch, director of the school.

806. LENASSI BIAGGIO Anthony, Spinner. *Goritz.*
Raw Silk.

807. MATUSIK John Nep. Piarist Teacher and Silk Producer. *Klausenburg, Transylvania.*
Reeled off silk.

808. RITTER von **ZAHONY** William. *Goritz.*
Floret Silk.
(See also Class 3.)

809. SCHEY Fr. Patented Wholesale Dealer. *Vienna.*
Agents: G. & A. Worms, London, 1, Austin Friars Old Broad Street.
Greggia (silk not thrown) of 3 cocoons, and Organzine $^{16}/_{20}$ of greggia from spinning establishments, in the Slavonian Military Borderland.

810. SCHWARZ John. *Vienna, Hundsthurm 90.*
Silk hatband.

811. SILK MILL *Gross-Zinkendorf, Hungary.*
Raw Silk in skeins and on reels.

812. SILK WORM REARING COMPANY at Styria. *Gratz.*
Silk.

813. SILK WORM REARING UNION in Austrian Silesia. *Troppau.*

Specimens of cocoons, raw silk, shute, thrown silk (of the last two sorts also dyed silk), and damask woven of this silk.

This Union, established since 1859, numbers already more than 1,300 members, possessing extensive plantations with a stock of nearly half a million of mulberry-trees, and producing silk that arose from the Milanese silk-worm race acclimated to that place and well preserved; the cocoons are tight and yield a rare length of thread, the silk having by many judges been estimated equal to the original Milanese. The exhibited specimens are the first products of numerous breeds of 1861.

814. STOFELLA D. A. Silk Establishment. *Roveredo, Tyrol.* Gold M. of the Lower-Austrian Trade Union 1843. Silver M. I. Cl. Vienna 1845. M. I. Cl. Munich 1854. M. I. Cl. Paris 1855. M. I. Cl. Triente 1857. Ware-house: Vienna, Stofella & Giacomozzi.

Various specimens of silk worked and unworked.

First inventor and introducer of silk à tours comptés (1841). Mechanical spinning according to the Stofella system. Yearly production 5,000 Kil. Twisting mill 6,000 Kil. That establishment receives orders for any article both in Roveredo and Vienna.

815. WIEDNER Francis, Weaver. *Raab, Hungary.*
Silk Counterpane.

816. WINKLER Theresa, Silk Producer. *Klausenburg, Transylvania.*
Raw Silk in skeins.

Class 21.

Woollen and Worsted, including Mixed Fabrics generally.

The yearly production of wool in Austria (700.000 Cwts) is almost totally worked up by home industry. The annual exports (191.000 Cwts averaging the years 1851—1860) are very little exceeding the imports (188.000 Cwts during the same period).

Any material difference is only owing to the circumstance that imports (from Russia and Turkey) comprise the coarse sorts in prevalence then, whereas the greater part of the exports concerns the superfine sorts produced from Austrian Merino sheep.

Coarse woollen manufactures and cloths of secondary quality make up the main production of Austrian weaving industry in this branch.

In the same manner as the production of linen, in the eastern portions of the empire, is exercised as a domestic industry for home wants, so woollen manufactures are produced here by the peasants. Wool is spun by the females of the household, and woven into coarse Halina-cloth in primitive looms. Besides these domestic looms, there is some cloth produced by regular weavers of Halina-cloth. Cloth of second quality in demand for the citizens of those parts must be procured from the cloth-manufactures of the western regions, with the exception of those little quantities produced by the cloth-making guilds of the Transylvanian Saxons.

Apart from the afore-named Transylvanian group, the whole Austrian cloth manufacture is almost limited to within four groups of industry viz. Reichenberg, Brünn, Iglau and Bielitz. The prevailing manufactures of Reichenberg and the surrounding country are those of the finer sorts of cloth and fashionable stuffs, of Iglau and Bielietz the common sorts, of Brünn almost exclusively the fashionable stuffs, such as peruvienne, toskin etc. Next to a considerable number of trades on a smaller scale, there have of late, in the western provinces

of the empire, been introduced manufactures of cloth and fashionable stuffs, uniting all the operations of cloth production from wool-scouring to cloth-finishing, and being provided with power looms. Spinning-mills for carding wool, worked by water power, are mostly in possession of some cloth-manufacturers or cloth-making corporations; jobbing spinning-trades of this kind are comparatively few in Austria. Moreover there is a great quantity of wool produced by hand-spinning. Upon the whole, there are computed more than 600.000 spindles for hand and machine spinning of carding wool; the number of looms (Halina-looms included) amounts to nearly 50,000.

Whereas the Austrian carding-wool spinning completely covers the wants of weaving, and exports even small quantities of that yarn (3.300 Cwts in 1860), those 9 worsted spinning mills with 30,000 spindles, existing at present do not suffice for the want of worsted; in 1860 there were imported nearly 35,000 Cwts of such yarns. The production of unfulled worsted weavings (partly mixed with cotton yarn) has its chief seats at Reichenberg and the surrounding country (for merino, thibet and orleans stuffs), and at Vienna (for shawls). The production of other mixed fabrics of carding-wool (trowsers and coat stuffs, etc.) is mainly exercised in north-western Bohemia. For the manufacture of the afore-named worsted and mixed stuffs, there are about 15.000 hand and power looms going.

The value of the whole production of woollen and mixed fabrics and yarns, woollen webs included, is estimated at 140 millions florins per annum; and more than 400,000 hands are employed.

Regarding foreign trade, there is a considerable export, having amounted to 57.200 Cwts in 1860, in comparison to an import of 8.600 Cwts.

817. AUSPITZ L's. Grand-children. Patented Cloth Manufactory. *Brünn*.
Woollen Manufactures.

818. BAUER Theodore. *Brünn*.
Woollen Manufactures.

819. BAUM Gustavus. Patented Cloth Manufactory. *Bielitz, Silesia*.
Woollen Manufactures.

820. BEUER Benjamin. *Reichenberg 115-1*.
Diverse Cloth Fabrics.

821. BLASCHKA & COMP. Manufactory of Woollen Fabrics. *Liebenau, Bohemia*. Ware-house: Vienna and Prague.
Plain Woollen Manufactures.
(See Class 23.)
This establishment, founded in 1836, comprises first the Liebenau weaving mill with finishing, dying, form-cutting and printing on 100 tables, and secondly the Katharinaberg weaving mill with 200 looms, 25 jacquard-looms, and 500 looms out of the house. The chief productions of these manufactories, where 500 hands and 2 steam-engines are working, consist in: Orleans and printed Thibet shawls, Paramattas, Lastings, Merinos, Ripse, Circas shawls, etc. etc.

822. BRÜNN, Commercial Board at.
Woollen Manufactures.

823. BUM Max. *Brünn*.
Woollen Manufactures.

824. CAFFIER & POHLENTZ. Commission-Merchants and Exporters. *Vienna, Neubau 194*. Representatives of several houses at New-York; branch business and stores of Austrian manufactures in Leipzig.
Collective Exhibition of Vienna shawl manufactures.

825. DÖPPER A. F. Present owner: Hosch Rainer. Manufactory of Woollen Fabrics. *Neutitschein, Moravia*.
Cloths of three different sorts.

826. ECKSTEIN Frederick, Director of the Moravian Upper Weaving-school. *Brünn*.
Linen table-cloth, 10/4 broad, of damask weaving in connection with gaze. Cartridge-bag for rifled cannons, of worsted thread, cylinder and round bottom woven in one piece.

827. ELGER John. *Reichenberg 41-3, Bohemia*.
Diverse Cloth Fabrics.

828. ELGER Joseph. *Reichenberg 41-3, Bohemia*.
Diverse Cloth Fabrics.

829. FELS Joseph. *Gratz*.
Machine-paper-felt without list.

830. FÖRSTER Charles. *Bielitz, Silesia*.
Cloth and Woollen Stuffs.

831. FÖRSTER John, jun. *Rumburg, Bohemia*.
Mixed Woollen Manufactures (Lama).
(See also Class 18.)

832. GÁCS, Woollen and Fine Cloth Manufactory at *Gács, Com. of Neograd, Hungary*. Silver M. Pesth 1842. Gold M. Pesth 1846. Bronze M. London 1851. Silver M. Kaschau 1857. Chief Ware-houses: Debreczin, Losoncz, Miskolcz, Pesth. Smaller Ware-houses: Kaschau, Leutschau, Lipto, Szt.-Miklos and Neusohl.
Cloth „Buskin" etc. in various qualities; Ladies shawls and Coverlets.
Established in 1762 by Count Anthony Forgách, since 1800 property of a share-holders' company, this manufactory at first produced only coarser stuffs, but indefatigably striving after improvement, it now keeps pace with those abroad. There are manufactured a great many articles, such as, cloth, mixed cloth, buskin, toskin, azore, kerchiefs, shawls, and coverlets etc. in all colours and qualities. Four medals were already awarded to its manufactures.

833. GINSKEY Ignatius. *Maffersdorf, Bohemia*.
Woollen Coverlets.

834. GINZEL Francis. *Reichenberg 14-1, Bohemia*.
Cloth.

835. GOEBEL C. *Jägerndorf, Silesia*.
Woollen Stuffs for coats and trousers.

836. GÖBL John. Manufactory of Silk Stuffs. *Vienna, Schottenfeld 257*.
Mixed Silk and Fashionable Gowns.

837. GÖCKEL, S. C. *Hermannstadt, Transylvania*.
Cloth.

838. GRUNER Francis. *Reichenberg 145-3, Bohemia*.
Diverse Cloth Fabrics.

839. HAAS Philip & Sons. Webbing Manufactory *Vienna*.
Woollen Furniture Stuffs.
(See Classes 20 and 22.)

840. HÄRTLT Henry. *Reichenberg 183-3, Bohemia*.
Diverse Cloth Fabrics.

841. HÄRTLT Severin. *Reichenberg 75-4, Bohemia*.
Diverse Cloth Fabrics.

842. HAYDTER Sebastian, Patented Manufactory of Shawls. *Vienna, Gumpendorf 328*. Silver M. Vienna 1845. Silver M. Leipzig 1850. Silver M. Munich 1854. Silver M. Paris 1855. Honourable Mention London 1851. Ware-house at New-York, Ladewig & Haydter
Long Shawls and Square Shawls.

843. HEIDE John, jun. *Jägerndorf, Silesia*.
Woollen Stuffs for coats and trousers.

844. HEIDE Ralph. *Jägerndorf, Silesia*.
Woollen Stuffs for coats and trousers.

845. HENNINGER Francis. *Reichenberg 157-1, Bohemia*.
Diverse Cloth Fabrics.

846. HENNINGER Joseph. *Reichenberg 336-1, Bohemia*.
Diverse Cloth Fabrics.

847. HERSCHMANN Henry, Woollen Manufacturer. *Brünn*.
Various Coupons of Woollen Manufactures.
The Circassian Coupon Nr. 1, 29,458 is worked on a simple treddle-loom without machine and out of one warp. It contains three different colours in two sorts of breadths, distributed so that one half of the piece, in double broad fabric, contains a larger square, the second half of the piece two different patterns in smaller squares (woven in one whole) besides each other, the whole to be divided into two simple broad coupons, each of which forms a pattern by itself.

848. HIEBEL Joseph. *Reichenberg 137-4, Bohemia*.
Diverse Cloth Manufactures.

849. HIEBEL Wenceslas. *Reichenberg 249-1, Bohemia*.
Diverse Cloth Manufactures.

850. HIRT Charles' Sons. *Wagstadt, Silesia*.
Fashionable Woollen Articles.

851. HLAWATSCH & ISBARY. Patented Manufactory of Shawls. *Vienna, Gumpendorf. 224 and 225*.

Agent in London: Messrs. Rawlinson & van der Beeck, 33, Bread Street; in Manchester: Mr. J. G. Fockish, 3, Milk Street. M. II. Cl. Paris 1855.
Shawls.

This manufactory produces all sorts of „Broché" shawls, not only of wool, silk and cotton, but also of real Indian cashmere wools, there being a great variety of the newest patterns. That manufactory has ten pattern designers of its own, in Vienna, does all the by-works belonging to the manufacture of shawls, and possesses a steam finishing-apparatus after the newest system. During the last five years the following quantities of yarn were worked up:

	Woollen yarns Engl. Pounds	Cotton yarns Engl. Pounds	Spunsilk Engl. Pounds	Woollen warps Engl. Pounds
1857	47.529	3.480	4.635	1.650
1858	28.006	6.765	3.036	730
1859	48.830	3.340	6.693	665
1860	83.066	3.365	9.826	3.140
1861	70.696	6.230	6.082	5.228
Total	278.127	23.380	30.272	11.413

Of these materials, there were manufactured:

in the year	by workmen	number of shawls	valued at fl.
1857	405	26.118	280.567
1858	347	23.511	230.000
1859	577	32.097	449.358
1860	804	47.998	713.000
1861	796	48.353	653.000
Total	2.929	178.077	2,325.925

The fourth part of these productions were sold in the Austrian empire, three-fourths of them to America, the Tariff Union States, Italy, Russia, England and the Orient.

852. HOFFMANN Anthony. *Reichenberg 88-2 Bohemia.*
Diverse Cloth Fabrics.

853. HOFFMANN Joseph. *Reichenberg 61-3, Bohemia.*
Diverse Cloth Fabrics.

854. HOFFMANN William. *Reichenberg 24-3, Bohemia.*
Diverse Cloth Fabrics.

855. HOFFMANN, GOENNER & Comp. *Klein-Baranau near Iglau, Moravia.*
Woollen Manufactures.

856. HOFMANN Aloysius & Comp. *Tischnowitz, Moravia.*
Satincloths.

857. HORN Adolphus. *Reichenberg 97-4, Bohemia.*
Diverse Cloth Fabrics.

858. HORN Christopher. *Reichenberg 97-4, Bohemia.*
Diverse Cloth Fabrics.

859. HÜBNER Francis. *Reichenberg 114-1, Bohemia.*
Diverse Cloth Fabrics.

860. HÜBNER Francis. *Reichenberg 128-1, Bohemia.*
Diverse Cloth Fabrics.

861. JAKOB Francis. *Reichenberg 126-3, Bohemia.*
Diverse Cloth Fabrics.

862. JAKOWITZ William. *Reichenberg 68-3, Bohemia.*
Cloths.

863. ILLEK Francis. *Brünn.*
Woollen Manufactures.

864. KAFKA Henry. *Brünn.*
Woollen Manufactures.

865. KAHL Francis. *Reichenberg 46-3, Bohemia.*
Diverse Cloth Fabrics.

866. KAHL Brothers. *Reichenberg, Bohemia.*
Diverse Cloth Fabrics.

867. KALLAB Francis & Sons. Patented Manufactory of Cloth. *Gross-Meseritsch, Moravia.* Prize-M. Vienna 1844. Honourable Mention: Munich 1854. Ware-house: Vienna, Köllnerhofgasse.
Woollen Stuffs.

868. KASPER Ferdinand. *Reichenberg 138-3, Bohemie.*
Diverse Cloth Fabrics.

869. KASPER Joseph. *Reichenberg 333-4, Bohemia.*
Diverse Cloth Fabrics.

870. KEIL'S Anthony Widow. *Reichenberg 167-4, Bohemia.*
Diverse Cloth Fabrics.

871. KELLER Joseph, Patented Manufactory of Wool-Spinning. *Brünn.*
Assortment of Carding Yarns.

872. KERN Henry, Army Cloth Manufactory at Altenberg. *Vienna, City 625.*
Cloth.

873. KIRSCH F. J. Sons. *Brünn.*
Woollen Manufactures.

874. KLAUS J. E. *Niedergrund, Bohemia.*
Mixed Woolen Stuffs for coats and trousers. (See Classes 18 and 19.)

875. KLEIBER A. Shawl Manufactory. *Vienna, Schottenfeld 336.* Agent: J. L. Meyer Sons & Comp. London, 21, 22, Iron monger Lane, Cheapside.
„Broché" Long Shawls and Square Shawls of various qualities, „Stella" and „Rayé" Shawls.

This manufactory is established since 1834, and produces Stella and Rayé Shawls in a variety of patterns, especially working for export. During this period of 28 years, that establishment has acquired the merit of having made important inventions and improvements of general introduction in the manufacture of shawls. Deserving of especial notice is the shawl Nr. 2997 (black plain ground worked with silk) which substitutes the Indian embroidered shawls, being manufactured in a highly artistic style. This manufactory has not yet taken part in any exhibition, but has, from 1840 to the present time, continually employed 120 and sometimes 200 hands, and is enabled to execute extensive orders at the shortest notice. Samples of borders and „Stella carré" are kept ready for inspection by his agent.

876. KLINGER Francis. *Reichenberg 335-1, Bohemia.*
Diverse Cloth Fabrics.

877. KOHN Max. Woollen Manufacturer. *Brünn.*
Woollen Manufactures.

878. KÖNIG Joseph I. *Reichenberg 160-3, Bohemia.*
Diverse Cloth Fabrics.

879. KÖNIG Joseph II. *Reichenberg 274-3, Bohemia.*
Diverse Cloth Fabrics.

880. KÖNIG Wenceslas A. *Reichenberg 167-3, Bohemia.*
Diverse Cloth Fabrics.

881. KRÄMER B. S., Manufactory of Fashionable Stuffs. *Vienna, Gumpendorf 570.*
Shawls.

882. KRECZY Brothers. *Brünn.*
Woollen Manufactures.

883. KRZYSZTOFOWICZ Francis. *Trybuchowce, District of Czortkow, Galizia.*
Coarse Cloth.

884. LANG John N. *Vienna, Mariahilf 45.*
Woollen Scarfs and Travelling Shawls. (See class 18.)

885. LARISCH Aloysius. Cloth Manufactory. *Jägerndorf, Silesia.*
Cloths.

886. LEIDENFROST Edward & Sons. Patented Wool Spinning Mill. *Brünn.*
Assortment of Carding Yarns.

887. LEUBNER Ferdinand. *Reichenberg 276-3, Bohemia.*
Diverse Cloth Fabrics.

888. LIEBIEG Francis, Woollen Manufacturer. *Reichenberg, Bohemia.* Ware-house: Vienna 767. Bronze Medal Leipzig 1850. M. I. cl. Paris 1855.
Woollen Gowns.
(See class 23.)

889. LIEBIG John & Comp. *Reichenberg, Bohemia.*
Mixed Woollen Manufactures, printed, dyed and plain.
(Also class 23.)

890. LÖW Adolphus & **SCHMAL**. Patented Manufactory of Woollen Vigogne Wares. *Brünn.* Branch establishment at Iglau, Moravia.
Several coupons of Woollen and Vigogne Manufactures.

This manufactory was founded in 1843 on a very small scale only as weaving concern with two looms. From this humble beginning the business gradually increased in extent, and in 1854 was first set up a steam-engine of 20 horse-power. In 1859, the branch establishment at Iglau was bought, and weaving, spinning and finishing of fashionable stuffs arranged therein. The manufactory at Brünn is worked with steam-power alone, the branch establishment at Iglau, both with steam and water-power; in connection with the spinning mill of both these establishments, there is also a special carding yarn spinning mill. At present there are working: machines of 80 horse-power of steam and 30 horse-power of water, 44 assortments of spinning, 700 looms, 500 of which are in the manufactories, and 200 are set up in the houses of country weavers; moreover complete finishing and dyeing apparatus are in connection with the establishment, where 15,000 pieces of fashionable articles are manufactured, per annum, both for home and foreign sale, much business being transacted with exporters of Hamburg, Berlin and Leipzig.

891. MAY Charles. Shawl Manufactory. *Vienna, Gumpendorf 502.*
Long and Square Shawls.

892. MAYER Charles. Woollen Manufacturer. *Brünn.* Agents: Ch. Nolde & Comp. London; Charles Mahler, Leipzig. Honorary Medal Munich 1854. Silver Medal Paris 1855.
Several coupons of Woollen Manufactures.

893. MAYER Edward & **SCHILLER**, Woollen Manufacturers. *Brünn.* Agents: William Doubleday, London; Charles Mahler, Leipzig.
Several coupons of Woollen Manufactures.

894. MICHEL Philip. Patented Manufactory of Woollen, Linen and Cotton Webbings. *Gärten near Schönlinde, Bohemia.*
Woollen Webbings of various fineness and size.
(See classes 18 and 19.)

895. MORO Brothers. Manufactory of Fine Cloth and Woollen Stuffs. *Klagenfurt, Carinthia.* Honourable Mention London 1851. Great Medal Munich 1854. M. I. cl. Paris 1855.
Woollen Stuffs.

This manufactory was founded in the second half of last century, and is since the beginning principally active in the production of cloth and woollen stuffs in white as well as all other colours, from the lightest to the darkest tints. The chief task proposed is, to produce the light and tender colours as clear and lively as possible, and they have completely succeeded in this task both in general and in particular regarding the white

and scarlet colours. Moreover, this establishment has turned its attention to supplies for the army, producing all sorts of woollen stuffs in demand for the uniforms both of Austrian and foreign corps. In these articles, that is, woollen stuffs of white or any other light colour, as well as in all cloths and stuffs suitable for regimentals, this manufactory carries on a brisk business not only in Austria but also abroad, a considerable number of foreign corps providing themselves with productions from this establishment.

896. MÜLLER Anthony Ed. *Reichenberg 323-3, Bohemia.*
Diverse Cloth Fabrics.

897. MÜLLER Anthony J. *Reichenberg 123-4, Bohemia.*
Diverse Cloths.

898. MÜLLER Anthony Lewis. *Reichenberg 145-4, Bohemia.*
Diverse Cloth Fabrics.

899. MÜLLER Wenceslas. *Reichenberg 338-4, Bohemia.*
Diverse Cloth Fabrics.

900. NAMIEST. Patented Fine Cloth Manufactory of Baron Charles Puthon & Comp. at. *Moravia.*
Woollen Manufactures.

901. OFFERMANN John Henry. Patented Fine Cloth Manufactory. *Brünn.* Silver and Gold Medal at several exhibitions.
Woollen Manufactures.

This manufactory founded under the firm of Henry Offermann, the father of the present owner Charles Offermann, exists since 1782, produces 10.000 pieces of wares in various qualities of cloths and fashionable stuffs, and is connected with a branch establishment at Fussdorf for the manufacture of army cloth. These manufactories are provided with all machines of the newest construction and are conducted on a very extensive scale. At the exhibitions of Vienna in 1835 and 1839, Berlin 1844, and Leipzig 1850, its productions were honoured with the Silver Medal, and in 1845 in Vienna with the large Gold Medal, whereas Charles Offermann having been Proxy Jury in 1851 at the London Exhibition, and Jury at Munich in 1854 and at Paris in 1855, consequently also excluded from receiving Medals, he received the knight's cross of the Imp. French légion d'honneur. Charles Offermann sen. was, on account of his merits in industry, distinguished by His Majesty the Emperor Francis with the middle Gold Medal for civil merit, by His Majesty the Emperor Ferdinand with the great honorary Medal for civil merit, and by His Majesty the Emperor Francis Joseph with the knight's cross of the Francis Joseph Order.

902. PFEIFER Giles, Patented, Manufactory of Cotton, Woollen and Linen Wares. *Rumburg, Bohemia.* Ware-house: Vienna, City 353.
Woollen Manufactures.

Woollen and mixed woollen stuffs for gowns are produced in great variety. Price ³/₄ Vienna ells broad, according to quality 18 cr., to 1 fl. upwards, per Vienna ell. The manufacture of these stuffs is but endennized in Rumburg since 15 years, yet is has attained an extent and perfection rendering protection-duty for this special article quite unnecessary. These articles have rather become articles of export, and are actually sent to Italy, to the Danubian Principalities, to Asia Minor and other provinces of Turkey, where they distance manufactures of the Tariff Union States. America, too, imports many of these productions. That manufactory produces also cotton stuffs for gowns in great variety, ⁹/₈ Vienna ells broad. With the present high prices of yarns, the price of a yard of export wares is 3¹/₄ Prussian Ngr. Better wares destined for the Austrian market, fast colours, are paid, loco Vienna, 20 cr. per Vienna ell. Besides Austria, Italy and in recent times also America, buy many of this articles; 20,000 pieces are annually furnished by that manufactory. The raw material, 40 medio warp and 50 mule warp, is imported from England.

903. PINTNER Wenceslas. Manufactory of Woollen Wares. *Brünn.*
Fashionable Wares.

904. PISKO Henry. *Brünn.*
Woollen Manufactures.

905. POPPER Brothers. Fine Cloth Manufactory. *Brünn.*
Woollen Manufactures.

906. POSSELT Francis A. *Reichenberg 342-3, Bohemia.*
Diverse Cloths.

907. QUITTNER James. Army and Fine Cloth Manufactory. *Troppau, Silesia.*
Army Cloths in the chief colours of the Austrian army. Woollen Stuffs for coats.

908. RASCHKA John jun., Cloth Manufactory. *Freiberg, Moravia.*
Cloths.

909. RATHLEITNER James & Son. *Gratz.*
Styrian Unfulled Cloth.

910. REDLICH Frederic. *Brünn.*
Woollen Manufactures.

911. REDLICH Maurice. Patented Cloth Manufactory. *Brünn.*
Woollen Manufactures.

912. REICHENBERG, Weaving-Scholl at. *Bohemia.*
Various Webbing Samples.

913. ROHN Joseph. *Reichenberg 136-4, Bohemia.*
Diverse Cloth Manufactures.

914. ROHN Charles. *Reichenberg 156-3, Bohemia.*
Diverse Cloth Manufactures.

915. SALOMON Charles. *Reichenberg, Bohemia.*
Diverse Cloth Manufactures.

916. SALOMON Francis. *Reichenberg 255-3, Bohemia.*
Diverse Cloth Manufactures.

917. SALOMON Ignatius. *Reichenberg 342-3, Bohemia.*
Diverse Cloth Manufactures.

918. SALOMON Joseph J. *Reichenberg 78-4, Bohemia.*
Diverse Cloth Manufactures.

919. SALOMON Joseph, *Reichenberg 21-2, Bohemia.*
Diverse Cloths.

920. SALOMON William. *Reichenberg 98-4, Bohemia*
Diverse Cloth Manufactures.

921. SCHMIEGER A. & Comp. Patented Manufactory of Worsted and Carding Yarns, and Woollen Fabrics. *Neudek, Bohemia.* Prize-M. London 1851. Honourable Mention Munich 1854. Paris 1855. Ware-houses at Vienna and Prague.
Various sorts of Worsted and Carding Yarns, and Woollen Manufactures.

This manufactory, in connection with dyeing, exists since 1848 and un interruptedly occupies 600 workmen. The whole establishment is worked with water-power and a steam-engine of 20 horse-power. The production of Hungarian and Russian woollens consists, per annum, in: 60,000 pounds of worsted, No. 10 to 100 upwards, 200,000 pounds of carding yarns, No. 6 to 40 upwards, as well as 3,000 pieces of webbings: cashmere, thibet, mousseline of any breadth, and 20,000 areas and long shawls.

922. SCHMITT F. *Böhmisch-Aicha, Bohemia.*
Printed and non-printed Woollen and Mixed Manufactures.
(See class 23.)

923. SCHÖLL Augustus. *Brünn.*
Woollen Manufactures.

924. SCHOELLER Adolphus. Manufactory of Woollen and Felt Wares, *Brünn.*
Felts.

925. SCHOELLER Brothers. *Brünn.*
Woollen Manufactures.

926. SIEBENEICHER Leopold jun., *Reichenberg, 151-3, Bohemia.*
Diverse Cloth Manufactures.

927. SIEBENEICHER Wenceslas. *Reichenberg 284-3, Bohemia.*
Diverse Cloth Manufactures.

928. SIEGMUND Wenceslas. *Reichenberg 84-4, Bohemia.*
Diverse Cloth Manufactures.

929. SIEGMUND William. *Reichenberg.* Branch establishment at Friedland, Bohemia.
Cloth.

930. SPITZ S. Woollen Manufacturer. *Brünn.*
Woollen Manufactures.

931. STOSSIMMEL Anthony. *Reichenberg, 198-3, Bohemia.*
Diverse Cloth Manufactures.

932. STRAKOSCH Brothers. Manufactory of Woollen Fabrics. *Brünn.* Bronze M. Vienna 1845. Gold M. Leipzig 1850. Gr. M. Munich 1854. Silver M. Paris 1855. Warehouses: Leipzig, Caffier & Pohlenz; Hamburg, C. Pfennig.
Various coupons of Woollen Manufactures.

933. STRAKOSCH Ralph. Patented Manufactory of Woollen Fabrics. *Butschowitz, Moravia.* Honourable Mention Munich 1854. Ware-house: Brünn, Ledergasse 38.
Various coupons of Woollen Manufactures.

934. STRAKOSCH Salomon Sons. Manufactory of Woollen Fabrics. *Brünn.*
Woollen Manufactures.

935. STRNISCHTIE Charles & Comp. *Brünn.*
Woollen Manufactures.

936. TEUBER Joseph & Sons. Patented Wool Spinning Mill. *Brünn.* Gold M. Vienna 1845. Gr. M. Munich 1854. M. I. cl. Paris 1855.
Diverse Carding Yarns.

This establishment was, in 1858, purchased by the present owners from a society that had, a year before, bought it from the firm H. F. and E. Soxhlet. Under the latter, this establishment was distinguished at the exhibitions in Vienna 1845, Munich 1854 and Paris 1855. At present this mill possesses 52 assortments for jobbing spinning with about 21,000 spindles, works up, per annum, about 12,000 Vienna Cwts of washed and dyed wool, producing from 6,500,000 to 7,000,000 skeins and occupying about 600 hands. Moreover there is a gas manufactory and workshop for machinery. The exhibited articles are: 30 different sorts of carding yarns, white, coloured, mingled, flamed and twisted, from 3 to upwards of 22 skeins per Vienna pound. One skein equal to 1,760 Vienna ells.

937. TRENKLER Adolphus. *Reichenberg 164-2, Bohemia.*
Diverse Cloth Manufactures.

938. TRENKLER Anthony & Sons. *Reichenberg 31-5, Bohemia.*
Cloths.

939. TRISZTINOVITS Michael, Galloon and Double Maker. *Pesth.*
Galloon and Double Wares.

940. TSCHÖRNER Joseph. *Reichenberg 96-4, Bohemia.*
Diverse Cloth Manufactures.

941. TUGEMANN Geoffrey. *Reichenberg 22-2, Bohemia.*
Diverse Cloth Manufactures.

942. ULLRICH Adolphus. *Reichenberg 164-3, Bohemia.*
Diverse Cloth Manufactures.

943. ULLRICH Andrew S. *Reichenberg 97-2, Bohemia.*
Diverse Cloth Manufactures.

944. ULLRICH Anthony jun. *Reichenberg 163-4, Bohemia.*
Diverse Cloth Manufactures.

945. ULLRICH Francis. *Reichenberg 48-3, Bohemia.*
Diverse Cloth Manufactures.

946. VONWILLER & Comp. Cloth Manufactory.
Ware-house: *Vienna, City 727.*
Stuffs for trousers and coats, Cloths of wool.
(See Classe 19.)

947. VÖSLAU, Worsted Yarn Shareholders' Company at. Lower-Austria. Gr. M. Munich 1854. Silver M. Paris 1855 (for spinning and dyeing) and M. II. Cl. for the Director C. L. Falk.

Worsted Yarn (shute, warp and twisted). Arras Yarn (twisted, lacemaker's, carpet-maker's etc.). Sampler of fourfold twisted zephyr-yarn in 1600 tints. Quality sampler of fourfold zephyr-yarn in some tints, 3 sorts. Samples of embroiderer's yarn. Small sampler of current tints.

That manufactory consumes, annually, about 9,000 Cwts of fleece combing wool from Hungary, Transylvania, the Danubian Principalities and Russia, occupying about 450 persons and producing, weekly, about 10.000 pounds of yarn on as many spindles. More than 5,000 lbs. of it are dyed in nearly 1.600 tints. The advantages offered by this establishment are variety of spinning (No. 6 to 52 upwards for various purposes) and particularly in dyeing. The establishment succeeded in distancing the previously powerful foreign competition in this branch by equal beauty of productions, and in opening a pretty considerable export-trade in dyed yarns to Italy, the Orient and South America. This manufactory is incessantly endeavouring to introduce improvements and has in Austria first practically brought to bear nearly all innovations in machinery relating to worsted yarn spinning and dyeing.

948. WEINLICH Francis. *Vienna, Gumpendorf 538.*
Mixed Woollen Shawls.

949. WESTHAUSER Joseph. *Vienna, Gumpendorf 345.*
Fashionable Stuffs.

950. WILSCH John. *Jägerndorf, Silesia.*
Woollen Stuffs for coats and trousers.

951. WOLFF Albert, Place of Manufactory: Gross-Siegharts. *Vienna, City 500.*
Furniture Stuffs.

Class 22.

Carpets.

Carpets of the commonest sort are manufactured in NW. Hungary together with coarse wool-tissues and horse-cloths This industry, with the exception of two rather extensive establishments, lies entirely in the hands of individual tradesmen and pedlars, these last serving as intermediaries between the producers and the home-consumers of the articles in question.

Table-covers of the common sort (generally with squared patterns) are produced at home and brought into commerce by the inhabitants of the Pusterthal (Tyrol).

A peculiar sort of Carpets, composed of Cloth-lists, is manufactured in an extensive manufacturing establishment at Vienna.

Similar establishments, converting either the waste, or more or less fine sorts of wool into patterned Carpets, are spread but in small numbers (in all, no more than seven), over Upper and Lower Austria and the environs of Reichenberg (Bohemia).

Table-covers of a somewhat refined quality, (to be mentioned under the article of „Printed tissues") are produced in the woollen-manufactories.

The manufacture of carpets of every description in Austria gives employment to about 3.000 persons and the average value of the articles thus brought into commerce every year may amount to about 2 millions florins.

---

952. (Omitted.)

953. HAAS Philip & Sons. Weaving Manufactory. *Vienna.*

(See Classes 20 and 21.)

954. HARBANDER Charles & Son, Weavers. *Freudenthal, Silesia.*
Table Cover.

955. ZUSCHIK Joseph, Pattern Designer and Painter. *Vienna, Gumpendorf 124.*
Painted Carpet.

Class 23.

Woven, Spun, Felted, and Laid Fabrics, when shown as specimens of printing or dyeing.

Dyeing (both of yarns, as red-yarn-dyeing, and of stuffs, as dyeing in high colours) and printing of cotton goods is, in Austria, almost exclusively the trade of special printing manufactories and small trades of printers and dyers. In Vorarlberg alone, there are some weaving establishments which at the same time print their goods. Whilst the minor printers and dyers are only jobbing, the more extensive cotton printing establishments, centred in Vienna, Prague, Pesth and Reichenberg, buy raw stuffs for their own account, in order to bring them into trade when manufactured. The greater part of laid fabrics is printed with the help of rollers and perotines, the rest, as well as all sorts of cotton-shawls, are left to hand-printing.

The same prevails with silk-goods, only that silk-dyeing and printing is almost exclusively exercised in Vienna and its neighbourhood. The dyeing of sewing and knitting silk is done by smaller trading-concerns in Verona and its neighbourhood.

Dyeing and printing of woollen and mixed wares (orleans etc.), in opposition to cotton and silk wares, is nearly all done in the respective weaving establishments of Bohemia. In Vienna and its neighbourhood alone, there are several printing establishments for woollen goods, and some of them united with weaving manufactories have branch printing houses.

The extent of dyeing and printing establishments is to be judged of by the fact that, in 1860, there were imported of colouring-matter: 163,600 Ctws of dyeing-wood in blocks, 34,000 Ctws of madder (besides home production amounting to 6,000 Cwts), 2,500 Cwts of cocheneal, 11.700 Cwts of indigo.

The number of people occupied in the Austrian dyeing and printing concerns amounts to about 100,000 men, women and children. The value of their production is included in the several branches of weaving industry, from the reason that, as before mentioned, the exercise of these trades mostly coincides with that of weaving.

---

956. BINDELES Joseph, Academical Artist and Teacher of Calligraphy. *Prague 118-5.*

Patterns for Embroidery.

The exhibited patterns serving for design-printing on stuffs, as well as those for letter-designs on stone monuments, to be made in a few minutes without proficiency in drawing, are invented by the exhibitor.

957. BLASCHKA & Comp., Manufactory of Woolen Fabrics. *Liebenau, Bohemia.* Ware-houses: Vienna and Prague.

Printed Woollen Manufactures.

(See also particulars about that manufactory Class 21.)

958. BOSSI Joseph, Manufactory of Printed Fabrics. *St. Veit near Vienna.* Ware-house: Vienna 648. Agent in London: Professor G. Hieser. Silver M. Vienna 1845. Leipzig 1850. Prize-M. London 1851. Bronze M. New-York 1853. Honourable M. Munich 1854. M. I. Cl. Paris 1855.

Shawls, Gowns and Stuffs of wool, Cashmere with silk, Barege.

This establishment produces printed stuffs partly with hand, partly with machine printing.

959. DORMITZER Leopold Sons, *Koleschowitz near Prague.*

Printed Cottons and other stuffs.

960. FRÖHLICH G. A. Son, Patented Manufactory of Printed Stuffs. *Warnsdorf, Bohemia.* Silver M. Vienna 1835. Leipzig 1850. Bronze M. London 1851. Honourable M. Munich 1854. Ware-house: Vienna, City 354.

Printed Cotton Wares.

This manufactory produces printed cotton wares, such as: printed, pressed and dyed velvets, printed piqués, orientals, cooper, satins, molesquins, tueffel; in general all articles belonging to the line of printed stuffs for trousers, coats and waistcoats. Its productions are, for the most part, intended to supply the wants of the working classes, and are not only distinguished for their cheapness, but also for their durability and fine appearance. This establishment exists since 1807, and was the first that, some years ago, performed, in oil-varnish and

with cylinders, the comparatively durable mineral-colour printing on velvet, piqués, molesquins and tueffel, by means of which velvets and piqués receive a peculiarly silklike, and molesquins and tueffels a woollenlike appearance. That manufactory possesses a cylinder-printing machine for 2, and one for 3 colours, 1 design calender, 2 mangles, 2 engraving machines, 1 steam dyeing and bleaching establishment, 4 boilers of 18, 20, 40 and 50 horse-power, 2 steam-engines of 10 and 20 horse-power. The production in 1861 amounted to 160,722 pounds of printed and cotton manufactures jobbing work, and 152,824 pounds of cotton wares export for the Tariff Union States, with an annual exchange of from 400,000 to 500,000 fl. This manufactory now works on 100 hand looms and employs a number of factors who work on 400 hand looms for the account of this establishment. At present there is a machine weaving building for 100 double looms. The sale of its productions is mediated through the manufactory at Warnsdorf and the ware-house at Vienna, whence they are sent to all the provinces of the Austrian empire and to the German Tariff Union States.

961. GANAHL Charles & Comp., *Feldkirch, Vorarlberg.*

Turkish red cotton yarn.

962. GETZNER & Comp., Cotton Spinning Mill, Turkish Red Dying Establishment, Machine Looms, Bleachery and Cotton-Ware Manufactory. *Feldkirch, Vorarlberg.* Ware-houses at Vienna, Linz and Verona. Agent: Messrs. William Bunge & Comp. London.

Turkish Red Cotton Yarns (Prime Colour).

A Packet of Nr. 4 seconds (from vaste),
   "   "   "   "   10 L. M. mule yarn,
   "   "   "   "   20 throstle yarn,
   "   "   "   "   40 L. mule yarn of pure Louisiana,
   "   "   "   "   30 Makó of pure Makó cotton,
   "   "   "   "   50   "   "   "   "   "

963. GOLDBERGER S. F. & Sons. *Pesth.*

Printed Cottons.

(See Class 18.)

964. HEINTSCHEL E. & Comp. *Heinersdorf, Bohemia.* Ware-house: Vienna, City 483.

Printed Woollen Manufactures.

965. HILLER Francis, Patented Manufactory of Woollen Stuffs. *Jungbunzlau, Bohemia.* Ware-houses at

Vienna and Prague. Agent in London: Augustus Koch & Comp. from Vienna.

Printed Cashmere, Merino and Wool Shawls.

This establishment, though not founded before 1855, yet already counting among the most extended and renowned of the empire, consists of 7 principal and 8 accessory buildings. Of the former are to be particularized: — the counting-house with stores of yarns, raw and finished manufactures; the goods-press and fringing locality; the engine-house, boiler-house, printing and weavings house, the building for hand and cylinder form engraving, besides studio for drawing and joiner's workshop. The accessory buildings contain the locksmith and forge houses, the sucking, steam contrivance for orleans, cashmere shawls etc. the stores for dyeing substances and building materials etc. As motors are used: 2 steam-engines with 50 horse-power each, 4 boilers, 3 graving machines, 6 washing machines, 7 shearing machines, 2 drying apparatus, 1 Dutch mangle, 2 hydroextractors, 1 embossing machine, 2 rouleaux, 2 perrotines, 2 dyeing-wood extractors, 1 lathe, 3 engraving machines, 1 hydraulic press, 45 machine printing tables, 108 hand printing tables, 120 regulator and 140 power looms, 13 weaver's aiding machines etc. The establishment occupies within the premises more than 900 persons, and out of them about 1,000 jobbing weavers, 220,000 fl. being paid for wages a year. There are produced all sorts of woollen and mixed manufactures, especially printed thibet shawls, cashmeres, printed cashmeres, circas and circas shawls, orleans and orleans shawls, plain patterned and printed coat stuffs, merinos, lastings etc. both for inland and export trade, for the latter of which the printed thibet shawls are peculiarly adapted, and actually go to the several Italian states and to South America. The yearly production of that establishment averages a value of 1,800,000 fl.

966. HOTTINGER Adolphus. *Vienna.*
Dyed Wool.

967. LEITENBERGER Francis. Owner: Leitenberger Frederick. Patented Cotton Printing Manufactory. *Cosmanos, Bohemia.* Ware-houses at Vienna, Prague and Linz. Gold. M. Vienna 1835, 1839, 1845. Prize-M. London 1851. Gr. M. Munich 1854. M. I. cl. Paris 1855.

Printed Cotton Manufactures.

This manufactory produces, annually, 30,000 pieces of about 60 Vienna ells of cottons, percalines, cambrics, coeper, croisé, brillantins, mixed woollen stuffs and kerchiefs.

968. LIEBIEG Francis. Woollen Manufactory. *Reichenberg, Bohemia.* Ware-house: Vienna 767. Bronze M. Leipzig 1850. Hon. M. Munich 1854. M. I. cl. Paris 1855.

Printed Woollen Manufactures.
(See Class 21.)

The exhibitor manufactures dyed and printed orleans, orleans shawls, dyed and printed thibets, printed furniture stuffs, printed table covers, dyed and printed thibet and mousseline shawls of all sizes with silk and without, printed thibet and mousseline long shawls with silk and without, variegated woven woollen shawls and long-shawls of various sizes, raw white and coloured tickens, as well as calicos for lining, manufactured for inland and foreign trade.

969. LIEBIEG John & Comp. *Reichenberg, Bohemia.*
Mixed Woollen Manufactures, printed, dyed and plain.
(See Class 21.)

970. MAYER V. & Sons. Owners of a Patented Cotton and Woollen Printing, and a Chemical Product Manufactory. *Guntramsdorf near Vienna.* Hon. Mention at the Industrial Exhibition in Vienna 1839.
Fifty printed Mousselines de laine.

The printing manufactory, founded in 1814 in Vienna, exists since 1834 in Guntramsdorf, and exclusively produces mousselines de laine, there being employed about 200 hand printers, 4 perrotines, 3 variegated rouleaux machines, which, like the rest of the mechanical motors, are worked partly with water-power, partly with a steam-engine. The establishment occupies 350 workmen, producing 60,000 pieces annually, which are sold at home and exported to Italy and South America. In the Manufactory of Chemical products, the chief production is tartaric acid for the American and Russian market, besides all chemicals requisite for printing manufactures. Samples and prices of the chemicals to be had at Mr. Herman Rubeck's, little Tower Street, City, London.

971. PORGES Brothers. *Smichow near Prague.*
Printed Cottons and other printed manufactures.

972. RÖDEL George, Professor at the Weaving School at *Brünn.*
Title page of a new ornamental work for artists and industrials, composed by the exhibitor. Brünn 1862. Distinctions received at Vienna, Dresden and Paris.

973. SCHMITT F. *Böhmisch. Aicha, Bohemia.*
Printed Woollen Manufactures.
(See Class 21.)

974. SEEBACH, Turkish Red Dyeing Establishment at. Ware-house: Vienna, City 901.
Red Yarns.

975. SPINNING MILL and DYEING in Adrianople Red. *Aidussina, Goritz.*
Cotton Yarns, red dye, for the use of Adrianople.
(See Class 18.)

976. THOMAS Leopold. *Graslitz, Bohemia.* Ware-house: Vienna, City 413.
Printed Woollens.

Class 24.

Tapestry, Lace and Embroidery.

Lace-bobbins have, since the sixteenth century, been handled in the unfertile high-land plains of the Bohemian Ore-mountains. By simplifying the method of Belgium, it was possible to produce common sorts of lace of linen-thread at cheap prices. In consequence of this circumstance, this branch of industry was speedily developed to a large extent; in the beginning of the present century, the number of hands employed with lace-making in the Bohemian Ore-mountains, amounted to more than 60.000 men, women and children.

Through the introduction of the bobbin-net-frame in Austria (1831), hand working was restricted to the use of cheaper cotton-thread, and such competition created as not only to reduce the number of bobbiners to a minimum, but also to make these, at times, suffer from want of occupation. At present, the number of those occupied with the production of common laces amounts to hardly 8.000 persons; joined to whom are 4.000 females producing valenciennes, point, and plate laces.

With the decrease of bobbin hand work, numbers of the females took to tambouring and embroidery. Since the beginning of this century the number of female embroiderers, in the western parts of the Bohemian Ore-mountains, has, up to the present moment, risen to nearly 8.000. Whilst there embroidery is solely exercised as a domestic industry, there are in Vorarlberg several manufactories in activity.

Besides embroidery in white, artistic embroidery is exercised only in Vienna, Prague and other principal cities of the empire.

The production of lace and works of embroidery employs in all near 40.000 persons (mostly women) and reaches yearly a value of about 8 millions florins, for the greatest part answering the demands of the inland trade.

---

977. BELLMANN Anne. *Prague.*
Embroidered Curtains.

978. BENKOWITS Maria, jun. *Vienna, City 1100.*
Artistic Embroidery, a picture.

979. DRAKULIC Smiljana. *Korlnica, Military Borderland.*
Embroidered Kirtle.
(See Class 27.)

980. FARKAS Caroline (11 years of age.) *Pesth.*
Embroidery.

981. FUCHS John, Sons. Manufactory at *Graslitz, Bohemia.* Ware-house: Vienna, City 441.
Laces and Embroidery.

982. GOSSENGRÜN, *Stad, Bohemia.*
Bone-lace.

983. HOFFMANN Leopoldine. *Freudenthal, Silesia.*
Embroidery.

984. HÜBER Joseph. *Hirschenstand, Bohemia.*
Tamboured Laces, Shawls.

985. MARIOTTI Catharine. *Trentino, Tyrol.*
Laces and Embroidery.

986. NOWOTNY Francis. Manufactory of Embroidery and Shawls. Ware-house of Fashionable Stuffs. *Vienna 1144.* Gold M. Vienna 1845. Hon. M. Munich 1854. M. I. Cl. Paris 1855.
Cashmere Shawls with Indian embroidery.

987. PIFFER Felicita. *Triento, Tyrol.*
White Embroidery.

988. PRAGUE, Central Committee for the promotion of industry in the Bohemian Ore and Giant mountains at.
Bone-lace, Embroidery and Crotchet-work.
(See Classes 27 and 29.)

989. SARTORELLI Zeffira. *Felve, near Trentino, Tyrol.*
Altar-cloths and Tapestry.

990. SCHMID Anne. *Vienna, Margarethen 123.*
Embroidered Long Shawl.

991. SOMMER Eliza. *Vienna, Alservorstadt. 14.*
Hon. Mention: Vienna 1845.
Embroidery.

992. SOUPPER Theresa. *Pesth, obere Donauzeile 9*
Embroidered Pictures.

993. SZOVATHY Samuel, Cloak Tailor. *Miskolcz, Hungary.*
Embroidered Guba.
(See Class 27.)

994. ULLMANN Anthony. *Hirschenstand, Erzgebirge, Bohemia.*
Tamboured Laces Manteau.

995. ULLMANN J. F. *Neudegg, Bohemia.*
Black silk bone-laces (Shawl, Voilet, Volant).

996. ZELGER Thomas. *Vienna, Schottenfeld 469.*
Embroidery begun and finished.

Class 25.

Skins, Furs, Feathers and Hair.

Sub-Class a) Exhibitors of Skins and Leathers.
  „ „ *b) Exhibitors of Feathers.*
  „ „ *c) Exhibitors of Manufactures from Hair.*

The use of Fur for male clothing has been adopted only in the Eastern parts of the Empire; in the Western parts, it has been nearly totally superseded by strong carding yarn tissues. The Hungarians and Transylvanians are generally dressed in sheep- and lamb-skins. In the western parts, furs of wild animals are in prevalent use for female winter-dress, foreign furs being imported in comparatively little quantities. The working out of these articles is in the hands of a special corporation of individual tradesmen.

The manufacture of Quills has suffered considerable restriction by the more and more extended use of Steel-pens. The use of imported feathers for female toilet, by plumassiers (especially at Vienna), is but of very little importance.

The Brush-making industry of Austria, is generally confined to the supply of local demands in provincial capitals and larger towns, from which these tradesmen frequent the markets, where the country-population is to supply their wants in such articles.

The branches of Industry, here in question, occupy about 25.000 hands, bringing into commerce a quantity of articles of the average value of 3 millions florins.

997. BANYAY L. Charles, Merchant. *Klausenburg, Transylvania.*

Lamb-skins.

(See Classes 2 and 3.)

998. BERGER Theresa. *Vienna, Mariahilf 64.*

Plumes.

999. CHLADEK John, Brush-maker. *Pesth, Palatingasse.*

Brush Manufactures.

1000. FLEISCHL S. D., Dealer in Produces. *Pesth, Felbergasse.*

Bed-feathers and Eider-downs.

1001. GOLDSTEIN & Son A., Fur Merchants. *Pesth.*

Furs.

1002. HUDOVERNIG Primus, Horsehair Sieve Maker. *Strassis near Laibach, Carniola.*

Horsehair Sieves.

1003. JONAS Michael, for the Furriers Guild at *Bistritz, Transylvania.*

Saxon Peasant's · Fur. — Saxon Boddice (for women).

1004. KAPICZKA Charles jun., Patented Perriwig Maker. *Prague 147-1* Honourable Mention at Munich 1854, Paris 1855.

Wigs.

The white crown-wig is entirely platted of human hair. The black wig all of human hair affords the advantage of not accepting any soil. The usual spring wig is of naturally curled human hair. The black stage wig is inaccessible to perspiration. The white stage wig is for representing an old man of 80. The exhibitor keeps a store of all French and English perfumes.

1005. KIBITZ George, Patented Manufactory of Dressed Furs. *Pilsen 25, Bohemia.* Agent G. II. Krause Leipzig.

Dyed bear, Russian lynx (dyed wheel-cat-like), Bohemian lamb-skin, dyed Istrian lamb-skin, plucked beaver, lightly dyed marten, lightly dyed fitchet, pole-cat skin dressed completely scentless, dyed oppossum.

The wheel-cat-like dye and dressing quite scentless of pole-cats is of the exhibitor's own invention. The manufacturer begs also to call attention to the fact that, in dyeing Istrian lamb-skins

black, neither the wool nor the hide is injured and that moreover the smell, usually occurring with lamb-skin of common dressing, is here completely removed. The latter are, according to the length of the wool, applicable for rugs in rooms or carriages, as also for lining fur-coats.

1006. PAIDLY Francis. *Lembery 137²/₄, Galicia.* Honourable Mention at Paris 1855. Agent during the Exhibition: Mr. Maurice Nirenstein.

10 numbers of cleaned and assorted bed-feathers under glass covers.

1007. PATTAK George. *Hermannstadt, Transylvania.*

Brush Manufactures.

1008. PERELES Maurice. *Prague 918.*

Hare's hair of Bohemian felts.

1009. PERELIS and **POLLACK**, Cleaning Establishment for bed-feathers and eider-downs. *Prague 1196-2* M. II. Cl. Paris 1855.

Bed-feathers and Downs.

This manufactory cleans bed-feathers and downs by means of steam, rendering them hereby scentless and safe from moths. Moreover the article acquires, by this manipulation, a beautiful appearance and elasticity not yet attained by any other method. Feathers are also sold uncleaned. 250 up 300 hands are occupied in this establishment, list of prices and samples gratis on application. (See inserted advertisement.)

1010. SCHINDLER Ralph jun., Wig Manufacturer. *Prague 953-1*

Wigs and Phytogens.

Wig of white hair upon stuff, out of human hair; wig of red hair, of one piece of gaze without seam, every hair separately set in; both for being pasted on with phytogens, warranted a complete imitation of nature. The exhibitor's manufactures of ladies and gentemen's wigs are expecially to be recommended for their lightness and natural likeness. Pasting on with phytogens is conducive to cleanness and safe sticking.

1011. SCHWER Joseph, Plumaiser to the I. R. Court. *Vienna, Neubau 256.*

Plumes and Tufts.

1012. STETTNER J., Brush-maker. *Pesth, drei Kronengasse.*

Brush Manufactures.

1013. TEXTORIS Albert, Merchant. *Bistritz. Transylvania.*

Goat's hair oats-sacks.

Class 26.

Leather including Saddlery and Harness.

Sub-Class a) Exhibitors of Leather.

   „ „ b) Exhibitors of Saddlery, Harness etc.

   „ „ c) Exhibitors of Manufactures generally made of Leather.

The production of Leather of every description and its mode of preparation in Austria, has hitherto nearly exclusively remained in the hands of Tanners and Tawers. It is only lately that in the western provinces (especially in and around Vienna and Prague, in Lower and Upper Austria, Lombardo-Venetia, Bohemia and Moravia), the Tanning of Leather has arrived to any importance. Tanners are exclusively employed in working out inland hides for local consumption; with the exception of the Sole and Russian Leather-manufacturers of Bolechow (Gallicia), the numerous Tanners of

Brzeczowa and Deutsch-Proben (Upper Hungary) and the Cordwain-makers of Kutty (Galicia), whose articles find their way through the whole Austrian Empire.

The inland production not being adequate to the extensive demands of home consumption, quantities of skins and hides are imported every year. In 1861, this import amounted to 188.000 Cwts, while the export scarcely reached to 17.000 Cwts. The chief articles of import are hides of horned cattle, either from South-America or (in a fresh state) from the neighbouring eastern countries. The Leather-manufactories provided with highly improved appliances, bring into circulation articles of superior quality, while those issued by individual tradesmen, especially Tanners, still remain inferior in quality, being nearly exclusively prepared with fir-bark, the recent improvements in this branch of industry remaining either unknown, or being purposely left aside. This circumstance may serve to explain, how that, in 1861, the import of Leather of every description (for the most part from the territory of the Zollverein) amounted to 52.000 Cwts, while the export only reached the insignificant amount of 14.000 Cwts.

In Vienna, only the trades of Saddlers and Harness-makers has extended beyond the local demands. Articles of this kind, produced at Vienna, are not only bought by home consumers, but also exported into the neighbouring eastern states.

The production of Leather fancy articles has, in the last few years, gained a conspicuous place among the branches of industry practised at Vienna, not only in several extensive establishments, but also among a great number of smaller tradesmen. Portfolios, Cigar-cases, Porte-monnaies and similar articles of Vienna make are exported in considerable quantities.

These branches of industry employ more than 150.000 persons, the average value of their yearly production amounts to 70 millions and with addition of fancy articles, quoted above, to nearly 80 millions florins.

1014. BRÜNNER and Sons. *Liebenau near Gratz, Stiria.*
Leather.

1015. CZERWENKA Francis, Tanner. *Chrudim, Bohemia.* Honourable Mention: Paris 1855.
Tanned upper-leather.

1016. EINHAUSER J. F., Saddler. *Uderns, Zillerthal, Tyrol.*
Saddler's Works.

1017. FOGÉS James, Leather Manufacturer. *Prague 1171-2.*
Various sorts of calf and cow leather, fronts, butts, cappings etc.

1018. GALLASCH Francis, Tawer. *Vienna, Hundsthurm 98.*
Pianoforte Hammer-leather.

1019. GIRARDET Charles. *Vienna, City 1100.*
Fancy Goods of Leather.

1020. GOLDSCHMIDT James J.. Manufactory of Leather and Lacquered Leather. *Smichow, Bohemia.*
Various sorts of Leather.

The establishment of this manufactory, meant to be extended on a large scale, was attempted in 1861 and working began but in the second half of that year. Not yet in full operation, the intention of the manufacturer is to produce all sorts of tanned leather and also sole-leather. 30 workmen are employed till now, the production of lacquered leather not beginning till summer 1862. The exhibited sorts are partially the first productions that were finished.

1021. HABENICHT Augustus, Bookbinder and Manufacturer of Fancy Leather Goods. *Vienna, Neubau 100.*
Leather Hangings.
(See Class 28.)

1022. KOTTBAUER Leopold, Leather Pipe Maker. *Vienna, Rossau 111.*
Leather Pipe and Bucket.

1023. KRAETSCHMAR C. A., Trader. *Rima-Szombath, Hungary.* M. and Dipl. London 1851, New-York 1853. M. II. Cl. and Dipl. Paris 1855. Agent Mr. Ernest Szumrák at Messrs. Williams and Norgate, London, 17, Henriette Street, Covent Garden.
Hungarian Saddle mounted with silver.
(See Classes 3, 4 and 27.)

1024. MANSCHÖN M. F., Leather Goods Manufacturer. *Pesth, Badgasse.*
Leather Goods.

1025. NORILLER Bernhard, *Roveredo 655, Tyrol.*
Skins and Hides.

1026. OTTENREITER Leopold, Manufacturer of Leather Fancy Goods, Veterinary Surgeon and Inventor of a Salve for the Gout. *Pesth.* Vienna at Wiltmann and Froyler's.
Leather Case. Liniment for horses.
(See Class 2.)

1027. RIEKH Francis, Manufactory of Leather. *Gratz, Stiria.* Ware-houses: Vienna, Wieden 513, Agram at Steiner and Račić's. Milan at F. C. Preyssl's.
Leather Wares.

The exhibitor produces all sorts of leather for harness-makers and saddlers, and begs to call attention to his lacquered calf-leather in particular. Samples of this, viz. lacquered shoe-leather, and coloured lacquered calf-skins, as well as common tanned, brown and black calf-leather are exhibited in the Palace.

1028. SCHMITT Brothers, Leather Manufacturers. *Krems, Lower-Austria.* Ware-house Vienna, City 1127.
Various sorts of calf-hides.

1029. SCHOEPPEL John, Glover. *Komotau, Bohemia.*
Dyed Leather.

1030. SEYKORA Joseph. *Kosteletz 233, Bohemia.*
Brown leather, black elastic leather.

1031. STAUDINGER Ignatius. *Marburg, Stiria.*
Horse-hide for roof.

1032. SUESS A. H. and Sons, Manufactory of Leather. *Sechshaus 114 near Vienna.* Ware-houses: Vienna 869, Pesth Neumarktsplatz. Gold. M. Vienna 1839 and 1845. Honourable Mention: Paris 1855.
Leathers.

This manufactory, founded in 1816 and now worked with steam, produces all current sorts of leather, such as: calf and cow leather, cylinder calf-skins, carding, lithograph, and harness-maker's hides, Glacé calf, kid and lamb leathers, Cordovan, Saffian, Furniture, Pocket-book, Chagrin and Comfort leathers, etc. in all kinds of finishing and colours. In the production of lacquered shoe-calf-leather, this manufacture is unique in Austria (100,000 pieces annually).

1033. SZEPESSY Anthony, Merchant. *Debreczin, Hungary.*
Whips.

1034. TOPERCZER Janos. *Vienna, Praterstrasse 513.*
Saddle-girths, stirrup-lashes and strap.

1035. TOSI Caesar. *Scariano near Goritz.*
Hides.

1036. WOLFF Frederick jun., Tanner. *Hermannstadt, Transylvania.* Silver M. Hermannstadt 1847. Bronce M. New-York 1853. Honorable Mention: Munich 1854.
Boar's hide, water-proof calf-leather, cordovan-leather.

Class 27.

Articles of Clothing.

Sub-Class a) Exhibitors of Hats and Caps.
    „   „   b) Exhibitors of Bonnets and General Millinery.
    „   „   c) Exhibitors of Hosiery, Gloves, and Clothing in general.
    „   „   d) Exhibitors of Boots and Shoes.

The trade of Cap-making is in itself of little importance as the adult portion of the Austrian population wear Felt or Silk Hats. The variety of forms of these hats, depending on national, and even personal habits and tastes, is the cause that the manufacture of Felt-hats lies entirely in the hands of Individuals, working merely for the supply of local demands. Only at Vienna, Prague and some other provincial capitals, the manufacture of Silk-hats exists on a somewhat larger scale.

Some years ago, the production of Linen was still a manual operation, exclusively performed by Sempstresses. Since that time, the introduction and rapid extension of Sewing-machines has given rise to manufacturing establishments at Vienna, in several provincial capitals and at Pilsen, for the most part, operating on cotton-tissues.

The same change has lately taken place, as to the manufacture of Clothes, Boots and Shoes, hitherto an exclusive attribute of the individual activity of the Tailors' and Shoemakers' trade-corporations. Now extensive establishments, organized at Vienna, Prague and Münchengrätz (Bohemia) according to the principle of division of labour, are producing these articles, not only for special demands, but also for large and diversified stores. In consequence of this, an extensive export of articles of clothing to Turkey and the Danubian Principalities has been opened within these few years.

The production of Gloves, Umbrellas, Parasols and artificial Flowers scarcely extends beyond the boundaries of Vienna but manufacture of Gloves in Prague is of some importance. The production of these goods at Vienna and Prague, not only supplies the total inland demands, but also takes a share in export trade.

Nearly 60,000 Tailors and 80,000 Shoemakers are subject to individual-trade-taxes. These, with the other relative branches of industry and the auxiliary working people, constitute a total of nearly 350,000 persons, employed in the manufature of Clothing articles. The value of the goods produced by them (Leather, Cloth, Linen etc.) have been brought under the head of the Classes, to which they specially belong.

1037. ADLER Valentine. *Vienna, City 1105.*
Ladies' Shoes.

1038. BACH Theodore. Manufactory of Boots and Shoes. *Vienna, Landstrasse 376.*
Boots and Shoes.

This manufactory produces all descriptions of half-boots and shoes for ladies and gentlemen, executing any order in a quality equal to the exhibited samples, and the fashion suitable to any country. All attention paid to elegant fashion, solid workmanship and the cheapest prices possible.

1039. BACHRICH Joachim. Patented Washing Crinoline and Gentlemen's Linen Manufactory, worked with Sewing Machines. *Vienna, Neubau 211.*
Shirts.

1040. BAUMANN Michael, Ladies' Shoemaker *Pesth.*
Ladies' Shoes.

1041. BIRNBAUM A. M., Gum-wares Manufactory. *Teplitz, Bohemia.*
Gum-wares.

1042. BODNÁR Joseph, Gentlemen's Tailor. *Pesth, Universitätsgasse 6.*
Gentleman's Attila. Waistcoat and Trousers. Lady's Riding Robe.

1043. BRUNNER F. F., Straw-bonnet Manufacturer and Toy Merchant. *Pesth, Waiznergasse.*
Straw-Bonnets.
(See Class 29.)

1044. BUDAN Joseph. *Prague 861-2.*
Leather Gloves.

1045. BUSCH James, Patented Screw-boot Maker. *Prague 376-1.* Ware-house: Vienna, Jägerzeile 524. When partner in the firm J. J. Pollak Sons, the exhibitor received: Great M. Munich 1854 and Great M. Paris 1855.

Boots and Shoes of various descriptions, the upper parts joined by means of sewing-machine, but the soles fastened to the upper-leather by means of screws.

The method of manufacturing Screw-shoes and the requisite machines, invented by the exhibitor, affords a more durable, cheaper and more speedily to be finished production, enjoying an extensive sale even in foreign parts. The exhibitor is also employed in considerable purveyances for the I. R. Army. The establishment is computed for the production of 600 pairds of boots a day, and may easily be enlarged, since any kind, even the finest, can be furnished within a few hours.

1046. CHRISTL Joseph. *Vienna, City 941.*
Boots and Shoes.

1047. DEPOLD Joseph, Shoemaker. *Pesth.*
Boots and Shoes.

1048. DRAKULIČ Smiljana. *Korenica, District of Ottocan, Military Border.*
Lady's Kirtle, Apron, Socks.
(See Class 24.)

1049. ELSINGER M. J. and Son, Manufactory of Water-proof Clothes. *Vienna.* Ware-house: Mariahilf 57.

Water-proof Clothes.

The exhibitors' productions will not grow clammy with 60 degrees Reaumur, nor when folded down in a wet state. The manufacture deviates from the usual method of tarring and gumming cloaks. That pleasing, supple article may be worn by any gentleman, and being of elegant fashion, joined to cheap-

E.

ness of price it has procured the exhibitors a yearly sale of more than 6,000 suits. In pursuance of a special permission of the Ministry of War, rain-cloaks of this kind are worn by Officers of the Austrian army. The exhibited tourist's costume consists of coat, trousers and hood with back-slouch, price 9 shillings.

1050. FALLER, TRITSCHELLER and Comp. *Vallonora near Vicenza.*
Straw Bonnets and Plaitings.

1051. FEDIGOTTI James, Manufacturer of Woollen Hats and Shoes. *Tiarno di Sotto (Val di Ledro), Tyrol.*
Three pair of boots of fulled wool.

1052. FRESE Anthony, Manufacturer of Gloves to the I. R. Court. *Prague 592-1.*
Gloves and Glover's Manufactures.

1053. FÜRTHWOLF and Co. Patented Manufactory of Oriental Caps (Feszes). *Strakonitz, Bohemia.* Silver M. Vienna 1845. Ware-houses: Vienna, Trieste, Alexandria, Cairo, Constantinople.
Oriental Feszes (Caps).

The production of this manufactory, founded in 1818, comprizes all sorts of feszes (caps), current in the Levant, Barbery and the western coast of Africa, and of late also the manufacture of negro-caps. Machines employed: 2 turbines, 24 horse-power, steam-engine 20 horse-power. The number of hands employed in that establishment amounts to from 500 to 600 persons, and there are besides 500 women and children of the surrounding country, in a circuit of 2 German miles, entrusted with domestic work. Yearly average of production: 1,200,000 pieces, estimated at 1,200,000 francs. The manufactory is, however, capable of enlargement and producing far more considerable numbers.

1054. GASTEIGER Anthony. *Wilten, near Innsbruck, Tyrol.*
Artificial Flowers of beetle-shards.
(See Classes 7 and 11.)

1055. GINTER, Brothers, Gentlemen's Tailors. *Pesth, Waiznergasse 17.*
Gentlemen's Dresses.

1056. GÜLCHER'S Theodore Son, Manufactory at *Neusteinhof.* Ware-house: Vienna, City 720.
Turkisch Caps (Feszes).

1057. HAAN, Brothers, Patented Manufactory of Shoes and Boots. *Vienna.*
Boots and Shoes.

1058 HAHN Leopold. Patented Manufactory of Boots and Shoes. *Vienna.* Ware-house: Vienna, City 737. 2. Warehouse: City 770. Agents for London and all England Messrs. L. Brandeis and Goldschmidt, 10, Love Lane, North Street, E. C.
A hundred pair of various kinds of shoes, according to the fashion requisite for exports to the different countries of Europe and America.

Existing since 1853 and occupying 200 persons annually, this manufactory is enabled to execute the most extensive orders at the shortest notice.

1059. HALLER Christopher. *Gratz, Stiria.*
Water-proof Stuffs for clothes, of silk, linen and wool, also ready-made surtouts and cloaks, easily to be stowed away in the coat pocket; and rain-cloaks of which tents may be made.
(See Class 10.)

1060. HIRSCH J. P., Hat and Cap Manufacturer. *Vienna, Wieden 923.*
Hats and Caps.

1061. HÖNIG Ignatius, Manufactory of Stocks and Cravats. *Vienna, Mariahilf 45.*
400 Stocks and Cravats.

1062. HUBER Anthony, Ladies' Tailor. *Pesth.*
Ladies' Dresses.

1063. HÜBSCH Joseph. *Prague 583.*
Hats.

1064. JAMBOR Andrews, Gentlemen's Tailor. *Pesth.*
Clothes.

1065. JANOWITZ S., Patented Manufactory of Hats. *Brünn, Moravia.*
Assortment of Hats.

1066. JAQUEMAR George, Glover. *Vienna.* Ware-house: City 251. Medals: Vienna 1839 and 1845, Berlin 1844, Leipzig 1850, Honorary M. Munich 1854.
Gloves of kid-leather, glacé and Swedish.

This manufactory, established in 1810, is the oldest in Austria, employing more than 100 persons, and exporting most of its productions. In consequence of the solidity of its manufactures, the reputation of its productions spread so far that the manufactory was honoured with orders by the Imperial Court at Paris, the Courts of Greece and several German Sovereigns. The exhibited gloves are not purposely made for the Exhibition, but such as are usually furnished in trade.

1067. KLEIN Charles, Milliner. *Pesth.*
Clothes.

1068. KRACH, Brothers, Manufactory of all sorts of Clothes and Mercantile Business. *Prague, Bohemia.* Warehouse: Vienna. Silver M. Vienna 1845. Bronze M. Leipzig 1850. Prize-M. London 1851. Bronze M. New-York 1853. Honorary-M. Munich 1854. Silver M. Paris 1855.

200 various pieces of apparel.

The exhibitors who exercise this business, are the third generation of the same family in this establishment. Having formerly turned their attention chiefly to the orders of higher ranks, they have in later times also applied to manufacturing articles of general use. At present their productions embrace all articles, both those of highly finished elegance and refined taste, and those of the humble wants of indigent journeymen. In 1859, proofs have been given of their capacities in furnishing Military Clothing, having, during 8 weeks, supplied complete fittings for 2,500 men of the Bohemian Volunteer Corps, an order executed to the entire satisfaction of the Committee entrusted with the outfitting. The exhibitors have ever been anxious to render themselves, in this line of business which is more than any other line subject to changes, independent of imitation not only respecting fashion and combination of various materials, colours and finishing, but also in the choice of stuffs as such. They have first started the invention of double sided stuffs introduced over all the world since. Most of the productions issued from their workshops, as well as those exhibited now, are made of stuffs and working materials furnished according to their own devices. Thus this establishment is enabled, always to offer novelties and not to hobble after known fashions, but rather to dictate it frequently, as is to be proved from journals for fashions (those of Paris not excepted) which regularly bring one or several figures designed after dresses made in this establishment. It is also acknowledged by competent authorities that the higher improvement in manufacturing of dresses in which Prague excels, is principally owing to the impulse given by this establishment. The attention of the exhibitors has also been mainly directed to producing suitable articles for any wants, proofs for which assertion may be found in numerous articles invented by them. The Exhibitors occupy in both their establishments at Prague and Vienna, an average number of 100 workmen; and besides there are, according to want, employed a number of masters in outhouse work. For their workmen in their house at Prague, there has, since 1860, been erected a sunday school in which instruction is given in geometry, drawing, counting, etc.

1069. KRAETSCHMAR C. A. *Rima-Szombath, Com. of Gömör, Hungary.* M. and Diploma London 1851, New-York 1853, M. I. Cl. Paris 1855. Agent: Mr. Ernest Szumrák at Messrs. Williams and Norgate, London, 17, Henriette Street, Covent Garden.
Three kinds of Cloaks, woven.
Three children's cloaks.
(See Classes 3, 4, and 26.)

1070. KRAUTZBERGER, Brothers. Patented Manufactory of Cloth Hats. *Marienbad, Bohemia.*
Cloth Hats and Caps.

That establishment produces, by means of sewing machines, hats of various stuffs after the newest and most tasteful fashions, besides being very solid and durable. Particularly to be recommended are the cloth-hats, manufactured in this establishment, and distinguished for lightness, elegance, elasticity, and for their being completely water-proof.

1071. KRISCHONIG Adam. *Vienna, Wieden 307 and City 939.*
Mignon-Bouquet.

1072. KRISTIAN Ignat., Patented HatManufacturer to the I. R. Court. *Vienna, Laimgrube 1.*
Hats of various forms.

1073. MADLBAUER Agnes, Sempstress. *Vienna, Fünfhaus 225-226.*
2 Shirts, 1 Chemise.

1074. MALATINSZKY Emerich, Halina Tailor. *Pesth, Landstrasse 17.*
Halina for ladies and gentlemen.

1075. MAUTHNER, Brothers. *Leippa, Bohemia.*
Clothes.

1076. MAYERHAUSER E. *Vienna, St. Ullrich 38.*
Straw-Flowers.

1077. MICSEI and Comp. Straw-Hat Manufactory. *Pesth 8.*
Straw-Hats trimmed after the national Hungarian fashion.

1078. MORTÁN Philip, Gentlemen's Tailor. *Pesth, Theaterplatz.*
Clothes.

1079. MOTTL'S M. Sons, Manufactory of Gentlemen's Clothes. *Prague, Bohemia.*
New Articles of Dress made of stuffs manufactured according to the exhibitors' designs. Mottl's Patented Reduction Machine for speedy and accurate execution of any drawing of cuts for clothes.

The firm, established since 1834, enjoys general credit for the good quality, comparative cheapness and elegance of its productions, with numbers of the most distinguished customers among whom gentlemen of the highest rank and even members of the Imperial House. All that is superior regarding novelty and beauty in tailoring industry, and all that is either indispensable or luxurious in such like fashionable creations, is to be found in the stores of Mottl's establishment which resembles a regular factory in its arrangement. The reviewer of „the Observator of Fashions" gives the following opinion about this establishment: „M. Mottl, Sons have not only founded the first tailoring establishment on a grand scale at Prague, but they have also materially promoted the progress of this branch of industry, by rousing competition into new life in all directions. In recognition of their merits in German tailoring industry and especially on account of improvements and inventions of new fashions, one of the partners, Aloysius Mottl, was named corresponding member of the German Tailor's Academy, and the other, Wendelin Mottl, director of the Austrian branch Academy of Fashion. By means of the Reduction Machine invented by the partner of the firm, Wendelin Mottl, the formerly necessary 36 reduction-measures for centimeter drawings of cuts are now useless. It was recommended by all Journals for Fashions as a useful and important invention, and is accepted by many for its practical utility. See particulars of the exhibited articles among the advertisement at the end of this volume.

1080. NITSCH Joseph. *Pesth.*
Articles of Dress.

1081. PAGET F., Manufactory of Waterproof Stuffs. *Vienna, Brigittenau 66.* Office: Vienna, City 487, and London, 37, Old Broad Street.
Elastic waterproof awning, waterproof cloak, sailor's hat.
(See Class 4.)

1082. PANESCH Anthony, Shoe Manufacturer and Patentee. *Vienna, City 1100.*
Boots and Shoes.
(See Class 2.)

1083. PERNER A., Manufactory of Wooden Pegs. *Budweis, Bohemia.*
Wooden Pegs.

1084. PETRY Maria. *Vienna, Strozzengrund 19.*
Bouquet of Florentine Straw.
Worked, free hand, of twisted Florentine straw. Price 20 francs. These articles are also made of glossy plates and picked straw. Samples of from 3 to 12 fl. per dozen on re-imbursement. (List of prices gratis.)

1085. POLLAK D. H., Manufactory of Ladies' and Gentlemen's Shoeing Articles. *Vienna, Windmühle 109.*
Ladies' and Gentlemen's Shoeing Articles.
This manufactory is continually occupying 500 persons, and can, therefore, satisfy demands to any extent. Its productions have in a very short time gained a far spread reputation, so that they are almost exclusively exported to the East and West Coasts of America, to Africa, Asia, and Australia. This establishment is, therefore, acquainted with the forms of all the world.

1086. PORFY Francis, Hatter. *Pesth, Waitznergasse 19.*
Hats.

1087. PRAGUE. Central Committee for the Promotion of Industry among the inhabitants of the Ore and Giant Mountains. Industry School at ses *Neudeck.*
Gloves.
(See Class 24 and 29.)

1088. RACZKOVICZ Cozma, Produceer. *Nagy-Kikinda, Comitate of Torontál, Hungary.*
Servian National Manufactures.

1089. RATTICH John and Son, Manufactory of Wooden Pegs. *Krumau, Bohemia.*
Wooden Pegs for Shoemakers.

1090. REGENSBURGER Michael-Angelo. *Predazio, District of Cavalese, Tyrol.*
Half-boots.

1091. REICH M. *Vienna, Gumpendorf 342.*
Ladies' and Gentlemen's Shoeing Articles.

1092. RICHTER Joseph. *Kaaden, Bohemia.*
Gloves.

1093. RÖMISCH Charles. *Prague 590-1.*
Clothes.

1094. ROTH Michael, Shoemaker. *Hermannstadt, Transylvania.*
Various Sorts of Boots.

1095. ROTHBERGER James, Tailor. *Vienna, City 627.*
Gentlemen's Clothes.

1096. SCHILLING Frederick, Patented Manufactory of Model Boots and Shoes.. *Vienna 638.*
Model Boots and Shoes.

1097. SCHMID Betty. *Vienna, Mariahilf 74.*
Patented Shirts.

1098. SCHMIDT Augustus. *Vienna, Kärntnerstrasse.*
Gentlemen's Dresses.

1099. SERVATIUS Andrew, Hatter. *Kronstadt, Transylvania.*
Woollen Felt Hats.

1100. SINGER Joseph, Tailor. *Vienna, City 1120.*
Dress Coat.

1101. SPITZMÜLLER Fred., Glover. *Vienna, Joseph-stadt 229.*

Six dozens of various gloves for ladies, gentlemen and children, at the price of from 6 to 12 florins a dozen.

The exhibitor's manufactory produces all sorts of glover's works. Orders are begged to be addressed to the afore-named direction, the sum in Austrian Money joined or a cheque upon an Austrian Commercial House. Prompt execution of orders warranted. Bandages and Fancy Articles are also manufactured.

1102. STEIDL Anthony, Shoemaker. *Vienna, City 819.*
Boots.

1103. STRASCHITZ BÄRMANN, Patented Manufactory of Gentlemen's Clothes. *Prague 147-1.* Warehouses: Vienna, Jägerzeile 516, Teplitz, Pilsen and Leitmeritz. Agent in London: Mr. Charles Netter, Cannon Street.

Various articles of dress.

Complete Travelling Dress of finest material, and peculiar fashion. — Black dress coat of particular cut and artistic finishing. — Fine Jaquet of blue peruvienne, new fashion and workmanship. — Hunting coat, double stuff new workmanship with invisible seams. Of the exhibited articles, No. 1 and 2 are made without there having previously been any measure taken, worked after a photographic picture, and suiting to a nicety. The

exhibitor is in possession of a patent for the invention of elastic lettings-in for trousers, by means of which buttons, buckles and braces are useless. This manufactory furnishes all sorts of gentlemen's clothes, boys' dresses, bonnets, cravats etc. The exhibitor has been established these seven years.

1104. SZÁBO Francis, Tailor. *Pesth, Schlangengasse.*
Gentlemen's Clothes.

1105. SZÓVÁTHY Samuel, Cloak Tailor. *Miskolcz. Comitate of Borsod, Hungary.*
Ladies' and Gentlemen's Szürs (Cloaks).

1106. SZURGENT Andrew, Ladies' Tailor. *Pesth, grosse Brückengasse 7.*
Lady's Fur-cloak (Menté).

1107. VADASZ Francis, Boot-maker. *Pesth, Herrengasse.*
Boots and Shoes.

1108. ZEIDLER & MENZEL. Patented Manufactory of Wooden Pegs. *Schönau near Schluckenau, Bohemia.*
Wooden Pegs.

This manufactory was founded in 1859, and furnishes yearly more than 4,000 Cwts of wooden pegs, produced with 29 machines worked by water-power, there being 100 workmen employed. The pegs are sold in Austria, Germany, Russia. The exhibitor was decorated by His Majesty the Emperor of Austria with the Golden Cross of Merit, in acknowledgment of his merits in promoting industry; and the manufactory received, in 1861, the Great Silver Medal of the Lower Austrian Trade Union, offered by that Society for this article.

Class 28.

Paper, Stationery Printing and Bookbinding.

Sub-Class a) *Exhibitors of Paper, Card, and Millboard.*

„   „   b) *Exhibitors of Stationery.*

„   „   c) *Exhibitors of Plate, Letterpress, and other modes of Printing.*

„   „   d) *Exhibitors of Bookbinding.*

The only articles (raw Silk excepted) charged with a high export-duty (4·20 florins per Cwt) in the Austrian tariff, are Rags, and as a normal consequence of this regulation, the export of this article is generally insignificant, amounting scarcely to 5,000 Cwt in 1861. It is, for the most part, worked out into Paper in establishments, spread over the whole of the Provinces of the Austrian Empire, with the exception of Dalmatia. Latterly, manual labor has gradually given way to mechanical proceedings, especially in the Western Provinces, the old proceeding being more and more confined to the Eastern parts of the Empire.

45 Mechanical Paper-manufactories, each employing from one to three engines, and 202 Paper-mills, each with between 2 and 3 dippers, are at present in activity. The rapid developement of mechanical Paper-manufacture has taken a conspicuous place in the improvement of Austrian international relations, as, in 1861, the export of Paper of every sort amounted to nearly 95.000 Cwt, whilst the import did not exceed an amount of 21.000 Cwt.

Steel-pens are, for the present, an exclusive feature in Vienna Industry. Their annual production of nearly 40.000 „Gross" is inadequate to the wants of inland-consumers. The home demands for Lead-pencils, Ink and other writing-materials are fully satisfied by inland manufacture.

Nearly 350 Printing-houses and 150 Lithographic establishments (with 1.500 hand-presses and 400 mechanical presses) are in activity for literary and artistical publications. There is scarcely in the W. Provinces any place, numbering above 5.000 inhabitants, without a printing-house of its own.

In the E. Provinces, these establishments are generally confined to the Capitals and to some few others of the larger towns. The greatest activity is displayed in the extensive establishments of Vienna, Prague, Pesth and Venice.

Book-binding lies, for the most part, in the hands of individual tradesmen, united into corporations; it is practised in a manufactural way only in the above named Capitals, the centres of intellectual improvement. There, this branch of Industry has made considerable progress, whilst, in other places, it has generally remained below the ordinary standard.

We must here mention the manufacture of Coloured Paper, as a branch of Vienna Industry, and the „Papiermaché" articles (snuff-boxes, matchet boxes etc.), which, in some parts of Bohemia, are not unimportant objects of domestic Industry.

The production of the articles here in question employing nearly 150.000 persons represents an annual average value of 120 millions of Austrian florins.

1109. BELLMANN Charles, Artistic Topographic Institute. *Prague.*
Pictures in Xylography.

1110. BOLDOG Lewis, Bookbinder. *Pesth.*
Various bookbinder's works.

1111. (Omitted.)

1112. GEROLD Charles' Son. *Vienna, City.*
Book-printer's Works.
(See Class 29, where particular about his printing office.)

1113. GIESSENDORF Charles von, Lithographer, Member of the I. R. State Printing Office. *Vienna, Ober-Meidling* 28. Studio: Vienna 508.
High-etchings on zinc-plates for the printer's press. Printing-plates and high-relief plates of old and new illustrations. Lithographic copy of a general-staff ground map of Wallachia together with high relief plate of the same map. Copy of a portrait in chalk manner on the printer's press; copies of all illustrations.
With the exhibited objects the proof is adduced that both lithographic or xylographic old pictures and new lithographs and xylographs may be transported upon zink and high-etched for the printer's press, without their wanting to be designed again.

1114. GUCKER Joseph, Bookbinder. *Pesth.*
Bookbinder's Works.

1115. HABENICHT Augustus, Bookbinder and Leather Fancy Goods Manufacturer. *Vienna, Neuhau* 100.
Various Bindings of books.
(See Class 26.)

1116. HARDTMUTH L. and Comp., Manufactory of Pencils and Elastic Writing and Ciphering Boards. *Budweis, Bohemia.* Ware-house: Vienna, am Peter. M. II. Cl. Paris 1855.
Pencils and Elastic Writing and Ciphering Boards.
Joseph Hardtmuth made, in the year 1797, the invention to produce artificial lead-pencils from inland graphite, which were henceforth, to the year 1848, manufactured in superior quality and in yearly increasing quantity by himself and his sons Lewis and Charles in Vienna, and afterwards by the latter in Budweis. At present the manufacture is carried on by his youngest son Charles Hardtmuth alone, under the firm of L. and C. Hardtmuth, and their productions enjoy high renown and a lively sale. There are about 200 people occupied, and 2 steam-engines of 40 horse-power as motors for 6 veneer-cutting-saws, 40 grooving, planing and top cutting machines, and 30 mills are in action. In 1861, there were produced more than a million dozen of pencils, considerable quantities of which were exported to Russia, Germany and America. Of raw material, this manufactory consumes a yearly quantity of about 600 Cwts of the best Bohemian graphite, and about 3,000 Cwts of cedar-wood (juniperus virginiana). The production of elastic writing and ciphering boards, another invention of J. Hardtmuth's, is likewise continually increasing, and there were, in 1861, produced about 20.000 dozens of all sizes, 20 workmen having been employed in it, and steam-power used for working mills and other machines.

1117. HARTINGER Anthony and Son., Patented Institute for Artistic Lithographic Productions. *Vienna, Mariahilf* 71. Bronze Medal: Munich 1854. Honourable Mention: New York 1853.
Colour-print pictures upon patent-linen.

1118. HASLINGER Charles, quondam Tobias, Music Publisher and Seller to the I. R. Court. *Vienna 281.* Sale of publications at Leipzig and Berlin. Commission Stores in all large towns of the empire, as well as of Europe and America. Silver Prize-Medal Vienna 1835.
Diverse Music-books.

This establishment, founded in 1809 and since 1826 in the sole possession of Tobias Haslinger (the father), is numbered among the most prominent Music Publishing Concerns of Europe, at present, undoubtedly, the most extensive in the empire, incessantly occupying 12 copper-plate and lithographic presses for music-printing. The present chief, Charles Haslinger, is possessed of the Golden Cross of Merit with the Crown, and is Honorary Member of several musical Societies. Among a profusion of publications the following may deserve particular mention: Beethoven's Pianoforte Concerts and Symphonies. Hummel's Pianoforte School. Spohr's Violin School; besides a considerable number of school works of all kinds. Moreover the Pianoforte Compositions of Beethoven, Hummel, Liszt, Schubert, Thalberg; the Song Compositions of Beethoven, Schubert, Weber and others. Particular mention may also be made of the publication of all the Dance Compositions of John Strauss, the father, and his sons. Exhibited are: Pianoforte Compositions of Bargiel, Haslinger and Tausig; Song Compositions of Francis Schubert; and Dance Compositions of John and Joseph Strauss.

1119. JOSEPHSTHAL, I. R. Patented, Manufactory of Paper, Oil and Dyeingwood at. *Josephsthal, near Laibach, Carniola.* Honorary Medal: Munich 1854. M. II. Cl. Paris 1855. Ware-houses: Vienna 579, Triest and Agram.
Paper.
(See Class 4.)
This manufactory has 3 paper machines of the best construction, working entirely with water-power, which puts in action 40 cylindrical paper-mills and other machines requisite for them. It produces daily about 8,000 Vienna pounds of fine and superfine paper which is much sought after on account of its excellent water of production. The supply of rags is drawn from Carniola, Carinthia, Croatia and Slavonia, and the sale of paper is one half to the Austrian provinces and in the other to foreign countries, viz. to Greece, Alexandria, Constantinople etc.

1120. KNEPPER W. and Comp. (Sole partner Francis Wertheim), Fancy Paper Manufactory. *Vienna,* Manufactory and chief ware-house: Wieden 348. Distinctions received at all great Exhibitions since 1844. Commission Stores in all large towns at home and abroad.
Fancy and Pressed Papers.
Existing since 1827, worked with steam and employing from 200 to 300 persons, this manufactory is the largest of its kind on the continent. Of particular extent, in recent times, is the production of cigaret-paper, so that this manufactory is capable of considerable trans-atlantic export which is very lively now. The contrivances made after the best models allow of prompt execution of the most extensive orders.

1121. LEIDESDORF Fr. and Comp., Patented Paper Manufactories at *Ebenfurth, Ober-Eggendorf and Wiener-Neustadt.* Ware-house: Vienna, City 427.
Machine Paper.
These manufactories produce all sorts of paper of any size and strength, viz. white and coloured letter paper and tinted paper, all sorts of writing, drawing, lithographing and printing paper, yarn-packing and other packing papers. Sugar, silk and cigaret papers, and so forth. Lists of prices and samples are given gratis on demand. This establishment, founded in 1840 gand really extended by enlarging that which existed and erecting new ones, is, according to want, worked by water or steam power. There are at present 200 workmen employed, who are, entirely, colonized there. 32 cylindrical paper mills with 5 paper machines are active, consuming a yearly quantity of 60,000 Cwts of rags, out of which nearly 40,000 Cwts of paper are produced, estimated at about 1,200,000 fl. These productions are not only sold at home, but also exported to the Orient. The raw material is collected at home, Hungary and Croatia being the best and most abundant places of collection. This industrial undertaking may be called the most extensive establishment in this branch in the Austrian monarchy, both with respect to its great extent and its considerable capacity of producing solid as well as beautiful manufactures.

1122. LORENZ Francis Sons., Patented Machine Paper Manufactory at *Arnau, Bohemia.* Ware-houses: Vienna Singerstrasse 896; Prague, Kolowratstrasse 578. Distinctions: Silver M. Berlin 1844. Silver M. Vienna, 1845. Bronce Medal Paris 1855. Agent in London: Mr. William Staats, 38, Newgate Hill Cannon Street.
Paper.

This manufactory produces especially: letter, writing and printing paper of any quality, from the finest vellum to common bull paper; as well as drawing, envelop, silk and packing paper. It works with two machines, is completely fitted out with water and steam-power and the corresponding secondary work, cylindrical mills and satinizing machines. and. therefore, enabled to undertake extensive supplies. On account of the cheapness and good qualities of its manufactures, it has extended its sale, besides the inland, to the Tariff Union States, England, Russia, and Turkey. Samplers, Lists of prices in German language, and business cards in German, French and English are at hand.

1123. LUIGI James and Comp. *Roveredo, Tyrol.*
Paper.

1124. LUKSICZ Abel, Owner of the Topographical Artistic Institute at *Karlstadt, Croatia.*
Printer's Productions.

1125. NEUMANN L. T., Printseller. *Vienna, City, Michaelerhaus.*
Frames, Albums, Portfolios.

1126. PATERNO F., Printseller. *Vienna 1064.*
Agent: Messrs. Droosten, Allan and Comp., London, 126, Strand.

Album of Costumes. Gauermann's Alpine Sceneries. Strassgschwandtner's Hunting Album. Portraits of Their Majesties the Emperor and Empress of Austria, after Schrotzberg. Taubinger's School of Figures.

The Album of the national costumes of Austria offers a faithful survey of the different races inhabiting the powerful monarchy. With respect to variety of costumes, it matches with the most considerable works of this kind, and is likely to supply a valuable addition to geographical and ethnographical illustrations exciting general interest in all circles. In a more special, but not less attractive manner, Gauermann's Alpine Sceneries afford a view of the picturesque life and pursuits of the Austrian Alps renowned for their beauties of nature; moreover, Strassgschwandtner in his characteristic Hunting Album has again proved a worthy successor of Ridinger. Austria's National Costumes, 140 numbers, coloured and bound 10 £. Gauermann's Alpine Sceneries, 26 numbers, coloured, bound, 20 £. Strassgschwandtner's Hunting Album, 50 numbers, coloured, bound, 10 £. Messrs. Droosten Allan and Co., London, 126, Strand, Gallery of Fine Arts, are so kind as to accept orders and answer inquiries.

1127. POSNER K. L., Patented Machine Ruler and Manufacturer of Commercial Books. *Pesth.*
Machine Ruler's Works and Book-bindings.

This establishment, existing since 1854, has gained great success. It has stores of its productions in all considerable towns of Hungary and Transylvania, carrying on a brisk business with the Danubian Principalities, with Turkey and Egypt. Through the mediation of the ware-house at Constantinople, this manufactory was, in 1856, entrusted with the supply of books and set forms required by the General Commissariate of the R. English Army in the Crimea, a work executed to their entire satisfaction. In 1857, the establishment was honoured with the expression of satisfaction of His Majesty the Emperor Francis Joseph. K. L. Posner composed and published, in 1854, a work under the title of „Book-keeping", greatly facilitating book-keeping as regulated by law, and having already appeared in the fifth edition. The net profit of these editions (more than 600 fl.) was devoted for purposes of schools. The owner of this manufactory was distinguished with laudatory diplomas by the I. R. Home Office, Trade Office, the Provincial Government at Buda, the Boards of Trade at Pesth, Buda, Pressburg, Kaschau, Debreczin, Klausenburg and Temesvar. Among the exhibited objects are the „Golden Book" and the „Stranger's Book," destined for gifts to the Hungarian Academy of Sciences.

1128. REIFFENSTEIN and **ROESCH**, Artistic Office. *Vienna, Leopoldstadt 482.*
Lithographic prints in frames and portfolios.

1129. REISSIG Joseph. *Vienna, Windmühle 79.*
Cartoon Works.

1130. ROLLINGER Brothers. Manufactory of ruled and bound mercantile books, and all sorts of bookbinder's works. *Vienna 644.* Commission stores: Prague, Leuthner J. and Comp. Agencies at Brünn, Ol-

mütz, Temeswar, Fiume, Trieste etc. M. I. Cl. Paris 1855.

Travelling Album, Account Books, and Machine-ruled papers.

This establishment, having existed these 62 years, was up to the year 1854 chiefly occupied with official works, but from that time it took to all descriptions of bookbinder's works and especially to the manufacture of mercantile account books. There are from 80 to 100 people occupied within the establishment and 10 out of it. By appropriately distributing work in the production of great numbers of all sorts of account books, from the smallest notice-tablets to the largest ledger, there is attained perfect solidity and adequate cheapness. At the Paris Exhibition, this establishment was awarded the above mentioned distinction for solid and cheap work. In 1860 the manufactory was entrusted with the manufacture of 800 splendid folio bindings for the festival of the Maria Theresa Order, when 400 volumes were to be furnished within a fortnight; unusually thick, they were bound in red gros grain Marocco leather and richly gilt. In 1861, 22.000 folio stitchings were supplied to the Vienna Municipality Office within 48 hours. The exhibited Gigantic Album (dedicated to the City of London) with original ornaments was invented and executed by Charles Rollinger. This Album (16 sq. ft.) was, on purpose, produced in such a gigantic size, in order to show the capacities of this establishment, and particularly to render the original ornaments plainly visible to every body. Besides, there are exhibited Account Books and Machine-ruled Papers of various kind. That manufactory is perfectly well provided with contrivances for all sorts of bookbinder's and Machine-ruler's Works, and orders of the greatest extents may be executed in the promptest, best and cheapest manner. A great store of its productions is kept in Vienna, City 644, purchasers of greater lots being allowed a considerable discount. Exports upon a large scale were already made to the Danubian Principalities, to Constantinople and Odessa etc.

1131. SIEGER Edward, Patented Lithographic Establishment. *Vienna, Leopoldstadt 642.* Medal Munich 1854. Silver M. Paris 1855.

Samplers of Lithographic Prints in black, bronze and colours. Assortment of official ruled paper and blank forms for all branches of commerce and trade business. Several Placards and Advertisements.

That establishment also produces, and keeps a store of, all sorts of manipulation paper and tables alike, for official use (viz. parish, military, provincial, tax offices, municipal boards etc.) and for private transactions. Of the whole of the exhibited articles, not a single one was purposely made for the Exhibition.

1132. SMITH and **MEYNIER**, Manufactory of Machine Paper. *Fiume.* Ware-houses: Trieste, Venice, Ancona, Messina, Constantinople etc. Gold M. Vienna 1845. Prize-Medal London 1851. Honorary Medal Munich 1854. M. I. Cl Paris 1855.

Machine Paper.

The Machine Paper Manufactory at Fiume, under the firm of Smith and Meynier, founder of the same, was erected in 1828, comprizing an area of 11,800 square fathoms and forming premises of 17 buildings; it is situated at the disemboguement of the Recina and the Fiumara into the sea, at a little distance from the city, and was, after the restriction of the free-port district, in 1855 incorporated with the inland. It introduced machine paper manufacture into the Austrian States, and, after having by its example and the general popularity of its productions called into existence several similar extensive enterprizes, it turned its attention almost exclusively to exports of its productions. It is in direct communication with all transatlantic staples where its manufactures are well known and enjoy a regular sale. That it is competent to enter the lists with any establishment abroad, is sufficiently proved by its chief purchasers at present being in London. Moreover there are ware houses of its productions in Trieste, Venice, Ancona, Messina, and Constantinople. There are from 600 to 800 persons, all natives, employed incessantly either for the internal management or for external transactions. Yearly consumption of rags about 30,000 Cwts; quantity of paper, produced of all sorts occurring in trade from 200,000 to 230,000 reams. The establishment is in possession of 7 honorary medals of 1st class, bestowed upon it by all Industrial Exhibition held till now.

1133. SOMMER Leopold, Patented Manufactory of Paper at *Guggenbach, Styria,* and Printing Office at Vienna. Domicil at Vienna, Alservorstadt 147.
Printer's Productions.

1134. SPINA C. A., Print and Music Seller and Publisher to the I. R. Court. *Vienna, City 1133.*
Printed Music.

1135. TRENTSENSKY Matthew, Patented Machine Ruling and Artistic Office. *Vienna, Leopoldstadt 642.*
Samples of ruled and bound papers.
(See Class 29.)

1136. WALDHEIM R., Chevalier., Xylographic Establishment. *Vienna 326.*
Two Tableaux., 12 Wood-cuts after designs of Charles Swoboda, to Geoffrey Kinkel's „Otto the Marksman."

1137. WINKLER Aloysius, Plate Printer. *Vienna, Gumpendorf 142.*
Samples of Plate Printing.

1138. ZAMARSKI L. and C. **DITTMARSCH,** Typographic Literary Artistic Office. *Vienna, Mariahilf 3.* Ware-house, City 24.
Literary Artistic Works.

This establishment embraces all branches of practical graphic art, occupying 18 fly, 20 hand and 14 lithographic presses, with a staff of 300 persons. It is especially adjusted for the printing of monetary-value papers: there have been produced, in this office during the last 12 years, for more than 500,000,000 fl. shares, lottery-tickets and such like value papers. Exhibited are: 21 Volumes of various printed works in quarto and octavo, in various languages. 2 volumes Talmud in folio; this newest, splendid edition of the Talmud will appear in 24 such volumes, on writing paper at the price of 40 Pr. dollars, on regal paper at the price of 50 Pr. dollars, on vellum paper at the price of 66 Pr. dollars (there have already been published 10 volumes). 10 plates for illustrating the progressive stages of typographical colour printing. Sampler with black and coloured wood-cuts (cut in the studio of that typographical office). Samples of Cards for playing, produced in typographical colour printing on the printer's press. This method is patented in Austria and in the Tariff Union States. These cards are distinguished for beauty, extraordinary cheapness, and durability of colours (oil-print) above all others. 8 oil-colour prints, produced in the chromolithographical manner, by means of the lithographic press. The publications of oil-colour prints of this establishments comprises about 60 pictures, of which the Saviour's Progress to the Cross in 14 stages is particularly prominent for its superior artistic execution. A complete copy of all the 14 stages costs 100 Pr. dollars; printed on painter's canvass, 120 Pr. dollars. See the illustrated catalogue of the establishment, Printsellers and exporters receive from 25 to 40% discount.

Class 29.

Educational Works and Appliances.

Sub-Class a) Publishers.
" " *b) Apparatus Makers.*
" " *c) Toy and Games Manufacturers.*
" " *d) Exhibitors of Specimens and Illustrations of Natural-History.*

The Chapter on intellectual culture in the previous General „Part" of this book contains particulars and introductory facts to this class, which are appropriately illustrated, with respect to the state of public educational institutions, in the Collective Exhibition of the I. R. State Ministry, as detailed farther below Nr. 1188.

There have already been given some data about book and print selling concerns (the latter frequently trading in maps and sometimes music) in the „General Part". The greater part of the text-books for primary schools have up to this time been furnished by the I. R. School-Book Selling Office, for the price of production. The printing of maps immediately upon weaved stuffs, instead of upon paper, will have great success, especially when promoted by the new invention of photographing maps upon stone and then etching them, hereupon to print off the etching, or galvano-plastically to transmute them into copper-plates fit for printing.

Lead-pencil manufacture has, of late, considerably diminished former imports of them. Fine water-colours are for the most part imported. Physical and optical (philosophical) instruments of which almost all Middle-Schools and Higher Institutes possess complete collections, are manufactured in superior quality; and the Precision Instruments of Vienna manufacture, in especial, enjoy a cosmopolitan reputation.

Imports are restricted to model-copies of new inventions.

Through the impulse given by the I. R. Institute of Geology, not only all educational institutions are enabled to provide themselves with collections of minerals at moderate prices, but commerce with minerals, too, has increased to a greater extent in general. In like manner, numerous Societies of Natural History influence the promotion of botanical and zoological collections.

The production of toys of wood, papier maché, iron plate etc. falls, in Austria, almost entirely within the province of domestic industry, which is of especial importance for the population of the Bohemian Ore Mountains as well as for the environs of the Traun lake in Upper Austria. [1])

Division of labour is followed out to the extreme in the production of wooden toys. This circumstance on the one hand, and the low pay for day-work in these parts on the other, account for the extraordinary cheapness of these productions, in consequence of which they find a ready sale in all directions abroad.

A particular sort of toys are the marbles which are manufactured on special marble mills in the environs of Salzburg.

This branch of Austrian industry occupies an average of 3.000 hands, but the value of the yearly production, on account of the cheapness of these articles, amounts scarcely to the sum of half a million florins.

---

[1]) Regarding their local extent, these two groups of the production of toys are represented on a map of wood industry exhibited under Nr. 1196 by the Direction of Administrative Statistics.

1139. AGRAM, Board of Commerce and Trade at. *Croatia.*
Publications.

1140. ARTARIA and Co. *Vienna, Kohlmarkt.*
Maps.

1141. BOTZEN, Board of Commerce and Trade at. *Tyrol.*
Publications.

1142. BRAUMÜLLER William, Bookseller to the I. R. Court. *Vienna.*

200 Publications in 250 volumes, most of which belong to the province of science and are illustrated with wood-cuts. 12 of these volumes treat of Chemistry and Pharmaceutics; 20 of Geography, History, Statistics; 32 of Agriculture and Forest-management; 15 of Mathematics; 50 of Medicine; 20 of Natural History; 23 of Law and Politics; 20 of Philology and Literature; 30 of Theology and Philosophy; 21 of Veterinary Surgery; 7 of diverse subjects.

The complete list of publications of this publishing-house, established since 1838, is contained in the catalogue printed in 1862. The guiding principles of this publishing-business are, to promote the progress and improvement of scientific literature in Austria, and, at the same time, by entering into connexions with the representatives of science at German universities, to draw Austrian publishing business from its seclusion and thus contribute to a more intimate union between Austria and Germany. The works registered in the catalogue embrace science in all its departments treated of by numbers of scientific celebrities in Austria and abroad, a series of publications upon which the publisher may look with just pride. In the province of Medicine, where Austria occupies a paramount position, Mr. Braumüller's list of works is distinguished by the productions of first-rate medical authorities, such as Bamberger, Engel, Fick, Hyrtl, Linhart, Rokitanski, Scanzoni, Scherer, Schroff, Sigmund, etc. Law and Politics are represented by works of Arndts, Damianitsch, Ellinger, Esmarek, Frühwald, Haimerl, Pachmann, Philipps, Riczy, L. Stein, Jos. Unger. Referring to the various branches of natural philosophy and history, we may call attention to works of Eichelberg, Haidinger, Hochstetter, Kunzek, Oscar Schmidt, C. Reichenbach, Francis Unger, Zippe. Historical works of celebrity are those of Aschbach, Gervinus, Hurter, Prokesch-Osten, J. K. Weiss. The list of theological works is graced by names of divines like: Auer, Seb. Brunner, Ginzel, Kutschker, K. Gellayer, Cardinal Rauscher, Veith, Werner. From the department of philosophy we mention: Larus, Günther, Zimmermann. The list of publications includes the majority of Austria's literary productions in the sciences of forest-husbandry and veterinary-surgery, besides a great number of the most approved and generally used text-books for schools. Lastly, the publisher considers himself entitled to mention that he has essentially exerted himself in embellishing the literary productions of Austria, and that his publications are, in this respect, worthy of being ranked with the most splendid performances of the German press.

1143. BRODY, Board of Commerce and Trade at *Galicia.*
Publications.

1144. BRÜNN Board of Commerce and Trade at. *Moravia.*
Works of the pupils of the Industrial School at Brünn (drawings, models, and modellings etc.). Works of the Moravian Upper Weaving School at Brünn (drawings and weavings.)
(See Class 21.)

1145. BRUNNER F. F., Straw-hat Manufacturer and Toy Merchant. *Pesth, Waiznergasse.*
Toys.
(See Class 27.)

1146. BUDWEIS, Board of Commerce and Trade at. *Bohemia.*
Publications.

1147. CRACOW, Board of Commerce and Trade at. *Galicia.*
Publications.

1148. CZERNOWITZ, Board of Commerce and Trade at. *Bukowina.*
Publications.

1149. EGER, Board of Commerce and Trade at. *Bohemia.*
Publications.

1150. ESSEGG, Board of Commerce and Trade at. *Slavonia.*
Publications.

1151. FINANCES, I. R. Ministry of. *Vienna.*
Statistic Publications.

1152. FIUME, Board of Commerce and Trade at. *Croatia.*
Publications.

1153. FRIESE F. M., Director of Mines. *Vienna.*
Statistic-geographical review of the total Austrian mining-industry and its progress these 36 years.

1154. GEOGRAPHICAL SOCIETY, I. R. *Vienna.*
Publications.

1155. a) GEOLOGICAL INSTITUTE of the Empire, I. R. *Vienna.*
Maps and Publications.
(See Class 1.)

1155. b) GEROLD's Charles Son, Printer, Bookseller and Publisher to the Imp. Academy of Sciences at *Vienna.* Agent in London, Baron Arthur Hohenbruck, of the Austrian Exhibition Commissioners.

Publications in various departments of science; text-books and manuals for school-instruction and autodidactic purposes.
(See Class 28.)

Gerold's publishing-house, established about the middle of last century and possessed of letters patent by the Emperor Joseph II. and the Emperor Francis I., pursued with unremitting assiduousness the tendency, taken by its founder, of attending to, and mediating, the progress and propagation of science, learning and letters. Though in the beginning limited to inland sale, the publications of this firm have, during the last ten years, not only been circulating in all German states, but have also been distinguished with honourable notice in other countries where German language and literature is appreciated and understood by the lettered public. Those branches of science and letters which this house particularly attended to, are:

1. Mathematics, from the simplest rudiments of the school up to the most abstruse investigations, with particular regard to practical purposes and application to technical pursuits.

2. Natural philosophy, especially Natural History, the several Kingdoms of which are pre-eminently represented by monographs like: Fauna Austriaca. (Coleoptera) Beetles, by L. Redtenbacher, 2 vol. (Diptera) Flies, by R. Schiner (Hemiptera) European hemiptera, by Fieber. Flora of Lower Austria, by A. Neilreich, 2 vol. Geographical Dissemination of Animals, by Schmarda. The mineralogical works of Mohs, Kenngott, Suess, etc.

3. Military Science.

4. Historiography, for which department of letters, the healthy impulse of new vitality in recent times has prompted and improved able writers of great celebrity.

5. Lastly, Educational Literature the promotion of which was the incessant task of that firm, and which it endeavoured to fulfill by producing cheap editions of Greek and Roman classics, dictionaries and commentaries, by publishing manuals of natural history, geography and statistics, some of which have had 12 and even 20 editions. The printing office, which has been gradually enlarging these 12 years, at present working with 16 small presses and 6 fly presses with a steam engine, and occupying 120 persons, is capable to answer all the demands of our time, directing its chief attention to all the improvement and perfection in typographical art hitherto attempted. In conclusion, there may be mentioned that this establishment keeps an extensive assortment of books and entertains connexions in all parts of the civilized world. The whole store together presents a picture of literature from all departments of human knowledge and life in the principal languages of civilization.

1156. GÖRZ, Board of Commerce and Trade at. *Littorale.*
Publications.

1157. GRATZ, Board of Commerce and Trade at. *Stiria.*
Publications.

1158. HOFMANN F. W., Counsellor of agriculture. *Vienna, Josephstadt 52.*
Books on Husbandry, 15 vols.
(See Class 9.)

1159. INNSBRUCK, Board of Commerce and Trade at. *Tyrol.*
Publications.

1160. KRONSTADT, Board of Commerce and Trade at. *Transylvania.*
Publications.

1161. LAIBACH, Board of Commerce and Trade at. *Carniola.*
Publications.

1162. LAUFFER and **STOLP,** Booksellers at *Pesth Waitznergasse, Hungary.*
Travelling Album.

1163. LECHNER Ralph, University Library. *Vienna.*
Books and Games for Children.

1164. LEMBERG, Board of Trade and Commerce at. *Galicia.*
Publications.

1165. LEOBEN, Board of Trade and Commerce at. *Stiria.*
Publications.

1166. LIEBICH Christopher, Counsellor of Forests and Lecturer at the Polytechnic Institute at Prague. Gold Medal for Arts and Sciences.
Tableau representing the advantages of producing mulberry-leaves and wood after the method of the exhibitor, in comparison with the usual method.

1167. LIEBSCHER Leopold and **KUNZ** John, *Vienna, City 258.*
Samples of Mechanical Toys and Automatons.

1168. LIHARZIK Francis Dr. *Vienna 1142.*
The Law of human growth, in plastic and pictorial representation, as theory of proportion for every age and both sexes.
Models by Müller Francis, Chased work by Sauter Francis, both in Vienna.

1169. LINZ, Board of Commerce and Trade at. *Upper-Austria.*
Publications.

1170. LOWER-AUSTRIAN, Board of Trade. *Vienna.*
Transactions of the Board since 1851. Trade Magazine.

1171. MECHITARIST'S College at Venice.
Publications.

1172. MILITARY GEOGRAPHICAL INSTITUTE, I. R. *Vienna.* Great London Medal of 1851.
Maps put together as tableaux for the wall: —

1. Environs of Vienna, southern part. (The northern was published for the London Exhibition of 1851.) The surveying at the scale of 1 : 14,400 and in the circumference of 112 sq. miles was undertaken in 1822 and 1823. The execution of engraving began at a time when lithography was still in its infancy, and no State had, till then, undertaken the application of it for maps in the way as shown in the tableau. Letters and framework were engraved upon a stone-plate, the ground having been drawn with chemical chalk upon a second plate. The cultures of the soil, viz: meadows, pastures, woods, and vineyards have each been taken down upon a stone and afterwards severally printed into the maps. It need not be mentioned how many experiments were to precede the execution, before a favourable

result was obtained. In order to obtain it, no larger size of paper was to be chosen than one of 10 sq. inches. The lithographic execution of this map consisting of 112 leaves lasted from 1825 to 1841. The price of the whole work, of which separate sections of 4 leaves each are also sold, is 56 fl.

2. Environs of Rome. This tableau composed of 20 leaves forms only one part of the large special map of Central Italy in connexion with that of Upper-Italy. Consisting of 52 leaves, the scale being 1 : 86,400, this map is the result of a military surveying executed in 1841—1843 by Austrian Staff-Officers under the direction of the present Lieutenant Field-marshal Baron Nagy, after an agreement having been entered into with the Papal and Tuscan governments. Although Italy is the land of classic antiquity, occupying such a high rank in all sciences and arts since centuries, it was, nevertheless, in point of geography, but inadequately represented. There was no want of descriptive geography, but of good surveying and maps founded upon a scientific basis, as is still the case in Naples. The map of Italy by Rizzi-Zanoni, and another smaller one by Inghirani were all the previous works worth mentioning. The Austrian government surmounted the great difficulties, incurred great pecuniary expenses, and presented to the scientific world the first accurate picture of the configuration of this beautiful country in the map designated here. The price of the complete map is 90 fl., separate parts, however, are also sold at different prices.

3. Special map of Bohemia, in 39 leaves, reduced to the scale of 1 : 144,000 from the extensive military surveying of the country, and engraved on copper. It is the latest production of the Institute and is sold at the price of 34 fl., separate leaves being also sold.

4. General Map of Europe, by Scheda, in 25 leaves, scale 1 : 2,592,000. The first map for which colour-printing linear was applied. Since that time the practice of colour-printing in this manner has been often made use of, and the editor has received marks of distinction from nearly all states. In the first edition, published for the London Industrial Exhibition of 1851, the ground was drawn on stone with chemical chalk, in the second edition it was executed in the engraved manner. Price 30 fl.

5. Tableau composed of photolithographic and other samples.

a) Photolithography: a topographic drawing or an engraving of this kind (ground plot), is copied in direct sunlight upon fine-grained lithographic limestone prepared with double chromate of potash and gum, and the stone rendered ready for printing. Besides the originals (drawing and copper-plate), there are exhibited three several lithophotographic prints, in order to show that the stones produce equally sharp proofs after a few hundred prints have been taken off.

b) Transferring an engraving from stone upon copper. By means of an affusion of a cold solution of resinous matter, a cast of the engraving in stone is taken off made electrically conducting, and hereupon the copper copy produced galvanically in the usual blue vitriol solution.

c) The engraving in stone being by far more shallow than the engraving on copper, wherefore the copper plate copies from the stone admits not of such a great number of proofs, the stone copy was, therefore, overcast with iron, and in this way the number of good proofs considerably augmented.

d) Samples of etching upon copper, with an improved method for etching.

6. A Case containing: a copy of the special map of Bohemia, in 39 leaves; a copy of Central Italy in 52 leaves; a copy of the general map of Hungary in 17 leaves; three leaves of a special map of Dalmatia, that are already published, and several that are engraving.

10 leaves of a general map of the Austrian Monarchy in course of engraving, by Scheda. Original Surveying Sections of Wallachia, scale 1 : 57,600. Original Surveying Sections of Hungary, scale 1 : 28,000, without being founded on terrar reductions. Original Surveying Sections of Dalmatia, founded upon reductions of previous terrar surveyings. The I. R. Military Geographical Institute, since 1840 a special state office, in Vienna, produces and multiplies the drawings, plans and maps requisite for service, hence the geodetical surveyings in the empire emanate from the same. Besides the direction, there are 10 divisions: 1. Office for topographical drawing; 2. Lithography; 3. Copper engraving; 4. Presses with bookbinding; 5. Electrotype; 6. Photography; 7. Proving; 8. Triangulation calculus; 9. Surveying — in all of which the respective service is performed by appointed military officers, technical officers, masters and assistants, and partly also by temporarily assigned commissioned and warrant officers.

Present Staff: 1 General, 14 Staff officers, 107 commissioned Officers, 40 warrant officers and privates, 62 technical officers, 4 masters, 34 assistants and servants; total 262.

The triangulation works of the empire are completed, save a few second surveyings; there were surveyed 9 bases; and in 18 stations, azimuth and latitude were determined. Of the empire, 8,200 square miles were surveyed in detail; there are still 3,619 sq. m. to be surveyed, viz: a small part of Hungary, half of Galicia and Bukowina, the larger part of Transylvania, then Croatia, Slavonia and the Military Borderland. Besides the empire, the whole of Central Italy and Wallachia was surveyed and triangulated. The surveying of the rest of the monarchy will be done in 9 years. The maps of the separate provinces of the empire, not published as yet, may be finished within 18 years. The yearly expenditure for all divisions of the Institute amounts to about 200,000 fl. The income amounts to 26,000—30,000 fl. The method to gain copper prints from engravings in stone, and the etching on copper with the new etching method are peculiar to the Institute.

1173. MINISTRY, I. R., of Trade and National Economy. *Vienna.*

Work on the State of Mining in Austria. Geographical Representation of the Austrian Rail-roads and Notices thereon.

Besides there are exhibited:

a) Drawing of a Suspension-bridge for railways;

b) Drawing of oblique cushion shafts of locomotive engines and carriages;

c) Drawing of a contrivance for doing away with the smoke of locomotive-engines.

These three drawings are designed by Mr. Martin Stirner, Imp. Counsellor and Railway Inspector. Moreover there are exhibited 14 lithographed views of several parts of the railway over the Semmering, during the time of building 1½ foot broad by 2 feet high, for hanging up in the engine-room; lastly the geographical longitudinal profile of the Empress Elizabeth Western Railway, from Vienna to Linz in 3 parts, broad 10 inches, high 2 feet designed by Mr. Henry Wolf I. R. Geologist.

Self-acting coupling for railway waggons, by G. Winter. 2 Plans. Description and Model.

Iron bridge. Plan, by the Civil-Engineer Schiffkorn.

1174. NAVY, I. R., Supreme Command of the. *Trieste.*

Novara Expedition. Some scientific results of the circumnavigation of the globe by the frigate Novara (Commodore B. v. Wüllerstorf-Urbair), undertaken by order of the Imp. Government in the years 1857, 1858, and 1859, under the auspices of His Imperial Highness the Archduke Ferdinand Maximilian, Commander in Chief of the Austrian Navy.

1. Chart of the track of the frigate Novara.

2. Maps of the isle of St. Paul in the South-Indian Ocean, and of the isles of the Nicobar Archipelagus in the gulf of Bengal, after the surveys of the officers of the expedition.

3. Plastic plan of the isle of St. Paul, after the surveys of the members of the Novara Expedition, executed by Captain J. Cybulz, I. R. Artillery.

4. Geological map of the isthmus of Auckland (New Zealand) and

5. of the province of Nelson on the southern island of New-Zealand, designed by Professor F. von Hochstetter, geologist of the expedition.

6. Circumnavigation of the Austrian frigate Novara in the years 1857, 1858, and 1859, under the command of Commodore B. v. Wüllerstorf-Urbair. Descriptive Part by Dr. Charles Scherzer. 3 vol. 1861—1862. I. R. Court and State printing-office (Original German Edition).

7. The same, translated into English. London 1861—1862. Saunders, Otley & Co, publishers.

8. The same, translated into Italian by Counsellor v. Bolza. Vienna 1862. I. R. Court and State printing-office.

9. Nautico-physical Part, embracing the astronomical, geodetical, magnetical, and meteorological observations made during the voyage by the Commander and the officers of the expedition, and edited by the Hydrographical Institute of the Imperial Navy at Trieste. Vol. I. Astronomy.

10. Medical Part, by Edward Schwarz, M. D., Vol. I.

11. Wood-cuts from the descriptive part of the voyage, executed in the xylographical establishment of J. von Berghof n Vienna.

12. Wood-cuts from the descriptive part of the voyage, executed in the xylographical establishment of R. von Waldheim in Vienna.

13. Prosopometer and Gnathometer, invented by Dr. Edward Schwarz, applied to two Indian skulls, besides a number of other anthropometrical instruments, used during the voyage, in order to determine the differential diagnosis of human races; accompanied by a pamphlet illustrating the system of investigation and some general results.

14. Cast of a skeleton of Palapteryx ingens Owen (Moa of the natives), extinct gigantic bird of New-Zealand. The original was brought home by Dr. F. von Hochstetter from the Moa cave of the Awatere valley of the province of Nelson (South Isle of New Zealand), and was set up in the Novara Museum at Vienna. Restored and moulded in gypsum by Dr. Gust. Jäger in Vienna. Casts of the Moa Skeleton may be had on application to the Austrian Commissioners.

15. Skull of Palapteryx ingens.

16. Specimens of Illustrations, intended to accompany the Zoological part, now in preparation.

17. Specimens of Illustrations, intended to accompany the publications on New Zealand.

18. Outline of the principal aims and general scientific results of the Novara Expedition.

1175. OLLMÜTZ, Board of Commerce and Trade at. *Moravia.*

Publications.

1176. PATRIOTIC ECONOMICAL SOCIETY, I. R. *Prague.*

History of the Society and its activity during the last ten years.

1177. PERINI A., Professor. *Trient, Tyrol.*

Work on the disease of silkworms.

1178. PILSEN, Board of Commerce and Trade at. *Bohemia.*

Books and Maps.

1179. PRAGUE, Central Committee for the Promotion of Industry in the Bohemian Forest-Districts at.

Annual Report of the Ore and Giant Mountains. Conditions of Trade in the Bohemian Ore Mountains.

Frequent distresses in the northern frontier mountains of Bohemia awakened charity, in consequence of which Committees were formed in the provincial metropolis and in those parts, which were visited by bad harvests and want of work, in order to convey to the poor mountaineers the donations flowing in from all quarters. The union existing at Prague was, in 1848, constituted as Central Committee. The end in view, viz. that of bestowing charity, was at a later period to give way to that of promoting industry. The establishment of schools for industry, therefore, soon became the chief aim of its activity. Hence there gradually arose the Industry Schools at Joachimsthal, Sonnenberg, Zinnwald, Trinkseifon, Kunau, Heinrichsgrün, Schmiedeberg, Graupen, Büringen, Pressnitz, Neudek, Katharinaberg, in the Ore Mountains, — as well as at Rochlitz, Hochstädt, Haida and Semil, in the Giant Mountains. About 2,000 pupils received instruction in these schools. Moreover, endeavours were made to promote industry by distributing materials and advancing money. The activity of the Central Committee has, of late, in a great measure been concentrated to the Ore Mountains, since, in the Giant Mountain-ranges more favourably adapted for the settlement of manufacturing establishments, there is an increasing extent of manufacturing industry, such as, the production of cotton, woollen, linen, and mixed stuffs, glass refining etc. whilst, in the high table-land of the Ore Mountains so little intersected with valleys, the inhabitants are chiefly limited to domestic industry, for reason of not only their metallic beds being exhausted, but also in consequence of calamities of war, religious intolerance, irrational and technically highly defective management of mining, in decay since the 17th century. Their domestic industry includes the production of: tinned iron-spoons (in the country about Platten and Neudek) occupying about 700 hands; looking-glasses framed in tin-plate (Platten) 18 hands; guns and bayonets (Weippert) 330 workmen; iron nails (Natsching and Heinrichsdorf) 300 hands; buckles and buttons (Peterswald) 1,000 hands; Karlsbad Manufactures (needles, cutlery, cabinet-work, objects of thermal tuff, tin-foundery, pottery); musical instruments (wind and stringed, Graslitz and Schönbach) 600 working people; wooden articles and toys (Upper-Leutensdorf, Katharinaberg) 2,000 hands; weaving (Weippert, Klostergrab, Katharinaberg) 1,500 hands; lacemaker's weaving (Weippert, Kupferberg) 400 hands; straw-plaiting and straw-border's weaving (Zinnwald, Graupen) 6,000—8,000 hands; Laces: a) bobbined, occupying several thousand females (spread over the greater portion of the middle mountain range), b) platted (Graslitz) 60 females, c) points (Gossen) grün, Bleistadt) 1,500 females, d) Valenciennes (Gottesgab-200 females, e) application (Gottesgab, Obertham) 300 females; Embroidery: a) with the crotchet-needle, tambouring, b) in flat stitch (Hirschenstand, Graslitz, Heinrichsgrün) several thousand females; gloves (Obertham, Neudek, Presnitz) 200 hands. — Wages for males: 35—70 kreutzers, for females 10—50 kr., children 5—25 kr. per day continually. Price of brown-coal at the pit: 8 kreutzers per Cwts. At present the Central-Committee supports also schools for industry at B l e i s t a d t (since 1854, in point-laces of linen and cotton thread, 239 who have served their apprenticeship), at N e u d e k (since 1859, in glove-sewing, 52 served out their apprenticeship), at S o n n e n b e r g (since 1861, in crotchet-work and stained embroidery, 10 served out their apprenticeship), at P r e s s n i t z newly opened Sept. 8. 1861, in cutting out and sewing of gloves, 28 females served their apprenticeship.

1180. PRAGUE, Board of Commerce and Trade at. *Bohemia.*

Publications.

1181. PRAGUE, Society for the encouragement of Bohemian trade at.

History and Annual Reports of the Society.

1182. PRÜFER Charles. *Vienna, Landstrasse 413.*

Crystal Models.

1183. PURGER John Bapt., Carver in wood and Sculptor. *Gröden, Tyrol.* Agent: Mr. William Meyerstein, 47, Friday Street, London, E. C. — Bronze Medal of Vienna 1845, and of Leipzig 1850. Honorary Medal of Munich 1854. Medal II. Cl. Paris 1855.

Wood-carvings, Toys.

The Gröden valley in Tyrol numbers about 3,500 inhabitants, and exports yearly about 6,000 Cwts of wooden articles, 3 fourths of which go abroad, even so far as America and Asia.

1184. REICHENBERG, Board of Commerce and Trade at. *Bohemia.*

Publications.

1185. REISS Henry, *Vienna, City 1167.*

Missale Romanum.

1186. ROVEREDO, Board of Commerce and Trade at. *Tyrol.*

Industry Map of Tyrol.

1187. SALZBURG, Board of Commerce and Trade at. Publications.

1188. STATES-MINISTRY, (I. R. Departement of instruction). *Vienna.*

Illustrative Exhibition of the Course, Progress and State of the present Public Instruction. See general part XXXIII.

(See Class 10.)

This collective Exhibition contains:

1 and 2. Graphical representation of the connecting links between the various educational institutions, designed by professor Stummer.

A. Primary Schools.

3. Model of the infant's asylum at Krems in Lower-Austria.
4. Plans (100) of primary schools.
5. Three maps showing the respective proportions between those children that attend schools and those that are bound to attend them; designed by Dr. Adolphus Ficker.
6 and 7. Models of school-forms, by counsellor J. Hermann in Vienna, and from Vicenza.
8. School-satchel with necessaries, from Brünn.
9. Intuitive method of instruction, by means of pictorial illustrations.
10. Composing-box for spelling according to the articulation method, by counsellor V. Prausek in Brünn.

11. German composing-box,
12. French composing-box,
13. Board of moveable letters,
14. Model of a black-board, together with a case of moveable letters,
15. Model-sheet for making moveable letters, besides wooden pens, } accompanied by directions for the use of them, by counsellor J. Herrmann in Vienna.

16. Composing-box and printing apparatus for the blind, by Schwarz in Brünn.
17. Copy-heads for instruction in writing, by Muck in Vienna.
18. Copy-heads and copy-books for instruction in writing, by Pokorny.
19. Copy-heads and copy-books, by Greiner in Vienna.
20. Copy-heads for calligraphy, by the teaching staff of the normal school at Prague.

21. Coloured boards,
22. Counting chains, with a board,
23. Arithmetical apparatus, } accompanied by directions, by counsellor J. Herrmann in Vienna.

24. Arithmetical apparatus with coloured beads, by Ziwný in Brünn.
25. Arithmetical apparatus with black beads, from the normal school at Troppau.
26. Arithmetical apparatus with red-white beads, by Krämer in Troppau.
27. Counting boards.
28. Copies for the first instruction in drawing by Lodl, teacher at Rockitzan, German and Bohemian.
29. Copies for elementary instruction in drawing, by Fr. Knappek at Znaim.
30. Maps for the blind, by the Institute for the Blind at Prague.
31. Model of a bee-hive, by the higher elementary school at Freudenthal in Silesia.
32. Metronome, by Matocha at Iglau.
33. Representation of the Euphorbia Lathyris, by Kopecky in Königgrätz.
34. Pictorial illustrations of venomous plants, by Hartinger in Vienna.
35. The edible mushrooms and the venomous fungi, by Becker and Hartinger.
36. Model of the gymnastic institute at Krems.
37. Prize drawing, by Fr. Ertl in Prague.
38 to 79 inclusive: Works and tasks of pupils of the following schools: Ursula girls' school at Laibach, school for English young ladies at Krems, girls' school at Verona, sunday-school at Rovigo, sunday-school at Venice, sunday-school at Belluno, institute for the blind at Brünn, institute for the deaf and dumb at Gratz, institute for the blind at Linz, Jews' institute for the deaf and dumb at Vienna, elementary school at Schwarzenau, elementary school at Mautern, elementary school at Ybbs, elementary school at Wagstadt, correctionary institute at Brünn, regimental military-school at Gospić, normal school at Brünn, civil ladies' institute at Vienna, school-masters' training institute at Vienna, school-masters' training institute at Königgrätz, school-masters' training institute at Brünn, industrial school at Schönlinde, lower realistic school (or higher elementary school) at Prague, lower realistic school (higher elementary school) St. Anne, at Vienna, higher elementary school at Freudenthal, and lower realistic school (higher elementary school) Jägerzeile, at Vienna.
80 to 91 inclusive: Text-books prescribed in elementary (primary) schools (German, Bohemian, Polish, Ruthene, Slovene, Servian, Croatian, Italian, East-Roman, Hungarian).

B. Middle-Schools.

a) Gymnasia.

92. Views and plans of gymnasia and seminaries, natural-philosophical cabinets and apparatus; plans of the Maria-Theresa Academy of the Nobility at Vienna.
93. Electrical machine (20 inches), by Ch. Winter in Vienna, price . . . . 65 fl.
94. Electrical machine (12 inches), by the same, price . 17 „
95. Electrical machine (6 inches), by the same, price . 8 „
96. Electrical mine firing machine, by the same, price 25 „
97. Hand electrical machine, by the same, price . . . 6 „
98. Laue's glass, by the same, price . . . . 6 „
99. Resinous cake under glass with Lichtenberg's figures, price . . . . 4 „
100. Hygro-electroscope, by Ch. Winter in Vienna, price . . . . 2 „
101. Universal rheometer, by Zenger, teacher at the gymnasium in Neusohl.
102. Impressions of butterflies and explanation (Freinberg, near Linz).
103 to 112. Maps for the wall, by F. Scheda, edited by the I. R. administration of text-books in Vienna.
113 to 119. Paulini's terminological relievo-maps, edited by the I. R. administration of text-books in Vienna, viz. Orteles, Tatra, Schneekoppe, environs of Adelsberg and Zirknitz.
120 and 121. Relievo of the country about Berchtesgaden and the Grossglockner, by Fr. Keil.
122. Apparatus for demonstrating the rise of trade-winds, by Pranghofer, executed by Hauck, machinist, in Vienna.
123 to 147 inclusive: Tasks and specimens of works done by pupils of the following gymnasia: S. Caterina at Venice, Josephstadt at Vienna, Schotten at Vienna, that of Gratz and of Marburg in Styria, that of Lemberg, Kremsmünster in Upper-Austria, Salzburg, Brünn, Teschen, Padua, Pisek, Troppau, and Klagenfurt.
148 to 174 inclusive: Text-books in the vernacular language.
175. Maps of Bohemia and Moravia, by Kořistka in Olmütz.

b) Realistic Schools.

176. Plans (views, ground-plans, etc.) of realistic schools, furniture and collections. Portfolio containing 73 pieces.
177. Photographs of models used in the higher realistic school Landstrasse, Vienna.
178. Models for the instruction in geometrical drawing, from the higher realistic school Landstrasse, Vienna (15).
179. School of figures, by L. Taubinger (60 leaves).

12*

180. Six portfolios of the higher realistic school at Gratz.

181. Box with numismatics, by Ign. Schrotter.

182. Synoptical representation of the instruction in geo-metrical drawing in all classes at the higher realistic school Schottenfeld, Vienna. 6 tableaux.

183 to 191 inclusive: Annual reports, 67 parts, of the higher realistic schools Schottenfeld, Landstrasse and Wieden in Vienna, Buda, Brünn, Prague, Gratz, Klagenfurt, Linz.

192 to 218 inclusive: Tasks done by pupils of the higher and lower realistic schools Schottenfeld, Landstrasse, Wieden, Rossau, and Gumpendorf in Vienna, Buda, Brünn, Troppau, Prague, Lemberg, Pressburg, Linz, and Olmütz.

219 and 220. Synoptical table of the known systems of short-hand and that of Gabelsberger in its development and application to various languages. The Austrian constitutional principles printed in one volume and written in short-hand in $1/_{61}$, by professor L. Conn, in Vienna.

221. Monographs (short-hand) by professor Heger in Vienna. (containing the civil code, the penal code, and judical and bankruptcy proceedings).

222. Short-hand types.

223. Printed and written short-hand works.

C. Universities and Polytechnic Institutes.

224. Professor Dr. Hyrtl's preparations of the human and comparative anatomy.

Explanatory list to these preparations.

When I was requested by the Imperial Committee for the London Exhibition at Vienna to place myselfamong the number of the Exhibitors in the Educational Department (Class 29), I thought it my duty, to form a collection of anatomical specimens in order to give a clear view, how and with what success I work in my professional science.

The objects, which I submit to the examination and impartial judgement of the distinguished Naturalists of Great Britain, represent the different tasks of Technical Anatomy, from those of the most unpretending character to those of the highest delicacy.

I should feel both proud and happy, if the introduction of my handiwork in this place is not considered as an intrusion by my learned English Colleagues, whose dissecting ability and exquisite taste in putting up anatomical preparations, I had ample opportunity to admire in the magnificent Museums of London and Edinburgh.

I have arranged my objects in such a style, that they may not strike the contemplator with that horror or disgust—those almost inseparable companions to every visitor of anatomical collections—but on the contrary, that they may be fit even for fair Ladies' eyes to feast upon.

Without further preliminary remarks I proceed to give a short explanation of the various objects numbered and ticketed in order to guide and facilitate closer inspection.

Nr. 1. This square case with rose wood-frame contains a collection of the innermost part of the Organ of Hearing, called Labyrinth, throughout all orders of the Mammalia.—10 series of 10 objects each, point out the very curious configuration of this important organ, beginning with Man, continuing down to the Whales and Monotremata. The Labyrinth is deeply buried and concealed in the strongest of all cranial bones, which by its stony firmness is called Petrosal.

To clear away the surrounding osseous matters from the delicate and whimsically shaped outlines of the Labyrinth, is justly thought by all Anatomists a very difficult task.

Good preparations of this kind are considered as treasures of anatomical cabinets.

Happily I succeeded in finding out a very commodious and almost never failing method [2]), which makes easy with all annoyances of cutting and digging out a Labyrinth, should it even be enclosed in petrosal bones of such thickness and strength that it may serve to strike fire with steel. A specimen of this latter sort belonging to a Whale, will be seen in front of the case, and direct the attention of the visitor to the Labyrinths of the Whale-tribe, ranged in the tenth line of the tableau.

The labyrinths of two antediluvian Mammalia (third line) are of all the huge Pachydermata, which, „to speak with Plinius, monstrorum ferox Africa in lucem prodit" of the rarest Edentata and Marsupialia, of the Monotremata and Cetacea worthy of peculiar attention. In the latter order the organ of hearing is of very imperfect development [3]), but certainly sufficient to the wants of those unwieldy monsters, living in the dreary and silent solitude of the arctic seas, never to be disturbed by other tunes, than the harsh cry of a hungry sea-bird, or the howling of a polar hurricane. We

are justly entitled, to wonder at the fictions of poets of older times, who let the Delphins be delighted by the lyre and songs of Arion, so that one of them had even the complacency to take the poor singer, when fallen over board in a storm, on his shoulders, and deposit him courteously safe and sound on the distant shore.

To catch a better glimpse of the smallest objects, I endeavoured, to bring them nearer to the glass, by fixing them on little tabourets of a Lilliputian drawing-room.

Those, who will be enabled distinctly to perceive the Incus and the Stapes of Pachyura and Nycteris in the third range, may boast of very sharp eyes. All orders and families of Mammalia have sent their representatives to the general meeting. At the bottom of the case lurks a very nice Mexican Lizard of the Scincoid tribe (Sphenops capistratus Wagl.), gracefully holding up between his jaws the chain of stately bones, belonging to the rarest of all animals, here collected: Elephas primigenius. Please to compare the ossicles with those of the Elephant of our days, last object in the range, labelled Pachydermata.

The cases No. 1 and No. 2, are well worth the honour of the costly and ornamental carving, which I have bestowed on them. Every Naturalist may appreciate, what difficulties I had to strive with, what expenses to defray, to gather in my dissecting room in a central town of the continent, the large hoard of rare and costly animals (some of them unique), living in the remotest parts of the globe, and what heroic decision I needed to destroy their skulls by forcing out the petrosal bones, with the special intention, to give a complete and most accurate morphology of an organ, hitherto but vaguely known. — Science has its weaknesses and fancies to satisfy at every cost.

Medical men will pay most attention to the first line of the strange, squared up battalion of Labyrinths. They will perceive, that the Human Labyrinth attains its full improvement already in the new-born child, and that it does not increase in size throughout the remainder of the lifetime of Man.

There is such intimacy between the sense of hearing and the intellectual education of the child, that the organ, which conveys the first sensitive impressions to the awakening soul, must be ready for action in the same moment, as the human embryo becomes an individual.

The Labyrinth of a deaf and dumb man, 40 years old, brings up the rear of the first line. These dimensions are equal to those of the leader of the file, an embryo, during the fifth month of pregnancy.

No. 2. A similar case to the former. It brings before the Spectator the chain of little bones, called Malleus Incus, and Stapes, which transmits the sonorous vibrations of the air, from the Membrana tympani to the sensible sphere of the Labyrinth.

Man and all orders of Vertebrate Animals are congregated in this little space. Among the 168 groups of bones, three bones each, there will be seen occasionally a Tympanic Membrane and its osseous encasement.

No. 3. A black round wooden case, covered with a large watch-glass, contains the Labyrinths of various families of Birds. The Labyrinths are taken out of the skull by pairs, to show the relative position of the right and left. A very young Siren bears in his mouth the Labyrinth of a Cassiowary.

No. 4. A similar case with birds-heads, whose Labyrinths are maintained in their natural position.

No. 5. (A blackened pedestal, with low glass-bell) holds the most curious skeleton, and the stuffed skin of Chlamydophorus truncatus. Skin and Skeleton belongs to the same individual. These few words will raise the attention of my brother Anatomists. They will certainly agree, that it is not of easy undertaking, to extract the skeleton, with all its dependencies, safe and unhurt, from the skin, and to get out even the minute bones of fingers and toes, without damaging the dermal covering and the horny shovels of the nails.

This little animal is the happy possessor of so many anatomical oddities, that ventured to publish a monography on its organization, compared with that of his nearest relation, the Dasypus gymnurus. [4])

There exist but two specimens of Chlamydophorus in Europe. One of them is exhibited in the anatomical Museum of the Zoological Gardens in Regents Park, by the care of the late highly esteemed London Zoologist, William Yarrell, the other, Reader, presents its compliments to you from under my glass-shelter, and assures men of my profession of his eagerness, to be on intimate terms with them.

Most unhappily the Chlamydophorus is doomed, never more to encounter any of his countrymen in the Museums of the world. The only spot, where the two above-mentioned specimens were found, viz. the town of Mendoza in the high-

---

[1]) We will let speak Prof. Dr. Hyrtl himself.

[2]) To be read, fully explained, in my „Handbuch der practischen Zergliederungskunst", Wien 1861. An English translation is prepared by Dr. John Barclay of Leicester.

[3]) In reference to the enormous size of the body.

[4]) Chlamidophori truncati cum Dasypode gymnuro comparatum examen Anatomicum, in the Transactions of the Academy of Sciences at Vienna, Vol. IX. 1855.

lands of Chili, has disappeared with all its inhabitants and surrounding territory from the surface of the earth. The dreadful catastrophe of an earth-quake in the Andes engulfed in the year 1861 a population of 14,000 souls, and an unknown number of Chlamydophorus in the boiling crater of a Volcano.

No. 6 and 7. The skeletons of O[r]nithorhynchus and Echidna. It may be thought superfluous, to make parade with animals, whose paradoxical organization has been fully explained by the admirable paper of Sir Richard Owen[1]), venerated by all anatomists as the first authority of our science. I brought them over to London, only to put forth, how the value of skeletons can be increased, by conserving the natural ligaments of the bones, instead of binding and fastening them together with wires.

No. 8. A large oblong wooden case, with natural skeletons of venomous vipers, and harmless snakes, set up in their graceful natural coilings. Some attention is due to the Embryo of a giant-snake (Python), coiled up in the egg and occupying the centre of the case, more to the right, and below the rattle-snake. The head of this little, but very good looking monster, which, when full-grown, may break the bones of a rhinoceros in his deadly embrace, is somewhat raised, to render visible for sharp sighted eyes the nimble cutting instrument of hard enamel, fixed at the top of the muzzle. This animal, when come of age, and anxious, to get free of his temporary confinement in the egg, cuts out with this chisel a little hole in the shell, which he breaks in fragments by boring in the hollow with his cone-shaped head. The rudiments of pelvic extremities in the Boa and Eryx are not forgotten.

Opposite the extremely light-boned venomous water-snake of the Sunda-Islands (Hyrophis) lies the terrible Aspis Haje, whose deadly poison killed the most unfortunate and heroic Queen of Egypt. For this atrocious crime the aspect of our snake is doubly odious and frightful to tender hearted persons. The Aspis seems to feel the curse, cast on its name for ever, and throws out his ribs, as if swollen with anger at the scorn of the world.

No. 9. A similar case with splendid pieces of fishes, all stately Lords of the Ganges, Nile and Mississippi. The accomplished elegance, the noble bearing, the snowy whiteness of these dignified personages, as well as their uncommon wealth and richness in undisturbed spines, can only be appreciated by anatomists, who command enough patience of Sisyphus, to engage themselves with osteological researches in Fishes The Hyoidean and Splanchnic Branchial Arches (Owen) of every individual are separately put up, to admit closer inspection. Geologists and Palaeontologists will not overlook the specimens of the extinct order of Ganoid Fishes, and the General Public will gaze with wonder at the Hindustan Cuchia, which fish is fond of rambles on dry land, where he is so frequently seen, that, according Hamilton, he is called a Snake in his country.

No. 10. A smaller case than the preceding, swarms with saltwater-fishes of tropical seas of lesser dimensions than the former, but of no less value and anatomical interest. They stare at each-other in utter amazement on account of the dry aërial element, in which they find themselves, where fins will do no more. The Kurtus indicus (Russel) shows the osseous cover of his air-bladder, the Corydoras, a little, but very pugnacious champion of the Amazon River, spreads out the powerful dentigerous spines of his pectoral fins. which enable our hero, to close with, and to rip up the belly of every valiant antagonist, the Pipe-Fish is proud of the delicate bony framework of his light and slender body, and the Labroid-Fish, Scarus, seems to have borrowed out the beak of a parrot, to look more interesting, and boasts of a very good grinding mill, united to his branchial department, which serves him to chew his rapidly swallowed food with comfortable deliberation. The blue fish of the Red Sea (cheilinus) tells the Visitor by means of a yellow ticket, that the strange colour of his bones is not a consequence of his (or his anatomists) dislike for cleanliness, but a natural prerogative of his own. I dare say, that this blue colour is produced by deposits of copper-salts, formerly contained in the blood of certain molluscs, on which the fish feeds with voracity. Although not subjected to dyspepsia, he can not digest the coloured metallic ingredient of his food, and lays it down in the tissue of the bones. [2])

No. 11. This case comprehends all known genera of tailed Batrachia of Northern America, together with their representatives living in Europe. The precious specimens are disposed of in such a manner, that each genus presents a double view, from above and from the under side, the complicated hyoid bones being fixed sideways to the individual. The stately Amphiuma, with its rudimentary and useless limbs, occupies the middle of the chest, having on each side a full grown, and a younger Siren, male and female. His next neighbour, down to the right, is the skeleton of the eyeless Proteus, inhabiting the eternal darkness of the Subterranean abysses of the grotto of Adelsberg in Carniola.

No. 12. In this case is exposed to full view the Osteology of the very singular families of Coeciliae and Amphisbaenae. The former, however snakes in respect of their elongated and cylindrical bodies, are in not a few osteological and splanchrological points allied to the non-tailed Batrachia (Frogs and Toads). The latter, by their rudiments of pelvic and scapular extremities, are regarded by Zoologists to link the Ophidian to the Saurian order. The magnificent specimen of Coecilia albiventris (Brazil) and the very rare Chirotes canaliculatus (Mexico) claim the particular notice of Visitors.

No. 13 forms the sitting room of a large and chosen society of foreign and indigenous Frogs and Toads. A venerable Mexican toad-patriarch presides and the meeting in the centre of the case. He has before him the most hideous of all female toads, but highly valued throughout the old and new world, on account of her tenderness towards her numerous descendents. She provides them with free lodgings on her back, in the shape of month-cells, where the eggs are hatched, and the young domiciliated, until the curiosity, common to young people, to see the world, raises their spirit of enterprise, and emboldens them, to make their own way through the swamps of Surinam.

Behind the Mexican grand-father, a large-headed Brazilian, half toad, half frog bears on its neck in sleepy indifference a four-panelled bony plate, — the only occurrence of dermal bones in the Batrachian order. The rest of the company, mostly youngsters of uncivilized countries, may'be told of in after time, when they will have merited our regard, on the praises of a pen.

No. 14. This case is the abode of a colony of Lizards, chosen out of the four quarters of the wide world. Even the enemies of Austria — and there are many in my days — will condescend to admit, that the long tailed lizard, with rudiments of scapular arch, and pelvic extremities, a countryman of mine, from the south-eastern parts of the empire[3]), is the prettiest among the number.

The flying dragon of Java expands his parachute, and the other specimens parade their heavy and short-legged bodies about or make show of their crest-adorned or horned heads or on the other hand, of their smart and nimble limbs, which enable them egregiously to the quickest movements, and very daring exploits in leaping when chasing their prey. All the said attributes of agility and alertness are totally wanting in the American Phrynosoma (first specimen to the left). This animal mourns in silence over the American war, or is perhaps ashamed of the strange ornament on his head, and the meaning of it in common life; we leave the poor toad like-sufferer to his unpleasant reflections on married life, and go on with our explanation to the following case.

No. 15. There are exposed to view the different forms of Turtles, inhabiting dry land, or living in fresh or salt-water. The Zoologist will be pleased by the presence of the very rare specimens of Staurotypus and Pelusios, — the latter brought home by the famous itinerant Lady, Madame Pfeiffer, from the marshy and pestiferous shores of Madagascar. A very young Chelys, and the ferocious Trionix delight in friendly intercourse with their peaceful and harmless neighbours of hot climates.

No. 16—20. Various organs of Man and Mammalia, injected and corroded. The fragility, precarious existence of corroded preparations is sufficiently known to all practical Anatomists. I tried, to obviate those very disagreeable deficiencies. and found out a mode of managing, to restore to the corroded ramifications of blood vessels the necessary firmness and durability. I immerse the corroded blood vessels repeatedly in a cooled solution of isinglass, and dry the wetted and dripping ramifications in a rapid draught of tepid air. The blood vessels become repossessed by this proceeding of their lost membranous coats and are at the same time provided with strength to such a degree, that they can bear rough treatment, which they happily will not experience from the hands of those Ladies and Gentlemen, who will honour them by closer inspection.

No. 21. A series of mahogany boxes, each containing 24 microscopical injections of the blood vessels of a single organ, beginning with Man, and descending to fishes. Every injection

---

[1]) „Monotremata" in the Cyclopaedia of Anatomy and Physiology.

[2]) In the tissue of fresh bones the copper can not be brought forth by chemical agents, as little as the iron in the human blood. But in bones, reduced to ashes, as well as in calcinated blood, you will not be disappointed, to prove the presence of the metal.

All the above are taken out of my private collection, counting 600 skeletons of fishes, all prepared in the same way, but set up, each separately in a cheaper style Neque ebur, neque aurum renites: I may be bold to say, that my collection of fish-skeletons is one of the most interesting scientific curiosities of Vienna.

[3]) The Morlachians in Dalmatia call the animal: Scheltopuzik, and rank it among the snakes, on account of the apparent want of extremities.

is put under glass in a little ebony case, and is provided with a ticket, bearing a special explanation of the character of the object.

I beg leave to observe, that the finest capillary vessels may be very well examined with a magnifying power of 20—30 diameter. A higher power does not let us see more of their peculiar arrangement and reticulation.

I hope and trust that all Anatomists and Microscopists, who visit the Exhibition, will make free with my specimens, and use them, as they may think proper for their purposes.

The injections of excretory ducts of Mamma, Liver, Kidneys and Salivary glands, — of lymphatic vessels and their origines in the mucous membrane of the intestines, and in the cellular ground work of some parenchymatic organs, — of precorneal and hyaloideal blood vessels of the eye of Reptiles, — of other tissure of difficult management, have met with due appreciation by all Anatomists on the continent, and I am sure, they are not unknown to some English Naturalists, to whom I presented them, for exchange, or for sale, in the London Natural History Review, July, 1861; the result of which were some applications, to which I paid attention.

No. 22. The important question of preserving anatomical objects in spirit, induced me, to add to the present collection of dry objects, some preserved in alcohol, in order to show, how these things are managed in the anatomical Museum under my care. The contents of the 18 glasses, will reward a nearer investigation. I point out especially: 1. the accessory branchial organ of Heterotis and Meletta, — 2. the inverted order, in which the branchial circulation goes on in ganoid fishes (Acipenser), compared with true osseous and cuntilagineous ones (Raja, Anguilla), — 3. the skeleton and subcutaneous pulmonary tack of Saccobranchus Singio (Bloch), — 4. the skeleton of the most curious turtle called Damatochelys, with true ribs and membranous intercostal pareties, — 5. the brain, spinal marrow, and organ of hearing of Raja batis, — 6. an eye, 20 years after the extraction of cataract (with reproduced lens), — and 7. the bloodvessels of the anterior half of an eye Camel and of its capsula lentis.

No. 23. The motorious and sensitive nerves ramifying in the human face and its cavities.

I thought it advisable, to exhibit this object, partly to encourage and reward the zeal of the young anatomist, who obliged his teacher by such a performance (Dr. Charles Pokorny, Assistant to the chair of Anatomy), partly to convince others, how well such dry preparations of nerves suit the exigencies of a crowded lecturing room, where the students are eager for a good sight of the Trigeminus and Facialis, which a wet preparation will never give.

Nr. 24. A porto-folio with ichthyotomical drawings. I began, years ago, to collect anatomical drawings of fishes, with the intention, to publish a Comparative Osteology of Fishes, for the special accommodation of the Palaeontologist, who needs very often the sight of skeletons of living genera. I gave up that idea with reluctance, my means not being sufficient, to meet the considerable expenses, which the engraving and printing of such a work would render indispensable.

No. 25. Some of my Anatomical Writings.

Among others 1. my Lehrbuch der Anatomie des Menschen, 8th adition preparing, 7 translations, — 2. my Handbuch der topographischen Anatomie, 5th edition preparing, 5 translations (an English one promised by my friend, Professor Percival Wright, Trinity College, Dublin), — 3. my Handbuch der praktischen Zergliederungskunst, 1861, translated already in Dutch and Italian, — an English translation promised by Dr. John Barclay of Leicester, under the care of the Sydenham Society; — 4. a host of various publications on Comparative Anatomy, edited by myself, or my pupils, who trod with success and honour the thorny path of anatomical authorship.

And now, gentle readers, I take my leave, half blushing at my incorrect and paltry conversation. My thoughts came with candour and respect to your hospitable threshold. And, if my humble efforts have met with your approbation, I shall be glad, indeed, not to have worked and toiled in vain for the honour of my country, but if, on the contrary, they are inferior to your expectations, I, as a stranger, claim your courtesy and forbearance.

225. Collection of calculi. Medico-surgical institute at Gratz.

226. Glacier-phenomena, represented by professor Fred. Simony in Vienna. That tableau intends giving a view of the general characteristics constituting the region of glaciers, and especially exhibiting the phenomena resulting from the rising and periodical mutations of the local extension of glaciers.

227. Artificial eye, by professor von Hasner in Prague.

228. Surgical-obstetric instruments, by Ign. Stelzig; price 85 florins.

229. Plans (Views, Sections etc.) of Universities and Polytechnic Institutes (54).

230. General and detailed plans of the polytechnic institute in Vienna, and statistic table (17).

231. Ten leaves for instruction in Botany. Johanneum at Gratz.

232. Six leaves for instruction in Zoology. Johanneum at Gratz.

233 to 241. Students' works of the Johanneum at Gratz and the Polytechnic Institute at Prague.

242. Works of students. Drawings of hydraulic and land architecture, practical and descriptive geometry and mathematics. Technical institute at Lemberg.

D. Special Schools.

a) Military Schools.

243 to 248 incl. Plans for military educational institutes military lower educational house, ground-plans, front-view, and details; military higher educational institutes, ground-plans, front-view, and details; gymnastic apparatus and implements (lithographed); military lower educational houses at Weiskirchen, Bruck on the Leitha, Prerau, Belluno and Fischau; military higher educational institutes at Kuttenberg, Kaschau, Kamenitz near Peterwardein, Güns in Hungary, and Strass; Infantry school companies at Olmütz and Hainburg; Artillery school companies at Olmütz, Liebenau, Prague and Lobzow near Cracow; Engineer school company at St. Pölten; Pioneer school company at Tulln; cadet institutes at Hainburg, Fiume, Marburg and Eisenstadt; Neustadt academy; Artillery academy at Weiskirchen; Engineer academy at Klosterbruck: — Writing copybooks with copy-heads; copy-books of the higher educational institutes; key for situation drawing (18), by J. Scheda.

Text-books of the military educational institutes 45 volumes. [1]

Progress made in learning shown in: Copy-books of lower educational houses; copy-books of higher educational institutes; translations and writings under dictation of the school companies; synoptical exhibition of the progress in technical drawing of the pioneer school company at Tulln; technical drawings (second and third annual course) of the pioneer school company at Tulln; geometrical and situation drawings of the engineer school company at St. Pölten; technical drawings of the artillery school companies at Prague, Olmütz, Cracow and Liebenau; free-hand drawings of the cadet institutes at Eisenstadt, Hainburg, Marburg and Fiume; free-hand drawings of the military teachers' training institute at Neustadt; descriptive geometry of the engineer academy at Klosterbruck (9); architectural drawings, third annual course, engineer academy at Klosterbruck near Znaim; hydraulic architectural drawings; fourth annual course, ornamental drawings, third and fourth annual course, engineer academy at Klosterbruck; architectural drawing of the higher engineer course at Klosterbruck; situation and machine drawings of the artillery academy at Weiskirchen; geometrical situation and free-hand drawings of the military academy at Wiener-Neustadt.

b) Special Civil Schools.

249 to 251. Tasks of pupils in the industrial school, Wieden (in Vienna), and builder's school, Jägerzeile (in Vienna).

252. Plans of agricultural schools.

253. Means of instructions in agricultural schools.

254. Pictorial illustrations of Austrian breeds of cattle. Published at the expense of the Government and edited by the I. R. State Printing office, after photographs. Text by professor Dr. Joseph Arenstein.

255. Shoeing of horses. I. R. Veterinary Institute in Vienna.

256. Universal instrument for forest-science, by professor Beymann in Maria-Brunn.

257 and 258. Various works of pupils and collections from schools for forestry and mining.

259 to 261. Tables for the wall, serving to illustrate lectures on machine construction. Mining and Forest academy at Schemnitz.

262. Fifteen volumes on agriculture, by Hoffmann.

263 to 268. History of Commerce, by Körner. Commercial Arithmetic, by Kaulich. Report of the Vienna Commercial academy. Art of surveying, by Beer. Earth-boring.

268 and 269. Manual of agriculture, and manual of cattle-breeding, by counsellor Pabst.

270. Ground-forms, plastically represented; a case with 25 and another with 43 models, by captain Czybulcz.

271. Dr. F. K. Hillardt's (Vienna) a) Patented apparatus for perspective drawing (Vienna, L. W. Seidel 1858), price £ 2. b) Geometrical tables for the wall.

272. Trimeter, by Ferd. Heissig, in Vienna. (Trimeter to a parallel-rule-perspective for stone masons, builders, architects, machinists engineers etc.; price 8 fl.)

273. Altimeter, by Amadeo Gentilli.

This instrument is destined for obviating the tedious process of levelling, where there is no other end in view but to know the differences of height of the most important points, that are influencing the projected work. It is founded on the trigonometrical method, but without requiring the surveying of a station line, because it gives both distances and angles of altitudes. In a pamphlet annexed to it, there is explained its simple theory and applicability to all summary surveys preceding some tracing, as well as to the solution of many special problems.

STATISTIC Office, Direction of, at *Vienna*.
(See No. **1196**.)

1189. STYRIAN Society of Industry and Trade at *Gratz*.
Publications.

1190. STUMMER Joseph, Dr., I. R. Professor, President of the Emperor Ferdinand Northern Railway, Vice-president of the Charles Lewis Railway, etc. *Vienna, Wieden 309*. Medal I. Cl. Paris 1855.
Historiographical Representation of the Emperor Ferdinand Northern Railway.

This historiographical representation contains in symbolical figures all events and results, in yearly succession, relative to the building, management and administration of the Northern Railway during its 25 years' existence, and was designed and executed by the exhibitor himself, who has been distinguished at the Paris Exhibition in 1855 with the Medal of the 1st Class, moreover, by His Majesty the Emperor of Austria, with the Gold Medal for Art and Science, and by the General Committee of the Emperor Ferdinand Northern Railway with an honorary gift of 1,000 ducats, the same Committee has also caused the multiplication of the work in colour-printing at their own expense.

1191. TRENTSENSKY Matthew, Artistic Office. *Vienna, Leopoldstadt 642*.
Pictures and other Objects for the instructive occupation of youth.

1192. TRIESTE, Board of Commerce and Trade at.
Statistic Reports.

1193. TROPPAU, Board of Commerce and Trade at. *Austria Silesia*.
Annual Report for 1851 and 1852; for 1853; for 1854—56; for 1857—60. Catalogue of the Industrial Exhibition occasioned by the Board in 1851. Report on wages for working people, and institutions existing for the support of traders in Silesia, 1852. Two volumes of Minutes on Sessions held in 1860 and 1861.

1194. UDINE, Board of Commerce and Trade at. *Lombardo-Venice*.
Publications.

1195. VIENNA, I. R. Central Commission for the preservation of architectural monuments at.
Books.

1196. STATISTIC OFFICE, I. R. Direction of *Vienna*.
Statistic and Cartographic Representation of the Austrian Empire.

1197. VIENNA, Board of Commerce and Trade at.
Annual and Statistic Reports. Cheap text-books for Trade Schools in Vienna.

1198. WEISS John Bapt. *Vienna, Wieden 667*.
Sampler of tools for wood-workers.
(Also Class 32.)

This work, designed for instructing artisans in wood, contains on 42 tables 700 illustrations of the tools used, in Austria, by joiners, instrument-makers, machinists, railway-workshops, carpenters, coopers, wheelwrights, stock-makers, chair-caners, matches-makers, book-binders, harness and belt-makers, glaziers and amateur workers. All these tools are manufactured in the Tool-Manufactory of John Weiss and Son, Vienna, Wieden 667, and are separately exhibited in 32d class which see. Price of the sampler 6 fl.

1199. WINTERNITZ Charles, emerited professor. *Vienna*.
Educational Aids for Children. Published at Lechner's, Vienna.

These popular Entertainment Aids comprise: Reading Aids for children of from 4 to 6 years of age; Writing, Counting, and Geography Aids for children of from 5 to 7 years of age; Aids for History fit for children of from 5 to 8 years of age; Entertaining Aids for learning French designed for children of from 8 to 10 years. — The end in view with these Aids is, to infuse into children the rudiments of knowledge by means of entertaining plays, without their undergoing regular schooling lessons. This method has become so extensively popular that 31,000 copies of these Aids were sold since 1854.

1200. WURZBACH Constantine, Dr. von. Biographical and bibliographical Works:
1. Austrian Literature, Reports for 1853, 1854, 1855; 5 vol. First Attempt statistically to exhibit the intellectual exertions of a polyglot state.
2. Bibliographical Central Organ of Austria, 1858. First special bibliographical organ of that empire.
3. Biographical Treasury of the Austrian Empire. 7 vol. A — Har. The first in the Monarchy, that took into account all the national races of the empire and exhibited distinguished persons of the last two centuries; containing many corrections of chronological data from numerous new sources some of which where quite neglected, and descriptions of coats of arms, medals, sepulchral and other monuments, portraits, taken from original sources, manuscripts and archivs.
4. Schiller Book. Gift for the centenary festivity of Schiller's birth. Comprizes the literature on Schiller, published till November 1859, both monographs and treatises found scattered in compilations and other publications. This review embraces all languages and nations having translations of Schiller's Works or scientific strictures on them. To this there are supplements of art, such as, portraits of Schiller, his parents, children, grandchildren, patrons and friends; the fac-similes of his monuments, medals, dwellings, his hand-writing and that of his relations.

1201. ZOOLOGICAL - BOTANICAL SOCIETY, I. R. at *Vienna*.
Books.

Class 30.

Furniture, and Upholstery, including Paper Hangings and Papier Maché.

Sub-Class a) Exhibitors of Furniture and Upholstery.

„   „   b) Exhibitors of Paper Hangings and General Decoration.

House-furniture, in the Eastern Provinces of the Empire, is generally manufactured by the house-owners themselves, the whole not going beyond a table, a bench and a trunk for keeping clothes and other property. The outsides of these Trunks are adorned with rude pictures of various colours, and are made by the Joiners in small towns and boroughs; those made at Kronstadt (Transylvania) form a conspicuous part of the export to the Danubian Principalities. The demands of the country inhabitants for ordinary articles of furniture are provided for by the joiners in small towns and boroughs, exclusively employed in satisfying local wants. Articles of the same sort of a more refined and ornamental description are made by joiners established in the provincial Capitals. Vienna is the chief seat of Furniture etc. not only for the whole of home consumption, but also for export into the neighbouring Eastern dominions. Furniture, made of **bent wood** and **Iron furniture** have but lately risen into commercial importance.

Next to Vienna, Prague, Venice and Pesth must be quoted, as being the seats of extensive and highly improved manufacture of furniture of refined description. Painting being still in Austria the prevailing way of ornamenting the inside of rooms, the want of **Paper hangings** is not large. Some few manufactories of this article, established at Vienna, Prague, Salzburg and Innsbruck, sufficiently provide for the inland consumption finding still time for producing coloured papers of all sorts, paper articles for book-binding etc.

These branches of Industry give employment to more than 60.000 persons and their total value falls but little short of 30 millions of Austrian florins.

1202. (Omitted.)

1203. BAUMANN L., Agent for the first Hungarian parquet floor manufactory, and steam sawing-mill. *Ghymes-Kosztolán, Com. of Neutra, Hungary.* Central Administration: Vienna, Neuwien 372. Ware-houses: Vienna, Pesth, Brünn, Temesvar.

Assortment of Parquets.

1204. FROYDA Francis. *Vienna, Schottenfeld 138.*

Joiner's Manufactures.

1205. GRÜNER Joseph, Cabinet Maker. *Vienna, Alservorstadt 27.* Agent J. L. Mayer Sons and Comp., London, 21, 22, Ironmonger Lane, Cheapside.

Fancy Cabinet Manufactures.

1206. KETTEL Henry, Manufactory of Sculptor's and Fancy Works of Stone-paste. *Vienna, Schottenfeld 83.*

Furniture and Fancy Works of Stone-paste and Antlers.

1207. (Omitted.)

1208. KLEJHONZ Joseph, Cabinet Maker and Inlayer. *Vienna, Wieden 704.* Honorary Mention Vienna 1845.

16 cases, 12 inlaid pianoforte labels; inlaid part of pianoforte top, pianoforte pedal and 2 legs; and inlaid veneer table-top.

Two of the cases and two of the pianoforte labels are ebonist-work, the rest of either sort inlaid work. The table-top is destined as veneering for joiners. The exhibitor's productions are mostly inlaid pianoforte labels, for which he recommends himself at home and abroad.

1209. KLOBASSER John, Paper-hangings Manufacturer. *Vienna, Alservorstadt 334.*

Paper Hangings.

1210. KNEPPER & SCHMIDT. *Vienna, Wieden 420.*

Frames spread with paper-hangings.

1211. KRONIG Charles. *Vienna, Landstrasse 94.*

Papier Mâché Manufactures. (See Classes 7 and 31.)

1212. KUHN Conrade, Carver and Gilder. *Olmütz 4. Moravia.*

Gothic pontifical chair of wood and gilt.

1213. (Omitted.)

1214. PODANY Francis & Matthew, Joiners. *Vienna, Schottenfeld 298.*

Samples of paper-thin veneers for fancy furniture, pannels and floors.

1215. RÖSNER Professor. *Vienna, City 435.*

Oratory of Her Imperial Highness the Archduchess Sophia with Paintings by Professor Kupelwieser; joiner's work by Leistler.

1216. SCHECK Ferdinand, Painter and Sculptor. *Linz 895.*

Carvings with joiner's and upholsterer's works.

1217. SCHMIDT Philip, Joiner to the I. R. Court. *Vienna, Gumpendorf 235.*

Buffet.

1218. SCHÖNTHALER F., Carver. *Vienna, Alte Wieden 213.*

Furniture and Carver's Works.

1219. SCHUBERT John, Patented Manufactory of Porcelain and Metal Nails, Knob and Box-rivets. *Vienna, Ottakring 413.* Ware-house: Vienna, Wieden 320 and 333.

Sampler of porcelain nails, knobs and boxrivets.

That manufactory produces, with 250 workmen, about 60,000,000 pieces a year continually increasing its productions, though only working upon orders given. Lists of prices and samplers are sent on demand and all articles belonging to this line are promptly executed Chief objects of production are: metal nails, knobs, nails for fastening stuffs, and buttons, golden, silver and steel flats, box-rivets of yellow brass with pegs or clamps, whip-knobs, livery-buttons, bell-wire-handles, and handles for drawers.

1220. SPOERLIN & ZIMMERMANN. Patented Manufactory, to the I. R. Court, of Paper Hangings and Fancy Paper. *Vienna.* Gold M. Vienna 1835, 1839, 1845. Prize-M. London 1851. Great M. Munich 1854. Prize M. Paris 1855. Ware houses: Vienna, City 1043 and Gumpendorf 368, and in all large provincial towns.

Drawing-room decorations, and ceiling decorations; assortment of paper-hangings, samples of fancy papers.

1221. THONET Brothers, Patented Manufactory of bent wood works at *Koritschan and Bistritz, Moravia.* Ware-houses: Vienna, Leopoldstadt 586. Hamburg, Paris, London. Prize-M. London 1851. Prize M. Munich 1854, M. I. Cl. Paris 1855.

Chairs, Arm-chairs, Sofas and Tables of bent wood. Wheels of the exhibitor's own construction.

Grounded upon an I. R. Patent, the exhibitors began, in 1850, their manufactures of furniture of bent wood, having soon become an article much sought after for their novelty, extraordinary lightness connected with durable strength. In consequence of the London Exhibition of 1851, where this furniture was distinguished with the Prize-Medal, these productions became known abroad, and export business commenced, which turned out very advantageous for reason of easy and light packing, because chairs and sofas are to be taken to pieces; in 1854 export business comprised one-third of the production. In 1854 the exhibitors were awarded the Prize-medal of Munich and in 1855 the Prize-medal 1. class of Paris, by means of which their manufactures gained a European reputation, and are since that time exported to all parts of the world. At present, more than 800 persons are occupied with the manufacture of these pieces of furniture at the establishments at Koritschan and Bistritz in Moravia; about 70,000 chairs are made, of which nearly 25,000 are sold in Austria, the rest being exported to all directions abroad, by means of the store-houses of the exhibitors at Hamburg, Paris, London. During the last two years that article was essentially improved. Whereas, formerly, the bents were consisting of several parts and glued together, now any bent whatsoever is made of one piece, hereby being adapted to every climate, which was not the case when there were glued joints. The exhibited wheels are of quite a new construction, varying from the usual, by the wooden nave being quite done away with, the spokes running together in the centre standing upon the axes and lying on a metal box, that has on the outside a lid to be removed and which is fastened with a nut. By means of this construction, which has proved considerably stronger than the usual one, it is possible to take out each spoke separately, without taking the wheel to pieces.

1222. VITTORELLI Peter. *Borgo di Valsugana, Tyrol.*

Presses.

(See Classes 4, 8 and 9.)

1223. WILDE Joseph, Joiner. *Zircz, Com. of Vesprim, Hungary.*

Chiffoniers.

1224. WOLF John, Joiner. *Vienna, Margarethen 22.*

Shop-front.

Class 31.

Iron and General Hardware.

Sub-Class a) Exhibitors of Iron Manufactures.

　　„　　„　　*b) Exhibitors of Manufactures in Brass and Copper.*

　　„　　„　　*c) Exhibitors of Manufactures in Tin, Lead, Zinc, Pewter, and General Braziery.*

The use of Iron for Railroads, Machinery and Implements (manufacturing and agricultural) for Building and Military Equipments etc. has been spoken of under the preceding Classes. Our present object is the extensive use of raw and forged Iron for the production of **Sheet Iron, Wire, Nails, Pins, Needles** etc. These objects are nearly exclusively produced in mills and other mechanical establishments, organized after the system of division of labour, and are sufficient for the demands of inland consumption. The once extensive production of **Nails** by hand-work, as it still exists at Zbirow (Bohemia), Losenstein (Upper Austria) and Kropp (Carniolia) is now evidently losing ground before the increased demands for Nails and Pins produced by machinery. Latterly the manufacture of **fire-proof safes** begun in Austria since 1852 and of **Kitchen-implements of cast or plate Iron** has considerably advanced. Vienna is the exclusive seat of these branches of Commerce, whilst the production of **enamelled cast-iron cooking vessels** belongs to the Ironworks in Bohemia, Moravia and Hungary.

The Corporations of Black-smiths, Lock-smiths, Spur-makers, Tinkers etc. are spread over the whole area of the Austrian Empire. These branches of Iron Industry employ nearly 40.000 persons (auxiliaries included) and the value of their annual productions amounts, in a round sum, to 20 millions of Austrian florins.

Next to Iron articles, those made of **Copper** hold a conspicuous place in Austrian Industry. While the insufficient production of raw Copper requires import (26.000 Cwts in 1861) for supplying the wants of inland Industry, hammered Copper in the shape of plane and concave plates and objects for immediate use, made of this metal, are the objects of an important foreign trade, especially with Turkey. The same is the case with **Zinc**. This metal must also be imported in a raw or half worked condition, to supply the demands of home Industry the export being of no importance (1.700 Cwts in 1861).

The import of raw **Lead** (6.500 Cwts in 1861) is nearly balanced by its export (6.400 Cwts). As to worked **Lead**, chiefly from Carinthia, the exports exceed the imports (in 1861 imports 753 Cwts, exports 2.700 Cwts).

In 1861, the import of **Tin** amounted to more than 12.000 Cwts, the quantity of this metal produced in Austria not being sufficient to supply the wants of home consumption. Although the use of Tin for dishes and table utensils had been nearly superseded by the use of China and earthenware, the application of this metal for chemical uses leaf-tin, metallic alloys, and above all, for Tin-salts, has essentially increased in the Austrian Empire.

Brass ranks first amongst **Metallic Alloys.** Plate brass is manufactured in rolling-works, chiefly in Lower-Austria, Tyrol and Lombardo-Venetia. The casting of brass-articles falls to the lot of minor brass-workers or of manufacturers, who work them out for their special purposes. **Packfong-** and **German-Silver articles,** and fancy **Bronze-articles** are almost exclusively confined to Austria, especially to Vienna and its environs, where they are produced either in extensive establishment or by tradesmen. Latterly they have succeeded in taking a conspicuous place among the objects of Austrian export.

Iron and Metal Industry (including the numbers mentioned above) gives employment to more than 70.000 persons, their productions being valued at 50 millions florins per annum.

1225. BARTELMUS Augustus & Comp. (Owner I. W. Jusa) Manufactory of Enamelled Hollow Ware. *Brünn.* Assortment of enamelled iron-plate hollow ware.

This manufactory produces iron-plate hollow ware enamelled inside and outside, either pressed out of one piece, or clinched and folded, as shown in the list of prices joined to the exhibited wares. The enamel is within light-grey, without dark-blue and free from any mineral hurtful to health, especially free from lead and zinc. The enamel is of superior lustre and, as it were, indestructible in use, since, when red-hot, it may be repeatedly plunged into cold water, without the inner or outer glazing being injured by it, hence it is, in fact, impossible that the enamel can crack off in the cooking operations, unless the vessel be hurt by rude violence. For reason of the outer enamel, our vessels are easy to be kept clean, and possess the advantage of being light and therefore convenient for being handled; besides, they are conducing to speedy boiling because of their thinness, hence considerable saving of fuel is attained.

1226. BARTELMUS Edward, Patented Manufactory of Enamelled Iron-plate Hollow Ware and Iron-foundery. *Neu-Joachimsthal, Bohemia.* Ware-house at Prague 741-1. Bronze M. of Vienna 1835, 1839, 1845. Bronze M. of Leipzig 1850. M. II. Cl. of Paris 1855.

Diverse enamelled and non-enamelled cast-iron cooking-vessels, water-tubs, basins, conduit-tubes, fodder-basins, water-closets, spitting-boxes, tobacco-boxes, ashes-boxes.

E.

That manufactory originally at Brünn, established in 1833, removed to Neu-Joachimsthal in 1840 and has existed these 29 years. It is the first establishment that, on a grand scale, produced the article denominated „cast-iron enamelled hollow ware" brought it into commercial intercourse and procured for it appreciation with the buying public. The productions of that manufactory are distinguished: a) regarding the cast, for softness, lightness of weight and pleasant form; b) respecting the enamel, for beautiful white appearance, bright lustre, equality and tenacity. The enamel is thinly overcast and resists rebounding shocks applied on the outside unhurt, that is, blows and strokes received outwardly rather effect the breaking of the iron than the cracking off of the enamel. This quality is preferable to the thick overcast enamel of English manufactures. This establishment in its present extent is, respecting the quantity of yearly productions in this article, likely to rank foremost among those that exist for this branch. The production amounted, in 1861, to 23,355 Cwts. Of these, 19,266 Cwts belong to enamelled iron pottery, and 4,089 Cwts to non-enamelled pottery and other cast-ware. Since November 1861 the manufactory received a greater extent through additional buildings, so that, since this period, 90 Cwts of enamelled pottery are daily manufactured and sold, which is equal to a yearly production of 27,000 Cwts in 300 working days. In case of favourable sales, the manufactory is already now enabled to attain an annual production of 33,000 Cwts. There were, in 1861, employed an average number of 290 workmen, and 3 steam-engines were working with 18 horse-power together. The sale attained amounted, in 1861, to 321,000 florins = £ 29,777. List of Prices gratis. Deduction, according to the lot 30 % and 35 %, of the noted.

1227. BAUER J. J. and Comp., Manufactory of Screws and Rivets. *Vienna, Goldschmiedgasse.*
Samples of Screws and Rivets.

1228. BAUMGARTEN FÜRST Michael, Hammer-work Owner. *Unterweissenbach, Upper-Austria.*
Iron-wares.
(See Classes 32 and 4.)

1229. BODE Fred. Max., Manufactory of Pocket Chemical Light Boxes. *Vienna, Wieden 704.*
Pocket Chemical Light Boxes without matchets, after a new patented invention.

These peculiarly convenient and safe chemical light boxes afford light 500 times without filling. Price 5 florins complete per dozen. Separate filling for 1,000 lights 25 kreutzers. Sample on receipt of the price noted.

1230. BREVILLIER and Comp. Patented Manufactory of Screws and Metal-wares. *Neunkirchen, Lower-Austria.* Ware-house in Vienna 803. Great M. of Munich 1854.
Sampler of Screws, Rivets and Nails.
(See Class 18.)
The sale of the productions of this manufactory does not only extend to all parts of the empire, but also to the States of the Tariff Union, Switzerland, Italy, Russia, and the Danubian Principalities. The works are at present carried on with water-power and employ 400 — 500 hands. Only inland (Styrian) iron is used for the manufacture.

1231. CHADT Joseph, Enameller and Engraver. *Vienna, St. Ulrich 157.*
Table in Bronze and Enamel.

1232. CHRISTALLNIGG Charles, Count; Foundery and Manufactory of Machines. *St. Johann am Brückl, Carinthia.* M. II. Cl. and Honourable Mention, Paris 1855.
Cast Wares.
(See Class 9.)

1233. COLLECTIVE EXHIBITION of Iron Wares of the following Steel and Iron Manufacturers at *Waidhofen and Ybsitz, Lower-Austria.*
Grossmann J. V. and Sons, Leik Francis, Plankh Lewis, Schweineker Brothers, Wagner Augustus,

Wertich Joseph, Windischbauer Francis.
Iron-Wares.
(See Class 32.)

1234. COLLECTIVE EXHIBITION of Iron - Wares produced by the following export-traders, manufacturers and iron-works in *Steyr and the adjoining country (Upper-Austria).*
Manufacturers:
Eigruber Matthew, Cutler,
Heindl Anthony,   „
Mitter John, Sword-smith,
Preitler Joseph,   „
Reichl Joseph,   „
Ries Francis,   „
Werndl Brother, Armourers.
Exporters:
Almeroth John,
Amort John,
Brandegsky Francis,
Engel M. A.,
Koller's Widow and Heirs,
Schönthan Francis,
Voith J. J.
Iron-Wares.
(See Class 32.)

1235. DIENER Charles, Manufactory of Zinc-ornaments. *Vienna, Landstrasse 34.*
Zinc Ornaments.
This establishment, the largest of its kind in Austria, manufactures building ornaments, plastic decorations of pressed, chased and drawn zinc-plate for the inner embellishment of castles, churches, theatres, ball and concert-rooms; also gas and taper candelabres, cast busts and sculptures. This manufactory keeps also stores of zinc-plates of the Silesian Share-holders' Company of Mining and Zinc-Smelting.

1236. DITMAR R., Patented Manufactory of Metal-Wares. *Vienna, Landstrasse 108.* Ware-houses: Vienna, City 939. Prague, grosser Ring. Pesth, Josephsplatz. Berlin, Firm A. Schubert & Comp. unter den Linden 20. Munich, Firm N. Bachmayer.
Moderator Lamps and Chandeliers of all descriptions. Mineral-Oil Lamps for heavy oils up to 36° B.
This manufactory exists since 1841, employs more than 500 workmen, and exports more than half of its productions to foreign countries, especially to Spain, Italy, Holland, Russia, Sweden, England, America, and the German States. All requisite component parts are made in the manufactory, only raw materials, copper, tin, lead, zinc etc. are taken from elsewhere. Lists of Prices are at service.

1237. DOMOKOSEN, I. R. Copper-Mining, Smelting and Hammerworks at *Balánbánya, Transylvania.*
Hammered Copper-plate.
(See Class 1.)

1238. EDER Francis, Exhibitor for the Patented Plate Manufacture Share-holders' Company at *Wöllersdorf, Lower-Austria.* Ware-house: Vienna 1096. Prize-M. London 1851. Honorary M. Munich 1854. M. I. Cl. Paris 1855.
English tinned planished plates, fine iron-plate, zinked iron-plates, and forms pressed of flat plates.
The above-named manufactory produces English tinned planished plates of various sizes and thickness, particularly distinguished for rich tinning, softness and rare ductility; furthermore dead-white tinned plates and fine iron-plates to the thinness of paper; also zinked iron-plates which, without wanting any paint, prove so completely free from rust, that they are, for reason of their superior durability, highly recommendable for roofing and eaves; lastly Forms pressed of flat plate.

1239. EGGER Nothburga, Countess, Owner of the Plate, Hoop and Rod-iron Rolling Works at Lippitz-bach, of the Puddling and Rolling Works at Freuden-

berg, of the Hammer and Wire-drawing Works at *Feistritz, Carinthia.* Agent A. Kreck, London. Gold M. Vienna 1845. Prize-M. London 1851. Great M. Munich 1854. Little Gold M. Paris 1855, and M. I. Cl. Paris 1855 for the Chief Director.
Wires, Lumps, Nails and Turf.

The lumps are from blomaries; the wires (from string-wire to No. 20) of hammered iron, each number of the same grain; coppered spring-wire, sley-wires in long grains. The material for all these wires is refined with char-coal, welded with wood and rolled out in equal heat; there are also: puddlings, lumps and flattings; an assortment of nails; iron in bars of puddling-iron; wires No. 20 and 21, black wire No. $\frac{000 \ 00 \ 0}{2, \ 2^{1}/_{2}, \ 3,}$ and fine, wires of puddling-iron, lastly dragged and pressed Frasser and Sprink turf as fuel for the puddling and rolling work at Freudenberg.

1240. FEIWEL Leopold, Locksmith. *Pesth.*
Locksmith's Works.

1241. FERNKORN Anthony, Knight. Sculptor and Brass-founder, born at Erfurt, March 17, 1813. Knight of the I. R. Austrian Leopold Order and the R. Bavarian Michael Order I. Cl.
Fountain. Monument (Archduke Charles).

In 1835, Fernkorn entered the studio and foundery of Stiglmaier at Munich, having been, at the same time, employed at Schwanthaler's. Since 1840, Fernkorn resides in Vienna, in the beginning only occupied with smaller works. The first work on a grand scale exciting attention and earning the approbation of connoisseurs, was the statue of St. George in combat with the dragon (11 feet high) erected in the Montenuovo house at Vienna. A smaller copy of it (zinc-cast) was purchased by the Austrian Art Union for the lottery. Of the remaining works of the artist we only quote: — „Statuettes from the Nibelungen“. — „The Statue of the Madonna“ (9 feet high) for the church in Foth. — Then the Statuettes: „Dancing“, „Idyl“, „National Song“, „Poetry“, „Tragedy“, „Music“; two of which, Music and Dancing, were executed by the artist in sandstone and in colossal size. — The „Hovering Angel“ for the Kopal monument. — 6 statues of Emperors for the cathedral at Speier. — The „Radetzky Bust“, for the city of Laibach — The „Colossal Lion“ in sandstone, for the monument of those fallen at Aspern and dedicated by the Archduke Albrecht to their memory in their cemetery — and recently „the Equestrian Statue of Archduke Charles“ in brass, on the Burgplatz in Vienna. For the present, there are the following works in preparation: — Colossal monument of Prince Eugen for the outer Burgplatz; Maria Theresa with 4 sitting figures, modelled by Gasser; sketches for the Jellachich monument at Agram; sketch for the Emperor Joseph for Brünn; to be cast in brass after the models of other sculptors; — Schwarzenberg monument, after Hänel in Dresden; Vörosmarty, after Baron Vay jun., and lastly the rich ornaments at the pedestal for Eugen.

Fernkorn's works bear the stamp of ingenious contrivance, bold conception and careful execution, extending even to the details. The localities of the former cannon-foundery were granted by His Majesty the Emperor Francis Joseph to the sculptor's artistic pursuits, and that institution bears the title of I. R. Art Foundery.

1242. FLORENZ Joseph, Scale and Weight-maker *Vienna, Leopoldstadt 249.*
Scales.

1243. GROSSAUER Francis, Mathematical Instrument Maker. *Steyer, Upper-Austria.*
Mathematical Instruments.

1244. HARRACH Francis Ernest, Count, Patented Manufactory of Iron Ware and Plate. *Janowitz near Römerstadt, Moravia.* Honorary Medal Munich 1854. M. I. Cl. Paris 1855.
White-iron-plate and iron-plate, wire-iron.

1245. HELLRIEGL Peter. *Innsbruck 249.*
Iron kitchen-utensils of all kinds, and Tools.

1246. HOLLENBACH David, I. R. Manufactory of Bronze Wares and Brass foundery. *Vienna, Josephstadt 167.* Silver M. Vienna 1845. Great M. Munich 1854.
Bronze Wares, viz: candelabres, door fittings, church lamps, chimney stands, etc. and a sampler.

This establishment was founded in 1839 by the exhibitor, occupying now from 50 to 60 workmen, besides artists and auxiliary hands out of the manufactory. Stores of clock-cases, chandeliers, candle-sticks, statuettes and groups after original designs and copies, inkstands etc., all works belonging to this line are accepted. All productions are originals of Vienna both regarding their designs and models. Main staple in Vienna and sale to the Austrian provinces, exports to Egypt, Turkey, Russia, etc. The exhibited red-bronzed candelabre is for 5 gas-flames with glass-balls, the gilt church-lamp is in byzantine style, the gilt lantern in Moorish, the chimney stands with utensils in rococco style.

The sampler contains in 125 leaves the photographic representations of the exhibitor's productions.

1247. KERL'S F. A. Heirs. *Platten, Bohemia.* Bronze M. Vienna 1845.
Tinned iron spoons.

The exhibitor produces tinned iron and plate spoons of any pleasing forms wished for, likewise curry-combs for cattle and horses, tinned plate looking-glasses, variously coloured and lacquered plate looking-glass boxes, and buckles for horses harness, for all of which he employs 200 workmen. He also keeps stores of iron, steel and mining products, such as, tin, arsenic, blood-stone, brown-stone, emery etc. moreover all sorts of hemp and Hungarian products. Lists of prices in the Exhibition palace.

1248. KOLBENHEYER E. Patented Manufactory of Metal Wares. *Vienna, Wieden 856.*
Metal Wares.

This manufactory is the only one in Austria, which produces English-brown oxydated copper tea-urns, samovars, steam-egg-boiling machines, censer-lamps etc. Britannia-metal tea-urns, tea, milk and coffee pots, sugar boxes, champagne cooling-vats, travelling cups to be taken to pieces, in cases, new-constructed cocks for wine and beer barrels, field bottles with metal cups and leather fillings etc. yellow brass samovars, tea-urns and chemical lamps. The same articles are also produced of pinchbeck and furbished copper. Great assortment of breakfast and dining necessaires of Britania metal, for 1 person up to 24 persons, in leather, Russin leather and oakwood cases adapted for journeys and voyages. Newest Patented self-acting machine for producing ices, advantageous for confectioners, owners of hotels and for a private household. The extraordinary preference to be given to this machine is founded upon the fact that the production of ices requires one-third less time with this machine, and that with equal quantities of substances the resulting quantity of ice is, in using this machine, stronger by the half of that produced in the usual method.

1249. KRONIG Charles. *Vienna, Landstrasse 94.*
Lacquered Plate.
(See Class 30.)

1250. KÜHN Edward and Charles, Patented Manufactory of Chemical Products. *Vienna, Sechshaus 46.*
Untinned and welded white-iron losses and tin.
(See Class 2.)

This manufactory produces of worthless waste of white-iron, with the help of 4 servants, per annum 300 Cwts of fine tin, 1,500 Cwts of weldable iron, 200 Cwts of ammoniac, 60 Cwts of Berlin blue. The Patented Invention, offering a net profit of 80 %, is to be sold for abroad.

1251. LANDERL Leopold, Manufactory of Machine Nails. *Steyer, Upper-Austria.*
Machine Nails.

1252. LEOBEN, Board of Commerce and Trade at. Collective Exhibition of that district:
Mining Products, Iron and Steel Manufactures.
FÜRST Ignace and Maria, Owner of iron-works. *Büchsengut and Thörl near Aflenz in Styria.*
(See Cass 1.)
SESSLER Victor Felix, Iron-works in *Krieglach.*
Iron wares.
(See Classes 1 and 32.)

1253. LUCKNER Anthony and **FABRICIUS** Charles. *Vienna, Wieden 48.*
Fire and rust-proof Safe.

1254. MARKY & GECMEN. *Komorau, Bohemia.*
Enamelled cast-iron hollow-ware.

13 *

1255. MONNIER Joseph, Manufacturer of Pianoforte Pegs. *Vienna, Wieden 1098.*
Diverse Pianoforte Pegs.

1256. MORGENTHALER Charles, Brass and Tin Plate Worker. *Pesth.*
Brass and Tin Plate Works.

1257. OETL John, Locksmith. *Pesth.*
Fire-proof Safe.

1258. OSTWALD & RITTIG. *Vienna, Wieden 752.*
Bronze Fancy Wares.

1259. PANLEHNER Francis. *Waidhofen, Lower-Austria.*
Iron Wares.
(See Class 32.)

1260. PFANNKUCHE G. & SCHEIDLER. *Vienna, Althann 1 and 2.*
Fire-proof Safes.

1261. PICHLER Joseph, Hammer and Tool Smith. *Steinbach, Upper-Austria.*
Chopping-blade, Shovels, Hoes and Hatchets.

1262. PLEISCHL Adolphus M., Manufactory for Enamelling hollow ware of iron-plate. *Vienna, Alservorstadt 109.* Prize-M. London 1851. New York 1853, Munich 1854, Paris 2 M. 1855.
Diverse enamelled healthy cooking-vessels of folded and clinched iron-plate. Cooking-vessels for ship-kitchens. Two field services for officers; bathing tubs, kettles.
(See Class 10.)
The advantages this manufactory is enabled to offer, are variety of productions, manufactures of very large vessels presenting particular difficulties, and completely inoffensive Enamel with which folded and clinched vessels of the largest size are to be durably enamelled.

1263. PRAGUE, Society of Iron Industry at.
Iron Wares.
(See Class 1.)

1264. REACH Fred., Manufactory of Decimal Weights and Metal Wares. *Karolinenthal near Prague 1049.* Store-houses: at Pilsen, Reichenberg, Bukarest.
Decimal Weights, Cattle Scales, and diverse Metal Wares.
(See Class 8.)

1265. REISS A., Brass and Tin Plate Worker to the I. R. Court. *Vienna, Laimgrube 87.*
Bronze M. Leipzig 1850. Honorary M. Munich 1854. 2 M. II. Cl. Paris 1855. Silver M. Vienna 1857.
Shower and Bath Apparatus, Diverse Tinman's Works.
(See Class 9.)
The principal productions of this establishment, existing since 1843, are all sorts of coffee and tea urns, a variety of shower and bath apparatus, Extraction-coffee-machines of yellow brass, pinchbeck, packfong, china-silver and porcellain, 30,000 of which were also sold Coffee machines non plus ultra, new invention, are till now the best both with respect to preparation of good coffee and easy management and comparative cheapness. Of shower-bath apparatus, the exhibitor produces: — uterus shower-baths, eye shower-baths, ear shower-baths, sitting-bath tubs with uterus showers, inhalation apparatus etc. used by the most celebrated professors and physicians in various diseases. Moreover the exhibitor furnishes a great variety of articles of tin and metal manufacture.
Besides the medals enumerated above the exhibitor received for horticultural utensils silver medals of the I. R. Horticultural Society in Vienna, in 1843, 1847, 1850, 1852 and in 1858 the silver medal of the Agricultural Union in Feldsberg, for domestic utensils.

1266. REWOLT Francis, Brass and Tin Plate Worker. *Vienna 604.*
Tinman's Works.

1267. REZNICEK A. M., Plate-Worker, *Smichow near Prague.*
Sugar-moulds of plate, Steam-pipes.
(See Classes 7 and 8.)

1268. ROTHSCHILD Baron, Iron works. *Wittkowitz, Moravia.* Silver M. Vienna 1845, London 1851, Paris 1855.
Rod-iron, Plates, and Puddling-steel.
(See Class 5.)

1269. RUBENZUCKER Michael, Machine-nail Manufactory. *Ramingsteg, near Steyer, Upper-Austria.*
Machine Nails.

1270. SCHOLTZ C. A., Manufactory of diverse tinned and non-tinned curry-combs and other similar productions of iron-plates. *Matzdorf, Hungary.*
Various Tinman's Works.

1271. SEIDAN Wenceslas, Medaller, Patentee for stamped superscription-boards. *Vienna, Gumpendorf 407.*
Medals; Superscription-boards in frames.

1272. SIKINGER Charles. Nail Manufacturer. *Steyer, Upper-Austria.*
Hammered and Machine Nails.

1273. THIRRING Charles, *Neu-Gaudenzdorf near Vienna 199.* Dyer and Patentee for untinning white-iron-plate.
Untinned white-iron-plate waste, tin and iron gained from it.

1274. WACHTLER Philip, Engraver. *Pesth, Waitznergasse.*
Diverse engraver's works.

1275. WAND Siegmund, Manufactory of Bronze Wares, *Vienna, Michelbeuern 56.*
Bronze Wares in connection with glass.
This establishment produces all descriptions of bronzed wares, especially finely gilt illumination requisites both for gas and candle lighting. Particularly to be recommended are hollow, drawn chandeliers making, when solidly constructed, quite the same effect as those of cast brass. In consequence of the light weight of these chandeliers, the import of them is particularly advantageous for those countries where the Tariff is regulated by weight. The fashion is of the manufacturer's own composition, novelties in this line apearing at short intervals. Moreover there are highly to be recommended: Amples (lamps hanging down from the ceiling) of the finest glass, for 3 or 6 lights, well adapted to be substituted for small chandeliers. The durability of the gilding is warranted.

1276. WERTHEIM F. & WIESE, Patented Manufactory of Fire-proof Safes not to be unlocked, for money, books and deeds. *Vienna, Wieden, Feldgasse.* Warehouse and Office in Vienna, City 435 and 437. Warehouses in all central provincial towns of the empire; then in Constantinople, Petersburg, Hamburg, Cairo, Alexandria, Athens, etc. Great-M. of Munich 1854. Silver-M. I. Cl. Paris 1855.
Fire-proof Safes.
This manufactory was established in 1852 and now enjoys universal renown. Upon order of his Excellency the Minister of Finances, several safes made in this manufactory were, February 19. 1853, under official surveyance and before a commission of the I. R. Polytechnic Institute in Vienna, publicly submitted to a proof in fire, when they furnished evidence of safety against danger from fire never exhibited before. Upon order of H. M, the Sultan, these safes were July 10. 1857, publicly submitted to a trial in fire at Constantinople, a trial crowned with the best success. In an other trial, 1858, in the enamelling oven

of the I. R. State Porcelain Manufactory in Vienna, when a common safe of the exhibiting firm was exposed to a heat exceding 1,000 degrees Celsius, its contents remained entirely unscathed. This manufactory has, during 10 years of its existence, finished 10,000 safes, and now, new-built, worked with steam and provided with the best engines, it produces, with a permanent number of 200 workmen, 1,000 and 1,200 safes annually, at the price of 200 rising to 1,000 florins, only using inland materials. More than half of the production is required for exports. Forty-six officially verified and frustrated attempts of burglary prove the power of resistance of these safes against violence, and far more numerous cases of danger from conflagration, when these safes were often buried for days in flame or embers, without their contents having suffered injury, speak for their fire-proof safety. This manufactory has, up to the present moment, provided about 700 I. R. Tax Offices and all monetary and credit institutions of the empire with safes, the constant store of which, amounting to 200 and 300 of all sorts, as well as the quantity of yearly production make this manufactory to be the most extensive and important in its kind on the Continent.

1277. WIESE Fred. *Vienna, Alservorstadt 188.*
Plate-Cooking-Vessels.

1278. WINIWARTER J. & G., Plate and Lead Ware Manufactory, *Gumpoldskirchen, Lower-Austria.* Warehouse, Vienna 816. Agent Mr. William Andrews (Mr. Royle and Comp.), London, 32, King-Street, Cheapside. Bronze-M. Vienna 1857.

Plate, Vessels, zinked wire and zinked nails, wire-pins, assortment of lead-tube fore-warmers etc.

In order to present an insight into the productiveness and manifold produces of this manufactory, a separate treatise with many illustrations was printed in English and German to be had for 2 1/2 s. both in the Exhibition palace and at booksellers'. This treatise contains informations on zinked iron-plate, its manufacture and further use for works of this manufactory, and illustrations of the most important machines and productions. A theoretical computation of the bearing capacity of arched roofs of zinked chamfered iron-plate is added. For the exhibition were only sent in some samples of material and smaller objects of zinked iron-plate, such as: *a)* zinked iron-plates, plain and chamfered; a bent chamfered zinked plate showing the superior material that this manufactory makes use of or their arched roofs, according to the annexed sketch.
b) Hollow shafts of zinked iron-plate for floors and construction of ceilings as shown by the drawing.
c) Employment of zinked oval plate tubes, enveloped with straw and loam, for erecting light fireproof partition-walls, of which complete houses may be built. The exhibitor's Office secured against burglary and fire, as shown in the machine room of the Exhibition, is partially built of such walls. *d)* Diverse hollow-ware of zinked iron-plate, half finished and quite finished. For the production of similar hollow-ware, this manufactory furnishes, at moderate prices, the zinked rims (without bottoms and handles) in the following form and size: for water buckets, for watering pots or usual water-cans (3 sizes), for coal-scuttle, for flour and corn tubs, for liquid vessels of 2 kilderkins capacity.

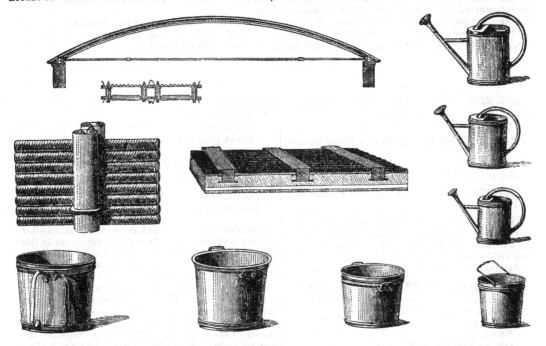

1279. WINKLER Michael. Manufactory of Combination Safety-Locks *Vienna, Gumpendorf 583, 589.* Ware-house, Vienna 260.

Combination Safety-Locks and Plastic Metal-Cast Sign-Boards.

The exhibitor's manufactory produces locks for doors, padlocks, chest-locks, drawer-locks, and all other descriptions of jutting and mortise locks, alike in forms of his own and also according to any desirable design and dimension, after his patented principle. Particularly worthy of mention are those keys, for doors of all kinds, to be worn as watch-guard breloques. The same key is fit locking with equal safety a collection of

locks from the housedoor to the case locks, whereas each separate lock has its own key not fitting any of the others.
Of Superscriptions for streets, lanes, way and mile sign-boards, commercial sign-boards, as well as stair and door numbers, the exhibitor has furnished hundreds of thousands for provincial governments, and was distinguished by His Majesty with the golden cross of merit and by the Trade Union with the great silver medal.

1280. ZIEGLER A. William. *Wilhelmshof, Post Klentsch, Bohemia.*
Tin-foil and Leaf-tin.

Class 32.

Steel and Cutlery, Edge Tools.

Sub-Class a) Exhibitors of Steel Manufactures.

　　„　　„　　b) Exhibitors of Cutlery and Edge Tools.

More than one half (140.000 Cwts) of the total quantity of Steel of every sort, manufactured in the Austrian Empire, is worked up by inland Industry. Deduction being made for the quantities of this raw material, used for the production of machines or worked out into scythes, sickles (see Class 9.), armoury, musical and surgical instruments, a yearly amount of about 45.000 Cwts remains for C u t l e r y.

The valley of the river Enns, offering means of transport for the steel-production of the Innerberg Mining-Company, and those of its affluents, Erlaf and Ybbs, lying partly in Upper-, partly in Lower-Austria, include a circumscribed group of establishments for the manufactures of Iron articles. The centres of these industrial regions are Waidhofen on the Ybbs and Scheibbs for Lower-, and Steyer and Losenstein for Upper-Austria.

The different sorts of Iron articles (Razors, Knives, Scissors, Awls, Drawing-knives etc.) are made by workmen united into Corporations, distinct from each other and never encroaching upon the speciality of any of its fellow-corporations.

Above half a million dozens of table knives and forks, 50,000 dozens of razors, 30 millions of awls, come every year out of these workshops, in good quality and at extremely moderate prices, being sold within the Empire or exported to the neighbouring Eastern dominions.

Refined Cutlery such as Razors, Pen-knives, Table-knives and forks etc. are manufactured in great quantities by the Cutlers of Vienna, Prague and other large towns. A circumstance, deserving special notice, is, that latterly Tungsten has come into extensive use for the production of Smiths' tools.

In Austria, the manufacture of Steel and Cutlery employs nearly 25,000 hands and produces articles, representing annual value of 12 millions florins.

1281. BACHNER Francis, Shoemaker's Tool Manufacturer. *Steyer, Upper-Austria.*
　　Various Shoemaker's Tools.

1282. BAUMGARTEN-FÜRST Michael. *Unter-Weissenbach, Upper-Austria.*
　　Steel Wares.
　　(See Classes 4 and 31.)

1283. BOSCH George. *Neuzeug, Upper-Austria.*
　　Table Knives.

1284. BRUNNER Anthony. Manufactory of Piercing-saws. *Vienna, Mariahilf 116.* Honourable Mention Vienna 1845 and Munich 1854, M. II. Cl. New York 1853 and Paris 1855.
　　Assortment of Piercing-saws.

1285. COLLECTIVE EXHIBITION of Steel-wares from the Steel and Iron Manufacturers. *Waidhofen and Ybsitz:*
　　Grossmann J. V. and Sons.
　　Leik Francis.
　　Plankh Lewis.
　　Schweinecker Brothers.
　　Wagner August.
　　Wertich Joseph.
　　Windischbauer Francis.
　　Steel and Iron Wares.
　　(See Class 31.)

1286. COLLECTIVE EXHIBITION of Steel-wares from the following Manufacturers. *Steyer and its environs, Upper-Austria.*
Manufacturers and Works:
　　Eigruber Matthew, Cutler,
　　Heindl A., Cutler,
　　Mitter John, Sword-maker,
　　Preitler Mathews, File-cutter,
　　Reichl Joseph, File-cutter,
　　Ries Francis,　　„　　„
　　Werndl Brothers, manufactory of armour.

Exporters:
　　Almeroth John,
　　Amort John,
　　Brandegsky Francis,
　　Engel M. A.,
　　Koller v. Widow and Heirs,
　　Schönthan Francis,
　　Voith J. J.
　　Steel-wares.
　　(See Class 31.)

Besides the above-named, there are still many other exhibitors of Steyer and its environs concerned in this collective exhibition. The production of iron-wares has flourished in these parts since several centuries, is now occupying from 12,000 to 15,000 workmen, and makes a capital of 5 upwards to 6 millions circulate, per annum. The sale is to all parts of the empire, considerable quantities, however, are exported to Italy, to the Danubian Principalities, the Levant, Russia, Germany, and North-America. Superior raw material (the best quality of steel and iron, besides the Swedish), as well as the ability of the working population inherited from father to son, have secured to the iron manufactures of Steyr a brisk sale in accordance with their long enjoyed reputation. The chief productions are: files, scythes, sickles, pocket-knives, razors, table-knives, hammered nails, machine nails, awls, tools, etc. In recent times, great success has been gained in the manufacture of arms, there being now produced considerable quantities of swords, guns barrels, bayonets, lances, pikes and component part-of muskets, etc. The bayonet manufactory of Werndl Brothers, ranks among the first on the continent, both on account of its power of producing great quantities and its exemplary organization. Its manufactures enjoy of late, a lively sale to Belgium and England.

1287. (Omitted.)

1288. EGGER Matthew. *Sterzing (Tyrol)*.
Scythes and Sickles.

1289. EISGRUBER Matthew, Cutler. *Steyer, Upper-Austria.*
Diverse Cutler's Manufactures.

1290. ERNST John, Tack Smith. *Steyer, Upper-Austria.*
Tacks.

1291. (Omitted.)

1292. GROSSAUER Joseph, Cutler. *Steyer, Upper-Austria.*
Table Knives.

1293. HACK Francis X. Cutler, *Steyer, Upper-Austria.*
Table Knives.

1294. HAIDER Edward. *Waidhofen an der Ybbs, Lower-Austria.*
Chopping Blades; Scythes.

1295. HASLINGER John, Tack Smith. *Garsten near Steyer, Upper-Austria.*
Tacks.

1296. HAUSER Joseph, Gimblet Smith. *Steyer, Upper-Austria.*
Various Gimblets.

1297. HEIDER Leopold, Saw Smith. *Königswiesen, Upper-Austria.*
Piercing Saws.

1298. HOLZINGER Joseph. *Neuzeug, Upper-Austria.*
Clasp and Table Knives.

1299. HUBER Joseph. *Sierning, Upper-Austria.*
Table Knives.

1300. KERBLER Joseph. *Gründberg, Upper-Austria.*
Table Knives.

1301. KERBLER Francis. *Sierninghofen, Upper-Austria.*
Table Knives.

1302. KERNREUTER Francis. Patented Manufactory of Tools. *Vienna, Hernals 205.*
Parallel vices, Hand Drilling-machine, Plate-shears, and various tools for Iron and Brass Workers.

1303. KRAUPA Anthony, Cutler. *Sierning, Upper-Austria.*
Various Knives.

1304. KRAUS Edward. Patented Manufactory of Steel Wares. *Ossegg, near Teplitz, Bohemia.*
3 steel-ribs for parasols, and 3 steel-ribs for umbrellas.

1305. LECHNER Matthew, File Cutter. *Steyer, Upper-Austria.*
Files.

1306. LEOBEN. Board of Commerce and Trade at. *Styria.* Collective Exhibition of the Board District, embracing Mining and Smelting Products, Iron and Steel Manufactures.
(See Classes 1 and 31.)
1. Weinmeister Christophus, Manufactory of Scythes; *Wasserleit near Knittelfeld.*
Scythes.
2. Zeilinger John Aloysius. Owner of Scythe Smithies. *Knittelfeld, Styria.* Silver M. Vienna

1845. Honourable Mention London 1851, Honourable Mention 1854. M. I. Cl. Paris 1855.
Various Scythes.

The scythe-forge „Hopfgarten" at Schottenberg near Knittelfeld, new-built since a twelvemonth, occupies 50 scythe-smiths and produces yearly about 90,000 scythes, going most of them to Russia, Turkey, Hungary, Germany and America. They amount, according to the present price, to 67,000 fl. The raw-steel is drawn from the I. R. Steel Works at Eibiswald, from Count Meran's Works at Krems, and from Prince Schwarzenberg's Works at Murau. The works at Hopfgarten have 10 fires, viz. 4 fires of charcoal and 6 fires of mineral-coal, and 7 hammer-strokes. The scythe-forge at Knittelfeld occupies 25 scythe-smiths and produces 45,000 scythes, amounting to 33,500 fl. at the present price. These works have 6 fire-, viz. 2 fires of charcoals 4 fires of mineral coals and 6 hammer-strokes. Prices of the exhibited scythes in Austrian Currency: 100 of 9 hands length, Russian, franco Vienna 78 fl. 100 of 9 hands length, Hungarian, franco Vienna, 78 fl., 100 of 7½ hands length common form for Germany and Switzerland, franco Salzburg 80 fl. 100 of 33 inches (French measure) large bearded, for France, franco Salzburg 100 fl. 100 of 10 hands length, American, franco Trieste 105 fl.

1307. LÖSCHENKOHL Joseph, Cutler. *Trattenbach, Upper-Austria.*
Pocket Knives.

1308. LOVREK Augustus. *Vienna, City 213.*
Tools and Tungsten Steel.

1309. MILLER'S Martin Son. *Vienna, Gumpendorf 351.*
Steel-wares. Saws, Clock-springs, Gold-plate Rollers, Gold-wire-drawing plate.
(See Class 16.)

1310. MITTER Joseph, Cutler. *Steyer, Upper-Austria.*
Knives.

1311. MOSDORFER Balthasar. *Weitz, Styria.*
Sickles.

1312. MOSER John. *Sierning, Upper-Austria.*
Pocket Knives.

1313. MOSER Joseph. *Grünburg, Upper-Austria.*
Various kinds of Knives.

1314. OFFNER J. M., Scythe Forge. *Wolfsberg, Carinthia.* Bronze M. London 1851. M. Berlin 1844. Honourable Mention New-York 1853.
8 grass scythes of refined Carinthian hammered and puddling steel, viz.

1 of 8	hands length,		broad Polish scythe	
1 „ 8'½	„	„	Russian	
1 „ 8½	„	„	middle broad Hungarian scythe	
1 „ 7½	„	„	narrow	„ „
1 „ 8	„	„	narrow Breslau	„
1 „ 8	„	„	broad Wallachian	„
1 „ 8	„	„	narrow „	„
1 „ 9	„	„	Turkish scythe.	

Store for sale in the manufactory. Prices according to quality and sort per 100, 60 to 80 fl. loco manufactory. The exhibitor produces also all other sorts of grass and corn scythes of the best Austrian steels.

1315. PACHERNEGG John, Patented Manufactory of Scythes. *Uebelbach, Styria.*
100 various scythes.

They are manufactured in 4 scythe and steel hammer-works, viz. 1. in Uebelbach, where a yearly average of 100,000 scythes are produced, which bear the mark of „7 stars" used since more than 100 years in these works. 2. in Feistritz where 80,000 scythes are made annually, bearing the mark of „a sword." 3. in Waldstein, where a yearly average of 50,000 scythes are made bearing the mark of „star-cross." 4. in Einöd, near Karpfenberg, where 80,000 scythes are annually produced with the mark of a „stag". Thus the yearly total production in all 4 hammerworks is 320,000 scythes.
Each scythe forge being at the same time a steel hammerwork where the steel for the former is produced, and great care being bestowed on the production of steel, the scythes are consequently of superior quality. The sale is mainly to Russia, Poland, Hungary, the Danubian Principalities, and Turkey.

1316. PANLECHNER Charles. *Vienna, Fünfhaus 142.*
Piercing Saws.

1317. PANLEHNER Francis. *Waidhofen, Lower-Austria.*
Iron Wares.
(See Class 31.)

1318. PESSL Aloysius. *Sierninghofen, Upper-Austria.*
Knives.

1319. PESSL Theophile. *Sierninghofen, Upper-Austria.*
Knives.

1320. PHILIPP Anthony, File Cutter. *Steyer, Upper-Austria.*
Files.

1321. PICKL M. J., Scythe Forge. *Himmelberg, Carinthia.* Honourable Mention Paris 1855.
Various Scythes, Sickles, Chopping Blades.

1322. PILS Charles. *Neuzeug, Upper-Austria.*
Table Knives.

1323. REICHENAU Matthew von, Widow. *Waidhofen an der Ybbs, Lower-Austria.*
Chopping Blades and Scythes.

1324. ROHRAUER Matthew, Jew's harp Maker. *Molln, Upper-Austria.*
Several kinds of jew's harp.

1325. SCHAFFENBERGER Francis, Awl Smith. *Steyer, Upper-Austria.*
Awls.

1326. SCHWARZ Ignatius, Jew's harp Maker. *Steyer, Upper-Austria.*
Jew's Harps.

1327. SCHWINGHAMMER Thad., Cutler. *Steinbach. Upper-Austria.*
Diverse Knives.

1328. STEEL-WORKS COMPANY I. R. Styrian Austrian, Patented. *Vienna,* for the I. R. Chief Steel Works of Innerberg at Weyer, Kleinreifling, Hollenstein and Reichming.
Files from their manufactory at Weyer.
(See Class 1.)

1329. STARKE Charles, Cutler. *Steinbach, Upper-Austria.*
Diverse kinds of Knives.

1330. STORNIGG Primus, Tool Smith. *Steyer, Upper-Austria.*
Tool-smith's Manufactures.

1331. VOLBERT Francis, Tool Smith. *Steyer, Upper-Austria.*
Various Tools.

1332. WEINMEISTER Joseph, Scythe Smith. *Leonstein, Upper-Austria.*
Scythes of various sorts.

1333. WEISS John and Son, First Patented Manufactory of Tools. *Vienna, Wieden 667.* Silver M. Vienna 1845. Honorary M. London 1851.

Assortment of tools for wood-workers, among which the patented new parallel-regulator-plane.

This manufactory founded in 1820 and newly built in 1854 is particulary adapted for the production of tools answering all the wants of our times. It is provided, besides the usual auxiliarly machines and circular saws, with several vice-pins, slitting-machines, boring-machines, square and vertical saws etc. put in motion with a steam-engine of 10 horse-power; there are more than 100 workmen active, and 1,500 planes besides other tools are produced in one week. In proof of the superiority of their productions may be mentioned that complete collections of the tools of their manufactory were purchased as samples for the I. R. technical cabinet at Vienna, for the higher school of industry at Hannover, and for the technological museum at Athens, and that the great number of their practical inventions and improvements were honoured with praises. Deserving of particular notice are his patented parallel-regulator-planes, invented by him, and consisting of a new construction by means of which all adjustable planes for wood-workers may, through a simple slide of the soles, be regulated parallel, in any desirable dimension, speedily and with full security. These regulator contrivances constructed of iron are to be made use of for other planes, when the first is worn out, wherefore the price is by no means higher for this plane than for others of the common construction.

The whole of the productions of this manufactory are contained in the „Atlas of Austrian Tools for wood-workers", illustrated with 700 figures, edited by the principal of that firm J. B. Weiss, and exhibited in the 29. class.

1334. WERTHEIM Francis, I. R. Court and Patented Tool Manufacturing Hammerworks at *Vienna and Scheibbs.* Principal Ware-house and Office: Vienna, Wieden 348. Silver M. Vienna 1845. Silver M. Petersburg 1849. Prize-M. London 1841. Great M. Munich 1854. Gold Honorary M. Paris 1855.
Assortment of Tools.

The exhibitor manufactures all kinds of tools of steel, iron, yellow brass and wood for joiners, carpenters, coopers, wheelwrights, turners, tanners, saddlers and machinists. Founded by the celebrated tool-smith Gruber, this manufactory exists since 1841, was new-built 1857 and provided with the best machines, steam-power is availed of and steel used which is produced in the manufacturer's hammerworks. The extent of the establishment may be seen by the fact that 60,000 to 80,000 planes are produced annually. The white beech-wood used for them is in no country surpassed in its superior quality, whilst the prices are very moderate; for instance, a smoothing plane, which costs 10 s. in London and 6 frs. in Paris, is furnished by this manufactory for 1 fl. Collections of samples of its tools are to be found in the Polytechnic Institute at Vienna, the Museum at Athens and Petersburg, the School of Industry at Fürth, and in the „Conservatoire des Arts et Métiers" at Paris. The drawings of these tools are everywhere used for being copied and are accompanied with an explanatory text. Besides the distinctions mentioned above, the exhibitor received: the silver medal at the exhibition of Laibach in 1844, the gold medal at Linz in 1847, and in 1850. His Majesty the Emperor of Austria conferred on him the golden medal for art and science and made him knight of the I. R. Austrian Francis Joseph Order, he is in possession of the R. Hannoverian Guelph Order, the great golden medal on the commander's ribbon, of the Imp. Russian St. Anne Order, and is officer of the I. Ottom. Medjidie Order.

1335. WINKLER Brothers, Iron Works. *Waidhofen an der Ybbs, Lower-Austria.*
9 Scythes.

1336. WINTER Wolfgang. *Neuzeug, Upper-Austria.*
Fire-steel Knives.

1337. ZEILINGER Francis. *Uebelbach, Styria.*
Scythes.

1338. ZEITLINGER Caspar, Scythe Smith. *Strub near Molln, Upper-Austria.*
Scythes of various kinds.

1339. ZEITLINGER'S Michael Son, Scythe Smith. *Blumau, near Kirchdorf, Upper-Austria.*
Scythes of various kinds.

Class 33.

Works in precious Metals and their imitations and Jewellery.

The Imperial Mint-Offices at Vienna, Kremnitz (Hungary), Carlsburg (Transylvania) and Venice work up a far greater quantity of Gold and Silver, than any other establishment within the Austrian Empire. In 1861, the coinage in them amounted to:

Silver coin . . . 26,200.000 pieces, in value of 21,500.000 A. fl.
Gold id. . . . . 700.000 „ „ „ „ 9,400.000 „ „
Silver, small coin 1,000.000 „ „ „ „ 100.000 „ „
Copper id. . . . 185,000.000 „ „ „ „ 2,400.000 „ „

Vienna, Venice and Prague are the only places, where the production of Gold- and Silver-Articles has reached importance; Pesth, Brünn, Trieste and Goricia being but second rate. The articles, made at Vienna by nearly 600 tradesmen, some of whom have given to their establishments the extent of manufactories, have succeeded in obtaining the particular favour of the public.

Vienna is the only place where the production of Leaf-gold and Leaf-silver, of Gold- and Silver-wire and of Laces, Strings etc. made out of these materials, has reached some importance. With the exception of some few establishments, which have attained the dimensions of manufactories, these branches of Industry lie wholly in the hands of tradesmen belonging to several Corporations.

Articles made of massive Gold and Silver are objects of exclusive home consumption. Laces and Strings, made of these metals, find their way into Turkey and the Orient in general.

Precious stones (with exception of the Garnets of Bohemia and the Noble Opals of Hungary) are generally imported in the condition of cut and polished Jewels. The Garnets are cut and polished in Bohemia near the localities where they are found; the Opals are fitted out for further use at Vienna.

„Pierres de Strass" artificial Rubies, Emeralds, Saphires and other Gems, are imitated in high perfection in special Glass-manufactures around Gablonz, and receive their finish from the hands of Glass-grinders.

The branches of Industry here in question (Mint-workmen included) give employment to 10.000 persons, and the annual value of their production (taking away the expenses for raw materials) is represented by a sum of 3 millions of Austrian florins.

1340. (Omitted.)

1341. BOLZANI Co. Patented Gold Chain Manufactory. *Vienna, Laimgrube 132.* Silver Medal New York 1853. Great Medal Munich 1854. Honourable Mention Paris 1855.
Works in Gold.

1342. BUBENICEK W. *Prague 549-1.*
Diverse Works in Gold.

1343. EGGER David, Goldsmith. *Pesth, grosse Brückengasse.*
Works in Gold.

1344. EGGER S., Trader in Antiques, Medals etc. *Pesth, Dorotheergasse 11.* Ware-house: Vienna, City 1134. Agent in London: Mr. Davidson, 32, Wilson Street, Finsbury.
Antique Jewelry.
(See Class 1.)

1345. FABER Charles Maria. Surgeon Dentist to His Imperial Highness the Archduke Maximilian. *Vienna, Graben.* Ware-houses: Berlin at Mr. Pappenheim's, London at C. Ash Sons.
Plastic weldable gold, in dendrite crystal forms, for stopping hollow teeth; teeth stopped with it; specific mouth-soap „Puritas".

1346. GLANZ Joseph, Manufactory of Bronze, Iron, Gold and Silver Cast-works, *Vienna, Wieden 508.* Gold Medals Vienna 1835, 1839, 1845. Honorary Medal Munich 1854.
Gilt Tankard with stand of silver.

The tankard (property of his Grace the Prince Colloredo-Mannsfeld) 12³/₄ inches high, weighing 9¹/₂ marcs in silver, is ornamented round about with 8 allegorical, free-standing figures, representing the activity of the departments of the Trade Union. The work in silver was executed in the establishment of the exhibitor, now existing for 30 years. and designed by Professors C. Rösner and J. Führich.

E.

1347. GROHMANN H., Jeweller to the I. R. Court and Patented Gold and Silver Smith. *Prague.*
Works in gold and silver.

1348. KOBEK Francis, Jeweller. *Vienna, Lugek 768.* Silver Medal Vienna 1845.
Diadem of brilliants and pearls; price 20,000 fl. Corsage with hangings of brilliants 10,800 fl. Sévigné of brilliants 7,800 fl. Feather of brilliants 6,000 fl. Loop of brilliants to be worn as brooch and as diadem 7,300 fl.

The exhibitor produces any kind of jeweller's work, after his own original designs, and has, during 20 years, designed and executed all jeweller's works for the I. R. Court and for the I. R. Treasury. In 1860, the exhibitor received the silver medal of the Austrian Art Union.

1349. LEONIC Manufactory, Management of J. Oberleitner, at *Stans, near Schwatz, Tyrol.*
Samples of leonic wires.

1350. LERL Gustavus. *Vienna, Strozzengrund 36.*
Bronze fancy-goods.

1351. LOBKOWITZ Ferdinand, Prince and Duke at Raudnitz. Administration of Mines at *Bilin, Bohemia.*
Sampler of raw and cut precious Bohemian garnets.

1352. NETZ Bernhard, Jeweller. *Vienna, St. Ullrich 152.* Agents: Michael Goldsmith's Sons, 40, Ely Place, Holborn, London.
132 rings in brilliants and coloured precious stones.

Extraordinary contrivances in working render it possible to that establishment to produce uncommon results with respect to cheapness of articles and fineness of workmanship.

1353. NEUSTADTL M. G. *Prague 703-1.*
Works in Gold.

1354. PAUL Charles, Bronze Worker. *Vienna. Schotenfeld 364.*
Bronze Fancy-works.

1355. PICHLER Lewis, Jeweller to the I. R. Court. *Prague 762-2.*

Works in Gold.

1356. PODEBRAD H. *Prague.*

Works in Gold.

1357. POPPE Ignatius G., Jeweller and Goldsmith. *Vienna, Schottenfeld 109.*

Golden Bracelet n o t t o b e l o s t.

The exhibitor manufactures all sorts of gold and jeweller's works, especially bracelets, not to be lost, of a new-invented construction. These bracelets of massive appearance, protected by their solid construction from breaking, have a mechanical contrivance facilitating putting on and off, and danger of being lost or snatched away by thieves is removed by doing away with the usual catch-springs and joints. They are made in the most fashionable forms and at the most moderate prices.

1358. SCHEIDL Thomas, Silver Smith. *Vienna, Neubau 258.*

Works in Silver.

1359. SCHMITT Adolphus. Manufactory of genuine Bohemian and Tyrolese String and Setting Garnets. *Deutschbrod, Bohemia.*

Genuine Garnets.
(See Classe 3.)

1360. SCHÖNBORN Ervin, Count. *Dlažkovic, Bohemia.* Ware-houses: Prague 365-3 and Skalken, near Lobositz. Honourable Mention Munich 1854. Medal II. Paris 1855.

Bohemian Garnets.

The garnets of Dlažkovic are distinguished among the Bohemian garnets by the dark fire of their colour, by the greater hardness, accepting, in a cut state, a more lively splendour and preserving the same much longer.

1361.ˑ ZLOCH, Dr. Anthony Ralph. *Prague.*

18 large crystals, artistically set in silver for a 5 feet high candelaber.

Class 34.

Glass.

Sub-Class a) Exhibitors of stained Glass, and Glass used n buildings and decorations.

  „   „   b) Exhibitors of Glass for household use and fancy purposes.

The rise and progress of glass manufacture in Austria dates from the beginning of the last century, and was materially promoted by letters patent granted by the empress Maria Theresa to the then immigrant glass-makers, securing to them encouraging protection The extensive forests of the Bohemian and the Giant mountains became, in consequence of their stores of pure quartz, the principal homesteads of the new-established glass manufactories. In process of time, glass manufacture spread southward and westward to the forests of the Alps and to the Carpathian mountains, so that now all provinces of the empire participate more or less in this branch of industrial activity. *)

All branches of glass production and its improvements are cultivated in Austria, since the manufacture of cast plate-glass has been introduced.

Though in most of the Austrian glass manufactories wood is still used as fuel, of late, coals and peat have been made extensive use of.

At present there are in the empire 215 glass-works with nearly 2.000 melting-pots producing 650.000 Cwt of rawglass of all sorts annually.

The Refining of raw-glass by grinding or cutting and staining is performed either in polishing mills in connection with glass-works, or is left to domestic industry. In respect to the last mentioned, there exist in Bohemia two special industrial groups, one of which (in Hayda and its neighbourhood) cultivate the refinement of hollow ware, the other (in Gablonz and the surrounding parts) attend to the production of glass fancy wares and polished glass-beads, whereas the production of enamel-beads is to this very day almost exclusively limited to Venice and Murano.

More than 60.000 hands are occupied with the production and refinement of raw glass the total value of the yearly production of glass ware of all sorts, calculated at the place of production, amounts to 18,375,000 florins. One half of this result falls to the share of the districts of the Boards of Commerce Reichenberg, Pilsen, Budweis, Eger and Prague (in Bohemia); the rest is distributed among Venice, Austria proper, Styria, Hungary, Transylvania and the remaining provinces of the empire, the Military Border-land alone excepted, glass not being produced there.

The Glass manufactures are of peculiar importance to Austria considering that it not only completely covers inland demands, but also, in increasing proportion, taking part in the exports which of late amount to nearly one-third of the whole quantity produced. Of glass wares there were exported

in 1841 . . . . 136.045 Cwt

„ 1851 . . . . 182.062 „

„ 1860 . . . . 210.532 „

Further consideration may be claimed for this increase, for reason of the exports being of refined ware only; polished and stained glass-ware, and of beads and plate glass which amounted

in 1841 to 20.090 Cwt

„ 1851 „ 58.840 „

„ 1860 „ 114.192 „

In comparison with these results of exports and their rise, the imports are insignificant. the latter amounted

in 1841 to 378 Cwt

„ 1851 „ 718 „

„ 1860 „ 6.520 „

The increase of imports since 1851 is almost exclusively to be put to the account of the favoured intercourse with the Zollverein comprising chiefly raw unpolished plate glass, common hollow glass ware and cast plate glass.

---

*) The local repartition of glass manufacture throughout the empire is represented on a special map, exhibited together with other maps of industry designed by the I. R. Directory of Administrative Statistics in Class 29.

1362. ADAM John Herm. and Co. Firm: „Brothers Janke". *Blottendorf, Bohemia.*
Hollow Glass Ware.

1363. BATTISTI John. *Innsbruck, Tyrol.*
Glass-pieces for strapping razors etc.

1364. CZECH And. Leop. *Haida, Bohemia.*
Hollow Glass Ware.

1365. FRIEDRICH Anthony, Glass Manufacturer. *Köflach, near Gratz, Styria.*
Tray with glass cover, produced with mineral coal fire.

1366. HARRACH, Count. Glass Manufactory. *Neuwelt, Bohemia.*
Fine Glass.

1367. HEGENBARTH Augustus. *Haida, Bohemia.*
Glass Ware.

1368. HOFFMANN William, Patented Glass Refiner and Trader to the I. R. Court. *Prague und Vienna.* Agents: Messrs. J. and R. M. Cracken, 7, Old Jewry, London. Honourable Mention London 1851. Honorary Medal Munich 1854.
Fine white and coloured Bohemian crystal glass; various vessels of glass.

The exhibitor first introduced into Austria Silvering after his peculiar method. All articles of this establishment are made, according to own designs, in the glass-houses and then refined by cutting, polishing, painting, gilding or silvering. That manufactory works for inland trade and export, also undertaking works and orders after designs and models sent in.

1369. KOSSUCH John, Glass-house Owner. *Pest, Palatingasse.*
Glass Ware.

1370. LOBMAYER J. and L. Refining Establishment. Furnishers to the Court. Ware-house: Vienna, City 940.
Glass Wares and diverse Chandeliers and Mirrors.

Lustre, Chandelier and Girandole. No. 1 in mediaeval manner of white crystal glass, in connection with gilt iron, ancient models having been used. No. 2 in rococo-style, of entirely cut white crystal glass, richly ornamented, in connection with gilt bronze. No. 3 of sevre-like painted alabaster glass, rich with gilt bronze. No. 4, in white, richly cut crystal glass, partially English manner, after own designs. Ampels (lamps hanging down from the ceiling) of various kinds; pillars for balustrades of white crystal glass; drinking vessels of white thin crystal glass with fine engraving; dessert services, in rococo style, of white, cut and engraved crystal glass with bronze ornaments; dessert services of sevre-like painted alabaster glass with gilt bronze. Rococo Mirrors with richly cut glass frames. Antique Goblets designed after ancient models. Fish and Flower stands; and various articles of glass with bronze ornaments. The exhibitors possess an extensive manufactory of lustres, chandeliers and girandoles in white and coloured crystal glass with and without bronze ornaments, likewise also a refining establishment producing all sorts of glass ware. The requisite raw glass is wholly drawn from inland manufactories (from their own branch manufactory at Blottendorf, Bohemia), the component parts of ornamentation are partially home productions, the composition, however, is executed in Vienna. All the exhibited objects are exclusively made after designs and models of their own, unless expressly mentioned as derived from elsewhere.

1371. MAYER J. Nephews. Manufactory of Glass. *Adolf- and Leonorenhain, Bohemia.*
Glass Ware.

1372. MOSER Lewis. *Karlsbad, Bohemia.*
Glass Goblet with 2 medaillons.

1373. PALME Ignatius, **KÖNIG** and Co. *Steinschönau, Bohemia.*
Glass Ware.

1374. PALME Reinhold, Glass Refiner and Lustre Manufacturer. *Haida, Bohemia.*
Two Crystal Plateaux.

1375. PALME Francis, **KÖNIG.** *Steinschönau, Bohemia.*
Bohemian white crystal and fancy Glass Wares.

1376. PELIKAN'S J. Nephews. *Haida, Bohemia.*
Glass.

1377. REICH Samuel and Co. Patented Glass Manufactory. *Krasna, Moravia.*
Diverse Glass Wares.

1378. SCHREIBER J. Patented Glass Manufactory. *Gross-Ullersdorf, Moravia.* Ware-house: Vienna.
Flower Plateau with fish vase. Component parts for oil and gas lamps.

1379. STÖLZLE Charles. Patented Glass Manufactories at *Alt and Neu Nagelberg, Eilfang, Ludwigsthal, Suchenthal and Georgenthal, Lower-Austria and Bohemia.* Ware-house Vienna, Wieden 1, and at Pressburg. Medal I. Class Paris 1855.
Glass Wares.

The exhibitor's manufactures are produced in Bohemia and Lower-Austria in 7 glass-houses with 9 glass-furnaces, some of which existed for a hundred years and longer, some 40 and 70 years. With these there are connected a number of grinding and stamping mills, worked with engines. In the manufactory Eugenia there is since 1861 a plate-glass furnace in activity, the first of this kind in the empire, heated with turf and furnishing superior fine plate-glass. These establishments produce white crystal glass, coloured glass, grinding, chalk, table and green glass, cylinder watch and common watch glasses, glass tubes, bells and covers. 584 persons are occupied, among whom many cutters, polishers, gilders, painters and engravers. The consumption of wood amounts to 14,000 cords a year, and the total sum of wages paid for work is 340,000 fl. The articles are sold in all parts of Austria and exported in great quantity to foreign countries. Among the exhibited objects, there are also some Iris and Transparent Glasses of the exhibitor's invention.

1380. UNGER F. and Co. United Manufactories of Glass, *Tiefenbach, Haida, Marschendorf, Bohemia.* Ware-houses: Vienna, Berlin; Dépôt: Paris, rue de haute ville 44. Agent: Mr. W. Bew, London, 26., Aldermanburgh.
Glass Wares.

That establishment produces hollow glass, crystal, half crystal, common, coloured and ornamented table-services, flower-vases, toilet-articles, fancy-articles, flacons, illumination articles for gas and oil lamps, finished chandeliers for gas and tapers, bottle-cases, glass beads hollow, massive, cut and plain, jewel-paste, glass-jewelry black and coloured, lustre-hangings (prisms), candle-trays, knife-reposers, salt-cellars, etc. etc besides all export articles, which are going mostly to America, Africa, the Indies, China, Japan. During the fair, there is also a ware-house at Leipzig.

1381. VOGEL Ignatius, Glass Refiner. *Jägersdorf, Bohemia.*
Glass Ware.

1382. WEIDLICH Sebastian. *Steinschönau, Bohemia.*
Glass Ware.

1383. WINTERNITZ Brothers. *Prague 1187-2 and Wiesenthal near Morgenstern.*
Glass Ware.

1384. ZAHN Joseph and Co. *Steinschönau, Bohemia.*
Glass Ware.

Class 35.

Pottery.

Brick and tile kilns are spread over all parts of the empire, but are erected to a larger extent in the neighbourhood of populous towns like Vienna, Pesth, Prague etc. The manufacture of hollow tiles has not been endenized till a few years ago near Vienna (In 4.300 kilns there are produced on an average 1.000 millions of bricks and tiles per annum. Besides there are, in some of these kilns 25 millions drainage-tubes pressed and baked.

Terra-cotta wares (building ornaments, statues etc.) are produced in three manufactories, belonging to the districts of the Commercial Boards at Vienna and Salzburg, to the amount of 60.000 Cwt yearly.

The production of common pottery for household use, and of clay stoves (from 850.000 to 900.000 Cwt) occupies about 8.500 potters in all parts of the empire. In some places where particularly serviceable material abounds, there is, a multitude of potteries to be found crowded together, but no extensive factory establishments have till now carried on this branch of industry. For the production of clay tobacco pipes (4.000 – 5.000 Cwt, 6 millions of pieces), there are some large manufactories in Lower Austria and Hungary, as well as small ones in Hungary, where most of these productions are consumed.

Fire-bricks are produced partly in stone-ware and porcelain manufactories, chamotte-stone being employed, partly in special establishments of smaller extent, employing a mixture of quartz; the quantity produced averaging from 15.000 to 20.000 Cwt.

Melting pots for glass-works are produced, in these establishments, of fire-proof clay, and under their own management. Graphite-pots, on the other hand, only few smaller establishments being occupied with their production, are therefore imported mostly from Bavaria. Retorts and stone vessels for conveyance of acids and mineral waters are partly produced in manufactories of chemicals under their own management, partly in particular establishments in Bohemia near Bilin, Eger and Brüx. The production of these articles amounts to at least 100.000 Cwt annually.

Terralithe and siderolithe wares are little sought in Austria itself. The yearly production of such wares, in which, added to

4 manufactories in the commercial board district of Reichenberg (on the Elbe), only few minor producers have a share, amounts to above 6.000 Cwt, most being sold abroad.

The production of stone-ware and wedgewood occupies 54 manufactories with 108 ovens, amounting from 35.000 to 40.000 Cwt. yearly; one half of which falls to the share of Bohemia where 10 manufactories of greater extent are active, at the same time also producing porcelain.

The Porcelain manufacture is centred in a group for the most part associated round about Carlsbad where kaolin-earth, felspar and quartz are found in superior quality, and fuel (brown-coal) is to be had at moderate prices. This group numbers 10 manufactories with 36 ovens. Besides there are porcelain manufactories at Prague and Tannowa (in Bohemia), at Vienna and Berend and Telkibanya (in Hungary). The total production reaches annually the quantity of more than 30.000 Cwt.

The total value of yearly production in the above named branches of earthen-ware industry in Austria is computed at nearly 30 millions florins. The number of hands employed herein amounts to 60.000 men, women and children.

With the enumerated productions of earthen-ware of all kinds, Austrian industry covers inland demands, without foreign trade in such manufactures having risen to any remarkable height. However, some increase of the imports in this trade is to be traced, as shown by the value of:

	Imports	Exports
	florins	
in 1841	181.400	413.800
„ 1851	250.300	541.100
„ 1860	782.000	957.200

The extraordinary rise since 1851 was almost entirely occasioned by the repeal of the prohibition of porcelain-wares; for whilst in 1851 the total trade (import and export) in porcelain-wares amounted to only 40.000 florins, the value of trade in these wares in 1860 amounted to nearly 1·3 million florins.

Far less variation is to be traced in the trade with the rest of the articles of this industrial branch; there is only left to be mentioned in particular an increase in the imports of pottery and melting-pots from 67.100 florins in 1851 to 177.100 florins in 1860.

1385. ASCAN Conrade. Patented Manufactury of Bricks, Tiles, Earthen-ware and Cement. *Charlotten-hütte, Salzburg.*

Assortment of Fayence Vessels, under the name of Hallein Vessels.

1386. BEHR Charles. *Karlsbad.* Ware-house: Prague, Rossmarkt 819.

Vases, Figures, Plants, Animals for Aquaria and Cabinets; Cut Stones and Mosaïc for brooches, Cases, etc., Relievos of figures, heads, pictures, and other objects of Karlsbad Thermal Tuff.

The application of Karlsbad thermal tuff for plastic representations is an invention of that firm, which was crowned with success after many expensive experiments. Quite new and interesting are the relievos of thermal tuff (Sinteroplastiques). These are neither cut, nor engraved, but are produced by deposits or precipitations of the carbonate of lime contained in the Karlsbad mineral waters, elastic forms being exposed to the continuous overflowings of the mineral water.

1387. ERNDT Francis, jun. Potter. *Vienna, Alser-vorstadt 237.*

Terra cotta Ware.

1388. FISCHER Maurice. Patented Porcelain Manufactory. *Herend, Com. of Veszprim, Hungary.* Great Gold Medal of the National Union of Industry 1846. Prize-Medal: London 1851, New York 1853. Prize-Medal I. Class: Paris 1855. Stores of samples: Pesth, Vienna, Paris.

Porcelain Services and Sets of antique and Chinese Forms.

This manufactory produces both articles for practical use and fancy objects; viz: vases, plateaux, candelabers, figures etc. exclusively in antique style, especially in old Saxon, vieux Sèvres of Chinese and Japan genre.

Established in 1839, this manufactory has in its beginning already, instigated partly by the unfavourable circumstances of the Austrian duties of those times partly by individual propensity of taste, taken a path contrary to that followed at present and so little characteristic, that is, it cultivated again the antique style such as that which reached its highest pitch in the last century. In these endeavours the establishment was at first readily supported by Hungarian noblemen, they having found a welcome opportunity of completing their ancient heirlooms of china and sets of plate, in superior executed fashion and at moderate prices. Richly provided with pleasing and very tasteful original pieces as models, the manufactory has in a very short time succeeded in aquiring confidence and custom for their achievements in all circles.

The activity of that establishment was particularly encouraged and engaged on the part of the Most High Imperial Court and several families of high nobility (like those of Liechtenstein, Schwarzenberg, Metternich, Schönburg, Esterhazy, Kinsky) as well as many embassies residing at various courts.

By means of taking part in several international exhibitions, that establishment succeeded in becoming known abroad in a most advantageous manner. Many high persons of England, Germany, Russia and almost without exception all the distinguished houses of Austria are provided with productions in the before-mentioned genre of this establishment.

In consequence of many years persevering experiments in the application of a fossil found in the Bakony forest (where the manufactory is situated), the manufactory succeeded in producing Chinese Kaolin both in its ground tint and its plastic. It carried the point in imitating to a nicety Chinese and Japanese porcelain manufactures so much valued and sought since the eldest times, the production of which was all along in vain attempted in Europe, and it likewise gained success in producing objects of a size not yet attained anywhere. It must not be left unmentioned that articles usually consisting of several parts, are not, as commonly done, joined with cements or bronze, but produced of one piece in the mass, which proceeding is also founded on a peculiar method invented in that establishment.

The productions of this manufactory have earned just acknowledgment from connoisseurs and competent authorities. The immortal Alexander von Humboldt expresses his highly favourable opinion on the Herend Establishment in a letter of Oct. 12., 1857, directed to the Owner. The great Natural Philosopher, after having in this letter explained his views on the manufacture of porcelain in general, utters the following sentiments respecting the endeavours of the Herend manufactory: „My old friendship with Alexander Brogniard, director of Sèvres, and with Stanislaus Julien, the translator of the Chinese works on porcelain manufacture, have also awakened a feeling for the importance of the generally diffused interest for your productions. In point of sublime work, tender thinness of forms, tint and taste of imitated ancient painting, I have never seen anything more similarly perfect than that which you produce in your great establishment at Herend: cups, plates, net-like perforation of vases, everything alike deceives the eye, and my gratitude is as intense as the esteem which is so largely due to the successful promoter of aesthetic industry"

1389. HUFFZKY'S Vinc. Widow. *Hohenstein near Teplitz, Bohemia.*

Terralith and Earthen Ware.

1390. MILLER and **HOCHSTETTER**. Earthen Ware Manufactory. *Hruschau, Silesia.*

Earthen Ware.
(See also Class 2.)

1391. PRAGUE, Patented Porcelain and Earthen Ware Manufactory at. *Smichow near Prague.* Warehouses: Prague and Vienna. Silver Medal: Vienna 1845. Silver Medal: Leipzig 1850.

Diverse Porcelain Ware.

The porcelain manufactory at Prague, established 25 years ago, produces all articles of common use as well as those that are ordered for purposes of luxury. Most materials being procured from the immediate neighbourhood and of superior quality, and those arriving from remoter places being procured very cheap, in consequence of the favourable site of that establishment near the water and rail-way, it has always been enabled to furnish excellent fine and particularly strong and cheap porcelain. For this reason the sale of its productions has extended, far beyond the boundaries of the Austrian monarchy, to Germany, Italy, Greece, Russia, the Orient and to America, and is continually upon the increase. Because of the scanty space for exhibition, the manufactory could not send more for the exhibiton of this year than 2 services for the dining-table, 4 coffee and tea-table services, from the quality and prices of

which, of course, only part of its power of production is to be seen.

1392. RESS Anthony. Clay Pipe Manufactory. *Wr. Neustadt, Lower-Austria.*

Clay Pipes.

1393. RICHTER Anthony, Königsaal, Earthen Ware and Sugar Manufactories. *Zabelitz, Bohemia.*

Earthen Ware.
(See Classe 3.)

1394. VIENNA, I. R. Porcelain Manufactory of the State at. *Vienna, Rossau 137.* Ware-house: City 1218. Commission Stores at Pesth: Engelbert Rorrich. Prize-Medal London 1851. Great Bronze Medal Munich 1854. Honourable Distinction Vienna 1835, 1839 and 1845.

Porcelain-pictures, vases, fruit-dishes, flower-baskets, plateaux, dining and tea services, groups, statuettes etc.

The present exhibiton of 964 pieces will prove that the I. R. Porcelain Manufactory, is devoted not solely to one particular branch of production, but comprehends, within the large province of that industry, the manifold application of porcelain as objects of its pursuit. The I. R. Manufactory of Porcelain exists since 1718, and is after that of Meissen (erected in 1710) the eldest in Europe. The raw materials are, with the exception of Passau china-clay, taken from inland products. There are consumed per annum:

China-clay	1,564 Cwts.
Quartz	559 „
Felspar	282 „
Gypsum	199 „
Marble	80 „
Fire-proof clay for saggars	14,481 „

Of fuel, the porcelain-kilns being heated with wood, the wants are: soft wood 2,172 cords (72 cubic feet each), coal and coke 1,516 Cwts.

A low-pressure engine of 10 horse-power works the mill with 4 stones, the stamping, searcing, and crushing mills. washing apparatus for fireproof clay and chamotte (burn saggars), crushingwork, sorting-sieve and kneading machine, besides a circular saw for wood cutting. The mass-washing has 17 washing-vats with 900 Cubic feet and 14 mass-holders with 900 Cubic feet, all built of cement. The 5 round kilns have 10½ feet width and 46 feet height with 4 stories. The capacity of the strong baking story is 478 Cubic feet. Each kiln has 5 chimneys for split wood. A chemical laboratory serves for the preparation of all the colours of the fluates and gold for painting (850 ducats were wanted in 1861) besides also for synthetic and analytic purposes. The Keramic Museum for clay and porcelain productions of ancient and modern times at home and abroad affords a picture of the historical development of pottery. The I. R. Porcelain Manufactory occupies about 200 workmen, among whom: 55 painters and gold-polishers etc. 53 white turners and washers, 36 saggar-turners and brickworkers etc., 10 coolers and burners, glaziers, 24 wood-workers, 10 house-servants, 8 store-house, ware-house and office servants, besides 15 clerks.

The object of manufacture is hard porcelain with felspar glazing. The usual mass is not generally subject to warping, wherefore articles with long straight lines and also large flat surfaces like picture-plates etc. are to be executed. Many objects which must be very thin, are cast, for instance: crucibles, evaporating vessels, tubes for laboratories etc. etc. Ornaments are produced with handwork or print. The latter comprizes gold and colour printing, according to a process peculiar to this I. R. Manufactory, joining an economical to the technical advantage. Its application was brought to bear in the execution of greater orders, especially a great table service for 500 persons. The sale of the productions of the I. R. Porcelain Manufactory is almost limited to the inland, there being exported only to some places (Dresden, Hamburg, Frankfort o. M., London), partly white, partly ornamented articles of luxury. The production of one year (1861) amounts, in raw productions of porcelain, to 277,230 pieces, which finished for sale, together with the fire-proof bricks formed of used up saggars, fetch the value of 94,700 fl. This establishment received Prize-Medals from the Austrian Art Union in 1860, and from the I. R. Horticultural Society at Vienna in 1861.

Orders of all sorts are executed.

Class 36.

Manufactures not included in previous Classes.

Sub-Class a) Exhibitors of Dressing Cases and Toilette articles.
„   „   b) Exhibitors of Trunks, and Travelling apparatus.

Vienna is nearly the only place in the Austrian Empire, where the articles to be mentioned under this Class are manufactured, partly in some few extensive establishments, partly by a numerous class of tradesmen. The productions of this industry, but recently established pass from the hands of the manufacturer into those of a class of shopkeepers, known under the general denomination of „traders in Nuremberg articles" who keep complete assortiments of them.

Case-makers, Book-binders, Harness-makers, Turners, Ebonists, Cabinet-makers and Wood-carvers are employed in producing the endless varieties of fancy-articles, sent in all directions from Vienna for inland consumption and, latterly also for export.

Fancy leather-articles (as porte-monnaies, Cigar-cases etc.) wooden chests of every description adorned and inlaid with bronze and carved wooden frames for photographic visit-cards have gained considerable importance.

The number employed in these branches of industry has increased to 5.000, producing every year au average value of nearly 4 millions florins.

1395. BREUL & ROSENBERG. *Vienna, City, Graben.*
Vienna Leather and Wooden Fancy Wares.

1396. ENDERS Felix. *Vienna, Gumpendorf 405.*
Leather Fancy Wares.

1397. GRIENSTEIDL Felix. *Vienna, City 746.*
Fancy Wares of Wood.

1398. HIRNER & SINGER. Manufactory of Turner's and Joiner's Works. *Vienna.* Agents: at Paris, M. Pappel and Cloitre St. Jacques; at London, Mr. W. Meyerstein, 47, Friday Street.
Buttons, Walking Sticks, Cases, Pipes and Articles for Smokers.

1399. JAFF Joseph. *Vienna, Leopoldstadt 4.*
Fancy Wares.

1400. KLEIN Augustus. Patented Manufactory of Leather, Wood and Bronze Wares. *Vienna, Mariahilf 37.* Ware-house: Vienna, City 282.
Fancy Wares of leather, bronze, wood, Portmanteau Maker's Works, Travelling Requisites.

In 1846, the exhibitor began his trade with a fund of 25 fl. With his new fashionable forms and solid workmanship, he soon excited the attention of all merchants of Vienna, and within the space of a year he occupied already 25 people. Perseveringly directing his energies to his trade, it soon enlarged and improved, so that he undertook journeys for the formation of direct intercourse and connexions. He had a share in the last exhibitions of Leipzig, Munich and Paris, and received the bronze medal of Leipzig and Munich, and the silver one of Paris. During his 2 months' stay at Paris, he got the conviction that his manufactures are apt to rival with all competitors, and exerted his powers to avail himself of that favourable circumstance. Since 6 years he has his own agent at Paris where the sale of his productions is yearly increasing, so that he can hardly satisfy the numerous orders which he receives. The exhibitor keeps also in Berlin a store of samples for Germany, sends a traveller to Russia, England, Belgium, and Holland for taking orders. In Milan he has a general agent who causes Italy, Greece, Spain, Portugal, Turkey, and South Russia to be traversed with the exhibitor's samples. The main business is export-trade. At present the exhibitor occupies 300 persons of various branches, viz: Leather workers, bookbinders, purse-markers, joiners, bronze-workers, sculptors, engravers, tin-man-workers, chasers, and steel-workers etc. etc. All articles of his manufacture emanate directly from his establishment. The exhibitor succeeded in selling thousands of his models, and it is flattering for him to know that in Vienna fancy wares are mostly imitated from his models, and that extensive sales are effected for his productions abroad.

1401. KOCH Augustus & Comp., Exporters. *Vienna.*
Representatives of the following firms, partaking in the Collective Exhibition, viz:

Bauer Matthew, Manufacturer of phisharmonica and accordions. *Vienna.*

Fischer Joseph, Pipe Carver. *Vienna.*

Haan Brothers, Patented Manufactory of Shoeing Wares. *Münchengrätz, Bohemia.*

Heimann Joseph, Manufactory of furniture to be taken to pieces. *Vienna.*

Mannstein, Furniture Manufactory. *Vienna.*

Nawratil Joseph and Lasitzka, Manufactory of Cabinet and Joiner's Works. *Vienna.*

Neiber and Breiter. Manufactory of Leather Fancy Wares. *Vienna.*

Schreiber Emanuel, Pipe Carver. *Vienna.*

Collective Exhibition of all sorts of mother-of-pearl buttons, and other specialities of Vienna fancy-ware manufacture.

The exhibitor undertakes to execute orders at the shortest notice for the above quoted manufacturers.

Bauer Matthew at Vienna produces phisharmonicas and accordions as well as all other musical instruments of this line. The manufactures of this firm so advantageously known since a long series of years need no further recommendation. The grand scale on which this manufactory is conducted, renders it possible to execute the most extensive orders, from the commonest to the finest sorts in the shortest time.

Fischer Joseph, Vienna, produces all kinds of pipes of real Meerschaum for exports to England and France.

Haan Brothers, Münchengraetz, Bohemia, Patented Manufactory of all sorts of high-low s for gentlemen, ladies and children, especially for exports to South-America, Australia, Batavia and the English Colonies in the East Indies. Deserving of particular mention are their newest kind of boots under the denomination of „Diggers" resisting every moisture. This manufactory disposes of considerable numbers of hands, so that extensive orders may be executed at the shortest notice.

Heimann Joseph, Vienna, produces furniture that is to be taken to pieces.

Mannstein, Manufactory of furniture, at Vienna, produces all descriptions of furniture from the simplest to the most elegant workmanship. Of particular popularity are his decomposable pieces of furniture which are not to be distinguished from those not to be taken to pieces, when the former are put together. The easy way of taking this furniture to pieces facilitates their being packed up in a very narrow space, and therefore, considerably diminishes the expenses of transport in removals or exports. For imports in remote countries this furniture is for this reason particulary advantageous.

Nawratil John, and Lasitzka, Vienna, Cabinet Maker, produces elegant wooden cases of all sizes, forms and kinds of wood, with trimmings and mountings of ivory, bronze and mosaïc.

Neiber and Breiter, Vienna, Patentees and inventors of a paper-frame album for photographic pictures, manufacture all kinds of portfolio works in fine and superfine manner and most solid workmanship, such as: travelling necessaires, travelling bags, simple and richly ornamented buvaras, porte-cahiers, photograph albums, ladies' necessaires, ladies' bags, porte-notes and porte-monnaies, pocketbooks, matchets-boxes, tobacco and cigar cases, of the newest fashions.

Schreiber Emanuel, Vienna, produces all sorts of pipes and cigar-pipes of meerschaum, imitations of meerschaum in all new fashions. Deserving of particular notice are his manufactures of pipes of pure meerschaum waste without admixture of foreign components. This production is distinguished from the usual meerschaum imitations by its being of the same quality and lightness and appearance as genuine meerschaum, is as apt to be smoked brown and indissoluble in water, which is not the case with other imitations.

1402. KRAMMER Hermann jun., Manufactory of Hunting and Travelling Requisites. *Vienna, Weissgärber 70.* Agent: Mr. Bergmann and Comp., 21, Queen Street, Cheapside, London.

Travelling-trunks in sailcloth, and other sorts. Lafaucheux - Cartouche. Dog-strings, Partridge - nets, Small-shot-bags.

1403. KREBS Anthony J., Exporter. *Vienna,* Wholesale Stores. *Vienna, City 771.* In Leipzig, during the fair, No. 5 near the alte Waage.
Fancy Goods.

The exhibitor keeps stores of: — fine veneered cabinet-maker's wares with elegant and tasteful ornaments of genuine gilt bronze and ivory, viz. ladies' cases, caskets, glove-cases, perfumery-cases, cigar-cases, tea-caddies, key boxes, letter-boxes, letter-holders, sugar-boxes, cover-boxes, book-shelves, étagères, boston-boxes, mark-plates, card-cases, card-presses, clock-cases, etc., in the finest sorts of wood and with the most fashionable mountings; then of massive oak, ash and tinged wood, turner's fancy works, viz. cigar-stands, cigar-cups, tobacco-boxes, watch-holders, inkstands, ashes-plates, table-candle-sticks, hand-candle-sticks, matchet-boxes, pipe-lighter-cups, liqueur-glass-stands, reading-desks, pipe-stands, savings-boxes, thermometers, toothpick-holders, etc., similar articles of cast and chased, genuine gilt bronze; also leather fancy wares mounted with gilt bronze, after the newest taste, as: writing-desk-port-folios, ladies' toilet-cases, caskets, travelling-necessaires, cigar-boxes, clock-cases, tea-caddies, cigar-cases, porte-monnaios etc.

The exhibitor's incessant endeavours to improve home industry in this branch have found general acknowledgment among the manufacturers of Vienna. Continually turning his attention to having many articles made at home, that were formerly only made abroad, and to improve taste and introduce variety of forms partly by his own contrivances, partly by making use of foreign models, the exhibitor has shunned neither pains nor costs to procure these articles a staple in all the considerable towns of the world.

1404. LOYSCH Albert. *Vienna, City 628.*
Fancy Wares.

Among the numerous articles of the exhibitor's productions, his wood, bronze and leather fancy articles are most deserving of notice, not only for their originality and elegance of forms, but especially for solidity and moderate prices.

1405. MÜLLNER G. & Co. *Vienna, Leopoldstadt 34.*
Leather Fancy Wares.

1406. REIF John. Manufactory of sounding-boards. sieve-hoops, and matchet-sticks. *Kuschwarda, Bohemia.*

Sounding-boards, Matchet-sticks, Pencil-casings, Sieve-hoops, pestled soft and hard Shavings, Frames, Covers, Pianoforte-key-wood.

The exhibitor produces sounding-boards, matchet-sticks, all sorts of soft pencil-casings, sieve-hoops of all sizes, pestled soft and hard shavings, key-frames and tops. The continuance and enlargement of these productions is secured by means of great forests in that region, and by the owner's (His Grace the Prince of Schwarzenberg) allowing his wood of renowned quality to be taken from these forests.

1407. RODEK, Brothers. *Vienna, City 1150.*
Fancy-Wares of Leather, Wood and Bronze.

1408. SCHMÖLL Fred. Ralph. *Vienna, Mariahilf 74.*
Fancy-Wares.

1409. STENZEL Charles, Manufactory of Fancy Wares of Wood. *Vienna, Wieden 947.*
Fancy Wares of Wood.

The exhibited goods are distinguished for originality and and solid workmanship. Attention is to be called to Leather-imitation articles of wood, a new kind invented by the exhibitor.

1410. THEYER Francis, Merchant and Patentee. *Vienna, City 905.*
Fancy Wares of Wood with bronze, Fancy Wares of Leather, and of Wood with inlaid paintings.

SECTION IV.

Fine Arts.

———

The English Program enounces expressly the purpose of the Exhibition in the following terms: „to set into light the progress and present condition of Modern Art, the period to be represented being left to the choice of any country, each of them being the best judge of what may best fit their own peculiar condition."

England has chosen to represent a period of hundred years, going back as far as to 1762. Austria would have adopted the same period, if the space, assigned to this Empire, had been in any way sufficient to fulfill the purpose of an Exhibition on such an extensive scale.

The space, however, originally allowed to Austria, together with its supplement, obtained only by repeated sollicitation, is so insufficient, that, from the first beginning, there could be no more thought of achieving, even approximatively, the problem enounced in the Program of the Exhibition.

It would be but idle self-delusion to believe Austria (even under more favourable conditions of space) to be enabled to rivalize with the greatest States of Europe; nevertheless, we must regret, that the uncommon narrowness of the space, set to our disposal, is an obstacle for asserting two facts, which had well deserved to be brought before the eyes of the enlightened portion of the English public. The first of these facts is the existence of a certain number of eminent Masters during the last gone period; together with a visible progress of Austrian Art within these last ten years.

Had circumstances allowed to bring to exhibition in some completeness the pictures of such an Artist, as Henry Füger in many respects equal, in many others superior to his British coëval Benjamin West, or those of Austrian „Genre" painters, such as Danhauser, Fendi, Ranftl, Schindler, Treml, Waldmüller, with all their talented successors belonging to a later period, such an exhibition would have excited interest among the English public, not only as a specimen of still imperfectly known tendencies of modern German Art, but also on account of the analogies between these tendencies and those of English Art.

In no other part of Germany (and this is the second fact, we have alluded to before), Sir Josuah Reynolds, Sir Henry Lawrence, Wilkie and Sir Edwin Landseer have found as many admirers, and, we me well say, successors as in Austria. Here, as in England, the „Genre", the pictures of Animals and the Portrait, are cultivated with conspicuous predilection, while the revival of Monumental Architecture and Casting in Metal, in contrast with the above-named more homely branches of Art, dates only from these last years, since the beginning of the reign of the present Emperor. The want of space, which prevented us from offering, by means of the objects to be exhibited, a complete comparison between Austrian and British Art, is also an obstacle against a full representation of the present state of monumental Architecture. The Vienna Architects had declared themselves willing to organize a collective exhibition of modern Architecture; the first practical essays, however, have proved the space allowed to be completely insufficient for bringing such a scheme to execution. A single model of the Great Arsenal, in the immediate vicinity of Vienna — undoubtedly the grandest and most artistical of all the military buildings achieved on the Continent during these last twenty years — or of the Church of Alt-Lerchenfeld, with the rich al fresco pictures decorating its interior, would have taken away half the space allowed to the totality of Austrian Exhibitors. The Vienna Association of Artists, together with the Committee intrusted with the arrangement of the Artistical section, have acknowledged this space to be absolutely insufficient. We reproduce here expressly this protest, as raised from more than one side, adressing it directly to the English public; so that nobody may suppose, that any competent person in Austria has thought it possible, under the exstant conditions, to fulfill — even with the best will — the real purpose of the intended Exhibition.

The artistical activity in Austria, during this last Century, took its point of departure from two establishments for superior artistical instruction: the Academies of Fine Arts of Vienna and of Venice.

The Academy of Vienna has been founded in 1704 under the reign of Leopold I.; the Academy of Venice dates from 1807. Both are State establishments in the full sense of the term, possessing some funds of their own, but being chiefly maintained at public expenses.

The Academy of Fine Arts of Vienna, reorganized in 1850, is governed by a Director (a practising Artist). There are three schools (called „Masterschools") for Historical Painting, a preparatory school for Picture and Sculpture, a school for smaller Plastics, a school for Landscape-painting and another for Architecture. The number of artistical Professors is sixteen, besides some Teachers of scientific branches. The average number of pupils, in these last years, was about 200. The Academy possesses an extensive Library and an interesting Gallery of Pictures, whose main portion is a legacy of Count Lamberg, formerly a Curator of this establishment.

The „Academia delle belle Arti" of Venice, has a Secretary, keepers of Galleries, eight Professors and four adjoined Professors. The number of pupils, in the course of the last years, varied between 230 and 240.

The Academical Gallery has acquired European celebrity. The teaching activity of the Academy is concentrated in the schools for Painting and for Sculpture. Both Academies dispose of certain pensions for sustentation and travelling expenses of talented pupils; special pensions of Imperial foundation for natives of Hungary are in the gift of the Vienna Academy. The sums spent out of Academical funds, or out of public funds for Academical purposes, are represented by the following numbers:

Academy of Vienna: in 1857: 75,100 florins; in 1862: 52,800 florins.

Academy of Venice: in 1856: 35,286 florins; in 1861: 34,143 florins.

Besides these State establishments, Austria possesses a number of private Institutes, among which the Academy of Prague, with five Professors, holds the first place. This Academy is a foundation of the Society of patriotic Friends of Art, who, in 1860, contributed a sum of 9,190 florins for the school attached to this establishment.

The school of Fine Arts in Cracovia numbers presently four Professors and two Assistants.

The Academy of the Styrian Estates at Gratz is sustained by an allowance of yearly 2,000 florins from the funds of the Joanneum. The Academy of Painters and Sculptors at Verona is defrayed by a Society established for this special purpose.

The admission of pupils into the private ateliers of Artists of renown has won less ground in Austria, than it did, and still does, in France and Belgium. Some few eminent Artists: Messrs. Frederick Amerling, Charles Rahl, Ferdinand Waldmüller etc. (besides the Professors of the academical „Masterschools", paid by Government) have succeeded in gathering around them a number of disciples. A number of Artists of note, whose works figure in the London Exhibition, have received their artistical education in the ateliers of the three above-named Painters.

Next to the above mentioned establishments for superior artistical education comes a number of private Drawing-schools for special purposes, as free-hand drawing, plan- and architectural drawing etc.

In all the Polytechnical Institutes, in Technical and even Primary schools, drawing is taught. In 1859, the Austrian Empire possessed 49 completely organized Technical schools with nearly 600 teachers (70 among them for drawing), frequented by 11,526 pupils. Lower Austria, in 1860, numbered 60 courses for teaching freehand drawing in Primary schools, frequented daily by 3,414, and on Sundays only, by 2,166 pupils. Information in drawing is more and more required for, especially in those parts, where Industry is progressing rapidly.

The Art Societies, founded on the principle of shares for the purpose of promoting Art, are essentially influent. Two Societies of this kind exist at Vienna and the later in date of them, known under the denomination of „Austrian Art Society" has established subsidiary organs in some Province towns. Other Societies of the same nature exist at Prague, Pesth, Linz, Salzburg, and Lemberg. Besides these Art Societies, the Artists themselves have formed into Associations, either for social intercourse or for direct promotion of Art. The Association of Vienna Artists, established in 1860 by fusion of the „Albert Durer Union" and the „Concordia" Association, holds incontestably the first rank among them.

The Archaeological Societies are of more or less direct influence in promoting the progress of Art. The I. R. Commission for preserving and investigating Architectural Monuments is a State establishment, publishing „Annals" and a monthly periodical under the title of „Mittheilungen" (Communications). Other Archæological Societies, as the Vienna Antiquarian Society, are of a private character, being sustained by the contribution of their fellows and publishing either regular periodicals or isolated archeological papers. Similar publications are issued by other Scientific establishments, as the Provincial Museum of Bohemia, the Archæological Society of Cracovia, the Archæological Sections of the Hungarian Academy and of the Moravian Society, the Museum Carolino-Augusteum of Linz, the Ferdinandeum of Innsbruck, the Historical Societies of Styria, Carinthia and Carniola etc.

Since the beginning of the present reign, the Fine Arts have occupied a more conspicuos situation in the intellectual life of the Austrian Empire, partly in consequence of the raising of an extensive series of monumental works, partly, because Government, throwing away the fetters of antiquated bureaucratical forms, resolutely acted on more liberal principles. Together with this governmental progress, powerful impulsion issued from privates, as from ecclesiastical and secular Corporations; nor can it be any more a subject of doubt, that Austria (provided external circumstances to be in any way favourable to the Arts of Peace) is progressing towards a grand artistical developement.

Space and time is wanting here for discussing in detail all these monumental works; we must satisfy ourselves with calling attention on a number of works and establishment of recent origin.

Among these establishments, the first place may be claimed by the Institute for Casting in Metal, raised to the rank of an Imperial establishment under the superintendence of the Sculptor Anthony Chevalier de Fernkorn.

The want of a great indigenous establishment for monumental Casts was, in former times, a subject of many complaints, and the necessity to entrust foreign establishments with the execution of such works (as the Monuments of King Charles IV. and Field-Marshall Radetzky at Prague, the Monument of the Archduke-Palatine at Pesth) was considered a national disgrace. Presently monumental Casting in Metal has found a fixed centre at Vienna, not merely in the private establishments of Messrs. Glanz & Hollenbach, but in Fernkorn's Imp. Casting establishment. The colossal equestrian Monument of Archduke Charles on the outer Residence-place is a first-rate work, both for casting and chiselling. In the first ten years of its existence, this establishment has executed a long series of Metal Casts. The Monuments for the Empress Maria Theresia and for Ressel—the inventor of the Steamboat-screw—are to be placed in the course of the present year; the first in the park of the Military Academy of Wiener-Neustadt, the second at Trieste.

Among the works of Monumental Architecture, raised since the beginning of the present Emperor's reign, the following may be made objects of special consideration.

1. The Imperial Arsenal in the next vicinity of Vienna, constructed by the Architects Prof. van der Nüll, A. de Siccardsburg, L. Förster, C. Roosner and Mr. Theophil Hansen. It is a brick-construction executed on artistical principles, comprehending, besides localities for confection and preserving Army implements of every description, a grand Museum of military objects, among which are mediæval armours of highest interest. The centre of the whole group of buildings is occupied by a ceremony-hall, overvaulted with a cupola and decorated with subjects chosen among the military history of Austria, whose execution al fresco has been entrusted to Mr. Charles Blaas, Professor of History-painting at the Academy of Venice.

2. The Parochial Church in the suburb of Alt-Lerchenfeld, achieved in 1861, after ten years of uninterrupted activity. The interior of this church is decorated with a cyclus of al fresco pictures, whose execution has been entrusted to Direct. Engerth, Prof. Blaas, Mayr, Schulz, Binder and Mr. Schömann; the splendid ornamentation of the inside of the walls, as also the arrangements and objects required for Divine service, have been executed by Prof. van der Nüll.

J. G. Müller, a native of Switzerland, who conducted the architectural execution, has deceased during the progress of the construction. The costs amounted to a total of 600,000 fl. Austrian money.

3. The Votive Church, a colossal cutstone construction in Gothic style, presently in work and probably not to be brought to end before a period of, at least, ten years. The Architect of this church, Henry Ferstel, whose plans are among the exhibited objects, belongs to the younger artistical generation of

E

Vienna. Mr. K r a n n e r, an able Architect, superintends the building-workhouse („Bauhütte").

4. The n e w B a n k at Vienna, executed at the expense of the National-Bank establishment by Mr. H. F e r s t e l, Architect of the Votive church, holds the first place among the civil buildings of the Capital, both for its monumental character and for the artistical ornamentation of its interior.

The buildings, coming next in importance after the first-rate constructions just-mentioned, are: the new Synagogue by Prof. L. F ö r s t e r, the Evangelic Technical School by Mr. Th. H a n s e n, the Lazarist-Church in Gothic style (presently in building) by Prof S c h m i d t, the Gothic Church next to the Belvedere by Engineer B e r g m a n n, the façade of the Greek non-united Church by Mr. Th. H a n s e n, the C o m m e r c i a l A c a d e m y by Mr. Ferd. F e l l n e r, the Francis Joseph-Ga e by Engineer R z i w n a t z, the Francis Joseph Barracks, the Central-Military-Riding-house, etc.

The building of the new Opera-House, on the open space before the lately demolished Carinthia Gate, is intended to begin in the course of the present year under the superintendence of Messrs E. van der N ü l l and A. von S i c c a r d s b u r g, whose plan has been adopted in the way of free concurrence. Plans of the Lazarist Church and of the new Opera - House figure among the exhibited objects.

The Provinces of the Austrian Empire have not remained insensible to the artistical impulsion issued from the Capital of the Monarchy.

An activity, hitherto unknown, has taken place· and many of its results highly deserve acknowledgement. Prague is indebted to the Society of patriotic Friends of Arts in Bohemia for a cyclus of al f r e s c o pictures in the Belvedere, entrusted to Director C h r. R u b e n and presently nearly brought to end, and for the monument of Field-Marshall Radetzky by Messrs M a x, of Prague, holding a dignified place aside the less recent monument of King Charles IV by Mr. K r a u n e r, the monumental fountain in commemoration of Emperor Francis I. by Mr. H ä h n e l. A stone-built Church in Roman style, planned and superintended by Prof. R œ s n e r, at Vienna, is rising in the suburb of Carolinenthal.

The Bishop of the Greek rite of C z e r n o w i t z (Bucowina) has entrusted Mr. H l a w k a with the construction of an extensive Episcopal Residence. Mr. Th. H a n s e n has ·made a plan for an Invalide Hospital at Lemberg. The new Synagogue of P e s t h is a work of the Viennese Architect Mr. L. F ö r s t e r; the House of the H u n g a r i a n A c a d e m y, in project for the same place, has been intrusted to a Prussian Architect, Mr. S t ü l e r. General H e n t z i has a monument in the Fortress of B u d e, in whose heroïc defence he had found his death. In the city of B u d e, the monument of the A r c h d u k e - P a l a t i n e Joseph has been achieved by a Sculptor of Munich, Mr. H a l b i g; another in commemoration of Count Stephan Szechényi is presently in preparation. At G r a t z, a monument, executed by Mr. Hans G a s s e r, has been erected in commemoration of General Welden. C r o a t i a is preparing a metal equestrian Monument of the late Banus (Governor) Count Jellačič.

The progress of Industry, enlivened by railroad-communications, has operated a visible change in the architectural features of our Austrian towns, nowhere as strikingly as (next to Vienna) at T r i e s t e, where the railroad-constructions and the immense Naval Arsenal, established by the Austrian Lloyd Society, have impressed a quite new physiognomy to the whole town.

The buildings at Pola and Verona stand first among the great military constructions in the southern Provinces of the Austrian Empire.

A special attention has been paid by Government to the o l d M o n u m e n t s of V e n i c e; since a long succession of years they have been made the object of incessant, though inobtrusive, activity, to which alone the civilized world is indebted for the preservation of these venerable relics. The Ducal Palace is in restoration down to the minutest details, and a great part of this work has already been brought to a successful end. In these last years, painters have been sent to Paris and Brussels with the mission of taking copies of pictures taken away during the French Revolution and not restored after the peace of Vienna. As the Ducal Palace, the Cathedral of St. Mark is an object of uninterrupted activity, all the expenses being supported exclusively out of Government funds. These restorations are not confined to the architectural department; they comprehend likewise the partly impaired Mosaïcs adorning the interior, as it may be seen by Prof. B l a a s' sketches in oilcolours, sent to the Exhibition. The Churches of S a n G i o v a n n i e P a o l o, ai S c a l z i, and several other Monuments of more or less importance are presently in restoration, either at the Government's or at the City's expenses.

Indifferent as the present generation of Venetians is to the progress of Fine Arts, their patronage has become incumbent to Government. Next to Prof. L u i g i F e r r a r i, to whom the execution of Marco Polo's monument has been entrusted, every Venetian Artist of some note has been and is still employed by Government.

The movement in S a c r e d A r t is a significative symptom of the present situation of Art in Austria. The interest taken by Government, as by Privates in public collection and in promoting Art within more restrained spheres deserves no less attention. Space allows us here but a succinct discussion of both these tendencies.

The progressive movement in S a c r e d A r t is manifested either by means of new constructions or by restoring enterprises. Several among these new constructions have been mentioned above. Mr. E s s e n w e i n brings to exhibition the plans and drawings of a Church in construction in Galicia. The artistical confection of smaller objects intended for ecclesiastical use, assisted by the progress of archæological studies and by the general improvement of taste, has made splendid progresses. Here we have another subject of regret about the narrowness of the space allowed to Austrian Exhibitors, as an obstacle against bringing into sight a great number of precious and splendid objects, executed in these last years at the expense of the Primate of Hungary, the Archbishops of Vienna, Prague, Ollmütz, Kalocsa, the Bishops of Raab, Diakovar, and other Dignitaries of the Church. The Prayer-Book, offered by the Artists of the Vienna Academy to Her Majesty the Empresson occasion of Her marriage is the only specimen of this branch of Art figuring at the Exhibition. Since the time that full autonomy and liberty of conscience has been imparted to acatholic Confessions, these have followed the artistical impulsion given by the example of the Catholic Church. Besides the restorations on a larger scale mentioned above, we must mention here the restoration of the Cathedral of St. Stephan at Vienna and the projected achievement of the Dome-church at Prague. Besides those in the Lombardo - Venetian Kingdom, the Austrian Empire numbers several G a l l e r i e s of P i c t u r e s and Arti s t i c a l C o l l e c t i o n s of high importance. The first rank among them belongs to those of Vienna, not only for their number, but also for the quantity of eminent pictures specially of the Old Dutch, Flemish and Venetian Schools preserved in them; so that Vienna, under this respect, among all the towns of the extra-italian Continent, takes without contest the first place after Paris. The I m p e r i a l G a l l e r y in the B e l v e d e r e and the A m b r a s C o l l e c t i o n possess objects of the highest value. The Austrian Nobility, like the Nobility and wealthy Gentry of the British Empire, possess large Galleries, as those of the Princes Liechtenstein and Esterházy, the Counts Harrach, Schönborn, Czernin, Zichy, Beroldingen at Vienna, Prince Rohan and Count Nostitz at Prague etc. Among Privates possessing valuable collections of ancient and modern objects of Art, partly represented in the Exhibition, we may quote here: Count N a k o, Baron Heekeren, Messrs de Arthaber, Fellner, Jäger, Ritter, von Galvagni, Koller, Engert, Böhm, Bühlmaier, Artaria, Goll, Kranner, Klinkosch jun., F. X. Mayer, Henry von Mayer, Zeppezauer, Nikotzky, Gsell-

This euumeration, incomplete as it is, shows how much interest for Art is alive among the population of Vienna and the nature and amount of the elements for artistical developement presently existing in Austria. The wish for refined enjoyment of life and for intellectual pleasures are not the only advantages Austrian Artists may boast of; they find also ample subjects for their activity in the characteristical physical and moral features of the manifold Nationalities inhabiting the Austrian Empire, who, for the most part, have hitherto preserved unaltered their original types.

With exception of the Lombardo-Venetian Artists, still adhering to the ancient historical traditions of Italian Art, all the other Austrian Artists (with extremely rare exceptions) follow the tendencies of the German School. National elements are manifesting themselves more in the choice of costumes and subjects than in artistical form in the strict sense of the term In these last years, the French and Belgian Schools of Painting have visibly influenced our Vienna Artists, as they did those of other countries. Vienna numbers presently some hundred Artists of every nationality and every religious confession. A fact-characteristic of this Association of Artists and highly creditable for them, is: that national elements, but too frequently in opposition, and even in hostile conflict, in the Provinces of the Austrian Empire, have never succeeded in disturbing the peace and concord among the Artists of Vienna.

Whoever views with unpartial eyes the intellectual movement and the artistical activity now alive in Austria, must come to the persuasion, that this movement is intense and follows an ascending direction. May the new era of civil and religious liberty, inaugurated by the own spontaneous will of our high-minded Emperor, promote the increase and progress of the peaceful Arts, assembled in the London Exhibition, as in a temple of Peace! !

Class 37.

Architecture. Architectural Models Designs.

1. ESSENWEIN Augustus, at Vienna, born at Carlsruhe 1831; Fellow of several Archæological Societies. Architect.

I. **Church and Monastery at Wielowice in Galicia,** Brick construction in Early Gothic style (to be constructed). Property of the Artist.

2. FERSTEL Henry, at Vienna, born 1830, Architect.

I. **Four views of the Votive Church at Vienna;** presently in construction. Property of the Artist. The Saviour Church (called „Votive Church") construction at Vienna in commemoration of the providentially fortunate preservation of H. M. the present Emperor on February 18th 1853, is a hewn stone building projected in Gothic style. Its foundations have been solemnly laid on April 24th 1856.

3. HENSZLMANN Emmerick, born 1813, ⎫
4. GERSTER Charles, born 1813, ⎬ Architects at Pesth.
5. FREY Louis, born 1813.   ⎭

I. **Projected Plan and drawings for the buildings of the Hungarian Academy at Pesth.** Three numbers. Property of the Hungarian Academy.

6. LIPPERT Joseph, born at Arad in Hungary, presently at Vienna. Architect.

I. **Reconstruction of the Dome of Ollmütz in Moravia;** five drawings. Property of the Artist.

7. 8. NÜLL Edward van der, born at Vienna 1812, Professor at the Imp. Rl. Academy of Vienna, Knight of the Imp. Rl. Austrian Order of the Iron Crown and of the Rl. Bavarian Order of St. Michael, Imp. Rl. Superior counsellor in the Building department. **SICCARDSBURG** Augustus von, born at Vienna 1813, Professor at the Imp. Rl. Academy of Vienna, Knight of the Imp. Austrian Order of the Iron Crown. Architects.

I. — IV. **Four architectural drawings of the new Opera-house to be built at Vienna.** Property of the Artists.

9. ROESNER Charles, born at Vienna 1805, Professor at the Architectural School of the Imp. Rl. Academy of Vienna, Knight of the Imp. Austrian Order of Francis Joseph and of the Papal Order of St. Gregory, possessor of the Imp. Rl. great Golden Medal of the Vienna Industrial Exhibition in 1845, of the great British Exhibition and Service Medal of the London Universal Exhibition in 1851 and of the great Rl. Bavarian Medal of the Munich Industrial Exhibition of 1854. Architect.

I. **Four Plans of the Church in Roman Style** under the invocation of St. Cyrillus and Methodius, presently in construction at Prague in the suburb of Carolinenthal. Property of the Artist.

10. SCHMIDT Frederic, born in Württemberg 1825. Professor of Architecture at the Imp. Rl. Academy of Fine Arts of Vienna, Knight of the Imp. Austrian Order of Francis Joseph and of the Rl. Prussian Order of the Red Eagle IV. Class

I. **The Church of the Lazarists at Vienna.** Architectural drawing.

II. **The Church at Fünfhaus.** Architectural drawing.

Class 38.

Paintings in oil, and Water-colours and drawings. Art Designs and Manufactures.

11. L'ALLEMAND Fritz, at Vienna, born 1812, Knight of the Rl. Order of the Netherland Oak-Crown. Painter of Battles.

I. **Scene from the Battle of Komorn,** April 26th 1849. Property of His Exc. Count Montenuovo.

12. ALT Rodolphus, born at Vienna 1812. Painter of Architecture and Landscape.

I. **Four Architectural „Intérieurs"** in watercolours Property of His Exc. Count Harrach.

II. **Tower of the bridge over the Moldaw at Prague.** In oil. Property of Mr. Bühlmayer.

13. AMERLING Frederick, born at Vienna 1803, Knight of the R. Bavarian Order of St. Michael, Member of several Academies. Portrait- and History-painter.

I. **A Mother with her child.** In oil. Property of Mr. Henry von Mayer.

II. **Portrait of the Artist.** In oil. Property of Mr. Fr. Xav Mayer.

III. **Portrait of the Sculptor Thorwaldsen.** In oil. Property of H. H. the Sovereign Prince John Liechtenstein.

IV. **A Girl with a rose.** In oil. Pperty of Mr. Kranner.

14. ANGELI J., of Vienna, presently at Munich. History-painter.

I. **Mary Queen of Scotland,** hearing her sentence of death read to her. In oil. Property of the Artist.

15. BLAAS Charles, born 1815 at Nauders (Tyrol), Professor at the Imp. Rl. Academy of Fine Arts of Venice (great Gold Medal at the Paris Exhibition 1855, Gold Medal at the Vienna Exhibition 1857) Portrait- and History-painter.

I. **Venetian Brides captured by Istrian Pirates** In oil, 1858. This event took place under the reign of the Doge Candiano II. The „Festa delle Marie" instituted for its commemoration, is still celebrated every year at Venice. (To be sold for 10.000 Francs.)

II. **Subjects taken from St. John's Apocalypsis;** painted 1861. Cartoons in oil for the Mosaïcs in St. Mark's Cathedral, whose execution has been ordered by H. M. the Emperor of Austria.

16. BRODSZKY Alexander, at Pesth, born 1819. Landscape-painter.

I. **Scenery from the Banks of Lake Balaton in Hungary** In oil Property of the Artist.

17. BRUNNER Joseph, born at Vienna 1826. Landscape-painter.

I. **Forest scenery.** Painted 1861. Property of the Artist. To be sold for. 150 Austrian florins.

18. CZERMAK Jaroslav, born at Prague 1830, presently at Paris; Knight of the Rl. Belgian Order of Leopold (great Gold Medals at the Paris and Brussels Exhibitions 1861).
I. Razzia of Baschi-Bojuks. Painted 1861 in oil. Property of Mr. Goupil at Paris.
II. Slavonian Raja. In oil. Property of Mr. Durand at Paris.
III. Mother with her child. In oil. Property of Mr. Peirere at Paris.

19. DAFFINGER M., born at Vienna 1790, † there 1849. Water colours- and Miniature-painter.
I. A Series of Miniature Portraits of eminent persons Property of H. H. Prince Metternich, Austrian Embassador at the French Court.

20. DANHAUSER John, born at Vienna 1805, † there 1845. History- and Genre-painter.
I. The Child amidst his world. In oil. Property of Mr. Federl.
II. The little Woman on the stairs („Stiegen-Weibchen"). In oil. Property of Mr. Bühlmayer.

21. EMLER Bonaventura, born at Vienna 1831. History-painter.
Three subjects taken from Dante. Paradise, Entorance to Paradise and Hell. Cartons in chinese ink. To be sold: each drawing for 50 Austrian florins.

22. ENDER Edward, born at Rome 1824, presently at Vienna. Portrait- and History-painter.
I. The Fugue, a Genre subject. In oil. Property of the Artist. To be sold for 80 £.

23. ENGERTH Edward, born at Pless 1818, Director of the Academy of the Estates of Bohemia, at Prague. Portrait- & History-painter.
I. Portrait of H. M. the Emperor of Austria. In oil. Property of the Committee of the Estates at Prague.

24. EYBL Francis, born at Vienna 1806. Portrait- and Genre-painter.
I. An old Woman In oil Property of Mr. Fr. Xav. Mayer.
II. A Peasant Woman. In oil. Property of H. M. the Emperor.

25. FENDI Peter, born at Vienna 1796, † there 1842. Genre-painter.
I. The Bride. In oil. Property of Mr. Bühlmayer.
II. The Seizure. In oil. Property of Mr. Bühlmayer.

26. FIEDLER J., born 1816 at Berlin, presently at Trieste, Knight of the Rl. Belgian Order of Leopold and of the Rl. Greek Order of the Saviour, Member of the Imp. Rl. Academy of Fine Arts at Venice. Landscape-painter.
I. The Granite quarries of the ancient Egypt in the Nubian desert. In oil. Property of the Artist. To be sold for 130 £

27. FÜGER Fr. Henry, born at Heilbronn 1751, † at Vienna, 1818. Portrait- & History-painter.
I. The death of Germanicus. In oil. Property of the Imp. Academy of Fine Arts of Vienna, whose member Füger had been from 1784 to his death.

28. FUEHRICH Joseph Knight of, born 1800 at Kratzau in Bohemia, Professor at the Academy of Fine Arts of Vienna, Knight of the Imp. Austrian Orders of the Iron Crown and of Francis Joseph, and of the Papal Order of St. Gregory. History-painter.
I. The Mourning Jews. In oil. Property of Count Nostiz-Rinek.
II. Three drawings in lead-pencil (Spring, the Resuscitation of Lazarus and Our Saviour appearing to St. Thomas). Property of Mr. Schulgen, Art-merchant at Düsseldorf.
III. A subject taken from a popular religious Song. St. Mary sheltering Her worshippers under Her Mantle. In lead-pencil. (To be sold for 1000 Austrian florins. Right of publication reserved to the author.)
IV. Cartoons for the Alt-Lerchenfeld-Church at Vienna. (Dome).

29. GAUERMANN Fred., at Vienna, born at Miesenbach 1807, Knight of the Rl. Netherland Order of the Oak-Crown, Member of the Academy of Fine Arts. Landscape- and Animal-painter.
I. Cows in a landscape. In oil. Property of Mr. Angeli.
II. A Bear. In oil. Property of H. Rl. Highness the Duke Augustus of Saxe-Coburg.

30. GAUL Gustavus, born at Vienna 1836. Portrait- and History-painter.
I. Portrait of the celebrated Actress Sophia Schröder. In oil.

31. GEIGER John Nep., born at Vienna 1801, Professor at the Imp. Rl. Academy of Fine Arts.
I., II., III. Three scenes from the Family history of Count Czernin. In water colours. Property of Count Czernin.

32. HALAUSCA L., born at Vienna 1829. Landscape-painter
I. A May-morning, Spring scenery. In oil. Property of the Artist. To be sold for 45 Austrian florins.

33. HANSCH Anthony, born at Vienna 1815. Landscape-painter.
I. The Glacier of Dux in Tyrol. In oil. Property of the Artist To be sold for 50 £.
II. The great Glacier of Oetzthal in Tyrol, with the Lake of Gurgl. In oil. Property of the Artist. To be sold for 60 £.

34. HAUSHOFER Maximilian, born at Munich 1811, Professor at the Academy of Prague. Landscape-painter.
I. Sunday-morning in the Bavarian Mountains (Chiemsee). In oil. Property of the Artist. To be sold for 70 £.

35. HOLZER J., born at Vienna 1824, Landscape-painter.
I. Forest-solitude, a Landscape. In oil. Property of the Artist. To be sold for 80 £

36. KOUDELKA Paulina Baroness, married to the Chevalier de Schmerling, born at Vienna 1806, † there 1840. Flower-painter.
I. Group of Flowers. In oil. Property of H. Exc. the Minister of State, Chevalier de Schmerling.

37. KRIEHUBER Joseph, born at Vienna 1800, Knight of the Imp. Austrian Order of Francis Joseph. Portrait- and Land-scape-painter.
I. and II. The Ducal Family of Saxe-Coburg and Countess Kolonicz. In water-colours. Property of H. R. H. the Duke Augustus of Coburg and of Countess Kolonicz.

38. KUPELWIESER Leopold, born at Vienna 1796, Professor at the Imp. Rl. Academy of Fine Arts, Knight of the Imp Austrian Order of Francis Joseph and of the Papal Order of St. Gregory, Member of the Academies of Munich and Milan. History-painter.
I. Vision of St. Altmann (afterwards Bishop of Passau), of St. Adelbert (subsequently Bishop of Würzburg) and of St. Gebhard, Bishop of Salzburg in 1030. In oil. Property of the Monastery of St. Florian.
II. Cartoons for the Alt-Lerchenfeld-Church at Vienna. (Dome).

39. LIBAY Louis, born in Hungary 1814, presently at Vienna, Knight of the Royal Danish Order of Danebrog III. Class, and of the Ottoman Medschid-Order IV. Class; possessor of the great Gold Medals for Art and Science of Austria, Wurttemberg, Saxony, Hannover, etc. Landscape-painter.
I. View of Pesth-Bude. In lead-pencil Property of Countess Szechényi.

40. LIGETH Anth., at Pesth, born 1823. Landscape-painter.
I. The desert of Sahara, seen at distance from the vicinity of Cairo. In oil. Property of the Artist.

41. LICHTENFELS E. de, born at Vienna 1833. Landscape-painter.
I. Autumn Scenery. In oil. Property of Mr. Mayer at Leoben in Stiria.

42. LOTZ Charles, at Pesth, born 1838. Painter of Animals.
I. Study on a pasture („Puszta"). In oil. Property of the Artist.

43. MARKO Charles, born in Hungary, † 1860 in the environs of Florence. Landscape-painter.
I. Fancy scenery. In oil.
II. Our Saviour baptized in the River Jordan by St. John. In oil. Property of the Hungarian National Museum.

44. MARASTONI Joseph, at Pesth, born 1834. Portrait- and History-painter.
I. Portraits of Deputies to the Hungarian Provincial Assembly. In oil. Property of the Artist.

45. MAYER Charles, born at Vienna 1810. Professor at the Imp. Rl. Akademy of Fine Arts. History-painter.
I. Cartoons for the Alt-Lerchenfeld-Church at Vienna. (Dome).

46. MOLNAR Joseph, at Pesth, born 1822. History-painter.
I. Abraham's Emigration. In oil. Property of the Artist.

47. NEUGEBAUER Joseph, born at Vienna 1810. Portrait-, Flower- and History-painter.
I. Group of inanimated objects („Stillleben"). In oil. Property of Mr. Bühlmayer.

48. NEUSTÆTTER Louis, born at Munich 1829, presently at Vienna. Portrait- and Genre-painter.
I. Genre-picture. In oil. Property of Mr. A. Zinner.

49. OECONOMO Aristides, Knight of the Rl. Greek Order of the Saviour. Portrait- and History-painter.
I. Portrait of Mr. Dumba. In oil. Private property.

50. ORLAI Samuel, born at Pesth 1822. History-painter.
I. Milton dictating his „Paradise Lost". In oil. Property of the Artist.

51. PETTENKOFEN Gustavus, born at Vienna 1826. Genre-painter.
I. Gipsies bathing. In oil. Property of Mr. Plach.
II. Scenes of Gipsey life in Hungary. Two pictures in watercolours. Property of Count Náko at Vienna.

52. PITNER Frantis, at Vienna, born 1826, Knight of the Ducal Parmese Order of St. Louis. Genre-painter.
I. **Procession of Pilgrims at Loretto.** In oil. Property of the Artist. To be sold for 1000 Austrian florins.

53. PRAYER BOOK, executed by Professors of the Imp. Rl. Academy of Vienna and offered to Her Majesty the Empress Elisabeth of Austria on occasion of Her Marriage in 1853.
This book contains the „Horarium Beatæ Mariæ Virginis" and a Calendar, written and painted on parchment and adorned with twenty Miniatures. The ornaments of the cover are in Enamelled and Relievo Work.

54. QUADAL M. F., born at Niemtschitz in Moravia 1736, † at St. Petersburg 1808. Portrait- and History-painter.
I. **Military Review at Minkendorf near Laxenburg** in the environs of Vienna. Emperor Joseph II, the Imp. Prince (afterwards Emperor) Francis, the Field-marshalls Count Haddik, Loudon and Lascy, Prince Eszterházy Captain of the Guards, Prince Liechtenstein, General of Cavalry, and Colonel Stipsicz are among the persons, represented on this picture. In oil. Property of H. Imp. H. Archduke Albert.

55. RAFFALT John, born at Murau in Stiria 1836, presently at Vienna. Genre-painter.
I. **Hungarian Market-place.** Property of Mr. Galvagni.

56. RAHL Charles, born at Vienna 1812, Court-painter to the Grand-Duke of Oldenburg, Knight of the Rl. Greek Order of the Saviour and of the Ducal Oldenburg Order of Peter Frederick Louis, Honorary Member of the Rl. Academy of Bavaria. Portrait- and History-painter.
I. **Persecution of Christians in the Catacombs.** In oil. Property of Dr Abendroth at Hamburg.
II. **Female Portrait.** In oil.

57. ROI Pietro, born at Vicenza, Member of the Academy of Fine Arts of Venice. History-painter.
I. **St. Mary on her Throne, the Saints Magdalen, Margareth and Vincent de Paula standing around her.** In oil. Painted 1857, by order of the Imp. Rl. Venetian Government for the Church of the Monastery „delle Convertite" at Venice.

58. RUBEN Christian, born at Treves 1806, Director of the Imp. Rl. Academy of Fine Arts of Vienna, Knight of the Imp. Austrian Orders of the Iron Crown and of Francis Joseph. History-painter.
I. **The heathen Prussians baptized by Ottokar of Bohemia.** A cartoon of one of the al fresco pictures intended to decorate the interior of the Belvedere of Prague.

59. SCHÆFFER Augustus, born at Vienna 1803, Landscape-painter.
I. **View of the lake of Königssee.** Property of Mr. Conradi.

60. SCHEFFER von LEONHARDSHOF, born at Vienna 1795, † there 1822. History-painter.
I. **Saint Cæcilia.** In oil. Property of H. M. the Emperor.

61. SCHINDLER A., born at Engelsberg in Silesia 1806, † at Vienna 1861. Genre-painter.
I. **The dying soldier.** In oil. Property of H. M. the Emperor.

62. SCHŒNN Aloys, born at Vienna 1826. Genre-painter.

I. **Coffee-house in the Herzegovina.** In oil. Property of the Artist.

63. SCHROTZBERG Fr., born at Vienna 1811. Portrait-painter.
I. **Portrait of H. M. the Empress of Austria.** Property of Countess Eszterházy.

64. STEINFELD Francis, born at Vienna 1787. Professor Emeritus at the Imp. Rl. Academy of Fine Arts of Vienna. Landscape-painter.
I. **Gastein in 1840.** In oil. Property of H. M. the Emperor.

65. STŒCKLER Emanuel, born at Vienna, presently at Venice. Painter in water-colours.
I. **Baptistery of the Cathedral of St. Mark at Venice.** To be sold for 28 £.
II. **A woman of Albano near Rome.** To be sold for 28 £.

66. SZEKELY J., born at Pesth, presently at Munich. History-painter.
I. **Finding of the dead body of Louis II., King of Hungary.** In oil. Property of the Artist.

67. THAN Maurice, at Pesth, born 1828. Portrait- and History-painter.
I. **Angelica and Medor from Ariosto's „Orlando furioso".** In oil.
II. **Female Head**, a study. In oil.
III. **Male Head**, a study. In oil.

68. THOREN Otto, Chevalier de, born at Vienna 1828, presently at Munich, Knight of the Imp. Russian Order of St. Wladimir with the swords, decorated with the Imp. Austrian Cross for Military Merit.
I. **Living Animals.** Property of the Artist. To be sold for 700 £.

69. VANDRAK Charles, at Pesth, born 1806. Genre-painter.
I. **Genre-picture.** In oil. Property of the Artist.

70. VŒSCHER Leopold, born at Vienna. Landscape-painter.
I. **Mountain Scenery.** In oil. Property of Countess Erdödy.

71. VOGLER J., at Vienna, died 1855. History-painter.
I. **The vision of the Riders at Jerusalem.** (Maccab. V. 1—4.)

72. WALDMÜLLER Ferdinand, born at Vienna 1793, Professor Emeritus, Knight of the Rl. Prussian Order of the Red Eagle IV. Class. Genre-painter.
I. **Christmass-Evening.** In oil. Property of Mr. Galvagni.
II. **Binding an Apprentice.** In oil. Property of the Rev. Father Dr. Emerich Gabely in the Scotch Monastery at Vienna.

73. WURZINGER Charles, born at Vienna 1817, Professor of the Imp. Rl. Academy of Vienna. History-painter.
I. **Emperor Ferdinand II. refusing the impetuous demands of the rebels** — the Cuirassiers of Dampierre arriving on the place before the Imp. Residence. In oil. Property of H. M. the Emperor.

74. ZONA Antonio, born at Venice, Member of the Academy of Fine Arts of Venice. Portrait- and History-painter.
I. **Titian and Paolo Veronese meeting on Ponte della Paglia next to the Ducal Palace.** In oil. Painted in 1861 by the Emperor's orders and presented by H. M. to the Academy of Fine Arts of Venice.

Class 39.

Sculpture by various processes; Models, Die Sinking and Intaglios. Fine Art in Repoussée Chasing etc.

75. DUNAINSKY Ladislav, at Pesth, born 1822. Sculptor.
I. **Buste of the Hungarian Poët Pétöfi.**
II. **Samson and Delila**, a group. Properties of the Artist.

76. FERRARI Antonio, born at Venice 1810, Professor of Sculpture at the Imp. Rl. Academy of Venice, Member of several Academies. Sculptor.
I. **Buste of Galileo Galilei.** Achieved 1861, in Carrara marble. Presented by H. M. the Emperor to the University of Padua, whose member Galilei had been for some years.

77. FERNKORN Anthony Chevalier de, born at Erfurt 1813 Knight of the Imp. Austrian Order of Leopold and of the Rl Bavarian Order of St. Michael. Sculptor and Caster in Metals
I. **Colossal Head of a Horse for the Monument of Archduke Charles.** Cast in plaister. Property of the Artist.
This head is equal in size to the head of the horse belonging to the Bronze-monument of Archduke Charles, erected,

in 1859, on the outer Palace-place at Vienna by order of His Imperial Majesty.
II. **St. George.** A statuette in bronze. Property of the Artist. To be sold for 600 Austrian florins.

78. KAEHSMANN Joseph, born at Vienna 1784, Prof. Emeritus at the Imp. Rl. Academy of Fine Arts of Vienna. Sculptor.
I. **A group: Jason, having overcome the Dragon, and Medea.** In Carrara marble. Property of H. M. the Emperor. (Dome.)

79. KISSLING Leopold, born at Schönegen in Austria 1770, † at Vienna 1827. Sculptor.
I. **A group: Mars and Venus with Amor.** In Carrara marble. Property of H. M. the Emperor. (Dome.)

80. LEVI Wenceslas, born at Kuglau in Bohemia, presently at Rome. Sculptor.

I. **Statue of St. Elizabeth.** In Carrara marble. Property of H. M. the Emperor.

81. MAX Emanuel, born at Bürgstein in Bohemia 1809 (?), presently at Prague. Sculptor.

I. **Statue of Vlasta.** In marble. Property of Prince F. Kinsky.

II. **Statue of a praying child.** In marble. Property of the Artist. To be sold for 110 £.

III. **Statue of David.** In marble. Property of the Artist. To be sold for 350 £.

82. RADNITZKY Charles, born at Vienna 1818, Professor of Sculpture at the Imp. Rl. Academy of Vienna. Medal-Engraver and Sculptor.

I. **Medals:** 1. Of the Cantatrice Jenny Lind. 2. Of the dramatic Artist Rachel. 3. Of the Historian and Erudite J. v. Görres. 4. Of the Poët Fr. Halm (Baron Münch). 5. Commemorative of His Majesty's tour through Hungary. 6. Jubilary Medal of the Joanneum of Gratz. 7. Prize-medal of the School of Arts of Trieste. 8. Medals commemorating the Motto of H. M. the Emperor Francis Joseph I.: „Viribus unitis“. 9. Commemorative of the Emperor's return from Hungary in 1852. 10. In honour of His Imp. H. Archduke Charles. 11. Of the Orientalist Hammer-Purgstall. 12. For Mozart's centenary jubilee at Vienna. 13. Of Count Zamoisky at Warsaw. 14. For the funeral Ceremony for Duke Frederick the Warlike. 15. Commemorative of the Meeting of German Naturalists at Vienna. 16. Prize-medal of the Academy of Fine Arts. 17. Cast Medal. 18. Medal in honour of J. Würth. 19. In honour of H. Em. Cardinal Scitovski, Primate of Hungary. 20. Of H. Em. Cardinal Landgrave Fürstenberg, Archbishop of Ollmütz.

II. **Portraits of King Mathias Corvinus and his Queen Beatrix,** Bassi-rilievi, cut in Steatite.

83. SEIDAN William, born at Prague 1817, presently a Vienna. Medal-Engraver.

I. **Commemorative Medals:** 1. Of the Battle of Temesvar 1849 (price: 4 Austr. florins). 2. Of the 500 years jubilee of the Rifle Corps of Prague. 3. Of the painter Overbeck 1847 (price: 5 Austr. florins). 4. Of the painter Cornelius (price: 5 Austr. florins). 5. Of Field-Marshall Count Radetzki 1858. 6. Of the inauguration of the Counsel of the Empire at Vienna, May 1st 1861 (price: 3 Austr. florins). 7. Prize-medal 1861. 8. Of Emperor Ferdinand 1848. 9. Of General Count Schlick 1848—49 (price: 4 Austr. florins). 10. Of the presence of Their Majesties at Prague 1854. 11. Of the secular jubilee of the Military Order of Maria Theresia 1857. 12. Of the erection of the monument in honour of General Hentzi at Bude 1861. 13. Portrait-medal of the Historiographer Palatzky 1861 (price: 5 Austr. florins).

Class 40.

Etchings and Engravings.

84. GREFE Conrad, born at Vienna 1823, possessor of the Gold Medal for Art and Science. Landscape-painter.

I. **Scenery with thunderstorm.** Copied from the Artist's own composition, etched.

85. MAYER Christian, at Vienna, born 1812. Engraver.

I. **The four parts of the World,** copied from Rubens picture. Property of the Artist.

86. POST Charles, at Vienna, born at Prague 1834, Member of the Polytechnic Institute of Paris. Engraver.

I. **Landscape,** from a picture of A. Achenbach, engraved in copper. Property of the Artist. To be sold for 50 Austrian florins.

87. SCHMIDT Leopold, born at Prague 1824, presently at Vienna. Engraver.

I. **Mozart's Apotheosis,** from a drawing of J. N. Geiger. Engraving on copper. Property of the Artist. To be sold for 40 Austrian florins.

88. SCHŒN Lawrence, born at Vienna 1817. Landscape-painter.

I. **Etched Landscapes.** Property of the Artist.

ALPHABETICAL INDEX.

Utility of the Maize-plant.

Where shall we in future get our paper from? is at the present time a stereotype question of the paper-manufacturers. And, indeed, they have reason to ask this question, for it is a wellknown fact, that the consumption of paper is in a rapidly augmenting progression in all civilized states. The explanation for this is not only the much increased productive activity in the field of the literature generally, and of the periodical especially, but also the quicker pulsation of public and private commercial life, caused by the freeer institutions of states, by the stimulus of competition, increased means of communication, etc. To this must yet be added, that a great quantity of the manufactured paper is used for quite different purposes than for printing and writing. So is, for inst., a great deal of the paper used for tapestry, cartonage-work, packing-paper, etc.

The consequences of this enormous paper-consumption are felt more and more, because the paper-manufacturers meet every day with greater difficulties to procure a sufficient supply of raw-material, necessary for the working of their factories. The rags, which are mostly used for the paper-pulp, cannot be produced at will, like other raw-material; the supply is confined, as well in regard to quality as quantity, to a certain limit, influenced by the activity of the rag-gatherers. It is therefore evident, that the moment must come, sooner or later, when it is absolutely impossible for the paper-manufacturers to keep pace with the paper-consumption, if they should not succeed to discover a satisfactory substitute for rags. Indeed, their exertions have been directed to this for years, and experiments, tried not without success, have proved the existence of many materials, containing fibres, which might serve as a surrogate for rags. But few are adapted for manufacturing purposes, partly because most of them are too costly, and partly because they cannot be produced in sufficient quantities. Only plants of culture are produced in great quantities. Of these the *Maize-plant* is best adapted for paper-manufacturing. This has already been discovered some time ago, and consequently it has been tried at different times. Already in the last century two maize-straw paper-manufactories were in existence in Italy, according to Dr. Joh. Christ. Schaeffer's „Sämmtliche Papierversuche", Regensburg 1772. But the procedure, applied by the manufacturers, seems to have got lost with the decay of the paper-mills.

A certain Moritz Diamant from Bohemia, directed recently again the attention to the importance of the Maize-plant as a surrogate for linen-rags, and indicated a process for the transformation of maize-fibres into paper-pulp. He presented already in the year 1856 to Baron Bruck, then Minister of the Finances, a project with regard to it. The Imperial Paper-manufactory Schlögelmühle near Gloggnitz, was consequently authorized to make, under Diamant's direction, out of a certain quantity of Maize-straw, paper. The produced paper was not satisfactory in regard to quality; also the cost of manufacturing proved to be considerably higher than that of rag-paper. In consequence of such results the Ministry of the Finances was obliged to stop further experiments.

Diamant tried now to get private individuals for his enterprise, to manufacture maize-straw-paper, but his endeavours did not meet with the desired success, for in the year 1859 he made, with recommendations from Trieste in his possession, a second application to the Austrian Minister of the Finances.

Baron Bruck consented, in consequence of a recommendation from judicious men, whose views had been consulted, to have a second trial made in the Imperial paper-mill under Diamant's direction. The mill was already at that time under my superintendence and I interested myself very warmly for it. Different kinds of paper were manufactured, partly writing-, partly printing-paper, which was not entirely satisfactory as far as quality was concerned; the cost of producing the paper was yet, in spite of all exertions to reduce the manufacturing-expenses, considerably higher than that of the rag-paper, consequently the Direction of the Imperial paper-mill could not recommend the manufacturing of maize-paper on a larger scale.

As the height of the expenses of manufacturing was principally caused by the great distance of transportation of the raw-material, it was proposed to undertake the manufacturing of maize-paper in such a locality, where Maize is raised in sufficient quantities and consequently the straw of the plant close at hand.

To bring the question of rentability nearer to its solution, it was resolved to take a middle-course, by erecting, in an experimental way, a factory for half-stuff. It was argued, that the cost of transportation, which had as yet increased the expenses of all experiments to make maize-paper, must be considerably reduced, when, instead of the voluminous, weighty straw, only the extract of it, the real substance adapted for manufacturing paper, should be delivered to the paper-mills.

The projected half-stuff Factory was erected at Román Szt.-Mihály near Temesvar, where the Maize-cultivation is extensive, and on the 6th day of March, 1860, it commenced to operate under Diamant's provisional direction. The length of time fixed for experiments, was one year. Diamant promised to manufacture in this period 4500 Cwts. half-stuff out of maize-straw, a quantum, of which not the seventh part was reached. The produced half-stuff was, in addition to this, so poor, that further experiments and the working of the Factory were suspended in consequence of Diamant's own suggestion, before the granted time had expired. Diamant was then released from his position, absented himself, and left the question undissolved. The experiments did cost more than 30,000 Florins, which had been advanced by the Imperial paper-mill, according to orders from the deceased Minister Baron Bruck.

This money had to be made up again, in case the Direction should be held responsible for it.

With this, the first stadium of the-maize-straw-fabrication was closed, as far as the experiments were conducted under Diamant's direction. Diamant did not participate in subsequent experiments.

The Imperial paper-mill had now to rely on itself.

The exertions of the Direction of the Schlögelmühl paper-mill, under whose superintendence the experiments were continued, aimed principally at two things: first, to reduce the cost of production through rational improvements in the mode of manufacturing paper; secondly, to investigate how the expenses would be, if, instead of the whole straw, only the fibre-stuff of the shucks (the leaves, enclosing the corn-ear), containing the fibres of the best and finest quality, would be used for manufacturing paper.

If those industriously continued experiments did not lead directly to the desired results, i. e.: to make paper as cheap out of maize-straw as out of rags, they led at least indirectly to it, and what is of greater weight, to a very important result: *the discovery of a new fibre, capable of being spun and woven*, which (the fibre) furnishes us, in its **Waste**, with a cheap paper.

The Genesis of this discovery is as follows:

The basis of all paper is vegetable fibre. The rags are but the fibres, produced out of the flax-, hemp-, or cotton-plant and used up by wearing. If those fibres would be used for paper, before they would be converted into textures, the paper would be certainly better, but also incomparatively more costly.

Paper of maize-straw, is paper of unworn plant-fibres. After the ideas had once run into this direction, the question was brought very near: Cannot the fibres of the maize-plant, before they are delivered to the paper-machine, just as well be worn, as the fibres of flax and hemp are worn first? In other words: cannot the maize-fibre be spun and woven? All what necessary, was a trial. It was made and succeeded. It appeared, that the maize-fibre could be extracted out of the plant in a form like flax, by a procedure very simple and at the same time requiring but little apparatus' and auxiliary means, that it could be spun like flax and be woven like the flax-thread. The process applied and invented by me, is not only in Austria, but also in all great European States protected by Patents, to secure for Austria the priority of the invention.

That the spinning and weaving of the maize-fibre is not yet as far advanced as manufacturing paper out of it, is not to be wondered, for it must be recollected, that the last named process has been tried for several years, while the invention of spinning and weaving has been experimented upon only during 6 months, and is consequently yet in its infancy. The textures of maize-flax will look quite different in a short time, especially when men of the profession will have taken it into their hands, and when the spinning- and weaving-machines will be adapted to the maize-fibre. Not one invention came out of the head of its author complete; every one had to have time for its perfection, why should it be otherwise in this case? But so much can be said already at present, that the detection of the capability of the maize-fibre to be spun and woven, is of the greatest consequence, and that the cultivation of this plant constitutes one of the most profitable branches of agriculture known.

The plant, not taking the corn-ear into consideration, which pays for itself already the cultivation, can be made useful in many different ways.

Through the process, applied for producing the maize-fibres, the components of the plant are separated into three different parts, to wit: *Fibres, Flour-dough* and *Gluten*. The fibres are spun and woven; the nutritive substance (flour-dough) which has the peculiarity to remain fresh for months in the open air, consequently to resist, unlike other organic substances, putrefaction—gives a pleasant-tasting, nutritive, healthy flour-dough. All the fibre- and gluten-**Wastes** of the maize-plant, which are precipitated during the process of extracting the fibres, are used for manufacturing paper. The catalogue of the Austrian Exhibition at London (1862) in German, French, and English, consists of such paper, viz. partly of pure maize-paper, partly of such, which is made out of maize-fibres, mixed with linen- or cotton-rags.

The maize-plant contains consequently not one particle, which cannot be used. It can be made useful from the corn-ear up to the head of the stalk. The ear and the maize-stuff-extract gives food to man, the fibres are woven into cloth, and the shorter fibre- and gluten-stuff is converted into beautiful paper. After the fibre has served for cloths, it is brought back as rags and manufactured into paper. Which plant can boast of such a general utility like the maize-plant?

The most remarkable thing in regard to the process, is its simplicity. The humblest labourer can learn the procedure, when he is but once instructed in writing or verbally, and is enabled to effect the production of the above named articles, on the maize-field itself, without the slightest expense. In want of wood, the lower parts of the stalks will supply him with fuel. Owners of large farms, or manufacturers can produce daily hundreds of Cwts. in steam-boilers. The shuck-gatherers effect through cash-payment the buying up from the smallest farmer to the largest planter, and bring the materials into the markets of the world. Austria will try first to acquire enough supply for its consumption and then realize a large foreign-export. The other countries, where maize is raised, will follow. The whole world will profit millions by this new branch of industry.

In conclusion it must be remarked, that I have succeeded to bring the invention in such a comparatively short time to its present degree of perfection, only through the industrious cooperation of Messrs. Pfob, Jahn, Jung, Marsanich in Schlögelmühle, as also through my inspection of the spinning-mills at Trummau, Pottendorf, Siegersdorf, in-der-Au, Pottschach, Stuppach, and Lambach. Besides, I am very much obliged to Mr. Roman Uhl, baker at Vienna, who, 'at the very first trials, undertaken with the nutritive substance, succeeded, by means of a certain quantity of common flour mixed with it, to convert it into excellent bread. Without his profound professional knowledge and his intuitive intelligence, as well as his rational procedures, I am sure, the problem would have been resolved neither in as short a time, nor in as satisfactory a manner.

Above all, I have here to acknowledge the merits of Moritz Diamant. As unsatisfactory as the results he effected, were, the merit belongs to him to have given the first impulse again to the idea of making paper out of maize-straw, after it had slumbered for so long a time, which led to the improvements now attained in manufacturing paper out of maize-straw.

Additional remarks.

1. 160 lbs. of rags valued at about 16 fl., are required for the production of 100 lbs. of foolscap.
2. 100 lbs. of the paper manufactured thereof sell for about 33 fl.
3. 4 fl. were paid up to this time for one hundredweight, loco Schlögelmühle.
4. From 3 to 3½ Cwts. of lischen (head-leaves) yield 100 lbs. of paper.
5. According to official minutes, there are in the Austrian dominions more than 2,800,000 yokes (1½ engl. acres = 1 yoke) of ground planted with maize; the produce of lischen (or head-leaves) may, therefore, be estimated at 2½ millions Cwts. of lischen at the lowest computation.
6. We may thus take it for granted that 1,200,000 Cwts. of rags are to be substituted by means of maize leaves.
7. 1 Cwt. of head-leaves yields, on an average, one-third substance for spinning, one-third for paper, and one-third for food; waste there is nearly none.
8. If the whole of the fibrous substance were worked up into paper, there would be produced about 1,500,000 Cwts. of paper from the lischen collected in the Austrian monarchy.
9. Regarding the virtues of papers manufactured from pure maize substance, there is no doubt whatever, that these papers by far surpass the best rag papers in strength, thoughness, durability and power of bearing. Experiments made in my own room and before my own eyes showed that one sheet of bleached maize-paper chosen from the portfolio, bore a weight of 460 Vienna pounds.
10. If the substance is ground short (of which circumstance the transparency of the paper is depending), maize-paper can probably be used as an excellent substitute for glass, owing partly to its natural transparency.
11. There is furthermore to be remarked that factories for the extraction of fibre and substance for bread, require no expensive machinery and but little adjutory materials.

Dr. Alois Ritter Auer von Welsbach,

Imp. Roy. Aulic Counsellor, Chief Director of the Imp. State-Printing Establishment in Vienna, and of the Imp. Paper-mill at Schlögelmühle, Member of the Imp. Academy of Sciences.

Imperial Printing Establishment, Vienna 1862.

Printed on maize-paper.

HANSE-TOWNS.

SOUTH-WEST TRANSEPT.

BREMEN.

CLASS II.

1.   SCHMIDT & ROHLAND, *Hemelingen, Bremen.*—White lead (dry, and as paint).

CLASS III.

2.   BOLLMANN, E. & M.—French, German, and malt vinegar.

3.   SCHOMBURG & Co.—Caraway-cordial.

4.   WALTJEN, E.—Flour.

CLASS VII.

5.   BRUNS, G. H.—Hand-blowing machine for forges.

CLASS VIII.

6.   WALTJEN, C. & Co.—Friction-balance for ascertaining the laws of friction, and testing lubricating substances.

CLASS XII.

7.   GLEISTEIN, G. & SON, *Vegesack, nr. Bremen.*—Machine-manufactured cordage for ships' use.

CLASS XIX.

8.   WALTJEN, F. W.—Kyanised linen.

CLASS XXIV.

9. EBEL, H.—Picture in embroidery.

10. ZIEGENHIRT, F.—Picture in embroidery.

CLASS XXVI.

11. MEYER, A.—Saddle and bridle.

CLASS XXVII.

12. ALBRECHT, J. H.—Dress-boots.

13. BÖDEKER, C. F., & KRÜGER, BROS.—Dress-coat.

14. BORTFELDT, C.—Silk and felt hats.

15. CORSSEN, J. F.—Boots.

16. DONOP, H. & Co.—Trousers and waistcoat.

17. HENNING, H.—Varnished leather riding-gaiters.

18. HOFFMEYER, C. H.—Jacket.

19. LEONHARDT, F. W.—Coat à deux mains.

20. SCHMIDT, J. H.—Coat.

21. SICK, C.—Silk hats.

22. WEBER & EEG.—Boots and shoes.

CLASS XXVIII.

23. GEFFKEN, D.—Ledger, cash-book, photograph album, portfolio.

CLASS XXX.

24. ARMERDING, J. H.—Secretaire.

25. BRUNS, G. H.—Iron work-table and embroidering tables.

26. FORQUIGNON, J.—Boudoir furniture.

27. LAMPE, C.—Sofa of basket-work.

28. SCHLEEF, J. H.—Sideboard and sofa table.

CLASS XXXI.

29. ASENDORPF, C.—Iron money-safe, with safety-locks and letter-combination.

CLASS XXXIII.

30.   KOCH & BERGFELD.—Silver tea and coffee service, German antique style ; silver tea tray.

31.   WILKENS, M. H. & SONS.—Silver table service, &c.

CLASS XXXVI.

32.   BIERMANN & FREVERT.—Glass pyramid with articles of cigar-manufacture.

33.   BÖDEKER, H. & SONS.—Brematin and alabaster candles, with the raw material and composition in blocks.

34.   DÖDING, F. H.—Cocks of tin composition, injured neither by corrosion nor friction.

35.   OETTING, D.—Tallow candles with prepared wicks.

36.   WALTJEN, F. W.—Kyanised wood. Fire-lighters.

HAMBURG.

CLASS I.

1.   KLEUDGEN & Co.—Hardened Portland cement.

2.   STAUB, J.—Grinding and sand stone, from the Deichbrüche Quarries, on the Bohemian frontiers.

CLASS II.

3.   BEHRMANN & V. SPRECKELSEN.— Mineral waters.

4.   DOUGLAS, SONS, J. J.—Soap, fancy shapes and slabs.

5.   FLÜGGER, J. D.—Lac varnish and dryers.

6.   GILLMEISTER, CROP, & Co.—Chemical preparations.

7.   HASPERG & SCHÄFER.—Concentrated non-ferruginous alum, from argillaceous cryolite.

8.   OBERDÖRFFER, A. & Co.—Artificial mineral waters.

9.   VÖLKLEIN, STEINERT, & Co.—Soap pomatum, tooth-powder, and perfumery.

CLASS III.

10.   DAMPFZUCKERSIEDEREI VON 1848.— Refined and raw sugar, and treacle.

11.   ECKHARD, A. F.—Sweetmeats and comfits.   Plastic   fancy   picture,   of sugar.

12.   FETT, J. M. & Co.—Smoked beef.

13.   GREEN, H. F.—Cherry cordial.

14.   MULSON, L. & Co.—Preserves.

15.   PETERS, J. J. W.—Liqueurs, brandy, and spirits.

16. REESE & WICHMANN. — Chocolate, cocoa, sweetmeats, and tea.

17. REESING, W. C.—White crystallized candy.

18. SCHRÖTER, G. — Fruit-drops and rocks.

19. TREDE, G. H. G.—Anti-cholera and stomachic bitters.

CLASS IV.

20. GOLDSCHMIDT & Co.—Merino-sheep wool.

21. KAEMMERER, A.—A glass containing colour.

22. PORTEN, I. V. D.—Cigars.

23. WERNER, P. O. E.—Splinter for upholstery work.

24. ZIPPERLING, KESSLER, & Co.—Pulverized and ground dye-wood.

CLASS VI.

25. SACHS, F. B. C.—A barouche.

CLASS VII.

26. BECKER, J. C.—Sewing machine.

27. CATHOR, M. & GRABAU, A. C.—Joiner's tool-case.

28. HEBERLING, H. P.—Gilder's press.

29. KOLTZAU, H. C.—Shoemaker's tools.

30. KÖNIGSLÖW, H. V.—Sewing machine.

31. WINTER, E.—Diamond pencils for lithographers, engravers, &c. Rubies for ruling.

CLASS VIII.

32. GÄBLER & VEITSHAUS, *Hamburg,* makers; DEWIT, W. C. & H., *Amsterdam,* patentees.—Manometer.

CLASS X.

33. PLATH, C. C.—Mathematical instruments.

CLASS XII.

34. STEINHAUS, C. F.—Model of a merchant frigate.

35. EISERN BOOT BAU Co.—Corrugated and galvanized metal articles.

CLASS XIII.

36. Krüss, A.—Dissolving-view apparatus. Photographical objective barometer.

37. Schröder, H.—Microscope, with minor apparatus. Optical articles.

CLASS XV.

38. Bröcking, W.—Double regulator.

39. Knoblich, T.—Astronomical constant pendulum clock.

40. Nieberg, J. L.—Astronomical constant clock, with free escapement.

CLASS XVI.

41. Albrecht, F. E. J.—Upright pianoforte.

42. Baumgardten & Heins.—Grand pianoforte.

43. Doll, J. & Kamprath.—Upright pianoforte.

44. Melhop, W.—Æolian harps.

45. Müller, L. W.—Upright pianoforte.

46. Mejer, A. W. A. & Co.—Mechanism for pianos.

47. Piering, T.—Bass-tube, bassoon, German silver cornet-à-piston.

48. Plass, C. H. L.—Pianoforte.

49. Rachals, M. F. & Co.—Upright pianoforte.

50. Rott, J. A. W.—Upright pianoforte.

51. Ruppach, R.—Upright pianoforte.

52. Schlüter, C. E. L.—Vertical grand pianoforte. Organ-box for children.

53. Schröder, C. H.—Piano, half oblique.

CLASS XVII.

54. Brüning, G. H.—Paralytic and valetudinarian's self-acting mahogany armchair, with wheels.

55. Matthias, B. & Co.—Artificial teeth.

56. Nebel, J. C. R.—Surgical wire leg and arm ligatures.

CLASS XXII.

57. Werner, P. O. E.—Floor-carpets of splint.

CLASS XXIII.

58. Mylius & Hasenohr.—Printed woollen shawls, table-cloths, and furniture-cloth.

CLASS XXIV.

59. FRANCK, GESCHW.—Gold and silver-embroidered table-cloth.

CLASS XXVI.

60. EISSFELDT, J. E. & Co.—Calf and goat-skins.

61. FEIDEL & Co.—Calf-skins, varnished knee and shade-leather, and horse-skins.

62. LAHRMANN, A. W. & Co.—Calf-skins and skivers.

63. MÖNCKEBERG, G. A. C.—Horse-collar, travelling-trunk, and box.

64. WAMOSY, D.—Calf-skins, varnished knee and shade-leather. Brown skins for saddlery, and black oil-cloth.

CLASS XXVII.

65. DOSSE, F. A.—Boots.

66. LADENDORF, H. W. F. — Boots and shoes.

67. PAPE, J. C. W.—Boots and slippers.

68. PREHN, G. C. W.—Samples of boots.

69. SANDER, A.—Various kinds of boots.

CLASS XXVIII.

70. ADLER.—Frame, containing samples of steel-engravings and lithographs.

71. ASPERN, W. M. v.—Counting-house books.

72. BADE, W.—Counting-house books.

73. BAHRS, T.—Chemical manifold copying-books.

74. BERENDSOHN.—Three albums.

75. KITTLER, R.— Dictionary, English and Portuguese.

76. MÖLLER, C. H. A.—Counting-house books.

77. SCHMIDT, C. O.—Watch-stand, cigar-salver, almanack, and stationery.

CLASS XXIX.

78. KÖHLER, A.—Writing-books, containing prototypes of alphabets.

CLASS XXX.

79. AHRENS, H.—Chairs, looking-glass, flower-stand, foot-stool, book-stand, sofas, &c.

80. BERNHARDT, J. H. A.—Upholstery.

81. BEYER, F.—Stuffed basket-work arm-chair, reed-sofa, &c.

82. BOCK, L. & Co.—Looking glass and table, with marble slab. Gilt pedestal.

83. BOCK, F. D.—Furniture.

84. DIÈBER, J. H.—Mahogany writing-desk.

85.—EHRENSMANN, R.—Osier arm-chairs, flower-stand, and basket-work.

86. FREESE, H. — Window, with wood-chipweft for blinds. Wood-chip hangings.

87. GRUPE, L.—Pattern for printed woollen table-cloth, curtains, furniture-cloth, and printed paper-hangings.

88. HANDWERKER-VEREINIGUNG v. 1853.—Furniture of walnut and mahogany.

89. HARDEN, G. H.—Tables with round and oval leaf. Tobacco-box.

90. HARDER, F. JUN.—Flower-vase and stand, chairs, work-baskets, &c.

91. HAUER, H. W.—Oak sideboard.

92. KLÜVER, D. H. J. — Walnut sideboard.

93. KÖPCKE, C. C. J.—Oak dining-table.

„ NIESS, A.—Walnut sideboard.

„ SCHINDLER, H.—Oak chairs.

„ STIEHR, J. W.—Oak sideboard.

„ WEHRSPOHN, H. G.—Walnut centre-table.

94. KOSTE, H. N. & Co.—Chairs, sofa, table, and piano-chair, in curved wood.

95. LAGERFELDT, J. G. A.—Gilt sofa-looking-glass.

96. LOOSE, J. R.—Jacquaranda centre-table.

97. MAACK, P.—Basket-work furniture, and furniture for children.

98. MÜLLER, J. H. & KRÖGER.—Oak chandelier, Gothic style.

99. NIESS, A.—Walnut centre-table.

100. PATEIN, J. F. M.—Black varnished tables.

101. PLAMBECK, C. F. M.—Church door, Gothic style : inlaid work.

102. PLÖNSKI, FRAU J.—Leaf of a table, painted.

103. RAMPENDAHL, H. F. C.—Fancy furniture of antlers, and articles of amber, &c.

104. SCHMIDT, E. — Wicker-work for door-plates ; round centre-table, with leaf in wicker-work.

105. SCHULTZE, C. F.—Cot, trimmed with wicker-work.

106. SCHULTZ, J. E. F.—Window-blinds.

107. TENGER, J.—Centre-table, with leaf inlaid in mosaic work and pedestal.

108. WEHRSPOHN, H. G.—Ladies' work-table.

109. WERNER, C. F. & PIGLHEIN.—Book-shelf and chimney-piece ; ebony etagere ; library table ; arm and other chairs.

110. WULF, E. & KELTER.—Field-beds for railway travelling ; railroad cushions.

111. ZIMMERMANN, F.—Rocking and other chairs, and perambulator.

CLASS XXXI.

112. BLECHER, I. H.—An iron safe.

113. CLASSEN, F.—Bird-cage.

114. KLEIN, T.—Spelter hip-bath, hexagonal lantern, hearth-blower in zinc.

115. KRAHNSTOVER, E. B.—Shower-bath, portable water-closets.

116. KRAMER, J. G. W., JUN.—Brass parrot-cage.

117. RACKE, F. W. — Spelter hearth-crown, brass bird-cage.

118. SCHULTZ, F. J.—Fire-screen, bird-cages.

119. TIMCKE, H. F.—Brass parrot-cage, coal-scuttle, German silver pendant kettle.

120. WEBER, J. F.—Varnished bird-cages.

121. WOLTERS, J.—Fancy figure, in embossed iron-plate ; iron-plate vase.

CLASS XXXII.

122.   WEBER, W.—Pocket-knives, razors, and carving knives.

CLASS XXXIII.

123.   PETERS, J. H. & Co.—Crystal spirit-gas lamps.

CLASS XXXIV.

124.   EIMBCKE, E.—A skylight for ship's use, in gutta percha, framed.

125.   MEYER, BROS.—Demijohns, covered with wicker-work.

126.   SCHULZ, A.—Transparencies, and various articles of glass.

CLASS XXXV.

127.   HAUTHAL, J. G. & Co.—Painted tea-cups in porcelain, flower-vase.

128.   ROTH, G. A.—Plaster-cast figure, chandeliers.

CLASS XXXVI.

129.   FEDERWISCH, G. L.—Kites.

130.   MEYER, H. C., JUN. — Hamburg Arms, composed of walking-sticks, figures, &c.

131.   NATHANSEN, W.—Coloured, stamped, and other paper.

132.   SCHNACK, H. F.—Sample card of mother-of-pearl escutcheons.

133.   WÖBCKE, H. — Meerschaum and cigar-tubes.

134.   ZUBER, J.—Brooches of pink and green shells, flower-basket, &c.

LÜBECK.

CLASS II.

1. FRITZ & BÖNING.—Eau de Cologne.

CLASS III.

2. CARSTENS, D. H.—Hermetically preserved victuals.

3. GRELL, A. F.—Hand-made marchpane.

4. HAHN, G. C. & Co.—Preserved alimentary substances in boxes.

5. JÜRGENSEN, F.—Aromatic stomachic liqueur.

6. PETERSEN, J. C. C.—Hand-made march-pane.

CLASS XI.

7. FISCHER, C. A. & SON.—Rifles and fowling-pieces.

CLASS XXIV.

8. AMANN, C.—Embroidered handkerchief.

CLASS XXVII.

9. FRITZ, J. H. & BÖNING.—Shirts.

CLASS XXVIII.

10. HEINRITZ, C. C. J.—Specimen of bookbinding.

CLASS XXIX.

11. SÖHLBRAND, J. M.—Caligraphical copy.

CLASS XXX.

12. LEDERHAUSEN, J. F. C.—Bed-screen of wicker-work.

CLASS XXXVI.

13. HERMBERG, H. M. C.—Cigar-boxes, portfolios, and portemonnaies.

MECKLENBURG-SCHWERIN.

SOUTH-WEST TRANSEPT

CLASS II.

Sub-Class A.

1. WINKLER, *Niegleve, near Gustrow.*—Bleached stuffs.

CLASS III.

Sub-Class A.

3. HILLMANN, *Scharstorf, near Rostock.*—Oats.

4. KLOCKMANN, *Harmshagen, near Wismar.*—Wheat and other grain.

5. PFANNENSTIEL, *Brusenback, near Kleinen.*—Groats and flour.

6. SCHLIEFFEN, COUNT, *Schlieffenberg, near Gustrow.*—Wheat, rye, and pease.

6*a.* WIRKEDE, C. VON, *Klein, Leukow, near Peuzlin.*—Wheat.

7. SCHUBART, *Gallentin, near Wismar.*—Roots of cereal plants.

Sub-Class B.

8. CLEVE, V. *Carow, near Plau.*—Cheese.

9. MICHAEL, V. *Gantzkow, near Neubrandenburg.*—Smoked goose-breast.

10. OERTZEN, V. *Woltow, near Tessin.*—Cheese.

11. SCHLIEFFEN, COUNT, *Schlieffenberg, near Gustrow.*—Cheese.

Sub-Class C.

12. SANITER & WEBER, *Rostock.*—Liqueurs.

12*b.* ENGELL & CO. *Wismar.*—Spirits and beer.

CLASS IV.

Sub-Class A.

13. BRUNNENGRÄBER, H. *Schwerin.*—Medical and other soaps.

Sub-Class B.

14. BEHR, V. *Rentzow, near Gadebusch.*—Wool.

15. SCHLIEFFEN, COUNT, *Schlieffenburg, near Gustrow.*—Wool.

16. CÖLLE, *Meetzen, near Gadebusch.*—Fleeces.

17. SCHACK, VON, *Retchendorf, near Schwerin.*—Fleeces.

18. SCHALBURG, *Herzberg, near Parchim.*—Fleeces.

19. HOFFSCHLAEGER, J. F. *Weisin, near Lübz, Estate.*—Fleeces and samples of wool.

CLASS VI.

24. FLORKOWSKY, *Schwerin.*—Carriage with swan's neck, and without axle-beam or box-seat.

CLASS VIII.

26. MEMMERT, *Schwerin.* — Machines for cutting angles for cabinet-makers, and for cutting almonds.

CLASS IX.

28. MEYER, *Schwaan.*—A Mecklenburg hook-plough.

CLASS X.

30. DAHSE, B. *Rostock.*—Roofing and isolating material of stone paste-board with asphalte.

CLASS XI.

33. SCHMIDT, *Schwerin.* — Lefaucheux changing double-barrelled rifle ; rifle for deer-stalking, with cast-steel barrel ; percussion needle-gun, after Tervy.

CLASS XII.

Sub-Class A.

36. DAHSE, B. *Rostock.*—Tightening material for paste-board with asphalte.

CLASS XIII.

37. BÖCKENHAGEN, *Gustrow.* — Spirit-gauge for distilleries.

38. DOLBERG, *Rostock.*—Reflecting-circle, after Borda, with repetition.

CLASS XIV.

39. DETHLEFF, *Rostock.*—Ambrotypes.

CLASS XV.

40. DREYER, *Kehna.*—Pendulum; second clock, with regulator.

CLASS XVI.

43. JENTZEN, W. *Grabow.*—Pianino.

CLASS XVII.

44. AHRENS, R. *Neubukow.*—Trusses.
45. MÖSSINGER, G. *Rostock.*—Models of artificial foot and hand.

CLASS XIX.

46. KRASEMANN, G. *Rostock.*—Covers for upholstery, woven from unspun Manilla hemp fibres.

CLASS XXIV.

Sub-Class A.

47. SIEGERT, C. L. *Rostock.*—Embroidery.

CLASS XXVI.

Sub-Class B.

48. BLIEFFERT, C. *Schwerin.* — Horse-collars, curved at the top, changeable.

Sub-Class C.

49. AHRENS, R. *Neubukow.* — Leather game-bag.

50. LEHMANN, C. *Schwerin.*—Gloves and dyed glove leather.

CLASS XXVII.

Sub-Class A.

51. SIEGERT, C. L. *Rostock.* — Straw hats.

Sub-Class C.

52. CRULL, E. *Rostock.*—Waterproof oil-cloth dress for mariners.

Sub-Class D.

54. MULLER, H. *Rostock.*—Boots.

CLASS XXVIII.

Sub-Class D.

56. DAHSE, B. *Rostock.*—Ledgers in leather and moleskin, with spring.

57. GARBE, *Rostock.*—Bookbinding articles.

57a. LAN, E. *Neu-Brandenburgh.*—Bookbinding articles.

CLASS XXIX.

Sub-Class B.

58. JANTZEN, F. *Schwerin.*—Drawing examples for the industrial schools of Mecklenburg, executed for the Government.

CLASS XXX.

Sub-Class A.

60. BEHR, H. *Rostock.*—Jewel-press.

63. HERMES, F.—Walnut drawing-room table.

65. PETERS, C. *Schwerin.*—Parquet floors, cloth press of oak, with carvings.

67. PETERS, C. — Portions of the inlaid floors in the saloons of the Grand Ducal Castle at Schwerin, and a folding-door.

CLASS XXXIV.

Sub-Class A.

71. GILLMEISTER, E. *Schwerin.*—Glass paintings.

CLASS XXXVI.

74. CRULL, E. *Rostock.*—Life-buoy.

75. JÜRSS & CROTOGINO, *Rostock.*—Garden table and seat of Portland cement; cement.

77. BERLING, L. *Rostock.*—Matches.

78. OPPERMANN, *Niegleve, near Gustrow.*—Knitted travelling stockings and woollen gloves.

79. BURMEISTER, F. *Schwerin.*—Lithographic stone with drawing, electrotype copy taken from this, second electrotype taken from the first, and print taken from the latter; electrotype copy, taken from a gutta-percha copy of the stone, and print obtained from it; lithographic print from the stone.

LONDON EXHIBITION 1862.

SPECIAL CATALOGUE

OF THE

ZOLLVEREIN - DEPARTMENT

EDITED

BY AUTHORITY

OF THE

COMMISSIONERS OF THE ZOLLVEREIN-GOVERNMENTS,

TOGETHER

WITH ADVERTISEMENTS, RECOMMENDATIONS AND ILLUSTRATIONS.

SECOND EDITION.

BERLIN. 1862.

R. DECKER, PRINTER TO THE COURT OF HIS MAJESTY THE KING OF PRUSSIA.

PREFACE.

The present catalogue has been composed in consequence of a desire to offer to the visitors a sketch of the Zollverein-contribution to the Exhibition, more extensive than the official catalogue affords and more complete, than the illustrated catalogue could have presented, as it contains a list of all the exhibitors, who have carried out their notified intentions. The compilation of the catalogue not having been begun until after the expiration of the time fixed for the delivery of the objects to be exhibited, such exhibitors only have been included, as have actually sent in their contributions or have applied for delay. Thus the catalogue will correspond with the actual Exhibition at least approximately.

Besides the name, condition, firm and residence of the exhibitors, the place where the manufactory is situated, when not identical with the latter, is given together with all prices; which have been awarded at the General German Exhibitions of Berlin and Munich in 1844 and 1854 and at the International Exhibitions of London and Paris in 1851 and 1855, and the names of the agents, by whom the exhibitors will be represented on the present occasion. These statements are followed by a specification of the objects exhibited and lists of the prices, where the exhibitors have wished to publish them, the statements of the exhibitors themselves having been taken as to price value, measure and weights. To facilitate reference to these statements, which from their diversity are somewhat complicated, a comparative table of the coins, measures and weights employed, reduced to English and French standards, is supplied pag. 179.

The arrangement is such, that the exhibitors in the States, to which they belong, and the States themselves follow in alphabetical order. The plan of division into classes, which Her Majesty's commissioners had decided on, has been observed as to the several States, but in the individual classes the list of the exhibitors is alphabetical. The numbers, which precede the name of every exhibitor, run through the whole of the catalogue corresponding with those of the official catalogue and the printed labels affixed to the exhibited articles.

The arrangement of the objects exhibited does not entirely agree with the plan laid down by Her Majesty's commissioners, the reason being the difficulty of assigning the true limits of each class by definite characteristics and also the circumstance, that the numbers in the special catalogue must correspond with those in the official catalogue.

The latter was constructed upon general statements made by the exhibitors of the nature of the articles, but when they were sent in, these did not prove accurate in every individual instance, so that the number had to be altered to bring the exhibitor into the proper class. It was preferred not to disturb this order, which materially facilitates the use of the catalogue.

A further division of the 36 principal classes into subclasses has not taken place. These were not determined on before the month of april, when the catalogue was already in the press. Owing to this circumstance it has been impossible to have any regard to the alteration in the division according to which perfumes are to belong to the fourth class, whilst formerly they were included in the second.

The difficulty of reference thus arising is obviated by referring to such objects as, being mentioned in one class belong to another, under the name and number of the exhibitor, and by noticing all the classes in which the exhibitor appears, under his name.

Of the objects, belonging to class I »Mining, Quarrying, Metallurgy and Mineral products« a special catalogue in English has been compiled, which besides enumerating the exhibitors, the specimens exhibited and the place whence taken, will afford a general view of the extent and importance of the mining and metallurgical resources of the Zollverein. The Commissioners will issue this catalogue to such visitors as take a particular interest in the subject. In consequence of that only a short list of the exhibitors and their products has been adopted.

The second division of the catalogue contains advertisements, recommendations and illustrations printed in the form desired by the exhibitors. The number, which the exhibitors bear in the official and in the present catalogue is printed in brackets at the head of every advertisement. In the special catalogue on the other hand reference is made to every exhibitor, mentioned in the appendix by giving the number affixed to his name there with the addition of the Roman numeral II.

The special catalogue is not intended as a guide for the occasional visitor. It will however on account of the precise and minute specification of the articles and the numerous prices current, be welcome to those persons, who consider the Exhibition as a field of instruction, and as a place, from which to derive practical advantage for trade and industry, by enabling them to form new connections and extend those already existing.

PREFACE TO THE SECOND EDITION.

In the second edition of this catalogue the corrections have been added which have been occasioned by comparing the contents with the exhibition and by a recent revision. Besides, the prize-medals and the honorable mentions awarded to exhibitors have been inserted between their names and those of their agents with »London 1862 medal« and »London 1862 honor. mention« respectively.

INDEX.

FIRST DIVISION.

SECOND DIVISION.

TABLE OF THE CLASSES.

CLASS I.
Mining, quarrying, metallurgy and mineral products.

CLASS II.
Chemical substances and products and pharmaceutical processes.

CLASS III.
Substances used for food, including wines.

CLASS IV.
Animal and vegetable substances used in manufactures.

CLASS V.
Railway plant, including locomotive engines and carriages.

CLASS VI.
Carriages not connected with rail or tram roads.

CLASS VII.
Manufacturing machines and tools.

CLASS VIII.
Machinery in general.

CLASS IX.
Agricultural and horticultural machines and implements.

CLASS X.
Civil engineering, architectural, and building contrivances.

CLASS XI.
Military engineering, armour and accoutrements, ordnance,
and small arms.

CLASS XII.
Naval architecture and ships' tackle.

CLASS XIII.

Philosophical instruments and processes depending upon their use.

CLASS XIV.

Photographic apparatus and photography.

CLASS XV.

Horological instruments.

CLASS. XVI.

Musical instruments.

CLASS XVII.

Surgical instruments and appliances.

CLASS XVIII.

Cotton.

CLASS XIX.

Flax and hemp.

CLASS XX.

Silk and velvet.

CLASS XXI.

Woollen and worsted, including mixed fabrics generally.

CLASS XXII.

Carpets.

CLASS XXIII.

Woven, spun, felted, and laid fabrics, when shown as specimens
of printing or dyeing.

CLASS XXIV.

Tapestry, lace, and embroidery.

CLASS XXV.

Skins, furs, feathers and hair.

CLASS XXVI.

Leather, including saddlery and harness.

CLASS XXVII.

Articles of clothing.

CLASS XXVIII.

Paper, stationery, printing and bookbinding.

CLASS XXIX.
Educational works and appliances.

CLASS XXX.
Furniture and upholstery, including paper-hangings and papier-maché.

CLASS XXXI.
Iron and general hardware.

CLASS XXXII.
Steel, cutlery and edge tools.

CLASS XXXIII.
Works in precious metals, and their imitations and jewellery.

CLASS XXXIV.
Glass.

CLASS XXXV.
Pottery.

CLASS XXXVI.
Manufactures not included in previous classes.

ABBREVIATIONS USED IN THE CATALOGUE.

Agt.	Agent.	kr.	Kreutzer.
manu.	manufacturer.	Zollz.	Zollzentner.
Berl. Ell.	Berlin or Prussian Ell.	Pfd.	Zollpfund.
Brab. Ell.	Brabant Ell.	Lth.	Loth.
Th.	Thaler.	cm.	centimètre.
sg.	Silbergroschen.	mm.	millimètre.
fl.	florin.		

FIRST DIVISION.

DUCHY OF ANHALT-BERNBURG.

CLASS III.

SUBSTANCES USED FOR FOOD, INCLUDING WINES.

1. BRUMME, A. F. & Co., manu. Bernburg, Dröbel. London 1851 honor. mention, 1862 medal. Agt. Lion M. Cohn, Phaland & Dietrich, 20 St. Dunstans-Hill, City.

Specimen of beet-root raw-sugar 12 Th. (1 £ 16 sh.) p. Zollz. (1 cwt.)

2. BRUMME, A. F. & Co., manu. Bernburg, Waldau. London 1851 honor. mention, 1862 medal. Agt. Lion M. Cohn, Phaland & Dietrich, 20 St. Dunstans-Hill, City.

Two loaves of unrefined beet-root sugar (saft-melis) 15¼ Th. (2 £ 5 sh. 6 d.); ground loaf-sugar 14¾ Th. (2 £ 4 sh.); moist sugar (first quality of afterproduct) 14 Th. (2 £ 2 sh.); moist sugar (second quality of afterproduct) 13½ Th. (1 £ 19 sh. 8 d.); all p. Zollz. (1 cwt.)

3. STENGEL, CHARLES (Cuny & Co.), manu., Bernburg. London 1862 medal. Agt. C. Trübner, represent. Lion M. Cohn and Phaland & Dietrich, 20 St. Dunstans-Hill, City.

2 loaves of beet-root sugar, the beet-roots grown by the exhibitors 15¾ Th. (2 £ 7 d.); crushed beet-root sugar No. 1. 15¾ Th. (2 £ 7 sh.); do. No. 2. 14½ Th. (2 £ 3 sh. 5 d.); do. No. 3. 14 Th. (2 £ 2 sh.); all p. 100 lb. (1 cwt.)

DUCHY OF ANHALT-DESSAU-CÖTHEN.

CLASS II.

CHEMICAL SUBSTANCES AND PRODUCTS, AND PHARMACEUTICAL PROCESSES.

6. ANHALT-ASSOCIATION FOR CHEMICAL MANUFACTURES represented by ADOLPH GRÄTZEL, Director, Rosslau o. E.

	pruss. lb.	Th.	sg.	gallon	sh.	d.
Brown-coal tar	100	—	5	1	—	11¼
Photogen of 0,765 spec. weight Baumé	100	—	16	1	3	—
Photogen of 0,790 do.	100	14	15	1	2	9
do. 0,805 do.	100	13	—	1	2	6
do. 0,815 do.	100	11	—	1	2	2½
Solar-oil 0,830 do.	100	8	15	1	1	9
do. 0,840 do.	100	8	—	1	1	7½
Paraffin-oil 0,870 do.	100	6	—	1	1	3
				lb.		
Cakes of paraffin I. Qual.	100	34	—	1	—	11
do. do. II. do.	100	20	—	1	—	6½
Paraffin candles I. do.	100	42	—	1	1	2
do. do. II. do.	100	38	—	1	1	½
				gallon		
Coal-tar	100	1	—	1	—	3
Benzin	100	15	—	1	2	6
Raw coal-tar oil	100	1	15	1	—	5
				lb.		
Raw phenylacid (creosot)	1	—	10	1	—	11
Purified phenylacid	1	—	20	1	1	10
Picrin-acid	1	2	15	1	6	10

(The named prices without packing.)

	pruss. lb.	Th.	sg.	lb.	sh.	d.
Coal-tar pitch	100	1	15	100	4	4
Coal-tar soot	100	6	—	100	16	6
Coal-tar oil soot	100	7	—	100	19	4

(The named prices with packing.)

CLASS III.

SUBSTANCES USED FOR FOOD, INCLUDING WINES.

8. HEIDEN, RICHARD, manu. and confectioner, furnisher to the court of His Highness the duke of Sax-Meiningen, Cöthen. London 1862 honor. mention. Agt. C. Trübner, London.

Dr. Arthur Lutze's homoeopathically medicated chocolate, approved by the faculty.

Richard Heidens improved aromatic vegetable lozenges for diseases of the chest, approved by the faculty.

CLASS IV.

ANIMAL AND VEGETABLE SUBSTANCES USED IN MANUFACTURES.

8a. KÄMMERER, C. G., manu. of soaps and perfumes, Dessau.

Imitation french toilet-soaps, cocoa-nut oil, grease, tallow and resin-soaps, perfumes; see price-current.

A

CLASS VIII.
MACHINERY IN GENERAL.

9. Achilles, W., Lithographer, Cöthen, inv., Heckert, H., Gröbzig, tin-man, maker, Agt. C. Trübner, London.

Beer preserving apparatus, to keep beer and wine drawn from the casks in perfectly good taste for a time of about 14 days, 1 £.

CLASS XI.
MILITARY ENGINEERING, ARMOUR AND ACCOUTREMENTS, ORDNANCE, AND SMALL ARMS.

11. Berger, Rudolph, Cöthen, gun-smith to His Highness, the duke of Anhalt, and to His Majesty the king of Prussia. London 1862 honor. mention.

A double barelled percussion gun; barrels of the best quality, Leclerc damask soldered with tin; the stock of spanish nut-wood, inlaid with german silver, 85 Th. (12 £ 15 sh.)

A double barrelled percussion rifle, Bernard damask; ranging to a distance of 150 paces, 75 Th. (11 £ 5 sh.)

A double barrelled needle-gun, system Berger; patented in Prussia, Anhalt and Coburg-Gotha; barrels of turkish damask, ornamented with hunting-scenes; stock of spanish nut-wood, 95 Th. (13 £ 25 sh.)

CLASS XIII.
PHYSICAL INSTRUMENTS AND PROCESSES DEPENDING UPON THEIR USE.

12. Oechelhäuser, director of the continental-gas-society, Dessau.

Gas-apparatus.

CLASS XX.
SILK AND VELVET.

13. Vollschwitz, William Francis (formerly Bachoven & Vollschwitz), merchant and manu. of silk-plush, Zerbst. London 1851 honor. mention; Paris 1855 bronz. medal; London 1862 honor. mention.

Remnant of silk hat-plush, 12 frs. p. m etre (8 sh 10 d. p. yard).

CLASS XXI.
WOOLLEN AND WORSTED, INCLUDING MIXED FABRICS GENERALLY.

14. Dessau-Wool-Spinning-Mill, Dessau. London 1862 honor. mention.

A box containing samples of woollen yarns.

15. Meinert, S., manu., Dessau. London 1862 medal.

No. 20890. Caschmir uni 6¼ ells, No. 20891. do. diagonal 18½ ells, No. 20233. do. jaspé 20¼ ells, No. 20236. do. à careaux 6¼ ells, No. 20857. do. uni 6½ ells, No. 20758. do. diagonal 11¼ ells, all p. ell 3 Th. (9 sh.)

16. Peukert & Körner, manu., Jessnitz.

1 piece of black cloth No. 8288. (28 yards) 1⅓ Th. p. ell. (4 sh. 5 d. p. yard); 1 piece of velvet for ladies cloaks No. 861. (16 yards) 1½ Th. p. ell (6 sh. 2 d. p. yard), 1 carpet ⁴⁄₄ striped No. 550. (38½ yards), 8 sg. p. ell (1 sh. 1 d. p. yard).

CLASS XXIII.
WOWEN, SPUN, FELTED, AND LAID FABRICS, WHEN SHOWN AS SPECIMENS OF PRINTING OR DYING.

17. Languth, H. (Friedrich Robitzsch' successors), dyers, Dessau.

A sheet of samples of dyed and printed silk and woollen stuffs, p. ell silks 2½ to 5 sg., p. lb. woollen stuff 3 to 12½ sg., dying and printing p. ell 3¾ to 5 sg.

18. Plaut & Schreiber, manu., Jessnitz. Paris 1855 honor. mention; London 1862 medal. Agt. Joseph Oxford & Co.

Printed woollen table covers: $\frac{12}{4}$ plain brown cloth with half silk border 7 Th. (1 £ 1 sh.), do. printed cloth table-covers No. 176. 177. 175. 205. B. 208. A. 6 Th. 15 sg. (19 sh. 6 d.), do. cachemire No. 217. A. 183. B.g.B. 189. B. 212. A. 221. A. 213. D.&E. 158. B. 165. B. 166. B. 204. A. 222. C. 6 Th. (18 sh.), do. lama (flannel) colored No. 214. 199. B. 182. A. 206. B. 4 Th. 15 sg. (13 sh. 6 d.), $\frac{11}{4}$ do. 2 colors No. 211. A. 218. C. 219. C. 220. E. 3 Th. 22½ sg. (11 sh. 3 d.), do. 4 colors No. 120. C. 153/36. 3 Th. 22½ sg. (11 sh. 3 d.), $\frac{12}{4}$ crimson red dyed, printed cloth No. 155. 4 Th. (12 sh.), $\frac{13}{4}$ do. lama (flannel) No. 209. 3 Th. 15 sg. (10 sh. 6 d.), $\frac{11}{4}$ anilin do. No. 150. 2 Th. 15 sg. (7 sh. 6 d.), $\frac{1}{4}$ green do. No. 193. 2 Th. 15 sg. (7 sh. 6 d.), $\frac{1}{4}$ crimson red do. II. No. 196. 2 Th. 2½ (6 sh. 3 d.) all p. piece.

CLASS XXVI.
LEATHER, INCLUDING SADDLERY AND HARNESS.

19. Spieler, Leopold, saddler and upholsterer, Dessau.

Two halters with muzzles for crib-biters, to prevent them from gnawing at the manger 4 Th. 20 sg. and 5 Th. (14 and 15 sh.)

CLASS XXVII.
ARTICLES OF CLOTHING.

20. Rosenthal, C., shoe-maker, Cöthen. London 1862 honor. mention.

A pair of riding-boots 5 £.

CLASS XXVIII.
PAPER, STATIONERY, PRINTING AND BOOKBINDING.

21. Katz, Brothers, printers, Dessau. Paris 1855 bronz. med.

36 printed books and specimens of printing.

CLASS XXIX.
EDUCATIONAL WORKS AND APPLIANCES.

21a. Achilles, W., lithographer, Cöthen, inv. Hobusch, H., Cöthen, joiner, maker. Agt. C. Trübner.

Apparatus for the instruction according to the method combining writing and reading, to be employed also as writing-copies in schools 2 £. Specimens of German letters 2 £.

CLASS XXXI.
IRON AND GENERAL HARDWARE.

22. Polysius, G., locksmith, Dessau.

A lock for money and book safes with invisible combinations, changeable 2,525,252,424 times 150 Th.

GRAND-DUCHY OF BADEN.

CLASS I.

MINING, QUARRYING, METALLURGY AND MINERAL PRODUCTS.

26. von Kilian, owner of a quarry, Waldshut. Agt. Ch. Trübner, 20 St. Dunstans Hill, City, and at the Exhibit. Building.
(s. II. No. 7.)
Two mill stones 70 fl. (5 £ 16 sh. 8 d.) and 67 fl. (5 £ 11 sh. 8 d.).

27. Zimmermann, teacher, Oppenau. Agt. No. 26.
Collection of minerals from the Renchthal 20 fl. (1 £ 13 sh. 4 d.).

CLASS II.

CHEMICAL SUBSTANCES AND PRODUCTS AND PHARMACEUTICAL PROCESSES.

28. Benckiser, manu., Pforzheim. Paris 1855 bronze medal; London 1862 medal. Agt. No. 26.
4 glass-jars with tartaric acid.

29. Clemm-Lennig, manu., Mannheim. London 1862 honor. mention. Agt. Amann, Roller & Co., 3 Love lane, East-cheap, London.
21 glass-jars with samples of chemical products: Heavy spar, chloride of barium and sulphide of manganese, hydrate of baryta, melted and crystallized, chloride of baryum, sulphate of baryta, acetate of baryta, nitrate of baryta, chloride of tin, arsenic acid, sulphate of copper, nitrate of mercury, chloride of mercury, do. for the preservation of railroad-sleepers, subchloride of mercury, pure bone black, raw gelatin, bone glue, superphosphate of lime, steamed bone dust, bone-dust for feeding.
Two pieces of kyanized railway sleepers having been 20 years in the ground.

30. Göhringer, Fr., Rippoldsau. Agt. No. 26.
Mineral waters; pastilles digestives.

32. Röther, H., manu., Mannheim. London 1862 honor. mention. Agt. No. 26.
One bottle with diamond color, one bottle with diamond cement.
One plate painted with diamond color, one do. and one piece of linen cloth painted with diamond color.

33. Ultramarine manufactory, Heidelberg. Munich 1854 medal of honor; Paris 1855 silver medal; London 1862 medal. Agt. G. Eckenstein, 150 Leadenhall-street, London E. C.
3 glass-jars with ultramarine, blue and green; 2 do. do. blue.

CLASS III.

SUBSTANCES USED FOR FOOD, INCLUDING WINES.

34. Bassermann, Herrschel & Dieffenbacher, manu., Mannheim. London 1862 medal. Agt. No. 26.
Macaroni, sago, starch and other pastes.

38. Hetzel, Burgomaster's adjunct, Willstedt.
Maize.

39. Grand ducal agricultural school, Carlsruhe. London 1862 medal. Agt. No. 26.
Maize.

40. Schollenberger, commercial gardener and seedsman, Carlsruhe. Agt. No. 26.
Maize.

41. Badish association for the cultivation and commerce of tobacco, Carlsruhe. London 1862 medal. Agt. No. 26.
Samples of Badish tobacco.

42. von Böcklin, Richard, Orschweier. London 1862 honor. mention. Agt. No. 26.
Badish tobacco.

43. Kübler, Burgomaster, St. Ilgen. London 1862 honor. mention. Agt. No. 26.
Badish tobacco grown in the Palatinate.

45. Seitz, Oftersheim. Agt. No. 26.
Tobacco grown in the Palatinate.

46. Bader, A. F., manu., Lahr. Munich 1854 medal of honor; London 1862 medal. Agt. Killy, Traub & Co., 52 Bread-street.
Cigars of tobacco grown in the Palatinate: Imperials, Regalia de la Reyna 15 fl. (1 £ 5 sh.); Cazadores, Prevas, Queen Manilla, Entractos, Chinese 12 fl. (1 £); Regalias, Cylindrados 11 fl. (18 sh. 4 d.); Trabucos, Operas, Manilla cheroots, Bayonetas 10 fl. (16 sh. 8 d.); Media Regalia, Regalia Londres, Conchas 9 fl. (15 sh.); Londres, Trabuquillos, Prensados, Panatelas, Napoleons 8 fl. (13 sh. 4 d.); Millars, Damas, Sultana, Queues de rats 7 fl. (11 sh. 8 d.); Rat tails 6 fl. (10 sh.); all p. mille.

47. LANDFRIED, J. P., merchant and manu., Rauenberg. London 1862 medal. Agt. No. 26. (s. II. No. 8.)

Cigars and tobacco grown in the Palatinate: Opera, $\frac{1}{1}$ Regalia, Imperials from 11 to 15 fl. (18 sh. 4 d. to 1 £ 5 sh.); $\frac{1}{2}$ Regalia, Trabucos, Entractos, Manilla from 9 to $10\frac{1}{2}$ fl. (15 sh. to 17 sh. 6 d.); Communos, Manilla round and square, Londres from 7 to $8\frac{1}{2}$ fl. (11 sh. 8 d. to 14 sh. 2 d.); all p. mille.

Cigars of any inferior quality of tobacco at lower prices.

A detailed price-current in the second section of the catalogue.

48. MAYER, BROTHERS, manu., Mannheim. Munich 1854 medal of honor; Paris 1855 silver medal; London 1862 medal. Agt. No. 26.

Cigars of tobacco grown in the Palatinate from $5\frac{1}{2}$ to 11 fl. (9 sh. 2 d. to 18 sh. 4 d.); of other tobacco: Regalia, Media Regalia, Queen Regalia, Regalia de Londres from 22 to 60 fl. (1 £ 16 sh. 8 d. to 5 £); Londres Habana 45 fl. (3 £ 15 sh.); Regulares from 16 to 32 fl. (1 £ 6 sh. 8 d. to 2 £ 13 sh. 4 d.); Londres from $16\frac{1}{2}$ to 26 fl. (1 £ 7 sh. 2 d. to 2 £ 3 sh. 4 d.); Trabuquillos from 15 to 20 fl. (1 £ 5 sh. to 1 £ 13 sh. 4 d.); Lady, Galanos, Goldposes, Prenzados, Caballeros from $8\frac{1}{4}$ to 18 fl. (14 sh. 6 d. to 1 £ 10 sh.); all p. mille.

50. SEELIG, DAV., manu., Mannheim. Agt. L. Leoni & Co., 153 Fenchurch-street, London E.C.

Assortment of cigars.

51. MAYER, JONAS, merchant, Heidelberg. London 1862 honor. mention. Agt. No. 26.

Hops grown in the Palatinate.

52. KÜBLER, Burgomaster, St. Ilgen. Agt. No. 26.

Hops grown in the Palatinate.

53. ALBRECHT, J., wine merchant, Freiburg. Agt. No. 26.

2 bottles of cherry brandy.

54. VON BABO, L., Weinheim. London 1862 honor. mention. Agt. No. 26.

3 bottles of wine grown in the neighbourhood of the Bergstrasse.

55. BLANKENHORN, BROTHERS, Müllheim. London 1862 medal and honor. mention. Agt. No. 26.

3 bottles of wine so called Markgräfler, year 1834; 3 do. Riesling picked out, year 1858; 3 do. of red wine, year 1858; 3 do., year 1859; 2 bottles of old cherry brandy; 2 do. of plum brandy, year 1859.

56. BOERSIG, landlord of the eagle inn, Oberkirch. London 1862 medal. Agt. No. 26.

Wines from the Renchthal and cherry brandy.

57. DILGER, O., merchant, Freiburg. London 1862 honor. mention. Agt. No. 26.

4 bottles of cherry brandy finest quality.

58. SEXAUER, Sulzburg. London 1862 medal. Agt. No. 26.

6 bottles of wine called Castelberger of the years 1857, 1859, 1860.

60. FISCHER, F. X., Offenburg. London 1862 medal. Agt. No. 26. (s. II. No. 59.)

3 bottles of cherry brandy 1859; 3 do. of Josephsberg wine; 3 do. of wine called Zeller red wine, Abtsberger Ausstich 1857; 2 do. Durbacher Clevner; 1 do. Durbacher Weissherbst.

61. HANOVER, AB., producer, Schmieheim. London 1862 medal.

Wine and cherry brandy.

62. KUENZER & Co., manu., Freiburg i. B. London 1862 medal.

Badish sparkling wines.

63. LANDFRIED, P. J., KÜNZLE & Co., merchants, Heidelberg. Agt. No. 26. (s. II. No. 30.)

12 bottles of Badish wine.

64. AGRICULTURAL OFFICE, Breisach. London 1862 medal. Agt. No. 26.

20 bottles of white and 4 bottles of red wine called Kaiserstühler.

66. LOTHER, TH. E., Eppingen. Agt. No. 26.

2 bottles of wine.

67. MERK, C., SONS, Bühl. Agt. No. 26.

6 bottles of wine.

68. SCHÜTT, F., Affenthal. London 1862 honor. mention. Agt. No. 26. (s. II. No. 88.)

Wine and cherry brandy.

69. SPITZMÜLLER, J. A., Biberach. London 1862 honor. mention. Agt. No. 26.

3 bottles of cherry brandy, 3 do. vinegar spirit.

CLASS VIII.

MACHINERY IN GENERAL.

74. MÜRRLE, GG. JAC., manu., Pforzheim. Paris 1855 honor. mention; London 1862 medal. Agt. No. 26.

Two steam apparatus for pharmaceutical purposes, one do. for chemical purposes.

74a. KÜHFUSS, taylor, Carlsruhe. London 1862 medal. Agt. No. 26.

Apparatus for firemen called the Carlsruhe fire cap.

CLASS XIII.

PHILOSOPHICAL INSTRUMENTS AND PROCESSES DEPENDING UPON THEIR USE.

75. SICKLER, C., mechanician and optician to the court, Carlsruhe. Munich 1854 honor. mention. Agt. No. 26.

Balance of iron for chemical analysis 60 fl. (5 £); do. of gilt brass 120 fl. (10 £); theodolite with multiplication 250 fl. (20 £ 16 sh. 8 d.); leveling instrument showing the differences of heights by per cent 100 fl. (8 £ 6 sh. 8 d.); do. small size with divided circle 36 fl. (3 £); instrument for measuring rectangles with telescope 18 fl. (1 £ 10 sh.).

CLASS XIV.

PHOTOGRAPHIC APPARATUS AND PHOTOGRAPHY.

76. LORENT, D., Mannheim. Munich 1854 medal of honor; Paris 1855 silver medal; London 1862 medal. Agt. No. 26.

19 photographs of monuments of the Lombardo-Venetian Kingdom.

CLASS XV.

HOROLOGICAL INSTRUMENTS.

77. JOINT-STOCK COMPANY FOR THE MANU-FACTURE OF CLOCKS, Lenzkirch. Munich 1854 great medal; London 1862 medal. Agt. Brugger & Straub, 79 High Holborn.

1 regulator, springs, 14 days, strike, clock-case of oak wood; 2 do. visible anchor movement, case of palisander; 1 do. of oak wood; 1 do. weight, 8 days, case of palisander; 1 do. strike, clock-case of palisander inlaid with metal; 1 do. clock-case of palisander, visible anchor movement; 1 do. springs, 14 days, strike, clock-case of palisander, visible anchor movement; 1 do. clock-case and dial of carved oak wood; 1 table-clock, round, 14 days, strike, case of carved palisander; 1 do. carved oak wood; 1 fancy clock, gilt bronze, 14 days, strike, with pedestal and glass-case, 20 raw clock-works, 24 finished and polished clock-works.

78. BEHA, J. A., manu., Eisenbach. London 1862 honor. mention. Agt. No. 26.

1 cukoo-clock with case and painting 36 fl. (3 £); 1 do. striking the quarters 42 fl. (3 £ 10 sh.); 1 8 days cukoo-clock, springs, clock-case 48 fl. (4 £); 1 do. of oak wood 90 fl.; 1 do. of carved oak wood and paintings 114 fl.

79. BOB, LORENZ, clock-maker, Furtwangen. Munich 1854 medal of honor; Paris 1855 honor. mention; London 1862 medal. Agt. No. 26.

1 regulator movement, going one year, with clock-case of nut-tree wood 200 fl. (16 £ 13 sh. 4 d.); 1 do., going one month, with clock-case of palisander 126 fl. (10 £. 10 sh.); 1 do., going eight days, with a do. case 112 fl. (9 £. 6 sh. 8 d.); 1 do. striking the quarters with a do. case 112 fl. (9 £ 6 sh. 8 d.)

79a. BOB, M., clock-maker, Tryberg. Agt. No. 26.

2 regulators, 2 night-clocks, iron frame, milk-glass dials, 8 days work, 2 do. 30 hours work, iron frame, 1 do. 8 days work, glass frame, 1 do. with pedestal and glass-case, 1 do. 30 hours work, with pedestal and glass-case, 1 travelling clock, silver plated with repetition, 1 do. of bronze, 1 fancy clock, 8 days work, 1 do. 30 hours work, 1 clock for hanging up against the wall, 8 days work, 5 do. 30 hours work.

80. BOB, V., clock-maker, Furtwangen. London 1862 medal. Agt. Beda Spiegelhalter, care of G. Spiegelhalter & Co., 6 Mount place, Whitechapel-road.

Several regulators.

80a. BÜHLER, C. H., clock-maker, Tryberg. Agt. No. 26.

1 night clock, entirely gilt, 8 days work; 2 do. iron, 8 days work; 1 do. bisquit porcellain, 30 hours work; 1 do. iron groupe; 1 pendule, gilt with pedestal and glass-case, 8 days work; 1 clock with crystal frame, black, cross-form, 8 days work; 1 fancy clock with pedestal and glass-case, entirely gilt, 8 days work; 1 do. 30 hours, with dial of porcelain; 1 do. case of gilt earthen-ware; 30 hours work; 5 do. 8 days works; 6 do. 30 hours works.

81. DILGER, O., clock-maker, Tryberg.

33 different Black-forest clocks.

81a. DOLD & HETTICH, manu., Furtwangen. Agt. No. 26.

11 pieces of paintings, with different frames, parts of Black-forest clock works.

82. FURTWÄNGLER, clock-maker, Gütenbach. London 1862 honor. mention. Agt. No. 26.

1 clock-work, going one month, with clock-case and painting 60 fl. (5 £); do. going 8 days 20 fl. (1 £ 13 sh. 4 d.); 1 do. 24 fl. (2 £); 7 clocks going 8 days from 4 to 12 fl. (6 sh. 8 d. to 1 £).

83. HEINE & DILGER, painters and photographers, Neustadt. Agt. No. 26.

4 paintings on zinkplate for clocks, figures in Black-forest costume No. 1. 14 fl. (1 £ 3 sh. 4 d.), No. 2. 8 fl. (13 sh. 4 d.); No. 3. 5 fl. (8 sh. 4 d.), hunting scene No. 4. 4 fl. (6 sh. 8 d.).

84. VON HERZER & STOCKER, clock-makers, Villingen. London 1862 medal. Agt. W. Stocker, 51a Mortimer-street, Cavendish square.

1 regulator, movement going one month, with clock-case of nut-tree wood 160 fl. (13 £ 6 sh. 8 d.); 1 do. of palisander 160 fl. (13 £ 6 sh. 8 d.); 1 do. going eight days with clock-case of palisander, richly inlaid with metal 33 fl. (24 £ 15 sh.); 1 do. with black polished clock-case 24 fl. (2 £).

85. KALTENBACH, L., clock-maker, Furtwangen. Paris 1855 honor. mention; London 1862 honor. mention. Agt. No. 26.

Regulator movement with clock-case 85 fl. (7 £ 1 sh. 8 d.); 2 small ones, going 8 days, 8 and 9 fl. (13 and 15 sh.); regulator, going 8 days, strike, with clock-case 45 fl. (3 £ 15 sh.).

86. KÄMMERER, S., clock-maker, Furtwangen. London 1862 medal. Agt. No. 26.

Clock-work, going 8 days, with clock-case; do. with clock-case to be attached to the wall; clock-work, going 2 days, with small English case.

86a. KETTERER, A., clock-maker, Furtwangen. Agt. No. 26.

Different parts of raw and finished clock works.

87. KÖRNER & HEILBOCK, manu., Villingen. Agt. No. 26.

Wooden clock-cases.

87a. KREUZER, R. & A., manu., Furtwangen. Agt. No. 26.

2 tableaux, samples of letters for firms, signboards etc. in Hyalophany with oval frames, each 10 fl. (16 sh. 8 d.); tableau with the Badish coat of arms, with square frame 12 fl. (1 £); tableau with samples of different writing 12 fl. (1 £); 2 pieces for decoration of clocks, one with a gilt, the other with a brown frame, each 3 fl. (5 sh.).

87b. LAULE, J., painter. Furtwangen. Agt. No. 26.

One 8 days clock work, springs, strike, with paintings 25 fl. (2 £ 1 sh. 8 d.); one small 8 days clock-work, springs, with paintings, frame of mahogany 25 fl. (2 £ 1 sh. 8 d.); one clock, weight, with a carved nut-tree wood case with paintings 30 fl. (2 £ 10 sh.).

88. MARTENS. J. H., & Co., manu., Furtwangen. London 1862 medal. Agt. No. 26.

Pocket chronometer and watches with anchor escapement and single parts of them.

88 a. MAURER, R., manu., Eisenbach. London 1862 honor. mention. Agt. No. 26.

Regulator movement, going one year, with a clock-case of nut-tree wood 200 fl. (16 £ 13 sh. 4 d.); do. going 14 days, springs, strike 48 fl. (4 £); do. going 8 days, weight 36 fl. (3 £); table clock, round, going 14 days, strike 32 fl. (2 £ 13 sh. 4 d.); fancy clock, going 8 days, 36 fl. (3 £).

89. METZGER, sculp., Carlsruhe. London 1862 honor. mention. Agt. No. 26.

Carved clock-case for regulators; 2 do. for Black-forest-clocks; 3 fancy-clocks with carved frames.

91. THOMANN, P., manu., Furtwangen. London 1862 medal. Agt. No. 26.

1 glass-case with an assortment of different clock-springs and rolled steel.

91 a. TRITSCHLER, S., manu., Schollach. London 1862 honor. mention. Agt. No. 26.

One 8 days work, springs, with case of mahogany 18 fl. (1 £ 10 sh.); one do. strike, with a do. case 27 fl. (2 £ 5 sh.).

92. WEHRLE, C., maker of cabinetwork, Neustadt. London 1862 honor. mention. Agt. No. 26.

Small gothic clock-case with a cukoo-clock 30 fl. 48 kr. (2 £ 11 sh. 4 d.); do. of oak wood with a cukoo-clock, springs 43 fl. 48 kr. (3 £ 13 sh); do. of nut-tree wood with a cukoo-clock, springs 34 fl. 48 kr. (2 £ 18 sh.); clock-case of polished mahogany with an 8 days clock-work 36 fl. (3 £).

93. WEHRLE, E., manu., Furtwangen. London 1862 honor. mention. Agt. No. 26.

Trumpet-clock 120 fl. (10 £).

93 a. MANUFACTORY OF MUSICAL CLOCK-WORKS OF THE BADISH BLACK-FOREST, Exhibitor M. WELTE, Vöhrenbach. London 1862 medal. Agt. No. 26.

Grand musical clock-work, called »Orchestrion«.

93 b. WINTERHALTER, CHARLES, manu., Vöhrenbach.

Two automatons.

93 c. SCHULTHEISS, BROTHERS, St. Georgen. London 1862 Med.

Parts of clock works of enamelled copper and iron plate.

CLASS XVI.

MUSICAL INSTRUMENTS.

94. PADEWET, J., musical instrument maker to the court, Carlsruhe. Munich 1854 medal of honor; Paris 1855 honor. mention; London 1862 honor. mention. Agt. No. 26.

Bass-viol 3500 fl. (300 £); 2 violins each 120 fl. (10 £).

CLASS XVIII.

COTTON.

95. HEUSS, FR., manu., Hassmersheim near Mosbach. Agt. No. 26.

Cotton sewing yarn.

98. WOLFF, J., SON, rope-makers, Mannheim. London 1862 honor. mention. Agt. No. 26.

Cotton ropes for cotton spinners and weavers.

99. ZÜRCHER, BROTHERS, manu., Lahr. Munich 1854 medal of honor; London 1862 medal. Agt. No. 26.

Summer and winter waist-coatings.

100. FINGADO, C., rope-maker, Mannheim. London 1862 medal. Agt. No. 26.

4 Ship's-ropes of Badish splitthemp 67 fl. (5 £ 11 sh. 8 d.), 142 fl. (11 £ 16 sh. 8 d.), 61 fl. (5 £ 1 sh. 8 d.) and 335 fl. (27 £ 18 sh. 4 d.).

102. FREI, J., rope-maker, Ettenheim. Agt. No. 26.

Hemp for spinning.

103. HANOVER, AB., producer, Schmieheim. Agt. No. 26.

Hemp for spinning.

104. HAUS, burgomaster, Altfreistett. London 1862 medal. Agt. No. 26.

Split-hemp and hemp for spinning.

104 a. HUTH & Co., producers, Neufreistädt. London 1862 honor. mention.

Samples of hemp.

105. SCHOCH, A. & F., Lichtenau. London 1862 medal. Agt. No. 26.

Raw-hemp and hemp for spinning; hempseed.

106. SCHOLLENBERGER, J., seedsman, Carlsruhe. Agt. No. 26.

Hempseed.

107. WAGNER, G., Emmendingen. London 1862 medal.

Samples of hemp.

CLASS XX.

SILK AND VELVET.

108. MEZ, BROTHERS, manu., Freiburg. London 1862 medal. Agt. Ch. Rumpf, Basinghall Chambers, 79 Basinghall-street, E. C.

Raw silk, double woof, dyed sewing-silk, sewing-machine-silk.

CLASS XXV.

SKINS, FUR, FEATHERS AND HAIR.

110. KAHN, M., SONS, manu., Mannheim. Agt. No. 26.

Samples of cleaned bed-feathers.

CLASS XXVI.

LEATHER, INCLUDING SADDLERY AND HARNESS.

112. HEINTZE & FREUDENBERG, manu., Weinheim. London 1851 bronce medal; Munich 1854 prize medal; Paris 1855 silver medal; London 1862 medal. Agt. No. 26.

Varnished and blackened calf's-skins.

113. KONSTANZER, A., tanner, Villingen. London 1862 honor. mention. Agt. No. 26.

Calf's-skins for coach-roofs, aprons, harness etc.

117. SCHWEICKHARDT & KURZ, manu., Lahr. London 1862 honor. mention. Agt. No. 26.

6 doz. of calf's-skins, different kind, and a few doz. of calf's-leather for shocs and boot legs.

CLASS XXVIII.

PAPER, STATIONERY, PRINTING AND BOOKBINDING.

122. BOHNENBERGER & Co., manu., Pforzheim. Munich 1854 honor. mention; London 1862 honor. mention. Agt. No. 26.

One large case and a small one containing paper, printing and copper-plate printing paper.

CLASS XXX.

FURNITURE AND UPHOLSTERY INCLUDING PAPER-HANGINGS AND PAPIER-MACHÉ.

124. HASSLINGER & Co., upholsterer to the court, Carlsruhe. Munich 1854 honor. mention; London 1862 honor. mention. Agt. No. 26.

Lady's writing table of ebony wood, richly ornamented with ivory, safety-lock 3000 fl. (250 £); chair (fancy) 72 fl. (6 £); chair (medaillon) 72 fl. (6 £).

CLASS XXXI.

IRON AND GENERAL HARDWARE.

126. HELMREICH, MOLL & Co., manu., Mannheim. Munich 1854 medal of honor; London 1862 medal. Agt. No. 26.

Nails, wire-tacks, rivets of iron, brass and copper.

127. SCHULTHEISS, BROTHERS, manu., St. Georgen. London 1862 honor. mention. Agt. No. 26.

Different objects of enamelled copper and sheet iron.

CLASS XXXII.

STEEL AND CUTLERY.

128. LACHMANN, cutler, Rastatt. London 1862 honor. mention. Agt. No. 26.

Cutlery.

CLASS XXXIII.

WORKS IN PRECIOUS METALS, AND THEIR IMITATIONS AND JEWELLERY.

PFORZHEIM JEWELLERS EXHIBITORS:

129a. BECKER, CH. London 1862 medal. Agts. A. Mönch & Co., 26 Lambeth hill, City, London.

Fancy-articles, bracelets, brooches, parures, earrings, medaillons, buttons for shirts and sleeves, chains.

129b. BENCKISER & Co. London 1862 medal. Agts. Vogl, Brothers, 1 Sambrook Court, Basinghall-street.

Bracelets, brooches, parures, ear-rings, pins, medaillons, chains, buttons, fancy-articles.

129c. SCHLESINGER&WEBER. Agt. No.129a.

Fancy-articles, bracelets, brooches, buttons, pins.

129d. GÜLICH, CH. London 1862 medal. Agt. No. 129a.

Bracelets, brooches, buttons, parures, medaillons, fancy-articles.

129e. NÖSGEN, A. Agt. No. 129a.

Fancy-articles, parures, ear-rings, pins.

129f. KIEHNLE, J. London 1862 honor. mention. Agt. No. 129a.

Fancy-articles, bracelets, brooches, buttons, medaillons, pins.

129g. GESCHWIND & Co. London 1862 honor. mention. Agt. No. 129a.

Fancy-articles, parures, medaillons, buttons.

129h. GRUMBACH, F. Agt. No. 129a.

Bracelets, brooches, buttons, pins.

129i. MAJER, G. London 1862 honor. mention. Agt. No. 129a.

Specimens of watch-keys, seals, breloques, medaillons.

129k. KÄMPF & Co. London 1862 medal. Agt. No. 129a.

Fancy-articles, bracelets, brooches, buttons, medaillons, pins, chains.

129m. MÜLLER, G. London 1862 honor. mention. Agt. No. 129a.

Chains and bracelets.

129n. ERHARDT & Co. Agt. No. 129a.

Fancy-articles, bracelets, buttons, brooches, pins.

129o. DILLENIUS & BOHNENBERGER. London 1862 medal. Agt. No. 129a.

Bracelets, brooches, buttons, pins, fancy-articles.

129p. TSCHOPP, FR. London 1862 honor. mention. Agt. No. 129a.

Rings, massy and hollow.

129q. HILLER, J. London 1862 honor. mention. Agt. No. 129a.

Rings, massy and hollow.

129r. BECKER, CH. Agt. No. 129a.

Specimens of all kinds of crosses.

129s. KELLER, H. London 1862 medal. Agt. No. 129a.

Rings, massy and hollow.

129t. ROHRECK, C. E. Agt. No. 129a.

Specimens of watch-keys, seals and breloques.

129u. ROLLER, ADOLF. Agt. No. 129a.

Chains and bracelets.

129v. STEINBRENNER, CH. Agt. No. 129a.

Rings, massy and hollow.

129w. SAACKE, G., & Co. London 1862 medal. Agt. No. 129a.

Parures, bracelets, brooches, buttons, medaillons, pins, buttons.

CLASS XXXIV.

GLASS.

133. MANNHEIM LOOKING GLASS MANUFACTORY, BRANCH OF THE JOINT STOCK COMPANY OF THE MANUFACTORIES OF ST. GOBEIN, CHAUNY & CIREY. The frames made by Raucourt & Co., Mannheim. London 1862 honor. mention. Agt. Grand-Ry, St. Pauls Wharf, 25 Upper Thames-street.

1 looking-glass, cut with facets, in a gold-frame and 1 plate glass in a gold-frame, each 800 fl. (68 £ 11 sh. 2 d.).

KINGDOM OF BAVARIA.

Apply to the R. B. Commissioner, Dr. Beeg for those Bavarian Exhibitors, who have no Agents particularly named.

CLASS I.

MINING, QUARRYING, METALLURGY AND MINERAL PRODUCTS.

141. FRÄNKEL, L. H., owner of mines, Massbach (Lower-Franconia).

Brown-coal, refractory clay, clay for Portland-cement.

142. KISSINGEN, R. B. establishment for the sale of Mineral-waters.

Natural Mineral-waters.

From Kissingen: Ragoczy, Pandur, Maxbrunnen, Bitterwater;

from Bocklet: Chalybeate-water;

from Bruckenau: Vernatzer, Sinnberger and Chalybeate-water.

Prices: 100 large stone-bottles 15 fl., 100 half stone-bottles 11 fl. 30 kr.

For transports to a great distance in glass-bottles: 100 large bottles 21 fl., 100 half bottles 18 fl.

Sold in London at A. Garden and Best & Son.

143. KLINGENBERG, municipality in Lower-Franconia. Agt. F. C. Claudius. 3 St. Helen's place, London.

Refractory clay, from the clay-pits belonging to the municipality. Price: 54 kr. (18 d.) p. 50 Kilo, on the banks of the Main at Klingenberg.

144. KROHER, ADOLPHUS, manu., Staudach (Upper-Bavaria).

Cement, roofing-plates and paving-stones of Cement.

145. MARTIUS, Dr. TH. W. C., Professor of Pharmacy and Pharmacognosy in the University of Erlangen.

Phosphorite, Apatite, Native Phosphate of lime from Amberg (Bavaria), discovered as detritus 1816 and found by the Exhibitor in large quantities 1855. Piece of 320 pounds weight.

146. SOLENHOFEN, joint-stock company of Solenhofen (Middle-Franconia). London 1862 medal. (s. II. No. 22.)

Lithographic lime-stones from the quarries of the company.

CLASS II.

CHEMICAL SUBSTANCES AND PRODUCTS AND PHARMACEUTICAL PROCESSES.

148. ADAM, J. M., manu., Rennweg near Nuremberg. London 1862 medal.

Ultramarine, Parisian blue, Prussiate of potash.

149a. GRAF & Co., manufactory of chemical products of gas-tar, Nuremberg. London 1862 honor. mention.

Benzol 100 fl., Benzin 28 fl., Tar-oil 26 fl., Camphin-oil for lighting 30 fl., Photogen do. 27 fl., Gas-ether do. 28 fl., Creosote, pure 120 fl., all p. 50 Kilo.

Creosote-oil for wood-preservation 3 fl., Naphthaline for caoutchuk manufactories 5 fl., both p. pound.

Asphaltum-lac or Nuremberg iron-lac 20 fl., Asphaltum-lac or Parisian metal-lac 25 fl., do. white for fine objects 80 fl., all p. 50 Kilo.

Nitrobenzol, chem. pure 2 fl. 42 kr., Nitro-Mirban-oil for perfumery 2 fl. 30 kr., Picrin-Acid, purest I a. 7 fl., Anilin, purest, for manufactories of colors for preparation of red, violet and blue Anilin colors 4 fl. 45 kr., Anilin sulphuric 4 fl. 40 kr., Anilin-red, violet, blue; Chloroform 1 fl. 45 kr., all p. pound.

149b. GROSBERGER & KURZ, manu., Nuremberg. London 1862 medal.

Yeast cake.

150. HEUFELD, Bavarian joint-stock-company for the manufactory of chemical and chemico-agricultural products of (established 1861) Heufeld (Station on the Munich-Salzburg-railroad). London 1862 medal. Agt. C. F. Claudius, 3 St. Helen's place, London.

Soda 85°. do. 98° Sulphate. Steamed ground bone. Super-phosphate of lime. Concentrated manure.

151. HOFFMANN, G., manu., proprietor of manufactories, Schweinfurt, Euerbach and Aura on the Saale. Schweinfurt (Lower-Franconia). Munich 1854 medal of honor; London 1862 medal. Agt. C. F. Claudius, 3 St. Helen's place, London.

Different sorts of ink for copper-printing, white-lead, Cremserweiss, oxydes, ochre from the raw state till the finest preparation, umber and other chemically prepared colors (Chrom-yellow, mineral-yellow, new-yellow etc.).

152. Kaiserslautern, Ultramarine-manufactory at (Palatinate), Director Dr. Wilkens. London 1862 medal. Agt. William Staats, Dowgate-Hill Chambers, 38 Dowgate-Hill, London. (s. II. No. 94.)

2 groups of raw Ultramarine, 24 sorts of pure and mixed Ultramarine, 4 sorts of blueing-balls of Ultramarine.

153. Lichtenberger, Carl, manufactory of argol and cognac-oil, Hambach near Neustadt on the Haardt (Palatinate). London 1862 medal. Agt. F. M. Pokorny, 15 Fish-Street Hill, City, London. (s. II. No. 1.)

1. Cognac-oil prime 5 £ 17 sh., 2. do. second 3 £ 15 sh., 3. do. rough 3 £ 7 sh., 4. Oenanth-acid, rectified 16 £ 14 sh., 5. do. rough 12 £ 10 sh. all p. ½ Kilo. 6. Wine-lees brandy 37 £ 10 sh., 7. Cognac-spirit 66 £ 14 sh. each p. 1000 litres, 8. Argol from the lees of wine (90 pCt. cont.) 5 £, 9. Tartar of lime (cont. 96 pCt.) 4 £ 4 sh., each p. 50 Kilo.

154. Meyer, H. (F. Mittler), manu. Augsburg. London 1862 honor. mention. Agts. G. Ingelbach & Wolffgang, London. (s. II. No. 47.)

Chromic acid-green, called: »Mittler's giftfreies Grün« (color free from poison) Patent; Price 20 fl. and more the 50 Kilo. Mittlers-green colored window-blinds, papers etc.

155. Martius, Dr. Th. W. C., Professor in the University at Erlangen. (s. No. 145.)

Zanzibar, Mogadore, Angola, Benguela, West-indian and Manilla-Copals solved in alcohol of 0,803 or spirit of 0,839 after a new, easy and simple process. As specimen: pieces of nut-wood polished with such Copal-solutions; Westindian Copal-solution as bookbinder's varnish.

156. Rösch, Friedrich, manu., Nuremberg. (s. II. No. 17.)

Nuremberg-Deckweiss (white color that covers well) for manufactories of paper-hangings, and to be mixed with all other covering colors. p. 50 Kilo 3 fl. free Nuremberg.

157. Sattler, W., manu., Schweinfurt. Berlin 1844 gold medal; London 1851 bronz. medal; Munich 1854 medal of honor; Paris 1855 bronz. medal; London 1862 medal. Agt. H. Kohnstamm, 39 Dowgate-Hill, Cannon-street, London. (s. II. No. 67.)

White-lead, black colors, earth and mixed colors for paper hangings, paper, for decoration-, fresco- and artistical painting; for lac- and oil-varnishes of every description; ink for typographic and lithographic printing.

A color-grinding-mill; Price: 18 fl.

158. Stollreither, C. P., Munich.

Blue color of the Exhibitor's own invention (substitute for Indigo).

159. Toussaint, G. F., manu., Fürth. London 1862 honor. mention.

Cyanide of potassium in cubes 130 fl., do. in bars 134 fl., Acid. borac. 76 fl., dextrine 15 fl., all p. 50 Kilo.

CLASS III.

SUBSTANCES USED FOR FOOD, INCLUDING WINES.

160. Barth, Steph., & Co., proprietors of vine-yards, and wine-merchants, Würzburg. London 1862 medal.

Fine Franconian wines from the most celebrated Franconian vine-yards: Stein and Leisten-wines 1811. 1822. 1846. 1857. 1858. 1659. Straw-wine 1857. Frozen Stein 1859. Wines from the Hill-sites: Harfe, Schalksberg, Pfülben etc. Sparkling Franconian wines of the manufactory of A. Kuhn in Würzburg: finest red Leisten-Mousseux, Mousseux Muscatel, Mousseux white.

The Exhibitor St. Barth will be present in London during the time of the Exhib.

161. Biffar, Andreas & Adam (Rheinische Früchte-Handlung), Deidesheim (Palatinate). Munich 1854 I. prize-medal; London 1862 medal.

Preserved Mirabelles, I. Qual. largest sort p. Kilo 3 sh.

Almonds; Nuts, white and black; Quince, red and yellow; Pears, red and white, all p. Kilo 2¾ sh.

Cherries, agriots, light-red; Apricots, whole, without stone; do. in halves; Peaches, all p. Kilo 3⅓ sh.

Glazed fruits ⅓ sh. p. Kilo more.

Elegant boxes with fruits, glazed or crystallised 1/1: 2 sh., ½: 1⅙ sh., ⅓: ⅔ sh.

Mixed fruits in brandy or sugar ½ Kilo 1⅙ sh., ¾ Kilo 1⅔ sh.

Dried fruits: Apples, pears, plums, cherries, mirabelles, reineclaudes and brunelles.

162. Doering, Ferdinand, manu. and merchant, Würzburg. Munich 1854 medal of honor. Agt. J. E. Guerra & Co., 14 America-Square, London.

German sparkling Wines. 4 different qualities, i. e. sorts, in whole and half bottles.

163. Eichhorn & Co., manufactory of Cigars. (Eichhorn & Blumröder), Speyer (Palatinate). London 1862 medal.

Cigars in original Habanese cedar-chests, manufactured for exportation: small Media, Regalia, Britannia, Rio del Norte, Exp.-cig. 30 fl., large Media Regalia, la Resolucion 34 fl., large Regalia A Fernandez 40 fl., very large Imperiales, Flor de Cabaños 50 fl., all p. mille.

164. Häutle, Theodor, manu., Munich. London 1862 medal.

Samples of Chocolate.

165. Leuchs, Joh. Carl (C. Leuchs & Co.), proprietor of an establishment for selling new inventions, Nuremberg. (s. No. 208.)

»Wine made of Water« prepared without grapes or other fruits and plants, without lees by a substance with permanent fermentation-power, more durable than wine from grapes. Price of the preparation 1—2 d. the bottle. Invention of the Exhibitor.

166. NEUBAUER, JACOB, manu., Winzingen (Palatinate). London 1862 medal.

Different sorts of Starch.

167a. OPPMANN, M., inspector of the royal wine-cellars, Würzburg. London 1862 medal.

From the royal wine-cellars:

Wines from the most celebrated Bavarian vineyards: 1. 1857. Steinwein, 2. 1857. Leisten, 3. 1858. Steinwein, 4. 1858. Leisten, 5. 1857. Hörsteiner.

167b. OPPMANN, M., inspector of the royal wine-cellars, Würzburg. London 1862 medal.

Sparkling Franconian wines from the Exhibitor's own manufactory.

CLASS IV.

ANIMAL AND VEGETABLE SUBSTANCES USED IN MANUFACTURES.

168. HECKEL, GABRIEL, manu., Munich.

Assortments of buttons of bone.

169. ROTH, municipality of, district of Spalt (Middle-Franconia). London 1862 medal.
(s. II. No. 44.)

Hops, not cured, of light green color.

170a. UHLMANN, S., merchant exporter, Fürth. London 1862 medal.   (s. II. No. 11.)

Samples of cured Bavarian hops.

170b. SECKENDORF, LE-VINO & Co., exporters, Nuremberg. London 1862 medal. Agt. L. U. Tuchmann, 9 Savage-Garden, E. C. London.

Uncured and cured hops.

CLASS V.

RAILWAY PLANT, INCLUDING LOCOMOTIVE ENGINES AND CARRIAGES.

171. KOLB, LUDWIG, manu., Bayreuth. Munich 1854 medal of honor; London 1862 honor. mention. Agt. C. F. Claudius, 3 St. Helen's place, London.

Elastic, waterproof tilt for railway-waggons. Samples of the raw-material.

CLASS VII.

MANUFACTURING MACHINES AND TOOLS.

172. DINGLER, JULIUS, & WOLFF, J. B. (Dingler'sche Maschinen-Fabrik), Zweibrücken (Palatinate). London 1862 medal.

Printing-press No. I. 450 fl., do. No. IV. 210 fl., Embossing-press No. I. 750 fl.

173. HEINTZ, DR. CARL FRIEDR. VON, Counsellor of State etc., Munich. London 1862 honor. mention.

Printing-machine, called »Tachytype« by which common letter-print may be performed without moveable types, after a new system, invented by the Exhibitor.

CLASS VIII.

MACHINERY IN GENERAL.

175. KLINGENFELD, C. A., Professor at the royal Polytechnical institution, Nuremberg. London 1862 honor. mention.

Nuremberg Patent weighing-scales. Invented by the Exhibitor.

176. PFANZEDER, GEORG, mechanician Munich.

Scale-beam on four edges, of particular construction. Patented invention of the Exhibitor.

CLASS IX.

AGRICULTURAL AND HORTICULTURAL MACHINES AND IMPLEMENTS.

177. DANZER, B., manu., Munich. London 1862 honor. mention.

Apparatus for watering delicate plants and for destroying pucerons.

CLASS X.

CIVIL ENGINEERING, ARCHITECTURAL AND BUILDING CONTRIVANCES.

179. ECKHARDT, P., Brick-factory. Grosshesselohe near Munich. London 1862 honor. mention.

Paving-stones of the Exhib. own invention.

180. KLETT & Co., Ironworks and manufactory of miscellaneous engines, railway-waggons, bridges etc., Nuremberg. London 1862 medal.   (s. II. No. 20.)

Model of a part of the railway-bridge across the Rhein at Mayence, constructed after the system of von Pauli. $\frac{1}{10}$ of the natural size.

181. TÖLZER, JOSEPH, carpenter, Tegernsee (Upper-Bavaria). London 1862 honor. mention.

Designs of wooden farm-houses, barns, stables etc. in the Bavarian highlands. Price of the original work consisting of 24 sheets, 30 £.

CLASS XI.

MILITARY ENGINEERING, ARMOUR AND ACCOUTREMENTS, ORDNANCE AND SMALL ARMS.

182. UTENDÖRFFER, H., manu., contractor to the royal Bavarian army, Nuremberg. London 1862 medal. Agt. A. G. Franklin & Co., 14 South-street, Finsbury-Square, London E. C.

Rich assortment of Percussion-Caps for every description of fire-arms.

CLASS XIII.

PHYLOSOPHICAL INSTRUMENTS AND PROCESSES DEPENDING UPON THEIR USE.

184. HAFF, BROTHERS, manu., Pfronten (province of Suavia). Munich 1854 I. prize-medal; London 1862 medal.

Cases of mathematical instruments.

186a. RIEFLER, CLEMENS, mechanician, Maria-Rhein (Suavia). London 1862 medal.

Case of mathematical instruments of German silver, with improved movement of the heads of the compasses 84 fl. (48 Thlr., 7 £.)

186b. RODLER, A., manu., Nuremberg.

Coal-cylinders and coal-plates for electro-galvanic apparatuses: Coal-cylinder No. 1. 4 fl. 48 kr., do. No. 2. 4 fl. 48 kr., do. No. 3. 3 fl. 36 kr., do. No. 4. 3 fl., do. No. 5. 2 fl. 24 kr., do. No. 6. 2 fl. 24 kr., do. No. 7. 1 fl. 48 kr., all p. doz.
Coal-plates 3 fl. 36 kr. p. doz.
Two coal-cylinders of particular electro-motoric power.

CLASS XIV.

PHOTOGRAPHIC APPARATUS AND PHOTOGRAPHY.

188. ALBERT, J., photographer to the court, Munich. London 1862 medal.

Photographic portraits size of life; reproductions of works of art.

189. GYPEN & FRISCH, publishers of religious works and prints. London 1862 honor. mention. (s. No. 222.)

Photographs of modern religious sculptural works; reproduction of original cartoons of modern artists.

CLASS XV.

HOROLOGICAL INSTRUMENTS.

190. MANNHARDT, J., manufactory, Munich 1854 I. prize-medal for engines, medal of honor. (II. prize-med.) for clocks; Paris 1855 I. and II. prize-medal for engines and clocks.

A clock-work of a novel, most simple kind, 8 days rate. Clock, striking quarters and hours, 8 days rate. Clock-work the rate not audible, without oiling, 8 days rate.
Model of a planing-machine with turning chisel.
Model of a grooving plane-machine with movable table.

CLASS XVI.

MUSICAL INSTRUMENTS.

191a. BÖHM, THEOBALD, musician, Munich. London 1851 council-medal; Munich 1854 I. prize-medal; Paris 1855 grande médaille d'honneur.

Wind-instruments of novel invention and construction.

191b. GSCHWENDTNER, manu., Oberstdorf near Lindau.

Sounding-boards.

192. HASELWANDER, JOHANN, instrument-maker, Munich. London 1862 medal.

Citherns.

193. HENTSCH, JACOB. manu., Lindberg (Lower-Bavaria). Paris 1855 bronz. medal; London 1862 medal.

Sounding-boards: 18 pieces of Bellywood, 2 do. Keywood, 2 packets of long Scaleboards, 1 do. short, 2 packets Splints, cornered, 4 do., round.

195. PFAFF, MICHAEL, wind-instrument-maker, Kaiserslautern (Palatinate). London 1851 honor. mention; Munich 1854 honor. mention; London 1862 medal.

Wind-instruments: 1 bombardoon (contra-base) in C, with 4 rotation-cylinders and B-piece 150 fl. 1 baryton-horn in B., with 4 rotation-cylinders with tune change to F 85 fl. 1 alto-horn in C, with 3 rotation-cylinders with tune-change to B 50 fl. 1 Flügelhorn in C, with 3 rotation-cylinders with tune-change to B 42 fl. 1 trumpet in C, with 3 rotation-cylinders and B-, A-, As- and G-piece 38 fl. 1 horn in G, with 3 rotation-cylinders and F-, E-, Es-, and D-piece 80 fl.

196. STEGMAIER, FERDINAND, wind-instrument-maker, Ingolstadt (Upper-Bavaria). Munich 1854 honor. mention; London 1862 honor. mention.

Wind-instruments:	fl.	frn.	sh.
Es-cornet with D-rivets and B-piece	40	85	69
C-Flügelhorn with B-piece	44	94	75
C-trumpet with B-, A-, As-piece	38	82	65
C-trumpet, base, with B-piece	42	90	72
F-alto-horn, with tune-change to Es and D	48	103	82
C-alto-horn with tune-change to B and A	52	111	89
B-baryton with 4 pieces	77	165	132
F-tenor-horn with tune-change to Es	77	165	132
F-bombardoon with 4 pieces	110	235	189
C-contra-base with 4 pieces	160	342	274

CLASS XVII.

SURGICAL INSTRUMENTS.

198. WOLFFMÜLLER, ALOIS, manu. Munich. Munich 1854 medal of honor; London 1862 medal.

Two pharmaceutical steam-apparatuses and utensils.

CLASS XVIII.

COTTON.

201. SALCHER, THOMAS, manu. Passau (Lower-Bavaria).

1 piece 39 Berlin ells, blue-white Penelope-canvas (for embroidery) ⅔ broad, No. 6. prime à 4¼ sg. 5 Th. 15¾ sg. 1 do. à 4¼ sg. 5 Th. 15¾ sg. 1 piece 39 Berlin ells Penelope No. 0. prime (4 corded cotton) à 4¼ sg. 5 Th. 15¾ sg. 1 piece Penelope for marking ⅓ ell in ☐ No. 5 1¼ sg. 1 do. ½ ell in ☐ 1½ sg.

CLASS XIX.

FLAX AND HEMP.

202. TEGELER, E., director of the »Company for linen-thread-manufacture«, Otterberg (Palatinate). London 1862 medal.

Linen:
White two-cord linen-thread as it comes from bleaching No. 30. 50. 60. 80. 100; each 8 pounds.
White two-cord linen-thread, ready packed for sale. No. 30. 50 60. 80. 100.; each 1 pound.
Black two-cord linen-thread half ready. No. 22. 30. 40. 60. 70.: each 8 pounds
Black two-cord linen-thread ready packed for

sale. No. 18. 20. 22. 25. 30. 35. 40. 50. 70. 80.; each 1 pound.

Prices of the exhibited linen-sewing-threads, First quality in English shillings and pence p. dozen, 1 dozen at 12 pounds Engl. W. with 10 pCt. discount on cash payment.

No.	18.	20.	22.	25.	30.	35.	40.
black at	22.9	23.6	24.3	25	25.6	30	35
white at	28	29	30	31.6	34	37	40

No.	50.	60.	70.	80.	100.	120.	140.
black at	42	49	56	66	78	—	—
white at	48	56	65	74	84	97	110

CLASS XX.

SILK AND VELVET.

203. Escales, Brothers, manu., Zweibrücken (Palatinate). Munich 1854 honor. mention; Paris 1855 honor. mention; London 1862 honor. mention.

Hat-plushes:

					frs.
Metres 5. 10	head-plush,	French breadth	4¼'''	at	12
» 5. 15	do.	» »	4¼	»	11
» 5. 10	do.	» »	4¼	»	10
» 5. 10	do.	» »	4½	»	8
» 5. 20	do.	» »	4¾	»	7
» 3. 60	do.	» »	4¾	»	5
» 5. 30	do.	» »	5	»	4½
» 5. —	rim-plush,	German »	3¼	»	8
» 5. 10	do.	» »	3½	»	6

204. Ritter & Thiel, manu., Kaiserslautern (Palatinate). London 1862 honor. mention. Agts. A. Heintzmann & Rochussen, 9 Friday-street, London.

	p. Stab of 115 cm. fl. kr.	p. metre of 1000 mm. frs. c.	p. yard of 914 mm. sh. d.
1 Piece of black Taffety Lit. F.	2 30	4 70	3 4
1 » idem » E.	2 40	5 —	3 7
1 » idem » D.	2 50	5 30	3 10
1 » idem » C.	3 —	5 60	4 —
1 » idem » B.	3 10	6 —	4 3
1 » idem » A.	3 20	6 30	4 6

82 cm. br. — Neat prices.

CLASS XXI.

WOOLLEN AND WORSTED, INCLUDING MIXED FABRICS GENERALLY.

206. Georg, Jacob, manu., St. Lamprecht (Palatinate). London 1862 honor. mention.

Broad-cloth and buckskins.

CLASS XXIII.

WOVEN, SPUN, FELTED AND LAID FABRICS, WHEN SHOWN AS SPECIMENS OF PRINTING OR DYEING.

208. Leuchs, Georg (C. Leuchs & Co.) establishment for the sale of new inventions, Nuremberg. (s. No. 165.)

Turkey-red cotton-cloth, dyed after the invention of the exhibitor, without oil and saving half of the costs and ¾ of the labour.

CLASS XXIV.

TAPESTRY, LACE AND EMBROIDERY.

209. Wünsch, J. B., manu. of spangles, gold and silver-embroideries etc., Nuremberg. Agt. A. Boden & Co., 33 Aldermanbury, London.

Gold- and silver-embroideries. The English arms 48 fl. p. piece.

No.	5. kr.	6. kr.	37. kr.	49. kr.	52. kr.	87½ kr.	99. fl. kr.	239. fl. kr.
Silver:	15	16	32	30	36	30	1 42	— 40
Gold:	22	24	48	45	54	45	2 24	1 —

p. Brabant. ell.

No.	186. kr.	191. kr.	194. kr.	196. kr.	198. kr.	199. kr.	248. kr.	280. fl. kr.	282. kr.
Silver:	16	12	18	18	14	15	9	2 48	28
Gold:	24	18	27	27	21	22	14	4 12	42

p. dozen.

No.	283. fl. kr.	284. fl. kr.	303. kr.	303½ kr.	420. fl. kr.	421. kr.	421½ kr.
Silver:	1 20	1 20	18	16	— 42	20	18
Gold:	2 —	2 —	27	24	1 3	30	27

p. dozen.

Ecclesiastical embroideries:

No.	328¼. fl.	328½. fl. kr.	329¼. fl. kr.	329½. fl. kr.	330¼. fl. kr.	330½. fl.	369. fl.
Silver:	2	3 —	3 —	9 —	3 —	12	7
Gold:	3	4 30	4 30	13 30	4 30	18	10

p. dozen.

CLASS XXVI.

LEATHER, INCLUDING SADDLERY AND HARNESS.

210. Mayer, Ignaz, leather-manufactory (Baron J. Eichthal's heirs), Munich. Berlin 1844 silver-medal; London 1851 honor. mention; Munich 1854 I. prize-medal; Paris 1855 silver-medal; London 1862 medal. Agt. William Baither, 47 Mark-Lane, City, London.

Leather: 4 coach-hides, 2 cow-hides, 2 horse-hides, 12 calf-skins for hunting-boots, 24 for shoes (patent calf), 6 calf-skins, light, 6 calf-skins heavy, each black varnished, 12 calf-skins japanned in various colors, 6 bronze japanned calf-skins, 2/2 black harness hides, 2/2 brown stirrup hides (backs), 12 curried calf-skins (russet calf), 12 waxed calf-skins.

211. Schwarzmann, Fr. X., leather-manufactory, Munich. London 1862 medal.

Leather of different tanning and dressing.

CLASS XXVII.

ARTICLES OF CLOTHING.

212. Greider, Georg, shoemaker, Tegernsee (Upper-Bavaria). London 1862 honor. mention.

Boots and shoes for Gentlemen: mountain-shoes 8 Th., Steigeisen to them 2 Th., hunting shoes for the plain 8 Th., elegant boots 7 Th., boots with side-lacing 7 Th.; for Ladies: mountain-boots 6 Th., high elegant leather boots 7 Th., boots 7 Th., boots of black satin with elastique and embroidery 6 Th.

213. Herbig, Carl, last-manu., Kaiserslautern (Palatinate). London 1862 medal.

Lasts: Boot-tree with mecanique 6 fl. 30 kr., do. common sort p. piece 2 fl. 30 kr., 3 pair of Gentleman's lasts, white-beech 6 fl. 48 kr., 3 pair do. red-beech 4 fl. 48 kr., each pr. doz. pair; 1 pair do. in the rough state, as they were cut after one pattern by the machine, constructed after the system of Beilich at Kaiserslautern, red-beech

3 fl. 48 kr.; 1 pair of lasts for enlarging, red-beech 4 fl., 1 pair do. white-beech 4 fl. 48 kr., all p. doz. pair; 3 pair of Ladies lasts, red-beech 2 fl., 3 pair do. white-beech 2 fl. 24 kr., each p. doz. pieces.

CLASS XXVIII.
PAPER, STATIONERY, PRINTING AND BOOKBINDING.

214. BECKER, AUGUST, publisher, Munich. London 1862 medal.

Lithographic prints in oil-colors without retouche, No. 1—4. The four seasons of the year, after C. Gugel; 5. Parting from the parental house after Eberhardt; 6. Wooing, after Mozet; 7. Beggars, after Murillo; 8. Evening, after Mozet; 9. Morning, after Mozet; 10. Madonna, after Murillo. Prices: No. 1—9. 18 sh. a piece without the goldframe. No. 10. 13 sh. 6 d. without the goldframe.

215a. BEISSBARTH, J. C. SON, manu., Nuremberg. Munich 1854 medal of honor; Paris 1855 silver medal; London 1862 medal.
(s. II. No. 52.)

Painters-brushes of every kind.

215b. R. B. GENERAL-BERGWERKS- AND SALINEN-ADMINISTRATION, Munich. London 1862 medal.

Geognostic description of the Bavarian Alps and lower Alps with two geognostic maps.

216. BEROLZHEIMER & ILLFELDER, lead-pencil-manufactory (established 1855), Fürth. London 1862 medal.

Lead-Pencils of all sorts.

217. BROWN & SCHNEIDER, publishers and proprietors of an establishment for xylography, Munich. London 1862 medal. Agt. H. Bender, 8 little Newport-street, Leicester-square.

1. The Munich pictorial sheets No. 1—312, black-print in loose sheets. 2. The same, colored, in loose sheets. 3. The same No. 1—312, 1—13 volumes, black print, paste-board-binding. 4. The same, colored, paste-board-binding. 5. The Munich Pictorial-books No. 1—12., beautifully colored and handsomely bound. (NB. The fornamed are continued to be published annually with 24 or more sheets.) — Christian catholic-pictures, after original drawings, 72 sheets, in 2 editions, black print and colored; Mr. Petermann's sporting-book, 3 volumes bound in 1 volume; Songs of the Bavarian Highlands, with their melodies. Published on particular request and with the munifizent support of His Majesty, the King, for the Bavarian mountain-people. Collected and edited by Fr. v. Kobell, with pictures by A. v. Romberg. A Natural History, in pictures, edited by Dr. A. Reinsch, with drawings by E. Froehlich. In two editions: black print and colored. A guide for the higher art of cooking etc., by J. Rottenhoefer; with drawings by E. Doepler. The Munich collection of Patterns for Artists etc, edited by L. Wind. Instruction for angling and flyfishing by W. Bischoff; with drawing. Pictures (12 of them) selected from the lives of Bavarian Sovereigns; executed at the express command of His Majesty, Maximilian II, King of Bavaria.

219. DESSAUER, ALOIS, stained-paper and glue-manufactory, Aschaffenburg. London 1862 medal.
(s. II. No. 50.)

Stained paper of every description.

220. FABER, A. W., lead-pencil-manufactory (established 1761), Stein near Nuremberg. London 1851 prize-med.; Munich 1854 I. prize-med.; Paris 1855 prize-medal; London 1862 medal. Agts. Heintzmann & Rochussen, 9 Friday-street, E. C. London.
(s. II. No. 53.)

Lead-pencils in gross and doz. colored pencils, black, white and red-chalk-pencils; slate-, ever-pointed- and artists-pencils; boxes with lead-pencils. Colored and everpointed pencils; drawing-requisites; new lead-pencils made of Siberian lead; boxes of the same etc.

221. GROSSBERGER & KURZ, lead-pencil-manufactory (established 1854), Nuremberg. London 1862 medal. Agt. C. E. Elliott, 5 Aldermanbury, London E. C.
(s. II. No. 45.)

Lead-pencils, colored oil-chalk-pencils (Creta laevis, Creta polycolor), red-pencils, crayons etc. of all sorts and kinds.

222. GYPEN & FRISCH, publishers of religious Works of Art, Munich.
(s. No. 189.)

6 copper-engravings, representing the life of St. Bonifacius, after the compositions of Heinrich Hess; dedicated to His Eminency, the Cardinal Wisemann in London. A series of 4 sheets 24 fl. (2 £). — Steel-engravings: Religious pictures of different kinds. — Lithographic oil-color-print: religious pictures.

223. HÄNLE, LEO, manu., Munich. London 1851 honor. mention; Munich 1854 medal of honor for gold and silver-paper; London 1862 medal. Store in Commission 93 Gt. Tower-street.

1 card with 47 sorts of gold- and silver-papers, real and imitation. These papers can be supplied varnished to protect them from the influence of the atmosphere.
1 patternbook with plain and perforated gold- and silver-borders, frames, corners, and other ornaments both real and imitation. These may also be protected against tarnishing by a varnish.

224. KATHAN, PETER, manu., Augsburg. London 1862 honor. mention.

Assortment of plain and perforated gold- and silver-papers, real and imitated.

225. KIMMEL, J. G. (C. G. Röser), manu. and merchant, Nuremberg. Berlin 1844 prize-medal; Munich 1854 medal of honor; London 1862 honor. mention. Agt. Frederick Rudolph, 188 Gresham-House, Old Broad-street, London E. C.

Real and half-real gold- and silver-borders, ornaments, lacquered colored paper-borders and ornaments resembling leather; relievoes real and half-real gold- and silver-paper, large cartoons and stained papers.

226. KITZINGER. J. G., artistical establishment for oil-color-printing (Patent), Munich. London 1862 honor. mention.

The transfiguration, after Raphael, oil-color-print on painters-canvas.

228. Mozet, Joseph, painter, Munich. Oil-color-prints. London 1862 honor. mention.

Neapolitan fisher-family, after A. Riedel in Rom. Departure and return; after original-pictures painted by the Exhibitor.

230. Städtler, J. G., pastel-and lead-pencil-manufactory (established 1784), Nuremberg. Munich 1854 medal of honor; London 1862 medal.   (s. No. 280 a.)

Pastels in 100 colors 8 fl., lead-pencils from 30 kr. to 9 fl., black-chalk, 4 cornered in cases 1½ fl. all p. gross.

231. Stern, Wilhelm, manu., Fürth. Stained-papers of different sorts.

232. Sussner, G. W., manu., Nuremberg. (H. Meusel jun. and L. Lammers). London 1862 medal. Agt. William Staats, Dowgate-Hill Chambers, 38, Dowgate-Hill, Cannon-street E. C. London.   (s. II. No. 33. 34.)

Rich assortment of Creta polycolor (colored oil-chalk) and crayons mécaniques, open, on cards, or packed in cases, drawings, performed with Creta polycolors, and color-sample-cards.

233. Topographic bureau of the General-quarter-master-staff of the R. B. army Munich. Munich 1854 I. prize-medal; London 1862 medal.

40 sheets of the topographic map of Bavaria.

CLASS XXIX.

EDUCATIONAL WORKS AND APPLIANCES.

235. Munich Association for industrial improvements, Munich. London 1862 medal.

Publications of the association 1851—1862; collection of drawings showing the methodical way used in the association's drawing-school.

236. Guembel, C. W., Royal inspector of mines. London 1862 medal.

Geological description of the Bavarian Alpes.

237. Müller, J., engraver and manu. of stencils, Nuremberg. London 1862 honor. mention.

New stencil-painting-plays, p. doz. cases of different size 5—25 fl.

238. Neussner, Lorenz, optician, Nuremberg. London 1862 honor. mention.

Large laterna magica with representations of »Schiller's Bell« in 49 pictures on 17 glasses 60 fl.; improved laterna magica with 12 glasses and 3 chromatropes 6 fl.

CLASS XXX.

FURNITURE AND UPHOLSTERY, INCLUDING PAPER-HANGINGS AND PAPIER-MACHÉ.

240a. Cnopf, Paul, merchant and manu. Nuremberg.

Articles of papier-maché.

240b. Gremser, Fr., sculp., Augsburg. London 1862 medal.

High-altar of the church at Blaubeuren, model, carved in ivory.

241a. Kübler, Brothers, manu., Munich. London 1862 medal.

Inlaid floor of differents kinds of wood.

241b. R. Kunstgewerbe-school, Nuremberg. London 1862 medal.

The show boxes of No. 182. 220. 221. 230. 232. suggested by professors of the establishment and in part executed there.

242a. Löhner, Johann, manu., Nuremberg. Cabinet-work, boxes and fancy-articles.

242b. Scheidig, Lorence, gilder, Furth. London 1862 honor. mention.

Looking glass- and picture-frames.

243. Trimborn, Christ., manu., Nuremberg. London 1862 honor. mention.

Articles of papier-mache; samples of lacquered and waxed child- and Lady-dolls-heads, child- and Lady-dolls-heads with glass-eyes, child and Lady-dolls-heads with dressed natural hair, doll-shapes with lacquered child-heads; noses; dominoes in paper, linen, velvet, silk; linen masks; rocking-horses, childs-helmets; figures of saints; models for dress-stores; figures of different sorts; heads of animals after nature, do. the same en miniature.

244. Zinn, Sam. & Thurnauer, M. (Sam. Zinn & Co.), manu., Redwitz (Upper-Franconia). London 1862 medal.

Fine wicker-wares (working-, fruit-, flower-, baskets etc.).

CLASS XXXI.

IRON AND GENERAL HARDWARE.

245. Bauer, Adolph (Joh. Paul Ammon), manufactory of gold- and silver-wire and lametta, Nuremberg. London 1851 honor. mention; Munich 1843 medal of honor; London 1862 medal. Agt. Louis Henlé, 9 Dowgate-Hill, Cannon-street, London.

1-, 2-, 3-, 4-times gilt-wire, ord. silver-wire, fine silvered wire, fine silvered and gilt-wires, different sorts of lametta.

246. Biberbach, J. C., manu., Nuremberg. London 1862 medal.

Brass in sheets and brass-wire.

247. Brandeis, J. jun., manu., Fürth. London 1851 prize-medal; Munich 1854 honor. mention; Paris 1855 bronz. medal; London 1862 medal. Agt. S. Brandeis (E. Brandeis & Co.), London.

Bronze-powder, slitted and rolled metals, beaten metal-leaf in paper, orsedew; beaten metal, half finished. Brocate.

248. Brunnbauer, Heinrich, gold-beater, Munich. London 1862 medal. Agt. Frederick Puckridge, 5 & 6 Kingsland-place, Kingsland, London.

Beaten gold- and silver-leaf. Mixed-gold-foil.

250a. Conradty, C., manu., Nuremberg. London 1862 medal.   (s. II. No. 63.)

Bronze-powder: Real gold p. ducate No. 4000. 9 fl. — Real-silver p. Mark Cologn No. 4000. 33 fl. — English silver p. 1 pound (½ Kilo) No. 4000. 38 fl., No. 2000. 36 fl., No. 1000. 34 fl. — Argentan

No. 4000. 20 fl., No. 2000. 16 fl., No. 1000. 14 fl.
— Pink and salmon No. 4000. 34 fl., No. 2000.
32 fl., No. 1000. 30 fl. — Flora and green-gold,
citron-flora No. 4000. 18 fl., No. 2000. 17 fl.,
No. 1000. 16 fl., No. 800. 15 fl., 600. 14 fl., No. 500.
13 fl., No. 400. 12 fl., No. 300. 11 fl., No. 200.
10 fl., No. 100. 7 fl. — English green-gold, citron-
and pale-citron No. 4000. 17 fl., No. 2000. 15 fl.,
No. 1000. 14 fl., No. 800. 13 fl., No. 600. 12 fl.,
No. 500. 11 fl., No. 400. 10 fl., No. 300. 9 fl.,
No. 200. 8 fl., No. 100. 6 fl. — Redbrown, car-
moisin, carmin, violet flesh-color, lila, richgold A. No. 4000. 16 fl., No. 2000. 14 fl., No. 1000.
13 fl., No. 800. 12 fl., No. 600. 11 fl., No. 500.
10 fl., No. 400. 9 fl., No. 300. 8 fl., No. 200.
7 fl., No. 100. 5 fl. — Richgold B., English white,
light-green, dark-green, grass-green, ducat-color
No. 4000. 14 fl., No. 2000. 13 fl., No. 1000. 12 fl.,
No. 800. 10 fl., No. 600. 9 fl., No. 500. 8 fl.,
No. 400. 7 fl., No. 300. 6 fl., No. 200. 5 fl.,
No. 100. 4 fl. — Pale-yellow, bright-yellow,
orange and silver-composition No. 4000. 12 fl.,
No. 2000. 11 fl., No. 1000. 10 fl., No. 800. 9 fl.,
No. 600. 8 fl., No. 500. 7 fl., No. 400. 6 fl.,
No. 300. 5 fl., No. 200. 4 fl., No. 100. 3 fl. —
Dead-bronze: green gold and copper p. 1 pound
(½ Kilo) 7 fl., yellow gold white p. 1 pound (½ Kilo)
4 fl.

Brocat for manufactories of paper-hangings.

	fine Quality	extraf. Quality	superf. Quality	Paper-bronze
	fl. kr.	fl. kr.	fl. kr.	fl.
No. 2½. 3. 4. und 7	3 15	3 45	4 30	6
Citron, highyellow, orange	3 30	4 —	5 —	7
Green- and german-silver	4 —	4 30	5 30	8
p. pound (½ Kilo).				

250 b. Eberle, J. N., manu., Augsburg.
London 1862 honor. mention.
          (s. II. No. 64. 65.)
Piercing-saws.

251. Eyermann, M. & Loewi, C., manu.,
Fürth. London 1862 medal. Agt. L. E. Werth-
heimer, 28 Basinghall-street, City, London.
Leaf-metal and bronze-powder.

253. Fuchs, G. L. and Sons, manu., Furth.
Munich 1854 medal of honor; Paris 1855 silver-
medal; London 1862 medal. Agts. Frauenknecht
& Stotz, 80 Bishops-gate St. within, London.
Leaf-metal, bronze-powder, orsedew.

254. Fürth, Society for promoting industrial progress at Fürth. Munich 1854 I.
prize-medal; London 1862 medal. Agts. Heintzmann & Rochussen, 9 Friday-street, Cheapside.
Representation of the leading and fancy-articles,
made at Fürth. Looking-glass, unfoliated, foliated
with tin and real silver; framed looking-glasses
of many sorts; optical articles; spectacles-frames
in steel, iron, brass, horn, eye-glasses; girdler-
wares; cabinet-maker's and turner's-articles; gold-
and silver-leaf; leaf-metal and bronze-powder;
boxes etc.; stained papers; playing-cards; tin-
wares; play things; lead figures, cutting in ivory,
bone, wood; porcellan-painting; bronzed-wood-
articles; pipes; combs; cotton and hosiery; candle-
yarns; night-candles wafers, chemicals etc.

255 a. Hänle, Leo. manu., Munich. London 1862 medal.   (s. No. 223.)
146 Samples of bronze-colors and brocate.

255 b. Heckel, J. Georg, heirs, manufactury
Allersberg near Nuremberg. London 1862 honor.
mention.
Plated wire and articles of plated wire.

256 a. Heininger, Anton, girdler, Passau
(Lower-Bavaria). London 1862 honor. mention.
Gilt monstranze or pix; after an old wooden model
of the XV century, found 1832 in the cathedral at
Freising.

256 b. Hofer & Schicketanz, manu., Furth.
London 1862 honor. mention.
Bronze colors.

257. Huttula, Wilhelm, goldbeater,
Nuremberg. London 1862 honor. mention.
Gold- and silver-leaf and mixed-gold-foil.

258. Issmaier, Jos. Mich., manu., Nuremberg. Berlin 1844 bronz. medal; Munich 1854
medal of honor; Paris 1855 bronz. medal.
1. Tin-figures as gas-burners, turning by the
pressure of the gas. 2. Tin-figures to be placed
on a jet of water. 3. Magnetic-play-things.

259. Linz, Joh. Leonh., metal-beater, Furth.
London 1862 honor. mention.
A Card, representing the fabrikation of beaten
white metal. White metal of No. 1½. 2½. 3 of
different size. Sample of white-metal-parings.

260. Meyer, J. C., manu., Furth. London 1862 medal.
Leaf-metal and bronze-powder.

261. Pauli, C. H., gold- and silver-leaf-
manufactory, Nuremberg. (established 1760.)
Munich 1854 medal of honor; London 1862
honor. mention.
Fine molten gold and silver. Rolled gold in
different colors. Rolled silver and aluminium.
Gold, beaten the first and second time in 10 different colors. Silver and aluminium. Thin beaten
gold in 20 different colors from 1 in. engl. to
7 in. engl. in □; thin beaten silver and mixed-
gold-foil.

263 a. Schätzler, Georg Ernst, gold-
beater, Nuremberg. Munich 1854 medal of honor;
Paris 1855 silver-medal; London 1862 medal.
Gold-leaf in different colors: white, green, citron,
yellow, orange, red; mixed-gold-foil; silver-leaf.

263 b. Scheiblein, Friedrich, manu., Weis-
senburg. London 1862 honor. mention.
Plated wire and articles of plated wire and lametta.

264. Schmidtmer, Ernst (E. Kuhn), manufactory, Nuremberg. London 1862 honor.
mention.
Gilt and silvered wire and lametta.

265. Schweizer, A., manu., Furth. Munich
1854 I. prize-medal; Paris 1855 bronz. medal;
London 1862 honor. mention. Agts. Heintzmann
& Rochussen, 9 Friday-street, E. C. London.
Steel-frames for spectacles and eye-glasses.

CLASS XXXII.
STEEL AND CUTLERY.

269. Birnböck, Thomas. engraver. to H.
M. the emperor of Russia, Munich. Munich
1854 medal of honor; Paris 1855 silver-medal;
London 1862 medal.
Impressions of seals, medals.

CLASS XXXIV.
GLASS.

270. ARNDT, BROTHERS, manu., Pirmasens (Palatinate). London 1862 medal.

Watch-glasses from 35 c. to 30 frs. p. Gross.

271. BACH, J., manu., Furth. London 1862 honor. mention.

Looking-glass-plates, unsilvered and silvered.

272. HILDEBRAND, CARL, manu., Munich. London 1862 honor. mention.

Muslin-glasses of different patterns.

273. KOCH's, C. W., looking-glass-manufactory Einbuch, near Ratisbonne. London 1862 honor. mention.

Glass-plates for photographs.

274a. STEIGERWALD, FRANZ, manu. and merchant, House in Munich and London. Munich 1854 I. prize-medal; Paris 1855 med. d'honneur; London 1862 medal.

Cut-crystal and ornamental-glass of every description.

274b. WINKLER, J. G., manu., Furth. London 1862 honor. mention.

Glasses coated with silver and mercury.

CLASS XXXV.
POTTERY.

275. GRUBER & RAUM, manu., Nuremberg.

Pressed graphite-crucibles (Patent).

276. SCHMIDT, C., establishment for painting on porcelain, Bamberg. Munich 1854 medal of honor; Paris 1855 bronz. medal; London 1862 medal. Agts. Schmitt & David, 102 Leaden-hall-street, London.

Paintings on porcelain: Shepherd, killed by lightning, after J. Becker, painted by L. Sturm. Greek widow, after Leonardo da Vinci, painted by C. Meinelt. Madonna after Raphael, painted by H. Schweizer. Family picture, after Georgoni painted by H. Kundmüller.

277. WIMMERS, HEINRICH, artistical-establishment, Munich. London 1862 medal.

Paintings on porcelain.

CLASS XXXVI.
MANUFACTURES NOT INCLUDED IN PREVIOUS CLASSES.

279. SCHWARZ, J. VON, Proprietor of several manufactories, Nuremberg. Munich 1854 medal of honor; London 1862 medal.

Steatite-gas-burners, with holes or cuts, with iron- or brass-fittings, without fittings (i. e. only of steatite), double-burners of novel construction in brass-fitting, Argant-burners; v. Buns's and v. Schwarz's pipes of different sorts.

280a. STÄDTLER, J. G., manu., Nuremberg. London 1862 honor. mention.

Gas-burners and buttons of a peculiar composition (Patented invention of the Exhibitor).

280b. SPRINGER, SPERL & DR. HAGEN, privileged wood dying establishment, Nuremberg. London 1862 medal.

Xylochromic and xyloplastic objects: dyed woods embossed and pressed objects of dyed wood (Patent).

DUCHY OF BRUNSWICK.

CLASS III.

SUBSTANCES USED FOR FOOD, INCLUDING WINES.

282. WITTEKOP, HERMANN (Wittekop & Co.), manu., Brunswick. London 1851 prize-medal; Paris 1855 bronze medal; London 1862 medal. Agts. Lion M. Cohn, Phaland & Dietrich, 20 St. Dunstans Hill, City.

Several samples of manufactured cacao and assortment of macaroni, vermicelli etc.

CLASS VIII.

MACHINERY IN GENERAL.

283. SEELE, FR., & Co., manu., Brunswick. London 1862 honor. mention. Agts. Lion M. Cohn, Phaland & Dietrich, 20 St. Dunstans Hill, City.

A centrifrugal sugar-drying apparatus driven by a cross belt 375 Th. excl. (60 £ incl.) packing-cases and wooden-frame.

CLASS XI.

MILITARY ENGINEERING, ARMOUR AND ACCOUTREMENTS, ORDNANCE, AND SMALL ARMS.

284. SEYDLITZ, W. A. E., gun-maker to the court, Brunswick. Agts. Lion M. Cohn, Phaland & Dietrich, 20 St. Dunstans Hill, City.

A percussion rifle with all the appurtenances, in a case: bullet-mould, cartridge-stoppers, screw-driver, breech- and nipple-drawer, bolt for making cartridges, oil-bottle, powder-measure, cleaning-stick, box for percussion-caps, cartridge apparatus and powder-flask, 250 Th. (37 £).

CLASS XIII.

PHILOSOPHICAL INSTRUMENTS AND PROCESSES DEPENDING UPON THEIR USE.

285. BORNHARDT, A., philosophical and mathematical instrument-maker, Brunswick. London 1862 honor. mention. Agt. L. Löffler, 3 Great George-street, Westminster.

Two balances for chemical analysis; a. for 500 Grammes weight 15 £. b. for 200 Grammes 12 £.

CLASS XVI.

MUSICAL INSTRUMENTS.

286. PAULUS, AUGUST, musical instrument-maker, Brunswick. Agt. Lion M. Cohn, Phaland & Dietrich, 20 St. Dunstans Hill, City.

Cornets à piston No. 1. 300 Th. (44 £), No. 2. 216 Th. (32 £), No. 3. 200. (29 £), all p. doz.

CLASS XXI.

WOOLEN AND WORSTED, INCLUDING MIXED FABRICS GENERALLY.

289. DEGENER & DABELSTEIN, manu., Brunswick. London 1862 honor. mention. Agt. Henry Hewetson, 25 Marklane.

Ladies' cloaking and coating: pure wool, cotton warp and woolen weft.

CLASS XXVIII.

PAPER, STATIONERY, PRINTING AND BOOKBINDING.

290. VIEWEG, FRIEDR. & SON, publishers, printers and manu. of paper, Brunswick. London 1851 prize-medal; Paris 1855 silver medal; London 1862 medal. Agts. Lion M. Cohn, Phaland & Dietrich, 20 St. Dunstans Hill, City.

Andriessen, Lehrbuch der unorganischen Chemie 5 sh. Bacmeister, Handbuch für Sanitäts-Soldaten 1 sh. 6 d. Beer, Einleitung in die höhere Optik 7 sh. 6 d. The same, Grundriss des photometrischen Calcüls 3 sh. Blasius, Fauna der Wirbelthiere Deutschlands I. vol. 8 sh. Bolley, Handbuch der chemischen Technologie I. 1., I. II. 1., II. II., III. 1. 12 sh. 5 d. Braun, the Ruins and Museums of Rome 9 sh. Bunsen, gasometrische Methoden 6 sh. The same, Gasometry 6 sh. Campe, Robinson der Jüngere, Prachtausgabe 6 sh. Curtius Rufus, ed. Zumpt, ed. major 12 sh. The same, ed. minor 3 sh. Dienger, Ausgleichung der Beobachtungsfehler 3 sh. 6 d. The same, Abbildung krummer Oberflächen 2 sh. Dufferin, Briefe aus hohen Breitengraden 5 sh. 6 d. Ecker, der feinere Bau der Nebennieren 6 sh. Feuerbach's nachgelassene Schriften. 4. vols. 14 sh. Fick, die medicinische Physik 9 sh. Frerichs, Klinik der Leberkrankheiten. 1 u. 2. vol. 18 sh. The same, Klinik, Atlas, 2 numbers 1 £ 12 sh. 6 d. Fresenius, qual. Analyse. 10. ed. 5 sh. 6 d. The same, quant. Analyse. 5. ed. 1 number 5 sh. Frick, physikalische Technik 7 sh. 6 d. The same, die Feuerspritze 8 sh. Gorup-Besanez, Lehrbuch der Chemie I. II. III. 1. 1 £ 6 d. Gottlieb, Lehrbuch der reinen und technischen Chemie. 2. ed. 6 sh. 9 d. Hamm, Grundzüge der Landwirthschaft. 2 vols. 2 £ 4 sh. The same, landwirthschaftliche Geräthe. 2. ed. 15 sh. Handwörterbuch der Chemie I. 1—8. II. I. 1—9. II. 1—10. III. ½ III. IV. V. VI. VII. VIII. 1—4. 6 £ 16 sh. Harting, das Mikroskop 15 sh. Hasselt, Handbuch der Giftlehre. 2 vols. 12 sh. Heinemann, Schmetterlinge. 1. vol. 10 sh. 6 d. Heinen, Rotationsapparate 2 sh. Henle, Handbuch der systematischen Anatomie I. 1—3. II. 1. 1 £ 2 sh. 6 d. Hettner, Literaturgeschichte. 1. u. 2. vol. bound 18 sh. Humboldt, ästhetische Versuche. 3. ed. 4 sh. Kolbe, organische Chemie

B °

I. 1—11. II. 1—4. 1 £. 2 sh. 6 d. Kopp, Geschichte der Chemie 4. vols. 1 £. 8 sh. 6 d. Liebig, Anleitung zur Analyse. 2. ed. 2 sh. The same, Grundsätze der Agriculturchemie. 2. ed. 2 sh. 6 d. The same, Ueber Theorie und Praxis 2 sh. 6 d. Meyer, Bellona orientalis, bound 5 sh. 6 d. Mohr, Titrirmethode. 2. ed. 9 sh. The same, Commentar. 2. ed. 2 vols. 16 sh. The same, pharmaceutische Technik. 2. ed. 7 sh. 6 d. Müller, Lehrbuch der Physik. 5. ed. 2 vols. 1 £ 3 sh. The same, kosmische Physik. 2. ed. with plates 12 sh. The same, Grundriss der Physik. 7. ed. 5 sh. 6 d. Nell, der Planetenlauf, with plates 3 sh. 6 d. Olshausen, Lehrbuch der hebräischen Sprache 8 sh. 6 d. Otto, Lehrbuch der Chemie I. II. ı—ııı. 3. ed. 2 £ 12 sh. 6 d. The same, Ausmittelung der Gifte. 2. ed. 2 sh. The same, Essigfabrikation. 2. ed. 4 sh. The same, landwirthschaftliche Gewerbe. 5. ed. 2 vols. 19 sh. 2 d. Prestel, Diagramm, with plates 11 sh. Regnault-Strecker, Chemie I. vol. 5. ed. 6 sh. The same, II. vol. 3. ed. 5 sg. 6 d. Reuleaux, Constructionslehre 1. vol., with plates 1 £ 18 sh. The same, Constructeur 7 sh. 6 d. Rosengarten, Stylarten, bound 11 sh. 6 d. Ruete, Lehrbuch der Ophthalmologie. 2. ed. 2 vol. 17 sh. 6 d. Scheerer, Löthrohrbuch. 2. ed. bound 4 sh. Scheffler, Eisenbahnwagenachsen 8 d. Schellen, der electromagnetische Telegraph. 3. ed. 6 sh. 8 d. Schödler, das Buch der Natur. 11. ed. I. II. 1. 2. 5 sh. Scholl, der Führer des Maschinisten. 5. ed. in boards 5 sh. 6 d. Schröder van der Kolk, Medulla spinalis 7 sh. Schrön, Logarithmen. 2. ed. bound 5 sh. 2 d. The same, not bound 5 sh. 2 d. The same, Hungarian edition 5 sh. 2 d. The same, Dutch edition 5 sh. 2 d. The same, Danish edition 5 sh. 2 d. Schunk, Handbuch der Pariser Feuerwehr 4 sh. 6 d. Semper, die vier Elemente der Baukunst 2 sh. The same, Wissenschaft, Industrie und Kunst 1 sh. 6 d. Siebold, Lehrbuch der Geburtshülfe. 2. ed. 8 sh. The same, geburtshülfliche Briefe 3 sh. Stöckhardt, Schule der Chemie. 12. ed. 6 sh. Süs, Swinegels Reiseabentheuer 2 sh. The same, Mähr von einer Nachtigall 2 sh. Tacitus Agricola ed. Wex 7 sh. 6 d. Valentin, Grundriss der Physiologie 12 sh. Vogt, Lehrbuch der Geologie, 2 vol. 2. ed. 15 sh. The same, Grundriss der Geologie 7 sh. The same, Schöpfungsgeschichte 2. ed. 5 sh. Voit, chemische Laboratorium zu München, with plates 15 sh. Vorträge, wissenschaftliche 9 sh. Walkhoff, Rübenzuckerfabrikant. 2. ed. 8 sh. 6 d. Weisbach, die neue Markscheidekunst. 2 vols. 2 sh. 4 d. The same, Lehrbuch der Mechanik. I. vol. 1. 2. number 4. ed. 3 sh. The same, II. vol. 3. ed. 17 sh. The same, III. vol. in 2 sections 1 £ 2 sh. 6 d. The same, Ingenieur. 3. ed. 1. and 1. number 4 sh. 8 d. Weltzien, systematische Zusammenstellung 11 sh. Wernicke, Lehrbuch der Mechanik 2 vols. 9 sh. Wiedemann, die Lehre vom Galvanismus I. II. 1. 19 sh. Wundt, Lehre von der Muskelbewegung 5 sh. Zernikow, Theorie der Dampfmaschinen 4 sh.

The above specified books are printed at the exhibitors' own office; the wood-cuts are engraved at their own establishment and the paper made in their own manufactory.

291. WESTERMANN, GEORGE, bookseller and printer, Brunswick. Paris 1855 bronze medal; London 1862 honor. mention. Agts. Lion M. Cohn, Phaland & Dietrich, 20 St. Dunstans Hill, City.

Books:

a. Illustrated works: Berg, the island of Rhodos (splendid work), the binding by Mr. H. Koch at Stuttgart. A collection of portraits, executed in wood-cut belonging to the English history. A collection of portraits, belonging to the universal-history. H. Lange, small school-atlas representing all the parts of the world. European-galery. A collection of the chief-works of painting in steel-engravings. The grand life of nature and nations. G. Westermann's Illustrated German monthly magazine, vol. I—XI. Müller, ophthalmology in the veterinary art. Winkler, the island of Iceland.

b. Printed with stereotype plates. Our days, glimpses from time to time, vol. I. II. Rost, a Greek and German dictionary, vol. I. II. Pfaff, German history, vol. I. II. Klotz, Latin and German dictionary, half-vol. I—IV. Elwell, dictionary of the English and German language. Macaulay, the history of England, translated by W. Beseler, vol. I—IV. Macaulay, select writings, vol. I—VIII.

c. Printed in movable type: Herzfeld, the history of Jisrael. Andree, North-Amerika. Künzel, the life and speeches of Sir Robert Peel, vol. I. II. Woodbury, compendium of the English language. Hoffmeister, letters from India. Hagen, the history of our own time, vol. I. II. The interior life of Russia. Schmarda, a voyage round the world, vol. I—III. Saunterings in London, vol. I. II. Hoefken, Flemish Belgium, vol. I. II. Salvador, the history of the Roman empire, vol. I. II. Barthold, Germany and the Hugenotts. Oelsner-Monmerqué, memorials. Plass, the tyrannis, vol. I. II. Schaumburg, the history of the counts of Valkenstein.

CLASS XXX.

FURNITURE AND UPHOLSTERY, INCLUDING PAPER-HANGINGS AND PAPIER-MACHE.

292. WALTER, ERNST, & SON, basketmaker to the court, Brunswick. London 1862 honor. mention. Agts. Lion M. Cohn, Phaland & Dietrich, 20 St. Dunstans Hill, City.

1 easy chair 1 £, 1 lady's work-table 12 sh., 1 foot-stool 2 sh. 6 d., 1 basket 9 sh.

CLASS XXXI.

IRON AND GENERAL HARDWARE.

292a. GOLDBERG & JACOBS (formerly C. Wried, Successor to Stobwasser, Brunswick), painter, Brunswick. London 1862 honor. mention. Agts. Lion M. Cohn, Phaland & Dietrich, 20 St. Dunstans Hill, City.

Oil-paintings, copies from works of ancient and modern masters, on canvas, metal-plates and wood-panels.

295. WILKE, AUGUST, locksmith, Brunswick. Agts. Lion M. Cohn, Phaland & Dietrich, 20 St. Dunstans Hill, City.

Machine for beating eggs 3 £ 10 sh.

FRANKFORT ON THE MAIN.

CLASS II.
CHEMICAL SUBSTANCES AND PRODUCTS AND PHARMACEUTICAL PROCESSES.

302. FRANKFORT-JOINT STOCK COMPANY FOR THE MANUFACTURE OF CHEMICAL PRODUCTS FOR AGRICULTURAL PURPOSES, Frankfort on the Main, manufactory Griesheim near Frankfort.

Specimens of: 1. calcined soda, 2. do. 98°, 3. crystalized soda, 4. chloride of calcium (bleaching-powder), 5. biphosphate of lime I., 6. do. II., 7. steamed bone dust, 8. artificial guano, 9. concentrated manure II., 10. concentrated manure for the cultivation of the vine.

303. ZIMMER, CONRAD, Dr. phil., Frankfort on the Main. London 1851 prize-medal, 1862 medal. Agts. Ammann Roller & Co,

100 chemical preparations of Peruvian bark.

CLASS III.
SUBSTANCES USED FOR FOOD, INCLUDING WINES.

304. ECKERT, WILL., & Co., manu. of cigars, Frankfort on the Main. Munich 1854 medal of honor; Paris 1855 bronze medal; London 1862 medal.

Jod-cigars, assortment of cigars for exportation made of tobacco from the Palatinate; Havannah-cigars.

305. GIORGI, DE, BROTHERS, chocolate manu., Frankfort on the Main. London 1862 medal.

Chocolate, cacao, cacaonut-oil in squares, preparations of cacao in boxes.

CLASS IV.
ANIMAL AND VEGETABLE SUBSTANCES USED IN MANUFACTURES.

308. MOUSON, J. G., & Co., manu. of soaps and perfumes, Frankfort on the Main. London 1862 honor. mention. Agts. S. Oppenheim & Sons, 4a Bread-street, Cheapside.

Transparent soaps in bars, balls and moulded pieces; various toilet-soaps in moulded pieces, extracts, pomatums.

309. RIEGER, WILLIAM, manu., Frankfort on the Main. London 1862 medal. Agt. William Rieger, 26 Lambeth Hill, Doctors Commons, London.   (s. II. No. 91.)

Several toilet-soaps and perfumes.

CLASS XIII.
PHILOSOPHICAL INSTRUMENTS AND PROCESSES DEPENDING UPON THEIR USE.

311. KNEWITZ, PH. J. F., mechanician and optician, Frankfort on the Main. London 1862 honor. mention.

Orthograph suggested by Dr. G. Lucae: Morphology of the skulls of races, Frankfort on the Main 1861. page 16. 24 fl. (2 £). Balance for chemical analysis, at 250 grms. weight, indicating 1 decimilligrm. 300 fl. (25 £). A like balance, at 100 grms. weight, indicating 1 decimilligrm. 126 fl. (10 £ 10 sh.). A set of gramm-weights 15 fl. (1 £ 5 sh.).

CLASS XIV.
PHOTOGRAPHIC APPARATUS AND PHOTOGRAPHY.

312. HAMACHER, G., Frankfort on the Main. London 1862 honor. mention.

Göthe-Gallery by William von Kaulbach, photographs, part 1—4. each 3 leaves 32 Th. (4 £ 16 sh.). Göthe-Gallery by William von Kaulbach with text by Adolf Stahr, copper-prints, part 1. 6 Th. (18 sh.). Kaulbach-Album, bound 3 Th. (9 sh.). Fresco paintings of the Landgrave-hall at the Wartburg, by Maurice von Schwind, bound 5 Th. (15 sh.). The style, architectonical manual by G. Semper I. with wood-cuts and 15 colored lithographic prints 6 Th. (18 sh.).

CLASS XVI.
MUSICAL INSTRUMENTS.

314. ANDRÉ, CHARLES AUGUST, merchant prop. of a manu., Frankfort on the Main. Munich 1854 honor. mention; London 1862 honor. mention. Agts. J. & R. Mac Cracken.

Mozart-pianoforte 840 fl. (70 £).

CLASS XXII.
CARPETS.

315. VACONIUS, JOHN JOSEPH, manu. of carpets, Frankfort on the Main.

Plush sofa-carpet, 3 colored 1 by 1¼ yards 15 sh., do. 3 colored 1¼ by 2 yards 1 £ 5 sh., do. 3 colored, crimson and grey 1½ by 2 yards 1 £ 5 sh., do. 5 colored, red, white, yellow 2¼ by 3 yards 4 £ 3 sh.; plush carpet, 3 colored 1¼ by 2 yards 5 sh.; plush sofa-carpet 3 colored 1 by 1¼ yards 15 sh.

CLASS XXIV.

TAPESTRY, LACE AND EMBROIDERY.

316. DANN, LEOPOLD, & Co., manu., Frankfort on the Main.

Curtain-bearers in wool and silk; furniture-fringes in wool and silk of different breadths and stoutness; curtain-gimps in wool and silk; curtain-tassels in wool and silk; several kinds of valance.

CLASS XXVI.

LEATHER, INCLUDING SADDLERY AND HARNESS.

317. HAUSMANN, BROTHERS, manu., Frankfort on the Main, manufactory Homburg vor der Höhe. London 1862 medal. Agts. Weintraud, Rumpff & Co., 4 King-street, Cheapside, London.

Colored sheep-skins, kid-skins, split and not split. and moscowy leather.

318. LANDAUER-DONNER, G. F., leather manu., Frankfort on the Main. London 1862 medal. Agts. Sprösser, Lorentz & Co., 41 Watling-street, E. C.

Differently dressed morocco-leathers; East-indian goat and sheep-skins; colored calf-leather; split sheep-leathers of various colors and dressings.

319. REGES, JOHANN ANDREAS BENJAMIN, bookbinder and portfolio maker, Frankfort on the Main. Agts. J. & R. Mac Cracken.

Miniature photograph-albums of different sizes and finished with photographs, covered with glass p. piece 12 fl. (1 £).

320. ROTH, JOH. ADAM, ROTH, JOH. PETER (John Ad. Roth, sons), leather manu., Frankfort on the Main. Manufactory Lorsbach near Frankfort. London 1862 honor. mention.

Shamoy-dressed buffalo-leather; calf-leather, varnished, waxed, brown, shamoy, cylinder.

321. ROTH, C. W., manu. Hausen near Frankfort on the Main. London 1862 medal. Agt. W. Baither, 47 Mark-Lane, City, London.

Leather for machine-straps, buffalo-leather, skins to be waxed, brown and black calf-leather, cylinder-skins, knapsak-skins.

322. SCHEIDEL, GOUDA, & Co., manu. Frankfort on the Main. Paris 1855 bronze medal; London 1862 honor. mention. Agts. Gouda, Scheidel & Co., 58 Watling-street, Cheapside, London.

Ladies' companions, necessaires, cigar-cases, albums' fine woodboxes.

CLASS XXVII.

ARTICLES OF CLOTHING.

323. QUILLING, JOHN FRED., Frankfort on the Main. (s. II. No. 24.)

Knitted shawl with embroidered border, knitted concert-cape, knitted capuchon, capuchon worked in the frame, slippers in crotchet-work, muff worked in the frame, child's-frock in crotchet-work.

324. SCHNAPPER, J., JUN., manu., Frankfort on the Main.

Silk drawers woven in different patterns marked with No. and price 36—72 fl. (3—6 £), woollen drawers No. 7. 23 fl. (2 £ 6 sh. 8 d.), do. No. 8. 40 fl. (3 £ 6 sh. 8 d.), cotton drawers No. 9. 8 fl. (13 sh. 4 d.), do. No. 10. 10 fl. (16 sh. 8 d.), do. No. 11. 12 fl. (1 £); silk camisols No. 12. 42 fl. (3 £ 10 sh.), woollen do. No. 13. 30 fl. (2 £ 10 sh.), do. No. 14. 40 fl. (3 £ 6 sh. 8 d.); knitted woollen socks No. 15. 6 fl. (10 sh.), do No. 16. 7 fl. (11 sh. 8 d.), do. No. 17. 3 fl. 18 kr. (5 sh. 6 d.); knitted woollen stockings for children No. 18. 5 fl. (8 sh. 4 d.), do. for ladies No. 19. 12 fl. (1 £); silk petticoats No. 20. 24 fl. (2 £); capuchons No. 20—27. 7 to 21 fl. (11 sh. 8 d. to 1 £ 15 sh.); all p. doz.

CLASS XXVIII.

PAPER, STATIONERY, PRINTING AND BOOKBINDING.

324a. FLINSCH, H. F. G. (Dressler's foundery, F. Flinsch), manu., Frankfort on the Main. Munich 1854 medal of honor; Paris 1855 silver medal; London 1862 medal. Agts. J. & R. Mac Cracken, London.

A book with specimens of printing; 410 letter-punches.

325. KNATZ, CHARLES, prop. of a typographic and lithographic printing office, Frankfort on the Main. Munich 1854 honor. mention; Paris 1855 honor. mention; London 1862 medal.

Works of typographic and lithographic printing.

CLASS XXIX.

EDUCATIONAL WORKS AND APPLIANCES.

326. KÖBIG, J., lithographer, Frankfort on the Main. Munich 1854 medal of honor; Paris 1855 bronze medal; London 1862 medal.

Specimens of lithographic printing.

327. KRUTHOFFER, CHARLES, prop. of a printing office for every kind of printing. Munich 1854 medal of honor.; Paris 1855 bronze medal.

Lithographic, typographic and embossed works for mercantile purposes; microscopic works for papers of value; new manner of combined typographic printing; colored prints for artistical and industrial purposes; imitated miniatures printed on parchement in the style of the middle ages; three typographic splendid editions; genre and religious paintings in colored lithographic prints published by the exhibitor.

CLASS XXX.

FURNITURE AND UPHOLSTERY, INCLUDING PAPER-HANGINGS AND PAPIER-MACHÉ.

328. BÖHLER, FRED., manu., Frankfort on the Main. London 1862 medal. Agts. J. & R. Mac Cracken, 7 old Jewry, E. C., Special agent for the exhibition Ferdinand André, 31 Grafton-street, Fitzroy Square, W.

A set of furniture of stag's horn: 1 closet for arms,

1 looking glass and console; 1 sofa, green plush, 2 armchairs do.; 6 chairs do., 1 table, oval, 1 small table, round; 2 chairs, doeskin; 2 tabourets, 1 stool, green plush; 2 flower stands; 1 pendule; 1 pair of candelabres; 1 lustre for 15 candles; 1 stand for hunting-implements; 3 decorations for curtains, 6 curtain-bracelets, 1 hunting-coat of armes. 34 pieces: 1250 £.

1 chair with doeskin, stag's horn pattern of another set of furniture.

Articles carved in ivory. Goblet: large H. 500 £, do. large L. 350 £, do. middlesized H. 80 £, do. small H. No. 35. 22 £, do. small H. No. 36. 25 £. Candle-screen H. No. 10. 30 £, do. H. No. 12. 27 £. Brooches: No. 75. TH. 4 £ 10 sh., No. H. 84. V. 4 £, No. 48. DH. 4 £, No. 18 1 £ 15 sh., No. 39. T. 2 £ 10 sh., No. 11½. H. 3 £ 10 sh., No. 32. DH. 2 £ 15 sh., No. 35½. DH. 2 £ 10 sh., No. 7. 2 £ 10 sh., No. 30½. D½. 2 £ 5 sh., No. 36. T. 2 £ 15 sh., No. 17. 2 £.

A set of ornaments carved in ivory consisting of a bracelet, a brooch and a pair of earrings 15 £.

329. JACQUET, H., SON, Frankfort on the Main.

1 house-altar 216 fl. (18 £), 1 press for smoking utensils 200 fl. (16 £ 15 sh.), 1 folding screen 150 fl. (12 £ 10 sh.), 1 etagère 84 fl. (7 £), 1 screen for arms 120 fl. (10 £), 1 carved round frame with looking glass 78 fl. (6 £ 10 sh.), 1 flower-case 30 fl. (2 £ 10 sh.).

CLASS XXXI.

IRON AND GENERAL HARDWARE.

330. LAUSBERG, CHARLES, merchant and manu., Frankfort on the Main. Agts. J. & R. Mac Cracken, 7 Old Jewry, London.

An assortment of various white and varnished metal-covers for corking bottles, pitchers, pots etc. The prices p. mille go by sizes and the manner of varnishing.

CLASS XXXIII.

WORKS IN PRECIOUS METALS, AND THEIR IMITATIONS AND JEWELLERY.

331. GOLDSCHMIDT, MAURICE, SON, manu. of jewellery, Frankfort on the Main. London 1851 honor. mention, 1862 honor. mention. Agt. Maurice Goldschmidt son. 3 Thavies' Inn Holborn, Representative J. Wertheim.

3 sets of jewellery, bracelet, brooch, earrings; 15 bracelets in a case; 12 brooches in a case; 3 colliers in a case; 6 etagère-articles in rococo-style.

332. FRIEDMANN, JOSEPH, manu. of jewellery, Frankfort on the Main. Paris 1855 silver medal; London 1862 medal. Agt. Adolf Friedmann, 17 Hatton-Garden, London. (s. II. No. 61.)

Parures, bracelets, brooches, earrings, pins, buttons, chains, charivaris, breloques and fancy-articles.

KINGDOM OF HANOVER.

Agent: G. W. Roese, 8 Trafalgar-square, Brompton, S. W.

CLASS I.

MINING, QUARRYING, METALLURGY AND MINERAL PRODUCTS.

341. Georg-Marien-Hütte and Beckeroder Hütte, near Osnabrück. London 1862 medal.

Raw materials: brown iron ore, calcined ochre, sparry iron ore; lump-coal, washed coal, coke from washed coal; limestone.

Pig-iron melted by coke and coal: specular iron, half-specular iron, radiated iron, grey foundry-pig, grey forge-pig.

Charcoal-iron: specular and radiated.

342. Egestorff, Georg, Linden near Hanover. Munich 1854 bronz. medal; Paris 1855 silver-medal.

Salt.

343. In der Stroth, H., Bentheim.

Jet-coal.

344. Grubenverwaltung am Piesberge near Osnabrück (Bergmeister Pagenstecher).

Glance-coal (anthracite) from the Piesberg:

lump-coal .	p. 125 lbs.	7 sg.	(p. 112 lbs. a. d. p.	$6\frac{4}{5}$ d.)
crumbs . . .	»	$5\frac{1}{2}$ »	(» »	$5\frac{1}{3}$ »)
small crumbs	»	5 »	(» »	$4\frac{9}{10}$ »)
sifted coal .	»	3 »	(» »	$2\frac{9}{10}$ »)
garbles . . .	»	1 »	(» »	1 »)

Paving stones from the Piesberg (conglomerate and carbonated sand-stone): 100 sq. ft. 7 Th. 10 sg., 100 sq. ft. engl. 24 sh.; 100 ft. borders 4 Th. 5 sg., 100 ft. engl. 13 sh. $7\frac{1}{3}$ d.

345. Mosqua, Carl; Hildesheim. London 1862 honor. mention.

	at Hildesheim Th.	at London £	sh.
A french mill-stone for grinding wheat (No. 234)	95	15	6
Do. (No. 235)	95	15	6
Do. for grinding rye (No. 236)	115	18	15
A Münden mill-stone for rye (No. 380)	27	5	2
A polishing stone for rice (No. 462)	130	21	15
A peeling stone for rice and barley (No. 463)	83	13	16
Do. for barley (No. 464)	20	4	1
A piece of sandstone, moulded (IV.), hard	—	—	6
Do. (V.)	—	—	4

CLASS II.

CHEMICAL SUBSTANCES AND PRODUCTS AND PHARMACEUTICAL PROCESSES.

346. Chemische Fabrik, Nienburg.

	p. 100 lbs. Th.	sg.	p. 100 lbs. a. d. p. sh.	d.
Carbonate of soda 50 pCt.	4	25	14	9
do. do. $52\frac{1}{2}$ »	5	—	15	3
do. do. $57\frac{1}{2}$ »	6	10	19	4
Crystal. soda	2	15	7	8
do. Glaubersalt	1	20	5	1
Chloride of lime 36 pCt.	5	$22\frac{1}{2}$	17	6
Acetate of lime	6	—	18	4
Sulphate of soda	1	20	5	1

347. Egestorff, Aug., Linden near Hanover. 9 different samples of Ultramarine.

348. Egestorff, Georg, Linden near Hanover.

Ultramarine.

349. Chemische Fabrik (E. F. Heins and C. Nöllner), Harburg. London 1862 honor. mention.

Tin-salt lb. 9 sg., 1 lb. a. d. p. $9\frac{5}{6}$ d.; saltpeter lb. 2 sg., 1 lb. a. d. p. $2\frac{1}{5}$ d.

350. Heuer, Aug., & Co., Lichtenstein near Osterode. London 1862 honor. mention.

	p. 100 lbs. Th.	sg.	p. 112 lbs. a. d. p. sh.	d.
White lead EF.	9	15	29	—
do. HB.	8	$22\frac{1}{2}$	26	8
do. FF.	7	$22\frac{1}{2}$	23	8
do. F.	7	—	21	4
do. MF.	6	—	18	$3\frac{1}{2}$
do. M.	5	—	15	3
do. O.	4	—	12	$2\frac{1}{2}$
do. OO.	3	—	9	2
Ground litharge	7	$7\frac{1}{2}$	22	1
Lichtenstein brown	2	15	7	8
Ground sulphuret of zinc	2	—	6	1
Ground heavy spar	1	—	3	$\frac{1}{2}$

351. Hilkenkamp & Co., Osnabrück.

	p. 100 lbs. Th.	p. 112 lbs. a. d. p. £	sh.	d.
Bremen green No. I.	50	7	12	6
do. No. II.	45	6	17	3
Parisian blue	78	11	17	10
Chrome-yellow, light	40	6	2	—
do. orange	40	6	2	—

353. Röhrig, Gebrüder, Linden near Hanover.

Ultramarine.

354. Schachtrupp, J. F., & Co., Osterode. London 1862 medal.

9 samples of white lead; 2 do. of litharge. Ground heavy spar. Sugar of lead.

355. Stackmann & Retschy, Lehrte.

Steamed Bone-dust 8 £ 3 sh.; superphosphate of lime 8 £ 15 sh.; nitrogenized superphosphate of lime, patent manure 9 £ 18 sh.; bone fat 42 £, all p. ton. Bone charcoal. Wool meal. Pig's hair meal. Horn meal.

CLASS III.

SUBSTANCES USED FOR FOOD, INCLUDING WINES.

356. Brede, C. L., Hanover, manu. of vermicelli and Klencke's proteïne-food. London 1862 medal.

	p. 100 lbs. Th.	sg.	p. 112 lbs. a. d. p. sh.	d.
Tubular maccaroni	14	—	42	8
Spelt vermicelli	11	—	33	6
Spelt groats with eggs	12	15	38	1½
Spelt lentils	12	—	36	7
Spelt groats	12	—	36	7
Fancy vermicelli paste	12	15	38	1½
Lasagne	11	—	33	6
Dented vermicelli with eggs	14	—	42	8
Ribbon vermicelli with eggs	14	—	42	8

	p. lb. Th.	sg.	p. 1 lb. a. d. p. sh.	d.
Proteïne groats	—	6	—	6½
do. grits	—	6	—	6½
do. powder	—	7½	—	8⅙
Ferruginous proteïne powder No. I.	—	10	—	11
do. do. No. II.	—	12½	1	1⅔
Salep groats	—	20	1	9¾

357. Breul, D., & Habenicht, tobacco and snuff-manu., Hanover. London 1862 medal.

	p. lb. Th.	sg.	pr. 1 lb. a. d. p. sh.	d.
Cavendish	—	12½	1	2
Ladyfinger	—	11	1	—
Negrohead	—	15	1	4½
Ladies twist	—	15	1	4½
Fine twist No. I.	1	—	2	9
do. No. II.	—	25	2	3½
Chewing tobacco Lit. B.	—	11	1	—
do. Lit. C.	—	10	—	11
do. Lit. C. fine.	—	12½	1	2

	p. 100 lbs. Th.	sg.	p. 112 lbs. a. d. p. sh.	d.
Fine cut Cavendish	40	—	122	—
Shag tobacco No. 1.	33	—	100	6
do. No. 2.	26	—	79	3
Birds-eye tobacco No. 1.	33	—	100	6
do. do. No. 2.	25	—	76	3
Chinese snuff	100	—	305	—
Granné do.	13	—	39	8
Râpé de Hanovre No. I.	32	—	97	7
do. do. No. II.	16	—	48	9
Au grand cardinal	46	—	140	2
Saint-Omer No. 1.	33	—	100	6
do. No. 2.	27	—	82	4
do. No. 3.	25	—	76	3
Râpé de Paris No. 1.	40	—	122	—
do. No. 2.	36	—	109	10
do. No. 3.	22	—	67	—
do. No. 4.	18	—	54	11

358. Grütter, Heinr., & Co., manu. of sparkling wines, Nienburg. Agts. Hinton Brothers & Co., 80 Old Broad-street. (s. II. No. 66.)

3 bottles Champagne 1. quality 8 Th. (24 sh.), do. for export 6 Th. (18 sh.), 3 bottles sparkling Hock 8 Th. (24 sh.), do. for export 6 Th. (18 sh.), 3 bottles sparkling Moselle 8 Th. (24 sh.), do. for export 6 Th. (18 sh.), 3 bottles sparkling Muscatel 8 Th. (24 sh.), do. for export 6 Th. (18 sh.), from 18 to 36 sh. and higher, all p. doz. bottles, franco London, in bond.

CLASS IV.

ANIMAL AND VEGETABLE SUBSTANCES USED IN MANUFACTURES.

359. Cohen, A., Vaillant & Co., Harburg. London 1862 medal. (s. II. No. 99.)

Large and complete assortment of all manufactures of vulcanized india-rubber:

A. India-rubber overshoes for men, women, boys, girls and children in most different shapes and linings.

B. Selfacting overshoes to be put on and out without using hands, varied shapes and linings.

C. Shoes and boots to be worn alone; long india-rubber boots in most different shapes; enameled india-rubber leather shoes, varnished india-rubber laced shoes, clothshoes, lasting and felt gaiter boots, buskins, felt shoes, felt cloth shoes, cotton velvet-shoes, canvass shoes.

D. Waterproof cloth, single and double texture of orleans, alpacca, woollen, silk and cotton in different colors and qualities.

E. Waaterproof clothing: overcoats, hoods, leggins, caps and hats of silk, twill, india cloth, alpacca, orleans, cashmere, percale, stout, moleskin, canvass and cloth for 3 classes of consumers.

F. Enameled india-rubber leather on stout and twill and all articles that are made of it. Cart and waggon covers, carriage laps, horse and gun-covers, carpetbags, pouches, mail and pilot-bags, mule-bags, fodder-bags, knapsacks, cartridge-boxes, umbrella cases, horsepads.

G. Airproof manufactures. Air and neck-cushions, pillows, railroad beds, mattresses, swimming-betts, heaters, washing and bathing tubs, pontoons, fishing and life-cloaks, life-coats, life-jackets.

H. Household-articles and manufactures for domestic use. Bracelets, sleevepads, nipples, thumb-stalls, stopples, kitchenpails, fire-buckets, tobacco-bags, urinals, eartubes, belts, clothes-rail, chest-expanders, doorsprings and billardcushions.

I. India-rubber mats and carpets. Entirely new specialities and invention.

K. India-rubber toys of every description and all shapes. Balls, ballons, flutes, rattles, figures, heads for dolls.

L. Articles for art. Drawing-models, complete assortment of the Basreliefs of the Partenons.

M. India-rubber for technical purposes. For railroads, marine, machinery and manufacturing purposes; india-rubber sheet and valves for steam-packing in 6 different qualities. Railroad-buffers, cart-covers, bands, hoses and tubing of every description in strength.

N. Surgical instruments of every description.

O. Manufactures for military and marine use. Tents, knapsacks, pouches, powder-bags, covers for baggage-carts, bottles, camp-beds, camp-blanckets, mule-bags, pontoons, buffers for guns, and sheet packing for gunboats.

P. India-rubber thread and elastic weavings made of it.

Specification, quantity and prices will be found in II. No. 99.

360. HARBURG INDIA-RUBBER COMB COMPANY, Harburg. London 1862 medal. Agt. Benda Brothers 96, Newgate-street. (s. II. No. 85.)

India-rubber combs: Dressing combs 3 sh. 3,6 d. to 14 sh., tail combs 5 sh. 3 d. to 9 sh. 6 d., back combs 4 sh. 6 d. to 1 £., neck combs 1—2 sh. side combs 1 sh. to 1 sh. 6 d., puff combs 1 sh. 6 d. to 2 sh., poll combs 2 sh. 8 d. to 7 sh., pocket combs 2—6 sh., etwee-combs 2—6 sh., dandrif-combs 2 sh. 6 d. to 6 sh., all p. doz. Discount as per agreement.

361. HENNECKE, JOH. HEINR., SÖHNE (Jac. Hennecke), Goslar.

	p. 100 lbs.	p. 112 lbs.
		a. d. p.
	Th.	sh. d.
Sesam oil	17	51 10
Palm-nut oil	18	54 11
Cocoa-nut oil	18	54 11

364. WILLHARM & MÜLLER, Melle.

Machine-made cork stoppers.

CLASS VII.

MANUFACTURING MACHINES AND TOOLS.

365. KNÖVENAGEL, A., machine-maker, Linden near Hanover. London 1862 honor. mention.

Sawing machine for cutting planks into boards of one quarter of an inch and upwards, working by 1, 2 or 3 blades.

366. VOIGTLÄNDER, F., machine-maker, Schladen. London 1862 honor. mention.

Tobacco spinning-machine, Breul & Habenicht's patent 18 £ 15 sh.

CLASS X.

CIVIL ENGINEERING, ARCHITECTURAL AND BUILDING CONTRIVANCES.

368. EGESTORFF, AUG., Linden near Hanover.

Asphalt.

369. HENNING, D. H., Limmer near Hanover. London 1862 medal.

	p. 100 lbs.	p. 112 lbs.
		a. d. p.
	Th. sg.	sh. d.
Ground asphaltum-stone	— 16	1 7½
Crude do.	— 10	1 —
Petroleum (earth oil) from Reitlingen	6 — 18	4
Mineral tar (goudron) id.	8 — 24	5
Mastic	1 2	3 3

370. HEYN, C. F., Patent Portland cement-manu., Lüneburg. London 1862 honor. mention.

Pillar-capital, cement casting, taken from St. Michaels church in Hildesheim, X. century; together with a pedestal of cement casting and a pillar of wood 120 Th. (18 £). Pillar-capital, cement casting, taken from the Wunstorf cathedral near Hanover, IX. century; together with a pedestal of cement casting and a wooden pillar 120 Th. (18 £).- Original cask of Portland cement, 400 lb. weight 3⅓ Th. (10 sh.).

371. MOSQUA, CARL, Hildesheim.

Roman cement. — Fine plaster.

CLASS XI.

MILITARY ENGINEERING, ARMOUR AND ACCOUTREMENTS, ORDNANCE AND SMALL ARMS.

372. EGESTORFF, GEORG, Linden near Hanover.

Percussion-caps.

373. HUISMANS, J. G. (Gerhard Lampen) Leer. London 1862 honor. mention.

Shot in 18 sorts or numbers: 0000. = BBBB., 000. = BBB., 00. = BB., 0. = B., 1., 2. to 14. 25 £ net p. ton, the shot in bags free delivered in the Thames or Humber, except duty, the expence of cask to the charge of buyer. — 60 frs. p. 100 Kilos, franc à bord.

374. KLAWITTER, TH., gun-maker to H. M. the King of Hanover, Herzberg.

A double-barrelled gun 80 Th. (12 £.).

375. SCHACHTRUPP, J. F. & Co., Osterode.

16 samples of patent shot.

377. WESTPHAL, L., gun-maker, Peine. London 1862 honor. mention.

Gun »Lefaucheux« with its appurtenances 118 Th. (17 £ 14 sh.).

Double-barrelled gun 65 Th. (9 £ 15 sh.).

CLASS XIII.

PHILOSOPHICAL INSTRUMENTS AND PROCESSES DEPENDING UPON THEIR USE.

378. LANDSBERG & PARISIUS, Hanover. London 1862 medal.

Kathetometer of the newest construction 180 Th. (27 £); theodolite with repetition, horizontal and azimuthal circle 150 Th. (22 £ 10 sh.); levelling instrument with a horizontal circle 150 Th. (22 £ 10 sh.), do. with fixed telescope on three feet 36 Th. (5 £ 8 sh.), do. of simple construction 26 Th. (3 £ 18 sh.); surveying table with stand and plate 44 Th. (6 £ 12 sh.); rule with telescope, compass and level 60 Th. (9 £); alhidade for levelling 14 Th. (2 £ 2 sh.); astrolabe with telescope, level and circle 25 Th. (3 £ 15 sh.); optometer of own construction 18 Th. (2 £ 14 sh.); ophthalmodiastimeter of own construction 7 Th. (1 £ 1 sh.); chemical balance 84 Th. (12 £ 12 sh.), small do. 40 Th. (6 £); essay-balance 25 Th. (3 £ 15 sh.); spectrum-apparatus after Bunsen and Kirchhof 35 Th. (5 £ 5 sh.), do. of new construction for comparing observation of two spectra 60 Th. (9 £); auxiliary apparatus for making experiments with spectra 15 Th. (2 £ 5 sh.); case of mathematical instruments and different articles of this kind 70 Th. (10 £ 10 sh.).

379. Löhdefink, D. A., manometer-maker, Hanover. London 1862 honor. mention.

12 different metallic manometers, viz: *a.* manometer with three-ways-cock and flange for the comptrolling manometer 15 Th. (2 £ 5 sh.), *b.* the same with maximum-index 18 Th. (2 £ 14 sh.), *c.* round manometer, 10 atmospheres (for locomotive engines) 14 Th. (2 £ 2 sh.), *d.* the same with three-ways-cock and stand 16 Th. (2 £ 8 sh.), *e.* a smaller one, 10 atm. (for locomotive engines) 13 Th. (1 £ 19 sh.), *f.* the same with three-ways-cock, stand and maximum-index 17 Th. (2 £ 11 sh.), *g.* smallest sort, 5 atm. 12 Th. (1 £ 16 sh.), do. 4 atm. 12 Th. (1 £ 16 sh.), *h.* round, 8 atm. divided in pounds 18 Th. (2 £ 14 sh.), *i.* vacuum-manometer 18 Th. (2 £ 14 sh.), *k.* small round manometer, 5 atm. divided in pounds 16 Th. (2 £ 8 sh.), *l.* comptrolling manometer 16 Th. (2 £ 8 sh.).

Boiler-armature containing water-gauge, magnetical swimmer and steam-gauge 75 Th. (11 £ 5 sh.).

Double water-gauge with gauge-cocks 35 Th. (5 £ 5 sh.).

Sewing-machine for solid and light work 100 Th. (15 £.).

CLASS XVI.

MUSICAL INSTRUMENTS.

380. Meyer, H. F. & Co., Hanover.

Instruments of German silver: fagott, bass clarinet, B-clarinet, flute, oboe with tubes.

B-clarinet of King's red ebony, keys and garniture of German silver; flute of King's red ebony, head piece of ivory, keys, cylinder and garniture of silver; oboe of King's red ebony, keys and garniture of German silver.

381. Riechers, Aug., Hanover. London 1862 honor. mention.

1 violin, guitar, cither.

CLASS XVIII.

COTTON.

382. Mechanische Weberei (Anonymous Society), Linden near Hanover. Munich 1854 medal of honor; Paris 1855 bronz. medal; London 1862 medal.

Common black velvet 1. and 2.; patent do. 1. and 2.; brown Victoria velvet; black do.; brown Fancy velvet; scarlet velvet.

CLASS XIX.

FLAX AND HEMP.

383. Meyerhof, J., Hildesheim.

6 table cloths and napkins p. set 7 Th. (21 sh.), 2 do. p. set 14 Th. (42 sh.).

1 doz. towels 8 Th. (24 sh.), 1 do. 7⅔ Th. (23 sh.).

2 pieces of linen cloth p. piece 31 Th. (93 sh.), 1 do. 32 Th. (96 sh.).

384. Schulze, D., linen-manu., Bodenteich. Berlin 1844, London 1851 and Munich 1854 honor. mention; London 1862 medal.

			p. ell.	p. yard.
			Th.sg.	sh.d.
69 Hanov. ells bleached linen			2 —	9 5
4 »	» do.		1 10	6 3
66 •	» do.		— 20	3 2
66 •	« do.		— 20	3 2
67 »	» do.		— 14	2 2½
35 »	» unbleached linen		1 25	8 7
45 »	» do.		— 13	2 —

Samples of bleached and unbleached linen, fine hand-spun linen yarn and finest flax.

385. Siemsen, Joh., Hanover. Munich 1854 medal of honor; London 1862 honor. mention.

7 bags of manila hemp 2—4 Th. (6—12 sh.), 3 manila-hemp cords (to raise one's self in bed) 1½ to 4⅓ Th. (4 sh. 6 d. to 13 sh.), 23 bell-ropes of manila hemp 1½—15 Th. (4 sh. 6 d. to 45 sh.).

CLASS XXII.

CARPETS.

389. Roskamp, Carl, Springe. London 1862 honor. mention.

3 pieces of fine Paris carpet 20 sg. (2 sh.), 1 do. striped twilled carpet 10 sg. (1 sh.), 1 do. fine Paris stair-carpet 14 sg. (1 sh. 8 d.), 1 do. Dutch stair-carpet 8 sg. (10 d.), 1 do. samples of striped Dutch carpet 13 sg. (1 sh. 7 d.), all p. flemish ell.

CLASS XXV.

SKINS, FUR, FEATHERS AND HAIR.

390. Leppien, Georg (Gottfried Leppien), Lüneburg. London 1862 honor. mention.

1 piece hair-seating, 26 in. wide, p. yard 2 sh. 3 d.; 1 do. 34 in., p. yard 4 sh.

CLASS XXVI.

LEATHER, INCLUDING SADDLERY AND HARNESS.

391. Goldschmidt, W., Wölpe. London 1862 honor. mention.

horse-leather: ½ doz. shoe-vamps 8 Th. (24 sh.); ⅓ doz. front parts of boots 10 Th. (30 sh.); ⅓ doz. back parts of boots 4 Th. (12 sh.); sample of black horse-leather 25 sg. (2 sh. 6 d.) all p. dozen.

392. Schritzmeyer, J. H., Harburg. London 1862 honor. mention.

Leather.

393. Wolff & Rohte, Walsrode. London 1862 medal.

	p. 100 lbs.		p. 112 lbs.		
			s. d. p.		
	Th. sg.	£	sh.	d.	
2 pieces of Buenos-Ayres sole-leather	52 —	7	18	6	
2 pieces dressed German cow-leather	— 19	—	1	9	
⅘ pieces dressed and blackened Buenos-Ayres horse-leather	— 27	—	2	5½	
			p. dozen		
⅙ doz. boot-vamps of dressed Buenos-Ayres horse-leather	10 —	—	30	—	
¼ doz. boot-vamps of dressed Buenos-Ayres horse-leather	7 15	—	22	6	
⅙ doz. shoe-vamps do. 4 Th.	4 —	—	12	—	
⅙ doz. boot-back-pieces do.	4 —	—	12	—	

CLASS XXVII.

ARTICLES OF CLOTHING.

394. Froeling, Carl, & Sohn, felt-manu., Münden. London 1862 honor. mention.

3 pairs of travelling boots, 1 do. boots, 1 do. overboots, 2 do. hose, 2 do. man's socks, 5 do. ladies' socks, 1 do. do. laced boots, 1 do. do. booteens, 2 do. young ladies' boots, 2 do. children's boots, 2 do. riding slippers, 3 do. booteens, 5 do. men's shoes, 12 do. ladies' shoes, 4 do. young ladies' shoes, 5 do. children's shoes, 2 pieces of felt, 1 filtering funnel, 2 saddle covers, 2 pieces printed felt-carpeting.

397. Hugo, Gebrüder (Wilhelm Hugo), Celle. London 1862 medal.

13 silk umbrellas, 12 do. en-tout-cas, 41 sticks for silk umbrellas, 30 do. for en-tout-cas, 3 silk parasols, 13 bundles patent-cane, 13 do. new-whalebone.

398. Keunecke, W., Peine.

3 pairs of gentleman's boots, $4\frac{2}{3}$, $4\frac{1}{2}$, $4\frac{1}{6}$ Th. (14 sh., 13 sh. 6 d., 12 sh. 6 d.), 2 do. ladies' cloth-booteens, pair 3 Th. (9 sh.), 2 do. do. leather-booteens, pair 3 Th. (9 sh.), 3 do. do. laced leather-booteens, pair $2\frac{2}{3}$ Th. (8 sh.), 3 do. do. cloth-booteens, pair $2\frac{3}{5}$, $2\frac{1}{3}$ and $2\frac{1}{3}$ Th, (7 sh. 10 d., and 7 sh.), 1 do. do. laced cloth-booteens, pair $2\frac{1}{3}$ Th. (7 sh.)

400. Pollmann, C. A., Hanover.

36 silk umbrellas 3 to 10 Th. (9 to 30 sh.), 7 cotton do. $\frac{2}{3}$ to $1\frac{1}{6}$ Th. (2 to $3\frac{1}{2}$ sh.), 4 alpaca do. 2 to $2\frac{1}{2}$ Th. (6 to $7\frac{1}{2}$ sh.), 30 silk en-tout-cas $1\frac{1}{3}$ to $5\frac{1}{2}$ Th. (4 to $16\frac{1}{2}$ sh.), 6 do. parasols $2\frac{5}{6}$ to 10 Th. ($8\frac{1}{2}$ to 30 sh.)

401. Voges, Carl, Peine.

Silk bell-rope 20 Th. (3 £); 3 gross silk waistcoat-buttons hand-made, 1 gross $2\frac{1}{2}$ Th. (7 sh. 6 d.), 2 of them each 2 Th. (6 sh.); 2 doz. sweepers for cleaning lamp-chimneys, p. doz. 3 sh.

CLASS XXVIII.

PAPER, STATIONERY, PRINTING AND BOOKBINDING.

402. Edler & Krische, Hanover. London 1862 honor. mention.

16 different account-books, viz: 1 ledger 90 Th. (13 £ 10 sh.), 1 account-current 17 Th. 15 sg. (2 £ 12 sh. 6 d.), 1 ledger 15 Th. (2 £ 5 sh.), 1 do. 6 Th. 10 sg. (19 sh.), 1 cash-account 17 Th. 15 sg. (2 £ 12 sh. 6 d.), 1 journal 12 Th. 25 sg. (1 £ 18 sh. 6 d.), 1 do. 14 Th. 15 sg. (2 £ 3 sh. 6 d.), 1 memorial 8 Th. 10 sg. (1 £ 5 sh.), 1 journal 6 Th. (18 sh.), 1 memorial 6 Th. 10 sg. (19 sh.), 1 Journal 2 Th. $27\frac{1}{2}$ sg. (8 sh. 9 d.), 1 memorial 2 Th. 10 sg. (7 sh.), 1 letters 24 sg. (2 sh. 5 d.), 1 do. 23 sg. (2 sh. 4 d.), 1 do. 22 sg. (2 sh. $2\frac{1}{2}$ d.), 1 do. $20\frac{1}{2}$ (2 sh.).

405. Jaenecke, Gebrüder, printers and lithographers to H. M. the King of Hanover, and type founders, Hanover. Munich 1854 medal of honor.

Specimens of prints made with the types from the exhibitor's own foundery; table printed in colors; a portfolio containing colored printings; 6 different typographic works; a large table containing the plan and 21 views of the town of Hanover in colored print.

406. Jaenecke, Gebrüder, & Schnee-mann, Fr., Hanover. Munich 1854 medal of honor; London 1862 medal. Agts. Stoer Brothers, Vulcan Wharf, Earl-street, London E. C.

Black ink for typography. *a.* For hand-presses: No. 000. extrafine ink for splendid printing p. lb. 1 Th. 18 sg. 0. fine do. p. lb. 1 Th. $2\frac{1}{2}$ sg.; A 00. finest ink for picture-printing 78 Th., A 1. fine ink for type-printing 38 Th., B 1. middle quality for do. 30 Th., C 1. common ink for do. 25 Th. p. 100 lb.

b. For machine presses: A 00. finest ink for picture-printing 78 Th., A 0. fine do. 53 Th., A 3. fine do. for type-printing 38 Th., A 5. do. 25 Th., B 4. middle quality for do. 23 Th., C 4 a. ink for newpapers 18 Th. all p. 100 lb.

Fine colored printing inks: No. 1. fine carmine-lake, 2 Th. 5 sg., 2. do. vermilion 1 Th. 15 sg., 3. ultramarine 2 Th. 5 sg., 4. French blue 1 Th. 15 sg., 5. chrome-yellow 1 Th. 10 sg., 6. Venetian white 20 sg., 7. blue ink for ruling books 15 sg., all p. lb.

Black lithographic inks: D. ink for pen-drawings 16 sg., E. do. for engraving 14 sg., F 1. for pencil-drawing I. 3 Th. 15 sg., F 2. do. II. 1 Th. 2 sg., G. do. for autographing 1 Th. 18 sg., H. do. for bronzing 1 Th. 2 sg., all p. lb.

Soots: finest lamp-black; fine soot from pine-wood.

A list of prices for English weight and money is exhibited.

407. Klindworth, F., printer to H. M. the King of Hanover. Hanover.

Publications printed in oil-colors: genealogical table of the Guelfs 18 Th. (54 sh.), the same, small edition 2 Th. (6 sh.); the arms of the King of Hanovre 20 sg. (2 sh.); the arms of the Duke, of Brunswick 1 Th. 24 sg. (5 sh. 5 d.), portrait of King Ernst August 15 sg. (1 sh. 6 d.); view of the spinning-manufactory at Hanover 20 sg. (2 sh.), the museum at Hanover 4 Th. (12 sh.), Christ-church at Hanover; Marienburg castle in colors 2 Th. 20 sg. (8 sh.); do. in sepia 1 Th. 20 sg. (5 sh.); Ernst-August-Album 5 Th. (15 sh.).

408. Koch, J. H. A., Freyburg a. E. London 1862 honor. mention.

Patent writing-ink for steel pens.

409. Koenig, J. C., & Ebhardt, account-book-manu., Hanover. Munich 1854 honor. mention; London 1862 medal.

Assortment of ruled and bound account-books, type-paged or folioed and printed headings:

Foolscap, $13\frac{3}{8}$ and $9\frac{1}{8}$ in.: 1 minute book 1 Th. 14 sg. (4 sh. 5 d.), 1 do. 1 Th. 4 sg. (3 sh. 5 d.). 1 memorandum 25 sg. (2 sh. 6 d.), 1 minute book 25 sg. (2 sh. 6 d.), 1 do. $12\frac{1}{2}$ sg. (1 sh. 3 d.), 1 cash book $27\frac{1}{2}$ sg. (2 sh. 9 d.), 1 ledger 4 Th. 5 sg. (12 sg. 6 d.), 1 private ledger 3 Th. 20 sg. (11 sh.)

Demy, $14\frac{1}{2}$ and $9\frac{1}{8}$ in.: 1 cash book 2 Th. 10 sg. (7 sh.), 1 memorandum 2 Th. 10 sg. (7 sh.), 1 ledger 3 Th. (9 sh.).

Medium, $16\frac{1}{4}$ and $10\frac{1}{4}$ in.: 1 journal 3 Th. 12 sg (10 sh. $2\frac{1}{2}$ d.), 1 memorandum 3 Th. 12 sg. (10 sh

$2\frac{1}{2}$ d.), 1 cash book 3 Th. 12 sg. (10 sh. $2\frac{1}{2}$ d.), 1 ledger 4 Th. $27\frac{1}{2}$ sg. (14 sh. 9 d.), 1 bill-book 3 Th. 12 sg. (10 sh. $2\frac{1}{2}$ d.), 1 correspondence-book 4 Th. 15 sg. (13 sh. 6 d.).

Royal, $17\frac{1}{2}$ and $10\frac{1}{2}$ in.: 1 account current 6 Th. 15 sg. (19 sh. 6 d.), 1 bill-book 6 Th. 15 sg. (19 sh. 6 d.).

Super royal, $19\frac{1}{2}$ and $12\frac{1}{2}$ in.: 1 account current 13 Th. 10 sg. (40 sh.), 1 numerical register of shares 15 Th. 25 sg. (47 sh. 6 d.).

Imperial, 22 and $15\frac{3}{4}$ in.: 1 general ledger 22 Th. (66 sh.).

Octavo: 1 rent-book 1 Th. (3 sh.), 1 travelling-ledger 25 sg. (2 sh. 6 d.), 1 do. 20 sg. (2 sh.), 1 travelling-adressbook 25 sg. (2 sh. 6 d.).

411. Schnell, Phil., paper-manu., near Münden. London 1862 honor. mention. Agts. Lion M. Cohn, Phaland & Dietrich of Berlin, 20 St. Dunstans-Hill, City. (s. II. No. 68.)

	Sizes, in.		Formats, cm.	
	20:24	17:22	51:61	$43\frac{1}{2}$:56
	p. ream		la rame	
	sh.	sh.	fr.	fr.
Middle glazed paper, different colors	18	14	$22\frac{1}{2}$	$17\frac{1}{2}$
Fine glazed paper, different colors	20	$15\frac{1}{2}$	25	$19\frac{1}{2}$
Extra fine glazed paper, as:				
bronze-brown	$25\frac{1}{2}$	$19\frac{1}{2}$	31	$24\frac{1}{2}$
steel-blue	31	23	39	$29\frac{1}{2}$
copper-blue	31	23	39	$29\frac{1}{2}$
Bremen-blue . . . . } Bremen-green . . . . }	$25\frac{1}{2}$	$19\frac{1}{2}$	31	$24\frac{1}{2}$

The ream contains 480 sheets weighing $25\frac{1}{2}$ and 20 lb. a. d. p. — Free delivered in any seaport of England or France.

CLASS XXX.

FURNITURE AND UPHOLSTERY, INCLUDING PAPER-HANGINGS AND PAPIER-MACHÉ.

412. Behrens, C., Alfeld (manu. of wooden ware at Kaierde on the Hils). London 1862 medal.

Assortment of shoe-lasts for gentlemen, ladies, youth and children in many different forms, from 2 sh. 3 d. to 11 sh. 6 d. p. doz. of pairs.

413. Herting, Carl, paper-hanging manu., Einbeck. London 1862 honor. ment. Agt. H. Beisner, 27 Duke-street, Grosvenor square, W. (s. II. No. 23.)

Decoration, containing: Flock paper-hanging with imitated mother of pearl 4 £ 10 sh., flock paper-hanging with gold-print 10 sh. 6 d., paper-hanging, imitation of wood 1 sh. 10 d., all p. piece; bordure 3 sh. p. yard; corner pieces and clasp, printed in gold and flock, p. clasp and 2 corners 3 sh. 6 d.

414. Schwikkard, Fr. (G. Schwikkard), Goslar.

Assortment of 110 different objects of alabaster-marbre from the Harz; as: vases, chests, cups, boxes, candlesticks, letter pressers, pin cushions, writing stands, match-boxes etc. etc. from 3 d. to 12 sh. p. piece.

CLASS XXXI.

IRON AND GENERAL HARDWARE.

416. Bernstorff, C. & Eichwede, manu. and statue-founders; Hanover. Berlin 1844 silver-medal; Munich 1854 medal of honor; London 1862 medal.

Bronze bust of General Count Alten 45 £.

A colossal lion of-Bronze in lying attitude, appointed for the Royal Welfen-palace at Hanover, 600 £.

417. Breitenbach, J., Nörten near Göttingen.

Sheet-iron ware, as: 1 coffee drum with frame 8 d., 1 do. for the economical hearth 8 d., 1 frying pan 6 d., 1 pail for ashes 5 d., 2 pot-lids, 2 fire covers, 2 fire shovels, 1 shovel for sweepings, 5 pans without legs $11\frac{1}{2}$ d. to 1 sh. 8 d., 3 do. for frying meat and sausages $7\frac{1}{2}$—$9\frac{3}{4}$ d., 3 do. for frying bacon $4\frac{1}{2}$—6 d., 6 do. with legs $4\frac{1}{2}$—$9\frac{3}{4}$ d., 6 coffee roasters 5 d. to 1 sh., 8 melting spoons $2\frac{1}{2}$—11 d.

418. Du Bois, J. E., pewterer to H. M. the King of Hanover, Hanover. Munich 1854 medal of honor; Paris 1855 honor. mention; London 1862 medal and honor. mention.

Steam-apparatus for pharmaceutic use 75 £.

Assortment of about 1400 different pewter figures in 12 tables and 3 chests 6 £ 12 sh.

419. Haendler & Natermann, lead manu., Münden. London 1862 medal.

$\frac{1}{2}$ roll of rolled lead, 8 ft. large, 5 mm. thick; $\frac{1}{4}$ do. 8 ft. large, 0,9 mm. thick; 13 tinned leaden tubes from $\frac{1}{8}$ in. to 5 in. wide, 4 coils of leaden tubes from $\frac{1}{8}$—$\frac{1}{2}$ in.; 3 leaden pump-chambers of 3, 4 and 5 in., 1 coil of tinned lead-wire, 7 mm. thick, 10 small coils of lead-wire, 19 bags of shot, No. $\frac{5}{0}$—14., 4 do. of small bullets, No. I.—V., 6 do. of leads for custom-houses, 6 sounding-leads, a book of samples of tin- and lead-foils.

CLASS XXXII.

STEEL AND CUTLERY.

420. Graboh, A., cutler to H. M. the King of Hanover, Hanover. London 1862 medal.

1 doz. table knives and forks in ivory with silver 6 £ 15 sh., 1 doz. dessert do. 5 £ 5 sh., 1 carving case in ivory with silver 1 £ 7 sh., 1 do. in hartshorn with silver 1 £ 7 sh., 1 hunting knife 9 sh., 2 breadknives in mother of pearl 18 sh., 2 large pruning knives in hartshorn 1 £ 4 sh., 2 small do. 1 £ 1 sh., 2 daggers in ivory 1 £ 10 sh., 2 nail-nippers 18 sh., $\frac{1}{2}$ doz. table knives and forks in deer-feet with silver 5 £ 8 sh., 1 carving case 1 £ 7 sh., 3 hunting knives 1 £ 16 sh.

421. Lage-Schulte, G., file-maker, Hanover.

6 arm-files 5 sh. 5 d., 6 hand-files 5 sh. 5 d., 6 bastard-files 2 sh. 3 d., 6 halfround files 3 sh., 6 three-square files 2 sh. 4 d., 6 round files 2 sh. 3 d., 6 flat files, smooth 3 sh. 6 d., 6 halfround do. 4 sh., 6 farriers rasps 2 sh. 3 d., all p. piece.

422. Pfuhl, H. C., file-maker; Linden near Hanover.

1 hand-file 9 sh., 1 do 5 sh., 4 machine-files, bastard 20 sh. 5 d., 2 do. smooth 14 sh. 3 d., 2 three-square, smooth, 15 in. 4 sh. 8 d., 1 round file, do. 16 in. 2 sh. 8½ d., 1 square do. 15 in. 2 sh. 4 d., 1 halfround do. 13 in. 1 sh. 8 d. 1 round do. bastard 17 in. 2 sh. 8 d., 1 square do. do. 16 in. 2 sh. 3 d., 1 halfround do. do. 12 in. 1 sh. 1½ d., 2 do. do. do. 10 in. 1 sh. 5 d., 1 do. saw-file 6 in. 4½ d., 3 triangular do. do. 5 in. 1 sh. 6 d., 3 files 11 d., 1 farrier's rasp. 1 sh. 6 d., 1 shoemaker's rasp. 2 sh., 1 do. do. 1 sh., 1 rasp for wood, rough, 13 in. 10½ d., 1 do. do. do. 10 in. 6 d., 1 do. do. fine 13 in. 1 sh. 4 d., 1 do. do. do. 10 in. 8 d., 1 tooth-rasp. 4 sh. 6 d., 1 pointing wheel for pin-maker's 15 sh., 1 riffler 1 sh. 6 d.

CLASS XXXIV.

GLASS.

423. Pezolt & Schele (G. W. Schele), glass-manu., Osterwald near Elze.

Assortment of more than 100 species of cheap hollow glass-ware, as wine-bottles, ale-bottles, drinking-glasses, medicine-glass, etc. etc.

The prices of bottles are as follows: Champagne-bottles $\frac{1}{1}$ 17 sh. 3 d., $\frac{1}{2}$ 12 sh. 6 d., $\frac{1}{4}$ 9 sh.; half-green Bordeaux-bottles $\frac{1}{1}$ 12 sh., Bordeaux-bottles $\frac{1}{1}$ 8 sh. 6 d., $\frac{1}{2}$ 6 sh.; English patent-bottles $\frac{1}{1}$ 9 sh. 3 d.; dark ale-bottles $\frac{1}{1}$ 9 sh. 3 d., Hock-bottles $\frac{1}{1}$ 9 sh. 6 d., brown Hock-bottles $\frac{1}{1}$ 9 sh. 9 d., $\frac{1}{2}$ 6 sh. 9 d. The prices refer to 100 bottles.

ELECTORATE OF HESSIA.

CLASS II.

CHEMICAL SUBSTANCES AND PRODUCTS AND PHARMACEUTICAL PROCESSES.

431. HABICHS, G. C., SONS, color-manu., Cassel. Berlin 1844 prize-medal; London 1862 honor. mention.

60 specimens of colors for dying linen cloths, for painters, for the manufacture of paperhangings and window-blinds.

CLASS IV.

ANIMAL AND VEGETABLE SUBSTANCES USED IN MANUFACTURES.

433. RUPERT, C., & Co., manu. of toilet-soaps and perfumes, Cassel. Agts. Phaland & Dietrich, 20 St. Dunstan's Hill, City.

Four assortments of toilet-soaps, eau de Cassel and eau de Cassel quintessence.

CLASS VIII.

MACHINERY IN GENERAL.

434. HENSCHEL & SON, machine-manu., Cassel. London 1862 medal. Agts. Phaland & Dietrich, 20 St. Dunstan's Hill, City.

1. Tubular steam-boilers, invented and constructed by Henschel and Son, rewarded by the société d'encouragement in Paris with a golden prize-medal and the sum of 6000 frs. 2. Revolving slide with variable expansion for steam-engines invented and improved by Henschel and Son.

CLASS XIII.

PHILOSOPHICAL INSTRUMENTS AND PROCESSES DEPENDING UPON THEIR USE.

436. BREITHAUPT, F. W., & SON, mathematical and mechanical institute, Cassel. Berlin 1844 bronze medal; London 1851 prize-medal, 1862 medal. Agt. G. Heydenreich, mechanician 11 Grove Road, Mile-End Road, London.

Repeating theodolite No. 453. with a graduated circle of 12 in., achromatic telescope with an aperture of 18 lines 59 £ 5 sh. 2 d.; repeating theodolite No. 533. with a graduated circle of 7½ in., achromatic telescope with an aperture of 16 lines 23 £ 14 sh. 1 d.; surveyors-apparatus with table and stand 6 £ 1 sh. 6 d.; rule of te-lescope with vertical circle 7 £ 9 sh. 7½ d.; large surveyor's table with stand 7 £ 17 sh. 1 d.; large levelling-instrument No. 615. with telescope and air-level 13 £ 6 sh. 8 d.; compensating-level No. 102. 8 £ 3 sh.; simple instrument for measuring angles 3 £ 8 sh. 2 d.; large miner's compass No. 738. with covering plate, hanging, graduated circle 12 £ 14 sh. 1 d.; table air-pump 10 £ 7 sh. 5 d.; two rules, each of one metre 6 £ 4 sh. 5 d.; rule consisting of a plate divided by hand 7 £ 8 sh. 2 d.

437. FENNEL, OTTO, mechanician, Cassel.

Large levelling-instrument with distance-measurer; small levelling-instrument; miner's surveying-instrument (compass, clinometer and plate belonging to it); pocket compass-apparatus (compass, pendant and plate belonging to it).

CLASS XXI.

WOOLLEN AND WORSTED, INCLUDING MIXED FABRICS GENERALLY.

439. BRAUN, BROTHERS, cloth-manu., Hersfeld. London 1851 prize-medal; Munich 1854 prize-medal. Agts. Phaland & Dietrich, 20 St. Dunstan's Hill, City.

Seven pieces of woollen cloth No. 98,444. 8 sh., No. 98,447. 9 sh., No. 98,439. 6 sh. 9 d., No. 98,523. 7 sh. 9 d., No. 98,497. 6 sh. 9 d., No. 98,528. 7 sh. 9 d., No. 98,486. 7 sh. 6 d. p. yard each.

440. BURKARD MÜLLER, manu., Fulda. Munich 1854 honor. mention; London 1862 medal and honor. mention. Agts. Lion M. Cohn, Phaland & Dietrich, 20 St. Dunstan's Hill, City, and Schunck, Souchay & Co.

An assortment of colored woollen zephir-yarns in 84 shades; 50 coupons of canvas for embroideries; 30 coupons of plush stuffs (velour d'Utrecht) for furniture, bags, waist-coats, shoes etc.; cards of samples of colored stramin-stuffs, plushes, lastings and serges de berry.

CLASS XXII.

CARPETS.

441. GEIBEL, FRITZ, designer, Kesselstadt near Hanau. London 1862 honor. mention. Agt. E. G. Zimmermann, 2 St. Paul's Buildings, Little, Carter Lane, Doctor's Commons.

Carpet-patterns.

CLASS XXV.

SKINS, FUR, FEATHERS AND HAIR.

442. ALLENDÖRFER, J. C., furrier, Cassel. Munich 1854 honor. mention; Paris 1855 silver medal; London 1862 medal. Agt. William Schnell, 63 Charlott-street, Fitzroy square, London.

Furs and hunting-produce of Germany undressed and dressed, viz: fur-coats, carpets, muffs, collars, skins etc.

443. RIEBE, CONRAD, furrier to the court, Cassel. Agts. Lion M. Cohn, Phaland & Dietrich, 20 St. Dunstan's Hill, City.

Suit of mink-skins 24 £; suit of fitchet-skins 8½ £; suit of rabbit-skins 2½ £; carpet of fox-skins 9 £; footstool 1½ £.

CLASS XXIX.

EDUCATIONAL WORKS AND APPLIANCES.

445. FISCHER, THEODOR, Cassel. Agts. Trübner & Co., 12 Paternoster-row, London.

Relief-plans: Sinai and Golgatha, relief of the holy land and its environs, with goldframe 10 £, without goldframe 6 £; relief-plans of Jerusalem, with goldframe 1 £, without goldframe 6 sh.

446. SCHELLER, WITTICH & SCHERB, manu. of toys, Cassel. London 1862 honor. mention.

Toys, viz: military accoutrements, guns, games, building-toys, furniture, doll-rooms, shops and kitchens, tool-chests in steel and iron, rocking horses, pianos for children.

447. WEBER, A., & ALLMEROTH, manu. of toys, Marburg. London 1862 medal. Agts. Krause & Auerbach, 68 Basinghall-street, E. C.
(s. II. No. 19.)

Toys: chromatic and diatonic glass-pianinos, furniture, tool-chests, muskets, rifles, pistols, crossbows, games etc.

CLASS XXXI.

IRON AND GENERAL HARDWARE.

449. ZIMMERMANN, E. G., manu., Hanau. London 1851 honor. mention; Munich 1854 prize-medal; Paris 1855 bronze medal; London 1862 medal. Agt. E. G. Zimmermann, 2 St. Pauls Buildings, Little Carter Lane, Doctor's Commons.

Artistic and useful objects in iron, zinc- and bronze-castings, as well as in marble: portraits busts and statues, figures, animals, clock-cases, inkstands, matchholders, candlesticks, cigarstands, watchstands etc.

CLASS XXXIII.

WORKS IN PRECIOUS METALS AND THEIR IMITATIONS AND JEWELLERY.

450. WEISHAUPT, C. M., SONS, manu. of jewellery, Hanau. Berlin 1844 gold medal; London 1851 council medal, 1862 medal. Agt. Philipp Wohack, 33 Frith-street, Soho.

Assortment of jewellery with and without precious stones number 1 to 75; five gold snuff-boxes, one of them with diamonds, the others with enamel; assortment of crosses of orders of knighthood 19 pieces.

CLASS XXXV.

POTTERY.

451. GOEBEL, J. P., & SONS, manu. of crucibles, Grossalmerode. Agt. C. Trübner, 20 St. Dunstan's Hill, City.   (s. II. No. 25.)

Plumbago-crucible of 600 kilos, already used, according to the testimony of the mint at Frankfort o. M. having been in fire 250 hours; plumbago-crucible of 600 kilos also already used, and according to the testimony of the royal Prussian mint at Berlin having suffered 36 meltings; 9 new plumbago-crucibles; 6 sets of Hessian crucibles; chest of clay-pipes.

452. PFEIFFER, C. H., manu., Cassel. Agts. Lion M. Cohn, Phaland & Dietrich, 20 St. Dunstan's Hill, City.

Cassel porcelain consisting of pitchers, vases, tableplates, flowerpots etc.

GRAND-DUCHY OF HESSE.

CLASS I.

MINING, QUARRYING, METALLURGY AND MINERAL PRODUCTS.

461. ADMINISTRATION OF THE GRAND DU-CAL COPPERFURNACE at Thal-Itter. London 1862 honor. mention.

Refined copper.

462. TASCHE, HANS, salt-work-inspector and surveyor of mines at Salzhausen in the Wetterau, Province of Oberhessen. London 1862 medal.

Collection of the technically important rocks of the Wetterau and of the Vogelsberg in 60 various specimens, arranged according to the system and scientific works published by the exhibitor. The collection is intended to serve in general as a model for technico-mineralogical collections and is therefore grouped after a peculiar idea of the exhibitor. The exhibitor wishes to dispose of the collection, but by particular agreement only.

CLASS II.

CHEMICAL SUBSTANCES AND PRODUCTS AND PHARMACEUTICAL PROCESSES.

463. ASSOCIATION »BLAUFARBWERK MARIEN-BERG« at Marienberg near Bensheim. Manager: Reinhold Hoffmann Ph. D. Established 1853. Paris 1855 bronz. medal; London 1862 medal.

Ultramarine-colors, consisting of: I. Those called Royal. 1 sample of blue raw ultramarine; 3 samples of finished ultramarine of a violet-blue shade, not affected by alum. II. Those denominated »Prin-cess«: 2 samples of finished ultramarine of true blue shade. III. Green: 1 sample of raw and 2 samples of finished green ultramarine. IV. samples of all the different sorts of ultramarine brought to the market with information of the prices and of the principal applications.

464. BÜCHNER, WILHELM, ultramarine-manu., Pfingstadt near Darmstadt. London 1851 prize-med.; Munich 1854 medal of honor; Paris 1855 silver medal.; London 1862 medal. Agt. H. Kohnstamm, 33 Dowgate Hill, Cannon-street.

1. Raw ultramarine such as it comes out of the kiln. 2. 8 glasses containing ultramarine in its finished state and of different shades and degrees, viz: 1. imperial extra; 2. imperial I.; 3. imperial II.; 4. imperial dark (1—4. for paper-making): 5. pure blue extra; 6. pure blue I.; 7. pure blue II.; 8. pure blue dark (5—8. for printing cloth and paper-hangings).

465. MERCK, EMANUEL, chemical works, Darmstadt. London 1862 medal. Agts. Ammann, Roller & Co., 3 Love-Lane, Eastcheap E. C.

Chemical manufactures.

466. OEHLER, K., chemical works, Offenbach. London 1851 honor. mention; Paris 1855 bronz. medal; London 1862 medal.

Anilin violet precipitated; anilin violet, tincture; anilin, red yellowish, syrup; anilin, red yellowish, tincture; anilin, red yellowish, cristals; anilin red, bluish, syrup; anilin red, bluish, tincture; anilin red, bluish, crystallized; benzole, cleansing liquid; coal-naphtha I.; coal-naphtha II.; naphthaline depurated; naphthaline crude; heavy naphtha; asphaltum-varnish; nitrate of iron for dyeing black; nitrate of iron for dyeing blue; nitrate of iron pure; acetate of iron; muriate of tin; oxymuriate of tin crystallized; oxymuriate of tin, liquid; pink salt; essence of rosin; oil of rosin; nitric acid; glauber-salt; compound of tartar for dyeing black; compound of tartar for dyeing in colors; black dyeing powder; Chromate of copper; Cuba wood, rasped; Campeachy wood, rasped; St. Martha wood, rasped; Pernambuco, rasped; Hungarian fustic, cut; Logwood, sliced; Logwood, hacked; Logwood, granulated; Logwood, groats; Logwood, powder; Campechay wood, powder; Lima wood, powder; calliatur wood, powder; red sanders wood, powder, fine; asphaltum; coal-tar pitch; pitch for shoemakers, yellow; pitch for shoemakers, black; grease, blue; grease, brown.

467. PETERSEN & Co., chemical works, Offenbach, represented by Theodor Petersen, Dr. phil., and Eduard Schmidt, civil-engineer. London 1862 honor. mention.

Products from pit-coal tar: pit-coal tar; heavy tar-oil; light tar-oil; benzole; poluole; xylole; cumole; cymole; nitrobenzole; aniline; aniline red; anilin violet; creosote; phenic acid; picric acid; sulphate of ammonia; naphthaline (crude); naphthaline (pure); pitch; asphalt-mortar; lacker; iron-lacker; asphalt-varnish; asphalt-plates; powder for preventing incrustation in steam boilers.

468. ALBRECHT. J., manu.. Mentz. Munich 1854 honor. mention; London 1862 honor. mention. Agts. Mittler & Eckhardt, 6 Grocer's Hall Poultry E. C.

Asphalt; amber and copal-, furniture-, carriage and tin-lacker etc.; spirit of wine varnish for pictures, leather. paper, metal and objects of wood etc.; polish of copal and lac: refined shell-lac, white, blond. red etc.; bottle-lac for lackering all sorts of bottles, only in fine colors.

469. MAYER, F. H., manu., Mentz. London 1862 honor. mention.

Varnish, shell-lac, sealing-wax.

470. MEHL & MOSKOPP, varnish-makers, Mentz. London 1862 honor. mention.

Bleached refined shell-lac with silver-lustre; do. mixed with colored do.; spirit-varnishes and all kinds of colored metal-varnishes; copal-, amber- and other kinds of oil-varnishes.

471. MELLINGER, DR. CARL (Carl Mellinger), manu. of lac and varnish, Mentz. London 1862 medal.

1. Lasting varnish for carriages (flowing) 170 fl.; 2. best wearing body varnish 170 fl.; 3. elastic carriage-varnish 140 fl.; 4. best hard body varnish 160 fl.; 5. best black Japan 130 fl.; 6., 7., 8. varnish for carriages No. 1., 2., 3., German manner, 130, 115, 100 fl.; 9., 10., 11. varnish for furniture No. 1. 120 fl., No. 2. 110 fl., No. 3. 100 fl.; 12. Lasting varnish for tin-goods (flowing) 120 fl.; 13. varnish for tin-goods No. 1. 110 fl.; 14. varnish of Damar No. 0. 70 fl., 15. No. 1. 60 fl.; 16. bleached linseed-oil 40 fl.; 17. oil for gilding 80 fl.; 18. painter's varnish No. 1. 250 fl., 19. No. 2. 200 fl.; 20. varnish for bookbinders 175 fl.; 21. varnish for sculptors 175 fl.; 22. white copal polish 175 fl.; 23. brown copal polish 150 fl.; 24. shell-lac polish 150 fl.; 25. varnish for metals, blue 200 fl.; 26. varnish for metals, red 200 fl.; refined shell-lac, white 160 fl., brown 140 fl., in other colors 160 fl. p. 50 kilogr.

472. CHRISTOPH SCHRAMM, chemikal works, Offenbach. London 1862 medal.

Oil varnish: best elastic carriage-varnish; railway-waggon-varnish No. 2. and 3.; grinding elastic carriage-varnish No. 1. and 2.; varnish for tin-wares; asphaltum varnish No. 1.; varnish for sugar-moulds; black varnish for tin-wares; varnish for sticks; Damar varnish No. 1.; West-indian copal-varnish white No. 00.; amber varnish of corals No. 1.; table and floor-varnish; copal furniture-varnish No. 1., 2. and 3.; mastix varnish for tapestry; fire-proof metal-varnish; black oil-cloth varnish; best black enamel-varnish; black Japan; Japan gold-size, clear; Japan gold-size, dark; polishing varnish.

Varnish of linseed-oil, double white; varnish of linseed-oil, white; varnish of linseed-oil for lithographers, very feeble No. 0., feeble No. 1., middle strong No. 2., 3., strong No. 4., double strong No. 5.; goldprinting varnish; mordant varnish for paperhangings.

Dry colors: American green No. 1.; fine chrome-yellow clear No. 0., do. No. 2.; chrome-yellow dark No. 1.; crimson color; improved chrome-green for fresco painting; chrome-orange No. 1.; new green No. 0.; Neuwied-green No. 0. yellow; Neuwied-green No. 0. blue; Neuwied-blue; Paris-blue; Schweinfurth-green, double distilled; violet color in oil, brown; cinnabar green A. 1. E. 1. S. 1. M. 1. 3. N. 1. 3.; do. prepared in oil; absolutely pure white lead in oil; floor ground, varnish-clear.

Black printing-ink: for printing-machines, strong and middle strong, for common work; strong and middle strong for fine works and illustrations; for manual printing: strong and middle strong for common work, strong for very fine editions, for very fine editions with woodcuts, for very fine editions with woodcuts, brown.

Printing-ink in colors: Paris blue, mahogany-color; snow-white; chrome-orange.

Black printing-ink for lithographs: for chalk-drawings No. 1.; with pens No. 1. and 2.; for engraving No. 1. and 2.

Printing-ink for lithographs in colors: miloriblue; cinnabar carmin; chrome-red; chrome-yellow, clear; chrome-yellow, dark; chrome-green, clear and dark. Lamp-soot, superior calcined; fine calcined.

Spirit-varnishes; best East-Indian copal, brilliant white; white copal for paintings in water-colors; white copal for furniture; East-Indian copal, red-brown; varnish for morocco leather.

473. VOLTZ, A., manu. of metal-cement, Mentz. Agt. Charles Sevin, 155 Fenchurch-street E. C.

Metal-cement for steam-engines, conduit-pipes etc. In place of minium.

CLASS III.
SUBSTANCES USED FOR FOOD, INCLUDING WINES.

474. WAGNER, JOH. PH., & Co., manu. of chocolate, Mentz.

Samples of Chocolate.

475. ENGELHARDT, F., manu. of chicory-coffee, Rüsselsheim near Mentz. London 1851 prize-medal, 1862 honor. mention.

Dried chicory-roots; fine chicory-flour; fine chicory-grits; homoeopathic coffee.

476. JUNGBLUTH, JOH. V., chicory-manu., Worms on the Rhine. London 1862 honor. mention.

No. 1. Moka-coffee; No. 2. cacao-coffee; No. 3. economical-coffee; No. 4. acorn-coffee; No. 5. substitute for coffee; No. 6. three different sorts of continental coffee.

477. CANTOR & SON, wine-merchant and manu. of sparkling Hock, Mentz. Manufactory Weisenau near Mentz. Agts. Cantor & Son, 11 St. Mary-at-Hill E. C.

Sparkling Hock 12 fl. (1 £); sparkling Rhine-wine 16 fl. 30 kr. (1 £ 8 sh.); sparkling Hock, finest 21 fl. 30 kr. (1 £ 16 sh.); sparkling Steinberg cabinet 28 fl. 30 kr. (2 £ 8 sh.); sparkling Moselle 12 fl. 30 kr. (1 £ 1 sh.); sparkling Moselle-wine I. Qual. 17 fl. (1 £ 9 sh.); sparkling Moselle, finest 21 fl. 30 kr. (1 £ 16 sh.); sparkling Scharzhofberger 28 fl. 30 kr. (2 £ 8 sh.); sparkling red Hock 17 fl. (1 £ 9 sh.); sparkling Asmannshäuser 21 fl. 30 kr. (1 £ 16 sh.); p. doz. bottles.

478. DAEL, GEORG, prop., grower and wine-merchant of Hock, Rhine- and Moselle-wines, Mentz and Geisenheim on the Rhine, established 1814. Establishment of sparkling Hock and Moselle-wines since 1832. Paris 1855 honor. mention; London 1862 medal. Agt. Fred. Class, Office 36 Crutched Friars E. C.

Sparkling Hock green label 27 fl.; black label 32 fl.; blue label 36 fl.; grand vin central 45 fl.; sparkling Hock Moselle, red label 32 fl.; 1857 Geisenheimer (own growth) 18 fl.; 1857 Kosackenberg (own growth) 30 fl.; 1857 Gräfenberger (own growth) 33 fl.; 1822 Geisenheimer Lickerstein 36 fl.; 1857 Geisenheimer Kosackenberg 39 fl.; 1857 Geisenheimer Rothenberg, gold-label 40 fl.; 1857 Gräfenberger picked 60 fl.; 1857 Geisenheimer Rothenberg picked cabinet 80 fl.; p. doz. bottles, free on board Rotterdam.

479. FRIDBERG & GUTTMANN, sparkling-wine-manu., Mentz.

Sparkling Hock.

480. GEORGE, ETIENNE, burgomaster, Rüdesheim near Bingen on the Rhine, prop. of vine-yards. London 1862 medal.

Rhine-wine: Scharlachberg of the years 1857, 1858 and 1859. One Stück à 600 Maas 2000 to 2500 fl., one quarter about 160 to 200 £.

481. HENKELL & Co., wine-merchant and sparkling Hock-manu., Mentz. London 1862 medal. Agt. J. F. Dupré, 5 Fowkes' Buildings, great Tower-street.

Sparkling Hock, Hochheimer, Moselle-Muscatel and Henkell and Co.

483. LAUTEREN, C., SON, wine-merchant and manu. of sparkling Hock and Moselle, Mentz. London 1862 medal. Agts. Cunliffe, Dobson & Co. 106 Fenchurch-street, London.

Rhine-wines and sparkling Hock and Moselle.

484. MÜLLER, JOHANN, JUN., wine-grower, Bensheim. Paris 1855 bronze medal.

Riessling of the year 1858; Roulant of the year 1858; mixed wine.

485. PABSTMAN, G. M., SON, wine-merchant, Mentz and Hochheim. London 1862 medal. Agt. French & German, vine-yard agency 47 Mark-Lane E. C. James Cowerdale 11 London-street, Fenchurch-street E. C.

Still Hocks, Queen Victoria-Hocks. These wines are to be had at moderate prices directly or by the above named agents to whom please apply for particulars.

486. PROBST, FRANZ ANTON, prop. of vine-yards and wine-merchant, Mentz and Rüdesheim. London 1862 medal. Agts. Verkrüzen & Co., 96 Hatton Garden E. C.

Wine-samples. For prices apply to the agents.

487. RAISER, CHRISTIAN, sparkling Hock-manu., wine-merchant and prop. of vine-yards, Oppenheim on the Rhine. Established 1857. Agt. Louis Raiser (Aug. & G. Fischer).

Sparkling Rhine-wine.

488. BARON VON RODENSTEIN, A., Bensheim. London 1862 medal.

Wine-samples: Bensheimer Kirchberg Riessling of 1834; Bensheimer Hochberg Riessling of 1858; Bensheimer Kirchberg Riessling of 1861.

489. VALCKENBERG, P. J., wine-merchant and wine-grower, sole prop. of the Liebfrauenmilch-monastery- and Luginsland vine-yards, Worms on the Rhine. Established 1786. Paris 1855 bronze medal; London 1862 medal. Agts. Roquette, Sillem & Co., 2 Crosby-square E. C.

Nierstein Glöck from the vine-yards of count Buol-Schauenstein of 1858; Liebfrauenmilch, Enclos Klostergarten of 1858 and Luginsland picked of 1857, both sole property; Johannesberg castle, prince Metternich's estate from 1857; Steinberg cabinet from 1857 and Steinberg cabinet No. 511., from 1846 both from the duke of Nassau's cellars at the convent Eberbach in the Rhinegau.

490. SICHEL, MORITZ, destiller, Mentz. London 1862 honor. mention.

Wine-oil p. english pound 7 £ 5 sh.; imitation-cognac of potato-spirit and wine-oil p. gallon 3 sh. 9 d.; tartar, from lees of wine p. 100 english pounds 3 £ 18 sh. 7 d.; tartrate of lime from lees of wine p. 100 english pounds 2 £ 2 sh. 1 d.

491. HEYL, LEONHARD, & Co., dealers in tobacco-leaves, Worms and Mannheim (Philipp Kloos). London 1862 medal. Agts. Mahler Brothers & Co., 58 Fenchurch-street.

1. Flat Tuetten; 2. flat Gundi, light color; 3. flat Gundi, brown color; 4. Gundi, folded; 5. stripped Tuetten covers, light color; 6. stripped Tuetten covers, brown; 7. stripped Gundi covers, light; 8. stripped bunchwrappers; 9. stripped fillers; 10. cutting tobacco, yellow; 11. cutting tobacco, butted; 12. ground leaves. All these sorts are German tobaccos for exportation.

492. WENCK, F. A., manu. of tobacco and cigars, Darmstadt. London 1862 medal.

No. 1. Havanna cigars, Nuremberg boxes 120 fl. (10 £); No. 2. La Semiramis Havanna regalia 100 fl. (8 £ 6 sh. 8 d.); No. 3. H. Upmann regalia 34 fl. (2 £ 16 sh. 8 d.); No. 4. Venus regalia, German cigars 10 fl. (16 sh. 8 d.); No. 5. la flor eminente Havanna 44 fl. (3 £ 13 sh. 4 d.); No. 6. la celebrada Havanna Londrés 40 fl. (3 £ 6 sh. 8 d.); No. 7. el Marinero 25 fl. (2 £ 1 sh. 8 d.); No. 8. Ugues 19 fl. (1 £ 11 sh. 8 d.); No. 9. Ambrosia Londres 25 fl. (2 £ 1 sh. 8 d.); No. 10. la flor eminente 26 fl. (2 £ 3 sh. 4 d.); No. 11. e pluribus unam Londres, German cigars 8 fl. (13 sh. 4 d.); No. 12. la Victoria 18 fl. (1 £ 10 sh.); No. 13. el Aguillas 14 fl. (1 £ 3 sh. 4 d.); No. 14. la Diana 12 fl. (1 £); No. 15. India 11 fl. (18 sh. 4 d.); No. 16. Guiana 12 fl. (1 £); No. 17. la real 12 fl. (1 £); No. 18. Sevilla 6 fl. (10 sh.); No. 19. demi Cabanas 5½ fl. (9 sh. 2 d.); No. 20. Kentucky 5¼ fl. (8 sh. 9 d.); (the cigars No. 17., 18., 19. and 20. are German cigars). No. 3. Macenilla ½ regalia 30 fl. (2 £ 10 sh.); No. 4. petit Londres façon 36 fl. (3 £); No. 5. petit Londres 19 fl. (1 £ 11 sh. 8 d.); p. 1000 cigars.

CLASS IV.
ANIMAL AND VEGETABLE SUBSTANCES USED IN MANUFACTURES.

493. KIEFER & PIRAZZI, importers, Offenbach on the Main. London 1862 honor. mention. (s. II. No. 27.)

Genuine pure Turkish rose-oil of superior quality drawn from the first producers at the original place.

494. MARTENSTEIN, J. D., SON, merchants, Worms on the Rhine. London 1862 honor. mention.

Very fine animal lubricating oil; very fine patent clarified lamp-oil; No. 1. very fine bleached animal lubricating oil; No. 2. unbleached animal lubricating oil; No. 3. imitated olive-oil; No. 4. very fine patent clarified burning-oil.

495. MIELCKE, CARL (formerly Betz & Mielcke), manu. of amber articles, Worms on the Rhine. Munich 1854 honor. mention; Paris 1855 honor. mention; London 1862 medal.

1. Assortment of different beads of amber in 24 strings mostly for Afrika; 2. assortment of different pipes and cigar-tubes, together 32 pieces; 3. 4 patterns of waste amber; 4. 2 pieces of raw amber.

497. GLÖCKNER, GEORGE, manu of soap, Darmstadt. London 1862 medal.

Specimens of soap.

498. SCHMITT, FRIEDRICH, manu., Darmstadt. Paris 1855 honor. mention; London 1862 honor. mention.

Specimens of soap.

499. MICHEL, J., ivory-cutter, Erbach.

Bracelets: No. 1. 1 £ 8 sh. 4 d.; No. 2. 11 sh. 8 d.; Brooches: No. 6. 1 £ 10 sh.; No. 7. 15 sh.; No. 8. 15 sh.; No. 10. 8 sh. 4 d.; No. 11. 8 sh. 4 d.; No. 13. 3 sh. 4 d.; No. 15. 2 sh. 2½ d.; No. 18. 3 sh. 4 d.; a wreath No. 19. 15 sh.; pendants, two pieces No. 4. 2 sh. 6 d., No. 12. 1 sh. 8 d.

500. HAMM, FRIEDRICH, manu. of glue, Offenbach. London 1862 honor. mention.

Glue of equal quality but of different color (dark and light). Price (the 6th March): 3 £ 6 sh. (82 frs. 50 c., 38 fl. 30 kr.) p. 110,232 lb., (50 ko. french 100 lb. Zollverein's weight) from Offenbach on the M.

501. HEYL, CORNELIUS, manu. of leather and glue, Worms. London 1862 medal. Agts. Oehlenschläger Brothers, 2 Fowke's Buildings, Great Tower-street.

Glue.

502. LYON, L., & SONS, manu. of glue, Michelstadt.

Cologne glue.

503. ZERBAN, CONRAD, glue-manu. Worms. London 1862 honor. mention. Agt. C. Trübner, 15 St. Dunstans Hill, City.

First rate yellow, colored, white, and liquid glue, the latter prepared without acid.

504. HEINRICH & HEYL, manu. of agricultural chemical products, Offenbach. Prop. Mrs. Elise Heinrich, widow, and Philipp Heyl. Agts. Dorgerloh, Sang & Co., 2 Savage Gardens.

Bone-dust and super phosphate.

505. GRAND DUCAL CHIEF-FORESTERY WALD-MICHELBACH. London 1862 medal.

Oak-barks for tanning p. 50 kilogr. 3—3¼ fl.

506. KELLER, HEINRICH, SON, seeds-establishment, Darmstadt. Agt. Georg Enes, 14 Cullum-street.

Forest, grass and clover-seeds.

507. HEIDEN-HEIMER & Co., hop-merchant, Mentz. London 1862 medal. Agts. Trier Mayer & Co., 7 Savage Gardens.

Samples of Bavarian hops.

508. SCHRÖDER-SANDFORT, hop-merchant, Mentz.

Hops.

CLASS VI.

CARRIAGES NOT CONNECTED WITH RAIL OR TRAM ROADS.

511. DICK & KIRSCHTEN, coach-makers and manu. of carriage-axletrees and springs, Offenbach. Prop. Carl Theodor Wecker. Berlin 1844 great gold medal; Munich 1844 great medal; London 1862 medal.

1. Coach on 8 springs, mounted in rough state by locksmith, smith and cart-wright; 2. different articles for coaches; 3. patent and common axletrees of different sizes; 4. all kinds of springs for coaches.

CLASS VIII.

MACHINERY IN GENERAL.

513. DE BARY, JULIUS, manu. of machines, Offenbach. London 1862 medal. Agts. Lamberg & Butler, Drury-Lane, London.

Machine for the manufacture of cigars, improved, patent in England, United-states of Amerika, France, Russia, Zollverein, Austria, Italy, Denmark, Sweden, Norway, Belgium, Holland.

514. HEIM, BROTHERS, manu. of machines, Offenbach. Munich 1854 medal of honor; Paris 1855 bronze medal; London 1862 medal. Agts. Morgan Brothers, 21 Bow-Lane, Cannon-street, City.

A large Heim lever-press; a small one; a large improved type stamping-press; a balancier; a press for gilding the inner side of hats; a seal-press; a litographic press for mercantile work; a lithographic press for mercantile work and chalk-printing; a roller-work for smoothing, for printing-offices and bookbinders; a paper cutting machine; a paste-board cutting machine.

515. JORDAN, J., & SON, manu. of pipe, tile and brick-machines, drainage-tools and implements, clay preparing-machines, horse mills, steam-engines, Darmstadt. Prop. J. Jordan, grand ducal architect, and J. Jordan Son, engineer. Munich 1854 medal of honor; Paris 1855 honor. mention; London 1862 medal.

A pipe, tile and hollow-brick making machine, Model No. 3. of the manufactory of J. Jordan & Son at Darmstadt, entirely of iron, with rack and pinion of wrought-iron, the box planed.

516. SCHULTZ, BROTHERS, engineers, Mentz. Machine for cutting toothed wheels of every diameter, invented by the exhibitor.

517. SCHRÖDER, J., manu. of sewing machines, Darmstadt. Agts. Jos. Myers & Co., 144 Leadenhall-street, London.

Sewing machines: 1. A. 220 fl. (16 £ 9 sh., 2. B. 150 fl. (12 £ 13 sh.), 3. B. 166 fl. (14 £); with table 190 fl. (16 £); 4. C. 157 fl. (18 £ 5 sh.); 5. D. 36 fl. (3 £); 6. D. 41 fl. (3 £ 10 sh.); 7. E. 100 fl. (8 £. 9 sh. 8 d.).

CLASS XI.

MILITARY ENGINEERING, ARMOUR AND ACCOUTREMENTS, ORDNANCE AND SMALL ARMS.

519. DELP, CARL, gun-smith to the court, Darmstadt.

A double barrelled gun.

CLASS XVI.

MUSICAL INSTRUMENTS.

524. Glück, C. L., piano-maker to his Royal Highness the Grand-duke of Hesse-Darmstadt, Friedberg near Frankfort.

A pianino demi-oblique, wholesale 450 fl. (38 £).

526. Seidel, Joseph, musical instrument-maker to the court of Hesse-Darmstadt, Mentz. Agt. G. André Augener, 33 Goswell-Road.

2 B-clarinets incl. cocoa-nut case, German silver keys and rings 10 £; an Es-clarinet do. 4 £ 10 sh.; a bassoon, maple-wood, brass-keys incl. case 7 £ 10 sh.; a piston cylinder machine, German silver mountings 4 £ 10 sh.; a B-trumpet, cylinder-machine do. 4 £ 10 sh.

CLASS XXI.

WOOLLEN AND WORSTED, INCLUDING MIXED FABRICS GENERALLY.

528. Wollengarnspinnerei Worms on the Rhine. Director Verhöven.

Carded yarn, weft-yarn and artificial wool.

529. List, Conrad & Theodor, manu., Lauterbach.

4 pieces of woollen damask table-cloth mixed with cotton, joint and woven together, surrounded with yellow woollen border viz: 1. one green 10 sh.; 2. one brown 10 sh.; 3. one red with firm 10 sh.; 4. one blue 10 sh.; 2 pieces of woollen damask table-cloth mixed with cotton, joint and woven together, surrounded above and below with yellow woollen border viz: 5. one black and scarlet with firm in yellow silk 9 sh.; 6. one black and green.

530. Lang & Hess, cloth-manu., Beerfelden i. O.

No. 5259. 1 piece 7 patterns 6½ D. E., demi-saison 4⅙ yard à 5⅔ sh. (1 £ 3⅔ sh.); No. 5331. 1 piece 8 patterns, 8 D. E. demi-saison 5⅓ yard à 5⅔ sh. (1 £ 10⅙ sh.); No. 5326. 1 piece 7 patterns, 6 D. E. demi-saison, 4 yards à 6⅑ sh. (1 £ 4½ sh.); No. 5318. 1 piece 4 patterns, 6½ D. E. d'hiver, 4⅓ yard à 8 sh. (1 £ 15 sh.).

CLASS XXVI.

LEATHER, INCLUDING SADDLERY AND HARNESS.

531. Dörr & Reinhart, leather-manu., Worms on the Rhine. Established 1839. London 1851 prize-medal; Munich 1854 great medal; Paris 1855 silver medal; London 1862 medal. Agts. Schmitt & David, 102 Leadenhall-street E. C.

Patent calf and goat-skins varnished in different colors; varnished cow-hides and calf-skins.

532. Hertz, Brothers, manu. of leather, Oppenheim on the Rhine. London 1862 honor. mention.

Half a doz. calf-skins and half a doz. waxed skins.

533. Heyl, Cornelius, Worms. London 1851 prize-medal; Munich 1854 great medal; Paris 1855 silver medal; London 1862 medal. Agts. Oehlenschläger, Brothers, 2 Fowkes'-Buildings, great Tower-street.

Black varnished leather; waxed calf-skins; russet calf-skins; boot-legs and vamps.

534. Mayer, Michel & Deninger, leather-manu., Mentz. Berlin 1844 ordre of the red eagle IV. Cl.; London 1851 prize-medal; Munich 1854 the great medal of honor; Paris 1855 great medal of honor; London 1862 medal.

Colored moroccos; skivers, roans and calf-skins in the different stages of dressing and finishing; black, white and bronze patent calf; japanned hides for coach and shoe-work; black and white patent hides for belts and saddlery; white patent and common buffle-hides; hides for straps and different other kinds of hides for saddlery.

535. Mayer, Paul, leather-manu., Mentz. Paris 1855 silver medal; London 1862 medal. Agt. Fr. Preyss, 20 St. Dunstans Hill, great Tower-street.

1. ½ skin of vache-leather 6½ kilogr.; 2. ½ skin for straps 12 kilogr.; 3. 6 curried white calf-skins 10 kilogr.; 4. 6 curried waxed calf-skins 8½ kilogr.; 5. 6 pair of white and fulled vamps 6 kilogr.; 6. 6 pair waxed fulled vamps 6 kilogr.; 7. 6 pair half white legs with hindparts 6 kilogr.; 8. 6 pair quite waxed legs with hindparts 6 kilogr.

536. Mayer, J., & Feistmann, manu., Offenbach. London 1862 honor. mention. Agts. Schlösser Brothers, 12 great St. Thomas-Apostle, E. C.

Calf-kid; black kid, glazed kid.

537. Melas, L., & Co., manu. of japanned calf-leather (patent-calf), Worms on the Rhine. Munich 1854 medal of honor; Paris 1855 bronze medal; London 1862 honor. mention. Agts. Schlösser, Brothers, 12 great St. Thomas-Apostle Cannon-street, E. C.

Japanned calf-leather (patent calf) and calf-leather to be japanned.

538. Müller, Gustav, leather-manu., Bensheim. Paris 1855 honor. mention; London 1862 medal.

⅔ hides for soles; 5 brown calf-skins; 5 waxed calf-skins.

540. Spicharz, Philipp Jacob, leather-manu., Offenbach. Munich 1854 medal of honor; London 1862 honor. mention.

Varnished cow-hides; grained varnished calf-skins; curried calf-skins, waxed and brown; black calf-kids.

541. Preetorius, Wilhelm, & Co., leather-manu., Alzey. Munich 1854 medal of honor; Paris 1855 bronze medal; London 1862 honor. mention. Agt. B. Cahn, 22 Cannon-street west.

½ doz. of waxed calf-skins; ½ doz. of waxed calf-leathers; ½ doz. of brown calf-leathers; ½ doz. of brown calf-skins; ½ doz. of calf-leather vamps; ½ doz. of calf-leather boot-legs; ½ doz. of calf-leather vamps; ½ doz. of calf-leather boot-legs.

542. Wormatia, leather-manufacturing-company, Worms. Munich 1854 medal of honor; Paris 1855 prize-medal; London 1862 medal. Agt. S. Frankel, 68 Queen-street, Cheapside.

Varnished calf-skins.

CLASS XXVII.

ARTICLES OF CLOTHING.

543. HERZ, OTTO, & NASS, manu. of every kind of first-class boots and shoes only whole-sale and for exportation, Mentz. London 1862 medal. Agt. Schmitt & David, 102 Leadenhall-street, City E. C.

1 doz. of elastic boots for gentlemen, varnished calf, kid-tops 140 fl.; 1 doz. of elastic boots for gentlemen, varnished neat's leather, kid-tops, 155 fl.; 1 doz. of curried elastic boots for gentl., satin kid 140 fl.; 1 doz. of elastic boots in colors for ladies, satin français, military heels, 87 fl.; 1 doz. of elastic boots for ladies, chevreaux mats, military heels, 95 fl.; 1 doz. of elastic boots with welts for ladies, chevreaux doré, military heels, 120 fl.

544. SCHUHMACHER, JOSEPH, SON, manu. of boots and shoes for ladies and gentlemen, Mentz. Store at Moskow, Troaba in Moskow: Troaba-street, Ossipoff-house and during the fair at Nischncy-Nowgorod. Berlin 1844 bronze medal; London 1851 honor. mention; Munich 1854 medal of honor; Paris 1855 bronze medal; London 1862 medal. Agts. Heintzmann & Ro-chussen, 9 Friday-street, Cheapside.

Assortments of fine shoes and boots for gentlemen and ladies.

545. WOLF, SIMON, Shoe-manu. and ex-porter, Mentz. London 1862 honor. mention. Agt. Fred. Preyss, 20 St. Dunstans Hill, Great Tower-street.

A. 1. 1 pair of hunting boots (patented) 30 fl. (2 £ 10 sh.); 2. 1 pair of varnished Wellington boots with morocco legs 22 fl. (1 £ 16 sh. 8 d.); 3. a pair of patent boots, high galoshed with kid-tops s. s. 12 fl. (1 £); 4. 1 pair of patent hide-boots with kid-tops 14 fl. (1 £ 3 sh. 4 d.); 5. 1 pair of patent boots with kid-tops s. s. 12 fl. (1 £); 6. 1 pair of slippers with gold patent vamps 10 fl. (16 sh. 8 d.); 7. a pair of patent calf-shoes to be laced 8 fl. (13 sh. 4 d.)

Ladies' boots: 8 1 pair of white satin boots with gold embroidery to be laced 12 fl. (1 £); 9. 1 pair of white satin shoes with gold embroidery to be laced 8 fl. (13 sh. 4 d.); 10. 1 pair of colored satin shoes with gold embroidery to be laced 12 fl. (1 £ 11 sh.); 1 pair of gold patent boots mock-eyelets ss. 10 fl. (16 sh. 8 d.); 12. 1 pair of gold patent shoes with gold embroidery 9 fl. (15 sh.); 13. 1 pair of kid-boots with patent vamps cork-soles 10 fl. (16 sh. 8 d.); 14. 1 pair of black French satin boots with embroidery 5 fl. (8 sh. 4 d.)

B. For exportation. For gentlemen: 15. oxonian calf-shoes to be laced, patented 144 fl. (12 £.); 16. patent calf-shoes s. s. 71 fl. (5 £ 11 sh. 8 d.); 17. calf-boots with kid-tops, mock-buttons, screwed 3 soles 88 fl. (7 £ 6 sh. 8 d.); 18. patent boots with kid-tops s. s. 80 fl. (6 £ 13 sh. 4 d.); 19. patent boots, high galoshed, with kid-tops s. s. 86 fl. (7 £ 3 sh. 4 d.); 20. patent boots s. s. (fulled) 78 fl. (6 £ 10 sh.); 21. patent hide-boots high s. s. (fulled) 2 soles 89 fl. (7 £ 8 sh. 4 d.); 22. patent hide-boots s. s. (fulled mock-eyelets) 98 fl. (8 £ 3 sh. 4 d.); 23. patent hide, high galoshed kid-tops, screwed s. s. 98 fl. (8 £ 3 sh. 4 d.); 24. patent boots with kid-tops s. s. 82 fl. (6 £ 16 sh. 8 d.); 25. patent boots with patent toe-caps (fulled) 81 fl. (6 £ 15 sh.); 26. French satin boots with patent calf galoshed 78 fl. (6 £ 10 sh.).

Ladies' boots: 27. black lasting boots, side-springs 42 fl. 24 kr. (3 £ 10 sh. 8 d.); 28. black lasting with patent toe-caps, springs 43 fl. 36 kr. (3 £ 12 sh. 8 d.); 29. black lasting with vamps, springs 46 fl. 24 kr. (3 £ 17 sh. 4 d.); 30. black lasting mock-eyelets, springs 47 fl. 24 kr. (3 £ 19 sh.); 31. black cashmere boots, springs 46 fl. 12 kr. (3 £ 17 sh.); 32. black cashmere boots, embroidery s. s. 51 fl. (4 £ 5 sh. 3 d.); 33. colored cashmere boots s. s. 48 fl. 12 kr. (4 £ 4 sh.); 34. col. cashmere boots with morocco toe-caps s. s. 49 fl. 24 kr. (4 £ 2 sh. 4 d.); 35. black French satin boots s. s. 46 fl. 12 kr. (3 £ 17 sh.); 36. black French satin boots with embroidery s. s. 51 fl. (4 £ 5 sh.); 37. col. French satin boots s. s. 48 fl. 24 kr. (4 £ 8 sh.); 38. col. French satin with embroidery and gold patent toe-caps s. s. 55 fl. 12 kr. (4 £ 12 sh.); 39. black kid-boots s. s. 50 fl. 24 kr. (4 £ 4 sh.); 40. black kid-boots with patent toe-caps s. s. 51 fl. 36 kr. (4 £ 6 sh.); 41. black kid-boots with patent vamps s. s. 54 fl. 36 kr. (4 £ 11 sh.); 42. black kid-boots with patent mock-eyelets s. s. 59 fl. 36 kr. (4 £ 19 sh. 4 d.); 43. white kid with patent s. s. 56 fl. (4 £ 13 sh. 4 d.); 44. gold patent boots, with gold patent camps, mock-eyelets s. s. 78 fl. (6 £ 10 sh.).

Children's boots: 45. black lasting boots s. s. 28 fl. (2 £ 6 sh. 8 d.); 46. black lasting boots with patent toe-caps s. s. 28 fl. 48 kr. (2 £ 8 sh.); 47. black lasting boots with patent vamps s. s. 30 fl. (2 £ 10 sh.); 48. black lasting boots with vamps, mock-eyelets 32 fl. (2 £ 13 sh. 4 d.); 49. black cashmere with s. s. embroidery 32 fl. (2 £ 13 sh. 4 d.); 50. colored cashmere with s. s. 33 fl. 36 kr. (2 £ 16 sh.); 51. colored cashmere with embroidery and gold patent toe-caps 36 fl. (3 £); 52. black French satin boots with embroidery s. s. 32 fl. (2 £ 13 sh. 4 d.); 53. col. French satin boots with s. s. 33 fl. 36 kr. (2 £ 16 sh.); 54. kid-boots with patent toe-caps 31 fl. 12 kr. (2 £ 12 sh.); 55. kid-boots with patent vamps 33 fl. 36 kr. (2 £ 16 sh.); 56. kid-boots with patent vamps, mock-eyelets 36 fl. (3 £).

546. SCHUCHARD, H., hat-manu., Darm-stadt. London 1862 medal.

Silk hats and bonnets for ladies, gentlemen and children.

547. KOLBE, L., & Co., button-manu., Bessungen near Darmstadt. Munich 1854 medal of honor; Paris 1855 honor. mention.

Metal and fancy buttons.

548. HAENLEIN & LESCHEDITZKY, lacker-ing-manufactory, Offenbach.

13 cap-visors; 8 cap-straps; 2 girdles for boys, 1 girdles for ladies; 1 piece of varnished stuff for cap-visors; 1 piece of varnished stuff; 1 piece of varnished stuff for cap-visors; 2 pieces of var-nished stuff of fine quality; 2 pieces of bathing-carps for ladies; 1 varnished cap-front shade stuff; varnished stuff of strong quality, of fine quality; varnished sheep's skin; 4 ruffles; 1 collar; var-nished stuff for cap-straps.

CLASS XXVIII.

PAPER, STATIONERY, PRINTING AND BOOKBINDING.

549. HEMMERDE & Co., paper-makers, Darmstadt, Manufactory at Bensheim.

Yellow and brown straw wrapping-paper and paste boards.

550. FREUND, E. A., manu. of colored paper, Offenbach. London 1862 medal and honor. mention.

Colored and pressed papers for labels, cartonage-work etc.; glazed boards for firm and visite-cards.

551. WÜST, BROTHERS, Darmstadt. Munich 1854 honor. mention; London 1862 medal.

Specimens of differently colored fancy paper as chagrin, moire, fancy, gold, carton, marble-paper.

552. SCHOTT's, B., SONS, music publishers, Mentz. Agt. Schott & Co., 159 Regent-street.

Six volumes music, partly printed in the usual manner and partly transferred from tin plates on lithographic stones and printed by the lithographic press.

553. WIRTZ, FERDINAND, litho-geographical office, Darmstadt.

Six maps from »Ewald's handatlas« printed in colors, viz: a map of Greece; a map of Germany geol. and high- and lowland; a map of the moon; a map of the distribution of plants; a map of variations of the surface of the earth; a map of Switzerland, printed in colors; a geological map of the environs of Darmstadt; a plan of Darmstadt, of Worms and of Offenbach, printed in colors; a large map of Europe (high- and lowland) printed in colors.

554. FROMMANN, M., manufactory of playing-cards, Darmstadt. Paris 1855 bronze medal; London 1862 medal.

A large sample-card. All kinds of German, English, French and Spanish playing-cards from the lowest to the most superior quality are manufactured at the price of 7 fl. and more per twelve doz. Samples and conditions will be given if desired.

555. REUTER, W., Darmstadt. Munich 1854 medal of honor; Paris 1855 bronze medal; London 1862 medal.

A frame with cards; 18 packs of Spanish, 9 of Portugese, 30 of French, 31 of English playing-cards.

CLASS XXIX.

EDUCATIONAL WORKS AND APPLIANCES.

559. LÖSSER, FRIEDRICH, drawing-master and modeller, Darmstadt.

Drawing-rules; prices p. doz.: No 2. 20 in. long 3 fl. 12 kr. (5 sh. 4 d.); No. 3. 25 in. long 4 fl. (7 sh.); No. 4. 30 in. long 4 fl. 30 kr. (7 sh. 6 d.); No. 5. 35 in. long 5 fl. 40 kr. (9 sh. 5 d.); No. 6. 40 in. long 7 fl. 48 kr. (13 sh.); No. 7. 45 in. long 8 fl. 30 kr. (14 sh. 2 d.); No. 8. 50 in. long 9 fl. (15 sh.).

45° angles; prices p. doz.: No. 1. (composed) cathete 5 in. 2 fl. 24 kr. (4 sh.); No. 2. cathete 6 in. 2 fl. 46 kr. (4 sh. 7 d.); No. 3. cathete 7 in. 3 fl. (5 sh.); No. 4. cathete 8 in. 3 fl. 20 kr. (5 sh. 7 d.); No. 5. cathete 9 in. 4 fl. 20 kr. (7 fl. 3 d.); No. 6. cathete 10 in. 4 fl. 45 kr. (4 sh.); No. 7. cathete 11 in. 5 fl. 54 kr. (9 sh. 10 d.); No. 8. cathete 12 in. 7 fl. 54 kr. (13 sh. 2 d.).

8 composed 60° angles with the same numbers and prices as before then.

45° angles (not composed): prices p. doz.: No. 1. cathete 5 in. 1 fl. (1 sh. 8 d.); No. 2. cathete 6 in.

1 fl. 20 kr. (2 sh. 3 d.): No. 3. cathete 7 in. 1 fl. 40 kr. (2 sh. 9 d.); No. 4 cathete 8 in. 2 fl. (3 sh. 4 d.); then 4 60° angles (not composed) the same numbers and prices as before.

Ellipsis-curve, prices p. doz.: No. 3. 2 fl. 24 kr. (4 sh.); No. 4. 2 fl. 24 kr. (4 sh.); No. 5. 3 fl. (5 sh.); No. 6. 3 fl. (5 sh.); No. 7. 3 fl. 20 kr. (5 sh. 7 d.); No. 8. 3 fl. 20 kr. (5 sh 7 d.). The prices are made for the place.

560. SCHRÖDER, J., polytechnisches Arbeits-Institut, Darmstadt. Berlin 1844 bronze medal; London 1851 prize-medal; Munich 1854 medal of honor; London 1862 medal. Agts. Jos. Myers & Co., 144 Leadenhall-street.

Models for teaching mechanics, mathematics etc. to be used in schools and colleges.

I. Geometrical bodies. Case No. 1—9. 5 sh. 10 d.; do. No. 1—12. 9 sh. 2 d.; do. No. 1—25. 1 £ 9 sh. 4 d.

II. Descriptive geometry. No. 26—45. first part, plate 1—20. fig. 1—62. 5 £ 17 sh.; No. 46—65. second part, plate 21—40. fig. 63—118. 12 £ 4 sh.

III. Crystals. Models of the chief forms of crystals No. 116. fig. 1—36. 1 £ 5 sh. 104 crystals from professor Koop's work of crystallization 5 £ 18 sh.; 6 models of crystals, showing the different systems of axes 2 £ 2 sh.

IV. Parts of machines as constructed by prof. Redtenbacher at Carlsruhe. No. 257—266. screw-joints 4 £ 16 sh.; No. 267—274. plates jointed by screws 5 £ 13 sh.; No. 287—296. axles and beams 5 £ 2 sh. 6 d.; No. 297—301. coupled axles 5 £ 1 sh. 8 d.; No. 302. 1 bearing 2 £ 7 sh. 5 d.; No. 304. 1 hanging bearing 2 £ 7 sh. 5 d.; No. 305. 1 foot-bearing 2 £ 7 sh. 5 d.; No. 328. part of a toothed wheel 18 sh. 8 d.; No. 329. part of a beveled wheel 18 sh. 8 d.; No. 343. connecting-rod of wrought iron 16 sh. 10 d.

V. Parts of machines according to the construction of professor F. Reuleaux at Zurich. No. 652—669. models of fastening screws 6 £ 10 sh. 6 d.: No. 670—677. screw-joints 6 £ 12 sh.; No. 697—704. axles and beams 5 £ 7 sh. 5 d.; No. 715. bearing to be fastened a the wall 2 £ 7 sh. 5 d; No. 716. hanging bearing 2 £ 7 sh. 5 d.

VI. Models of mechanisms for changing motion by professor Redtenbacher at Carlsruhe. No. 441. beveled-wheels 3 £ 16 sh. 3 d.; No. 446. screw-bevels 4 £ 4 sh. 8 d.; No. 215. tangent screw 4 £ 4 sh. 8 d.; No. 455 chain of wheels 6 £ 7 sh; No. 463. counter 6 £ 15 sh. 6 d.: No. 464. two pulleys in different plaines 3 £ 1 sh.; No. 465. pulley fastened upon Dr. Hooke's universal joint 4 £ 4 sh. 8 d.; No. 466. a combination of pulleys 3 £ 1 sh.; No. 474. crank and wheel 2 £ 10 sh. 10 d.; No. 475. sliding crank 3 £ 8 sh.; No. 476. oscillating crank 2 £ 10 sh. 10 d.; No. 481. excentric 3 £ 8 sh.; No. 490. triangle for moving the expansion-slide 3 £ 16 sh. 3 d.; No. 495. engine-beam and radius-rod 3 £ 1 sh.; No. 497. Watt's parallel motion 3 £ 16 sh. 3 d.: No. 498. the same for marine engines 3 £ 16 sh. 3 d.: No. 504. mandrel and excentric 3 £ 8 sh.; No. 507. mangle-wheel 6 £ 7 sh.; No. 521. disengaging coupling 6 £ 7 sh.: No. 526. governor 6 £ 7 sh.; No. 533. slide-geer for Woolfe's steam-engine 7 £ 4 sh.: No. 534 b. paddle-wheel 9 £ 6 sh. 4 d.: No. 537. overshot water-wheel 8 £ 9 sh. 8 d.

VIb. Different mechanisms No. 148 windlass for ships 8 £ 9 sh. 8 d.; No. 852. guide for conchoids by Reuleaux 2 £ 19 sh. 4 d.

VII. Models of building No. 1266—1272. models of different size and joints at the corner 2 £ 11 sh. 10 d.; No. 1273—1274. openings in inclined and round walls 12 sh.; No. 1278—1288. walls with straight and arched openings 9 £ 5 sh. Roofing: No. 1623. roofing of the corn-market at Mentz 4 £ 16 sh.; No. 1895—1896. wood-beams 2 £ 8 d.

VIII. Agricultural models: No. 2020. a grate 1 £ 9 sh. 8 d.; No. 2046. double sluice 2 £ 7 sh. 5 d.; No. 2047. sluice 1 £ 9 sh. 8 d.; No. 2104—2105. boilers for breweries 3 £ 17 sh.; No. 2068. water-wheel for raising water 9 £ 8 sh.

IX. Drawing-instruments No. 1—12. set of rulers; No. 1—14. drawing rulers; No. 1—30. sets of squares; No. 1—12. palettes; No. 1—40. curves; No. 1—100. circular curves; No. 1. a desk for originals; No. 1. pantograph.

CLASS XXX.

FURNITURE AND UPHOLSTERY, INCLUDING PAPER-HANGINGS AND PAPIER-MACHÉ.

561. KNUSSMANN, WOLFGANG, manu. of furniture, Mentz. Berlin 1844 bronze medal; Munich 1854 medal of honor; London 1862 medal. Agt. J. & R. Mac Cracken, 7 Old Jewry.

Sideboard; dining-table for 24 persons; chairs for dining-rooms; armchair; model for inlaid floors; all the furniture are of oakwood, gilt and gothic style.

562. NILLIUS, F. C., & SON, manu. of furniture, purveyor to His Highness the Grand Duke of Hesse, Mentz.

A sofa of nut-wood fine workmanship; easy chairs of the same style and workmanship and a candelled table (plate of grey marble) with one leg. The above mentioned articles, sofa and six easy chairs are covered with fine aniline plush superior quality, and stuffed with good horse-hair. Price of sofa and six chairs 420 fl., of the table 80 fl.

563. REITMEYER, ANDR., furniture-manu., Mentz.

A causeuse with silkstuff, and two armchairs with silkstuff; a bed with complete bedding and curtains.

564. REINHARDT, JOHN MARTIN, chair-manu., Mentz.

Cane-chair No. 5. of nut-wood 23 fl. (2 £); straw-chair No. 3. of nut-wood 6 fl. 30 kr. (11 sh.); cane-chair No. 1. of nut-wood 6 fl. (10 sh. 3 d.); cane-chair No. 2. of nut-wood 5 fl. (8 sh. 6 d.); cane piano-chair of cherry-wood 5 fl. (8 sh. 6 d.); 1 straw piano-chair of nut-wood 3 fl. 40 kr. (6 sh. 6 d.); straw-chair No. 1. of nut-wood 5 fl. 30 kr. (9 sh. 6 d.); straw-chair No. 3. of nut-wood 2 fl. 45 kr. (4 sh. 9 d.).

567. HOCHSTÄTTER, FELIX, Darmstadt. London 1862 medal. Agt. H. Kohnstamm, Dowgate-Hill, 33 Cannon-street.

Samples of tapestry, principally made with machines, in satinet with the prices marked on them, from 12 kr. (4 d.) upwards.

568. BUSCH, LOUIS, foundry of bronze work, Mentz. London 1862 honor. mention.

Schiller's statue cast of bronze with pedestal price 25 £.

569. FISCHER, JOH., & Co., manu., Offenbach.

Bronzed and black enamelled fine cast-iron wares.

570. FOUNDRY AND MANUFACTORY OF GAS-APPARATUS, Mentz. London 1862 medal.

Lustres for gas and wax-candles of pure bronze and zinc.

571. SEEBASS, ALFRED RICHARD, & Co., manu. of cast-iron wares, Offenbach. London 1851 honor. mention; Munich 1854 medal of honor; Paris 1855 bronze medal; London 1862 honor. mention. Agt. Angenman, 192 Bishopsgate-street.

Cast-iron wares; prices p. doz.:

Ash-cups: No. 2667. 2 Th. (6 sh.); No. 2665. 2 Th. 22½ sg. (8 sh.); No. 2666. 2 Th. 22¼ sg. (8 sh.); No. 2673. 2 Th. 15 sg. (7 sh. 6 d.); No. 2674. 2 Th. (6 sh.); No. 2678. 3 Th. (9 sh.).

Letter-pressers: No. 2751. 10 Th. 15 sg. (1 £ 11 sh. 6 d.); No. 2760. 5 Th. 15 sg. (16 sh. 6 d.); No. 3434. 9 Th. (1 £ 7 sh.); No. 3417. 7 Th. 15 sg. (1 £ 2 sh. 6 d.); No. 3435. 4 Th. 15 sg. (13 sh. 6 d.); No. 3436. 5 Th. (15 sh.); No. 3437. 6 Th. (18 sh.); No. 3411. 7 Th. (1 £ 1 sh.); No. 3429. 5 Th. (15 sh.); No. 2815. 6 Th. 15 sg. (19 sh. 6 d.).

Flower-stands: No. 1459. 18 Th. (2 £ 14 sh.); No. 1457. 40 Th. (6 £).

Crucifixes No. 3557. 14 Th. 15 sg. (2 £ 3 sh. 6 d.); No. 2826. 15 Th. (2 £ 5 sh.); No. 3567. 12 Th. (1 £ 16 sh.); No. 3566¼. 18 Th. (2 £ 14 sh.); No. 3561. 6 Th. 15 sg. (19 sh. 6 d.); No. 2837. 8 Th. 15 sg. (1 £ 5 sh. 6 d.); No. 3573. 2 Th. (6 sh.); No. 3582. 13 Th. 15 sg. (2 £ 6 d.); No. 3583. 19 Th. (2 £ 17 sh.).

Cigar-stands: No. 3622. 7 Th. 15 sg. (1 £ 2 sh. 6 d.); No. 3661. 13 Th. 15 sg. (2 £ 6 d.).

Thermometer No. 8028¼. 9 Th. 15 sg. (1 £ 8 sh. 6 d.).

Cigar-stands: No. 3676. 13 Th. 15 sg. (2 £ 6 d.); No. 3688. 15 Th. (2 £ 5 sh.); No. 3696. 22 Th. (3 £ 6 sh.); No. 3706. 12 Th. 15 sg. (1 £ 17. sh. 6 d.); No. 3720. 5 Th. 15 sg. (16 sh. 6 d.); No. 4040. 8 Th. 15 sg. (1 £ 5 sh. 6 d.); No. 4027. 10 Th. 15 sg. (1 £ 11 sh. 6 d.); No. 4042. 7 Th. (1 £ 1 sh.); No. 4047. 16 Th. (2 £ 8 sh.); No. 4043. 16 Th. (2 £ 8 sh.); No. 4028. 11 Th. (1 £ 13 sh.); No. 4048 9 Th. 15 sg. (1 £ 8 sh. 6 d.); No. 3735. 17 Th. (2 £ 11 sh.); No. 3733. 9 Th. (1 £ 7 sh.); No. 4035. 12 Th. 15 sg. (1 £ 17 sh. 6 d.); No. 3700. 9 Th. (1 £ 7 sh.); No. 4030. 15 Th. (2 £ 5 sh.); No. 3692. 7 Th. (1 £ 1 sh.); No. 4049. 7 Th. (1 £ 1 sh.).

Tinder-boxes: No. 3805. 3 Th. 15 sh. (10 sh. 6 d.); No. 3828. 5 Th. (15 sh.); No. 3815. 4 Th. (12 sh.); No. 3816. 5 Th. 15 sg. (16 sh. 6 d.); No. 3783. 6 Th. 15 sg. (19 sh. 6 d.); No. 3824. 5 Th. (15 sh.); No. 3822. 7 Th. 15 sg. (1 £ 2 sh. 6 d.).

Penholders: No. 3432. 9 Th. 15 sg. (1 £ 8 sh. 6 d.). Hasps: No. 2945. 16 Th. (2 £ 8 sh.); No. 4051. 17 Th. (2 £ 11 sh.); No. 4054. 12 Th. 15 sg. (1 £. 17 sh. 6 d.); No. 2950. 17 Th. (2 £ 11 sh.); No. 4057. 12 Th. (1 £ 16 sh.).

Candle-sticks: No. 4103. 7 Th. 15 sg. (1 £ 2 sh. 6 d.); No. 4688. 18 Th. (2 £ 14 sh.); No. 4111. 4 Th. 15 sg. (13 sh. 6 d.); No. 4110. 5 Th. 15 sg. (16 sh. 6 d.); No. 1758. 5 Th. 15 sg. (16 sh. 6 d.); No. 4108. 5 Th. (15 sh.); No. 4112. 5 Th. (15 sh.) No. 4094¼. 4 Th. 15 sg. (13 sh. 6 d.); No. 4106. 5 Th. 15 sg. (16 sh. 6 d.).

Lamp-screens: No. 3048. 17 Th. (2 £ 11 sh.); No. 3052. 18 Th. (2 £ 14 sh.).

Pair of candle-sticks: No. 4197. 14 Th. 15 sg. (2 £ 3 sh. 6 d.); No. 4215. 5 Th. 15 sg. (16 sh. 6 d.); No. 4217. 9 Th. (1 £ 7 sh.); No. 4196. 14 Th. (2 £ 2 sh.); No. 4203. 15 Th. (2 £ 5 sh.); No. 4209. 22 Th. (3 £ 6 sh.); No. 4211. 9 Th. (1 £ 7 sh.); No. 4219. 28 Th. (4 £ 4 sh.); No. 4225. 5 Th. (15 sh.); No. 4158½ 28 Th. (4 £ 4 sh.); No. 4207. 12 Th. (1 £ 16 sh.)

Night-lamps: No. 4340. 24 Th. (3 £ 12 sh.); No. 4311. 12 Th. (1 £ 16 sh.); No. 4344. 13 Th. (1 £ 19 sh.); No. 4346. 12 Th. 15 sg. (1 £ 17 sh. 6 d.); No. 4356. 16 Th. 15 sg. (2 £ 9 sh. 6 d.); No. 8071. 9 Th. (1 £ 7 sh.); No. 8072. 9 Th. 15 sg. (1 £ 8 sh. 6 d.); No. 8074. 17 Th. 15 sg. (2 £ 12 sh. 6 d.); No. 8075. 8 Th. 15 sg. (1 £ 5 sh. 6 d.); No. 8076. 21 Th. (3 £ 3 sh.); No. 4328. 13 Th. (1 £ 19 sh.); No. 4327. 8·Th. (1 £ 4 sh.).

Pin-cushions: No. 4423. 3 Th. 22½ sg. (11 sh.); No. 4432. 3 Th. 22⅔ sg. (11 sh.); No. 4425. 4 Th. 15 sg. (13 sh. 6 d.); No. 4429. 3 Th. 22½ sg. (11 sh.).

Night-watches: No. 4368. 66 Th. (9 £ 18 sh.); No. 3117. 66 Th. (9 £ 18 sh.); No. 4369. 68 Th. (10 £ .4 sh.).

Pendulum-clocks: No. 7000. 62 Th. (9 £ 6 sh.). Paper-crotches: No. 1988. 2 Th. (6 sh.); No. 1984. 2 Th. 22⅔ sg. (8 sh.); No. 1985. 3 Th. (9 sh.); No. 1983. 1 Th. 22½ sg. (5 sh.).

Knitting-basket No. 2195. 7 Th. 15 sg.; (1 £ 2 sh. 6 d.).

Inkstands: No. 4640. 6 Th. 15 sg. (19 sh. 6 d.); No. 4644. 28 Th. (4 £ 4 sh.); No. 4659. 9 Th. 15 sg. (1 £ 8 sh. 6 d.); No. 4658. 14 Th. (2 £ 2 sh.); No. 4662½. 10 Th. (1 £ 10 sh.); No. 4696. 10 Th. (1 £ 10 sh.); No. 4690. 21 Th. (3 £ 3 sh.); No. 4627. 21 Th. 15 sg. (3 £ 4 sh. 6 d.); No. 3226½. 10 Th. (1 £ 10 sh.); No. 4594. 9 Th. 15 sg. (1 £ 8 sh.); No. 4650. 26 Th. (3 £ 18 sh.); No. 4651. 19 Th. (2 £ 17 sh.); No. 4616. 6 Th. 15 sg. (19 sh. 6 d.); No. 4677. 7 Th. (1 £ 1 sh.); No. 4620. 9 Th. (1 £ 7 sh.)

Dressing-glasses: No. 4781. 19 Th. (2 £ 17 sh.); No. 4782. 18 Th. (2 £ 14 sh.); No. 4760½. 18 Th. 15 sg. (2 £ 15 sh. 6 d.); No. 4764. 14 Th. 15 sg. (2 £ 3 sh. 6 d.).

Pin-cushions: No. 4428. 6 Th. 15 sg. (19 sh. 6 d.). Dressing-glasses: No. 4783. 48 Th. (7 £ 4 sh.); No. 4762. 30 Th. (4 £ 10 sh.); No. 4785. 15 Th. (2 £ 5 sh.).

Thermometers: No. 8020. 12 Th. 15 sg. (1 £ 17 sh. 6 d.); No. 8023. 18 Th. (2 £ 14 sh.); No. 8028. 11 Th. (1 £ 13 sh.); No. 8029. 18 Th. (2 £ 14 sh.); No. 8030. 16 Th. (2 £ 8 sh.); No. 8031. 11 Th. 15 sg. (1 £ 14 sh. 6 d.); No. 8034. 9 Th. 15 sg. (1 £ 8 sh. 6 d.); No. 8036. 15 Th. (2 £ 5 sh.); No. 3516. 18 Th. (2 £ 14 sh.); No. 3523. 13 Th. (1 £ 19 sh.); No. 8010. 19 Th. (2 £ 17 sh.); No. 8014. 14 Th. (2 £ 2 sh.); No. 3535. 44 Th. (6 £ 12 sh.).

Watch-hooks: No. 5022. 7 Th. 15 sg. (1 £ 2 sh. 6 d.); No. 5025. 8 Th. (1 £ 4 sh.); No. 5031. 7 Th. (1 £ 1 sh.); No. 4832. 6 Th. (18 sh.); No. 4843. 10 Th. (1 £ 10 sh.); No. 5032. 5 Th. 15 sg. (16 sh. 6 d.); No. 4822. 7 Th. (1 £ 1 sh.); No. 4836. 10 Th. 15 sg. (1 £ 11 sh. 6 d.); No. 5021. 9 Th. (1 £ 7 sh.); No. 4080. 7 Th. 15 sg. (1 £ 2 sh. 6 d.); No. 5030. 4 Th. 15 sg. (13 sh. 6 d.).

Wax-stand: No. 2483. 7 Th. 15 sg. (1 £. 2 sh. 6 d.).

Tooth-pick stands: No. 1409. 3 Th. 15 sg. (10 sh. 6 d.). — 2 figures with the firm; 1 book of patterns, containing lithographic copies of fine iron-wares adress-cards and feather-brooms.

573. DOLLMANN, A., & Co., Offenbach. London 1862 honor. mention. Agts. partners of Riesbeck, Becker & Co., 51 Aldermanbury, E.C.

Travelling and ladies' bags of leather; photograph-albums; reticules; ladies' companions; workboxes; scent-cases; workbags; card-cases; cigar-boxes; tobacco-boxes; elastic city-purses; portemonnaies; needle-books.

574. KNIPP, J. F., & Co., prop. J. F. Knipp & Henry Fuchs, manu., Offenbach. Munich 1854 medal of honor; London 1862 medal. Agt. Killy Traub & Co., 52 Bread-street.

Assortment of purses; cigar-cases; letter-cases; match-boxes; pocket-books; travelling-cases; ladies' companions; photograph-books; wood-boxes such as tobacco, cigar, tea, glove, enveloppes, writing-stand boxes with brass and packfong-mounting.

575. POSEN, EDUARD, & Co., manu., Offenbach. London 1862 medal. Agt. Krausse & Auerbach, 68 Basinghall-street.

Portemonaies; cigar-cases; reticules; albums; bags.

576. SPIER, PHILIPP, portfolio-manu., Offenbach. Agt. A. H. Merzbach, 40 St. Mary's Axe, E. C.

Portemonnaies No. 102., 105., 107., 112½., 115.; porte-bourses No. 129., 139., 141.; portemonnaies No. 145., 146.; portemonnaie-pockets No. 160½.; portemonnaies No. 217., 223½., 258., 259., 301.; portemonnaie-pockets No. 325., 326.; portemonnaies No. 332., 337., 389., 390., 391., 393b., 399.; porte-bourse No. 401.; portemonnaies No. 404.; portemonnaie-pockets No. 411., 412., 418.; portemonnaies No. 419., 426., porte-cigars No. 1338., 1338½., 1352., 1384., 2320., 2329., 2332., 2335., 2341., 2342., 2343., 2344., 2349.; photograph-albums No. 1612., 1614., 1617., 1628., 1629., 1630., 1631.; Cabas No. 6004., 6005., 6006., 6007., 6008., 6009., 6010., 6011., 6012.; necessaries No. 6013., 6014—6020.; cabas No. 6021—6026.

578. STEINHART & GÜNZBURG, manu., Offenbach. Agt. A. Oppenheim, 32 Basinghall-street, E. C.

6 ladies' bags No. 54., 56., 59., 60., 70., 188.; 10 travelling-bags and porte-manteaux; No. 351 c., 353 B., 355 B., 356 D., 358., 360 B., 361., 364., 367., 368.; 3 travelling-bags No. 526., 528., 533.; 3 money-bags No. 658., 659., 660.

579. WEINTRAUD, CHRISTIAN, JUN., & Co., Offenbach. London 1862 medal. Agt. Weintraud, Rumpff & Co., 4 King-street, Cheapside.

Money-bags: portemonnaies; cigar-cases; elastic purses, ladies' bags; ladies' companions and albums of every kind.

580. KALTENHÄUSER, D., & Co., Schotten. London 1862 medal.

No. 1. a flower-table with 6 sides in open work, 3 chains 18 fl. (1 £ 10 sh. 4 d.); No. 2. a flower-stand with 6 sides in open work 3 fl. 50 kr. (6 sh. 5 d.); No. 3. a flower-stand with solide pot 3 fl. (5 sh.); No. 4. a fancy table with carved top 7 fl. (11 sh. 8 d.); No. 5. a tea-canister with 2 bottles 8 fl. 36 kr. (14 sh. 4 d.); No. 6. a flagon-chest with 2 flagons à odeur 4 fl. 30 kr. (7 sh. 6 d.); No. 7. a porcelain-plate with wood-ornaments 7 fl. (11 sh. 8 d.): No. 8. a liqueur-chest with 2 bottles and 12 glasses 25 fl. (2 £ 3 sh. 7 d.); No. 9. a letter-box 6 fl. 36 kr. (11 sh.); No. 10. a cigar-box with cigar-bundles on the cover 7 fl.

(11 sh. 8 d.); No. 11. an other cigar-box 18 fl. (1 £ 10 sh. 4 d.); No. 12. a glove-box, rustique with ribbons and nails 1 fl. 56 kr. (3 sh. 1 d.); No. 13. an other glove-box 1 fl. 56 kr. (3 sh. 1 d); No. 14. a whist-chest 4 fl. 50 kr. (8 sh. 1 d.); No. 15. an ink-stand with glass 3 fl. 30 kr. (6 sh.); No. 16. an ink-stand with two glasses 2 fl. 30 kr. (4 sh. 2 d.); No. 17. a folding-chair (ordinary) 4 fl. 30 kr. (7 sh. 6 d.); No. 18. a folding-chair of middle size with clouts and carved work 8 fl. 30 kr. (13 sh. 4 d.); No. 19. a large folding-chair with clouts, arms and carved work 10 fl. 12 kr. (17 sh.); a small folding-chair with reed basked-work 8 fl. (13 sh. 4 d.); No. 21. two glass-chests 12 fl. (1 £); No. 22. a tea-chest twisted with two straps for two bottles 5 fl. (8 sh. 4 d.).

SUPPLEMENT.

581. The central board for measures and weights for the Grand-Duchy of Hesse.

1 metre, divided in centimetres and inches; dry-measures: 1 bushel, 1 peck, 2 litres, $\frac{1}{2}$ litre. Fluid-measures: 1 gallon, 1 pottle, 1 quart, 1 pint; weights: $\frac{1}{4}$, $\frac{1}{2}$, 1, 2, 3, 4, 5, 10, 25, 50 pounds; description of the Hessian system of measures and weights.

582. Heumann, Otto, laboratory for technical chemestry, Ober-Ramstadt. London 1862 honor. mention.

Drying-oil for the manufacture of copal-varnish representing a new method; green color of baryt as representing the method of pruducing a color for oil-painting free of arsen, lead and copper and not altered by light.

583. Dolles, Joseph, wine-merchant and prop. of exquisite vine-yards, Bodenheim on the Rhine. Paris 1855 honor. mention.

Bodenheim Riesling Ausbruch, label white and gold; 1 qual. label violet and silver, both of the year 1857; Bodenheimer Riesling Ausbruch, label white and gold; 1 qual. label violet and silver both of the year 1858; Bodenheimer Riesling Ausbruch, label white and gold; 1 qual. label violet and silver both of the year 1859; all wines grown by the exhibitor.

585. The administration of Ludwigsbrunnen near Gross-Karben. General-Depositor Philipp Forster, Frankfort on the Main.

12 stone-bottles of mineral water 13 fl. p. 100 bottles.

586. Eckhard, Dr., Darmstadt.

A sailing vessel of the smallest resistance

PRINCIPALITY OF LIPPE.

CLASS II.

CHEMICAL SUBSTANCES AND PRODUCTS AND PHARMACEUTICAL PROCESSES.

591. SCHIERENBERG, G. A. B., manu., Meinberg. Agt. F. W. Koch, 6 Olive Terrace, Camberwell New-road, Kensington.

Samples of Cremnitz - white in paste (en pâte) 100 pounds zoll-weight 10 th. 112 lbs. 1 £ 10 sh.

592. WIPPERMANN, F. W., manu., Lemgo Agt. H. Landwehr from Bielefeld, exhibition-buildings.

Spirit of vinegar.

CLASS IV.

ANIMAL AND VEGETABLE SUBSTANCES USED IN MANUFACTURES.

593. HOFFMANN, F., manu., Salzuflen (Starch manufactory near Salzuflen). Agts. Blow & Raeke, 14 London-street, E. C.

Samples of fine kiln-dried starch in sticks, air-dried do., patent in pipes do. hair powder do.; all extremely white and clear.

Fine kiln-dried in sticks, patent in pipes do., both blued and very clear.

Roasted wheat starch (wheat-gum) of three qualities for technical purposes, of every color from light to dark brown according to ordre.

For prices apply to the above named agents.

CLASS XIX.

FLAX AND HEMP.

594. KOTZENBERG, F. L., manu., Salzuflen. Munich 1854 honor. mention. Agt. H. Landwehr from Bielefeld, exhibition-buildings.

Bleached linen thread No. 50. I. a. 38 sg. (3 sh. 10 d.); do. No. 150. I. a. 85 sg. (8 sh. 6 d.); do. No. 30. II. a. 25 sg. (2 sh. 6 d.); do. No. 60. II. a. 36 sg. (3 sh. 7 d.); black and glazed No. 40. I. a. 34 sg. (3 sh. 5 d.); do. not glazed No. 80. I. a. 43 sg. (4 sh. 3 d.); do. not glazed No. 30. II. a. 20 sg. (2 sh.); do. not glazed No. 60. II. a. $32\frac{1}{2}$ sg. (3 sh. 3 d.); do. not glazed No. 25. III. a. $17\frac{1}{2}$ sg. (1 sh. 9 d.); black and glazed No. 35. III. a. 20 sg. (2 sh.); grey No. 35. II. a. $22\frac{1}{2}$ sg. (2 sh. 3 d.); in various colors No. 35. II. a. 25 sg. (2 sh. 6 d.). Prices refer to 1 lb.

CLASS XXXVI.

MANUFACTURES NOT INCLUDED IN PREVIOUS CLASSES.

595. REHTMEYER, GOTTFRIED, manu of tobacco-pipes, Lemgo.

Genuine Meerschaum tobacco pipe-bowls, boiled in wax and mounted with silver, and Meerschaum-cigar-pipes with amber-tips.

GRAND-DUCHY OF LUXEMBOURG.

CLASS I.

MINING, QUARRYING, METALLURGY AND MINERAL PRODUCTS.

601. Godin, David, engineer, director of the society of mines of Stolzenburg, Liege.

Various specimes of copper-ore, 500 kilo 200 frs. The ores such as they are taken out of the lodes contain 23 pCt. of copper, by an analysis recently made at Swansea.

603. de la Fontaine, Edmund, prop., Remich on the Moselle. Agts. Lion M. Cohn, Phaland & Dietrich, 20 St. Dunstans Hill, City. (s. No. 606.)

2 pieces of alabaster, 3 pieces of paving-stones (gris de Luxembourg), 1 piece of lime-stone.

CLASS XVIII.

COTTON.

606. de la Fontaine, Edmund, prop., Remich on the Moselle. Agts. Lion M. Cohn, Phaland & Dietrich, 20 St. Dunstans Hill, City. (s. No. 603.)

Cotton goods woven by power-looms.

607. Charles, Aug., & Co., Bonnevoie near Luxembourg. Berlin 1844 bronze medal; London 1851 honor. mention; Munich 1854 honor. mention; Paris 1855 silver medal; London 1862 medal. Agts. Ch. Oppenheimer & Co., $79\frac{1}{2}$ Watling-street, Cheapside, for. gloves. — John W. Münch, 11 Union Court, Old Broadstreet, for leather.

66 pair of gloves: 4 p. habit-gloves embroidered; 11 p. ladies-gloves, laced, Louise, Dutchess, Eugenie, Josephine, double stitched; 3 p. gentlemengloves castor, dog-skin; 2 p. Sweden-glowes; 8 p. kid-gloves for gentlemen, ladies, gentlemen, embroidered (brodes au crochet); 4 p. kid-gloves for ladies, embroidered (brodes au crochet); 4 p. kidgloves for gentlemen stitched (piques $\frac{1}{2}$ anglais); 4 p. kid-gloves for ladies stitched (piques $\frac{1}{2}$ anglais); 12 p. ladies-gloves arlequin, fancy, $\frac{1}{4}$ long à festons; 1 p. gentlemen-gloves lined with flannel; 1 p. ladies-gloves unsewed; 12 p. children's-gloves; 190 dyed kid-skins; 12 black calf-hides.

CLASS XXXI.

IRON AND GENERAL HARDWARE.

611. Neuen-Therer, locksmith, manu. of iron furniture, inv., Luxembourg.

Two bed-steads, one 40 frs.; wholesale price 34 frs.

DUCHY OF NASSAU.

CLASS I.

MINING, QUARRYING, METALLURGY AND MINERAL PRODUCTS.

617. DÖPPENSCHMIDT, FR., manager of the slate-mines of His Highness the Duke of Nassau, Caub.

Slates, prepared in English square shape and German rounded shape. Cut plates, polished and varnished.

619. HAAS, L., & Co., prop. of mines and smeltingworks, Dillenburg.

Nickel-ores: iron- and copper-pyrites with nickel-pyrites.

Nickel-smelting products: nickel-copperrust, concentrated nickel-copperrust; do. freed from iron.

620. HEUSLER, C. L., prop. of mines and smeltingworks, Dillenburg. London 1862 medal.

Nickel-ores: iron- and copper-pyrites with nickel-pyrites.

Nickel-smelting-products: roasted nickel-ores, nickel-copperrust, roasted nickel-copperrust; concentrated nickel-copperrust, do. freed from iron; refined alligation of copper and nickel.

621. LOSSEN, A., SONS, prop. of iron-smeltingworks and iron forges, Michelbach. London 1851 and 1862 honor. mention.

Pig-iron: cast iron in bars, for trying the degree of its tenacity and elasticity.

Screw-apparatus: to try the tenacity and elasticity of the cast-iron bars.

Rod-iron: refined charcoal-iron, bent, twisted and perforated in cold temperature, to show the tenacity and quality of the iron.

622. SCHÄFER, lawyer, prop. of mines, Diez.

Slates, prepared in English and German shape.

623. THE GOVERNMENT ENGINEERS OF MINES at Diez, Dillenburg, Weilburg, Wiesbaden — London 1851 prize-medal, 1862 medal — in the name of the prop. of mines and smeltingworks:

Collection of the products of mines and smeltingworks.

MINING COMPANY HOLZAPPEL at Holzappel and FRIEDRICHSEGEN at Oberlahnstein; REMY, HOFMANN & Co. at Bendorf (Ems).

Silver- and lead-ores: leadglance, rich in silver, do. with copper-pyrites; carbonate of lead and copper: cerusite and malachite; pure copper-ores: copper-pyrites without admixture of lead-ore.

HEYL, burgomaster at Weyer; HEYMANN, burgomaster at Niedertiefenbach; HEYMANN & Co. at the same place; LAUX, BROTHERS, at Diez; MAI, HILF & Co. at Limburg; REPPERT, L., at Cologne (Niedertiefenbach); REUSS & SONS at Heilbronn (Geissenheim); SCHWARZ II., G., & Co., at Niedertiefenbach; UNKELBACH at Limburg.

Manganese-ore: peroxyde of manganese, pyrolusite, psilomelan.

THE MANAGERS OF MINES of His Highness the Duke of Nassau at Dillenburg and Weilburg; MINING COMPANY CONCORDIA at Eschweileràue (Schaumburg, Villmar); MINING COMPANY DEUTSCH-HOLLAND at Johannishütte, Duisburg (Balduinstein); MINING COMPANY DILLINGEN, ironworks at Dillingen (Weinbach); C. HERBER at Wiesbaden; MINING COMPANY HÖCHST, ironworks at Höchst; JACOBI, HANIEL & HUYSSEN at Ruhrort (Oberneisen); MINING COMPANY NIVERN, ironworks at Nivern; REMY, HOFMANN & Co. at Bendorf (Diez, Ems); STRAUSS, J., & Co., at Wiesbaden; MINING COMPANY PHÖNIX at Laar near Ruhrort (Villmar, Ahausen).

Iron-ores: red-hematite, do. with calcspar, superior flux for iron-smelting; do. with iron-foam; brown hematite; clay iron-stone; spathase-iron.

COMPANY OF HÖCHST, ironworks at Höchst.

Pigiron: grey charcoal-pigiron; specular-iron: oligiste-iron.

MARPOW, DE, GERVAIS, AUG., & Co., at Puy (Nauroth).

Heavy-spar.

KOCH, C. & L., at Dillenburg; SLATEMINING COMPANY LANGHECKE at Langhecke.

Slates.

ALEFF at Ebernhahn; SCHRÖDER, schultheis at the same place; COMPANY OF JUG-MANUFACTORERS WIRGES & HILLSCHEID at Wirges & Hillscheid; KOCH, C. & L., at Dillenburg; LADE, G. & F., at Geissenheim.

Fire-clay; china-clay; fire-bricks.

BURGER, burgomaster, HAIBACH, BROTHERS, SCHMIDT, F., SEELBACH, BROTHERS, at Allendorf.

Fullers-earth.

MARBLE MANUFACTORY of the house of correction at Diez.

Marble of varied colors, grey, red, black.

RÜBSAMEN, BROTHERS, at Hof.

Brown-coal, lignite, bituminous wood.

CLASS II.

CHEMICAL SUBSTANCES AND PRODUCTS AND PHARMACEUTICAL PROCESSES.

624. DIETZE, H. & Co., manu., Mayence, manufactory Lorch. London 1862 honor. mention.

Verdigris, extra dry: subacetate of copper 50 fl. (4 £ 5 sh. 9 d.), do. refined: acetate of copper 58 fl. (4 £ 18 sh. 9 d.); chemically pure acetic-acid 7° B. 18 fl. (1 £ 10 sh. 10 d.); do. 8° B. 21 fl. (1 £ 16 sh.); do. 9° B. 24 fl. (2 £ 1 sh. 1 d.); acetic-acid for technical use, first quality 7° B. 12 fl. 5 kr. (1 £ 1 sh. 5 d.); do. 8° B. 14 fl. 5 kr. (1 £ 4 sh. 10 d.); do. 9° B. 16 fl. 5 kr. (1 £ 8 sh. 8 d.); do. for the manufacture of anilin 7° B. 12 fl. (1 £ 7 d.); do. 8° B. 14 fl. (1 £ 4 sh.); do. 9° B. 16 fl. (1 £ 7 sh. 10 d.); concentrated refined pyrolignous acid, vinegar of wood 32 pCt. A. H. W. 11 fl. (18 sh. 11 d.); 39 pCt. A. H. W. 13 fl. (1 £ 2 sh. 4 d.); 40 pCt. A. H. W. 15 fl. (1 £ 5 sh. 8 d.); kreosot from beechwood-tar 175 fl. (15 £.).

CLASS III.

SUBSTANCES USED FOR FOOD, INCLUDING WINES.

624a. Aschrott, H. S., manu., Hochheim. London 1862 honor. mention.
Hock.

625. Dietrich & Ewald, manu. of sparkling hock and Moselle, Rüdesheim on the Rhine. London 1862 medal. Agts. Lightly & Simon, 123 Fenchurch-street, E. C.
Sparkling Hock and Moselle six different sorts: No. 1. sparkling Rüdesheim (brown Hock), No. 2. and 3. sparkling Hock nonpareil; No. 4. sparkling Moselle nonpareil non muscatel, No. 5. and 6. Moselle nonpareil.

626. Dilthey, Sahl & Co., prop. and growers, Rüdesheim, on the Rhine. Agt. W. Foster, Newton, Falconhall cellars and 11 Silver-street, Falcon square, E. C.
Hock: wine of Rüdesheim in twelve different sorts from the seasons: 1846, 1857, 1858 and 1859.

627. Müller, Math., manu. of sparkling Hock and Moselle, Eltville on the Rhine. Berlin 1844 honor. mention; Munich 1854 great medal; London 1862 medal. Agts. Camphausen & Co., 6 Billiter square.
Sparkling Hock and Moselle in seven different sorts: Nr. 1. sparkling Hock, No. 2. do. superior, No. 3. do. nonpareil, No. 4. flower of sparkling Johannisberg, No. 5. sparkling Moselle, No. 6. do. superior, No. 7. do. nonpareil.

628. Siegfried, G. W., prop. and grower, Rauenthal, Rheingau. London 1862 medal. Agts. Charles Wooley & Co., 66 Mark-lane.
Hock, wine from Rauenthal (Berg) in three different sorts from the season of 1858.

629. Wiesbaden, central administration of the domains of His Highness the Duke of Nassau, Wiesbaden. London 1862 medal.
1. Hock, cabinet-wines: Hochheim season of 1839 and 1858, Neroberger season of 1858 and 1859, Marcobrunner season of 1858, Steinberger season of 1846, 1857 and 1858, Rudesheimer 1839 and 1857.
2. Mineralwaters in their original jugs and bottles from Selters, Fachingen, Geilnau, Ems-Kranchen, Ems-Kesselbrunnen, Schwalbach-Paulinenbrunnen, Schwalbach-Stahlbrunnen, Schwalbach-Weinbrunnen.
3. Pectoral-comfit of Ems, mixed with the dry salts of the Ems-mineralwaters, exhibited by the administration of the Baths of Ems.

630. Wiesbaden, the directory of the society for agricultural and forest matters of Nassau.   (s. No. 632.)
Red-wheat from the district of the Aar, largely exported to Holland and England; samples of flour from this wheat; early-barley and early-oats, the latter grown 1500 ft. above the sea, on the high Westerwald.

CLASS XVI.

MUSICAL INSTRUMENTS.

631. Schellenberg, A., manu., Wiesbaden.
Guitar with inlaid adornments 75 fl. (6 £ 8 sh.), case 13 fl. (1 £ 2 sh.); bass-guitar of palixanderwood 36 fl. (3 £ 1 sh. 8 d.), case 6 fl. (10 sn.); guitar of palixander-wood 32 fl. (2 £ 13 sh. 6 d.), case 6 fl. (10 sh.); guitar ordinary 21 fl. (1 £ 15 sh.); case with four dozen twined strings for bass-guitars 5 fl. 2 kr. (8 sh. 8 d); case with four dozen strings for guitars 4 fl. 8 kr. (8 sh.).

CLASS XIX

FLAX AND HEMP

632. Wiesbaden, the directory of the society for agricultural and forest matters of Nassau.   (s. No. 630.)
Flax, grown on the high Westerwald, 1500 ft. above the sea.

CLASS XXI.

WOOLLEN AND WORSTED, INCLUDING MIXED FABRICS GENERALLY.

633. Stirn, Math., Sons, shoddy and mungo manufacturers, Oberursel near Frankfort on the Main.
Restored (artificial) wool, free from remains of rags, long of hair, and pure in color. coarse Mungos 2 to 2½ d.; fine do. 5 to 12 d.; fine cachemire and tibet 10 to 18 d.; shoddy 3½ to 9½ d.; all p. lb. (50 kilogr.).

CLASS XXVIII.

PAPER, STATIONERY, PRINTING AND BOOKBINDING.

634. Kreidel, Chr. Will., publisher, Wiesbaden.
Printed books of his own publishing, mostly with illustrations: Braun, history of art etc., 2 vol. 10 fl. 40 kr. (17 sh. 9 d.); Hartwig, the high north 4 fl. 2 kr. (7 sh.); do., the tropics 5 fl. 25 kr. (8 sh. 9 d.); do., the isles of the Great Ocean 5 fl. 25 d. (8 sh. 9 d.); Heusinger v. Waldegg and Wilh. Clauss, representation and description of the locomotive engine 17 fl. 5 kr. (1 £ 9 sh. 2 d.); Heusinger, organ for the progress of railroad-engineering, continued by Dr. H. Scheffler, 1857—1861 a year 7 fl. (11 sh. 8 d.), Neubauer, Dr. C. and Dr. J. Vogel, quant. and qual. chemical analysis etc., 3. edition 4 fl. (6 sh. 8 d.); Reden, Dr. von, Germany and the other parts of Europe 12 fl. 8 kr. (1 £ 1 sh. 4 d.); Sandberger, Dr. Frid., fossil shells of the tertiary formation of the Mayence basin 36 fl. 75 kr. (3 £ 1 sh. 9 d.); Sandberger, Dr. Guido and Dr. Frid., fossil shells of the Rhenish devonian strata 48 fl. (4 £ 1 sh. 8. d.); Schenkel, Dr. Daniel, christian dogmatik, 2 vol. 16 fl. (1 £ 6 sh. 8 d.); Technological dictionary, English, German, French, by Prof. Dr. Franke 4⅔ fl. (7 sh. 9 d.); annual review of analitical chemistry, by Dr. R. Fresenius 5 fl. 25 kr. (8 sh. 9 d.).

GRAND-DUCHY OF OLDENBURG.

CLASS I.

MINING, QUARRYING, METALLURGY AND MINERAL PRODUCTS.

636. Lürssen, B. H., manu. of lighting stones, Delmenhorst. Agt. Albert Wodecki, 10 St. Mary's Axe. (s. No. 638.)

Lighting stones 100 Pfd. 2 Th. gold (112 lb. engl. 7 sh.).

CLASS III.

SUBSTANCES USED FOR FOOD, INCLUDING WINES.

637. Hoyer & Son, Oldenburg. Agts. Loeder, Krapp & Co., 1 New Broad-street, City. (s. No. 639.)

Strongbeer in bottles p. doz. 15 sg. (1 sh. 6 d.); pale ale p. doz. bottles 20 sg. (2 sh.).

CLASS IV.

ANIMAL AND VEGETABLE SUBSTANCES USED IN MANUFACTURES.

Cassebohm, H. (s. No. 645.)

638. Lürssen, B. H., manu., Delmenhorst. London 1862 medal. Agt. Albert Wodecki, 10 St. Mary's Axe. (s. No. 636.)

Corks: homoeop. I. quality No. 0. 40 grotes (1 sh. 10 d); No. 1. 45 gr. (2 sh. 1 d.); No. 2. 55 gr. (2 sh. 6 d.); No. 3. 65 gr. (3 sh.); No. 4. 75 gr. (3 sh. 6 d.); No. 5. 90 gr. (4 sh. 2 d.); No. 6. 105. gr. (4 sh. 10 d.); No. 7. 140 gr. (6 sh. 6 d.); No. 8. 170 gr. (8 sh. 4 d.); No. 9. 205 gr. (9 sh. 6 d.); No. 10. 250 gr. (11 sh. 6 d.); do. II. quality No. 1. 20 gr.; (11 d.) No. 2. 25 gr. (1 sh. 2 d.); No. 3. 30 gr. (1 sh. 5 d.); No. 4. 37½ gr. (1 sh. 9 d.); No. 5. 47⅓ gr. (2 sh. 2 d.); No. 6. 57½ gr. (2 sh. 8 d.); No. 7. 72 gr. (3 sh. 4 d.); No. 8. 90 gr. (4 sh. 2 d.); No. 9. 110 gr. (5 sh. 1 d.); No. 10. 135 gr. (6 sh. 3 d.); all in grotes (gold) p. 1000 pieces free on board Bremen.

½ long corks for bottles I. quality 6 Th. (1 £); do. II. quality 4 Th. (13 sh. 4 d.): ⅓ long do. I. quality 4 Th. (13 sh. 4 d.); do. II. quality 3 Th. (10 sh.); corks for champagne bottles I. quality 18 Th. (3 £); do. II. quality 9 Th. (1 £ 10 sh.); all p. 1000 pieces free on board Bremen in Thaler (gold).

639. Hoyer & Son, Oldenburg, manufactory at Donnerschwa. Medals Munich and Paris. Agts. Loeder, Krapp & Co., 1 New Broad-street, City. (s. No. 637.)

Stearic acid 35 Th. (5 £ 4 sh. 6 d.); stearine-candles 35 Th. (5 £ 4 sh. 6 d.); cocoa-oil soap 11 Th. (1 £ 13 sh.); oleine-soap 12½ Th. (1 £ 17 sh. 6 d.); all p. 100 lb. (cwt.)

CLASS VII.

MANUFACTURING MACHINES AND TOOLS.

640. Brader & Co., Zwischenahn.

Thrustle bobbins 2⅓ Th. (7 sh.); roving bobbins 6 Th. (18 sh.); roving tubes 2⅓ Th. (7 sh.); slubbing tubes 4⅓ (13 sh.); warping bobbins 6 Th. (18 sh.); p. Gross.

CLASS XI.

MILITARY ENGINEERING, ARMOUR AND ACCOUTREMENTS, ORDNANCE, AND SMALL ARMS.

641. Wichmann, J. H., captain of arms, Oldenburg. Agts. Philipps, Graves & Philipps, London.

A musket (with description) 25 Th. (4 £); two empty cartridges.

CLASS XVII.

SURGICAL INSTRUMENTS.

642. Kirchmeyer, Ph., saddler and truss-maker, Ovelgönne. Agts. Philipps, Graves & Philipps, London. (s. No. 643.)

A double truss with elastic pads to which a regulator is attached. The screws of the regulator being lifted up the pads are found to be moveable in order to be fixed properly to the port of the rupture 9 Th. (1 £ 10 sh.); a single truss for an inguinal hernia with a rather narrow port 3 Th. 10 sh.

CLASS XXVI.

LEATHER, INCLUDING SADDLERY AND HARNESS.

643. Kirchmeyer, Ph., saddler and truss-maker, Ovelgönne. London 1862 honor. mention. Agts. Philipps, Graves & Philipps. London. (s. No. 642.)

A quilted gentleman's saddle with stirrup-straps 50 Thl. Gold (6 £); elastic saddle-tree entirely of leather with whalebone and steel springs 9 Th. Gold (1 £ 10 sh.).

CLASS XXXVI.

MANUFACTURES NOT INCLUDED IN PREVIOUS CLASSES.

645. Cassebohm. H., cook to H. R. H. the Grand Duke of Oldenburg. London 1851 honor. mention; Munich 1854 honor. mention: London 1862 honor. mention. Agts. Philipps, Graves & Philipps. London.

Phelloplastic. Model of the Heidelberg castle, northside 210 Th. (30 £); epitaph of Anton Gunther last count of Oldenburg 77 Th. (11 £); choragic monument of Lysicrates at Athens 28 Th. (4 £): the Heidelberg castle in relievo (not to be sold).

KINGDOM OF PRUSSIA.

CLASS I.

MINING, QUARRYING, METALLURGY AND MINERAL PRODUCTS.

Concerning the Special Catalogue of the objects exhibited in Class I. the reader is referred to the notice in the preface.

Of the numbers on the labels of this class those in the left-hand corner refer to the before mentioned special catalogue, those in the right-hand corner to the present and to the official or shilling-catalogue.

651. ACTIEN-GESELLSCHAFT FOR MINING AND THE MANUFACTURE OF LEAD AND ZINC, Stolberg a. in Westphalia.   London 1862 medal.

Ores and metallurgical products.

652. ACTIEN-GESELLSCHAFT FOR MINING AND SMELTING, Stolberg and Nordhausen.

Raw and dressed copper-ores.

652a. ALAUNWERK MUSKAU, director general Rieloff.

Alum and vitriol.

653. ALTENBEKER HÜTTENWERK.

Ironstone.

654. AMALIA, colliery near Minden.

Coal and ironstone.

655. AM SCHWABEN, colliery near Dortmund.

Fire-clay.

656. ANANIAS, ironstone-mine near Prussian Oldendorf.

Ironstone.

657. ATZROTT & Co., Cologne.

Table-top of roofing-slate.

658. BACHEM & Co., and SPINDLER & Co., Königswinter near Bonn.

Trachyt from Stenzelberg, basaltic lava from Hannebach, blue trachyt from Wolkenburg.

659. BAUTENBERGER EINIGKEIT, iron-mine near Unterwilden.

Ironstone.

BEERMANN, Riesenbeck, freestone.
(s. No. 803.)

660. BELGISCH-RHEINISCHE GESELLSCHAFT FOR MINING UPON RUHR, Düsseldorf.

Coal.

661. BERGER & Co., Witten.

Cokes.

662. BERG-INSPECTION, ROYAL, Ibbenbüren. London 1862 honor. mention.

Coal and beds from the coal measures.

663. BERG-INSPECTION, ROYAL, Königshütte (Silesia).

Coal and clay-ironstone.

664. BERG-INSPECTION, ROYAL, Tarnowitz.

Raw and dressed lead-ores.

665. BERG-INSPECTION, ROYAL, Wettin.

Coal of the collieries of Wettin and Loebejün; cokes.

666. BERG-INSPECTION, ROYAL, Zabrze.

Coal.

667. BERG- UND SALINEN-INSPECTION, ROYAL, Stassfurt near Magdeburg.   London 1862 medal.

Rock-salt; products of potassium; minerals.

668. BERGISCHER GRUBEN- UND HÜTTEN-VEREIN, Hochdahl. London 1862 honor. mention.

Common pig-iron, pig-iron for steelmaking, lead-ore.

670. BLEIBTREU, DR. H., director of mines and works at Obercassel near Bonn.

Patent lime-cokes.

671. BOECKING, BROTHERS, Asbacher-Hütte near Kirn.

Ironstones.

672. BÖRNER, M.; Siegen.   London 1862 medal.

Ores and products of iron-works.

673. BOLLENBACH, iron and manganese-mine near Neunkirchen.

Iron and manganese-ores.

674. BONNER BERGWERKS- UND HÜTTEN-VEREIN, Bonn.   (s. No. 1339.)

Iron-ores brown-coal, and products of it.

675. Boltze, Salzmuende near Halle.
Fire- and china-clay.

676. Bonzel, J., Mecklinghausen near Olpe.
London 1862 honor. mention.
Marble-slabs.

677. Borussia, colliery near Dortmund.
Coal and cokes.

678. Braut, colliery near Werden.
Coal.

679. Broicher Eisenstein-Bergwerk Muehl-
heim upon Ruhr.
Ironstone.

680. von Brewer, Xaver, Nieder-Mendig
(Coblenz).
Volcanic tuf, roofing-slate.

682. Buderus, G. W., Sons, rolling-mill
Germania near Neuwied.
Tinplates.

683. Buderus, J. W., Wetzlar.
Ironstone.

684. Burgholdinghauser Huttenwerk,
near Siegen.
Ironstone and products of blast-furnaces.

685. Camp & Co., Wetzlar.
Ironstone.

686. Capitain, Theodor, Obercassel near
Siegen.
Fine and coarse-grained basalt in flags from Ober-
cassel.

687. Cappeln, quarry, proprietor Scheer
at Ibbenburen.
Jurassic sandstone from the quarries on the Ibes-
knapp and at Westerbeck near Mettlingen.

689. Caroline Erbstollen, colliery near
Unna.
Coal.

690. Charbon, Pieter Eduar, Friedrichs-
Glück near Lintorf.
Iron-pyrites and lead-ore.

691. Charlotte, colliery near Hattingen.
Coal and beds from the coal-measures.

692. Christian Gottfried, colliery near
Tannhausen.
Antracite.

693. Churfurst Ernst, lead-mine near
Bonkhausen.
Lead-ores.

694. Coln-Müsener Bergwerks-Verein,
Müsen near Siegen. London 1862 medal.
Iron, lead, copper and silver-ores and smelting-
products.

695. Concordia, company of Eschweiler
for mining and smelting, Ichenberg near Eschweiler. London 1862 medal.
Iron-ores and blast-furnace-products.

696. Conradine, ironstone-mine near Porta
(Minden).
Ironstone.

697. Constantin der Grosse, colliery
near Bochum.
Coal and cokes.

698. Dahlbusch, colliery near Gelsenkirchen.
Coal.

699. Deutsch - Holländischer Actien-
Verein for smelting and mining, Johannishütte
near Duisburg. London 1862 honor. mention.
Coals, ironstone, blast-furnace and rolling - mill-
products.

700. Direction, Royal, of mines, Saarbrucken. London 1862 medal.
Collection of coals, cokes, maps, models of coal-
veins.

701. Doinet, proprietor of mines at
Zuelpich.
Iron and manganese-ores, brown-coals and pro-
ducts from them.

701a. von Dobschütz, Friedersdorf near
Greifenberg (Silesia).
Brown-coal from the pit Heinrich near Langenöls.

702. Dorstfeld, colliery near Dortmund.
Coal.

703. Dreckbank, colliery near Werden.
Ironstone.

704. Dressler, J. H., sen., Siegen. London 1862 medal.
Iron, lead, zinc, copper, nickel-ores, blast-furnace
and rolling-mill-products.

705. Eckardt, Julius, merchant, Gefell
district of Ziegenrück.
Washed ochre.

707. Einigkeit, colliery near Steele.
Fire-clay, ironstone.

708. Eisenzeche, iron-mine near Siegen.
Ironstone and manganese-ore.

709. Emonds, H. J., Bergisch-Gladbach.
Brown-coal.

710. Endemann, Wilhelm, & Co., Bochum.
Cokes.

711. Eschweiler Bergwerks - Verein,
Eschweiler. London 1862 medal.
Coal, cokes, maps.

712. Eschweiler Gesellschaft for mining and smelting, Stolberg near Aachen. London 1862 medal.
Lead, zinc, silver-ores, smelting-products.

713. Eunicke, Leopold (Firma Leopold
Kayser & Co.), Naumburg upon Bober.
Nickel in cubes.

714. Fanny, colliery near Kattowitz.
Specimens of calcined works of the coal-measures.

715. Felthaus, H., Wetzlar.
Ironstones and quicksilver-ores.

716. Fina, colliery near Mettmann.
Fire-clay, ironstone.

717. FLEITMANN, DR., Iserlohn. London 1862 medal.

Cobalt and nickel-metal and products from them.

718. FRAUENBERGER EINIGKEIT, iron-mine near Neunkirchen.

Ironstone.

719. FRIEDERIKE, colliery near Bochum.

Ironstone.

720. FRIEDRICH WILHELM, iron-mine near Siegen.

Ironstone.

721. FRIEDRICH-WILHELMS-HUTTE, joint-stock company for mining at Mühlheim upon Ruhr.

Pig-iron and cast-iron work.

722. FRIEDRICH-WILHELMS-HUTTE near Siegburg. London 1862 medal. (s. No. 1262, 1320.)

Iron-ores, products of blast-furnace.

723. FUCHS, colliery near Waldenburg.

Coal and specimens of intermediate rocks.

724. GENERAL No. I. TIEFBAU, colliery near Dahlhausen.

Coal.

725. GERLACH, GABRIEL & BERGENTHAL, Germaniahutte near Grevenbrück upon Lenne. London 1862 honor. mention.

Ironstones and products of blast-furnaces.

726. GIBRALTAR ERBSTOLLEN, colliery near Dahlhausen.

Coal

727. GLÜCKAUF TIEFBAU, colliery near Brüninghausen.

Coal.

728. GLÜCKHILF, colliery near Waldenburg.

Coal.

729. GOTTESSEGEN, concession on Bleiberg near Commern.

Knobby copper-ore.

730. GRAF BEUST, colliery near Essen.

Coal.

731. GREVEL, iron-mine near Unna.

Ironstone.

732. GRÜNEBERGER CONSOLIDIRTE BRAUNKOHLENGRUBEN (United brown-coal collieries), Gruneberg.

Brown-coal of different kinds.

733. GUETTLER, Reichenstein.

Products of arsenic and gold-works.

734. GUTER TRAUGOTT, colliery near Myslowitz.

Coal.

735. HAEUSLINGS-TIEFE, iron-mine near Siegen.

Ironstone.

736. HARPEN, MINING SOCIETY OF, Dortmund.

Coal

737. HASENWINKEL, colliery near Hilgenstock.

Beds from the coal-measures.

738. HECTOR, ironstone-mine near Ibbenbüren.

Ironstone.

739. HEINRICH GUSTAV, colliery near Bochum.

Cokes.

740. HEINRICHSSEGEN, lead-mine near Müsen.

Grey copper-ore.

741. HELENE AND AMALIE, colliery near Essen.

Coal.

742. HELLE, iron-mine near Sundwig.

Ironstone.

743. COUNT HENCKEL OF DONNERSMARCK, HUGO, Siemanowitz (Silesia). London 1862 medal.

Coal; ironstone; products of zinc and iron-works.

744. HENRICHSHÜTTE, aministration of the Henrichshütte near Hattingen. London 1862 medal.

Pig-iron; rolling-mill products and lead ores.

745. HIEPE, S., Wetzlar.

Specimens of marble.

746. HOERDER BERGWERKS- UND HUTTEN-VEREIN, Hoerde. London 1862 medal. (s. No. 1258.)

Coal; beds from the coal-measures; ironstone; phosphate.

747. HOLLAND, colliery near Bochum.

Coal.

748. HOLLERTERZUG, iron-mine near Dermbach.

Manganese-ores.

749. HORHAUSEN, mine near Hamm upon Sieg.

Ironstone.

750. HOYER, GUSTAV, & Co., Schonebeck.

Salt licking-stones and cribs for cattle.

751. HUEBNER, DR., BERNARD, manufactory of mineral oil and parafine, Rehmsdorf near Zeitz. London 1862 medal.

Brown coal and products from it.

752. HÜTTENAMT, ROYAL, Creuzburgerhutte (Silesia).

Ironstones and products of iron-works.

754. HÜTTENAMT, ROYAL, Konigshütte (Silesia). London 1862 medal.

Calamine and products of zinc-works, iron-ores and products of iron-works.

755. HÜTTENAMT, ROYAL, Malapane. London 1862 honor. mention.

Iron-ores and products of iron-works.

756. HÜTTENAMT, ROYAL, Messingwerk near Neustadt-Eberswalde.

Products of yellow and common copper.

757. HÜTTENWERK, ROYAL, Rybnik.

Rolled and hammered iron.

757 a. HÜTTEN-INSPECTION, ROYAL, Friedrichshütte near Tarnowitz. London 1862 honor. mention.

Lead and silver-ores and products from them.

758. HUYSSEN, DR., Royal Director, Breslau. London 1862 medal.

Maps of the coal measures of Lower-Silesia.

759. JACOBI, HANIEL, HUYSSEN & CO., Oberhausen. London 1862 honor. mention.

Coal, cokes, ores, products of blast-furnace and rolling-mill.

760. IBBENBÜREN, proprietor of a quarry.

Coal-sandstone.

761. JUNG, C., Dahl near Hagen.

Steel-bars.

762. JUNGE SINTERZECHE, iron-mine near Siegen.

Ironstone.

763. KAERGER, C. H. L., Willmannsdorf near Jauer.

Red iron-ores.

764. KARL GEORG VICTOR, colliery near Alt-Laessig.

Coal.

765. KELLER, KÜPPERS, RAABE & CO., proprietors of quarries at Aix-la-Chapelle.

Sandstone and crystal-sand.

766. KLOSE, captain of the united copper-mines, Kupferberg.

(Copper-ores; products of copper-smelting works.

767. KOLLMANN, Clotten (Coblenz).

Roofing-slate from Müllenbach.

767 a. KRAMSTA, Freiburg.

Coal, ironstones, zinc-plates.

768. KRAUSE, Berlin. (s. No. 2064.)

Iron-ores and products of Silesian iron-works.

769. KRAUSS, representative of the Westerwald mining-company at Bensberg.

Lead, silver, zinc-ores, raw and separated.

770. KREUTZ, JACOB, owner of mines and smelting-works at Siegen. London 1862 honor. mention.

Ironstone and products of blast-furnaces.

770 a. KRUPP, FR., Essen. (s. No. 1308.)

771. KULMIZ, counsellor of commerce, Saarau (Liegnitz).
(s. No. 998, 1104, 1353, 2171.)

Granite from the »Streitberg« near Striegau (Silesia). Raw and dressed coal; cokes, brown-coal.

772. LAMARCHE, company, Velbert.

Alum-shale, ironstone.

773. LANDAU, SALOMON, proprietor of mines, Coblenz and Andernach upon Rhine. London 1862 medal.

Lava-millstones and trass.

774. LANDESKRONE, lead-mine near Siegen.

Lead-ores.

775. COUNT OF LANDSBERG-VELEN, Velen near Münster.

Iron-pyrites and iron-ores.

777. LAURA, colliery near Minden.

Coal.

778. LENNE - RUHR, joint-stock company for mining and smelting near Altenhundem. London 1862 medal.

Iron-ores, iron-pyrites, iron and steel-products.

779. LIMBURGER FABRIK UND HÜTTEN-VEREIN, Limburg upon Lenne. London 1862 medal. Agt. Heintzmann & Rochussen, 9 Friday-street, Cheapside.

Steel-products.

782. LOHMANNSFELD, lead-mine near Alten-Seelbach.

Lead-ores.

783. LOMMERSDORF, iron-mine near Adenau.

Ironstones.

783 a. LÜSCHWITZ, ADMINISTRATION OF MINES AND SMELTING WORKS, Breslau.

Ores and products of arsenic.

784. MÄRKISCH - WESTPHÄLISCHER BERG-WERKS-VEREIN, Iserlohn.

Zinc- and lead-ores and metallurgical products.

785. MAIKAMMER, iron-mine near Wülfrath.

Ironstone.

786. MANSFELD SCHE GEWERKSCHAFT, COMPANY FOR MINING OF COPPER-SLATE, Eisleben. London 1862 medal. (s. No. 2090.)

Brown-coal; copper-slate and products of smelting-works; maps.

787. MARIA, colliery near Höngen in the Worm-district. London 1862 medal.

Coals, cokes, maps.

788. MARIE-LOUISE, colliery near Dahlhausen.

Ironstone.

789. MECHERNICHER BERGWERKS-ACTIEN-VEREIN, Mechernich in Rhenish Prussia.

Raw and separated lead-ores and products.

790. MECKLINGBÄNKER ERBSTOLLEN, colliery near Steele.

Ironstone.

791. MERKELL, master stone-cutter, Halle on Saale.

Motley sandstone from Nebra near Querfurt.

792. MEURER, WILHELM, merchant producer of pig-iron and proprietor of mines, Cologne.

Ironstones, lead and copper-ores, pig-iron and smelting products.

793. MINERVA, joint-stock company (director general Conrad), Breslau. London 1862 medal.

Products of the iron-works of Minerva, and steel.

794. MORGENSTERN, vitriol-work near Landeshut.

Iron-pyrites, products of sulphur, vitriol and alum-works.

795. VON MÜLMANN, ALBERT, proprietor of mines, mine Plato near Siegburg.
(s. No. 2174 a.)

Brown-coal, clay, ironstone.

D •

796. Müllensieffen, Gustav, Crengeldanz near Witten
Coal

797 Neu-Altstädde III., iron-mine near Ibbenbüren.
Ironstone.

798. Neu-Düsseldorf, colliery near Dortmund.
Coal and stones from the coal-measures.

799. Neumann & Zimmermann, Lindlar near Wipperfürth (Rhenish province)
Sandstone.

800. Neu-Schottland, joint-stock company for mining and smelting, Steele.
Pig-iron.

801. Niederfischbacher Gesellschaft for mining and smelting, Niederfischbach, district of Siegen.
Ores; products of separating and smelting-works.

802. H. R. H. Princess Marianne of the Netherlands.
Marble from Seitenberg near Habelschwerdt (Silesia).

803. Niemann, Bevergern, and Beermann, Riesenbeck (Münster).
Sandstone (Hils).

804. Nordstern, ironstone-mine near Siegen.
Ironstone.

805. Oberbergamt, royal, Bonn. London 1862 medal.
Geological map of Rhenish-Prussia and Westphalia by Mr. von Dechen.

806. Oberbergamt, royal, Breslau. London 1862 medal.
Geological maps and maps of the coal-measures. Specimens of rocks and ores.

807. Oberbergamt, royal, Dortmund. London 1862 medal.
Maps of the Westphalia coal-districts.

808. Oberbergamt, royal, Halle. London 1862 medal.
Geological maps and sections.

810. Ostermann, A., & Co., Bochum.
Cokes.

811. Philippshoffnung, cobalt-mine near Siegen.
Cobalt-ores.

812. Phönix, joint-stock company for mining and smelting, Laar near Ruhrort. London 1862 medal.
Coals, iron-ores, blast-furnace and rolling-mill-products.

813. Pörtingssiepen, colliery near Werden.
Beds from the coal-measures.

814. Porta Westphalica, Hüttengesellschaft near Porta.
Ironstone.

815. Prang, Jean Baptiste, Münster, owner of marble-quarries at Alme near Brilon. London 1862 honor. mention.
Marble-slabs.

816. Primavesi & Co., Gravenhorst.
Ironstone.

817. Prinz Wilhelm, mine near Langenberg.
Raw and dressed lead and zinc-ores.

818. Raab, Ludwig, Wetzlar.
Ironstone.

819. Redenshütte near Zabrze (Silesia).
Products of iron-works.

820. Rehm, Heinrich, Wetzlar.
Specimens of marble.

821. Reicher Trost, mine near Reichenstein.
Arsenic-ores and manufactured arsenic.

822. Remy, Brothers, Werdener Hütte, District Olpe. London 1862 honor. mention.
Ironstones and products of blast-furnaces.

823. Rhonard, mine near Siegen.
Copper-pyrites.

824. Ritter, Friedrich, Bochum.
Cokes.

825. Roemel, iron-mine near Neunkirchen.
Ironstone.

826. Roland, colliery near Mülheim upon Ruhr.
Coal and stones from the wal-measures.

827. Ruffer, Breslau. London 1862 medal. (s. No. 1261.)
Products of iron and zinc-rolling-mills.

828. Ruge (parafine fabric of Berlin-Wildschütz), Wildschütz near Hohenmölsen. London 1862 medal.
Brown-coal and products from it.

829. Rumpe, Johann Casp. & Wilhelm, Altena.
Products of antimony smelting-works.

830. Runge, royal mining engineer, Breslau. London 1862 medal.
Geological sections of the Giant-mountains (Riesengebirge) and the basin of Waldenburg-Nachod.

831. Sächsisch-Thüringische Actien-Gesellschaft (joint-stock company for products of browncoal) Halle. London 1862 medal.
Products of brown-coals.

832. Sack, mining-director at Sprockhövel. London 1862 honor. mention.
Coal, ironstones and cokes.

833. Saelzer & Neuack, colliery near Essen.
Coal.

834. Salzamt, royal, Schönebeck.
Common salt.

835. Saturn, Rhenish joint-stock mining-company, Cöln.
Lead, copper and zinc-ores of the mines Bluecher and Katzbach.

836. Scheuer, iron-mine near Siegen
Copper-ores.

837. Schmiedeberg, iron-mine near Siegen.
Ironstone.

838. Schneider, H. D. F., Neunkirchen near Betzdorf. London 1862 honor mention.

Ironstones and products of blast-furnaces.

839. Schran, Franz, Gleidorf near Oberkirchen.

Ironstones and specimens of roofing-slate of the slate-company at Fredeburg.

840. Schreckendorfer Hütte near Landeck (Silesia).

Iron-ores and products of iron-works.

841. Schruff, Herbst & Eisleb, Call near Commern.

Separated ores from old heaps and lead-smelting products.

842. Schulten, Friedrich, metal-foundry, Duisburg.

Metal-tewels for blast-furnaces.

843. Schultz & Wehrenbold, Justushütte near Gladenbach.

Ironstone.

843a. Schwidtal (mining-director), Bromberg.

Pressed brown-coal.

844. Shamrock, colliery near Bochum.

Coal and cokes.

845. Sieg, Bergwerks-Gesellschaft (manager Mr. Rütgers), Cologne.

Crystallized galena.

846. Silberart, lead and zinc-mine near Altenseelbach.

Blende.

847. Societé des mines de l'Eifel (Fournier, de Reiset & Co.), Heistert near Call.

Knobby lead-ores.

848. Solms, mining-administration of prince Solms-Braunfels, Braunfels near Wetzlar.

Iron-ores.

849. Stadtberger Gewerkschaft, Altena. London 1862 medal.

Copper-ores, products of the copper-works, vitriols.

850. Steinhauser Hütte, Witten.

Products of rolling-mill and stamped iron-work for waggons.

851. Steinheuer, C. H., Lindlar near Wipperfürth.

Sandstone.

852. Stinnes, Gustav, Mülheim upon Ruhr.

Separated coal.

853. St. Josephsberg, copper-mine near Linz.

Copper-ores.

854. Stoecklein, quarrier, Sangerhausen.

Brown-coal sandstone from quarries near Sangerhausen. (Witzschel, Brothers, merchants, Sangerhausen, propr.)

855. Count of Stolberg-Wernigerode's Factory Ilsenburg (Magdeburg). (s. No. 2053.)

Rolled wire, card-wire.

856. Storch & Schöneberg, iron-mine near Siegen.

Ironstone.

857. Die Strothe, iron-mine near Tecklenburg.

Ironstone.

858. Stumm, Brothers, Neunkirchen.

Iron and manganese-ores.

859. Teutonia, Hüttengewerkschaft near Willebadessen.

Ironstone.

860. von Thiele-Winkler, Kattowitz

Iron-ores, coals and specimens of intermediate rocks, products of iron-works.

861. Thisquen, Paul Val., manu. Neu-Hattlich near Montjoie.

Artificial char-coal made from peat.

862. Tremonia, colliery near Dortmund.

Coal.

863. Duke of Ujest, Hohenlohehütte.

Clay iron-stone.

864. Ulrich, Theodor, owner of mines and smelting-works at Bredelar.

Ironstone and products of blast-furnace.

865. Uslar, Rudolph, manager of the company W. Gessner & Co. at Nuttlar. London 1862 honor. mention.

Specimens of roofing-slate.

866. Vereinigte Pörtingssiepen, colliery near Werden.

Coal.

867. Vereinigter Hannibal, colliery near Bochum.

Coal.

868. Vereinigter Präsident, colliery near Bochum.

Coal and cokes.

869. Vereinigte Sellerbeck, colliery near Mülheim upon Ruhr.

Coal.

870. Vereinigte Wiesche, colliery near Mülheim upon Ruhr.

Brick-coal.

871. Vereinigungs-Gesellschaft for coal-mining in the Worm-district, Kohlscheid near Aix-la-Chapelle. London 1862 medal.

Coals, Cokes, maps.

872. Victoria-Mathias, colliery near Essen.

Coal and cokes.

873. Victorsfeld, lead-mine near Lippe.

Galena.

874. Vollmond, colliery near Bochum.

Coal and cokes.

875. Vorwärts-Hütte near Waldenburg. Magnetic iron-ore of Schmiedeberg.

877. Werschen-Weissenfels sche Actien-Gesellschaft, joint-stock company at Weissenfels. London 1862 medal.

Brown-coal and products from it.

878. Werther, colliery near Halle (Westphalen).

Beds from the coal-measures.

879. Wetzlar, joint-stock company, Wetzlar.

Ironstone.

880. Wiesehahn, Julius, merchant, Dortmund.

Cokes.

881. Wiesmann, A., & Co., manu., Bonn.

Brown-coal, tar, oil, candles from it.

882. Wildberg, lead and iron-mine near Siegen.

Lead-ores and copper-ores.

883. Wilde-Wiese, mine near Arnsberg.

Ironstone.

884. Wilhelmine Victoria, colliery near Gelsenkirchen.

Coal.

885. Winkler, Ernst, representative of the joint-stock company of Altonrath at Overath.

Copper-ores.

886. Winter, G., & Co., proprietor of the mine Weidenstamm near Braunfels at Frankfurt on the Main.

Black manganese.

Witzschel.   (s. No. 854.)

887. Wolff, Gollenfels (Coblenz).

Marble from Stromberg.

888. Zollverein, colliery near Essen.

Coal and beds from the coal-measures.

CLASS II.

CHEMICAL SUBSTANCES AND PRODUCTS AND PHARMACEUTICAL PROCESSES.

951. Amende, R., manu., Berlin, Müllerstr. 146. 147.

Manure-powder; bone-dust; crushed horn; phosphate of lime; blood in powder.

952. Andrae & Grüneberg, manu., Stettin, manufactory Alt-Damm. London 1862 honor. mention. Agt. Lion M. Cohn, Phaland & Dietrich, represented by Ch. Trübner, 20 St. Dunstans Hill, City, and at the Exhib. Building.

Nitrate of potash pulverized and crystallized 13 Th. (39 sh.); muriate of potash 6 Th. (18 sh.) p. cwt.

953. Behrend, Gustav (G. Behrend), apothecary and chemist, Hirschberg. London 1862 honor. mention. Agt. s. No. 952.

15 divers sorts of the finest vermilion and cinnabar 25 to 27½ sg. (2 sh. 5 to 8 d.); borate of manganese for the preparation of wine yellow varnish and various lackers 1 Th. (3 sh.); 10 bottles of divers copal-lackers 20 sg. to 1 Th. (2 to 3 sh.); all p. lb.; 1 bottle of fair amber-lacker 32 Th. (4½ £.); 1 bottle of dammar-lacker 26 Th. (3⅖ £.) p. Zollz. each; 13 divers sorts of sealing-wax p. lb. 10 to 30 sg. (1 to 3 sh.).

Benneke & Herold, Berlin, lacker and varnish.   (s. No. 1163.)

954. Beringer, A., manu. of colors for colored and room-papers, Charlottenburg near Berlin. Munich 1854 medal of honor.; London 1862 medal. Agts. Ed. Ehrensperger & Co., 4 Laurence Pountney Place, Cannon-street.

Colors; book with 42 leaves on which the application of the colors is shown Several rolls of paper hangings and typographical prints in colors.

Beyrich, F., Berlin, chemical agents for photography.   (s. No. 1419.)

955. Boden, Rud. Th., merchant and manu., Cologne. Agts. Morgan Brothers, 21 Bow-Lane, Cannon-street, City.

I. Sugar of lead, crystallized in different ways 36 to 38 sh. p. cwt. free London; II. spirits of vinegar (concentrated vinegar 15 pCt. of free acetic acid), to be used as first rate table vinegar by mixing it with 3—4 vol. of water, p. gallon 2 sh. to 2 sh. 6 d. all free London; p. 100 litres 40 to 44 frs. f. o. b. Antwerp or Rotterdam.

Braune, B., Danzig, chemical products from amber.   (s. No. 2183.)

957. Bredt, Otto (Otto Bredt & Co.), manu., Barmen. London 1862 honor. mention

Chemical products in 10 bottles: persio d'aniline, red 1 £ 6 sh.; persio d'aniline, violet 6 £; red aniline, liquid 4 sh.; violet aniline, liquid 7 sh. 6 d.; picric acid 11 sh.; indigo carmine, blue 3 sh; indigo carmine, red 3 sh.; cudbear 10½ d.; bichlorid of tin 10 d.; stannate of soda 9½ d.: all p. lb. engl.

958. Bühring, Constantin, manu., Berlin Wilhelmsstr. 100.

Nine bottles of crystal-water, for cleanig all stuffs' and for washing kid-gloves 4 Th. (12 sh.) p. doz

960. Cohn, Dr. Wilhelm (Berlin manufactory of steamed bone-dust), Berlin, Martiniquefelde near Moabit. Agt. M. D. Lasker, 15 Fish-street Hill.

Samples of: bone-dust 3 Th. (9 £); sulphate of bone-dust 3¼ Th. (9 £ 10 sh.); superphosphate of lime 2¼ Th. (7 £ 10 sh.); foddering bone-dust 3¾ Th. (11 £); crushed bones for preparing animal charcoal 2¾ Th. (8 £ 1 sh.); solid bonefat 14 Th. (42 £ 18 sh.) all p. Zollz. (p. ton.); some biscuits of bone-dust for foddering 2¼ Th. (9 sh.) p. 100 pieces. Circular-letters in German, English and French language are annexed.

962. Cuntze, Eberhard, distiller of eau de Cologne, Cologne. London 1862 honor. mention. Agt. s. No. 952.

25 bottles of genuine eau de Cologne; (label with the name of the manufacturer and exhibitor) p. 100 doz. 52 £ 10 sh.

963. CURTIUS, JULIUS, Duisburg. London 1851 prize-medal; Munich 1854 medal of honor; Paris 1855 bronze medal.; London 1862 medal. Agts. Oehlenschlager Brothers, 2 Fowke's Buildings, Great Tower-street.

10 samples of ultramarine.

966. ENGEL & VON SCHAPER, apothecaries of I. ordre and technical chemists, Berlin, Mohrenstr. 17 and Wrietzen on the Oder. Agts. R. Schomburg & Co., Cannon-street, City.

12 bottles of concentrated restitution - fluid 6 Th. (18 sh.).

EUNICKE (Leopold Kayser), Naumburg upon Queiss, Nickel. (s. No. 713.)

967. FARINA, JOHANN MARIA (Johann Maria Farina, Jülichs-Platz 4), merchant, Cologne. London 1851 honor. mention, 1862 medal. Agt. John Dodds, 12 Coopers Row, Crutched Friars.

Eau de Cologne.

968. FARINA, JOHANN MARIA (opposite the Jülichs-Platz), distiller of eau de Cologne, and purveyor to TT. MM. the King of Prussia, the Queen of England, the Emperor of Russia, the King of Hanover, the King of Denmark, the King of Saxony, Cologne. London 1851 prize-medal; Paris 1855 honor. mention; London 1862 medal. Agency 7 Old Jewry, E. C.

Eau de Cologne; for prices apply to the named agency.

969. FARINA, JOHANN MARIA (23 Rheinstrasse), manu., Cologne. Agt. Johann Maria Farina, 2 Salters-hall Court, Cannon-street, E. C.

Extrait d'eau de Cologne and eau de lavande ambrée.

970. FASSBENDER, GERHARD (Johann Maria Farina, opposite the Altenmarkt 54, or vis-à-vis le Marche 54), merchant and manu., Cologne. London 1862 medal. Agt. Ed. Loysel, 92 Cannon-street, E. C.

1. Eau de Cologne, in common short angular, or long ½ bottles 2 Th. (6 sh.); 2. eau de Cologne double, in like bottles 3 Th. (9 sh.); 3. extrait d'eau de Cologne, in like bottles 4 Th. (12 sh.); all p. doz.

971. FARINA, JOHANN MARIA (opposite the New-Markt), manu., Cologne. London 1851 honor. mention, 1862 honor. mention. Agt. Moritz Mathias, 38 Dowgate Hill, Cannon-street.

105 bottles of eau de Cologne; prices p. doz. Bottles with fine straw: 10 Th. (1 £ 10 sh.); 6 Th. (18 sh.); 5 Th. 15 sg. (16 sh. 6 d.); 5 Th. 10 sg. (16 sh.).

Bottles in straw plaiting 20 Th. (3 £); 10 Th. (1 £ 10 sh.); 10 Th. 15 sg. (1 £ 11 sh. 6 d.); 6 Th. 20 sg. (1 £); 10 Th. 15 sg. (1 £ 11 sh. 6 d.); 7 Th. 15 sg. (1 £ 3 sh. 6 d.); 6 Th. (18 sh.); 5 Th. 15 sg. (16 sh. 6 d.); 5 Th. 10 sg. (16 sh.); 6 Th. 20 sg. (1 £); 13 Th. 10 sg. (2 £).

Bottles without straw-plaiting 16 Th. (2 £ 8 sh.); 8 Th. (1 £ 4 sh.)

972. FARINA, F. MARIA, Glockengasse No. 4711, distiller, Cologne. London 1862 honor. mention. Agt. J. W. Green (E. G. Zimmer-

mann), 2 St. Paul's Buildings, Little Carter-Lane, Doctor's Commons.

Eau de Cologne, 12 doz. bottles; vinaigre oriental, 3 doz. bottles.

973. FARINA, JOHANN ANTON, Stadt Mailand, Hochstrasse 129, manu., Cöln. Paris 1855 honor. mention; London 1862 medal. Agts. Alexander Mönch & Co., 26 Lambeth Hill, Doctor's Commons, E. C.

Eau de Cologne in bottles.

FASSBENDER, GERHARD, Cöln. (s. No. 970.)

GAILLARD, C. F., Berlin, perfumes. (s. No. 1181.)

975. GASSEL, RECKMANN & Co., manu., Bielefeld. Agt. H. Landwehr, Exhibition Building. (s. No. 1345.)

2 bottles of benzine 5 £ 5 sh.; bottle of white oil 2 £ 5 sh.; p. cwt. each; 4 bottles of benzin (scouring drops) all four 2 sh.

976. GEISS, FRANZ GUSTAV, DR. PHIL., chemist, druggist and prop. of a destillery of essences at Aken on the Elbe. Berlin 1844 honor. mention; Munich 1854 medal of honor.; London 1862 honor. mention. Agt. Lion M. Cohn, Phaland & Dietrich, represented by Ch. Trübner, 20 St. Dunstans Hill, City, and at the Exhib. Building.

Fourty bottles containing as many kinds of essences distilled from wild-growing as well as from cultivated vegetables, grown in Germany, in the lump 200 Th. (30 £).

977. GEORGSHÜTTE NEAR ASCHERSLEBEN, mineral-oil and paraffin-works, Aschersleben. London 1862 medal. Agt. s. No. 976. (s. II. No. 12.)

1. A plate of paraffin, 30 Rhenish in. long, 28 in. broad, 1½ in. thick, with the Prussian arms painted in oil, set in a gold-frame. 2. A box containing 4 plates of paraffin, melting point 58° Cels. (136,4° Farenh.) p. cwt. 34 Th. (5 £). 3. A bottle of photogen, clear, specific gravity 0,795, boiling-point 140—250° C. (284—482° F.) p. cwt. 13½ Th. (2 £). 4. A bottle of paraffin-oil (solar oil), light-yellow; spec. grav. 0,830, boiling-point 250—330° C. (482—626° F.) p. cwt. 8 Th. (1 £ 4 sh.).

These articles are made of brown-coal from the Aschersleben mines.

Notice: The above stated prices are now current, but they are dependent on junctures of time.

979. GRAHN, ANNA, Ww., manu., Halle on the Saale. Agt. s. No. 976.

Two specimens of clay, p. Ctr. 10 Th. (1 cwt. 1 £ 10 sh.); two specimens of size for gilding wood, p. Ctr. 13⅓ Th. (1 cwt. 2 £).

980. GRASS, MARGARETHA, distiller, Cologne. Domhof 2.

Eau de Cologne double 4 Th. (12 sh.); Spanish Carmelite-spirit 4 Th. (12 sh.); prices p. doz. bottles incl. packing.

GROSSMANN, C. E., Berlin, perfumes. (s. No. 1187.)

982. HARTMANN, WM., merchant, Cologne, manufactory at Rolle. (s. II. No. 5.)

Argols; prices p. cwt., freight paid London, cash. The pCt. numbers indicate the quantity of the bitartrate of potassa. Rhenish I. a. 81½ pCt. 90 sh.

do. II. a. 76½ pCt. 84 sh.; crystallized I. a. 95 pCt. 100 sh.; do. II. a. 87 pCt. 92 sh.

These crystallized argols are made from lees of wine.

983. HERMANN, O. (Chemical works), Schönebeck. Berlin 1844 prize-medal; London 1851 prize-medal; Munich 1854 medal of honor; Paris 1855 medal of honor; London 1862 medal. Agt. Lion M. Cohn, Phaland & Dietrich, represented by Ch. Trübner, 20 St. Dunstans Hill, City, and at the Exhib. Building.

	prussian. Th. sg.		engl. £ sh. d.
Nitrate of baryta	Pfd. — 9	lb.	— — 10
Chloride of Barium ...	» — 7½	»	— — 9
Crystallized phosphoric acid	» 1 20	»	— 4 6
Boracic acid	» — 12	»	— 1 2
Cyanide of potassium ..	» — 27½	»	— 2 6
Magnesia	» — 15	»	— 1 5
Carbonate of potassa ..	» — 20	»	— 1 10
Nitrate of Bismuth ...	» 5 —	»	— 13 9
Hydrate of potassa ...	» — 20	»	— 1 10
Acetate of potassa	» — 25	»	— 2 3
Phosphoric acid spec. grav. 1,300	» — 16	»	— 1 5
Fuming nitric acid spec. grav. 1,550	» — 12	»	— 1 2
Strong acetic acid spec. grav. 1,060	» — 25	»	— 2 3
Hyposulphite of soda ..	Zollz 10 —	Cwt.	1 10 6
Crystallized nitrate of strontia	» 19 —	»	2 18 —
Oxide of zinc	» 68 —	»	10 7 6
Bicarbonate of potassa .	» 33 —	»	5 1 —
Tartrate of antimony (emetic tartar)	» 56 —	»	8 11 —
Nitrate of strontia.....	» 19 —	»	2 18 —
Golden sulphur of antimony	» 50 —	»	7 12 —
Bicarbonate of soda...	» 9 20	»	1 9 6
Crystallized tartrate of potassa	» 78 —	»	11 18 —
Acetate of soda	» 23 —	»	3 10 —
Sulphate of iron and of ammonia	» 6 25	»	1 1 —
Chloride of barium ...	» 14 —	»	2 3 —
Phosphate of soda ...	» 19 —	»	2 18 —
Sulphate of magnesia ..	» 5 10	»	— 16 4
Nitrate of soda	» 10 —	»	1 10 6
Crystallized nitrate of soda	» 7 —	»	1 1 6
Crystallized bicarbonate of soda	» 18 —	»	2 15 —
Carbonate of magnesia .	» 17 —	»	2 12 —
Oxide of tin	» 60 —	»	9 3 —
Carbonate of soda	» 11 —	»	1 13 6
Sulphate of soda	» 4 10	»	— 13 3
Hydrochloric acid spec. grav. 1,195	» 9 —	»	1 7 6
Nitric acid spec. grav. 1,420	» 22 —	»	3 7 —
Sulphuric acid spec. grav. 1,840	» 12 15	»	1 18 —

984. HERSTATT, CHRISTOPH, & Co., manu., Cologne, 17 Fettenhennen. Agt. J. M. Robson, 32 Lawrence Lane, Cheapside.

12 bottles of eau de Cologne double p. doz. 2⅔ Th. (8 sh.).

985. HEYL, RICHARD, & Co., Berlin, manu. Charlottenburg. London 1862 medal. Agt. Gottfr. Steinhoff, Kirstall near Leeds.

Colors for fancy papers and paper hangings; do. for painters; varnishes; acetic acid; ice acetic acid;

essence of vinegar; acetate of alumin, soda, baryta, lime, iron, strontian; minium; litharge.

986. HIRSCH, DR. G., and Magistrate, Königsberg. Munich 1854 honor. mention. Agt. s. No. 983.

1. Several samples of granulated animal charcoal for sugar-refineries; 2. powdered bone-black; 3. pure bone-dust for manure; 4. bones dissolved with 33⅓ pCt. of sulphuric acid; superphosphate of lime; 6. powdered gypsum of French stones; 7. do. of stones from the river Saal; 8. powdered magnesite of Silesian magnesite-stones; 9. shoe-blacking; 10. do. in boxes.

988. HOFFMANN & Co., manu., Müngersdorf near Cöln. Agt. Schwann & Co., 4 Gresham-street.

Steamed bone duct; bone earth; superphosphate.

HOYER, G., & Co., Schönebeck, salt licking-stones. (s. No. 750.)

990. HUGUENEL, C. P., Breslau. London 1862 honor. mention. Agt. s. No. 983.

Specimens of madder; of two years, 1860 first quality; do. 1861; second quality 1861; third quality 1861; of one year 1861; of six months 1861; garancine, first quality; do. second quality; flower-wood; flowers of madder; quercitrine; sumacine.

991. HUTTER, CARL LUDWIG EDUARD (Hutter & Co.), merchant, Berlin, Alte Jakobsstrasse 51. Agt. s. No. 983.

a. 4 bottles of esprit des cheveux, vegetable hair-balm; b. 4 bottles of lenticulosa, cosmetic water; c. 2 bottles of extrait Japonais, for dyeing the hair; d. 2 bottles of scurf-water. The contents of each bottle 8 loth (4 ounces); price of a. to c. p. bottle 3 sh., p. 100 bottles 15 £; of d. p. bottle 1 sh. 6 d., p. 100 bottles 7 £ 10 sh.

992. JÄGER, CARL, manu., Barmen. Munich 1854 honor. mention; Paris 1855 bronze medal; London 1862 medal. Agt. Gerh. Andreae & Co., 6 Mincing Lane.

Extract of safflower, liquid, p. Qrt. 16 sh., do. dryed, p. lb. 36 £. Rosaïn (red of aniline), liquid, p. gallon 1 £ 7 sh. 6 d., do. in crystals p. lb. 6 £. Mauwe of aniline, liquid, p. gallon 1 £ 10 sh., do. in paste, p. lb. 15 sh. Samples of silk dyed with these colors.

KAYSER, LEOPOLD, jun., & Co., Naumburg upon Queiss. (s. No. 713.)

994. KOBER, THEODOR, manu., Sömmerda. Agt. s. No. 983.

Indigodine; carmine of indigo 1. quality, fine light-blue; do. 2. quality, fine; extract of indigo 1. quality; do. 2. quality; do. 3. quality.

995. KOCH, ANDREAS, merchant, Zell. Agt. s. No. 983.

Lactucarium (dried milky juice of lactuca virosa), 2 lb. 6 Th. (18 sh.).

996. KÜDERLING, H. F., manu., Duisburg. London 1862 medal. Agt. William Schultz, 4 Lawrence Pountney Place, Cannon-street.

Samples of prussiate of potash.

997. KUNHEIM, DR., & Co., manu., Berlin. Berlin 1844 silver medal; London 1851 prize-medal; Munich 1854 medal of honor.

Muriate of ammonia; alum; chromium-alum; iron-alum; sulphite of soda; cyanide of potassium; crystallized verdigris; soda-ashes; crystallized soda; caustic soda; bicarbonate of soda; sulphate of copper; silicate of soda; pink salt; super-phosphate; hyposulphite of soda; nitrate of strontia; stannate of soda; nitrate of lead; oxide of tin; oxide of uranium; protoxide of tin; protoxide of copper; carbonate of copper; carbonate of nickel.

998. Kulmiz, Carl (C. Kulmiz), counsellor of commerce, Ida- and Marienhütte near Saarau (Breslau). London 1862 honor. mention. Agt. Lion M. Cohn, Phaland & Dietrich, represented by Ch. Trübner, 20 St. Dunstans Hill, City, and at the Exhib. Building.

Calcined high-degreed soda 14 sh.; crystallized soda 8 sh.; calcined glauber-salt 6 sh; crystallized glauber-salt 3 sh. 6 d.; crystallized nitrate of soda 1 £ 3 d.; chloride of lime 16 sh.; sulphate of baryta (Blanc fixe) 10 sh.; muriatic acid free of sulphuric acid 4 sh.; nitric acid free of chlorine 1 £ 1 sh. 6 d.; phosphate of lime precipitated 10 sh.; all p. cwt., casks incl.

Kunzmann, H., Berlin, chemical agents for photography. (s. No. 1427.)

999. Lehmann, J. R., manu., Labagienen near Königsberg.

Bone-dust; fish-guano.

1000. Leverkus, Dr. C., manu., Wermelskirchen (Düsseldorf). Berlin 1844 silver medal; London 1851 prize-medal; Paris 1855 silver medal; London 1862 medal. Agt. Frederick Rudolph, 188 Gresham House, Old Broad-street, E. C.

Eight bottles of ultramarine, blue of different sorts, marked Ia. I—VIII.

1002. Lindgens & Sons, manu., Mülheim near Cologne. Agt. Friedr. Osterroth, 1 Bell-Yard, South Side, St. Pauls, E. C.

I a red lead, absolutely pure; do. for glassmakers; do. orange; litharge for apothecaries; do. fine powdered and sifted; do. in flakes, I. quality; do. ordinary; sugar of lead, white 3 fold crystallized; do. crude; white lead absolutely pure; do. powdered No. II. and III.

1004. Lucas, Moritz, apothecary and chemist, Kunersdorf near Hirschberg (Liegnitz). London 1851 honor. mention; Munich 1854 medal of honor; London 1862 honor. mention. Agts. s. No. 998.

China-cinnabar; carmine-cinnabar; patent-cinnabar No. 0. 2. b. (dark); patent-cinnabar No. 1. (light); jaune brillant; mangan extract (borate of manganese); siccative-powder (free from lead).

1006. Marquart, Dr. Ludwig Clamor, manu. of chemical product, Bonn on the Rhine. London 1851 honor. mention; Munich 1854 medal of honor; Paris 1855 silver medal; London 1862 medal. Agts. Morgan Brothers, 21 Bow-Lane, Cannon-street, City.

Acid. acetic. glac.; acid. butyric. oleos.; acid. malicum; acid. phosphor. glac.; acid. succinic.; acid. valerianic. trihydr.; acid. valerianic. monoh.; aether butyric.; bromamyl; bromoform; cadmium jodat.; calcar. bimalic.; cognac-oil; aether valerian.; amyloxyd. acetic.; atropinum; calcar. hypophosphor; cerium oxalic; chlorclayl; coniin; digitalin; extr. filic. mar.; glycerin; jodamyl; jodoform; nicotin; propylamin; santonin; phosphor pentachlorid; filicin. Prices according to quantities.

1007. Martin, Mary Clementine, nun, Cologne on the Rhine. London 1851 prize-medal; Paris 1855 honor. mention; London 1862 medal. Agts. Th. Foster & Co., 45 Cheapside.

Perfumes; prices p. doz., deduction 15 pCt. and 5 pCt. discount cash. Eau de cologne double: full size champaign bottles a 6 single ones 27 Th. 18 sg. (4 £ 2 sh. 10 d.); half size do. à 3 single ones 13 Th. 24 sg. (2 £ 1 sh. 5 d.); double wicker-bottles à 2 single ones 11 Th. (1 £ 13 sh.); single do. 5 Th 24 sg. (17 sh. 5 d); double size glassbottles à 2 single ones 9 Th. 6 sg. (1 £ 7 sh. 8 d.); single do. 4 Th. 18 sg. (13 sh. 10 d.); half do. 2 Th. 9 sg. (6 sh. 11 d.); single long glass-bottles 4 Th. 18 sg. (13 sh. 10 d.); half do. 2 Th. 9 sg. (6 sh. 11 d.).

Melissa carmelite-spirit: single bottles 4 Th. (12 sh.); half do. 2 Th. (6 sh.).

1008. Matthes, E., & Weber, manu. of chemical products, Duisburg on the Rhine. Berlin 1844 bronze medal; London 1851 prize-medal; Munich 1854 medal of honor; Paris 1855 silver medal; London 1862 medal. Agts. Heintzmann & Rochussen, 9 Friday-street, Cheapside E. C.

Soda ash, 58,2 pCt. soda, 99,5 pCt. carbonate of soda; crystallized soda; caustic soda, 80 pCt. hydrate of soda; sulphate of soda 98 pCt.; bleaching powder (choride of lime) 35 pCt.

1009. Nachtweyh, Carl, manu., Dziewenttine near Militsch (Breslau). Agts. s. No. 998.

Potatoestarch sugar.

1010. Neudorff, W. (W. Neudorff & Co.), manu., Königsberg. Agt. Dr. Krause, 35 Alfred-street, Bedford-square, W. C.

Dr. Scheibler's sulphur-soap 1⅓ Th. (4 sh.), containing brome and jod, for preparing artificial baths of Aix la chapelle, p. lb. 1⅓ Th. (4 sh.), p. ½ lb. ¾ Th. (2 sh. 3 d.).

1013. Oster, J. B., druggist, Königsberg. London 1862 honor. mention.

Acid. succinic. sublim. 4⅓ Th. (11 sh. 7 d.); do. dep. alb. 5½ Th. (14 sh. 4 d.); do. pur. albiss. 5⅚ Th. (15 sh. 7 d.); all p. lb. Prus. resp. Engl.

1015. Pommerensdorf, manu. of chemical products, joint-stock company, Stettin. London 1862 honor. mention. Agts. s. No. 998.

	Zollz. Th.	ton. £	sh.	d.
Bone-black, granulated III. . . .	4⅔	14	4	3
Bone-black, granulated IV. . . .	3⅓	10	3	—
Ivory-black, ground	2⅙	6	12	—
Bitter salt	2⅓	7	2	—
Borate of soda	20	60	18	—
Sulphate of soda, free from iron	2	6	1	9
Roll-sulphur	3⅔	11	3	3
Hyposulphite of soda	9	27	8	—
Ferro-prussiate of potash	40	121	16	2

1016. Prager, Louis, manu., Erfurt. Agt. William Schultz, 4 Lawrence Pountney Place. Cannon-street, and during the exhibition also C. Trübner, 20 St. Dunstans-Hill, City. (s. No. 1222 a.)

3 packets of blacking for exportation free of acid à 10 boxes 3 £; 3 packets à 10 wood boxes of military-blacking 2 £; 3 packets à 10 boxes blacking for exportation 1 £ 10 sh.; 3 tin boxes of steam-blacking free of acid, four years old, containing ½ Pfd. 22 £. Prices p. 1000 boxes.

1017. RATHKE, G. (Kaumann's Apotheke), apothecary, Berlin, Alexandrinenstr. 41. Agts. Lion M. Cohn, Phaland & Dietrich, represented by Ch. Trübner, 20 St. Dunstans Hill, City, and at the Exhib. Building.

Capsules gelatineuses; trochisci; cigars with jod etc., sticks of lunar caustic (nitrate of silver).

1019. RUFFER & Co., manu., Breslau. London 1862 honor. mention.

White zinc.

1019a. RUNGE, DR. FRIEDLIEB FERDINAND, professor of technology, Oranienburg near Berlin. Paris 1855 bronze medal. (s. No. 1961.)

Manure-dust of waste wool, leather and horn.

SCHERING, E., chemical agents for photography, Berlin. (s. No. 1433.)

1020. SCHERR, J., SON, manu. of artificial mineral waters, Triers. (s. No. 1130.)

20 bottles of artificial mineral water.

1021. SCHINDLER & MÜTZEL, manu., Stettin. London 1862 honor. mention. (s. No. 1237.)

Perfumes.

1022. SCHOENFELD, DR. F. (St. Schönfeld), chemist, Düsseldorf. Agt. s. No. 1017.

Oil-colors in tin-tubes; moist water-colors in tubes, dry colors for painting, japanning and printing.

1023. SCHÜR, OTTO, DR. PHIL., apothecary, proprietor of a manu. for artificial mineral-waters and of glass-works, Stettin. Paris 1855 honor. mention.

A siphon and thirty eight bottles of different artificial mineral-waters.

1023a. SCHÜR, DR. OTTO, representative of the Stettin joint-stock company for the manu. of manure, Stettin, manufactory at Bredow near Stettin.

manure for grain 2 Th. (6 sh.); for rapes 2 Th. (6 sh.); for meadows 1½ Th. (4 sh. 6 d.); poudrette ⅚ Th. (2 sh. 6 d.), all p. Zollz. (cwt.).

1023b. SCHULTZE, J. C., manu., Berlin. Charlottenstr. 82. London 1862 honor. mention.

Specimens of lac-varnish.

1024. SCHUSTER DR., & KAEHLER, Danzig. London 1862 honor. mention.

Acid. succinic. white 9⅓ Th.; acid succinic depurated 8 Th.; acid. succinic. sublimated 6 Th ; rosin of amber (colophony of amber) 5 sg. all p. lb.; Aix-la-Chapelle bathing-soap p bottle 1⅓ Th.; 30 bottles of mineral waters.

1027. SPENDECK, JOHANN, PETER (J. P. Spendeck & Co.), merchant, Cologne. Agts. Morgan Brothers, 21 Bow-Lane, Cannon-street, City.

1 large bottle of extract of eau de Cologne (⅛ Quart) 12 sh.; 2 half bottles do. (½ Quart) 12 sh.; 6 bottles of double eau de Cologne (1 Quart) 12 sh.; 24 bottles do. (2 Quart) 1 £ 10 sh.; 12 bottles do (1 Quart) 12 sh.; 12 bottles of genuine eau de Cologne (1 Quart) 9 sh.; 12 bottles do. 2nd quality (½ Quart) 3 sh.

1028. SPIRO, PAUL, manu. of bronze-powder, Berlin. Agts. S. Henschel & Co., 18 Ironmonger-Lane Cheapside, E. C.

50 different bronze-powders along with the schabin, (Dutsh metal-parings) of which they are made, p. lb. 1½—5 sh.

STETTIN-KRAFTDÜNGER-FABRIK; joint-stock company for the manuf. of manure. (s. No. 1023a.)

1030. SZITNICK, OTTO, merchant & manu., Königsberg. Agts. s. No. 1017.

I. Bone-dust (fat removed) prepared according to the instruction of professor Stoeckhardt at Tharand. Analysis of Dr. Dullo: 5,5 water, 34,0 organic substances, 51,0 phosphate of lime, 7,0 carbonate of lime, 2,5 sand, per Zollz. 2⅔ Th. p. ton 7 £ 10 sh. free on bord.

II. Bone-dust prepared with sulphuric acid according to the instruction of professor Stoeckhardt at Tharand. Analysis of Dr. Dullo: 6,0 water, 27,5 organic substances, 10,8 soluble phosphate of lime, 19,0 insoluble phosphate of lime, 35,2 sulphate of lime, 1,5 sand, p. Zollz. 4 Th. p. ton 11 £ free on bord.

1031. THURN, JOHANN PAUL, WIDOW (Joh. Maria Farina, opposite the cathedral), manu., Cologne, 17 Fettenhennen. Munich 1854 medal of honor. Agt. J. M. Robson, 32 Lawrence-Lane, Cheapside.

12 bottles of eau de Cologne p. doz. 2⅓ Th. (7 sh. 6 d.)

1032. VILTER, W., manu., Berlin, kleine Waldemarstr. 4. Agt. s. No. 1017.

Specimens of glue of bones 4½ Th. (13 sh. 6 d.); steamed bone-dust 2⅔ Th. (8 sh.); animal oil 14 Th. (2 £ 2 sh.); sulphate of ammonia 5⅔ Th. (17 sh.), all p. Zollz.; ferro-prussiate of potash 1 lb. 14 sg. (1 sh. 5 d.).

1033. VOIGT & HAVELAND, manu., Breslau.

Garancine of Silesia 40 Th. (6 £); milled madder of Silesia 12 Th. (1 £ 16 sh.), p. Zollz. (cwt.) each; albumin of eggs p. lb. 1 Th. (3 sh.).

1034. VORSTER & GRÜNEBERG, manu., Cologne manufactory Kalk near Cöln. London 1862 honor. mention. Agts. G. Dahlke, Morgan Brothers, 21 Bow-Lane, Cannon-street, City.

Chemical products; prices p. 100 pounds; nitrate of potash, absolutely pure, powdered 13⅔ Th. (41 sh.); do. crystallized 13⅓ Th. (40 sh.); soda-ash of 95 pCt. or 55 pCt. engl., waste product from the manufacture of saltpeter 4⅔ Th. (14 sh.); refined nitrate of soda, crystallized, 6⅔ Th. (20 sh.); refined carbonate of potash 93 pCt., free of soda 12 Th. (36 sh.).

1035. WEISS, J. H., & Co., manu., Mühlhausen (Erfurt). Berlin 1844 bronze medal; London 1851 prize-medal; Munich 1854 medal of honor; London 1862 medal. Agt. s. No. 1017.

1 Colors of madder for house-painting and for printing paper hangings, p. pound or ½ kilogr. ½—4 sh.; 2. pure, splendid, light-resisting colors of madder, soluble in oil, for artists and printing in colors, p. pound or ½ kilogr. 9 sh. to 2 £.

1036. WEITZE, DR. C. G., manu., Stettin. (s. No. 1248.)

Perfumes.

CLASS III.

SUBSTANCES USED FOR FOOD, INCLUDING WINES.

1040. APPEL, HEINRICH, & Co., manu., Schwedt on the Oder. London 1862 honor mention.

Samples of tobacco and cigars.

1041. AXMANN, R., manu. of vermicelli, Erfurt. London 1862 medal. Agts. Lion M. Cohn, Phaland & Dietrich, represented by Ch. Trübner, 20 St. Dunstans Hill, City, and at the Exhib. Building.

An assortment of vermicelli, samples of different kinds. For ¼ boxes or 100 lbs. net weight: No. 1—6. 13½ Th ; No. 7—12. 9¾ Th.; No. 13—20. 9⅔ Th ; No. 21—34. 8¼ Th.; No. 35—38. 9¾ Th.; boxes incl.

1042. BAEVENROTH, C. F., manu., Stettin. Agts. s. No. 1041.

	pr. 192 Quart p. Imperial-Gallon	
	Th.	sh. d.
Finest rectified spirits of wine from potato-spirits, strength 90 a 91 pCt. Tralles equal to 59 à 60 over proof english	36	2 3
Stettin rum, strength 80 pCt. Tralles equal to 40 over proof	35½	2 2½
Vinegar, one ounce of which will neutralize 67 à 68 grain of dry carbonate of potash	12	— 9

Casks incl. free on board Stettin.

Extrait d'absinthe from potato-spirits per bottle, package incl., 13½ sg. (1 sh. 4 d.).

1043. BARRE, E., manu., Lübbecke (Minden). London 1851 honor. mention, 1862 honor. mention.

Wheat-starch of different kinds: 1. air-dried starch 8 Th.; 2. kiln-dried starch 9 Th.; 3. patent-starch 10 Th.; 4. hair-powder 9 Th.; all p. cwt.

1044. BAUTE & Co., merchants and manu., Camen (Arnsberg). London 1862 medal. Agt. William Schultz, 4 Lawrence Pountney-place, Cannon-street, E. C.

Westphalian smoked hams of the finest quality.

Bitter extract of Dr. Kortum (author of the Jobsiade) called »Jobs«

1045. BECKHARD & SÖHNE, wine-merchants and manu. of sparkling wines, Kreuznach. Agts. Daniel Taylor & Sons, 27 Leadenhall-street.

Rhenish and Palatinate wines. Sparkling Hock and Moselle.

1046. BEISERT, ADOLPH (A. Beisert), prop. of a corn mill, Sprottau. London 1862 medal. Agts. s. No. 1041.

Wheaten flour and grits, rye-flour.

1047. BIESCHKY, A., manu., Danzig.

Bitter cordial.

1048. BLUME, C., manu., Berlin, Königsstr. 51. London 1862 medal.

Wine made of honey.

1049. BONNE, WILHELM, manu., Rheda (Minden). Agt. Gustav Weiss, 6 Commercial street, Whitechapel. London N. E.
(s. II. No. 57.)

12 tin-boxes of saveloy preserved in lard, 44 Pfd. 14 Lth. Prus. 16½ Th., 49 lb. p. lb. gross-w. 1 sh. loco Rheda.

1051. BROMBERG, ADMINISTRATION OF THE ROYAL MILLS, Bromberg. London 1862 medal. Agt. Ferdinand Pickert, 79 Mark-Lane E. C.

5 samples of wheat-flour No. 1. 2. 3. 14 sh.; wheat-pollard, wheat-bran 4 sh. 6 d.; 5 samples of rye-flour No. 1. 2. 3. 11 sh.; rye-pollard, rye-bran 4 sh. 6 d.; all p. Zollz. after the price-current.

1052. BUDDE, ERNST, manu., Herford. London 1862 honor. mention. Agt. Heinrich Landwehr, Exhibition Building.

1 jar of mustard, 15 Pfd. 1 Th (3 sh.).

1053. BUEHL, ALBERT, & Co., wine-merchants and manu., Coblenz. London 1862 medal. Agt. M. Edersheim, 45 Eastcheap E. C.

6 bottles sparkling Johannisberg 2 £ 8 sh.; 6 do. sparkling Scharzberg 2 £ 8 sh.; 6 do. sparkling Moselle muscadel 18 sh.; 6 do. sparkling Hock 18 sh.; all p. doz. bottles.

1054. CARSTANJEN, A. F., SONS, manu. of tobacco and cigars, Duisburg on the Rhine. Berlin 1844 honor mention; London 1851 honor. mention; Paris 1855 bronze medal; London 1862 medal. Agt. Eduard Rhodius, countinghouse by Killy, Traub & Co., 52 Bread-street.

Tobacco 3 d. to 4 sh. 6 d., tobacco for chewing 11 d. to 3 sh. p. pound; carrots, p. 100 pounds 4 £ 9 sh. to 6 £; snuff, p. pound 7 d. to 1 sh. 11 d.; cigars, p. thousand 1 £ 2 d. to 5 £ 7 sh.

1055. CARSTANJEN, CARL & WILHELM, manu., Duisburg, Munich 1854 prize-medal; Paris 1855 silver medal; London 1862 medal. Agts. Heintzmann & Rochussen, 9 Friday-street, Cheapside.

6 barrels of snuff, each 8 pounds net; prices p. pound: No. 1. 13 sg.; No. 2. 10 sg.; No. 3. 8½ sg.; No. 4. 7½ sg.; No. 5. 7 sg.; No. 6. 6½ sg.
3 boxes cigars; prices p. mill.: No. 7. 18 lb. 10 Th.; No. 8. 18½ lb. 6 Th.; No. 9. 16 lb. 7 Th.
3 boxes tobacco; prices p. pound: No. 10. 8 lb. 14 sg.; No. 11. 9 lb. 5 sg.; No. 12. 6 lb. 8½ sg.

1056. CRESPEL & SONS, manu., Ibbenbüren (Westphalia). London 1862 honor. mention.

Wheat-starch: ff. refined patent-starch, white in 1 lb. packets; do. in barrels; do. blue in 1 lb.-packets; do. in barrels; do. starch, white, dryed by 157 pCt. Fahrenheit in 4 and 5 lb.-packets; do. in barrels; do. blue in 4 and 5 lb.-packets; do. in barrels; ff. refined starch No. 1. air-dried; f. do. middling, air-dried; do. No. 2. air-dried; do. scraping-starch No. 3. air-dried; ff. refined hair-powder No. 1. in 1 lb.-packets; do. in barrels; do. No. 3. in barrels; ff. refined British gum (roasted starch) No. 1. in barrels.

1057. DELIUS, RUD., & Co., manu., Bielefeld, New steam-mill near Bielefeld. London 1862 medal. Agt. H. Landwehr, Exhibition Building.

Wheat, wheat-grit, wheat-flour, wheat-pollard, wheat-bran; loco-prices on te 4. of march p. 100 lb. engl.: wheat 11 sh. 2 d.; wheat-grit No. 1/3. 16 sh. 4 d.; wheat-flour No. 00. 16 sh. 4 d.; do. No. 0. 15 sh.; do. No. I. 13 sh. 7 d.; do. No. II. 11 sh. 6 d.; do. No. III. 9 sh. 6 d.; wheat-pollard No. 1. 5 sh. 10 d.; do. No. II. 5 sh.; wheat-bran No. I. 4 sh. 2 d.; do. No. II. 3 sh. 8 d.

1058. DITTRICH, HEINRICH, landed-prop., Seitendorf near Frankenstein (Breslau). London 1862 medal.

One Prussian Scheffel of white wheat, 90 lb., (1 bushel 2 pecks 3 gills).

1059. DOMMERICH & Co., manu., Magdeburg. Agts. A. Bodens & Co., 33 Aldermanbury, London E. C.

1. Prepared chicory in parcels 15 sh.; 2. chicory and beet-root meal 15 sh.; 3. dried chicory and beet-roots 10 sh. 6 d.; all p. cwt.

1060. Drebs, Otto, landed prop., Ottomin near Danzig.

6 Samples of:

	p. Schfl.	p. Quarter.	
	sg.	£	sh.
wheat	125	3	15
rye	80	2	10
barley	67	2	—
oats	50	1	10
pease	80	2	10
gray pease	100	3	—

1061. Drucker, Jacob, distiller and wine-merchant, Coblenz. Agt. John Mohun, 36 St. Mary at Hill E. C.

4 Bottles Rhenish brandy (eau de vie du Rhin);

	in	sh.	d.
No. 1. pale, 57 pCt. Tralles, proof,	hogshead	3	10
	casks	4	—
No. 2. pale, 52 pCt. Tralles, 10—12 under proof,	hogshead	3	6
	casks	3	8
No. 3. brown, 57 pCt. Tralles, proof,	hogshead	4	—
	casks	4	2
No. 4. brown, 52 pCt. Tralles, 10—12 under proof,	hogshead	3	8
	casks	3	10
all p. gallon.			
Of No. 2. proof in cases of 12 bottles (2 gallons)	case	12	6
of No. 3. do.	case	13	—

Free on board, Rotterdam.

1062. Eisenmann, R., manu., Berlin, 23 Alexanderstr. London 1862 medal.

Potato-spirit of 95 pCt. (Prussian standard).

Elsner von Gronow. (s. No. 1082.)

1063. Engelbrecht & Veerhoff, manu., Herford. London 1862 honor. mention. Agt. Landwehr of Bielefeld.

White air-dried wheat-starch 8 Th. (24 sh.); do. blue 8½ Th. (25½ sh.); air-dried, bright wheat-starch 8½ Th. (25½ sh.); white steam-dried wheat-starch 8½ Th. (25½ sh.); do. blue 9 Th. (27 sh.); patent wheat-starch, white, 9 Th. (27 sh.); do. with high luster 9½ Th. (28½ sh.); wheat-starch powder 9 Th. (27 sh.); half dextrine for dressing warps 5⅔ Th. (17 sh.); dextrine, white 10 Th. (30 sh.); do. light brown 9 Th. (27 sh.); do. brown 9 Th. (27 sh.). Prices p. Zollz.

1064. Ermeler, W., & Co., manu., Berlin, Breitestr 11.

Cigars, tobacco, snuff and tobacco for chewing.

1065. Fasquel, D., manu., Berlin, Lindenstr. 119.

Arrac, French brandy (cognac), rum, essence of absinthium and cumin-oil.

1067. Fier, Joseph (Joseph Fier), prop. of vine-yards, merchant, Zell on the Moselle (Coblenz). Agts. Lion M. Cohn, Phaland & Dietrich, represented by Ch. Trübner, 20 St. Dunstans Hill, City, and at the Exhib. Building.

1 bottle Zeltinger 1857er No. 1. G. 16 £; 1 bottle Brauneberger 1857er No. 2. G. 23 £; 1 bottle Piesporter 1857er No. 135. 27 £; 1 bottle Josephshöfer 1857er No. 125. 30 £; 1 bottle Zeller-Petersborner 1857er No. 123. 36 £; 3 bottles red wine 1857er No. 115, 117. u. 139. of Bordeaux vines 15 £; all p. 140 Qrt. (35 gallons).

1069. Flatau, Jos. Jac., merchant, agent of the cultivators of hops at Neutomysl, Berlin. Paris 1855 honor. mention. Exhibitor himself present.

2 packets of hops.

1070. Neumann, Brothers, merchants, Grünberg. London 1862 honor. mention Agts. s. No. 1067.

Samples of dried fruits; prices p. lb: pears 1. quality peeled 7½ sg. (9 d.); do. 2 quality 6 sg. (7⅕ d.); do. 3. quality, not peeled 4 sg. (4⅘ d.); apples, peeled and free of kernels 7½ sg. (9 d.); do. curled 5 sg. (6 d.); do. not peeled 4 sg. (4⅘ d.); plums, peeled 6 sg. (7⅕ d.); do. without stones 6 sg. (7⅕ d.); do. old sugar-plums 4 sg. (4⅘ d.); do. new ones 3 sg. (3⅗ d.); cherries, sour 5 sg. (6 d.); peaches, peeled 20 sg. (2 sh.); chalk of plums 4 sg. (4⅘ d.); chalk of cherries 5 sg. (6 d.); walnuts p. threescore 5 sg. (6 d.);

Preserved fruits; prices p. bottle: straw-berries 20 sg. (2 sh.); rasp-berries 15 sg. (1 sh. 6 d.); currants 15 sg. (1 sh. 6 d.); Ostheim-cherries 20 sg. (2 sh.); plums, peeled 12½ sg. (1 sh. 3 d.); do. Reine-claude 15 sg. (1 sh. 6 d.); quinces, peeled 12½ sg. (1 sh. 3 d.); apricots, peeled 20 sg. (2 sh.); peaches, peeled 20 sg. (2 sh.); pine apples, peeled 1 Th. 15 sg. (4 sh. 6 d.); walnuts 15 sg. (1 sh. 6 d.); heps 20 sg. (2 sh.); bilberries 5 sg. (6 d.); rasp-berry juice 15 sg. (1 sh. 6 d.); cherry-juice 15 sg. (1 sh. 6 d.).

1071. Förster & Grempler, growers, Grünberg (Liegnitz). Munich 1854 honor. mention; Paris 1855 bronze medal; London 1862 medal. Agts. s. No. 1067.

An assortment of sparkling and other wines; prices p. bottle: 1859er white and red-wine 8 sg. (10 d.); 1852er white, red 10 sg. (1 sh.); 1846er white, red 12 sg. (1 sh. 2 d.); 1858—1859er sweet wine 8 sg. (10 d.); sparkling wine 1 Th. (3 sh.).

1073. Franke, Johann Heinrich (J. H. Franke), merchant and manu., Magdeburg. Agts. s. No. 1067.

16 bottles of different German liqueurs 6 Th. 15 sg. 6 pf. (2 £ 19 sh. ½ d.) 8 quarts (2 gallons incl. the bottles).

1074. Frenzel, E., grower, Baugsch Korallen (Königsberg).

Grey pease, beans.

1076. Gerten, J. H., alias Brüggemann, grower, Obrighoven (Wesel) London 1862 medal.

Corn; field and garden-fruits; jellies.

1077. Geyger, Adolph, & Co., wine-merchants, manu. of sparkling wines, Creuznach. Agt. William Schultz, 4 Lawrence Pountney-place, Cannon-street.

Sparkling Hock and Moselle 1. qual. 1 £ 16 sh., 2. qual. 1 £ 10 sh., p. doz. bottles, free on board, Rotterdam.

1078. Gilka, Carl Joseph Aloys (J. A. Gilka), manu. of spirits and liqueurs, Berlin. Berlin 1844 bronze medal; Munich 1854 medal of honor; Paris 1855 honor. mention; London 1862 medal. Agts. s. No. 1067.

Samples of 40 different sorts of liqueurs and spirits; spirit of beet-root molasses à 96 pCt. Tralles; potato-spirit à 96 pCt. Tralles; cognac of beet-root molasses à 50 pCt. Tralles.

1079. Graff, H., grower, Janischken (Königsberg).

Wheat; species of pease called »capuciner«

1080. GRASHOFF, MARTIN, horticulturist, farmer of a royal domain, Quedlinburg and domain Westerhausen. Paris 1855 bronze medal; London 1862 medal. (s. II. No. 49.)

1. Seeds of vegetables to be grown in the open-ground; 2. seeds of vegetables, to be grown in hot-houses; 3. vegetables; 4. root-plants; 5. lettuces; 6. sorts of roots; 7. sorts of rapes, turnips etc.; 8. pod-plants with and without pods, in the green state; 9. vegetables to be dryed and preserved; 10. subsitutes for coffee and beet-roots; 11. beet-roots for the food of cattle; 12. root-plants, cabbage, food for cattle and sheep; 13. carrots food for horses, sheep and cattle; 14. sorts of plants and roots, food for sheep; 15. seeds for the produce of food, sorts of grass for meadows and parks; 16. legumes; 17. oil-seeds; 18. seeds of dyers' plants; 19. seeds of officinal plants; 20. seeds of poisonous plants; 21. kinds of corn in grains and ears; 22. seeds of park and wood-shrubs; 23. potatoes; 24. potatoes from China, James potatoes; 25. flower-seeds; 26. finer flower-seeds, chamber and pot-flowers; 27. seeds of ornamental plants for open ground; 28. seeds of climbers for the embellishment of walls and bowers; 29. ornament-gourds; 30. eatable gourds.

The prices are calculated on the produce of the crop. Catalogues for special information are to be had.

1081. GREVE PHILIPP (Philipp Greve-Stirnberg), merchant, Cologne London 1862 honor. mention. Agts. Heintzmann & Rochussen, 9 Friday-street.

Six bottles of bitter cordial »Alter Schwede«, p. bottle $\frac{1}{2}$ Th. ($1\frac{1}{2}$ sh.).

1082. ELSNER VON GRONOW, MARTIN, member of the royal board of agriculture, Captain R. E. etc., Kalinowitz near Gogolin. London 1862 Juror for Classe III. Agts. Lion M. Cohn, Phaland & Dietrich, represented by Ch. Trübner, 20 St. Dunstans Hill, City, and at the Exhib. Building.

$1\frac{1}{2}$ bushel Correns rye. A variety of rye grown by the exhibitor; $1\frac{1}{2}$ bushel of Kalina barley, grown by the exhibitor, the bushel weighs 49 lb. av. dupois; $1\frac{1}{2}$ bushel of St. Foin grown by the exhibitor, the bushel weighs 26 lb. av. d. p.

The prices go by the state of the market; seeds 10 sg. a bushel over and above the market price.

1083. GRÜNEBERG, JOH. H., cook and manu. of preserved vegetables, fruits, meat, broth, milk etc., Berlin, Oranienburgerstr. 56. London 1862 medal. Agts. s. No. 1082.

In pots beef preserved in its own broth ($\frac{4}{1}$ gross weight 9 lb., neat weight 8 lb.) a $2\frac{2}{3}$ Th.; field-fares $\frac{2}{2}$ a 1 Th.; condensed broth 8 pots a 20 sg.; asparagus $\frac{2}{1}$ a $1\frac{5}{6}$ Th., $\frac{2}{2}$ a $1\frac{1}{4}$ Th.; green peas $\frac{2}{1}$ a $1\frac{5}{6}$ Th., $\frac{2}{2}$ a $17\frac{1}{2}$ sg.; asparagus cut in pieces $\frac{2}{1}$ a $1\frac{1}{4}$ Th., $\frac{2}{2}$ a 20 sg.; beans $\frac{2}{1}$ a 18 sg.; carrots $\frac{2}{2}$ a $12\frac{1}{2}$ sg. morils $\frac{2}{4}$ a 12 sg.; pine-apples in sugar, three whole fruits in 2 glasses $\frac{1}{1}$ 12 Th., $\frac{1}{2}$ 8 Th. In bottles: pine-apples in slices 2 a $1\frac{1}{2}$ Th.; apricots 2 a $1\frac{1}{2}$ Th.; peaches 3 a 2 Th.; raspberries 2 a 10 sg.; ananas-strawberries 2 a 25 sg.; agriots 2 a 1 Th.; amorellas 2 a 25 sg.; gooseberries 2 a 1 Th.; plums (Reine claude) 2 a $1\frac{1}{6}$ Th.; muscadel-pears 2 a 1 Th.; heps 2 a $1\frac{1}{6}$ Th.; currants without kernels 4 a 10 sg.; cherries (Bernstein-Kirschen) 2 a 1 Th.; cow-milk 12 a 4 sg.; mashed onion-soup in four pieces 4 lb. a $7\frac{1}{2}$ sg.; crab-tails in 4 bottles à $22\frac{1}{2}$ sg.; apple-jelly in glasses $\frac{2}{1}$ a 1 Th.; $\frac{2}{2}$ à 12 sg., $\frac{2}{4}$ à $7\frac{1}{2}$ sg.

The prices go every year by the market-price.

1084. GURADZE, A., counsellor of commerce, Tost (Oppeln). London 1862 honor. mention Agts. s. No. 1082. (s. No. 1192.)

3 small bags containing wheaten flour No. 0., No. II., No. III.; 3 small bags of rye-flour No. I., No. II., No. III.; each 50 pounds.

1085. HAEUSLER, MATHILDE, WIDOW (Carl Samuel Haeusler), Hirschberg. London 1862 medal. Agts. s. No. 1082. (s. No. 1347.)

Sparkling wine from Silesian grapes, 6 bottles, each 20—25 sg.; raspberry-juice p. lb. 6 sg.; Cherry-juice p. lb. 6 sg.; cider sweet and harsh p. 192 quart 22—28 Th.

1086. HALLE, ASSOCIATION OF THE BEET-ROOT SUGAR MANUFAKTURERS OF THE ZOLLVEREIN. London 1862 medal.

Samples of raw and refined sugar.

1087. VON HANDEL, F. C., Triers, representing the following exhibitors. London 1862 medal. Agts. s. No. 1082.

ALFF, FRANZ, Prum near Triers.

1. 1857 Scharzhofberg Saar-wine produce of Müller.

VON BEULWITZ, MRS., Triers.

2. 1857 Scharzhofberg Saar-wine; 3. 1857 Casel Moselle-wine.

BOCKING, A., prop., Trarbach on the Moselle.

4. 1857 Neuberg-choice vintage Moselle-wine; 5. 1857 Oligsberg do; 6. 1858 Brauneberg do.; 7. 1858 Neuberg do; 8. 1859 Brauneberg do.; 9. 1859 Neuberg do

BREUNING, HEIRS, Zeltingen on the Moselle.

10. 1857 Schlossberg Moselle-wine; 11. 1858 Schlossberg do.

GEBERT, C., prop., Temmels near Triers.

12. 1859 Bockstein Saar-wine.

GORTZ, prop., Triers.

13. 1857 Ayll Herrenberg Saar-wine; 14. 1858 do.; 15. 1857 Pisport Moselle-wine.

GRACH, JOSEPH, prop., Triers.

16. 1857 Agritius Saar-wine.

JULIUS, vicar, Waldrach near Triers.

17. 1859 Waldrach red Moselle-wine.

KOCH, APOLLINAR, prop., Wiltingen near Triers.

18. 1858 Scharzhofberg Saar-wine; 19. 1859 do.

LINTZ, FRIEDR., bookseller, Triers.

20. A. 1857 Wavern Herrenberg Saar-wine; 21. B. 1857 do.; 22. 1857 do. red do.

VON NELL, J. P., merchant and prop., Triers.

23. 1857 Thiergarten Moselle-wine.

NIESSEN, PETER CHR., & SON, Mülheim on the Moselle.

24. 1857 Elisenberg choice vintage Moselle-wine; 25. 1857 Brauneberg do.; 26. 1857 Oligsberg do.; 27. 1857 Neuberger do., the last both produce of Bocking.

RAUTENSTRAUCH, VALENTIN, prop., Triers.

28. 1857 Eitelsbach Moselle-wine; 29. 1859 do.; 30. 1859 do. Burgund do.

REVERCHON, ANTON, banker, Triers.
31. 1857 Conen red Saar-wine.

RHEINART, MRS., prop., Saarburg near Triers.
32. 1859 Geisberg Saar-wine.

BARON VON SOLEMACHER, prop., Grünhaus near Triers.
33. 1859 Maximin Grünhaus Moselle-wine.

VOM STEIN, F. W., prop., Triers.
34. 1859 Paulinsberg Moselle-wine.

WELLENSTEIN, prop., Triers.
35. 1859 Wiltingen kupp Saar-wine.

WILCKENS, rendant, Triers.
36. 1857 Scharzhofberg do. produce of A. Koch;
37. 1857 Grach Moselle-wine prod. of the seminary.

ZELL, F., Triers.
38. 1857 Pisport Moselle-wine; 39. 1857 Josephs-hof' do., from count Kesselstadt's estates.

DEY, A., & Co., Coblenz. London 1862 medal
40. Sparkling Moselle: 41. do. Hock; 42. 1857. do. Scharzberg; 43. 1857 do. Ehrenbreitstein.
The wines specified from 2 to 25 and from 28 to 35 are the exhibitors' own produce.

1088. HARTMANN, F. A., Münster. Paris 1855 bronze medal. Agts. Lion M. Cohn, Phaland & Dietrich, represented by Ch. Trubner, 20 St. Dunstans Hill, City, and at the Exhib. Building.
1 Spirit of vinegar 160 grain of pure acid 16⅔ Th. (2 £ 10 sh.); 2. do. 80 grain do. 8⅓ Th. (1 £ 5 sh.); 3. do. 60 grain do. 6 Th. (18 sh.); 4. do. vinegar 40 grain 4 Th. (12 sh.); 5. do. 20 grain 2 Th. (6 sh.), all p. 120 quart (30 gallons).

1089. HEERS, F. W., manu., Telgte. London 1862 honor. mention. Agts. s. No. 1088.
6 bottles of bitter cordial p. quart 1 sh.

1090. VAN HEES, WILHELM, distiller, Cleve. Agts. s. No. 1088.
One bottle of corn-brandy and one bottle of genever, called »Schiedamer« 1 anker or 30 quart Prus. (1 quart 1 pint 1,56 gills) 9 Th. (1 £ 7 sh.).

1092. HURTER, HUB., & SON, purveyors to His Majesty the King of Prussia, wine-merchants, Coblenz. Agt. Francis Chubb, 61 Moorgate-street.
Sparkling Hock and Moselle-wines of the exhibitors own manufacture p. doz. bottles free London 28 to 45 sh.

1093. COUNT ITZENPLITZ, manu., Kunersdorf near Wrietzen on the Oder. London 1862 honor. mention.
Beet-root raw sugar, crushed lump and moist sugar.

1094. JODOCIUS, BROTHERS, & Co., wine-merchants, Coblenz. London 1862 medal. Agt. Fr. Mecke, 28 Great Tower-street.
Sparkling Hock and Moselle-wines.

1095. JOSSMANN, J., merchant, Berlin.
Collection of such seeds and products as are exported from Prussia.

1096. KANTOROWICZ, E., manu., Posen. London 1862 honor. mention.
Brandy, made of rye and flavoured with cumin.

1097. KEILER, J. S., merchant and manu., Danzig. Agts. s. No. 1088.
Liqueurs: Danzic goldwater; cherry (ratafia); maraschino; curaçao; rose; vanilla; (luft) peppermint; double ginger; double bitter orange; double goldwater; double Danzic goldwater; finest do.; Russian crême de allasch; finest Danzic rum punch-essence.

1099. KOBKE, FRIEDRICH (Köbke & Bergener), merchant, Magdeburg. London 1862 honor. mention. Agts. s. No. 1088.
Twelve crystal bottles of German liqueurs, p. Prussian quart without bottle 16½ sg. (p. gallon without bottle 6 sh. 8 d.)

1100. KRAFORST, P. H., seedsman, Leichlingen (Dusseldorf). London 1862 medal. Agts. Pet. Lawson & Son, 27 Great George-street, Westminster.
Seeds of grasses, clover and herbs for fodder; Herbarium of grasses and meadow-plants.

1101. KRAUSE, KARL HEINRICH, butcher, Langensalza.
German smoked saveloy, p. lb. 1 sh. 3 d.

KRAUSE, F. W.   (s. No. 1121a.)

1102. KREUTZBERG, MATHIAS JOSEPH, prop., Ahrweiler   Agts. s. No. 1088.
Six bottles of red Ahr-wine, 60 bottles, glass and pakage not included, 34 Th. (5 £.)

1103. KREUZBERG, PETER JOSEPH, & Co., merchant and wine-grower, Ahrweiler. London 1862 medal. Agt. Edward Scheller, 21 Wellington-street, Strand, W. C.
Ahr-wines, Walporzheim-Domley of 1857, 1858 and 1859.

KRUSE, A. T.   (s. No. 1106.)

1104. KULMIZ, C., sugar-refiner, Berlin, Adalbertsstr. 39.
    (s. No. 771, 998, 1353, 2171.)
3 loaves of sugar.

KUPFERSCHMIDT, A. H.   (s. No. 1151.)

1106. KRUSE, A. T., merchant and manu., Stralsund. London 1851 honor. mention, 1862 medal   Agts. C. A. Jonas, care of Lane, Hankey & Co., 23 Old Broad-street, E. C.
12 loaves of wheat-starch weighing 4 to 5 lb. a piece, 100 lb. Prus. weight 12⅔ Th. (cwt 38 sh.); 1 box cont. a loaf of starch, broken, to show the interior; ¹/₁, ²/₂, ⁴/₄ English pounded wheat-starch, 100 lb. 12⅔ Th. (38 sh.); 1 box cont. a sample of hair-powder 100 lb. 12⅔ Th. (cwt. 38 sh.); ¹/₁, ²/₂ english pounded hair-powder, 100 lb. 12⅔ Th. (38 sh.). Prices without duty, free on board, Stralsund.

1108. LANDER & KRUGMANN (formerly A. Werth & Co.), manu. and prop. of a corn-mill, Beuel near Bonn. London 1851 honor. mention; Paris 1855 bronze medal; London 1862 medal. Agt. Morgan Brothers, 21 Bow-Lane, Cannon-street, City, E. C.
17 sorts of potato and wheaten starch; starch-gum and liquid dressing.

1109. LANGGUTH & KAYSER, merchants and owners, Trarbach on the Moselle. London 1862 medal. Agt. M. F. Andres, 12 Mark-Lane.
1 bottle sparkling Moselle 1 £ 5 sh.; 24 bottles do. Hock 2 £ 8 sh.; 24/1 bottles do. Moselle

2 £ 8 sh.; 18/2 and 6/3 bottles do. 2 £ 12 sh.; 24 bottles do. 1 £ 10 sh.; 24 bottles do. 1 £ 12 sh.; 18/2 and 6/3 bottles do. 1 £ 14 sh.; 24 bottles do. 1 £ 5 sh.; 24 bottles do. 1 £ 6 sh.; 18/2 and 6/3 bottles do. 1 £ 9 sh.; 1 bottle do. 1 £ 10 sh., prices p. dozen ¼ bottles.

1110. LAUE, H., manu., Wehlau (Königsberg).

Wheat-flour; lin and rape-seed cakes.

1111. LEHMANN, ROBERT, landed prop., Nitsche (Posen). Agts. Lion M. Cohn, Phaland & Dietrich, represented by Ch. Trübner, 20 St. Dunstans Hill, City, and at the Exhib. Building. (s. No. 1208a, 1516a.)

A sample of rye p. Prussian bushel or 90 pounds 2⅔ Th. (98 pounds 8 sh.). A sample of barley p. Prussian bushel or 75 pounds 2¼ Th. (82½ pounds 7 sh.).

1112. LEHMANN, JOHANN KARL, LEHMANN, JOHANN ALBERT (J. C. Lehmann), purveyor to His Majesty the King of Prussia, manu. of preserved fruits and vegetables etc. London 1862 medal. Agts. s. No. 1111.

No. I. 2 tin-boxes with preserved pine-apples each 2 Pfd. (2 lb. 4 ounces) 2 Th. (6 sh.); No. II. 2 do. each 1 Pfd. (1 lb. 2 ounces) 1 Th. (3 sh.); No. III. 2 do. each ½ Pfd. (9 ounces) 17 sg. (1 sh. 8 d.), all p. box, net weight; No. IV. 1 glass-jar with pine-apples gross-w. 36½ Pfd. net 26 Pfd.; No. 1. 2 glass-bottles with preserved pine-apples, each 2 Pfd. (2 lb. 4 ounces) 2½ Th. (6 sh. 8 d.); No. 2. 2 do. each 1⅓ Th. (4 sh.); No. 3. 2 do. each 1 Pfd. (1 lb. 2 ounces) 1 Th. (3 sh.); No. 4. 2 do. each ⅔ Pfd. (12 ounces) 21 sg. (2 sh. 1 d.); No. 5. 2 do. each ½ Pfd. (9 ounces) 17 sg. (1 sh. 8 d.); No. 6. 2 do. each ⅓ Pfd. (6 ounces) 11 sg. (1 sh. 1 d.), all p. bottle, net weight; No. 7. 2 glass-bottles with peaches; No. 8. and 9. 2 and ²/₂ glass-jars with apricots; No. 10. 2 do. with hips; (No. 7—10 preserved 1860); No. 11. 2 bottles with essence of pine-apple, each net 1⅓ Pfd. (net 1 lb. 8 ounces) p. bottle 22½ sg. (2 sh. 2½ d.); No. 12. 2 vessels with mixed pickles both btt. 14⅓ Pfd. (btt. 15 lb. and 13 ounces); No. 13. 2 bottles with royal punch-essence, each ¾ quart (1½ pints) 22 sg. (2 sh. 2 d.); No. 14. ²/₂ bottles do. each ⅜ quart (¾ pints) 12 sg. (1 sh. 2 d.).

1114. LUDORFF, FRANZ, manu., Münster. Agts. s. No. 1111.

Dr. C. von Bönninghausen's homoeopathic coffee, p. pound 2¼ sg. (3 d.) loco Münster.

1116. MEYER, H. G., manu., Herford (Minden). London 1862 medal. Agts. s. No. 1111.

1 barrel of mustard p. 100 pound (20 St.) 6½ Th. (19 sh. 4 d.); 1 bottle of vinegar p. 180 quart 8 Th. (gallon 4 £ 7 sh. 5 d.).

1117. RAVENSBERG STEAM-CORN-MILL, Bielefeld, manufactory Porta. London 1862 medal.

An assortment of wheaten flour, grits, pollard, bran in 12 bags: wheaten flour known as emperor's flour No. 00., No. 0., No. 1., No. 2., No. 3.; wheaten grits No. 1., No. 2., No. 3., No. 4.; pollard and bran of wheat.

1118. MOLLARD, E. O. PH., manu., Gora (Posen).

Preserved green pease, carrots partridges.

1119. NERNST, HERMANN, farmer and captain, Taplacken (Königsberg). London 1862 honor. mention. Agts. s. No. 1111. (s. No. 1216.)

Cigars 1000 pieces 4—16 Th. (12 sh. to 2 £ 8 sh.); tobacco 100 pounds 15—30 Th. (2 £ 5 sh. to 4 £ 10 sh.)

NEUMANN, BROTHERS. (s. No. 1070.)

1121. VON NIESSEN, A., manu. of liqueurs, Danzig. London 1862 honor. mention.

6 bottles of double goldwater-liqueur 6 Th. 15 sg. (19 sh. 6 d.); 2 bottles of double stomach-liqueur 6 Th. (18 sh.); prices p. 12 bottles.

1121a. KRAUSE, F. W. (Ohlsen-Bagge successor), spirit-manu., Frankfurt on the Oder. London 1862 medal. Agts. s. No. 1111.

Pure spirit of wine from raw potatoe-spirit. Prices according to the prices for raw spirit at the exchange of Berlin.

POMERANIAN PROVINCIAL SUGAR REFINERY. (s. No. 1142.)

1122. POPPELSDORF, ROYAL AGRICULTURAL-COLLEGE, by Hartstein, director, Poppelsdorf near Bonn. (s. No. 1222.)

Nine varieties of wheat from originally English seed, cultivated the last 5 years. (harvest of 1861.)

1122a. PRAGER, LOUIS, manu., Erfurt. London 1862 medal. Agt. William Schultz, 4 Lawrence Pountney-Place, Cannon-street, E. C., and during the exhibition also C. Trübner, 20 St. Dunstans Hill, City. (s. No. 1016.)

Samples of barley, raw and manufactured, showing, the whole of the manufacturing of pearled barley, unto the highest perfection, as: Thuringian barley; peeled barley; bran; split barley No. 2., No. 1., No. 0.; groats of barley; forage; half pearled barley No. 3., No. 2., No. 1., No. 0., No. 00.; dark flour; pearled barley N. 3., No. 2., No. 1., No. 0., No. 00., No. 000., No. 0000.; white flour.

Potatoe-sago and flour; wheat and wheat-groats; oats, raw and manufactured.

Millet, red, yellow, white and peeled.

Legumes: horse-beans; white beans; yellow, green and peeled peas; lentils; vetches.

Oil seeds: dodder; rape; poppy, blue, grey and white.

Clover seeds: melilot, yellow, trefoil; lucern, white sainfoin.

Seeds for trade: fenugreek, white mustard, anise, canary, coriander, dill, fennel, hemp, caraway, nigella.

Dried fruits: plums, cherries and juniper-berries. 3 packets of pearled barley, superior quality 1 cwt. 3 £.

1122b. PRITZKOW, HERRMANN (George Broche), distiller, Berlin. London 1862 medal. Agts. s. No. 1111.

Berlin caraway-brandy 1 bottle 1 sh.; Berlin caraway-liqueur ½ bottle 1 sh. 6 d.; mulled-wine extract ½ bottle 1 sh. 6 d.; raspberry-limonade extract ½ bottle 1 sh. 6 d.

1123. PROCHNOW, J. F., merchant, Rügenwalde (Cöslin).

Smoke-dryed geese.

1123a. COUNT RACZYNSKI, ROGER, grower, Woynowice (Posen).

Hops.

1124. Radicke, Adalbert, Leffkowitz, Wolf (A. Radicke & Co.), merchants and distillery-prop., Berlin and Grunberg. Agts. Lion M. Cohn, Phaland & Dietrich, represented by Ch. Trubner, 20 St. Dunstans Hill, City, and at the Exhib. Building.

6 bottles of imitation French brandy or cognac produced 1860 in the distillery at Gruenberg from Gruenberg grapes.

1125. vom Rath, H., manu., Lauersfort (Crefeld) (s. No. 1225.)

Samples of flour.

1126. Raussendorff, Hugo, merchant and distiller, Berlin. London 1862 honor. mention. Agts. s. No. 1124.

Essence of rum $22\frac{1}{2}$ sg. (2 sh. 3 d.); essence of arrack $22\frac{1}{2}$ sg. (2 sh. 3 d.); essence of French brandy $1\frac{1}{4}$ Th. (3 sh. 9 d.) all p. lb. (for the manufacture of rum, arrack and French brandy 2 pounds are sufficient for 180 quart of 54 pCt. spirit)

Ravensberg, steam corn-mill. (s. No. 1117.)

1126a. Reiffen, Gustav, merchant, inv., Saarbrücken, manufactory Reiffenburg near Saarbrucken. Agts. s. No. 1124.

Cordials 1 bottle and $\frac{1}{2}$ bottle of Magenwurze, 1 bottle and $\frac{1}{2}$ bottle of Biedermann, 1 bottle and $\frac{1}{2}$ bottle of Boonekamp, p. $\frac{10}{16}$ quart 14 sg. ($\frac{11}{16}$ quart 1 sh. 6 d.)

1127. Roeder, Johann Adam, manu. of liqueurs, purveyor to His Royal Highness Prince Frederick of Prussia Paris 1855 bronze medal; London 1862 medal.

Punch-essences: arrack; rum; vanilla; portwine; burgundy; table-liqueurs: double anisette; double Curacao; crême de gingembre; creme de vanilla; creme de mocca.

Rohloff, C. G. (s. No. 1140.)

1129. Sachs & Hochheimer, manu. of sparkling Hock and Moselle-wine, Coblenz. Agts. J. & R. Mac Cracken, 7 Old Jewry.

6 bottles sparkling Hock: 6 bottles sparkling Hock sup. quality; 6 bottles sparkling muscadel Moselle; 6 bottles sparkling Moselle of Scharzberg I. quality.

1130. Scherr, J., Son, manu. of mineral-waters and liqueurs, Triers. London 1862 honor. mention. (s. No. 1020.)

6 bottles of liqueurs.

1131. Schiffer, Carl, distiller, Düsseldorf.

Punch-essence of portwine; punch-essence of burgundy-wine; royal punch; p. bottle 25 sg. (2 sh. 6 d.) wholesale 23 sg. (2 sh. 3 d.)

1134. Schmellitscheck, W., manu., Wesel. London 1862 honor. mention.

Chocolate.

1135. Schmidt, A., manu., Labes (Stettin).

Compressed vegetables.

1136. Schmitz, J. A., manu., Hübsch near Wesel. London 1862 honor. mention.

Fruit and beet-root jelly.

1137. Schoeller, Leopold, privy counsellor of commerce, prop., Schwieben (Oppeln). London 1862 medal. Agt. Charles Nolda, 2 Church-Court, Old Jewry. (s. No. 1238.)

$\frac{1}{4}$ of a bushel of flower-wheat 3 Th. (9 sh.); $\frac{1}{4}$ of a bushel of lemon-peas $2\frac{1}{2}$ Th. (8 sh.); $\frac{1}{4}$ of a bushel of Siberian spring-barley $1\frac{2}{3}$ Th. (5 sh.); $\frac{1}{4}$ of a bushel esparcet (sainfoin) all p. Prus. Scheffel.

1138. Schultze, J. C., manu., Berlin, Charlottenstr. 82. (s. No. 1240.)

Home-made French brandy (cognac).

1140. Rohloff, C. G. (G. F. A. Steiff), merchant, brewer and distiller, Danzig. Berlin 1844 bronze medal.

A bottle rectified potatoe spirits of 95 pCt. tralles 100 quarts of 90 pCt. 22 Th. (100 gallons of 356 pCt. (13 £ 1 sh 6 d.).

Simple brandies: cummin, juniper, orange 20 Th. (9 £ 18 sh.); kernel-water (Persico) half double brandy 32 Th. (15 £ 17 sh.); Ratafia (cherry brandy) a double brandy 60 Th. (29 £ 14 sh.); all p. 120 quart (100 gallons).

Double spruce-beer $\frac{1}{8}$ barrel $2\frac{1}{2}$ Th. ($3\frac{1}{7}$ gallon 7 sh. 6 d.)

1142. Stettin, Pomeranian Provincial Sugar-Refinery, Stettin. London 1862 honor. mention. Agts. s. No. 1124.

One loaf and a half of fine beet-root sugar refined $13\frac{1}{2}$ Th. (2 £); one loaf and a half of beet-root sugar, melis $12\frac{1}{2}$ Th. (1 £ 17 sh. 6 d.); fine white beet-root moist sugar 13 Th. (1 £ 19 sh.); fine yellow do. $11\frac{3}{4}$ Th. (1 £ 15 sh. 6 d.); yellow do. $10\frac{3}{4}$ Th. (1 £ 12 sh. 6 d.); brown do. $9\frac{3}{4}$ Th. (1 £ 9 sh. 6 d.) all p. Zollz. (cwt.), free on board.

1143. Stettin steam-mill company, Stettin, mill at Züllchow near Stettin. Agt. Geo. Cartellieri, 3 New London-street.

Samples of wheatflour, ryeflour, bran and pollard.

1145. Uhlendorff, Louis Wilhelm, prop. of a corn-mill, Hamm on the Lippe. Paris 1855 bronze medal; London 1862 medal. Agts. s. No. 1124.

Samples of corn, legumes, flour and bran.

1146. Underberg-Albrecht, H., inv. and sole distiller of the celebrated Boonekamp of Maagbitter, known under the device: »Occidit qui non servat«; purveyor to the courts of Their Majesties the King of Prussia and the King of Bavaria, His Royal Highness the Prince Frederik of Prussia and His Royal Highness the Prince of Hohenzollern-Sigmaringen, Rheinberg on the Rhine. London 1862 medal. Agts. Morgan Brothers, 21 Bow-Lane, Cannon-street, City, and Hermann Rübeck, 12 Little Tower-street, E. C. (s. II. No. 90.)

Boonekamp of maag-bitter a quart bottle 2 sh; a pint do. 1 sh. $1\frac{1}{4}$ d; a small one 6 d; from Rheinberg, package excl.

1148. Visser, Brothers, manu., Cologne. Agts. Alexander Mönch & Co., 26 Lambeth-Hill, Doctor's Commons, E. C.

Starch, samples of different kinds: white and blue patent wheat-starch; white and blue radiated wheat-starch; white and blue air-dried wheat-starch.

1149. Volcker & Rickmann, merchants, Schwedt on the Oder. Agts. s. No. 1124.

$\frac{1}{10}$ box of cigars of Prussian growth 3—5 Th. (9 to 15 sh.) p. 1000; 1 sample of tobacco in rolls do. 6—9 Th. (18 sh. 4 d. to 1 £ 7 sh. 5 d.); 2 samples of tobacco in leaves do. 8—10 Th.

(1 £ 4 sh. 4 d. to 1 £ 10 sh. 5 d.); 1 sample of tobacco out do. 10—12 Th. (1 £ 10 sh. 5 d. to 1 £ 16 sh. 6 d.) all p. Zollz. (cwt.).

1150. WAGENER, LOUIS, Royal counsellor of rural economy and administrator of the domain Proskau (Oppeln). Agts. Lion M. Cohn, Phaland & Dietrich, represented by Ch. Trubner, 20 St. Dunstans Hill, City, and at the Exhib. Building. (s. No. 1530.)

Twenty pounds of self-grown hops p. cwt 30 to 100 Th.

1151. KUPFERSCHMIDT, HEINRICH ADOLPH (Isaac Wedling Wwe. & Eydam Dirck Hekker, called »the Lachs«), distiller, Danzig. Agts. s No. 1150.

Liqueurs: bottle-case No. 1. wine-cinamom 1 Th. 25 sg.; essence of bitter oranges 1 Th. 20 sg.; wine gold-water 1 Th. 15 sg.; liliencomfolgen 1 Th. 10 sg; cordial 1 Th. 10 sg.; cumin 1 Th. 5 sg : 6 bottles containing together 7½ Berlin quart 10 Th. 25 sg ; 6 bottles 12 pints 1 £ 12 sh. 8 d.; bottle-case No. 2 double gold-water 15 sg . double cordial 13 sg.; kurfurstl. Magen 13 sg.; double persico 13 sg.; bitter orange 12 sg.; cumin 11 sg.; 6 bottles containing 7⅕ Berlin quart 4 Th. 17 sg.; (6 bottles or 12 pints 13 sh. 8 d.); each bottle-case 2 Th.

WELLMANN, A., Stettin. (s. No. 1248 a.)

1156. WETTENDORFF, JOHANN WILHELM, prop., Triers (Balduinhouse near Triers). Paris 1855 bronze medal; London 1862 honor. mention. Agts. s. No. 1150.

Hops of 1861 grown by the exhibitor.

1157. WIGANCKOW, OTTO, merchant and manu., Berlin, Badstr. 93. Agts. s. No. 1150.

Specimens of cigars; the list of prices is annexed.

1158. WILCKE, F. H. W. (Franz Wilcke), merchant and distiller, Berlin, Mohrenstr. 16. London 1862 medal. Agts. s. No. 1150.

3 bottles of rectified-sprit 7½ sg. (4 sh. 10 d.); 2 bottles of orange-liqueur 20 sg. (12 sh. 10 d.); 2 bottles of extract of ginger-liqueur 20 sg. (12 sh. 10 d.) all p. quart (p. gallon).

1160. WITTEN, STEAM-MILL JOINT-STOCK COMP. Witten on the Ruhr (Arnsberg). London 1862 medal. Agt. A. L. Cross, 29 Norfolk-street, Strand.

Two qualities of flour from red wheat; two qualities of flour from white wheat; two qualities of flour from rye; two qualities of grit; different sorts of macaroni, vermicelli etc.; at the time of exhibition extra price-currents will be issued.

CLASS IV.

ANIMAL AND VEGETABLE SUBSTANCES USED IN MANUFACTURES.

1161. ADENAU, XXII. a. local-department of the Rhenish-Prussian agricultural society. London 1862 medal.

Oak-bark for tanning 100 lb. or 50 kilogr. 2⅓ Th. (6 sh. 7 d.).

BALTIC ASSOCIATION FOR THE PROMOTION OF AGRICULTURE AT ELDENA. (s. No. 1173.)

BENDER, W., Bleicherode, Flax. s. No. 1500.

1163. BENNEKE & HEROLD, lacker-manu., Berlin, manufactory Schöneberg near Berlin. London 1862 honor. mention. Agts. Dorgerloh. Sang & Co., 2 Savage-Gardens.

20 bottles of lacker and varnish.

1164. BLANCKE, JULIUS C. W., manu. Magdeburg. Agt. E. G. Zimmermann, 2 St. Pauls Buildings, Little-Carter-Lane, Doctor's Commons.

India-rubber goods: 1 piece of hose with a spiral wire; 2 pieces of canvas-hose; 2 pieces of gas-pipes; 4 pieces of packing-cords; 2 pieces of sheet with cloth-inserting; 1 piece of sheet with cloth outside; 2 pieces of pure sheet, grey; 2 pieces of ring-packings for flanges; 1 piece of gutta-percha hydraulic ram-packing; 1 piece of gutta-percha pump-bucket; 1 piece of hemp-hose; 1 paper-box with solid balls; do. with hollow balls; do. with plain inflated balls; do. with colored inflated balls; 2 paper-boxes with nipples.

BRAUNSBERG AGRICULTURAL ASSOCIATION. Flax. (s. No. 1502.)

1167. VON BRUNNECK, burgrave, and VON BRUNNECK, SIEGFRIED, landowners, Belschwitz near Rosenberg (Westpreussen). London 1851 prize-medal, 1862 honor. mention.

4 fleeces of electoral-wool p. Zollz. 120 Th. (110 lb. 18 £).

1168. BÜRGERS, WOLLER JOSEPH, merchant, Cologne. London 1862 honor. mention. Agts. Morgan Brothers, 21 Bow-Lane, Cannon-street, City.

Cologne-glue.

1169. BURRMEISTER, H., manu., Stettin. London 1862 honor. mention.

Rapeseed oil.

1170. DEHMEL, HEINRICH, purveyor to H. M. the King of Prussia, Quaritz (Liegnitz). Agts. s. No. 1150.

Soaps; prices p. doz.: 1 brown marbled soap cake; 6 pieces of savon des deux mondes 3 sh.; 4 pieces of piever-soap 1 sh. 6 d.; 24 pieces of soap with horse-heads 7½ d.; 6 pieces of soap with stag heads 7⅕ d.; 8 pieces marbled barrel-soap 7⅕ d.; 8 pieces of soap with flower-baskets 4½ d. 8 pieces of soap with palm-trees 3⅗ d.; prices p. doz. 34 pieces of small soap-figures.

1171. DOUGLAS, ANTON, prop., Amalienan near Konigsberg. (s. II. No. 29.)

No. 470. fleece of a ram, born 1859; No. 112. fleece of a ewe do.

1172. DYHRENFURTH, LUDWIG, landed prop., Jacobsdorf (Breslau). London 1862 honor. mention. Agts. s. No. 1150.

2 ewe-fleeces; 1 yearling ewe-fleece; 1 yearling ram-fleece.

1173. BALTIC ASSOCIATION FOR THE PROMOTION OF AGRICULTURE AT ELDENA, province of Pomerania. Agts. s. No. 1150.

Three chests containing samples of wool from different sheep-farms in the district of the Baltic association.

1174. VAN ELS, C. & Co., manu., Düsseldorf.

Oil for lubricating machinery.

1175. Engel & von Schaper, manu., Berlin, 17 Mohrenstr.

Perfumed soaps.

1177. Foerster, Ferd. Siegfr., manu., Halbmeilmühle near Grunberg. Agts. Lion M. Cohn, Phaland & Dietrich, represented by Ch. Trübner, 20 St. Dunstans Hill, City, and at the Exhibit. Building.

(s. No. 1071, 1343, 1605, 1606, 1871 a.)

Prices p. Zollz.: Raw rape-seed oil 12⅝ Th.; do refined 13⅓ Th ; lubricating oil 15 Th; rape-seed cakes 1⅔ Th ; rape-seed cake powder 2 Th.

1178. Fonrobert & Reimann, india-rubber and gutta-percha manu., Berlin. Berlin 1844 silver medal; Munich 1854 medal of honor; Paris 1855 silver medal; London 1862 honor. mention. Agt. Oliver Kirkmann, at Messres. George Spill & Co., 149 Cheapside.

I India-rubber-Material: 7 pieces Para-rubber in sheets; 2 blocks and 4 balls East-India-rubber; cleaned rubber; 4 blocks of masticated rubber; 1 roll do ; 10 pieces and 4 rolls of manufactured and mixed rubber; 10 sheets vulcanized rubber.

II. Manufactured articles: 6 packing-rings; 9 sugar loaf funnels; valve; 6 pump valves; 23 buffer-rings (8 iron plates); 4 packing-rings; 15 samples of tubing; 7 do. of delivery-hose; locomotive-hose; suction-hose with spiral wire and gutta-percha sucker; 10 different thickness of cord; billard-cushion; bed-pan; 4 breast drawers with glass and horn-mountings; 2 clyster syringes; enema syringe with double acting valve; urinal syringe; 4 pessaries; 10 corks; 13 bottle-caps with and without tubes; 2 nipples with shields; 2 inflaters; 1 urinal, 1 cupping-glass with horn mounting; do. with glass; truss-pads; 1 pair of gloves; 2 dress-protectors; 11 bottle-nipples; tobacco pouch; 2 pair trowser straps; 12 doz. armlets; 4 cards with penholders and with erasers; 2 sticks; large comb; 30 half doz. combs; 2 syringes; 2 pipe-mouthpieces; guard to trigger; 2 brush-cases; insulating sheet; 4 insulating cups; rolls of copper-wire covered with gutta-percha; roll do. covered with India-rubber; telegraphic apparatus with battery and bell; battery-chamber of gutta-percha; 2 gutta-percha drinking-cups; 3 gutta-percha funnels; 2 gutta-percha flower-pot stands; large shoe; 2 waterproof coats; 4 caps; 2 air-cushions; 5 different gas-tubing; 3 pyramids with 119 figures, animals and heads; pyramids with 165 grey and colored balls; 144 grey and colored balls; large football with figures and landscapes 3 ft. in. diameter.

Price-lists in English, French and German will be published.

1180. Friedenthal, C. P., landed prop., Domslau near Breslau. Agts. J. G. Marc & Co., 21 Mark-Lane.

Teazels p. 1000 40—60 sg. (4—6 sh.)

1181. Gaillard, Carl Friedrich (C. F. Gaillard), merchant and manu. of perfumes, Berlin. Agts. A. G. Franklin & Co., 14 South-street, Finsbury-Square. (s. II. No. 18.)

Fine perfumes and toilet soaps of all sorts, own manufactures, particularly soaps like mosaik work with figures.

1183. Giessler, Nicolaus Heinrich, farmer, Trochtelborn near Erfurt.

Thuringian woad in small balls; p. Zollz. 6 Th. (18 sh. 6 d.)

1184. Goebel, Eberhard (Jacob Goebel), manu., Siegen. London 1862 medal.

Glue.

1185. von Goldfuss, A., Prussian major, prop. of Gross-Tinz near Nimptsch (Breslau). Agts. s. No. 1177.

4 Fleeces of electoral escurial wool 20 Th. (3 £), from the flock of sheep at Niklasdorf near Strehlen.

1187. Grossmann, C. E., manu., Berlin, Alexanderstr. 38 a. London 1862 honor. mention. Agts. s. No. 1177.

Soaps and perfumes; prices pr. doz.; Soaps: No. 102. royal de chridace 3 Th.; No. 80. sweet rose 3 Th ; No. 81. sweet violet 2 Th ; No. 82. sweet bouquet 2 Th ; No. 64. ponce 1⅓ Th ; No. 68. rice-flour 1⅓ Th ; No. 110. imperial 1½ Th ; No. 124. fine rose 1½ Th ; No. 125. violet 1½ Th ; No. 132. fancy 1 Th.; No. 160. India 4 Th ; No. 162. bitter almond 1⅓ Th.; No. 163. finest bitter almond 2 Th ; No. 180. fine glycerine 2 Th ; No. 204. honey 1½ Th ; No. 177. king 1 Th ; No. 360. king-coronation ⅚ Th.; No. 245. fine bitter almond ½ Th.; No. 312. transparent mosaik ⅔ Th.; No. 24. Windsor 1 Th.; No. 188. guimauve 1⅓ Th.; No. 142. musk 2 Th.; No. 72. vegetable ⅔ Th.; No. 104. sulphur 1 Th ; No. 103. tar 1 Th.; No. 66. gall ½ Th.; No. 288. shaving-oil ½ Th.; No. 305. transparent shaving 1½ Th.; No. 298. Victoria 1½ Th.

Perfumes: No. 406. pomatum in pots 8 Th : No. 482. glycerine-cream 4 Th. No. 456. Borsdorf apple-pomatum 2 Th.; No. 463. eggoil-pomatum 2 Th.; No. 421. brillant pomatum 2⅔ Th.; No. 470. pearl pomatum 1½ Th.; No. 489. philocome pomatum 4 Th.; No. 411. tooth-soap 1 Th.; No. 497. eau de Cologne triple 3 Th ; No. 498. eau de Cologne double 1½ Th.; No. 510. finest imitated extracts 4 Th.

Soaps: No. 201. omnibus- ⅔ Th ; No. 202. eagle- ⅔ Th.; No. 200. shell- ¼ Th.; No. 199. oval ¼ Th.; No. 90. transparent soap-hands 2 Th.; No. 91. transparent coronation 2 Th.; No. 92. transparent oval 1½ Th ; No. 93. transparent small oval 1 Th.; No. 94. dove-girl 1⅓ Th.; No. 95. cupide 1½ Th.; No. 96. flower-girl 1½ Th.; No 12. fruits of soap 1½ Th.; No. 13. soap-cake 1 Th.

Perfumes: a glass-box containing a large bouquet of violets and camellias 36 Th.; No. 2. violet bouquet 6 Th.; box containing a bouquet of violets and roses 24 Th.; violet-bouquet 4 Th.; No. 511. perfumes in china-pots 3 Th.

1191. Guradze, Siegmund, prop., Kottlischowitz (Oppeln). Agts. s. No. 1177.

3 washed fleeces of merino sheep of the electoral-negretti breed.

1192. Guradze, A., royal counsellor of commerce, Tost Munich 1854 honor. mention; London 1862 medal. Agts. s. No. 1177.

(s. No. 1084.)

Wool-fleeces.

1194. Haller, J., manu., Halle on the Saale.

Starch.

1196. Herz, S., manu., Berlin, Dorotheenstr. 1. London 1862 medal.

Rape, lin-seed and »dotter« (camelina sativa) cake, whole and crushed and oil of the same.

1197. HEYL, J. F., & Co., manu., Berlin, Leipzigerstr. 75.

Line, rape and hemp-seed and oils; castor, raddish and cumin oil.

1198. HOFMANN, J. G., machine and oil-manu., »Koinonia« (Breslau). London 1862 honor. mention.

Three bottles, containing 1. raw rape-seed oil, obtained not by pressure but in a new and patented apparatus by means of sulphuret of carbon; 2. rape-seed oil free of water and acid, refined in a new manner; 3. rape-seed oil, refined in a new manner (lampoil).

HOHLSTEIN, B., Bollstedt. Flax. (s. No. 1514.)

1199. HOLTZ, H., landowner, Wogenthin near Koslin (Bruckenkrug). Agts. Lion M. Cohn, Phaland & Dietrich, represented by Ch. Trübner, 20 St. Dunstans Hill, City, and at the Exhib. Building.

Bark of oak, dryed and cleaned 1. best quality 2 Th. (6 sh.); II. 1½ Th. (4½ sh.); III. 1 Th. (3 sh); all p. Zollz. (cwt.).

1200. HOMEYER, FRIEDRICH, landowner, Ranzin near Mökow (Stralsund). Agts. Nicolson Brothers, 5 Jeffrey's-Square.

Wool fleeces perfectly free of grease.

VON HUHN, N., Ober-Gerlachsheim. Flax. (s. No. 1511.)

1201. JANSSEN, MICHELS & NEVEN, manu., Cologne, manufactory, Nippes, suburb of Cologne. London 1862 medal. Agts. Robinson & Fleming, 21 Austin Friars.

Column of candles and blocks of stearine, moulded; oleïne.

1202. KAMPER, THEODOR, merchant and manu., Cologne. Agts. Morgan Brothers, 21 Bow-Lane, Cannon-street, City.

A bottle of oil for all sorts of machinery, newest invention.

KAMPHAUSEN, N. W., Bendorf. Silk-cocoons. (s. No. 1536.)

1203. KIRSTEIN, C., merchant, Hirschberg. Agts. Heintzmann & Rochussen, 9 Friday-street, Cheapside. (s. No. 1516.)

Officinal vegetables from the Riesengebirge: radix angelica 9⅓ Th. (1 £ 8 sh.); radix elebori albi 10⅔ Th. (1 £ 13 sh.); lichen Islandic. 3⅓ Th. (10 sh.); baccae mirtylli 9 Th. (1 £ 7 sh.); prices p. Zollz. free Hamburgh.

1204. KISTENMACHER & GUERKE, merchants, Sprottau. Agts. s. No. 1199.

Palm-soap 14 Th. (2 £ 2 sh.); soap from palm-oil and cocoa-nut oil 12 Th. (1 £ 16 sh.); resin-soap 9 Th. (1 £ 7 sh.); cocoa-soda soap 20 Th. (3 £); prices p. Zollz.

1205. KLEIST, apothecary to the Prussian army, Berlin, inventor HILLER L., Berlin, Jerusalemerstr. 26. Agts. s. No. 1199.

Oil for watches and chronometers: a. in holes in a brass piece; b. in 4 small black bottles; c in a bottle for inspection; d. in glas tubes to be tested; p. doz. 18 sh.

1207. LÉDER, AMADEUS, THEODOR (Leder Brothers), apothecary of 1. order, technical chemist and manu. of perfumes, Berlin. Agts. s. No. 1199.

Cocoa-nut oil soda-soap 20⅓ Th. (3 £ 1 sh.); earth-nut oil (peanut-) balsamic-soap.

Honey-soap 24⅓ Th. (3 £ 13 sh.); toilet-soap, one colour and marbled 22⅓ Th. (3 £ 7 sh.); all p. 1 Zollz. 1½ lb. (1 cwt.).

Bathing-soap, light, all in blocks No. 1 to 14.

Figure-soaps No. 1. to 25., 65., 66., 84., 113. 2½ to 12 Th. (7 sh. 6 d. to 1 £ 16 sh.); toilet-soaps with rounded angles, partly with the firm of the exhibitor, partly with decorations (sujets en bas-relief) No. 26 to 33., 37., 38., 40., 43., 44., 45., 81., 82., 85., 86., 87., 93., 95., 96. 2 to 15 Th. (6 sh. to 2 £ 5 sh.); p. gross each.

The following prices p. dozen: toilet-soap in various scents with and without labels, rose, orange, musk, violet etc. No. 39., 42., 48., 52., 54., 60., 94., 103., 104., 105., 106. 1 to 4 Th. (3 to 12 sh.); almond-soap No. 34., 36., 112. ½ to 1⅔ Th. (1 sh. 6 d. to 5 sh.); glycerine yolk-soap No. 50. 2 Th. (6 sh.); earth-nut oil balsamic soap No. 57. 1 Th. (3 sh.); cocoa-nut oil-soap, highly refined, No. 73. 12½ sg. (1 sh. 3 d.); ball-soaps No. 49., 51., 113 a., 113 b. 20 sg. to 1 Th. (2 to 3 sh.); shaving-soap, balsamic, No. 58. 1⅓ Th. (4 sh.). transparent soaps No. 46., 88., 89 to 92., 98., 100 to 102., 108. 12½ sg. to 3⅓ Th. (1 sh. 3 d. to 10 sh.); medicinal soaps No. 47., 53., 55., 56., 59. 20 sg. to 1½ Th. (2 sh. to 4 sh. 6 d.); almond-cream No. 78.; 2½ Th. (7 sh. 6 d.); cold cream. No. 76 2⅓ Th. (7 sh.); oriental skin-water No. 107. 3 Th. (9 sh.); chilblain-paste No. 79. 1½ Th. (4 sh. 6 d.); cosmetic paste No. 75. 1⅓ Th. (4 sh.); complexion-balsam No. 72. 1⅓ Th. (4 sh.); pomatum (cosmetiques) No. 35., 97., 99. 15 sg. to 1 Th. (1 sh. 6 d. to 3 sh.); spermaceti wax-pomatum No. 83. 1½ Th. (4 sh. 6 d.); balsamic oil of eggs No. 68. 3 Th. (9 sh.); balsamic egg-oil pomatum No. 64. 3 Th. (9 sh.); remedies for promoting the growth of hair: komaphition forte, komaphition mite, komaphylacticon, lepielasia and oil of herbs of the Alps No. 65., 66., 67., 70., 77. 2 to 7 Th. (6 sh. to 1 £ 1 sh.); extraits d'odeurs No. 61., 63., 69. 3 to 4 Th. (9 to 12 sh.); perfumery No. 41. 71. 20 sg. to 1½ Th. (2 sh. to 4 sh. 6 d.); remedy for the teeth 48 a., 53 a., 62. 10 sg. to 1 Th. (1 to 3 sh.); smelling-cushions (sachets) 114., 115., 116. 5 to 7 Th. (15 sh. to 1 £ 1 sh.); paints 74., 80. 1 Th. (3 sh.).

1208. LEHMANN, ROBERT, landed prop., Posen. London 1851 honor mention; Paris 1855 silver medal; London 1862 medal. Agts. s. No. 1199. (s. No. 1111. 1516a.)

8 washed wool-fleeces p. 100 lb. 130 Th. (19 £ 10 sh.).

1209. PRINCE LICHNOWSKY, CARL, Kuchelna (Silesia). Agts. s. No. 1199.

Washed wool-fleeces; samples of wool.

1211. LÜBBERT, EDUARD, landed prop., Zweybrodt near Breslau. London 1851 prize-medal; München 1854 medal of honor.; Paris 1855 silver medal; London 1862 medal. Agts. Lion M. Cohn Phaland & Dietrich, represented by Ch. Trübner, 20 St. Dunstans Hill, City, and at the Exhib. Building.

Wool-fleeces.

VON LUTTWITZ, R., Simmenau. Flax. (s. No. 1517.)

1212. LUYKEN, RIGAUD & REMY, manu. of stearine candles, Wesel. Agts. Mess & Co., E. C.

Stearine candles and stearic acid.

1214. MAHR, BROTHERS, manu. of ivory- and boxwood-combs; Naumburg on the Saale. London 1862 honor. mention.

Ivory bastard-combs of 1¼ to 3¾ in. engl. measure 1 sh. 3 d. to 4 sh. 6 d.; dandrif, of 2¼ to 4 in. engl. measure 3 sh. 6 d. to 8 sh. 6 d.; super-fine, of 2⅔ to 4 in. engl. m. 5 sh. 6 d. to 12 sh.; superfine deep, of 3 to 4 in. engl. m. 11 sh. to 15 sh. 6 d.; super-superfine, of 2⅔ to 4 in. engl. m. 8 to 15 sh.; super-superfine deep, of 3 to 4 in. engl. m. 12 sh. 6 d. to 20 sh.; imperial, of 3 to 4 in. engl. m. 14 to 20 sh.; imperial deep, of 3¼ to 4 in. engl. m. 18 sh. to 1 £ 6 sh.; all p. doz.

Boxwood-combs bastard of No. 1. to No. 6. 4 sh. 3 d. to 10 sh. 9 d.; dandrif of No. 1. to No. 6. 5 sh. 5 d. to 13 sh. 3 d.; superfine of No. 0. to No. 3. 18 sh. to 1 £ 7 sh.; super-superfine of No. 0. to No. 2. 48 sh. to 72 sh.; all p. gross.

MEVISSEN, GERH., Dulken. Flax.
(s. No. 1518.)

1215. MOTARD, DR. A., manu., Berlin, Bruderstr. 11. London 1862 medal.

Elaïne, samples of soap, stearine and paraffin-candles.

1216. NERNST, HERMANN, farmer of crown-land and captain, Taplacken (Königsberg). Agts. s. No. 1211. (s. No. 1119.)

Turnsol-oil p. lb. 2 sh.; Turnsol-seed cakes.

1217. COUNT VON OPPERSDORF, EDUARD, prop. of Ober-Glogau (Silesia).

3 fleeces of rams and 3 fleeces of ewes.

1218. OTTO, JOHANN, ALBRECHT, THEODOR (J.F.Otto), manu. of paraffine-candles and soaps, Frankfurt on the Oder. London 1862 medal. Agts. s. No. 1211.

Paraffine-candles, 1a. quality 42 Th. (6 £ 8 sh.); peat-paraffine 35 Th. (5 £ 6 sh. 6 d.); turf-paraffine 30 Th. (4 £ 11 sh. 6 d.); tallow-soap 14½ Th. (2 £ 4 sh. 3 d.); olive-oil soap 16½ Th. (2 £ 10 sh. 3 d.); cocoa-nut oil-soap 16 Th. (2 £ 8 sh. 9 d.); palm-oil soap 14 Th. (2 £ 2 sh. 9 d.); rape-seed oil-soap 12 Th. (1 £ 16 sh. 9 d.); oleïne-soap 10 Th. (1 £ 10 sh. 6 d.), all p. Zollz. (cwt.).

Bleached rape-seed oil, palm-oil, cocoa-nut oil; market-prices with the addition of 6 to 9 sh. for bleaching.

VON PANNWITZ, W., Burgsdorf. Flax.
(s. No. 1521.)

1220. POHL, GEORGE, Instructor, Canth. Paris 1855 bronze medal. Agts. s. No. 1211.

Silesian teasels, p. Zollz. 1½ Th. (4 sh. 6 d.).

1221. POLLACK, L., manu., Liegnitz.

Rape-seed oil and rape-seed cake.

1222. POPPELSDORF, ROYAL AGRICULTURAL COLLEGE, by Hartmann, director, Poppelsdorf near Bonn. London 1862 honor. mention.
(s. No. 1122.)

Althaea rosea, mallow for dyeing.

1223. PRINGSHEIM, EMANUEL, landed prop., Ober-Schonau (Breslau). Agts. s. No. 1211.

4 wool-fleeces from the breeding-stock of Ober-Schönau near Oels.

1225. VOM RATH, H., grower, Lauersfort (Crefeld). (s. No. 1125.)

Oak-bark for tanning.

1226. DUKE OF RATIBOR, VICTOR, Rauden (Oppeln), Niedane near Ratibor. London 1862 medal Agts. s. No. 1211.

9 washed wool-fleeces and 120 samples of not washed wool.

1226a. VON REIBNITZ, J., breeder, Holzkirch.

Fleeces of wool.

VON REISWITZ, B., Wendrin. Flax.
(s. No. 1524.)

1227. RISTOW, C., farmer, Repkow (Cöslin). (s. No. 1558.)

Fleeces of wool.

1228. ROMER, CARL (Romer & Hackenberg), manu., Cologne. London 1851 and Paris 1855 honor. mention. Agts. Morgan Brothers, 21 Bow-Lane, Cannon-street, City.

Machine-oil 17 Th. (2 £ 11 sh 9 d.); spindle-oil 16 Th. (2 £ 8 sh. 9 d.) p. Zollz. (cwt.) each.

ROMPLER & TOLLE, manu., Barmen.

Articles of caoutchouc.

1229. VON RUDZINSKI-RUDNO, CARL, prop. of Liptin (Oppeln). Munich 1854 medal of honor; Paris 1854 silver medal; London 1862 medal. Agts. s. No. 1211.

12 fleeces of superfine electoral sheep: 6 ram-fleeces, 6 ewe-fleeces p. Zollz. 150 Th. (111,1 lb. 22½ £).

1231. SCHROEDER, THEODOR CARL & SIERMANN, LOUIS (H. Sarre & Co.), soap-manu., Berlin, Stralauerstr. 16. Agts. s. No. 1211.

Tallow-soap 14 Th. (2 £ 2 sh.); marbled soap 18 Th. (2 £ 14 sh.); mosaic-soap 15 Th. (2 £ 5 sh.), all p. Zollz.

1232. SARRE, HEINRICH, JUN., manu., Moabit near Berlin, Friedrichsstr. 135. Berlin 1844 bronze medal; London 1851 prize-medal; Munich 1854 honor. mention; Paris 1855 honor. mention; London 1862 medal. Agts. Schomburg & Co., 90 Cannon-street.

Stearine-candles and oleïne of waste greasy parts of rape and cotton-seed oil from oil-mills, of yolk and soap-suds from spinning-establishments and cloth-manufactories, of used clouts from machine-manufactories.

As secondary product of inodorous asphaltum for the manufacture of roofing-paper.

1234. COUNT SAURMA, EUGEN, by primogenitureship prop. of Ruppersdorf (Breslau). London 1862 medal. Agts. Lion M. Cohn, Phaland & Dietrich, represented by Ch. Trübner, 20 St. Dunstans Hill, City, and at the Exhib. Building.

4 wool-fleeces from the breeding-stock of Zülzendorf near Nimptsch.

1235. Scheidt, Victor, comb-maker, Cologne. Agts. s. No. 1234.

Four dressing-combs of a new-invented construction (patent), in 4 sorts: A. of ivory with teeth of tortoise-shell 48 Th. (7 £ 4 sh.); B. and C. of buffle-horn with teeth of ivory 22 Th. (3 £ 6 sh.); D. do. 18 Th. (2 £ 14 sh.); prices p. doz.

1237. Schindler & Mutzel, manu., Stettin. London 1862 medal. (s. No. 1021.)

Soaps, grease.

1238. Scholler, Leopold, privy counsellor of commerce and prop., Schwieben (Oppeln). London 1862 honor. mention. Agt. Charles Nolda & Co., 2 Church-Court, Old Jewry. (s. No. 1137.)

4 washed fleeces of wool 120 Th. (20 £) p. Zollz. (cwt.).

Scholler, Mevissen & Bücklers, Düren. Flax. (s. No. 1526 a.)

1239. Schram, Peter & Joseph, Brothers (P. J. Schram), starch-manu., Neuss (Düsseldorf) Munich 1854 medal of honor; Paris 1855 bronze medal.

Starch in the different stages of manufacture; dressing-starch; starch in sticks.

1240. Schultze, J. C., manu., Berlin, Charlottenstr 82. (s. No. 1023 b.)

Schulze, Fr. A., Berlin. Key-hole lids of bone. (s. No. 2008.)

1241. Siegen association for culture and industry, Siegen. Agts. s. No. 1234. (s. No. 1369.)

1 bundle of oak-bark for tanning.

1243. Thaer, A. P., counsellor of agriculture and prop., Möglin near Wrietzen on the Oder (Berlin). London 1851 honor. mention; Paris 1855 silver medal and cross of the legion of honor; London 1862 medal.

Wool-fleeces.

1244. Vilter, W., manu., Berlin, kleine Waldemarstr. 4. Agts. s. No. 1234. (s. No. 1032.)

Oil for lubricating machines 20 Th. (3 £) p. Zollz.

1245. Vogt, August Gottfried, manu. of horn-articles, Mühlhausen (Erfurt). Agts. s. No. 1234.

14 pair of weighing scales, prices p. pair: a. oval scales for the weight of 1 lb. 3 sh.; of ½ lb. 2 sh. 6 d.; of ¼ lb. 1 sh. 9½ d.; of $\frac{1}{10}$ lb. 1 sh. 3½ d.; of $\frac{1}{15}$ lb. 11 d.; of $\frac{1}{30}$ lb. 7 d.; of ¼ oz. 6 d. b. round scales, diameter of 6 in. 1 sh. 9½ d., of 5 in. 1 sh. 5 d., of 4 in. 11 d., of 3 in. 7¾ d., of 2 in. 5 d., of 1 in. 2½ d., of 2¾ in. (with spouts) 7¾ d.;

A shaving-basin 1 sh. 6 d.

4 Knives 1 sh. 9½ d., 2 sh. 1¼ d., 2 sh. 1¼ d., 6 sh.; 5 pair of knives and forks 5 sh., 9 sh., 7 sh., 6 sh., 9 sh.; all p. doz.

Two spoons joined like a pair of scissors p. piece 4 sh.; a spoon and fork joined like a pair of scissors p. pair 2 sh. 9 d.

5 spoons with spatula; prices p. doz.: length of 9½ in. 6 sh. 7 d., of 8½ in. 5 sh., of 6½ in. 3 sh. 9½ d., of 5½ in. 3 sh., of 4½ in. 2 sh. 2½ d.; 6 spoons with pointed handles: length of 9½ in. 5 sh. 9½ d., of 8 in. 4 sh., of 7 in. 3 sh. 2½ d., of 6½ in. 2 sh. 9½ d., of 3½ in. 1 sh. 3½ d., of 3 in. 1 sh. 2½ d.; 5 spoons with straight handles: length of 9 in. 4 sh. 9½ d., of 6 in. 2 sh. 3½ d., of 5½ in. 1 sh. 11 d., of 4½ in. 1 sh. 5 d., of 3½ in. 1 sh. 2½ d.; 6 spoons for snuff or spices: length of 6½ in. 4 sh. 6 d., of 6 in. 3 sh. 9½ d., of 5½ in. 3 sh. 3½ d., of 4 in. 2 sh., of 3½ in. 1 sh. 9½ d., 3 in. 1 sh. 7¼ d.; a mustard-spoon 2 sh.; a taring-plate 6 in. diameter 1 sh. 9½ d.; 4 plates: diameter of 6 in. 1 sh., of 5 in. 9½ d., of 3¾ in. 6 d., of 3 in. 4½ d.

A sieve for pills, p. piece 1 sh. 2½ d.

4 round cases, diameter of 6 in. 3 sh. 7¼ d., of 5 in. 2 sh. 8½ d., of 4 in. 2 sh. 1¼ d., 3 in. 1 sh. 7¼ d.; 4 open cases of 6 in. 3 sh. 7¼ d., of 5 in. 2 sh. 8½ d., of 4 in. 2 sh. 1¼ d., of 3 in. 1 sh. 7¼ d.; 8 spatulas: length of 10 in. 5 sh. 6 d., of 9 in. 4 sh. 7¼ d., of 8 in. 3 sh. 9½ d., of 7 in. 3 sh., of 6 in. 2 sh. 5 d., of 5 in. 1 sh. 9½ d., of 4 in. 1 sh. 2½ d., of 3 in. 1 sh. Prices p. doz.

A machine for making pills, p. piece 7 sh.

1246. Wächter, Johann, Tilsit. Agts. G. H. & J. F. Wulff, Mincing-Lane.

Two linseed-cakes 70 sgr. (7 £); two rapeseed-cakes 50 sgr. (5 £); 1 bottle containing raw linseed-oil 12 Th. (36 £); 1 bottle containing linseed-varnish 13⅓ Th (40 £); 1 bottle containing refined rapeseed-oil 15 Th. (45 £), prices p Zollz. (p. ton) free on board Tilsit. 1 case containing soaps and perfumes.

Wagener, Proskau. Flax. (s. No. 1530.)

1247. Wehowski, Joseph, farmer, Graase (Oppeln).

Washed wool-fleeces; 100 samples of not washed wool.

1248. Weitze, Dr. C. G., manu., Stettin. (s. No. 1036.)

Oil; soap.

Wellhauser, E. L., Elberfeld. Articles of caoutchouc. (s. No. 2201.)

1248 a. Wellmann, Albert, merchant, Stettin, flour-mill at Dittersbach. Agts. de Wolff, Schuck & Co., 7 Idol-Lane, City.

Starch-meal 5⅝ Th. (17 £ 15 sh.); starch 5¾ Th. (17 £ 10 sh.); p. Zollz. (p. ton) casks incl.

The same, flour-mill at Barwalde.

Wheat-flour 00. 6½ Th. (19 £ 10 sh.): do 0. 6 Th. (18 £); do 1. 5½ Th. (16 £ 10 sh.); do. 2. 5 Th. (15 £); rye-meal 0 4⅓ Th. (13 £); rye-meal 1. 3¾ Th. (11 £ 5 sh.); bags incl.; wheat-bran 1¼ Th. (3 £ 17 sh. 6 d.); rye-bran 1$\frac{5}{12}$ Th. (5 £ 15 sh.); rye-bran 1$\frac{7}{12}$ Th. (4 £ 17 sh. 6 d.); bags excl. prices p. Zollz. (p. ton), free on board Stettin.

1249. von Wiedebach, Friedrich, landed prop. of Beitzsch near Jesnitz (Frankfort on the Oder). London 1862 honor. mention. Agts. s. No. 1234.

6 unwashed wool-fleeces from the exhibitor's sheep-flock 66⅔ Th. (10 £).

Willmann, A. & Sons. Patschkey. Flax. (s. No. 1531.)

1250. Wunder, L., soap-manu. and purveyor to H. M the King of Prussia, Liegnitz. London 1851 prize-medal; Paris 1855 bronze medal; London 1862 medal. Agt J Chr. Rein, 108 Strand.

Household-soap, of tallow of the exhibitor's own melting and Hungarian hogslard 15 Th. (2 £ 6 sh.); fuller's soap, made of the soap-suds and fat-refuse from the cloth-manufactories 10 Th. (1 £ 11 sh.); p. Zollz. (cwt.) each; hereto a sample of this fat from the cloth-manufactories.

2 sapophanies (transparent-soap) representing both the sides of the prize-medal of the great exhibition in London 1851, framed in blocks of cocoa-soap in the English and Prussian national colors p. piece 10 Th. (1 £ 10 sh.)

Wunder's patent shaving-soap in bars usual bars 3 Th. (9 sh.); 4 monster-bars hereto 36 Th. (5 £ 8 d.) p. doz. each one block of this shaving-soap, ou which the firm of the manufacturer is engraved and filled with white cocoa-soap; prices from Liegnitz.

1251. Zecher, August, manu. of horn-articles, Mühlhausen (Erfurt). London 1862 medal. Agts. Lion M. Cohn, Phaland & Dietrich, represented by Ch. Trübner, 20 St. Dunstans Hill, City, and at the Exhib. Building.

A pattern-card of apothecaries' utensils of horn and bone, and some other articles; as: 3 pill-machines, 1. 2. 3.; pillsieves and plates; silvering-boxes for pills; round and oval scales; scales with a spout; powder-boxes, open on one and two sides· spatulas of horn and bone; small plates and spatulas for the cleaning of mortars; spoons with spatulas and pointed handles double-spoons; spoons marked: poison; spoons of bone; a cup of bone for taring; a pair of tongues to be used in poisonous liquids. Some salad-scissors; salad-spoon and fork; spice-spoon; salt- and mustard spoon; fruit-knife; bon-bonnières; color-spatulas; shoeing horns.

CLASS V.

RAILWAY PLANT, INCLUDING LOCOMOTIVE ENGINES AND CARRIAGES.

1252. Berlin, Joint-stock company for the manufacture of railway requisites (Actien-Gesellschaft für Fabrikation von Eisenbahn-bedarf). London 1862 medal.

Railway passenger-carriages I. and II. class, do. III. and IV. class.

1253. Bochum company for mining and cast-steel manufacturing, Bochum (Arnsberg). Paris 1855 great medal of honor; London 1862 medal. Agts. Heintzmann & Rochussen, 9 Friday-street, E. C.

2 cast-steel tires, 9¾ ft. diameter; cast-steel tire, bent cold; a set of railway carriage disk-wheels, entirely polished, axle and both the wheels of cast-steel; an other one, ordinary workmanship; cast-steel driving-wheel of a locomotive engine, 5 ft. diameter; cast-steel bell, weight 11 tons.

1254. Borsig A., manu., Berlin. Berlin 1844 golden medal; Paris 1855 great medal of honor; London 1862 medal.

Locomotive engine with tender.

1255. Cologne machine making com-

pany, Cologne. Manufactory at Bayenthal near Cologne. Agts. s. No. 1251. (s. 1294, 2029.)

Part of a railway-crossing; ten samples of the same. (Frogs.)

1258. Hoerde, society of mines and iron works, Hoerde near Dortmund. Berlin 1844 golden medal; London 1851 prize-medal; Paris 1855 great medal of honor; London 1862 honor. mention. Agts. Heintzmann & Rochussen, 9 Friday-street, E. C.

Disc-wheel of wrought-iron for railway-waggons forged with the rim in one piece; disc-wheel of wrought-iron for railway-waggons, forged with the rim in one piece and with a turned puddled-steel tire; disc-wheel of wrought-iron for railway-waggons, rolled with the rim in one piece; disc-wheel of wrought-iron for railway-waggons with a peculiar fastening of tire; centred locomotive-tire of puddled steel; telegraph-post of wrought-iron for electro-magnetic telegraph-wires; plates for boiler-bottoms rolled with the rim in one piece.

1258a. Krupp, Fr., Essen. London 1862 medal. (s. No. 1308.)

Wheels, axles, springs for railway-carriages, tires.

1259. Lehrkind, Falkenroth & Co., Haspe (Arnsberg). London 1851 prize-medal; Paris 1855 silver medal; London 1862 medal. Agts. s. No. 1251.

Set of wrought-iron railway-wheels with axle-tree and tires of puddled-steel; centred tire of puddled-steel; case with samples of broken rails, switch-rails, axle-trees and tires of puddled steel, as well as of the raw material they are made of, partly by means of an improved process with a refine l grain.

1261. Ruffer, G. H., privy coun e lor of commerce and manu., Breslau. Iron-worl s Piala-hütte near Rudzinitz (Oppeln) London 1862 honor. mention. (s. No. 827.)

Two tender-tires; two driving-wheel tires; carriage axle-tree; an axle-tree; a boiler-plate, all of fine-grained iron; fracture-proofs of iron; a cast-iron railway carriage-wheel; a half one do. for showing the fracture.

Schulz, Knaudt & Co. Parts of locomotive engines. (s. No. 1318.)

1261a. Schwartzkopff, Berlin, Chaussée-str. 20. (s. No. 1282, 1319.)

Railway carriage-wheels, case-hardened casting.

1262. Sieg-Rhine mining and smelting joint-stock company, society for mining and the management of iron-furnaces, Cologne on the Rhine. Iron works Friedrich-Wilhelms-Hütte at Siegburg near Cologne. Agts. Morgan Brothers, 21 Bow Lane, Cannon-street, City. (s. 722, 1320.)

Railway carriage-wheels, case-hardened casting; tram-road truck for the conveyance of iron-ores etc., 100 Zollz. load; tram-road waggon for 100 Zollz. load.

CLASS VI.

CARRIAGES NOT CONNECTED WITH RAIL OR TRAM ROADS.

1264. Karwiese, Gust. Ad. (A. G. Karwiese), manu., Graudenz. Agts. s. No. 1251.

A lady's phaeton of plate-glass panes; the frame consisting entirely of iron-panes composed of three single parts of wrought iron.

1265. Neuss, Jos., carriage-manu. to His Majesty the King of Prussia, Berlin. London 1862 medal. Agt. Jos. Neuss Son, Basinghall-street 22.

State-coach for His Majesty the King; coupé d'Orsay, smallest size; double caleche.

1267. Schran, Franz, manu., Gleidorf near Schmallenberg (Arnsberg). Agts. Lion M. Cohn, Phaland & Dietrich, represented by Ch. Trübner, 20 St. Dunstans Hill, City, and at the Exhib. Building.

Refined charcoal-iron 38 Th. (11 £ 8 sh.); axle-tree with stamped arms 48 Th. (14 £ 8 sh.); axle-tree for a heavy waggon, finished 62 Th. (18 £ 12 sh.); axle-tree for a phaeton 84 Th. (25 £ 4 sh.), all p. 1000 lb. (p. 20 cwt.)

CLASS VII.

MANUFACTURING MACHINES AND TOOLS.

Bialon, J., Berlin.   (s. No. 1286.)

1269. Braun, Johann Michael, manu. of cards, Düren near Aachen. Paris 1855 honor. mention. Agts. s. No. 1267.

Samples of cards with felt, invented by the exhibitor 1844; 2. samples of cards of the exhibitor's invention, working in a straight line; 3. cards, having been used for 10 years.

Gaertner, Theyson & Ede, Borgholz-hausen. Patent cork-cutting machine.
(s. No. 1299.)

1271. Hamann, August, machine-manu., Berlin. Berlin 1844 golden medal; London 1851 prize-medal.

A small iron turning-lath, 4 ft. long, height of the centres 6 in., 45 £.

1274. Kamp, Johann, reed-manu., Berlin, Krautsstr. 56.

Sleys, prices p. 100 reeds: for woollen fabrics 5⅖ d.; for cotton fabrics 4⅕ d.; for silk fabrics, 2 pieces 5⅖ d.

1276. Kühnen, Fr., cardmaker, Wesel. Agts. s. No. 1267.

Assortment of cards for carding cotton and wool as well as for dressing cotton and woollen cloth.

1278. Müller, Julius, manu. of wire-gauze, Berlin, Mühlenstr. 59. London 1862 honor. mention. Agts. s. No. 1267.

4 pieces of wire-gauze for paper-machines: laid first pressing-roller for paper-manufactories 85 Th. (13 £); sieve-roller with different water-marks for paper manufactories; pattern of deckle-straps for paper-machines.

1279. Offenhammer, Ferd., merchant and manu., Berlin.

Show-card with needle-files 1 sh.; knife-files 3 sh. 2 to 6 d.; gravers 1 sh. 8 d.; square gravers 1 sh. 5 to 10 d.; broaches 1 sh. to 3 sh. 5 d.; smooth crossing-file 3 sh. 6 d.; steel-collars 12 sh.; steel wristbands 9 sh.; prices p. doz.

1280. Sauer, Fr., Lennep. Agts. Charles Nolda & Co., 2 Church Court.

Tools for cloth-weaving: a leaf with steel-wire heddles; a double leaf do.; 2 sleys with steel-reeds; a steel-wraith for dressing-machines. For prices apply to the agents.

1281. Schulze, Wm., Vellinghausen & Co., manu., Stockum near Witten (Arnsberg). Agt. Heintzmann & Rochussen, 9 Friday-street, Cheapside.

A case-hardened roller, hard only to 1 in. of depth, p. lb. 3 sg. (3½ d.).

1282. Schwartzkopff, L., manu., Berlin, Chausséestr. 20. London 1862 honor. mention. (s. No. 1261a., 1319.)

Patent steam-hammer.

1283. Sigl, Georg (G. Sigl), machine-manu., Berlin. Berlin 1844 bronze medal; Munich 1854 medal of honor; Paris 1855 bronze medal; London 1862 medal. Agts. s. No. 1267.
(s. II. No. 46.)

A litographic printing-press 2000 Th. (300 £).

Thomas, H., Berlin, machines for shearing and frizzling.   (s. No. 1322.)

1284. Zeyland, J., joiner, Posen. Agts. s. No. 1267.   (s. No. 1376.)

Three planes with double irons and without wedges, each 1 Th. 25 sg.; four planes with single irons and without wedges, each 1 Th. 17½ sg.

CLASS VIII.

MACHINERY IN GENERAL.

1286. Bialon, J. (C. Hummel), engineer and prop. of the machine-works under the firm: C. Hummel, Berlin. Paris 1855 juror; London 1862 medal. Agts. s. No. 1267.

Steam-drying cylinder-machine; cylinder printing-machine for four colors; block-printing machine for two colors; double pantograph-machine for cotton-printers; calender.

1287. Bleyenheuft, J. H., tanner, Aachen. Munich 1854 medal of honor; London 1862 honor. mention. Agts. s. No. 1267.

Driving belts: 2 in. after the French method 16⅔ Th. (2 £ 10 sh.); 3 in. after the German method 27½ Th. (4 £ 2 sh. 5 d.); 1 double strap, 3 in. after the English method 55 Th. (8 £ 5 sh.); 1 rivetted strap 2 in., own method of the exhibitor 16⅔ Th. (2 £ 10 sh.); 1 rivetted strap 4 in. do. 36⅔ Th. (5 £ 10 sh.); 1 double strap of 6 in. improved by the exhibitor 140 Th. (21 £); 1 double strap of 3 in. with invisible joints 55 Th. (8 £ 5 sh.); all p. 100 ft. (97¼ ft.).

Improved leather-ropes: fourfold 33⅓ Th. (5 £); twofold 23⅓ Th. (3 £ 10 sh.). Leather-cords: 1 in. 10⅓ Th. (1 £ 10 sh.); ⅝ in. 6⅔ Th. (1 £); ⅜ in. 5 Th. (7 sh. 5 d.); ½ in. 4 Th. (6 sh.); ¼ in. 3¼ Th. (5 sh.); ⅓ in. up to the thinnest 2¼ Th. (3 sh. 3¾ d.); all p. 100 ft. (97¼ ft.); 1 rivetted fire-bucket, p. piece 2 Th. (6 sh.).

1287a. Bleyenheuft-Milliard, M. F., manu., Eupen. London 1862 medal.

Machine-straps; fire-buckets; hose for fire-engine; machine-strap leather.

1288. Boecke, J., manu. of sewing and knitting-machines, Berlin. London 1862 honor. mention. Agts. s. No. 1267.

Sewing-machines: shuttle-machine for heavy fabrics 90 Th.; do. for fine fabrics 80 Th.; shuttle-machine for shoe-makers (system Thomas) 90 Th.; tambour-work machine with one thread 45 Th.; machine of Wheeler-Wilson for shirts 80 Th.; do. 60 Th.; do. for heavy fabrics (cloth, leather) 70 Th.; machine for house-hold use 18 Th.; for gloves and cotton fabrics 60 Th.; shuttle-machine, improved Groove and Baker-system 75 Th.; cord-stitch machine, Groove & Baker, 60 Th; cord-stitch machine 60 Th.

1290. BROSOWSKY, W. A., manu., Jasenitz near Stettin.

Machine for cutting peat, working to a depth of 10 ft. 144 Th. (21 £ 15 sh.).

1291. BUCHNER, C. W., copper-smith, Pyritz, Pomerania. Agts. Lion M. Cohn, Phaland & Dietrich, represented by Ch. Trübner, 20 St. Dunstans Hill, City, and at the Exhib. Building.

Distilling apparatus.

1292. CADURA, HEINRICH, merchant and manu. of driving-belts, Breslau. Agts. s. No. 1291.
(s. No. 1503, 1807, 1938.)

Leather driving-belts for machines.

1293. CAHEN, A., LEUDESDORFF & Co. (Rhenish manufactory of leather for machines and straps), Mülheim on the Rhine. Agts. s. No. 1291.

Double strap, 30½ ft. long, 4 in. broad, 3 sh. 2 d.; single strap, 21 ft. long, 6 in. broad, 2 sh. 5 d.; prices p. ft.

1294. COLOGNE MACHINE MAKING COMPANY, Cologne, manu. Bayenthal near Cologne. Agts. s. No. 1291.   (s. No. 1255, 2029.)

Transportable steam-engine; steam-pump; 5 water slide-valves; five taps with cast-iron cases; six taps with brass-cases.

1295. DREWITZ, EDUARD, manu. of machines, Thorn. Paris 1855 silver medal; London 1862 honor. mention. Agts. s. No. 1291.

Apparatus to be used in distillation for measuring the quality and quantity of spirit.

1297. EGELLS, F. A., manu., Berlin, Chausséestr. 2. London 1862 medal.

Woolf's steam-engine of 30 horse power.

1298. FRAUDE, W. O., & Co., manu., Berlin, Auguststr. 68. London 1862 medal and honor. mention. Agts. s. No. 1291.

I. Soda-water apparatus with pump, constructed after the combined system of Struve, Soltmann and Brahma with a drawing-off screw-valve and improved machine for corking bottles 75 £; without corking-machine 70 £.

II. Apparatus for boiling, digesting, and distilling with high-pressure steam, with double cylindre, cooling apparatus after the system of Mitscherlich, with all the vessels appertaining to the apparatus 55 £.

1299. GÄRTNER, THEYSON & EDE, machine-manu., Borgholzhausen (Minden) and Hannover. London 1862 honor. mention. Agt. Gärtner.

Patent cork-cutting machine 1000 Th. (150 £).

1301. GRESSLER, JULIUS, & BUSCH, ARNOLD (J. Gressler & Co.), Berlin, Königsstr. 34. London 1862 honor. mention. Agts. s. No. 1291.

2 machines for preparing aërated liquids; prices pr. piece. Self preparing apparatus for 25 quarts contents of the condensing vessel 200 Th. (30 £); for 60 quarts contents 336 Th. (50 £ 8 sh.); pump-apparatus for 20 quarts contents of the condensing vessel 400 Th. (60 £).

1 corking-machine belonging to the above apparatus 34 Th. (5 £ 2 sh.).

1303. HECKMANN, C. (Berlin copper and brass-works), merchant and manu., Berlin, Schlesische Str. 18 and 19. Berlin 1844 order of the red eagle IV. class; London 1851 council medal; Paris 1855 bronze medal; London 1862 medal. Agts. s. No. 1291.

A vacuum-apparatus for refining sugar 10 ft. diam, and 12 ft. high with fittings complete 1150 £. package included.

HEMPEL, O. M., manu., Berlin, manometer. (s. No. 1404.)

1307. KROPFF, OSCAR, & Co., manu. of mechanic apparatus, Nordhausen. Paris 1855 honor. mention. Agts. s. No. 1291.

1. A model of a mash-cooling apparatus as used for distillation 3 to 400 Th. (50 £). 2. A model of a heating-apparatus for smiths' forges 14 Th. (2 £). 3. A safety-manometer for beer-casks 1 Th. (3 sh.).

1308. KRUPP, ALFRED (Fried. Krupp), cast-steel works, Essen. Berlin 1844 golden medal; London 1851 council medal; Munich 1854 great medal; Paris 1855 great medal of honor; London 1862 medal. Agt. Alfred Longsdon, 9 Chepstow place, Camberwell New Road.
(s. II. No. 13.)

Cast-steel manufactures:

Propeller-shaft with two cranks, ordered by the Bremer Lloyd, weighing in the finished state, 9 tons, 2400 £; paddle-wheel shaft with one crank in the forged state, 16 tons, 2250 £; propeller-screw, 10 cwt., 112 £; anchor.

2 finished and 1 forged locomotive crank-shaft, 2½ tons, 750 £; 2 sets of wheels and axles for railway-carriages, and one for American street-railways, 1¾ tons, 112 £; 6 railway-carriage and locomotive-engine springs, 7½ cwt., 26 £; locomotive-engine crankpin, 2¼ cwt., 37 £; 24 patent tires without welding, of different sizes; one of them measuring 8 ft. in diam. and being highly polished, 7¾ tons, 670 £.

3 pairs of hardened and highly polished rolls, one of them being 16 in. long and 10 in. diameter 645 £.

6 pieces of ordnance of various bores, the largest 9 in. diam. and weighing when finished 8 tons, (some of them are completely finished, the others are only finished in the bore), 20 tons, 5625 £; 1 gun, cut in two parts in its longitudinal direction, to show the fracture; 1 gun, split in its longitudinal direction, to show the tenacity of the metal. Gun-barrels. 6 cylinders with differently shaped riflings in them.

Pump-rod for mines, 30 ft. long, with coupling, 1¼ tons, 150 £.

Block of cast-steel, cast and broken into two parts to show the fracture, 20 tons, 1200 £; forged bar, broken into 3 parts to show the fracture, 12 tons, 840 £; block of cast-steel, partly in the cast and partly in the forged state, with fractures at both ends, 4 tons, 140 £; different samples of fractured and bent cast-steel.

Values of every position are approximative.

LIEBERMANN & MESTERN.   (s. No. 1321.)

1311. LORENZ, H., & VETTE, Th. (Plastic-coal Factory), manu., Berlin, Engel-Ufer 15. Agts. Charles Egan, 9 Tower Hill, East.
(s. II. No. 14.)

Filtering-machines, called reservoir filtering-balls, in the shape of porous hollow charcoal-balls of 4,

6 and 8 in. diam., used by means of a tin-siphon or of an India-rubber hose as siphon for rendering bad water drinkable (prices p. piece): 4 in. with 2 ft. hose in japanned box for travelling 5 sh.; 4 in. with 2 ft. and pewter-cock 4 sh. 7 d.; 4 in. in canel-netting with pewter-siphon 5 sh. 6 d.; 6 in. with 4 ft. hose and pewter-cock 8 sh. 7 d.; 6 in. in cane-netting with pewter-siphon 12 sh.; 8 in. with 4 ft. hose and pewter-cock 11 sh. 7 sh.; 8 in. in cane-netting with pewter-siphon 15 sh. 6 d.

1312. UNITED STEAM NAVIGATION CO., manufactory at Buckau near Magdeburg. London 1862 medal.
30 horse power horizontal steam-engine with Corliss, self adjusting expansion-gear 3000 Th. (428 £); 1 centrifugal-machine for sugar-manufactories 400 Th. (57 £).

1314. PHILIPPSON, F. C., & Co., engineers, Berlin. Agts. Lev. Jametel & Co., Brazennose-street, Manchester.
3 iron valves with oval flanges, diam. ½, 1 and 1½ in.; 3 brass valve with oblong flanges, diam. ½, ¾ and 1 in.; 2 do. with screws and nuts, diam. ½ and 1 in.; mercury vacuum-gauge; double lever spring-balance for locomotive-engines; spring-balance for portable steam-engines and common boilers; alarum for preventing too high or too low a water-level in steam boilers; 3 water-indicators, with guard for the glass; do. for transportable engines; water-indicators, without guard; boiler-cock; do. smaller size; lubricator; valve grease-box; steam-whistle for transportable engines; injector; 7 patent steam-gauges in brass of divers pressures and fastenings; 6 patent steam-gauges in iron of divers pressure and fastening; patent steam gauge in brass, small size; condensing box; lower part of mercury steam-gauge.

1315. RULAND, WILHELM, manu. of strap-leather. London 1862 honor. mention. Agts. Morgan Brothers, 21 Bow-Lane, Cannon-street, City.
3 pieces of leather for straps.

1316. SCHÄFFER, BERNARD, BUDENBERG, CHRIST. FRIEDRICH (Schäffer & Budenberg), Buckau-Magdeburg. Branches: Manchester, 92 George-street; New-York 37 Chatam-street. Paris 1855 bronze medal; London 1862 honor. mention.
Pat. steam-pressure-gauges: No. 3. 60 lb. pressure, No. 36,750. No. 2. 100 lb. pr., No. 36,754; No. 1. 100 lb. pr., No. 36,753; No. 1. 300 lb. pr., No. 33,969; 4 in. dial, 100 lb. pr., No. 36,752; 3 in. dial, 100 lb. pr., No. 36,751. Pat. vacuum-gauge, No. 2., 30 in. engl., No. 36,755; do. No. 1., 30 in. engl., No. 36,756. Pat. hydraulic gauge No. 2., 200 atmospheres, No. 30,837. Pat. testing-gauge, 2 in. dial, 180 lb. pr., No. 24,805; mercurial gauge, 2 atmospheres, No. 3362; engine-counter. running up to 10,000 No. 1662; do., running up to 100,000 No. 1776; do., running up to 1,000,000 No. 1713; do., running up to 10,000 No. 1660. for centrifugal machines; do., running up to 1,000,000 No. 1236 (arranged to be replaced again to zero); engine-counter, with visible work running up to 1,000,000; water-gauge No. 1808; pat. water-gauge with alarum; Giffard's patent self-acting water-injector, affords 150 pounds water p. minute No. 355; do., affords 100 pds. water p. min. No. 357; do., affords 45 pds. water p. min. No. 327; do., affords 25 pds. water p. min. No. 345; do., affords 15 pds. water p. min. No. 422; pat. safety-valve 4 in. dial; 2 selfacting apparatus for removing the condensed water from the steam-cylinder; conduit-pipe for the condensed water; level with iron-plate 12 in. length No. 576; do., 6 in. lth., No. 569; double-

level do., 18 in. lth., No. 600; 3 grease-cups for steam-cylinder; 12 grease-boxes of divers construction; dynamometer, 20 tons, No. 112; signal-apparatus for mines; counter-apparatus with watch attached; tableau: consisting of allegorical figures, patent steam-pressure gauges, patent vacuum-gauge, engine-counter and clock.

1317. SCHLICKEYSEN, C., manu., Berlin. The exhibitor himself present. London 1862 medal. (s. II. No. 3.)
Patent combined brick, tile, pipe and peat (turf)-machine with complete transmission and elevator for bricks and drying platform, forming from 15 to 20 thousand bricks a day. With horse- or steam-power from 30 to 500 £.

1318. SCHULZ, KNAUDT & Co., prop. of puddling-furnaces and rolling-mills, Essen. London 1862 honor. mention.
Iron back-plate for the fire-box of a locomotive engine with 7 in. inside and outside flanges; locomotive-engine dome, 24 in. diam., 14 curved in., with 6 in. flanges; welded boiler tube, 3 ft. high, 2 ft. diam., with 6″ flanges; proofs of broken and bent pieces of the iron-plates used for the manufacture of the above named objects.

1319. SCHWARTZKOPFF, L., manu., Berlin, Chausséestr. 20. (s. No. 1261a, 1282.)
Caloric engine.

1320. SIEG-RHINE MINING AND SMELTING JOINT-STOCK COMPANY, society for mining and the management of iron-furnaces, Cologne. Foundry Friedrich-Wilhelmshütte at Siegburg near Cologne. London 1862 medal. Agts. Morgan Brothers, 21 Bow-Lane, Cannon-street, City. (s. No. 722, 1262.)
A smoke-consuming and fuel-saving grate for steam-boiler furnaces, patented to Eugen Langen.

· 1321. LIEBERMANN, B., & MESTERN, A. (machine manufactory and iron works Wilhelmshütte), machine-manufactory Wilhelmshütte near Sprottau (Liegnitz). London 1851 honor. mention; Munich 1854 medal of honor; London 1862 medal. Agts. Semenza, Mazini & Comber.
A horizontal high-pressure steam-engine of 20 horse power after Corliss' system with self-adjusting expansion-valve.

1322. THOMAS, H. (engineer and iron-founder), Berlin, Grabenstr. 31. London 1851 prize-medal; Paris 1855 bronze medal; London 1862 juror for Cl. VIII. (s. II. No. 86.)
Longitudinal shearing-machine with two shearing-cylinders; machine for frizzling woollen goods in five different directions.

1323. UHLHORN, HEINRICH (D. Uhlhorn), machine-maker. Grevenbroich. London 1851 council-medal; Munich 1854 prize-medal; London 1862 medal. Agts. Morgan Brothers, 21 Bow Lane, Cannon-street, City.
A large coining-machine for stamping medals in the Exhibition Building 576 £.

1324. WAGNER, F. G., JUN., mechanician to His Majesty the King of Prussia, Berlin, Mauerstr. 72. Berlin 1844 silver medal; Paris 1855 bronze medal; London 1862 medal.
Relievo-machine for copying medals and surfaces in relief-lines. It copies in the size of the original, it also magnifies and reduces the copies, 500 Th. (75 £); panto- or micrograph, machine for engraving

copies of initials, letters and ornaments in reduced size 1000 Th. (150 £); number-printing machine applicable both to the hand- and engine-press, for numbering banknotes, shares etc. 500 Th. (75 £); printing proofs of the relievo and the guilloshee-machines constructed by the exhibitor.

1325. WATREMEZ & KLOTH, manu., Aix-la-Chapelle. London 1862 honor. mention. Agt. Leopold Stiebel, Brook's Wharf, Upper Thames-street, E. C.

3 patent safety apparatus for steam-boilers; improved by Black.

WILHELMSHÜTTE.   (s. No. 1321.)

CLASS IX.

AGRICULTURAL AND HORTICULTURAL MACHINES AND IMPLEMENTS.

1329. CEGIELSKY, H., manu., Posen. London 1862 honor. mention.

Ploughs; thrashing machine; straw-cutter; sowing machine; roller.

1330. ECKERT, H. F., manu., Berlin, Kleine Frankfurterstr. 1. London 1862 medal. Agt. J. G. Dalke (Morgan Brothers), 21 Bow-Lane, Cannon-street.

Self-sharpening Ruchadlo swingle-plough drawn by two horses (Eckert's patent) 12¼ Th. (1 £ 17 sh. 6 d.); the same, drawn by one horse only 11 Th. (1 Th. 13 sh.); the same, drawn by two horses with Eckert's patent ploughcart 19½ Th. (2 £ 18 sh. 6 d.); the same, with double-share and Eckert's patent ploughcart 21 Th. (3 £ 3 sh.); plough for clayland 16 Th. (2 £ 8 sh.); the same with Eckert's patent ploughcart 23 Th. (3 £. 9 sh.); plough for miners 10 Th. (1 £ 10 sh.); heaping-plough with movable ears 8 Th. (1 £ 4 sh.); plough for sowing with three shares and Eckert's patent plough-cart 20 Th. (3 £); improved Tennant-grubber 45 Th. (6 £ 15 sh.); shoveling and heaping-plough 10 Th. (1 £ 10 sh.); broad-sowing machine with paddle-wheels 80 Th. (12 £); thrashing-machine drawn by three horses with winch 300 Th. (45 £); Boston winnowing-machine 40 Th. (6 £); chain-pump for filthy fluids 17 to 30 Th. (2 £ 11 sh. to 4 £ 10 sh.): earth-bore, stick-shaped 3 Th. (9 sh.); machine for washing potatoes and removing stones from them 110 Th. (16 £ 10 sh.); earth-bore with a diameter of 3, 6 or 9 in. 3½ to 6 Th. (10 sh. 6 d. to 18 sh.).

1332. PINTUS, J., & Co., iron-founders and machinists, Brandenburg on the Havel near Berlin. Paris 1855 two bronze medals. J. PINTUS, London 1862 juror for Cl. IX. Agts. A. Heintz-mann & Rochussen, 9 Friday-street, Cheapside.

(s. No. 2082 a and II. No. 9.)

Subsoil-plough 15 Th. (2 £. 5 sh.); Bohemian chain-harrow 40 Th. (6 £); centrifugal sowing-machine 10 Th. (1 £ 10 sh.); new general purpose broadcast sowing-machine 85 Th. (13 £ 5 sh.); improved two horse mowing-machine including all extras 140 Th. (21 £); improved two horse combined mowing and reaping-machine 200 Th. (30 £); improved combined reaping and mowing-machine 250 Th. (35 £): small combined thrashing, shaking and riddling-machine for two horse power 300 Th. (46 £); two churns, a. 22 Th. (3 £ 6 sh.), b. 12 Th. (1 £ 10 sh.).

1335. TONNAR, A., manu., Eupen. London 1862 honor. mention.

Drying and winnowing-machine.

CLASS X.

CIVIL ENGINEERING, ARCHITECTURAL AND BUILDING CONTRIVANCES.

1336. ADLER, MARCUS, architect, Berlin. Agts. Lion M. Cohn, Phaland & Dietrich, represented by Ch. Trübner, 20 St. Dunstans Hill, City, and at the Exhib. Building.

An oval smoke-consumer of cast-iron, small size, with grate from 1 Th. upwards according to size.

1337. BARHEINE, RUDOLPH ROBERT (R. Barheine), purveyor to TT. MM. the King and the Queen-dowager of Prussia, manu. of marble-wares, Berlin. Paris 1855 honor. mention; London 1862 honor. mention.

Table of green marble (verde di Genoa) 120 Th. (18 £).

Monument of gray Silesian marble in gothic style 250 Th. (37 £ 10 sh.).

Cross-monument of gray Silesian marble in gothic style 175 Th. (26 £ 5 sh.).

Chimney piece of white Carrara marble 500 Th. (75 £).

1338. THE ROYAL PRUSSIAN MINISTRY OF TRADE AND PUBLIC WORKS, Berlin. Place of manufacture: Royal engine-factory at Dirschau. London 1862 medal.

I. A model of part of the great bridge over the river Vistula near Dirschau, in the Berlin-Königsberg railway.

This bridge is for the passage of railway-trains, carriages and passengers.

It contains two endpiers, five middlepiers and six openings. The whole length of the bridge is 2747 ft. 3 in. (2668 ft. Prus. meas.); each opening is 397 ft. 6 in. (386 ft. Prus. meas.) in the clear. The superstructure consists of 3 wrought iron girders, reaching each over two openings.

The model shows but one endpier, one middle-pier, and $\frac{7}{23}$ of the length of one of the girders. It is a true copy of the general construction of the bridge, and of the joinings of the different parts of the iron in particular.

The model is in $\frac{1}{12}$ size = 1 in. to the ft.

II. Model of one of the inclined planes in the Elbing-Oberland-canal.

The upper tract of the canal, which unites the lakes of the Prussian Oberland, communicates with the lower line of canal, conveying trough the river Elbing into the »Frische Haff«, by 4 inclined planes of a respective lifting height of 67 ft., 61 ft. 9 in. 80 ft. 3 in., 72 ft. = (65, 60, 78, 70 ft. Prus. meas.).

Propulsive motion is derived from a water-wheel of 27 ft. 9½ in. (27 ft. Prus. meas.) diameter and transmitted to a rope-barrel of 12 ft. 4 in. (12 ft. Prus. meas.) diameter by means of shafts, wheels and a reversing coupling-box.

The aforsaid water-wheel and the transmission-parts are not copied in the model; all other parts concerning the general shape of the incline, the construction of the mechanism and the carrying of the boats are shown.

By one passage of 15 minutes duration two boats of 242,500 lbs. (220,000 pounds Prus. weight) loading or 286,600 lbs. (260,000 pounds Prus. weight) gross weight each, are conveyed over the incline in opposite direction, the loss of time for carrying the boats in and out of the waggons being included.

The model is in $\frac{1}{32}$ size = $\frac{3}{8}$ in. to the ft.

1339. BONN MINING AND SMELTING COMPANY by Dr. Hermann Bleibtreu, Bonn. London 1862 med. Agts. Morgan Brothers, 21 Bow Lane, Cannon-street, City. (s. No. 674.)

Portland-cement of Bonn in an original cask and a bottle, 1 cask of 400 pounds gross w. or 375 pounds net 6 Dutch fl. free on board Rotterdam; a block of Portland-cement hardened under water, showing the crystalline structure of the Portland-cement of Bonn; a piece of house-door stairs, built in december 1858 of Portland-cement of Bonn; and since that time continually very much; used paving-plates and ornaments of Portland-cement of Bonn; (the ornaments are moulded by Jos. Hartzheim, modeller at Cologne).

1341. BÜSSCHER & HOFFMANN, manu., Neustadt-Eberswalde and Berlin. London 1862 medal. Agts. Lion M. Cohn, Phaland & Dietrich, represented by Ch. Trübner, 20 St. Dunstans Hill, City, and at the Exhib. Building.

1. Carton-pierre for flat and fireproof roofing 1 □ ft. 11 pf. (1 d.).

2. Asphaltum plates for covering vaults of tunnel, insulating houses etc. 1 □ ft. 2½ sg. (3 d.).

3. Gas and water-tubes of asphaltum-paper p. ft. (length of the building) 2 in. wide 4⅓ sg. (4⅓ d.); 3 in. wide 7 sg. (7 d.); 4 in. wide 10 sg. (1 sh.); 5 in. wide 14 sg. (1 sh. 4 d.); 6 in. wide 18⅔ sg. (1 sh. 8⅔ d.); 7 in. wide 23⅓ sg. (2 sh. 3⅓ d.); 8 in. wide 28 sg. (2 sh. 8 d.); 9 in. wide 33 sg. (3 sh. 3 d.); 10 in. wide 38 sg. (3 sh. 8 d.); 11 in. wide 43⅓ sg. (4 sh. 3⅓ d.); 12 in. wide 50 sg. (5 sh.).

4. Model and drawings of a circular kiln for burning bricks, tiles, limestone and other substances. (Hoffmann & Licht patent.) Price according to agreement.

1342. CARSTANJEN, JULIUS, manu. of asphalted objects principally for roofing, Duisburg on the Rhine. London 1862 medal. Agts. Eduard Rhodius, counting house of Killy, Traub & Co., 52 Bread-street.

A roll of asphalted roofing paste-board 3 ft. broad, 200 ft. length p. □ ft. 1 d.; a roll of asphalted roofing paste-board 28 in. broad, 14 ft. length p. piece 2 sh. 1 d.; a packet of asphalted roofing paste-board in sheets of 29 and 33 in. p. 100 pieces 1 £ 13 sh.; a model of a roof to show the manner of covering with asphalted roofing paste-board, as well in long rolls as in small sheets.

1343. FÖRSTER, FERD. SIEGF. (Förster's paper-manufactory Krampe near Grünberg). Agts. s. No. 1341.
(s. No. 1071, 1177, 1605, 1606, 1871a.)

Roofing-paper in sheets and rolls, p. □ Ruthe 3⅓ Th. Model of a roof variously covered and coated.

1343a. GÄNICKE, L., & SCHÖNDUVE, B. (L. Gänicke), manu., Wittenberge. Agts. s. No. 1341. (s. II. No. 21.)

1 roll asphaltum-paper for roofing.

1344. GALOPIN, JOS., founder of ornaments and sculp., Aix-la-Chapelle. Agts. s. No. 1341. (s. No. 2186.)

Chimney-piece of white marble 200 Th.; wash-hand-stand top 24 Th.

1345. GASSEL, RECKMANN & Co., manu., Bielefeld. Agt. H. Landwehr, Exhib. Building. (s. No. 975.)

4 rolls of felt (asphaltum-felt for roofing and sheathing walls and ships), each 35 yds. by 32 in. 1 £ 19 sh.; 9 sheets of felt, each 20 in. by 30 in. 1 sh.

1347. HAEUSSLER, MATHILDE, WIDOW (Carl Samuel Haeusler), Hirschberg. London 1862 honor. mention. Agt. s. No. 1341. (s. No. 1085.)

A model of a roof covered with »wood cement« invented by Charles Samuel Haeusler, which shows the roofing with this material in the beginning as well as finished. Instructions for carrying out this roofing with price-currents are annexed to the model. The fire-proof quality of this roofing is testified by the police.

1348. HAURWITZ, L., & Co., manu., Stettin.

Asphalt floor-slabs; asphalt-paper; roofing pasteboard.

1352a. KRZYZANOWSKI, A., manu., Posen. London 1862 honor. mention. Agt. s. No. 1343.

Sixtin madonna of moulded stone, 4½ ft. high 65 Th.; Bust of Lelewel of moulded stone, 1½ ft. high 4 Th.

1353. KULMIZ, CARL (C. Kulmiz), counsellor of commerce and manu., Ober-Streit near Striegau (Breslau). London 1862 medal. Agts. s. No. 1341.
(s. No. 771, 998, 1104, 2171.)

Granit capital of a pillar for the battle-monument near Liegnitz, ½ of its natural size.

A basis-stone for the ice-breakers of the Vistula-bridge near Warsaw; ⅙ of its natural size.

A granit step for a free standing stair-case.

A granit slab with polished surface.

A sepulchral-monument of polished Silesian granit 350 Th. (50 £).

1355. MEISSNER, WILH. (Pomeranian manufactory of tar-paper and carton-pierre), manu., Stargard (Pomerania). Agts. s. No. 1341.

Model of a stable-building roofed with asphalted carton-pierre. Rolls of fireproof sheets coated with stone-lac. Each roll 150 □ ft. Prus. 3 Th. or 154 □ ft. engl. 9 sh.

1356. MEWS, HERRMANN, architect prop. of a manufactory of artificial stones, Stettin.

A Corinthian column with socle and capital, crowned with an eagle of artificial cement-stone 53 Th. 10 sg. (8 £).

1357. MICHELI, BROTHERS, manu., Berlin, Jägerstr. 52. London 1862 medal.

Plaster of Paris and marble-work.

1358. MÖHRING, L. (the Berlin manufactory of asphaltic paste-board for roofing), manu., Berlin. Manufactory Moabit.

1 □ Ruthe asphalt paste-board for roofing 3 Th. (9 sh.); ½ Ctr. asphaltum lacker for roofing 1 Th. (3 sh.).

1359. POHL, H. (H. Pohl & Co.), Berlin, Alte Jakobsstr. 21. London 1862 medal. Agt. J. F. Chinnery, 67½ Lower Thames-street.
(s. No. 2083.)

6 Corinthian and Ionian capitals of raw cast zinc p. piece 26⅔ Th. (4 £).

1360. QUISTORP, merchant and consul, Stettin, owner of the »Pommersche Portland-Cementfabrik« at Lebbin. Agts. Lion M. Cohn, Phaland & Dietrich, represented by Ch. Trübner, 20 St. Dunstans Hill, City, and at the Exhib. Building.

1 pillar of cement, with eagle 54 Th. (8 £); 1 table of cement 7 Th. (1 £); samples of Pomeranian Portland-cement, p. cask of 400 lb. 3 Th. (9 sh.).

1362. SCHLESING, F., asphaltum-manu., Berlin. Paris 1855 honor. mention; London 1862 honor. mention.

Asphalt-goudron sec. 18 sh.; asphalt-cement for telegraphs 1 £ 13 sh. p. 100 lb.

1363. SCHOLZ, THEODOR (Johann Scholz & Son), manu., Klitschdorf and Siegersdorf (Liegnitz). Agts. s. No. 1360.

Fire and water-proof paste-board for roofing, 1 Zollz. (150 ☐ ft. rheinisch) 3⅓ Th. free Stettin.

1364. SCHOTTLER & Co., paper-manu., Lappin near Danzig. Agt. Ferd. Pickert, 79 Mark-Lane.

Paste-board, p. Zollz. 5 Th. (15 sh.); asphaltum paste-board, 1 ☐ Ruthe 4 Th. (12 sh.).

1365. SCHRÖDER & SCHMERBAUCH, manu., Stettin. Manufactory Finkenwalde near Stettin. Agts. s. No. 1360.

A water or gas-pipe of tar-paper, 3 ft. long, ½ ft. diameter; p. ft. 10 sg. or 1 sh.

1367. SCHÜTTLER, R. J., manu., Moabit (Berlin).

Cement-manger and cement-tiles.

1368. SCHULZE, HERMANN, & WILHELM, ANTON (Schulze & Wilhelm), manu., Nordhausen. Agts. s. No. 1360.

Works in marble: Chimney-piece with iron frame 150 Th. (22 £ 10 sh.); table with plate of mosaic work and pedestal of marble 300 Th. (45 £). The following prices p. dozen: bowl 30 Th. (4 £ 10 sh.); tobacco-box 15 Th. (2 £ 5 sh.); cigar-tub 4 Th. (12 sh.); do. 4 Th. (12 £); ash-cup 2 Th. (6 sh.); tinderbox with plate 2 Th. (6 sh.); 2 tinderboxes 1¼ Th. (4 sh.); tinderbox with plate 2 Th. (6 sh.); tinderbox in the form of a tub 2 Th. (6 sh.).

Works in rose-colored alabaster: Smoking-necessary 48 Th. (7 £ 4 sh.); ink-stand 24 Th. (3 £ 12 sh.); office candle-stick with garniture 12 Th. (1 £ 16 sh.); letterpresser with a statuette 8 Th. (1 £ 4 sh.); bowl 120 Th. (18 £); 2 twisted candlesticks 15 Th. (2 £ 5 sh.); ash-cup 8 Th. (1 £ 4 sh.); watch-case 8 Th. (1 £ 4 sh.); sewing-stone, twisted 12 Th. (1 £ 16 sh.); wafer-box 2 Th. (6 sh.).

Works in white marble: thermometer 15 Th. (2 £ 5 sh.); ash-cup 4 Th. (12 sh.); clew-holder 4 Th. (12 sh.).

1369. SIEGEN, ASSOCIATION FOR CULTURE AND INDUSTRY, Siegen. Agts. s. No. 1360. (s. No. 1241.)

Model in relievo of an artificial meadow and of the environs of Siegen.

1370. STETTIN PORTLAND-CEMENT WORKS by Wm. Lossius and Dr. Delbrück, Directors, Stettin. Paris 1855 bronze medal; London 1862 medal. Agts. s. No. 1360.

Samples of Stettin Portland-cement No. 1.

quickly setting, No. 2. slowly setting, p. barrel of 375 Pfd. Prussian weight net 3 Th., p. barrel of 415 lb. Engl. weight net 9 sh.

Specimens of Stettin Portland-cement see No. 1356. H. Mews.

1372. COUNT OF STRACHWITZ, HYACINTH CARL, Gross-Stein (Oppeln). Agts. s. No. 1360.

Fire-bricks: *a.* for blast-furnaces: 1 hearth-brick 15 sg.; 8 bricks of boshes, p. 100 pieces large size 4 Th.; small size 3 Th.; 2 shaft-bricks p. 100 pieces 2 Th.; 1 brick as used for puddling furnaces p. 100 pieces 1⅓ Th.; *b.* for heating by air: 1 cover-brick, 1 side-brick p. piece (bent plate) 1 Th., p. piece (straight plate) ½ Th.; *c.* 1 plate for a baker's oven 2¼ sg.

1373. TARNOWITZ CEMENT MANUFACTORY by E. Metke, director, Tarnowitz. Agts. s. No. 1360.

1. Roman-cement p. ton 3¾ hdwght. 2⅓ Th. (3,37 hdwght. 7 sh. 9 d.); 2. two floor-flags, hexagonal, ornamented, diameter 12 in. by 1 in. thick, p. piece 2½ sg. (4 d.); 3. two floor-flags, hexagonal, plain, diameter 12 in. by 1 in. thick, p. piece 2½ sg. (4 d.).

1374. WEIMAR, JOHANN PH., sculptor, Berlin. (s. II. No. 26.)

Objects of cast marble (new invention):

Gladiator (antique); frieze by Fischer, high relievo 4 ft.; 3 relievo's with scenes from the English history, modeled by Her Royal Highness the Princess Royal of Prussia; an ornamented glass-goblet; an ornamented porcelain-vessel; several ornamented plates of different metals.

1375. WOLFFHEIM, W., merchant and manu., Stettin. Agt. Henry Schirges, 9 Frith-street, Soho-square.

Asphalt-paste-board for roofing 627 ☐ ft. à 11 pfg. (engl. 646 ☐ ft. 2 £ 17 sh. 8 d.); asphalt-paper 1 sh. 8 d.

1376. ZEYLAND, J., joiner, Posen. Agts. s. No. 1360. (s. No. 1284.)

A window with four leaves and espagnolettes 3 £ 13 sh. 5 d.; a double-window with espagnolettes to be hermetically shut by means of india-rubber packing 3 £ 7 sh. 3 d.

CLASS XI.

MILITARY ENGINEERING, ARMOUR AND ACCOUTREMENTS, ORDNANCE, AND SMALL ARMS.

1377. ALBRECHT, F., gunsmith, Bromberg. Double barrelled gun.

1378. BEERMANN, B., gunsmith, Münster (Westphalia). Paris 1855 bronze medal; London 1862 honor. mention. Agts. s. No. 1360.

1. A double barrelled gun, one barrel rifled; 2. a double barrelled gun (system Lefaucheux); 3. a pair of pistols for firing at a target. Each of them with all appurtenances. Price-list at the agents'.

1379. BERGER & Co., manu. of cast steel, Witten òn the Ruhr. London 1862 medal. Agts. Heintzmann & Rochussen, 9 Friday-street, Cheapside.

A rifled piece of ordnance of cast steel with carriage. 20 Gun and rifle-barrels of cast steel in the different states of manufacturing. Models of guns as used in Prussia, England, France, Russia, Austria, Hannover, Württemberg and Switzerland.

1383. HAENEL, C. G., manu., Suhl. London 1862 honor. mention. Agts. Lion M. Cohn, Phaland & Dietrich, represented by Ch. Trübner, 20 St. Dunstans Hill, City, and at the Exhib. Building.

Infantry minié-musket after the Baden model 16⅓ Th. (2 £ 9 sh.); infantry-rifle, cast-steel barrel, Podewills' system 17⅔ Th. (2 £ 13 sh.); light infantry minié-musket, with brass-mountings 13 Th. (1 £ 19 sh.); artillery-rifle, Württemberg model 12 Th. (1 £ 16 sh.); rifle carabine 12 Th. (1 £ 16 sh.); free Hamburg, cases incl.

1384. HÖSTEREY, J. P., Barmen. London 1862 honor. mention. Agts. Morgan Brothers, 21 Bow-Lane, Cannon-street, City.

Percussion copper-caps, prices p. 1000 pieces: 400 cartouches métalliques or Flobert caps silvered 9m/m 5 Th. (15 sh.); 6m/m 1 Th. 18 sg. (4 sh. 9 d.); gilt 9m/m 5 Th. (15 sh.); 6m/m 1 Th. 18 sg. (4 sh. 9 d.); 200 military caps 1 Th. 3 sg. (3 sh. 4 d.); silvered 1 Th. 3 sg. (3 sh. 4 d.); 200 hunting-caps, gilt 16 sg. (1 sh. 7 d.); silvered 16 sg. (1 sh. 7 d.); 1600 hunting-caps, plain, silvered 8 sg. (9½ d.); rifled, gilt 9 sg. (10½ d.); plain waterproof copper 11 sg. (11 d.); rifled do. 12 sg. (11⅘ d.); 1000 large oeillets 8 sg. (9½ d.); 1000 small do. 5½ sg. (6½ d.).

1386. KIRSCHBAUM, CARL REINH., manu. of side-arms, Solingen. London 1862 medal. Agts. Morgan Brothers, 21 Bow-Lane, Cannon-street, City. — C. F. W. Rust, 129 London-Wall, E. C., City.

30 side-arms as used by the British (East-India-Comp.), Prussian, Dutch, French, American and other armies at the price of 6 sh. to 1 £ 10 sh.; 20 pieces of fancy steel arms with gilt and electro-plated mountings at the price of 3 £ to 225 £; 30 blades for small arms of various sizes and qualities at the price of 9 sh. to 9 £; 20 Arkansas knives with gilt and electro-plated mountings at the price of 10 to 30 sh.

1386a. KRUPP, FR., Essen. London 1862 medal. (s. No. 1308.)

Pieces of ordnance, guns.

1387. LAUTE, GUSTAV, manu. of side-arms, Berlin, Brüderstr. 32. Agts. s. No. 1383.

Arms, whole-sale prices p. piece: garde du corps sword 15 Th. (2 £ 5 sh.); curassier sword 15 Th. (2 £ 5 sh.); 2 cavalry drawing-room swords 8 Th. (1 £ 4 sh.); ancient Prussian sword with Toledo-blade 18 Th. (2 £ 14 sh.); dragoon sword with steel-basket and damask blade 18 Th. (2 £ 14 sh.); artillery sword 12 Th. (1 £ 16 sh.); husar sword with damask blade 30 Th. (4 £ 10 sh.); ulane sword with damask blade 18 Th. (2 £ 14 sh.); sword of honor, model of the one for prince Charles of Prussia 60 Th. (9 £); battle-sword with basket for infantry 18 Th. (2 £ 14 sh.); fusilier officer's sword 12 Th. (1 £ 16 sh.); do. with basket to be unscrewed 18 Th. (2 £ 14 sh.); 2 swords for the court 10 Th. (1 £ 10 sh.); sword for marine-officers 14 Th. (2 £ 2 sh.); dagger for marine-officers 10 Th. (1 £ 10 sh.); hunting-knife 12 Th. (1 £ 16 sh.); do. oval shape 10 Th. (1 £ 10 sh.); do. old fashioned 25 Th. (3 £ 15 sh.); 2 old fashioned daggers, knight-figures 10 Th. (1 £ 10 sh.); hunting-knife with oak-branches 12 Th. (1 £ 16 sh.); 2 long daggers with ivory-handle 15 to 30 Th. (2 £ 5 sh. to 4 £ 10 sh.); dagger with skeleton-handle 25 Th. (3 £ 15 sh.); 3 Oriental daggers with damask blades 20 Th. (3 £).

1389. LEONHARDT, J. E., manu., Berlin, Wilhelmsstr. 46. Agt. Heintzmann & Rochussen, 9 Friday-street, Cheapside.

Machine for casting bullets to minie rifle-guns.

1390. LÜNESCHLOSS, P. D., manu., Solingen. London 1862 medal.

Sabre made on the occasion of His Prussian Majesty's coronation.

1391. PRETZEL, AUGUST, gun-maker, Gross-Glogau (Liegnitz). Agts. s. No. 1383.

A brace of target-pistols with hair-grooves, ranging at a distance of 25, 120 and 200 paces, Turk-damask barrels, finely engraved, the firm inlaid with gold, in a case with all appurtenances. The partitions for bullets etc. are covered with finely engraved silver-plates; priee 135 Th. (20 £).

1392. RÖDDER, MATHIAS, manu. and merchant, Cologne. London 1862 honor. mention. Agts. Morgan Brothers, 21 Bow-Lane, Cannon-street, City.

6 double barrelled guns: No. 1. 6 £ 12 sh.; No. 3. 6 £ 13 sh.; No. 4. 5 £ 2 sh.; No. 6. 5 £ 17 sh.; No. 7. 3 £ 15 sh.; No. 9. 3 £ 15 sh.; 1 double barrelled gun, one barrel rifled No. 11. 6 £ 11 sh.; 5 rifled-guns No. 2. 7 £ 4 sh.; No. 5. 4 £ 4 sh.; No. 8. 5 £ 3 sh.; No. 10. 2 £ 18 sh.; No. 12. 3 £ 18 sh.; 1 Tesching No. 13. 1 £ 7 sh.

Needle fire-arms: rifled musket No. 14. (new needle-gun) 3 £ 1 sh.; rifled carabine with pike No. 15. (needle-gun) 3 £ 1 sh.; rifled carabine No. 16. (needle-gun) 6 £ 14 sh.; rifled carabine No. 17. (new needle-gun) 4 £ 1 sh.: brace of pistolets No. 18½. (needle-syst.) 4 £ 1 sh.; 4 powder-flasks and powder-horns No. 46. 5 sh.; No. 47. 4 sh. 9 d.; No. 48. 6 sh.; No. 49. 2 sh. 5 d.

24 dog-whistles, fox-angles, patent percussion-cap box, nipple-filling, screw-driver, shot-charger, No. 19. 3 sh.; No. 20. 3 sh. 4 d.; No. 21. 2 sh.; No. 22. 6½ d.; No. 23. 2 sh.; No. 24. 7 d.; No. 25. 9 d.: No. 26. 1 sh. 3 d.; No. 27. 3 sh.; No. 28. 6 d.; No. 29. 9½ d.; No. 30. 2 sh. 6 d.; No. 31. 5 d.; No. 32. 5 sh.; No. 33. 6½ d.; No. 34. 9 d.; No. 35. 8½ d.; No. 36. 1 sh. 3 d.; No. 37. 5 d.; No. 38. 9 d.; No. 39. 3¾ d.; No. 40. 9 d.; No. 41. 8 d.; No. 42. 10 d.

3 revolvers and pocket-revolver No. 43 2 £ 8 sh.; No. 44. 1 £ 15 sh.; No. 45. 2 £. 8 sh.

No. 50. cap with movable head-piece to a Lefaucheux gun 8½ d.; apparatus belonging to it 2 sh; No. 51. gun-lock p. piece 3 sh.; No. 52. pricker 10 sh.; No. 53. and 54. a brace of gun-locks, p. brace 8 sh. 6 d.; 6 gun-crochets (animals' heads) No. 55. 56. 57. 58. 59. 60. p. piece 5 sh. 6 d.

1392a. SCHALLER, A., manu., Suhl.

Revolver.

1392b. SCHILLING, VAL. CHR., manu. of arms, Suhl. Berlin 1844 bronze medal; Paris 1855 silver medal; London 1862 medal. Agts. s. No. 1383.

Infantry gun, grand-ducal Hessian model 16 Th. (2 £ 8 sh.): Enfield gun 17⅓ Th. (2 £ 12 sh.); rifle with yatágan, Württemberg model 22⅔ Th. (3 £ 8 sh.); Swedish breech-loading rifle with Whitworth bore 20 Th. (3 £): pistol with butt-end, Hamburgh model 11 Th. (1 £ 13 sh.); pistol for a customs officer, with Enfield bore 6⅔ Th. (1 £): buck double barrelled rifle, No. 10,596. with damask (Turk) barrels 56⅔ Th. (8 £ 10 sh.): percussion rifle-gun, No. 11,001. with damask Leclerc barrels 46⅔ Th. (7 £); percussion rifle No. 11,011. 18⅓ Th. (2 £ 15 sh.)) do. double barrelled gun No. 11,004. 16⅔ Th. (2 £ 10 sh.): double barrelled gun, system Lefaucheux with all the appurtenances in a box 153⅓ Th. (23 £).

A brace of pistols No. 10,110. with cast-steel barrels and all the appurtenances in a box 73⅓ Th. (11 £); revolver, system Lefaucheux No. 598. 18⅔ Th. (2 £ 16 sh.); military revolver, systemAdams 16 Th. (2 £ 8 sh.); do. system Beaumont 18⅓ Th. (2 £ 15 sh.); revolver, system Adams middle model 13 Th. (1 £ 19 sh.); do. system Baumont 15⅓ Th. (2 £ 6 sh.).

CLASS XII.

NAVAL ARCHITECTURE AND SHIP'S TACKLE.

1392 c. Krupp, Alfred, (Friedrich Krupp), Essen. (s. No. 1308.)

Propeller-shaft; propeller-screw.

CLASS XIII.

PHILOSOPHICAL INSTRUMENTS AND PROCESSES DEPENDING UPON THEIR USE.

1393. Belle, Richard, optician and mechanician, Aix-la-chapelle. Agts. Lion M. Cohn, Pha land & Dietrich, represented by Ch. Trübner, 20 St. Dunstans Hill, City, and at the Exhib. Building.

Scale-beams 24 in. long of red-brass No. 1. 4 £ No. 2. 3 £.; both these beams for 50 kilogr. weight.

1394. Bredemeyer, Julius, mechanician, Frankfurt on the Oder. Paris 1855 honor. mention. Agts. s. No. 1393.

Theodolite 125 Th. (18 £ 15 sh.); levelling-instrument 54 Th. (8 £ 2 sh.); 3 miner's compasses with additional instruments in three sizes 70 Th. (10 £ 10 sh.); 50 Th. (7 £ 10 sh.); 40 Th. (6 £); pocket miner's compass 19 Th. (2 £ 17 sh.); hydraulic press 50 Th. (7 £ 10 sh.)

1395. Elster, Siegmar, gas-engineer, Berlin, Neue Königsstr. 67. London 1862 medal. Agt. W. Siemens, 7 Adelphi-street.

Experimental gas-apparatus: 1. new air-mixing photometer, transforming the hydrocarbon of the flame into carbonic oxide and apparatus for ascertaining the specific gravity of gases 26⅔ Th. (4 £); 2. pressure-photometer indicating the pressure required by a set of definite height 20 Th. (3 £); 3. multiplying gauge and pressure-photometer 20 Th. (3 £); 4. Bunsen control-photometer with experimental gasmeter, governor, multiplying pressure-gauge, Argand test-burner and jet-burner 73⅓ Th. (11 £); 5. the consumer's instructive gasmeter 14 Th. (2 £ 2 sh.); 6. selfregistering pressure-indicator 73⅓ Th. (11 £); 7. alarm pressure-gauge 20 Th. (4 £).

1397. Fessel, Friedrich, mechanician & optician, Cologne. London 1862 medal. Agts. s. No. 1393.

Myographion, an apparatus for registering the contractions of irritated muscles, and for measuring the velocity of propagation of the nervous agent 31 £.

1400. Greiner, A. & C. (J. C. Greiner sen. & Son.), manu. of meteorological instruments, Berlin, Kurstr. 15. Agts. s. 1393.

12 Prussian standard alcoholimeters p. piece 8 Th. (1 £ 4 sh.); 2 cylinders p. piece 1 Th. (3 sh.)

1401. Greiner, F. F., manu. of meteorological glass-instruments and apparatus, Stutzerbach (Erfurt). London 1862 medal. Agt. John J. Griffin, 119 Bunhill Row E. C.

Meteorological glass-instruments and chemical apparatus; 311 numbers.

1402. Gressler, Julius, Busch, Arnold (J. Gressler & Co.), Berlin, Königsstr. 34. London 1862 honor. mention. Agts. s. No. 1393. (s. No. 1301.)

Galvanic apparatus. Coal-zinc batteries, of several constructions:

A. Transportable batteries, as used in rooms, of 4 elements upon wood-stand with pillar and gyrotrope p. piece 14 Th. (2 £ 2 sh.)

B. Six batteries, each containing 4 elements of hollow coal-cylinders, with strong-removable copper-setting, in six sizes: of 3 in. height of the cylinder 3⅔ Th. (11 sh.) p. batterie; 4¼ in. height of the cylinder 5 Th. (15 sh.) p. batterie; 5 in. height of the cylinder 5⅔ Th. (17 sh.) p. batterie; 6½ in. height of the cylinder 6⅓ Th. (19 sh.) p. batterie; 7 in. height of the cylinder 7⅔ Th. (1 £ 3 sh.) p. batterie; 8 in. height of the cylinder 8⅔ Th. (1 £ 6 sh.) p. batterie; 12 in. height of the cylinder 23⅓ Th. (3 £ 10 sh.) p. batterie.

C. Five batteries, each of 4 elements of solid coal-rolls, with removable copper-setting of 3½ in. height of the rolls 2⅔ Th. (8 sh.) p. batterie; 5½ in. height of the rolls 4⅓ Th. (13 sh.) p. batterie; 7 in. height of the rolls 5⅔ Th. (17 sh.) p. batterie; 8 in. height of the rolls 8 Th. (1 £ 4 sh.) p. batterie; 12 in. height of the rolls 19⅓ Th. (2 £ 18 sh.) p. batterie.

D. Three batteries, each of 4 elements of coal-cylinders with setting protected against the acid by gutta-percha coating: 5 in. height of the cylinder 5⅔ Th. (17 sh.) p. batterie; 7 in. height of the cylinder 7⅔ Th. (1 £ 3 sh.) p. batterie; 8 in. height of the cylinder 8⅔ Th. (1 £ 6 sh.) p. batterie.

E. Two batteries, each of 4 elements, with lead-setting: 5 in. height of the coal-cylinder 4 Th. (12 sh.) p. batterie; 7 in. height of the coal-cylinder 5⅔ Th. (17 sh.) p. batterie.

F. Three batteries, each of 4 elements, of gas-retort plates, with removable brass-setting: of 4½ in. height of the plate 4⅔ Th. (14 sh.) p. batterie; 7 in. height of the plate 7⅓ Th. (1 £ 2 sh.) p. batterie; 8¼ in. height of the plate 9⅓ Th. (1 £ 8 sh.) p. batterie.

Apparatus: 1 machine, moved by electro-magnetic power 30 Th. (4 £ 10 sh.); 1 apparatus for electric light (self regulating) slide-apparatus 60 Th. (9 £); 1 do. screw-apparatus 22½ Th. (3 £ 7 sh. 6 d.); 1 assortment of 28 samples of copper-wire covered with silk or cotton (not for sale); 1 telegraph of Morse's system, improved construction 40 Th. (6 £); 1 finger-telegraph 22½ Th. (3 £ 7 sh. 6 d.).

1403. Hager, F., mechanician, Stettin. Agts. s. No. 1393.

A transparent compass with arms 25 Th. (3 £ 15 sh.); an amplitude-compass which, held with the hand, may serve in sounding. By means of a screw it may be fastened to any wooden object serving as a stand. The figures are read off by the medium of a prism. 20 Th. (3 £); an octant of brass, the divisions on silver 32 Th. (4 £ 16 sh.); an octant of brass, the divisions on ivory 32 Th. (4 £ 16 sh.)

1404. Hempel, O. M., steam-gauge manu., Berlin. London 1862 honor. mention. Agts. s. No. 1393.

Assortment of spring steam-gauges.

1406. Lüttig (C. Lüttig), mechanician, Berlin. Berlin 1844 honor. mention; London 1851 honor. mention. Agts. s. No. 1393.

1. a large levelling-instrument with a thirty-fold

magnifying achromatic telescope 16 in. long, which may be turned and laid into its bearings. The air-level indicates an inclination of five seconds at a deviation of a line; the horizontal circle indicates the angles to thirty seconds by the nonius; 160 Th. (24 £); 2. a like instrument, middle-sized, 120 Th. (18 £); 3. do. a smaller one 90 Th. (13 £ 10 sh.); 4. a large case of mathematical instruments, containing 55 pieces 140 Th. (21 £).

1407. ROHRBECK, W. J. (J. F. Luhme & Co.), manu. of chemical, pharmaceutical and physical apparatus, Berlin. Berlin 1844 honor. mention; London 1851 prize-medal; Munich 1854 honor. mention; London 1862 medal.

1. Fine balance for chemical purposes for 1 kil. weight indicating $\frac{1}{10}$ mgrm. with arrangement for stopping the beam and the scales and for attaching the weights on the divided beam, in a glass case, 300 Th.

2. A set of weights of brass and platinum with corn-tongs in a mahogany case 42 Th.

3. A fine balance for chemical analysis in a glass case, at a charge of 150 grm. indicating $\frac{1}{10}$ mgrm., divided beam, arrangement for attaching the centigrm. weights and for ascertaining the specific gravity of bodies, 85 Th.

4. A set of chemical weights of brass and platinum with corn-tongs 16 Th.

5. A fine balance for chemical purposes in a glass case; 2nd. quality, at a charge of 60 grm. indicating $\frac{1}{2}$ mgrm.; divided beam arranged to be stopped, middle axis on agate, scales plated with platinum hanging by platinum wire 50 Th.

6. A set of weights of german silver and platinum with corn-tongs from 100 grm. to 1 mgrm. 11 Th.

7. 2 Apothecary's balances with brass-pillars, 8 in. high, one with Prussian eagle, one with Russian eagle, each with brass-beam, 12 in. long, on mahogany case with two boxes for weights, 20 and 22 Th.

8. A balance for technical laboratories and gauge-offices on mahogany-stand with stopping and brass-scales in a glass-case 16 Th.

9. Assortment of apothecary's vessels of glass and porcelain with burnt labels, besides an assortment of labels of porcelain, milk-glass and enamel for drawers and cases in apothecaries' shops; prices according to the annexed catalogues.

10. A Plattner blow-pipe apparatus with balance, weights and all the appurtenances for blow-pipe investigations in a case 78 Th.

11. 2 test-chests of mahogany with drawers for analysis-requisites: a. with 30 ground bottles with burnt labels and chemical formulae on the ground stoppers 20 Th.; b. a like chest with 48 ground bottles 30 Th.

12. Chemical lamps: a. Berzelius lamp with porcelain plate $4\frac{5}{12}$ Th.; b. Mitscherlich lamp with Plattner blowing-apparatus of 5 jets 8 Th. (both for alcohol); c. Rammelsberg lamp for coal-gas with 5 Bunsen burners and mounting for wire-gauze with rings and supports for alembics on porcelain plate $8\frac{1}{3}$ Th.; d. 2 different Luhme lamps with tripods, ring and handles $4\frac{1}{2}$ Th.

13. Assortment of chemical glass-apparatus: a. beakers, 1 set of 12 pieces $1\frac{1}{2}$ Th., of 6 pieces with spouts 1 Th., of 6 pieces with 2 spouts $1\frac{1}{2}$ Th.; b. 2 funnels $3—4\frac{1}{2}$ sg.; c. 2 retorts $5—10$ sg.; d. sulphureted hydrogen-apparatus after Kipp $3\frac{1}{2}$ Th.; e. 2 alcali metric apparatus after Geisler and Schrötter, each 2 Th.; f. new apparatus for investigating the quality of coal-gas by ascertaining quantitatively the carbonic acid and other gases $8\frac{1}{2}$ Th.

14. a. India-rubber blowing-apparatus and air-vessel for coal-gas and air with selfacting blow-pipe on stand with porcelain plate $12\frac{3}{4}$ Th.; b. Mitscherlich melting-apparatus for coal-gas and atmospheric air on porcelain plate 14 Th.

15. Mahogany-case containing 4 different areometers of Baumé, 1 alcoholometer, 5 thermometers on milkglass with divided tube, 1 thermometer divided in $\frac{1}{10}$ of degrees for physicians to ascertain the heat of human bodies 15 Th.

16. Stands with different burettes and pipettes graduated after Mohr, Erdmann and Geisler with glass-cocks after Rammelsberg 18 and 10 Th.

17. Apparatus for evolving fluoric acid consisting of platinum-crucible with tight fitting cover, cooling-pipe and platinum-receiver, all without soldering in case 120 Th.

18. 10 platinum-basins and 6 crucibles with covers and pipes; weight of the platinum 12 Loth, p. Loth 6 Th.

19. 2 glass-gasometers after Schrötter with glass-cocks 18 Th., after Mitscherlich with brass-cocks $8\frac{1}{2}$ Th.

20. Large air-pump with two glass-clyinders of $2\frac{1}{3}$ in. diam., Grassmann cock, barometer and glass-plate with cover 150 Th.

21. Polarization-apparatus after Dove on stand complete with prisms arranged for applying an open telescope, microscopes, circular polarization-instruments with case in which the annealed glass, the Nicol and double refracting prisms as well as the crystals of polarization are put. This apparatus has been improved by Dove by adding a dichroiscope which serves for interference-experiments and for showing dichroisms 130 Th.

22. Case with a Mitscherlich polarization-apparatus with requisites for testing sugar for sugar-manufacturers and physicians 40 Th.

23. Model of a hydraulic press with glass-cylinder of 800 lb. pressure 45 Th.

24. 1 apparatus for spectral analysis after Kirchhoff and Bunsen with a tube after Mousson and prism 24 Th.; 1 do. with 2 prisms, large apparatus 36 Th.

25. 1 microscope with 2 ocular and 3 object-lenses, 600 fold magnifying, complete in case 32 Th.

1408. MEISSNER, ALBRECHT, mechanician, Berlin, Friedrichsstr. 71. Agts. Lion M. Cohn, Phaland & Dietrich, represented by Ch. Trübner, 20 St. Dunstans Hill, City, and at the Exhib. Building. (s. II. No. 81.)

St. Schmeisser's patent hemispheric sundials in two different constructions; the one to be used between the 20. and 42. degree, the other between the 42. and 63. degree of north latitude p. doz. 84 Th. (12 £ 12 sh.) at Berlin.

1410. NOBERT, F. A., mechanician and optician, Barth (Pommern). Berlin 1844 prize-medal; London 1851 prize-medal; Paris 1855 bronze medal; London 1862 medal. Agt. Thomas Ross, optician, 2 and 3 Featherstone Buildings, High Holborn.

A compound microscope optically separating 8000 parts of a Par. line and directly measuring 50000 parts of a Par. line, 500 Th. (75 £); 6 test-plates for the microscope, with divisions: a. 20 degrees $\frac{1}{1000}'''$ to $\frac{1}{6000}''''$ one 15 Th. (2 £ 5 sh.); b. 30 degrees, $\frac{1}{1000}'''$ to $\frac{1}{8000}'''$ one 30 Th. (4 £ 10 sh.); c. 19 degrees, $\frac{1}{1000}'''$ to $\frac{1}{10000}'''$ one 30 Th. (4 £ 10 sh.); 3 diffraction-plates, plain-parallel, with divisions: a. 1600 lines in 8 Par. lines $21\frac{1}{2}$ Th. (3 £ 4 sh.); b. 1800 lines in 6 Par. lines 24 Th. (3 £ 16 sh.); c. 2000 lines in 4 Par. lines $26\frac{2}{3}$ Th. (4 £).

1411. OERTLING, AUGUST, optician and mechanician, Berlin, Oranienburgerstr. 57. Berlin 1844 gold medal; London 1851 prize-medal.

Agt. L. Oertling, 13 Store-street, Bedford-square.

Goniometer for optical purposes in measuring angles; it consists of a circle of 15 in. diam. with divisions on silver of $\frac{1}{12}$ degree. By means of a micrometrical apparatus it is allowed to ascertain the angles to a second. The circle can be directed vertically as well as horizontally. The vertical position is necessary, if working with liquids, whose horizontal surface serves as plain of reflection. It is to be employed to measure in crystals the alterations of the angles by heat; also to measure the angles of very small crystals as well as those formed by the faces of large crystals. All the investigations and measurings which relate to spectral analysis and polarization can be made with it.

1411a. Pintus, J. & Co. (s. No. 2082a.)

1412. Reichel, C., mechanician, Berlin, Alte Jakobstr. 65. Agts. Lion M. Cohn, Phaland & Dietrich, represented by Ch. Trübner, 20 St. Dunstans Hill, City, and at the Exhib. Building.

6 spirit-levels of different sizes indicating different angles. Lists of prices in English, German and French are annexed.

Reimann, L., Berlin. Chemical scales.
(s. No. 2086.)

1413. Siemens & Halske, Berlin. London 1851 council medal; Munich 1854 great medal; Paris 1855 silver medal; London 1862 medal. Agt. C. W. Siemens, 3 Great George-street, Westminster.

a. 4 recording apparatus of different constructions worked by transmitting currents, connected in circuit for an overland line; b. 2 submarine recording apparatus, connected in circuit for submarine lines; c. 4 recording apparatus of different constructions worked by interrupting a continuous current, connected in circuit for a line; d. line of alarm-bells including 1 clockwork for increasing strength of current, 1 do. for interrupting current, 1 inductor and 1 contactbreaker; e. 1 mechanical transmitting apparatus with types; f. 2 ink recording apparatus of different constructions worked by magnet-key instead of battery; g. 2 dial apparatus, in circuit; h. 1 bell-telegraph for houses; i. 3 induction-coils of different sizes; k. 1 resistance-bridge for determining resistances and places of faults; l. electrical indicator for lengths of cables paid out, and electrical log for measuring ship's way; m. 2 current commutators, one self-acting, the other with clockwork for measuring speed of currents in submarine wires; n. medico-galvanic apparatus; o. thermo-electrical pyrometer; p. apparatus for developing ozone by electrical induction.

1415. Schmidt, Franz, Berlin, Alexandrinenstr. 74. Agts. s. No. 1412.

1 Soleil-saccharometer (polarization-apparatus) with all appurtenances as: the complete apparatus with magnifying lens, 3 observing tubes of 200, 100 and 50 mm. length. lamp with glass and clay-cylinders, filtring-stand for 2 coal-cylinders, 2 filtring-cylinders, 2 glasses, areometer, cylinder with brass-foot, Celsius' thermometer, 2 sized cylinders, pipette, 144 Th. (21 £ 12 sh.).

1417. Winckler, G., Dr. phil., Quilitz, H. (Warmbrunn, Quilitz & Co.), merchants and manu., Berlin, manufactories Jemmlitz and Tschornow (Niederlausitz). London 1862 medal. Agts. A. G. Franklin & Co., 14 South-street, Finsbury-Square E. C

Test-chest with 35 labeled bottles and all the requisite apparatus; medicine-chest for domestic use; test-tubes with stand; 12 burettes with stand; 12 pipettes with stand; model of a hydraulic press; two-barrelled air-pump with manometer; hand-air-pump; model of a high and low-pressure steam-engine; model of a chain-pump; Nörremberg's polarization-apparatus; Seebeck's polarization-apparatus with Nicol's brass-mounted prism and 6 pieces unannealed glass; Heron's-fountain; Galileo's incline; assay-balance and weights in palisander-case; Mohr's balance in palisander-case; 6 prisms on brass-stand; Newton's color-rings; 219 bottles with labels of white, milk-white, hyalith and blue Jemmlitz glass; 37 porcelain bottles with labels; mahogany-tablet with milkwhite labels; mahogany-tablet with porcelain labels; 4 separating-funnels with cover, stop-cock and stopper; 8 glass-mortars with spouts and pestles; Mitscherlich's chlorine-apparatus; Pugh's sulphuretted hydrogen-apparatus; Kipp and Mohr's sulphuretted hydrogen-apparatus; 3 Berzelius flasks for evolving gases; 3 alembics with movable heads and stoppers; 3 sets of glass-beakers, plain and with spouts; 2 sets of glass semi-globular evaporating-dishes; 10 glass-cylinders graduated to cub. cent.; 5 burettes, Gay-Lussac and English; 3 stoppered flasks graduated to $\frac{1}{2}$, $\frac{1}{4}$ litre; 6 agate-mortars with pestles; thousand-grain bottle with thermometer; 4 flasks, graduated to grammes; 3 sets of milk-white glass speculums; 3 do. with black reflectors; 6 gum-bottles, English and cylinder-shape; 8 test-glasses with feet; 6 cylinder and bell-shaped glass-measures; 15 flasks for filtring collodion with covers; 22 photographic baths of white Jemmlitz glass; 51 white glass pressed photographic dishes; 6 alcalimetric apparatus; specific gravity-apparatus; three-neck Woolfe-bottles with stop-cock.

CLASS XIV.

PHOTOGRAPHIC APPARATUS AND PHOTOGRAPHY.

1419. Beyrich, Ferd., manu. of chemical preparations etc., Berlin, Friedrichsstr. 101. London 1862 medal and honor. mention. Agts. s. No. 1412.

Chemical preparations and papers for photographical purposes.

1419a. Burchard, A., Berlin, 39 Neue Friedrichsstr.

Photolithographs after a new process invented by the exhibitor: Copy of a drawing made with the pen by Schinkel (the Waterfall of Gastein); Copies of wood-cuts by Dürer.

1420. Busch, Emil, optician manu. of photographic apparatus, Rathenow. Berlin 1844 bronze medal; London 1851 bronze medal, 1862 medal. Agts. s. No. 1412.

1 photographic apparatus for visit-cards with 4 double object-glasses of 30''' (Parisian lines) in diameter 206 Th. (30 £ 18 sh.); 1 stand 26½ Th. (4 £); 1 camera obscura (with elastic piece to lengthen it) with an universal object-glass (double object-glasses and dispersing lenses) of 36''' (Parisian lines) 109 Th. (16 £ 7 sh.); 1 stand 26½ Th. (4 £).

Duncker, A., publisher, Berlin, Französische Str. 21. Photographs. (s. No. 1868.)

1422. Fessler & Steinthal, manu., Berlin, Französische Str. 48.

Photographs.

1424. Hammerschmidt, W. A., manu., Neu-Schöneberg near Berlin.
Photographs.

1425. von Klitzing glass-manufactory, Bernsdorf (Liegnitz), by W. Barth. (s. No. 2152.)
12 cuvettes for photographical purposes in English measures; prices p. piece: 3½ to 5½ in. 12 sg. (1 sh. 2⅔ d.); 4 to 6 in. 12½ sg. (1 sh. 3 d.); 5 to 7 in. 20 sg. (2 sh.); 5½ to 8½ in. 22½ sg. (2 sh. 3 d.); 10 to 12 in. 1½ Th. (4 sh. 6 d.); 9 to 11 in. 1⅓ Th. (4 sh.); 8 to 10 in. 1 Th. (3 sh.); 7 to 9 in. ⅚ Th. (2 sh. 6 d.); 6 to 8 in. ⅚ Th. (2 sh. 6 d.).

1427. Kunzmann, H., manu., Berlin, Friedrichsstr. 218. London 1862 honor. mention.
Chemical agents and paper for photography.

1428. von Minutoli, Alexander, Royal Prussian counsellor of the provinzial regency, Liegnitz. London 1862 honor. mention. Agts. A. Asher & Co., care of D. Nutt, 270 Strand.
Minutoli. Four thousand (4000) works of antiquity, photographed from the originals and intended as models for manufacturers and artisans. Liegnitz 1855—1862, 24 sections, 7 volumes gr. folio 800 Th. (120 £).

1429. Moser & Senftner, stereoscope-manu. and publishers of stereoscopic pictures, Berlin, Unter den Linden 44. Agts. Lion M. Cohn, Phaland & Dietrich, represented by Ch. Trübner, 20 St. Dunstans Hill, City, and at the Exhib. Building.
13 different stereoscopes: 2 Moser's drawing-room-stereoscopes (mahogany-case) the exhibitor's own new system, each with 50 simple or transparent pictures; 2 drawing-room-stereoscopes each with 50 pictures on glass or paper (1 caryed antic polixander-case); 6 cases for stereoscopic pictures.
The 112 pictures on paper in the stereoscopes are published partly by the exhibitors, partly by others, made partly by German, partly by French or English photographers.
Those of the 100 glass-pictures which represent the interior views of the palaces at Berlin and Potsdam, as well as the other views, have by permission, granted to the exhibitor, been photographed by Mr. Soulier.

1430. Nicolai publisher (G. Parthey), Berlin. Agts. s. No. 1429. (s. No. 1913.)
Photographs: Wilhelm von Kaulbach: The death of Julius Caesar. Photographed after an original cartoon by Joseph Albert at Munich. 1861. Edition I. 23½ by 33½ in. 8 Th. (1 £ 4 sh.); edition II. 18¾ by 24¾ in. 5 Th. (15 sh.). Shakespeare-album. Photographs after the artist's own drawings by Joseph Albert at Munich. fol. 1862. Part. I. Macbeth, 3 plates; part II. the tempest, 2 plates; part. III. King John, 3 plates; all 10 Th. (1 £ 10 sh.).

1431. Oehme, G., & Jamrath, F., photographers to His Majesty the King of Prussia, Berlin, Jägerstr. 19. London 1862 medal.
No. 1. Portrait of His Majesty the King of Prussia; No. 2. portrait of His Royal Highness the Prince Royal of Prussia; No. 3. portrait of Mr. de Huelsen, superintendent of the royal theatres; No. 4. portrait of Miss Pellet, of the King's theatre; No. 5. Messrs. Hendrichs, Karlowa, Grua, Doering, Mrs. Kierschner and Miss Doellinger of the King's theatre; No. 6. portrait of a gentleman (all from life), 24 in. high by 18 in. broad, each 40 Th. (6 £); No. 8—13. appartments in the palace of His Royal Highness the Prince Royal of Prussia, each 5 Th. (10 sh. without frame), size 12 and 11 in.; No. 14. portrait of the general count de Nostiz 40 Th. (6 £).

1432. Schauer, G., photographer, Berlin, Friedrichsstr. 188. London 1862 honor. mention.
Specimens of photography.

1433. Schering, E., chemist, Berlin, Chausseestr. 21. Paris 1855 honor. mention; London 1862 medal. Agts. s. No. 1429. (s. II. No. 95.)
Acid. aceticum glac. pur.; acid. pyrogallic. albiss. leviss.; ammonium hydrobromat.; ammonium hydryodat. pur. albiss.; ammonium hydrofluorat.; argent. nitric. fus. albiss.; do. griseum; auro-kalium chlorat. cryst.; aurum chlorat. cryst.; cadmium bromatum; cadmium fluoratum; cadmium iodat. cryst. albiss.; calcium bromatum; ferrum sulphuric. cryst. pur.; glycerin chem. pur. albiss.; iodum resubl. opt.; kalium bromat. opt.; kalium chlorat. pur.; kalium cyanat. fus. albiss.; do. in bacill.; kalium fluoratum pur.; kalium iodatum puriss.; do. fus. in bacill. pharm. Svecic.; lithion hydrobromat.; lithion hydrochlorat.; lithion hydroiodat.; natrium cloratum puriss.; natrium iodatum; natrum hyposulphurosum; platinum chlorat. siccum; pyroxylin; strontium chloratum sicc.; uranium nitric. cryst.; zincum bromatum; zincum chlorat. sicc. albiss.; zincum iodat. albiss.

Winkler & Quilitz, Berlin. Photographic baths and dishes. (s. No. 1417.)

1435. Wothly, Jacob, photographer to the court, Aix-la-Chapelle. London 1862 medal. Agts. s. No. 1429.
1. Portrait of Mr. Distery photographer to His Majesty the Emperor of the French; 2. portrait of the exhibitor; 3. a group.; 4. portrait of madame de Koehnen; 5. a milk-maid; 6. portrait of an officer.

CLASS XV.

HOROLOGICAL INSTRUMENTS.

1437. Becker, Gustav, manu. of regulators, Freiburg (Breslau). London 1862 honor. mention. Agts. s. No. 1429. (s. II. No. 96.)
A regulator with clock-work, the dial 7 in., flat and enamelled, I. quality, in a case of nut-wood (carver's work) 31 Th. (4 £ 13 sh.); a regulator, pendulum-clock of I. quality, the case of nut-wood (carver's work) 24 Th. (3 £ 12 sh.); a regulator, pendulum-clock, the dial 8 in., enamelled, second, 80 part. I. quality in a nut-wood case (silvered garniture) 25 Th. (3 £ 15 sh.); a regulator, pendulum-clock, the dial 7 in., enamelled II. quality (the case of carved polixander) 14 Th. (2 £ 2 sh.); a regulator, pendulum-clock, the dial 8 in. enamelled I. quality (the case of polixander-wood) 11½ Th. (1 £ 14 sh. 6 d.); a regulator, a small pendulum-clock, the dial 5 in.. enamelled, deepened II. quality (a gothic case of nut-wood), 2 pieces, each 14 Th. (2 £ 2 sh.); a regulator-work, the dial 7 in., enamelled, I. quality 8 Th. (1 £ 4 sh.); 2 regulator-works the dial 6 in. and deepened II. quality 7 Th. (1 £ 1 sh.); 2 do., 5 in. 6 Th. (18 sh.); a clock-work (ebauche) scheme 11½ Th. (1 £ 14 sh. 6 d.); a pendulum-clock do., I. quality 5½ Th. (16 sh. 6 d.): do., II. qual. 4½ Th. (13 sh. 6 d.); some wheels.

1438. Eppner, Albert, & Co., manu. of

watches and watch-makers to His Majesty the King and His Royal Highness the Prince Royal of Prussia, Berlin, Behrenstr. 31. Manufactory Lähn in Silesia. London 1862 honor. mention. Agts. Lion M. Cohn, Phaland & Dietrich, represented by Ch. Trübner, 20 St. Dunstans Hill, City, and at the Exhib. Building. (s. II. No. 70.)

Watches: gold-chronometer No. 16,236. 21 lines 200 Th. (30 £); gold anchor-watches: No. 15,247. 17 lines, savonette, to be wound up by button 90 Th. (13 £ 10 sh.); No. 15,741. 18 lines, to be wound up by button 78 Th. (11 £ 14 sh.); No. 16,133. 20 lines, 19 rubies 74⅔ Th. (11 £ 4 sh.); No. 16,179. 12 lines, 19 rubies 44 Th. (6 £ 12 sh.); 15 lines, 15 rubies, gold dial 38 Th. (5 £ 14 sh.); No. 15,548. 18 lines, 15 rubies, brass cuvette 36⅔ Th. (5 £); No. 15,637. 16 lines, 15 rubies, gold cuvette 33⅓ Th. (5 £ 9 sh.); No. 15,551. 18 lines, ¾ P., 15 rubies 33 Th. (4 £ 19 sh.); No. 15,547. 15 lines, 15 rubies, brass cuvette 29 Th. (4 £ 7 sh.).

Gold cylinder-watches: No. 15,553. 13 lines, 10 rubies, gold cuvette 24 Th. (3 £ 12 sh.); No. 13,554. 15 lines, 6 rubies, latten cuvette 22 Th. (3 £ 6 sh.).

Silver anchor-watches: No. 15,839. 18 lines, 15 rubies 15 Th. (2 £ 5 sh.); No. 15,550. 16 lines, 11 rubies 12 Th. (1 £ 16 sh.); No. 15,638. 17 lines, 15 rubies 16 Th. (2 £ 8 sh.).

German-silver anchor-watch No. 15,953. 17½ lines, ¾ P., 7 rubies 8 Th. (1 £ 4 sh.).

Silver cylinder-watches: No. 15,555. 18 lines, 10 rubies 9 Th. (1 £ 9 sh.); No. 15,639. 16 lines, 10 rubies 8 Th. (1 £ 4 sh.); No. 15,940. 19 lines, 6 rubies 7 Th. (1 £ 1 sh.); No. 15,556. 17 lines 6⅔ Th. (1 £).

German-silver cylinder-watch No. 15,954. 19 lines, 6 rubies 5⅔ Th. (17 sh.).

Gold anchor-watch No. 73,973. with repetition, 20 to 30 rubies, savonette, to be wound up by button 220 Th. (33 £).

1439. Felsing, Conrad, watch- and clock-manu. to the court, Berlin. London 1862 honor. mention. Agts. s. No. 1438.

A regulator-clock in polixander-case, going one month, indicating seconds and fifth parts of seconds, the periodical change of the moon, the months and days with barometer and thermometer 500 Th. (75 £); seven different dials.

1442. Tiede, Fr., chronometer- and watch-maker to His Majesty the King of Prussia, to the Royal observatory and to the academy of sciences, Berlin. Berlin 1844 golden medal; London 1862 medal. Agt. Th. Tiede, 38 Holford-square, Pentonville, W. C.

1 marine two-day chronometer with Hartnup's compensating balance made of one piece, without binding screws; 2 marine two-day chronometers, having an auxiliary compensation-arrangement on a new principle; 2 Hartnup compensating-balances made of one piece, without binding-screws (rough); 3 small regulators with steel and zinc-gridiron compensating-pendulum: one in gilt ornamental case, two in plain rose-wood cases.

1443. Weiss, C., watch-manu., Gross-Glogau. Paris 1855 honor. mention; London 1862 honor. mention. Agt. Emil Rauscher, 62 Regent-street.

A church-clock going eight days and striking the quarters and hours 236 Th. (35 £).

1444. Wiese, Robert, watch-maker, Landsberg on the Warthe. Agts. s. No. 1438.

A clock.

CLASS XVI.

MUSICAL INSTRUMENTS.

1445. Adam, Gerh., manu., Wesel. London 1862 medal.

Pianoforte; pianino; square pianoforte.

1446. Bechstein, Friedr. Wilh. Carl (C. Bechstein), pianoforte-maker to His Majesty the King of Prussia, Berlin. London 1862 medal. Agts. s. No. 1438.

Two large concert-pianofortes without ornaments No. 1. 700 Th. (105 £); No. 2. 566⅔ Th. (85 £).

1450. Engel, F. A., teacher, Danzig.

1. New constructed tuning-apparatus for tuning pianofortes, p. piece with information 5 sh.; p. 6 pieces 28 sh., p. doz. 55 sh.; 2. tuning-forks of the usual chambre-tune as well as of that of Paris; p. piece 3 d., p. 6 pieces 1 sh. 5 d., p. doz. 2 sh. 9 d.

1451. Esslinger, C. W., merchant and manu. of musical instruments, Berlin, Jerusalemerstr. 53. Agt. A. Vincent, 11 Panton-square, Regent-Circus, W.

D-flute with drawer and 8 keys of German silver 3 £; D-flute with drawer and 6 keys 1 £ 4 sh.; D-piccolo-flute with drawer and 4 keys of German silver 1 £ 1 sh.; flageolet (English form) 9 sh.; aeolian-harp 18 sh.; violin-bow 17 sh.; 4 mouth-pieces for bassoon p. doz. 6 sh. 9 d.; 3 mouth-pieces for contra-bassoon 3 sh.; 4 mouth-pieces for oboe, p. doz. 9 sh.; clarinet-cane, p. doz. 2 sh.; 2 signal-whistles for fire-men, p. doz. 15 sh.; 1 pair of castanets with laces 17 sh.; do. without laces not yet finished, p. doz. pair 1 £ 16 sh.; 4 ornamented violin-screws, p. doz. 7 sh.; violin string-holder, p. doz. 7 sh. 6 d.; Violin-bridge, p. doz. 3 sh.; violoncello-bridge, p. doz. 4 sh. 6 d.; violin-sordet, p. doz. 7 sh.; sets of hair for violin-bows, p. doz. 7 sh.; violoncello-sordet 9 d.; colophony in boxes, p. do. 8 sh. 6 d.; flute-sweeper, p. doz. 9 sh.; pocket music-desk 2 sh.; music-pen, p. doz. 4 sh.; 3 rings for guitars (zither), p. doz. 4 sh. 9 d.; 2 complete sets of strings for guitars (zither), p. set 9 sh.

1453. Grimm, Carl, instrument-maker to His Majesty the King of Prussia, Berlin, Kurstr. 15. London 1862 medal. Agts. s. No. 1438. (s. II. No. 79.)

One cello, two violins, one viola, together 150 £.

1455. Hartmann, W., pianoforte-maker, Berlin.

Pianinos: No. 1. large size, jacaranda-wood, gothic ornaments, 7 octaves, iron frame and iron brace 220 Th. (33 £); No. 2. small size, jacaranda-wood, 6¾ octaves, iron frame and iron brace, 180 Th. (27 £); these instruments keep remarkably well in tune, four months at least. Prices in cash at the place; for packing 1 £ more.

1456. Ibach, Adolph, Sons, pianoforte- and organ-makers, Barmen. London 1862 honor. mention. Agts. Morgan Brothers, 21 Bow-Lane, Cannon-street, City. (s. II. No. 84.)

A pianoforte with repetition-mechanism 105 £; a piano with iron for a damp and hot climate 52 £.

1457. Kloss, Ernst, chanter and violin-maker, Bernstadt. Agts. M. Caspari & Co., 7 Cullum-street, City.

Violin and violin-stick 24½ Frd'or. (20 £).

1458. KNACKE, BROTHERS, pianoforte-makers, Münster. London 1862 medal. Agts. Lion M. Cohn, Phaland & Dietrich, represented by Ch. Trübner, 20 St. Dunstans Hill, City, and at the Exhib. Building.

A grand pianoforte for concerts 600 Th. (90 £); a pianino 350 Th. (52½ £).

1459. KÜNTZEL, L., instrument-maker to the court, Berlin, Kronenstr. 75. Agts. s. No. 1458.

1 violoncello, 2 violins, 2 bass-viols for a quintetto 2000 Th. (300 £).

1463. MAHLITZ, E., manu., Berlin, Kloster-str. 82. London 1862 honor. mention.

Patent grand pianoforte; pianino.

1464. MANN, THEOPHILUS, manu., Bielefeld. Agt. H. Landwehr, Exhib. Building.

A Pianino 4 ft. long, 4½ ft. high, 2 ft. broad 30 £.

1465. OBERKRÜGER, FRIEDRICH, Piano-manu., Cologne. Agts. s. No. 1458.

A piano oblique in polixander-wood 220 Th. (33 £).

1466. OECHSLE, A., SONS, manu. of drums, Berlin, Wallstr. 86. London 1862 honor. mention. Agts. s. No. 1458.

1. A large janissary-drum with braces, sticks and strap, English arms, 5 £; 2. a large janissary-drum with screw (of small height), with sticks, key, strap and cushion, English arms 6 £; 3. a military-drum of brass of small height with drum-strap, drum-sticks, hook (eagle) with loop. 2 £ 10 sh.; 4. and 5. a pair of kettle-drums with stand and drum-sticks 12 £ 10 sh.

1467. OTTO, LUDWIG, merchant and manu. of musical instruments, Cologne. London 1862 honor. mention. Agts. Morgan Brothers, 21 Bow-Lane, Cannon-street, City.

Double-bass 25 £; violoncello 20 £; alto-viola 10 £; 3 violins, each 10 £; 2 double bass-bows, each 15 sh.; violoncello-bows 1 £ 10 sh.; 4 violin-bows, each 15 sh.; 3 guitars, No. 1. 5 £; No. 2. 3 £ 10 sh.; No. 3. 2 £.

1468. SCHMIDT, FRIEDR. ADOLPH, brass wind instrument-maker, Cologne. London 1862 medal. Agts. Morgan Brothers, 21 Bow-Lane, Cannon-street, City.

Trumpet in B (piston); euphonio; C-bass.

1469. SCHWECHTEN, G., pianoforte-manu., Berlin, Lindenstr. 40. London 1862 honor. mention. Agts. s. No. 1458.

Polixander-pianino 400 Th. (60 £).

1471. SPANGENBERG, WILHELM, pianoforte-manu., Berlin, Charlottenstr. 60. London 1862 medal. Agts. s. No. 1458.

Concert-pianoforte 125 £; pianino 90 £.

1472. WILLMANNS, G. (Westermann & Co.), manu., Berlin. Berlin 1844 honor. mention; London 1851 honor. mention; Munich 1854 medal of honor; Paris 1855 bronze medal. Agts. s. No. 1458.

Grand pianoforte 600 Th. (90 £).

CLASS XVII.
SURGICAL INSTRUMENTS.

1473. GOLDSCHMIDT, S., mechanician and truss-maker to His Majesty the King of Prussia, manufactory and magazine of Prussian patent trusses, surgeons machines, bandages and instru-ments, as well as every object, required for nursing sick persons, Berlin. Berlin 1844 honor. mention; London 1862 medal.

Surgical bandages, machines, instruments and trusses.

1474. IMME, JULIUS, & Co., manu., Berlin.

Volta-electric brush.

1475. KAESTNER, FR., dentist, Cologne.

Instruments for dentists, as: hand and mouth-reflectors; a drilling engine; hand-drills; tooth-saws; English and American tooth-tongs, straight and crooked, with round and angular bits; instruments for filling and cleaning; caoutchuc and leaf-gold; an operation-chair; a turning-lathe and a grinding-stool; an apparatus for vulcanizing; a pump for injecting the caoutchuc into the basins; an inductive apparatus; a galvano-caustor for destroying the nerves; teeth of hippopotamus, gold and caoutchuc.

1478. LUTTER, AUGUST, manu. of surgical instruments and bandages, Berlin. London 1862 medal. Agt. Henry Griebel, 3 Crown-street, Walworth.

Surgical instruments and bandages.

1479. PISCHEL, E., manu., Breslau.

A box containing 4 Groove-cells; a case with galvano-caustic instruments.

1480. ROETTGEN, F., manu., Bonn.

Hearing-tubes; carriage speaking-pipes; respirators; life-exciters, called abductors; flesh-brushes.

1481. TRESCHINSKY, F. W. G., manu., Berlin, Krausenstr. 62.

Orthopaedic instruments and apparatus.

1482. WINDLER, H., surgical instrument-maker, Berlin, Mittelstr. 64. London 1862 medal.

Surgical instruments.

CLASS XVIII.
COTTON.

1484. BORNEFELD, W., manu., Gladbach (Düsseldorf). London 1862 medal.

Cotton and worsted canvas.

1485. COLSMANN, JOH. FRIEDR. (Colsmann & Co.), manu., Barmen. Agts. Morgan Brothers, 21 Bow Lane, Cannon-street, City.

(s. No. 2030 a.)

Match-cords No. 1. 2. 3. 18 sg. p. Pfd.

1486. ERMEN & ENGELS, cotton-spinners, manu. of cotton sewing and knitting-yarns, prop. of bleaching and dyeing establishments, Barmen. Manufactory Engelskirchen. Berlin 1844 silver medal; Munich 1854 prize-medal; Paris 1855 bronze medal.

Knitting-yarns, 13 sorts, each ½ bundle; prices p. lb. Engl. at which having been sold from January to Juli 1861; at present 10—15 pCt. more: 6 fold Estremadura No. 7. white 2 sh. 1 d.; do. No. 4. grey 1 sh. 7 d.; 4 fold superior No. 30. white 2 sh.; do. No. 36. grey 1 sh. 11 d.; 4 fold best No. 24. white 1 sh. 9½ d.; do. No. 20. grey 1 sh. 6½ d.; 3 fold secunda No. 18. white 1 sh. 3½ d.; do. No. 16. grey 1 sh. 1 d.; 4 fold do. No. 14. mixed colors 1 sh. 4½ d.; 8 fold best No. 30. marbled 2 sh. 10 d.; 4 fold imitation of merino No. 12. grey 4. 1 sh. 5 d.; do. No. 12. drab 3. 1 sh. 5 d.; do. No. 12. brown drab 2. 1 sh. 5 d.

Sewing-yarns: 2 fold salmon tie ¼ oz. skeins No. 24. white 1 sh. 11 d.; do. secunda (tambour) No. 18. white 1 sh. 7 d.

F °

(Eisengarn) Diamond-thread: $\frac{1}{2}$ bundle 2 fold $\frac{1}{4}$ oz. skeins No. 20. white 2 sh. 2 d.; do. No. 20. black 1 sh. 10 d.

Twist: No. 30. prima mule; No. 18. do. water.

Also: samples of white, brown, mixed colors, flamy and marbled knitting-yarns of white, brown and colored sewing-yarns, of black, white and colored diamond-thread (Eisengarns) and of black, white and colored diamond-thread bobins.

1487. FLEISCHER, LUDWIG, Mühlhausen (Erfurt). Agts. Lion M. Cohn, Phaland & Dietrich, represented by Ch. Trübner, 20 St. Dunstans Hill, City, and at the Exhib. Building.

Silk canvas No. 6. 20$\frac{1}{2}$ Elle 16 Th. 12 sg. (15 yards 2 £ 11 sh. 4 d.); imitated silk canvas of cotton No. 5. 22 Ellen 4 Th. 12 sg. (16 yards 13 sh. 2 d.); cotton canvas No. 4/0. 22 Ellen 3 Th. 10 sg. (16 yards 10 sh.); do. No. 2. 22 Ellen 3 Th. 10 sg. (16 yards 10 sh.); do. No. 5. 22 Ellen 3 Th. 10 sg. (16 yards 10 sh.)

1488. GRUNSFELD, JOSEPH, SONS, Heiligenstadt (Erfurt). London 1862 medal: Agts. s. No. 1487.

a. Bedticking: No. 367. 12/4 broad 59$\frac{1}{2}$ Ellen (43$\frac{1}{2}$ yards) 9$\frac{2}{3}$ sg. (1 sh. 4 d.); No. 577. 12/4 broad 58 Ellen (42$\frac{1}{2}$ yards) 9$\frac{1}{3}$ sg. (1 sh. 3 d.); No. 300. 6/4 broad 60 Ellen (44 yards) 5$\frac{1}{4}$ sg. (9 d.); No. 387. 6/4 60$\frac{1}{2}$ Ellen (44 yards) 4$\frac{5}{6}$ sg. (8 d.); No. 531. 6/4 broad 60 Ellen (44 yards) 4$\frac{2}{3}$ sg. (7$\frac{1}{2}$ d.); No. 364. 6/4 broad 59 Ellen (43 yards) 4$\frac{1}{4}$ sg. (7 d.); No. 570. 6/4 broad 59$\frac{1}{2}$ Ellen (43$\frac{1}{2}$ yards) 4$\frac{1}{4}$ sg. (7$\frac{1}{2}$ d.); No. 327. 6/4 60$\frac{1}{2}$ Ellen (44 yards) 4$\frac{2}{3}$ sg. (7$\frac{1}{2}$ d.); all p. Berl. Elle (p. yard).

b. Négligé-stuffs: No. 1000. 6/4 broad 57$\frac{1}{2}$ Elle (41 yards) 4$\frac{1}{3}$ sg. (7 d.); No. 1001. 6/4 broad 58 Ellen (41$\frac{1}{2}$ yards) 4$\frac{1}{3}$ sg. (7 d.); p. Berl. Elle (p. yard).

c. Stuffs for aprons and ladies' dresses: No. 714. 7/4 broad 29$\frac{3}{4}$ Ellen (21$\frac{1}{2}$ yards) 6 sg. (9$\frac{3}{5}$ d.); No. 718. 7/4 broad 30$\frac{1}{4}$ Ellen (22 yards) 5$\frac{2}{3}$ sg. (9 d.); No. 870. 7/4 broad 29$\frac{1}{2}$ Ellen (21$\frac{1}{2}$ yards) 5$\frac{3}{4}$ sg. (9 d.); No. 784. 7/4 broad 30$\frac{1}{4}$ Ellen (22 yards) 5$\frac{5}{12}$ sg. (8$\frac{1}{2}$ d.); all p. Berl. Elle (p. yard).

1490. KLEMME, GUST. (Klemme & Co.), manu., Crefeld. London 1862 medal. Agts. Gd. Ingelbach & Wolffgang, 11 Staining-Lane, Gresham-street.

Embossed ribbons of silk, silk and cotton velvet.

1492. MITSCHERLICH, FRIEDRICH AUGUST, & FERDINAND (F. A. Mitscherlich), manu., Eilenburg. Manufactories at Eilenburg, Halle and Görlitz. Munich 1854 honor. mention; London 1862 medal.

Satin-striped No. 1.; quilting-raised No. 2.; do. not raised No. 3. 4. 5.; quilting-ribs No. 6.; do. rayé No. 7.; quilting raised No. 8.; damasc No. 9.; mock-quilting No. 10.; not raised quilting with border for petticoats No. 11.; dimitty raised No. 12.: imitated quilting No. 13.; raised quilting with border for petticoats No. 14.; mock-quilting No. 15.; each one piece.

6 pieces of measured quilting-petticoats raised and not raised.

1493. PFERDMENGES & SCHMÖLDER, Rheydt; cotton spinning and weaving establishment, Grevenbroich. London 1862 medal.

Cotton cloths, printed (beaver, calmuc and chinchilla) 6—8 d. p. yard.

1494. ROLFFS & Co., calico-printers, Cologne. Manufactory at Siegfeld near Cologne.

10 pieces of glazed calicoes for furniture 6/4 wide: 4 pieces do. twill; cotton handkerchiefs, printed in madder and steam-colors, 2 doz. 5/4, 23 doz. 1 3/8; 22 doz. 6/4; 1$\frac{1}{2}$ doz. 7/4.

1495. SENNE, JOH. FRIEDR., wick-manu., Erfurt. Agts. s. No. 1487.

Chemically prepared lamp-wicks; prices p. 12 doz.:
a. Shade lamp-wicks No. 1. 13 sg. (1 sh. 3$\frac{3}{5}$ d.); No. 2. 16 sg. (1 sh. 7$\frac{1}{5}$ d.); No. 3. 20 sg. (2 sh.); No. 4. 24 sg. (2 sh. 4$\frac{4}{5}$ d.); No. 5. 28 sg. (2 sh. 9$\frac{3}{5}$ d.); No. 6. 32 sg. (3 sh. 2$\frac{2}{5}$ d.); No. 7. 36 sg. (3 sh. 7$\frac{1}{5}$ d.); No. 8. 40 sg. (4 sh.).
b. Cylindrical-wicks No. 1. 21 sg. (2 sh. 1$\frac{1}{5}$ d.); No. 2. 25 sg. (2 sh. 6 d.); No. 3. 29 sg. (2 sh. 10$\frac{4}{5}$ d.); No. 4. 33 sg. (3 sh. 3$\frac{3}{5}$ d.); No. 5. 37 sg. (3 sh. 8$\frac{2}{5}$ d.); No. 6. 41 sg. (4 sh. 1$\frac{1}{5}$ d.); No. 7. 45 sg. (4 sh. 6 d.); No. 8. 49 sg. (4 sh. 10$\frac{4}{5}$ d.).
c. Flat-wicks No. I. from 1—8. in. Leipsic m. No. 1. 36 sg. (3 sh. 7$\frac{1}{5}$ d.); No. 2. 48 sg. (4 sh. 9$\frac{3}{5}$ d.); No. 3. 60 sg. (6 sh.).

1496. STERNENBERG, J. H., & SONS, manu., Schwelm (Arnsberg). Paris 1855 bronze medal; London 1862 honor. mention. Agts. Heintzmann & Rochussen, 9 Friday-street, Cheapside, E. C.

10/4 Brab. Ell or 68$\frac{3}{4}$ in. 1 a. bed-ticking, No. 2. 40 yards 7 £ 10 sh.; 10/4 Brab. Ell or 68$\frac{3}{4}$ in. do. No. 715. 36$\frac{1}{2}$ yards 5 £ 17 sh.; 10/4 Brab. Ell or 68$\frac{3}{4}$ in. stuff for blinds No. 2. 40 yards 4 £. 10 sh.; 10/4 Brab. Ell or 68$\frac{3}{4}$ in. do. No. 22. 40 yards 4 £ 19 sh.; 9/4 Brab. Ell or 61$\frac{3}{4}$ in. 1 a. bed-ticking, No. 2. 38$\frac{1}{4}$ yards 6 £ 7 sh. 6 d.; 9/4 Brab. Ell or 61$\frac{3}{4}$ in. do. No.831. 36$\frac{1}{2}$ yards 4 £ 5 sh. 6 d.; 9/4 Brab. Ell or 61$\frac{3}{4}$ in. stuff for blinds No. 79. 40$\frac{1}{2}$ yards 4 £ 4 sh.; 9/4 Brab. Ell or 61$\frac{3}{4}$ in. do. No. 111. 34$\frac{1}{2}$ yards 3 £ 12 sh.

1497. and **1498.** WOLFF, SCHLAFHORST & BRÜEL (formerly Wolff, Schlafhorst & Brüel, and Wolff & Schlafhorst), manu., prop. of establishments for spinning, weaving, dyeing, printing, dressing and finishing, M. Gladbach. Paris 1855 bronze med.; London 1862 medal. Agts. Morgan Broth., 21 Bow-Lane, Cannon-street, City.

70 coupons of cotton goods: dark brown and white calmuc 16 Brab. Ell (12 yards) 4$\frac{1}{4}$ to 4$\frac{5}{8}$ sg. (7$\frac{1}{2}$ to 7$\frac{5}{8}$ d.); cotton beaver, plain colored 32 Brab. Ell (24 yards) 3 to 3$\frac{1}{2}$ sg. (4$\frac{3}{4}$ to 5$\frac{1}{4}$ d.); fustian 16 Brab. Ell (12 yards) 2$\frac{11}{12}$ to 3 sg. (4$\frac{3}{4}$ to 4$\frac{7}{8}$ d.); different kinds of printed cotton beaver 296 Brab. Ell (222 yards) 3$\frac{1}{8}$ to 3$\frac{11}{12}$ sg. (5 to 6$\frac{1}{4}$ d.); different kinds of cotton for trousers 200 Brab. Ell (150 yards) 5$\frac{1}{2}$ to 6$\frac{3}{4}$ sg. (8$\frac{3}{4}$ to 10$\frac{3}{4}$ d.), all p. Brab. Ell (p. yard).

CLASS XIX.

FLAX AND HEMP.

1499. ADLER, ELKAN (Adler Brothers), merchant and manu., Neustadt (Silesia). London 1862 medal. Agts. Lion M. Cohn, Phaland & Dietrich, represented by Ch. Trübner, 20 St. Dunstans Hill, City, and at the Exhib. Building.

Linen damask table-cloth, napkins of colored linen, with silk mixture, towels etc.

No. 1. 1 damask table-cloth, pure linen, 11/4 broad, 3 Ells long, 6 napkins, 6/4 □, 9 sh.; No. 2. and 3. 2 do. fine, each 12 sh. 9 d.; No. 4. and 5. do. with 12 napkins, fine, each 1 £ 4 sh.; No. 6. do. 5$\frac{1}{2}$ Ells long, 11/4 Ells broad, 17 sh. 3 d.;

No. 7. diaper table-cloth, 3 Ells long, 11/4 Ells broad, 6 napkins, 9 sh.; No. 8. do., 5½ Ells long, 11/4 Ells broad, 12 napkins, 17 sh. 6 d.; No. 9. damask table-cloth, 12/4 broad, pure linen, 12 napkins, 1 £; No. 10. do. 18 napkins, 1 £ 10 sh.; No. 11 do. 24 napkins, 2 £ 2 sh.; No. 12. 1 doz. damask towels, ⅞ broad, 10/4 long, 12 sh.; No. 13. lilac silk napkins, 11/4 □, 1 £ 5 sh. 3 d.; No. 14. yellow do. 1 £ 4 sh.; No. 15. pink do. 1 £ 5 sh. 3 d.; No. 16. red do. 1 £ 9 sh.; No. 17. black and white do. 1 £ 2 sh. 6 d.; No. 18. unbleached linen do. 3 sh. 6 d.; No. 19. chamois linen do. 4 sh.

1500. Bender, sen., Wilhelm, manu. of linen damask, Bleicherode (Erfurt). London 1862 honor. mention.

1. Sample of flax prepared in the common way with yarn spun from it and tow.

2. Sample of flax (originally of the same kind as No. 1.) prepared without chemical means after the process improved by the exhibitor, with yarn spun from this flax and tow.

1501. Linen-manufactories of Bielefed, represented by H. Landwehr at the Exhibition Building. London 1862 medal.

1. Delius, E. A. & Sons. Berlin 1844 silver medal.

34 pieces of bleached linen, hand and machine-spun; handkerchiefs, 92 £ 4 sh. 3 d.

2. Krönig, F. W. & Sons. London 1851 honor. mention; Paris 1855 bronze medal.

20 pieces of bleached linen, 101 £ 14 sh.

3. Kisker, A. W. (formerly F. Lüder & Kisker), purveyor to the court. Berlin 1844 silver medal; Paris 1855 silver medal.

114 pieces of diaper, damask, table-cloths, towels, napkins (grey and bleached) and bleached linen, 172 £ 11 sh. 6 d.

4. Wittgenstein, H. M.

16 pieces of bleached linen 66 £ 12 sh.

5. Gante, C. F. & Sons.

Bleached linen, handkerchiefs and shirt-fronts, 106 £ 18 sh. 8½ d.

6. Lueder, F. (formerly F. Lüder & Kisker). Berlin 1844 silver medal; Paris 1855 silver medal.

Bleached linen, diaper and damask table-cloths, napkins (grey and bleached) quilted diaper and towels, 91 £ 15 sh. 9 d.

7. Bertelsmann & Son.

17 pieces of bleached linen 56 £ 12 sh. 6 d.

8. Krönig & Jung (formerly Krönig & Bökemann). Munich 1854 medal of honor.

18 pieces of bleached linen and handkerchiefs 48 £ 9 d.

9. Rabe & Consbruch.

Bleached linen, handkerchiefs and shirt-fronts, 67 £ 15 sh. 3 d.

10. Potthoff, C. H. Munich 1854 medal of honor.

Bleached linen and handkerchiefs 33 £ 4 sh. 6 d

11. Piderit, F.

Bleached linen, handkerchiefs and shirt-fronts, 48 £ 8 sh. 6 d.

12. Goldbeck & Vieler.

Bleached linen, handkerchiefs and shirt-fronts, 94 £ 13 sh. 4 d.

13. Colbrunn, C.

1 bleached linen, handkerchiefs for ladies and gentlemen 103 £ 17 sh. 6 d.

14. Heidsick, L.

Bleached linen, handkerchiefs and shirt-fronts, 37 £ 4 sh. 6 d.

15. Meyer, S. & Co.

Shirt-fronts and shirts 29 £ 8 sh.

1502. Braunsberg agricultural association, Boehmenhoefen near Braunsberg. London 1862 honor. mention. Agts. Lion M. Cohn, Phaland & Dietrich, represented by Ch. Trübner, 20 St. Dunstans Hill, City, and at the Exhib. Building.

Samples of Varmian flax.

1503. Cadura, Heinrich, merchant and manu. of driving-belts, Breslau. Agts. s. No. 1499. (s. No. 1292, 1807, 1938.)

Hemp-hoses.

1504. Delius, Conrad Wilhelm & Co., manu., Versmold (Minden). Berlin 1844 silver medal; London 1862 medal.

Sail-cloth of hemp yarns, hand-spun and hand-woven, four pieces at 37½ yards: No. 00. 15 d.; No. 0. 14 d.; No. I. 13 d.; No. II. 12 d.; all p. yard.

1505. Eichelbaum, S. (S. Eichelbaum), merchant, Insterburg, house of correction Insterburg. Munich 1854 honor. mention. Agts. s. No. 1502.

6 packs of fishing-nets from No. 1. to 6., every pack 10 cords long and 60 couplings broad, p. 1 cord or 6 ft.: No. 1. 2½ sg. (3 d.); No. 2. 2⅔ sg. (3⅖ d.); No. 3. 3 sg. (3⅗ d.); No. 4. 3⅔ sg. (4⅖ d.); No. 5. 4¼ sg. (5⅒ d.); No. 6. 6½ sg. (7⅘ d.); 1 windlass-cabel one inch in diameter, worked with 6 strands and a heart, twenty pounds weight p. lb. 7 sg. (8⅖ d.); 1 carriage-line for two horses, of 6 strands and worked with a heart, 3¼ pounds weight 22½ sg. (2 sh. 3 d.); 2 Hungarian halters p. piece 6 sg. (7⅕ d.);

1506. Engel, F. E., rope-maker and manu., Görlitz. Agts. s. No. 1502.

Travelling-bag 12 Th. (1 £ 16 sh.); game-bag net decorated with a hart 5 Th. (15 sh.); fowling-bag with girdle 3 Th. (9 sh.); two game-bag nets p. piece 1 Th. (3 sh.); clothes-line, made of Manilla hemp 3 Th. (9 sh.); bell-rope of Manilla hemp 4 Th. (12 sh.); 1500 Ells of sewing-thread 1 Th. (3 sh.).

1507. Erdmannsdorf flax yarn-spinning-mill and weaving-establishment, Erdmannsdorf near Hirschberg. London 1862 medal. Agts. s. No. 1502.

a. ½ cut of creas No. 150. breadth 16/4 Ells, length 30 Ells (Berl.) 20½ Th. (p. yard 2 sh. 9⅓ d.); b. 1 cut of creas No. 90. breadth 12/4 Ells, length 60 Ells (Berlin) 26⅔ Th. (p. yard 1 sh. 9⅞ d.); c. 1 cut of creas No. 130. breadth 12/4 Ells, length 60 Ells (Berlin) 34 Th. (p. yard 2 sh. 3⅞ d.); d. 1 cut of linen creas No. 105. breadth 7/4 Ells, length 60 Ells (Berlin) 25½ Th. (p. yard 1 sh. 8⅞ d.); e. 1 piece do. No. 70. breadth 6½/4 Ells, length 50 Ells (Berlin) 12 Th. (p. yard 1 sh.); f. ½ length of creas à la morlaix No. 50. length 45 Ells (Berlin) 9⅓ Th. (p. yard 9⅝ d.); g. do. No. 55. ½ length 45 Ells (Berlin) 9$\frac{5}{12}$ Th. (p. yard 10⅛ d.); h. do. No. 60. length 45 Ells (Berlin) 10⅔ Th. (p. yard

11⅛ d.); i. do. No. 65. length 45 Ells (Berlin) 11⅝ Th. (p. yard 1 sh. ¼ d.); k. do. No. 70. length 45 Ells (Berlin) 13⅓ Th. (p. yard 1 sh. 1⅔ d.).

1508. Exner & Stockmann, manu., Schweidnitz (Breslau). London 1862 honor. mention. Agts. Heintzmann & Rochussen, 9 Friday-street, Cheapside.

Indigo-dyed and printed linen-cloth.

1509. Fränkel, S., manu., Neustadt (Oppeln). London 1862 medal.

Plain linen, table-cloths and napkins.

1509 a. Goszlau, C., manu., Nitsche (Posen).

Damask towels.

1511. von Huhn, Nicolai, prop., Ober-Gerlachsheim. London 1862 honor. mention. Agts. Lion M. Cohn, Phaland & Dietrich, represented by Ch. Trübner, 20 St. Dunstans Hill, and at the Exhib. Building.

Specimens of lin-seed, p. ton. 15 Th.; flax p. Ctr. 25—30 Th.

1512. Helling, Fritz, Sail-cloth manu., Borgholzhausen (Minden). London 1862 medal.

Sail-cloth, 15 pieces.

1513. Herford company for the manufacture of hand-spun linen, Herford (Minden). Paris 1855 silver medal; London 1862 medal. Agt. Heinr. Landwehr, at the Exhibition Building.

6 pieces of white linen, 37½ yards length, 34 in. breath, every piece: No. 9984. 30 Th. (4 £ 10 sh.); No. 9471. 40 Th. (6 £); No. 515. 50 Th. (7 £ 10 sh.); No. 9988. 60 Th. (9 £); No. 3553. 85 Th. (12 £ 15 sh.); No. 2269. 100 Th. (15 £); 5 doz. of white handkerchiefs, 25 in. square No. 2992. 50 Th. (7 £ 10 sh.). Samples of warps and wefts, hand-spun.

These articles will be followed by a very fine piece of grey linen, the finest ever made, 37½ yards length, 34 in. breadth 240 Th. (36 £).

1514. Hohlstein, Bernhard, flax-cultivator, Bollstedt near Mühlhausen (Erfurt).

Samples of flax in three assortments, together with samples of the tow obtained in treating it.

von Huhn. (s. No. 1511.)

1516. Kirstein, C., manu. of linen-goods, Hirschberg. Berlin 1844 bronze medal; London 1851 honor. mention, 1862 medal. Agts. Heintzmann & Rochussen, 9 Friday-street, Cheapside. (s. No. 1203.)

Pure linen: table-cloth with twelve napkins, damask 14 Th. (2 £ 2 sh.); do. Jaquard 11 Th. (1 £ 13 sh.); tea table-cloth, damask 3⅓ Th. (10 sh.); bleached linen No. 60. 52 Berl. Ells (37½ yards) 11 Th. (1 £ 13 sh.); do. No. 90. 16 Th. (2 £ 18 sh.); estopilles claires unies, 9½ Berl. Ells (7 yards) 2 Th. (6 sh.).

Half linen: wharp of cotton-yarn, weft of linen-yarn, the figures of cotton-yarn; 5 pieces white, blew, red, violet and lilac, estopilles claires à fleurs, 10¼ Berl. Ells (7½ yards) 2½ Th. (7 sh. 6 d.).

1516 a. Lehmann, Robert, landed prop., Nitsche (Posen). London 1862 honor. mention. Agts. s. No. 1511. (s. No. 1111, 1208.)

Sample of flax, 24 Pfd. 6 Th. (18 sh.).

1517. Baron von Lüttwitz, Rudolph, landed prop., Simmenau near Creutzburg (Oppeln). Berlin 1844 red eagle 3. class; London 1851 prize-medal; Munich 1854 medal of honor; Paris 1855 bronze medal and from the academy of Paris the great gold medal 1. class; London 1862 medal. Agts. s. No. 1511.

8 packets of flax, water-retted p. Zollz. 25 Th. (100 lbs. 3 £ 15 sh.).

1518. Mevissen, Gerh., manu., Dülken (Düsseldorf).

Flax; linen yarn; thread.

1519. Müller, Melchior Heinrich, manu., Münster. London 1862 medal. Agt. Andr. O. Brückmann, 3 Rood-Lane, City.

Sail-cloth, tilt and tent-cloth, caoutchouc-tilts for rail-road cars without seam.

List of prices at the agent's.

1520. Nitzsche, J. H., Berlin, Frankfurter Thor. London 1862 honor. mention.

Hemp driving-straps, hose for fire engine.

1521. von Pannwitz, W., farmer of the Royal domain Bürgsdorf, Creutzburg (Silesia). London 1862 medal. Agts. s. No. 1511.

3 packets of flax, p. 100 pounds 23 Th.

1522. Pohl, August, damask-weaver, Stralsund. London 1862 honor. mention. Agts. s. No. 1511.

Six tea table-cloths, brown and white, pattern representing: »christmas tree« 6 Th. (18 sh.); six tea table-cloths, brown and white, pattern representing: »the hunting-seat of prince Putbus in the isle of Rügen« 5 Th. (15 sh.); two damask table-cloths, each with 12 napkins 18 Th. (2 £ 14 sh.); two doz. damask towels 10 Th. (1 £ 10 sh.). The above articles are woven of English yarn, partly bleached, partly grey, just as taken from the loom and without finish.

One dozen towels, home-made and afterwards bleached 18 Th. (2 £ 14 sh.).

1523. Ravensberg Spinning Establishment. London 1862 medal.

45 bundles of wet and dry spun linen and tow-yarn No. 2¾—80. 19 £ 16 sh. 3 d.

1524. Baron Reiswitz, gentleman farmer, Wendrin near Sausenberg (Silesia). London 1862 honor. mention. Agts. s. No. 1511.

Several samples of flax.

1525. Rösner, C., manu., Wüste-Waltersdorf (Breslau).

Linen cloth.

1526. Schoeller, Mevissen & Bücklers, manu., Düren. Munich 1854 medal of honor.

Machine-spun linen and tow-yarn.

1529. Stoltenburg, Eduard (E. Stoltenburg), linen drell- and damask-manu., Stralsund. Agts. s. No. 1511.

One damask linen table-cloth with 24 napkins 5 £; do. with 12 napkins 2 £ 8 sh.; one doz. damask towels 28 sh.; one doz. diaper towels 18 sh.

1529a. Spinning Establishment „Vorwärts". Paris 1855 silver medal; London 1862 honor mention.

81 bundles of wet-spun linen and tow-yarn No. 14—70. 32 £ 6 sh. 1½ d.

1530. Wagener, Louis, Royal counsellor of rural economy and administrator of the domain of Proskau, prod. Agts. Lion M. Cohn, Phaland & Dietrich, represented by Ch. Trübner, 20 St. Dunstans Hill, City, and at the Exhib. Building. (s. No. 1150.)

25 pounds of flax, prepared for the market, p. Zollz. 18—25 Th.

1531. Willmann, A., & Sons, manu., Patschkey (Breslau). London 1862 medal.

Specimens of flax.

Wipprecht, E., Berlin. Linen horse-clothes. (s. No. 1711.)

1532. Zöllner, A., manu., Stralsund. Paris 1855 honor. mention; London 1862 honor. mention. Agts. s. No. 1530.

One damask table-cloth with 12 napkins No. 50., 12/4 broad., satin-worked with eight threads 17 Th. (2 £ 10 sh.); one damask table-cloth with 12 napkins No. 40., 10/4 broad, satin-worked with five threads 12 Th. (1 £ 15 sh.); one ticking table-cloth with 12 napkins No. 25., 10/4 broad 10 Th. (1 £ 10 sh.); one doz. damask towels No. 50., 8/4 long, satin-worked with eight threads 8 Th. (1 £ 4 sh.); one doz. ticking-towels No. 25., 8/4 long 6 Th. (17 sh. 9 d.).

CLASS XX.

SILK AND VELVET.

Adler, Elkan, Brothers, Neustadt (Oppeln). Silk napkins. (s. No. 1499.)

1534. Andreae, Christoph, silk and velvet-manu., Mülheim on the Rhine. London 1851 prize-medal; Paris 1855 great medal of honor; London 1862 medal. Gen. Agt. Eduard Rhodius, care of Killy, Traub & Co., 52 Bread-street, City. Spec. Agt. for furniture-goods D. Walters & Sons, 43 Newgate-street.

Plain silk velvets, black and colored; fancy-velvet for waist-coats; plain velvet-ribbons, black and colored; figured velvet-ribbons; velvet-scarfs; furniture-plushes of mohair or wool; figured furniture-velvets; table-covers; taffeta-ribbons, plain and figured.

1535. vom Baur, J. H., Son, manu., Ronsdorf. Paris 1855 bronze medal; London 1862 honor. mention. Agts. A. G. Franklin & Co., 14 South-street, Finsbury-square E. C.

Ribbons and galloons of silk, silk-mixture and wool; assortment of hatbands.

1536. Camphausen, Nicol. Wilh., as representative of the central silkworm and winding-establishment of the Rhenish Prussian agricultural society, Bendorf on the Rhine, manufactoryEngers near Coblenz. London 1862 honor. mention.

5 lbs. raw silk; 1 oz. Balkan cocoons; ½ oz. Japan cocoons; ¼ oz. silkworm-gut.

1537. Bielefeld Silk - Manufacturers. London 1862 medal. Represented by H. Landwehr, Exhibition Buildings:

1. Delius, E. A. & Sons.

Different kinds of black and colored silks, 24 pieces, 130 £ 15 sh. 11 d.

2. Krönig, C. & Th.

Different kinds of black silks and colored velvets, 12 pieces, 16 £ 7 sh. 9½ d.

3. Bökemann & Wessel.

Different kinds of black silks and black velvet, 8 pieces, 36 £ 6 sh. 1 d.

4. Wittgenstein, C. H., & Son.

Different kinds of black and colored silks, 33 pieces.

5. Bertelsmann & Son.

Different kinds of black and colored silks, 15 pieces, 47 £ 2 sh. ½ d.

6. Wertheimer, M.

Different kinds of black velvet, 5 pieces, 16 £ 5 sh. 8 d.

7. Bartels, Brothers, Gütersloh.

Different kinds of black and colored silks, 10 pieces, 43 £ 6 d.

1538. vom Bruck, H., Sons, silk-manu., Crefeld. Berlin 1844 silver medal; London 1851 prize-medal; London 1862 medal.

Plain black and colored velvets and velvet-ribbons; fancy velvet-ribbons; velvet waist-coatings and scarfs.

Cleff, Brothers, Barmen. Silk slips. (s. No. 1593.)

1540. Dappen, Böckner & Schnütgen, silk-manu., Crefeld.

Silk fancy-stuffs for gentlemen; cravats, slips and scarfs; those articles also ready made.

1541. Baron von Diergardt (Friedrich Diergardt), privy counsellor of commerce, silk-goods manu., Viersen (Düsseldorf). Berlin 1844 the slip to the order of the red eagle 3. class; London 1851 prize-medal; Munich 1854 knight of the order of merit of St. Michael; Paris 1855 officer of the legion of honor.; London 1862 juror for Cl. XX. Agt. H. Borckenstein, 8 Moorgate-street.

Plain silk-velvets, black and colored, in several qualities and widths to 72 in., in English, German and French make. Fancy waistcoat-velvets and terry-velvets. Plain and fancy silk velvet-ribbons, black and colored, with cut and woven edges, several qualities in narrow and wide numbers.

1542. Draemann, widow, Elvira, & Dellmann, Hugo (Draemann & Dellmann), manu. of silk and silk-mixtures, Crefeld. Agts. Jordan & Parker, 26 Noble-street, E. C.

A doll dressed with the exhibitor's silks and silk-mixtures 2 £; 4 dresses 20 in. popeline unie; 3 dresses 20 in. popeline cannelée □; 6 shawls of silk-mixture.

1543. Engelmann & Bohnen (Ch. Engelmann & Son), Crefeld. Paris 1855 bronze medal: London 1862 honor. mention. Agts. Morgan Brothers, 21 Bow-Lane, Cannon-street. City.

Pure silk-stuffs, colored and black for ladies' dresses: silk shawls.

1544. Fränkel. S., manu., Neustadt(Oppeln). Silk table covers.

1547. Gressard & Co., manu., Hilden. London 1862 medal. Agts. A. Stenger & Co., 4 Gresham-street.

Printed silk corahs and bandana-handkerchiefs: printed silk bandana-dresses. For prices apply to the agents.

1548. HAMERS, A., Crefeld. (s. No. 1724.)

1549. HEIMENDAHL, GUSTAV, manu. and merchant, Crefeld. Munich 1854 honor. mention; Paris 1855 bronze medal; London 1862 medal. Agt. M. Dahlke.

Cocoons and silks of the exhibitor's own breeding: raw silk; 22/26. organzine, fort apprêt; 22/24. organzine moyen apprêt; 26/30. trame; 40/44. China organzine dop. strafato; 26/30. organzine dop. strafato; 40/44. China trame à tours comptés; 50/60. Japan trame à tours comptés; 50/60. Japan trame à 3 bouts; crêpe-silk; sewing-silk; silk for reeds or slays, grenadine; 20/22. organzine excellent; 140 points filato; 130 points torto; Bengal 26/30. organzine.

1550. HIPP, H. G., & BETTER, silk-manu., Crefeld. Paris 1855 silver medal; London 1862 honor. mention.

Plain and fancy-velvets; plain and fancy velvet-ribbons; silk shawls; silk veils and grenadine-stuffs.

1551. JACOBS, J. H. & Co. (formerly Jacobs & Bering), silk-manu., Crefeld. London 1851 honor. mention; Paris 1855 bronze medal; London 1862 honor. mention. Agt. John B. Taylor, 1 and 2 Mumford-Court, Milk-street E. C.

Silks for parasols- and umbrellas.

KAMPHAUSEN, N. W. (s. No. 1536.)
KLEMME, G., Crefeld. Embossed ribbons of silk, silk-velvet. (s. No. 1490.)

1553. KÜPPERS, L., & Co., silk-manu., Crefeld. London 1862 honor. mention.

Figured and plain silks; satins and taffetas.

1554. KÜPPERS & KNIFFLER, silk-manu., Crefeld. London 1862 honor. mention. Agt. Wm. Meyerstein, 47 Friday-street, Cheapside.

Popeline; satin (silk and cotton); gold-brocade, do.; silver-brocade, do.; moiré, do.

1556. MAEHLER & TRAPPEN, silk-manu., Crefeld. London 1862 honor. mention. Agts. Fr. Bennoch & Co., 80 Wood-street.

Plain black taffetas, colored; umbrella-silks; plain and figured velvets.

1557. OEHME, C. W., manu., Berlin, Spandauerstr. 74. London 1862 medal.

Black hat-plushes.

1558. RISTOW, C., farmer, Repkow (Cöslin). (s. No. 1227.)

Raw silk.

1559. SCHEIBLER & Co., silk-manu., Crefeld. Berlin 1844 gold medal; London 1851 prize-medal; Paris 1855 medal of honor; London 1862 medal.

Plain and fancy-velvets; velvet-ribbons; silks; silk ribbons.

1561. SCHRÖDER, W. & Co., manu., Crefeld. London 1862 honor. mention.

Taffeta and satin.

1562. SCHROERS, G. & H., manu. of silk-goods, Crefeld. Paris 1855 silver medal; London 1862 medal. Agt. Albert Schroers, at Messrs. Desgrand's père & fils, 9 d. New Broad-street, E. C.

Plain and fancy velvet-ribbons; fancy ladies' velvet-scarfs; pieces of plain velvet and bonnet-velvets; fancy silk and velvet waist-coatings.

1563. SCHUMACHER & SCHMIDT, ribbon-manu., Wermelskirchen. London 1862 medal. Agt. Henry Borckenstein, 8 Moorgate-street.

Black taffeta-ribbons; do. with colored edge.

1564. SEYFFARDT & TE NEUES, silk-manu., Crefeld. London 1862 medal. Agts. Charles Oppenheimer & Co., 79½ Watling-street, Cheapside.

Fancy dress and parasol-silks; fancy-shawls.

1565. TOEPFFER, GUST. AD., merchant and breeder of silk-worms, Stettin. London 1862 medal. Agt. James Wm. Green (E. G. Zimmermann), 2 St. Paul's Buildings, Little Carter-Lane, Doctors Commons.

Cocoons: No. 1 a. Libanon race; No. 1 b. cross-breed of the Libanon and Milan races; No. 1 c. Milan race; No. 1 d. cross-breed of the Lyons and Milan races; No. 1 e. Brianza race; No. 1 f. Brianza and Lyons races; No. 1 g. Lyons race; No. 1 h. Sina race; No. 1 i. cross-breed between the Sina and China races; No. 1 k. China race; No. 1 l. Japan race. The silkworm-eggs of this last race were sent by the Royal Prussian expedition to Japan. Price of each kind 2 sh. 6 d. p. pound.

A piece of pasteboard No. 1 m. with cocoons out of which the moths have crept; a box No. 1 n. with cocoons of an autumn-breed of the Brianza race 2 sh.: a paper-box No. 2. with cocoos of the American birch-silkworm; a paper-box No. 3. with cocoons of the Ricinus-silkworm; No. 4. a spinning-hurdle after M. d'Avril with Pomeranian cocoons 4 sh.; No. 5 a. two climbing-ladders after d'Avril with Pomeranian cocoons, p. piece 2 sh. 6 d.; No. 5 b. two do. with Japanese cocoons p. piece 2 sh. 6 d.; No. 6. a brooding-machine for silkworm-eggs made in Stettin 1 £ 10 sh.; No. 7. a silk coverlet, from cocoons out of which the moths have crept 2 £ 4 sh. 6 d.; No. 8. three pieces of silk coteline of Pomeranian cocoons p. yard 5 sh. 9 d.; No. 9. a piece of pasteboard with raw silk from Pomeranian cocoons p. Pfd. 1 £ 16 sh.; No. 10. three plants of morus Lhou two years old, out of the plantation of the exhibitor p. piece 6 d.; No. 11. a printed report containing directions for the culture of the mulberry tree, with drawings by the exhibitor 9 d.; No. 12. several reports of the culture of silkworms in Pomerania; No. 13. two red silk handkerchiefs made of Pomeranian cocoons p. piece 5 sh.; No. 14. two yellow do. p. piece 5 sh. All these cocoons were produced in the establishment of the exhibitor and reared upon the leaves of mulberry-plants of seven years' growth. The silk-stuffs are manufactured by J. A. Heese in Berlin from Pomeranian silk.

1566. VON DEN WESTEN, velvet-manu., Crefeld. Paris 1855 silver medal; London 1862 medal.

Velvets and velvet-ribbons.

CLASS XXI.

WOOLLEN AND WORSTED, INCLUDING MIXED FABRICS GENERALLY.

ANDREAE, CHR., Mühlheim on the Rhine, worsted shag. (s. No. 1534.)

1570. ARENDT, EDUARD, manu., Zielenzig and Berlin. Paris 1855 bronze medal. Agts. Lion M. Cohn, Phaland & Dietrich, represented by Ch. Trübner, 20 St. Dunstans Hill, City, and at the Exhib. Building.

Glass-case containing the following specimens of yarn of carded wool for fulling stuffs:

No.		1 lb. = pieces.	p. lb. sh.	d.
185.	warp, zebra	3,6	4	—
269.	weft, medley	5,4	3	5½
399.	do. do.	5,4	3	5½
543.	warp, black	6,3	4	3
543.	weft, do.	6,3	4	—
543.	do. do.	2,7	2	4
559.	do. medley	5,4	3	8½
747.	do. mode	5,4	3	7
780.	do. do.	5,4	3	7
791.	do. medley	2,7	3	9
795.	do. mode	5,4	3	7
836.	do. medley	2,4	3	9
846.	do. do.	5,4	3	8½
873.	do. mode	5,4	3	7
917.	do. dotted	5,4	4	8
919.	warp, medley	7,2	4	7
919.	weft do.	5,4	3	7
921.	do. do.	5,4	3	8½
922.	do. do.	5,4	3	8½
926.	do. do.	2,7	3	10
924.	do. dotted	5,4	4	5
927.	do. do.	5,4	4	5
928.	do. do.	5,4	4	5
930.	do. do.	5,4	4	5
934.	warp, mode	7,2	4	7
936.	weft, dotted	5,4	4	5
939.	do. medley	5,4	3	7
944.	warp, do.	7,2	4	7
949.	do. do.	7,2	4	7
951.	do. do.	7,2	4	7
952.	do. do.	7,2	4	7
953.	do. do.	7,2	4	7
959.	weft, do.	5,4	3	8½
960.	do. do.	5,4	3	8½
—	white warp	5,4	3	4
—	do. do.	6,3	3	7
—	do. do.	7,2	4	—
—	white weft	7,2	3	10½
—	white warp	9	4	8

An entire piece of yarn 1640 yards or 2250 Prussian ells long.

1571. ARON, J., manu., Berlin, Dorotheen-str. 9.

Woollen and mixed shawls.

1572. AX, HEINRICH, Rheydt. Munich 1854 medal of honor; Paris 1855 silver medal; London 1862 honor. mention.

Buckskins of cotton and wool.

1573. BECKER & AUERBACH, manu. of worsted fabrics, Berlin. Agts. Krause & Auerbach, 68 Basinghall-street, E. C.

Woollen shawls.

1574. BELLINGRATH, C. H., & LINKENBACH, Barmen. London 1862 medal. Agt. Otto Linkenbach, 2 Milk-street, second floor.

Fancy ribbons of silk, wool and cotton; tailors bindings, trimmings and fancy articles.

1575. BERGER, MORITZ, manu. of fancy cassimeres and tenant of the formerly Royal iron-foundery and forge at Peitz, Frankfurt on the Oder. Agts. Lion M. Cohn, Phaland & Dietrich, represented by Ch. Trübner, 20 St. Dunstans Hill, City, and at the Exhib. Building.

77¼ a ⅝ Berl. ells of fancy summer-buckskins in twelfe remnants as: No. 9851., 9685., 9926., 9916., 9962., 9877., 9970., 10,016., 10,021., 10,033., 10,019., 10,006., p. Berl. ell 1½ Th. (p. yard 6 sh. 1 d.).

1575 a. BERNSTEIN & LICHTENSTEIN, manu., Königsberg.

Shoddy and mungo wool.

1576. BERTELSMANN & NIEMANN, manu., Bielefeld. Agt. H. Landwehr, Exhibition Building.

Different kinds of velours d'Utrecht, worsted velvets, worsted shags and damasks, 18 pieces 62 £ 9 d.

1577. · BIEGER, RUDOLPH, & BIEGER, REINHOLD (G. M. Bieger), cloth-manu., Finsterwalde (Frankfurt on the Oder). Agts. s. No. 1575.

3 pieces of woollen cloth. No. 17,003. 24¾ ells à 57½ sg. (18 yards à 8 sh.); ¾ cloth No. 15,861. 23½ ells à 45 sg. (17 yards à 6¼ sh.); twilled No. 16,910. 21¼ ells à 55 sg. (15½ yards à 7½ sh.).

1578. BLECHER & CLARENBACH, cloth-manu., Hückeswagen. London 1862 honor. mention. Agt. Charles Nolda, Church Court, Old Jewry.

7 pieces of woollen cloth; prices p. Brab. ell of 69½ cent. and p. yard; black doeskin 67 sg. (8 sh. 11 d.); do., 73 sg. (9 sh. 8 d.); woolblue spring-coating 66 sg. (8 sh. 10 d.); darkblue esquimau 89 sg. (11 sh. 11 d.); marineblue ratiné 99 sg. (13 sh. 2 d.); winter-coating 94 sg. (12 sh. 6 d.); ratiné, dotted, 99 sg. (13 sh. 2 d.).

1579. BLEISSNER, FRIEDRICH & CARL (J. G. Bleissner), cloth-manu., Neudamm (Frankfurt on the Oder). Agts. s. No. 1575.

4 coupons of woollen cloth: brown melange tricot 5½ Berl. ells 45 sg. (8 sh. 3 d.); gray croisé 6 Berl. ells 40 sg. (5 sh. 6 d.); black croisé 5 Berl. ells 45 sg. (6 sh. 2 d.); gray 5¼ Berl. ells 45 sg. (6 sh. 2 d.); all p. ell (p. yard).

1580. NOBILING, LUDWIG, & ZUELZER, JULIUS (S. J. Bluhm), Haynau (Liegnitz). Agts. Charles Nolda, 2 Church Court, Old Jewry.

Billiardcloth 2⅝ Th. (11 sh. 6 d.); red cloth 2⅓ Th. (9 sh. 5 d.); do., yellow 2 Th. (8 sh. 2 d.); do., blue 1½ Th. (6 sh. 1 d.); all p. Berl. ell (p. yard).

1581. BOCKHACKER'S, CARL, SUCCESSORS, cloth-manu., Hückeswagen. Munich 1854 great medal; London 1862 medal. Agt. Charles Nolda, Church Court, Old Jewry.

12 pieces of woollen cloth and coatings: black drap-supra; woolblack drap-double; black drap-esquimau; black winter-doeskin; black winter-tricot; dahlia mixed granite-coating; Oxford mixed do.; black tricot-coating; woolblue mixed tricot-coating; dahlia piqué-coating; mulberry silk mixed-coating; woolblue, mixed, chinchilla-coating. For prices apply to the agents.

1582. BOCKHACKER, FRIEDRICH, & SON, Hückeswagen. London 1862 honor. mention. Agt. Charles Nolda, 2 Church Court, Old Jewry.

Samples of woollen yarn, grey not washed 30th., 10th., 8th., 7th., 6th., 5th., 4th., 3rd. and samples of woollen yarns of mixed and different colors washed and not washed, 6th. washed or 7th. not washed 5th., 4th., 3rd.

1583. BOCKMÜHL, FRIEDRICH, SONS, spinners, Dusseldorf. Paris 1855 bronze medal;

London 1862 medal. Agt. Friedrich Osterroth, 1 Bell-Yard South side, St. Pauls E. C.

An assortment of worsted yarns, combed and spun without oil and without silk; prices p. engl. pound. Weft No. 144. 11 sh.; No. 130. 10 sh.; No. 115. 9 sh.; No. 100. 8 sh.; No. 86. 6 sh. 6 d.; No. 72. 5 sh. 10 d. Warp No. 100. 8 sh. 8 d.; No. 86. 8 sh.; No. 72. 6 sh. 6 d.; No. 56. 5 sh. 10 d.; No. 43. 5 sh. 5 d.; No. 28. 4 sh. 10 d. Threefold No. 72. 7 sh. 6 d. Twofold, blue and white No. 50. 7 sh. 5 d.; No. 115. 10 sh.; No. 86. 7 sh. 2 d.; crimson No. 44. 6 sh. 4 d. Threefold No. 56. 5 sh. 10 d. Tops 6 sh. 6 d.

1584. Böhmer & Ercklentz, manu., M.-Gladbach. London 1862 medal. Agts. Morgan Brothers, 21 Bow Lane, Cannon-street, City.

90 coupons of cloth; prices p. yard, free to port. Madeira 10$\frac{1}{2}$ d.; pointé B. 14 d.; Russian III. 14$\frac{1}{2}$ d; do. III B. 13$\frac{1}{2}$ d.; Augustins 9$\frac{1}{4}$ d.; Canada B. 9$\frac{3}{4}$ d.; drap imperial 10 d.; Palmerston 21 d.; silk warps 17$\frac{1}{2}$ d.; Brésil 13$\frac{3}{4}$ d.; Turino 16 d.; Américains 19 d.; dandies 16$\frac{1}{2}$ d.; castors 15 d.; Italiens 15 d.; Panama B. 9$\frac{1}{4}$ d.; do., W. 16 d.; Russian II. 20 d.; Germania 12$\frac{1}{2}$ d.; gentlemen 16$\frac{1}{2}$ d.; Flora 16 d.; Panama I. 12 d.; Laplata 14$\frac{1}{2}$ d.; buckskin à soie V. 18 d.; do. WU. 25 d.; diagonal à soie II. B. 10$\frac{3}{4}$ d.; royal à soie 21 d.; galon C. 13 d.; do. 13 d.; diagonal double 14 d.; cannelé 20 d.; $\frac{8}{4}$ galon C. 26 d.

1586. Brach & Co., manu., Berlin, Spandauerstr. 76. London 1862 honor. mention. Agts. Rawlinson & van der Beeck, 33 Bread-street, Cheapside.

Woollen shawls, do. square shawls, do. goods.

1587. Bruck, Brothers, manu. of worsted goods, Berlin. Agt. D. Born, 2 Tower-Royal, Watling-street.

7/4 Tartan à satin $\frac{1}{4}$ doz. 5 Th. (15 sh.); 7/4 zéphir $\frac{1}{6}$ doz. 5$\frac{1}{4}$ Th. (15 sh. 9 d.); 8/4 plaids $\frac{1}{6}$ doz. 6 Th. (18 sh.); 9/4 zéphir $\frac{5}{12}$ doz. 8$\frac{3}{4}$ Th. (1 £ 6 sh. 3 d.); 9/4 do., à satin and tartan à satin 1$\frac{5}{12}$ doz. 9 Th. (1 £ 7 sh.); 9/4 tartan and chiné à satin $\frac{1}{2}$ doz. 10 Th. (1 £ 10 sh.); 9/4 do., à satin noir $\frac{1}{12}$ doz. 10 Th. (1 £ 10 sh.); 9/4 chiné à soie $\frac{1}{12}$ doz. 10$\frac{1}{2}$ Th. (1 £ 11 sh. 6 d.); 9/4 zéphir écossais $\frac{1}{3}$ doz. 12 Th. (1 £ 16 sh.); 9/4 do., cannelé à soie $\frac{1}{4}$ doz. 13 Th. (1 £ 19 sh.); 9/4 Victoria 1 doz. 14 Th. (2 £ 2 sh.); 10/4 zéphir $\frac{5}{12}$ doz. 11$\frac{1}{2}$ Th. (1 £ 14 sh. 6 d.); 10/4 tartan à satin $\frac{1}{6}$ doz. 12 Th. (1 £ 16 sh.); 9/4 tartan cachenez $\frac{1}{12}$ doz. 7$\frac{5}{6}$ Th. (1 £ 3 sh. 6 d.); 9/4 zéphir do., $\frac{7}{12}$ doz. 8$\frac{1}{4}$ Th. (1 £ 4 sh. 9 d.); 10/4 cachenez à soie $\frac{5}{12}$ doz. 11 Th. (1 £ 13 sh.); 10/4 Cabyles à coin 4/ $\frac{1}{12}$ doz. 12 Th. (1 £ 16 sh.); 12/4 do., sans coin 4/ $\frac{1}{12}$ doz. 14 Th. (2 £ 2 sh.); 12/4 do., à coin 4/ $\frac{1}{12}$ doz. 15 Th. (2 £ 5 sh.); 14/4 do., à coin 2/ $\frac{1}{12}$ doz. 17 Th. (2 £ 11 sh.); 14/4 do., 4/ $\frac{1}{6}$ doz. 19 Th. (2 £ 17 sh.); 16/4 do., 4/ $\frac{1}{6}$ doz. 23 Th. (3 £ 9 sh.); 16/4 tartan $\frac{1}{12}$ doz. 18 Th. (2 £ 14 sh.); 16/4 crêpe à satin $\frac{1}{12}$ doz. 21 Th. (3 £ 3 sh.); 16/4 zéphir à fleurs $\frac{7}{12}$ doz. 32 Th. (4 £ 16 sh.); 4/4 tartan and camara $\frac{1}{6}$ doz. 34 Th. (5 £ 2 sh.); 4/4 tartan imprimé, tartan écossais and mélange à cannelé $\frac{5}{12}$ doz. 38 Th. (5 £ 14 sh.); 4/4 chiné satiné $\frac{1}{3}$ doz. 39 Th. (5 £ 17 sh.); 4/4 chiné à soie, mosaïque à satin and à soie, crêpe à satin $\frac{5}{12}$ doz. 42 Th. (6 £ 6 sh.); 4/4 zéphir II. $\frac{1}{3}$ doz. 45 Th. (6 £ 15 sh.); 4/4 do., à soie $\frac{1}{6}$ doz. 56 Th. (8 £ 8 sh.); 4/4 do., à fleurs $\frac{1}{12}$ doz. 62 Th. (9 £ 6 sh.); 4/4 do., cachemir $\frac{1}{6}$ doz. 69 Th. (10 £ 7 sh.).

1588. Brügmann, Wm., & Co., wool-spinner, Burtscheid (Aix-la-Chapelle). Agts. Lion M. Cohn, Phaland & Dietrich, represented by Ch. Trübner, 20 St. Dunstans Hill, City, ad at the Exhib. Building.

1. Washed weft; free from grease; each $\frac{1}{15}$ lb. Prus. or $\frac{1}{16}$ lb. Engl.: 7 pincops (noppenyarns) 55 sg. (5 sh.); No. 1/33. weft No. 6., a. light mixed 46 sg. (4 sh. 2 d.); b. dark colors 49 sg. (4 sh. 4$\frac{1}{2}$ d.).

2. White weft with grease; each $\frac{1}{8}$ lb. Prus. or $\frac{1}{10}$ lb. Engl.: No. 44. No. 11. 50 sg. (4 sh. 6$\frac{1}{2}$ d.). No. 45. No. 10. 45 sg. (4 sh. 1 d.); No. 46. No. 8. 40 sg. (3 sh. 8 d.); No. 47. No. 6. 32 sg. (2 sh. 11 d.); No. 48. No. 5. 28 sg. (2 sh. 7$\frac{1}{2}$ d.); No. 49/50$\frac{1}{4}$. No. 4. 25 sg. (2 sh. 3$\frac{1}{2}$ d.); No. 49$\frac{1}{2}$/50. No. 3$\frac{1}{2}$. 24/22 sg. (2 sh. 3 d.); No. 51. No. 3. 18 sg. (1 sh. 8 d.).

3. Colored weft; each $\frac{1}{15}$ lb. Prus. or $\frac{1}{16}$ lb. Engl.: No. 35/38. No. 5. 40 sg. (3 sh. 8 d.); No. 34. No. 3. 21 sg. (1 sh. 11$\frac{1}{2}$ d.); No. 40/43. No. 4. 35 sg. (3 sh. 2$\frac{1}{2}$ d.).

4. Cashmere weft; each $\frac{1}{6}$ lb. Prus. or $\frac{1}{8}$ lb. Engl.: No. 52/63. 74. No. 2$\frac{1}{2}$: dark colors 50/55 sg. (4 sh. 6$\frac{1}{2}$ d.), light mixed 40/45 sg. (3 sh. 8 d.).

5. Alpacca weft; each $\frac{1}{15}$ lb. Prus. or $\frac{1}{16}$ lb. Engl.: No. 68. No. 1$\frac{3}{4}$ 40 sg. (3 sh. 8 d.); No. 69. No. 2. 35 sg. (3 sh. 2$\frac{1}{2}$ d.); No. 70. No. 1$\frac{1}{4}$. 30 sg. (2 sh. 9 d.); No. 71. No. 1. 28 sg. (2 sh. 7 d.); No. 72. No. 1. 28 sg. (2 sh. 7 d.); No. 73. No. 1. 25 sg. (2 sh. 3$\frac{1}{2}$ d.); No. 104. No. $\frac{3}{4}$. alpacca-extract weft 11 sg. (1 sh.); No. 105. No. 1$\frac{1}{4}$. do. 13 sg. (1 sh. 2 d.); No. 106. No. 1$\frac{1}{4}$. do. 15 sg. (1 sh. 4$\frac{1}{2}$ d.); No. 107. No. 2. do. 20 sg. (1 sh. 10 d.); No. 108. No. 1$\frac{1}{2}$. 25 sg. (2 sh. 3$\frac{1}{2}$ d.); No. 109. No. 2. 30 sg. (2 sh. 9 d.); No. 110. No. 1$\frac{1}{2}$. 32 sg. (2 sh. 11 d.).

6. Mohair mixed weft; $\frac{1}{7}$ lb. Prus. or $\frac{1}{8}$ lb. Engl.: No. 64/67. No. 2$\frac{1}{2}$. 30/34 sg. (2 sh. 9 d.).

7. Shoddy weft; each $\frac{1}{15}$ lb. Prus. or $\frac{1}{16}$ lb. Engl.: No. 111. No. $\frac{3}{4}$. 9 sg. (10 d.); No. 112. No. 3. 20 sg. (1 sh. 10 d.).

8. 3 thread woolen yarn: No. 100. 3 thread blue 28 sg. (2 sh. 7 d.); No. 101. do. 26 sg. (2 sh. 4$\frac{1}{2}$ d.); No. 102/3. do. 24 sg. (2 sh. 3 d.).

9. 2 thread woolen list-yarn: No. 81. No. 40/ 24 sg. (2 sh. 3 d.); No. 83. à 20/, No. 84 à 24/ 20 sg. (1 sh. 10 d.); No. 85 à 40/. No. 93. à 60/ 25 sg. (2 sh. 3$\frac{1}{2}$ d.); No. 87 à 40/, No. 98 à 50/, No. 99. à 36/ 27 sg. (2 sh. 6 d.); No. 88/89. à 20/ 11/10 sg. (1 sh.); No. 90. à 32/ 14 sg. (1 sh. 3$\frac{1}{2}$ d.); No. 91/97. à 36/ 17 sg. (1 sh. 7 d.); No. 92. à 70/ 30 sg. (2 sh. 9 d.); No. 94. à 50/ 21 sg. (1 sh. 11$\frac{1}{2}$ d.).

10. 2 thread mohair mixed list-yarn: No. 75. à 60/ white 35 sg. (3 sh. 2$\frac{1}{2}$ d.); No. 76. à 50/ do. 30 sg. (2 sh. 9 d.); No. 77. a 45/ do. 26 sg. (2 sh. 4$\frac{1}{2}$ d.); No. 78. à 40/ do. 24 sg. (2 sh. 3 d.); No. 79. à 32/ do. 21 sg. (1 sh. 11$\frac{1}{2}$ d.); No. 80. à 26/ do. 18 sg. (1 sh. 8 d.); all p. Zollpound, (p. lb. Engl.)

1589. Budde, Carl (Budde & Münter), manu., Herford. Agt. Heinrich Landwehr, Exhibition Building.

Woollen toilet-cover, 50 Berl. ells 5$\frac{3}{4}$ Th. (36$\frac{1}{2}$ yards 16 sh. 10 d.).

1591. Camphausen, J. P., & Küppers, manu., M.-Gladbach. Paris 1855 bronze medal. Agts. Heintzmann & Rochussen, 9 Friday-street, Cheapside, E. C.

Goods for coats and trousers: cotton goods; cotton mixed with wool; cotton mixed with silk; cotton mixed with wool and silk; cotton mixed with flax from 6$\frac{1}{4}$ to 11 sg. p. Brab. ell, p. yard from 10 to 17$\frac{5}{8}$ d.

1592. CARO & ROSENHAIN, merchants and manu., Berlin. Agt. William Oelrichs, 13 George-street, Mansion-house.

Long and double shawls, wool and wool and silk-mixture.

1593. CLEFF, BROTHERS, manu. of trimmings (cords, braids, bindings and ties), Barmen. London 1862 medal. Agts. A. Stengert & Co., 4 Gresham-street.

Taylor's trimmings, ladies' dress trimmings, ribbon-ties.

1596. DAVID, LEOPOLD (David & Co.), plush and woollen goods manu., Berlin. Agts. Max Nanson & Co., 48 Watling-street, E. C.

4 pieces of plush for furniture; checkered 36th.; garnet 46th.; brown 5/4 Ells broad 36th.; 1 piece of cap-plush; 2 pieces of brown and black trimming-plush; 2 pieces do. frizzled; 8 pieces of coating, 2 Ells and 8/4 Ells broad; 1 piece of cloth for trousers with borders.

1598. DELLMANN, W., & Co., Elberfeld. Agt. E. van der Beeck, 33 Bread-street, Cheapside, E. C.

Worsted goods: plaids; square and long-shawls; gentlemen mufflers and cravats; flannels, colored.

1599. DEUSSEN, JUL., manu. of woollen cloth, Sagan. Paris 1855 bronze medal. Agts. Heintzmann & Rochussen, 9 Friday-street, Cheapside, E. C.

50/51 in. royal, cloth, black 22½ sg. (3 sh. 1 d.); 52/53 in. blue 30 sg. (4 sh. 2 d.); 52/53 in. mixture grey 28¾ sg. (4 sh.); 54/55 in. imperial, black 40/42½ sg. (5 sh. 7 d. to 5 sh. 11 d.); 53 in. electoral do. 37½/40 sg. (5 sh. 3 d. to 5 sh. 7 d.); 52/54 in. croisé do. 60/52½ sg. (6 sh. 7 d. to 7 sh. 4 d.); all p. Berl. Ell (p. yard).

1600. DEUTZ, JUN., & STROM, cloth-manu., Aix-la-Chapelle. Agt. Charles Nolda, 2 Church Court, Old Jewry.

Black satin, croisé (lustred): No. 18,534. 4/4 4 sh. 3 d.; No. 18,640. do. 4 sh. 7 d.; No. 18,517. do. 5 sh. 4 d.; No. 18,420. do. 5 sh. 1 d.; No. 18,858. 8/4 11 sh. 10 d.; No. 18,432. and 18,635. do. 10 sh. 2 d.; No. 18,775. and 18,851. do. 11 sh. 1 d.; all p. yard.

1602. ERCKENS, JOHANN, SONS, cloth-manu., Burtscheid near Aix-la-Chapelle. Paris 1855 silver medal; London 1862 honor. mention. Agt. Charles Nolda, 2 Church Court, Old Jewry.

14 pieces of cloth: 9/4 scarlet No. 95,477. 110 sg. (14 sh. 8 d.); 9/4 orange No. 102,397. 110 sg. (14 sh. 8 d.); 9/4 Prus. hussar-scarlet No. 102,396. 110 sg. (14 sh. 8 d.); 4/4 blueish white satin No. 108,957. 38 sg. (5 sh.); 4/4 blueish white cassimere No. 103,625. 36 sg. (4 sh. 10 d.); 8½/4 black twilled cloth (without lustre) No. 104,229. 90 sg. (11 sh. 10 d.); 8½/4 do. No. 103,935. 75 sg. (9 sh. 11 d.); 8/4 do. No. 106,378. 50 sg. (6 sh. 8 d.); 4/4 black satin No. 102,803. 42 sg. (5 sh. 7 d.); 4/4 do. cassimere No. 104,238. 50 sg. (6 sh. 8 d.); 4/4 do. satin (lustred) No. 102,958. 40 sg. (5 sh. 4 d.); 4/4 do. No. 109,151. 32 sg. (4 sh. 3 d.); 4/4 black satin (without lustre) No. 102,853. 36 sg. (4 sh. 10 d.); 4/4 do. (lustred) No. 108,925. 32 sg. (4 sh. 3 d.); net, loco Burtscheid; all p. Berl. Ell (p. yard).

1603. ESCHENHAGEN, FR. ADOLPH, cloth-manu., Cottbus. Agts. Lion M. Cohn, Phaland & Dietrich, represented by Ch. Trübner, 20 St. Dunstans Hill, City, and at the Exhib. Building

Black woolly orange-white silky summerstuff, No. 700. 13½ yards, p. yard 1⁵⁄₁₂ Th. (5 sh. 11 d.); violet woolly white silky summerstuff; No. 560., 12¾ yards p. yard 2¹⁄₁₆ Th. (6 sh. 2 d.)

1603a. FELLER, A. (J. G. Feller & Son), cloth-manu., Guben (Frankfort on the Oder). Berlin 1844 bronze medal; Paris 1855 honor. mention. Agts. s. No. 1603.

5 pieces of woollen cloth: black ¾ cloth; 20¾ Ells or 15 yards 38 sg. (5 sh. 3 d.); croisé, 12¼ Ells or 8¾ yards 47 sg. (6 sh. 6 d.); 4/4 cloth, 8½ Ells or 6¼ yards 56 sg. (7 sh. 9 d.); tricot, 8 Ells or 5¾ yards 59½ sg. (8 sh. 3 d.); indigoblue 4/4 cloth, 11 Ells or 8 yards (dyed in wool) 72 sg. (10 sh.); all p. Ell (p. yard). An assortment containing 18 specimens of different cloth as manufactured by the exibitors for the Orient.

1604. FEULGEN, BROTHERS, cloth-manu., Werden on the Ruhr. Munich 1854 medal of honor; Paris 1855 silver medal. Agts. Heintzmann & Rochussen, 9 Friday-street, Cheapside.

Woollen cloth: No. 7951. 22½ yards black satin; No. 6807. 16½ yards bleumourant; No. 7984. 10⅕ yards black; No. 7868. 8⅗ yards black.

1605. FÖRSTER, FERD. SIG., cloth.-manu., Grünberg. Berlin 1844 titled counsellor of commerce; London 1851 prize-medal; London 1862 medal. Agts. s. No. 1603.

(s. No. 1071, 1177, 1343, 1606, 1871a.)

An assortment of woollen cloth-samples. The prices are for black cloth where nothing else is said; for colored fabrics they are raised according to the price of the color 1—3 sg. p. Ell or 2 to 5 d. p. yard; the breadths are noted in English inches.

A. Export-goods for North-America, West-Indies and the Brasils. Royal No. 29. 48—49 in. broad 3 sh. 2 d.; No. 30. 50—51 in. broad 4 sh. 2 d.; No. 31. 52—53 in. broad 4 sh. 8 d.; No. 32. 53—54 in. broad 5 sh. 4 d. Electoral, 54 in. broad No. 33. 6 sh. 1 d.; No. 34. 6 sh. 10 d.; No. 35. 7 sh. 6 d.; No. 36. 8 sh. 2 d.; No. 37. 9 sh. 4 d.; all p. yard.

B. Exported to England, Holland, Switzerland and Australia. Cloth No. 8. 51—52 in. broad 5 sh. 10 d.; No. 9. 52—53 in. broad 6 sh. 7 d.; No. 10. 53—54 in. broad 7 sh. 2 d.; No. 11. 54 in. broad 7 sh. 10 d.; No. 12. do. 8 sh. 6 d. Croisé No. 23. 53 in. broad 8 sh. 2 d.; No. 24. 53—54 in. broad 8 sh. 11 d.; No. 25. 54 in. broad 9 sh. 7 d.; all p. yard.

C. Exported to the La Plata-States, the western coast of South-America and Canada. Satin No. 90. 52—53 in. broad 6 sh. 7 d.; No. 100. 53—54 in. broad 7 sh. 2 d.; No. 110. 54 in. broad 7 sh. 10 d.; No. 120. 8 sh. 6 d.: No. 130. 9 sh. 2 d.; No. 140. 9 sh. 11 d.; No. 150. 10 sh. 7 d.; No. 160. 11 sh. 4 d.: all p. yard.

D. Exported to the Orient and Egypt. No. 10a/k. 54. in. broad wool-dyed cloth, 10 coupons, average price 7 sh. 6 d.

E. Exported to the East-Indies. China and Japan. No. 40. 60 in. broad Spanish stripes, in the usual sortment, 4 sh.

F. Manufactures for Germany, France, Italy, Scandinavia, soft lustred. Electoral No. 320 d. 52 in. broad 6 sh. 7 d.; No. 330 d. 53—54 in.

broad 7 sh. 2 d.; No. 340d. 54 in. broad 7 sh. 10 d.; No. 350d. 8 sh. 11 d.; No. 360d. 9 sh. 5 d.; No. 370d. 9 sh. 11 d. Croisé No. 23d. 53 in broad 8 sh. 6 d.; No. 24d. 53—54 in. broad 9 sh. 2 d.; No. 25d. 54 in. broad 9 sh. 11 d. Cloth 54. in. broad No. 11d. 7 sh. 10d.; No. 12d. 8 sh. 6 d.; No. 13d. 9 sh. 2 d.; No. 14d. 9 sh. 7 d. Satin, 54 in. broad No. 110d. 8 sh. 2 d.; No. 120d. 8 sh. 6 d.; No. 130d. 9 sh. 2 d.; No. 140d. 9 sh. 11 d.; No. 150d. 10 sh. 7 d.; No. 160d. 11 sh. 4 d.; all p. yard.

1606. Förster, Ferd. Sigf. (Ferd. Sigf. Förster, wool spinning establishment), Grünberg. Manufactory Luckau near Polkwitz. Agts. Lion M. Cohn, Phaland & Dietrich, represented by Ch. Trübner, 20 St. Dunstans Hill, City, and at the Exhib. Building.

(s. No. 1071, 1177, 1343, 1605, 1871a.)

Samples of vigogne, prices marked on them.

1607. Fränkel, Isaac, manu., Berlin.

20 pieces of long-châles, interwoven.

1608. Fremerey, Joh. Pet., manu., Eupen. London 1862 honor. mention. Agts. s. No. 1606.

Cloth, prices p. yard: No. 21. 11¼ yards brown double 7 sh. 5 d.; No. 22. 13½ yards do. 7 sh. 8 d.; No. 23. 10½ yards black ribs 5 sh. 1 d.; No. 24. 13½ yards blue double 8 sh.; No. 25. 12¾ yards black double ribs 6 sh.; No. 26. 9 yards tricot 7 sh.; No. 27. 9 yards blue ratiné 8 sh. 9 d.; No. 28. 13 yards black travers 6 sh.; No. 29. 12¾ yards black double 7 sh. 4 d.; No. 30. 13¼ yards black tricot 6 sh. 9 d.; No. 31. 9 yards black ratiné 8 sh. 9 d.; No. 32. 9 yards black ratine 8 sh. 6 d.

1609. Friedheim, S. M., Sons, manu., Berlin, Spandauerstr. 18. Paris 1855 bronze medal; London 1862 honor. mention.

Orleans, plain and figured.

1610. Wirth, Wilhelm (Gebhardt & Wirth), manu., Frauenmühle near Sorau. Paris 1855 bronze medal; London 1862 honor. mention. Agt. Charles Nolda, 2 Church Court, Old Jewry.

Six remnants of black woollen cloth consisting of zephirs, half-cloth and three quarter-cloth: quality A. 50 in. broad 21¼ sg. (2 sh. 11 d.); B. 52 in. broad 25 sg. (3 sh. 5 d.); C. 52/53 in. broad 28¾ sg. (3 sh. 10½ d.); D. 53/54 in. broad 32½ sg. (4 sh. 4½ d.); E. 54 in. broad 36¼ sg. (4 sh. 11 d.); F. 54 in. broad 40 sg. (5 sh. 5 d.); all p. Berl. Ell (p. yard).

1611. Geissler, Ernst, cloth-manu., Görlitz. Manufactories at Görlitz, Köslitz and Nieda. Munich 1854 medal of honor; Paris 1855 bronze medal. Agt. Charles Nolda, 2 Church Court, Old Jewry.

No. 29,715. cinnamon-colored broad-cloth, dyed in the piece; No. 29,242. black broad-cloth, dyed in the piece; No. 28,349. mulberry broad-cloth, dyed in the wool, with yellow selvages; No. 35,353. green broad-cloth, dyed in the wool, with three-colored selvages; No. 32,626. light-colored broad-cloth, dyed in the wool, with yellow selvages; No. 30,520. scarlet broad-cloth, with broad black selvages; No. 27,695. black summer-satin, with lustrous finish; No. 28,799. black fine satin, finished without lustre; No. 27,239. black summer-tricot, with lustrous finish; No. 28,479. black winter-tricot, with lustrous finish; No. 28,804. black fine croisé, finished without lustre; No. 30,086. black summer-diagonal, with lustrous finish; for prices apply to the named agents.

1612. Gevers & Schmidt, manu., Görlitz. London 1862 honor. mention. (s. No. 1716.)

Woollen cloth.

Goetz & Jahn. (s. No. 1636.)

Gohr, A. (s. No. 1685.)

1613. Graeser, Brothers & Co., manu., Langensalza (Erfurt). Berlin 1844 bronze medal; Munich 1854 honor. mention; London 1862 honor. mention. Agt. Frederic Osterroth, 1 Bell Yard, South Side St. Paul's, E. C.

80 pieces of summer-buckskin, 54 in. broad: 4 pieces 5¼ yards 7 sh.; 18 do. 24¾ yards 7¼ sh.; 18 do. 25⅜ yards 7½ sh.; 36 do. 62$\frac{3}{16}$ yards 7½ sh.; 4 do. 5½ yards 8 sh.; all p. yard.

1614. Greiff & Co., manu., Barmen. London 1862 medal.

Samples of coat-bindings of silk, mohair and cotton.

1615. Gröschke, C. A., cloth-manu., Forst (Frankfort on the Oder). Berlin 1844 bronze medal; Paris 1855 bronze medal. Agts. s. No. 1606.

4 pieces of woollen cloth: summer-coating, 10¾ yards 6 sh.; Germania coating 3½ yards 8 sh.; buckskin 4/4 broad 4 yards 4 sh.; buckskin (velour) 7 yards 8 sh.; all p. yard.

1617. Gruender, T., manu. of woollen goods, Peitz near Cottbus (Frankfort on the Oder). Paris 1855 bronze medal. Agts. s. No. 1606.

No. 1—7. winter-pantaloons stuff, 7 coupons, 2$\frac{6}{11}$ yards 70 sg. (9 sh. 7½ d.); No. 8. winter-paletot cloth 2$\frac{4}{11}$ yards 75 sg. (10 sh. 3½ d.); No. 9. do., 2$\frac{3}{11}$ yards 82½ sg. (11 sh. 6 d.); No. 10. do., 2$\frac{5}{11}$ yards 90 sg. (12 sh. 5 d.); No. 11. do., 2$\frac{4}{11}$ yards 100 sg. (13 sh. 9 d.); No. 12. do., 2$\frac{1}{11}$ yards 90 sg. (12 sh. 5 d.); No. 13. do., 2$\frac{4}{11}$ yards 90 sg. (12 sh. 5 d.); No. 14. do., 2$\frac{6}{11}$ yards 90 sg. (12 sh. 5 d); No. 15—21. 7 coupons of various summer-stuffs, different meas. 45 sg. (6 sh. 2½ d.); No. 22—23. 2 coupons of summer-stuff, different meas. 50 sg. (7 sh.); No. 24/25. 2 coupons of stuffs for summer-cloaks à 2$\frac{2}{11}$ yards 37½ sg. (5 sh. 7½ d.); all p. Berl. Ell (p. yard).

1619. Haberland, G. Aug., cloth-manu., Finsterwalde. Berlin 1844 bronze medal; London 1851 prize-medal.

2 pieces of woollen cloth: black quilted, No. 9738. 14¾ Ells, p. Berl. Ell 70 sg. (p. yard 9 sh. 8 d.); black twilled, No. 9557. 12 Ells, p. Berl. Ell 72½ sg. (p. yard 10 sh.).

1621. Hahn & Huldschinsky, merchants and manu. of shoddy- and mungo, Berlin. London 1862 honor. mention. Agts. s. No. 1606.

Samples of shoddy and mungo wool 240 lbs. 1 £ 10 sh. to 20 £.

1622. Halbach, Wilhelm, Wolferts, Ludwig (Halbach & Wolferts), manu., Barmen.

London 1862 medal. Agt. John B. Taylor, 1 and 2 Mumford Court, Milk-street, E. C.

Silk and halfsilk taffeta ribbons: silk and halfsilk sashes and hat-bands.

1623. Heegmann & Mesthaler, manu., Barmen. Agt. E. Brassert, 9 Bow-Lane, Cheapside.

Samples of buttons in metal, silk and other stuffs.

1625. Hendrichs, Franz, cloth-manu., Eupen. Berlin 1844 silver medal; London 1851 prize-medal; Munich 1854 medal of honor; Paris 1855 silver medal; London 1862 medal. Agt. Charles Wilson (C. Candy & Co.), Watling-street.

Woollen cloth: No. 3209. 11 yards, 8½/4 electoral noir, velouté à demi lustre, 61 sg. (8 sh. 1 d.); No. 21,225. 11 yards, 8/4 croisé do. 66 sg. (8 sh. 9 d.); No. 21,204. 11 yards, 8½/4 croisé fin, noir do. 73 sg. (9 sh. 9 d.); No. 21,367. 11 yards, 9/4 supra croisé do. 85 sg. (11 sh. 4 d.); No. 21,438. 11 yards, 9/4 supra cachemir do. 95 sg. (12 sh. 8 d.); No. 21,530. 10¾ yards, 9/4 supra electoral croisé do. 105 sg. (14 sh.); No. 21,635. 11 yards, 9/4 drap de cour noir 105 sg. (14 sh.); No. 21,740. 11 yards, 9/4 cachemir de cour do. 110 sg. (14 sh. 8 d.); No. 21,874. 10½ yards, 8½/4 supra satin do. 85 sg. (11 sh. 4 d.); No. 21,975. 10½ yards, 8½/4 satin de cour do. 95 sg. (12 sh. 8 d.); prices neat p. Brab. Ell (p. yard).

1626. Herrmann, Philipp, cloth-manu., Bromberg. (s. No. 1717, 1750.)

Woollen yarn 15 sg. (1½ sh.).

1629. Hilger, Brothers, manu. of woollen cloth, Lennep. Paris 1855 silver medal; London 1862 medal. Agt. Charles Nolda, 2 Church Court, Old Jewry.

10 coupons of woollen cloth, plain, twilled and satin; for prices apply to the agents.

1630. Hirnstein, Friedrich, manu., Meschede (Arnsberg).

Woollen knitting-yarns in various qualities and colors 6½ lbs.

1631. Hoffmann, Erdmann, cloth-manu., Sorau. Berlin 1844 bronze medal. Agt. M. Teubner.

No. 73,385. cloaking 8 sh.; No. 74,000. winter-buckskin 9 sh.; summer-stuff: No. 74,446. 5 sh. 10 d.; No. 74,470. 5 sh. 2 d.; No. 74,525. 5 sh. 2 d.; No. 74,507. 5 sh. 2 d.; No. 74,402. 5 sh. 2 d.; No. 73,677. black tricot 4 sh. 5 d.; No. 72,145. black ¾ cloth 4 sh. 8 d.; No. 72,101. black zephyr 2 sh. 10 d.; No. 73,546. fancy colored zephyr 3 sh. 6 d.; No. 73,627. scarlet ¾ cloth 5 sh 2 d. Prices p. yard.

1632. Hoffmann, Goenner & Co., manu., Görlitz. Paris 1855 bronze medal. Agts. Streckeisen, Bischoff & Co.

13 pieces of cloth, each 4½ yards, prices p. yard: Black royal No. 319,159. 6 sh. 9 d.; No. 319,162. 6 sh. 9 d.; No. 318,864. 7 sh.; No. 319,158. 7 sh. 3 d.; black Brazil No. 319,640. 8 sh. 3 d.; No. 319,700. 8 sh. 3 d.; black cloth No. 319,695. 7 sh. 3 d.; No. 319,787. 7 sh. 6 d.; No. 319,068. 8 sh.; No. 319,218. 8 sh. 3 d.; No. 163,723. 10 sh. 6 d.; purple cloth No. 319,684. 7 sh. 6 d.; scarlet cloth No. 318,125. 7 sh. 6 d.

1634. Hüffer, E., & Thelosen, J. (Hüffer & Morkramer), manu., Eupen. Berlin 1844 bronze medal; Paris 1855 bronze medal. Agt. Charles Wilson (C. Candy & Co.), Watling-street.

12 pieces of cloth etc., viz: No. 43,997. black cloth 14½ ells (10⅝ yards) 52 sg. (7 sh.); No. 40,565. 13 ells (9⅝ yards) 67 sg. (9 sh.); No. 39,370. 13¼ ells (9⅞ yards) 75 sg. (10 sh.); No. 40,249. black Segovienne 14¾ ells (11 yards) 82½ sg. (11 sh.); No. 40,917. 14¼ ells (10⅝ yards) 85 sg. (11⅓ sh.); No. 44,023. 14¼ ells (10⅝ yards) 90 sg. (12 sh.); No. 42,920. black cachemir 25¾ ells (19⅜ yards) 68 sg. (9 sh.); No. 42,487. black Moskowa 17¼ ells (12⅞ yards) 88 sg. (11⅔ sh.); No. 44,589. black Segovienne, dyed in wool 11 ells (8⅜ yards) 75 sg. (10 sh.); No. 46,545. blue do. 13½ ells (10⅛ yards) 72 sg. (9½ sh.); No. 46,593. 14¼ ells (10⅝ yards) 72 sg. (9½ sh.); No. 42,385. billards-cloth 6¼ ells (4⅝ yards) 92½ sg. (12¼ sh.); all p. Brab. ell (p. yard).

1635. Itzigsohn, Marcus, cloth-manu., Neudamm. Berlin 1844 bronze medal, London 1851 prize-medal. Agts. Lion M. Cohn, Phaland & Dietrich, represented by Ch. Trübner, 20 St. Dunstans Hill, City, and at the Exhib. Building.

5 pieces of woollen cloth: No. 3897. military-cloth, gray melange, ready for sewing 32½ ells or 24¼ yards 33¾ sg. (4½ sh.); No. 3898. do. 23 ells or 17¼ yards 37½ sg. (5 sh.); No. 3814. do. without finish 35½ ells or 26½ yards 30 sg. (4 sh.); No. 3800. common cloth, blue melange, ready for sewing 22¾ ells or 17 yards 30 sg. (4 sh.); No. 3321. cloth, Russian gray, ready for sewing 23¼ ells or 17½ yards 41 sg. (5½ sh.); all p. Berl. ell (p. yard).

1636. Goetz, Gustav, counsellor of commerce, Jahn, Carl, manu., Jahn, Louis, manu. (Carl Gotthilf Jahn), Neudamm. London 1862 medal. Agts. s. No. 1635.

12 coupons of cloth, each 12 or 12½ ells: No. 90,320. 9/4 fine black cloth 2⅓ Th. (9 sh. 4 d.); No. 90,500. 9/4 fine black croisé 2⅙ Th. (8 sh. 7 d.); No. 90,826. 8/4 black tricot 2⅓ Th. (9 sh. 4 d.); No. 90,867. 8½/4 dark blue cloth 2⅙ Th. (8 sh. 7 d.); No. 88,172. 8/4 light blue cloth 2$\frac{1}{12}$ Th. (8 sh. 4 d.); No. 89,501. 8/4 drap colored doeskin 2⅓ Th. (9 sh. 4 d.); No. 89,528. 8/4 light blue do. 2⅓ Th. (9 sh. 4 d.); No. 91,137. 8/4 dark blue do. 2$\frac{7}{12}$ Th. (10 sh. 3 d.); No. 91,100. 8/4 variously colored summercloth 1$\frac{7}{12}$ Th. (6 sh. 4 d.); No. 90,383. 8/4 dark do. 1⅚ Th. (6 sh. 7 d.); No. 91,405. 8/4 fine drap colored cloth for railway-waggons 1$\frac{11}{12}$ Th. (7 sh. 7 d.); No. 91,495. 8/4 drap colored cloth for railway-waggons 1⅚ Th. (6 sh. 7 d.); all p. Berl. ell (p. yard).

Specimens of yarns.

1637. Jansen, J. W., manu., Montjoie. London 1862 medal.

Trouseries and coatings.

1638. Kauffmann, Meyer, manu., Tannhausen near Waldenburg (Breslau). London 1862 honor. mention. Agts. s. No. 1635.

Dresses of barége worsted and cotton mixture No. 1. 2. 3. 4. 5. 3½ sg. (7 sh.); do. with silk No. 6. 5 sg. (10 sh.); mozambique worsted and cotton-mixture No. 7. 8. 5 sg. (10 sh.); do. No. 9. 5½ sg. (11 sh.); do. Ia. No. 10. and 11. 6½ sg. (13 sh.); do. IIa. with silk No. 12. 6¾ sg. (13 sh. 6 d.); do. Ia. No. 13. 8¼ sg. (16 sh. 6 d.); poil do. No. 14. 7 sg. (14 sh.); crêpe do. IIa. No. 15. 8¼ sg. (16 sh. 6 d.); do. Ia. No. 16. 17. 18. 9¼ sg. (18 sh. 6 d.); poil de chèvre, worsted and cotton-mixture No. 19. 20. 5 sg. (10 sh.); all p. Berl. ell (p. dress of 14½ yards).

1639. KAYSER, ALFRED, cloth-manu., Aix-la-Chapelle, manufactory Burtscheid. Paris 1855 bronze medal; London 1862 medal. Agt. Charles Nolda, 2 Church Court, Old Jewry.

15 pieces of black cloth: 5 doeskin 6/4; 1 summer do.; 2 do. 3/4; 3 plain cloth 6/4; 2 twilled do.; 2 without lustre.

1640. KEBEN, SIEGFRIED (Keben & Co.), manu., Berlin, manufactories Brieg and Berlin. London 1862 medal. Agts. S. Oppenheim & Sons, 4a. Bread-street, Cheapside.

4 pieces of trimming plush, fine mohair, in different colors 1½ Th. (6 sh. 3 d.); 2 pieces of stuff like crimea fur fine mohair, in different colors 1½ Th. (6 sh. 3 d.); bouclé façonné, 52 in. broad, curled plush in fine mohair, two colors 4 Th. (16 sh.); toile amiantine (silk cloth), stuff for rifled canon cartridges; sole manufacturers for the kingdom of Prussia 7½ sg. (1 sh. 6 d.); bouclé changeant, in fine black woollen plush, with red ground 3½ Th. (14 sh. 6 d.); all p. Berl. ell.

Portiere of fine double plush, best mohair, 52 in. broad on the upper part the Prussian eagle in a red field, the lower part in fine violet plush 1000 Th. p. 6 ells (170 £). Instead of the Prus. eagle any other coat of arms may be inserted.

1641. KELLER & STROETER, manu., Barmen. London 1862 medal.

Trimming-lace and ribbons.

1644. KLEMM, ROBERT, GUSTAV & JULIUS, BROTHERS (G. Klemm), cloth-manu., Forst (Frankfurt on the Oder). Agt. Friedrich Osterroth, 1 Bell Yard, South Side, St. Paul's, E. C.

3 pieces of summer-buckskin No. 12,483 1⅚ Th. (5 sh. 10 d.); No. 12,495. 1½ Th. (6 sh. 2 d.); No. 12,534. 1½ Th. (6 sh. 2 d.); 5 pieces of winter-buckskin No. 12,417. 1$\frac{11}{12}$ Th. (7 sh. 10 d.); No. 12,446. 1$\frac{11}{12}$ Th. (7 sh. 10 d.); No. 12,545. 2$\frac{1}{12}$ Th. (8 sh. 6 d.); No. 12,361. 2⅛ Th. (9 sh.); No. 12,409. 2⅛ Th. (9 sh.); all p. Berl. ell (p. yard) commission excl.

1645. KRAGE, PAUL HEINRICH, manu. of woollen cloth Quedlinburg. Berlin 1844 bronze medal; München 1854 honor. mention. Agts. Lion M. Cohn, Phaland & Dietrich, represented by Ch. Trübner, 20 St. Dunstans Hill, City, and at the Exhib. Building. (s. No. 1742.)

1. Winter-paletot, cloth wool-dyed, stout, fine 52 in. 77½ sg. (10 sh. 5 d.); 2. duffel F. mulberry 52 in. 65 sg. (8 sh. 6 d.); 3. do. 3. 52 in. 43 sg. (5 sh. 8 d.); 4. do. 6. fast-blue for seamen 52 in. 39 sg. (5 sh. 2 d.); 5. velour 4½. wool-dyed half-stout 52 in. 40 sg. (5 sh. 3 d.); 6. lady 5. light 55 in. 32 sg. (4 sh. 3 d.); all p. Berl. ell (p. yard). All pure wool.

1646. KRAUSE, C. F., & SONS, manu., Görlitz. Agts. Elkan & Co., 4 Gresham-street.

6 pieces black cloth: Royal prima 25¼ yards 5 sh.; superfine ¾ 25 yards 6 sh.; imperial 20¾ yards 7 sh.; drap elastique 20¼ yards 7 sh. 6 d.; SSF. 1/1 16⅞ yards 7 sh. 6 d.; electoral délustré 17¼ yards 9 sh. 6 d., all p. yard.

1647. KRUGMANN & HAARHAUS, manu. of upholstery-goods, plushes, table-covers, curtains etc., Elberfeld. Paris 1855 bronze medal; London 1862 medal. Agt. W. Meyerstein, 47 Friday-street, Cheapside.

Furniture fabrics: silk brocate and damask, plain and figured ribs: curtains of wool and half wool; plushes of different kinds and widths, to 72 in., for furniture, carriages and shoes; table-covers of different qualities; cotelines of silk.

1648. LANGENBECK & WEX, manu., Barmen. Agts. S. Oppenheim & Sons, 4 Bread-street, Cheapside.

Fancy and plain lady-dress buttons; stuff-buttons for tailors.

1649. LEHMANN, D. J., LEHMANN, ANTON, LEHMANN, LUDWIG (D. J. Lehmann), manu., Berlin, Spandauerstr. 64., manufactory Nowawess, Kloster Zinna. Berlin 1844 honor. mention; London 1851 honor. mention; Munich 1854 medal of honor; Paris 1855 silver medal; London 1862 medal. Agts. Eichholz, Morris & Co., 16 Gresham-street, E. C.

Velours d'Utrecht: 4/4 wide, prices p. yard, 9 pieces 00., 36¾ yards; 11 sh. 2 d.; 6 do. 0., 29 yards, 9 sh. 9 d.; 13 do. I., 55½ yards 6 sh. 10 d.; 7 do. I a., 55¾ yards, 6 sh. 1 d.; 6 do. II., 30¾ yards, 5 sh. 5 d.; 4 do. III., 32 yards, 4 sh. 9 d.; 2 do. IIIB., 17 yards, 3 sh. 2 d.; 1 do. IV., 7 yards, 2 sh. 5 d.

Damask, 4/4 wide, 4 pieces 22¾ yards; embossed velvet, 4/4 wide, 2 pieces 4 yards; do. 5/4 wide, 1 piece 5 yards; printed velvet, 4/4 wide, 5 pieces 25 yards; velvet for dressing gowns, 4/4 wide, 3 pieces 2⅜ yards; figured velvet, 4/4 wide, 3 pieces 6 yards; I. crimea-fur, 4/4 wide, 8 pieces 46 yards; do. frisé, 4/4 wide, 3 pieces 8½ yards; Camilla française, 4/4 wide, 6 pieces 71⅜ yards; embossed velvet, 6/4 wide, 1 piece 10¾ yards; prices p. yard from 2 sh. 9 d. to 8 sh. 5 d.

Knickerboker, 8/4 wide, 4 pieces 25½ yards, p. yard 3 sh. 10 d.

Camilla, 8/4 wide, 4 pieces 14 yards; do. française, 4 pieces 28 yards, each p. yard from 9 sh. 9 d. to 12 sh. 7 d.

Double alpacca, 8/4 wide, 1 piece, 3½ yards, p. yard 7 sh. 9 d.; Wellington and velvet, 8/4 wide, 6 pieces 63 yards, p. yard from 2 sh. 9 d. to 4 sh. 2 d.; table-covers, 8/4 wide, 3 pieces each 13 Th. (1 £ 19 sh.); shawls, 8/4 wide, 240 pieces, p. piece 3—10 Th. (from 9 sh. to 1 £ 10 sh.); seadog, 4/4 wide, 2 pieces 14 yards, p. yard 5 sh. 3 d.; I. crimea-fur, 4/4 wide, 6 pieces 58½ yards, p. yard 5 sh. 7 d.

1650. LENDER, CARL (Lender & Co.), manu., Rheydt. Agts. Morgan Brothers, 21 Bow-Lane, Cannon-street, City.

24 pieces of cloth for coats and trousers: wool, cotton and silk 12—13 sg. (18—20 d.); wool, and cotton 8½—13 sg. (13—20 d.); cotton and linen 8½ sg. (13 d.), all p. Brab. ell (p. yard) Prices p. yard marked on each piece.

1651. LEVY, MAGNUS, ARON, JOSEPH (Levy & Aron), Berlin, Magazinstr. 17. Munich 1854 prize-medal; Paris 1855 bronze medal; London 1862 medal. Agts. Loewe & Co., 5 Queen-street, Cheapside, E. C.

Woollen shawls, worked by Jacquard apparatus and common-looms.

1654. LOEWEN, S., & HILDESHEIMER, cloth-manu., Brandenburg on the Havel. Agts. s. No. 1645.

1 remnant of twisted silk-hosiery 1½ Th. (6 sh. 3½ d.); 1 do. ribs 1⅔ Th. (7 sh.); 3 remnants of buckskin, of twisted silk and woollen yarn 1¾ Th. (7 sh. 3½ d.); 1 remnant of buckskin, pure wool 1⅔ Th. (7 sh.); each remnant 12 Berl. ells (8⁴⁄₇ yards).

1655. Lütgenau & Wiehager, cloth-manu., Hückeswagen. Agts. Charles Nolda & Co., 2 Church Court, Old Jewry.

Woollen cloth and coatings: No. 23. black Esquimau; No. 24. black cloth; No. 25. black doeskin; No. 26. blue, friezed; No. 27. and 28. mixed, friezed. For prices apply to the agents.

1656. Marggraff, Brothers, cloth-manu., Schwiebus. Agts. Lion M. Cohn, Phaland & Dietrich, represented by Ch. Trübner, 20 St. Dunstans Hill, City, and at the Exhib. Building.

4 pieces of black cloth: thick-cloth, No. 8. 54 in. 43¾ sg. (6 sh.); No. 4. 54 in. 38¾ sg. (5 sh. 4 d.); No. 2. 52 in. 36¼ sg. (5 sh.); ¾ cloth No. 7. 52 in. 33¾ sg. (4 sh. 7½ d.); all p. Berl. ell (p. yard).

1657. Marx, N., & Sons, cloth-manu., Aix-la-Chapelle. Agts. Charles Nolda & Co., 2 Church Court, Old Jewry.

No. 36,686. satin de laine 45 sg. (6 sh.); No. 36,163. satin electa 48 sg. (6 sh. 5 d.); No. 36,497. tricot ribs 38 sg. (5 sh. 1 d.); No. 36,567. tricot Sedan 43 sg. (5 sh. 9 d.); No. 36,701. tricot double 48 sg. (6 sh. 5 d.); No. 36,295. drap de Russie 55 sg. (7 sh. 4 d.); No. 36,933. Siberienne 57 sg. (7 sh. 7 d.); all p. Brab. ell (p. yard).

1658. Matthesius, F. L., & Son, manu., Cottbus. Berlin 1844 bronze medal; Paris 1855 bronze medal. Agts. s. No. 1656.

3 pieces of black cloth: croisé, 22¾ ells (16¼ yards), p. Berl. ell 50 sg. (p. yard 7 sh.); tricot, 24¾ ells (17¾ yards), p. ell 57¼ sg. (p. yard 8 sh.); cloth, 20 ells (14¼ yards), p. ell 64 sg. (p. yard 9 sh.).

1659. Mayer, Ferdinand, merchant and manu. of lasting, Cologne, manufactories at Neuss and Frechen. London 1862 medal. Agts. Heintzmann & Rochussen, 9 Friday-street, Cheapside.

Four pieces of lasting, lustred stuffs of wool and cotton, each piece 42 Berl. ells or 30 yards, qual. No. 1. 32 Th. (4 £ 16 sh.): No. 2. 29 Th. (4 £ 7 sh.); No. 3. 27½ Th. (4 £ 1 sh. 6 d.); No. 4. 25½ Th. (3 £ 16 sh. 6 d.).

1660. Mayer, J. F., cloth-manu., Eupen. Paris 1855 bronze medal; London 1862 medal. Agt. Charles Wilson (C. Candy & Co.), Watling-street.

11 pieces of cloth, superfine: No. 60,160. scarlet 20½ yards; No. 60,159. gentianblue 20¾ yards; No. 60,162. black 20 yards; prices p. yard 6 sh. 10 d.

Spanish stripes, superfine: No. 59,622. purple 18½ yards; No. 59,586. orange 18 yards; No. 59,562. green 17½ yards; No. 59,260. brown 18¾ yards; No. 59,520. scarlet 18½ yards; No. 59,292. gentianblue 18¾ yards; No. 59,517. darkblue 18¾ yards; No. 60,167. white 18½ yards; prices p. yard 4 sh.

1661. Mebus & Rübel, Barmen. London 1862 medal. Agts. Heintzmann & Rochussen, 9 Friday-street, Cheapside.

Coat-bindings.

1662. Meyer, M., & Co., cloth-manu., Aix-la-Chapelle. Paris 1855 bronze medal; London 1862 medal. Agts. Charles Nolda & Co., 2 Church Court, Old Jewry.

Cloth, width 8/4 Brabant ells, novelties for winter: No. 10,207., 10,218., 10,913., 10,913½., 10,224., 10,224½., p. Berl. ell 75 sg. (p. yard 10 sh.).

No. 10,203., 10,204., 10,205., 10,214., p. Berl. ell 80 sg. (p. yard 10 sh. 8 d.).

Talma diagonal: No. 15,591., 15,595., p. Berl. ell 65 sg. (p. yard 8 sh. 8 d.).

Novelties for summer: No. 9895. 134., 9614. 195., 9770. 204., 9785. 176., p. Berl. ell 55 sg. (p. yard 7 sh. 3 d.).

1663. Nausester, Wilhelm, manu., Lötmaringhausen near Meschede (Arnsberg).

Woollen wowen, colored jackets, 12 pieces; woollen knitting yarns of different qualities and colors, 3 Pfd.

1664. Nettmann, H. D., & Son, cloth-manu., Limburg on the Lenne (Arnsberg). Paris 1855 bronze medal; London 1862 honor. mention. Agts. Heintzmann & Rochussen, 9 Friday-street, Cheapside.

5 pieces of woollen cloth; prices p. yard: black No. 12,879. 22 yards 10 sh. 4 d.; Mulberry No. 13,027. 20½ yards 9 sh. 7 d.; black diagonal-tricot No. 13,063. 20½ yards 11 sh.; colored summer-tricot No. 13,078. 23¾ yards 8 sh. 4 d.; colored frizzled cloth No. 13,118. 10¼ yards 12 sh. 6 d.

1665. Niederheitmann & Buchholz, cloth-manu., Aix-la-Chapelle, new establishment. Agts. Charles Nolda & Co., 2 Church Court, Old Jewry.

4 pieces of winter fancy cloth; 2 pieces frizzled; 3 pieces woollen velvet; 1 piece black tricot diagonal; 2 pieces summer fancy cloth.

1666. Niemann & Gundert, manu., Barmen. London 1862 medal. Agts. Charles Oppenheimer & Co., 79½ Watling-street, Cheapside.

Worsted coat-bindings and braids.

1667. Noss, C., Cologne. Paris 1855 silver medal; London 1862 honor. mention. Agt. Wm. Meyerstein, 47 Friday-street.

6 pieces of velours d'Utrecht: No. 22,324. light granate; No. 22,307. dark granate; No. 22,117. green; No. 22,308. rose des Alpes; No. 22,018. violet and yellow; p. Brab. ell 45 sg. (p. yard 6 sh.) with 4 pCt. discount.

No. 6689. velours d'Utrecht frisé, black, p. Brab. ell 52 sg. (p. yard 6 sh. 10 d.) with 4 pCt. discount; table-cover of Utrecht-velvets crimson, p. piece 12½ Th. (37 sh.); do. printed 15 Th. (45 sh.).

1668. Offermann, F. W., manu., Imgenbroich (Aix-la-Chapelle). Agt. s. No. 1664.

Coatings and trouserings.

1669. Osterroth, Wm. & Son, manu., Barmen. Paris 1855 silver medal.

72 pieces of 21½ yards 65 th. Alpaca braids 2d. Qual. colored; 16 pieces of 36 yards superfine pure mohair braids: 12 pieces of 36 yards fonyàs B.; 12 do. rifle-braids: 12 do. fonyàs S.; 32 pieces of 27 yards galloons mohair nouveauté bindings; 20 do. galloons mohair de Paris; 24 pieces of 41 yards soutaches: 16 pieces of 27 yards merinos picté bindings: 6 do. merinos renforçé surfin bindings: 12 pieces of 36 yards galloons mohair figuré bindings: 16 do. superfine galloons mohair bindings; 12 do. galloons mohair diagonal binding; 16 pieces of 27 yards Alpaca braids first Qual.; 12 do. Milan braids MF.

1671. Pastor, Godf., prop. of a spinning establishment, Aix-la-Chapelle. Munich 1854 prize-medal; Paris 1855 silver medal: London 1862 medal. Agts. s. No. 1656.

Woollen yarns (carded): No. 1. warp No. 11. (common quality) 23 sg. (2 sh. 1 d.): No. 2. warp No. 14. do. 26 sg. (2 sh. 4 d.); No. 3. warp No. 14.

29 sg. (2 sh. 6 d.); No. 4. warp No. 18. (common quality) 29 sg. (2 sh. 6 d.); No. 5. weft No. 18. do. 29 sg. (2 sh. 6 d.); No. 6. warp No. 18. 32 sg. (3 sh.); No. 7. warp No. 20. (common quality) 32 sg. (3 sh.); No. 8. weft No. 20. do. 31 sg. (2 sh. 11 d.); No. 9. warp No. 20. 35 sg. (3 sh. 2 d.); No. 10. warp No. 24. 40 sg. (3 sh. 6 d.); No. 11. weft No. 24. 38 sg. (3 sh. 5 d.); No. 12. warp No. 28. 47 sg. (4 sh. 2 d.); No. 13. warp No. 32. 52 sg. (4 sh. 9 d.); No. 14. warp No. 45. 75 sg. (6 sh. 9 d.); No. 15. ½ warp (weft) No. 84. 95 sg. (8 sh. 6 d.); No. 16. do. No. 18. mixed 35 sg. (3 sh. 2 d.); No. 17. do. No. 18. mixed 35 sg. (3 sh. 2 d.); No. 18. do. No. 18. blue 38 sg. (3 sh. 5 d.); No. 19. do. No. 18. brown 35 sg. (3 sh. 2 d.); No. 20. warp No. 24. pensée 50 sg. (4 sh. 6 d.); No. 21. warp No. 36. brown 85 sg. (7 sh. 8 d.); No. 22. warp No. 40. blue 88 sg. (7 sh. 11 d.); No. 23. doubled yarn No. 32. 90 sg. (8 sh. 2 d.); No. 24. do. No. 45. 110 sg. (9 sh. 11 d.); No. 25. do. No. 32. 95 sg. (8 sh. 6 d.); No. 26. do. No. 34. 97 sg. (8 sh. 8 d.); No. 27. do. No. 34. 97 sg. (8 sh. 8 d.); No. 28. do. No. 40. 110 sg. (9 sh. 11 d.); No. 29. do. No. 28. (with silk) 90 sg. (8 sh. 2 d.); No. 30. knickerbocker yarn ⅔ warp No. 16. 42 sg. (3 sh. 10 d.); all, with the grease; prices p. Zollpfd. (p. lb. engl.).

1672. PETERS, LEONHARD, cloth-manu., Eupen. Agts. Rawlinson & van der Beeck, 33 Bread-street, Cheapside.

Cloth, 12 pieces: Spanish stripes, yellow No. 64,440. 20¼ yards 4 sh.; Spanish stripes, scarlet No. 64,571. 20¼ yards 4 sh. 2 d.; zephircloth, black No. 6000. 5¾ yards 4 sh. 8 d.; twilled cloth, black No. 12,000. 5¾ yards 7 sh. 10 d.; doeskin, black No. 8000. 5¾ yards 4 sh.; ribs, black No. 33,000. 5¾ yards 7 sh. 2 d.; diagonal No. 35,000. 5¾ yards 4 sh. 6 d.; fancy cassimere No. 230. 5⅝ yards 5 sh. 4 d.; do. No. 247. 5½ yards 5 sh. 8 d.; do. No. 217. 5¾ yards 5 sh. 8 d.; double cloth, black No. 47,000. 5⅛ yards 8 sh. double diagonal, black No. 35604. 5¾ yards 7 sh. 3 d.; all p. yard.

1673. PFERDMENGES, J. H., & SON, manu., Rheydt. Paris 1855 silver medal; London 1862 honor. mention. Agts. Heintzmann & Rochussen, 9 Friday-street, Cheapside.

Twelve pieces of buckskin of wool, mixed with cotton and silk 112¾ Berl. ells 60 Th. (84½ yards 9 £).

1675. POLZIN CLOTH WEAVERS GUILD, Cöslin.

White and blue flannels.

1677. RICHTER, AUGUST, manu., Forst and Muskau (Frankfort on the Oder). Paris 1855 bronze medal. Agt. Friedrich Osterroth, 1 Bell-Yard, South Side, St. Paul's, E. C.

8 pieces of figured woollen cloth I a. 1⅝ Th. (8 sh.); 4 pieces do. II a. 1 5/12 Th. (7 sh.); p. Berl. ell (p. yard) each incl. 5 pCt. provision.

1679. RITTINGHAUS & BRAUNS, manu., Kettwig. Paris 1855 bronze medal. Agt. Friedrich Osterroth, 1 Bell-Yard, South Side, St. Paul's, E. C.

Figured woollen stuffs for trousers. For prices apply to the agents.

1680. RITZ & VOGEL. cloth-manu., Aix-la-Chapelle. London 1862 honor. mention. Agt. Charles Nolda, 2 Church Court, Old Jewry.

(s. II. No. 62.)

17 pieces of plain and twilled woollen cloth in black and fine colors: No. 4406. 1 piece 4/4 broad Summer doeskin, black, with lustre 21 sg. (2 sh. 11 d.); No. 4040. 1 do. 26 sg. (3 sh. 7 d.); No. 2792. 1 do. 37 sg. (5 sh. 1 d.); No. 3162. 1 do. electa, black, with lustre 45 sg. (6 sh. 2 d), No. 3161. 1 do. without lustre 45 sg. (6 sh. 2 d.); No. 3127. 1 piece 4/4 broad Winter doeskin, black, without lustre 45 sg. (6 sh. 2 d.); No. 1885. 1 piece 8½/4 broad croisé electoral, black (twilled cloth) do. 85 sg. (11 sh. 8 d.); No. 5502. 1 piece 8½/4 broad Spanish stripes, light green 28 sg. (3 sh. 10 d.); No. 1627. 1 piece 8/4 broad drap royal (plain cloth), chemical blue 40 sg. (5 sh. 7 d.); No. 1042. 1 piece 8/4 broad drap electoral (plain cloth), white 56 sg. (7 sh. 8 d.); No. 1058. 1 do. yellow 56 sg. (7 sh. 8 d.); No. 2234. 1 piece 8½/4 broad do. scarlet 77 sg. (10 sh. 8 d.); No. 2110. 1 piece 4/4 broad Marocco (plain cloth), scarlet 50 sg. (6 sh. 11 d.); No. 2101. 1 do. amaranth 50 sg. (6 sh. 11 d.); No. 2133. 1 do. sky-blue 50 sg. (6 sh. 11 d.); No. 1454. 1 piece 4/4 broad cassimere, white 36 sg. (5 sh.); No. 1174. 1 piece 8/4 broad pianoforte cloth, scarlet 90 sg. (12 sh. 5 d.); all p. Berl. ell (p. yard).

1681. RUFFER, CARL, privy counsellor of commerce, RUFFER, HEINRICH, counsellor of commerce (S. B. Ruffer & Son), Liegnitz. Munich 1854 honor. mention; Paris 1855 bronze medal. Agts. Lion M. Cohn, Phaland & Dietrich, represented by Ch. Trübner, 20 St. Dunstans Hill, City, and at the Exhib. Building.

(s. II. No. 28.)

6 pieces of woollen cloth, satin and croisé p. Berl. ell 2—3 Th. (p. yard. 8 sh. 3 d. to 12 sh. 4 d.)

1685. GOHR, A. (Scheiffchen & Son), cloth-manu., Güntersberg near Crossen on the Oder. Berlin 1844 bronze medal. Agts. s. No. 1681.

10 pieces of woollen cloth, 8/4 ells or 53 in. width: No. 4860. satin d'hiver 42½ sg. (5 sh. 8 d.); No. 4863. do. 45 sg. (6 sh.); No. 4866. do. double 47½ sg. (6 sh. 4 d.); No. 4649. do. d'hiver 50 sg. (6 sh. 8 d.); No. 4191. do. prima 55 sg. (7 sh. 4 d.); No. 4000. do. brillant 60 sg. (8 sh.); No. 4124. do. de cour 52½ sg. (7 d.); No. 4675. do. Victoria 47½ sg. (6 sh. 4 d.); No. 4546. tricot 50 sg. (6 sh. 8 d.); No. 5153. do. 60 sg. (8 sh.); all p. Berl. ell (p. yard).

1686. SCHLIEF, E. P., manu., Guben (Frankfort on the Oder). Paris 1855 bronze medal. Agts. Charles Nolda & Co., 2 Church Court, Old Jewry.

7 coupons of woollen cloth, each 10¼ Berl. ells or 7½ yards; No. 7406. 8/4 black cloth, 52 in. broad 42 sg. (6 sh.); No. 7793. 8½/4 fine 7/8 black cloth, 53 in. broad 45 sg. (6 sh. 2 d.); No. 7100. 8½/4 fine black cloth, 55 in. broad 65 sg. (9 sh.); No. 7597. 8/4 tricot with galloon, 54 in. broad 65 sg. (9 sh.); No. 7129. 8/4 woollen satin, 53 in. broad 58 sg. (7 sh. 11 d.); No. 7379. 8½/4 double elastique, 54/55 in. broad 70 sg. (9 sh. 6 d.); No. 7601. 8/4 double travers 53 in. broad 58 sg. (7 sh. 11 d.); prices p. Berl. ell, 4 pCt. discount incl.

1686 a. SCHLIEF, CARL (Samuel Schlief), manu., Guben. Berlin 1844 silver medal; Paris 1855 bronze medal. Agts. s. No. 1681.

Black woollen cloth: No. 5404. black diagonal 13 yards 6 sh.; No. 5328. do. 13 yards 5 sh. 11 d.; No. 5117. do. 13 yards 6 sh. 8 d.; No. 5422. tricot 13 yards 7 sh. 10 d.; No. 5357. do. 13 yards 7 sh. 6 d.; No. 9341. cloth 17¼ yards 5 sh. 10 d.; No. 5319. do. beaver 13 yards 6 sh. 2 d.; all p. yard.

1687. Schmidt, Wilhelm & Carl (Friedrich Schmidt & Co.), Sommerfeld (Frankfort on the Oder). Paris 1855 bronze medal. Agts. Lion M. Cohn, Phaland & Dietrich, represented by Ch. Trübner, 20 St. Dunstans Hill, City, and at the Exhib. Building.

No. 400. 24⅛ yards 60 in. broad Prussian blue zephyr 38 sg. (5 sh. 3 d.); No. 300. 25⅝ yards 60 in. broad scarlet do. 35 sg. (4 sh. 10 d.); No. 53,694. 26⅞ yards 52 in. broad black do. 32 sg. (4 sh. 4 d.); No. 53,660, 22⅛ yards 54 in. broad do. ¾ cloth 40 sg. (5 sh. 5 d.); all p. Berl. ell (p. yard).

1689. Schneider, August, manu. of shawls and cloths, Berlin, Kaiserstr. 30. Agts. s. No. 1687.

10 pieces of zephyr-shawls 54 Th. (8 £ 2 sh.); 3 pieces of cachemir-shawls 66 Th. (10 £.); 2 pieces of Victoria-shawls 60 Th. (9 £); all p. doz.

1690. Schoeller, Johann Peter, cloth-manu., Düren. Paris 1855 silver medal; London 1862 medal. Agts. Charles Nolda & Co., 2 Church Court, Old Jewry.

Cloths: No. 124,901. jaquette blue veneziano; No. 124,920. invisible cloth; No. 124,921. black cloth; No. 124,924. black cachemir; No. 124,932. black croisé; No. 124,939. black satin double; No. 124,936. black crêpe de laine; No. 124,974. blue gentry; No. 124,955. darkblue moutonné.

1691. Schoeller, Leopold, & Sons, manu., Düren. Berlin 1844 silver medal; London 1851 prize-medal; Munich 1854 the order of merit of St. Michael; Paris 1855 medal of honor; London 1862 medal. Agts. Charles Nolda & Co., 2 Church Court, Old Jewry.

Woollen cloths, stuffs and fancy goods.

1692. Schürmann, Petfr, & Schröder, Hermann (Peter Schürmann & Schröder), cloth-manu., Lennep. Berlin 1844 silver medal; London 1851 prize-medal; Paris 1855 silver medal; London 1862 medal. Agts. Charles Nolda & Co., 2 Church Court, Old Jewry.

Eleven pieces of woollen cloth, viz: No. 34,555. 19¾ yards billard-cloth, breadth 78½ in., 96 sgr. (13 sh. 2 d.); No. 33,185. 7½ yards green drap de brésil, breadth 54½ in., 67 sg. (9 sh. 2 d.); No. 35,319. 7½ yards melange tricot, breadth 54½ in.; 81½ sg. (11 sh. 2 d.); No. 34,527. 7½ yards black tricot, breadth 54½ in., 80 sg. (10 sh. 11½ d.); No. 34,537. 25¼ yards black satin, breadth 27¼ in., 38½ sg. (5 sh. 3 d.); No. 35,336. 7½ yards Marengo-tricot, breadth 54½ in., 57½ sg. (7 sh. 10½ d.); No. 32,072½. 7½ yards lightblue satin, breadth 54½ in., 78½ sg. (10 sh. 9 d.); No. 34,065. 7½ yards black drap de Brésil, breadth 58 in., 69 sg. (9 sh. 5½ d.); No. 34,615. 7½ yards black twilled, breadth 58 in., 86 sg. (11 sh. 9½ d.); No. 33,960. 7½ yards darkblue drap de Brésil, breadth 58 in., 86 sg. (11 sh. 9½ d.); No. 34,560. 9¾ yards bleu grec cloth, breadth 58 in., 91 sg. (12 sh. 5½ d.); all net prices p. Berl. ell (p. yard).

1693. Schwamborn, Neuhaus & Krabb, cloth-manu., Aix-la-Chapelle. Agts. Charles Nolda & Co., 2 Church Court, Old Jewry.

Fourteen pieces of fancy woollen cloth of different weight and quality.

1695. Souter & Alt, manu., Eupen.

Woollen yarns.

1696. Sterken, H., manu., Aix-la-Chapelle. London 1862 honor. mention.

Coatings, trouseries and cloakings.

1697. Sternickel & Gülcher, cloth-manu., Eupen. London 1862 honor. mention. Agt. Charles Wilson (C. Candy & Co.), Watling-street.

Seven pieces of woollen cloth: No. 5129. 5 sh. 4 d.; No. 5159. 5 sh. 7 d.; No. 5222. 5 sh. 1 d.; No. 9611. 6 sh. 8 d.; No. 9659. 5 sh. 10 d.; No. 9731. 6 sh. 3 d.; No. 9805. 6 sh.; all p. yard.

1699. Teschemacher & Kattenbusch, manu., Werden (Düsseldorf). London 1862 honor. mention.

Plain and twilled woollen cloth.

1703. Ulenberg & Schnitzler, manu. of worsted yarns, Opladen near Cologne. Berlin 1844 silver medal; Paris 1855 bronze medal; London 1862 honor. mention. Agts. s. No. 1687.

A. Hosiery (woollen) yarns. 8 bundles, each 5 lb. Prus. or 5½ lb. Engl. weight, dyed and twisted woollen yarns in different colors, varying in price from 3 to 5⅓ Th. (9 to 16 sh. p. bundle).

B. Knitting worsted yarns. 12 bundles, each 5 lb. Prus. weight or 5½ lb. engl. weight dyed and twisted worsted yarns (carded and combed yarns) in different colors, varying from 3⅝ to 6⅔ Th. (11½ to 20 sh.) p. bundle.

1705. Voss, Carl & Julius, cloth-manu., Kettwig on the Ruhr. Agts. Heintzmann & Rochussen, 9 Friday-street, E. C.

Cloth and buckskin p. Berl. ell 41½ to 73 sg. (p. yard 5 sh. 8½ d. to 10 sh. 1 d.).

1708. Weigert, Brothers, manu., Berlin Manufactories Berlin and Ratibor. Munich 1854 medal of honor; Paris 1855 bronze medal. Agts. A. Schröder & Co., 9 Gresham-street, East.

Stuffs for mantelets, dresses, coatings, tweeds etc., for ladies, gentlemen and children, 52 coupons, each 52 in. broad: Tricot 24 sg. (3 sh. 4 d.); divers satins 22 sg. (3 sh. 1 d.); miltons 21 sg. (2 sh. 11 d.); duffs 13½ sg. (1 sh. 10 d.); Doffites 18 sg. (2 sh. 6 d.); Wellingtons 18 sg. (2 sh. 6 d.); shipmans 25 sg. (3 sh. 6 d.); cotton zephyr 8½ sg. (1 sh. 3 d.) plush seadog 2⅓ Th. 12 sh.; castorins, 23 in. broad (cotton plush) 5¾ sg. (10 d.); p. Berl. ell (p. yard).

1710. Werner, Heinrich, manu., Forst (Frankfort on the Oder). Paris 1855 honor. mention. Agts. s. No. 1687.

Three pieces of buckskin, new patterns, p. Berl. ell 50 sg. (p. yard 7 sh.).

1711. Wipprecht, E., manu., Berlin, Mauerstr. 76.

Woollen and linen horse-cloths.

Wirth, W. (s. No. 1610.)

CLASS XXII.

CARPETS.

1714. Burchardt, B., & Sons, manu. of painted window blinds and oil-cloth, Berlin, Brüderstr. 19. Berlin 1844 bronze medal; London 1851 prize-medal, Paris 1855 honor. mention;

London 1862 honor. mention. Agts. Lion M. Co m, Phaland & Dietrich, represented by Ch. Trübner, 20 St. Dunstans Hill, City, and at the Exhib. Building.
Oil-cloth for tables; floor cloth; painted window-blinds.

1715. DINGLINGER, A. F., manu. of carpets and worsted yarns, manufactories at Berlin and Hirschberg. Berlin 1844 bronze medal; London 1851 prize-medal; Munich 1854 medal of honor; Paris 1855 bronze medal; London 1862 honor. mention. Agts. s. No. 1714.
Carpets, prices p. piece: 2 Brussels 3¾ by 3 yards 22 Th. (3 £ 6 sh.); 2 do. 3 by 2¼ yards 13 Th. (1 £ 19 sh.); 2 do. 2¼ by 1½ yards 6½ Th. (19 sh. 6 d.); patent velvet sofa carpet 2¼ by 1¼ yards 7 Th. (1 £ 1 sh.); tapestry do. 2¼ by 1¼ yards 5 Th. (15 sh.).
Undressed yarns, prices p. lb. engl.: 3 bundles of knitting yarn 3 and 4 thread 19½ sg. (1 sh. 11¼ d.); 1 bundle carded yarn single 19 sg. (1 sh. 10¾ d.); 1 bundle do. 2 thread 19½ sg. (1 sh. 11¼ d.); 1 bundle do. 2 by 2 thread cord 20 sg. (2 sh.).

1716. GEVERS & SCHMIDT, Görlitz, manufactory Schmiedeberg. London 1862 honor. mention. Agts. Eichholz, Morris & Co., 16 Gresham-street, E. C. (s. No. 1612 and II. No. 93.)
Drawing-room-carpet No. 2491. contents 80 ☐ Berl. ells or 40⁴⁰⁄₄₉ ☐ yards 26 £; Drawing-room-carpet No. 2502. contents 42 ☐ Berl. or 21²¹⁄₇₉ ☐ yards 13 £ 10 sh.; Uschak-carpet No. 2517. contents 30 ☐ Berl. ells or 15¹¹⁄₄₉ ☐ yards 11 £.

1717. HERRMANN, PHILIPP, cloth-manu., Bromberg.   (s. No. 1626, 1750.)
1 carpet of cow's hair, long 45 ells, broad 26 in., 7½ sg. p. Berl. ell (9 d.); 1 do. 1 Th. 10 sg. (4½ sh.).

1718. KÜHN, THEODOR, & Co., manu. of Turkish carpets, Cottbus (Frankfort on the Oder). London 1862 honor. mention. Agts. s. No. 1714.
An imitation Turkish carpet, 44 ☐ ells 88 Th. (13 £ 4 sh.); by whole-sale, 4 pCt. discount.

1719. LEHMANN, M., manu., Berlin, Brüderstr. 16.
Oil-cloth; oil floor-clooth.

1720. PRAETORIUS & PROTZEN, carpet-manu., Berlin. Munich 1854 honor. mention; Paris 1855 silver medal. Agts. Charles Nolda & Co., 2 Church Court, Old Jewry.
An assortment of 6 patent velvet sofa carpets: an assortment of 6 3frame and 5frame Brussel sofa carpets; 3 5frame velvet sofa carpets.

1721. STEIDEL, C. F., SOMMER, FERD. (Steidel & Sommer), manu. of carpets, Berlin, Weberstr. 8. Agts. s. No. 1714.
Carpets, prices p. piece: 3¾ yards long, 3 yards broad, 30 Th. (4 £ 10 sh.); 3 yards long, 2⅖ yards broad, 18 Th. (2 £ 14 sh.); 3 yards long, 2⅖ yards broad, 15 Th. (2 £ 5 sh); 2⅖ yards long, 1⅗ yards broad, 6¼ Th. (18 sh. 9 d.); 1¾ yards long, 1⅜ yards broad, 6 Th. (18 sh.).
The following prices p. doz.: Scotch carpets, 2⅖ yards long, 1⅗ yards broad, 30 Th. (4 £ 10 sh.); bed carpet 1½ yards long, 8/11 yard broad, 30 Th. (4 £ 10 sh.); ribs carpet 1½ yards long, 8/11 yard broad, 18 Th. (2 £ 14 sh.); travelling bags 8/11 yard long, 6/11 yard broad 18 Th. (2 £ 14 sh.); do. 8/11 yard long, 6/11 yard brod, 23 Th. (3 £ 9 sh.); do. 14/22 yard long, 16/33 yard broad, 14½ Th. (2 £ 2 sh.); do. 14/22 yard long, 16/33

yard broad, 18½ Th. (2 £ 15 sh. 6 d.); do. 6/11 yard long, 16/33 yard broad, 12 Th. (1 £ 16 sh.); do. 6/11 yard long, 16/33 yard broad, 15¼ Th. (2 £ 6 sh. 6 d.); do. 5/11 yard long, 4/11 yard broad, 7¼ Th. (1 £ 1 sh. 9 d.); do. 5/11 yard long, 4/11 yard broad, 9 Th. (1 £ 7 sh.).
Small bags, No. 1. 14/22 yard long, 4/11 yard broad, 4⅖ Th. (13 sh. 6 d.); No. 2. 6/11 yard long, 4/11 yard broad, 4 Th. (12 sh.); No. 3. 6/11 yard long, 7/22 yard broad, 3¼ Th. (10 sh. 6 d.); No. 4. 16/33 yard long, 7/22 yard broad, 3½ Th. (9 sh. 6 d.).

1722. TOEPFFER, GUST. AD., BLEUDORN, HUGO (Gust. Ad. Toepffer & Co.), merchants and manu., Stettin. London 1862 medal. Agt. James Wm. Green (E. G. Zimmermann), 2 St. Pauls Buildings, Little Carter-Lane, Doctors Commons.   (s. No. 1565.)
16 coir mats and rugs, 36 yards each (prices p. yard): plain a. 22 in. wide 9 d.; b. 26 in. wide 11 d.; c. 35 in. wide 1 sh. 2½ d.; border I. d. 22 in. wide 11 d.; do. III. e. 26 in. wide 1 sh. 1 d.; do. IV. f. 35 in. wide 1 sh. 4½ d.; do. V. g. 26 in. 1 sh. 3 d.; V. h. 35 in. wide 1 sh. 7 d.; figured 18. i. 26 in. wide 1 sh. 2½ d.; do. 19. k. 35 in. wide 1 sh. 7½ d.; do. 20. l. 26 in. wide 1 sh. 3 d.; do. 19. m. 35 in wide 1 sh. 7½ d.; 4 red stripes n. 35 in. wide 1 sh. 7 d.; plain extra o. 35 in. wide 1 sh. 8 d.; extra border V. p. 35 in. wide 2 sh. 2 d.; extra pattern 10. q. 35 in. wide 2 sh. 2 d.; Japan plain r. 26 in. wide of 19 yards lgth. 2 sh. 3 d.
Prices p. doz.: 6 coir double mats No. 1. 12 d.; No. 2. 15 sh.; No. 3. 20 sh.; colored border No. 1. 14 sh.; No. 2. 17 sh.; No. 3. 22 sh.; 6 Manilla double mats No. 1. 27 sh.; No. 2. 37 sh.; No. 3. 45 sh.; 3 coir lattice mats No. 1. 12 sh.; No. 2. 15 sh.; No. 3. 20 sh.; 3 Manilla lattice mats No. 1. 24 sh.; No. 2. 35 sh.; No. 3. 45 sh.; 6 coir Japan mats No. 1. 15 sh.; No. 2. 18 sh.; No. 3. 23 sh.; No. 1. 18 sh.; No. 2. 25 sh.; No. 3. 36 sh.; 3 Manilla Japan mats No. 1. 20 sh.; No. 2. 27 sh.; No. 3. 37 sh.; 3 coir frame double mats No. 1. 28 sh.; No. 2. 37 sh.; No. 3. 51 sh.; 3 skeletons No. 1. 21 sh.; No. 2. 27 sh.; No. 3. 36 sh.; 4 Manilla skeletons No. 1. 27 sh.; No. 2. 42 sh.; No. 3. 45 sh.
2 cocoa skeletons 15 ☐ ft. 13 sh.; colored 15 ☐ ft. 15 sh.; cocoa frame double 12 ☐ ft. 13 sh.; prices p. piece.

CLASS XXIII.

WOVEN, SPUN, FELTED AND LAID FABRICS, WHEN SHOWN AS SPECIMENS OF PRINTING OR DYEING.

1723. BERGMANN & Co., manu., Berlin, Krausenstr. 39. London 1862 medal.
Dyed Berlin wool.

1724. HAMERS, ANTON, hire establishment for dressing, tabbying, embossing, silk-dyeing and weaving, Crefeld. London 1862 medal.
Dyed silk in strings; silk and half-silk atlas (satin) finished; taffeta dyed and treated after the French method; tabby, silk and half-silk goods; piece-goods and ribbons embossed; pattern designs for Jaquard weaving.

1725. LAUEZZARI, CARL, manu., Barmen. Munich 1854 medal of honor.; Paris 1855 silver medal; London 1862 medal.-
Genuine Turkish red marking yarn wound on bottoms.

1728. NEUHAUS, H. J., prop. of a silk-dyeing establishment, Crefeld. Berlin 1844 honor. mention; London 1862 medal. Agts. Morgan Brothers, 21 Bow-Lane, Cannon-street, E. C.

Dyed silk.

PFERDMENGES & SCHMOELDERS, Grevenbroich. Printed cotton fabrics. (s. No. 1493.)

1729. REUISHAGEN, F. W. (J. P. Rittershaus), manu., Dusseldorf. Agts. Wattenbach, Hulger & Co., Mincing-Lane.

One bundle of Turkish red, dyed yarn, middling color No. 20. water twist, 10 Pfd. 8 Th., p. Engl. lb. 2 sh. 6 d.

ROLFFS & Co., Cologne. Printed cotton fabrics. (s. No. 1494.)

1730. SPINDLER, W., dyer, silk and woollen-printer, manu. of aniline colors and safflower extract, Berlin. London 1862 medal.

Divers dyed and printed silks for weaving, as well as floss-silks, cordonnet (of the manufactory of Mr. C. A. Hovemann, Berlin) and sewing silks; printed Berlin wool and printed wefts; divers stuffs (old and worn clothes that have been partly dyed again, partly printed or watered).
Aniline-red; aniline-purple; aniline-violet; aniline-blue and safflower extract.

1731. WOLFF, JOH. FR., Elberfeld. Berlin 1844 bronze medal; London 1862 medal. Agts. Morgan Brothers, 21 Bow-Lane, Cannon-street, City.

Turkish red and pink colored cotton yarn 3 packs (30 pounds) 5 £.

CLASS XXIV.

TAPESTRY, LACE AND EMBROIDERY.

BELLINGRATH, C. H., Barmen. Ribbons. (s. No. 1574.)

1732. BESSERT - NETTELBECK, PAULINE, gold, silver and silk embroidery establishment, Berlin, Kronenstr. 52. London 1862 honor. mention. Agt. Louis Andrae, Care of Wm. Klein, 3 Fowke's Buildings, Great Tower-street, E. C. (s. II. No. 10.)

1 genuine gold embroidered cross for altar-cloth 140 Th. (21 £); the arms of Her Majesty the Queen of England, embroidered in gold and silk 66 Th. 20 sg. (10 £); a sketch cut in cardboard which was used in making the cross; 41 patterns for uniform embroideries in gold and silver for military- marine- and civil-officers.

CLEFF, BROTHERS, Barmen. Ribbons. (s. No. 1593.)

1733. FRIEDBERG, LOUIS, lace-maker, Berlin. Agt. Julius Friedberg, 19 Beaumont-square, E.

Patterns of chenille, chenille-tassels, galloons, fringes, buttons; especially for ladies' mantles and dresses.

GREIFF & Co., Barmen. Ribbons for coat trimmings. (s. No. 1614.)

HALBACH, W., & WOLFERTS, Barmen. Ribbons. (s. No. 1622.)

1736. HANCKE, CLARA, prop. of an establishment for embroidery. Agts. Morgan Brothers, 21 Bow-Lane, Cannon-street, City.

Cushion embroidered on ribs 9 Th. (1 £ 7 sh.); embroidery on straw for an étagère 8 Th. (1 £ 4 sh.); slippers on cloth 3 Th. 10 sg. (10 sh.); cushion in crossstitch 8 Th. 10 sg. (1 £ 5 sh.); étagère in crossstitch 8 Th. (1 £ 3 sh.); slippers on velvet 4 Th. (12 sh.).

KELLER & Co., Barmen. Ribbons. (s. No. 1641.)

1738. KRISTELLER, HERRMANN, manu. of white wares, Berlin. Agts. Lion M. Cohn, Phaland & Dietrich, represented by Ch. Trübner, 20 St. Dunstans Hill, City, and at the Exhib. Building.

Undress caps of plain stuffs and lace; prices according to the annexed price current.

LANGENBECK & WEX, Barmen. Buttons covered with worsted. (s. No. 1648.)

MEBUS & RÜBEL, Barmen. Trimming lace and ribbons. (s. No. 1661.)

NIEMANN & GUNDERT, Barmen. Ribbons, lace. (s. No. 1666.)

OSTERROTH, W. & SON, Barmen. Lace. (s. No. 1669.)

1739. PAREY, C. F. W., tapestry-manu., Berlin. Leipzigerstr. 39. London 1851 honor. mention. Agts. s. No. 1738.

A chimney-screen with embroidery in relievo on canvass with Berlin wool representing the eagle of the Prussian arms 100 Th. (50 £).

1740. PLASMAN, F. J. (F. J. Plasman), manu., Barmen. Agts. T. S. & C. Wyche, 13 Sise Lane, E. C.

Lace and ribbons, 12 pieces of each kind; nuns' lace No. 2. to No. 28. 81 yards 9—44 d.; cotton lace No. 2. to No. 38. 86 yards 16—64 d.; German ribbons No. 1. to No. 12. 207 yards 41—92 d.; figured ribbons No. 6. to No. 28. 72 yards 20 to 48 d.; cotton braids No. ¼ to No. 2. 96 yards 34—43 d.

1741. SCHÄRFF, ROBERT, manu., Brieg (Breslau). Berlin 1844 bronze medal; London 1851 prize-medal; Paris 1855 bronze medal; London 1862 medal. Agts. P. & F. Schäfer, 6 Golden-square.

Patterns of every kind of lace-work: A. for coach makers: coach-laces, tuffs, tassels, agreements, fringes, two models of hammer-clothes in ⅓ of the usual size.
B. for harness-makers: every kind of saddle-cloth: girths for halters, bridles and guns; borders for saddle-cloths and horse-cloths; decorations for harnesses, halters, cords, head-nets etc.; wool- and linen stuffs for horse-cloths and housings.

1742. STEINER, J., lace-manu., Breslau. London 1862 medal.

Various patterns of laces, seam cords, girths, bridles, tassels etc., especially for coach-makers and saddlers.

1744. WECHSELMANN, JOHANN, JACOB, manu., Berlin, manufactory Hirschberg (Silesia).

London 1862 medal. Agts. Lion M. Cohn, Phaland & Dietrich, represented by Ch. Trübner, 20 St. Dunstans Hill, City, and at the Exhib. Building.

Volant 6 mètres long 75 cm. high in points à l'aiguille, gauze 720 Th. (108 £); 12 cols and manchettes in points à l'aiguille and in points d'Alençon: No. 2. 14 Th. (2 £ 2 sh.); No. 3. $13\frac{1}{3}$ Th. (2 £); No. 4. $14\frac{1}{2}$ Th. (2 £ 3 sh. 6 d.); No. 5. 6 Th. (18 sh.); No. 6. $5\frac{1}{2}$ Th. (16 sh. 6 d.); No. 7. $5\frac{1}{2}$ Th. (16 sh. 6 d.); No. 8. $5\frac{1}{4}$ Th. (15 sh. 6 d.); No. 9. $4\frac{5}{6}$ Th. (14 sh. 6 d.); No 10. $4\frac{1}{2}$ Th. (13 sh. 6 d.); No. 11. 4 Th. (12 sh.); No. 12. $3\frac{1}{3}$ Th. (10 sh.); No. 13. $3\frac{3}{4}$ Th. (11 sh. 3 d.); 1 garniture worked on horse hair No. 14. 36 Th. (5 £. 8 sh.); 2 do. in application No. 15. 18 Th. (2 £. 14 sh.), No. 16. $12\frac{1}{2}$ Th. (1 £ 17 sh. 6 d.); 1 do. in point gauze No. 17. $35\frac{1}{3}$ (5 £ 6 sh.); 3 pocket handkerchiefs in gauze and application No. 18. 19 Th. (2 £ 17 sh.); No. 19. $13\frac{1}{3}$ Th. (2 £); No. 20. $12\frac{1}{2}$ Th. (1 £ 17 sh. 6 d.); 1 barbe gauze No. 21. $26\frac{2}{3}$ Th. (4 £.); 2 barbes guipure No. 22. $6\frac{2}{3}$ Th. (1 £.); No. 23. $5\frac{1}{3}$ Th. (15 sh. 9 d.).

CLASS XXV.

SKINS, FUR, FEATHERS AND HAIR.

1745. Becke, C. G., manu., Berlin, Wassmannsstr. 33. Agts. s. No. 1774.

Horse-brushes, p. doz. 14 Th. (2 £ 2 sh.).

1747. Engeler, H. M., & Son, brush manu. and furnishers to the court, Berlin, Behrenstr. 36. Berlin 1844 bronze medal; London 1851 prize-medal; Munich 1854 medal of honor; Paris 1855 bronze medal; London 1862 medal. Agts. s. No. 1744. (s. II. No. 32.)

Assortment of hair- and clothes-brushes of plain Jacaranda wood, 1. quality; assortment of hair-, clothes-, hat-brushes and brushes for table-cloth, inlaid with mother of pearl; assortment of ivory hair-brushes: a. of plain ivory, b. of ivory with carved figures; various hair- and flesh-brushes, with steel bristles, simple, magnetic and electro-galvanic; skin rubbing brushes, in the shape of gloves; assortment of shaving-brushes; assortment of horse-brushes, mane-brushes etc.

1748. Friesecke, Wilhelm, brush-maker, Wittenberg.

60 specimens of hair- and clothes-brushes, hat- and oil-brushes. Numbers and prices are marked on one brush of every sort; p. doz. or 12 pieces.

hair-brushes:					£	sh.	d.
	£	sh.	d.	No. 18.	—	18	—
No. 100.	1	16	—	» 103.	1	4	—
» 96.	1	16	—	» 51.	—	15	—
» 92.	1	19	—	» 55.	—	18	—
» 54.	1	10	—	» 93.	—	18	—
» 101.	1	19	—	» 19.	—	9	9
» 91.	1	10	—	» 48.	—	15	—
» 5.	1	4	—	» 4.	—	15	—
» 80.	1	10	—	» 29.	—	12	—
» 53.	1	16	—	» 31.	—	7	6
» 56.	1	10	—	» 14.	—	15	—
» 52.	1	1	—	» 3.	—	13	6
» 94.	1	7	—	» 26.	—	9	—
» 61.	1	1	—	» 2.	—	9	—
» 1.	—	9	—	» 23.	—	6	—
» 50.	—	19	6	» 12.	—	6	—
» 95.	1	7	—	» 47.	—	13	6
» 102.	1	7	—	Litr. H.	—	4	6
» 88.	1	4	—	» I.	—	6	—

hair-brushes:					£	sh.	d.
	£	sh.	d.	No. 7.	—	9	—
Litt. A.	—	5	—	» 8.	1	1	—
» M.	—	6	9	» 10.	1	10	—
» E.	—	4	6	» 1.	—	9	—
No. 2.	—	4	6	» 2.	—	7	—
Litr. C.	—	3	9	hat-brushes:			
No. 9.	—	2	—	No. 1.	—	9	—
Litr. G.	—	4	6	» 2.	—	6	—
» K.	—	4	—	» 3.	—	4	6
» F.	—	4	6	» 0.	—	9	—
» B.	—	2	6	oil-brushes:			
clothes-brushes:				No. 3.	—	8	—
No. 5.	—	18	—	» 2.	—	4	6
» 6.	—	12	—				

1749a. Hegewald, Heinrich, wigmaker and hairdresser, Bromberg.

A gentleman's wig 96 Th. (14 £ 8 sh.); a lady's wig 48 Th. (7 £ 4 sh.); each p. doz.

1750. Herrmann, Philipp, cloth-manu., Bromberg. (s. No. 1626, 1717.)

Cowhair yarn 2 Th.

1751. Hornemann, A., manu., Goch (Dusseldorf). Paris 1855 bronze medal; London 1862 honor. mention. Agts. s. No. 1744.

Brushes in iron rings: Lit. A. 20 sg.; B. $24\frac{1}{2}$ sg.; C. $28\frac{2}{3}$ sg.; D. 35 sg.; E. 42 sg.; F. 50 sg.; G. 56 sg.; H. 65 sg.; I. 76 sg.; K. 86 sg.; L. 96 sg.; M. 108 sg.; N. 116 sg.; O. 124 sg.; P. 132 sg.; Q. 140 sg.; R. 152 sg.; S. 164 sg.; in copper casing: Lit. 00. 30 sg.; A. 37 sg.; C. 48 sg.; E. 62 sg.; G. 74 sg.; I. 100 sg.; L. 132 sg.; N. 178 sg.; badger's hair: No. 1. $3\frac{1}{5}$ Th.; No. 2. $4\frac{1}{5}$ Th.; No. 3. $5\frac{1}{5}$ Th.; No. 4. $7\frac{1}{3}$ Th.; No. 5. $8\frac{1}{5}$ Th.; No. 6. $9\frac{1}{5}$ Th.; No. 7. $10\frac{1}{5}$ Th.; No. 8. $11\frac{1}{5}$ Th.; No. 9. $12\frac{3}{5}$ Th.; No. 10. $13\frac{3}{5}$ Th.; No. 11. $14\frac{3}{5}$ Th.; No. 12. 16 Th., all p. dozen.

1752. Keller, Franz (J. C. Keller & Son), furrier, Weissenfels. London 1862 medal. Agts. Johann Moritz Oppenheim & Co., Cannon-street, West, E. C.

800 pieces of meniver, 100 pieces 8—$35\frac{1}{3}$ Th. (1 £ 4 sh. to 5 £ 6 sh. 6 d.); 100 pieces of backs and bellies 24 Th. (3 £ 12 sh.); 8 breadths meniver bellies for lining, 1 piece $5\frac{1}{2}$—$10\frac{1}{2}$ Th. (16 sh. 6 d. to 1 £ 11 sh. 6 d.).

1753. Koch, Carl Ferd. (C. F. Koch), manu., Zeitz. Agts. Heintzmann & Rochussen, 9 Friday-street, Cheapside.

83 different hair-brushes; 1 hair-brush with pearl embroidery; 28 several clothes-brushes; 2 clothes-brushes with pearl embroidery; 3 hat-brushes; 3 hat-brushes with pearl-embroidery; 3 plush brushes; 1 furniture-brush; 2 table-brushes; 21 pocket-brushes; 6 horse-brushes; 4 polishing-brushes; 3 chess-boards to be unrolled; 2 table-covers to be unrolled; every article is marked with the price.

1754. Krafft, Georg, manu., Wetzlar London 1862 honor. mention.

Wigs in double silk 96 Th.; do. silk gauze (angora-hair) 72 Th.; toupet in silk gauze 36 Th.; ladies' tops in silk gauze 36 Th.; curled tops 60 Th.; crowns in silk No. 1. 2 Th.; do. in silk net No. 2. $2\frac{1}{3}$ Th.; do. silk gauze, No. 4. 3 Th.; tops for gentlemen's wigs in silk and silk gauze 6 in. 10 Th.; do. in silk net 7 in. $10\frac{1}{2}$ Th.; skin tops for ladies 7 in. 18 Th.; ladies' tops in silk gauze 5 in. 9 Th.; do. $4\frac{1}{2}$ in. $10\frac{1}{4}$ Th.; do. $4\frac{1}{2}$ in. 7 Th.; do. (large)

5½ in. 12 Th.; do. in hair gauze 4½ in. 18 Th.; prices p. doz.

Short hair prepared 2½ Th.; do. fine 5 Th.; top-hair do. 15 in. 5½ Th.; 17 in. 6 Th.; 18 in. 8½ Th.; 21 in. 11 Th.; hair for braids, 25 in. 15 Th.; 30 in. 20 Th.; 36 in. 30 Th.; 40 in. 36 Th.; 42 in. 45 Th.; 115 cm. 60 Th.; white 50 cm. 30 Th.; gray 45 cm. 15 Th.; fair 57 cm. 14 Th. Ringlet-hair, 18 and 20 in. 12 Th.; 9 and 13 in. common 3⅓ Th.; 9 and 13 in. fine 5 Th. Curled hair for gentlemen's wigs, fine 5 Th.; common 3⅓ Th.; gray 5 Th.; mixed 15 Th.; half white 20 Th.; white 30 Th.; prices p. pound.

Gentleman's wig weft, brown, red, 2¼ sg.; white 3½ sg.; p. mètre each.

White Angora hair 10 Th.; white goats hair 6 Th.; each p. pound.

1755. LANGE, JOHANN TRAUGOTT, manu. fo feather-bunches, Berlin, Jüdenstr. 51.

Feather-dusters, dusters for feather-bunches p. piece 5 sg. to. 3 Th. (6 d. to 9 sh.).

1758. NANNY, AUGUST, manu. of bristle-wares, Königsberg. London 1862 medal and honor. mention. Agts. Lion M. Cohn, Phaland & Dietrich, represented by Ch. Trübner, 20 St. Dunstans Hill, City, and at the Exhib. Building.

Fifteen species of artificially prepared badger's hair and bristles in four natural and four arti-ficial colors; prices p. pound. Soft bristles: No. 1. gray 35 sg. (3 sh. 6 d.); No. 2. pale yellow 37½ sg. (3 sh. 9 d.); No. 3. black 40 sg. (4 sh.); No. 4. white 50 sg. (5 sh.). Hard bristles No. 5. gray 50 sg. (5 sh.); No. 6. pale yellow 55 sg. (5 sh. 6 d.); No. 7. black 60 sgr. (6 sh.); No. 8. white 90 sg. (9 sh.). Hard and soft bristles No. 9. blue 85 sg. (8 sh. 6 d.); No. 10. red 85 sg. (8 sh. 6 d.); No. 11. yellow 85 sg. (8 sh. 6 d.); No. 12. green 85 sg. (8 sh. 6 d.); No. 13. pale yellow mixed shoe-makers bristles a. 75 sg. (7 sh. 6 d.); b. 90 sh. (9 sh.); c. 105 sg. (10 sh. 6 d.); No. 14. extra fine white mixed bristles d. 120 sg. (12 sh.); e. 150 sg. (15 sh.); f. 180 sg. (18 sh.); No. 15. badger's hair from 1½ to 4½ in. long 17½ sg. (1 sh. 9 d.).

II. Brushes and painters' brushes; prices p. piece: No. 1. 1 foot-brush, mahogany, 30 Th. (4 £ 10 sh.); No. 2. cylinder foot-brush 12 Th. (1 £ 16 sh.); No. 3. a velvet-brush 5 Th. (15 sh.); No. 4. a clothes-brush with embroidery 3 Th. (9 sh.); No. 5 do. to be regulated hard and soft 3 Th. (9 sh.); No. 6. do. serving as a case 3½ Th. (9 sh. 6 d.); No. 7. do. a larger one 4 Th. (12 sh.); No. 8. hair brush 1⅔ Th. (5 sh.); No. 9. do. 2 Th. (6 sh.); No. 10. do. with embroidery 3 Th. (9 sh.); No. 11. do. to be regulated hard and soft 2½ Th. (7 sh. 6 d.); No. 12. a bath-brush (leather) 1⅓ Th. (4 sh.); No. 13. a flesh-brush 3 Th. (9 sh.).

No. 14. badger hair-brushes for painters; No. 15. badger hair-brushes in tin; No. 16. do. round, with, red band.

1760. ROEGNER, C. H., Striegau. London 1862 honor. mention. Agts. s. No. 1758.

Hair-, clothes-, velvet-, pocket-brushes.

1761. SAMTER, LOUIS, merchant, Lissa (Posen). Agt. D. Born, 2 Tower Royal, Cannon-street, West, E. C.

5 bundles of hog's-bristles with price current.

1763. SCHULZE, RUDOLPH (Heinrich's successor), hair-dresser and wig-maker. Gross-Glogau. Agts. s. No. 1758.

A gentleman's fine periwig 1 7/10 Loth 12 Th. (1 £

16 sh.); a gros-de-Naples top 5 Loth 2 sh. 1½ d.; do. to sit fast on the skin 5½ Loth 3 sh. 7½ d.

1764. STANGE, FRIEDRICH-DAVID, manu., Aschersleben. London 1862 honor. mention. Agts. s. No. 1758.

Spun and sodden horse-hair.

CLASS XXVI.
LEATHER, INCLUDING SADDLERY AND HARNESS.

ALVES, H., Berlin, whips.   (s. No. 2181.)

1765. BARTSCH, FRIEDRICH, SONS, leather-manu., Striegau (Breslau). London 1862 honor. mention. Agts. E. Beckh & Co., 19 St. Dunstans Hill.

1 sole-leather 37 Pfd. (41 lb. Engl.) 55 Th. (8¼ £); 1 inner sole-leather 25 Pfd. (27 lb. Engl.) 50 Th. (7½ £); 6/2 harness-leathers 58 Pfd. (64 lb. Engl.) 60 Th. (9 £); 1 hide of neat's leather 13 Pfd. (14 lb. Engl.) 60 Th. (9 £); all p. Zollz. (p. cwt.). 8 calf-leathers 14 Pfd. (15 lb. Engl.) 2⅓ Th. (⅓ £); 1 roof-leather 11 Pfd. (12 lb. Engl.) 8 Th. (1¼ £); each p. piece. 1 machine-strap 8 Pfd. (9 lb. Engl.) 100 Th. (15 £ p. cwt.).

1766. BENJAMIN, M. A., tanner, Cologne. Agts. Page & Welch, 118 Bishopsgate-street.

Twelve calf-skins 40 Pfd. 2 sh. 9 d.; twelve Cal-cutta hides 50 Pfd. 1 sh. 10 d.; p. lb. each.

BLEYENHEUFT, J. H., Aix-la-Chapelle, ma-chine-straps, leather ropes, fire-buckets.
(s. No. 1287.)

BLEYENHEUFT, M. F., (MILLIARD), Eupen, machine-straps, fire-buckets, fire-engine hose; leather for machine straps.   (s. No. 1787a.)

1770. EHRHARDT, C. T., saddler, Nord-hausen. Agts. s. No. 1758.

A trunk 30 Th. (5 £); a hunting pouch 12 Th. (2 £).

1771. GAMMERSBACH, FRANZ WILHELM, manu., Roisdorf near Cologne. Agts. Charles Dittges & Co., 41 Basinghall-street, E. C.
(s. No. 1816.)

Calf-leather: brown, varnished, smoothed, grained calf-skin; goat-skin.

GUNDLACH, W., Breslau. Fancy articles or-namented with leather flowers.   (s. No. 1879.)

1772. HARFF, P. J., tanner, Cologne. Lon-don 1862 medal. Agts. Heintzmann & Rochussen, 9 Friday-street, E. C.

4/2 cow-hides for soles (Laplata-hides) 52 Pfd. 26 Th. (3 £ 18 sh.); 1 piece of sole-leather for packing pump pistons 6 Pfd. 4 Th. (12 sh.); 6 pair of soles 3 Pfd. 2 Th. (6 sh.).

1773. HARTMANN, FR., saddler to His Royal Highness the Prince Royal of Prussia, Berlin, Friedrichsstr. 166. Berlin 1844 bronze medal: London 1862 medal. Agts. s. No. 1758.

1 pair of harnesses 100 £.; 2 saddles 9 £.

1776. HÜTTENHEIN, H., manu., Hilchen-bach (Arnsberg). London 1862 medal.

Sole-leather.

1777. JACOBI, BERNHARD, JUN., leather-manu., Weissenfels. London 1862 honor. mention. Agts. Lion M. Cohn, Phaland & Dietrich, represented by Ch. Trübner, 20 St. Dunstans Hill, City, and at the Exhib. Building.

A hide for machine-straps; a black hide for harnesses; a colored hide for riding-equipage, as used by the Prussian cavalry; p. 100 Pfd. Zollgew. 57 Th. (100 Kil. 428 frs., 112 lb. Engl. 9 £ 6 sh.).
Two brown calf-hides, p. 100 Pfd. Zollg. 100 Th. (100 Kil. 750 frs., 112 lb. Engl. 16 £ 12 sh.).

1779. KÄRNBACH, C., manu., Berlin, Louisenstr. 46.

riding-equipage.

1780. KLEINSCHMIDT, CARL CHRISTOPH, tanner, Mühlhausen. London 1862 honor. mention. Agts. s. No. 1777.

German lamb-skins: 10 pieces tanned like kid-leather, marked No. 1.; 10 smaller ones marked No. II.; p. 100 pieces 50 Th. (7 £ 10 sh.).

1781. KLEIST, apothecary to the Prussian army, Berlin, inv. Agts. s. No. 1777.
(s. No. 1205.)
Sole-leather and boots, the former prepared and of the double durability of common sole-leather; 2 common and 1 prepared leather-sole, to prove the double durability of the latter; 3 pieces of prepared leather, tanned in different ways; 1 prepared sole from the neck of the skin; 1 pair of high waterproof boots with the front soles and heels of prepared leather.

1782. KORN, RICHARD & AUGUST, manu. of varnished leather, Saarbrück. Paris 1855 honor. mention; London 1862 medal. Agts. Richard & August Korn, 11 Union-Court Old Broad-street. — Joh. Wm. Münch, Agent.
2 doz. of black varnished calf-skins.

1783. KORNFELD. L. (L. Kornfeld & Co.), Berlin. Agts. s. No. 1777.
Sheeps-leather; prices p. dicker: red, fine and gross-grained 19 sh. 6 d.; green, fine-grained 19 sh. 6 d.; gross-grained 18 sh.; yellow, fine- and gross-grained 19 sh. 6 d.; black, dressed with sumac-bark 19 sh. 6 d.; brown, fine- and gross-grained 19 sh. 6 d.; pink 1 £ 1 sh. 6 d. and 18 sh. 6 d.; bronzed 1 £ 1 sh., 19 sh. 6 d., 16 sh. 6 d.; tanned black embossed 14 sh. 6 d.; red, yellow, green, black, dressed with sumac-bark 16 sh. 6 d.; pink 18 sh. 6 d. and 15 sh. 6 d.; red 12 sh.
The following prices p. dozen: may-green, pink 1 £ 10 sh.; in diverse colors 1 £ 10 sh. and 1 £ 2 sh. 6 d.; pink and in diverse colors 1 £ 1 sh.; red without gloss 1 £ 8 sh. 6 d.; buck-leathers: gold-bronzed, do. lissé 2 £ and 1 £ 17 sh.
Sheep-leathers: black, tanned, pr. dicker 9 sh.
Split Solferino 1 £ 14 sh.; brown shagreen 1 £ 4 sh.; grosgrain morocco 3 £ 19 sh.; shagreen do. 3 £; fine-red 2 £ 5 sh.; shammy-leathers 2 £ 8 sh.; red morocco 2 £ 17 sh. 6 d.; brown do. 2 £ 12 sh. 6 d.; gold-bronzed, grained morocco 2 £ 17 sh. 6 d.; all p. doz.
Shagreen-morocco, Solferino, Magenta and red p. dicker 2 £ 17 sh. 6 d.
White calf-leathers 3 £ 3 sh.; black morocco-leathers 2 £ 8 sh.; gold-bronzed lissé morocco-leathers 2 £ 17 sh.; all p. doz.
Calf-leather, brown 11½ pounds, à 3 sh. 9 d.; o 4 £ 2 sh.; kid 2 £ 17 sh. 6 d.; glacé 4 £ 6 d., 3 £ 15 sh., 3 £ 3 sh.; all p. doz.; black, rough 3 £ 9 sh. p. dicker; varnished 3 £ 12 sh.,

3 £ 6 sh., 2 £ 11 sh., 2 £ 5 sh., 1 £ 6 sh. 3 d., all p. doz.; brown, 15 pounds 2 sh. 9 d. p. pound.
Varnished neat's leathers 1 £ 19 sh., 2 £ 6 d., 18 sh. 6 d.; varnished calf-leathers 11 sh. 9 d.; all p. skin; do. 4 £ 8 sh.; calf-leather legs with backparts 2 £ 10 sh.; calf-leather vamps 1 £ 5 sh.; all p. doz.
Horse-leather vamps (unfulled) 1 £ 5 sh.; horse-leather backparts 12 sh.; p. doz. each.

KULLRICH, F. F., Berlin, Book-binding leather-articles.
(s. No. 1898.)

1784. KÜHLING, AUG., Düsseldorf. Agts. s. No. 1777.
Elastic saddle with leather tree 6 £ 10 sh.

1785. LAMM, J. H., manu., Berlin, Rossstr. 29. Agts. s. No. 1777.
Leather; prices p. doz. Calf-skins: 2 blacked, plain, gloss of spirit 24 Th. (3 £ 12 sh.); 1 blacked plain, sap-color 24 Th. (3 sh. 12 d.); 1 do. gloss of wax 22 Th. (3 £ 6 sh.); 2 blacked quadrated, gloss of spirit 28 Th. (4 £ 4 sh.); 2 blacked, plained on the flesh-side 26 Th. (3 £ 18 sh.); 1 unblacked, plained on the flesh-side 30 Th. (4 £ 10 sh.); 1 do. 22 Th. (3 £ 6 sh.); 1 unblacked shagreened, without grease 26 Th. (3 £ 18 sh.); 1 blacked shagreened do. 26 Th. (3 £ 18 sh.); 1 blacked plain calf-skin do. 26 Th. (3 £ 18 sh.); goat-skins, 1 blacked shagreened 16 Th. (2 £ 8 sh.); 1 unblacked shagreened 16 Th. (2 £ 8 sh.); 2 blacked buck-skins 13 Th. (1 £ 19 sh.).
The following prices p. doz. pairs: 2 pair of shagreened legs of neat's-leather, for hunters, waterproof 60 Th. (9 £.); 1 pair of chagreened legs of neat's-leather 18 Th. (2 £ 14 sh.); 1 pair of flesh-plain legs of neat's-leathers 17 Th. (2 £ 11 sh.); 1 pair of flesh-plain vamps of neat's-leather 9 Th. (1 £ 7 sh.); 1 pair of shagreened vamps of neat's-leather 9 Th. (1 £ 7 sh.); 2 pair of hairy legs of calf-leather 36 Th. (5 £ 8 sh.); 1 pair of hairy vamps of calf-leather 18 Th. (2 £ 14 sh.); 5 pair of legs of calf-leather, blacked and unblacked 17 Th. (2 £ 11 sh.); 4 pair of blacked vamps of calf-leather 9 Th. (1 £ 7 sh.).

1786a. MEYER, A., manu., Gumbinnen. Agts. s. No. 1777.
Two horse-collars, two bridles, both of twisted work, a line 24 £; a snaffle 2 £ 2 sh.

1789. PASCHEN, W., manu., Königsberg. London 1862 honor. mention.
Saddles with elastic saddle-trees.

1791. ROLKE, FERDINAND, saddler, Breslau Agts. s. No. 1777.
A large travelling-trunk of brown horse-leather and lined with red ticking 1 £ 4 sh. 5 d.; a smaller one 19 sh.; a travelling-pouch of figured plush with brass-handle 15 sh. 8 d.; a smal reticule for ladies of brown calf-leather, lined with red silk 9 sh.

1792. ROSENBAUM, F. W., manu., Breslau. Agts. s. No. 1777.
A pair of harnesses with white garnishment 110 Th. (16 £ 6 sh.); do. with black garnishment 40 Th. (6 £).

RULAND, W., Bonn, leather for machine straps.
(s. No. 1315.)

1796. STERNEFELD, Brothers, manu., Goch (Düsseldorf.)
Calf leather.

1797. STRATHMANN, H., manu., Berlin, Hirschelstr. 44. London 1862 honor. mention.

Roof-leather without varnish.

1799. WIEHR, FRANZ LOUIS, WIEHR, FRIEDRICH (Fr. Wiehr Sons), whip-makers, Berlin, Fischerstr. 10 and 22. Berlin 1844 bronze medal. Agts. Lion M. Cohn, Phaland & Dietrich, represented by Ch. Trübner, 20 St. Dunstans Hill, City, and at the Exhib. Building.

Assortment (55 pieces) of all kinds of whips; for prices apply to the agents.

CLASS XXVII.

ARTICLES OF CLOTHING.

1803. VAN BERLO, J. & A., glove-manu., Aix-la-Chapelle. Paris 1855 bronze medal; London 1862 medal. Agts. S. Oppenheim & Sons, 4 Bread-street, Cheapside.

Kid-gloves for ladies: 1 doz. light colored 1 £ 4 sh.; 1 doz. black do. 1 £ 4 sh. 6 d.; 1 doz. colored 1 £ 4 sh. 6 d.; gloves for gentlemen: 1 doz. piqué, black 1 £ 16 sh.; 1 doz. with embroidery, colored 1 £ 10 sh. 6 d.; 1 doz. without embroidery 1 £ 8 sh. 6 d.; 18 pair fancy gloves at different prices.

Fine cloth-gloves; prices p. doz.: No. 1. fancy colored 16 sh. 6 d.; No. 2. black 16 sh. 6 d.; No. 3. white 16 sh. 6 d.; No. 4. lined with flannel 1 £ 4 sh. 9 d.; No. 8. fancy, embroidered 17 sh. 3 d.; No. 43. mixed, with border 1 £ 3 sh.; No. 44. fancy colored 18 sh.; No. 55. do. embroidered with border 1 £ 2 sh. 6 d.; No. 56. do. embroidered and edged 1 £ 3 sh. 6 d.; No. 57. do. with double border 1 £ 1 sh. 3 d.; No. 16. do. 15 sh.; No. 62. do. 12 sh.; 12 pair of fancy gloves at different prices.

1806. BRESLAU ASSOCIATION OF SHOEMAKERS »WEINTRAUBE«, Breslau. London 1862 medal. Agts. R. Schomburg & Co., 90 Cannon-street, City.

15 pair of Wellington boots with single soles 4 Th. 20 sg., 5 Th., 5 Th. 10 sg.; 15 pair do. with double soles 5 Th. 15 sg., 6 Th., 6 Th. 10 sg.; 4 pair of gaiters of calf-leather with single soles 4 Th. 10 sg.; 4 pair do. with double soles 4 Th. 20 sg., 5 Th.; 4 pair do. of kid with calf-leather facings with single soles 4 Th. 20 sg.; 4 pair do. with double soles 5 Th. 15 sg.; 8 pair do. with neat's leather facings and double soles 6 Th.; 6 pair of gaiters of kid with varnished facings and single soles 5 Th.; 4 pair of varnished leather with single soles 5 Th.; 7 pair of varnished neat's leather with double soles 5 Th. 15 sg.; 1 pair of English lacing gaiters with double soles 7 Th. 15 sg.; 4 pair do. with double soles and toe-piece 8 Th.; 3 pair of shoes of varnished leather 4 Th.; 6 pair do. of calf-leather 3 Th. 15 sg., 4 Th. 5 sg.; 2 pair of waterproof boots with swellings and double soles 10 Th.; 3 pair of common waterproof boots with double soles 7 Th.; 1 pair of high boots 12 Th.; prices p. pair.

BÜLOW, C., Görlitz, masks. (s. No. 2183.)

1807. CADURA, HEINRICH, merchant and manu. of leather driving belts, Breslau. Agts. s. No. 1799. (s. No. 1292, 1503, 1938.)

Woollen India-rubber coats; cotton India-rubber coats.

1808. CIESIELSKI, AUGUST, shoemaker, Bromberg.

A pair of »Prince Albert«- boots, sewed 84 Th. (12 £ 12 sh.); a pair of calfskin boots do. 60 Th. (9 £) p. doz. pair; a shoe-shaped leather snuff box, silver rimmed 96 Th. (14 £ 8 sh.) p. doz.

1809. CLASSEN-KAPPELMANN, manu. of woollen yarns and hosiery, Cologne. Agts. Morgan Brothers, 21 Bow Lane, Cannon-street, City.

Woollen yarns, jackets and drawers of wool, cotton, silk etc.

1811. DOMBROWSKY, STANISLAUS, shoemaker, Posen. London 1862 honor. mention. Agts. s. No. 1799.

A pair of varnished neat's leather boots 2 £ 2 sh.

DELLMANN, W., & Co., Elberfeld, cravats, slips. (s. No. 1598.)

1812. DODECK, ROBERT, manu. of gloves, Burg near Magdeburg. Agts. s. No. 1799.

Tamboured ladies' gloves, dyed 12 pair 5¼ Th. (15 sh. 9 d.); do. light color. 12 pair 5 Th. (15 sh.); gentlemen's gloves of doe-skin 12 pair 6 Th. (18 sh.); do. white 12 pair 5 Th. (15 sh.); black gloves 6 pair 4½ Th. (13 sh. 6 d.); dyed do. 6 pair 4½ Th. (13 sh. 6 d.); doe-skin gloves for gentlemen 12 pair 6¼ Th. (18 sh. 9 d.); tamboured gloves for gentlemen, dyed 12 pair 6 Th. (18 sh.); tamboured kid-gloves for ladies 12 pair 7 Th. (21 sh.); do. black 12 pair 5½ Th. (16 sh. 6 d.); 17 pair of gloves black or dyed in light colors and ornamented with different embroidery works at the wholesale prices of 2¾, 3¼, 5, 5¼, 5½, 6, 7, 8½—11 Th. p. doz. — Besides 1 pair of gloves merely cut. Two samples of colors.

1813a. EIGEL, JOH., manu. and merchant, Cologne. Agt. Hermann Kerkhoff, Esq.

12 pair of gentlemen's boots and shoes. The prices p. pair in Prus. money, p. doz. in Engl. money, freight paid London. No. 1. Wellington boots, calf leather, double soles, sewed 3 Th. 7 sg. 6 pf. (6 £ 7 sh.); No. 2. do. single soles 2 Th. 27 sg. 6 pf. (5 £ 8 sh.); No. 3. do. double soles 3 Th. 2 sg. 6 pf. (5 £ 14 sh.); No. 4. and 5. do. single soles 2 Th. 22 sg. 6 pf. (5 £ 2 sh.); No. 6. do. 2 Th. 15 sg. (4 £ 13 sh.); No. 7. do. with elastic 2 Th. 12 sg. 6 pf. (4 £ 10 sh.); No. 8. do. neat's leather double soles 3 Th. 2 sg. 6 pf. (5 £ 14 sh.); No. 9. do. single soles 2 Th. 17 sg. 6 pf. (4 £ 16 sh.); No. 3—9. pegged; if sewed 5 sg. or 6 d. more p. pair; workmen's shoes: No. 10. 1 Th. 18 sg. (3 £); No. 11. 1 Th. 14 sg. (2 £ 14 sh.); No. 12. 1 Th. 16 sg. (2 £ 17 sh.).

1814. ELSTER, ALEXANDER, straw-hat-manu., Berlin. Paris 1855 honor. mention; London 1862 medal. Agt. Edward Stadler, 2 Lawrence Lane. E. C.

24 white straw-hats of 7 halms and indented plaits, plain and fancy sorts; made by applying a peculiar bleaching method; prices affixed to the hats. Straw-plaits, 6 parcels; straw-halms, 2 parcels.

1816. GAMMERSBACH, FRANZ WILHELM, manu., Roisdorf near Cologne. Agts. Charles Dittges & Co., 41 Basinghall-street, E. C. (s. No. 1771.)

Vamps ready for the last and legs of boots; rosettes for shoes; cap-front-shades; cap-straps; silk hats for gentlemen.

1818. HACKENBERG, F., manu. of buttons, Elberfeld. Agts. Lion M. Cohn, Phaland & Dietrich, represented by Ch. Trübner, 20 St. Dunstans Hill, City, and at the Exhib. Building.

Samples of patent buttons for gentlemen and ladies; the gross from 4 d. to 1 £.

1819. HERRMANN, EDUARD, & Co., manu., Berlin, Breite Str. 7. Agts. s. No. 1818.

Silk umbrellas: No. 1. 1 piece 24 in. D. steel-rods, wood-stick 1 doz. 22 Th.; No. 2. 1 piece 26 in. C. artificial whale-bone 1 doz. 36 Th.; No. 3. 1 piece 26 in. A. steel-rods, woodstick 1 doz. 40 Th.; No. 4. 1 piece 26 in. raw Croisé Paragon stick 1 doz. 54 Th.
Silk parasols: litt b. 1 piece 16 in. entoutcas w. horn handle 1 doz. 15 Th.; litt. n. 1 piece 17 in. do. wood stick 1 doz. 28 Th.; litt. s. 1 piece 17 in. do. 1 doz. 33 Th.; litt. w. 1 piece 16 in. do. w. fine bone handle 1 doz. 50 Th.; No. 5. 1 piece Knicker fond jaspé à volant 1 doz. 40 Th.

1820. HESSTHAL, W. J., manu., Aix-la-Chapelle. London 1862 medal and honor. mention.

Worsted gloves.

1821. HOLLAENDER, B., manu., Leobschütz (Oppeln.)

Jackets and tippets for children and ladie's caps.

1823. KIPPMEYER, G., Crefeld. Agts. Morgan Brothers, 21 Bow Lane, Cannon-street, City.

Shoes and boots.

1825. KÜHN, J. W., Crefeld. Agts. Morgan Brothers, 21 Bow Lane, Cannon-street, City.

Shoes and boots.

1826. LANGETHAL, GOTTLOB, Erfurt. Paris 1855 honor. mention; London 1862 honor. mention. Agt. William Schultz, 4 Lawrence Pountney Place, Cannon-street.

Boots and shoes, 18 pair; prices p. doz.: boots calf leather 7 £ 4 sh.; buskins, double soles. elastic 7 £ 4 sh.; do. varnished leather, elastic 6 £ 18 sh.; shoes, patent neat's leather, double soles 5 £ 14 sh.; buskins, calf leather, elastic 5 £ 8 sh.; boots, calf leather, 2. quality 5 £ 8 sh.; shoes, elastic 5 £ 2 sh.; buskins for ladies, elastic and varnished leather, borderings 3 £ 18 sh.; do. of another kind 3 £ 12 sh.; boots, neat's leather, 2. quality 3 £ 12 sh.; buskins for ladies, of serge with elastic and heels 3 £ 1 sh.; do. of another kind 2 £ 15 sh.; do. 2 £ 9 sh.; do. 2. quality, with heels 2 £ 8 sh.; do. 2 £ 2 sh.; buskins for children of serge 1 £ 1 sh.; do. with rosettes 1 £ 1 sh.; Victoria shoes, of varnished leather 1 £ 1 sh.

1827. LANGMEIER, C. A., hat-manu., Berlin, Grosse Friedrichsstr. 106. Agts. s. No. 1818.

18 hats of woollen textures and felt for gentlemen, ladies, boys and girls, p. doz. 6½ bis 30 Th. (1 £ to 4 £ 10 sh.).

1828. LAUFFER, EMIL, & Co. (C. d'Heureuse), Berlin. London 1851 prize-medal.

Ladies' bonnets of sewed plaiting: fine fine Leghorn, 11 halms; do. Brussels, 7 halms; middle fine Brussels, 7 halms; fine fine German, 7 halms; white Lisse Swiss plaiting; white horse-hair plaiting. Round girls' and boys' bonnets and hats fine fine English, German, Swiss whip edge plaiting, German, 7 halms plaiting.
Gentlemen's hats of colored English plaiting.
Boys' caps of fine fine English whip edge plaiting.

1829. LENZ, CARL FEDOR, Berlin. London 1862 medal. Agt. David Nau, 4 Queen-street, Cheapside.

43 pieces of woollen hosiery and fancy goods, caps etc.

1834. MÜLLER, JOHANN LUDWIG, shoemaker, Berlin, Schützenstr. 76.

No. 1. A boot for a lady, whose one leg is eleven inches shorter than the other 13 Th. (50 frs.) (2 £); No. 2. boot lined eleven inches high with cork and wood for a gentleman, whose legs have the same difference in length 13 Th. (50 frs.) (2 £); No. 3. four boots with last for persons with differently deformed feet 13 Th. (50 frs.) (2 £); No. 4. a pair of varnished boots for a person with a large flat foot p. pair 13 Th. (50 frs.) (2 £).

NAUSESTER, W., Lötmaringhausen, jackets and caps. (s. No. 1663.)

1837. NOACK, EDUARD, manu. of felt-wares, hat-maker to His Majesty the King of Prussia, Berlin. Manufactories at Berlin and Brandenburg on the Havel in the Royal penitentiary. Agts. s. No. 1818.

18 pair of felt-shoes with soles of felt and of leather, at prices of 5 to 22½ sg. (6 d. to 2 sh. 3 d.), p. pair.

OFFENHAMMER, Berlin, Steel collars, and wriss bands. (s. No. 1279.)

1838. OPPERMANN, E. F., manu. of boots and shoes, Berlin, Unter den Linden 60. London 1862 honor. mention. Agts. s. No. 1818.

For gentlemen: No. 1. varnished cannon boots 25 Th., with spurs 30 Th. (4 £ 10 sh.); No. 2. leather top boots, with tops 12 Th.. (1 £ 16 sh.); No. 3. varnished boots 8 Th. (1 £ 4 sh.); No. 4. grained hunting boots 10 Th. (1 £ 10 sh.); No. 5. vachette elastic spatterdashes, with strings at the top 10 Th. (1 £ 10 sh.); No. 6. spatterdashes of cloth with buttons and cork soles 8 Th. (1 £ 4 sh.); No. 7. glacé elastic bottines, with strings, vachette border, double soles 8 Th. (1 £ 4 sh.); No. 8. chevreaux elastic bottines 7 Th. (1 £ 1 sh.); No. 9. drap de soie elastic bottines, varnished 7 Th. (1 £ 1 sh.); No. 10. varnished elastic shoes 5 Th. (15 sh.); No. 11. fur slippers 4 Th. (12 sh.).
For ladies: No. 12. bronzed leather elastic spatterdashes 6 Th. (18 sh.); No. 13. Balmoral spatterdashes with cork bottoms varnished 8 Th. (1 £ 4 sh.); No. 14. gray drap de soie spatterdashes, with strings and chevreaux borders 6 Th. (18 sh.); No. 15. brown drap de soie spatterdashes with strings and bronze border 4 Th. (12 sh.); No. 16. silk wadded half boots 6 Th. 18 sh.); No. 17. red velvet slippers with genuine gold embroidery and pearl 60 Th. (9 £); No. 18. blue velvet shoes 6 Th. (18 sh.); No. 19. bronze shoes 4 Th. (12 sh.); No 20. gray silk shoes 3 Th. (9 sh.).

1840a. SALKOWSKI, JOHANN, tailor, Posen. Agts. s. No. 1818.

An overcoat of black cloth (czamarka), with the appurtenant undercoat (Żupan), buttoned to the former, Polish national costume 60 Th. (9 £).

1842. SCHMIDT, C. R., manu., Breslau.
Boots and spatterdashes.

1843. SCHRÖDER, CHRISTIAN, shoemaker to the court, Laasphe near Wittgenstein (Arnsberg). London 1862 honor. mention. Agts. Heintzmann & Rochussen, 9 Friday-street.
8 pair of boots: top-boots with tops of varnished leather; long-boots with nails; vachette-boots; varnished leather boots with legs; boots of calf skin; boots with India-rubber border of vachette and double soles; boots with elastic clocks and varnished leather borders; lace boots with double soles.

1844. SEELIG, S., manu., Berlin, Alexanderstr. 53.
Clothes made of different kinds of hair.

1846. SOMMERFELD, LOUIS, merchant, Berlin, Königsstr. 30. Agts. Lion M. Cohn, Phaland & Dietrich, represented by Ch. Trübner, 20 St. Dunstans Hill, City, and at the Exhib. Building.
A great-coat containing another coat and waistcoat; a pair of riding and walking pantaloons; a paletot of double cloth.

1847. SONDERMANN, F. W., manu., Gummersbach near Cologne. Agt. Victor Bauer, 1 Ironmonger-Lane, Cheapside.
Samples of knit woollen and worsted hosiery.

1848. STEGMANN, C. A. F., manu., Berlin, Scharrnstr. 8. Agts. s. No. 1846.
31 silk and cotton umbrellas.

1849. STEINBERG, E., manu., Berlin, Markgrafenstr. 40.
Bodices.

1850. TEICHMANN, M., manu., Leobschütz (Oppeln). London 1862 honor. mention.
Worsted shawls, jackets, caps, sleeves and hoods.

1851. TESCHEMACHER, RUD., & KATTENBUSCH, AUG., manu., Werden on the Ruhr. London 1862 honor. mention. Agt. Friedrich Osterroth, 1 Bell-Yard, South side St. Pauls, E. C.
Felt shoes; prices p. doz. of pair; the numbers refer to size.
1 quality with felt soles 1. No. 4. 2½ Th. (7 sh. 6 d.); 2. No. 8. 3½ Th. (10 sh. 6 d.); 3. No. 12. 4½ Th. (13 sh. 6 d.); 4. No. 15. 5½ Th. (16 sh. 6 d.); 5. No. 19. 6½ Th. (19 sh. 6 d.); soles of plaited felt 6. No. 13. 4½ Th. (13 sh. 6 d.); 7. No. 17. 6½ Th. (19 sh. 6 d.); double felt soles 8. No. 14. 7 Th. (1 £ 1 sh.); 9. No. 18. 8 Th. (1 £ 4 sh.); leather soles 10. No. 2. 4 Th. (12 sh.); 11. No. 10. 5½ Th. (16 sh. 6 d.); 12. No. 13. 7 Th. (1 £ 1 sh.); 13. No. 16. 8 Th. (1 £ 4 sh.); 14. No. 19. 9 Th. (1 £ 7 sh.); leather soles and set with leather 15. No. 11. 8½ Th. (1 £ 5 sh. 6 d.); 16. No. 14. 9½ Th. (1 £ 8 sh. 6 d.); upper leathers 17. No. 15. 11 Th. (1 £ 13 sh.); 18. No. 18. 12 Th. (1 £. 16 sh.); double soles felt and leather 19. No. 13. 8½ Th. (1 £ 5 sh. 6 d.); 20. No. 17. 10½ Th. (1 £ 11 sh. 6 d.); Russian or half-boots with leather soles 21. No. 12. 9 Th. (1 £ 7 sh.); do. set with leather 22. No. 14. 11½ Th. (1 £ 4 sh. 6 d.); upper leathers entire 23. No. 16. 13 Th. (1 £. 19 sh.); plain plush with leather soles 24. No. 11. 10 Th. (1 £ 10 sh.); do. set with leather 25. No. 15. 12 Th. (1 £ 16 sh.); black with frieze lining 26. No. 15. 12 Th. (1 £ 16 sh.); pattern plush with leather soles 27. No. 13. 11 Th. (1 £ 13 sh.); checkered do. 28 No. 14. 12 Th. (1 £ 16 sh.); do. set with leather 29. No. 9. 10½ Th. (1 £ 11 sh. 6 d.); 30.

No. 18. 14 Th. (2 £ 2 sh.); pattern do. 31. No. 18. 14 Th. (2 £ 2 sh.); colored felt, Indian rubber soles and upper leathers entire 32. No. 15. 11 Th. (1 £ 13 sh.); black felt do. 33. No. 17. 12 Th. (1 £ 16 sh.); ice galoshes 34. No. 14. 9 Th. (1 £ 7 sh.); felt spatterdashes with buttons without lining 35. No. 7. 8 Th. (1 £ 4 sh.); 36. No. 13. 10 Th. (1 £ 10 sh.); plush do. elastic sides and lining with rosettes 37. No. 12. or 36. 19 Th. (2 £ 17 sh.); without rosettes 38. No. 14. or 38. 20 Th. (3 £.); carpet shoes set with leather 39. No. 13. 13 Th. (1 £ 19 sh.); 40. No. 16. 14 Th. (2 £ 2 sh.); leather parlour slippers 41. No. 11. or 34. 10 Th. (1 £ 10 sh.); 42. do. No. 13. or 37. 10 Th. (1 £. 10 sh.); stramin shoes (canvass) 43. No. 11. 5 Th. (15 sh.); 44. No. 14. or 38. 6 Th. (18 sh.); second quality felt shoes, felt soles 45. No. 9. 3¼ Th. (9 sh. 9 d.); 46. No. 13. 4 Th. (12 sh.); 47. No. 18. 6 Th. (18 sh.); do. leather soles 48. No. 2. 3¼ Th. (10 sh. 6 d.); 49. No. 13. 6 Th. (18 sh.); do. 50. No. 20. 8 Th. (1 £. 4 sh.); do set with leather 51. No. 15. 8 Th. (1 £ 4 sh.); do. upper leathers entire 52. No. 19. 10 Th. (1 £ 10 sh.); do. double soles felt and leather 53. No. 16. 8½ Th. (1 £ 5 sh. 6 d.); Russian or half-boots, leather soles 54. No. 12. 7 Th. (1 £ 1 sh.); do. upper leather entire 55. No. 17. 11 Th. (1 £. 13 sh.); Lastings boots: children's boots 56. No. 0. 6 Th. (18 sh.); 57. No. 4. 7 Th. (1 £ 1 sh.); girls' boots 58 No. 6. 8 Th. (1 £ 4 sh.); 59. No. 10. 13 Th. (1 £ 19 sh.); 60. No. 5. 10 Th. (1 £ 10 sh.); 61. No. 8. 12 Th. (1 £. 16 sh.); children's button-boots 62. No. 1. 9 Th. (1 £ 7 sh.); girls' do. 63. No. 9. 15 Th. (2 £ 5 sh.); welted boots 64. No. 6. 14 Th. (2 £ 2 sh.); do. 65. No. 10. 18 Th. (2 £. 14 sh.); do. 66. No. 3. 12 Th. (1 £ 16 sh.); do. 67. No. 7. 16 Th. (2 £ 8 sh.); French satin children's boots 68. No. 7. 12 Th. (1 £ 16 sh.); do. girls' boots 69. No. 27. 18 Th. (2 £ 14 sh.); lasting ladies' boots without heels 70. No. 12. 14 Th. (2 £. 2 sh.); on welts 71. No. 16. 16 Th. (2 £ 8 sh.); with heels 72. No. 14. 16 Th. (2 £ 8 sh.); elastic sides 73. No. 12. 18 Th. (2 £ 14 sh.); 74. No. 14. 18 Th. (2 £ 14 sh.); 75. No. 13. 21 Th. (3 £ 3 sh.); set with leather 76. No. 13. 19 Th. (2 £ 17 sh.); upper leathers entire on welts 77. No. 15. 20 Th. (3 £); French satin ladies' boots 78. No. 36. 26 Th. (3 £ 18 sh.); leather boys' boots 79. No. 1. 13 Th. (1 £ 19 sh.); 80. No. 7. 16 Th. (2 £ 8 sh.); 81. No. 12. 19 Th. (2 £ 17 sh.); gentlemen's boots 82. No. 18. 30 Th. (4 £. 10 sh.); 83. No. 17. 32 Th. (4 £ 16 sh.); 84. No. 19. 36 Th. (5 £ 8 sh.); 85. No. 16. 39 Th. (5£ 17 sh.); 86. No. 18. 40 Th. (6 £.); 87. No. 15. 40 Th. (6 £); with double soles 88. No. 19. 44 Th. (6 £ 12 sh.); 89. No. 17. 44 Th. (6 £ 12 sh.); gentlemen's shoes (wax leather) 90. No. 17. 34 Th. (5 £ 2 sh.); (patent leather) 91. No. 18. 36 Th. (5 £ 8 sh.); (vachette) 92. No. 16. 36 Th. (5 £ 8 sh.); first qual. gentlemen's leggings 93. No. 18. 10 Th. (1 £ 10 sh.); 94. No. 17. 11 Th. (1 £. 13 sh.); 95. No. 18. 12 Th. (1 £ 16 sh.); 96. No. 18. 16 Th. (2 £ 8 sh.); 97. No. 15. 21 Th. (3 £ 3 sh.); 98. No. 17. 14 Th. (2 £ 2 sh.); 99. No. 18. 21 Th. (3 £ 3 sh.); 100. No. 19. 24 Th. (3 £ 12 sh.); 101. No. 18. 15 Th. (2 £ 5 sh.); 102. No. 16. 15 Th. (2 £ 5 sh.); 103. No. 17. 22 Th. (3 £ 6 sh.); 104. No. 18. 24 Th. (3 £ 12 sh.); 105. No. 19. 21 Th. (3 £ 3 sh.); 106. No. 18. 22 Th. (3 £ 6 sh.); 107. No. 19. 23 Th. (3 £ 9 sh.); vamps: wax leather 108. No. 17. 18 Th. (2 £ 14 sh.); varnished leather 109. No. 18. 19 Th. (2 £ 17 sh.); 110. No. 16. 20 Th. (3 £); leggings for gentlemen: second qual. 111. No. 18. 8 Th. (1 £. 4 sh.); 112. No. 17. 16 Th. (2 £ 8 sh.); for ladies: leather first qual., elastic sides 113. No. 14. 18 Th. (2 £ 14 sh.);

with rings 114. No. 14. 16 Th. (2 £ 8 sh.); lasting first qual., with lacing holes 115. No. 12. $5\frac{2}{5}$ Th. (16 sh.); 116. No. 14. $5\frac{2}{3}$ Th. (17 sh.); 117. No. 16. 6 Th. (18 sh.); with metallic rings 118. No. 15. 6 Th. (18 sh.); second qual. with metallic rings 119. No. 1. 3 Th. (9 sh.); with lacing holes 120. No. 8. $4\frac{1}{2}$ Th. (13 sh. 6 d.); 121. No. 16. $5\frac{3}{5}$ Th. (17 sh.); with buttons 122. No. 00. $3\frac{1}{2}$ Th. (10 sh. 6 d.); 123. No. 6. $4\frac{1}{2}$ Th. (13 sh. 6 d.); first qual. elastic sides 124. No. 13. 8 Th. (1 £ 4 sh.); 125. No. 14. 10 Th. (1 £ 10 sh.).

Saddle cloths: prices p. doz.: gray, 126. 127. 12 Th. (1 £ 16 sh.); white, 128. 13 Th. (1 £ 19 sh.); colored, 129. 130. 131. 15 Th. (2 £ 5 sh.); fine, 132. 133. 134. 135. 136. 22 Th. (3 £ 6 sh.); jockey, 137. 138. 139. 22 Th. (3 £ 6 sh.); silk embroidered saddle cloth, 140., p. piece $13\frac{1}{3}$ Th. (2 £); gold do. 141. $26\frac{2}{3}$ Th. (4 £.); silver do. 142. $26\frac{2}{3}$ Th. (4 £).

Carriage saddle cloths, 143. p. doz. 12 Th. (1 £ 16 sh.).

Gray sole felt, 144. p. piece 1 Th. (3 sh.); white do. 145. p. piece 1 Th. (3 sh.); 1 pair of felt soles to be put into the boots 146. No. 18. p. doz. pair $1\frac{1}{6}$ Th. (3 sh. 6 d.); 1 pair of rolled leather soles 147. No. 20. p. doz. pair $3\frac{2}{3}$ Th. (11 sh.).

Lining socks, stitched, 148. No. 14. p. doz. pair $\frac{3}{4}$ Th. (2 sh. 3 d.); do. 149. No. 15. p. doz. pair $\frac{3}{4}$ Th. (2 sh. 3 d.); milled 150. No. 16. p. doz. pair $\frac{3}{4}$ Th. (2 sh. 3 d.); do. 151. No. 17. p. doz. pair $\frac{5}{6}$ Th. (2 sh. 6 d.); 5 ells plush binding 152. p. ell $\frac{5}{12}$ Th. (1 sh. 3 d.); 12 ells felt for shoe uppers 153. p. Berl. ell $1\frac{5}{12}$ Th. (4 sh. 3 d.); 12 ells woven shoe uppers 154. p. ell 1 Th. (3 sh.); felt for rubbers 155. p. lb. $3\frac{1}{3}$ Th. (10 sh.); felt for pumps 156. 157. p. lb. $\frac{2}{3}$ Th. (2 sh.); filter funnels 158. p. piece $\frac{5}{12}$ Th. (1 sh. 3 d.); do. 159. p. piece $\frac{7}{12}$ Th. (1 sh. 9 d.); do. 160. p. piece $\frac{2}{3}$ Th. (2 sh.); 3 printed beer glass waiters 161. p. doz. $\frac{2}{3}$ Th. (2 sh.); machine felt 162. p. lb. $\frac{1}{6}$ Th. (6 d.).

1853. Wechselmann & Co., manu. and merchants, Berlin, Leipzigerstr. 42.

6 morning-caps viz: No. 6. 20 Th. (3 £); No. 7. 14 Th. (2 £ 2 sh.); No. 8. 20 Th. (3 £); No. 9. 14 Th. (2 £ 2 sh.); No. 10. 20 Th. (3 £); No. 11. 20 Th. (3 £); all p. doz.

1 tippet and sleeves in muslin (mull) No. 1. 3 Th. (9 sh.); 1 tippet and sleeves, in bobbin-net No. 2. 6 Th. (18 sh.); 1 muslin (mull) smock-frock No. 3. $2\frac{5}{6}$ Th. (8 sh. 6 d.); 1 pair of shoulder bandeaux in muslin No. 4. $1\frac{5}{6}$ Th. (2 sh. 6 d.); 1 berthe in silk lace (tull) $3\frac{1}{2}$ Th. (10 sh. 6 d.).

1854. Weidling, August, shoe-manu., Erfurt. Agts. Lion M. Cohn, Phaland & Dietrich, represented by Ch. Trübner, by Ch. Trübner, 20 St. Dunstans Hill, City and at the Exhib. Building.

Boots and shoes, 8 pair, prices p. pair: laced ladies' boots of lasting 1 Th. 25 sg. (5 sh. 6 d.); do. of satin 2 Th. 15 sg. (7 sh. 6 d.); do. of prunello 1 Th. 25 sg. (5 sh. 6 d.); ladies' kid boots with elastic 2 Th. 20 sg. (8 sh.); do. (à la visite), of silk 2 Th. 5 sg. (6 sh. 6 d.); varnished boots for gentlemen with elastic 3 Th. 10 sg. (10 sh.); gentlemen's half-boots with elastic and varnished leather borders 3 Th. 20 sg. (11 sh.); gentlemen's half-boots with elastic varnished neat's leather borders and English soles 4 Th. 10 sg. (13 sh.).

1855. Weissler, S., merchant, Leobschütz. Agts. s. No. 1854.

25 pieces of woollen caps for ladies and children, jackets, sleeves. The price per dozen marked on the label of every piece.

1856. Włościborski, Ludwig, shoemaker, Posen. Agts. s. No. 1854.

A pair of ladies' shoes, white satin, $2\frac{1}{2}$ ozs. wt. 5 sh.; a pair of black satin ladies' boots with heels, 6 ozs. wt. 8 sh.

1859. Zarrad, F. A., Crefeld. Agts. Morgan Brothers, 21 Bow Lane, Cannon-street, City.

Shoes and boots.

1860. Zarrad, J. A., Crefeld. Agts. Morgan Brothers, 21 Bow Lane, Cannon-street, City.

Shoes and boots.

CLASS XXVIII.

PAPER, STATIONERY, PRINTING AND BOOKBINDING.

1861. Behrend, Bernhard, paper-manu., Cöslin. London 1862 medal.

1. Specimens of printing paper sized and unsized. 2. Samples of paper sheets for writing and printing telegraphs. 3. Specimens of tinted paper. 4. Specimens of vegetable parchment of 4 thicknesses as substitute for bladder, tracing paper etc. The prices are noted on the specimens.

1862. Patent-paper-manufactory, Berlin. London 1862 medal. Agts. s. No. 1854. (s. No. 1890.)

Different samples of ordinary and fine account-book- and letter-papers.

1863. Bornefeld, H. W., manu., Barmen.

Wafers.

1864. Braselmann & Vorster, manu. of paper, Stennert near Eilpe (Arnsberg). London 1862 honor. mention.

52/4 ream of letter-paper; $\frac{1}{4}$ ream of every sort; weight and price p. ream of 960 lettersheets:

No.	lb.	sh.	d.	No.	lb.	sh.	d.
103.	10	7	—	135.	12	9	11
104.	10	7	—	135.	15	12	5
105.	11	7	8	136.	12	9	11
105.	14	9	10	136.	15	12	5
106.	11	7	8	137.	13	10	8
106.	14	9	10	138.	13	10	8
107.	12	8	5	143.	10	9	—
108.	12	8	5	144.	10	9	—
113.	11	8	—	145.	11	9	11
114.	11	8	—	145.	15	13	6
115.	12	8	8	145.	18	16	2
115.	15	10	11	146.	11	9	11
116.	12	8	8	146.	15	13	6
116.	15	10	11	146.	18	16	2
117.	13	9	5	147.	12	10	10
118.	13	9	5	148.	12	10	10
123.	10	8	—	153.	11	10	2
124.	10	8	—	154.	11	10	2
125.	11	8	10	155.	12	11	1
125.	14	11	2	155.	15	13	11
126.	11	8	10	155.	18	16	8
126.	14	11	2	156.	12	11	1
127.	12	9	7	156.	15	13	11
128.	12	9	7	156.	18	16	8
133.	11	9	1	157.	13	12	—
134.	11	9	1	158.	13	12	—

24/4 ream of writing-paper; $\frac{1}{4}$ ream of every sort; weight and price p. ream of 480 sheets:

No.	lb.	sh.	d.	No.	lb.	sh.	d.
42.	8	4	5	62.	8	4	10
43.	9	4	11	63.	11	6	7
43.	11	6	1	67.	26	15	7
44.	11	6	1	69.	42	25	2
47.	26	14	4	72.	9	5	8
49.	42	23	1	73.	12	7	6
52.	9	5	2	77.	26	16	3
53.	10	5	9	79.	42	26	3
53.	12	6	11	82.	8	5	5
Stempel.	13	7	6	83.	11	7	5
57.	26	14	11	92.	9	6	4
59.	42	24	2	93.	12	8	5

Prices free at Hagen; the freight hence to London $\frac{3}{10}$ d. p. lb. ($\frac{1}{2}$ ream).

1867. CRAMER, JULIUS, manu., Cologne. Agts. Morgan Brothers, 21 Bow-Lane, Cannon-street, City.

4 bottles of aleppo copying-ink, 1 bottle (1 litre, $\frac{1}{4}$ gallon) 20 sg. (2 sh.).

1867 a. DECKER, RUDOLPH (Königliche Geheime Ober-Hof-Buchdruckerei), printer to His Majesty the King of Prussia, type founder, publisher and paper-manu.; paper-manufactory Eichberg near Hirschberg (Silesia). London 1851 prize-medal. Great gold medal for services in the promotion of industry; London 1862 medal. Agts. Lion M. Cohn, Phaland & Dietrich, represented by Ch. Trübner, 20 St. Dunstans Hill, City, and at the Exhib. Building.

Monumenta Zollerana. Records of the house of Hohenzollern. Imp. 4o. with many woodcuts 7 vol.

The works of Frédéric the Great. 30 vol. with a chronological index, maps and plans. (In French language.) Imp. 4o. with many woodcuts 34 vol. The same. Imp. 8o. unbound 34 vol. 55 Th. (8 £ 5 sh.).

Dante Allighieri, la divina Commedia. With a photograph of Dante's bust. One of the two only existing copies struck off on parchment. Large 4. 1333$\frac{1}{3}$ Th. (200 £). The same on vellum for engravings, in marocco cloth 16$\frac{1}{2}$ Th. (2 £ 9 sh.). The same do. in cloth 13$\frac{1}{3}$ Th. (2 sh.). The same (edizione minore) in 8o. unb. 2 Th. (6 sh.).

Novum Testamentum Graece ad fidem codicis vaticani recensuit Philippus Buttmann. Splendid edition in 4o. on vellum, unb. 4$\frac{1}{2}$ Th. (13 sh. 6 d.). The same, in 8o. unb. 1$\frac{1}{3}$ Th. (4 sh. 6 d.).

The Botanical Results of the travels of H. R. H. the late Prince Waldemar of Prussia, during the years 1845—1846; edited by the late Dr. F. Klotsch and Dr. Aug. Garcke. In Fol. with 100 lithographic plates, unb. 20 Th. (3 £).

A. v. Witzleben, the life and actions of Prince Frederic Josias of Coburg-Saalfeld, duke of Saxony etc. 3 vol. in 8o. with portrait and a map with 17 plans, unb. 4 vol. 13$\frac{1}{3}$ Th. (2 £).

Waldemar, Prince of Prussia. Travels in India during the years 1844—1846. Imp. in 8o. with portrait, 4 maps and 4 plans, unb. 3$\frac{1}{4}$ Th. (10 sh.).

The Marble Statues on the »Schlossbrücke« at Berlin. 16 woodcuts and explanatory letterpress. Elephant 8o., unb. $\frac{1}{2}$ Th. (1 sh.). The same. Splendid edition. Royal in fol. bound 10 Th. (1 £ 10 sh.). The same do. in French, bound 10 Th. (1 £ 10 sh.).

Dr. Hubert. Prussian spelling-book. 2 vol.

in 8o. on vellum with 60 woodcuts, unb. $\frac{1}{2}$ Th. (1 sh. 6 d.).

Dr. Hubert. Prussian school rom spelling-lessons. 44 plates in royal with 60 woodcuts 1$\frac{1}{2}$ Th. (4 sh.).

The Holy Bible with the apogrypha, translated by Dr. Martin Luther. Splendid edition, large in 4o. with border, printed in two colors, vellum for engravings, unb. 15 Th. (2 £ 5 sh.). The same, without border, 4o., vellum unb. 3 Th. (9 sh.); do. white print. pap. unb. 2 Th. (6 sh.); do. ordinary print. pap. unb. 1$\frac{1}{4}$ Th. (4 sh.); The same, 8o. with large types, vellum unb. 1$\frac{1}{4}$ Th. (3 sh. 6 d.); do. white print. pap. unb. $\frac{3}{4}$ Th. (2 sh.); do. ordinary print. pap. unb. $\frac{1}{2}$ Th. (1 sh. 6 d.). The same, 8o. with types middle-sized, vellum unb. $\frac{5}{6}$ Th. (2 sh. 6 d.); do. white print. pap. unb. $\frac{1}{2}$ Th. (1 sh. 6 d.); do. ordinary print. pap. unb. $\frac{1}{3}$ Th. (1 sh.). The same, 8o. with small types, vellum unb. $\frac{3}{4}$ Th. (2 sh. 3 d.); do. white print. pap. unb. 11$\frac{1}{4}$ sg. (1 sh. 2 d.); do. ordinary print. pap. unb. 8$\frac{3}{4}$ sg. (9 d.).

The New Testament and Psalms. Splendid edition, large in 4o. with border, unb. 4 Th. (12 sh.). The same, 4o. without border, vellum unb. $\frac{3}{4}$ Th. (2 sh. 3 d.); do. white print. pap. unb. $\frac{1}{2}$ Th. (1 sh. 6 d.); do. ordinary print. pap. unb. $\frac{1}{3}$ Th. (1 sh.). The same, 8o. with large types, vellum unb. 9 sg. (9 d.); do. white print. pap. unb. 5 sg. (5 d.); do. ordinary print. pap. unb. 4 sg. (4 d.). The same, 8o. with types middle-sized, vellum unb. 7 sg. (7 d.); do. white print. pap. unb. 4 sg. (4 d.); do. ordinary print. pap. unb. 2$\frac{1}{2}$ sg. (3 d.). The same, 8o. with small types, vellum unb. 6 sg. (6 d.); do. white print. pap. unb. 3 sg. (3 d.); do. ordinary print. pap. unb. 2 sg. (2 d.).

Book of hymns as used for divine service in evangelical churches with prayer-book and liturgy, large in 8o. vellum unb. 1 Th. 2 sg. (3 sh. 2 d.); do. white print. pap. unb. 20$\frac{1}{2}$ sg. (2 sh.); do. ordinary print. pap. unb. 13$\frac{3}{4}$ sg. (1 sh. 4 d.). The same, 8o. vellum unb. 14$\frac{1}{4}$ sg. (1 sh. 5 d.); do. white print. pap. unb. 7$\frac{1}{2}$ sg. (8 d.); do. ordinary print. pap. unb. 5$\frac{3}{4}$ sg. (6 d.). The same, miniature edition, vellum unb. 12 sg. (1 sh. 2 d.); do. white print. pap. unb. 6$\frac{1}{2}$ sg. (7 d.); do. ordinary print. pap. unb. 5 sg. (5 d.).

Railway-Post- and Steam-Navigation-Guide. Official edition, unb. 12$\frac{1}{2}$ sg. (1 sh. 3 d.).

v. Wolzogen. From the posthumous papers of Schinkel. Diary of travels letters with aphorisms. 2 vol. in 8o. with photographic pictures and portraits. 2 vol. 5$\frac{2}{3}$ Th. (17 sh.).

Proofs of printing types from the type-foundery of R. Decker.

The Saviour, St. Matthew, St. Luke, St. Mark, St. John, woodcuts by Unzelmann, after drawings by Kaulbach, white paper $\frac{1}{3}$ Th. (1 sh.), chinese paper $\frac{1}{2}$ Th. (1 sh. 6 d.); all p. piece.

Chapter XXI. of the apocalypse. The new Jerusalem. Woodcut by Unzelmann after a composition by Cornelius, white paper $\frac{1}{3}$ Th. (1 sh.), chinese paper $\frac{1}{2}$ Th. (1 sh. 6 d.).

The telegraphic Map of Europe. Galvanoplastic reproduction from printing-types in copper.

Impression of the same.

1868. DUNCKER, ALEXANDER, bookseller to His Majesty the King of Prussia and publisher, Berlin. Paris 1855 bronze medal; London 1862 honor. mention. Agts. s. No. 1867a.

A. Menzel: Aus König Friedrich's Zeit. Splendid edition with wood-cuts 47 Th. (7 £); C. Merkel: Biblische Geschichten, nach den vier Evangelien. Splendid edition with wood-cuts, black and colored 6⅔ Th. (1 £); A. Duncker: Die ländlichen Wohnsitze und Schlösser der ritterschaftlichen Grundbesitzer in der Preussischen Monarchie, with lithographic prints in colors 33⅓ Th. (5 £); Die Irrlichter, with photographs 6⅔ Th. (1 £).

1869. Ebart, Brothers, merchants and paper-manu., Berlin, manufactories Spechthausen and Weitlage near Neustadt-Eberswalde. Berlin 1844 gold medal; London 1851 prize-medal; Munich 1854 great medal; Paris 1855 bronze medal; London 1862 medal. Agts. Lion M. Cohn, Phaland & Dietrich, represented by Ch. Trübner, 20 St. Dunstans Hill, City, and at the Exhib. Building.

Samples of letter, writing, printing and other machine and hand-made papers, and glazed boards.

1870. Ebbinghaus, Friedr. Wilh., paper-manu., Letmathe near Iserlohn (Arnsberg). Berlin 1844 silver medal; Munich 1854 medal of honor; Paris 1855 silver medal; London 1862 medal. Agts. s. No. 1869.

4⅝ ream of letter-paper, 14 quires of writing-paper, 2 quires of drawing-paper 108 lb. (54 kilo.) value 4 £ 5 sh. 6 d.

1871. Ehlert, Heinr. & Co. (type-founding and engraving-establishment), Berlin.

12 printing-proofs of typographical ornaments and ornamental types.

1871 a. Förster, Ferd. Siegfr. (Förster's paper-manu.), Krampe near Grünberg (Liegnitz). London 1862 honor. mention. Agts. s. No. 1869. (s. No. 1071, 1177, 1343, 1605, 1606.)

Packing-paper 1 £ 6 sh. p. 100 lb.; sugar-paper 3⅝ d; covers for acts, deeds etc. 4⅓ d.; printing-paper 4 d.; do., pressed 4⅓ d.; ord. writing-paper No. 1. 4⅝ d.; do., No. 2. 4⅞ d.; writing-paper 6⅛ d.; champaign-paper, gray 6½ d.; blue, do. 6½ d.; red, do. 6½ d.; 4 sorts of tinted wrapping-paper 6½ d.; all p. lb.

1872. Friedlaender, Julius, book- and musik-seller, manu. of stereotype-plates, Berlin, Friedrichsstr. 217. London 1862 medal. Agts. Ewer & Co., Regent-street.

4 stereotype-plates, p. plate 9 to 18 sh.

1873. Friedlaender, Dr. J. (R. Friedländer & Son), bookseller, Berlin, Kurstr. 9. London 1862 honor. mention. Agt. Bernard Quaritch, bookseller, 15 Piccadilly.

Books printed in facsimiles from the original editions by way of a new chemical process: Fermati opera. Tolosae, 1679. 1 vol. fol. 10 Th. (1 £ 10 sh.); Taylor, Methodus incrementorum. Londini, 1715. 1 vol. 4. 3⅓ Th. (10 sh.); Gergonne, Annales des mathématiques. Nimes, 1814—23. 9 vols. 4. 90 Th. (13 £ 10 sh.).

1874. Friederich, Theodor, manu. of straw-paper and straw-pasteboards, Camen (Arnsberg). Agts. Heintzmann & Rochussen, 9 Friday-street, Cheapside, E. C.

Straw-pasteboard; 15 sheets, yellow 24/28 in.; 9 sheets, brown 24/28 in.; straw-paper, 4 half reams, yellow 13/16 and 24/28 in.; brown 13/16 and 24/28 in.

1875. Gädicke, Johann, chemist, Berlin, Lindenstr. 34., manufactory Treuenbrietzen. London 1862 honor. mention. Agt. Samuel Copping, 27 Tokenhouse-Yard, City.

Straw and maize-paper, made without any admixture of rags or other material, partly used for type, steel-engraving, lithograph and wood-cut printing p. pound 2⅔ sg. (2⅕ d.).

1877. Glüer, Louis, patern-painter to His Majesty the King of Prussia, academical artist, Berlin. Paris 1855 silver medal.

Embroidery-patterns.

1878. Gogler, L., manu., Berlin, Friedrichsstr. 65.

Embroidery-patterns.

1879. Gundlach, W., manu., Breslau. London 1862 honor. mention. Agts. s. No. 1869.

Fancy articles ornamented with leather-flowers; prices p. dozen: 3 work-baskets 1 £ 14 sh. and 2 £; 4 ash-cups 17 sh., 17 sh. 6 d., 18 sh. and 19 sh.; 1 letterpresser 1 £ 2 sh. 6 d.; 1 letter-holder 2 £; 2 cigarholders 1 £ 4 sh. and 1 £ 19 sh.; 2 cigarboxes 2 £ and 2 £ 15 sh.; 2 cigarbaskets 1 £ 6 sh. and 1 £ 14 sh.; 2 consoles 13 sh. and 1 £ 15 sh.; 5 matchboxes 16 sh., 1 £ and 1 £ 6 sh.; 1 spellbox 1 £ 7 sh.; 1 glovebox 1 £ 12 sh.; 1 journalholder 2 £ 16 sh.; 2 almanacks 1 £ 15 sh. and 2 £ 4 sh.; 1 visiting-cardsholder 1 £; 1 key-holder 1 £ 6 sh.; 2 inkstands 2 £ 2 sh. and 2 £ 4 sh.; 1 knittingcup 1 £; 2 thermometers 1 £ 14 sh. and 2 £ 11 sh.; 7 watchstands 13 sh. 6 d., 14 sh. 6 d., 17 sh. 6 d., 1 £, 1 £ 2 sh., 1 £ 8 sh., 1 £ 14 sh.; 2 wallbaskets 1 £ 3 sh. and 2 £ 16 sh.; 1 foothpickcase 12 sh. 6 d.

1882. Hieronimus, Wilh. (Wilh. Hassel), printer and bookseller, Cologne. Agts. Morgan Brothers, 21 Bow-Lane, Cannon-street, City.

The holy bible, in imp. 4., in do. 8., in 24.

1883. Hayn, A. W., counsellor of commerce, publisher, printer and type-founder, Berlin.

A work entitled: The book of the royal order of the red eagle.

1884. Hendler, F., paper-manu., Alt-Friedland near Waldenburg (Breslau). London 1862 honor. mention. Agts. s. No. 1869.

Samples of paper; prices p. pound free Hamburgh: Letter-paper No. 1., No. 1½., No. 2.; middle-sized No. I., No. I½., No. II.; plate-paper; tinted papers.

1885. Heymann, Carl, publisher and bookseller, counsellor of commerce, Berlin.

Borussica: An illustrated album or catalogue of 480 works and maps concerning the kingdom of Prussia and published by the exhibitor.

Hieronimus, W. (Hassel), Cologne. (s. No. 1882.)

1887. Hoesch, Ludwig & Emil (Hoesch & Sons). paper-manu., Düren near Aix-la-Chapelle, manufactory Krauthausen near Düren. London 1851 prize-medal; Paris 1855 silver medal; London 1862 medal. Agts. G. Ingelbach & Wolffgang, 11 Staining-Lane, E. C., and 6 Lyllipot-Lane.

Superfine cream glaced note paper; do. blue and

colored; ordinary letter-paper in cream and blue; superfine cream writing-paper; ordinary writing-paper; printing-paper; ruled note-paper; note-paper with water-mark; cream and colored tissue paper; copying-paper; paper for the manufacture of papier trempé.

1888. HOESCH, WILHELM EDMUND, HOESCH, EDUARD (Hoesch, Brothers), paper-manu., Düren near Aix-la-Chapelle; manufactories Friedenau and Kreuzau. Berlin 1844 silver medal; London 1851 prize-medal; Paris 1855 silver medal. Ed. Hösch London 1862 juror for Cl. XXVIII. Agt. G. A. Seeger, Old-Trinity-House, Water-Lane.

Weight p. ream uncut of 480 sheets in plano.

First quality of white wove and laid glazed letter-paper: wove ecu 4o. No. 103¾. 13¾ lbs.; coquille 4o. No. 102½. 13¼ lbs.; thick 8o. No. 106½. 21½ lbs.; large medium 4o. No. 103. 16½ lbs.; 8o. No. 105. 21 lbs.; thick 8o. No. 106. 33 lbs.; laid ecu 4o. No. 140. 13¼ lbs.; large medium 4o. No. 142. 15½ lbs.; 8o. No. 143. 21 lbs.; thick 8o. No. 144. 23½ lbs.

First quality of blue, wove and laid glazed letter-paper: wove large medium 4o. No. 108. 12⅝ lbs.; thick 8o. No. 110. 21 lbs.; laid ecu 4o. No. 145. 13¼ lbs.; coquille 4o. No. 146½. 12 lbs.; large medium 4o. No. 147. 15½ lbs.; thick 8o. No. 148. 21 lbs.

Extra thin wove and laid, white and blue letter-paper: large medium wove, white 4o. No. 130. 6½ lbs.; blue 4o. No. 131. 6½ lbs.; laid, white 4o. No. 155. 8¼ lbs.; blue 4o. No. 156. 8¼ lbs.

Second quality of white and blue, wove and laid letter-paper 4o.: large medium blue, wove No. 112. 12⅝ lbs.; laid No. 112 G. 12⅝ lbs.; wove, white No. 116. 12⅝ lbs.; No. 118. 16½ lbs.; blue No. 122. 12⅝ lbs.; No. 124. 16½ lbs.; coquille (medium) No. 127. 11 lbs.; large medium, white laid No. 151. 16½ lbs.

First quality of colored wove and laid letter-paper: large medium wove 8o. No. 160—163. 15½ lbs.; thick do. No. 164—167. 21 lbs.; laid 8o. No. 168—171. 15½ lbs.; thick do. No. 172—175. 21 lbs.; colored, blue and white No. 160—167., with lines 4o., No. 109. and 103. 15½ lbs.

Cigarette- (papyros) paper, folio: wove white No. 1., rosé No. 1 R.; laid white No. 1 G.; rosé No. 1 GR.; wove dark rosé No. I R.; rum-colored in 4 shades No. 49.; dark rosé No. 2⅝.; gray No. 30.; rum-colored laid No. 49 G.; bulle in 2 shades wove; each ream 5½ lbs.

Tissue-paper, folio: extra fine rosé No. 6 c., 6 B., 6 A., 6.; dark rosé No. 6¼., 6½., 6¾., 6⅞.; blue No. 18. dark blue, No. 18½.; each ream 6½ lbs; black No. 22. 7½ lbs.; gray No. 22½. 7½ lbs.; wove white No. 1.; blue No. 1½., white laid No. 1 G.; each ream 6½ lbs.; sixty five sorts of tissue-paper in different colors and shades folio, each ream 6½ lbs.

Copying letter-paper, folio: raisin white No. 132.; blue No. 133.; p. ream 5¾ lbs.

Second quality, large medium colored letter-paper No. 176. 12½ lbs.

First quality of wove and laid, white and blue writing and account-book paper: foolscap white wove No. 202.; laid No. 202 G.; blue laid No. 202 BG., each ream 13¼ lbs.; white wove No. 204.; laid No. 204 G., each 15½ lbs.; laid thick No. 206 G.; blue wove No. 206 B., each 17½ lbs.; medium: white wove No. 207.; laid No. 207 G.; blue wove No. 207 B., each 21 lbs.; white large laid No. 208 G. 28½ lbs.

Second quality of wove and laid, white and blue writing and account-book paper: white medium wove No. 225. 21 lbs.; blue laid foolscap with pro-patria-arms No. 255 B. 10½ lbs.; white laid foolscap No. 259. 12 lbs.; do. medium No. 263. 21 lbs.; do. large No. 264. 28½ lbs.; blue laid foolscap No. 269. 12 lbs.

1889. HOFERDT, JULIUS (Julius Hoferdt & Co.), merchant and manu. of account-books and portfolios, Breslau, manufactories of the Royal houses of correction at Brieg, Ratibor and Breslau. London 1862 honor. mention. Agts. Lion M. Cohn, Phaland & Dietrich, represented, by Ch. Trübner, 20 St. Dunstans Hill, City, and at the Exhib. Building.

Account-books, portfolios, note-books and pencases. Wholesale prices marked on the articles.

1890. PAPER-MANUFACTORY HOHENOFEN near Neustadt on the Dosse. London 1862 honor. mention. Agts. s. No. 1889.

(s. No. 1862.)

Different samples of ordinary writing-, account-book and letter-papers.

1890 a. HÜTTENMÜLLER, PHILIPP, Lorenzdorf near Bunzlau (Liegnitz). Paris 1855 bronze medal; London 1862 medal. Agts. s. No. 1889.

27 pieces of pressing-boards of divers qualities and thickness, 30 in. long 18 in. broad (Engl. measure); pressing-board, 72 in. long 36 in. broad, for pressing large shawls; 1 piece of paste-board, 72 in. long 36 in. broad and ½ in. thick, for the like use; a card of patterns of all the specimens mentioned above.

1891. KATZSCHKE, REINHOLD, bookbinder Weissenfels. London 1862 honor. mention Agts. s. No. 1889.

1. Gigantic fashion-lantern, Rhenish meas. 3 ft. 3. in. high, 2 ft. 4 in. diam. 4 Th. (1 yard 3 in. high, 28 in. diam. 12 sh.); 2. gigantic drawing-lantern with firm 3 ft. long, 22 in. diam. 2½ Th. (1 yard high, 22 in. diam. 7½ sh.); 3. crown with arms 17 in. high, 25 in. diam. 2 Th. (6 sh.); 4. gigantic balloon-lantern 21½ in. diam. 1 Th. (3 sh.); 5. a Chinese fashion-lantern 20½ in. high, 15 in. diam. 1⅓ Th. (4 sh.); 6. a star-lantern 34 in. high, 8 in. diam. 1⅓ Th. (4 sh.).

1892. KOCH, CARL AUGUST, Köppenmühle, Berg.-Gladbach near Cologne. London 1862 medal. Agts. s. No. 1889.

Specimens of different kinds of paper, as letter-paper, drawing-paper, card-paper etc.

1895. KÜHN, CLEON (Carl Kühn & Sons), merchant and manu., Berlin. London 1851 honor. mention, 1862 medal. Agts. s. No. 1889.

Account-books in different bindings, with printing in English, French, Spanish and German language; large portfolio for engravings of genuine morocco-leather, with paste-board samples.

1897. KUHLHOFF, W., manu., Neheim. Agts. s. No. 1889.

Straw-paper, brown, yellow and bleached.

1898. KULLRICH, F. F., bookbinder and portfolio-manu., Berlin. London 1862 medal. Agts. Gebhardt, Rottmann & Co., 24 Lawrence-Lane, Cheapside, E. C.

Albums for 25 to 200 photographic cards, porte-monnaies, cigarcases, pocket-books, portfolios with and without writing-materials, inkstands, sewing-cases, ladies' bags with and without sewing-materials, workbaskets, etc.

1899. Lamberts, Wilhelm, merchant, M.-Gladbach. Paris 1855 honor. mention; London 1862 honor. mention.

Account-current 1 £ 10 sh. journal 11 sh.; letter-book 2 sh. 3 d.; account-current 1 sh. 3 d.

1900. Lehmann & Mohr, manu., Berlin, Linienstr. 114.

Forms of bills; types for the blind; types for bills posted up, military and war-game.

Lenz, J., publisher, Berlin. Lithographic prints and works on gymnastics. (s. No. 1956.)

1901. Lesimple, Ad. (Langen's printing-office), printer, Cologne. Agts. Morgan Brothers, 21 Bow-Lane, Cannon-street, City.

Specimen-products of the printing-presse 1. printing on zinc 100 pieces 40 Th. (6 £); 2. printing in colors 50 pieces 250 Th. (41 £).

1905. Lunge, Dr. Georg, chemist, Breslau. Agts. Lion M. Cohn, Phaland & Dietrich, represented by Ch. Trübner, 20 St. Dunstans Hill, City, and at the Exhib. Building.

Half-stuff, manufactured entirely out of wheat-straw, for white writing and printing-paper, p. cwt. 4 Th. (12 sh.); 2. white paper, made of the same half-stuff, without artificial glue, but fit for writing, without any addition of rags, or rag-stuff, pure straw-mass p. cwt. 8 Th. (1 £ 4 sh.).

1906. May, C. F., manu., Berlin, Stralauerplatz 5.

Cards and parchment.

1907. Meissner, C. F., & Son, paper-manu., Raths-Damnitz near Stolp (Pomerania). Agts. s. No. 1905.

Proof-sheets of different sorts of letter, writing and printing-paper.

No.		Size Rhenish. in.	Weight p. ream.	Price p. Pfd.	Price p. ten reams.		
	I. Letter-paper.		Pfd.	d.	£	sh.	d.
1.	extra superfine cream ribbed . .	11 —17¼	20	10	8	6	8
2.	do.		16	10	6	13	4
3.	do. vellum		15	9	5	12	6
4.	do. ribbed		15	10	6	5	—
5.	superfine cream vellum		14	8½	4	19	3
6.	do.		12	8½	4	5	—
7.	fine cream vellum		10	8	3	6	8
8.	fine bluish ribbed		16	8½	5	13	4
9.	do.		15	8½	5	6	3
10.	extra superfine azure ribbed . .		14	10	5	16	8
11.	do.		10	10	4	3	4
12.	superfine azure vellum		12½	9	4	13	9
13.	fine bluish		10	8	3	6	8
	II. Writing-paper.						
14.	f. f. ribbed . . .	13¾—16¾	14	7¼	4	4	7
15.	f. f. vellum . . .	13¼—16¼	14	7¼	4	4	7
16.	f. f. ribbed . . .	13 —16	8	7¼	2	8	4
17.	f. ribbed		8	7	2	6	8
18.	f. vellum	13⅜—16¼	11	6¾	3	1	10½
19.	middle fine vellum	13½—16½	12	6¾	3	7	6
20.	ordinary	12⅜—15	6½	6	1	12	6
21.	m. f. vellum . .	17½—21	22	6¾	6	3	9
22.	do.	18 —21	13	6¾	3	13	1½
23.	m. f. cartoon . .	22 —23½	79	6¾	22	4	4½

No.		Size Rhenisch in.	Weight p. ream.	Price p. Pfd.	Price p. ten reams.		
			Pfd.	d.	£	sh.	d.
24a.	m. f. for tarts .	18 —18	11	6	2	15	—
24b.	do.	13 —13	6½	6	1	12	6
	III. Printing-paper.						
25.	f. sized satinet .	17½—20½	24	6⅓	6	6	8
26.	m. f. halfsized satinet		15	6	3	15	—
27.	do.	16 —19½	13	6	3	5	—
28.	m. f. unzised satinet	18 —22	14	5½	3	4	2
29.	do.	16¼—20	11	4⅚	2	4	3⅔
30.	ordinary	15¾—20½	10	4½	1	17	6
	Different sorts.						
31.	rose-colored satinet sized for printing	13 —16	8½	5½	1	18	11½
32.	rose-colored filtering-paper . . .	13¾—16¾	8½	4⅚	1	14	2⅚
33.	white filtering-paper	14 —17	15	6¾	4	4	4½
34.	colored envelop	24 —29	48	6	12	—	—

Prices p. cash with 2 pCt. discount free on board Stettin or Danzig on receipt of the bill of lading. Package in bales 6 d. p. Zollz. gross weight; do. in cases ⅙ d. p. Zollz. gross weight.

1908. Meister, Th. Wilh. (formerly A. Todt), merchant, Berlin. London 1851 honor. mention.

Three embroidery-patterns: fire-screens, with the portraits of Shakespeare, of Milton, each 15 sh.; carpet 1 £ 1 sh.

1909. Möser, W., manu., Berlin, 34 Stallschreiberstr.

Illustrated typographic tableau.

1910. van der Moolen, manu., Geldern. Agts. s. No. 1905.

150 specimens of sealing wax of different qualities and in all colors; 40 specimens of writing and copying-ink; lists of prices annexed.

1911. Müller, W., manu., Kettwig.

Paper.

1912. Muench, C. H., bookbinder to the university of Königsberg. Agts. s. No. 1905.

Letter-portfolio for ladies, violet shagreen (Magenta) with relievo and gilding made by the hand 25 Th. (3 £ 15 sh.).

1913. Nicolai (G. Parthey), publisher, Berlin. Agts. s. No. 1905. (s. No. 1430.)

Becker, W. A. (Royal architect in Berlin), practical instruction on the application of »coment« for architectural, industrial, agricultural purposes and objects of art. Parts 1 to 3. 18 plates chromolithographed. 28 sheets of letterpress in folio. 1860—61. 10 Th. (1 £ 10 sh.). Journal for architects and artisans, containing information of all the most recent inventions and improvements in architecture and the industrial arts, principally with respect to constructions in stone, cement, wood and iron. Edited by G. Töbelmann and H. Kaemmerling. First year. 6 parts 24 plates in chromolithography and copper-engravings. 1860. Folio. 4 Th. (12 sh.). The same 2. year. Edited by W. A. Becker (Royal architect in Berlin), 12 parts with 40 engravings on copper. 1861. Folio. 8 Th. (1 £ 4 sh.). Kaemmerling, H. (architect in Berlin), The civil architect, a collection of designs and sketches for private buildings in town and country; containing plans, sections and elevations with details for architects, masons and carpenters. 37 plates chromolithographed and engraved. Parts 1—6. Fol. 1860—61. 10 Th. (1 £ 10 sh.). Tietz, E. (architect in Berlin), designs of executed public and private buildings. Part I. The Hôtel d'Angleterre in Berlin. 6 plates. Folio. 1859. 2 Th. (6 sh.). Tietz, E., Kroll's garden in Berlin. 12 plates chromolith. and engraved. Folio. 1861. 5 Th. (15 sh.). Tietz, E., the Victoria-theatre in Berlin. 25 plates chromolith. and engraved. Folio. 1861. 9 Th. (1 £ 7 sh.).

1916. REIMER, D., Berlin. (s. No. 1960.)

1917. RHEINEN, H. J., HEIRS, Elberfeld. Munich 1854 honor. mention; London 1862 honor. mention. Agts. Heintzmann & Rochussen, 9 Friday-street, Cheapside, and R. Schomburg & Co., 90 Cannon-street, E. C.
10 pieces of tracing-paper.

1918. ROSENTHAL, J., manu., Berlin, Neue Friedrichsstr. 45. London 1862 honor. mention. Commercial account-books.

1920. SCHMIDT, G. A., bookbinder, Halle on the Saale.
Two photograph-albums.

1921. SCHMITZ, BROTHERS, paper-manu., Düren, manufactory Menken near Düren. London 1862 medal. Agts. Ingelbach & Wolffgang
Specimens of tinted paper.

1922. SCHNEIDER, FRIEDRICH, manu. and academical artist, Berlin, Linkstr. 9. London 1862 medal. Agts. Lion M. Cohn, Phaland & Dietrich, represented by Ch. Trübner, 20 St. Dunstans Hill, City, and at the Exhib. Building.
An assortment of embossed Bristol-boards, especially for photographical purposes according to the annexed price current; a polyscope (an angular looking-glass); p. 100 pieces 112 Th. 15 sg. (16 £ 17 sh. 6 d.).

1923. SCHOELLER, H. A., & SONS, paper-makers, Düren. Berlin 1844 silver medal; Paris 1855 silver medal; London 1862 medal. Agts. Louis Henlé, 9 Dowgate-Hill, Cannon-street. (s. II. No. 6.)
Different sorts of paper.

1924. SCHULZE, H., Royal bookkeeper and retired Lieutenant, Berlin, Ritterstr. 49. Paris 1855 honor. mention. Agts. s. No. 1922.
A book, containing: the statutes and ordinances of foundation of all known orders of knighthood and badges of honour bestowed by sovereigns and governments, with an album of 45 plates with more than 1200 engravings of the badges of orders in colored lithographic prints most of them of natural size 9 £ 12 sh. 6 d.

1925. SCHULZE, F., Berlin, 68 a. Leipziger-street.
Pictures printed in oil-colors: Christ's head after Correggio; Christ on the Olive-mountain; Christ blessing after Kaselowski.

1926. STANGE, EDUARD, merchant and manu. of fancy papers, Berlin. Munich 1854 honor. mention; London 1862 honor. mention.
Superfine letter-papers; envelopes; letter-cases fully assorted; ornamental wafers; albums; lamp-screens; colored pictures; cards of congratulation; reading-marks; decorations for cotillons etc.

1928. STERN, ADOLPH, wholesale merchant and manu., Berlin. (s. II. No. 58.)
A table with samples of ruled and printed account-books; a table with samples of ruled music-paper; a table with samples of ruled paper for copy-books.

1929. TENGE, paper-manu., Dalbke near Bielefeld. London 1862 medal. Agts. s. No. 1922.
1. blue hand-made paper ($\frac{1}{2}$ ream, 240] sheets) 7$\frac{1}{2}$ d.; 2. yellow do. 7 d.; 3. black do. 6$\frac{1}{2}$ d.; 4. chamois do. 7 d.; 5. fine blue do. 8$\frac{1}{2}$ d.; 6. 12 sheets of glazed paste-board 7$\frac{1}{2}$ d.; all p. pound.

1930. TROWITZSCH, EUGEN (Trowitzsch & Son), publishers, printers and letter-founders, Berlin. London 1862 medal. Agts. s. No. 1922.
Exhibition of the printing-office: Bibles and testaments in various editions and languages (printed for the British and foreign bible-society in 954,000 copies within the last five years); a variety of other publications etc.
Exhibition of the type-foundery: specimens of printing-types, steel-punches, machine-types, types in embossed characters for the blind, stereotype- and electrotype-plates, stereotype-blocks, brass-rules, galleys etc.

1933. WEISS, EWALD & Co., manu. of books for mercantile business, M.-Gladbach. Agts. Morgan Brothers, 21 Bow-Lane, Cannon-street, City.
Elegant ledger 40 Th. (6 £); journal 5 Th. (15 sh.).

1935. ZECHENDORF & BERTHOLD, manu., Berlin, 1 Wilhelm-street. London 1862 honor. mention.
Electro metallurgic plates and blocks for copperplate and typographic printing; typographic samples.

CLASS XXIX.
EDUCATIONAL WORKS AND APPLIANCES.

1936. BRENNECKE, DR., principal of the commercial and military school, Posen.
Trigonometry for the use of higher schools. Introduction into modern Geometry. The problem of tactions for the circle and the sphere in sixfold geometrical method. Classbooks for the study of the English language in German schools. 5 parts in 1 Vol. elegantly bound by Niklaus, bookbinder at Posen. 1. Grammar for the use of schools; 2. themes to translate from the German into the English language; 3. selections from the poetical works of Shakespeare; 4. English reading book with division of syllables, accentuation and pronunciation.

1937. BURO, LUDWIG, modeller, Berlin, Gipsstr. 6. London 1862 honor. mention. Agts. s. No. 1922.
Relievo-map of Europe 20 Th. (3 £).

1938. CADURA, HEINRICH. merchant and manu. of leather driving belts, Breslau. Agts. s. No. 1922.
Colored India-rubber balls.

1939. DÜMMLER, F., publisher, Berlin, Mohrenstr. 26. London 1862 medal.
Engravings and books, educational works illustrated according to Fröbel's ideas.

1940. FLEMMING, CARL, Glogau. London 1862 medal. Agt. S. Borkheim. 27 Crutched Friars, City.
Reymann's map of Central-Europe, 300 sheets

18 £ 16 sh.; Ungewitter, sketches of buildings of wood 1 £ 1 sh.; do. of buildings in stone 1 £ 1 sh.; do. of gothic furniture 16 sh. 6 d.; do of tombstones 1 £. 1 sh.; do. of buildings for town and country 1 £ 19 sh.; Lipsius, sketches of shop-windows 12 sh.; Sonntag, sketches of decorations for the garden 10 sh.; Brehm, Dr., Das Leben der Vögel 1 £ 2 sh.; Hartwig, Dr. G., Das Leben des Meeres 17 sh.; Rhode, historical school-atlas 6 sh.; Hammer, G., Hubertusbilder 11 sh.; Corrodi, Deutsche Räthsel und Reime 6 sh.; Georgens, Bildewerkstatt 5 sh. 6 d.; the same, Ausschneideschule 4 sh.; Gumpert, Töchteralbum 8 sh. 6 d.; Masius, Jugend Lust und Lehre 8 sh. 6 d.; Gumpert, Herzblättchens Zeitvertreib 6 sh. 3 d.; Smidt, H., Zu Wasser und zu Lande 6 sh. 3 d.; Wagner, Buch der Natur 4 sh.; Süs, Nussbäumchen 3 sh.; Fröhlich, Fabeln, 2 vols, 3 sh.; Ruhkopf, Zur Grossmutter 3 sh. 3 d.; Gumpert, Schlosspeterchen und Bauerhänschen 2 sh. 3 d.

1941. Franz, Rector, Oranienburg near Berlin.

Four tables with diagrams for the first intuitive instruction in geometry. The equality of the diagrams shown by cutting and placing them on each other. The Pythagorean proposition. Textbook: »Geometry by Franz«.

1942. Grunert, Carl, lithographer, Berlin, Wollankstr. 20.

Representation of the German (of Arends) shorthand as applied to the English language 20 £.

1943. Guttentag, J., bookseller, Berlin. Agts. Lion M. Cohn, Phaland & Dietrich, represented by Ch. Trübner, 20 St. Dunstans Hill, City, and at the Exhib. Building.

The works of the building-joiner for the instruction and practical use of architects and workmen, sketched and elaborated by F. A. W. Strauch architect, with XXXIX. copper-plates, bound 6½ Th. (1 £).

1945. Henning, August, painter and manu., Halle on the Saale. Agts. s. No. 1943.

2 puppets, fine No. 1., 22 in. high, 9 Th. (1 £ 7 sh.); 2 do. No. 2., 17 in. high, 5½ Th. (16 sh. 6 d.); 2 do. ord. No. 3., 15 in. high, 3 Th. (9 sh.); 2 do. No. 4., 15 in. high, 2¼ Th. (6 sh. 9 d.); 2 do. No. 5., 12 in. high, 1⅓ Th. (5 sh.); all p. gross.

Figures for playing at the ball: fine, No. 1. 5½ ft. high, 7 Th. (1 £ 1 sh.); do. No. 2., 4 ft. high, 5¼ Th. (15 sh. 9 d.); do. No. 3. 3 ft. high, 3¼ Th. (10 sh. 6 d.); 1 head for playing at tennis, 2 ft. high, 3¼ Th. (10 sh. 6 d.), all p. doz.

1946. Hermes, Wilhelm, book- and print-seller, Berlin, Königsstr. 26. London 1862 honor. mention. Agts. Joseph, Myers & Co., 144 Leadenhall-street.

1 copy of Wilh. Hermes »Berlin systematic drawing-school« for teachers and for self-instruction, number 1. to 250, arranged by series and bound in 7 albums. 24 various numbers of the above school, as they are published. 1 copy of »Wilh. Hermes Berlin drawing-master.« A collection of exercises for advanced pupils. Number 1. to 120., arranged by series, bound in 5 albums; 24 various numbers of the »drawing-master«, as they are published. 4 placards in a frame, containing specimens of the different drawing-schools.

Herzig, G., manu., Hermsdorf. Toys. (s. No. 2189.)

1947. Hildebrandt, Carl, inspector, Burg.

Prospectus of Pieschel's establishment for the education of poor children with drawings of the house, front-elevation, back-elevation and ground-plan. Specimens of drawing and writing of the pupils.

1948. Hirsch, Dr. M., author and publisher, Berlin, Schützenstr. 65.

Educational works; the doctrine of fractures (osteodolis); travels in Africa.

1950. Hupp & Wülfing, manu., Dusseldorf. Agts. s. No. 1943.

An assortment of track-copy-tablets (metal-tablets with deep engraved letters to be traced with the slate-pencil). The German writing is executed in four courses: on one tablet with copies on both sides, on two tablets each with copy on one side only, large size p. piece 6 sg.; on three tablets each with copy on both sides, on six tablets each with copy on one side only, small size p. piece 4 sg.

1952. Köhler, August, (Institution for geographic and plastic works of Albert and Julius Abelsdorff), Berlin, Unter den Linden 8. London 1862 honor. mention. Agt. Franz Thimm, 3 Brook-street, Grosvenor Square.

Relief-globe 16 in. diameter in English language. The mountains brown, the plains green, and the deserts in yellow colors, on carved wood-stand with compas, constructed by August Koehler 20 Th. (3 £); the same upon a plain stand without compas in German, French or Russian language 12 Th. (1 £ 18 sh.).

Mont-blanc, executed from the best materials and with the most recent notices given by the naturalist Doctor Pitschner, constructed in relief by August Koehler 1862, 100 Th. (15 £). Upon a dimension of 13 sq. ft. the surrounding country is extended north-west to the St. Bernhardt, south-east to the Valley of Sixt, north-east to the Valley Montjoie and south-west to the gorge d'Etalons.

1953. Krantz, Dr. phil. August (Rhenish mineral office), Bonn. Paris 1855 silver medal; London 1862 medal. Agts. Morgan Brothers, 21 Bow-Lane, Cannon-street, City.

A collection of 100 minerals, 100 rocks and 100 fossils, systematical arranged for lecturers and elementary instruction 40 Th. (6 £); a collection of 300 choice minerals in mahogany-cabinet 40 Th. (6 £); a collection of 100 minerals for blowpipe-experiments 4 Th. (12 sh.); a scale of hardness 3 Th. (9 sh.); a collection of 675 most carefully executed models of mineral crystals made in maple-wood in 6 drawers with printed catalogue 126 Th. (18 £ 18 sh.); a smaller collection of 114 of the same in 1 drawer with catalogue 17 Th. (2 £ 11 sh.)

1954. Kühn, Bernhard (Gustav Kühn), bookseller, printer and prop. of a lithographic establishment, purveyor to the court of His Royal Highness the Prince Royal of Prussia, Neu-Ruppin. Munich 1854 honor. mention; London 1862 honor. mention. Agts. s. No. 1943.

Ordinary sheets of paper filled with figures.

1955. LANGENSCHEIDT, JOHANN LUDWIG GUSTAV, man of letters and philologist, Berlin. Agts. Lion M. Cohn, Phaland & Dietrich, represented by Ch. Trübner, 20 St. Dunstans Hill, City, and at the Exhib. Building.

3 copies of French lessons per letter after the method Toussaint-Langenscheidt (by Charles Toussaint, prof. of the French language, and G. Langenscheidt) complete; 3 copies of English lessons per letter, after the method Toussaint-Langenscheidt (by Dr. C. v. Dalen, professor H. Lloyd and G. Langenscheidt) letters 1—8.; p. course 17 sh. (each language 2 courses).

LEHMANN & MOHR, Berlin, military and war-game. (s. No. 1900.)

1956. LENZ, G. F., prints eller, Berlin, Gertraudtenstr. 24. London 1862 honor. mention. Agt. Novra, 95 Regent-street.

Prints in sheets: in folio p. 1000 sheets: No. 1. 3 sh. tinted 9 £; No. 2. 9 sh. colored 12 £; No. 3. 4 sh. do. on black ground 24 £; No. 4. 2 sh. do. decorated 30 £; No. 5. 2 sh. do. on black ground 45 £.

Large folio, price p. 1000 sheets: No. 6. 5 sh. tinted 20 £; No. 7. 10 sh. colored 30 £; No. 8. 9 sh. do. on dark ground 50 £; No. 9. 2 sh. do. decorated 55 £; No. 10. 2 sh. do. on dark ground 75 £. No. 11. Portrait of F. L. Jahn, professor of gymnastics 4 sh.

Samples of forms of bills, labels and cards.

1957. MOHR, C. W., & Co., Berlin, Linienstr. 114. London 1862 honor. mention.

Apparatus for teaching to read German, English and Russian; juvenile library.

1960. REIMER, DIETRICH, publisher, Berlin, Anhaltstr. 11. München 1854 medal of honor; Paris 1855 bronze and silver medal; London 1862 medal. Agts. Williams & Norgate, booksellers, 14 Henrietta-street, Covent-Garden.

Kiepert's new handatlas, 40 maps, bound; Kiepert's atlas antiquus, 10 maps, bound; terrestrial globes, 13½ in. diam., on pedestal, German, English, Russian edition; with brass half-meridian, English edition; gilt frame, brass meridian, quadrant and compass, English edition. Celestial globe, 13½ in. diam. polished black frame, brass meridian quadrant and compass; terrestrial globe, 31 in. diam. polished black frame, meridian, quadrant and compass; celestial globe, in same size and execution as the terrestrial globe; six wall-maps, mounted and varnished; Kiepert, the world in Mercator's projection, Palestine; Graecia antiqua, new map of Tropical America north of the Equator; general map of the Turkish empire; Wetzell, wall-map for the instruction of mathematical geography.

1961. RUNGE, DR. FRIEDLIEB FERDINAND, Oranienburg near Berlin. Paris 1855 bronze medal; London 1862 medal. (s. No. 1019 a.)

A book in folio: »Paintings, produced by chemical action in approaching different salts (dissolved in water) upon blotting paper«; two volumes in octavo: »Chemistry for technical schools«, illustrated by the chemical combinations in natura.

1963. SCHNABEL, DR. CARL, director of the higher school. named »Realschule erster Ordnung«. Siegen. Paris 1855 honor. mention; London 1862 honor. mention.

51 specimens of crystal-models made of glass-

plates and paste-work, with visible axes. Every piece is paid by the number of the glass and paste-plates. For further particulars see the price current adjoined to the models. Every glass-plate is accounted for two sg., every paste-plate and axis for one sg.

1965. STOLLE, A., manu., Erfurt.

Toys (horse, sheep, goat).

1966. THAERMANN, G. HUGO, teacher and pen-man, Königerode near Harzgerode. Agts. s. No. 1955.

Model of penmanship with writing-copies by which a systematic and rational instruction in calligraphy is given. By careful imitation of these copies people will acquire a fine hand-writing in 16 lessons.

1967. VON WARNSDORF, L., publisher, Berlin, Köthenerstr. 33.

Jacobi's letters for the study of the English, Italian and French languages.

CLASS XXX.

FURNITURE AND UPHOLSTERY, INCLUDING PAPER-HANGINGS AND PAPIER-MACHE.

ALVES, H., Berlin, gold borders. (s. No. 2181.)

1969. ARON & JACOBI, manu., Berlin, Brüderstr. 29.

Window-blinds.

1970. BASCH, W., & Co., manu., Berlin, Neue Friedrichsstr. 9 and 10.

Safety-lock.

1972. BISSING, FR., joiner, Lissa (Posen).

Chess-board table and stools.

1973. BEMBE, ANTON, manu., Cologne. London 1862 medal. Agts. s. No. 1955.

A sample of parquetry, composed of several woods The foot of it shows different patterns of $\frac{1}{16}$ in. of natural size. These patterns are marked with prices p. Engl. square foot.

1974. CARL, J. F., Spandow.

Various pieces of gilt and polished cornices in a frame of gilt cornices.

1975. DREYKLUFT, A., furniture-maker, Merseburg. Agts. s. No. 1955.

A carved table for jewels, wholly of nut-tree 140 Th. (20 £).

1976. FERRENHOLTZ, G. J. (formerly Weyersberg & Co.), manu., Wesseling near Cologne. Paris 1855 bronze medal; London 1862 honor. mention. Agts. Heintzmann & Rochussen, 9 Friday-street, Cheapside.

Gold-cornices and cornices of imitated woods, as: polixander, nut-tree, maple and ebony. Prices p 110 ft. Engl.: No. 26. R. verr. 3 £ 15 sh.; No. 21. N. verr. 3 £; No. 26. Q. 3 £ 3 sh.; No. 14. N. 1 £ 16 sh.: No. 13½. N. 1 £ 16 sh.; No. 11½. Q. 1 £ 13 sh.; No. 26. RR. verr. 3 £ 18 sh.: No. 21. NN. verr. 3 £ 3 sh.; No. 26. QQ. 3 £ 6 sh.; No. 13½. NN. 1 £ 18 sh.: No. 14. NN. 1 £ 18 sh.; No. 13½. MM. 1 £ 13 sh.: No. 13½. US. verr. 2 £ 8 sh.; No. 27. UM. 3 £: No. 13½. Ur. 2 £: No. 9. U. 1 £ 7 sh.; No. 21. UM. verr. 3 £; No. 9½. US.

2 £ 2 sh.; No. 13½. B. verr. 2 £ 9 sh.; No. 22. B. 2 £ 14 sh.; No. 17. B. 2 £ 2 sh.; No. 13½. B. 1 £ 13 sh.; No. 7½. B. 1 £ 1 sh.; No. 11½. B. 1 £ 13 sh.; No. 28. 2 £ 18 sh.; No. 27. 3 £ 12 sh.; No. 22. 3 £; No. 9½. 2 £ 2 sh.; No. 10. 1 £ 13 sh.

1977. FRANZ, OTTO, sculptor in wood, Berlin, Wilhelmsstr. 88. Agts. Lion M. Cohn, Phaland & Dietrich, represented by Ch. Trübner, 20 St. Dunstans Hill, City, and at the Exhib. Building.

Carved wood-furnitures; prices p. doz.: No. 1. a. watch-case 13 Th. (1 £ 19 sh.); b. console to it 9 Th. (1 £ 7 sh.); No. 2. a. watch-case 13 Th. (1 £ 19 sh.); b. console to it 10 Th. (1 £ 10 sh.); No. 3. a. watch-case 14½ Th. (2 £ 3 sh. 6 d.); b. console to it 11 Th. (1 £ 13 sh.); No. 4. watch-hook 4½ Th. (13 sh. 6 d.); No. 5. do. 6½ Th. (19 sh. 6 d.); No. 6. frame 22 Th. (3 £ 6 sh.); No. 7. a. writing-stand with almanack, 2 ink-stands with lids 34 Th. (5 £ 2 sh.); b. stand for a thermometer 7½ Th. (1 £ 2 sh. 6 d.); c. watch-stand 7½ Th. (1 £ 2 sh. 6 d.).

1978. FRILING & Co., Cologne. Agts. F. Leonards & Co., 2 Salter's Hall-Court, Cannonstreet.

Gilt and polished cornices.

1979. FUHRBERG, F., manu., Berlin, Friedrichsstr. 190.

Household-furniture of wicker-work and work-baskets.

1980. GERIKE, HUGO, & PFITZNER, manu., Berlin, Prinzessinnenstr. 21. Agt. Jos. van Oudenhoven, Esq., 37 St. James' Place, New Cross, E. C., 3 Barge-Yard, Bucklersbury.

Writing-desk of artificial wood, with turned and carved work, and different samples of artificial wood-plates and veneers, surrounded by natural mahogany-wood to show the difference. Six frames of artificial wood.

GESELL, Görlitz, looking-glass frames. (s. No. 2187.)

1983. HEINRICH, C., dealer in bed feathers and downs. Nordhausen. Agts. s. No. 1977.

2 quilted coverlets, filled with downs, p. piece 6 Th. (18 sh.).

1984. HEINRICH, JOH. (formerly Black & Grumm), manu. of cornices, Bonn. Munich 1854 medal of honor; Paris 1855 bronze medal. Agts. S. Oppenheim & Sons, 4a. Bread-street, Cheapside.

An assortment of gold cornices for frames waved an smooth in imitation of polixander, ebony and mahogany-wood; an assortment of cornices for papered rooms in all wood-colors, with steel mounting and in gold.

1985. HERBST, AUGUST, basket-maker, Bonn. Agts. s. No. 1977.

Brown chair of wicker-work 20 Th. (3 £).

HERZIG, G., Hermsdorf, wooden ware for household use. (s. No. 2189.)

1987. KELTERBORN, ROBERT, wood-sculptor, Berlin. London 1862 honor. mention. Agts. s. No. 1977.

Carved furniture and utensils: arme-chair with cushions 40 Th. (6 £); table, square 24 Th. (3 £

12 sh.); do., small round 7 Th. (1 £ 1 sh.); do., marble-top 8 Th. (1 £ 4 sh.); all antique style.

The following prices p. doz.: Wine-cooler 78 Th. (11 £ 14 sh.); bowl with plate of china 72 Th. (10 £ 16 sh.); cigar-box 90 Th. (13 £ 10 sh.); do. (like-cudgels) 78 Th. (11 £ 14 sh.); boston-case 50 Th. (7 £ 10 sh.); tea-caddy 26 Th. (3 £ 18 sh.); card-case 18 Th. (2 £ 14 sh.); frame 46 and 38 Th. (6 £ 18 sh. and 5 £ 14 sh.); do. with ivy-leaves 40 Th. (6 £); do. with oak-leaves 40 Th. (6 £); do. with china-picture 60 Th. (9 £); do., small 22 Th. (3 £ 6 sh.); 2 small frames for photographic cards 6 Th. (18 sh.); 1 do. 5 Th. (15 sh.); card-press 26 Th. (3 £ 18 sh.); candle-stick 20 Th. (3 £); writing-stand 26 Th. (3 £ 18 sh.); do. with vine-leaves 32 Th. (4 £ 16 sh.); pen-holder 6 Th. (18 sh.); paper-folder 6 Th. (18 sh.); almanack 28 Th. (4 £ 4 sh.); album for photographic cards 34 Th. (5 £ 2 sh.); flower-vase, glass-chalice 27 Th. (4 £ 1 sh.); do. chinese 44 Th. (6 £ 12 sh.); cabinet-clocks, repeaters 144 and 192 Th. (21 £ 12 sh. and 28 £ 16 sh.).

1988. KEMPEN, WILHELM, & Co., manu., Emmerich. Agts. s. No. 1977.

Patterns of gilt and polished cornices.

1989. KILIAN, GABR., manu., Bonn. London 1862 medal.

Brooms and brushes of straw.

1990. KLEIN, JULIUS, manu. of gold cornices, Berlin. Agts. s. No. 1977.

Patterns of gilt mouldings and cornices and imitated cornices of wood-color for picture and lookingglass-frames.

1993. KOKSTEIN, ADOLPH, manu. of sculptures and joiner's work, Berlin, Wollanksstr. 1a.

A lookingglass-frame carved of American nut-wood 93 Th. 10 sg. (14 £); wholesale price 80 Th. (12 £).

KOEHLER, W. G., Zeitz, wood-consoles and ornamental wood-work. (s. No. 2193.)

1997. LÜDEKE, A., merchant, Brandenburg on the Havel. Agts. A. Boden & Co., 33 Aldermanbury.

Samples of gilt and polished cornices.

1998. LÖVINSON, LOUIS & SIEGFRIED (Renaissance, Commandit-Gesellschaft für Holzschnitzkunst), merchants and manu. of carved furniture, contractors for the instruction and employment of prisoners in wood-carving, Berlin, Unter den Linden 8, manufactories Berlin, Spandau, Moabit. London 1862 honor. mention. Agt. Julius Jacoby.

Side-board 1400 Th. (216 £); bureau 250 Th. (37 £ 10 sh.); smoking-chair 38 Th. (5 £ 14 sh.); drawing-room table 80 Th. (12 £); looking-glass 75 Th. (11 £ 5 sh.); 2 chandeliers, each 27½ Th. (4 £ 2½ sh.); bread-plate 4 Th. (12 sh.); water-plate 10 Th. (1 £ 10 sh.); wine-cooler 30 Th. (4 £ 10 sh.); 3 stools, each 8 Th. (1 £ 4 sh.); chess-table 30 Th. (4 £ 10 sh.); paper-basket 38 Th. (5 £ 14 sh.); turning-chair for writing-desks 24 Th. (3 £ 12 sh.); woodbox, in shape of a chopping block 25 Th. (3 £ 15 sh.); flower-table, small 25 Th. (3 £ 15 sh.); screen for stoves 36 Th. (5 £ 8 sh.); flower-table, large 68 Th. (10 £ 4 sh.); table, design of oak-leaves in vegetation 50 Th. (7 £ 10 sh.); 2 drawing-room chairs, stuffed and with carved helmets on the backs, each 15 Th.

(2 £ 5 sh.); fauteuil 40 Th. (6 £); eagle-chair with cushion for dining-rooms 18 Th. (2 £ 14 sh.); cane-chair with escutcheon 12 Th. (1 £ 16 sh.); cane-chair with helmet 10 Th. (1 £ 10 sh.); round table 28 Th. (4 £ 4 sh.); 2 small chandeliers, each 14 Th. (2 £ 2 sh.); 2 relievoes, portraits of Victoria and Frederic William of Prussia, each 30 Th. (4 £ 10 sh.); chandelier with globe and clock 90 Th. (13 £ 10 sh.); cigar-case, in shape of a travelling trunk 45 Th. (6 £ 15 sh.); case for private papers or money 40 Th. (6 £).

2000. Meller, Friedr., & Co., mánu. of gold and polished cornices, Bonn. London 1862 honor. mention. Agts. Charles Dittges & Co., 41 Basinghall-street, E. C.

Gold and polished cornices. For prices apply to the named agents.

2001. Methlow, Eduard, & Co., manu., Berlin. London 1862 honor. mention.

Frames, prices p. doz.: I. Plain, black polished oval frames ornamented with metal or gold-border, for photographs or other pictures, No. 1 P., 1 M., 2 P., 2 M., 3 P., 3 M., 4., 5., 6 P., 6 M., 7 P., 7 M., 7 A., 8., 9., 10., 11., 12., 13., 14., 15., 17. 1—7 Th. (3—21 sh.).

II. Brown oval frames with baroque decoration for photographs etc., No. 33., 34., 36., 37., 39., 47., 49. 1⅓—8 Th. (4 to 24 sh.).

III. Gilt oval frames for photographs etc., No. 40., 46. 3—18 Th. (9—54 sh.).

IV. Small frames pressed in relief with glass and box for photographs of the card-size shape, No. 50., 51., 52., 53., 54., 55., 56., 57., 59., 63., 64., 65., 66., 67., 68., 69., 71., 72. 1—1½ Th. (3—4½ sh.).

V. Square brown frames with baroque decoration for photographs etc., No. 101., 102., 102½., 104., 106. 1½ to 4½ Th. (4½—13½ sh.).

2002. Nees, A. F., manu. of gold cornices and frames, Cologne.

A frame of oak-wood, containing: samples of gilt and polished cornices (plain and waved) in various wood-colors, for frames and ornaments of walls; samples of curtain rests; samples of gilt and fancy-wood oval frames for the use of photographers, with passepartouts; samples of fancy-wood oval frames for looking-glasses.

2003. Paschen, Ernst, joiner, Stendal. Agts. Lion M. Cohn, Phaland & Dietrich, represented by Ch. Trübner, 20 St. Dunstans Hill, City, and at the Exhib. Building.

Travelling-furniture, consisting of: 1. a wardrobe with two doors, serving as a bed-screen when doors and sides are opened, 2. a writing-table, 3. a fancy-table, at the same time washhand-stand, card- and dressing-table, 4. a sofa-table, 5. four chairs, 6. a couch, serving at the same time as a packing-chest, for the above named objects, 600 lb. 500 Th. (80 £).

2004. Reichardt, Johann Friedrich, basket-maker, Erfurt. Agts. s. No. 2003.

One flower-stand; four arm-chairs of different kinds; two children's chairs; two paper-baskets; one bottle-basket; thirteen fancy-baskets of different kinds.

2005. Schievelbein, J. F. E., cabinet-maker, Berlin, Spandauerstr. 46, & Baudouin, E., & Co., manu. of silk-ware, Berlin, Manufactory Züllichau. Munich 1854 bronze medal. Agts. s. No. 2003.

2 arm-chairs.

2006. Schirow, C. A., & Co., manu., Berlin, Leipzigerstr. 68a.

Cart for invalids; fountain; arm-chair; chairs.

2007. Schmidt, Ludwig, joiner, Culm. Agts. s. No. 2003.

A bureau of polixander-wood mounted with metal-work (120 £).

2008. Schulze, Friedrich August, turner, Berlin, Neue Grünstr. 6. Agts. s. No. 2003.

Sample-card of bone key-hole lids. Prices p. gross.

	No.	Th.	sg.	sh.	d.	frs.	c.
	1—9.	3	12½	10	3	12	81¼
	10.	3	5	9	6	11	87½
	11.	2	27½	8	9	10	93¾
	12.	3	12½	10	3	12	81¼
	13.	3	5	9	6	11	87½
	14.	2	27½	8	9	10	93¾
	15.	3	5	9	6	11	87½
	16.	2	25	8	6	10	62½
	17.	2	5	6	6	8	12½
	18.	3	15	10	6	13	12½
	19.	3	10	10	—	12	50
	20.	3	5	9	6	11	87½
	21.	2	20	8	—	10	—
	22.	2	15	7	6	9	37½
	23.	2	10	7	—	8	75
	24.	2	7½	6	9	8	43¾
	25.	2	7½	6	9	8	43¾
	26—27.	2	15	7	6	9	37½
	28.	2	10	7	—	8	75
	29.	3	20	11	—	13	75
	30.	3	15	10	6	13	12½
	31.	3	—	9	—	11	25
	32.	2	15	7	6	9	37½
	33.	2	10	7	—	8	75
	34.	3	15	10	6	13	12½
	35.	3	—	9	—	11	25
	36.	2	15	7	6	9	37½
	37.	3	15	10	6	13	12½
p. in.	38.	3	—	9	—	11	25
1¾	39.	2	20	8	—	10	—
1⅝	40.	2	15	7	6	9	37½
1½	41.	2	12½	7	3	9	6¼
1⅜	42.	2	10	7	—	8	75
1¼	43.	2	7½	6	9	8	43¾
1⅛	44.	2	5	6	6	8	12½
1	45.	2	2½	6	3	7	81¾
⅝—1	46—51.	3	15	10	6	13	12½
⅝—1	52—55.	1	10	4	—	5	—

Notice. From No. 39—45. the prices are calculated by the size, without regard to pattern.

No. 46—51. are boxes with escutcheons.

2010. Sonnenburg, direction of the royal house of correction, (Frankfurt on the Oder). London 1862 honor. mention. Agts. s. No. 2003.

A chair carved in wood 340 Th. (49 £ 16 sh.).

2014. Unger, Julius, merchant, Erfurt. London 1862 honor. mention. Agts. s. No. 2003.

Railway-pillow with frame 11⅔ Th. (1 £ 15 sh.); iron bedstead, spring-bottomed after the American system of Mr. Tucker, p. doz. 60—75 Th. (9 to 12 £).

2015. Vallentin & Schaefer, merchants and manu., Schweidnitz (Breslau). Agts. s. No. 2003.

Sundry articles of wood; prices p. gross. Standermill 2 £ 12 sh.; needle-case (umbrella) 1 £ 3 sh.; back-scraper 1 £ 9 sh.; mushroom (instrument for darning) 15 sh.; articles for winding thread on 1 £ 15 sh.; a tape-measure 1 £ 15 sh.; pair of nut-cruckers 2 £ 3 sh.; pounce-box 15 sh.; perspective - glass 1 £ 6 sh.; dog - whistle 12 sh.; knitting-gland 1 £ 9 sh.; knitting-ball 3 £ 9 sh.; humming-top 2 £ 12 sh.; pepper-box 1 £ 15 sh.; instrument for darning 15 sh.; jou-jou 1 £ 15 sh.; sewing-cushion 2 £ 12 sh.; tumbler 15 sh.; rattle 1 £ 15 sh.; pen-case 15 sh.; bilboquet 1 £ 6 sh.; egg-cup stand 17 £ 3 sh.; sugar-hammer with knife 3 £ 9 sh.; 2 yarn-reels 8 £ 12 sh.; pocket looking-glass 1 £ 15 sh.; cigar-pipe 18 sh.; cuckoo-whistle 1 £ 3 sh.; ordinary-needle-case 9 sh.; clew basket 2 £ 12 sh.; machine for catching fleas 1 £ 9 sh.; napkin-ring 15 sh.; money-box 15 sh.; spell-cup 2 £ 12 sh.; egg for darning 1 £ 3 sh.; drinking-cup 1 £ 15 sh.; spice - box 10 £ 6 sh.; roulette 8 £ 12 sh.; cigar-stand 13 £ 15 sh.; rattle (for children) 15 sh.; 2 pop-guns 1 £ 6 sh.; screw pin-cushion with looking-glass 3 £ 9 sh.; do. without looking-glass 2 £ 3 sh.; squirt 1 £ 15 sh.; shaving-box 15 sh.; whip 1 £ 6 sh.; ruler 10 sh.; do. 10 sh.; box No. 1. 1 £ 15 sh.; do. No. 2. 1 £ 9 sh.; instrument for waxing thread. 1 £ 3 sh.

2016. WERKMEISTER, A., manu., Berlin, Michaelkirchstr. 11.
Gilt fillets for picture-frames.

2017. WIEDEMANN, D. P., manu., Berlin, Jerusalemerstr. 54. London 1862 medal.
Household-furnitures of wicker-work.

2019. WINCKLER, HERMANN, manu. of willow baskets, Berlin. Berlin 1844 bronze medal. For particulars apply to Wm. Payne & Son, 33½ Liquor pond-street, Gray's Inn-Lane, Holborn.
Furniture of basket - work: chair, do. for children, flower-stand, 2 work-tables, 2 paper-baskets, toilet-basket, 2 bottle-baskets, chair for dolls, 21 different baskets, 32 pieces 6 £ 2 sh.

CLASS XXXI.

IRON AND GENERAL HARDWARE.

2020. ARNHEIM, S. J., manu., Berlin, Rosenthalerstr. 36. London 1862 honor. mention.
Iron safe; improved locks.

    BASCH, W., & Co., Berlin. Safety lock.
        (s. No. 1970.)

2022. BERG, W., manu., Lüdenscheid (Arnsberg). London 1862 medal. Agt. Heintzmann & Rochussen, 9 Friday-street, Cheapside.
Metal buttons.

2023. BERLIN ROYAL PRUSSIAN IRON FOUNDRY. London 1851 council medal and prize-medal; Munich 1854 medal of honor; Paris 1855 bronze medal; London 1862 medal. Agt. August Pils, warehouse-man of the Royal porcelain - manufactory at Berlin.
No. 1. A monument, representing King Frederic William III. of Prussia. On a socle, richly ornamented by hautreliefs, arabesques, pearl-bars etc. the statue of the King stands by the side of a pedestal, on one side of which is a basrelief representing the Queen Louise. The wreath which the King holds in his hand over the basrelief is intended for his consort, the Queen. The original model is by professor Drake, sculptor 600 Th. (90 £).
No. 2. and 3. Two busts. Each of these busts stands on a richly ornamented pedestal; they are copied at the Royal iron foundry on a reduced

scale from busts of the sculptor professor Hagen. Busts each 85 Th. (12 £ 15 sh.); pedestal 275 Th. (41 £ 5 sh.).
No. 4. A vase. The figures below the handles represent war and peace. Between the two and amidst the basrelief running round the vase there is on one side, as the principal figure, Borussia on a throne, to whom the arts and sciences etc., are presented by a genius and on the other side warriors by an other genius. On the opposite side there is Eris. The design is by Mr. Stüler, Geheimer Baurath. 460 Th. (69 £).
No. 5. A candelabre for a drawing-room. On the top of a column lies Phoenix in his nest. The flame which is bursting out of the nest carries a boy who holds in his hand a golden ball surrounded by candlesticks. The design is by Mr. Hesse, Hofbaurath. 580 Th. (87 £).
No. 6. A lamp - stand. On a richly festooned base is the middle piece, a round perforated decoration, and then a group of boys playing on musical instruments. From amidst them rises a richly ornamented shaft crowned by two eagles bearing the plate on which the lamp is to be placed. The design is by Mr. Strack, Professor and Hofbaurath. 570 Th. (85 £ 10 sh.).
No. 7. Two altar - candlesticks. On the twelve sides of these candlesticks the twelve apostles are represented on consols and under canopies. 150 Th. (22 £ 10 sh.).
No. 8. A font. Upon an octagonal base, composed of columns and pointed arches, rests the main piece of the font, covering this base. As a continuation of the columns, the corners of the main piece are formed by small ornamented turrets which present eight fields and which are filled up alternately by the four evangelists and flowers. The design is by Mr. Stüler, Geheimer Baurath. 600 Th. (90 £).
No. 9. Candelaber for churches. From a richly decorated pedestal arises an octagonal trunk out of which extend in three parts towards both sides three branches richly ornamented with leaves each branch being ornamented at the end by tulips for the candles. A similar tulip is placed upon the mainstand. The design is likewise by Stüler. 750 Th. (112 £ 10 sh.).
The models (with the exception of No. 1.), the casting, chasing and inlaying with silver of all these objects have been executed at the Royal iron foundry.

    BOCHUM (ARNSBERG) MINING AND CAST-STEEL MANUFACTORING-COMPANY, Arnsberg. Cast-steel, bell.
        (s. No. 1253.)

2025. VON BOGDANSKI, JOSEPH, locksmith and machine-maker, Posen. Agts. Lion M. Cohn, Phaland & Dietrich, represented by Ch. Trübner, 20 St. Dunstans Hill, City, and at the Exhib. Building.
Case-lock 1 £; safety-lock for doors 1 £.

2029. COLOGNE MACHINE-MAKING COMPANY, Cologne, manufactory at Bayenthal near Cologne. Agt. s. No. 2025. (s. No. 1255, 1494.)
1 water-gauge; 6 socket-pipes.

2030. COHN, E. J., & Co., manu., Berlin. Agts s. No. 2025.
Hermetically closing doors for stoves; prices p. piece: Steel - polished with outside-beam and ornaments of German silver 2 £ 8 sh.; steel-polished with outside-beam 1 £ 16 sh.; bronze-painted, with inside - beam and large ornamented brass latch (6 sh. 6 d.); bronze - painted, with brass levers and a covering ornamental brass-plate 9 sh.

2030a. Colsman, Joh. Friedr. (Colsman & Co.), manu., Barmen. Agts. Morgan Brothers, 21 Bow Lane, Cannon-street, City. (s. No. 1485.)

Sandboxes and inkstands, No. 1. 5 sg. 3 pf.; No. 2. 5 sg. 9 pf.; No. 3. 6 sg. 6 pf.; No. 4. 7 sg. 10 pf.; No. 5. 10 sg. 6 pf.; all p. pair.

Hunting and railway whistles No. 1. and 2. 8 Th.; No. 3. and 4. 10 Th. 15 sg.; p. gross.

Carving-knife-handles 1 Th. 10 sg.; Table-knife-handles No. 0. 1. 3. 5. 7. 24 sg.; dessert-knife-handles No. 00. 2. 4. 6. 19 sg.; all p. doz.

Syringes, No. 1. (14 ounces) 14 Th. 24 sg.; No. 2. (12 ounces) 13 Th. 6 sg.; No. 2½. (10 ounces) 12 Th.; No. 2¾. (8 ounces) 11 Th. 6 sg.; No. 3. (5 ounces) 9 Th. 18 sg.; No. 4. (3 ounces) 8 Th. 24 sg.; No. 4½. (2 ounces) 8 Th.; No. 5. 6. 7. 6 Th. 24 sg.; No. 8. 6 Th.; No. 9. 4 Th.; No. 10. and 11. 2 Th. 12 sg.; No. 12. 2 Th. 24 sg.; No. 13. and 14. 2 Th. 12 sg.; No. 15. 2 Th.; No. 16. 16 Th. 6 sg.; all p. doz.

Spout-pipes, No. 1. 4 Th.; No. 2. 2 Th. 12 sg.; No. 3. 3 Th.; No. 4. 4 Th. 24 sg.; No. 4½. 2 Th. 12 sg.; No. 5. 4 Th. 12 sg.; No. 6. 4 Th.; all p. doz.

Ladles (with wooden handle) 1⅓ to 2¼ Th.; do. (with a sharp metal-handle) 2¼ to 2½ Th.; do. (with a blunt metal-handle) 1⅝ to 3 Th.; soup-ladles (small size) 1⅓ to 2⅝ Th.; milk-spoons 1 to 1⅕ Th.; all p. doz.

Table-spoons 5⅓ to 12⅓ Th.; do. (small, for children) 3⅝ to 5⅝ Th.; tea-spoons 1⅝ to 3½ Th.; forks 4 to 7⅝ Th.; all p. gross.

Coffee-mills No. 1. 18 Th.; No. 2. 20 Th.; No. 3. 24 Th.; all p. doz.

Thimbles No. 2/0. 0. 1. 2. 3. 4. 5. 6. 7. 8., No. 2/0. 1 Th. 8 sg.; each following No. 2. sg. less, therefore No. 8. 20 sg. p. gross; do., with colored rims, 2 sg. more; tailor-thimbles, 3 sg. less; all p. gross.

Cocks (Litt. I. No. 1—5.) 1 9/10 to 6⅔ Th.; do. (Litt. II. No. 0—5.) 1⅛ to 4½ Th.; do. (Litt. III. No. 0—5.) 1⅓ to 6⅓ Th.; all p. doz.

Tinder-boxes with cotton match-cord and agate-stone, No. 1. 17 Th.; No. 1⅛. 23 Th.; No. 1¼. 20 Th.; No. 1⅓. 23 Th.; No. 1½. 21 Th.; No. 1¾. 22 Th.; No. 1⅔. 22 Th.; No. 1⅕. 32 Th.; do. with a thinner pipe, No. 2—2⅓, cheaper 2 Th.; do. with a still thinner pipe, No. 3—3⅓, again cheaper 2 Th.; (No 1. 17 Th. p. gross, No. 2. 15 Th., No. 3. 13 Th); do. with flint less than with agate-stone 2 Th.; do. with silk match-cords more than with cotton 3 Th.; all p. gross.

2031. Cosack & Co., prop. of iron-works, Hamm (Arnsberg). London 1862 honor. mention. Agts. Lion M. Cohn, Phaland & Dietrich, represented by Ch. Trübner, 20 St. Dunstans Hill, City, and at the Exhib. Building.

8 various axle-trees, p. 1000 lb. 60 Th.; 10 rings of iron-wire (telegraphs, angular, half-round, round), p. 100 lb. 4½ Th.; 2 straps of iron-wire, p. lb. 3⅓ sg.; 8 ropes of iron-wire, p. lb. 2¾ sg.; 3 doz. of smooth spiral-springs, p. lb. 1⅔ sg.; 2 doz. spiral-springs, galvanic coppered, p. lb. 1⅓ sg.; 1 sample-card of rivets, p. lb. 1⅔ sg.; 2 sample-cards of forged-nails, p. lb. 1½ sg.

The prices of the axle-trees, iron-wire, rivets and nails refer only to the heavier sorts; the finer ones are proportionally dearer.

2032. Dahm, Knödgen & Kirchner, manu., Fraulautern near Saarlouis. London 1862 honor. mention. Agt. E. G. Zimmermann, 2 St. Pauls Buildings, Little Carter Lane, Doctors Commons.

An assortment of kitchen-utensils of sheet-iron, stamped, folded, polished, tinned.

2033. Diebitsch, C. von, manu., Berlin, 4 Hafen-place. London 1862 medal.

Zinc vase.

Count Einsiedel. (s. No. 2067.)

2035. Epstein, Ludwig, manu., Lublinitz, iron-works Mochala near Lublinitz. London 1862 medal. Agts. s. No. 2031.

Various spoons of tinned-iron. The prices p. 100 doz. are specified on the cover of the box in English, French and Prussian value.

2036. Fabian, M., locksmith and manu., of fire-proof iron-safes, Berlin, Spandauerstr. 75. Munich 1854 honor. mention; Paris 1855 bronze medal; London 1862 medal. Agts. s. No. 2031.

1 fire-proof iron-safe 1000 Th. (150 £).

Galopin, J., Aix-la-Chapelle. Model of an iron spiral stair-case and ornaments. (s. No. 2186.)

2037. Geck, August Theodor, merchant, Iserlohn. Paris 1855 bronze medal; London 1862 medal. Agt. August Theodor Geck, 22 College Hill, Cannon-street, E. C. (s. II. No. 16.)

Brass window cornices, curtain-bands, curtain-pins, ceiling ornaments etc. Gilt cornices for picture frames and room decorations. Hardware of every kind.

2038. Geiss, M., zinc-works, Berlin. Berlin 1844 silver medal; London 1851 prize-medal; Munich 1854 medal of honor; Paris 1855 silver medal; London 1862 medal.

Juno, statue, 7 ft. high, galvanized zinc-casting after the antique marble in the Royal museum at Berlin 160 Th. (24 £); Niobe, statue, 7 ft. 2 in. high in white marble-color after the antique marble in the gallery of Florence 144 Th. (21 £ 11 sh.); shepherd-group in full length, model by J. Franz in Berlin, galvanized zinc-casting 400 Th. (60 £).

2 figures, 7 ft. high, fighting a duel, standing upon a pedestal with basso-relievoes, models by professor Molin in Stockholm, galvanized zinc-casting, (exhibited in the Swedish division).

2039. Gladenbeck, Herrmann, manu., Berlin, Münzstr. 10—12. London 1862 medal. Agts. s. No. 2031.

1. Group, a horseman fighting with a lion after the original of Albert Wolff, zinc-casting electro-plated with bronze. 2. A colossal bust of Schiller after Dannecker's original, executed in bronze. 3. Two statues of Russian soldiers after Mehnert's models.

2041. Hasemann, M. A., manu., Berlin, Friedrichsstr. 154. London 1862 honor. mention.

Scales with columns.

2042. Hauschild, C., manu., Berlin, 3 Neander-street. London 1862 medal.

Iron safe.

2043. Heckel, Georg, metal-string-manu., St. Johann near Saarbrücken. Paris 1855 bronze medal: London 1862 medal. Agts. s. No. 2031.

Iron and cast steel-wire-strings for musical instruments.

Heegmann & Mesthaler, Barmen, metal buttons. (s. No. 1623.)

2044. HEINTZE & BLANKERTZ, manu., Berlin, Fliederstr. 4. London 1862 medal.

Steel pens.

2045. HENRIETTENHÜTTE, Ducal Sleswick-Holstein-Sonderburg-Augustenburg iron and enamel-work by William Guetzloe, chief director, Henriettenhütte near Primkenau (Liegnitz). London 1862 medal. Agts. Lion M. Cohn, Phaland & Dietrich, represented by Ch. Trübner, 20 St. Dunstans Hill, City, and at the Exhib. Building.

Enamelled cast iron ware: kettle No. 24. 10 sh. 6 d.; 2 horse-mangers No. 1. à 6 sh.; water-case No. 1. 3 sh.; do. No. 4. 5 sh. 6 d.; cylindrical pot No. 24. 3 sh. 6 d.; stew-pot No. 4. 9 d.; do. No. 8. 1 sh. 3 d.; frying-pan D. No. 3. 8 d.; do. No. 9. 1 sh. 6 d.; sauce-pan with feet No. 0. 3 d.; do. No. 2. 5 d.; milk-bowl No. 3. 1 sh. 3 d.; bellied pot No. 22. 3 sh.; pots with feet: No. 4½. 3 d., No. 5. 4 d., No. 5½. 5 d., No. 6½. 6 d., No. 7½. 7 d., No. 8½. 9 d., No. 9. 10 d., No. 9½. 11 d., No. 10. 1 sh., No. 10½. 1 sh. 2 d., No. 11. 1 sh. 4 d., No. 11½. 1 sh. 5 d., No. 12. 1 sh. 6 d., No. 14. 2 sh. 6 d., No. 15½. 3 sh. 4 d., prices p. piece.

2046. HERRMANN, C., bronzer, Danzig.

A gothic chandelier for thirty six candles.

2048. HITSCHLER, JOHANN, locksmith and manu. of patent weighing machines, Crefeld. Agts. Morgan Brothers, 21 Bow Lane, Cannon-street, City.

Iron-safe to be locked with or without a key.

2049. HOBRECKER, WITTE & HERBERS, manu., Hamm. London 1862 medal. Agts. Heintzmann & Rochussen, 9 Friday-street, E. C. (s. II. No. 31.)

Samples of rod-iron and rolled round iron. All sorts and sizes of iron-wire, round, half-round, flat, angular and square shaped. Wire for card cloth. Iron telegraph-wire, annealed, lackered with oil, and galvanized. Submarine telegraph wire of cast-steel. Tinned iron-wire. Galvanized and soft iron-wire for fastening telegraph-wires. Springs, smooth and coppered. All kinds of wire-tacks, round, square, plain, with hooks, curved etc. from 16 in. in length, ½ in. in diameter, to the smallest dimensions. For prices apply to the above named agents.

2051. HUPP, CARL, engraver, Düsseldorf. Agts. s. No. 2045.

Engraved and chased parts of a hunting-gun representing on one of the lock-plates: fighting stags, on the other: a wild boar pursued by hounds; upon the throat-band: the stag of St. Hubertus, upon the cocks ornaments 258 Th.; the finished hunting-gun 360 Th.

2052. HÜSTEN TRADE-COMPANY, Heister near Arnsberg. Munich 1854 medal of honor; Paris 1855 silver medal. Agt. J. W. Green, 2 St. Pauls Buildings, Little Carter Lane, Doctors Commons.

Tinned and black iron-plate.

2053. ILSENBURG, Count Stolberg-Wernigerode factory. Berlin 1844 gold medal; Paris 1855 silver medal; London 1862 medal. (s. No. 855.)

1. Bower of malleable cast iron and wire 213 Th. 10 sg. (32 £.); 2. 1 chandelier Mauresque 60 Th.

(9 £); 3. 2 lustres each, 4½ lbs. weight, 56 Th. 20 sg. (8 £ 10 sh.).

4. Ancient weapons, 13 pieces viz: 4 swords each ½ £ (3 Th. 10 sg.), 13 Th. 10 sg. (2 £); 1 morning-star (holy water sprinkle) 1 Th. 20 sg. (5 sh.); neck piece 3 Th. 10 sg. (10 sh.); a buckler 3 Th. 10 sg. (10 sh.); 1 do. with lions 1 Th. 20 sg. (5 sh.); 4 halberds each ½ £ (1 Th. 20 sg.), 6 Th. 20 sg. (1 £); 1 dagger 3 Th. 10 sg. (5 sh.).

5. Bronzed cast fancy objects: basin with satyr 3 Th. 10 sg. (10 sh.); do. with angel 3 Th. 10 sg. (10 sh.), do. rococo 3 Th. 10 sg. (10 sh.); do. antique 1 Th. 20 sg. (5 sh.); ash-cup with ape and dog 3 Th. 10 sg. (10 sh.); plate »our daily bread« 3 Th. 10 sg. (10 sh.); do. with »nuptials« 6 Th. 20 sg. (1 £); do. with »prodigal son« 6 Th. 20 sg. (1 £); 6. frieze from the Parthenon 13 Th. 10 sg. (2 £); do. with frame 20 Th. (3 £); 7. coat of armour as stove 86 Th. 20 sg. (13 £); 8. table with antique vase (prodigal son) 15 Th. (2 £ 5 sh.); 9. do. with »nuptials« 20 Th. (3 £).

10. Book covers and albums: a pair of book covers, gothic 3 Th. 10 sg. (10 sh.); do., imitated wood 3 Th. 10 sg. (10 sh.); album, imitated wood 6 Th. 20 sg. (1 £); album, Alhambra 13 Th. 10 sg. (2 £.); portfolio with flowers 6 Th. 20 sg. (1 £).

11. Cast fancy objects: 2 fruit-baskets each ½ £ 3 Th. 10 sg. (10 sh.); crucifix 1 Th. 20 sg. (5 sh.); tortoise, inkstand 3 Th. 10 sg. (10 sh.); letter presser, leaf-shaped 1 Th. 20 sg. (5 sh.); dressing-glass with boys lying at the foot 3 Th. 10 sg. (10 sh.); plate, (prodigal son), rough casting 3 Th. 10 sg. (10 sh.); do. with »nuptials« 1 Th. 20 sg. (5 sh.).

12. Galvanized cast fancy objects viz. goblet with hunting scenes 28 Th. 10 sg. (4 £ 5 sh.); do., gothic, lined with silver 60 Th. (9 £); 3 lamps each ¼ £ 5 Th. (15 sh.); boar 3 Th. 10 sg. (10 sh.); folding-stick 25 sg. (2 sh. 6 d.); lizard as letter presser 25 sg. (2 sh. 6 d.).

IRON FOUNDERY ROYAL, Berlin. (s. No. 2023.)

2056. KAHN, J., & Co., manu., Cologne. Paris 1855 bronze medal; London 1862 medal. Agts. Morgan Brothers, 21 Bow Lane, Cannon-street, City.

40 pieces of higly polished iron chains.

2057. KISSING & MÖLLMANN, manu. and merchants, Iserlohn. London 1862 honor. mention. Agt. J. W. Green (E. G. Zimmermann), 2 St. Pauls Buildings, Little Carter Lane, Doctors Commons. (s. II. No. 92.)

Brass church-lustre in gothic Style, 90 in. high and 75 in. diam. for 25 candles; brass lustre 30 in. high and 25 in. diam. for 8 candles; brass gas-lustre, 30 in. high and 25 in. diam., 8 candles; brass sconce for 1 candle; do. in gothic style for 6 candles; set of gilt brass chandeliers consisting of, 1 chandelier for 9 candles; 2 do. for 7 candles; 1 pair of brass chandeliers, each for 3 candles; 2 pair of brass table candlesticks for 1 candle, brass inkstand; 5 hand bells; assortment of brass cornices; curtain bands; paters; rosettes; watch ornaments; assortment of brass cast-ware; assortment of thimbles; brass rings; iron wire-tacks; samples of brass-wire; copper wire and web-wire; proof sheet of double polished latten brass.

2058. KNOLL, L., brass-founder, Berlin, Linienstr. 113. Munich 1854 honor. mention; Paris 1855 honor. mention; London 1862 honor. mention. Agts. s. No. 2045.

Bronze statue, the divine child, 150 £.

2059. Koch & Bein, manu., Berlin, Neue Friedrichsstr. 49. London 1862 medal: Agt. J. O. Bieling, 19 Gloucester-street, Queen Square, W. C. (s. II. No. 2.)

Tableau of metal and glass-letters, the initials in the national colors of the countries of Europe; coat of arms of the King of Prussia, zinc, with socle 170 Th. (27 £).

2060. Koeppen, J. W., manu., Berlin, Friedrichsstr. 235. Agts. Lion M. Cohn, Phaland & Dietrich, represented by Ch. Trübner, 20 St. Dunstans Hill, City, and at the Exhib. Building.

Alabaster-glass-lustre with 24 branches 38 Th.; do. shell-shaped, 9 branches $18\frac{1}{2}$ Th.; do. like a pearl-brooch, 9 branches $14\frac{1}{2}$ Th.; do. chandelier, 9 branches 15 Th.; bronze chandelier with 6 branches $5\frac{1}{2}$ Th.; do. Alhambre, 4 branches $4\frac{1}{4}$ Th.; do for cigars, 2 branches 3 Th.; do. with flower-vase, 6 branches 12 Th.

2061. Körner, H., manu., Berlin, Zimmerstr. 91.

Assortment of lamps.

2062. Kolesch, Heinrich, lock-smith and manu., Stettin. London 1851 honor. mention; Paris 1855 honor. mention; London 1862 medal. Agts. s. No. 2060.

A fire and thief-proof iron-safe, 46 cwts. 800 Th. (120 £).

2064. Krause, Friedrich Wilhelm, Royal counsellor of commerce, merchant, Berlin, iron-foundry and enamelling-works, Neusalz on the Oder. London 1862 honor. mention. (s. No. 768.)

Enamelled iron pots and utensils; Silesian pots: No. 4. 3 quart ($\frac{3}{4}$ gallon) $8\frac{1}{4}$ sg. ($9\frac{9}{10}$ d.); No. 5. $3\frac{1}{2}$ quart ($\frac{7}{8}$ gallon) $10\frac{1}{4}$ sg. (1 sh. $\frac{3}{10}$ d.); No. 6. 4 quart (1 gallon) $11\frac{1}{2}$ sg. (1 sh. $1\frac{4}{5}$ d.); No. 8. 5 quart ($1\frac{1}{4}$ gallons) 15 sg. (1 sh. 6 d.); Hambg. stewing pans No. 8. $4\frac{3}{4}$ quart ($1\frac{3}{16}$ gallons) $15\frac{1}{2}$ sg. 1 sh. $6\frac{3}{5}$ d.); No. 12. $6\frac{1}{2}$ quart ($1\frac{5}{8}$ gallons) 21 sg. (2 sh. $1\frac{1}{5}$ d.); Berlin stewing-pans No. 6. 4 quart (1 gallon) $9\frac{3}{4}$ sg. ($11\frac{7}{10}$ d.); frying-pans No. 3. $2\frac{1}{4}$ quart ($\frac{9}{16}$ d.) 10 sg. (1 sh.); No. 4. $3\frac{1}{2}$ quart ($\frac{7}{8}$ gallon) 13 sg. (1 sh. $3\frac{3}{5}$ d.); No. 5. $4\frac{1}{4}$ quart ($1\frac{1}{16}$ gallons) 15 sg. (1 sh. 6 d.); No. 10. $1\frac{1}{2}$ quart ($\frac{3}{8}$ gallon) $7\frac{3}{4}$ sg. ($9\frac{3}{10}$ d.); flat pans No. 9. $3\frac{1}{4}$ quart ($\frac{13}{16}$ gallon) $11\frac{1}{2}$ sg. (1 sh. $1\frac{4}{5}$ d.); No. 10. $4\frac{1}{2}$ quart ($1\frac{1}{8}$ gallons) 14 sg. (1 sh. $4\frac{4}{5}$ d.); water-pan No. 15. 16 quart (4 gallons) $42\frac{1}{2}$ sg. (4 sh. 3 d.); bellied pots No. 4. $2\frac{3}{8}$ quart ($\frac{5}{8}$ gallon) $8\frac{1}{4}$ sg. ($9\frac{9}{10}$ d.); No. 6. $3\frac{1}{2}$ quart ($\frac{7}{8}$ gallon) $11\frac{1}{4}$ sg. (1 sh. $2\frac{1}{10}$ d.); camp-dish $\frac{3}{4}$ quart ($\frac{3}{16}$ gallon) 5 sg. (6 d.); 2 plates, each 20 sg. (2 sh.); pattern of solid cast-iron.

2065. Krieg & Tigler, manu., Wesel on the Rhine. Agts. s. No. 2060.

Iron-wires and wire-ropes.

Krzyzanowski, A., prop. of a stone-casting establishment, Posen. (s. No. 1352 a.)

2067. Iron-works of count Einsiedel at Lauchhammer near Mückenberg in Prussian Saxony. Berlin 1844 council-medal; London 1851 prize-medal; Munich 1854 council-medal; Paris 1855 bronze medal; London 1862 medal. Agts. Bauerrichter & Co., 41 Charter-house. (s. II. No. 4.)

No. 1. Chimney enclosed in a mantel-piece 10 £ 4 sh., No. 2. stove-chimney 10 £ 10 sh.; No. 3. stove-chimney 5 £ 2 sh.; No. 4. stove-chimney 6 £ 6 sh.; No. 5. cylindrical stove-chimney 4 £ 10 sh.; No. 6. hall-stove 4 £ 4 sh.; No. 7. fender 1 £ 7 sh.; No. 8. fender 2 £ 4 sh.; No. 9. fender 2 £ 7 sh.; No. 10. fender 2 £ 13 sh.; No. 11. stand for fire-irons 8 sh.; No. 12. do. for fire-irons 9 sh., all p. piece; No. 13. set of 3 fire-irons 15 sh.; No. 14. do. of 3 fire-irons 1 £ 4 sh.; No. 15. do. of 3 fire-irons 1 £ 4 sh., all p. set.

No. 16. Group of a stag and dog 1 £ 8 sh.; No. 17. group of a horse and wolf 1 £ 8 sh.; No. 18. console-table 2 £ 11 sh.; No. 19. fisher-boy 2 £ 5 sh.

No. 20. Watch-stand 2 sh. 3 d.; No. 21. ash-cup 3 sh. 6 d.; No. 22. tinder-box 4 sh. 6 d.; No. 23. tinder-box 2 sh.; No. 24. candlestick 3 sh. 6 d.; No. 25. candlestick 3 sh. 6 d.; No. 26. letter-presser 3 sh.; No. 27. inkstand 6 sh.; No. 28. do. 5 sh.; No. 29. do. 1 sh. 6 d.; No. 30. do. 4 sh.; No. 31. do. 3 sh.; No. 32. do. 3 sh.; No. 33. toilet-glass 6 sh.; No. 34. do. 4 sh. 6 d.; No. 35. candlestick with movable arm 5 sh.; No. 36. fire-pan 8 sh.; No. 37. fire-pan 15 sh.; No. 38. plate for fruits 4 sh. 6 d.; No. 39. console-glass and table 12 sh.; No. 40. console-table 11 sh.; No. 41. chandelier 12 sh.; No. 42. looking-glass frame 1 £ 16 sh.; No. 43. photograph-frame 6 d.; No. 44. do. 8 d.; No. 45. cast-iron frame with a looking-glass 1 sh. 7 d.; No. 46. photograph-frame 2 sh.; No. 47. do. 7 d.; No. 48. table 4 £; No. 49. match box 1 sh. 3 d.; No. 50. do. 1 sh. 3 d.; No. 51. chandelier 2 sh. 6 d.; No. 52. do. 2 sh. 6 d.; No. 53. hand-candlestick 2 sh.; No. 54. gas-arm 7 sh.; No. 55. stand for lamps 2 £.

No. 56. Portrait of Professor Rietschel, in relievo 9 £; No. 57. statue of Luther 21 £; No. 58. statue of Lessing 18 £.; No. 59. statue of Holbein 19 £; No. 60. group of Schiller and Goethe 35 £; all p. piece.

2068. Lenzmann, Caspar Wilhelm (C. W. Lenzmann), manu., Hagen. Agt. Heintzmann & Rochussen, 9 Friday-street, Cheapside.

A. Locks for cupboards; prices p. doz.: 1. black 2 in., 1 Th. 2 sg. (3 sh. 1 d.); 2. rough $2\frac{1}{4}$ in., 1 Th. 6 sg. (3 sh. 7 d.); 3. halffine $2\frac{1}{2}$ in., 1 Th. 12 sg. (4 sh. 2 d.); 4. with fancy key-bit, highly finished $2\frac{3}{4}$ in., 1 Th. 24 sg. (5 sh. 5 d.); 5. double locking, brass $3\frac{1}{4}$ in., 3 Th. 19 sg. (10 sh. 10 d.); 6. as usual in Switzerland 3 in., 2 Th. 26 sg. (8 sh. 7 d.); 7. do. rivetted 4 in., 3 Th. 25 sg. (11 sh. 6 d.); do. with brass wheel and key $3\frac{1}{2}$ in.; 5 Th. 13 sg. (16 sh. 4 d.).

B. Locks with free main-plate: 0. black 3 in., 1 Th. 7 sg. (3 sh. 8 d.); 1. rough $1\frac{3}{4}$ in., 26 sg. (2 sh. 7 d.); 1. fancy key-bit 2 in., 1 Th. 6 sg. (3 sh. 7 d.); 1. brass ledges $2\frac{1}{4}$ in., 1 Th. 8 sg. (3 sh. 9 d.); 2. do fine $2\frac{1}{2}$ in., 1 Th. 5 sg. (3 sh. 6 d.); 2 fancy key-bit $2\frac{3}{4}$ in., 1 Th. 15 sg. (4 sh. 6 d.); 2. do. with pin-keys $2\frac{3}{4}$ in., 1 Th. 8 sg. (3 sh. 9 d.); 2. with brass ledges, fancy key-bits 3 in., 1 Th. 24 sg. (5 sh. 5 d.); 3. brass, common key-bit $1\frac{1}{2}$ in., 1 Th. 15 sg. (4 sh. 6 d.); 3. fancy key-bit, brass 2 in., 2 Th. 4 sg. (6 sh. 5 d.).

C. Mortise locks, black, 1. fancy key-bit $2\frac{1}{4}$ in., 1 Th. (3 sh.); 1. common key-bit $1\frac{3}{4}$ in., 1 Th. 4 sg. (3 sh. 5 d.); 1. brass ledges 3 in., 1 Th. 12 sg. (4 sh. 2 d.); 2. with fine polished key, rough $2\frac{1}{2}$ in., 2 Th. (6 sh.); 3. fine 2 in., 1 Th. 3 sg. (3 sh. 4 d.).

D. Locks for chests of drawers, prices p. set: 1. black 5 in., $9\frac{1}{2}$ sg. ($11\frac{1}{2}$ d.); 2. rough $4\frac{1}{2}$ in., $11\frac{3}{4}$ sg. (1 sh. 2 d.); 3. half fine 4 in., $11\frac{1}{4}$ sg., (1 sh. $1\frac{1}{2}$ d.); 4. fancy key-bit, fine $3\frac{1}{2}$ in., $12\frac{1}{2}$ sg. (1 sh. 2 d.);

E. Locks with free main plate for chests of dra-wers: 1. rough $4\frac{1}{4}$ in., 9 sg. ($10\frac{3}{4}$ d.); 2. fine $4\frac{1}{4}$ in., $9\frac{1}{2}$ sg. ($11\frac{1}{2}$ d.);

F. Mortise-locks for chests of drawers: 1. black $4\frac{1}{2}$ in., 9 sg. ($10\frac{3}{4}$ d.); 2. rough 4 in., $9\frac{1}{2}$ sg. ($11\frac{1}{4}$ d.); 3. with brass ledges, fancy key-bit, fine $3\frac{1}{2}$ in., $11\frac{5}{8}$ sg. (1 sh. $2\frac{1}{4}$ d.).

G. Pianoforte locks, prices p. doz.: 1. with brass ledges, black $1\frac{1}{2}$ in., 1 Th. 22 sg. (5 sg. 2 d.); 2. rough $1\frac{1}{4}$ in., 1 Th. 24 sg. (5 sh. 4 d.); 3. with fancy key-bit fine 1 in., 2 Th. 3 sg. (6 sh. 3 d.).

H. 1. trunk lock varnished $3\frac{1}{2}$ in., 1 Th. 12 sg. (4 sh. 2 d.); 2. do. chest-locks 4 in., 1 Th. 15 sg. (4 sh. 6 d.).

I. Locks for travelling trunks: 1. varnished 3 in., 1 Th. 15 sg. (4 sh. 6 d.); 2. do. fine brown var-nished $2\frac{1}{4}$ in., 1 Th. 27 sg. (5 sh. 8 d.); 3. do. brass 2 in., 3 Th. 4 sg. (9 sh. 4 d.); 3. do. $1\frac{3}{4}$ in., 2 Th. 20 sg. (8 sh.).

K. 2. brass cramps for trunks 1 Th. 3 sg. (3 sh. 3 d.).

L. Berlin pad-locks with wheels: 1. varnished No. 5/0. 1 Th. 4 sg. (3 sh. 4 d.); 2. with fancy key-bit, fine No. 0. 2 Th. 11 sg. (7 sh. 1 d.); 3. with half thick pin No. 6/0. 1 Th. 18 sg. (4 sh. 10 d.); 4. quite thick pin No. 2. 2 Th. 25 sg. (8 sh. 6 d.); 5. common key-bit No. 0. 2 Th. 20 sg. (8 sh.); 6. complete No. 1. 4 Th. 4 sg. (12 sh. 5 d.); 7. heavy magazin lock, fine varnished No. 2. 5 Th. 15 sg. (16 sh. 6 d.).

M. 1. street-door lock, French, 7 in., 1 Th. 7 sg. (3 sh. 8 d.).

N. Door-lock with cornice-latch 4 in., 15 sg. (1 sh. 6 d.); 2. clapper latch $4\frac{1}{2}$ in., $16\frac{3}{4}$ sg. (1 sh. 8 d.);

O. 1. mortise door-lock, Hamburgh $3\frac{1}{2}$ in., 20 sg. (2 sh.); 2. Danzick $3\frac{1}{2}$ in., 24 sg. (2 sh. 5 d.); 5. Berlin $3\frac{1}{2}$ in., 1 Th. (3 sh.); 6. common with 1 bolt, without key 2 in., 10 sg. (1 sh.).

P. 1. cellar-door-lock, without latch $3\frac{1}{2}$ in., $7\frac{1}{2}$ sg. (9 d.).

Q. Rail-way waggon door-locks: 1. Thuringia 20 sg. (2 sh.); 2. do. with screw 20 sg. (2 sh.); 3. do. with level-bolt 20 sg. (2 sh.); 4. Hannover 15 sg. (1 sh. 6 d.); 5. Berlin with key 15 sg. (1 sh. 6 d.); 6. do. without key $14\frac{1}{4}$ sg. (1 sh. 5 d.); 6. do. with pin $14\frac{1}{4}$ sg. (1 sh. 5 d.).

R. Hinge for waggon doors $14\frac{1}{4}$ sg. (1 sh. 5 d.).

S. Locks for desks and hunting trunks, p. doz.: 1. $2\frac{1}{2}$ in., 6 Th. 5 sg. (18 sh. 6 d.); 2. $1\frac{3}{4}$ in., 3 Th. 24 sg. (11 sh. 5 d.); 3. $2\frac{1}{4}$ in., 4 Th. 28 sg. (14 sh. 10 d.).

T. Keys, prices p. gross.: 1. 4 rough 26 sg. (2 sh. 7 d.); 2. 24 fineshed 3 Th. (9 sh.); 3. 3 fine po-lished with fancy ring 13 Th. 10 sg. (2 £).

U. Window-bolts, p. gross: 1. 1 Th. 5 sg. (3 sh. 6 d.); 2. do. 1 Th. 14 sg. (4 sh. 4 d.); 3. do. 1 Th. 14 sg. (4 sh. 4 d.); 4. do. 2 Th. 2 sg. (6 sh. 2 d.).

V. Hooks for hats and cloaks, varnished, p. doz. 17 sg. (1 sh. 8 d.).

W. Bedstead mortise-hooks, black p. 12 sets each 8 pair, $2\frac{1}{4}$ in., 24 sg. (2 sh. 4 d.); 2 do. white, p. 12 sets 26 sg. (2 sh. 7 d.).

X. Handles, p. doz. pair 5 in., 2 Th. 2 sg. (6 sh. 2 d.).

2069. LIEB'L, A., copper-smith, Berlin. Gollnowstr. 19. Agts. Lion M. Cohn, Phaland & Dietrich, represented by Ch. Trübner, 20 St. Dunstans Hill, City, and at the Exhib. Building.

A plated dyeing copper No. 134. to be heated by steam 500 Th. (75 £); a plated dyeing copper No. 135. to be heated by fire 300 Th. (45 £).

2071. LUDWIG, BROTHERS, manu., Breslau. London 1862 honor. mention.
Curry-combs.

2072. MACIEJEWSKI, STANISLAUS, lock-smith. Agts. s. No. 2069.
Mortise-gate-lock 18 Th.

MANSFELD ASSOCIATION. (s. No. 786, 2090.)

2074. MERTINS, C. P., zinc-founder, Ber-lin, Lindenstr. 90. London 1862 honor. mention. Agt. Oliver Kerkman, per Adr. G. Spill & Co., 149 Cheapside.
2 lying lions of cast zinc and bronzed 60 £; 1 Eng-lish coat of arms with crown and motto, cast zinc, gilt 15 £.

2075. MEVES, A., manu., Berlin, Chaussée-str. 86. London 1862 medal.
Group of cast zinc, chased; silver chess-board and figures, chased.

2076. MULACK, HEINRICH, tin-man, Ber-lin, Kurstr. 21. London 1862 medal. Agts. s. No. 2069.
Large gothic church-window 140 Th. (21 £); go-thic cross-flower, ornament for the spire of a steeple 20 Th. (3 £); pressed vase with socle 12 Th. (1 £ 16 sh.).

2077. NELCKE, OTTO, engraver, Berlin, Krausenstr. 41. London 1862 honor. mention. Agts. s. No. 2069.
Five engraved steel stamps No. 1. 32 Th. (4 £ 16 sh.); No. 2. 30 Th. (4 £ 10 sh.); No. 3. 28 Th. (4 £ 4 sh.); No. 4. 15 Th. (2 £ 5 sh.); No. 5. 12 Th. (1 £ 16 sh.); 1 engraved brass-plate No. 6. 25 Th. (3 £ 15 sh.); No. 2. and 4. and No. 3. and 5. belong together.

2078. NEUMANN, L., manu., Königsberg. London 1862 honor. mention. Agts. s. No. 2069.
Fire-proof iron safe 650 Th. (100 £).

2079. NOELLE, BROTHERS, manu., Lüden-scheid. London 1862 honor. mention. Agts. Heintz-mann & Rochussen, 9 Friday-street, Cheapside.
Spoons, soup-ladles, forks, tobacco-boxes, cocks, squirts of »Britannia-metal«.

2080. OHLE'S, E. F., HEIRS, manu., Breslau. Latches and escutcheons.

2080 a. PASCHMANN, J., turner, Hörstgen near Moers (Düsseldorf).
Sample card of key-hole lids, drawer knobs and door-handles.

2081. PETERS, F., manu. of zinc-ornaments, Berlin. Berlin 1844 bronze medal; Munich 1854 honor. mention; Paris 1855 honor. mention; London 1862 medal. Agts. s. No. 2069.
A gothic church-window in embossed zincwork 15 £.

2082 a. PINTUS, I., & Co., iron-founders and mechanicians, Brandenburg on the Havel near Berlin. Paris 1855 2 bronze medals; London 1862 honor. mention. Agts. A. Heintzmann & Ro-chussen, 9 Friday-street, Cheapside. (s. No. 1332.)
4 Royal Prussian post office weighing scales viz: for letters $2\frac{1}{2}$ Th. (7 sh. 6 d.); for small parcels 10 Th. (1 £. 10 sh.); for middling do. 18 Th. (2 £ 14 sh.); for large do. 26 Th. (3 £ 18 sh.); Pro-fessor Schönemann's patent platform weighing ma-chine 22 Th. (3 £ 6 sh.).

2083. POHL, H. (H. Pohl & Co.), Berlin, Alte Jacobstr. 21. London 1862 medal. Agt. J. F. Chinnery, 67½ Lower Thames-street. (s. No. 1359.)

Table of cast zinc with marble top surrounded with small statues and groups of galvanized cast zinc; gothic lustre for gas, galvanized 133 Th. 10 sg. (20 £); 2 figures 5 ft. high, harvest-maid and fisher-maid, galvanized, p. piece 80 Th. (12 £); 2 figures, 5 ft. high, bearing candle-sticks for gas, galvanized, p. piece 80 Th. (12 £); 1 large vase, cast zinc, colored, 600 Th. (90 £); 2 large stags, galvanized, p. piece 666 Th. 20 sg. (100 £).

2084. POKORNY, J. A., mechanician, Berlin, Oberwallstr. 17. Agts. Lion M. Cohn, Phaland & Dietrich, represented by Ch. Trübner, 20 St. Dunstans Hill, City, and at the Exhib. Building.

Show-card containing: 29 different scale-beams; 1 fine balance for 100 lb. weight, the tongue, pointing down-wards 5 £ 5 sh.; 1 do. with brass beam for 40 lb. 2 £ 14 sh.; 1 tareing balance with chest and holder 1 £ 10 sh.; 1 tareing balance with chest, holder and stopper, the tongue pointing downwards 2 £ 2 sh.

2085. PUTH, H., manu., Blankenstein (Arnsberg). London 1862 honor. mention. Agts. Heintzmann & Rochussen, 9 Friday-street, Cheapside, E. C.

Rope of cast-steel wire.

2086. REIMANN, L., mechanician, Berlin. Berlin 1844, London 1851, Munich 1854 honor. mention.

Chemical balance, when loaded with 500 grammes, indicating ½ milligramme 80 Th. (12 £, 300 frs.); chemical balance, indicating $\frac{2}{10}$ milligramme, when loaded with 100 grammes 60 Th. (9 £, 225 frs.); chemical balance, indicating $\frac{2}{10}$ milligramme, when loaded with 50 grammes 40 Th. (6 £, 150 frs.); set of gramme-weights of German silver and platina, from 500 grammes to 1 milligramme, p. set 16 Th. (2 £ 8 sh., 60 frs.); set of gramme-weights of brass and platina, from 100 grammes to 1 milligramme, p. set 9 Th. (1 £ 7 sh., 33¾ frs.); set of gramme-weights of brass and platina from 50 grammes to 1 milligramme, p. set 8 Th. (1 £ 4 sh., 30 frs.); 20 pCt. discount to the trade.

2090. RÖTHENBURG ON THE SAALE, ADMINISTRATION OF THE COPPER WORK of the Mansfeld-Association whose domicil is Eisleben. Berlin 1844 silver medal; London 1862 medal. Agts. s. No. 2084. (s. No. 786.)

a. A vacuum pan 8 ft. 10 in. wide, 30⅞ in. deep, with a flange of $3\frac{2}{16}$ in., weight 1724 pounds 48⅓ Th. (6 sh. 11 sh.); b. a vacuum pan 8 ft. 10 in. wide, 25⅞ in. deep. with a flange of $3\frac{2}{16}$ in., weight 1833 pounds, 48⅓ Th. (6 £ 11 sh.); c. a plate 33 ft. long, 5 ft. 6 in. broad, weight 1290 pounds 40⅓ Th. (5 £ 9 sh. 6 d.); d. a caldron 41⅛ in. wide, 27⅞ in. deep, weight 143,3 pounds, 43⅓ Th. (5 £ 18 sh. 9 d.); e. a caldron, 41⅛ in. wide, 27⅞ in. deep, weight 143,86 pounds, 43⅓ Th. (5 £ 18 sh. 9 d.); prices p. Zollz. (p. 100 pounds). All the articles made of Mansfeld-copper.

2091. SCHAEFFER & WALCKER, manu., Lindenstr. 19. London 1862 honor. mention. Agts. R. Schomburg & Co., 90 Cannon-street E. C.

Lusters with six branches No. 323. 91½ Th. (14 £ 10 sh.); sconce No. 340. 40 Th. (6 £); chandelier No. 353. 45 Th. (7 £ 15 sh.); lyre No. 316. 22⅚ Th. (3 £ 6 sh.); do. No. 327. 20⅓ Th. (3 £ 2 sh.); do. No. 328. 10⅓ Th. (1 £ 15 sh.); luster with six branches No. 307. 93⅓ Th. (14 £); do. with ten burners No. 331. 86 Th. (13 £); do. with six branches No. 318. 50 Th. (7 £ 10 sh.); do. with five branches No. 319. 90⅓ Th. (13 £ 10 sh.); large luster No. 350. 550 Th. (82 £ 10 sh.).

2095. SCHMALZ & SIMSON, Magdeburg, manufactory Suhl. London 1862 medal. Agts. s. No. 2084.

Specimens of hardwares.

2097. SZWARE, ROBERT, locksmith, Posen Agts. s. No. 2084.

Safety-lock for doors 10 Th.

2100. SOMMERMEYER & Co., manu., Magdeburg, manufactory Neustadt-Magdeburg. London 1851 prize-medal; Munich 1854 medal of honor; Paris 1855 silver medal; London 1862 medal. The exhibitors own agent at the Exhibition Building.

No. 6a. fire-proof iron-safe, the doors to be locked double hermetically with improved bramah- and newly constructed chub-locks, 1800 Pfd. 50 £; No. 6b. fire-proof iron-safe with partitions, for cash and books, construction the same as above, 2000 Pfd. 60 £.

2101. SPINN, J. C., & SON, manu., Berlin. London 1862 honor. mention.

Chandeliers, 64 candles, varnished bronze, with crystal-glass, Greek style, 5 ft. diam., 604 Th. (90 £ 12 sh.); 30 candles, varnished bronze, style rococo, 3½ ft. diam., 62 Th. (9 £ 6 sh.); 30 candles, gilt bronze, Chinese style, 3¼ ft. diam. 317 Th. (47 £ 10 sh.); 18 candles, varnished bronze, gothic style, 2⅔ ft. diam., 59 Th. (8 £ 17 sh.); 16 candles, varnished bronze, rococo style, 2½ ft. diam., 27⅔ Th. (4 £ 3 sh.); 12 candles, varnished bronze, rococo style, 2½ ft. diam., 23½ Th. (3 £ 10 sh.); gas-chandelier, 6 candles, varnished bronze, with drops etc., Greek style, 2⅔ ft. diam., 85 Th. (12 £ 15 sh.); 3 candles, zinc, gilt and painted, Alhambra style, 2¼ ft. diam., 42 Th. (6 £ 6 sh.); all p. piece.

2102. STOBWASSER, GUSTAV (C. H. Stobwasser & Co.), manu., Berlin. London 1851 prize-medal; Munich 1854 great medal; Paris 1855 silver and bronze-medal; London 1862 juror for Cl. XXXI. Agt. E. G. Zimmermann, 2 St. Pauls Buildings, Little Carter Lane, E. C.

A three-folded door screen with oil paintings in plates of imitation-malachite and gilt frames 1500 Th. (225 £); 139 various colza- and paraffine-oil-lamps, table-, wall- and hanging lamps, bronzed and gilt; 34 lackered, brass and german-silver-waiters and trays of different patterns; see printed price-books in the hands of the above-named agent.

COUNT STOLBERG-WERNIGERODE, Ilsenburg. (s. No. 2053.)

2103. STROBEL, GUSTAV, modeller, Frankenstein (Breslau). Agt. R. Schomberg, 90 Cannon-street, City.

A crucifix of metal after Benvenuto Cellini; original found A. D. 1861 in the monastery-church of Heinrichau, 8 £.

2105. TURCK, WILHELM & JULIUS (P. C. Turck, widow), manu., Lüdenscheid. London 1851 honor. mention; Munich 1854 medal of honor; Paris 1855 honor. mention; London 1862 medal. Agts. Schrage & Renninger, 30 Iron-monger Lane, Cheapside.

Samples of coat-, waistcoat-, cuff-, solitaire- and all other sorts of buttons of metal partly with different ornaments in mother-of-pearl, tortoise-shell, beads, glass-stones, enamel and other fancy-stuffs; samples of brooches, girdle- and garter-clasps, buckles in different sorts of metal partly with fancy-decorations; samples of match-boxes in metal partly with leather and calico; samples of upholsterers-nails, trunk-rivets, key-hole-lids and picture-nails wall- and furniture-buttons partly with inlaid stones; the prices are annexed to the articles.

2106. UHLHORN, GERHARD (D. Uhlhorn), counsellor of commerce and manu., Greven-broich. Berlin 1844 bronze medal; Munich 1854 medal of honor; Paris 1855 bronze medal; London 1862 honor. mention. Agts. Morgan Brothers, 21 Bow-Lane, Cannon-street, City.

198 samples of cards for combing wool, shoddy, cotton, silk, flax and tow, for teasling, dressing and finishing cloth, for cutting hare-wool, which cards are made of leather, artificial leather and patent card cloth set with round, triangular tinned brass-wire.

2107. VARENKAMP, FRIEDRICH, blacksmith, Düsseldorf. London 1862 medal. Agts. Lion M. Cohn, Phaland & Dietrich, represented by Ch. Trübner, 20 St. Dunstans Hill, City, and at Exhib. Building.

Horseshoes with two explanatory descriptions, 26 pieces 15 £.

2108. WILD & WESSEL, lamp-manu., Berlin. (s. II. No. 48.)

Lamps and burners:

Moderator-lamps: bronze No. 99. 1 £ 7 sh.; No. 60. 18 sh.; No. 58. 7 sh. 6 d.; No. 6. 10 sh. 6 d.; porcelain No. 106. 1 £ 5 sh. 6 d.; bronze No. 102. 15 sh.; porcelain No. 43. 13 sh.; glass No. 109. 10 sh.; bronze No. 100. 8 sh. 6 d.; porcelain No. 101. 11 sh. 6 d.; glass No. 101. 6 sh. 9 d.; bronze No. 4. 3 sh. 6 d.

Sliding-lamps: No. I. 4 sh. 10 d.; No. II. 4 sh. 6 d.; No. III. 4 sh. 2 d.; No. IV. 3 sh. 9 d.; No. V. 3 sh.; No. I.V. 5 sh. 10 d.; No. 10. 3 sh. 2 d.; No. 93. 3 sh.

Paraffine-lamps: No. 46. 9 sh.; No. 90. 2 sh. 5 d. Paraffine-burners: universal 4 sh. 9 d.; round No. 14. 2 sh. 9 d.; No. 14. 2 sh. 6 d.; No. 10. 3 sh.; No. 10. 1 sh. 8 d.; flat No. 10. 1 sh. 10 d.; No. 10. 1 sh. 6 d.; No. 7. 1 sh. 8 d.; No. 7. 11 d.; No. 7. 1 sh. 1 d.; improved No. 7. 8 d.; No. 5. 6 d.; prices p. piece.

CLASS XXXII.

STEEL AND CUTLERY.

2109. BEISSEL, STEPHAN, WIDOW AND SON, Aix-la-Chapelle. London 1851 prize-medal; Paris 1855 bronz. Med.

Needles; according to price-current.

2110. BÖLLING & VON DER CRONE, manu., Haspe. London 1862 honor. mention. Agts. s. No. 2107.

36 pieces of scythes, sickles, straw-knives.

2111. CORTS, GOTTLIEB, Remscheid. Munich 1854 prize-medal; Paris 1855 silver medal; London 1862 medal. Agts. s. No. 2107.

190 files and rasps.

2112. DICKERTMANN, BROTHERS, manu., Bielefeld. London 1862 honor. mention. Agt. H. Landwehr, Exhibition Buildings.

18 files of different kinds p. lb. 9½ d.

2113. EDELHOFF, J. W., & Co., manu., Remscheid. Manufactory Bremen near Remscheid. London 1862 honor. mention. Agts. s. No. 2107.

Saws, files and a machine straw-knife.

2115. JUNG, CARL, merchant and manu., Dahl near Hagen. Paris 1855 honor. mention. Agts. Heintzmann & Rochussen, 9 Friday-street, E. C.

Card I. 2 small cast-steel scythes for Germany, 5½ and 6 spans 8 Th. (24 sh.); card II. 2 cast-steel corn scythes for Russia, 5 and 5½ spans 8 Th. (24 sh.); card III. 2 polished and varnished cast-steel scythes for Denmark and America, 5½ and 6 spans 9 Th. (27 sh.); card IV. 2 scythes for Holland and Belgium, 36 and 30 in. 6 Th. (18 sh.); card V. 1 straw-knife of cast-steel 8½ Th. (25 sh. 6 d.); 1 do. of mint-steel for Germany, 24 in. 7 Th. (21 sh.); card VI. 1 polished and varnished cast-steel straw knife for Denmark, 21 in. 9 Th. (27 sh.); card VII. 2 cast-steel reaping sickles for Germany, Holland and Belgium; prices I.—VII. p. doz. 7 Th. (21 sh.)

Card VIII. 2 blue steel scythes for Germany, 4½ and 5 spans 46⅔ Th. (7 £); card IX. 2 do. for Russia, 6½ and 8 hands 30 Th. (4 £ 10 sh.); card X. 1 do. for Spain, 30 in. 38⅓ Th. (5 £ 15 sh.); 1 yellow cast-steel scythe for France, 34 in. 76⅔ Th. (11 £ 10 sh.); card XI. 2 do. for Italy and Switzerland, 34 in. 80 Th. (12 £); card XII. 1 do. for America, 37 in. 80 Th. (12 £); 1 blue steel scythe for America, 37 in. 46⅔ Th. (7 £); prices VIII.—XII. p. 100 pieces.

1 anvil with 2 horns and 1 small jumping anvil for rail-way workhouses 100 pound 8½ Th. (1 £ 8 sh. 3 d.)

2116. KLEB, W., manu., Allenbach (Arnsberg).

Hatchet; hoe; axe.

2116a. KRUPP, FR., Essen. (s. No. 1308.)

2117. LAMMERTZ, LEO, needle-manu., Aix-la-Chapelle. London 1862 medal. Agts. s. No. 2107.

An assortment of sewingmachine and other needles, containing: sewing machine needles, quality 32., 300 sg. (1 £ 10 sh.); do. 51., 600 sg. (3 £); do. 41., 900 sg. (4 £ 10 sh.); do. 61., 1200 sg. (6 £); drill'd ey'd needles, quality 104., 40 sg. (4 sh.); do. 107., 30 sg. (3 sh.); do. 120., 20 sg. (2 sh.); do. 123., 13½ sg. (1 sh. 4 d.); for gloves, quality 953., 40 sg. (4 sh.); do. 954., 30 sg. (3 sh.); do. 955., 20 sg. (2 sh.); do. 956., 15 sg. (1 sh. 6 d.); darning needles, quality 1057., 50 sg. (5 sh.); all p. thousand.

2120. MANNESMANN, A., manu., Remscheid. London 1851 prize-medal; Munich 1854 great medal; Paris 1855 gold medal of honor; London 1862 medal. Agts. Heintzmann & Rochussen, 9 Friday-street, Cheapside, E. C.

Files in every size and cut from double smooth to rough rough, from two in. upwards to twenty four in. length and from two pounds to sixty pounds weight; rasps in all used forms, sizes and cuts.

OFFENHAMMER, Berlin, files and edge-tools. (s. No. 1279.)

2122. Printz, Georg, & Co., manu. of needles, Aix-la-Chapelle. Munich 1854 prize-medal; Paris 1855 silver medal; London 1862 honor. mention. Agts. J. & S. W. Green (E. G. Zimmermann), 2 St. Pauls Building, Little Carter Lane, E. C.

A card with samples of needles; a case containing 251 thousand of different needles in 60 parcels. For prices apply to the agents.

2124. Reinshagen, G., manu., Remscheid. London 1862 medal. Agts. Heintzmann & Rochussen, 9 Friday-street, Cheapside, E. C.

Files and rasps.

2125. Schleicher, Carl, needle-manu., Düren. Manufactory at Schönthal near Aix-la-Chapelle. London 1851 prize-medal; Paris 1855 silver medal and cross of the legion of honor; London 1862 medal. Agt. James Wm. Green, 2 Little Carter Lane, Doctors Commons.

Samples of patented needle points; needles, fixed on cards and in papers, each containing 25 pieces.

2126. Scmumacher, F., & Co., manu., Aix-la-Chapelle. London 1862 honor. mention. Agts. s. No. 2124.

Glass breast-pins; glass-buttons, steel and brass-pins; needles.

2129. Werninghaus, J. C., merchant and manu., Hagen. Agts. s. No. 2124.

Two knives for a rag engine 20 Th. (2 £ 14 sh.), 18 Th. 10 sg. (2 £ 9 sh. 6 d.); each p. 100 lb.

2130. Witte, Stephan, & Co., needle-manu., Iserlohn. Berlin 1844 silver medal; Paris 1855 silver medal; London 1862 medal. Agts. s. No. 2124.

Samples of needles: 1. best gold gutter needles double drill'd sharps, blunts and betweens 6 sh.; 2. silver eye-needles, sharps, blunts and betweens, 7 d. p. thousand each.

CLASS XXXIII.

WORKS IN PRECIOUS METALS AND THEIR IMITATIONS AND JEWELLERY.

2131. Beckmann, A., manu., Elberfeld. London 1862 honor. mention.

Gold and silver ware.

2132. Brandt, E. F. & H. (Friedrich & Brandt), manu., Stettin.

Silver drinking-cup; butter-box; salt-cellar; pepper-box; cane-head and knop for whip-handle.

2134. Friedeberg, S., Söhne. goldsmiths, jewellers etc. to their Majesties the King and the Queen and His Royal Higness the Prince Royal and the Princess Royal of Prussia. Berlin, Unter den Linden 42. Paris 1855 bronze-medal; London 1862 medal. Agts. Lion M. Cohn, Phaland & Dietrich, represented by Ch. Trübner, 20 St. Dunstans Hill, City, and at the Exhib: Building.

A table-service in silver with small statues of the heroes Dessauer, Zieten, Seydlitz, Blücher, Bülow, Scharnhorst, York, Gneisenau, Kleist, Schwerin.

A baptismal-basin with the figures of St. John and Christ 1200 Th. (180 £).

2135. Gosche, H., manu., Berlin, Stall-schreiberstr. 9.

Chased silver work.

2136. Graeger, Klug & Hartung, manu., Mühlhausen. Agts. s. No. 2134.

A picture, silvered cast-iron, representing the Lord's supper 13 Th. 10 sg. (2 £); a cushion mounted with silvered iron nails; a box with samples of silvered iron bullen-nails p. mill No. 0. 1 Th. 5 sg. (3 sh. 6 d.); No. 1. 1 Th. 2 sg. (3 sh. 2⅔ d.); No. 2. 29 sg. (2 sh. 10½ d.); Nr. 3. 26 sg. (2 sh. 7⅓ d.); No. 4. 23½ sg. (2 sh. 4⅓ d.).

2137. Köhler, August, manu. of silver-wares, Liegnitz. Munich 1854 medal of honor. Agts. s. No. 2134.

Laurel wreath of silver; assortment of silver- and gilt silver-jewellery.

2138. Löwenthal, A. M., & Co., manu., Cologne.

Plated table-service.

2139. Mosgau, F., manu., Berlin, Markus-str. 49.

Assortment of silversmith's ware.

2140. Schütz & Hoffmeister, manu. of hard ware and carriage-lanterns, Berlin, Linden-str. 112. Agts. E. G. Zimmermann, 2 St. Pauls Buildings, Little Carter Lane, Doctors Commons, E. C.

Silver-plated carriage-lanterns; prices p. pair; dodecagonal 20 Th. (3 £); hexagonal 18 Th. (2 £ 14 sh.); oval with glass corners 18 Th. (2 £ 14 sh.).

2141. Schwartz, C., jeweller, Berlin, Mohrenstr. 26. Agt. Zimmermann, 5 Claremont-Square.

Bracelets, watch-chains and other objects in gold; busts of eminent men in fine silver.

2142. Sy, Louis, Wagner, Albert (Sy & Wagner, form. G. Hossauer), goldsmiths to His Majesty the King of Prussia, Berlin. London 1851 council-medal, 1862 medal. Agts. s. No. 2134.

1. In the possesion of their Royal-Highnesses the Prince and Princess Royal of Prussia: a. a shield wrought in frosted silver with arms inlaid in enamel, a wedding present from the nobility of the Rhenish provinces. The four reliefs represent the four Rhenish provinces historically depicted offering their congratulations to the Royal couple. The arms are those of the donors; b. a silver dish with silver enamelled; c. a. jug and two goblets in silver; d. a large goblet in silver and gilt.

2. Wedding presents from the 27 departements of Hessia, in the possession of the Grand-duke of Hesse-Darmstadt, consisting of a dinner-service, viz: a. a table-service; b. four chandeliers with thirteen branches; c. four silver dishes; d. two goblets with stand.

3. A gilt tankard with stand, ornamented with stag-heads, in the possession of the Prince of Lippe Schaumburg.

4. A table-service (silver dish); 5. a letter-presser; 6. four communion-cups; 7. two jugs for the altar; 8. a gilt drinking-cup; 9. a tea- and coffee-service in arabesque; 10. a table-service ornamented with hunting scenes; 11. a vase; 12. all the Prussian orders of knighthood.

2144. Vollgold, D., & Son, manu. of silver-ware, Berlin, Kommandantenstr. 14. Paris 1855 silver medal and great medal of honor; London 1862 medal.

Present of the city of Berlin to their Royal Highnesses the Prince and Princess Royal of Prussia in commemoration of their marriage.

2145. WINKELMANN, H. F., manu. of German silver, Berlin. Agts. Lion M. Cohn, Phaland & Dietrich, represented by Ch. Trübner, 20 St. Dunstans Hill, City, and at the Exhib. Building.

An assortment of corks with animals and figures; prices p. doz.: simple 5 sh. 9 d.; double 9 sh.; threefold 12 sh.; oxydized simple 6 sh.; gilt simple 7 sh. 6 d.; knife-banks in several forms (balls) 5 sh.; (the same) in several forms (trunk of a tree) 6 sh.; rings for napkins 13 sh. 6 d.; salt- and pepper-box 2 £ 8 sh.; breast-pins (brooches) 7 sh. 6 d.; salt-shovels 5 sh.

CLASS XXXIV.

GLASS.

2147. PLATE-GLASS COMPANY ST. GOBAIN, CHAUNY & CIREY, by Biver, director, Aix-la-Chapelle, manufactory Stolberg near Aix-la-Chapelle. Paris 1855 silver medal; London 1862 medal. Agt. Aug. de Grand-Ry, Pauls Wharf, 25 Upper Thames-street, E. C.

Not silvered framed plate-glass, height 438 cm. or 172 in. Engl., breadth 298 cm. or 117 in. Engl., 13 sq. mètres or 140 Engl. sq. ft. 930 frs. (37 £); silvered framed plate-glass, height 345 cm. or 136 in. Engl., breadth 219 cm. or. 86 in. Engl., $7\frac{1}{2}$ sq. mètres or 81 sq. ft. Engl. 760. frs. (30 £).

2148. HAARMANN, SCHOTT & HAHNE, manu., Witten (Arnsberg). Agts. Heintzmann & Rochussen, 9 Friday-street, Cheapside.

Plate-glass.

2149. HECKERT, CARL FERDINAND UND FRIEDRICH WILHELM (Carl Heckert), academical-artists and manu., Berlin. Paris 1855 bronze-medal; London 1862 medal. Agt. E. G Zimmermann, 2 St. Pauls Buildings, Little Carter Lane, Doctors Commons.

1. 1 large lustre for 36 candles with leaves and flowers of plain cut crystal glass; 2. 6 small lustres, the same species differing only in size and pattern; 3. 1 large lustre for 42 candles with colored leaves and flowers of glass; 4. 5 small lustres, the same species differing only in size and pattern; 5. 4 suspended lamps in plain cut crystal glass; 6. 2 suspended lamps with flowers and leaves of colored glass; 7. 1 pair of branch candle-sticks for 6 candles in plain cut crystal glass; 8. 1 pair of branch candle-sticks with leaves and flowers of colored glass; 9. 2 pair of sconces in plain cut crystal glass; 10. 4 pair of sconces with flowers and leaves of colored glass; 11. 4 table-services with flowers and leaves of colored glass; 12. 1 looking-glass with flowers and leaves of colored glass; 13. 2 looking-glasses in silvered plain cut crystal glass; 14. 1 looking-glass in milk-glass with beads of colored glass; 15. different articles in plain cut crystal glass combined with bronze, as: candle-sticks, clocks, inkstands, thermometers, letter-pressers, vases, albums, caskets, casters etc.; 16. the same articles in plain cut crystal imitated malachite; 17. the same articles in plain cut crystal black imitated wood etc. with gold; 18. the same articles in milk-glass with beads of colored glass; 19. vases, cups, candle-sticks etc. in Greek style in painted milk-glass combined with gilt bronze; 20. the same articles, as specified under No. 15. with leaves and flowers of colored glass; 21. 1 chess-board in crystal; 22. 1 large looking-glass frame and console in Mauresque style; 23. 1 large looking-glass in silver crystal glass, renaissance-style, the plate of the console in crystal, imitated lapis lazuli; 24. samples of letters in glass for firms; 25. do. of glass slates with leather frames; 26. do. of curtain-holders in painted, cut and gilt glass.

For prices and drawings apply to the agent.

2150. HECKERT, EDUARD, manu., Halle on the Saale. Agt. J. W. Green (E. G. Zimmermann), 2 St. Pauls Buildings, Little Carter Lane, St. Pauls Churchyard, South Side.

1 window with stained and figured panes of glass. NB. All the colors are burnt into the glass. 13 specimens and panes of figured glass, p. □ ft. 8 to 30 sg. ($9\frac{2}{3}$ d. to 3 sh.).

2152. VON KLITZING'SCHE GLASS-MANUFACTORY, by W. Barth, Bernsdorf (Liegnitz). (s. No. 1425.)

Concave glass; prices p. doz.: Geweck lamp-chimneys No. 1., 2., 3., 4., 5., $12\frac{1}{2}$ sg. (1 sh. 3 d.); Benkler chimneys $1\frac{7}{16}$ in. $1\frac{7}{8}$ in. 9 sg. ($10\frac{4}{5}$ d.); lantern chimneys 6 in. long, $1\frac{5}{8}$, $1\frac{13}{16}$, $1\frac{3}{4}$, $2\frac{1}{8}$ in. wide 8 sg. ($9\frac{3}{5}$ d.); French chimneys No. 1., 17., 21. 9 sg. ($10\frac{1}{5}$ d.); camphin-chimney $\frac{7}{12}$ in. burner $7\frac{3}{4}$ in. long, $1\frac{1}{4}$ in. diam. 9 sg. ($10\frac{4}{5}$ d.); oblong belly-chimney $10\frac{3}{4}$ in. long, $1\frac{11}{16}$ in. wide 10 sg. (1 sh.); gas-chimney annealed 7 in. long, $1\frac{7}{8}$ in. wide 12 sg. (1 sh. $2\frac{2}{5}$ d.); do. 8 in. long, $1\frac{3}{4}$ in. wide 12 sg. (1 sh. $2\frac{2}{5}$ d.); do. bluish 9 in. long, $1\frac{7}{8}$ in. wide 12 sg. (1 sh. $2\frac{2}{5}$ d.); gas-chimney smoothed 9 in. long, 2 in. wide 10 sg. (1 sh.); do. 9 in. long., $1\frac{7}{8}$ in. wide 10 sg. (1 sh.); bent chimneys No. 10., 15., 20. 9 sg. ($10\frac{4}{5}$ d.); Franconian milkwhite lamp shades, plain $5\frac{7}{8}$ in. wide $1\frac{2}{3}$ Th. (5 sh.); do. $5\frac{3}{8}$ in. wide $1\frac{1}{2}$ Th. (4 sh. 6 d.); do. $4\frac{7}{8}$ in. wide $1\frac{1}{3}$ Th. (4 sh.); do. high $5\frac{7}{8}$ in. wide $1\frac{2}{3}$ Th. (5 sh.); do. pressed $5\frac{7}{8}$ in. wide 2 Th. (6 sh.); pressed sugar-loaf-shaped shades $4\frac{1}{2}$ Th. (13 sh. 6 d.); do. with rim 6 Th. (18 sh.); camphin-vases, $1\frac{1}{2}$ pound $1\frac{1}{3}$ Th. (4 sh.); do. milkwhite belly 6 in. diam. $1\frac{2}{3}$ Th. (5 sh.); do. 1 pound $1\frac{1}{2}$ Th. (4 sh. 6 d.); large lantern ball $1\frac{2}{3}$ Th. (5 sh.); do. middle sized $1\frac{1}{2}$ Th. (4 sh. 6 d.); pear-shaped do. 1 Th. (3 sh.).

The following prices p. 100 pieces. White medicin glasses $\frac{1}{2}$ Loth $1\frac{1}{6}$ Th. (3 sh. 6 d.); 1 Loth $1\frac{1}{6}$ Th. (3 sh. 6 d.); 6 Loth $1\frac{5}{6}$ Th. (5 sh. 6 d.); 10 Loth 2 Th. (6 sh.); 14 Loth $2\frac{1}{2}$ Th. (7 sh. 6 d.); 24 Loth $3\frac{1}{3}$ Th. (10 sh.); perfume-glasses No. 10., 11. $1\frac{2}{3}$ Th. (5 sh.); No. 34. $1\frac{1}{2}$ Th. (4 sh. 6 d.); No. 47. 2 Th. (6 sh.); No. 61. $1\frac{5}{6}$ Th. (5 sh. 6 d.); No. 63. $1\frac{1}{3}$ Th. (4 sh.); No. 66. $1\frac{2}{3}$ Th. (5 sh.); No. 84. $1\frac{2}{3}$ Th. (5 sh.).

Hyalite stand bottle with stopper and bottom-ball p. piece 4 sg. ($4\frac{1}{5}$ d.); do. white without bottom-ball p. piece $3\frac{1}{2}$ sg. ($4\frac{1}{5}$ d.).

KOCH & BEIN, Berlin, metal- and glass-lettres. (s. No. 2059.)

2154. MÜLLENSIEFEN, BROTHERS, manu., Crengeldanz (Arnsberg). Agts. s. No. 2148.

Plate-glass.

2155. OIDTMANN, DR. HEINRICH, & Co., prop. of a glass-painting and printing establishment, Linnich (Aix-la-Chapelle). Agts. s. No. 2145.

A mosaic window printed en grisaille in encaustic colors, representing the portraits of the King and Queen of Prussia and the statue of Frederic I., height 12 ft. Rhenish, breadth. 6 ft., price p. sq. ft. 10 sg. or 1 sh.

2155 a. COUNT SCHAFFGOTSCH, manu., Josephinenhütte.

Fancy glass of every kind.

2156. SCHMIDTBORN, R., manu., Friedrichsthal near Saarbrücken. London 1862 honor. mention.

Window-glass; wine-bottles; large bottles for oil of vitriol.

2157. SCHULZE, THEODOR OSCAR (Th. O. Schulze), manu., Rauscha (Liegnitz). Agts. s. Lion M. Cohn, Phaland & Dietrich, represented by Ch. Trübner, 20 St. Dunstans Hill, City, and at the Exhib. Building.

1. A pane of doubled glass, partly red colored for windows, 32 in. sq. $2\frac{1}{2}$ Th.; 2. do. $1\frac{1}{2}$ as thick, 33 by 30 in. 4 Th.; 3. do. of yellow glass, 33 by 30 in. $1\frac{1}{2}$ Th.; 4. do. milk-white $1\frac{1}{2}$ as thick, 33 by 30 in. 6 Th.; 5. do., 33 by 31 in. 6 Th.; 6. do., 32 by 26 in. 4 Th.; 7. do. single thick, 29 by 31 in. $2\frac{2}{3}$ Th.; 8. do. green, 24 by 28 in. 1 Th.; 9. do., 23 by 28 in. 1 Th.; 10. 2 panes of violet-red glass, 22 by 28 in. $1\frac{5}{6}$ Th.; 11. 1 do. of violet-blue glass, 23 by 29 in. $\frac{3}{5}$ Th.; 12. 1 do., 22, by 28 in. $\frac{3}{5}$ Th.; 13. 2 panes of doubled glass partly blue colored, 29 by 23 in. $2\frac{2}{3}$ Th.; 14. a pane of blue glass, 29 by 22 in. $\frac{11}{15}$ Th.; 16 pieces together 5 £ 2 sh.

Twelve different milkwhite lamp-shades 3 Th.; six different lamp-shades doubled and partly milkwhite, green and blue colored $1\frac{1}{2}$ Th.; a yellow-green lamp-shade, sugar loaf-shape $\frac{5}{6}$ Th.; twelve different lamp chimneys $12\frac{1}{2}$ sg.; 31 pieces 17 sh. 5 d.

A milkwhite stained lamp-shade $1\frac{1}{3}$ Th. (4 sh.).

2158. SEILER, A., manu., Breslau.

Encaustic painted glass.

2159. COUNT OF SOLMS-BARUTH, FRIEDRICH HEINRICH LUDWIG (Count Solms glass-works administration), Baruth near Potsdam. London 1862 medal. Agts. s. No. 2157.

Show-board with samples of colored table glass and pictures, escutecheons etc. cut in double glasses; 28 pieces of large colored panes viz: 3 simple and 1 double opaque milk-glass pane, 7 lilac and violet from light to dark, 3 blue, 3 green, 3 yellow, 3 red, 1 aquamarine, 1 flesh-colored, 1 flesh-colored by lilac and yellow doubling, 2 scaled and one common pane; 34 common lamp chimneys of different shapes for differently constructed lamps and different lighting materials; 13 colored: green, light and dark blue, red, milkwhite vases for photogen lamps; 11 pressed and ribbed milk-glass screens of different sizes and shapes for gas-, photogen-, and oil-lamps; 5 for sinumbre lamps, 2 Wageman screens, 4 plain screens for Franken- and sliding lamps, 3 balls for moderator-lamps, 1 gasmoon, 1 Franken- and Wageman doubled green screen, 2 pressed sugar loaf-shaped gas screens, green and opaque, three milkwhite lamp stands, 5 gas pipes with mouthpieces, 2 lamp stands of filigree glass in different colors and 2 glasses candle-stick-shaped of Venetian work.

Adress cards and price lists for colored, table glass and lamp glasses are laid out.

2160. STRILACK, A., manu., Waitze (Posen).

Glass.

CLASS XXXV.

POTTERY.

2164. AUGUSTIN, A., earthen-ware-manu., Lauban. Agts. Lion M. Cohn, Phaland & Dietrich, represented by Ch. Trübner. 20 St. Dunstans Hill, City, and at the Exhib. Building.

Conduit pipe of clay, 6 in. diam., 4 ft. long 1 Th. 4 sg. (3 sh. $4\frac{4}{5}$ d.); do. 16 in. diam. $2\frac{1}{2}$ ft. long. 2 Th. 20 sg. (8 sh.) p. piece each.

2166. BERLIN ROYAL PORCELAIN-MANUFACTORY. Berlin 1844 prize-medal; London 1851 prize-medal; Munich 1854 great medal; Paris 1855 great medal of honor; London 1862 medal. Agt. August Pils, warehouse-officer of the Royal porcelain-manufactory. (s. II. No. 100.)

I. Vases and decorating vessels: 1. two vases, amphora-shape, about 5 feet high, with painted groups of elves after Steinbrueck; 2. one vase, do. with a picture of the »Borussia«; 3. one vase model Medici, with the triumphal procession of king Wine, after Schrötter; 4. one do., with »Nausikaa and Ulysses«, after Bendemann; 5. two very large vases with female figures in the reed; 6. three vases with griffin-handles, with views of Sanssouci and its environs, seladon ground; 7. three Urbino-vases with views of Potsdam, blue ground; 8. one do. with the portrait of Her Royal Highness Victoria Princess Royal of Prussia; 9. five do. with hovering figures, after Kaulbach; 10. two do. with mermaids, platinum lustre; 11. four do. painted with views of Babelsberg castle, glazed colored ground; 12. one vase with handles like serpents, with »Goethe, Humboldt, Grimm«, after Kaulbach; 13. five Dorian vases, with colored arabesques and figures; 14. two do. ornamented with gold; two wine-coolers with children after Rubens, painted in Majolika-manner; 16. two vases, fauns as handles; 17. two do. female heads as handles, marble ground; 18. five Weimar-vases with Watteau groups, royal-blue ground; 19. four Brunswic-vases, with flowers and Cupids; 20. three Chinese-vases relievo, with painted birds; 21. two do. with views of Babelsberg in blue; 22. Chinese potpourries of different sizes and ornaments; 23. two rococo vases, light-blue with white flowers; 24. two do. biscuit with colored flowers; 25. two rococo urns with heads, light-blue and white; 26. two rococo flower-vases with Watteau groups; 27. four vases Indo-Chinese, 2 blue, 2 red and gold ornamented; 28. two Voltaire-vases with a figure on the cover; two Chinese ones, brown glazed; two flower-bottles with stoppers; eight flower-vases with different ornaments; 29. Florence-service after Benvenuto: can with swan-handle; cup with mask; cup with harpies as handles; 30. ribbed plates and lamps on stands: imitations in Giallo, Rosso, Verde antico; 31. goblet with Centaurs, ivy in relievo; do. with groups of gnomes, antique tankard with goats; 32. very large dish, style renaissance, with views of Babelsberg in the 5 fields; 33. chalice with stand, gilt.

II. Pictures on porcelain: 34. Madonna after the original by Raphael in the Royal Museum at Berlin; 35. portrait of the King Frederic II., original by Graff.

III. Plastic works in biscuit: 36. statue of Borussia with side-figures: peace and prosperity with gilding; 37. monument of King Friedrich Wilhelm III. in the park of Berlin; 38. statues of King Wilhelm I. and the Prince Royal of Prussia; 39. busts of the Prince and Princess Royal of Prussia; 40. copy of Moses of Michel Angelo in Rome; 41. copy of the fountain-monument in Kissingen: Hygiaca, Pandur and Ragoczy: 42. a Corinthian capital with figures.

IV. Fountains: 43. Boy holding a serpent upon a plate on dolphins, biscuit with gilding: 44. the same colored, on a stand of zinc; 45. girl holding a fish upon a plate on dolphins, biscuit with gilding; 46. the same colored, on a stand of zinc; 47. the gooseman of Nuremberg, biscuit.

V. Furniture: 48. a large lustre with figures and flowers; 49. a small lustre do.; 50. two sconces with flowers; 51. four candlesticks with flowers; 52. a pier-glass with medaillon and Cupids; 53. a dressing-glass with a bust and flowers; 54. a clock-case do.; 55. two consoles with flowers (all the pieces of white porcelain ornamented with colors); 56. a wooden table with porcelain leaf: Cupid and water-fairy; 57. several round table-leaves of painted porcelain; 58. two Chinese stools, ornamented in blue and yellow.

VI. Utensils for use, as: 59. ink-stands, match-boxes, watch-stands, snuff-boxes.

VII. Several figures and groups: 60. glazed and colored, partly after ancient models.

VIII. 61. Table-ornaments: Cupid and Psyche on a shell, carried by sea-deities, 2 side-pieces: Horn of plenty with nymphs and cupids, colored; the same three pieces in biscuit with gold; 62. Cupid on a plate carried by Nereids, biscuit with gold, two Tritons Barberini on a pedestal do.; 63. a large fruit plate on three female figures: Agriculture, Culture of the vine, Navigation; two plates on hippopotamus No. 1., two fruit-plates on Caryatides; four fruit-baskets on 1 figure, model Pomona, two wine-coolers with figures as handles, all ornamented blue and gold; 64. plate on hippopotami No. 2., white and golden; 65. two plates on Caryatides, green and gold; 66. four baskets, model of Pomona, green and gold; 67. four fruit-plates on groups of children, ornamented pedestals 68. two fruit-baskets carried by five boys; 69. two do. with female Chinese; 70. two do. with Flora; 71. four do. palm-tree with children No. 1.; 72. four do. No. 2., four do. No. 3.; 73. four do. No. 4. with pedestals; 74. four small fruit-baskets with wrestling children; 75. four fruit-baskets on female figures; ad 67. to. 75. colored, in rococo-style.

IX. Table-service; 76. Plain-English table-service for 12 persons; every course differently ornamented; 77. an assortment of ornamented table-services: 1. imitation of the flower-service of Frederic II.; 2. do. of the blue scales-service of Frederic II.; 3. model Flora with blue flowers; 4. model Neuosier with colored flowers; 5. model rocaille with flowers of three colors; 78. twelve various plates with fine paintings; 79. twelve plates as patterns differently ornamented; 80. twelve small plates model rocaille, with Cupids; 81. two cabarets in order to change the plates.

X. Coffee- and tea-services: 82. three Japanese coffee-services No. 1. (thin) painted: Pompeyan figures; green lustre ground; yellow lustre and gold net; 83. a coffee-service (thin) with views of Babelsberg-castle; 84. two Japanese coffee-services: with flowers on blue ground, with Indo-Chinese decoration; 85. a coffee- and tea-service, blue with gold fuchsias; 86. one do. with Watteau-figures; 87. one do. with views of Berlin, turkois-blue ground; 88. one do. with birds on blue ground; 89. one do. with flowers on chamois ground; 90. one do. with flowers and gold; 91. one do. with fuchsias and gold; 92. one do. with birds, royal-blue ground; 93. fifty cups and saucers, differently painted and ornamented; 94. coffee- and tea-urns, coffee- and tea-canisters, cake-baskets and plates of different fashion and ornament.

XI. White porcelains: 95. an assortment of pharmaceutical and chemical vessels; 96. an assortment of telegraphic insulators; 97. an assortment of lithophanies.

2167. Berlin Royal Gesundheits-Geschirr-Manufactory, Thiergarten (Berlin). London 1862 medal. Agt. August Pils, store-keeper of the Royal porcelain-manufactory at Berlin.

A. Sanitary-Chinaware (porcelain); utensils for chemists and apothecaries: 13 evaporating dishes No. 0—12. 1 £ 3 sh. 6 d.; 9 do. 6—14 in. 1 £ 1 sh. 6 d.; 8 do. 8—15 in. 1 £ 13 sh. 6 d.; 10 do. small, with spout 1 sh. 9 d.; 8 do. flat 1 sh. 3 d.; do. with wooden handle No. 1. 9 d.; do. No. 3. 1 sh.; do. No. 5. 2 sh.; jar with cover for leeches 1 sh. 6 d.; do. with lid 1 sh. 6 d.; spoon for leeches 1 sh.; skillet with handle No. 00. 4 d.; No. 0. 5 d.; No. 1. 6 d.; No. 2. 9 d.; No. 3. 1 sh.; No. 4. 1 sh. 6 d.; skillet with wooden handle: No. 00. 9 d.; No. 0. 9 d.; No. 1. 1 sh.; No. 2. 1 sh. 3 d.; No. 3. 1 sh. 6 d.; infusion-jar with wooden handle: No. 1. 2 sh. No. 2. 2 sh. 3 d.; No. 3. 2 sh. 6 d.; medicin-spoons: No. 1. and 2. each 4 d.; No. 4., 5., 6., 7. each 6 d.; measure-jugs: for 3 oz. 2 d.; for 6 oz. 5 d.; for 12 oz. 7 d.; for 24. oz. 11 d.; for 32. oz. 1 sh.; 8 mixture-mortars with spout 4 sh.; 6 do. flat with spout 5 sh. 6 d.; 17 do. with spout 2 £ 2 sh. 9 d.; powder spoon 4 d.; 2 pestles, p. piece No. 00. and 0. 1 d.; 3 do. No. 1., 2., 3. 2 d.; 3 do. No. 4., 5., 6. 3 d.; 3 do. No. 7., 8., 9. 4 d.; 3 do. No. 10., 11., 12. 6 d.; quicksilver-jar 1 sh.; pipes, 2 ft. long, ½ in. bore 2 sh. 6 d.; do. ¾ in. bore 3 sh.; do. 1 in. bore 3 sh. 6 d.; retorts for 2 oz. 6 d.; for 4 oz. 1 sh.; for 8 oz. 1 sh. 3 d.; for 16 oz. 2 sh.; do. with tubulus for 8 oz. 1 sh. 6 d.; for 24 oz. 2 d.; salve-boxes p. doz. No. 0. 4 d.; No. 6. 1 sh. 3 d.; No. 12. 3 sh.; 5 crucibles with cover 2 sh. 6 d.; crucible with lid (biscuit) for 1 oz. 2 d.; for 2 oz. 3 d.; for 3 oz. 4 d.; for 4 oz. 5 d.; for 6 oz. 6 d.; spattles, 4 in. long 5 d.; do. 9 in. 10 d.; do. 18 in. 3 sh.; dish for drying sulphuric acid 1 sh. 3 d.; filtring-funnel No. 1. 1 sh. 3 d; No. 3. 1 sh. 6 d; with large holes No. 2. 1 sh. 3 d.; filtring-funnel small No. 1. 4 d.; No. 2. 6 d.; funnel No. 3. 1 sh.; do. No. 1. 7 d.; apothecary's-jar, fig. G. 3 in. 3 sh.; 4 in. 5 d.; 5 in. 6 d.; 6 in. 10 d.; 7 in. 1 sh.; fig. B., with inscription 2½ in. 6 d.; 3 in. 7 d.; 3½ in. 8 d.; 4 in. 8 d.; 4½ in. 10 d.; 5 in. 1 sh.; 5½ in. 1 sh. 2 d.; 6 in. 1 sh. 9 d.; with ⊙ med. 6½ in. 2 sh. 3 d.; 7 in. 1 sh. 6 d.; fig. A. with insciption 2½ in. 6 d.; with ⊙ med.; 3 in. 8 d.; 3½ in. 8 d. 4 in. 8 d.; 4½ in. 10 d.; 5 in. 1 sh.; 5½ in. 1 sh.; 6 in. 1 sh. 9 d.; with ⊙ med. 6½ in. 2 sh. 3 d.; 7 in. 1 sh. 6 d.

8 specula, p. piece 6 d.; dish for Photographers 3 sh. 6 d.; cuvettes for Photographers 6 sh.; 3 letters small p. piece 6 d.; do. large p. piece 9 d. plates with inscription p. piece No. 33. 4 d.; No. 35. 6 d.; No. 30. 8 d.; No. 36. 8 d.; No. 88. 5 d.; No. 101. 6 d.; No. 103. 8 d.; No. 98. 8 d.; No. 97. 10 d.; brown edged No. 109. 3 sh.; No. 110. 9 d.; Argand burners 3 d.; air-basket to regulate the gas-light 6 d.; burner for heating 1 sh. 6 d.; insulators VIII.B. 3 sh., XV. 3 sh., U. 3 and 4 d., H. 4 d.; B. 3 d., X. 4 d., O. 1 sh., C. 4 d., K. 7 d., J. 8 d., W. 8 d., Ab. 6 d., F. 4 d., L. 6 d., Aa. 6 d., R. 6 sh., P. 3 d., XIV. 3 d.; plates with 48. holes 6 d., with 72 holes 9 d.

Different articles; prices p. piece: 2 large fruit-baskets 15 sh.; 2 baskets 9 sh.; 2 do. with cover 9 d.; 4 dessert-plates 3 sh.; 2 plates 3 sh.; basket, small 3 sh.; large barrel 4 £; punch-bowl with cover and stand 2 £; punch-pail with cover 1 £ 10 sh.; foot-bath 1 £ 10 sh.; jar and cover 1 £ 10 sh.; barrel and cover 1 £ 10 d.

B. Marble-porcelain, prices p. piece: 2 oval bathing-pans 3 £; 2 chandeliers 2 £; 4 candlesticks 10 sh.; 2 eagles (Rauch) 8 £; 2 Echo (Dankberg) 1 £; A. v. Humboldt (L. Drake) 1 £; Beuth (Kiss) 1 £; 2 pedestals (L. Drake) 6 sh.; Bachus on the panther (F. Friedrich) 1 £; Amor on the lion (Friedrich) 1 £; setting-dog (Mêne) 15 sh.; 2 Russian soldiers (Mehnert) 1 £; 2 brackets (Gerber) 12 sh. The following articles invented by Heidel: 2 goblets 2 £ 5 sh.; 2 Jugs 12 sh.; Cup 9 sh.; Jug with ivy 9 sh.; 2 sugar-basins 1 sh. 4 d. 2 match-boxes 4 sh.

2168. DRYANDER & Co., manu. of earthen-ware, Saarbrücken (Triers). Paris 1855 honor. mention. Agts. Lion M. Cohn, Phaland & Dietrich, represented by Ch. Trübner, 20 St. Dunstans Hill, City, and at the Exhib. Building.

Common earthen ware; prices in centimes p. dozen: plates, common No. 0. 70 c.; do. No. 1. 80 c.; do. deep and flat No. 3. 120 c.; do. bordered No. 2. 90 c.; do. colored with sentences No. 00. 80 c.; do. No. 2. 160 c.; bowls painted No. 2. 18 c.; breakfast-cups painted No. 1. 13 c.; do. with stand painted and marbled No. 1. 14 c.; do. plain and ribbed, white No. 1. 10 c.; do. with stand No. 0. 10 c.; spitting-box grooved 100 c.; sauce-boat No. 1. 25 c.; salad dish, round and ribbed No. 5. 43 c.; do. English No. 4. 36 c.; plates: round No. 2. 20 c., oval No. 4. 27 c., oblong flat No. 3. 50 c., deep with garland No. 6. 37 c., fruit dish, oval No. 4. 18 c.; bowl with large rim No. 1. 16 c.; do. with round rim No. 6. 48 c.; coffee-pot, English No. 5. 80 c.; tea-pot do. No. 3. 42 c.; milk-pot do. No. 3. 19 c.; sugar bowl, vase-shaped No. 3. 33 c.; coffee-cup English 105 c.; milk-pot barrel-shaped No. 1. 20 c.; do. No. 4. 40 c.; pie-dish marbled No. 2. 35 c.; pie-tureen No. 4. 40 c.; preserve-tureen 40 c.; sugar dishes green and blue No. 1. 20 c.; fruit plates do. 45 c.; fruit dishes do. 80 c.; coffee-pot, red, pyramidal No. 4. 32 c.; soup dish No. 2. 45 c.; do. vase-shaped No. 4. 110 c.; do. oval No. 2. 75 c.; pap dish, yellow No. 0. 22 c.; breakfast cups, printed No. 1. 20 c.; plates, printed No. 0. 140 c.; do. No. 1. 160 c.

Fine earthen ware: coffee-pot Greek f. colored No. 1. 90 c.; milk-pot do. No. 1. 60 c.; sugar-pot do. No. 1. 60 c., all p. piece; coffee-cup do. 300 c.; plates, ribbed No. 0. 110 c.; No. 1½. 135 c.; do. deep No. 2. 150 c.; do. deep and flat No. 3. 180 c., all p. doz.; ribbed plates round No. 3. 65 c.; do. square No. 1. 45 c.; No. 3. 100 c.; square mustard-pot 70 c.; square salad dish No. 2. 100 c.; do. oval No. 0. 20 c.; do. round No. 4. 75 c.; square ribbed fruit dish No. 1. 25 c.; No. 3. 50 c.; square ribbed sauce-pot 120 c.; soup dish with plate No. 4. 525 c.; ragout dish with cover No. 2. 180 c., all p. piece; ribbed printed plates: No. 0. 180 c.; No. 1½ 220 c.; No. 2. 250 c.; No. 2⅓ 280 c.; do. flowing No. 1½. 280 c.; do. even No. 0. 170 c.; No. 1. 190 c., all p. doz.; ribbed flowing: washing basin 230 c.; ewer 225 c.; chamber-pot 190 c.; soap-box 140 c.; tooth brush-box 140 c.; sponge-bowl 50 c.; pomatum-box 50 c.; tooth powder-box 50 c.; sugar dish, printed No. 3. 23 c., all p. piece; tea-service round, rococo form, consisting of tea-pot, milk-pot, sugar-pot, 2 cups and saucers and 2 plates 350 c.; inkstands oval 290 c.; do. round, ribbed 160 c.; do. barrelshaped 175 c.; do. shell-shaped 200 c.; fruit dishes with stands 300 c.; square fruit-plates with relievo ornaments 300 c.; flower-bases, printed 500 c.; rococo No. 3. 200 c.; fruit baskets with saucer No. 1. 260 c.; do. round with stand 150 c.; do. oval with garlands 400 c., all p. piece; tea-service consisting of tea-pot, milk-pot, sugar-pot,

2 cups and saucers, plateau, printed and plated, together 1880 c.; candle-sticks p. piece 120 c.; plate, perforated No. 1. 450 c.; do. rococo No. 1½ 280 c.; do. Greek No. 1½ 350 c., all p. doz; flower-pot No. 8,0 18 c.; No. 6/0 24 c.; No. 4/0 33 c., all plated, prices p. piece. India coffee-pot No. 4. 500 c.; milk-pot No. 4. 250 c.; sugar-pot No. 4. 325 c., all. p. piece; coffee-cups No. 2. 550 c.; plates No. 2½. 800 c., p. doz., all printed and gilt. — Letter-presser painted and gilt 190 c., p. piece.

Fire proof kitchen-pot No. 0. 27 c.; No. 2. 42 c.; do. oval No. 1. 42 c.; No. 3. 80 c.; do. high No. 1. 80 c.; coffee-pot No. 2. 35 c.; kitchen-pot with handle No. 1. 12 c.; No. 3. 20 c.; eggcodler No. 2. 25 c.; No. 3. 30 c.; flower-vase, plated No. 200. c., all p. piece; coffee-cups and saucers rococo, printed, plated and lustred, p. pair 120 c.

2171. KULMIZ, CARL (C. Kulmiz), counsellor of commerce and manu., Ida-Marienhütte near Saarau (Breslau). Agts. s. No. 2168.
(s. No. 771, 998, 1104, 1353.)

Fire-bricks p. 1000 pieces 3 £; some crucibles p. piece 1 sh.; enamelled dutch tiles p. piece 4 d.; a large form-piece of fire-clay p. Zollz. 1 sh. 3 d.

2172. LINDEN, GUSTAV, manu. and proprietor of an estate at Ratingen near Düsseldorf.

Two pieces of blue-glazed right-hand tiles 20 Th.; two left-hand do. 20 Th.; one do., double hook 40 Th.; one do., ridge or covering-pan 40 Th.; four do. drain-pipes of different size, each 1 ft. long 7—17 Th.; all p. thousand.

2173. MARCH, ERNST, manu. of terra cottas, stone-ware, chemical apparatus and pipes, Charlottenburg near Berlin. Berlin 1844 great golden medal; London 1851 honor. mention; Munich 1854 great medal; London 1862 medal. Agts. s. No. 2168.

Figures: No. 1—4. evangelists St. Mathew, St. Mark, St. Luke, St. John, each 7¼ ft. high. Relievoes: No. 5. angel of peace; No. 6. Prussian heraldic eagle. Figures: No. 7. Flora 5 ft. 2 in. Engl.; No. 8. John the baptiste 2 ft. Engl.; No. 9. prophet Hezekiel 2 ft. Engl.; No. 10. do. Jeremia 2 ft. Engl.; No. 11. 12. 13. 3 consoles or brackets for these figures. Vases: No. 14. a. 15. 16. a. 17. 18. a. 19. in 3 different shapes. No. 20. 21. 2 corner-ornaments; No. 22. baptismal font; No. 23—34. a variety of architectural ornaments; No. 35. 36. 2 conduit pipes; No. 37. vessel for chemical purposes; No. 38. 39. 40. 41. stone-cocks for chemical purposes.

2174. MÜLLER, J. H., porcelain-painter, Berlin, Mohrenstr. 61. Agts. s. No. 2168.

1 dejeuner, yellow with chalk-pencil drawings (genre), composed of: 1 tray, 2 cups and saucers, 1 sugar-basin, 1 cream-pot, 1 coffee-pot, 1 slop-basin, 2 bread-plates, 1 cake-plate 35 Th. (5 £ 5 sh.); 1 dejeuner, gray with chalk-pencil drawings, landscapes, Harz-views, 28 Th. (4 £ 4 sh.); 1 vase, 1 ft. high in Pompeyan style 30 Th. (4 £ 10 sh.); 1 cup and saucer in the same style 5 Th. (15 sh.)

2174a. VON MÜLMANN, ALBERT, merchant, Zeche Plato near Siegburg. Paris 1855 silver medal. Agts. s. No. 2168. (s. No. 795.)

Clay-manufactures; prices p. 1000 pieces. Fire-bricks 1st. qual. No. 1. 3 £ 7 sh. 6 d.; No. 2. conical in length 3 £ 12 sh.; No. 3. conical in breadth 3 £ 12 sh.: 2d. quality: No. 4. 2 £ 14 sh.; No. 5. 2 £ 9 sh. 6 d.; 3d. quality: No. 6. 1 £ 4 sh.; No. 7. 1 £ 16 sh.; No. 8. hollow-bricks

2 £ 5 sh.; No. 9. glazed 2 £ 11 sh.; No. 10. paving-bricks 1 £ 7 sh.; No. 11. pantiles 1 £ 16 sh. Assortment of drainpipes No. 1. 12 sh.; No. 2. 15 sh.; No. 3. 18 sh.; No. 4. 1 £; No. 5. 1 £ 10 sh.; No. 6. 3 £; No. 7. 5 £ 8 sh.

2175. Rothenbach, W. (W. Rothenbach & Co.), merchant and prop. of an establishment for porcelain-painting, Breslau. Agts. Lion M. Cohn, Phaland & Dietrich, represented by Ch. Trübner, 20 St. Dunstans Hill, City, and at the Exhib. Building.

2 large vases with genre-pictures, portraits of girls 233 Th. 10 sg. (35 £); 2 vases with butterflies (an Indian) 6 Th. (18 sh.); 2 do. (a couple of dancers) 6 Th. (18 sh.); 2 do. (Watteaus) 6 Th. (18 sh.); cake-dish (lion hunt) 20 Th. (3 £); 2 cake dishes („the first step", „the first care") 20 Th. (3 £); cake dish (fleurs animées, ball of flowers) 5 Th. (15 sh.); 1 doz. of dessert-plates with fleurs animées 18 Th. (2 £ 14 sh.); 1 dejeuner with butterflies, 10 pieces 25 Th. (3 £ 15 sh.); 2 cake-dishes with nosegays and gold baroque 7 Th. (1 £ 1 sh.); 2 do. with nosegays in fond 7 Th. (1 £ 1 sh.); 1 coffee- and tea-service with black gold and yellow lustre-edges, 31 pieces, 18 Th. (2 £ 14 sh.); 1 cake-plate with vine-branches in gold 9 sg. 2 pf. (11 d.); 3 bouillon cups and saucers with gold branches, German, French and English inscription p. pair 8 sg. 4 pf. (10 d.); 3 cups and saucers with gold-branches, German, French and English inscription p. pair 3 sg. 4 pf. (4 d.); 3 bouillon cups and saucers with elegant genre-pictures in medaillon, blue fond, gold ornaments and nosegay p. pair 4 Th. 10 sg. (13 sh.); 1 dejeuner with white edge in yellow lustre fond 10 pieces 9 Th. (1 £ 7 sh.); 2 cake-baskets with nosegay and gold ornaments 3 Th. 20 sg. (11 sh.).

2177. Schomburg, H., & Co., manu. of fire proof china, Moabit near Berlin. Agts. R. Schomburg & Co., 90 Cannon-street. City. (s. II. No. 87.)

Coffee-machines of different sizes (new invention) p. piece 1⅔—10 Th. (5—30 sh.); insulators for telegraphs 30 different models and sizes; clay tubes (porous) for telegraphs in any desirable shape; both p. piece ½—10 sg. (5—12 d.); Argand gas-burners from 1 to 40 holes p. doz. 10—30 sg. (1—3 sh.); door and window mountings p. piece 7½—40 sg. (9 d. to 4 sh.); bell rope handles p. doz. 1—2½ Th. (3—7½ sh.); rings for curtains or blinds p. gross 20—40 sg. (2—4 sh.); knife sharpeners (new invention) p. doz. ⅔—2 Th. (2—6 sh.); gold for lustre painting on porcelain, earthenware and glass p. Loth 3½ Th. (10 sh. 6 d.).

2178. Strahl, Otto, counsellor of commerce and manu., Frankfurt on the Oder. London 1851 prize-medal; Paris 1855 bronze medal.

A stove of white glazed dutch tiles 27 Th. (4 £).

2179. Tielsch, Carl (C. Tielsch & Co.), counsellor of commerce, Altwasser (Breslau). London 1851 honor. mention, 1862 medal. Agts. W. Adolph & Co., 9 Bury-Court, St. Mary Axe.

2 chandeliers for a church, gilded; punchbowl with stand, gilt; vase on stand with Christ and St. John of biscuit porcelain; vase, ornamented; 2 idem of green biscuit porcelain; idem mauresque, decorated; idem, painted; 2 scent-boxes with animals of white and green biscuit porcelain; break-fast service, decorated with oak-leaves; idem, decorated with flowers; fruit-basket, painted with violets and lily of the valley; coffee- and tea-service, with painted ivy and gold; idem, decorated with green lustre and gold; idem, decorated with lustre and flowers; idem, with painted corn-ears and wild flowers; cake-basket, gilt and with lustre; idem, decorated; jewel-box, painted; cake tray with a painted bird; plate with a painted bouquet of flowers; cup with flowers; idem, gilt; idem, decorated; Tureen, oval dish, 2 plates and dish with cover, gold and green borders; crucifix of white biscuit porcelain.

2180. Vygen, H. J., & Co., manu. of fire clay bricks, retorts and crucibles, Duisburg. Agts. s. No. 2175.

Oval and D shaped clay-retorts p. 1000 pound 20 Th. (p. ton 6 £); tymp-brick for blast-furnaces; hearth brick; boshes-brick p. 1000 pound 10 Th. (p. ton 3 £); fire-room-brick p. 1000 pound 8 Th. (p. ton 2 £ 8 sh.); tunnel brick and covering brick for coke-ovens; p. 1000 pound 6 Th. (p. ton 1 £ 16 sh.); common-sized bricks for steel, iron, gas and chemical-works p. 1000 pieces from 20 to 40 Th. (3—6 £).

a., b., c. Crucibles for smelting copper; d. do, for cast-steel p. 100 mark capacity 5⅔, 3⅓, 1¼. 3⅔ Th. 200 pounds capacity (17 sh., 11 sh., 3 sh. 9 d., 11 sh.); e. crucible for ultramarine, f. do. for glass-houses p. 1000 pound 20 Th. (p. ton 3 £).

CLASS XXXVI.

MANUFACTURES NOT INCLUDED IN PREVIOUS CLASSES.

2181. Alves, Heinrich, merchant & manu., Berlin, manufactory Crayne. London 1862 honor. mention. Agt. W. Meyerstein, 47 Friday-street.

Patterns of gilt mouldings; whips.

2182. Braune, Bernhard, merchant and manu. of chemical producs, and of essential oils, Danzick. London 1862 honor. mention.

Raw amber scrapings for the manufacture of varnish, acid etc: bead-chips 1 sh. 5 d.; prime bright shelled 1 sh. 1½ d.; fine yellow-bright 9 d; fine red-bright 6½ d.; varnish in sticks 3¾ d.; shreds 5½ d.; scrapings, chopped-of 4¼ d.

Chemical preparations: a. of amber: succinic acid, sublimate 11 sh. 3 d.; crystallized yellow 13 sh. 6 d.; do. half-white 15 sh. 3 d.; do. superfine snow-white 17 sh. 6 d.; succinic oil, raw 3¾ d.; do. rectified 1 sh. 4 d.; succinic rosin 3¾ d.; all p. pound; b. pine rosin-oil, rectified, p. cwt. 2 £ 5 sh.

2183. Bülow, Carl, manu., Görlitz.

10 whole masks of wire-gaze partly with painted beards 4 Th. (12 sh.); 3 whole masks with natural beards 9 Th. (1 £ 7 sh.); 2 ladies' dominoes with curtains 3 Th. 10 sg. (10 sh.); all p. doz.

2186. Galopin, Jos., founder and sculp., Aix-la-Chapelle. Agts. s. No. 2175. (s. No. 1344.)

Model of an iron spiral stair case; two dogs-heads, ornaments for stoves; different specimens of architectural ornaments.

2187. GESELL, F., gilder and manu. of frames for looking glasses and pictures, Görlitz. Agts. Lion M. Cohn, Phaland & Dietrich, represented by Ch. Trübner, 20 St. Dunstans Hill, City, and at the Exhib. Building.

Genuine gilt oval looking glass frame, 35 in. broad, 22 in. high, 4 £.; quadrangular looking glass frame not genuine gilt; dimensions of the glass, 30 in. high by 18 in. broad, 1 £ 3 sh.

2188. GOLDSCHMIDT, J. P., manu., Berlin, Mittelstr. 35. Agts. s. No. 2187.

1 doz. of razor-straps A. M., p. doz. 9 Th., p. piece 1½ Th.; 1 doz. do. A. K., p. doz. 7½ Th., p. piece 1¼ Th.; 1 doz. do. B., p. doz. 7½ Th., p. piece 1⅛ Th.; 1 doz. do. C., p. doz. 5 Th., p. piece 22 sg.

2188 a. VON HAANEN, GEORG, artist, Cologne. Agts. Heintzmann & Rochussen, 9 Friday-street, Cheapside.

1 picture with artificial marble, 1 frame with divers patterns of artificial marble.

2189. HERZIG, GUSTAV, manu. of fancy wood articles, Hermsdorf (Liegnitz), manufactory Agnetendorf. Agts. Heintzmann & Rochussen, 9 Friday-street, Cheapside.

Fancy wood articles in an original-style, household-articles, writing utensils, wood-salve-boxes for apothecaries.

2192. KADE, EDUARD, Berlin, Gertraudtenstr. 8. London 1862 honor. mention. Agt William Meyerstein, 47 Friday-street.

18 photograph-albums, 3 ladies' necessaries, 2 writing-portfolios.

2193. KOEHLER, W. GUSTAV, sculp. and manu. of cornices, Zeitz. Agts. s. No. 2187.

1. A board on which wood-carvings and ornaments for pianos are exhibited 10 Th. (1½ £ 4 sh); 2. do. with the like articles 9 Th. (1⅓ £); 3. do. with patterns of cornices and firms mounted in metal 7 Th. (1 £); 4. a top-piece for a piano 3 Th. (⅜ £); 5. a pair of piano-consoles of rosewood 6 Th. (1 £).

2194. LAUE, JOHANN CHRISTIAN FRIEDRICH (C. F. Laue), turner, Naumburg on the Saale. Agts. s. No. 2187.

Chess board and chessmen of ebony and boxwood, 1 piece 15 £.

2196. NIESE, FERD., Danzick. Paris 1855 honor. mention; London 1862 honor. mention. Agts. s. No. 2187.

No. 1. 1 packet of cut beads of amber (corallen) 30 Th. (4 £ 10 sh.); No. 2—8. 7 packets do. 25 Th. (3 £ 15 sh.), 20 Th. (3 £), 18 Th. (2 £ 14 sh.), 16 Th. (2 £ 8 sh.), 15 Th. (2 £ 5 sh.), 13 Th. (1 £ 19 sh.), 11 Th. (1 £ 13 sh.); No. 9. to 13. 5 packets of bastard-beads 36 Th. (5 £ 8 sh.), 30 Th. (4 £ 10 sh.), 22 Th. (3 £ 6 sh.), 18 Th. (2 £ 14 sh.), 15 Th. (2 £ 5 sh.); No. 14. pure and polished 32 Th. (4 £ 16 sh.); No. 15—17. 3 doz. do. 18 Th. (2 £ 14 sh.), 10 Th. (1 £ 10 sh.), 8 Th. (1 £ 4 sh.); No. 18. cut beads of amber (olives) 25 Th. (3 £ 15 sh.); No. 19. 1 collier of bastard do. 15 Th. (2 £ 5 sh.); No. 20. do. 28 Th. (4 £ 4 sh.); No. 21. mouth piece of a Türkish pipe 30 Th. (4 £ 10 sh.); No. 22. cigar-pipe 12 Th. (1 £ 16 sh.); No. 23. a pair of bracelets and buttons 14 Th. (2 £ 2 sh.); No. 24. a brooch 7 Th. (1 £ 1 sh.); No. 25—28. 4 pieces of raw amber, 1⅓ lb. 90 Th. (13 £ 10 sh.), 27 Lth. 90 Th. (13 £ 10 sh.), 18 Lth. 80 Th. (12 £), 16 Lth. 65 Th. (9 £ 15 sh.); No. 29. 10 pattern-cards of mouth pieces and tips, for tobacco- and cigar-pipes 30 Th. (4 £ 10 sh.).

2197. PERLBACH, H. L., manu. of amber-beads merchant, Danzick. London 1862 medal. Agts. Meyer Levin, 17 South-street, Finsbury Square. (s. II. No. 15.)

I. Raw-amber; a piece of amber, most rare in point of size form and color, weighing 12 lb. for 600 £.; 2 pieces of amber weighing 4 lb. and 2 lb.; 5 samples à 1 lb. of differents sorts of raw-amber.

II. Amber-beads: an assortment of fine bastard-beads No. 1., 2., 3. à 2 lb.; an assortment of fine Leghorn amber beads No. 1., 2., 3. à 2 lb.; 4 strings of fine olive beads (bastard); 5 do. (transparent); 2 doz. of round bastard beads; 3 strings of transparent round cut beads of 2 oz.; 3 do. of 1¼ oz.; 3 do. of 1 oz.; 4 do. of ½ oz.; ½ doz. amber-colliers.

2198. RÖMPLER & TOELLE, manu., Barmen, Agts. Schrage & Renninger, 30 Ironmonger Lane, Cheapside.

India-rubber articles: elastic cords and braids, shoewebs and all other sorts of elastic webs, braces, belts, garters and watch-guards.

2200. VITÉ, F., manu., Berlin, Kommandantenstr. 7. London 1862 honor. mention. Etwees, bags, albums.

2201. WELLHÄUSER, E. L., manu., Elberfeld. Articles of caoutchouc.

2202. WESTPHAL, CARL AUGUST, manu. of articles of amber, Stolp (Pommern). Paris 1855 honor. mention; London 1862 medal. Agts. Weintraud, Rumpf & Co., 4 King-street, Cheapside.

1. A fine set of opaque amber in Oriental fashion; 2. a fine set of cut pale transparent amber; 3. three strings of opaque beads for the African trade; 1, 2 and 3 to 1 lb., 100 strings assorted 1530 Th. (230 £); 4. two strings of transparent cut beads, 8 strings to 1 lb., p. string 20 Th. (3 £), 19 do. 9 Th. (1 £ 7 sh.); 5. specimens of cigar-pipes and smoking utensils, of different shape and size, also of transparent cut olive-beads and of transparent smooth-beads.

KINGDOM OF SAXONY

AND

PRINCIPALITY OF REUSS Y. B.

The Exhibitors of Reuss are marked with *.

CLASS I.

MINING, QUARRYING, METALLURGY AND MINERAL PRODUCTS.

2301. Company of the marble quarries of Fürstenberg, Grünhain (presid. Capt. Naundorff, Schneeberg). London 1862 honor. mention.

Articles of white marble: 1 block 43 in. long, 18 in. broad and high 30 Th. (4½ £); 1 slab 38 in. long, 18 in. broad, 7 Th. (21 sh.); 1 tabletop 33 in. long, 18 in. broad, 5⅔ Th. (17 sh.); 1 do. 36 in. long, 18½ in. broad, 6 Th. (18 sh.); 1 polishing slab 5 Th. (15 sh.); 2 chimney-tops à 2½ Th. (7½ sh.); 1 whetstone ⅚ Th. (2½ sh.); 1 do. ⅔ Th. (2 sh.); various letterstones etc. à ⅓ Th. (1 sh.).

2302. Millstone-manufactory, Johnsdorf near Zittau. London 1862 honor. mention.

Specimens of millstone.

2303. Zwitterstocks-Factory, Altenberg.

Fine, chemically tried metallic tin in blocks, balls and rods, p. cwt. 6 £. 8 sh.

CLASS II.

CHEMICAL SUBSTANCES AND PRODUCTS AND PHARMACEUTICAL PROCESSES.

2304. Duvernay, Peters & Co., Chemnitz. London 1862 medal.

30 bottles with preparations of orchilla, aniline etc.; 1. orchilla in paste, red; 2. do. blue; 3. orchilla extract, simple; 4. do. blue; 5. do. double; 6. orchilla violet lac (laque de Parme); 7. cudbear blue-violet; 8. do. red-violet; 9. do. red; 10. indigo extract; 11. indigo carmine; 12. do. dryed; 13. substitute for tartar; 14. aniline red (fuchsine); 15. do. free from resin; 16. do. roseine; 17. do. in paste; 18. do. cudbear; 19. aniline red in crystals; 20. do. violet (mauve) in paste; 21. do. in solution; 22. do. dryed; 23. aniline, violet-blue in solution; 24. do. dryed; 25. aniline, blue in solution; 26. do. dryed; 27. orchilla weed from Lima; 28. do. Madagascar; 29. do. Benguela; 30. do. Cap Verd.

2305. Heine & Co., Leipzig. Paris 1855 honor. mention; London 1862 medal and honor. mention. Agts. J. L. Pfungst & Co., 32 Crutched Friars.

Ethereal oils and essences: oils of mustard seed 21½ Th. (3 £ 4½ sh.); angelica 6 Th. (18 sh.); parsley 9¼ Th. (1 £ 8 sh.); camomile vulgar 45 Th. (6 £ 15 sh.); aniseseed 3⅔ Th. (11 sh.); fennel 1½ Th. (4½ sh.); caraway, twice rectif. 2⅓ Th. (7 sh.); coriander 5⅔ Th. (17 sh.); cognac (grape) 35 Th. (5 £ 5 sh.); calamus 3½ Th. (10½ sh.); valeriana 5½ Th. (16½ sh.); juniper berries 3⅔ Th. (11 sh.); camomille, Roman, 10 Th. (1 £ 10 sh.); essence for Jamaica rum ⅔ Th. (2 sh.); do. for cognac 1 Th. (3 sh.); ether of apples 2 Th. (2 sh.); do. of pine apples 2 Th. (6 sh.); do. of rasp berries 2 Th. (6 sh.); all prices p. ½ kilogr.

2306. Krause, Heinrich Hermann, apothekary, Freiberg. Agt. C. Trübner, 20 St. Dunstans Hill, City.

Collections of chemical preparations for the use of instruction, two sizes at 5 Th. and 20 Th. (15 sh. and 3 £).

2307. Pommier & Co., Neuschönefeld near Leipzig. London 1862 honor. mention. Agts. Hecht Brothers.

Orchil and other chemical products, viz: orchilla-paste p. 100 Pfd. (à ½ kil) 9 Th. (18 sh.).

The following prices p. 1 Pfd.; orchilla extract 6½ sg. (8 d.); do. double 7½ sg. (9 sh.); cudbear 10 sg. (1 sh.); lacs, Parme 20 sg. (2 sh.); cochineal 20 sg. (2 sh.); do. ponceau 7 sg. (8½ d.); of weld yellow 7 sg. (8½ d.); fustic yellow 7 sg. (8½ d.); do. orange 1 Th. (3 sh.); ammoniac cochineal 1 Th. 25 sg. (5 sh. 6 d.); acid. picric crystal. 3 Th. 22 sg. (11 sh. 3 d.); do. paste 1 Th. 10 sg. (4 sh.).

The following prices p. 100 Pfd.: alum 3 Th. 10 sg. (10 sh.); do. powder 3 Th. 15 sg. (10 sh. 6 d.); do. purified 4 Th. (12 sh.); sulphate of alumine 3 Th. 12 sg. (10 sh. 3 d.); do. purified 4 Th. (12 sh.).

2308. Sachsse, E., & Co., Leipzig. London 1862 medal. Agt. Ludolf Kindler, 3 Dunster Court, Mincing Lane, E. C.

Ethereal oils and essences; prices p. ½ kilogr. (p. lb. avoir du poids free on board Hamburg).

Oils: angelica 6 Th. (18 sh.); anise-seed, Saxon 3 Th. 10 sg. (10 sh.); arnica flowers 270 Th. (40 £ 10 sh.); birch-tar, rectif. 2 Th. 20 sg. (9 sh.); botryos-leaves 20 Th. (3 £); bucco-leaves 20 Th. (3 £); calamus 3 Th. 10 sg. (11 sh.); caraway rectif. 2 Th. 12 sg. (7 sh. 2 d.); do. chaff 1 Th. 2 sg. (3 sh. 2 d.); cascarilla 20 Th. (2 £ 15 sh.); chamomile roman 15 Th. (2 £ 5 sh.); do. pure 45 Th. (6 £ 15 sh.); oil for cognac 33 Th. (5 £); oils, coriander 5 Th. 25 sg. (18 sh.); cumin 5 Th. (15 sh.); dill 5 Th. 15 sg. (16 sh. 6 d.); elder flowers 96 Th. (14 £ 8 sh.); fennel, sweet 1 Th. 12 sg. (4 sh. 2 d.); galbanum 9 Th. (1 £ 7 sh.); ginger 21 Th. (3 £ 3 sh.); hyssop 16 Th. (2 £ 8 sh.) juniper berries 3 Th. 25 sg. (12 sh.); juniper wood 15½ sg. (1 sh. 6 d.); laurel 10 Th. (1 £ 10 sh.); levistic 11 Th. (1 £ 11 sh.); majoram, German 9 Th. (1 £ 7 sh.); mirrh 16 Th. (2 £ 8 sh.); nigella 30 Th. (4 £ 10 sh.); onions 12 Th. (1 £ 16 sh.); parsley 9 Th. 15 sg. (1 £ 8 sh. 6 d.); pepper 12 Th. 15 sg. (1 £ 17 sh. 6 d.); peppermint, German 6 Th. 20 sg. (1 £); rue 2 Th. 15 sg. (7 sh 6 d.); sage 3 Th. (9 sh.); spearmint, German 4 Th. (12 sh.); valeriana 6 Th. 15 sg. (19 sh. 6 d.; vetiver 60 Th. (9 £); water-fennel 7 Th. 15 sg. (1 £ 2 sh. 6 d.); weedseed 7 Th. (1 £); oils for absynth, Swiss 6 Th. 15 sg. (19 sh. 6 d.); eau de Cologne (12 Th. (1 £ 16 sh.); English bitter 7 Th. (1 £ 1 sh.); gin 4 Th. 20 sg. (14 sh.); Maraschino di Zara 12 Th. (1 £ 16 sh.).

Essences: apple 2 Th. 15 sg. (7 sh. 6 d.); pear 2 Th. 15 sg. (7 sh. 6 d.); pine apple 2 Th. 15 sg. (7 sh. 6 d.); quince 2 Th. 15 sg. (7 sh. 6 d.); raspberry 2 Th. 15 sg. (7 sh. 6 d.); strawberry 2 Th. 15 sg. (7 sh. 6 d.).

Essences: for arrack 1 Th. 15 sg. (2 sh. 3 d.); for Boonekamp 2 Th. (6 sh.); for cognac 2 Th. 6 sg. (6 sh. 7 d.); for rum, Jamaica 1 Th. 7½ sg. (3 sh. 8 d.); flavouring for claret 1 Th. (3 sh.); coloring do. 1 Th. (3 sh.).

2309. SCHIMMEL & Co., distillers of essential oils and spirituous essences, Leipzig. Munich 1854 medal of honor; Paris 1855 bronze medal; London 1862 medal. Agt. C. F. Claudius, 3 St. Helens Place, Bishopsgate-street.

Spirituous essences for arrack, cognac (brandy), gin and rum.

Oils of angelica, anise-seed (Saxon), arnica flowers, Balm (Melissa), birch-tar rectif., botryos, calamus, chamomiles Roman, chamomiles vulgar with lemon oil, do. pure, caraway-seed twice rectif., cardamom Ceylon, cascarilla, cedarwood, celery, cognac (grape), coriander, cumin, dill, elder flowers, elecampane, fennel, fernwort with ether, galanga, galbanum, ginger, juniper berries, juniper wood, laurel, lovage, majoram, milfoil, mirrh, mustard-seed, nutmeg, parsley, pepper, peppermint German, pimento, rhodium, sage, sevin, sambalus (maskwort), tansy, valeriana, wax rectif., wormwood, wormseed levantine.

2310. SCHÜTZ, AUGUST, Wurzen near Leipzig. London 1862 honor. mention.

Dyed wool-dust for the use of paper-hanging-manufacturers.

2311. THEUNERT & SON, manufactory of ultramarine, Chemnitz. London 1862 honor. mention.

16 specimens of ultramarine.

2312. WÜRTZ, TH. (formerly Mottet & Würtz), Leipzig. London 1851 prize-medal;

Paris 1855 silver medal; London 1862 medal. Agt. C. Trübner, 20 St. Dunstans Hill, City.

Aniline, roseine, orchilla paste, orchilla extract, cudbear, different log-wood pastes (laques).

CLASS III.

SUBSTANCES USED FOR FOOD, INCLUDING WINES.

2313. JORDAN & TIMAEUS, Dresden. Berlin 1844 silver medal; Munich 1854 medal of honor; London 1862 medal. (s. II. No. 78.)

Chocolates, fancy articles in chocolate, confectionary. Prices affixed to the objects.

2314. STENGEL, WILHELM, Leipzig. London 1862 medal.

Specimens of distilled potatoe-sprit of the exhibitors own manufacture, 95—96° Tralles.

CLASS IV.

ANIMAL AND VEGETABLE SUBSTANCES USED IN MANUFACTURES.

2315. VON BURCHARDI, FRIEDRICH, Hermsdorf near Königstein. Paris 1855 bronze medal; London 1862 medal.

Bee-hives with movable combs and double glass sides; No. 1. an observation hive 13 Th. 10 sg. (2 £); No. 2. a stand hive 13 Th. 10 sg. (2 £), with 2 small honey bells 3 Th. 10 sg. (10 sh.); No. 3. a lying and travelling hive 10 Th. (1 £ 10 sh.), with 2 small honey bells 3 Th. 10 sg. (10 sh.); No. 4—11. eight large honey bells, each 6 Th. 20 sg. (1 £); No. 12. three pieces of combs, built by the bees out of melted and artificially pressed wax, each 1 Th. (3 sh.). Prices inclus. packing franco railway-station Koenigstein.

2316. KIND, JOHANN CHRISTIAN HEINRICH, owner of the original Negretti-sheep-farm Gleina near Bautzen. Paris 1855 bronze medal.

A case with three fleeces of wool and a stuffed sheep 300 Th. (45 £).

2317. VON SCHÖNBERG, ARTHUR, Rothschönberg. Munich 1854 medal of honor; Paris 1855 silver medal; London 1862 medal.

Fleeces of superfine wool.

2318. STEIGER, ADOLPH, prop. of the original Merino-sheep-farm Leutewitz and Löthayn near Meissen. Agt. Otto Neuhaus p. F. M. Schwarz, 9 Scott's Yard, Cannon-street.

15 samples of wool from rams; 15 do. from ewes. The living beasts are to be exposed in Battersea-park 25. Juni.

CLASS V.

RAILWAY PLANT, INCLUDING LOCOMOTIVE ENGINES AND CARRIAGES.

BURSCHE. (s. No. 2381.)

2319. HARTMANN, RICHARD, Chemnitz. Berlin 1844 silver medal; Munich 1854 great medal

J*

and cross of the order of merit; Paris 1855 silver medal; London 1862 medal. (s. II. No. 77.)

1. Locomotive engine with movable fore-axle-tree for mountainous countries and narrow curves 14,200 Th. (2130 £); 2. steam engine with condensation and expansion, 20 horse power 3300 Th. (495 £).

3. Seven working tools: a. patent machine for turning nuts and taking off the edges 140 Th. (21 £); b. patent machine for turning nuts and screws 280 Th. (42 £); c. patent double nut-shaping machine, for shaping squares and working simultaneously the two opposite sides of nuts or screw-heads 600 Th. (90 £); d. selfacting traversing drilling machine and selfacting boring machine 730 Th. (109½ £); e. selfacting radial drilling machine and selfacting boring machine 800 Th. (120 £); f. slotting machine, with 3 selfacting movements of the table and movable tool-holder for planing edges 800 Th. (120 £); g. slide lathe for turning and cutting screws, with selfacting longitudinal and traversing motion 630 Th. (94½ £).

4. A set of wool carding machines: a. first scribbling machine with cleaning rollers 520 Th. (78 £); b. second scribbling machine without cleaning rollers 450 Th. (67½ £); c. slubbing machine, 1 doffing cylinder with 2 rovings each of 20 slubbings 640 Th. (96 £); d. iron grinding roller for the carding machines above named 30 Th. (4½ £).

QUAST. (s. No. 2463.)

CLASS VII.

MANUFACTURING MACHINES AND TOOLS.

GROSSMANN. (s. No. 2338.)

HARTMANN. (s. No. 2319.)

2320. PURSCH, THEOBALD, Dresden. Agts. Müller & Schreiber, Crescent-Place, Blackfriars.

Artificial grinding tools; prices p. piece: Grinding wheels 1¼—4¼ in diameter 6 d. to 1 sh. 6 d.; grinding stones with layer of emeril 6 d. to 10 d.; grinding stones on slate 10 d. to 1 sh. 3 d.; conical buts 4 d. to 6 d.; files of various shapes, on wood 6 d. to 1 sh. 8 d.

2321. SAUER, JULIUS, Plauen.

A girth-rope, length 37½ yards, breadth 7¾ in. weight 58⅗ pounds 26⅔ Th. (4 £.); thread for Jacquard-looms p. pack a 5 lbs. ⅓ Th. (1 sh.).

2322. SONDERMANN & STIER, Chemnitz. London 1862 honor. mention.

A machine for cutting boiler tubes 425 Th. (63 £ 15 sh.).

2323. ZIMMERMANN, JOHANN, manu. of tools, Chemnitz. Munich 1854 honor. mention; London 1862 medal. (s. II. No. 69.)

Machines for working in metal and wood: A. Selfacting slide-lathe for surfacing and screw-cutting; top driving apparatus 400 Th. (60 £); F. large selfacting slide-lathe (12 ft. between the centres, 19½ ft. length of bed), top drivingapparatus 1580 Th. (237 £); H. centring apparatus for shafts etc. 200 Th. (30 £); V. selfacting planing machine 420 Th. (63 £); CA. large selfacting planing machine (for articles 20 ft. long, 6½ ft. broad, 6½ ft. high) for planing in all dirrctions, with a second tool-holder, the tools independent and selfacting in all directions, each tool-holder with a second tool 3750 Th. (562½ £); GA. planing-apparatus for planning the valve-faces of locomotive steam engine cylinders in their places 200 Th. (30 £); JA. selfacting universal shaping-machine, working in all directions, with top driving apparatus 450 Th. (67½£); NA. slotting and shaping machine; round table, circular motion, top driving apparatus 750 Th. (112½ £); TE. selfacting vertical drilling machine, top driving apparatus 280 Th. (42 £); ZD. portable hand-drill 75 Th. (11¼ £); BB. selfacting drilling machine for long holes, serviceable also as common drilling machine, with complete driving apparatus 600 Th. (90 £); WD. selfacting nut-cutting and facing machine, with top driving apparatus 450 Th. (67½ £); MA b. selfacting slotting and nut-shaping-machine 400 Th. (60 £); QE. machine for cutting boiler-tubes 400 Th. (60 £); BC. noiseless ventilator 180 Th. (27 £); DD a. small vertical-saw for curved work 150 Th. (22½ £); CD. endless band-saw, fitted up for circular sawing railway breaks 320 Th. (48 £); GD. roller planing machine for timber 650 Th. (97½ £); DE moulding and planing machine for the manufacture of doors and window-frames 800 Th. (120 £); LE. tenoning and mortising machine do. 450 Th. (67½ £); JE. wood-drilling and mortising machine 450 Th. (67½ £); parallel iron-vice, 7 in. opening 26 Th. (3 £ 18 sh.); do. 9½ in. opening 30 Th. (4½ £); do. for planing, shaping, slotting etc. machines, 4 in. opening 20 Th. (3 £); do. 30 in. opening 60 Th. (9 £).

2324. LINDNER, WILHELM FERDINAND, manu. of spindles etc., Chemnitz.

Tools for planting trees: with 4 knives 12 in. circle 4 Th. 5 sg. (12 sh. 6 d.); with 4 knives 10½ in. circle 4 Th. (12 sh.); with 3 knives 8 in. circle 3 Th. 15 sg. (10 sh. 6 d.).

CLASS VIII.

MACHINERY IN GENERAL.

HARTMANN. (s. No. 2319.)

MILLSTONE-MANUFACTORY. (s. No. 2302.)

2326. PFITZER, ERNST, manu. of weighing machines, Oschatz. London 1862 honor. mention.

1 decimal-weighing machine, with mahogany stand to weigh 50 lb. 16 Th. (2 £ 8 sh.); 2. patent decimal weighing machine, with unlimited horizontal weighing board, to weigh 100 lb. 13 Th. (1 £ 19 sh.); 3. counter weighing machine, square brass scale to weigh 25 lb. 8½ Th. (1 £ 5½ sh.). Wholesale prices.

CLASS XI.

MILITARY ENGINEERING, ARMOUR AND ACCOUTREMENTS, ORDNANCE AND SMALL ARMS.

2328. BÖSENBERG, C. H., Leipzig.

A percussion-needle double barelled gun, new invention 96 Th. (14 £ 8 sh.).

CLASS XIII.

PHILOSOPHICAL INSTRUMENTS AND PROCESSES DEPENDING UPON THEIR USE.

2330. KOOSEN, J. H., Burkhardswalde near Pirna.

Two galvanic chronometers with centrifugal regulator and isochronic motion. New invention. Not to be sold.

2331. RUETE, DR. TH., professor, Leipzig. London 1862 medal.

Model of a human eye, for the use of instruction, executed after Dr. Ruete's directions by M. Tauber, Leipzig, 25 Th. (3 £ 15 sh.); Ophthalmotrope, or apparatus for demonstrating the action of the eye-muscles after Dr. Ruete's directions constructed by Emil Stoehrer at Dresden 33⅓ Th. (5 £).

2332. SCHICKERT, HUGO, mechanical instrument-maker, Dresden.　　　(s. II. No. 73.)

A balance for physical purposes 160 Th. (24 £).

2333. SCHNEIDER, K. H. E., manu. of mathematical-instruments and gold balances, Leipzig.

30 cases of mathematical instruments from 20 sg. (2 sh.) to 19 Th. (3 £); 9 gold-scales from 11½ sg. (1 sh. 2 d.) to 2 Th. 17½ sg. (7 sh. 9 d.); a pair of compasses 4 Th. (12 sh.); a scale beam 28½ sg. (2 sh. 10 d.); do. 1 Th. 19 sg. (5 sh.); do. 1 Th. 5 sg. (3 sh. 6 d.).

CLASS XIV.

PHOTOGRAPHIC APPARATUS AND PHOTOGRAPHY.

2335. MANECKE, FR., Leipzig. London 1862 medal.

Photograph, portrait, full size; a do. smaller one; photographic apparatus.

2336. BROCKMANN, F. & O., Dresden.

A collection of photographs, taken from designs of Prof. Schurig after pictures of the Dresden museum.

CLASS XV.

HOROLOGICAL INSTRUMENTS.

2337. ASSMANN, JULIUS, watch-manu., Glashütte near Dresden. Munich 1854 medal of honor; London 1862 honor. mention. Agt. Louis Höber, watch- and clockmaker, 31 Duke-street, Grosvenor-square.

An assortment of 15 lever-watches in gold and silver, open face and hunting; some movements and parts of watches.

2338. GROSSMANN, M., watch- and clock-manu., Glashütte near Dresden. London 1862 honor. mention. Agt. C. Trübner, 20 St. Dunstans Hill, City.

Assortment of watches and chronometers and movements for d.; movement of a marine-chronometer; astronomical clock with steel and zinc-pendulum; two dead-beat-seconds, pocket instrument; micrometer and tenth-of-millimeter-gauge; assortment of parts of watches in different states of finishing; turn for watch-makers; lathe, 20 in. long, prismatic bar; do. 12 in. long, rectangular bar; screw die with plates and cutters.

2339. LANGE, A., & Co., founders of the Saxon watch-manu. (1845), Glashütte near Dresden. Berlin 1844 silver medal; Munich 1854 great medal; London 1862 medal. Agt. C. Trübner, 20 St. Dunstans Hill, City.

5 gold open faced lever pocket watches, improved construction; 2 do. to be wound up in the pendant;

3 do. hunting; 3 do. to be wound up in the pendant; 1 do. half hunting; 3 silver open faced watches; 1 do. to be wound up in the pendant; 2 do. hunting; 1 do. to be wound up in the pendant; 1 do. half hunting; 6 lever-watch movements of diff. size and quality especially for the American market.

2340. SCHNEIDER, ADOLPH, manu. of lever-watches, Glashütte near Dresden. Agt. C. Trübner, 20 St. Dunstans Hill, City.

Gold watches: hunting watch, to be wound up in the pendant; do open faced, to be wound up in the pendant, with the day of the month; hunting watch; open faced watch with golden dial; do. with enamelled dial; do. to be wound up in the pendant. Silver watches: hunting watch; open faced watch; do. to be wound up in the pendant; lever-watch-movement, not yet gilt; do. gilt.

CLASS XVI.

MUSICAL INSTRUMENTS.

2341. BREITKOPF & HÄRTEL, Leipzig. Berlin 1844 silver medal; London 1851 prize-medal; Munich 1854 medal of honor; London 1862 medal.

I. Pianofortes: full sized grand piano, 7 octaves, double action, in rosewood, 650 Th. (97 £ 10 sh.); cover 9 Th. (1 £ 7 sh.); short grand piano, 7 octaves, double action, in rosewood, 400 Th. (60 £); cover 8 Th. (1 £ 4 sh.); symmetric square-piano, 7 octaves, double action, in wal-nut, 280 Th. (42 £); cover 6 Th. (18 sh.); large oblique upright pianino, 7 octaves, double action, in rosewood, five ironbars above the strings, 270 Th. (40 £ 10 sh.); cover 8 Th. (1 £ 4 sh.). The case and the sounding-board of the first named piano are made after new patent construction; the double action of all instruments is patented.

II. Music prints: 1 copy of J. Seb. Bachs works, 10 vol. 50 Th. (7 £ 10 sh.); 1 copy of G. F. Haendel's works, 10 books, 35 Th. (5 £ 5 sh.); 1 copy of Palestrina's motettes 5 Th. (15 sh.); 1 copy of Beethoven works. New complete edition (in score No. 1. 2. 37—42. 54. 65. 124—126.; in parts No. 1. 54.) 13 Th. (1 £ 19 sh.); 1 copy of Mendelsohn's music to Athalia, score 12 Th. (1 £ 16 sh.); 1 copy of Mendelssohn's opera: die Heimkehr, score 10 Th. (1 £ 10 sh.); 1 copy of Mendelssohn's music to the summernights-dream, score 10 Th. (1 £ 10 sh.).

III. Typographic prints: 1 copy of the holy bible. Illustr. by Julius Schnorr von Carolsfeld. Splendid edition. 30 Th. (4 £ 10.).

IV. Published by the exhibitors: 1 copy of: Bildnisse berühmter Deutschen (portraits of celebrated Germans) 30 engravings on steel 15 Th. (2 £ 5 sh.).

2342. GLÄSEL, CARL WILHELM, Markneukirchen.

A guitar, new model, with case, 60 Th. (9 £).

2344. IRMLER, ERNST, Leipzig. Munich 1854 medal of honor; London 1862 medal.　(s. II. No. 83.)

Grand piano, English construction, in rosewood, 82 £; cottage piano in rosewood 52 £

2345. KAPS, ERNST, pianomaker to the court, Dresden. London 1862 honor. mention.

Grand piano.

2347. OTTO, FRIEDRICH AUGUST, manu. of violin-bows, Markneukirchen. London 1862 honor. mention.

10 violin-bows; 6 cello-bows; 1 bass-bow.

SCHLESSIGER & LUMMER. (s. No. 2488.)

2348. SCHUSTER, BROTHERS, Markneukirchen. London 1862 medal. Agt. Heintzmann & Rochussen, 9 Friday-street, Cheapside.

Violins, ord. doz. 9 sh. and upwards, fine p. piece up to 4 £; violoncello 7 £ 10 sh.; bows for violins and cellos, ord. p. doz. 2¼ sh. and upwards, fine p. piece up to 1⅓ £; guitars, 3 sh. to 7⅓ £; Flutes, fine, syst. Boehm, up to 4½ £ p. piece; wind-instruments of wood and metal of all kinds; patterns of trills, stringholders, pitch-pipes, music-desks etc.

2349. SCHUSTER, MICHAEL, JUN., Markneukirchen. Munich 1854 honor. mention; London 1862 medal. Agt. s. No. 2348.

Musical-instruments and music-wares of all kinds: 7 div. violinos; 4 Spanish guitars; 1 French do.; 4 div. concertinos; 8 piccolo-flutes; 4 flutes; 4 drum-fifes; 6 flageolets; 2 clarinets; 6 horns of German silver, bell backwards American system, with rotary system; 5 div. cornets; 1 chromatic horn; 1 tenor-trombone; 1 bass-trombone; 1 bugle; 4 hunting-horns; 33 violin-bows; 1 cello-bow; 1 bass-bow; bridges, pegs, tailpieces for violins, guitars and banjo's; capo d'astros; spring-locks; resin-boxes; an assortment of strings for violins, guitars etc.

2350. SEYFERT, FRIEDRICH WILHELM, Chemnitz. Agt. Victor Bauer, 1 Ironmonger Lane, E. C.

A pianino, built by Jul. Graebner at Dresden 300 Th. (45 £).

*2351. WAGNER & Co., Gera. London 1862 medal. Agt. D. Winkler, 29 Queen-street, Cheapside.

7 accordions of diff. quality; 1 concertino; 9 mouth-harmonicas of diff. quality.

CLASS XVIII.
COTTON.

2355. BAUMGAERTEL, C. E., & Son, Lengenfeld i. V.

1 p. 8/4 spotted cambric 12 yards 4 Th. 20 sg. (14 sh.); 1 p. 10/4 mull ramage bordered 24 yards 7 Th. 15 sg. (1 £ 2 sh. 6 d.); 1 p. 6/4 gauze ramage curtains 24 yards 6 Th. (18 sh.).

2356. CHALYBAEUS & MÜHLMANN, Chemnitz.
Patterns of cotton thread, raw, bleached and colored.

2358. FÖRSTER, OTTOMAR, Chemnitz. Munich 1854 honor. mention.

Cotton yarns, threads, strings, wicks: yarns for knitting; vicognia estremadura, supra, greentie, raw, bleached or colored, col. marbles 2 twisted; sewings 2 and 4 thread, bleached; canva, raw or bleached; bowyarns for manufacturers of curtains; kitchen wicks, from rowing, raw or colored; do. plaited, raw or bleached; wick yarns, wound in balls, twisted, raw or bleached; do. plaited, bleached, for stearine and paraffine candles; plaited strings for spinning-machines; patent strings twisted out of prepared knitting-yarns, for mule-machines and selfactors.

Threads for weaving, raw or bleached, cotton, or cotton mixed with silk, with vicognia, with wool, also combed with silk; threads of cotton and flax for lace-manufactures, raw, bleached or colored.

2360. GRUNER, HERRMANN, Ebersbach near Löbau. London 1862 medal.

Colored cotton woven goods for the Oriental market; dimitys, croisés, mouscacelles, mouchés, Stambul atlas, Stambul Chalys, Atagias etc.

2361. HÄBLER, BROTHERS, Grossschönau.

Cotton and half-linen drills: union drill (linen and cotton), coup. 1—6. 7¼ d.; coup. 7—9. 8¼ d.; coup. 10. 10⅝ d.; coup. 11—14 11½ d.; coup. 15. 10⅝ d.; coup. 16. 7⅝ d.; coup. 17. 5½ d.; Jacquard-drill, cotton, coup. 18. 8 d.; linen drill coup. 19. 9 d.; coup. 20. 17½ d.; prices p. yard; free Hamburgh payable in cash.

2362. HERZOG, H. W., Neugersdorf near Löbau. London 1862 medal. Agt. Victor Bauer, 1 Ironmonger Lane, E. C.

Cotton and mixed stuffs for gentlemen's coats and pantaloons; prices p. yard: Jacquard 1 sh. 10 d.; tricot 1 sh. 3⅓ d.; Georgia 1 sh. 1 d.; Victoria 1 sh. ½ d.; cord P. P. 11½ d.; Klapka 10 d.; cord I. 9¾ d.; Hercules 10¼ d.; trapet 9¾ d.; small cloth I. 8½ d.; adlas II. 8⅛ d.; crêpe P. P. 7⅓ d.; crêpe II. 5⁴⁄₁₁ d.; struck 6½ d.; Virginia 8⅛ d.; rips 4½ d.; small cloth 7 d.

2363. HEYDENREICH, R., Witzschdorf near Zschopau.

Patterns of cotton twists, mule and water, for hosiery and for weaving, No. 16—40. p. lb. 11¾ to. 19 d.

2364. HOFFMANN, C. G., Neugersdorf near Löbau. London 1862 honor. mention.

Cotton and mixed stuffs for gentlemen's coats and pantaloons: cords, velours, cassinets, tricot, rips, Bristol, diagonal, Klapka, Hercules etc.

2365. HUETTIG, BENJ., & Co., Leutersdorf near Eybau. Munich 1854 medal of honor; London 1862 honor. mention.

Cotton and mixed stuffs for gentlemen's coats, pantaloons and waistcoats.

2366. KELLER & GRUBER, Chemnitz. London 1862 honor. mention.

Patterns of raw bleached and colored cotton 4 and 6 thread knitting and crotcheting-yarns.

2367. KLEMM, ROBERT, Plauen. London 1862 honor. mention. Agt. Barnes, Marggraf & Co., 52 Bread-street, Cheapside.

White stitched filosh curtains; do. mull curtains; colored stitched filosh curtains; colored brillantines (for ladies' dresses).

2368. MARX, H. R., Seifhennersdorf near Löbau. Munich 1854 honor. mention; London 1862 honor. mention.

Cotton stuffs for gentlemen's coats and pantaloons.

2370. SCHÖNE, JOHANN GOTTFRIED, Gross-Röhrsdorf. Munich 1854 medal of honor; London 1862 honor. mention.

Girt-ribbons for suspenders, saddle-girths etc.; plush- and velvet-ribbons, border-laces etc., of cotton, linen and half-wool; Penelope-canevas and green-gauze for windows.

2371. Seyfert, Gustav, Auerbach. London 1862 medal.

White cotton curtain-stuffs; prices p. 24 yards. No. 1. 32 in. broad 13 sh.; No. 2. 42 in. br. 18 sh.; No. 3. 52 in. br. 25 sh.; No. 4. 52 in. br. 24 sh.; No. 5. 52 in. br. 27 sh.; No. 6. 42 in. br. 19 sh.; No. 7. 52 in. br. 27 sh.; No. 8. 52 in. br. 27 sh.; No. 9. 52 in. br. 27 sh.

2372. Tetzner, C. A., & Son, Burgstädt, Paris 1855 honor. mention.

Cotton yarns for knitting, crotcheting and weaving: raw and bleached Estremadura, 6 thread; do. supra 3 and 4 thread, do. secunda 4 thread; woolblue, wool-gray and woolbrown, Turkish red, blue, brown and gray colored, variegated, flamed etc. knitting yarns; single merino yarn for weavers, in gray, blue and brown.

2374. Waentig, & Co., Zittau. Munich 1854 great medal; Paris 1855 silver medal; London 1862 medal. Agt. W. Meyerstein, 47 Friday-street.

Cotton, union and linen-drills from 6½ to 19½ d. p. yard, for home trade and exportation.

2375. Zschimmer & Grimm, Plauen. Agt. C. Trübner, 20 St. Dunstans Hill, City.

Cotton gauze-curtains p. piece 23 yards 1 £ 13 sh.

CLASS XIX.
FLAX AND HEMP.

2381. Bursche, Johann Gotthelf, Pulsnitz. London 1862 honor. mention.

1 piece of double sail-cloth, 178 in. br., p. yard 7 sh. 6 d.; 1 coup. linen-atlas, 105 in. br., 1 do. twill, 105 in. br., for railway-carriages etc., p. yard 4 sh. 3 d.; a complete railway-lowry-cover, p. yard 2 £ 11 sh.

2383. Meyer, Joseph (au petit Bazar), Dresden. London 1862 medal. Agt. C. Trübner, 20 St. Dunstans Hill, City.

Linen and silk-damasks, woven at Gross-Schönau for the exhibitors account: Table-cloths 3½ by 2½ yards, with 12 napkins 33 in. □: No. 1. 17 Th. 15 sg. (2 £ 12 sh.); No. 2. 25 Th. (3 £ 14 sh.); No. 3. 26 Th. (3 £ 17 sh.); No. 4. 27 Th. (4 £); No. 5. 30 Th. (4 £ 9 sh.); No. 6. 32 Th. (4 £ 15 sh.); No. 7. 35 Th. (5 £ 4 sh.); No. 8. 35 Th. (5 £ 4 sh.); No. 9. 39 Th. (5 £ 16 sh); No. 10. 42 Th. (6 £ 4 sh.); tea table-cloth of white silk; do. of white and yellow silk; 12 doilies do., 6 do. of white and gray linen-damask; 2 table-cloths do.

2384. Neumann, C. F., jun.. Eybau. Munich 1854 great medal; Paris 1855 bronze medal; London 1862 medal.

Linen and half-linen goods for exportation. Prices franco Hamburgh. No. 1—5. 5 pieces pure linen creas, à 70 Varas (66 yards), No. 36: 38 sh. 6 d.; No. 40. 40 sh.; No. 50. 46 sh.; No. 56. 50 sh. 6 d.; No. 60. 55 sh.; 5 pieces Listados de hilo No. 160., à 48 var. (45 yards), No. 6. blue ⅔ □ half-linen, 19 sh. 3 d.; No. 7. blue clear-ground, ¼ lin., do. 20 sh.; No. 8. varieg. □ holanda, ½ lin., do. 20 sh.; No. 9. blue and red ⅔, ⅓ ||| ¼ lin., do. 20 sh.; No. 10. blue div. ||| ¼ lin., do. 20 sh.; No. 11. 1 piec. varieg. □ ¼ lin. Listados imperiales No. 220. 24 sh. 6 d.; No. 12. 1 piece blue ⅔ □, ½ lin., Listados primera No. 160. 20 sh.; No. 13. and 14.

⅚ pieces Listados, à 24 Varas (22½ yards), ⅜ □, ½ lin., No. 150. blue 8 sh. 6 d., red 9 sh. 7⅓ d.; No. 15. and 16. ¾ □, ½ lin., Listados legitimos No. 160. blue 10 sh., red 10 sh. 4½ d.; No. 17. blue clearground, ¼ lin., Listados Arabias No. 220. 12 sh. 7½ d.; No. 18. varieg. □ do. No. 260. 13 sh.; No. 19. ⅓ piece blue ⅔ □ pure linen Arabias No. 260. 14 sh. 9 d.

Sauer.   (s. No. 2321.)

CLASS XXI.
WOOLLEN AND WORSTED, INCLUDING MIXED FABRICS GENERALLY.

2385. Albrecht, Robert, Chemnitz. Munich 1854 medal of honor; Paris 1855 bronze medal; London 1862 honor. mention.

Woollen and mixed damasks and other stuffs for furniture, curtains, carriages etc., table-covers: 6 coup. Carola-portières; 6 coup. cord-travers do.; 4 pieces do. Ponchos; 2 pieces do. with-silk; 6 pieces rips do.; 5 pieces diff. saddle-cloths; 2 pieces table-covers of velour-rips façonné: 1 heft half-wollen damask; 1 heft do. rips uni et façonné; 1 heft do. striped; 1 heft do. velour-rips façonné; 1 heft woollen-damask; 1 heft do. rips uni et façonné.

2386. Bauch, Carl August, Rosswein.

Calmucks: 1 piece 35 in. (30½ yards) 1 sh. 2¼ d. 1 piece 26 in. (44½ yards) 9¼ d.

2387. Bauch, Traugott, Rosswein.

Calmucks: 1 piece 35 in. br. 22¾ yards long p. y. 1 sh. 2¾ d.; 1 piece 26 in. br. 43¼ yards long, p. y. 9¼ d.

2388. Beckert, Richard, Chemnitz. London 1862 honor. mention. Agt. Ferd. Drewes, 67 Newgate-street, E. C.

50 coup. mixed woollen-damasks for furniture; 6¼ doz. Ponchos of woollen and silk mixture; 4 doz. half-woollen Imperial-blankets.

2389. Bleyl, Fr., Camenz. Munich 1854 medal of honor.

Woollen cloth: No. 1. green, 21⅔ yards, 7 sh. 2 d.; No. 2. blue, 19⅓ yards, 7 sh. 2 d.; No. 3. woolblue, 22⅔ yards, 9 sh. 1 d.; No. 4. sai, 21 yards, 7 sh. 2 d.; No. 5. amranth, 22⅓ yards, 8 sh. 5½ d.; all p. yard.

2391. Böttger.Brothers. Leisnig. London 1862 honor. mention.

1 piece 21¼ yards summer-coat-stuff p. yard 7 sh. 6 d.; 1 piece 18⅜ yards winter do. 12 sh. 10 d.

2392. Brodengeyer. Fr., & Co.. Annaberg. Agt. Killy, Traub & Co., 52 Bread-street, Cheapside.

Chenille scarfs and collars of cotton, wool and silk.

2393. Caspari, J. F., Grossenhain.

Woollen cloth: 1 p. black cloth 16¼ yards à yard 8 sh. ¾ d.; 1 do, finished 17½yards à 8 sh. 10½ d.; 1 p. do. croisé 11½ yards à 8 sh. ¾ d.; 1 p. do. satin 20⅓ yards à 8 sh. 5½ d.

2394. Claus. R., & Co., Schedewitz nea Zwickau. London 1862 medal. Agt. C. G. Zimmermann, 2 St. Pauls Buildings, Little Carter Lane, Doctors Commons.

Halfsilk stuffs for ladies' dresses; prices p. piece of

28/29 yards: 4 dresses Alpaca with silk, 31 in. 19$\frac{1}{2}$ Thlr. (48$\frac{3}{4}$ sh.); Paramattas do. 27 in., 13$\frac{1}{2}$ Th. (33$\frac{3}{4}$ sh.); mixed lustre do., 27 in., 12$\frac{3}{4}$ Th. (31$\frac{1}{2}$ sh.); Orleans do. 27 in., 12 Th. (30 sh.); 13 dresses div. cords with silk, 23 in., 13$\frac{1}{2}$ Th. (33$\frac{3}{4}$ sh.); 1 dresses cord embroidered with gold, 23 in., 18$\frac{1}{2}$ Th. (46$\frac{1}{4}$ sh.).

2395. Dieterich, Heinrich, Meerane. London 1862 medal.

Woollen, mixed and halfsilk stuffs for ladies' dresses: 12 coup. crêpe imperial 35 in.; 1 coup. do. chiné 35 in.; 5 coup. do. à soie 35 in.; 2 coup. do. brodé 35 in.; 13 coup. poil de chêvre 23$\frac{1}{2}$ in; 5 coup. Mozambique 23$\frac{1}{2}$ in.; 2 coup. do. à soie 23$\frac{1}{2}$ in.; 1 coup. do. épinglé 23$\frac{1}{2}$ in.; 3 coup. do. brodé 23$\frac{1}{2}$ in.; $\frac{1}{6}$ doz. châles Mozambique à soie; $\frac{1}{12}$ doz. do. brodé; $\frac{1}{12}$ doz. do. crêpe à soie.

2396. Eckhardt, Brothers, Grossenhain. Munich 1854 medal of honor; Paris 1855 bronze medal. Agt. C. Trübner, 20 St. Dunstans Hill, City.

Woollen stuffs; prices p. yard: 1 piece winter stuff □, 8 yards, 10 sh. 7 d.; 9 pieces summer- do. diag. 9$\frac{1}{2}$ yards, 7 sh. 10 d.; do. □, 9$\frac{7}{8}$ yards, 7 sh. 4 d.; do. □, 9$\frac{3}{4}$ yards, 7 sh. 4 d.; do. velvet 8 yards, 7 sh. 4 d.; do. 9$\frac{3}{4}$ yards, 7 sh. 4 d.; do. 8$\frac{3}{4}$ yards, 7 sh. 4 d.; velv. striped 9$\frac{3}{4}$ yards, 7 sh. 4 d.; do. 9$\frac{1}{8}$ yards, 7 sh.; do. 9$\frac{3}{4}$ yards 7 sh.

2397. Facilides & Wiede, Plauen. Munich 1854 honor. mention; London 1862 medal. Agt. W. & W. Paterson & Co., Glasgow. (s. II. No. 75.)

Vicognia or Angola yarns of different qualities, mixtures and plain colors, No. 12/18. and 16., prices from 1 sh. 6 d. to 3 sh. 2 d.

2399. Frohberg, C. G., Rosswein. London 1862 honor. mention.

1 piece cylinder-cloth 19$\frac{7}{8}$ yards, p. yard 10 sh. 9 d.

2400. Frohberg, Brothers, Rosswein.

1 piece black cloth 17$\frac{3}{4}$ yards, p. yard 6 sh. 3 d.; 1 do. satin 18$\frac{7}{8}$ yards p. yard 6 sh. 9 d.

2402. Herrmann, F. G., & Son, Bischofswerda. Berlin 1844 silver medal; Munich 1854 great medal; Paris 1855 silver medal; London 1862 medal.

10 pieces cloth, twilled and satin; the prices are affixed to the pieces.

2403. Hertel & Büchelen, Meerane. Agt. Otto Sprösser, 41 Watling-street.

Woollen, halfwoollen and half silk stuffs.

Herzog. (s. No. 2362.)

Hoffmann. (s. No. 2364.)

2404. Hüffer, Heinrich, Crimmitzschau. London 1862 honor. mention. Agt. Chr. Höhs by F. Buth & Co.

Angola yarns from 1 sh. 8 d. to 2 sh. 9 d. p. lb. Engl.

Hüllig. (s. No. 2365.)

2406. Krah, Carl, Camenz.

2 pieces of woollen cloth: black 23$\frac{8}{10}$ yards p. yard 6$\frac{3}{10}$ sh.; brown 19$\frac{1}{2}$ yards p. yard 6$\frac{4}{10}$ sh.

2407. Krämer & Markendorf, Glauchau. London 1862 honor. mention. Agt. Philipp Beaufort, 31 Bush Lane, Cannon-street.

Woollen, mixed and halfsilk fancy goods for ladies; dresses and châles with bordures.

2408. Kratz & Burk, Glauchau. Paris 1855 silver medal; London 1862 medal.

Fancy goods for ladies' dresses: made of wool mixed with cotton; of do., decorated or embroidered with silk; of worsted entirely (namely worsted popelines); of do. decorated or embroidered with silk.

2409. Kürzel, Heinrich Ferdinand, Crimmitzschau. London 1862 medal.

Yarns of wool and cotton mixed (Vicognia or Angola).

2410. Lange, Adolph, Camenz.

3 pieces of woollen cloth: boil-blue 17$\frac{2}{3}$ yards p. yard 8$\frac{1}{3}$ sh.; black 22$\frac{9}{10}$ yards p. yard 6$\frac{6}{10}$ sh.; black 22$\frac{1}{2}$ yards p. yard 6$\frac{4}{10}$ sh.

2411. Lehmann, F. G., Böhrigen near Rosswein. Paris 1855 silver medal; London 1862 honor. mention.

An assortment of woollen and halfwoollen flannels, Lama-flannels, Domets, Boys, moltons; Spanish stripes; Circassiennes; Zephyrs; double velours; jupes d'Espagne.

2412. Leonhardt, F., & Son, Hainichen. Agt. A. T. Katsch, 29 Basinghall-street.

Three pieces of 27 in. twilled molleton.

2413. Leonhardt, G. F., Hainichen, am Markt. London 1862 medal. Agt. Victor Bauer, 1 Ironmonger Lane, E. C.

Div. white and colored flannels.

2415. Lohse, Eduard, Chemnitz. Berlin 1844 silver medal; London 1851 prize-medal; Munich 1854 medal of honor; Paris 1855 silver medal; London 1862 medal.

6 pieces reps plain, striped and fancy, union and worsted for furniture, curtains etc.; 1 piece reps cotelaine, to quilt carriages; 6 pieces worsted damasks with reps and satin ground; 8 pieces union and 4 pieces cotton damasks fancy and striped for furniture, curtains etc.; 4 pieces table-covers, all-wool and union.

2417. Meissner, Friedrich Traugott, Grossenhain. London 1851 prize medal, 1862 honor. mention. Agt. C. Trübner, 20 St. Dunstans Hill, City.

4 pieces of woollen cloth: blue $\frac{3}{4}$ cloth 25$\frac{1}{3}$ yards 8 sh. 3 d.; black do. 21$\frac{5}{8}$ yards 9 sh. 6 d.; wool-black 21$\frac{3}{8}$ yards 8 sh. 7$\frac{1}{2}$ d.; black twilled 18$\frac{7}{8}$ yards 8 sh. 7$\frac{1}{2}$ d.; prices p. yard.

2418. Metzler, Gottlieb, Rosswein. Munich 1854 honor. mention; London 1862 honor. mention.

2 pieces of calmucks 24$\frac{3}{4}$ and 24$\frac{1}{2}$ yards, p. yard 1 sh. 7$\frac{1}{3}$ d.

2419. Metzler, Wilhelm, Rosswein.

Two pieces of black cloth 21$\frac{5}{8}$ and 22 yards, p. yard 6 sh. 10 d.

2420. Minckwitz, Adolph, Camenz.

1 piece black twilled cloth 23 yards, p. yard 7 sh. 8 d.; 1 piece black plain cloth 23$\frac{1}{2}$, p. yard 6 sh. 10$\frac{1}{4}$ d.

***2421.** Morand, & Co., Gera. Berlin 1844 silver medal; London 1851 prize-medal; Munich 1854 great medal and knight of the order of St. Michael; Paris 1855 silver medal; London 1862 medal. Agts. Heintzmann & Rochussen, 9 Friday-street, Cheapside.

½ piece 74 in. cachemire d'Ecosse FFF.; ½ piece 48 in. tissu cachemire; ⅔ piece 36 in. grosgrain uni I.; ⅔ piece 36 in. grosgrain cannelé; ½ piece 47 in. cuir de laine No. 81.; ½ piece 47 in. drap d'été No. 84.; ½ piece 39 in. Zanella electoral; ½ piece 27 in. imperial rayé; 1 piece 39 in. popeline SPF.; ⅔ piece 44/45 in. thibet SSSF.; ½ piece 44/45 in. thibet SSSS.; ⅔ piece 44/45 in. thibet SSSA.; ½ piece 44/45 in. cach. d'Ecosse FFF.

2422. Mörbitz, C. G. E., Bautzen.

1 piece black woollen cloth 10 sh. 6 d.; 1 piece black woollen tricot 11 sh. 4 d.; 1 piece black woollen satin 11 sh. 4 d.; 1 piece boil-blue woollen cloth 12 sh. 11 d.; 1 marine blue woollen satin 12 sh. 11 d.; 1 piece boil-blue woollen satin 12 sh. 11 d.; all p. yard.

2423. Müller & Co., Crimmitzschau, Agts. S. Oppenheim & Sons, 4a. Bread-street, Cheapside.

7 pieces summer-buckskin 7 sh. 8 d.; 3 pieces do. with silk 8 sh. 2 d.; 2 pieces summer-tricot 8 sh. 4 d.; all p. yard.

2424. Nosske, Ernst, Camenz.

2 pieces of black woollen cloth 16 and 15⅔ yards, p. yard 6 sh. 10¼ d.

2425. Nosske, F., Camenz.

1 piece black cloth, mat, ready for the needle, p. yard 7 sh. 7½ d.; 1 piece black twilled cloth, do. p. yard 7 sh. 6 d.

2426. Nosske, Wilhelm, Camenz.

2 pieces blue woollen satin p. yard 8 sh. ⅔ d.

2430. Reichel, Christ. Fried., Rosswein. Munich 1854 medal of honor; London 1862 honor. mention.

1 piece black cloth 60 in.; 1 piece mixed drap royal; 8 pieces mixed drap d'été; 1 piece black ¾ cloth, each 52/53 in.; 1 piece black twilled cloth 51/52 in.; 2 pieces do. 54 in.

2431. Reissmann & Träger, Spinning-mill near Reichenbach.

Soft worsted yarn No. 60. 3 d. woof.

2432. Resch & Co., Meerane.

1 piece half-woollen poplin with silk; 2 dresses do. crêpe.

2433. Sievers & Engell, Meerane. Agts. Barnes, Marggraf & Co., 52 Bread-street, Cheapside.

Fancy stuffs for ladies of cotton and wool, of cotton, wool and silk mixed, prices p. Berl. ell (p. yard): gros d'Aumale, breadth 90 cm. (35½ in.), 12 sg. (1 sh. 8 d.); gros d'Aumale uni do. 11 sg. (1 sh. 6 d.); gros d'Aumale à soie do. 15 sg. (2 sh. 5 d.); brillantine, breadth 57 cm. (22½ in.), 5¾ sg. (9½ d.); brillantine chinois do. 6 sg. (10 d.); brillantine à oie, breadth 90 cm. (35 in.) 12½ sg. (2 sh. 1 d.).

2434. Solbrig, C. F., Harthau near Chemnitz. London 1851 prize-medal; Munich 1854 great medal; London 1862 medal.

Soft worsted yarns: aaa: merino twist No. 60/30.; electa-merino weft No. 76.

2435. Söllheim, G. F., Chemnitz. Munich 1854 honor. mention; London 1862 honor. mention.

Damasks for furniture etc., prices p. yard: Damask, wool and silk 55 in. 7 sh. 6 d.; woollen damask, 26 in., first qual. 1 sh. 9 d.; red 1 sh. 10 d.; extra first 1 sh. 10 d.; royal, silk, wool and cotton 55 in. 1 sh. 10 d.; royal, wool and cotton 55 in. 1 sh. 7 d.; imperial wool and cotton 25 in. I. 2 sh. 5 d.; do. II. 2 sh. 4 d.; cotton damask, red 23 in. I. 8½ d.; do. II. 6 d.

2436. Spengler, Carl, Crimmitzschau. London 1851 prize-medal.

4 coup. woollen buckskin, 36 in., p. yard 6 sh.; 2 coup. do. with silk, 36 in., p. yard 6 sh. 8 d.

2438. Straff & Son, Meerane. London 1862 honor. mention. Agt. Will. Meyerstein, 47 Friday-street, E. C.

Fancy stuffs for ladies' dresses in wool, union and silk: popeline 23¼ in. 10 d.; grenadine 46½ in. 1 sh. 7 d.; marbré do. 1 sh. 11 d.; Mozambique a soie do. 2 sh. 1 d.; Valencia do. 2 sh. 3½ d.; Mozambique mohair do. 2 sh. 4½ d.; Argentine 37¼ in. 2 sh. 4¼ d.; mantelet 50¼ in. 3 sh. 2¼ d.; all p. yard. Châles Mozambique 67 in., p. doz. 75 sh.

2439. Strübell & Müller, Meerane. London 1862 medal. Agt. Ch. Oppenheimer, 79½ Watling-street, Cheapside.

Half-wooll and half-silk fancy stuffs for ladies' dresses: 1 coup. poil de chêvre II a. 22½ in.; 3 coup. barège 23½ in.; 2 coup. chaly do.; 4 coup. poil de chêvre fin do.; 1 coup. grenadine do.; 1 coup. grenadine 47 in.; 1 coup. grenadine 66 in.; 2 coup. royal 35 do.; 3 coup. royal à soie do.; 2 coup. florentine à soie do.; 1 coup. camayeux à soie do.; 3 châles.

2440. Thieme & Co., Meerane. Agts. Barnes, Marggraf & Co., 52 Bread-street, Cheapside.

Woollen, halfwoollen and halfsilk fancy stuffs for ladies: 2 pieces Victoria prima 35 in.; 2 pieces do. secunda 36 in.; 2 pieces reps 23½ in.; 2 pieces lasting 23½ in.; 3 pieces Carolina 26 in.

2441. Unger, C. G., Kirchberg. Munich 1854 honor. mention.

Woollen cloth: blue p. yard 2 sh. 9 d.; checkered p. yard 2 sh. 7 d.; gray 1 sh. 10 d.; red p. yard 5 sh. 2 d.

2442. Vorwerg, Ephraim, Camenz.

Woollen cloth: 1 piece brown cloth p. yard 6 sh. 10¼ d.; 1 piece drap twilled cloth p. yard 6 sh. 10¼ d.

***2444.** Weissflog, Ernst Fr., Gera. London 1851 prize-medal; Munich 1854 medal of honor; London 1862 honor. mention. Agts. Morgan Brothers, 21 Bow-Lane, Cannon-street, City.

Soft worsted goods, woven on power-looms:

	Breadth.			p. Brb.ell	mtr.	yard	
	in. Sax.	cm.	in. Engl.	sg.	frs.	sh.	d.
Thibet qual. A.	46	108	43	10	1,78	1	3
» » 22.	45	106	42	12	2,13	1	7
» » 32.	45	106	42	16	2,88	2	2
» » 36.	45	106	42	17	3,06	2	3½
» » 42.	45	106	42	18	3,25	2	5
Lasting	45	106	42	20	3,56	2	8
Woven on hand-looms:							
Cachemire d'Écosse qual. VII. . .	49	116	46	24¼	4,31	3	3
do. qual. 40. . .	72	170	66	30½	5,43	4	1
rips façonné . .	41	96	38	22	3,94	3	1

Prices franco Gera netto.

2445. Wippern, C., & Wiehe, Crimmitzschau.

Angola yarns, prima and extra prima p. lb. 2 sh. 8 d. and 3 sh.

***2446.** Wittmer & Schönherr, Gera. Agts. Barnes, Marggraf & Co., 52 Bread-street, Cheapside.

Soft worsted goods:

		p. Brb. ell.	mètre	yard.	
	in.	sg.	frs.	sh.	d.
1 p. cachemire imprimé genre I. .	22/23.	10½	1,90	1	4¾
1 p. do. genre II. .	22/23.	13½	2,47	1	9¾
2 p. satin de laine façonné.	33.	14½	2,62	1	11¼
3 p. velours de laine façonné.	34/35.	23½	4,25	3	1½
1 p. velours de laine uni qual. 60. . .	37.	20	3,62	2	8
1 p. do. qual. 40. .	38/39.	22¾	4,12	3	½
1 p. electa	36/37.	19¾	3,58	2	7½
1 p. thibet	37/38.	16	2,90	2	1½
		p. piece:			
1 thibet □ shawl .	78 □	110	13,75	11	—
1 do. long-shawl. .	66/132.	147½	18,45	14	9

2447. Wolf, J. G., sen., Kirchberg.

4 pieces of woollen cloth, each 37 in.; black 14 yards 1 sh. 10 d.; light gray 18 yards 3 sh. 9 d.; black 21 yards 4 sh.; red 17 yards 4 sh. 2 d.; all p. yard.

2448. Wolf, Jul. Herm., Burgstädt near Chemnitz. London 1851 prize-medal, 1862 honor. mention.

All wool carded yarns, and cotton and wool mixed vicognia (angola-) and merino-yarns; both also with colored spots. Vicognia (angola) No. 12. dark blue mixed 1517. 2 sh. 2 d.; do. blue mixed 1520½ 2 sh. 4 d.; do. blue gray mixed 1519. 2 sh. 6 d.; do. black olive mixed 1523. 2 sh. 3 d.; do. No. 14/16. white washed 1437. 2 sh. 4 d.; do. No. 12. gray silver 1528. 1 sh. 8 d.; do. 12 black with diff. spots 1417. 2 sh. 10 d.; do. brown 1442. 3 sh. 2 d.; all wool carded yarn, rough No. 8—10. 1249. 3 sh. 4 d.; do. spotted, No. 10. diff. blacks 1426. 4 sh. 5 d.; do. No. 8. diff. silver gr. 1536. 4 sh. 5 d.; do. No. 10. diff. gray mix. 1425 4 sh. 6 d.; all p. lb.

***2449.** Zetzsche & Münch, Gera. Agt. J. C. Geiselbrecht, 8 Leadenhall-street.

Soft worsted goods, viz.: satin façonné, tartan laine, satin de laine, mousseline de laine, thibet, cachemire d'Écosse, reps tartan, reps roubaix, popeline, coteline. Shawls of thibet, cachemire and popeline.

2450. Zschille, Fedor, & Co., Grossenhain. Munich 1854 medal of honor; Paris 1855 silver medal.

Woollen stuffs: summer elastic velours 7 sh. 8 d.; do. with silk 8 sh. 4 d.; winter elastic velours 10 sh.; do. with silk 11 sh. 4 d.; do. paletot 11 sh. 4 d.; all p. yard.

2451. Zschille, Brothers, Grossenhain. Munich 1854 honor. mention; Paris 1855 silver medal; London 1862 medal.

Woollen cloth; 5 pieces cloth 7 sh. 7 d. to 10 sh.; 5 pieces ¾ cloth for exportation 8 sh. to 11 sh. 3 d.; 8 pieces elastic cloth, ready for the needle 8 sh. to 12 sh.; all p. yard.

CLASS XXII.

CARPETS.

2463. Quast, Friedrich, Leipzig. Munich 1854 honor. mention; London 1862 medal. Agts. P. & F. Schäfer, 6 Golden Square.

Printed oil-fustians for the lining of railway-waggons; floor-cloth.

2464. Röller & Huste, Leipzig. London 1851 prize-medal; Munich 1854 medal of honor; Paris 1855 silver medal; London 1862 medal.

Articles of oil-cloth: floorcloth of various qualities and sizes; do. for stairs; oil-cloth-canvas; canvas for painters, linen; do. ticken; window-screens in frames. 5 geographical maps printed on oil-cloth; 14 geographical skeleton maps on varnished paper, for the use of schools; conf. No. 2512.; oil-cloth-girdles; do. measures; do. collars and cuffs; oil-cloth-linings for slippers; moleskin piano-cover; printed moleskins, wood and marble imitation and var. designs; japanned muslin plain black; cotton, japanned, embossed, printed; japanned cotton for panotypes; printed gauze for window-screens; printed table-covers on moleskin; do. table-mats; oil-cloth-apron.

2465. Schäfer, Joh. Heinrich, Chemnitz. Munich 1854 and Paris 1855 honor. mention; London 1862 honor. mention.

Oiled floor-cloth, inlaid floor imitation; do. fustians, wood and marble imitation; do table-covers, wood imitation.

2466. Schumann, Alexander, Leipzig. London 1862 honor. mention. (s. II. No. 98.)

1 double oiled oil-cloth-carpet upon linen stuff, 73 in. wide, 10 yards long; 2 pieces of oiled muslin each 40 in. wide, 8¾ yards long, printed, *a.* gray ground, *b.* yellow do.; 2 pieces of oiled fustian, each 43 in. wide, 8¾ yards long, *a.* gray ground, *b.* yellow do.

2467. Wäntig, Ernst Ferd., Leipzig.

Imitations of leather and substitutes for leather, and fancy articles etc.

CLASS XXIII.

WOVEN, SPUN, FELTED AND LAID FABRICS, WHEN SHOWN AN SPECIMENS OF PRINTING OR DYEING.

2468. Chevalier, L., & Son, Leipzig. London 1862 medal. Agt. C. Trübner, 20 St. Dunstans Hill, City.

Shawls; prices p. dozen: 11 pieces cachemire Ima. 66 in. with 12 in. fringes 6 £; 12 pieces IIda. 5 £ 11 sh.; with do. 5 pieces circassienne 68 in. w. 9 in. fringes 7 £ 4 sh.; 6 pieces satin 66 in. w. 12 in.

fringes 6 £ 12 sh.; 8 pieces cachemire Ima. 33 in. 1 £ 7 sh.

2471. Römer, Brothers, Hainsberg near Dresden. Munich 1854 honor. mention.

Turkish red yarn No. 40. medio; pink do. No. 20. water.

Schütz.   (s. No. 2310.)

2474. Unger & Co., Schönheide. Agt. W. Meyerstein, 47 Friday-street.

Woollen shawls, 66 in., prices p. piece: 33 pieces printed, with wool fringes 9 sh.; 4 do. with silk stripes and woollen fringes 11¼ sh.; 5 do. application of col. velvet mohair-fringes 12 sh.; 1 do. 13½ sh.; 6 do. embroidered with silk (florissants) and mohair-fringes 16½ sh.

2 edges of a table-cover, application of col. velvet 3 £.

2475. Winter, Wilhelm, Chemnitz. London 1862 medal.

Woollen printed shawls, cravats, brooches, points and table-covers of all qualities and sizes.

CLASS XXIV.

TAPESTRY, LACE AND EMBROIDERY.

2478. Hietel, J. A., manu. of embroideries and tapestry, Leipzig. London 1851 prizemedal; Munich 1854 medal of honor; Paris 1855 bronze medal; London 1862 medal.

Embroideries: Portraits embroidered on white silk, with crape and hair: No. 1. of the Queen dowager of Saxony 15 £; No. 2. of the late King of Saxony 10 £; No. 3. of King John of Saxony 10 £; No. 4. of Prince Albert of Saxony 10 £; No. 5. of Prince George of Saxony 10 £.

No. 6. Portrait of count Chambord 20 £; No. 7. Germania guarding the Rhine 15 £; No. 9 to 12. bouquets embroidered in silk on white silk à 1 £ 13 sh.; No. 15 to 25. div. embroideries in wool and pearls on royal stramin, variegated bouquets (relief) the leaves in cross-work· and pearls from 1 £ 1 sh. to 6 sh.; No. 51 to 62. 12 embroidered linen shirts with collars, fronts and cuffs. The shirts have been washed on account of the embroidery, which is very difficult to be obtained clear and white on handspun-linen from 1 £ 4 sh. to 1 £ 10 sh.

2479. Klemm, Herrmann Th., Lössnitz. Agt. H. F. Goblet, 20 Milk-street, E. C.

Embroideries: 6 muslin-collars from 1 d. to 6 d.; 1 do. cravat 1 sh. 6 d.; 1 ribbs do. 2 sh. 5 d.; 2 linen-sets 4 sh. 6 d. and 5 sh. 6 d.; 4 muslin-sets 1 sh. 7 d. to 10 sh. 6 d.; 1 do. with entre-deux 9 sh. 6 d.; 4 do. with lace 8 sh. 3 d. to 13 sh. 6 d.; 3 do. with sleeves 10 sh. 6 d. to 12 sh.; 1 ff. pillow French linon with ff. lace 12 £; 1 small sewing-frame.

2480. Reim, C. G., Buchholz. Paris 1855 bronze medal. Agt. L. Walpole, Great Sutton-street, City.

Laces of hand-bob-work, galloons, trimmings etc.: Silk laces: p. piece à 15 yards. No. 1013. 32 sh. 8 d.; No. 1020. 41 sh.; No. 1024. 58 sh. 9 d.; No. 1026. 30 sh. 9 d.; No. 1027. 30 sh.; No. 1054. 14 sh. 9 d.; No. 1055. 30 sh.; No. 1056. 30 sh.

Cotton trimmings for curtains, p. piece à 15½ yards: No. 461. 5 sh.; No. 463. 3 sh.; No. 470. 2 sh. 4 d.; No. 471. 4 sh. 5 d.; No. 472. 5 sh.; No. 473. 5 sh. 8 d.; No. 474. 8 sh. 6 d.; No. 475. 7 sh. 3 d.; No. 476. 15 sh.; No. 477. 15 sh.; No. 478. 18 sh.; No. 479. 20 sh. 5 d.; No. 480. 20 sh. 6 d.

Negligé-trimmings, p. 15½ yards: No. ⅛. 2 sh.; No. ⅛. 1 sh. 10 d.; No. 1. 2 sh. 7 d.; No. 1. do. double 4 sh. 8 d.

Silk agrements, p. 12½ yards: No. 2549. 3 sh. 4 d.; No. 2798. 6 sh. 9 d.; No. 2816. 2 sh. 5 d.; No. 2823. 8 sh. 8 d.; No. 2824. 7 sh. 10 d.; No. 2825. 6 sh. 8 d.; No. 2826. 6 sh. 8 d.; No. 2827. 8 sh.; No. 2828. 13 sh. 8 d.; No. 2829. 34 sh.; No. 2830. 7 sh. 10 d.; No. 2831. 19 sh.; No. 2832. 48 sh.; No. 2833. 9 sh. 8 d.; No. 2834. 32 sh., No. 2835. 32 sh.; No. 2836. 6 sh. 9 d.; No. 2837; 8 sh. 9 d.; No. 2838. 13 sh. 5 d.; No. 2839. 20 sh.; No. 4712. 12 sh.; No. 4713. 12 sh.; 10 pCt. scontro, in draughts for three months.

2481. Schubert, Ernestine, Annaberg. London 1862 honor. mention. Agt. C. Trübner, 20 St. Dunstans Hill, City.

Two dolls completely attired and dressed; diff. articles of lace; crotchet-work and embroideries.

2482. Unger, Max, Johanngeorgenstadt. Agt. Eugen Herz, 59 Basinghall-street.

A ff. embroidered mantilla of silk with velvet 22 £.

Wimmer & Dietrich.   (s. No. 2547.)

CLASS XXVI.

LEATHER, INCLUDING SADDLERY AND HARNESS.

2484. Beck, Daniel. Doebeln. London 1862 medal. Agts. M. Sichel & Co.

1 lacquered neats-hide for coaches 33 sh.; 2 lacquered calf-skins 4½ sh.; 2 lacquered kid-skins 4½ sh.; all p. piece.

2 pair of legs for boots 27 sh.; 2 pair of vamps for boots 21 sh.; 2 pair of upper-leathers for boots 12 sh.; all p. doz.

2486. Lange, Ferd.. Oschatz. London 1862 honor. mention.

Two dickers of col. morocco leather 22 sh.; one dicker do., black 19 sh. p. dicker.

2487. Neubert, Amalie, Leipzig. (s. II. No. 71.)

Articles of patent plaited leather-work with leather flowers: 1. waste-paper basket 3 £ 15 sh.; 2. do. Chinese fashion 2 £ 17 sh.; 3. basket to hang on the wall 1 £ 18 sh.; 4. small work-basket with cover, white morocco 2 £; 5. do. 1 £; 6. tray for visiting-cards do. 14 sh.; 7. lamp-stand 10 sh.; 8. card-holder, white morocco 4 sh.

*2488. Schlessiger & Lummer. Gera. London 1862 medal.

Leather for piano-makers: 5 p. hammer-leather, of American deerskins; 3 p. brown leather for English mechanism do.; 3 p. yellow buck-leather for English mechanism, of English deerskins: 2 p. black leather for English mechanism of German calf skins; 1 p. hopper-leather, yellow, for German mechanism of English deerskins; 1 p. brown ground leather do.

CLASS XXVII.

ARTICLES OF CLOTHING.

2489. Braunsdorf. Wilh., Leipzig. Agt. Victor Bauer, 1 Ironmonger-Lane, E. C.

Samples of shoemaker's ware; prices p. doz. pair: Lasting-boots for women No. 121. 42 sh.; No. 122. 37 sh. 6 d.; leather boots for women No. 125. 72 sh.; No. 126. do. for men 108 sh.;

leather legs for men No. 127. 48 sh.; No. 128. 57 sh.; upper parts for men's boots No. 129. 24 sh.; No. 130. 25 sh. 6 d.; leather legs for men's boots No. 131. 47 sh.; No. 132. boots for men 96 sh.; No. 134. shoe-parts for men 45 sh.; lasting-legs for women No. 135. 24 sh.; No. 136. do. with leather 30 sh.; No. 138. do. without leather 15 sh.; No. 139. 15 sh. 6 d.; No. 140. 15 sh.; No. 141. 16 sh.; with elastic No. 142. 19 sh. 6 d.; No. 143. 24 sh.; lasting-legs for misses No. 144. 11 sh. 3 d.; No. 145. 11 sh. 3 d.; lasting-legs for children No. 146. 8 sh.; No. 147. 9 sh.; lasting-boots for women No. 148. 54 sh.; No. 149. 40 sh. 6 d.; No. 150. 45 sh.; slippers No. 151. 18 sh.; No. 152. 28 sh. 6 d.; No. 153. 30 sh.; No. 154. 33 sh.

2490. Haugk, Herrmann, Leipzig. London 1862 medal.

Hats of velvet and felt and other hat-maker's articles.

2491. Hausding, L., manu. of umbrellas, Chemnitz. London 1862 medal. Agt. C. Trübner, 20 St. Dunstans Hill, City. (s. II. No. 72.)

8 umbrellas of silk 4 £ 1 sh. to 7 £ 8 sh.; 18 parasols and en-tous-cas 1 £ 19 sh. to 4 £ 16 sh. p. dozen.

2492. Hecker, Gottlieb, & Sons, Chemnitz. Berlin 1844 silver medal; Munich 1854 great medal; Paris 1855 silver medal; London 1862 medal. Agts. Cusel & Co., 62a. Cannon-street, West.

All kinds of cotton and linen hosiery.

2493. Herrmann, Friedr. Gotthold, Oberlungwitz near Chemnitz. Paris 1855 bronze medal; London 1862 honor. mention.

Hosiery, viz: ladies' hose, men's socks and gloves of cotton, lisle thread and silk; ladies' hose, men's socks; jackets and drawers of merino; jackets of wool and silk.

2494. Hiller, Carl, & Son, Chemnitz. London 1862 honor. mention. Agt. John B. Taylor, ½ Mumford Court, Milk-street, E.C. (s. II. No. 89.)

Cotton hosiery; prices p. dozen.

No. 2. 6 pairs white ladies' hose 2 sh. 1 d.; No. 3. 6 do. 2 sh. 2 d.; No. 25. 6 do. 2 sh. 8 d.; No. 27. 6 do. 2 sh. 9 d.; No. 28. 5 pairs pink ladies' hose 3 sh. 4 d.; No. 29. 6 pairs white ladies' hose 3 sh.; No. 342. 7 pairs pink ladies' hose 3 sh. 6 d.; No. 37. 7 pairs brown ladies' hose 3 sh. 9 d.; No. 38. 6 do. 4 sh. 1 d.; No. 36. 6 do. 3 sh. 11 d.; No. 39. 6 pairs mode ladies' hose 4 sh.; No. 9. 6 pairs slate ladies' hose 3 sh. 1 d.; No. 13. 5 pairs white ladies' hose 3 sh. 3 d.; No. 12. 5 do. 3 sh. 6 d.; No. 22. 3 do. 4 sh.; No. 51. 6 do. 5 sh. 2 d.; No. 64. 6 do. 4 sh. 2 d.; No. 44. 5 do. 3 sh. 8 d.; No. 76. 6 do. 5 sh. 9 d.; No. 81. 6 pairs brown ladies' hose 6 sh. 9 d.; No. 83. 6 do. 7 sh. 3 d.; No. 82. 6 do. 7 sh. 6 d.; No. 84. 5 pairs white ladies' hose 7 sh. 7 d.; No. 87. 5 do. 8 sh.; No. 92. 4 pairs brown ladies' hose 7 sh.; No. 94. 5 do. 7 sh.; No. 53. 4 pairs white ladies' hose 5 sh.; No. 61. 7 do. 4 sh. 4 d.; No. 142. 6 do. 9 sh.; No. 122. 6 do. 8 sh.; No. 121. 6 do. 7 sh. 7 d.; No. 153. 6 pairs brown ladies' hose 6 sh. 8 d.; No. 157. 6 pairs white ladies' hose 6 sh. 7 d.; No. 162. 7 do. 7 sh. 7 d.; No. 233. 6 do. 7 sh.; No. 179. 7 do. 16 sh.; No. 172. 5 pairs brown ladies' hose 8 sh. 10 d.; No. 194. 5 pairs mixed ladies' hose 2 sh. 6 d.; No. 198. 5 do. 3 sh. 2 d.; No. 202. 4 do. 4 sh.; No. 211. 6 do. 3 sh. 7 d.; No. 242. 3 do. 4 sh. 9 d.; No. 265. 4 do. 5 sh. 8 d.; No. 312. 5 pairs black ladies' hose 8 sh. 7 d.; No. 9. 6 pairs white ladies' hose 3 sh. 8 d.; No. 20. 5 do. 4 sh. 7 d.

No. 10. 7 pairs men's white hose 3 sh. 9 d.; No. 41. 3 pairs men's brown hose 6 sh. 4 d.; No. 72. 6 do. 8 sh. 5 d.

No. 3. 8 pairs men's brown half hose 1 sh. 6 d.; No. 5. 8 do. 2 sh.; No. 9. 7 do. 2 sh. 5 d.; No. 12. 5 do. 3 sh.; No. 15. 4 do. 3 sh. 1 d.; No. 18. 3 do. 3 sh. 5 d.; No. 20. 6 do. 3 sh. 5 d.; No. 22. 8 do. 3 sh. 3 d.; No. 25. 5 do. 4 sh. 7 d.; No. 27. 8 do. 4 sh. 3 d.; No. 31. 6 do. 4 sh. 5 d.; No. 35. 6 do. 2 sh. 11 d.; No. 37. 6 do. 2 sh. 9 d.; No. 40. 4 do. 3 sh. 9 d.; No. 44. 6 do. 3 sh. 7 d.; No. 54. 6 do. 4 sh.; No. 57. 6 do. 4 sh. 5 d.; No. 60. 6 do. 5 sh. 4 d.; No. 63. 5 pairs men's white half hose 5 sh. 11 d.; No. 72. 4 pairs men's brown half hose 4 sh. 5 d.; No. 80. 5 do. 8 sh. 1 d.; No. 84. 6 do. 9 sh.; No. 96. 5 pairs men's white half hose 2 sh. 2 d.; No. 115. 6 do. 2 sh. 9 d.; No. 118. 9 do. 3 sh. 9 d.; No. 128. 6 do. 2 sh. 2 d.; No. 130. 6 pairs men's mixed half hose 1 sh. 9 d.; No. 133. 6 do. 2 sh. 5 d.; No. 136. 5 do. 3 sh.; No. 142. 5 do. 3 sh. 4 d.; No. 151. 5 do. 2 sh. 6 d.; No. 160. 8 do. 2 sh. 5 d.; No. 164. 6 do. 3 sh. 1 d.; No. 192. 6 do. 4 sh. 6 d.; No. 212. 6 do. 6 sh.; No. 36. 6 pairs men's fancy half hose 2 sh. 11 d.; No. 197. 6 do. 4 sh. 5 d.; No. 198. 6 do. 4 sh. 5 d.; No. 202. 7 do. 3 sh. 4 d.

No. 167. 4 pairs children's mixed hose No. 4. 1 sh. 7¾ d.; No. 53. 3 do. 2 sh. 10½ d.; No. 62. 6 do. 2 sh. ¼ d.; No. 58. 4 do. 2 sh. 4 d.; No. 23. 5 pairs children's white hose No. 1. 2 sh. 2 d.; No. 15. 6 do. No. 5. 2 sh. 1 d.; No. 31. 4 do. No. 3. 4 sh. 11 d.; No. 42. 6 do. No. 1. 2 sh. 5 d.; No. 10. 6 do. No. 2. 1 sh. 4½ d.; No. 2. 5 do. No. 5. 2 sh. 7 d.; No. 72. 5 pairs children's fancy hose No. 1. 1 sh. 2 d.; No. 77. 6 do. No. 3. 2 sh. 4 d.; No. 84. 4 do. No. 4. 5 sh. 1½ d.; No. 85. 5 do. No. 4. 4 sh. 8¾ d.; No. 92. 5 do. No. 3. 4 sh. 4½ d.; No. 91. 6 do. No. 3. 2 sh. 1 d.; No. 91. G. 6 do. No. 3. 2 sh. 3 d.; No. 80. G. 6 do. No. 1. 1 sh. 9 d.; No. 97. 6 do. No. 3. 3 sh. 7 d.; No. 95. 6 do. No. 1. 1 sh. 11 d.; No. 101. 6 do. No. 3. 2 sh. 4 d.

No. 2. 8 pairs boys' white half hose No. 5. 1 sh. 9 d.; No. 22. 6 pairs boys' brown half hose No. 4. 2 sh. 2 d.; No. 34. 6 do. No. 5. 6 sh. 5 d.; No. 44. 5 pairs boys' mixed half hose No. 5. 3 sh. 2 d.; No. 38. 5 do. No. 4. 1 sh. 8½ d.; No. 59. 5 pairs boys' fancy half hose No. 5. 5 sh. 2 d.; No. 54. 7 do. No. 1. 2 sh. 7 d.; No. 60. 6 do. No. 1. 1 sh. 7 d.; No. 61. 6 do. No. 4. 2 sh. 5 d.; No. 62. 5 do. No. 2. 1 sh. 9½ d.; No. 63. 5 do. No. 3. 1 sh. 10½ d.; No. 17. 1 piece men's cotton shirts 10 sh. 10 d.; No. 20. 1 piece men's woollen shirts 37 sh. 6 d.

2495. Muehle, Aug., Pirna. Munich 1854 honor. mention.

Diff. sorts of shoes, slippers, galoches and boots of felt.

2498. Reichel, H. H., Dippoldiswalde near Dresden. Munich 1854, Paris 1855 honor. mention; London 1862 medal.

Straw-plaitings, white and colored, of different degrees of fineness, all worked by the direction of the exhibitor.

2499. Rudloff, H., Leipzig.

Shoemaker's ware: 1 pair of varnished riding-boots 12 £; 1 pair varnished boots with morocco legs 1 £ 4 sh.; p. pair.

1 pair varnished gaiters with double soles 10 £ 16 sh.; 1 pair do. single 9 £; 1 pair varnished dancing shoes for gentlemen 10 £ 16 sh.; 1 pair varnished shoes 9 £; 1 pair varnished shoes with chagrin and buttons 9 £; 1 pair gaiters of calf's leather with double soles 9 £; 1 pair do. single 7 £ 10 sh.; 1 pair shoes of calf's leather with double soles 8 £ 8 sh.; 1 pair do. single 7 £ 4 sh.; all p. doz. pair.

2500. SCHMIDT & HARZDORF, Hartmannsdorf near Chemnitz. London 1862 honor. mention. (s. II. No. 74.)

Cotton hosiery, particularly for the South-American market.

2501. UHLE & Co., Neustadt near Chemnitz. London 1862 honor. mention. Agts. Heintzmann & Rochussen, 9 Friday-street, Cheapside. Cotton hosiery.

2502. VOECKLER, TH., & Co., Meissen. Munich 1854 medal of honor; London 1862 medal. Agts. Wolf & Baker, 1 Sambrook-Court, Basinghall-street. (s. II. No. 82.)

Umbrella-ribs of Wallosine (Vöckler's pat. artificial whalebone). Prices p. bundle of 96 ribs: 16 in. 40 d.; 18 in. 45 d.; 20 in. 51 d.; 22 in. 58 d.; 24 in. 64 d.; 26 in. 72 d.; 28 in. 80 d.; 30 in. 88 d.; 32 in. 96 d.; 34 in. 110 d.; 36 in. 130 d.

2503. WEX & SONS, Chemnitz. London 1851 prize-medal; Paris 1855 silver medal; London 1862 medal.

Cotton hose for exportation.

2504. WOLLER, FRIEDR. EHREGOTT, Stollberg near Chemnitz. London 1862 medal.

Cotton hosiery for exportation.

CLASS XXVIII.

PAPER, STATIONERY, PRINTING AND BOOKBINDING.

2505. BACH, J. G., Leipzig. Munich 1854 medal of honor.

Portrait of His Maj. King John of Saxony, after Gonne's original, printed in oil colors 3 £; a specimen-book of litographic works.

2506. BRANDSTETTER, FRIEDR., bookseller and publisher, Leipzig.

Books and engravings on steel: Blätter und Blüthen Deutscher Poesie und Kunst, ein Album, bound 1 £; the steel-engravings from the same, each 3 sh.; Oesers Briefe an eine Jungfrau über Gegenstände der Aesthetik, 7 edit., bound 11¼ sh.; v. Heyden, das Wort der Frau, 10. edit., bound 6 sh.

BREITKOPF & HÄRTEL. (s. Nr. 2341.)

2507. BROCKHAUS, F. A., Leipzig. London 1851 prize-medal; Munich 1854 medal of honor; Paris 1855 silver medal; London 1862 medal. Agts. C. Trübner & Co., 60 Paternoster Row E. C.

Books and prints: Schlagintweit, Results, vol. 1. 2. with atlas, part. 1. 2. à 26 Th. 20 ngr., 53 Th. 10 ngr. (8 £); Goethe-Gallerie, illustrated edition, part. 1. 2. à 2 Th. 12 ngr., 4 Th. 24 ngr. (14 sh.); Schiller-Gallerie, illustrated edition, bound 30 Th. (4 £ 10 sh.); Schulze, Bezauberte Rose, illustrated edition, bound 6 Th. (18 sh.);

Lange, Atlas von Sachsen, bound 5 Th. 20 ngr. (17 sh.); Brockhaus' Reise-Atlas, bound 7 Th. (1 £ 1 sh.); Arendts, Naturhistor. Schulatlas 1 Th. 5 ngr. (3½ sh.); Arendts, Histoire élément. natur. 1 Th. 5 ngr. (3½ sh.); Götz von Berlichingen, bound 7 Th. (1 £ 1 sh.); Bunsen, Bibelwerk, vol. 1. 2. 5., bound 8 Th. (1 £ 4 sh.); do., Bibel-Atlas 1 Th. (3 sh.); Schödler, Chemie der Gegenwart, boards 2 Th. 10 ngr. (7 sh.); Vogt, Künstliche Fischzucht 1 Th. 10 ngr. (4 sh.); Heussi, Geodäsie 3 Th. 20 ngr. (11 sh.); Richter, Kinderleben 1 Th. (3 sh.); d'Alquen, Angelfischerei, bound 1 Th. 18 ngr. (4½ sh.); Illustrirter Handatlas, part. 1—4. 6 Th. 12 ngr. (19 sh.); Gruner, Dom zu Orvieto 30 Th. (4 £ 10 sh.); Schulz, Denkmäler, 4 vols with atlas 120 Th. (18 £); Hafis, bound 30 Th. (4 £ 10 sh.); kleines Brockhaus'sches Conversations-Lexikon, vol. 1., bound 1 Th. 27½ ngr. (6 sh.); illustrirtes Haus-Lexikon, vol. 1. 2., bound 5 Th. 18 ngr. (17 sh.); Byron, poetical works, 3 vols., bound 4 Th. (12 sh.); do. sewed 3 Th. (9 sh.); Scott, poetical works, 3 vols., bound 4 Th. (12 sh.); do., sewed 3 Th. (9 sh.); Lewes, Selections etc., 2 vols., bound 2 Th. 20 ngr. (8 sh.); do., sewed 2 Th. (6 sh.); Froude, hist. of England, vols. 1. 2. 3., sewed à 1 Th., 3 Th. (9 sh.); Paton, Researches, 2 vols., sewed 3 Th. (9 sh.); Rilyiff, Russische Bibliothek (Werke), bound 3 Th. (9 sh.); do. sewed 2 Th. 20 ngr. (8 sh.); Caballero, Clemencia, bound 1 Th. 10 ngr. (4 sh.); do. sewed 1 Th. (3 sh.); the same Cuentos, bound 1 Th. 10 ngr. (4 sh.); do. sewed 1 Th. (3 sh.); Trueba, El Cid, bound 1 Th. 10 ngr. (4 sh.); do. sewed 1 Th. (3 sh.); the same, Las Hijas, bound 1 Th. 10 ngr. (4 sh.); do. sewed 1 Th. (3 sh.); Caballero, la Gaviota, bound 1 Th. 10 ngr. (4 sh.); do. sewed 1 Th. (3 sh.); Cervantes, Don Quixote, 2 vols., bound 2 Th. 20 ngr. (8 sh.); do. sewed 2 Th. (6 sh.); Herrmann, Composiciones, bound 1 Th. 10 ngr. (4 sh.); do. sewed 1 Th. (3 sh.); Trueba, de los cantares, bound 1 Th. 10 ngr. (4 sh.); do. sewed 1 Th. (3 sh.); Caballero, la familia de Alvareda, bound 1 Th. 10 ngr. (4 sh.); do. sewed 1 Th. (3 sh.); Garczynskiego, Poe-zye, bound 1 Th. 10 ngr. (4 d.); do. sewed 1 Th. (3 sh.); Slowackiego, Pisma, 4 vols., bound 5 Th. 10 ngr. (16 sh.); do. sewed 4 Th. (12 sh.); Gordona, Obrazki, bound 1 Th. 10 ngr. (4 sh.); do. sewed 1 Th. (3 sh.); Borkowskiego, Pamietrik, bound 1 Th. 10 ngr. (4 sh.); do. sewed 1 Th. (3 sh.); Leopardi, opere, vol. 1 bound 1 Th. 10 ngr. (4 sh.); do. sewed 1 Th. (3 sh.); Manzoni, promessi sposi, bound 1 Th. 10 ngr. (4 sh.); do. sewed 1 Th. (3 sh.); Dias, Cantos, bound 2 Th. (6 sh.); do. sewed 1 Th. 15 ngr. (4½ sh.); 6 prints Schiller-Gallerie, 6 prints Goethe-Gallerie 6 Th. (18 sh.); Rottner, Comtoirwissenschaft, 2 vols., sewed 4 Th. 15 sh. (13½ sh.); 2 cuts (resurrection and crucifixion) in frames; Tischendorf, novum testamentum, sheet 1 to 10 as specimen. Catalogue. Zur Erinnerung an das 50jähr. Jubiläum. Proofs of the typefounding establishment, with price current. 6 prices current of the mechanical establishment. 2 cards of the engraving-establishment.

2508. FISCHER, C. FR. AUG., Bautzen. Berlin 1844 silver medal; London 1851 prize-medal. 1862 medal. Agt. James Watt. Esq., 47a. Moorgate-street, City.

An assortment of plate-papers for copper-plate, steel-plate and lithographic printing qual. I. 8¼ d., qual. II. 8 d., qual. III. 7 d.; prices p. lb. engl.

free on board London — cases to be paid. Each quality sized ¼ d. more.

2509. Flinsch, Ferd., Leipzig. Berlin 1844 silver medal; London 1862 honor. mention.

Stained papers for title-sheets etc.; do. card-board papers; white wrapping paper for manufactures; plate-papers of chinese tint.

2510. Giesecke & Devrient, letter-press printing-office, stereotype-foundery, lithographic, copper and steel-plate printing-office, engraving and guilloche-office, galvanoplastic and mechanical workshops for typographical purposes. Munich 1854 medal of honor; Paris 1855 silver medal; London 1862 medal. Agt. J. W. Green (E. G. Zimmermann).

Technical and statistical report on the establishment, exhibiting its gradual development and affording some insight into its arrangements, together with a glance at its performances.

Works: a part of the codex Sinaiticus Petropolitanus, ed. Tischendorf, with lithographic facsimile of the original; Fragmenta Sacra Palimpsesta, ed. Tischendorf, vol. I—III.; Bibliotheca Tamulica, ed. Graul, vol. I—III.; Theologia Thetica, ed. Ziegenbalg (Tamil); Buch der Sachsen, by Böttger; Grundriss der Akiurgie, by Ravoth; Esquisses de la Vie populaire en Hongrie, by Prónay, etc. Scientific, belletristical, illustrated and elegant editions of works.

Specimens of miscellaneous printing; specimens of printing-types, cut by the exhibitors; Palimpsesta from the »Monumenta Sacra Palimpsesta«, ed. Tischendorf; typographical Tableau of artistic printing in black, gold, silver, and colored letter, and embossing; plates produced by the galvanoglyphotipic process, printed by the common printingpress or machine; productions of the pantographic, relief, and the exhibitors guilloche - machines.

Three tableaux of artistic printing in thirteen compartments; a collective representation of the various modes of printing in all branches of typography. Specimens of all the Oriental founts employed in the establishment.

2511. Grumbach, C. (E. Kretzschmar), Leipzig. London 1862 honor. mention.

Samples of wood-cut printing in the letter-press, in black, and in colors; specimens of letter-printing: Blätter und Blüthen Deutscher Kunst und Poesie (the same work exhibited under No. 2506. by the publisher).

2512. Hinrichs, J. C., publisher, Leipzig.

Books and maps published by the exhibitor; Theology, Archaeology Statistics (Theologie, Alterthumswissenschaft, Statistik) Monumenta sacra inedita. Nova collectio. Nunc primum eruit atque ed. Prof. Dr. Aenoth. Frid. Const. Tischendorf. Volumina V. et appendix. 4. maj. Each volume 16 Th.; Index. I. fragmenta sacra palimpsesta sive Fragmenta cum Novi tum Veteris Testamenti. (XLVIII. and 278. p. with lithographs) 1855.; II. Fragmenta evangelii Lucae et libri Genesis. (XLVII. and 322 p. with 1 lithograph.) 1857. III. Fragmenta Origenianae octateuchi editionis cum fragmentis evangeliorum palimpsestis. (XL. and 300. p. 2 litographs.) 1860.; IV. Psalterium Turicense purpureum (to be published shortly); V. Reliquiae textus sacri utriusque (to be published shortly); Appendix, codex Laudianus actuum apostolorum graece et latine (to be published shortly). Patrum aposto-

licorum opera. Textum ad fidem cod. et graecorum et latinorum ineditorum copia insignium, adhibitis praestantissimis editionibus recensuit atque emendavit Alb. Rud. M. Dressel. Accedit Hermae pastor ex fragmentis Lipsiensibus ed. Const. Tischendorf. large 8. (LXII and 672 p.) 1857. 3 Th.; Clementinorum epitomae duae, altera edita correctior, inedita altera nunc primum integra ex cod. romanis et excerptis Tischendorfianis cura Alb. Rud. Max. Dressel. large 8. (IX. and 344 p.) 1859. 2 Th.; Brugsch, Dr. H., geographische Inschriften altägyptischer Denkmäler, gesammelt während der auf Befehl Sr. Maj. des Königs Friedrich Wilhelm IV. v. Preussen unternommenen wissenschaftlichen Reise in Aegypten, erläutert und herausgegeben. 3 vol. large 4. 42 Th.; Index. I. Die Geographie des alten Aegyptens. With 57 lithographs and 2 maps. (XI. and 304 p.) 1857. 25 Th.; II. Geographie der Nachbarländer Aegyptens. With 23 lithographs and 2 maps. (XI. and 96 p.) 1858. 8¼ Th.; III. Die Geographie nach den Denkmälern aus den Zeiten der Ptolomaeer und Römer. Nebst einem Nachtrag zum 1. u. 2. Band u. Register zu dem ganzen Werke. With 17 lithographs and 1 map (XII. and 125 p.) 1859. 8⅔ Th.; the same Histoire d'Egypte dès les premiers temps de son existence jusqu'à nos jours. Ouvrage dédié à Son Altesse le Vice-Roi d'Egypte Mohammed - Saïd - Pascha. Accompagné de planches lith. et d'un altlas de vues pittoresques et de cartes. I. Partie. L'Egypte sous les rois indigènes. large 4. (VI. and 295 p. with 19 lithograps) 1859. 8 Th. The complete work will consist of 3 volumes and 1 atlas at the price of 32 Th.; the same Recueil de monuments égyptiens dessinés sur lieux et publiés sous les auspièces de Son Altesse le Vice-Roi d'Egypte Mohammed-Saïd-Pascha. 1. Partie. imp. 4. (IV. and 56 p. with 50 lithographs) 1862. 8 Th.; the 2. (last) vol. in the press; Overbeck, J., Geschichte der Griechischen Plastik für Künstler und Kunstfreunde. With illustr. by H. Streller, engr. by J. G. Flegel. 2 Vol. impr. 8. (XVIII. and 690 p. with woodcuts in the text and 16 wood-cuts) 1858. Splend. bound, gilt edge, 9 Th. 10 ngr.; Corpus legum ab Imperatoribus Romanis ante Justinianum latarum, quae extra constitutionum codices supersunt. Accedunt res gestae Imperatorum, quibus Romani juris historia et imperii status illustratur. Ex monumentis et scriptoribus Graecis Latinisque collegit, indicibus instruxit Dr. Gust. Haenel. imp. 4. (XXVIII. and 935 p.) 1860. 18 Th.; Wappäus, Prof. Dr. J. E., allgemeine Bevölkerungsstatistik. Vorlesungen. 2 parts imp. 8. (XXVIII. and 539 p.) 1859 and 1861. 5 Th. 20 ngr.; the same Handbuch der Geographie und Stastistik von Nord-Amerika. Nebst einer allgemeinen Uebersicht von Amerika. A. u. th. t.: Handbuch der Geographie und Statistik, begründet von Stein und Hoerschelmann. 1. vol. 2. part. imp. 8. (XIV. and 828 p.) 1855. 3 Th. 12 ngr.; Brachelli, Prof. Dr. H. F., Handbuch der Geographie und Statistik des Kaiserthums Oesterreich. Nebst einer Einleitung: der deutsche Bund im Allgemeinen. A. u. th. t: Handbuch der Geographie und Statistik begründet von Stein und Hoerschelmann. 4 vols. 1. part. imp. 8. (X and 676 p.) 1861. 2 Th. 18 ngr.

Icelandic and oldnorthern Literature (Isländische und Altnordische Literatur): Maurer, Prof. Dr. K., Isländische Volkssagen der Gegenwart, vorwiegend nach mündlichen Ueberlieferungen gesammelt und verdeutscht. imp. 8. (XII and 372 p.) 1860. 1 Th. 25 sgr.; Arnason, J., ilenskar p jodsögur og œfintyri. 1. Bindi. imp. 8. (XXXIV and 666 p.) 1862. 4 Th. 10 ngr.; Edda Saemundar hins Fróda. Mit einem Anhang zum Theil bisher ungedruckter Gedichte, ed. by Th. Möbius. 8. (XVIII. and 302 p.) 1860. 2 Th.; Fornsögur, Vatnsdalasaga, Hallfredarsaga, Flóamannasaga. Ed. by Gudbrandr. Vigfússon and Th. Möbius. imp. 8. (XXXII and 239 p.) 1860. 1 Th. 20 ngr.; Analecta Norroena. I. Auswahl aus der Isländischen und Norwegischen Literatur des Mittelalters, ed. by Th. Möbius. imp. 8. (XV and 319 p.) 1859. 2 Th.

Fischer, J. D., der praktische Baumwollspinner. Ein Hand- und Hülfsbuch für Spinnereibeflissene. Bevorwortet v. G. Bodemer. Nebst einem Nachtrag: Die Fortschritte in den Jahren 1855 bis 1861. l. 8. (XII and 354 p. with 70 wood-cuts printed in the text and 37 lith. in gr. Folio) 1862. 5 Th. 6 ngr.

School-maps and atlasses (Schul-Karten und atlanten): Vogel's and Delitsch's, mural-maps in oil-colors on cloth. (These maps are also exhibited by the manufacturers Rölle & Huste, No. 2464.) 1. 2. The two Hemispheres. Size 65 in. by 65 in. in brown colors showing the elevations above 500, 1500, 4000 and 8000 ft.; 2 mural maps on black oil-cloth with rollers 14 Th. 25 ngr.; 2 mural maps on blue oil-cloth with rollers 16 Th. 5 ngr.; 1 mural blank-projection on black oil-cloth with rollers 3 Th. 27½ ngr.; 3. Europe, size 55 in. by 61 in. in 4 brown colors showing the elevations above 300, 1500 and 4000 ft.: 1 mural map on black oil-cloth with rollers 6 Th. 12½ ngr.; 1 mural-map on blue oil-cloth with rollers 7 Th. 2½ ngr.; 1 mural blank-projection on black oil-cloth with rollers 3 Th. 12½ ngr.; 4. Central-Europe, size 52 in. by 61 in. in 5 brown colors showing the elevations above 300, 1000, 2200 and 4000 ft.: 1 mural map on black oil-cloth with rollers 7 Th. 22½ ngr.; 1 mural map on blue oil-cloth with rollers 8 Th. 12½ ngr.; 1 mural blank-projection on black oil-cloth with rollers 3 Th. 27½ ngr. An explanation to this maps is to be found in Delitsch's Mittel-Europa, orographisch, hypsometrisch und hydrographisch. in 8. 55 p. 7½ ngr.; Vogel's series of blank-projections (Netz-Atlas) on black oil-paper for map-drawing in schools. 6th ed. 7 sheets, size 12 in. by 14 in. 14 ngr.; the same, Schulatlas mit Randzeichnungen. 9th ed. in 22 sheets. Umgearbeitet, verbessert und mit Höhenschichten versehen von O. Delitsch. Imp. 4. 1861. bound 1 Th. 16 ngr.; the same, kleiner Schul-Atlas der reinen Elementargeographie. Mit Randzeichnungen und deren Erklärung. 2. ed. 6 sh. l. 4. 1855. 16 ngr.; the same and Delitsch's elementary altas of modern geographiy. Six maps in brown shadowings 12 ngr.; do., the same atlas with illustrations in natural history 16 ngr.; Delitsch's series of blank-projections (Netz-Atlas) on black oil-paper for map-drawing in schools. 15 sheets. Size 15½ in. by 18½ in. 1 Th.; Atlas, neuer, der ganzen Erde, für die Gebildeten aller Stände und höhere Lehranstalten. 31st ed. Auswahl in 24 Karten. Mit Berücksichtigung der geographischstatistischen Werke von Dr. C. G. D. Stein u. A. entworfen und gezeichnet von J. M. Ziegler, Dr. H. Lange, G. Heck u. A. boards 4 Th. 20 ngr.

2513. HOFMEISTER, FRIEDRICH, publisher of music. Leipzig.

Handbuch der musikalischen Literatur etc. 3 d. edit reaching unto 1844. Publ. by Ad. Hofmeister 22 sh.; first continuation of the »Handbuch«: 1844—1851 12 sh.; second continuation 1852—1859 15 sh.; a catalogue of all music works etc. published in Germany and the neighbouring countries in 1860 and 1861, in alphabetical order 5⅓ sh. Musikalisch-literarischer Monatsbericht, year 1829—1862, year 1834, 2 sh.; Verdi, la Traviata — for piano with German and Ital. words 16½ sh.; do. for piano 4 hands, without words 12½ sh.; do. for piano 2 hands, do. 9 sh.; 32 diff. musical pieces.

2514. KRAETZSCHMER, FRIEDR., lithogr. institut, Leipzig. Munich 1854 honor. mention; Paris 1855 bronze medal.

A volume of lithographic title-pages for musical works.

2516. MEINHOLD, C. C., & SONS, printers to the court, lithogr. and copper-plate printing-office, type and stereotype-foundery, galvanoplastic office, publishers, Dresden. Berlin 1844 bronze medal; Munich 1854 medal of honor.

The German history in pictures. 3 vol. with 232 original woodcuts. Splendidly bound 2 £ 13 sh.; Festive Ode for Prince George of Saxony and His consort Princess Mary 4½ sh.; Ruprecht, atlas of natural history 40 sheets of illum. and color prints 1 £ 13 sh.; 13 div. books for the youth, illustrated and bound in boards; each vol. 2 to 4½ sh.

2517. MEISSNER & BUCH, manu. of fancy-papers, printed oil-paintings, carved covers, papeteries, patent envelopes, lithogr. printers and stationers, Leipzig. London 1862 honor. mention.

A large pattern-book and six frames with specimens of colored printings, fancy papers, envelopes, covers etc.; a glass case wit specimens of different papeteries, envelopes etc.

2518. ROEDER, C. G., musical engraver and printer, lithogr. establishment, Leipzig.

Specimens of musical engraving and printing.

2519. SIEGEL, C. F. W., publisher of music Leipzig.

D. Krug, Op. 141., Melodienschatz f. Pianof. 12 sh.; Ch. Mayer, Op. 330. Album f. Pianof. 3 sh.; S. Rietz, Op. 36. »Lied vom Wein«, arrangirt f. Piano 6 sh.

2520. ROYAL STENOGRAPHIC INSTITUTION, Dresden.

Printed stenographic works: A Manual of Gabelsberger's German Shorthand-Writing. Composed with the consent of the Home-departement by H. Raetzsch, professor of stenography. 4. edition, augmented. Dresden 1861. 3 sh. 6 d.; The Reader to the compendium (prize-essay) of Gabelsberger's stenography. Written according to what is agreed upon by the committee of shorthand-writers at Dresden. 9th. edition. Dresden 1862. 1 sh. d.; Gabelsberger, shorthand-writers' almanack, published under the care of Dr. J. Zeibig, for the year 1861. 1 sh.; the same for the year 1862. 1 sh.

CLASS XXIX.

EDUCATIONAL WORKS AND APPLIANCES.

2530. HAWSKY, ADALBERT, Leipzig. Munich 1854 honor. mention; London 1862 honor. mention.

Playthings; prices p. doz.: magazine of game 34 Th. (5 £ 2 sh.): spirotonton 4 Th. (12 sh.); rackets,

Engl. form, No. 1. 8 Th. (1 £ 4 sh.); do. No. 2. 9 Th. (1 £ 7 sh.); spiralifer 1 Th. 6 sg. (3 sh. 7 d.); puzzle-knots: cross-form 12 Th. (1 £ 16 sh.); bullet-form 10 Th. (1 £ 10 sh.); cubical 8 Th. (1 £ 4 sh.); case-form, jacar. 11 Th. (1 £ 13 sh.); do. mahog. 10 Th, (1 £ 10 sh.); do. ordy. 7 Th. (1 £ 1 sh.); star-shaped mahog. 5 Th. (15 sh.); do. ordy. 3 Th. (9 sh.); sling-form 5 Th. (15 sh.); sixfold 2 Th. 15 sg. (7 sh. 6 d.); ordy. No. 3. 24 sg. (2 sh. 5 d.); No. 2. 12½ sg. (1 sh. 3 d.); No. 1. 7½ sg. (9 d.); little cock-game 3 Th. (9 sh.); do. 1 Th. 7½ sg. (3 sh. 9 d.); arrow-shot 1 Th. 15 sg. (4 sh. 6 d.); do. 3 Th. (9 sh.); humming-top 22½ sg. (2 sh. 3 d.); bullet-game 27½ sg. (2 sh. 9 d.); running joujoux 27½ sg. (2 sh. 9 d.); humming joujoux 1 Th. 5 sg. (3 sh. 6 d.); embroide-ring-frame 3 Th. 22½ sg. (11 sh. 3 d.); wardrobe-swivel 7 Th. (1 £ 1 sh.); do. without frame 4 Th. (12 sh.); key-swivel 3 Th. (9 sh.); foot-stool, mahog. 11 Th. (1 £ 13 sh.); do. ordy. 6 Th. (18 sh.); pen-case No. 1. p. gross 4 Th. (12 sh.); No. 3. p. doz. 15 sg. (1 sh. 6 d.); No. 4. 1 Th. 20 sg. (5 sh.); No. 5. 2 Th. 25 sg. (8 sh. 6 d.); No. 6. 2 Th. 5 sg. (6 sh. 6 d.); No. 7. 6 Th. (18 sh.); kitchen-utensils 8 Th. (1 £ 4 sh.); stable-utensils 6 Th. (18 sh.); shops 2 Th. (6 sh.); poli-chinels 2 Th. 22½ sg. (8 sh. 3 d.); cloth-puppet 6 Th. (18 sh.); bayonet-gun No. 1. 14 Th. (2 £ 2 sh.); No. 2. 15 Th. (2 £ 5 sh.); No. 3. 16 Th. 15 sg. (2 £ 9 sh. 6 d.); percussion-gun No. 1. 4 Th. 20 sg. (14 sh.); No. 2. 6 Th. (18 sh.); No. 3. 6 Th. 20 sg. (1 £); percussion-pistol 4 Th. 20 sg. (14 sh.); standard, fine No. 3. 5 Th. 20 sg. (17 sh.); No. 4. 6 Th. 20 sg. (1 £); ordy. No. 3. 3 Th. (9 sh.); cuirassier-helm 5 Th. (15 sh.); head-piece 5 Th. (15 sh.); English helmet 5 Th. (15 sh.); French do. 2 Th. 25 sg. (8 sh. 6 d.); game at nine-pins 12 Th. (1 £ 16 sh.); hunting-bag No. 1. 5 Th. (15 sh.); do. No. 2. 8 Th. (1 £ 4 sh.); hobby-horse No. 1. 3 Th. 22½ sg. (11 sh. 3 d.); No. 3. 5 Th. 20 sg. (17 sh.); cartridge-box 1 Th. 20 sg. (5 sh.); cartouch-box with sward belt 3 Th. (9 sh.); leather pouch for children 5 Th. (15 sh.); pouch hanging near the sabre 12 Th. (1 £ 16 sh.); wool-figure 10 Th. (1 £ 10 sh.); wool-dog 10 Th. (1 £ 10 sh.); wool-bird 8 Th. (1 £ 4 sh.); wool-butter-fly 3 Th. (9 sh.); doll in glass-case 5 Th. (15 sh.); all p. doz.

Puppet-head with natural hair:- No. 7. 5 Th. 18 sg. (16 sh. 10 d.); No. 8. 7 Th. (1 £ 1 sh.); No. 9. 7 Th. 6 sg. (1 £ 1 sh. 7 d.); No. 10. 8 Th. 20 sg. (1 £ 6 sh.); No. 12. 12 Th. 15 sg. (1 £ 17 sh. 6 d.); No. 14. 15 Th. 6 sg. (2 £ 5 sh. 7 d.); crying dolls: No. 3. 4 Th. 20 sg. (14 sh.); No. 4. 5 Th. 18 sg. (16 sh. 10 d.); No. 9. 5 Th. (15 sh.); No. 10. 2 Th. 15 sg. (7 sh. 6 d.); No. 12. 12 Th. 15 sg. (1 £ 17 sh. 6 d.); No. 15. 1 Th. 18 sg. (4 sh. 10 d.); dolls with porcelain head: 3/0. 2 Th. 15 sg. (7 sh. 6 d.); 2/0. 3 Th. 6 sg. (9 sh. 7 d.); 0. 3 Th. 25 sg. (11 sh. 6 d.); 1. 4 Th. 15 sg. (13 sh. 6 d.); dolls, fine; 3 in. 1 Th. 10 sg. (4 sh.); 4 in. 1 Th. 15 sg. (4 sh. 6 d.); 5 in. 1 Th. 20 sg. (5 sh.); 6 in. 2 Th. (6 sh.); dolls, ordy.: 2 in. 18 sg. (1 sh. 10 d.); 2½ in. 20 sg. (2 sh.); 3 in. 22½ sg. (2 sh. 3 d.); 4 in. 25 sg. (2 sh. 6 d.); 5 in. 1 Th. 2½ sg. (3 sh. 3 d.); paper-lanterns: No. 1. 8½ d.; No. 2. 1 sh. 2½ d.; No. 3. 1 sh.; No. 4. 1 sh. 6 d.; No. 5. 7 sh. 8 d.; No. 6. 2 sh.; No. 7. 3 sh.; No. 8. 3 sh. 7¼ d.; No. 9. 6 sh. 7¼ d.; No. 10. 3 sh. 6 d.; No. 11. 4 sh. 6 d.; No. 12. 7 sh. 10 d.; No. 16. 2 sh. 6 d.; No. 17. 4 sh. 2½ d.; No. 18. 1 sh. 6 d.; No. 19. 6 sh.; No. 20. 1 sh. 9 d.; No. 21. 6 sh.; No. 24. 12 sh.; No. 25. 7⅛ d.; No. 26. 9¾ d.; No. 27. 3 sh. 9 d.; No. 28. 6 sh.; No. 29. 1 sh. 6 d.; No. 30. 2 sh. 6 d.; No. 31. 4 sh. 2½ d.; No. 32. 6 sh.; harle-

quins: No. 1. 1 Th. 27 sg. (5 sh. 8½ d.); No. 2. 2 Th. 10 sg. (7 sh.); do. ordy. 20 sg. (2 sh.).

Hinrichs. (s. No. 2512.)

2531. Hülse, Ernst R. A., Dresden. Paris 1855 honor. mention; London 1862 honor. mention. Agts. Müller & Schreiber, 4 Crescent-place Blackfriars.

Military toys: English general staff 39 men. 1 £ 16 sh.; standard platoon of Coldstream-guards with music-band 107 men 4 £ 1 sh.; English life-guards 24 men 18 sh.; 11th. of hussars 24 men 18 sh.; highlanders, infantry 24 men 18 sh.; dragoon-guards 24 men 15 sh.; dragoons 24 men 15 sh.; lancers 24 men 15 sh.; English infantry 24 men 12 sh.; French do. 24 men 12 sh.: camp of the Coldstream-guards 29 men 1 £ 10 sh.

Krause. (s. No. 2306.)

Meinhold. (s. No. 2516.)

Röller & Huste. (s. No. 2464.)

CLASS XXX.

FURNITURE AND UPHOLSTERY, INCLUDING PAPER-HANGINGS AND PAPIER-MACHÉ.

2535. Guenther, Julius, Waldheim. Agt. Victor Bauer, 1 Ironmonger-Lane, E. C. (s. II. No. 80.)

Parts of a complete furniture in the same style: cupboard of walnut-wood 180 £; lady's library do. 30 £; dressing-glass do. 4½ £; all objects richly ornamented with sculptures and worked by the prisoners of the penitentiary at Waldheim.

2536. Madack, Rudolph, jun., basket-maker, Leipzig. London 1862 honor. mention.

Sofa No. 1. 2 £ 11 sh.; 5 armchairs No. 2. 1 £ 7 sh.; No. 3. 18 sh.; No. 4. 16 sh.; No. 5. 15 sh.; No. 6. 13 sh. 6 d.; stool No. 7. 12 sh.; 2 flower-stands No. 8. 1 £ 4 sh.; No. 9. 13 sh.; 2 flower-stands each 6 sh. No. 10. and 11. 12 sh.; 2 carriages for children No. 12. 2 £ 11 sh., No. 13. 1 £; 3 carriages for dolls No. 14. 5 sh., No. 15. 4 sh. 6 d., No. 16. 7 sh. 6 d.; sofa for children No. 17. 7 sh.; table for children No. 18. 5 sh.; 2 chairs for children No. 19. and 20. à 4 sh. 6 d.; 2 do. No. 21. and 22. each 4 sh.; 3 do. No. 23. 3 sh. 6 d., No. 24. 3 sh. 6 d., No. 25. 2 sh. 6 d.; 2 footstools No. 26. 3 sh., No. 27. 2 sh. 6 d.; 2 baskets for ladies No. 28. 2 sh. 6 d., No. 29. 2 sh.; 1 besom for carpets No. 30. 1 sh.

2537. Merz, Oscar, jun., basket-maker, Dresden. London 1862 medal.

Arm-chair of cane-work 2 £ 5 sh.

2538. Schmidt, T. F. C., manu. of varnished wood-imitation paper hangings, Leipzig.

15 pieces of paper-hangings: light deal paper-hangings (dull) 1 sh. 6 d.; do. maple (varnished) 1 sh. 6 d.; do. oak (dull) 1 sh. 9 d.; dark oak (varnished) 2 sh.; 3 nut (various) à 2 sh., 6 sh.; mahogany 2 sh.; 3 granite (brilliant) à 1 sh. 9 d., 5 sh. 3 d.; 2 granite (varnished) à 1 sh. 9 d., 3 sh. 6 d.; 2 large pannelling (different) à 7 sh. 6 d., 15 sh.

2539. Tuerpe, A., cabinet-maker, Dresden. Munich 1854 medal of honor.

Bool clock with high postament; do. with console;

small black cabinet with bronze ornaments; ebony-cabinet with ornaments of painted china; bool writing table.

2540. WOELLFERT, CARL, Dresden. Paris 1855 honor. mention; London 1862 medal.

An assortment of tooth-picks of diff. quality; prices p. mille: No. 6. 2 sh.; No. 7½. 3 sh.; No. 10. 4 sh.; No. 15. 4½ sh.

CLASS XXXI.
IRON AND GENERAL HARDWARE.

2541. HERRMANN, JULIUS, Dresden. London 1862 medal.

A specimen of wire-plaiting; for malt-kilns.

2542. LENK, CHARLES, manu., Dresden.

Steel pens to be filled with ink without dipping.

2543. MUENNICH, A., & Co., manu. of machines and wire-work, Chemnitz. London 1862 medal.

No. 1—3. woven of brass-wire; No. 4—13. woven of iron-wire for var. purposes; No. 14. do. patent malt-dryer; No. 15—19. do. for var. purposes; No. 20. do. pat. chicory-dryer; No. 21—23. woven brass-wire for various purposes; No. 24. centrifugal kettle; No. 25. vessel of iron-wire; No. 26—31. vessels of brass-wire for var. purposes.

*2544. WEISKER, A., & Co., Schleiz.

Lamps, candlesticks and other turned and spun articles of brass and German silver.

2545. WINKELMANN, JULIUS, establishment for galvanoplastic, Leipzig. London 1851 honor. mention; Paris 1855 silver medal; London 1862 honor. mention.

Galvanoplastic sculptures: the christmas-angel 90 £; King Ludwig of Bavaria 15 £; a gazel 5 £ 5 sh.; Frederic the great of Prussia 75 £.

CLASS XXXII.
STEEL AND CUTLERY.

2546. SAXON CAST-STEEL WORKS, Döhlen near Dresden. London 1862 medal and honor. mention. Agt. Adolph von Kanig, 11 High Row, Knights-bridge. (s. II. No. 76.)

Frog with cast-iron armature; locomotive-crank, finished; locomotive-connecting rod, finished; axletree for steam-engines, forged and turned for the journals; 2 tension-springs with welded ears; 2 supporting springs with rolled ears and groovings; 5 volute-springs for buffing and drawing-apparatus: rolling-mill with hardened rollers; 2 rollers, hardened and highly polished; 2 anvils with cast-iron head-pieces; 2 knives for paper-cutting machines: 2 circular knives for paper-machines; knife, two-edged, for cylindrical paper-mills: do. for the ground-works of the same; 2 knives for rag-cutting machines; knife for a leather-splitting machine; div. fractures of raw cast-steel.

CLASS XXXIII.
WORKS IN PRECIOUS METALS AND THEIR IMITATIONS AND JEWELLERY.

2547. WIMMER & DIETRICH, gold and silver manu., Annaberg. London 1862 honor. mention.

Articles of gold and silver-plated wire and tinsel, colored tinsel and spangles: laces, galloons, fringes, trimmings, purl, tresses, parts of ladies' head-dresses etc.

CLASS XXXV.
POTTERY.

2549. BUCKER, HEINRICH, painter on china, Dresden. London 1851 prize-medal, 1862 honor. mention.

Paintings on china in gold frames 6 £ 6 sh. to 22 £ 10 sh.; do. smaller ones without frames 6 sh. to 4 £ 10 sh.

2550. FIKENTSCHER, FR. CHRIST., Zwickau.

Apparatus for developing chlorine gas, of stone ware, 270 gallons 8 £.

2551. FISCHER, CHRIST., Zwickau. Munich 1854 great medal.

1 dinner service of china for 12 persons, bordered with gold and green lustre, containing: 36 dinner plates; 12 soup plates; 12 pie plates; 12 cheese plates; 2 round flat dishes No. 3.; 2 do. No. 2.; 1 oblong flat dish. No. 10.; 2 do. No. 8.; 2 do. No. 5.; 1 oblong fish-dish; 2 vegetable-dishes and covers; 2 salad-bowls No. 4.; 2 do. No. 2.; 2 bowls on pedestal; 2 oblong salad-bowls No. 2.; 2 sauce-bowls on stands, obl.; 1 sauce-terreen and spoon; 1 mustard-pot; 2 salt-boxes; 2 soup-terreens with stand No. 3. and 2. 20 £.

2 cups, with gold p. piece 4 sh.; 2 do. lustre and gold p. piece 3 sh.; 1 large dish, 5 partite, with lustre and gilt 21 sh.; 1 oblong flat dish No. 11. white 18 sh.

2552. KRAMER, HEINRICH EDUARD, Leipzig.

Specimens of the new invented »Kramers« China-chromatype: 1 coffee set 6 p. 18 sh.; 8 plates 7 sh.; 1 leaf with 6 col. pictures 3 sh.; 4 different do. 2 sh.

2553. KÖNIGLICH SÄCHSISCHE PORZELLAN-MANUFAKTUR (Royal Saxon china-manufactory), Meissen near Dresden. Berlin 1844 silver medal; London 1851 prize-medal; Munich 1854 great medal; London 1862 medal. Agts. Rittener & Saxby, 41 Albemarle-street, Piccadilly.

No. 1. 1 complete fireplace with mantle-piece 4 ft. 8 in. high, 5 ft. 8 in. broad 180 £; No. 2. 1 large mirror with 2 branches for 8 candles 6 ft. 4 in. high, 3 ft. 8 in. broad 60 £; No. 3. 1 clock with movement 20 £; No. 4. 1 stand for the above clock 6 £ 6 sh.; No. 5. and 6. 2 vases, painted figures after Watteau, p. piece 8 £ 5 sh.; No. 7. and 8. 2 square stands, p. piece 2 £ 16 sh.; No. 9. and 10. 2 candelabers 25 in. high, for 7 candles each, p. piece 11 £ 5 sh.; No. 11. 1 console-table 34 in. high, with china plate, painted with »the garden of love« after Rubens 90 £; No. 12. 1 mirror 54 in. high, 34 in. broad 35 £; No. 13. and 14. 2 lustres 43 in. high, for 28 candles each, p. piece 52 £; No. 15. 1 large lustre 64 in. high, for 18 candles 76 £; No. 16. 1 round table 31 in. high, with china plate, darkblue glazed, painted thereon »the pianiste« after Netscher 60 £; No. 17. 1 do., lightblue, glazed, painted thereon figures after Watteau 52 £; No. 18. and 19. 2 wooden oval tables 30 in. high, with china plates, painted thereon fruits and flowers, p. piece 15 £; No. 20. and 21. 2 vases à la Majolika painted thereon figures, p. piece 75 £; No. 22. 1 vase 39 in. high, with china handles, lightblue, glazed, painting thereon »the golden age« after Hübner 135 £; No. 23. and 24. 2 vases 30 in. high, light-blue, glazed, painting thereon »the seasons« after Hübner, p. piece 45 £; No. 25. 1 vase 39 in. high, with china handles, darkblue, views of Dresden and Meissen, painted thereon 90 £; No. 26. and 27. 2 vases 30 in. high, views of Pillnitz, Wesenstein painted thereon, p. piece 45 £; No. 28. 1 large vase 60 in. high, with china handles, figures painted thereon Diana and Aktäon after Albano 180 £; No. 29. 1 large vase 60 in. high, with china handles, colored and gilt 84 £; No. 30. 1 vase, embossed flowers and leaves 40 in. high 45 £; No. 31. 1 vase, flowers painted under the glaze 40 in. high 45 £; No. 32. and 33. 2 vases 25 in. high, painted with figures after Watteau thereon, p. piece 52 £; No. 34. and 35. 2 vases 35 in. high, painted with figures and flowers after Watteau thereon, p. piece 19 £; No. 36. and 37. 2 square stands for vases, p. piece 6 £; No. 38. and 39. 2 vases 23 in. high, pierced work, painted thereon figures after Watteau and landscapes, p. piece 29 £; No. 40. and 41. 2 vases 24 in. high, darkblue, glazed, painted thereon views of Dresden, Pillnitz and Wesenstein, p. piece 42 £.; No. 42. and 43. 2 vases 24 in. high, painted thereon flowers, p. piece 23 £; No. 44. and 45. 2 vases 18 in. high, do., p. piece 11 £; No. 46. and 47. 2 vases 22 in. high, painted thereon flowers and fruits, p. piece 13 £; No. 48. and 49. 2 vases 28 in. high, embossed flowers and leaves, p. piece 15 £; No. 50. 1 complete fountain 66 in. high 75 £;

No. 51. 1 large candelabre 68 in. high, branches for 27 candles 55 £; No. 52. and 53. 2 candelabres 40 in. high, branches for 18 candles each, p. piece 36 £; No. 54. and 55. 2 do. 36 in. high, branches for 6 candles each, p. piece 15 £ 10 sh.; No. 56. and 57. 2 do. 30 in. high, branches for 6 candles each, p. piece 18 £; No. 58. and 59. 2 do. 26 in. high, branches for 5 candles each, p. piece 13 £ 10 sh.; No. 60. and 61. 2 do. 19 in. high, branches for 5 candles each, p. piece 4 £ 18 sh.; No. 62. and 63. 2 do. 20 in. high, branches for 4 candles each, p. piece 4 £ 3 sh.; No. 64. to 67. 4 candlestick-groups, branches for 3 candles each, p. piece 3 £ 15 sh.; No. 68 to 71. 4 do. branches for 2 candles each, p. piece 3 £ 17 sh.

No. 72. 1 clock with movement 17 £; No. 73. 1 stand for it 3 £ 13 sh.; No. 74. 1 clock with movement 15 £ 12 sh.; No. 75. 1 stand for it 3 £ 6 sh.; No. 76. clock with movement 15 £; No. 77. 1 stand for it 4 £.

No. 78. and 79. 2 large centre-baskets 24 in. long, 16 in. high, p. piece 25 £ 15 sh.; No. 80. 1 large oval centre-basket 22 in. long, 11 in. high 18 £; No. 81. square stand for it 6 £; No. 82. 1 plat de menage 22 in. high, with 12 cups and covers and oval basket 23 £ 5 sh.; No. 83. and 84. 2 groups with baskets 21 in. high, p. piece 12 £; No. 85. and 86. 2 square stands for them, p. piece 3 £ 8 sh.; No. 87. and 88. 2 groups with baskets 17 in. high, p. piece 6 £ 10 sh.; No. 89. and 90. 2 square stands for them, p. piece 2 £ 12 sh.; No. 91. and 92. 2 round fruit-stands, p. piece 6 £ 4 sh.

No. 93. and 94. 2 jewel-boxes and covers, pierced work, richly decorated, p. piece 14 £; No. 95. 1 large centre-group, 8 fig., branch for 7 candles, 33 in. high 18 £ 10 sh.

No 96—175, a collection of 80 different groups painted and decorated: No. 96. 1 large group, 6 fig. the olympus, 22 in. high, 28 £; No. 97. 1 group of 4 fig. 10 £.; No. 98. and 99. 2 do. of 6 fig. each, p. piece 6 £ 10 sh.: No. 100. 1 do. of 7 fig. 8 £; No. 101. 1 do. of 6 fig. 8 £ 12 sh.; No. 102. 1 do. of 7 fig. 5 £ 15 sh.; No. 103. 1 do. of 7 fig. 6 £ 15 sh.; No. 104. 1 do. of 6 fig. 6 £ 6 sh.; No. 105. 1 do. of 4 fig. 5 £ 17 sh.; No. 106.

1 do. 5 fig. 4 £ 10 sh. No. 107. and 108. 2 do. of 3 fig. each, p. piece 3 £ 7 sh.; No. 109. and 110. 2 do. of 4 fig. each, p. piece 3 £ 8 sh.; No. 111. and 112. 2 do. of 5 fig. each, and pot-pourri, p. piece 3 £ 13 sh.; No. 113. and 114. 2 do. of 3 fig. each, p. piece 3 £ 16 sh.; No. 115. and 116. 2 do. of 4 fig. each, p. piece 3 £ 4 sh.; No. 117. 1 do. of 4 fig. 4 £ 15 sh.; No. 118. 1 do. of 4 fig. 3 £ 10 sh. No. 119. 1 do. of 4 fig. 4 £ 5 sh.; No. 120. 1 do. of 3 fig. 3 £ 12 sh.; No. 121. 1 do. of 3 fig. 2 £ 18 sh.; No. 122. 1 do. of 2 fig. 2 £ 17 sh.; No. 123. 1 do. of 2 fig. 3 £; No. 124. 1 do. of 2 fig. 4 £ 2 sh.; No. 125. 1 do. of 2 fig. 3 £ 7 sh.; No. 126. 1 do. of 3 fig. 3 £; No. 127. 1 do. of 3 fig. 3 £ 6 sh.; No. 128. 1 do. of 3 fig. 3 £ 3 sh.; No. 129. 1 do. of 3 fig. 2 £ 3 sh.; No. 130. 1 do. of 4 fig. 2 £ 4 sh.; No. 131. 1 do. of 4 fig. 2 £ 10 sh.; No. 132. 1 do. of 4 fig. 2 £; No. 133. 1 do. of 4 fig. 2 £ 17 sh.; No. 134. 1 do. of 5 fig. 2 £ 4 sh.; No. 135. 1 do. of 5 fig. 3 £ 8 sh.; No. 136. 1 do. of 5 fig. 3 £ 9 sh.; No. 137. 1 do. of 2 fig. 2 £ 9 sh.; No. 138. 1 do. of 4 fig. 2 £ 9 sh.; No. 139. 1 do. of 2 fig. 2 £; No. 140. 1 do. of 2 fig. 1 £ 18 sh.; No. 141. 1 do. of 3 fig. 4 £ 10 sh.; No. 142. 1 do. of 3 fig. 2 £ 6 sh.; No. 143. 1 do. of 3 fig. 2 £ 11 sh.; No. 144. and 145. 2 do. of 2 fig. each, p. piece 2 £ 11 sh.; No. 146—149. 4 do. of 4 fig. each, p. piece 2 £ 10 sh.; No. 150—153. 4 do. of 2 fig. each, p. piece 1 £ 13 sh.; No. 154—157. 4 do. of 2 fig. each, p. piece 1 £ 4 sh.; No. 158—161. 4 do. of 1 fig. and emblems, p. piece 17 sh.; No. 162. and 163. 2 do. of 2 fig. each, p. piece 1 £ 2 sh.; No. 164. and 165. 2 do. of 2 fig. each, p. piece 15 sh.; No. 166. and 167. 2 do. of 1 fig. and geese, p. piece 14 sh.; No. 168. and 169. 2 do. of 2 fig. each, p. piece 1 £ 9 sh.; No. 170. 1 do. of 2 fig. 1 £ 10 sh.; No. 171. 1 do. of 2 fig. 1 £ 12 sh.; No. 172. 1 do. of 2 fig. 1 £ 11 sh.; No. 173. 1 do. of 2 fig. 1 £ 11 sh.; No. 174. 1 do. of 2 fig. 1 £ 15 sh.; No. 175. 1 do. of 2 fig. 1 £ 10 sh.

No. 176—198. a collection of 23 groups and other objects, decorated white and gold: No. 176. and 177. 2 groups of 6 figures each, p. piece 6 £ 2 sh.; No. 178. and 179. 2 do. of 4 fig. each, p. piece 3 £ 15 sh.; No. 180. 1 do. of 4 fig. 3 £ 9 sh.; No. 181. 1 do. of 5 fig. 3 £ 4 sh.; No. 182. 1 do. of 5 fig. 3 £ 6 sh.; No. 183. and 184. 2 do. of 4 fig. each, p. piece 2 £ 18 sh.; No. 185. 1 lace-figure 7 £ 6 sh.; No. 186. 1 do. 9 £ 10 sh.; No. 187. and 188. 2 square stands for them, p. piece 3 £ 12 sh.; No. 189. and 190. 2 groups with baskets 21 in. high, p. piece 12 £ 15 sh.; No. 191. and 192. 2 square stands for them, p. piece 3 £ 16 sh.; No. 193. and 194. 2 groups with baskets, 17 in. high, p. piece 7 £ 4 sh.; No. 195. and 196. 2 square stands for them, p. piece 3 £; No. 197. and 198. 2 candelabres with branches for 4 candles each, p. piece 4 £ 3 sh.

No. 199—264. collection of 66 figures with lace: No. 199. 1 figure with veil 3 £ 12 sh.; No. 200. 1 do. 3 £ 14 sh.; No. 201. 1 do. 2 £ 15 sh.; No. 202—206. 5 do., p. piece 2 £ 8 sh.; No. 207. 1 do. 2 £ 15 sh.; No. 208. 1 do. 2 £ 17 sh.; No. 209. 1 do. 1 £ 8 sh.; No. 210. 1 do. 1 £ 18 sh.; No. 211. 1 do. 2 £ 5 sh.; No. 212. 1 do. 2 £.; No. 213. 1 do. 1 £ 11 sh.; No. 214. 1 do. 1 £ 1 sh.; No. 215—231. 17 do., p. piece 1 £ 8 sh.; No. 232—250. 19 do., p. piece 18 sh.; No. 251—264. 14 do., p. piece 1 £ 3 sh.

No. 265—318. collection of 54 figures without lace: No. 265—270. 6 figures, p. piece 1 £ 17 sh.; No. 271—272. 2 do., p. piece 1 £ 18 sh.; No. 273—276. 4 do., p. piece 2 £ 7 sh.; No. 277—280. 4 do., p. piece 1 £ 11 sh.; No. 281—282. 2 do., Chinese, p. piece 2 £ 5 sh.; No. 283—284. 2 do., p. piece 1 £ 18 sh.; No. 285—286. 2 do., piece 15 sh.; No. 287—302. 16 do., band of musicians, p. piece 14 sh.; No. 303—318.

16 do. band of Cupids disguised, 1/16, p. piece 14 sh.; one of them No. 12. 19 sh.

No. 319—324. collection of 6 groups and figures, biscuit China: No. 319. 1 group of 3 figures 5 £ 8s h.; No. 320. 1 do. of 2 fig. 3 £ 15 sh.; No. 321. 1 do. of 2 fig. 5 £ 6 sh.; No. 322—323. 2 do. of 2 fig. each, p. piece 1 £ 10 sh.; No. 324. 1 figure, Ganymedes, 1 £ 6 sh.

No. 325. and 326. 2 oval toilet-mirrors, p. piece 5 £ 12 sh.; No. 327. and 328. 2 do., p. piece 3 £ 15 sh.; No. 329. and 330. 2 hand-mirrors, p. piece 1 £ 6 sh.; No. 331. and 332. 2 round waiting-plates on 4 feet 1 size, painted thereon figures after Watteau, p. piece 4 £ 17 sh.; No. 333. and 334. 2 do. 2 size, painted thereon figures after Watteau, p. piece 3 £ 10 sh.; No. 335. and 336. 2. do. 3 size, painted thereon figures after Watteau, p. piece 2 £ 18 sh.; No. 337. and 338. 4 round baskets, pierced border, do., p. piece 2 £ 4 sh.; No. 341. 1 large oval plateau, darkblue, glazed, do. 9 £; No. 342. 1 do. pink, glazed, painted thereon view of Dresden 8 £ 14 sh.; No. 343. 1 do. yellow, glazed, painted thereon view of Dresden 8 £ 14 sh.; No. 344. 1 do. thereon fruits 8 £ 17 sh.; No. 345. 1 large square do., lightblue, glazed, painted thereon figures after Watteau 6 £ 10 sh.; No. 346. 1 do. yellow do. 6 £ 10 sh.; No. 347. and 348. 2 cups, covers and stands, darkblue, glazed, cups with 2 handles, painted thereon figures after Watteau and flowers, p. piece 1 £ 14 sh.; No. 349. to 352. 4 do. 2 yellow, 2 green, do. 1 £ 13 sh.; No. 353. and 354. 2 do. darkblue, glazed, cups with 1 handle, painted thereon figures after Watteau and flowers, piece 1 £ 10 sh.; No. 355. and 356. 2 do. yellow, do., p. piece 1 £ 9 sh.; No. 357. and 358. 2 do. white, do., p. piece 1 £ 2 sh.; No. 359—364. 6 do. with 2 handles, green, glazed, painted thereon figures after Berghem, p. piece 1 £ 13 sh.; No. 365. 1 dejeuner, coffee-pot, milk-pot, sugar-box and cover, 2 cups and saucers, plateau, darkblue, painted thereon figures and landscapes after Berghem, complete 12 £ 12 sh.; No. 366. 1 do. same pieces, green, glazed, painted, thereon figures after Watteau and flowers, complete 8 £ 5 sh.; No. 367. 1 do. same pieces, darkblue, do. Wouvermann, complete 12 £ 15 sh.; No. 368. 1 same pieces, lightblue, do. Berghem, complete 12 £ 6 sh.; No. 369. do. landscapes and flowers, complete 15 £; No. 370. 1 do. painted and decorated after Mr. Gruner's design, complete 23 £; No. 371. do. complete 24 £; No. 372. and 373. 2 mounted dessert-plates, yellow glazed border, painted thereon figures after Watteau, p. piece 6 £ 16 sh.; No. 374. and 375. 2 do., pierced border, do. Berghem and Boucher, p. piece 4 £; No. 376—379. 4 do., 2 green, 2 yellow, do. Watteau and Boucher, p. piece 5 £; No. 380—391. 12 dessert-plates, pierced border, painted thereon figures after Watteau, p. piece 1 £ 13 sh.; No. 392—403. 12. do. birds and insects, p. piece 1 £ 2 sh.; No. 404—415. 12 do., p. piece 18 sh.; No. 416—427. 12 do., darkblue, glazed thereon landscapes, p. piece 1 £ 4 sh.; No. 428—439. 12 do. pink, glazed, thereon lupids, p. piece 1 £ 9 sh.; No. 440—451. 12 do. fruits, p. piece 1 £; No. 452—463. 12 do., 6 pink, 6 yellow, glazed, thereon figures after Watteau, p. piece 2 £ 10 sh.: No. 464—475. 12 do. yellow, glazed, thereon fowl, p. piece 1 £; No. 476—487. 12 do. fruits, p. piece 15 sh.; No. 488.

to 499. 12 do. flowers, p. piece 12 sh.; No. 500—515. 16 patterns of a dinner-sets, painted thereon figures after Watteau and flowers, richly gilt 43 £ 13 sh.; No. 516—529. 14 do. flowers and insects, richly gilt 21 £ 2 d.; No. 530—542. 13 do. 15 £ 12 sh.; No. 543—557. 15 do. darkblue and gold-border 33 £ 6 sh.; No. 558—569. 12. do. painted and gilt à la chinoise 7 £ 12 sh.; No. 570—583. 15. do. darkblue, glazed border, painted thereon flowers and insects 19 £; No. 584—597. 14. do. lightblue, do. 19 £ 6 sh.; No. 598—613. 16 do. green, do. 24 £ 7 sh..

No. 614—682. Collection of 69 birds and other animals: No. 614. 1 cock 10 £ 5 sh.; No. 615. 1 eagle 5 £ 10 sh.; No. 616. 1 bustard 3 £ 12 sh.; No. 617. and 618. 2 pheasants (1. 5 £ 12 sh., 1. 5 £ 18 sh.) 11 £ 10 sh.; No. 619. and 620. 2 herons, p. piece 4 £; No. 621. and 622. 2 cockatoos, p. piece 2 £ 17 sh.; No. 623. and 624. 2 do., p. piece 1 £; No. 625. 1 hawk 2 £ 9 sh.; No. 626. 1 do. with squirrel 3 £ 4 sh.; No. 627. and 628. 2 hoopoos, p. piece 1 £ 6 sh.; No. 629. and 630. 2 parrots, p. piece 1 £ 11 sh.; No. 631. and 632. 2 parrots, p. piece 7 sh.; No. 633. 1 do. 1 £ 2 sh.; No. 634. and 635. 2 yellow thrushes, p. piece 1 £; No. 636. and 637. 2 snipes, p. piece 1 £ 5 sh.; No. 638. and 639. 2 fieldfares, p. piece 1 £; No. 640. 1 dove 16 sh.; No. 641. 1 group of 2 doves 10 sh.; No. 642. and 643. 2 larks, p. piece 8 sh.; No. 644. and 645. 2 bullfinches, p. piece 8 sh.; No. 646. and 647. 2 titmice on branch, p. piece 6 sh.; No. 648. 1 do. 4 sh; No. 649. 1 thistle finch 6 sh.; No. 650. and 651. 2 flaxfinches, p. piece 12 sh.; No. 652. and 653. 2 goldhammers, p. piece 14 sh.; No. 654. and 655. 2 finches, p. piece 8 sh.; No. 656. and 657. 2 swallows, p. piece 5 sh.; No. 658. and 659. 2 canarybirds, p. piece 5 sh.; No. 660. and 661. 2 small swans, p. piece 6 sh.; No. 662. 1 group of 2 goats 1 £ 13 sh.; No. 663. 1 do. of 5 cats 2 £ 6 sh.; No. 664. 1 do. of 2 dogs 2 £ 14 sh.; No. 665. 1 do. of 3 dogs 1 £ 8 sh.; No. 666. and 667. 2 stags, p. piece 1 £ 5 sh.; No. 668. and 669. 2 cats, p. piece 10 sh.; No. 670. and 671. 2 dogs, p. piece 1 £ 6 sh.; No. 672. and 673. 2 do., p. piece 12 sh.; No. 674. 1 do. 1 £ 3 sh.; No. 675. and 676. 2 pug-dogs, p. piece 1 £.; No. 677. and 678. 2 do., p. piece 8 sh.; No. 679—682. 4 do., p. piece 6 sh.; No. 683—695. 13 patterns of a dinner-set with painted flowers under the glaze, colored and gilt 10 £.; No. 696. 1 dejeuner, consisting of 1 coffee-pot, 1 milk-pot, 1 sugar-box, 2 cups and saucers with plateau 19 £ 5 sh.

2554. Thorschmidt, C. L., & Co., Pirna. Agts. Heintzmann & Rochussen, 9 Friday-street, Cheapside.

Vases, figures, drinking vessels etc. of a kind of stone-ware called »siderolithe« or »lava«.

2555. Wenzel, Mich., teacher of the Royal modelling-school, Dresden.

Three vases of Saxon serpentine, turned and ornamented, with engravings and gilding.

2556. Harkort, Charles & Gustave, manu., Altenbach near Leipzig.

Earthenware vessels for keepingcold butter and water.

GRAND-DUCHY OF SAXONY.

CLASS XXVII.

ARTICLES OF CLOTHING.

2602. ZIMMERMANN, CHRISTIAN, & SON, manu., Apolda. London 1862 medal. Agt. Henry Kreiter, merchant of Apolda.

Woollen and worsted hosiery and fancy articles: 125 comforters; 4 scarfs; 14 admirables; 2 boas; 8 nets; 14 spencers; 16 caszabaicas; 7 capes; 2 handkerchiefs; 1 hunting waistcoat; 1 twine; 1 travelling waistcoat; 2 ladies' waistcoats; 1 child's frock; 5 colliers; 5 ladies' caps; 14 hoods; 11 fanchons; 15 children's caps; 8 pair of gaiters; 12 caps; 18 pair of menottes; 11 pair of wristbands; 21 pair of children's shoes; 5 pair of cuffs; 38 pair of sleeves; 9 neck handkerchiefs; 1 jacket; 4 pair of gloves; 87 camisoles; 8 pair of pantaloons; 6 ladies' shirts; 6 spencers; 1 pair of drawers; 2 petticoats; 6 girdles; 30 pair of ladies' stockings; 33 pair gentlemen's socks; 3 pair gentlemen's stockings; 17 pair children's stockings; 1 pair boys' stockings; 1 travelling shirt; 1 pair hunting stockings.

For prices and other particulars apply tho the above named H. Kreiter, Exhibition Building.

CLASS XXIX.

EDUCATIONAL WORKS AND APPLIANCES.

2602 a. SCHREINER, O., maker, Weimar.

Durable artificial caterpillar bodies for the instruction in natural sciences.

CLASS XXXI.

IRON AND GENERAL HARDWARE.

2603. BARDENHEUER, CHRISTIAN, & Co., manu., Ruhla.

a. Clasps for porte-monnaies, cigar etuis, ladies' reticules, purses etc. of brass and gilt brass, polished steel and polished German silver; b. gilt settings for hair and cloth brushes; c. ornaments for album covers; d. humming tops.

Prices current at wholesale in Prussian, French and English currency are sent with the samples.

DUCHY OF SAXE-ALTENBURG.

CLASS II.

CHEMICAL SUBSTANCES AND PRODUCTS AND PHARMACEUTICAL PROCESSES.

2606. KÜHNEMANN, GOTTHOLD, chemist, Kahla.

Potassium; sodium; phosphoric acid.

CLASS XI.

MILITARY ENGINEERING, ARMOUR AND ACCOUTREMENTS, ORDNANCE, AND SMALL ARMS.

2607. HEU. AUGUST, fire-engine maker, Altenburg. Agts. Lion M. Cohn, Phaland & Dietrich, 20 St. Dunstans Hill, City.

1. A cross-bow with bow of steel, besides a lever and a bullet-mould 2 £ 5 sh; 2. three bows of steel with three arrows 9 sh.; 3. a bow for a cross-bow 12 sh.

CLASS XXI.

WOOLLEN AND WORSTED, INCLUDING MIXED FABRICS GENERALLY.

2611. MÜNZER, CARL HEINRICH, & SON. (formerly Damsch & Münzer Sons), woollen-yarn-spinners and manu. of flannels, Ronneburg. Berlin 1844 silver medal; Munich 1854 prize-medal; London 1862 medal. Agt. W. Meyerstein, 47 Friday-street, E. C.

1. F. white woollen gauze shirt-flannels; 2. f. colored woollen shirt-flannels; 3. plain and twilled colored shirt-flannels; 4 woollen and cotton mixed shirt-flannels; 5. f. striped and checkered lama shirt-flannels; 6. 2¼ doz. different woollen colored shirts for gentlemen.

CLASS XXV.

SKINS, FURS, FEATHERS AND HAIR.

2612. Meuschke, Joh. Aug. (J.C. Meuschke & Son), brush-maker, Altenburg. Berlin 1844 honor. mention; Munich 1854 medal of honor; London 1862 honor. mention. Agts. Lion M. Cohn, Phaland & Dietrich, 20 St. Dunstans Hill, City.

1. 35 hair-brushes; 2. 19 clothes-brushes; 3. 15 pocket-brushes; 4. 2 sopha-brushes and 1 hat-brush; 5. 3 cards of 3 different sorts of painters' brushes.

CLASS XXVI.

LEATHER, INCLUDING SADDLERY AND HARNESS.

2614. Ranniger, J. L., & Son, manufactory of gloves, Altenburg. London 1851 honor. mention, 1862 medal. Agt. J. G. Jockisch, Manchester, 4 Milk-street.

1. Assortment of kid-gloves made of horse, goat- and lamb-skins of divers colors and sorts; 2. assortment of 100 pieces of dyed lambskins.

The prices are marked on the different articles.

CLASS XXX.

FURNITURE AND UPHOLSTERY, INCLUDING PAPER-HANGINGS AND PAPIER-MACHÉ.

2615. Springer, Julius, basket-maker to the court, Altenburg. Agts. Lion M. Cohn, Phaland & Dietrich, 20 St. Dunstans Hill, City.

1. A. book-case, inside lined green, to be locked 1 £. 4 sh.; 2. a folding chair 12 sh.; 3. a folding arm-chair 1 £ 4 sh.; 4. a folded chair of the same kind 18 sh.

DUCHY OF SAXE COBURG-GOTHA.

CLASS II.

CHEMICAL SUBSTANCES AND PRODUCTS AND PHARMACEUTICAL PROCESSES.

2621. Holzapfel, C. F. & S. F., manu., Grub near Coburg. London 1862 honor. mention. Paris- and steel-blue, yellow prussiate of potash, preparations of manganese.

CLASS XI.

MILITARY ENGINEERING, ARMOUR AND ACCOUTREMENTS, ORDNANCE, AND SMALL ARMS.

2624. Kley & Barthelmes, Zella near Gotha. Agts. Lion M. Cohn, Phaland & Dietrich, 20 St. Dunstans Hill, City.

A double-barrelled gun.

CLASS XIII.

PHILOSOPHICAL INSTRUMENTS AND PROCESSES DEPENDING UPON THEIR USE.

2625. Ausfeld. Herman. mechanician and optician. Gotha. London 1851 honor. mention; Munich 1854 prize-medal; Paris 1855 honor. mention; London 1862 medal. Agts. Lion M. Cohn, Phaland & Dietrich, 20 St. Dunstans Hill, City.

An astronomical and electrical self-registering apparatus, constructed and executed according to the direction of Mr. Hansen, astronomer and privy counsellor at Gotha 280 Th. (42 £).

CLASS XIX.

FLAX AND HEMP.

2626. Burbach, Brothers, & Co., manu. Gotha, manufactory Hoerselgau near Gotha London 1851 honor. mention; Munich 1854 prize-medal; London 1862 medal. Agt. A. Milne, 11 Great St. Helens.

a. 1 varnished fire-bucket No. 1.; b. 1 do. No. 16.; c. 1 unvarnished fire-bucket No. 12. pliable; d. 1 piece 50 ft. Engl. patent double-stout fire-engine hose No. 6.; d. 1 piece 56 ft. Engl. patent fire engine hose No. 6.; f. 1 piece 103 ft. Engl. hempen hose No. 8.; g. 1 piece 428 ft. Engl. one length hose No. 000½.; h. 1 piece 170 ft. Engl. flaxhose No. 3.; i. 1 piece 6 ft. Engl. gray hempen hose No. 31.: k. 1 piece 6 ft. Engl. white double-stout hose No. 31.; l. 1 piece 6 ft. Engl. do. No. 21. (for mining purposes).

CLASS XXIII.

WOVEN, SPUN, FELTED, AND LAID FABRICS. WHEN SHOWN AS SPECIMENS OF PRINTING OR DYEING.

2627. Fischer. Georg Johann. manu. Coburg. Munich 1854 honor. mention. Agts.

Lion M. Cohn, Phaland & Dietrich, 20 St. Dunstans Hill, City.

22 doz. pair of printed plush shoe-shapes (cotton and wool) p. doz. 2 Th. 20 sg. (8 sh.).

CLASS XXVI.

LEATHER, INCLUDING SADDLERY AND HARNESS.

2628. ARNOLDT, WILHELM, manu., Gotha. London 1862 medal. Agts. Lion M. Cohn, Phaland & Dietrich, 20 St. Dunstans Hill, City.

Black smooth leather; light colored bridle-leather; brown calf-leather; waxed do.

CLASS XXVII.

ARTICLES OF CLOTHING.

2629. RABUS, CARL, manu., Gotha. London 1862 honor. mention. Agts. Lion M. Cohn, Phaland & Dietrich, 20 St. Dunstans Hill, City.

Gentlemen's varnished boots with elastic $4\frac{1}{2}$ Th.; ladies' leather boots with eyes and heels $2\frac{4}{6}$ Th.; white satin-boots $2\frac{1}{3}$ Th.; satin français boots with elastic and heels $2\frac{1}{3}$ Th.; lasting-boots without heels $1\frac{1}{3}$ Th.; satin français boots with lace $2\frac{1}{3}$ Th.; all p. pair.

CLASS XXIX.

EDUCATIONAL WORKS AND APPLIANCES.

2630. BENDA, G. (G. Benda & Co.), merchants and manu., Coburg. London 1862 medal. Agts. G. Benda & Co., 79 Basinghall-street.

An assortment of toys, dolls etc.

2631. COMMERCIAL-COMPANY NEUSTADT near Coburg, represented by L. G. Köhler, Nöthlich & Otto Richard Falk, Anton Fromann, Christoph Wittbauer. London 1862 medal. Agts. N. L. D. Zimmer, 20 Bevis Mark, City, E. C.; L. Rindskopf, 59 Basinghall-street; Loewe & Co., 5 Queen-street, Cheapside, E. C.; Louis Ph. Falk, 18 Basinghall-street; Z. Steiner & Co., 11 London Wall, City, E. C.

Thuringian toys of papier-mâché and dolls. For prices apply to the agents mentioned.

2632. HELM, OTTILIE, WELLHAUSEN, GEORG (Helm & Wellhausen), manu., Friedrichroda near Gotha. Munich 1854 prize-medal; London 1862 honor. mention. Agts. Killy, Traub & Co., 52 Bread-street, Cheapside.

Toys of papier-mâché, porcelain and wood of different qualities, descriptions and numbers.

2633. KRAUSE, THEODOR, pewterer, Gotha. London 1862 honor. mention. Agt. W. Meyerstein, 47 Friday-street.

Specimens of tin-toys of different species.

2634. PERTHES, JUSTUS, publisher, Gotha. Munich 1854 prize-medal; Paris 1855 silver medal; London 1862 medal. Agts. Williams & Norgate, booksellers.

Maps; atlasses, books; glyphographic plates.

CLASS XXX.

FURNITURE AND UPHOLSTERY, INCLUDING PAPER-HANGINGS AND PAPIER-MACHÉ.

2635. HOFFMEISTER, TOBIAS, GRASSER, MORITZ, (T. Hoffmeister & Co.). upholsterer to the court and manu. of furniture, Coburg. London 1851 honor. mention. Agts. Lion M. Cohn, Phaland & Dietrich, 20 St. Dunstans Hill, City.

1 Side-board of unpolished wall-nutt tree 310 Th.; 1 chair to match 15 Th.; 1 sofa, 2 arm-chairs and 6 easy chairs of polished black wood with bronze-ornaments, covered with gobelin-fabric of Aubusson, 1 sofa-table to match together 475 Th.

CLASS XXXI.

IRON AND GENERAL HARDWARE.

2636. GRÜNEWALD, WILH., pewterer, Coburg. Agts. Lion M. Cohn, Phaland & Dietrich, 20 St. Dunstans Hill, City.

1 Wine-tankard, 3 ft. high, 14 Pfd. 54 fl. (4 £ 10 sh.); set of coffee- and tea-things, oval-shaped: 1 coffee-pot, 2 Pfd. 23 Lth. 7 fl. 30 kr. (12 sh. 6 d.); 1 tea-kettle with stand, 4 Pfd 8 Lth. 13 fl. 48 kr. (1 £ 3 sh.); 1 cream-pot, 1 Pfd. 10 Lth. 3 fl. 36 kr. (6 sh.); 1 sugar-box 1 Pfd. 7 Lth. 3 fl. 36 kr. (6 sh.). 1 coffee-pot, crown-shaped, 2 Pfd. 9 Lth. 6 fl. (10 sh.); 1 tea-pot do. 1 Pfd. 23 Lth. 6 fl. (10 sh.); 1 tea-kettle with stand, melon-shaped, 3 Pfd. 27 Lth. 12 fl. (1 £); 1 cream-pot do. 20 Lth. 2 fl. 24 kr. (4 sh.); 1 cup with lid 1 Pfd. 23 Lth. 6 fl. (10 sh.); 1 cut glass-tankard with lid, 1 Pfd. 16 Lth. 2 fl. 24 kr. (4 sh.); 1 coffee-pot, pear-shaped, 1 Pfd. 27 Lth. 2 fl. 24 kr. (4 sh.).

CLASS XXXV.

POTTERY.

2637. DORNHEIM, HEINRICH, manu. of earthen ware, Gräfenrode near Gotha. Agts. Lion M. Cohn, Phaland & Dietrich, 20 St. Dunstans Hill, City.

Boar's head 9 Th.; stag's head with horns 10 Th.; head of a hunted roe-buck 4 Th.; roe-buck's head, turned to the left hand 3 Th.; roe-buck's head middling size 25 sg.; roe-buck's head, smallest size 10 sg.; chamois' head, turned to the right hand 3 Th.; fox-head 2 Th.

All the horns are made of clay.

2638. HENNEBERG, AUGUST (F. E. Henneberg & Co.). manu., Gotha. Berlin 1844 gold prize-medal; London 1851 and 1862 honor. mention. Agts. Lion M. Cohn, Phaland & Dietrich, 20 St. Dunstans Hill, City.

1. A gilt porcelain dressing-table with picture of the Madonna of Raphael 93 Th. 10 sg. (14 £); 2. a gilt porcelain vase with flowers in relievo, showing a view of Coburg 60 Th. (9 £); 3. a gilt porcelain coffee-board with picture (Palicaren by Jacobs) in case 60 Th. (9 £); 4. two gilt cake-trays with perspective views of Coburg, Gotha, in one case 53 Th. 10 sg. (8 £); 5. a gilt coffee-service of muslin-porcelain in case 10 Th. (1 £ 10 sh.); 6. a porcelain plate with flowers in gilt-

frame in case 73 Th. 10 sg. (11 £); 7. a porcelain plate with fruits in gilt frame in case 73 Th. 10 sg. (11 £); 8. a porcelain plate with the firm of the exhibitor, in gilt frame; 9. a porcelain bucket 4 Th. (20 sh.).

CLASS XXXVI.

MANUFACTURES NOT INCLUDED IN PREVIOUS CLASSES.

2639. ARNOLDI, H., counsellor of commerce, Gotha. Agts. Mittler & Eckhardt, 20 St. Dunstans Hill, E. C.

Collection of artificial fruits, consisting of 42 species of apples, 31 species of pears, 1 species of peaches, 15 species of plums and 1 species of damask-plums, published till now in 15 series; each series 2 Th.

2640 a. BEER, C., manu., Coburg. London 1862 honor. mention.

Tress-work of ruítes.

2641. WENIGE, ERNST. EMIL. merchant, Ohrdruff. Agts. Lion M. Cohn, Phaland & Dietrich, 20 St. Dunstans Hill, City.

Specimens of shirt-buttons.

DUCHY OF SAXE-MEININGEN.

CLASS II.

CHEMICAL SUBSTANCES AND PRODUCTS AND PHARMACEUTICAL PROCESSES.

2646. MERLET & Co., manu., Sophienau near Eisfeld. Agts. Belger & Engelhardt, 8 Sise Lane, E. C.

Different sorts of ultramarine.

2647. ORTLOFF, FRIEDRICH, DR. PHIL., Eisfeld. Munich 1854 honor. mention. Agt. C. Trübner, 20 St. Dunstans Hill, City.

A card of 52 specimens of encaustic colors burnt on small tables of porcelain representing a picture.

CLASS IV.

ANIMAL AND VEGETABLE SUBSTANCES USED IN MANUFACTURES.

2648. SCHMIDT, C., soap-boiler to the court, Pösneck.

Toilet-soaps.

CLASS XXI.

WOOLLEN AND WORSTED, INCLUDING MIXED FABRICS GENERALLY.

2649. DIETRICH, J. F., & SON. manu., Pösneck. Agts. Lion M. Cohn, Phaland & Dietrich, 20 St. Dunstans Hill, City.

10 pieces woollen flannels: No. 1. 6 yards 32 in. genuine blue navy-molton 1 sh. 7 d.; No. 2. 6 yards 32 in. black spagnolet 1 sh. 6 d.; No. 3. 6 yards 32 in. gentian spagnolet 1 sh. 7 d.; No. 4. 6 yards 32 in. scarlet spagnolet 1 sh. 8 d.; No. 5. 6 yards 37 in. light blue printed twilled flannel 1 sh. 6 d.; No. 6. 6 yards 37 in. scarlet do. 1 sh. 8 d.; No. 7. 6 yards 37 in. crimson do. 1 sh. 8 d.; No. 8. 6 yards 45 in. black and white checked flannel

1 sh. 9 d.; No. 9. 6 yards 45 in. green do. 1 sh. 10 d.; No. 10. 6 yards 45 in. gentian do. 1 sh. 10 d.; all p. yard.

CLASS XXVI.

LEATHER, INCLUDING SADDLERY AND HARNESS.

2650. DIESEL & WEISE, merchants and manu., Pösneck. London 1862 honor. mention.

15 pieces of curried calf-skins.

CLASS XXIX.

EDUCATIONAL WORKS AND APPLIANCES.

2651. HUTSCHENREUTHER & Co., manu., Wallendorf.

Dolls of papier mâché; slates: slate pencils; stone, porcelain and glass marbles.

CLASS XXXI.

IRON AND GENERAL HARDWARE.

2652. DEHLER, BROTHERS. & Co., merchants, Saalfeld. London 1862 medal. Agts. Belcher & Engelhardt, 8 Sise Lane. E. C., and Trübner (Phaland & Dietrich), 20 St. Dunstans Hill, City.

1 roll blue iron gauze, 100 sq. ft., 6 Th. 20 sg. (1 £): 1 roll brown do., 6 Th. 20 sg. (1 £); 1 roll gray do.. 6 Th. 20 sg. (1 £); 1 roll white do. 6 Th. 20 sg. (1 £); 1 roll yellow do., 6 Th. 20 sg. (1 £): 1 roll green do., 6 Th. 20 sg. (1 £); 1 roll pink do., 6 Th. 20 sg. (1 £); 1 roll transparent gauze, 58 sq. ft., 6 Th. 14 sg. (19 sh. 4 d.).

17 pieces wire window-blinds, with genre-paintings. 8 Th. (1 £ 4 sh.): 4 do., portraits, 8 Th. (1 £ 4 sh.); 6 do., hunting-scenes, 8 Th. (1 £ 4 sh.); 8 do.. landscapes. 7 Th. (1 £ 1 sh.); 1 fire-screen, with genre painting, 48 Th. (7 £ 4 sh.); 1 do., landscape, 48 Th. (7 £ 4 sh.): all p. doz.

4 pieces horse-hair textures, for sieves, No. 1. 24 ☐ in., 1 Th. (3 sh.); No. 2. 24 ☐ in., 15 sg. (1 sh. 6 d.); No. 3. 24 ☐ in., 10 sg. (1 sh.); No. 4. 24 ☐ in., 10 sg. 10 pf. (1 sh. 1 d.); 3 pieces brass-wire textures for sieves No. 1. 19½ ☐ in., 20 sg. (2 sh.); No. 2. 19½ ☐ in., 21 sg. 8 pf. (2 sh. 2 d.); No. 3. 19½ ☐ in., 24 sg. 2 pf. (2 sh. 5 d.); all p. piece.

5 sample-cards of transparent iron gauze and wire-textures: 1 piece brass-wire texture, No. 65. 77 sq. ft., 9 sg. 7 pf. (11⅓ d.); do. No. 90. 144⅔ sq. ft., 14 sg. 2 pf. (1 sh. 5 d.); do. No. 50. 65⅓ sq. ft. 7 sg. 9 pf. (9⅓ d.); do. No. 120. 100⅓ sq. ft., 1 Th. (3 sh.); 1 piece of copper-wire texture, No. 40. 28½ sq. ft., 8 sg. 4 pf. (10 d.); 1 piece of iron-wire texture, No. 42. 60¼ sq. ft., 5 sg. 10 pf. (7 d.); 1 piece of metallic cloth, No. 60. 178⅛ sq. ft., 9 sg. 10 pf. (11⅕ d.); 1 piece of iron-wire texture, No. 32. 47 sq. ft., 3 sg. 9 pf. (4½ d.); do. No. 15. 89⅔ sq. ft., 1 sg. 5 pf. (1 7/10 d.); do. brome-grass, 52 sq. ft., 1 sg. 8 pf. (2 d.); do. No. 10. 87½ sq. ft., 1 sg. 9 pf. (2 1/10 d.); do. No. 6. 29⅓ sq. ft., 10 sg. (1 sh.); do. for iron garden-chairs, 28 sq. ft., 6 sg. 8 pf. (8 d.); all p. sq. ft.

1 silver-wire-basket, French fashion, No. 3. 11 Th. (1 £ 13 sh.); do. No. 2. 10 Th. (1 £ 10 sh.); do. No. 1. 9 Th. (1 £ 7 sg.); do. No. 0. 8 Th. (1 £ 4 sh.); do. No. 00. 7 Th. (1 £ 1 sh.); 1 do. German fash. No. 3. 9 Th. (1 £ 7 sh.); do. No. 2. 8 Th. (1 £ 4 sh.); do. No. 1. 7 Th. (1 £ 1 sh.); do. No. 0. 6 Th. (18 sh.); do. No. 00. 5 Th. (15 sh.); do. No. 000. 4 Th. 15 sg. (13 sh. 6 d.); do. No. 0000. 4 Th. (12 sh.); 1 money basket, No 1. with set and lock 16 Th. (2 £ 8 sh.); do. No. 2. 14 Th. (2 £ 2 sh.); do. No. 1. 6 fold with cover 8 Th. 25 sg. (1 £ 6 sh. 6 d.); do. No. 1. with 6 fold large set 10 Th. (1 £ 10 sh.); do. No. 1. with 5 fold large set 9 Th. 16 sg. (1 £ 8 sh. 7 d.); do. No. 1. with 4 fold large set 9 Th. 12 sg. (1 £ 8 sh. 3 d.); d. No. 1. with 3 fold large set 8 Th. 15 sg. (1 £ 5 sh. 6 d.); do. No. 1. with 3 fold small set 8 Th. 6 sg. (1 £ 4 sh. 6 d.); do. No. 1. with 2 fold small set 8 Th. (1 £ 4 sh.); do. No. 1. 6 fold 6 Th. 24 sg. (1 £ 5 d.); do. No. 1. 5 fold 6 Th. 12 sg. (19 sh. 2 d.); do. No. 2 a. 5 fold 6 Th. (18 sh.); do. No. 2 a. 4 fold 5 Th. 24 sg. (17 sh. 5 d.); 1 knife basket No. 1. with brass-garniture 8 Th. 15 sg. (1 £ 5 sh. 6 d.); do. No. 2. with tinplate-garniture 6 Th. (18 sh.); do. No. 3. do. 5 Th. 15 sg. (16 sh. 6 d.); 1 tea-spoon basket 4 Th. (12 sh.); 1 key-basket No. 1. 5 Th. (15 sh.); do. No. 2. 3 Th. 20 sg. (11 sh.); 1 lady's work-box No. 397. 12 Th. (1 £ 16 sh.); 1 postage-stamp basket 6 Th. (18 sh.); 1 round dish-cover, 12 in. diam. 4 Th. 24 sg. (14 sh. 5 d.); 1 oval do, 12 in. diam. 4 Th. 24 sg. (14 sh. 5 d.); all p. doz.

CLASS XXXIV.

GLASS.

2654. Böhm, F., manu., Ernsthal.

Glass-cup, glass-flowers, cigar-pipes, glass-figures.

CLASS XXXV.

POTTERY.

2655. Conta & Böhme, manu., Pösneck.

43 fancy articles of china: figures No. 1. and 2. Scotch hunters 2/0. size; No. 3. and 4. do. I ; No. 5. and 6. hunters and lady 2/0.; No. 7. and 8. Marquis à la samis 0.; No. 9. and 10. minuet-dancers I.; No. 11. and 12. Mirabeau and Lisette I.

Boxes: No 13. boars-head; No. 14. Germania; No. 15. the first address 0.; No. 16. evening-rest; No. 17. the shared repast; No. 18. a covey of hens; No. 19. sleeping child in chair II.; No. 20. casket; No. 21. imitated wood; No. 22. jewel-box, antique III.; No. 23 do. with cherries; No. 24. do. with gold flowers; No. 25. and 26. do. with white roses.

No. 27. Watch-stand, Spanish lady; No. 28. and 29. 1 pair of figures, flagons 1413.; No. 30. group, feast of roses 0.

Inkstands No. 31. spring-amusements 0.; No. 32. sisterly love 0.; No. 33. boy with sheep 0.; No. 34. girl with sheep 0.; No. 35. chess-players 0.; No. 36. hunter with hound 0.; No. 37. rural 0.; No. 38. autumn-amusements I.; No. 39. the blessings of house and field 0; No. 40. group of Zouaves II.; No. 41. pen-case II.; No. 42. box (case) like wood; No. 43. watch-stand (Germania).

2656. Eberlein, Joh. Chr., manu. of china, Pösneck. Agts. Lion M. Cohn, Phaland & Dietrich, 20 St. Dunstans Hill, City.

Cigar-stand, match-stand, boxes for matches, jewels, pins, groups, figures, letter-pressers etc., p. doz. from 3 sh. to 7 £.; according to the special price current.

2657. Heubach, Kämpfe & Sontag, manu., Wallendorf near Saalfeld. Agts. Fr. Haas & Co., 34 Jewin-street.

No. 995. 1 group of figures, Luther and family on Christmas-eve after Schwerdgeburth; No. 996. 2 statues, Ariadne on the panther; No. 998. 1 christmas-tree; No. 974. 977. 922. 981. 975. 972. 973. 921. 8 statues, Maria and several saints; No. 950. 947. 4 statues, Zouave musicians and dancers; No. 889. 931½. 932½. 914. 930½. 992. 6 inkstands with figures; No. 933. 855. 912. 3 large boxes with figures; No. 954. 965. 968. 994. 966. 952. 937. 971. 8 attrapes, colored and gilt, all in biscuit china. 6 sets of tea-things for children, each consisting, of 1 tea-pot, 1 cream-jug, 1 sugar-box and 6 cups and saucers, colored and gilt; No. 1000. 1 garland of roses in biscuit china; No. 1001. 1 nosegay in biscuit china, 10 dolls' heads in natural colors; 7 nanking dolls with china heads, legs and arms.

2658. Müller, Fr., & Strasburger, manu. and merchants, Sonneberg. Agts. Petit & Wilborn of Paris, care of H. B. & G. Lang, 5 Dunstans Hill, City.

An assortment of paper mâché and china dolls' heads, leather and cotton-dolls, crying babies.

2660. Unger, Schneider & Co., Gräfenthal, formerly Taubenbach.

1 pair of vases, height 3½ ft., diam. 1½ ft., with gold-decoration; 1 dinner-service for children, consisting of 36 pieces. 3 sets of tea-things for children, No. 1. A. 1 fl. 12 kr. (2 sh.); No. 293. C. 20 fl. (1 £ 13 sh 4 d.); No. 26. C. 42 fl. (3 £ 10 sh.); 1 small dinner-service, No. 119. A. 1 fl. 12 kr. (2 sh.); prices p. doz.

PRINCIPALITY OF SCHWARZBURG-RUDOLSTADT.

CLASS XXIX.
EDUCATIONAL WORKS AND APPLIANCES.

2667. SPECHT, B. (B. Specht & Co.), manu., Rudolstadt. London 1862 honor. mention. Agt. Fred. Preyss, 20 St. Dunstans Hill, Great Tower-street.

23 boxes of water-colors, prices p. doz.: No. 1. 18 colors 4 d.; No. 2. 18 colors 6½ d.; No. 3. 12 colors 11 d.; No. 3a. 12 colors with 4 small cups 35 d.; No. 4. 12 colors 18 d.; No. 5. 24 colors 9 d.; No. 6. 24 colors 12 d.; No. 7. 24 colors 15 d.; No. 9. 24 colors 34 d.; No. 9a. 18 colors 21 d.; No. 11. 24 colors with 2 small cups 52 d.; No. 12. 24 colors 52 d.; No. 14. 18 colors with 2 small cups 55 d.; No. 30. 18 colors with 2 small cups 72 d.; No. 30a. 18 colors with 4 small cups 8 sh.; No. 30b. 24 colors with 4 small cups 11 sh.; No. 33. 12 colors 6 sh. 6 d.; No. 34. 18 colors with 2 small cups 10 sh.; No. 34a. 24 colors with 2 small cups 18 sh.; No. 40. 24 colors 2 small cups 53 d.; No. 11a. 15 colors with 4 small cups 5 d.; No. 14a. 15 colors with 4 small cups 5 sh. 9 d.; No. 38a. 45 colors with 6 small cups 6 £.

PRINCIPALITY OF SCHWARZBURG-SONDERSHAUSEN.

CLASS I.
MINING, QUARRYING, METALLURGY AND MINERAL PRODUCTS.

2670. THE LUTHERSTEUFE COMPANY, Ilmenau in Saxe-Weimar. Manufactory Oehrenstock in Schwarzburg-Sondershausen. Agts. Steinthal, Löwenthal & Co., 14 Little Tower-street.

Crystallized manganese in large lumps; crystallized manganese in smaller lumps, crystallized manganese, powdered p. 100 lb. weight 13 sh.

CLASS XXXI.
IRON AND GENERAL HARDWARE.

2671. BROEMEL, AUGUST, manu. of weighing-machines, Arnstadt. Berlin 1844 bronze medal; London 1862 honor. mention.

Weighing machines: No. 8539. for 10 to 15 Zollz. weight 35 Th. (5 £ 5 sh.); No. 9272. for 2 to 4 Zollz. weight 16 Th. (2 £ 8 sh.); No. 9271. for 1 to 2 Zollz. weight 12 Th. (1 £ 16 sh.); No. 9270. for 1 Zollz. weight 10 Th. (1 £ 10 sh.); prices current annexed; p. doz. 25 pCt. discount.

CLASS XXXV.
POTTERY.

2672. SCHIERHOLZ, C. G., & SON, manu., Plaue.

Figures and utensils; crucifixes and lithophanies of china; part of them combined with wood and metal.

PRINCIPALITY OF WALDECK.

CLASS IV.
ANIMAL AND VEGETABLE SUBSTANCES USED IN MANUFACTURES.

2676. BACKHAUS, BROTHERS, brewers, Affoldern (Waldeck). Agts. Lion M. Cohn, Phaland & Dietrich, 20 St. Dunstans Hill, City.

A bottle with liquid brown pitch for stopping the beer-casks; 100 lb. 16 Th. (2 £ 11 sh.). A small barrel pitched with it.

CLASS VIII.
MACHINERY IN GENERAL.

2677. DUNCKER, ED., manu., Friedensthal. Bread-cutting machine with adjusting screws.

KINGDOM OF WÜRTTEMBERG.

2681. BÖHRINGER, C. F., & SONS, manu. of quinine, santonine etc., Stuttgart. London 1862 medal. Agt. the Württemberg Trading-Company at Stuttgart, represented by Sam. Sessing, 10 Huggin-Lane, Wood-street, Cheapside, E. C.

No. 1. sulphate of quinine; No. 2. muriate of quinine; No. 3. sulphate of cinchonine; No. 4. santonine.

2682. BÜRKLE, J. F., manu. of chemical products, Gross-Seppach. Agt. s. No. 2681.
(s. II. No. 55.)

Chemical products: No. 1. aromatic sulphur barrel-brand, free from arsenic, for wine-coopers, prepared after a peculiar method of the exhibitor's own invention p. 100 lb. 5 £ 16 sh. 8 d.; No. 2. do., without aromatics p. 100 lb. 3 £ 12 sh. 3 d.; No. 3. sealing-wax in different sorts and colors, p. 100 lb. from 4 £ 3 sh. 4 d. to 33 £ 6 sh. 8 d.; No. 4. tooth-powder p. 100 boxes 1 £ 13 sh. 4 d., 3 £ 6 sh. 8 d. and 5 £; No. 5. aromatic lotion 2 £ 1 sh. 8 d. and 4 £ 3 sh. 4 d.; No. 6. oil to destroy bugs 3 £ 6 sh. 8 d. and 6 £ 13 sh. 4 d.; No. 7. hair-oil 8 £ 6 sh. 8 d.; No. 8. tooth-tincture 4 £ 3 sh. 4 d. and 8 £ 6 sh. 8 d.; all p. 100 glasses.

2683. FRANKEN, J. H., manu. of the »Stuttgart Water«, Stuttgart. London 1862 medal. Agt. s. No. 2681.

»Stuttgart-water« or superfine eau de Cologne in different flasks of different form and size viz $\frac{1}{4}$ and $\frac{1}{2}$ flasks, whereof 4 and 8 = 1 pint; further: in form of wine-bottles whereof 8, 4, 2 = 3 pints.

2685. HÄCKER, C., manu. and merchant, Stuttgart. London 1862 medal. Agt. s. No. 2681.

Bronze-colors: prices in florins in the 52½ fl.-standard p. pound Zollgew., in one ounce-packets with 25 pCt. Disct. and 15 pCt. Disct. in one pound-packets. Fine silver 64 fl. p. ½ Kil. (netto.); English silver 30 fl. p. ½ Kil.; German silver 12 fl. p. ½ Kil.; English green gold, pale flora, lemon-flora: No. 50. 4 fl., No. 60. 5 fl., No. 80. 6 fl., No. 90. 7 fl., No. 100. 8 fl., No. 200. 9 fl., No. 300 ... fl., No. 400. 12 fl., No. 500. 14 fl., No. 600. 15 fl., No. 800. 17 fl., No. 1000. 19 fl., No. 2000. 21 fl. Lemon-gold, rich gold, flesh-color, carmine, crim-

son, bright red, violet, purple, dark brown: No. 50. 2 fl., No. 60. 3 fl., No. 80. 4 fl., No. 90. 5 fl., No. 100. 6 fl., No. 200. 8 fl., No. 300. 9 fl., No. 400. 10 fl., No. 500. 12 fl., No. 600. 13 fl., No. 800. 15 fl., No. 1000. 16 fl., No. 2000. 18 fl. Rich gold B., lemon B., English-white: No. 50. 2 fl., No. 60. 3 fl., No. 80. 4 fl., No. 90. 5 fl., No. 100. 5½ fl., No. 200. 6 fl., No. 300. 7 fl., No. 400. 8 fl., No. 500. 9½ fl., No. 600. 10 fl., No. 800. 12 fl., No. 1000. 13 fl., No. 2000. 15 fl. Pale yellow, deep yellow, orange, green, silver-composition: No. 50. 1½ fl., No. 60. 2 fl., No. 80. 3 fl., No. 90. 3½ fl., No. 100. 4 fl., No. 200. 5 fl., No. 300. 6 fl., No. 400. 7 fl., No. 500. 8 fl., No. 600. 9 fl., No. 800. 11 fl., No. 1000. 12 fl., No. 2000. 13 fl.; all p. ½ Kil.

2687. KNOSP, RUDOLPH, manu. of Aniline-colors, Indigo-carmine and Archil, Stuttgart. Paris 1855 bronze medal; London 1862 medal. Agt. Fr. Köbel, partner of W. Jameson & Co., London.

Aniline-colors, red, violet and blue; indigo-carmine, archil and cudbear.

2689. RUND, GG. FRIED., manu. of sugar of lead and white lead, Heilbronn. Agt. s. No. 2681.

2 bottles containing crystallized sugar of lead, three times refined p. 100 lb. 21 fl. 30 kr.

2690. SCHWEIKHARDT, DR. EDUARD, manu. of artificial manure, Tübingen. Agt. s. No. 2681.

2 bottles containing artificial manure: No. 1. I. quality 6 sh. 4 d., No. 2. II. quality 4 sh. 3 d. p. cwt.

2691. SIEGLE, HEINRICH, manu. of colors and flower-papers, Stuttgart. London 1851 prize-medal; Munich 1854 great medal; Paris 1855 silver medal; London 1862 medal. Agt. s. No. 2681.

No. 1. Assortment of lakes- and other colors; especially fine colors, carmine, carmine-lakes, red wood-lakes, madder-lakes, pink-violet, brilliant-lakes, pensée and violet lakes, groseille-lakes, chrome-green for water and oil-paints, emerald-green, zinc-green, green, blue, yellow and brown lakes, cadmium-yellow, chrome-yellow and orange, zinc-yellow, kings-yellow, yellow and brown ochres, white colors in drops etc.; No. 2. colors for confectioners; No. 3. fancy-paper for artificial flowers and carmine-cloth; No. 4. samples of the application of the colors for paper-hangings, fancy-papers, lithography, oilcloth, lackering, painting in oil and water.

2692. Veit-Weil, manu., Oberdorf near Bopfingen. Munich 1854 medal of honor; Paris 1855 honor. mention. Agt. the Württemberg Trading Company at Stuttgart, represented by Sam. Sessing, 10 Huggin-Lane, Wood-street, Cheapside, E. C. (s. No. 2717.)

Artificial manure; prices p. Zollz. Bone-dust containing 4½ pCt. of nitrogen and 52 pCt. of phosphate of lime, by-product of the manufacture of bone-glue 5 sh. 1 d.; bone-dust containing 6 pCt. of nitrogen and 60 pCt. of phosphate of lime 9 sh. 4 d.; prepared bone-dust for making colorless jellies deprived of taste and odour 3 £ 2 sh. 8 d.; phosphate of lime 5 sh. 1 d.; artificial manure made exclusively from animal offal 6 sh. 10 d.; bone grease 2 £ 10 sh. 10 d.

2693. The Württemberg glue and manure manufacturing company, Reutlingen. Agt. s. No. 2692. (s. No. 2718, 2732.)

No. 6. artificial guano 7½ sh.; No. 7. bone superphosphate 6 sh.; No. 8. super-phosphate of lime 8½ sh.; all p. cwt.

2693 a. The Württemberg trading company, Stuttgart. Agt. s. No. 2692. (s. No. 2724 a, 2771 a, 2822, 2849.)

1 case containing powder for hardening forged iron p. ½ Kil. 10 d.

2694. Ziegler, Ernst, manu., Heilbronn. London 1862 honor. mention. Agt. s. No. 2692.

Sample of patent substitute for animal char-coal in grains, employed for discoloring liquids as for instance in the refining of sugar, discoloring and filtering of raw citric and tartaric acids 5 sh.; sample of patent substitute for animal charcoal in powder, employed as coloring matter, viz: for the manufacture of inks, paints etc. for letter-press and copper-plate printing 6 sh. p. cwt. each.

CLASS III.

SUBSTANCES USED FOR FOOD, INCLUDING WINES.

2695. The Royal board of agriculture, Stuttgart, on behalf of the following Württemberg wine-growers, viz: Albrecht (Gemeinderath) at Heilbronn; Bröm (Gerichtsnotar) at Weinsberg; Brüselle, Baron von, at Kleinbottwar; Brunner (Gutsbesitzer) at Neckarsulm; Burrer (Gutsbesitzer) at Gündelbach; Burkert, Friedrich, at Ingelfingen; Closs, F., at Heilbronn; Drauz, Heinrich, at Heilbronn; Egner, Friedrich, at Ingelfingen; Ellrichshausen, Baron von, at Assumstadt; Fehleisen (Apotheker) at Reutlingen; Feyerabend, sen., (Rechtsconsulent) at Heilbronn; Friz (Rathschreiber) at Fellbach; Fröhlich, Christ., at Bönnigheim; Gurrath, Adam, at Heilbronn; Häfner (Schultheiss) at Markelsheim; Heimsch (Gemeinderath) at Stuttgart; Herrmann (Oekonom) at Fellbach; Hild & Sohn at Weinsberg; Höchheimer & Sohn at Mergentheim; Hof-Domainenkammer, Königliche, at Stuttgart; Huber at Heilbronn; Hügel, Baron von, at Eschenau; Kayser, Christ., at Obertürkheim; Koch & Stierlen at Esslingen; Fürstlich Hohenlohesche Domainen-Kanzlei; Lobmüller at Reutlingen; Mayer, Julius, at Stuttgart; Marquardt (Hôtelbesitzer) at Stuttgart; Meiss, E. von, at Lichtenberg; Metz, Carl, at Heilbronn; Mittler & Eckhardt at Stuttgart; Müller, Johann, at Ingelfingen; Münzing (Fabrikant) at Heilbronn; Neipperg's Gräfliche Kellerei at Schwaigern; Nickel (Waldinspektor) at Heilbronn; Niclas, Ludwig, at Griesbach; Oehringen (Fürstliches Rentamt); Palm, Baron von, at Messbach; Pfander, Friedrich, at Fellbach; Romig, Friedrich, at Griesbach; Scheuerlen, Louis, and Scheuerlen, August, at Erligheim; Siber (Postverwalter) at Vaihingen; Stählen (Notar) at Heilbronn; Stückle, Julius, at Stuttgart; Sturmfeder, Barou von, at Oppenweiler; Thorn, Wilhelm, at Bönnigheim; Volz, Heinrich, at Heilbronn; Weitzel (Gutsbesitzer) at Sonnenberg; Wiedersheim (Hof-Kammerverwalter) at Stetten; Wölz, Wittwe, at Ingelfingen. London 1862 medal. Agts. Mittler & Eckhardt, 10 St. Dunstans Hill, E. C. (s. No. 2707.)

No. 1—139. Assortment of Württemberg wines from 51 growers, arranged into a whole by the Royal board of agriculture. It contains in all 139 different sorts, of the years 1811, 1842, 1846' 1857, 1858 and 1859, viz: 75 white and 61 red wines. The best ones, derived from but one species of wine (that is to say by using either the blue »Clevner« or the white »Clevner« alone; or the »Trollinger«-, »Traminer« or »Riessling«-grapes each for themselves), will be sold at prices from 40 kr. to 1 fl. 30 kr. p. bottle, whilst the ordinary sorts, being likewise very recommendable, are selling at from 18 to 36 kr.

2696. Daur, Heinrich, manu. of pearl-barley, Ulm. London 1862 medal. Agt. s. No. 2692.

Barley; prices p. Zollz. Ulm pearl-barley No. 000. 18 fl., No. 00. 17 fl., No. 0. 16 fl., No. 1. 15½ fl., No. 2. 14½ fl., No. 3. 13 fl., No. 4. 12 fl., No. 5. 10 fl., No. 6. 8 fl.

Dry-peas, shelled and freed from the husks 9½ fl. Barley-meal, by-product of the manufacture of pearl-barley 5 fl.

2697. Engelmann & Co., manu. of sparkling wines and liquors, Stuttgart. London 1862 medal and honor. mention. Agt. s. No. 2692. (s. II. No. 54.)

Wines and liqueurs; prices p. dozen bottles, package not charged free in port London: 9 bottles sparkling-wine 28 sh.; 2 bottles Maraschino, 2 bottles extrait d'Absynthe, 2 bottles crème de vanille, 1 bottle liqueur aux herbes aromatiques (Chartreuse) each 9 sh. 4 d.; each of the following liqueurs in 2 bottles: anisette, crème de cumin (alash), crème de menthe, curaçao, eau de noyaux, maagbitter, parfait d'amour, each 8 sh. 6 d.: double cumin 6 sh.; arrackpunch-essence, rumpunch-essence, wine-punch-essence, grog-essence 11 sh. 8 d.

2698. Administration of the farms of His Majesty the King of Württemberg (K. Hof-Kameral-Amt), Stuttgart. London 1862 medal. (s. No. 2695, 2710, 2773.)

Wines.

2700. Kirsner, W., merchant and manu. of essential oils, Rottweil. Agt. s. No. 2692.

Finest essential oil of cumin, twice rectified, p. Zollpound 6 sh. 3 d.

2701. Laiblin, Eduard, & Co., manu. of Champagne, sparkling Hock, Neckar- and Moselle wines, Stuttgart. London 1862 medal. Agt. Robert Selby, 31 Eastcheap. (s. II. No. 35.)

Sparkling-wine: 32 bottles of different size say: »Magnums«, quarts, pints and ½ pints, to show the different labels and qualities of the manufactures viz: champagne, sparkling-Neckar-, Hock- and Moselle-wines.

2702. Ludwig, J. F., & Co., manu. of candied and dried fruits and preserved vegetables, Stuttgart. London 1862 medal. Agt. the Württemberg Trading Company at Stuttgart, represented by Sam. Sessing, 10 Huggin-Lane, Wood-street, Cheapside, E. C.

Candied and dried fruits, prices p. lb.: No. 1. candied cherries 1 fl. 6 kr.; No. 2. white nuts 1 fl. 12 kr.; No. 3. candied apricots, halves 1 fl. 12 kr.; No. 4. various fruits 1 fl.; No. 5. apricots, halves 1 fl. 18 kr.; No. 6. apricot-paste 54 kr.; No. 7. peach-paste 54 kr.; No. 8. plum-paste 1 fl.; No. 9. candied peach-paste 1 fl.; No. 10. candied apricot 1 fl.; No. 11. candied damask-plums 1 fl. 12 kr.; No. 12. white pears 1 fl. 6 kr.; No. 13. red pears 1 fl. 12 kr.; No. 14. candied plums 1 fl. 12 kr.; No. 15. candied cherries without stones 48 kr.; No. 16. paradise-pears 1 fl. 30 kr.; No. 17. plums 1 fl. 12 kr.; No. 18. paradise-apples 1 fl. 12 kr.; No. 19. goose-berries 1 fl. 6 kr.; No. 20. Mirabelles 1 fl. 18 kr.; No. 21. apricots, entire 1 fl. 18 kr.; No. 22. damask-plums 1 fl. 6 kr.; No. 23. peaches 1 fl. 18 kr.; No. 24. quinces in slices, red and yellow 1 fl. 6 kr.; No. 25. plums 1 fl. 48 kr.

Prices p. bottle: 10 bottles, containing cherries, under water 48 kr.; green peas 1 fl. 45 kr.; plums 48 kr.; goose berries 48 kr.; various fruits in jelly 2 fl.; green beans under water 48 kr.; rasp berries, sweet, prepared without heating 54 kr.; sweet peas 48 kr.; carviol 1 fl. 24 kr.; rasp berries 54 kr.

2703. Mittler & Eckhardt, wine growers, large establishment for sparkling wines, own production, Stuttgart. London 1862 honor. mention. Agts. Mittler & Eckhardt, 20 St. Dunstans Hill.

Prices p. doz. quart bottles: No. 1. sparkling Neckar-wine 1859. label: »Neckar Champagne« 24 sh.; No. 2. do. 1857. label: »Moussirender Wein« 26 sh.; No. 3. do. 1857. Riesling grape, label, green: »sparkling Hock, extra superior« 35 sh.; No. 4. do. 1857. Riessling and Muscatel grape, label, green: »sparkling Moselle, extra superior« 30 sh.; No. 5. do. 1859. for exportation, label, white: »sparkling Hock, finest quality 20 sh.; No. 6. do. 1859. Muscatel-wine for exportation, label white: »sparkling Moselle, finest quality« 20 sh.

No. 7. red sparkling Neckar-wine 1857. Burgundy grape, label pink: »Moussirender Wein« 35 sh.

2704. Renner, J. A., manu. of starch, Schw. Hall. London 1862 medal. Agt. s. No. 2702.

No. 1. 1 crystallized starch; No. 2. starch in lumps; No. 3. starch in powder.

2705. Schöllkopf, Johann, manu., Ulm. London 1862 medal. Agt. s. No. 2702.

Starch: No. 1. finest quality; No. 2. middling fine quality.

2706. Seelig, Emil, manu. of finest dry chicory coffee, Heilbronn. London 1862 medal. Agt. s. No. 2702.

Finest chicory-coffee, prepared after a peculiar process of dry fermentation.

CLASS IV.

ANIMAL AND VEGETABLE SUBSTANCES USED IN MANUFACTURES.

2707. The Royal board of agriculture, Stuttgart, on behalf of the Württemberg woolgrowers. Munich 1854 honor. mention for a simular collection. (s. No. 2695.)

Collection of samples of wool from 80 of the Wurttemberg wool-growers, arranged in order to show the different qualities produced in Wurttemberg according to their degree of fineness and the use they are intended for. A list of the names of the growers is annexed.

2708. Gruner, Friedrich, soap-manu., Esslingen. Paris 1855 silver medal; London 1862 medal. Agt. s. No. 2702.

Soaps of different sorts: No. 1. bust of His Majesty the King of Württemberg; No. 2. bust of Her Majesty the Queen of Great-Britain; No. 3. bust of His Royal Highness the late Prince consort; No. 4. the arms of Württemberg; No. 5. 2 pedestals for the busts; No. 6. 2 bases for the bust 2. and 3.; No. 7. a base for the bust No. 1.; No. 8 a. 1 piece of medical soap with stamp and form of pieces, as they are usually sold; No. 8 b. piece of soap for woollen manufacture, with stamp of the arms of Württemberg; No. 8 c. 1 piece of olive-oil-soap, marked by stamp; No. 9. 1 glass containing potash-soap; No. 10. 1 glass containing washing soap for woollen manufacture.

2710. Administration of the farms of His Majesty the King of Württemberg (K. Hof-Kameral-Amt), Stuttgart. Agt. s. No. 2702. (s. No. 2698, 2773, 2773 a.)

Raw silk and stuffs made from it; the silk is from the breeding-establishment of the Royal domaine »Weil«, the stuffs are woven in the manufactory of Gessler & Co., Tettnang.

Wool for drapery from the merino-flock of the Royal domain »Achalm«.

Wool for worsteds from the merino-flock of the Royal domain »Seegut«.

Angora-goats-hair, raw and prepared; Cashmeregoats-down, raw and prepared; hairs of the white horn-bearing Yak, and hairs of the black Yack without horns, from the Royal establishment of acclimatisation at »Seegut«.

2711. Kauzmann, Brothers, manu. of articles of ivory, hartshorn and bone, Geislingen. Munich 1854 medal of honor. Agt. Fr. Haas & Co., 34 Jewin-street.

Carved work of ivory and hartshorn, viz: drinkingcup, screen, garnitures for the scrutoire, carved tops for walking-sticks, umbrellas and riding-whips, bracelets, brooches, ear-pendants, shawl-pins for ladies and gentlemen, shirt and cuff-buttons of ivory and hartshorn, pocket, note and card-books with covers of ivory, hartshorn and chamois-horn.

Small fancy-articles of ivory, hartshorn and bone, viz: small furniture of various forms, étagères, altars, saints' figures, finely carved baskets, thermometers etc.

Turned articles of bone, viz: needle-boxes, screw-cushions, thimbles, measure bands, needle-cushions, knitting-boxes, children's rattles, pipes, boxes in various fashions containing play-things, chess and domino-plays, tooth picks, watch-boxes, egg-cups, spice-boxes, spoons for salt and spices, spoons for children etc.

2712. KÖLLE, THOMAS, manu., Ulm. London 1862 medal. Agt. the Württemberg Trading-Company at Stuttgart, represented by Sam. Sessing, 10 Huggin-Lane, Wood-street, Cheapside, E. C.

Tinder (spunk), own manufacture; glue in different sorts.

2713. LINSE, JOH., manu. of glue, Bopfingen. Munich 1854 honor. mention; London 1862 honor. mention. Agt. B. Bamberger from Crailsheim.

1. Noerdlingen glue; 2. Cologne glue; 3. Calw glue.

2717. VEIT-WEIL, manu., Oberdorf near Bopfingen. Munich 1854 medal of honor; Paris 1855 honor. mention. Agt. s. No. 2712.
(s. No. 2692.)

Prices p. cwt.: No. 1—4. colorless gelatine, produced at a temperature of 50° R. in 4 different sorts from 7 £ 4 sh. to 10 £ 3 sh. 3 d.; No. 5—8. fine bone glue, extremely serviceable for fine work as well as in dressing and finishing fabrics from 6 £ 7 sh. to 7 £ 4 sh.; No. 9. fine bone glue 4 £ 13 sh. 3 d.; No. 10. bone glue I. perfectly soluble and very strong 2 £ 5 sh. 9 d.; No. 11. bone glue II. do. 2 £ 4 sh. 1 d.; No. 12. size I e., of great cementing strength 3 £ 2 sh. 8 d.; No. 13. and 14. size black specially adopted for treating woollen goods 2 £ 2 sh. 4 d.; No. 15. bone glue, white 4 £ 4 sh. 8 d.

2718. THE WÜRTTEMBERG GLUE AND MANURE MANUFACTURING COMPANY, Reutlingen. London 1862 honor. mention. Agt. s. No. 2712.
(s. No. 2693 and 2732.)

4 packets of finest glue; prices p. cwt.: No. 1. perfectly transparent 2 £ 14 sh.; No. 2. in broad tablets 2 £ 14 sh.; No. 3. Cologne fashion, transparent 3 £ 4 sh.; No. 4. do. opaque 3 £ 4 sh.; No. 5. bottle of liquid glue (stiffening) p. cwt. 12 sh.

CLASS VII.

MANUFACTURING MACHINES AND TOOLS.

2720. DOERTENBACH & SCHAUBER, card-manu., Calw. Berlin 1844 bronze medal; Munich 1854 honor. mention; Paris 1855 bronze medal. Agt. s. No. 2712.

1. Case with 3 show-glasses containing: 1 cards for cotton: No. 1. 1 sheet No. 22. 31/3 in., woollen card cloth; No. 2. 1 sheet No. 24. do. caoutchouc natural; No. 3. 1 sheet No. 26. do. plain cloth 6 plies; No. 4. 1 sheet No. 22. do. double leather; No. 5. 2½ ft., fillet No. 26. $\frac{20}{12}$ in. leather; No. 6. 2½ ft., fillet No. 24. $\frac{20}{12}$ in. plain cloth 6 plies.

2. Cards for wool: No. 7. 1 sheet for main cylinders No. 28. 29,4½ in. leather; No. 8. 1 sheet for main cylinders No. 24. do.: No. 9. 1 sheet for main cylinders No. 22. do.; No. 10. 2½ ft. sheet for workers No. 28. $\frac{21}{12}$ in. leather; No. 11. 2½ ft. sheet for workers No. 26. do.; No. 12. 2½ ft. sheet for workers No. 24. do.; No. 13. 2½ ft. sheet for workers No. 22. do.; No. 14. 2½ ft. sheet for main cylinders No. 22. $\frac{25}{12}$ in. leather; No. 15. 2½ ft. sheet for doffers No. 24. do.; No. 16. 1 sheet fly No. 26. 29 in. to 5 in. in calf leather; No. 17. 1 sheet fly No. 24. do. vulc. caoutchouc.

All the sizes in Paris measure.

2722. LANCASTER, KLEEMANN & Co., machine-makers, Obertürkheim. Agt. s. No. 2712.
(s. II. No. 56.)

1 powerloom 12 £.

2723. STEINER, JOS., manu. of tools for joiners and carpenters, Laupheim. Agt. s. No. 2712.

No. 1. trying plane, South-Germany-fashion 2 sh. 6 d.; No. 2. double plane do. 1 sh. 6 d.; No. 3. long plane do. 10 d.; No. 4. round, nose plane do. 9⅓ d.; No. 5. »Putzhobel« 1 sh. 5 d.; No. 6. toothing plane do. 1 sh.; No. 7. compass plane do. 1 sh.; No. 8. do. double 1 sh. 8 d.; No. 9. fillister to change side, double 3 sh. 1⅓ d.; No. 10. plough with boxwood screws 4 sh. 1 d.; No. 11. plough with iron screws, North-Germany-fashion 6 sh. 8 d.; No. 12. do. for glaziers 2 sh. 11 d.; No. 13. notching plane 3 sh. 4 d.; No. 14. dove-tail plane 2 sh. 3 d.; No. 15. skew rabbit plane 8½ d.; No. 16. moulding plane 8⅓ d.; No. 17. rabbit plane 1 sh. 7 d.; No. 18. do. 2 sh.; No. 19. double moulding plane 2 sh.; No. 20. hollow plane 8 d.; No. 21. round plane 8 d.; No. 22. round plane, North-Germany-fashion 8 d.; No. 23. cornice do. 8 d.; No. 24. trying plane, North-Germany-fashion 2 sh. 8 d.; No. 25. double plane do. 1 sh. 7 d.; No. 26. long plane do. 11 d.; No. 27. round nose plane do. 10 d.; No. 28. »Putzhobel« do. 1 sh. 6 d.; No. 29. toothing plane do. 1 sh. 1 d.; No. 30. fillister for glaziers 9⅓ d.; No. 31. do. South-Germany-fashion 9⅓ d.; No. 32. fillister for glaziers 11 d.; No. 33. double moulding plane, North-Germany-fashion 2 sh.; No. 34. compass double do. 1 sh. 9 d.; No. 35. notching plane, to changed, wooden garniture North-Germany-fashion 3 sh. 11 d.; No. 36. plough with boxwood screws, North-Germany-fashion 6 sh.; No. 37. grooving plane to be changed, with boxwood screws 6 sh. 6 d.; No. 38. »Gargelhobel«, Rhenish fashion 3 sh. 11 d.; No. 39. do. to be changed 3 sh. 11 d.; No. 40. »Einschnitthobel« for coopers 3 sh. 11 d.; No. 41. notching plane 2 sh. 10 d.; No. 42. plough, Dutch fashion 5 sh.; No. 43. fillister for glaziers 5 sh.; No. 44. »Zahnleistenhobel« 4 sh.; No. 45. »Jalousiehobel« to be changed 7 sh.; No. 46. saw for cutting grooves fa. A. 8 d.; No. 47. marking gauge 4⅓ d.; No. 48. »Faustsäge« 3 ft. 1 sh. 5 d.; No. 49. »Absatzsäge« 2½ ft. 1 sh 5 d.; No. 50. »Lochsäge« with polished handle 4⅔ d.; No. 51. »Fidibushobel« 1 sh.; No. 52. hold fast 3 in. 4⅔ d.; No. 53. do. 4 in. 5 d.; No. 54. do. 5 in. 5⅔ d.: No. 55. shooting board 5 sh. 10 d.: No. 56. »Zündhölzchenhobel« 5 sh. 10 d.; No. 57. round hobel for cartwrights 4 sh. 2 d.: No. 58. stock and bit with springs I a. 8 sh. 4 d.; No. 59. stock and bit with conical hafts 2 sh. 11 d.; No. 60. »Lederhobel«, to be changed 3 sh. 4 d.: No. 61. pastrain, sweep plane, Paris fashion 2 sh. 11 d.; No. 62. notching-plane 8⅔ d.: No. 63. hollow-plane No. 7. 10 d.: No. 64. router plane with 2 irons 2 sh. 2 d.: No. 65. »Putzhobel« 4 sh.; No. 66. »Putzhobel« 3 sh. 8 d.; No. 67. hand-saw 10 d.; No. 68. »Schittersäge« for wood cutters 2 sh. 1 d.: No. 69. »Schweifsäge« 1 sh. 3 d.; No. 70. 71. 2 pieces of cornice planes,

Engl. fashion p. piece 1 sh. 3 d.; No. 72. round nose plane, Engl. fashion 1 sh.; No. 73. long plane 1 sh. 1 d.; No. 74. double plane 1 sh. 9 d.; No. 75. compass double 3 sh.; No. 76. toothing plane 1 sh. 3 d.; No. 77. trying plane, double 2 sh. 8 d.; No. 78 rabbit plane 1 sh. 8 d.; No. 79. grooving and tongue-plane 3 sh.; No. 80. »Winkelhobel« 2 sh. 8 d.; No. 81. plough, Engl. fashion 16 sh. 8 d.; No. 82. notching plane 14 sh. 10 d.; No. 83. marking gauge f. A. 4½ d.; No. 84. do. f. B. 4½ d.; No. 85. Bogen-Zirkel, for coopers 2 ft. 2 sh. 6 d.; No. 86. compass to be changed 4 sh.; No. 87. plough for cart wrights 8 sh. 4 d.; No. 88. »Putzhobel« 6 sh. 8 d.; No. 89. a pair of grooving and tongue planes 9 sh. 4 d.; No. 90. »Salztenne« 1 sh. 5 d.; No. 91. »Mehltenne« 2 sh. 9 d.; No. 92. saw-frame 6 d.; No. 93. 3 pieces of squares p. 3 pieces 5 d.; No. 94. 1 do. p. piece 2 d.; No. 95. 1 do. p. piece 3 d.

12 mitre-squares p. doz. 3 sh. 4 d.; 10 do. 2 sh. 4 d.; ²⁄₁₂ doz. do. 5 sh. 4 d.; ⅛ do. 7 sh. 4 d.; ⅛ do. 7 ,sh.; ⅓ do. 4 sh. 8 d.; ⅓ do 4 sh.; ¹⁄₁₂ do.; 2 sh. 4 d.; all p. doz.

2724. VOGEL, MORITZ, manu. of tools, Ulm, manufactory at Laupheim. Agt. the Württemberg Trading-Company at Stuttgart, represented by Sam. Sessing, 10 Huggin-Lane, Wood-street, Cheapside, E. C.

Tools for joiners and carpenters: No. 1. trying plane South-Germany-fashion 1 sh..5 d.; No. 2. double plane do. 7½ d.; No. 3. long plane do. 6½ d.; No. 4. round-nose-plane do. 6 d.; No. 5. rabbit plane do. 1 sh. 3 d.; No. 6. side-fillister do. 1 sh. 5 d.; No. 7. notching plane 3 sh. 3 d.; No. 8. dove-tail plane 2 sh. 3 d.; No. 9 double moulding plane 2 sh. 1 d.; No. 10. moulding plane 6½ d.; No. 11. hollow plane 9 d.; No. 12. round plane 9 d.; No. 13. cornice plane 9 d.; No. 14. do. 9 d.; No. 15. plough with wooden screws 3 sh. 8 d.; No. 16. plough with boxen screws 5 sh. 10 d.; No. 17. grooving plane to be changed, with boxen screws 6 sh. 3 d.; No. 18. plough with iron screws 7 sh.; No. 19. plough with boxen screws North-Germany-fashion 7 sh. 4 d.; No. 20. trying plane do. 1 sh. 5 d.; No. 21. double plane do. 7½ d.; No. 22. long plane do. 6½ d.; No. 23. round-nose plane do. 6 d.; No. 24. tooth plane do. 7 d.; No. 25. side-fillister to be changed do. 2 sh. 10 d.; No. 26. notching plane to be changed, wooden garniture 4 sh. 2 d.; No. 27. dove tail plane to be changed 2 sh. 6 d.; No. 28. 29. 1 pair of grooving and tongue planes 2 sh. 9½ d.; No. 30. cornice plane North Germany-fashion 9 d.; No. 31. round plane do. 9 d.; No. 32. cornice plane do. 9 d.; No. 33. stock and bit with conical hafts 2 sh.; No. 34. stock and bit with springs 2 sh. 9½ d.; No. 35. pastrain, sweep plane, Paris fashion 2 sh. 9½ d.; No. 36. router plane with one iron 1 sh. 10 d.; No. 37. saw for cutting grooves 6 d.; No. 38. hold fast 5½ d.; No. 39. do. 6 d.; No. 40. marking gauge 4½ d.; No. 41. do. 4½ d.; No. 42. 43. 2 cutter-hafts p. piece 6 d.; No. 44. 45. saw-frame 5½ d.; No. 46. 47. 48. 3 squares p. piece 2 d., No. 49. 50. 51. 3 inlaid angles p. piece 2 d., No. 52. 1 do. 2½ d.; No. 53. 54. 2 do. p. piece 3 d.; No. 55. mitre-square 3 d.; No. 56. model 2½ d.; No. 57. 58. 2 do. p. piece 3 d; No. 59. 1 do. 3½ d.; No. 60. do. 4 d.

2724 a. THE WÜRTTEMBERG TRADING-COMPANY, Stuttgart. Agt. s. No. 2724.
(s. No. 2693 a., 2771a., 2822, 2849.)
1 sewing machine 2 £ 9 sh.

CLASS VIII.

MACHINERY IN GENERAL.

2725. FOUQUET & FRAUZ, manu. of circular knitting machines, Stuttgart. Munich 1854 medal of honor. London 1862 honor. mention. Agt. s. No. 2724.

No. 1. 1 circular knitting-machine 14 in., 27 course, with 2 »mailleuses« (courier-wheels), system Fouquet à coulisse, contrivance for plush and »chaineuse« (lining wheel); No. 2. 1 circular knitting machine 14 in., 28 fine, with 2 »mailleuses« (lining wheels) new system Fouquet for worsteds, silk and linen; No. 3. 1 circular knitting-machine with peculiar contrivance, for right and right-hand meshes, (ribbed frame), pearl meshes, as well as for seams; No. 4. 1 circular knitting-machine new invention for left and right-hand meshes.

2726. KURTZ, HEINRICH, manu. of fire-engines, Stuttgart. London 1862 honor. mention. Agt. s. No. 2724.

Portable fire-engine with suction-pump, on two-wheeled cart, water reservoir of linencloth 50 £.

CLASS IX.

AGRICULTURAL AND HORTICULTURAL MACHINES AND IMPLEMENTS.

2727. DITTMAR, BROTHERS, manu., Heilbronn. London 1862 medal.   (s. No. 2863.)
Garden tools.

2728. RAU, DR. L., professor of the Royal agricultural institution at Hohenheim. London 1862 honor. mention. Agt. s. No. 2724.

The history of the plough, illustrated by a hundred of models.
A. Originating from the hoe.
First class. Share inclined, no mould-board. No. 1. Ancient-Syracuse; No. 2. Ancient-Greece; No. 3. Ancient-Etruria; No. 4. Morocco; No. 5. Ceylon; No. 6. Ancient-Egypt; No. 7. Germany (Schwartzwald); No. 8. Morlachia; No. 9. Russia (Oesel-Island); No. 10. Arabia.
Second Class. Share horizontal, no mould-board. No. 11. Ancient-Rome; No. 12. Ancient-Greece; No. 13. Georgia; No. 14. East-Indies; No. 15. Norway; No. 16. France; No. 17. Switzerland; No. 18. Italy (Rome); No. 19. France; No. 20. Persia.
Third Class. Two narrow mould-boards. No. 21. Spain; No. 22. China; No. 23. Italy; No. 24. Austria; No. 25. France; No. 26. Ancient-Greece; No. 27. Sweden; No. 28. Italy; No. 29. Old-Norman; No. 30. France (South); No. 31. France; No. 32. France; No. 33. Germany (Dresden); No. 34. Sweden; No. 35. Germany (Tirol); No. 36. Portugal and Algiers; No. 37. Abyssinia.
Fourth Class. Share inclined, mould-board in the middle. No. 38. Anglo-Saxonian; No. 39. China; No. 40. Germany; No. 41. Switzerland; No. 42. France; No. 43. Italy; No. 44. Italy; No. 45. Germany (Bohemia); No. 46. Greece.
Fifth Class. Mould-board at the side. No. 47. Russia; No. 48. Germany (Thuringia); No. 49. England; No. 50. France; No. 51. Germany; No. 52. France; No. 53. East-Indies; No. 54. Italy (Milan); No. 55. Sweden; No. 56. Belgium (Flanders); No. 57. France; No. 58. France; No. 59. England; No. 60. England; No. 61. United-States; No. 62. Germany; No. 63. Germany.

B. Originating from the spade. No. 64. Germany; No. 65. Germany (Bohemia); No. 66. Germany (Saxony); No. 67. Germany (Nassovia); No. 68. Germany (Lunenburgh); No. 69. Germany (Saxony); No. 70. Belgium; No. 71. Belgium; No. 72. Belgium.

C. Originating from the two-pronged hoe. No. 73. Russia; No. 74. China; No. 75. Finland; No. 76. Livonia; No. 77. Courland; No. 78. Russia; No. 79. Russia; No. 80. Poland; No. 81. East-Prussia.

D. Turn-wrest ploughs, side-hill ploughs, swivel-ploughs, one way ploughs. No. 82. France; No. 83. Germany; No. 84. France; No. 85. France; No. 86. England; No. 87. Nassovia, Transylvania; No. 88. Germany; No. 89. Germany; No. 90. France; No. 91. Germany; No. 92. Germany; No. 93. Germany; No. 94. United-States; No. 95. Germany; No. 96. United-States; No. 97. Germany; No. 98. England; No. 99. Germany; No. 100. Germany.

2729. THE SOCIETY FOR THE REARING OF BEES IN WÜRTTEMBERG, by its president Eduard Weitzel of Sonnenberg near Waiblingen. Agt. the Württemberg Trading-Company at Stuttgart, represented by Sam. Sessing, 10 Huggin-Lane, Wood-street, Cheapside, E. C.

Model of an apiary, containing 26 doublebee-hives, improved after the system of the society; $\frac{1}{10}$ of real size.

CLASS X.
CIVIL ENGINEERING, ARCHITECTURAL AND BUILDING CONTRIVANCES.

2730. CHAILLY, J., manu. of cement, Kirchheim and Teck. London 1862 honor. mention. Agt. s. No. 2729.

No. 1. 1 barrel of Roman cement; No. 2. 1 cask of Portland cement 50 kilogr. net.; slabs No. 3. composed of 2 parts of Roman cement and 1 part of sand; No 4. 1 part of Roman cement and 1 part of sand; No. 5. of Portland cement; No. 6. of 2 parts of Portland cement and 1 part of sand; No. 7. of 1 part of Portland cement and 1 part of sand; No. 8. of 2 parts of Portland cement and 3 parts of sand; No. 9. of 1 part of Portland cement and 2 parts of sand; No. 10. of 2 parts of Portland cement and 5 parts of sand; No. 11. of 1 part of Portland cement and 3 parts of sand; No. 12. 2 slabs of Roman cement 10/12 in. thick; No. 13. 2 do. 12/12 in. thick; No. 14. 2 pieces of vaulted ceilings; No. 15. 6 pieces of ornaments for building-purposes made of Roman cement; No. 16. 4 do. of Portland cement.

2731. LEUBE, BROTHERS, manu. of cement, Ulm. Munich 1854 honor. mention; Paris 1855 silver medal: London 1862 medal. Agts. s. No. 2729.

1. An underground waterpipe made of cement. sand and bricks. Price of pure cement about 1 sh. 5 d. p. 50 kilogr.; 2. a piece of a similar object, cut off; 3. a table-top made of 2 parts of cement and 1 part of sand, smoothed by means of a mason's trowel; not immersed into water; 4. a sample-piece of a mixture of cement with a large quantity of gravel, having been under water for two months; 5. a small ball of pure cement having been under water for three months, broken in two by a heavy blow; 6. another ball of pure cement having been four months under water; price of the cement p. 50 kilogr. 1 sh. 5 d.

2732. WÜRTTEMBERG GLUE AND MANURE MANUFACTURING COMPANY, Reutlingen.
(s. No. 2693, 2718.)
Roman cement.

2733. ZIEGLER, CHRISTIAN, manu. of swimming bricks, Heilbronn. London 1862 honor. mention. Agt. s. No. 2729.

Prices p. 1000 pieces. No. 1. brick, large size for filling up the space between the joists 5 £; No. 2. do. small size 1 £ 10 sh.; No. 3. swimming (porous) brick, fire-proof 2 £ 10 sh.; No. 4. swimming and hollow side-wedge 3 £ 5 sh.; No. 5. swimming side or arch-wedge 1 £ 10 sh.; No. 6. porous arch-brick made of ordinary clay 2 £; No. 7. swimming chimney or partition-brick 2 £; No. 8. do. hollow 2 £; No. 9. do. halfround 1 £ 10 sh.; No. 10. guard for gas-retorts fire-proof 2 £ 10 sh.

CLASS XIII.
PHILOSOPHICAL INSTRUMENTS AND PROCESSES DEPENDING UPON THEIR USE.

2735. ENGLER, RUDOLPH, editor of Eble's horoscopes, Ellwangen. Agt. s. No. 2729.
(s. II. No. 51.)

Eble's horoscopes or indicators of time as mentioned by Zech, Reuschle, Littrow, Grunert etc.; 1. edition for the elevation of the pole of 45 to 55° 7 sh. 6 d.; 2. do. 45 to 60° 10 sh.; 3. do. 35 to 50° 11 sh. 8 d.

2736. SAUTER, AUGUST, manu. of balances, Ebingen. London 1862 medal. Agt. s. No. 2729.
(s. II. No. 97.)

No. 1. chemical balance with glass case, 15 in. beam to carry 100 grammes in each scale, contrivance for moving the »rider« as also for steadying the scales connected with the same axis which moves the beam 8 £ 7 sh.; No. 2. simple balance, with glass case 1 £ 14 sh.; No. 3. balance with 14 in. beam and 18 in. stative 1 £.

2737. WOLFF, F. A., & SONS, manu. of pharmaceutical apparatus, Heilbronn. London 1851 honor. mention; Munich 1854 great medal; Paris 1855 silver and bronze medal; London 1862 medal.

2 steam distilling-apparatus for pharmaceutical purposes of the newest construction, each with a contrivance for stirring No. 1. 38 £; No. 2. 49 £.

CLASS XV.
HOROLOGICAL INSTRUMENTS.

2739. BENZING, RAPP & Co., manu. of dutch-clocks, Stuttgart. Agt. Fred. Preyss, 20 St. Dunstans Hill. Gt. Tower-street, E. C.

Clocks, prices p. doz.: No. 104. with alarm in oak case 3 £ 18 sh.; No. 112. case of dork wood 4 £ 4 sh.; No. 112a. strike-work 5 £ 4 sh.; No. 111. wheel-work 3 £ 16 sh.; No. 111a. do. and alarm 4 £; No. 100. with thermometer 6 £ 12 sh.; No. 187. tin case, varnished 4 £ 16 sh.; No. 36. with alarm, porcelain, 5½ in. 1 £ 7 sh.; No. 33. wheel-work 5½ in. 1 £ 4 sh.; No. 32. do. 4½ in. 1 £ 2 sh.; No. 169. spring-work, case in gothic style 16 £; No. 47. cuckoo-clock 19 £; No. 35. with alarm, porcelain 4¼ in., 1 £ 5 sh.; No 38. strike-work 4¼ in. 2 £ 3 sh.;

No. 39. do. 5¼ in. 2 £ 7 sh.; No. 26. do. 8¼ in. 3 £ 8 sh.; No. 25. do. 7¼ in. 2 £ 16 sh.; No. 26½. 9 in. 3 £ 12 sh.; No. 31. wheel-work, bronze ornaments 1 £; No. 32½. do., oval frame 2 £ 4 sh.; No. 31¼. do., cast-iron ornaments 2 £ 4 sh.; No. 34. with alarm, bronze ornaments 1 £ 2 sh.; No. 99. tableau with gilt frame and clock, on the back an oil - painting 9 £ 6⅔ sh.

2740. SAUTTER, AD., manu., Ravensburg.

Watch-movements.

2742. BÜRK, J. (»Württemberg clock-manufactory«), Schwenningen. London 1862 medal. Agt. the Württemberg Trading-Company at Stuttgart, represented by Sam. Sessing, 10 Huggin-Lane, Wood-street, Cheapside, E. C. (s. II. No. 36.)

I. Time - pieces of 30 hours'run, clock-work allmetal: 1. No. 88. dial - frame, brown and green moiré 12 sh. 3 d.; 2. No. 89. do. 11. sh. 8 d.; 3. No. 89. do., oak 11 sh. 8 d.; 4. and 5. No. 89. do., brown of guttapercha 2 samples, each 12 sh. 8 d.; 6. No. 100. do., black with tortoise - shell 12 sh.; 7. No. 100. do. 11 sh. 6 d.; all with striking work; 8. No. 100. dial-frame, oak, without striking work 8 sh. 9 d.; 9. No. 118. do., black, inlaid with brass 17 sh. 6 d.; 10. No. 125. do., brown, 11 sh. 4 d.; 11. No. 127. do., black 12 sh.; 12. No. 130. do. 15 sh.; 13. No. 134. do., brown with thermometer 14 sh. 6 d.; all with striking work; 14. travelling-clock with alarm-work, without striking work 14 sh.

II. Time-pieces of 8 days'run, clock - work all metal: 15. No. 130. black, strike-work 1 £ 10 d.

III. Dutch-clocks with springs of 30 hours'run, clock-work all metal: 16. No. 39. dial-frame brown gilt 11 sh. 8 d.; 17. No. 40. do. ☐ 11 sh. 8 d.; 18. No. 48. do. 14 sh. 6 d.; 19. No. 48. do. 13 sh. 6 d.; 20. No. 53. do. ☐ blue inlaid with tin, gilt 14 sh.; 21. No. 53. do. ☐ brown 14 sh.; 22. No. 104. do. ☐ gilt ornaments and painting 16 sh. 8 d.; all with striking work; 23. No. 111. dial-frame black, with thermometer, without striking work 10 sh. 10 d.; 24. and 24½. No. 112. 2 do. black, Turkish ciphers, without striking work each 15 sh. 8 d.; 25. and 26. No. 119. 2 do. brown, gilt, oval à 7 sh. 10 d., and 13 sh. 8 d., with striking work 1 £ 7 sh. 4 d.; 27. No. 131. do. 16 sh.; 28. do. with painting and striking work 17 sh. 2 d.; 29. 132. do. 11 sh. 8 d.; 30. No. 138. do. 18 sh.

IV. Dutch - clocks of 8 days'run with springs, clock-work all metal: 31. No. 101. dial-frame, oval brown and gilt, petalophany, striking work 1 £; 32. No. 104. do. gilt, with oil - painting, striking work 1 £ 1 sh. 8 d.; 33. No. 106. do. with tableau 2 £ 15 sh. 8 d.; 34. No. 116. do. round, black 18 sh. 2 d.; 35. No. 137. do. oval, brown, gilt, with painting 1 £ 6 sh.; 76. No. 99. do. oval, brown 1 £ 8 sh. 4 d.

V. Dutch clocks with weights (regulator-fashion): 36. nut and polixander-case, of 8 days'run 1 £ 15 sh.; 37. do. with marqueterie 1 £ 15 sh. 6 d.

VI. Dutch-clocks (Black forest fashion): 38. trumpet-clock, with carved case of oak 3 £ 7 sh.; 39. cuckoo-clock do. No. III. L. 1 £ 3 sh.; 40. do. polished case 1 £ 1 sh. 8 d.; 41. with polished case and painted dial-frame 9 sh.; 42. with carved case and painted dial-frame No. 101. L. 6 sh. 8 d.; 43. with oak case and painted dial-frame No. 105. L. 8 sh. 4 d.

Middle - sized clocks, Scotch: No. 44/45. dial-frame 7 in. of porcelain, 2 samples, each 4 sh.; No. 46. do. 8½ in. 4 sh. 8 d.; No. 47. do. 10 in. 5 sh. 8 d.; No. 48. do. earthen No. 83. R. 4 sh.

8 d.; No. 49. do. No. 82. R. 4 sh. 6 d.; No. 50. do. 84. R. 4 sh. 4 d.; No. 51. do. No. 81. R. 3 sh. 10 d.; dial-frame, oval, brown, No. 52. Cir. 3 L. 4 sh. 10 d.; No. 53. Cir. 2. M. 4 sh. 10 d.; No. 54. No. 975½ L. 5 sh. 2 d.; No. 55. gilt No. 954¼ L. 5 sh. 7 d.; No. 56. round do. No. 994/2. L. 5 sh. 10 d.; No. 57. round do. R. E. g. D. 4 sh. 10 d.

VII. Clock-works. *a.* with springs: works of 30 hours' run No. 58. rough with striking work 4 sh. 10 d.; No. 59. varnished do. 6 sh. 5 d.; No. 60. polished do. 6 sh. 10 d.; No. 61. rough 2 sh. 8 d.; No. 62. varnished, with alarm 6 sh. 5 d.; 8 days' spring-works No. 63. rough, striking work 9 sh.; No. 64. varnished 12 sh. 6 d.; No. 65. polished 13 sh. 2 d.; 8 days' works No. 66. rough 4 sh. 2 d.; No. 67. polished 6 sh. 5 d.; No. 68. 30 hours' travelling-work with alarm 8 sh. 8 d. *b.* with weights, Black forest works: No. 69. Scotch fashion with door, polished 2 sh. 8 d.; No. 70. do. with case 2 sh. 4 d.

VIII. Telegraph-works and others: No. 71. writing telegraph-work 2 £ 10 sh.; No. 72. gas-controlling work 2 sh. 6 d.; No. 73. do. 14 sh.

IX. Portable alarm and controlling clock, No. 74. for 6 stations, patent, 3 £ 12 sh. 11 d.; 1 one year's bulletin 1 sh.; 1 control-book 2 sh. 4 d.; 6 key-boxes with chaines à 1 sh. 8 d.

X. Parts of clocks No. 75. 1 case containing wheels, movements and screws 1 £.

CLASS XVI.

MUSICAL INSTRUMENTS.

2746. HARDT & PRESSEL, manu. of pianofortes, Stuttgart. London 1862 medal. Agt. s. No. 2742.

No. 1. pianino oblique, with 7 octaves, of polixander-wood, keys of ivory 42 £.; No. 2. do. 37 £ 10 sh.

2747. HUNDT, FR., & SON, manu. of pianofortes, Stuttgart. London 1851 prize-medal, 1862 medal. Agt. Fr. Hundt, 21 Ebury-street, Pimlico.

1 cottage-piano with 7 octaves of best nut-wood, ornamented with rose-wood 42 £.; 1 square-piano with 6¾ octaves and the newest improvements, of polixander wood 25 £.

2748. MISSENHARTER, CARL, JUN., manu. of musical instruments, Ulm. London 1862 medal. Agt. s. No. 2742.

Wind-instruments: No. 1. cornet in B. 2 £. 10 sh.; No. 2. cornet in B., German silver 3 £ 10 sh.; No. 3. trumpet in C. 2 £ 7 sh.; No. 4. trumpet in C. 2 £ 2 sh.; No. 5. trumpet in F. 2 £. 10 sh.; No. 6. trumpet in F. 2 £. 10 sh.; No. 7. treble-horn in C. 2 £ 18 sh.; No. 8. treblehorn in C. German silver, 2 mouth-pieces with case 4 £ 5 sh.; No. 9. altohorn in C., brass, 4 £ 5 sh.; No. 10. altohorn in C., German silver, 5 £ 2 sh.; No. 11. baroxyton in F., 5 £ 8 sh.; No. 12. basstrombone in B. 3 £ 7 sh.; No. 13. concerthorn in F., German silver, 7 £ 15 sh.; No. 14. counterbass in Fis, 4 valves, 6 £ 12 sh.; No. 15. counterbass in Fis, 6 valves, 8 £. 4 sh.; No. 16. helikon in B., round fashion, 10 £ 17 sh.; No. 17. signalhorn 14 sh.; No. 18. signal-trumpet 12 sh.; No. 19. posthorn, 2 valves, 1 £ 12 sh.; No. 20. speaking-trumpet 5 sh.; No. 21. fire-trumpet, small, 1 £; No. 22. fire-trumpet, large, 5 sh.; No. 23. fire-whistle, small, ½ sh.; No. 24. fire-whistle, large, 2 sh. All these instruments are furnished with mouth-pieces.

2749. Oehler, Christian, manu. of pianofortes and pianinos, Stuttgart. London 1862 honor. mention. Agt. the Württemberg Trading-Company at Stuttgart, represented by Sam. Sessing, 10 Huggin-Lane, Wood-street, Cheapside, E. C.

Square piano-forte with 6¾ octaves, nut-wood 400 fl.

2750. Pross, Gschwind & Co., manu. of harmoniums, Stuttgart. London 1862 medal. Agt. B. Baur & Co., 1 Catharine Court, Seething-Lane, City.

No. 221. harmonium with 5 octaves, 4 plays, 2 manuals and 15 stops, oak wood with gothic ornaments 70 £; No 222. harmonium with 5½ octaves, 2 plays and 10 stops of polixander wood 20 £.

2751. Schiedmayer, J. & P., manu. of pianos, harmoniums, harmonicordes and church-organ stops, Stuttgart. Munich 1854 great medal; Paris 1855 bronze medal. Agts. Sevin, Chinery & Co., 155 Fenchurch-street.

No. 1. harmonicorde: 5 octaves, 18 stops, rose-wood 56 £; No. 2. cottage-piano, 7 octaves, rose-wood 30 £; No. 3. harmonium, 5 octaves, 8 stops, with percussion-action, rose-wood, 21 £; No. 4. harmonium 5½ octaves, 3 stops with percussion-action, rose-wood, 14 £.; 1 tableau containing 3 church-organ stops: No. 5. clarinette, 4½ octaves, 4 £ 10 sh.; No. 6. hautbois, 4½ octaves 4 £ 10 sh.; No. 7. seraphine, 4½ octaves 4 £ 10 sh.

2752. Schiedmayer & Sons, Stuttgart. London 1851 prize-medal; Munich 1854 great medal; Paris 1855 silver medal; London 1862 medal. Agts. Sevin, Chinery & Co., 155 Fenchurch-street.

No. 6130. grand piano, 7 octaves, rose-wood, 60 £; No. 6131. cottage-piano, 7 octaves, oak-wood, gothic, 42 £; No. 6132. square piano, 7 octaves, nut-wood, 33 £.

2753. Trayser, Ph. J., & Co., manu. of harmoniums, Stuttgart. Munich 1854 medal of honor; London 1862 medal. Agt. J. W. Green (E. G. Zimmermann).

No. 1. harmonium with 5 octaves, 4 plays, 15 stops and 2 manuals, nut-wood 40 £.; No. 2. harmonium with 5 octaves, 2 plays and 8 stops, solid oak-wood 18 £.

CLASS XVIII.

COTTON.

2755. Baumann, Carl, manu., Leonberg. (s. No. 2774, 2772, 2802.)

Cotton fabrics.

2755 a. Faber, Carl, Stuttgart. (s. No. 2765.)

White cotton coverlets.

2756. Gutmann, Brothers, manu., Göppingen. Munich 1854 honor. mention; Paris 1855 honor. mention. Agt. s. No. 2749.

Cotton and half-linen goods; prices p. yard: No. 1. breadth 9/4 colored half-linen twills (tickings) 34½ pounds, 46⅜ yards 16 d.; No. 2. breadth 8/4 colored half-linen do. 31½ pounds, 45½ yards 14 d.; No. 3. do. 30½ pounds, 44⅜ yards 14 d.; No. 4. do. 33½ pounds, 46⅘ yards 14 d.; No. 5. do. 33¼ pounds, 46 yards 14 d.; No. 6. do. 31½ pounds, 45½ yards 14 d.; No. 7. do. blue 30½ pounds, 46⅜ yards 14 d.; No. 8. do. colored checks 32 pounds, 45⅝ yards 16 d.; No. 9. do. half-linen 32 pounds, 45¾ yards 14 d.; No. 10. do. 30½ pounds, 47 yards 14 d.; No. 11. breadth 8/4 red and gray half-linen curtain-twill 31 pounds, 46⅜ yards 16 d.; No. 12. do. red and white half-linen ticking 31½ pounds, 46⅜ yards 16 d.; No. 13. do. blue 32½ pounds, 45⅝ yards 18 d.; No. 14. breadth 1⅞ gray check trousers 31 pounds, 47 yards 12 d.; No. 15. do. 8/4 gray striped trousers 31½ pounds, 46½ yards 12 d.; No. 16. do. 11/8 blue striped satin-ticking 22½ pounds 46⅜ yards 13½ d.; No. 17. do. red striped satin-ticking 22 pounds, 46 yards 16 d.; No. 18. do. colored twilled ticking 16½ pounds, 46⅜ yards 10 d.; No. 19. do. white half-linen bodice-twill 22 pounds, 47½ yards 13 d.; No. 20. do. red cotton ticking 11½ pounds, 46⅛ yards 9 d.; No. 21. do. colored ordinary gingham 11½ pounds, 45⅝ yards 8 d.

2757. Kaufmann & Sons (Max Kaufmann, Julius Kaufmann, formerly Kaufmann & Kaufmann Brothers), manu. of cotton and linen goods, Göppingen. Paris 1855 bronze medal. Agt. s. No. 2749.

Prices p. yard: No. 1. 9/4 rouleau half-linen twill 1 sh. 6½ d.; No. 2. 9/4 half-linen furniture-twill, red and gray 1 sh. 6½ d.; No. 3. and 4. do. mattress-ticking, variously colored and flammé 1 sh. 4 d.; No. 5. 8/4 twill for window-blinds 1 sh. 4 d.; No. 6—8. do. ticking, variously colored, each 1 sh. 2 d.; No. 9. do. flammé 1 sh. 2 d.; No. 10. do. flammé blue 1 sh. 1 d.; No. 11. 8/4 twill for window-blinds yellow and gray 1 sh. 2 d.; No. 12. 8/4 trouser-twill 1 sh.; No. 13. 8/4 cotton vari-colored 11½ d.; No. 14. do. ticking blue and white 11½ d.; No. 15. 8/4 bodice-twill with yellow stripes 1 sh. 1 d.; No. 16. bleached plumeau 1 sh. 2 d.; No. 17—19. 11/8 satin bed-fustian, red, each 1 sh. 4½ d.; No. 20—21. do. blue, each 1 sh. 2 d.; No. 22. do. bleached 1 sh. 2 d.; No. 23. 11/8 half-linen bodice-twill 1 sh. 2½ d.; No. 24. do. 1 sh. 1½ d.; No. 25. 11/8 cotton do. 9 d.

2758. Kolb & Schüle, manu. of cotton-fabrics of different kinds, Kirchheim u. T. Munich 1854 medal of honor; London 1862 medal. Agt. s. No. 2749.

Cotton fabrics; prices p. yard: No. 1. 20 yards blue bed-fustian, 11/8, 11 d.; No. 2. do. satin bed-fustian, 11/8, 14 d.; No. 3. do. 14 d.; No. 4. do. red satin bed-fustian, 11/8, 16 d.; No. 5. do. half-linen ticking, 11/8, 12 d.; No. 6. 23⅓ do. gray white fustian for lining, 5/4, 9 d.; No. 7. do. bleached tricot, 5/4, 10 d.; No. 8. 33⅓ do. croisé variously-colored, 5/4, 7½ d.; No. 9. 38 do. jacket-stuff, blue, 11/8, 6 d.; No. 10. 22 do. gray fustian for lining, 5/4, 6 d.; No. 11. 21⅓ do. bleached double fustian, 11/8, 18 d.; No. 12. 20 do. black jacquet-fustian, 11/8, 12 d.; No. 13. 32 do. ticking, cottonet, 11/8, 9 d.; No. 14. do., 11/8, 8 d.; No. 15. do. 11/8, 7½ d.; No. 16. do. ticking fustian, 11/8, 9½ d.; No. 17. do., 11/8, 7½ d.; No. 18. 31⅓ do. clothing fustian, 5/4, 7½ d.; No. 19. 31⅓ do. clothing fustian, 5/4, 8 d.; No. 20. 32⅔ do. lustre, 5/4, 7½ d.; No. 21. 32⅔ do. figured gingham, 5/4, 7½ d.; No. 22. 31⅓ do. gingham. 5/4, 7 d.; No. 23. 21⅓ do. valencias, 4/4, 10 d.; No. 24. 22 do., 4/4, 10 d.; No. 25. 24⅔ do. quilting, 15/16, 9 d.; No. 26. 20 do., 15/16, 9 d.; No. 27. 30⅔ do. flammé cottonet, 4/4, 6 d.; No. 28. 32⅔ do. croisé cottonet, 4/4, 6½ d.;

No. 29. 31½ do. jaspé cottonet, 4/4, 6 d.; No. 30. 20 do. checks 4/4, 7½ d.; No. 31. 20 do. 4/4, 7½ d.; No. 32. 30 do. croiséumbrella-stuff, 9/8, 8½ d.; No. 33. 2 pieces of bed-covers, filet and satiné, 5 sh. 5 d.

2758 a. KISSEL, AUGUST, manu., Böblingen. Munich 1854 medal of honor; Paris 1855 silver medal, under the firm Kissel & Krumbholtz. London 1862 medal. Agt. the Württemberg Trading-Company at Stuttgart, represented by Sam. Sessing, 10 Huggin-Lane, Wood-street, Cheapside E. C.   (s. No. 2769.)

Half-linen stuffs for trousers; prices p. yard: No. 6535. and 6537. 2 sh. 6 d.; No. 6625. 6626. 6630. 2 sh. 5 d.; No. 6551. and 6552. 2 sh. 2 d.; No. 6509. 6986. 6988. 6989. 6992. 6236. 6425. 6428. 6802. 6799. 4466. 4468. 4465. 2 sh. 1 d.; No. 6518. 6530. 6701. 6896. 6887. 2 sh.; No. 6729. 1 sh. 10 d.

Cotton and wool mixed stuffs for trousers: No. 6369. 6370. 2 sh. 6 d.; No. 6380. 6379. 6150. 2 sh.; No. 6203. 1 sh. 10 d.

2759. KRUMBHOLZ, LOUIS, manu. of fancy drill etc., Böblingen. Munich 1854 medal of honor; Paris 1855 silver medal; London 1862 medal. Agt. s. No. 2758 a.

Fancy twill for trousers; prices p. yard: 6266. twill, white and gray checks 30 d.; 6164. and 6129. do. gray and red checks 30 d.; 6131. and 6160. do. brown 30 d.; 5848. do. diagonal 30 d.; 6265. twill, white and brown checks, diagonal 26⅝ d.; 5725. do. fancy colored 26⅝ d.; 6291. do. brown with black checks 26⅝ d.; 6290. do. brown with red and black checks 26⅝ d.; 6251. do. fancy colored gray and red 26⅝ d.; 6248. do. white, gray and red 26⅝ d.; 6489. do. gray diagonal with black stripes 26⅝ d.; 5844. do with red stripes 26⅝ d.; 5704. twill, white with black and red checks 25⅓ d.; 6209. do. brown checks 25⅓ d.; 6490. do. blue stripes 25⅓ d.; 6491. do. red stripes 25⅓ d.; 6186. twill brown with black checks 22⅓ d.; 6185. twill, white 21⅝ d.; 6226. do. naturel 21⅝ d.

2760. LANG & SEIZ, manu. of linen and cotton fabrics, Stuttgart. Munich 1854 medal of honor; London 1862 medal. Agt. s. No. 2758 a.   (s. No. 2778.)

Cotton fabrics: 36 Württ. ells 8/4 plumeaux, blue, No. 221. à 40 kr.; 36 do. red, No. 9461 à 45 kr.; 2 ticking, yarn bleached No. 1. p. piece 4 fl. 24 kr.; 2 do. No. 2. p. piece 4 fl. 24 kr.; 2 ticking, blue and white No. 1. p. piece 4 fl. 48 kr.; 2 do. red and white No. 1. 5 fl. 24 kr. Prices p. Württ. Ell.

2761. LEVINGER, LEOP., manu., Ulm. Cotton fabrics.

2762. OTTENHEIMER & DETTELBACH, manu., Jebenhausen.   (s. No. 2771.) Cotton ticking for beds and matresses.

2763. VAIHINGER, A., & Co., manu. of cotton and cotton-mixed fabrics, Göppingen. Agt. s. No. 2758 a.

28 pieces of cotton and mixed fabrics: No. 1. 8/4 half-linen ticking, blue striped Ia.; No. 2. 8/4 variously colored ticking Ia.; No. 3. 8/4 flammé do. IIa.; No. 4. 8/4 do. II.; No. 5. 8/4 variously colored III.; No. 6. 8/4 do. III.; No. 7. 8/4 red and gray III.; No. 8. 8/4 do. blue III.; No. 9. 8/4 variously colored furniture-twill, checks; No. 10. 8/4 blue furniture-twill, checks; No. 11. 8/4 trousers, striped; No. 12. 8/4 do. naturel; No. 13. 8/4 do. checks; No. 14. 8/4 red cotton twill;

No. 15. 8/4 cotton twill, flammé; No. 16. 8/4 do. blue; No. 17. 11/8 satin bed-fustian, blue; No. 18. 11/8 do. red; No. 19. do. blue; No. 20. do. red; No. 21. 11/8 twilled fustian, variously colored; No. 22. do. red; No. 23. 11/8 flower-stuff, blue; No. 24. do. red; No. 25. do. yellow; No. 26. 11/8 bodice stuff, white; No. 27. do. naturel; No. 28. do. nanking.

CLASS XIX.
FLAX AND HEMP.

2765. FABER, CARL, manu. of linen goods, Stuttgart, establishment Beuren, Kirchheim u. T., Vaihingen, Leinzell. Munich 1854 great medal; London 1862 medal. Agt. s. No. 2758 a.

Linen fabrics: No. 1. 1 tabel-cloth with 18 napkins 9 £ 10 sh.; No. 2. 4 napkins 1 £; No. 3. 1 table-cloth with 12 napkins 3 £ 5 sh. 6 d.; No. 4. 1 do. with 6 napkins 1 £ 17 sh. 9 d.; No. 5. 1 do. with 6 napkins 1 £ 10 sh. 5 d.; No. 6. 1 do. with 6 napkins 1 £ 5 sh.; No. 7. 1 do. with 12 napkins 1 £ 17 sh. 11 d.; No. 8. 1 do. with 8 napkins 1 £ 6 sh. 7 d.; No. 9. 1 do. with 6 napkins 1 £ 5 d.; No. 10. 1 do. with 12 napkins 1 £ 9 sh. 4 d.; No. 11. 1 do. with 6 napkins 15 sh. 10 d.; No. 12. 1 do. with 6 napkins 14 sh. 7 d.; No. 13. 1 do. with 12 napkins 1 £ 4 sh. 7 d.; No. 14. 1 do. with 8 napkins 17 sh. 3 d.; No. 15. 1 do. with 6 napkins 12 sh. 8 d.

3 tea-cloths: No. 16. 10 sh. 5 d., No. 17. 8 sh. 4 d., No. 18. 9 sh. 7 d.; No. 19. 12 dessert-napkins 5 sh. 10 d.

Towels: No. 20. 1 doz. 1 £ 2 sh. 6 d.; No. 21. 1 doz. 15 sh. 5 d.; No. 22. 2 doz. each 2 £.; No. 23. 1 doz. 18 sh. 4 d.; No. 24. 1 doz. 12 sh. 11 d.; No. 30. 2 pieces of tea-cloths 1 £ 16 sh. 8 d.; No. 31. 2 doz. dessert-napkins 1 £ 10 sh.

Cotton goods: white coverlets, No. 25. 2 pieces 1 £ 6 sh. 8 d.; No. 26. 6 pieces 3 £ 10 sh.; No. 27. 4 pieces 2 £; No. 28. 4 pieces 1 £ 3 sh. 4 d.; No. 29. 4 pieces 19 sh. 4 d.

2766. GUTMANN, BROTHERS, manu., Göppingen. London 1862 honor. mention.   (s. No. 2756.)

2767. KAUFMANN & SONS (Max Kaufmann, Julius Kaufmann), manu. of cotton and linen-fabrics, Göppingen. London 1862 medal.   (s. No. 2757.)

2768. LANG, A. F., manu., Blaubeuren. Berlin 1844 silver medal; Munich 1854 great medal; Paris 1855 silver medal; London 1862 medal. Agt. s. No. 2758 a.

Linen cloth: No. 6. 9. 11. 13. 16. 21. 6 pieces of heavy white linen, finished, breadth 11/8, length 65 to 66 Württ. ells, in different packings p. yard 1 sh. 1½ d. to 7 sh. 6 d.; No. 3. 7. 10. 3 pieces of the same p. yard 1 sh. to 1 sh. 7 d.; No. 5. 10. 2 pieces of heavy white linen, breadth 12/4, length 65 to 66 Württ. ells, p. yard 2 sh. 4 d. to 3 sh. 6 d.; No. 2. to 18. 15 doz. white linen pocket-handkerchiefs, breadth 3/4, 4/4, 9/8, in different packings, p. doz. 4 sh. 7 d. to 32 sh.

2769. LANG & SEIZ, manu. of linen and cotton fabrics, Stuttgart. Munich 1854 medal of honor; London 1862 honor. mention. Agt. s. No. 2758 a.   (s. No. 2760.)

Linen fabrics; prices p. Württemberg ells: linen 65 W. ells, 11/8 No. 517/45. 26 kr.; 65 do. 5/4 No. 530/60 27 kr.; 64 do. 11/8 525/60. 29 kr.;

65 do. 12/4 No. 9618/65. 1 fl. 6 kr.; linen for
towels 70 do. 3/4 No. 9737/10. 17 kr.; 70 do.
3/4 No. 9598/20. 17 kr.; 67 do. 7/8 No. 462/65. 24 kr.
67 do. 7/8 No. 278/100. 24 kr.; 2/2 doz. 8/4 towels
No. 270. p. doz. 9 fl. 30 kr.; 2/2 doz. 8/4 towels
No. 100. 10 fl.; ½ doz. napkins No. 20. 7 fl. 30 kr.;
table-cloth No. 60. 2 fl. 54 kr.; do. No. 70. 3 fl.
15 kr.; 2/2 doz. napkins No. 65. p. doz. 8 fl.;
large table-cloth No. 65. 6 fl. 30 kr.; 2/2 doz.
napkins No. 90. p. doz. 8 fl.; 2 table-cloths each
3 fl. 15 kr.; 2/2 doz. napkins No. 100. p. doz. 8 fl.;
2 table-cloths each 3 fl. 15 kr.; 2/2 doz. napkins
No. 200. p. doz. 8 fl. 36 kr.; table-cloth No. 200.
3 fl. 30 kr.; large table-cloth 7 fl.; 2/2 doz. napkins No. 250. p. doz. 8 fl. 30 kr.; large table-
cloth 6 fl. 48 kr.; half-linen diaper: 36 W. ells
8/4 No. 22. 23 kr.; 36 do. No. 38. 24 kr.; linen
diaper 61½ do. 4/4 No. 30. 15 kr.; 36 do. 8/4
No. 35. 27 kr.; 36 do. No. 40. 30 kr.; satin 67 do.
4/4 No. 35. 20 kr.; 65 do. No. 45. 24 kr.; 64 do.
No. 65. 28 kr.

2771. OTTENHEIMER & DETTELBACH, manu.,
Jebenhausen. Agt. the Württemberg Trading-
Company at Stuttgart, represented by Sam. Ses-
sing, 10 Huggin-Lane, Wood-street, Cheapside,
E. C.   (s. No. 2762.)
Linen ticking for beds and matresses; linen twills
for window-blinds; linen twills for trousers.

2771 a. THE WÜRTTEMBERG TRADING-COM-
PANY, Stuttgart. Agt. s. No. 2771.
(s. No. 2693 a., 2724 a., 2822, 2849.)
Mats of seagrass: 1 piece, oval No. 1. 2 sh. 5 d.;
1 piece, long 1 sh. 9 d.

CLASS XX.
SILK AND VELVET.

2773. ADMINISTRATION OF THE FARMS OF
HIS MAJESTY THE KING OF WÜRTTEMBERG (Kö-
nigliches Hof-Kameral-Amt, Stuttgart). London
1862 medal.   (s. No. 2710, 2698.)
Raw silk.

2773 a. GESSLER & Co., manu. of silk-
stuffs, Tettnang. London 1862 honor. mention.
Agt. s. No. 2771.   (s. No. 2710.)
Plain silk stuffs, black and other colors, made of
Württemberg silk. Prices p. French aune: No. 729.
12½ aunes ribs azulin 30 in. 4 fl. 24 kr.; No. 736.
32 au. rayés, pensée 22 in. 2 fl. 24 kr.; No. 736.
27 au. rayés gris 22 in. 2 fl. 24 fl.; No. 733. 20 au.
gros du Rhin 32 in. 3 fl. 42 kr.; No. 738. 25 au.
poult de soie 30 in. 4 fl. 30 kr.; No. 734. 23 au.
gros du Rhin 27½ in. 3 fl. 42 kr.

CLASS XXI.
WOOLLEN AND WORSTED, INCLUDING MIXED
FABRICS GENERALLY.

2776. HARTMANN, BROTHERS, cloth-manu.,
Esslingen a. N. Berlin 1844 silver medal; Paris
1855 bronze medal; London 1862 medal. Agt.
s. No. 2771.
Woollen fabrics; prices p. yard: No. 1. 8 4/16 yards
satin, blue-mixed, for Swiss and Ital. army 8 sh.;
No. 2. 8¼ yards satin blue, wool-dyed, for Württemberg army 12 sh.; No. 3. 8 4/16 yards satin black
10 sh.; No. 4. 8½ yards sibérienne brown wool-
dyed 9 sh. 8 d.; No. 5. 8¼ yards pilot brown

mixed 8 sh. 8 d.; No. 6. 10½ yards do. 8 sh. 8 d.;
No. 7. 6½ yards tricot black 10 sh.; No. 8. 11½ yards
fancy cloth, twilled, dess. 106. 9 sh. 2 d.; No. 9.
7 yards do., dess. 364., 10 sh.; No. 10. 11¾ yards
do., dess. 357. 10 sh.; No. 11. 8 yards do., dess.
373. 10 sh. 3 d.; No. 12. 13 yards mille raies for
summer 5 sh. 6 d.; No. 13. 8 yards ribs for summer
6 sh. 6 d.

2777. KAUFFMANN, S. & J., merchants and
manu., Stuttgart, manufactory of wool and cotton mixed stuffs for gentlemen's clothing, Sindel-
fingen. London 1862 medal. Agt. s. No. 2771.
Stuffs with cotton warp; prices p. yard: No. 1.
for summer clothing 3 sh. 6 d.; No. 2. demi sai-
son 3 sh. 6 d.; No. 3. do. 4 sh. 6 d.; No. 4. for
winter clothing 6 sh.; No. 5. do. 6 sh. 4 d.

2778. KISSEL, AUGUST, manu., Böblingen.
London 1862 medal.   (s. No. 2758 a.)
Cotton and wool mixed stuffs for trousers.

2780. MÜLLER, J. G., JUN., manu. of drapery, Metzingen. Munich 1854 honor. mention;
Paris 1855 bronze medal; London 1862 medal.
Agt. s. No. 2771.
Woollen cloth; prices p. W. Elle: No. 6728. peau
d'agneau 5 fl. 24 kr.; No. 6430. black cloth 3 fl. 36 kr.;
No. 6443. blue cloth 3 fl. 30 kr.; No. 6278. dark-
green cloth 3 fl. 24 kr.; No. 6510. black »cuir«
3 fl.; No. 6531. ribs 3 fl. 12 kr.; No. 6532. diagonal 3 fl. 12 kr.

2781. RAIFSTÄNGER, M., manu., Metzingen.
Woollen cloths.

2782. SCHILL & WAGNER, manu. of woollen
goods, Calw. Berlin 1844 silver medal; Munich
1854 great medal; Paris 1855 silver medal; London 1862 medal. Agt. s. No. 2771.
a. Woollen stuffs for trousers; prices p. yard:
No. 1. and 2. 5 sh. 1 d.; No. 3. and 4. 5 sh. 9 d.;
No. 5. 5 sh. 11½ d.; No. 6. and 7. 6 sh. 9½ d.;
No. 8. and 9. 5 sh. 9 d.; No. 10. and 11. 6 sh.
4½ d.; No. 12. 5 sh. 11½ d.; No. 13. 7 sh. 8 d.;
No. 14. 6 sh. 9½ d. b. Velours: No. 15. and 16.
p. yard 6 sh. 7 d.

2784. ZOEPPRITZ, BROTHERS, manu. of
woollen and halfwoollen flannels, blankets of
various kinds, Heidenheim. Berlin 1844 bronze
medal; Munich 1854 great medal; Paris 1855 silver
medal; London 1862 medal. Agt. s. No. 2771.
Blankets and flannels.

No.	piece.	Blankets:	breadth in centim.	lenght in English in.:	£	sh.	d.
1.	2	twilled, white and pink-checks D.	151/205	60/82	—	15	1
2.	1	twilled, white with scar-let stripes G.	164/219	65,86	—	18	7
3.	2	twilled, white with small scarlet stripes J.	151/205	60,82	—	19	3
4.	1	twilled, white with small scarlet stripes H.	206/233	82/91	1	7	5
5.	2	plain, white with small scarlet stripes E.	151/205	60,82	—	9	10
6.	2	twilled, white with small scarlet stripes Bel.	151/205	60,82	—	8	5
7.	2	twilled, scarlet with black stripes G.	151,205	60,82	—	19	5
8.	1	twilled, white and blue checks G.	151 205	60 82	—	17	1
9.	2	twilled, scarlet with black stripes C.	151/205	60,82	—	14	7

L

Blankets:

Co.	piec.		breadth in centim.	lenght in English in.:	p. piece £	sh.	d.
10.	2	twilled, white with scarlet stripes C.	164/219	65/86	—	13	—
11.	1	twilled, roe-brown with red and blue stripes B.	151/178	60/70	—	9	5
12.	2	twilled, white with scarlet stripes D.	151/205	60/82	—	12	5
13.	1	twilled, lightbrown with black and light dahlia stripes EF.	151/178	60/70	—	9	—
14.	1	twilled, brown checks dess. 50., for travelling purpose EF.	151/178	60/70	—	9	9
15.	2	twilled, white with scarlet stripes, cotton warp BC.	137/192	54/75	—	6	1
16.	1	twilled, brown with red and black stripes, for military-purposes and for horses A.	151/178	60/70	—	6	2
17.	1	twilled, black and fancy checks, dess. 90. B. .	137/158	54/63	—	6	10

Flannel:

No.	piec.			breadth in cm.	lenght in English in.:	p. yard d.
18.	1	white, plain, halfwool (flannel)	Eb. . .	69	27/28	12⅓
19.	1	do. (domet)	G. . . .	69	27/28	10¾
20.	1	do.	H. . . .	77	31	12½
21.	1	do.	H. . .	77	31	13½
22.	1	do.	Hb. . .	69	27/28	13½
23.	1	white, twilled halfwool (domet)	Ga. . .	69	27/28	13½
24.	1	do.	Ha. . .	69	27/28	14½
		Welsh flannel:				
25.	1	white, plain all wool (imitation of Welsh-flannel)	J. . . .	69	27/28	23¾
26.	1	swanskin, white twilled all wool (Espagnolet)	F. . . .	77	31	23
27.	1	do.	Ja. . .	86	34/35	34
		Flannel:				
28.	1	white twilled all wool (finet)	M. I.	77	31	38¼
29.	1	do.	Kb. I.	69	27/28	27¾
30.	1	do.	J. I. .	69	27/28	23½
31.	1	do.	Ha. I.	69	27/28	24
32.	1	do.	Hb. II.	69	27/28	22¼
33.	1	do.	Ga. II.	69	27/28	19¼
34.	1	do.	Fa. . .	69	27/28	18¼
35.	1	white plain all wool (flannel)	Eb. . .	94	38/39	27
36.	1	do.	Ga. . .	86	34/35	26½
37.	1	white plain all wool (finet)	M. I.	69	27/28	31½
38.	1	do.	L. I. .	69	27/28	29
39.	1	do.	Jb. I. .	69	27/28	24
40.	1	do.	Ja. II.	77	31	23
41.	1	do.	Hb. I.	69	27/28	21½
42.	1	do.	Ha. II.	77	31	21¼
43.	1	do.	Ga. II.	69	27/28	17
44.	1	do.	Fa. . .	77	31	18
		Swanskin:				
45.	1	white twilled halfwool (Espagnolet) .	Ga. . .	77	31	19¼
46.	1	do.	Fa. . .	69	27/28	16
47.	1	do.	Ca. . .	60	24	13
48.	1	flannel scarlet, plain all wool (finet) . . .	Ga. II.	77	31	23½
49.	1	flannel royal blue, plain all wool (finet)	Fa. . .	69	27/28	18¼
50.	1	flannel white, plain all wool (flannel) . .	Fa. . .	259	104	72⅘

CLASS XXII.

CARPETS.

2785. Erlenbusch, J., teacher at the Royal weaving school, Stuttgart. Paris 1855 bronze medal. Agt. the Württemberg Trading-Compagny at Stuttgart, represented by Sam. Sessing, 10 Huggin-Lane, Wood-street, Cheapside, E. C.

Tapestry of the high warp: No. 1. carpet executed after a new method invented by the exhibitor 180 fl.; No. 2. do. for foot stools 6 fl. 15 kr.; No. 3. do. 6 fl. 15 kr.; No. 4. do. 6 fl. 15 kr.

CLASS XXIV.

TAPESTRY, LACE, AND EMBROIDERY.

2786. Bek & Salzmann, manu. of white embroidered stores, plain and embroidered curtains, stuffs for furniture and clothing, Ulm. London 1862 medal. Agt. s. No. 2785.

(s. II. No. 60.)

White embroidered stuffs: No. 1. 1 store, on »tulle« 18/4; No. 2. 1 pair of embroidered curtains 16/4; No. 3. 1 drapery, embroidered on »tulle«; No. 4. 2 curtain-holders on »tulle«; No. 5. 6. bordures, embroidered 10/4 19 au., No. 5. p. piece 46 frs., No. 6. p. piece 45 frs.; vitrages embroidered 7/4 19 au.; No. 7. 1 p. 50 frs.; No. 8. 1 p. 31 frs.

5 pair of ready made curtains: No. 9. p. pair 25 frs.; No. 10. p. pair 31 frs. 50 c.; No. 11. p. pair 19 frs. 50 c.; No. 12. p. pair 18 frs. 50 c.; No. 13. p. pair 26 frs.; No. 14. 1 »echarpe tambour« p. piece 4 frs. 50 c.

2787. Deffner, Otto, manu. of curtain-stuffs, Ravensburg. Munich 1854 great medal. Agt. J. W. Green (E. G. Zimmermann), 2 St. Pauls Buildings, Little Carter Lane, Doctors Commons.

Curtain-stuffs, double-shade: 6 pieces of gauze: No. 1. 6 sh. 4 d.; No. 2. do. 6 sh. 8 d.; No. 3. single shade 5 sh.; No. 4. do. 5 sh.; No. 5. double shade 6 sh. 4 d.; No. 6. do. 6 sh. 4 d.

5 pieces figured muslin: No. 7. à jour 1 £ 7 sh.; No. 8. do. gauze 1 £ 14 sh.; No. 18. do. 1 £ 8 sh.; No. 19. do. à jour 1 £ 6 sh.; No. 20. do. gauze 1 £ 7 sh.

9 pieces of muslin: No. 9. store embroidered 14 sh.; No. 10. store curtain, network 20 sh.; No. 11. tulle rideau network 17 sh. 6 d.; No. 12. do. 12 sh.; No. 13. do. 1 £ 3 sh.; No. 14. do. 1 £ 5 sh.; No. 15. muslin-curtain 14 sh. 6 d.; No. 16. tulle curtain, network 15 sh.; No. 17. muslin-store 14 sh. 6 d.

2788. Hummel, Sophie, superintendent of a private industrial school, Stuttgart. Agt. s. No. 2785.

Fine silk embroidery, executed after a new method invented by the exhibitor, representing His Royal Highness, the late Prince Consort 50 £.

2789. Neuburger, H., Sons, manu. of linens, shirtings and embroidery, Stuttgart and Ulm. Munich 1854 medal of honor; Paris 1855 bronze medal. Agt. s. No. 2785.

Embroideries: 12 embroidered petticoats; 6 do. »pleine-petite«; 9 do. B.; 3 do.; 2 piqué; 6 piqué covers; 5 doz. shirts; 12 pieces of organdis; 5 pieces of jaconet; 5 pieces of flower jaconet; 5 pieces of nainsook; 4 pieces of cambric; 2 doz. button-collars.

2790. Weiss, W. A., manu. of embroidered and damask-stuffs for curtains, Ravensburg. Berlin 1844 silver medal, Munich 1854 great medal; London 1862 honor. mention. Agt. the Württemberg Trading - Company at Stuttgart, represented by Sam. Sessing, 10 Huggin - Lane, Wood - street, Cheapside, E. C.

Embroideries: 1. embroidered drapery on muslin (921) 5 sh.; 2. do. on guipure (922) 7 sh. 6 d.; 3. do. store (1000) 13 sh. 4 d.; 4. do. table - cloth (1100) 13 sh. 4 d.; 5. 14/4 do. curtain (1002), 6. do. (1004) p. pair 1 £; 7. sample of embroidered curtain (989) 8 sh. 4 d; 8. do. on tulle (1052) 6 sh. 8 d.; 9. do. on guipure (1054) 6 sh. 8 d.; 10. do. on tulle (1074) 8 sh. 4 d.; 11. 7/4 embroidered vitrage on tulle 4 aunes, 6 sh. 8 d.; 12. do. 5 sh.; 13. (424), 15. (432), 18. (460), 19. (463), 20. (466), 21. (471) 12/4 damask, each 3¼ aunes, each 3 sh. 11 d.; 14. (425), 17. (454) do. 4 sh. 2 d.; 16. (452) do. 4 sh.; 22. (456), 23. (459) 14/4 damask each 3¼ aunes 5 sh. 12/4 damask on gauze 3¼ aunes; 24. 4 sh. 2 d.; 25. 4 sh. 4 d.; 26. 4 sh. 6 d.

2791. von Zwerger, Franz, manu of jaconets and muslins by hand and power-loom, Ravensburg. Munich 1854 great medal, under the firm von Zwerger, Deffner & Weiss; London 1862 medal. Agt. s. No. 2790.

Embroideries: No. 1. 18/4 broad 1 store on guipure 3½ au., 3 £ 6 sh. 8 d.; No. 2. 14/4 broad 1 coupon damassé a gaze double ombré 3½ au., p. piece of 19 aunes 2 £; No. 3. do 2 £; No. 4. 1 pair of embroidered tulle curtains 14/4 broad, 3¼ au. 2 £ 11 sh. 8 d.; No. 5. do. 2 £ 10 sh.; No. 6. 1 pair of curtains damassé a gaze double ombré 14/4 broad, 3½ aunes 14 sh. 2 d.; No. 7. 1 flower - atlas 1¾ W. ells 4 sh. 4 d.; No. 8. 1 flower - jaconet 8 aunes 8 sh. 9 d.; No. 9. 1 power - loom jaconet 16 aunes 9 sh. 2 d.; No. 10. 1 cambric 8 aunes 1 £ 1 sh. 3 d.; No. 11. 1 Nainsooc 8 aunes 1 £ 5 sh.; No. 12. 1 moll elastique 8 aunes 1 £ 5 sh.; No. 13. 1 powerloom moll 16 aunes 5 sh. 5 d.; No. 7—13. each 10/4 broad.

CLASS XXVI.

LEATHER, INCLUDING SADDLERY AND HARNESS.

2793. Dittmann, Brothers, leather-manu., Stuttgart, manufactory Gablenberg near Stuttgart. London 1862 honor. mention. Agt. s. No. 2790.

No. 1. and 2. 2/2 smoothed-vache-hides.

2794. Eckart, J. M., manu. of varnished leather and oilcloth, Ulm. London 1862 honor. mention. Agt. s. No. 2790. (s. II. No. 42.)

No. 1. 2 pieces of double side varnished »nettle« stuff 16 frs. 50 c.; No. 2. 1 piece double varnished stuff for military-caps 9 frs. 70 c.; No. 3. 3 pieces superfine oilcloth, each 5 frs. 53 c.; No. 4. 3 pieces oilcloth for military, each 5 frs. 53 c.; No. 5. 4 pieces sheepskins. varnished, black, p. doz. 13 frs.; No. 6. frame with samples of cap - peaks stripes and varnished leather sashes.

2795. Kiderlen & Marius, leather-manu., Ulm. London 1862 honor. mention. Agt. s. No. 2790.

Leather: No. 1. Calcutta, brown, No. 2. Calcutta, waxed p. lb. 2 sh. 1 d.; No. 3. and 4. 2 calfskins, brown, No. 5. and 6. 2 calfskins, waxed, p. lb. 3½ sh.

2796. Klemm, Theodor, manu. of leather, Pfullingen. Munich 1854 honor. mention: London 1862 honor. mention. Agt. s. No. 2790.

Samples of leather of extraordinary strength and durability, prepared after a new method of tanning, invented by the exhibitor and patented in Württemberg, Great-Britain and the United states of America. No. 1. ½ skin for harness; No. 2. 7 pieces of straps for testing the strength of the leather; No. 3. 4 pairs of boot-legs.

2797. Linse & Co., manu. of japanned leather, Crailsheim. London 1862 medal. Agts. Reiss & Co., 27 Watling-street, and the Württemberg trading-company at Stuttgart, represented by Sam. Sessing, 10 Huggin-Lane, Wood-street, Cheapside, E. C.

4 grained cow-hides: No. 1. japanned black for coverings 30 fl.; No. 2. do. for shoes 30 fl.; No. 3. japanned red 33 fl.; No. 4. do. blue 33 fl.; 4 grained horse - hides: No. 5. japanned black for coverings 20 fl.; No. 6. for shoes 20 fl.; No. 7. japanned green 24 fl.; No. 8. japanned yellow 24 fl.

No. 9. Cow-hide, japanned smooth black on the grain side 30 fl.; No. 10. horse-hide, japanned smooth black on the grain side 20 fl.

No. 11. Split flesh-side, japanned smooth, black 10 fl.; No. 12. do. 10 fl.; No. 13. do. green 12 fl.; No. 14. do. blue 12 fl.

2799. Möllen & Co., manu. of varnished leather, Bopfingen. Munich 1854 medal of honor; Paris 1855 bronze medal; London 1862 medal. Agt. s. No. 2790.

2 varnished neat's skins, black in the grain for carriage roofs; 1 piece of a neat's skin varnished on one side; 2 pieces of varnished calf's skins, black in the grain.

2800. Roser, Carl. Fr., leather - manu., Stuttgart. London 1862 medal. Agt. s. No. 2790.

No. 1. One light bridle hide; No. 2. one light bridle hide used for trunk, straps and cigar cases etc. etc.; No. 3. two fine light swine - hides for saddles.

2801. Schmid, Christian, leather manu., Stuttgart. Munich 1854 honor. mention; London 1862 medal. Agt. s. No. 2790.

2/2 Cow hides, p. lb. 17 d.

CLASS XXVII.

ARTICLES OF CLOTHING.

2803. Binder, Friedr. Wilh., manu. of hosiery, Ebingen. London 1862 medal. Agt. s. No. 2790.

Worsted hosiery; ½ dozen of each No.; prices p. dozen: gentlemen's jackets, 1. knitted, bluish gray No. 8. 2 £ 11 sh. 8 d.; 2. do. No. 10. 3 £; 3. do. dark-gray No. 8. 2 £ 10 sh.; 4. woven do. No. 8. 2 £ 13 sh. 4 d.; 5. do. No. 5. 2 £ 5 sh.; 6. do. No. 10. 2 £ 15 sh.; 7. woven bluish - gray No. 10. 3 £: 8. do. No. 3. 2 £; 9. do. No. 2. 1 £ 15 sh.; 10. short - stockings, dark - gray No. 5. 8 sh. 4 d.; 11. bluish - gray No. 3. 6 sh. 3 d.; 12. do. No. 1. 5 sh.; 13. ladies' stockings: black No. 60. 14 sh. 2 d.; 14. dark-blue No. 60. 15 sh. 10 d.; 15. do. gray No. 60. 14 sh. 2 d.; 16. bluish - gray No. 30. 8 sh. 4 d.; 17. do. No. 40. 9 sh. 2 d.; 18. gentlemen's gloves gray No. 4. 6 sh. 3 d.; 19. green No. 4. 7 sh. 1 d.; 20. ladies' shoes No. 3. 7 sh. 6 d.; 21. do. No. 3. 17 sh. 6 d.

2804. FALKENSTEIN, G., manu. of boots and shoes, Balingen. London 1862 medal. Agt. the Württemberg Trading-Company at Stuttgart, represented by Sam. Sessing, 10 Huggin-Lane, Wood-street. Cheapside, E. C.

Shoes and boots; prices p. dozen: No. 1. Wellington boots varnished 13 £; No. 2. gentlemen's boots 9 £; No. 3. English boots with double soles 12 £; No. 4. varnished boots gold stitched 11 £; No. 5. do. silver stitched 11 £; No. 6. varnished boots with wooden pegs 10 £; No. 7. lacing-boots with elastic 6 £.

2806. GROEBER, FERD., manu. of hosiery, Riedlingen a. D.

Hosiery; prices p. dozen; 1. Hoods: Juno 18 fl.; Ceres 24 fl.; Emma 12 fl.; Adeline 14 fl.; Lidia 11 fl.; Frieda 8 fl. 30 kr.; Mathilde 10 fl.; Bertha 11 fl.; Albertine 15 fl.; Blanda 1. 11 fl.; do. 2. 15 fl.; do. fancy 16 fl.; Antonette 16 fl.; Flora 15 fl.; Selma 12 fl.; Cera 11 fl. 30 kr.; Perpetua 11 fl. 30 kr.; Rezia 12 fl. 30 kr.; Anna 1. 2 fl. 36 kr.; do. 2. 3 fl. 24 kr.; do. 3. 4 fl. 12 kr.; do. 4. 5 fl.; Clara 1. 2 fl. 48 kr.; do. 2. 3 fl. 36 kr.; do. 3. 4 fl. 24 kr.; do. 4. 5 fl. 12 kr.; Marie 1. 3 fl.; do. 2. 4 fl.; do. 3. 5 fl.; do. 4. 6 fl.; Lina 2. 3 fl. 24 kr.; do. 3. 4 fl. 12 kr.; do. 4. 5 fl.

2. Sleeves, woollen: No. 80. 3 fl. 24 kr.; No. 116. 8 fl. 30 kr.; do. 6 fl. 30 kr.; No. 114. 6 fl. 30 kr.; No. 110. 9 fl.; No. 120. 12 fl.; No. 124. 11 fl.; No. 125. 16 fl.

3. Sleeves for summer: No. 260. 5 fl. 12 kr.; No. 264. 5 fl.; No. 264. 5 fl.; No. 275. 5 fl. 48 kr.; No. 276. 4 fl. 48 kr.; No. 277. 5 fl.; No. 278. 6 fl. 52 kr.; No. 279. 2 fl. 11 kr.; No. 280. 9 fl. 23 kr.; No. 281. 4 fl. 54 kr.; No. 282. 4 fl. 54 kr.; No. 283. 2 fl. 16 kr.; No. 284. 2 fl. 12 kr.; No. 285. 2 fl. 23 kr.; No. 286. 5 fl. 46 kr.; No. 287. 4 fl. 58 kr.; No. 288. 5 fl. 15 kr.; No. 289. 1 fl. 56 kr.

4. Menottes: No. 500. 54 kr.; No. 501. 39 kr.; No. 502. 54 kr.; No. 503. 2 fl. 48 kr.; No. 504. 2 fl. 20 kr.; No. 505. 3 fl. 16 kr.; No. 506. 2 fl. 34 kr.; No. 507. 42 kr.; No. 508. 2 fl. 2 kr.; No. 509. 3 fl. 9 kr.; No. 510. 3 fl. 27 kr.; No. 511. 3 fl. 44 kr.; No. 512. 2 fl. 16 kr.; No. 513. 1 fl. 19 kr.; No. 514. 1 fl. 27 kr.; No. 515. 1 fl. 24 kr.; No. 516. 44 kr.; No. 517. 1 fl. 2 kr.; No. 518. 1 fl. 38 kr.; No. 519. 1 fl. 10 kr.; No. 520. 1 fl. 27 kr.; No. 600. net gloves 1 fl. 38 kr.; No. 601. do. 1 fl. 38 kr.; No. 602. do. 6 fl. 11 kr.; No. 30. collars. white 3 fl.; No. 31. do. black 3 fl.; No. 32. variously colored 3 fl.

2807. HAAS, J. P., & Co., manu. of Leghorns and ordinary straw bonnets etc., Schramberg. Berlin 1844 bronze medal; London 1851 prize-medal; Munich 1854 great medal; London 1862 medal. Agt. s. No. 2804. (s. Il. No. 38.)

Ladies' bonnets, prices p. piece: No. 1. fancy 290. No. 242. 3 sh. 8 d.; No. 2. do. 289. 2 sh. 11 d.; No. 3. do. 290. extrafine 2 sh. 11 d.; No. 4. do. 290. Florentine 14. 3 sh.; No. 5. do. 268. 4 sh. 8 d.; No. 6. do. 290. No. 1. palm. 3 sh. 4 d.; No. 7. do. 293. Panama 6 sh.; No. 8. do. 292. No. 417. 2 sh. 6 d.; No. 9. do. 293. No. 1. palm. 2 sh. 4 d.; No. 10. do. 293. No. 1. palm. 2 sh. 2 d.; No. 11. do. 288. No. 421. 3 sh.; No. 12. do. 288. No. 4. Florentine 3 sh.; No. 13. do. 292. extrafine 2 sh. 11 d.; No. 14. do. 291. extrafine 2 sh. 11 d.; No. 15. do. 292. No. 19. Florentine 4 sh. 9 d.; No. 16. do. 291. No. 16. Florentine 3 sh. 7 d.; No. 17. do. 288. No. 417/42.

2 sh. 10 d.; No. 18. do. 293. extrafine palm. 5 sh.; No. 19. do. 292. No. 15. Florent. 3 sh. 4 d.; No. 20. do. 293. extrafine 2 sh. 11 d.; No. 21. do. 287. T. E. 2 sh.; No. 22. do. 285. No. 15. white Flor. 3 sh. 1 d.; No. 23. do. 281. No. 17. white Flor. 3 sh. 11 d.; No. 24. do. 281. No. 16. white Flor. 3 sh. 7 d.; No. 25. do. 285. No. 1. palm. A. 2 sh. 6 d.; No. 26. do. 285. No. 15. white Flor. 3 sh. 1 d.; No. 27. do. 262. No. 13. black Flor. 3 sh.; No. 28. do. 293. white 497. 4 sh. 2 d.; No. 29. do. 293. blue 5 sh.; No. 30. do. 281. No. 18. black Flor. 3 sh. 4 d.; No. 31. do. 293. extrafine Flor. 2 sh. 11 d.; No. 32. do. 293. extrafine palm. 5 sh.; No. 33. do. 293. No. 491. 6 sh. 6 d.; No. 34. do. 293. No. 506. 7 sh. 6 d.; No. 35. do. 293. No. 1. palm. black 3 sh. 4 d.; No. 36. do. 262. No. 15. black Flor. 3 sh. 4 d.; No. 37. girl's fancy bonnet 285. extrafine 2 sh. 2 d.; No. 38. children's fancy bonnet 272. black 1 sh. 8 d.; No. 39. do. palm. 1 sh. 10 d.; No. 40. do. 293. black Florent. 15. 3 sh. 4 d.; No. 41. Jerome, black palm. gentleman's hat 7 sh. 6 d.; No. 42. do. brown 7 sh. 6 d.; No. 43. sailor's, extrafine palm. 4. 4 sh. 2 d.; No. 44. do. 1. 2 sh. 3 d.; No. 45. do. 2 sh. 3 d.; No. 46. do. extrafine black 5 sh.; No. 47. gentleman's hat, oval, palm. extrafine 6 sh. 4 d.; No. 48. do. 6 sh. 4 d.; No. 49. do. 6 sh. 4 d.; No. 50. do. 6 sh. 4 d.; No. 51. Schiller 5 sh. 2 d.; No. 52. whale-bone-gentleman's hat 5 £; No. 53. white gentleman's hat, palm. 4 sh. 4 d.; No. 54. do. 4 sh. 4 d.; No. 55. gentleman's hat, Panama No. 40. 10 sh.; No. 56. do. No. 32. 6 sh.; No. 57. do. No. 31. 5 sh. 10 d.; No. 58. do. No. 35. 9 sh. 2 d.; No. 59. do. No. 33. 6 sh. 8 d.; No. 60. boy's hat, black, palm. 254½. No. 1. 1 sh. 3 d.; No. 61. do. 254. No. 1. 1 sh. 3 d.; No. 62. do. 254. No. 1. 1 sh. 3 d.; No. 63. do. Flor. 254. No. 11. 3 sh. 1 d.; No. 64. do. white 254. No. 41. 1 sh. 4 d.; No. 65. do. black No. 1. 1 sh. 3 d.; No. 66. ladies' bonnet fac. 279. A. 1. palm. 2 sh. 6 d.; No. 67. do. extrafine 4 sh. 8 d.; No. 68. do. black, palm. 4 sh. 8 d.; No. 69. do. 294. 4 sh. 8 d.; No. 70. do. 4 sh. 8 d.; No. 71. gentleman's hat 282. extrafine palm. A. 4 sh. 2 d.; No. 72. boys hat fac. 282. No. 8. black Florent. 1 sh. 8 d.; No. 73. do. 237. palm. 1. 1 sh. 8 d.; No. 74. do. extrafine 2 sh. 4 d.; No. 75. do. Scotch, black 2 sh. 2 d.; No. 76. ladies' bonnet 446. black 3 sh. 8 d.

Baskets: No. 77. No. 2. 5 sh.; No. 78. tress 21. 2 sh. 8 d.; No. 79. do. 2 sh. 8 d.; No. 80. palm. 2 sh. 8 d.; No. 81. No. 18. 3 sh. 4 d.; No. 82. do. 2 sh. 4 d.; No. 83. No. 12. 2 sh. 6 d.; No. 84. No. 14. 1 sh. 4 d.; No. 85. No. 27. 1 sh. 10 d.; No. 86. No. 76. 1 sh. 6 d.; No. 87. No. 75. 2 sh. 3 d.; No. 88. No. 66. 2 sh. 2 d.; No. 89. No. 92. 1 sh. 11 d.; No. 90. No. 116. 1 sh.; No. 91. violet 3 sh. 4 d.; No. 92. No. 23. 1 sh.

No. 93. 1 doz. pair of ordinary soles 2 sh. 8 d.; No. 94. do. double 2 sh. 10 d.; No. 95. 2 pair of palm. shoes 1 sh.; No. 96. ½ doz. cigar-cases 7 sh.

2808. HAAS, CLEMENS, manu., Schramberg. Agt. s. No. 2804.

Knitted, woollen jackets: No. 1. brown paletot machine-made 334 frs. 30 c.; No. 2. blue do. 282 frs. 88 c.; No. 3. gray do. 154 frs. 28 c.; No. 4. brown knitted jackets hand-made 72 frs. 88 c.; No. 5. blue do. 75 frs.; No. 6. blue gentlemen's waist-coat 70 frs. 74 c.; No. 7. blue boy's waist-coat 51 frs. 40 c.; No. 8. blue machine-made jackets, plain, 70 frs. 74 c.; No. 9. gray machine-made jackets, striped, 72 frs. 88 c.; prices p. doz.

2809. KIENZLE, BERNHARD, manu. of hosiery, Balingen. Agt. the Württemberg Trading-Company at Stuttgart, represented by Sam. Sessing, 10 Huggin-Lane, Wood-street, Cheapside, E. C.

Hosiery: hoods No. 1. 4 sh; No. 2. 3 sh. 4 d.; No. 3. 3 sh.; shawls No. 4. 3 sh. 8 d.; No. 5. 3 sh. 2 d.; No. 6. 2 sh. 8 d.; jackets No. 7. brown 5 sh. 8 d.; No. 8. colored 5 sh. 10 d.; No. 9. do. 5 sh. 6 d.; No. 10. 1 pair of drawers with stripes 2 sh. 6 d.; No. 11. 1 pair of drawers, woven 2 sh. 4 d.; No. 12. ¼ doz. of ladies' stockings, p. doz. 18 sh. 8 d.; No. 13. ⅓ doz. of gentlemen's stockings, p. doz. 1 £ 3 sh. 4 d.; No. 14. ¼ doz. of socks p. doz. 12 sh.; No. 15. ½ doz. of ladies' shoes, colored 16 sh. 8 d.; No. 16. ¼ doz. of ladies' boots 1 £ 3 sh. 4 d.; No. 17. 1 pair of gentlemen's boots, p. pair 3 sh.; No. 18. 1 pair of long boots, p. pair 5 sh. 4 d.; No. 19. 1 pair of spatterdashes 4 sh. 2 d.; No. 20. 1 ladies' jacket 6 sh. 1 d.; No. 21. 2 pair of children's shoes 2 sh.

2811. KISPERT & STICHLING, manu. of worsted knitted and crotchet articles, Ulm a. D. London 1862 medal. Agt. s. No. 2809.

Hoods: 1. No. 311. 1 £ 19 sh. 1 d.; 2. No. 451. 1 £ 12 sh. 11 d.; 3. No. 455. 1 £ 17 sh.; 4. No. 457. 1 £ 4 sh. 8 d.; 5 No. 460. 2 £ 5 sh. 3 d.; 6. No. 461. 2 £ 3 sh. 2 d.; 7. No. 463. 2 £ 3 sh. 2 d.; 8. No. 464. 1 £ 16 sh.; 9. No. 471. 1 £ 19 sh. 1 d.; 10. No. 472. 2 £ 10 sh. 9 d.; 11. No. 473. 2 £ 2 sh. 2 d.; 12. No. 475. 2 £ 3 sh. 3 d.; 13. No. 476. 2 £ 13 sh. 6 d.; 14. No. 478. 1 £ 18 sh. 5 d.; 15. No. 479. 3 £ 1 sh. 9 d.; 16. No. 481. 2 £ 9 sh. 4 d.; 17. No. 482. 2 £ 19 sh. 8 d.; 18. No. 484. 2 £ 11 sh. 5 d.; 19. No. 487. 1 £ 8 sh. 10 d; 20. No. 506. 1 £ 11 sh. 7 d.; 21. No. 510. 1 £ 17 sh.; 22. No. 512. 1 £ 12 sh. 11 d.; 23. No. 513. 2 £ 3 sh. 3 d.; 24. No. 514. 2 £ 2 sh. 2 d.; 25. No. 517. 1 £ 9 sh. 10 d.; 26. No. 518. 1 £ 16 sh.; 27. No. 519. 2 £ 1 sh. 2 d.; 28. No. 305. 1 £ 6 sh. 1 d.; 29. No. 322. 1 £ 5 sh. 9 d.; 30. No. 348. 1 £ 10 sh. 10 d.; 31. No. 346. 2 £ 4 sh. 7 d.; 32. No. 452. 1 £ 19 sh. 1 d.

Sleeves: 33. No. 368. 1 £ 15 sh.; 34. No. 369. 1 £ 10 sh. 2 d.; 35. No. 531. 1 £ 8 sh. 1 d.; 36. No. 357. 1 £ 7 sh. 9 d.

Cuffs: 37. No. 542. 10 sh. 4 d.; 38. No. 543. 14 sh. 5 d.; 39. No. 544. 1 £ 3 sh. 8 d.

Socks: 40. No. 447. 12 sh. 4 d.; 41. No. 448. 11 sh. 4 d.

Cape 42. No. 547. 2 £ 19 sh. 8 d.

2812. KNAPP, BENJAMIN, manu. of corsets, Reutlingen. London 1862 honor. mention. Agt. s. No. 2809.

Corsets; prices p. doz.: 1 white, with clasp 1 £ 11 sh.; 3. do. without clasp 1 £ 10 sh.; 7. do. with clasp 1 £ 18 sh.; 27. gray do. 1 £ 14 sh.; 28. do. 1 £ 15 sh.; 31. white do. 1 £. 11 sh.

2813. LIEB, FRIEDRICH, manu. of hosiery, Ulm. London 1862 honor. mention. Agt. s. No. 2809.

Hosiery: 1. 2. 2 large dolls; 3. small doll, harlequin No. 201.; 4. do. peasant No. 201.; 5. do. Krethel No. 202.; 6. woollen ladies' cape No. 132.; 7. do. jacket No. 133.; 8—11. woollen children's jackets No. 134.; No. 1.; No. 2.; No. 3.; 12. 1 pair of ladies' sleeves No. 112.; 13. do. No. 152.; 14. 1 pair of ladies' gaiters No. 142.; 15. 1 pair of children's gaiters No. 254.; 16. 1 do. stockings

No. 343.; 17. do. cape No. 346.; 18. do. No. 437.; 19. stocks No. 570.; 20. 1 pair ladies' cuffs No. 356.; 21. gentlemen's cap. No. 337.; 22. children's cap. No. 51.; 23. woollen children's cap. No. 128.; 24. do. bonnet No. 127.; 25. do. theater hood No. 329.; 26. do. No. 32.; 27—32. woollen ladies' hoods No. 82.; No. 21.; No. 40.: No. 393.; No. 97.; No. 19.; 33—38. woollen children's hoods No. 83.; No. 46.; No. 14.; No. 22.; No. 98.; No. 45.

2814. MUNDORFF & MÜLLER, manu. of hosiery, Calw. London 1862 honor. mention. Agts. Sprösser, Lorenz & Co., London. For the time of the exhibition Agt. s. No. 2809.
(s. II. No. 39.)

Fine woollen hosiery; prices p. doz.: 1. fine woollen gentleman's waist coat double, white No. 5. 57 sh. 6 d.; 2. do. No. 4. 58 sh.; 3. do. scarlet No. 3. 51 sh. 6 d.; 4. do. white No. 3. 45 sh.: 5. ladies' doublet, white No. 4. 55 sh.; 6. gentleman's do. No. 5. 55 sh.; 7. do. No. 5a. 48 sh. 6 d.; 8. gentleman's doublet scarlet No. 5. 62 sh.; 9. ladies' doublet, white No. 5. 50 sh.; 10. do. No. 5. 52 sh.

2815. OCHS, J. F., manu. of hosiery, Reutlingen. Agt. s. No. 2809.

Knitted and crochet goods; one piece of each kind; prices p. doz.: No. 1. white knit. children's cap with flowers of beads 8 sh. 8 d.; No. 2. do. with garland of beads 8 sh. 4 d.: No. 3. do. 5 sh. 4 d.; No. 4. do. with garland of oak leaves in beads 5 sh. 4 d.; No. 5. fine children's cap with embroidered flowers 8 sh. 4 d.; No. 6. fine children's cap 6 sh. 8 d.; No. 7. do. with white flowers 6 sh.; No. 8. fine children's knit. cap 6 sh. 8 d.; No. 9. white knit. collar 16 sh. 8 d.; No. 10. do. 15 sh.; No. 11. do. 16 sh.; No. 12. white woven collar 1 £ 3 sh. 4 d.; No. 13. do. 1 £ 3 sh. 4 d.; No. 14. fine white knit. collar 2 £ 10 sh.; No. 15. white knit. children's frock of fine quality 1 £ 6 sh. 8 d.; No. 16. do. 10 sh.; No. 17. do. of wool 16 sh. 8 d.; No. 18. white knit. children's collar 8 sh.; No. 19. do. 8 sh. 4 d.; No. 20. do. 9 sh. 4 d.; No. 21. white knit. gloves with beads 16 sh.; No. 22. do. 5 sh. No. 23. silk gloves without fingers, network 8 sh. 8 d.; No. 24. do. 13 sh. 4 d.; No. 25. silk fanchon network 2 £; No. 26. head dress twisted of brillant yarn 16 sh. 8 d.; No. 27. do. of silk yarn 1 £ 3 sh. 4 d.; No. 28. white knit. children's cap 2 sh.; No. 29. white fine knit. children's cap 4 sh. 8 d.

2817. OTTENHEIMER, J. M., & Sons, manu. of corsets without seam, Stuttgart. London 1862 medal. (s. II. No. 43.)

3 corsets with clasps: No. 4. white, damask; No. 6. do. common quality; No. 5. drab, do. 3 bodices: No. 3. white, damask, embroider'd; No. 1. do. matelassé; No. 2. drab, damask.

2818. ROSENTHAL, D., & Co., manu. of corsets without seam, Göppingen. London 1862 honor. mention. Agt. William Meyerstein, 47 Friday-street.

Corsets without seam and drill: No. 1. corset L. G. with buttons, white; No. 2. do. Olga superfine, with busk, white; No. 3. do. G., with busk, flammé: No. 4. bodice with buttons L. Jeanne d'Arc, yellow; No. 5. bodice with buttons, English color; No. 6. bodice with buttons La. E. P., white; No. 7. bodice with busk La., white: No. 8. waist à ocillet, Jenny Lind: No. 9. waist with buttons white; No. 10. corset Olga superfine with silver clasps, white: No. 11. bodice Victoria with buttons, white; No. 12. 13. 14. suit of clothes of half linen twill, striped (coat, trousers and waist-coat); No. 15. 16. 17. do.

2819. SCHUMM, FR., manu., Calw.
Various articles of hosiery.

2820. STEINHART, HERZ & Co., manu. of corsets, without seam, Göppingen. London 1862 medal. Agts. Krauss & Auerbach, 68 Basinghall-street, E. C.　　　　(s. II. No. 37.)

Corsets without seam: No. 1. embroidered cour EPB. 01. 3 fl. 36 kr.; No. 2. do. 2. 2 fl. 48 kr.; No. 3. do. 3. 2 fl. 24 kr.; No. 4. Norma à boutons 1 fl. 42 kr.; No. 5. Bellona à boutons, with red stripes 1 fl. 42 kr.; No. 6. do., couvert 1 fl. 36 kr.; No. 7. Victoria do. 1 fl. 36 kr.; No. 8. Olga with busk 1 fl. 24 kr.; No. 9. Bellona do. 1 fl. 21 kr.; No. 10. Adelaide à boutons 1 fl. 30 kr.; No. 11. flammé Bellona do. 1 fl. 36 kr.; No. 12. flammé Lucretia do. 1 fl. 24 kr.; No. 13. Hell-flammé do. 1 fl. 27 kr.; No. 14. white Giulia do. 2 fl. 45 kr.; No. 15. white Alliance do. 2 fl. 33 kr.; No. 16. flammé Olga do. 1 fl. 42 kr.; No. 17. flammé Olga bod. do. 1 fl. 30 kr.; No. 18. Hell-flammé do. 1 fl. 33 kr.; No. 19. flammé GB. do. 1 fl. 27 kr.; No. 20. flammé AB. do. 1 fl. 21 kr.; No. 21. flammé SB. do. 1 fl. 15 kr.; No. 22. flammé S. do. 1 fl. 9 kr.; No. 23. white Eugenie bod. à boutons of silk 5 fl.; No. 24. Amelje with busk 57 kr.; No. 25. Waist G. grand 1 fl. 18 kr.; No. 26. do. 1 fl. 6 kr.; No. 27. Lucretia à boutons. 1 fl. 24 kr.; No. 28. SB. do. 1 fl. 15 kr.; No. 29. Amelje do. 1 fl. 9 kr.; No. 30. Olga do. 1 fl. 30 kr.; No. 31. HN. à boutons camb. 1 fl. 12 kr.; No. 32. white GB. Union couvert 1 fl. 27 kr.; No. 33. white GNB. Union couvert camb. 1 fl. 18 kr.; No. 34. flammé GNB. do. 1 fl. 18 kr.; No. 35. white extension Olga à boutons 1 fl. 42 kr.

2821. VOTTELER, BROTHERS, manu. of articles of hosiery in cotton, wool and silk, Reutlingen. London 1862 honor. mention. Agt. the Württemberg Trading-Company at Stuttgart, represented by Sam. Sessing, 10 Huggin-Lane Wood-street, Cheapside, E. C.

Knitted and crochet-work-articles; prices p. doz. a. caps, (shuttle-sewing): No. 627. No. 80. 3 fl. 30 kr. No. 628. do. 4 fl. 36 kr.; No. 629. do. 5 fl. 33 kr. No. 630. do. 6 fl. 36 kr.; No. 631. No. 100. 4 fl. 3 kr.; No. 632. do. 5 fl. 3 kr.; No. 633. do. 6 fl. 6 kr.; caps, embroidered: No. 634. 2 fl. 54 kr.; No. 635. 3 fl. 18 kr.; No. 636. 4 fl. 3 kr.; No. 624. 3 fl.; No. 625. 3 fl. 48 kr.; No. 626. 4 fl. 12 kr.; No. 797. cap, crochet-work 2 fl. 36 kr.; caps knitted: No. 608. 1 fl. 30 kr.; No. 609. 1 fl. 48 kr.; No. 610. 2 fl. 6 kr.; No. 611. 2 fl. 24 kr.; No. 612. 2 fl. 36 kr.; No. 613. 3 fl.; caps, with lace: No. 615. 2 fl. 6 kr.; No. 617. 2 fl. 48 kr.; No. 619. 3 fl. 36 kr.; caps, knitted tight No. 1. 2 fl. 18 kr.; No. 2. 3 fl. 3 kr.; No. 3. 3 fl. 48 kr.; ordinary yarn-caps No. 601. 42 kr.; No. 602. 45 kr.; No. 603. 54 kr.; No. 604. 1 fl.; No. 605. 1 fl. 12 kr.; No. 606. 1 fl. 24 kr.

b. Children's napkins: No. 657. 5 fl. 54 kr.; No. 683. 4 fl. 48 kr.; No. 685. 6 fl.; No. 686. 6 fl.; No. 682. 5 fl. 36 kr.; No. 656. 8 fl. 12 kr.; No. 659. 5 fl. 6 kr.; No. 660. 6 fl. 6 kr.; No. 687. 4 fl. 24 kr.; No. 684. 4 fl. 24 kr.; No. 688. 2 fl. 48 kr.; No. 270. 4 fl. 36 kr.

c. Jackets: No. 230. 11 fl.; No. 229. 10 fl.; No. 228. 9 fl. 36 kr.; No. 227. 9 fl.; No. 226. 8 fl.

d. Waists: No. 232. 11 fl.; No. 233. 12 fl. 48 kr.; No. 236. 13 fl. 12 kr.; No. 237. 19 fl.; No. 235.

12 fl.; No. 222. 7 fl. 36 kr.; No. 223. 8 fl. 36 kr.; No. 224. 9 fl. 36 kr.; No. 225. 10 fl.

e. Hoods, crochet-work: No. 229. 3 fl. 36 kr.; No. 507. 11 fl. 36 kr.; No. 141. 8 fl. 24 kr.; No. 512. 10 fl.; No. 515. 3 fl. 24 kr.; No. 516. 10 fl. 24 kr.; No. 504. 8 fl.; No. 517. 7 fl. 36 kr.; No. 111. 2 fl. 48 kr.; No. 516. 7 fl. 12 kr.; No. 518. 7 fl. 36 kr.; No. 511. 3 fl. 12 kr.; No. 130. 8 fl. 24 kr.; hoods, lace-work: No. 519/1. 14 fl. 24 kr.; No. 519/2. 14 fl. 24 kr.; No. 519/3. 14 fl. 24 kr.; No. 521/4. 7 fl. 12 kr.; No. 521/5. 7 fl. 12 kr.; No. 519/6. 7 fl. 12 kr.; hoods, crochet-work No. 520. 24 fl.; No. 509. 5 fl. 24 kr.; No. 523. 12 fl. 36 kr.; No. 524. 8 fl.; No. 532. 22 fl. 48 kr.; No. 522. 31 fl. 12 kr.; No. 525. 3 fl. 12 kr.; No. 109. 48 kr.; No. 526. 4 fl. 48 kr.; No. 527. 9 fl.; No. 505. 15 fl.; No. 128. 8 fl. 24 kr.; No. 530. 5 fl.; No. 531. 3 fl. 24 kr.

f. Silk-fanchons: No. 2261. 28 fl. 48 kr.; No. 2259. 29 fl.; No. 2258. 31 fl. 12 kr.

g. Hoods: No. 797. 31 fl. 12 kr.; No. 799. 18 fl. 36 kr.; No. 800. 18 fl. 36 kr.; No. 801. 16 fl.; No. 802. 18 fl.; No. 803. 17 fl. 24 kr.; No. 804. 21 fl.; No. 805. 21 fl. 36 kr.; No. 806. 21 fl.; No. 807. 22 fl. 48 kr.; No. 808. 18 fl.; No. 810. 19 fl. 12 kr.; No. 811. 16 fl. 12 kr.; No. 812. 14 fl. 12 kr.; No. 813. 19 fl. 12 kr.; No. 815. 12 fl.; No. 816. 10 fl. 12 kr.; No. 817. 32 fl.; No. 819. 25 fl. 12 kr.; No. 820. 34 fl. 48 kr.; No. 821. 31 fl. 12 kr.; No. 824. 24 fl.; No. 826. 20 fl. 12 kr.; No. 827. 25 fl. 12 kr.; No. 818. 25 fl. 12 kr.

2822. THE WÜRTTEMBERG TRADING-COMPANY, Stuttgart. Agt. s. No. 2821.
(s. No. 2693a, 2724a, 2771a, 2849.)

Clothes, corsets boots and shoes: 14 coats, prices p. coat: No. 1014. 13 sh.; No. 1011. 10 sh. 2 d.; No. 1008. 6 sh. 10 d.; No. 969. 1 £ 1 sh. 6 d.; No. 993. 6 sh.; No. 848. 3 sh. 5 d.; No. 934. 3 sh. 2 d.; No. 1017. 6 sh. 8 d.; No. 1002. 5 sh.; No. 999. 4 sh. 11 d.; No. 857. 3 sh. 4 d.; No. 851. 3 sh. 8 d.; No. 1005. 5 sh. 8 d.; No. 860. 3 sh. 10 kr.

9 pair of trousers; prices p. pair: No. 1024. 5 sh. 3 d.; No. 1021. 5 sh.; No. 1009. 4 sh. 6 d.; No. 1030. 6 sh. 3 d.; No. 997. 3 sh.; No. 849. 2 sh. 2 d.; No. 855. 2 sh. 3 d.; No. 926. 4 sh. 6 d.; No. 1027. 3 sh. 4 d.

Vests: No. 936. 1 sh. 6 d.; No. 850. 1 sh. 6 d.; No. 1010. 2 sh. 2 d. p. piece.

17 stays without seam; prices p. piece: No. 3077. 2 sh. 6 d.; No. 3078. 2 sh. 9 d.; No. 3079. 3 sh. 2 d.; No. 3900. 3 sh.; No. 3080. 2 sh. 11 d.; No. 3081. 3 sh. 7 d.; No. 4768. 3 sh. 8 d.; No. 3084. 7 sh. 3 d.; No. 3086. 3 sh. 2 d.; No. 4769. 2 sh. 10 d.; No. 4770. 3 sh.; No. 4771. 3 sh. 2 d.; No. 4772. 3 sh. 4 d.; No. 4773. 2 sh.; No. 4600. 2 sh. 6 d.; No. 4601. 2 sh. 6 d.; No. 4602. 2 sh. 6 d.

18 pair of ladies' boots; prices p. pair: No. 3805. 7 sh. 5 d.; No. 4122. 6 sh.; No. 4120. 6 sh. 6 d.; No. 3799. 6 sh.; No. 4121. 5 sh. 4 d.; No. 3499. 5 sh. 4 d.; No. 4123. 4 sh. 9 d.; No. 3500. 4 sh. 9 d.; No. 3276. 4 sh. 11 d.; No. 3607. 4 sh. 4 d; No. 4123. 4 sh. 9 d.; No. 4715. 3 sh. 5 d.; No. 4686. 1 sh. 6 d.; No. 4686. 1 sh. 6 d.; No. 4686. 1 sh. 6 d.; No. 4225. 2 sh. 5 d.; No. 4716. 2 sh. 2 d.; No. 4717. 2 sh. 6 d.

8 pair of girls' boots; prices p. pair: No. 3496. 4 sh. 4 d.; No. 3494. 4 sh. 9 d.; No. 3498. 4 sh. 4 d.; No. 3492. 4 sh. 4 d.; No. 3491. 4 sh.; No. 3493. 4 sh. 4 d.; No. 3495. 4 sh.; No. 3497. 4 sh.

2 pair of children's boots: No. 4764. 4765. 1 sh.
7 d. p. pair.

18 pair of men's boots; prices p. pair: No. 3453.
10 sh. 8 d.; No. 4763. 12 sh. 4 d.; No. 4710. 10 sh.
9 d.; No. 3334. 10 sh. 4 d.; No. 4655. 10 sh. 2 d.;
No. 6058. 9 sh. 4 d.; No. 6058½ 8 sh. 10 d.; No.
4709. 10 sh. 4 d.; No. 4654. 9 sh. 9 d.; No. 3897.
7 sh. 2 d.; No. 3464. 8 sh. 3 d.; No. 3885. 6 sh.;
No. 4573. 8 sh. 4 d.; No. 3902. 6 sh. 8 d.; No.
4762. 7 sh.; No. 4036. 3 sh. 6 d.; No. 3465. 12 sh.
6 d.; No. 3329. 9 sh. 4 d.

1 pair of slippers: No. 3941. 2 sh. 2 d.

2823. ZEILE, J. P., manu., Reutlingen. Agt.
the Württemberg Trading-Company at Stuttgart,
represented by Sam. Sessing, 10 Huggin-Lane,
Wood-street, Cheapside, E. C.

Hosiery; prices p. doz.; a. 16 caps of woollen yarn.
No. 1. 12 sh.; No. 2. 1 £ 4 sh; No. 3. 1 £ 3 sh.:
No. 4. 12 sh.; No. 5. 1 £ 2 sh. 4 d.; No. 6. 18
sh.; No. 7. 1 £ 4 sh.; No. 8. 1 £ 3 sh. 4 d.;
No. 9. 1 £ 14 sh.; No. 10. 1 £ 18 sh.; No. 11.
1 £ 10 sh.; No. 12. 1 £ 18 sh.; No. 13. 1 £
16 sh. 8 d.; No. 14. 1 £. 4 sh.; No. 15. 1 £
8 sh.; No. 16. 2 £.

b. 7 woollen fanchons: No. 1. 1 £ 8 sh.; No. 2.
1 £ 10 sh.; No. 3. 1 £ 6 sh. 8 d.; No. 4. 1 £
8 sh.; No. 5. 1 £ 8 sh.; No. 6. 1 £ 6 sh. 8 d.;
No. 7. 1 £ 6 sh. 8 d.

c. 5 pair of woollen sleves: No. 1. 1 £ 8 sh.:
No. 2. 1 £ 5 sh.; No. 3. 1 £ 2 sh.; No. 4. 1 £
6 sh. 8 d.; No. 5. 1 £ 1 sh.

d. 6 pair of shoes for children: No. 1. 11 sh.;
No. 2. 11 sh.; No. 3. 12 sh. 8 d.; No. 4. 12 sh.;
No. 5. 11 sh.; No. 6. 11 sh.

e. 3 bonnets for children of woollen yarn: No. 1.
1 £ 6 sh. 8 d.; No. 2. 1 £ 6 sh. 8 d.; No. 3.
1 £.

f. 2 hats for children: No. 1. 2 sh.; No. 2. 1 £
10 sh.

g. frocks for children: No. 1. 1 £ 4 sh.; jacket
No. 1. 1 £ 15 sh.; do. No. 1. 2 £ 12 sh.; glove
do. No. 1. 1 £; girl do. No. 1. 1 £ 4 sh.; harlequin do. No. 1. 1 £. 4 sh.

h. 30 caps of cotton-yarn: No. 1. 1 sh. 8 d.;
No. 2. 2 sh. 8 d.; No. 3. 4 sh.; No. 4. 2 sh. 8 d.;
No. 5. 3 sh. 4 d.; No. 6. 5 sh.; No. 7. 5 sh. 4 d.;
No. 8. 5 sh. 4 d.; No. 9. 4 sh. 8 d.; No. 10.
5 sh. 8 d.; No. 11. 5 sh.; No. 12. 4 sh. 4 d.;
No. 13. 5 sh. 8 d.; No. 14. 4 sh.; No. 15. 7 sh.
8 d.; No. 16. 9 sh. 4 d.; No. 17. 11 sh.; No. 18.
7 sh.; No. 19. 4 sh. 4 d.; No. 20. 5 sh. 8 d.;
No. 21. 4 sh. 8 d.; No. 22. 4 sh. 4 d.; No. 23.
1 £ 4 sh.; No. 24. 16 sh.; No. 25. 4 sh. 8 d.;
No. 26. 4 sh.; No. 27. 5 sh.; No. 28. 5 sh.; No. 29.
4 sh. 4 d.; No. 30. 5 sh.

i. 7 aprons for children of cotton yarn: No. 1.
11 sh. 4 d.; No. 2. 10 sh. 8 d.; No. 3. 7 sh. 4 d.;
No. 4. 7 sh. 4 d.; No. 5. 8 sh. 8 d.; No. 6. 8 sh.
8 d.; No. 7. 8 sh.

k. 3 frocks for children of cotton yarn: No. 1.
1 £.; No. 2. 1 £ 6 sh. 8 d.; No. 3. 1 £ 6 sh. 8 d.

l. 6 collars for ladies of cotton yarn: No. 1.
14 sh.; No. 2. 15 sh. 8 d.; No. 3. 14 sh.; No. 4.
12 sh.; No. 5. 15 sh.; No. 6. 9 sh. 4 d.

m. Head-dress, silk, No. 1. 2 £ 10 sh.

n. Gloves, silk, No. 1. 16 sh. 8 d.; do. No. 2.
10 sh. 8 d.

o. 6 nets, silk cord: No. 1. 1 £ 4 sh.; No. 2.
18 sh.; No. 3. 18 sh. 8 d.; No. 4. 1 £ 6 sh. 8 d.;
No. 5. 1 £ 10 sh.; No. 6. 2 £ 4 sh.

p. 8 under-sleeves, half silk: No. 1. 10 sh. 4 d.;
No. 2. 12 sh. 4 d.: No. 3. 10 sh. 4 d.; No. 4.
11 sh. 4 d.; No. 5. 11 sh. 8 d.: No. 6. 12 sh.
4 d.; No. 7. 11 sh.; No. 8. 10 sh.

CLASS XXVIII.

PAPER, STATIONERY, PRINTING AND
BOOKBINDING.

2824. ENSLIN & CLOSTERMEYER, manu. of
fancy leather-articles, Kirchheim u. Teck. London 1862 honor. mention. Agt. s. No. 2823.

Ladies' bags: No. 159. 1 £ 11 sh. 8 d.; No. 160.
3 £ 12 sh.; No. 162. 2 £ 18 sh. 4 d.; No. 163.
2 £ 10 sh.; No. 164. 2 £ 10 sh.; No. 165. 3 £.
13 sh. 4 d. Necessaires: No. 200. 4 £; No. 201.
3 £ 6 sh. 8 d. Inkstand No. 203. 5 £ 6 sh. 8 d.
No. 206. Travelling-necessaire 18 £. Necessaires:
No. 214. 3 £; No. 225. 3 £; No. 227. 7 £ 13 h.
4 d.; No. 229. 3 £ 15 sh.; No. 230. 3 £ 4 sh.
8 d.; No. 232. 2 £ 10 sh. No. 235. Papeterie
6 £. Necessaires: No. 236. 5 £ 13 sh. 4 d.; No.
238. 3 £; No. 239. 1 £ 13 sh. 4 d.; No. 240.
4 £; No. 242. 2 £ 10 sh.; No. 244. 1 £ 10 d.;
No. 246. 1 £ 10 sh.; No. 247. 1 £ 9 sh. 2 d.;
No. 252. 4 £ 13 sh. 4 d.; No. 253. 2 £ 16 sh.
8 d.; No. 254. 1 £ 15 sh.; No. 255. 1 £ 8 sh.
4 d.; No. 258. 3 £ 1 sh. 4 d.; No. 259. 2 £. 10 sh.;
No. 260. 4 £; No. 261. 5 £.; No. 262. 3 £ 5 sh.;
No. 263. 3 £ 1 sh. 8 d.; No. 264. 2 £ 1 sh. 8 d.;
No. 265. 2 £ 8 sh. 4 d.; No. 266. 2 £ 6 sh. 8 d.;
No. 267. 2 £ 16 sh. 8 d.; No. 268. 3 £ 3 sh. 4 d.;
No. 273. 2 £ 16 sh. 8 d.; No. 275. 2 £ 10 sh.;
No. 276. 1 £. 12 sh. 6 d.; No. 277. 2 £ 1 sh.
8 d.; No. 280. 2 £ 5 sh. 10 d.; No. 281. 1 £
16 sh. 8 d.; No. 284. 3 £; No. 285. 2 £ 15 sh.;
No. 288. 1 £ 6 sh. 8 d. Jewel-boxes: No. 290.
5 £; No. 291. 4 £. Necessaires No. 293. 2 £;
No. 294. 6 £.; No. 295. 2 £. Photograph-albums:
No. 450. 4 £ 3 sh. 4 d.; No. 451. 3 £ 6 sh. 8 d.;
No. 452. 3 £ 3 sh. 4 d. Pocket-books: No. 500.
2 £ 10 sh.; No. 500a. 1 £ 13 sh. 4 d.; No. 501.
1 £ 13 sh. 4 d.; No. 514. 1 £ 10 sh.; No. 527.
2 £; No. 535. 1 £ 10 sh. 10 d.; No. 536. with
necessaire 2 £ 6 sh. 8 d. Pocket-books: No. 540.
1 £ 8 sh. 4 d.; No. 544. 1 £ 5 sh.; No. 545. 1 £
13 sh. 4 d.; No. 548. 8 sh.; No. 550. 18 sh. 4 d.;
No. 550a. 1 £ 1 sh. 8 d.; No. 551. 17 sh. 6 d.;
No. 551a. 1 £; No. 552. 16 sh. 8 d.; No. 552a.
19 sh. 2 d.; No. 553. 1 £ 6 sh. 4 d. No. 556.
Visites 11 sh. 4 d. No. 557. Pocket-book with
necessaire 2 £ 10 sh.; pocket-books: No. 559. 1 £
6 sh. 8 d.; No. 564. 1 £ 7 sh. 6 d. No. 567. Note-
book 1 £ 5 sh. Pocket-books: No. 583. 1 £ 8 sh.
4 d.; No. 586. 4 £; No. 587. 1 £ 8 sh. 4 d.; No.
588. do. with port-cigar 1 £ 13 sh. 4 d. Pocket-
books: No. 589. 1 £ 10 sh. 10 d.; No. 590. 1 £;
No. 591. do. with necessaire 2 £ 1 sh. 8 d. Pocket-
books: No. 592. 1 £ 3 sh. 4 d.; No. 596. 1 £ 2 sh. 6 d.;
No. 597. 16 sh. 8 d.; No. 598. 1 £ 18 sh. 4 d.;
No. 599. 1 £ 6 sh. 8 d.: No. 1000. 1 £ 6 sh. 4 d.;
No. 1004. 8 sh. 4 d.; No. 1005 1. 10 sh. 10 d.;
No. 1005 3. 7 sh. 6 d.; No. 1005 4. 6 sh. 8 d.;
No. 1007. do. with necessaire 1 £ 6 sh. 8 d.; No.
1008. 15 sh. 8 d.; No. 1009. 13 sh. 4 d.; No. 1010.
1 £ 13 sh. 4 d.; No. 1011. do. with port-cigar
15 sh.; No. 1012. 1 £. 10 sh.: No. 1013. card-etui
15 sh. Porte-monnaies No. 1. 1 £. 9 sh. 2 d.;
No. 14. 1 £: No. 31. with port-folio 1 £ 6 sh.
8 d.; No. 55. 1 £ 5 sh.; No. 60. 1 £ 10 sh.;
No. 64. 2 £: No. 67. 1 £ 2 sh. 6 d.: No. 74. 1 £
16 sh. 8 d.; No. 75. 1 £ 5 sh.: No. 77. 1 £; No.
78. 12 sh. 6 d.; No. 81. 1 £ 6 sh. 8 d.; No. 82.
1 £ 2 sh. 6 d.; No. 608. 1 £ 3 sh. 4 d.; No. 624.
1 £ 13 sh. 4 d.: No. 625. 1 £ 3 sh. 4 d.; No. 626.
1 £; No. 631. 1 £ 16 sh. 8 d.; No. 633. 1 £;
No. 634. 1 £ 10 sh.; No. 635. 1 £ 13 sh. 4 d.;
No. 651. 16 sh. 8 d. Cigar-etuis: No. 704. 1 £
8 sh. 4 d.: No. 716. 2 £; No. 717. 2 £ 6 sh. 8 d.;

No. 719. 1 £ 15 sh.; No. 720. 1 £ 12 sh. 4 d.; No. 723. 1 £; No. 726. 1 £ 6 sh. 8 d.; No. 728. 2 £; No. 735. 2 £; No. 39. 1 £ 12 sh. 6 d.; No. 51. cigar-etui with book 2 £. 4 sh. 2 d. Tinder-boxes: No. 6. 12 sh. 6 d.; No. 11. 9 sh. 8 d.; No. 13. 11 sh. 8 d.; No. 16. 19 sh. 2 d.; No. 17. 10 sh. 8 d.; prices p. doz.

2825. Faber, G. F., manu. of black-lead pencils and creta polycolor, Crailsheim. Agt. the Württemberg Trading-Company at Stuttgart, represented by Sam. Sessing, 10 Huggin-Lane, Wood-street, Cheapside, E. C.

Samples of black-lead pencils and creta polycolor.

2826. Ade, Eduard, drawing and wood-engraving establishment, Stuttgart. London 1862 honor. mention. Agt. s. No. 2825.

a. Copies of different kinds of engravings; the prices of the drawings, and engravings are as follows: No. 2., 5., 8. drawing 3 £ 4 sh., engraving 8 £ 7 sh.; No. 1., 3., 7., 9. drawing 2 £ 5 sh. engraving 6 £ 5 sh.; No. 6. drawing 1 £ 17 sh., engraving 5 £ 10 sh.; No. 4. drawing 1 £ 10 sh., engraving 3 £ 7 sh.

b. Portraits: drawing 1 £ 14 sh., engraving 4 £ 12 sh.

2827. Müller & Richter, manu. of photograph-albums and fancy leather-goods, Stuttgart. London 1862 medal. Agt. s. No. 2825.

Photogragh-albums: No. 1. very large, p. piece 10 £; No. 3016. quart, p. piece 2 £ 10 sh. The following prices p. dozen: No. 3041. ⅙ doz. 4 £ 10 sh.; No. 3041. ⅙ doz. 3 £ 5 sh. 10 d.; No. 3041 K. ⅙ doz. 5 £ 10 sh.; No. 3041 a. ⅙ doz. 6 £ 10 sh.; No. 3008. ⅙ doz. (a., b., c., d.) 10 £ 13 sh. 4 d.; No. 3020. ⅙ doz. 10 £.; No. 3019. ⅙ doz. 13 £ 16 sh. 8 d.; No. 3005/50. ⅙ doz. 7 £; No. 3005/30. ⅙ doz. 6 £ 10 sh.

2828. Rometsch, Carl, manu. of patent metal-slates, Stuttgart. London 1851 prize-medal; Munich 1854 medal of honor; Paris 1855 bronze medal; London 1862 medal. Agt. Fred. Haas & Co.. 34 Jewin-street, E. C. (s. II. No. 41.)

Patent metal-slates: 4 pieces of slates for household purposes; school-slates: ruled for music; for geography; ruled for ciphering; ruled for writing; 2 with linen frame; white, for lead pencil; 2 black, for slate pencil; 2 white, for lead pencil, with ornamented frames; 1 ruled school-slate; 4 door-slates, with leather frames; 1 »Klapptafel« ruled; 1 sign-board; 1 »Klapptafel« in leather; 2 school-slates; 1 wood-board for schools; 8 slates for offices; 8 note-books; 4 school-portfolios for girls; 4 slates with writing-patterns.

2829. Schäuffelen, Gustav, paper-manu., Heilbronn. Berlin 1844 silver medal; London 1851 prize-medal; Munich 1854 great medal; Paris 1855 silver medal; London 1862 medal. Agt. s. No. 2825.

Fine papers, such as: letter-paper, white and colored, carton and satin-paper, white and colored, telegraph-paper etc.

2830. Schwenk, Christian, Ludwigsburg. Agt. s. No. 2825.

Patent pergamentan: No. 1. ¼ sheet patent pergamentan-paper, p. sq. ft. 5 kr.; No. 2. ¼ do. carton-white 14 kr.; No. 3. ¼ do. red 14 kr.; No. 4. ¼ do. pale-yellow 14 kr.; No. 5. ¼ do. dark-yellow 14 kr.; No. 6. ¼ do. tracing-paper 14 kr.

No. 7. 1 map of Europe with pergamentan cover 10 kr.; No. 8. 1 map mounted on canvas, part of the Rhenish countries 28 kr.; No. 9. 1 map mounted on canvas 1 fl. 24 kr.; No. 10. 1 wood-cut, pergamented 1 fl. 24 kr.; No. 11. 1 map mounted on canvas, part of Darmstadt 2 fl. 20 kr. Some specimens of a map, pergamented.

2831. Völters, Heinrich, Sons, paper-manu., Heidenheim a. B. and Gerschweiler. Munich 1854 great medal; Paris 1855 bronze medal; London 1862 medal. Agt. s. No. 2825. (s. II. No. 40.)

Divers assortments of paper from wood, the wood-pulp being used and prepared according to the process invented by Heinrich Voelter; prices p. lb. No. 1. to 3. printing-papers (No. 1. »Kölnische Zeitung« etc.) 4⅘ to 5⅗ d., containing from 20 to 40 pCt. wood. The jury of the Munich exhibition 1854, in possession of official proofs as to the actual employment, several years since, of paper of »wood-pulp« (for news-papers etc. etc.) spoke in the highest terms of approbation of the invention. Compare report-book 11 page 21—25. No. 4. to 6. writing-papers, containing from 20 to 30 pCt. wood, 4⅗ to 6 d.; No. 7. to 9. wrapping-paper (English brown, glazed, skip etc.) containing from 30 to 50 pCt. wood, 5⅕ to 6⅕ d.; No. 10. to 14. papers for paper-hangings, containing from 30 to 80 pCt. wood, 4⅕ to 4⅕ d.; No. 15. to 22. bill-papers (also adapted for covers and cigar-boxes) in different colours, containing 40 pCt. pine-wood, 5⅗ d. NB. All these divers species from No. 1. to 22. contain the ordinary admixture of china clay. No. 23 to 32. tissue-papers for wrapping fruits, bottles etc., white and colored, containing from 40 to 50 pCt. of pine-wood 5⅗ to 8 d.

CLASS XXIX.

EDUCATIONAL WORKS AND APPLIANCES.

2832. Benz, August, teacher of drawing at the High-schools, Ellwangen. Agt. s. No. 2825.

Guide for drawing linear ornaments: No. 1. Greek ornaments: a. Maeander, b. »Knotenverzierungen«; No. 2. Mauresque traceries; No. 3. patterns for embroidery, twist and crotchet-work; No. 4. patterns for inlaid floors and wainscot; No. 5. continuation; No. 6. pattern, having the rhomboid for basis, Mauresque and Greek traceries formed by equal tracings; No. 7. appendix.

2833. Blumhardt, H., & Co., manu. of varnished tin-toys, Stuttgart. London 1851 honor. mention; Munich 1854 medal of honor; London 1862 honor. mention. Agts. Mittler & Eckhardt, 6 Grocers Hall Court, Poultry.

Tin toys; one sample of each; prices p. doz.: No. 4. basket 1 fl. 45 kr.; No. 63½. knife and fork-basket 6 fl. 12 kr.; No. 72. flower-vase 1 fl. 18 kr.; No. 73. do. 2 fl. 12 kr.; trays: No. 89. 54 kr.; No. 90. 1 fl. 18 kr.; No. 91. 1 fl. 33 kr.; No. 92. 2 fl.; No. 93. 2 fl. 39 kr.; No. 104. coffee-pot 3 fl. 45 kr.; No. 111. coffee-mill 4 fl. 24 kr.; No. 136. pen-case 2 fl.; No. 146. bottle-basket 2 fl. 6 kr.; No. 147. glass-basket 2 fl. 6 kr.; No. 148. do. 3 fl. 6 kr.; No. 150. vinegar and oilstand 2 fl. 45 kr.; No. 152. do 3 fl. 6 kr.; No. 159. watering-pot 1 fl. 45 kr.; No. 161. do. 3 fl. 6 kr.; No. 174. hand washing-tub 3 fl. 45 kr.; No. 181. bird cage with bird 1 fl. 54 kr.; No. 183. do. with parrot 3 fl. 45 kr.; No. 184.

do. with fish-glass 4 fl. 24. kr.; No. 186. broom 40 kr.; No. 188. do. of wood 1 fl. 18 kr.; No. 192. do. of bone 1 fl. 18 kr.; No. 231. moderateur-lamp 2 fl. 51 kr.; No. 232. do. for oil 5 fl. 6 kr.; No. 235. lantern 3 fl. 6 kr.; No. 240. washing-basin 1 fl. 18 kr.; No. 240½. do. 1 fl. 45 kr.; No. 308. menagère 4 fl. 51 kr.; No. 321. fruit-basket 1 fl. 6 kr.; No. 322. do. 1 fl. 39 kr.; No. 323. do. 2 fl. 12 kr.; No. 324. do. 2 fl. 45 kr.; No. 362. salt-box 2 fl.; No. 368. saving box, round 3 fl. 18 kr.; No. 373. do. 3 fl. 45 kr.; No. 392. duster 54 kr.; No. 393. do. 1 fl. 18 kr.; No. 394. do. 1 fl. 45 kr.; No. 458. clock without glass 1 fl. 45 kr.; No. 459. do. 2 fl.; No. 461. clock with glass 5 fl. 18 kr.; No. 466. balance 4 fl.; No. 495. water-pail 54 kr.; No. 497. do. 1 fl. 18 kr.; No. 550/1. parasol 5 fl. 30 kr.; No. 550/2. do. 4 fl. 24 kr.; No. 550/p. flower-stand 7 fl. 42 kr.; No. 634. work-table, empty 3 fl. 6 kr.; No. 635. work-table, filled 4 fl. 24 kr.; No. 637. bedstead 3 fl. 45 kr.; No. 638. do. 4 fl. 24 kr.; No. 639. do. 5 fl. 18 kr.; No. 640. do. 6 fl. 24 kr.; No. 652. cradles, empty 2 fl. 52 kr.; No. 653. do. 4 fl. 24 kr.; No. 658. flower-table, 3 pots 3 fl. 51 kr.; No. 663. do. with fish-glass 7 fl. 42 kr.; No. 666. side-table 4 fl. 12 kr.; No. 667½. do. 6 fl. 36 kr.; No. 669. grand pianoforte 21 fl. 6 kr.; No. 671. pianoforte-stool 2 fl. 6 kr.; No. 674. étagère 4 fl. 52 kr.; No. 679. rocking-chair 5 fl. 3 kr.; No. 680. children's table 2 fl. 39 kr.; No. 681. do. 3 fl. 45 kr.; No. 682. child' stool 2 fl. 24 kr.; No. 685. cloth-chest 10 fl. 36 kr.; chests of drawers: No. 687. 4 fl.; No. 688. 4 fl. 51 kr.; No. 689. writing-desk 10 fl. 36 kr.; No. 694. night-stand 2 fl. 52 kr.; No. 695. do. 3 fl. 18 kr.; No. 696. stove, round 3 fl. 6 kr.; No. 698. do. 5 fl. 18 kr.; No. 700. stove, square 8 fl. 12 kr.; No. 701. stove-utensils 5 fl. 3 kr.; No. 719. looking-glass 4 fl. 36 kr.; No. 721. couch 4 fl. 39 kr.; No. 723. writing-desk, filled 6 fl. 24 kr.; No. 725. do. 14 fl. 51 kr.; No. 736½. square table 3 fl. 18 kr.; No. 737. washing-stand, filled 4 fl. 24 kr.; No. 738. do. 8 fl. 48 kr.; No. 739. linen-chest 11 fl. 54 kr.; No. 742. washing-stand 6 fl. 48 kr.; No. 744. washing-stand, filled 10 fl. 12 kr.; No. 744b. book-shelf, empty 11 fl. 54 kr.; No. 744f. side-board 10 fl. 48 kr.; furniture, prices p. box: No. 752. 1 fl. 27 kr.; No. 753. 1 fl. 39 kr.; No. 754. 2 fl. 33 kr.; No. 755. 3 fl.; No. 757½. 5 fl. 12 kr.; No. 758. 5 fl. 6 kr.; No. 759. 1 fl. 33 kr. The following prices p. doz.: fountain No. 794. 8 fl. 48 kr.; No. 802. steamboat 19 fl. 24 kr.; No. 805. boat with sails 9 fl. 54 kr.; No. 815. gun 4 fl. 24 kr.; No. 819. doll-cart 5 fl. 42 kr.; No. 820. do. 7 fl. 42 kr.; No. 828. swing 5 fl. 30 kr.; No. 829. do. 5 fl. 6 kr.; No. 837a. doll-cart 9 fl. 54 kr.; No. 859. christmass-tree 1 fl. 45 kr.; No. 860. dressing-box 7 fl. 3 kr.; No. 861. glass with cover 1 fl. 39 kr.; No. 862. fruit-stand 3 fl. 18 kr.; No. 863. fish-glass on stand 3 fl. 18 kr.; No. 864. candlestick, gold 1 fl. 45 kr.; No. 865. do. 3 fl. 18 kr.; No. 866. chandelier 6 fl. 24 kr.

2834. THE ROYAL COMMISSION FOR THE INDUSTRIAL SCHOOLS, Stuttgart. London 1862 medal.

Plaster-of-Paris-models and copies for the instruction in drawing.

2836. ENGLER & LUTZ, manu. of japanned tin-toys, Ellwangen. London 1862 honor. mention. Agt. Jos. W. Green (E. G. Zimmermann), 2 St. Paul's Buildings, Little Carter-Lane, Doctors Commons, E. C.

Japanned tin-toys, one sample of each, prices p. piece: No. 23. milk-waggon 1 fl. 24 kr.; No. 26.

flower do. 1 fl. 36 kr.; No. 27. fruit do. 1 fl. 36 kr.; No. 32. brougham and pair of horses 2 fl.: No. 33. park-phaeton 2 fl. 12 kr.; No. 41. small Paris omnibus 3 fl. 36 kr.; No. 42. large do. 6 fl. 24 kr.; No. 43. small English children's cart 54 kr.; No. 114. railway-train No. 5. 8 fl.; No. 131. boat 38 kr.; No. 134. sloop No. 1. 55 kr..; No. 138. sloop with 2 sails No. 2. 1 fl. 50 kr.; No. 142. steamboat No. 1. 1 fl. 15 kr.; No. 155. steamboat with screw-propeller 4 fl. 30 kr.; No. 152. steam-boat No. 8. with propelling-apparatus 12 fl.; No. 204. cottage with pond 1 fl. 45 kr.; No. 206a. pigeon-house and poultry 2 fl. 24 kr.; No. 219. castle 8 fl.; No. 220. railway-station with swan-pond and moving railway-train 9 fl.; No. 221. railway-tunnel with moving railway-train 14 fl.; No. 223. Stirian landscape with forge 45 fl.; No. 242. fortress with detached works 5 fl.; No. 258. carrousel 1 fl. 45 kr.; No. 261. sliding-course 2 fl. 36 kr.; No. 264. bombardment-game 4 fl.; No. 306. night-stand No. 2. 14 kr.; No. 307. side-board No. 1. 15 kr.; No. 316. square table 16 kr.; No. 320. cradle No. 2. 20 kr.; No. 322. bed-stead No. 1. 18 kr.; No. 324. do. No. 3. 26 kr.; No. 332. swathing table 31 kr.; No. 334. flower-stand 32 kr.; No. 335. writing-table 40 kr.; No. 338. linnen-drawer No. 1. 48 kr.; No. 343. organ with action 6 fl. 24 kr.; No. 372. set of furniture varnished 1 fl. 36 kr.; No. 373. do. à medaillon 2 fl. 36 kr.; No. 379. do. antique furniture, best quality 4 fl.; No. 381. do. furniture, stuffed 3 fl.; No. 436. fountain 1 fl. 4 kr.; No. 437. do. 1 fl. 12 kr.

2837. GROSS, CARL, toymaker, Stuttgart. Munich 1854 medal of honor; Paris 1855 bronze medal; London 1862 medal. Agts. Mittler & Eckhardt, 6 Grocer's Hall-Court, Poultry.
(s. No. 2843.)

Assortment of toys, one sample of each; prices p. doz. if nothing else is remarked: building boxes No. 4 aa. 4 fl. 48 kr.; No. 4b. 16 fl.; No. 4g. 20 fl. 48 kr.; No. 4h. 6 fl. 48 kr.; draught-board No. 7l. 32 fl.; humming-tops No. 8e. 2 fl.; gun, brass-canon No. 9a. 4 fl. 30 kr.; gun, wood-canon No. 9½ 3/a. 8 fl.; game of devil on 2 sticks No. 12a. 7 fl. 30 kr.; cart, horse and barrel No. 18a. 6 fl. 30 kr.; do. 5 barrels No. 18e. 20 fl.; Engl. beer carthorse No. 18½ c. 23 fl.; No. 18½ f. 28 fl. 30 kr.; cart-horse and case No. 20. 3/a. 5 fl. 36 kr.; cart, 2 horses and case No. 20 aw. 9 fl. 48 kr.; small cart, assorted No. 23 o. 2 fl. 36 kr.; doll-coach with handle No. 30 bb. 18 fl. 24 kr.; cart with bricks, 1 horse No. 33½ a. 5 fl. 12 kr.; Engl. dust-cart, 1 horse No. 35 a. 34 fl.; No. 35 c. 29 fl.; No. 35 e. 25 fl.; railway cart, horse No. 36 c. 25 fl. 24 kr.; games of skittles, polished: No. 42½ a. 4 fl.; No. 42½ e. 9 fl.; basket-waggon with handle No. 46 a. 5 fl.; game of German balls, varnished No. 48 o. 6 fl. 12 kr.; games of German balls, polished: No. 48 gg. 13 fl.; No. 48 k. 28 fl.; No. 48 l. 33 fl. 36 kr.; game of German balls, striped No. 48 n. 19 fl. 12 kr.: cabriolet-horse No. 49 3/a. 11 fl. 30 kr.; do. No. 49 a. 24 fl.: 2 ball-heads No. 51 a. 16 fl.; rattling hoops: No. 56 d. 3 fl. 40 kr.; No. 56 g. 9 fl. 30 kr.; rattling hoop with handle No. 56 ee. 8 fl. 24 kr.; chess-boards: No. 58 k. 10 fl. 48 kr.; No. 58 u. 18 fl.; 1 long skittle-game No. 59 a. 15 fl. 48 kr.: 1 swing with leather-cloth No. 60 b. 29 fl.: targets: No. 62 s. 9 fl.; No. 62 n. 7 fl. 12 kr.; popingjay No. 62 ff. 33 fl.; turning figure, varnished No. 64 a. 3 fl. 12 kr.: turning figure, white No. 64 b. 2 fl. 12 kr.; turning figure, polished No. 64 d. 4 fl.: do. No. 64 e. 5 fl. 24 kr.: tivoli-board No.

69 b. 28 fl. 45 kr.; bird-waggon with stick No. 71 b.
2 fl. 36 kr.; set of washing-tubs No. 73. 4 fl. 24
kr.: wind-wheels: No. 78 d. 1 fl. 42 kr.; No. 78 ff.
3 fl. 15 kr.; cup and ball, small No. 82 a. 1 fl.
24 kr.; do. large No. 82 b. 3 fl. 48 kr.; do. middle
No. 82 c. 2 fl. 12 kr.; set of 3 trunks No. 84. 50 fl.
24 kr.; shaking figures No. 89 c. 2 fl. 24 kr.; No.
89 d. 2 fl. 6 kr.; No. 89 g. 4 fl. 48 kr.; tower-game
No. 90 b. 16 fl.; wood-egg, striped No. 97 a. 32 kr.;
do. triple No. 97 e. 1 fl. 36 kr.; do. with hares No.
97 f. 36 kr.; do. colored No. 97 g. 1 fl. 12 kr.;
game of patience No. 99 a. 8 fl. 15 kr.; omnibus 2 horses No. 104 b. 48 fl.; whistling figure,
white No. 106 a. 1 fl. 36 kr.; do. polished No.
106 b. 3 fl. 12 kr.; do. ladies, No. 106 c. 4 fl. 30
kr.; steeple-chase board No. 115 b. 42 fl.; do. varnished No. 115 c. 22 fl.; French dancer, striped
No. 117 a. 1 fl. 54 kr.; do. varnished No. 117 c.
1 fl. 24 kr.; chess-board with figures No. 118 a.
28 fl.; No. 118 b. 21 fl.; collodion gas-balloon
No. 119 a. 7 fl. 30 kr.; set of 6 straw-chairs No.
121. 2 fl. 30 kr.; target with bell and puff and
darts No. 123 a. 22 fl.; spillekin No. 124 g. 9 fl.
30 kr.; roulettes: No. 126 e. 16 fl. 30 kr.; No. 126 i.
38 fl.; roulette-board No. 127,a. 9 fl.; percussion-
guns: No. 128 a. 9 fl.; No. 128 c. 14 fl.; percussion-
gun with bajonet No. 128 d. 21 fl.; do. large No.
128 e. 42 fl.; the building globe No. 129 b. 60 fl.;
alphabetical game No. 132 e. 36 fl.; coffee-mill No.
136. 3 fl.; grocer's shop No. 148 cc. 27 fl.; castle-
bombardment No. 156 b. 28 fl.; do. No. 156 f. 12 fl.;
bombardment with balls No. 158 b. 9 fl. 30 kr.;
do. No. 158 c. 32 fl.; rattles, polished: No. 167 a. 3 fl.
36 kr.; No. 167 c. 2 fl. 24 kr.; No. 167 d. 1 fl. 48
kr.; No. 167 e. 2 fl. 30 kr.; tourniquet-game No.
168 a. 23 fl.; hen-house No. 173 b. 35 fl. 15 kr.;
puff and dart-games: No. 175 b. 12 fl. 36 kr.; No.
175 dd. 3 fl. 48 kr.; No. 175 ff. 4 fl. 12 kr.; boxes
of gymnastic games: No. 177 bb. 28 fl. 30 kr.; No.
177 d. 5 fl.; Japanese tivoli No. 180 b. 29 fl.; lawn-
billard No. 185 b. 48 fl.; game of croquet No. 186.
138 fl.; cube-games: No. 189 a. 6 fl.; No. 189 d.
21 fl.; No. 189 g. 63 fl.; No. 189 h. 26 fl.; No.
189 i. 32 fl.; No. 189 m. 16 fl. 30 kr.; No. 189 n.
43 fl.; No. 189 p. 13 fl. 36 kr.; No. 189 s. 27 fl.;
percussion-pistol No. 191a. 8 fl. 30 kr.; roulette-tivoli
No. 194 c. 19 fl.; railway No. 194 f. 42 fl.; waterfall
No. 194 g. 38 fl. 30 kr.; castle No. 194 h. 34 fl.;
nine-pins No. 194 k. 24 fl.; Mexican No. 194 l. 18 fl.;
Russian No. 194 m. 28 fl.; windmill No. 194 n. 28 fl.;
tree with monkey No. 194 o. 20 fl.; skittle-game with
rings No. 195. 24 fl.; skittle-roulette No. 197. 22 fl.
30 kr.; knitted clown No. 198 a. 12 fl. 30 kr.;
do. bird No. 198 c. 12 fl. 30 kr.; figure-waggons
No. 199 a. 2 fl. 36 kr.; do. with handle No. 199 c.
3 fl.; No. 199 f. 11 fl. 15 kr.; with handle No. 199 h.
10 fl. 30 kr.; parachute-top No. 402. 6 fl. 30 kr.;
monkey-dancer No. 403. 5 fl.; 2 games of figures
No. 414. 8 fl. 30 kr.; stairs' game No. 421. 20 fl.;
humming-top with figure No. 422. 4 fl. 36 kr.;
Tell's boy, ball-figure No. 424 b. 19 fl. 12 kr.;
2 ball-figures No. 426. 8 fl. 48 kr.; roulette with
pins No. 427. 8 fl. 30 kr.; steeple-chase game No.
429. 11 fl.; mouse-trap No. 430 a. 5 fl.; do. colored No. 430 b. 8 fl.; dancing-saloon with bell No.
431 c. 15 fl.; do. with music No. 431 d. 45 fl.; target-
board No. 432. 15 fl.; Roman ball-game No. 433 a.
18 fl. 30 kr.; triple dancers: No. 434 a. 2 fl. 36 kr.;
No. 434 b. 3 fl. 12 kr.; rifleman, turning figure
No. 436. 13 fl.; new bullet percussion-gun No.
438 28 fl.; harmonica-popguns No. 439. 4 fl. 12 kr.;
harmonica humming-top No. 440 a. 3 fl. 12 kr.;
Chinese ball-game No. 441. 7 fl. 36 kr.; stable with
2 horses No. 442 b. 53 fl.; do. 4 horses No. 442 e.
75 fl.; colored dancer No. 443. 1 fl. 42 kr.; the

decimal puzzle No. 446. 9 fl.; screw-dancer No.
448. 5 fl. 30 kr.; kettle-holder No. 217. 1 fl. 15 kr.;
table-mats, square: No. 221 oo. 3 fl. 48 kr.; No.
221 a. 6 fl. 36 kr.; No. 221 c. 11 fl. 24 kr; No. 221 e.
21 fl. 36 kr.; No. 221 g. 29 fl. 24 kr.; table-mats,
round: 235 o. 1 fl. 6 kr.; No. 235 aa. 2 fl. 12 kr.;
No. 235 bb. 4 fl. 48 kr.; No. 235 d. 9 fl.; coat-
holders: No. 237 a. 13 fl.; No. 237 g. 9 fl. 36 kr.;
table-mats, oval: No. 248 a. 3 fl.; No. 248 c. 4 fl.
12 kr.; No. 248 e. 5 fl. 24 kr.; No. 248 g. 9 fl.

2839. MALTÉ, FRANZ, artistical establishment, Stuttgart. London 1862 honor. mention.
Agt. the Württemberg Trading-Company at Stuttgart, represented by Sam. Sessing, 10 Huggin-
Lane Wood-street, Cheapside, E. C.

Published works: No. 1. »Geologische Karte von
Central-Europa von H. Bach«, printed in colors
(Verlag von E. Schweizerbarth in Stuttgart). No 2.
Mappa topographica da Cidade de S. Salvador e seus
suburbios levantada por C. A. Weyll e publ. por Ferd.
Glocker, Escala: 1/15,000. No. 3. Carte du Can-
ton de Neufchâtel d'après M. d'Osterwald par A.
de Mandrot, Echelle: 1/50,000. No. 4. Erdglobus
von Imle und Baur, revidirt von Heinr. Berghaus.
Farbendruck. Durchmesser 6″. With semi-meri-
dian of brass on a wooden foot. No. 5. Neuer
Schulatlas über alle Theile der Erde, bearbeitet
von Bach, Baur, Gross und Imle; revidirt von
Heinrich Berghaus; 27 plates printed in 3 colors.

2840. NITZSCHKE, WILHELM, bookseller,
Stuttgart.

Books and Atlasses: Mo. 1. W e i s s e r, Bilderatlas
zur Weltgeschichte. Nach Kunstwerken alter und
neuer Zeit. Mit erläuterndem Text von Dr. Heinr.
Merz und Hermann Kurz, I. vol. I. division,
History of the ancient times, 50 plates 17 fl. 36 kr.;
No. 2. I. vol. II. division. Representations of life
from classical antiquity, gods and heroes, 43 plates
16 fl. 33 kr.; No. 3. F r. B r a u n's celestial-atlas in
transparent maps, 30 asterisms and 1 large map
of the heavens 10 fl.; No. 4. W e n d e, Dr. Edmund,
Atlas der Naturgeschichte der drei Reiche, 52
plates taken from nature, finely colored, 8 fl. 24 kr.

2841. NÖRDLINGER, DR. H., professor of
the science of forest-cultivation at the Royal
academy Hohenheim. London 1851 prize-medal,
1862 medal. Agt. s. No. 2839.

1. Collection of 100 different sorts of woods with
their bark, which can be arrayed in form of a
library, containing all the European forest and
bosket-woods, which are of importance in a technical point of view, with their correct botanical
names, intended to show the distinguishing marks
of these woods as regards hardness, weight, color.etc.

2. Very thin cross-cuts of 100 different woods,
vol. I. containing woods of the forest and garden-
trees, and of the most common exotic bosket
trees of Germany, France and England. Published
by J. G. Cotta, Stuttgart 1852.

3. Vol. II. containing another hundred of cross-cuts
of European and exotic woods. Published by J.
G. Cotta, Stuttgart 1855.

4. Vol. III. An additional 100 specimens of European and exotic woods. Published by J. G. Cotta 1861.

5. 50 cross-cuts of the principal woods used as
timber and fuel, which are growing in Germany.
For men engaged in forest and technical concerns and
in wood-work. 1858. School-edition.

6. Collection of 60 cross-cuts of the principal
forest-woods, for the use of the students of the imperial academy of Nancy. 1855. Grosjean, printer

and bookseller. School-edition. Each collection forms a book of the size of a small dictionary, and is furnished with exact anatomical descriptions, in order that the different general and mostly even the species of a genus may be distinguished by the naked eye or by the aid of a magnifying glass.

2842. ROCK & GRANER, merchants and manu., Biberach. London 1851 prize-medal; Munich 1854 medal of honor; Paris 1855 silver medal; London 1862 medal. Agt. Charles Mittler (Mittler & Eckhardt), 6 Grocer's Hall Court, Poultry.

Assortment of varnished tin-toys; one sample of each; prices pr. doz.: No. 46. round pump No. 1. 6 fl. 12 kr.; No. 116. steamboat No. 3. 36 fl.; No. 132. fire-engine No. 2. 26 fl; No. 157. watering-pot, painted No. 2. 4 fl. 12 kr.; No. 181. gun No. 3. 5 fl 36 kr.; No. 952. railway-train No. 1. 25 fl. 12 kr.; No. 960. oval pump No. 0. 3 fl. 24 kr.; No. 1020. railway-train No. 1⅖. 42 fl.; No. 1030. do. No. 3. 74 fl. 24 kr.; No. 1174. sutler's barrel No. 1. 11 fl.; No. 1283. railway-train No. 0. 15 fl.; No. 1290. monkey-waggon 12 fl.; No. 1291. cab No. 1. 12 fl. 24 kr.; No. 1292. cab No. 2. 23 fl.; No. 1298. hunting-tilbury No. 1. 14 fl.; No. 1300. do. No. 3. 48 fl. 36 kr.; No. 1302. hunting-gig No. 3. 24 fl.; No. 1304. carriage No. 3. 23 fl.; No. 1343. garden-furniture 15 fl.; No. 1352. Chinese cabriolet 6 fl. 36 kr.; No. 1353. Berlin fire-engine No. 1. 29 fl.; No. 1384. steamboat with clock-work No. 6. 96 fl.; No. 1389. small boat No. 1. 4 fl. 24 kr.; No. 1391. do. No. 3. 5 fl.; No. 1392. do. No. 4. 7 fl. 48 kr.; No. 1394. sloop No. 1. 12 fl.; No. 1397. Thames boat with clockwork 24 fl.; No. 1399. steamboat do. No. 4. 49 fl.; No. 1400. London omnibus No. 1. 34 fl.; No. 1403. garden-phaeton No. 1. 33 fl.; No. 1408. corn-mill 25 fl.; No. 1410. Venetian boat No. 1. 8 fl. 12 kr.; No. 1418. sailing boat 18 fl. 36 kr.; No. 1419. lugger No. 3. 29 fl.; No. 1429. cab with clock-work No. 1. 41 fl.; No. 1431. cabriolet do. No. 4/0. 37 fl.; No. 1433. hunting-cabriolet with clock-work No. 1. 37 fl.; No. 1434. hunting-gig do. No. 1. 41 fl.; No. 1435. Parisian carriage do. No. 1. 55 fl.; No. 1458. cabriolet No. 7/0. 6 fl.; No. 1460. do. No. 5/0. 7 fl.; No. 1462. do. No. 3/0. 7 fl. 36 kr.; No. 1468. bathing-room No. 1. 21 fl.; No. 1471. doll-room No. 2. 36 fl.; No. 1481. perambulator No. 2. 42 fl.; No. 1482. do. No. 3. 50 fl.; No. 1483. steamboat No. 0. 12 fl; No. 1485. Indian boat 21 fl.; No. 1494. cottage with fountain 36 fl.; No. 1497. lugger with clock-work 58 fl.; No. 1501. perambulator No. 0. 11 fl.; No. 1506. carrousel with tunnel 68 fl.; No. 1507. do. with steeple chase 38 fl.; No. 1514. perambulator No. 2/0. 7 fl.; No. 1517. covered, phaeton No. 3. 112 fl.; No. 1521. water pails in sets No. 1. 6 fl.; No. 1522. railway-train No. 2/0. 14 fl. 24 kr.; No. 1525. Thames boat, with clock-work No. 2. 72 fl.; No. 1536. magic bottle No. 1. 9 fl. 36 kr.; No. 1541. carnival carriage No. 1. 10 fl.; No. 1550. hoop with clown, handle No. 1. 9 fl. 24 kr.; No. 1551. do. No. 2. 14 fl.; No. 1556. Italian cottage with fountain 51 fl.; No. 1559. fir-tree No. 2. 50 fl.; No. 1560. swan- and pigeonhouse with 2 fountains 27 fl.; No. 1564. pigeonhouse No. 2. do. 56 fl.; No. 1570. gardenhouse with fountain 54 fl.; No. 1580. swing No. 2. 16 fl.; No. 1582. sledge with handle No. 3. 54 fl.; No. 1586. sutler's barrel No. 2. 22 fl.; No. 1593. game at nine-pins No. 3. 42 fl.; No. 1594. hoop with Franconi 24 fl.; No. 1596. Swiss mill 42 fl.; No. 1598. railway-train No. 1⅕. 33 fl. 36 kr.; No. 1600. water-cart No. 1. 12 fl.; No. 1602. sand-cart No. 1. 13 fl.;

No. 1606. railway-train with rails 31 fl. 12 kr.; No. 1607. garden-phaeton No. 3. 72 fl.; No. 1609. children's carriage with clock-work 41 fl.; No. 1611. steamboat with sails and clock-work 96 fl.; No. 1612. Leipzig omnibus No. 1. 44 fl.; No. 1616. Hamburg do. No. 2. 66 fl.; No. 1617. Berlin do. No. 1. 46 fl.; No. 1620. castle Babelsberg No. 2. 180 fl.; No. 1625. gun boat No. 1. 72 fl.; No. 1626. do. No. 2. 102 fl.; No. 1627. screw gun-boat 102 fl.; No. 1628. screw-steamer No. 1. 100 fl.; No. 1630. glass-carriage with handle No. 1. 92 fl.; No. 1631. do. No. 2. 112 fl.; No. 1633. Spanish bark No. 2. 30 fl.; No. 1634. saving-box, gun boat 13 fl.; No. 1635. do. locomotive engine 12 fl.; No. 1636. carriage, 1 horse 76 fl.; No. 1639. saving-box, pail 3 fl.; No. 1640. Queen's phaeton, 2 horses No. 1. 84 fl.; No. 1641. do. with handle No. 2. 93 fl.; No. 1642. Berlin omnibus 180 fl.; No. 1643. cab, 1 horse No. 3. 96 fl; No. 1644. London omnibus No. 2. 66 fl.; No. 1645. Munich pump No. 1. 17 fl.; No. 1649. fountain No. 5. 45 fl.; No. 1651. roulette No. 1. 14 fl.; No. 1652. do. No. 2. 18 fl.; No. 1653. screw carrousel 18 fl.; No. 1654. carrousel No. 2/0. 5 fl.; No. 1655. screw dancer No. 1. 8 fl.

CLASS XXX.

FURNITURE AND UPHOLSTERY, INCLUDING PAPER-HANGINGS AND PAPIER-MACHÉ.

2843. GROSS, CARL, toymaker, Stuttgart. (s. No. 2837.)

Fancy wooden articles; prices p. doz.: 5 square table-mats: No. 221. 00. 3 fl. 48 kr.; No. 221. a. 6 fl. 36 kr.; No. 221. c. 11 fl. 24 kr.; No. 221. e. 21 fl. 36 kr.; No. 221. g. 29 fl. 24 kr. 4 round table-mats: No. 235. 0. 1 fl. 6 kr.; No. 235. aa. 2 fl. 12 kr.; No. 235. bb. 4 fl. 48 kr.; No. 235. d. 9 fl. 2 coat holders: No. 237. a. 13 fl.; No. 237. g. 9 fl. 36 kr.; 4 oval table-mats: No. 248. a. 3 fl.; No. 248. c. 4 fl. 12 kr.; No. 248. e. 5 fl. 24 kr.; No. 248. g. 9 fl.

2844. KIENLE, E., manu. of furniture, Stuttgart. London 1862 honor. mention. Agt. the Württemberg Trading-Company at Stuttgart, represented by Sam. Sessing, 10 Huggin Lane, Wood-street, Cheapside, E. C.

Furniture: No. 1. side-table 25 £; No. 2. sofa-frame of walnut 1 £ 13 sh. 4 d., p. doz. 18 £; No. 3. arm-chair-frame of walnut 1 £ 3 sh. 4 d., p. doz. 13 £; No. 4. chair-frame of walnut 10 sh., 6 doz. 28 £.

Frames with carvings: No. 5. sofa-frame of walnut 3 £ 6 sh. 8 d., p. doz. 36 £; No. 6. arm-chair-frame of walnut 1 £ 15 sh., pr. doz. 19 £; No. 7. chair-frame of walnut 18 sh. 4 d., 6 doz. 60 £; No. 8. chair-frame of walnut 15 sh., 6 doz. 48 £.

2845. VETTER, CARL, manu. of cornices and frames, gilt and varnished, Stuttgart. Munich 1854 medal of honor; Paris 1855 bronze medal; London 1862 honor. mention. Agt. s. No. 2844.

Tableau with samples of gilt cornices and frames.

2846. WAIDELICH, K., manu. of gilt cornices, Ulm. Agt. s. No. 2844.

Samples of gilt cornices of different kinds, prices p. 100 ft. Prus.: No. 1. profil No. 0. 3 fl.: No. 2. do. No. 0b. 3 fl. 12 kr.; No. 3. do. No. aa. 3 fl. 30 kr.: No. 4. do. No. 1. 4 fl. 30 kr.: No. 5. do.

No. 1 a. 5 fl. 30 kr.; No. 6. do. No. 1 c. 5 fl. 36 kr.; No. 7. do. No. 1 c. 5 fl. 36 kr.; No. 8. do. No. 2. 5 fl. 30 kr.; No. 9. do. No. 2 a. 6 fl. 36 kr.; No. 10. do. No. 3. 6 fl.; No. 11. do. No. 3 a. 6 fl. 30 kr.; No. 12. do. No. 3 b. 6 fl. 30 kr.; No. 13. do. No. 3 c. 8 fl.; No. 14. do. No. 3½. 9 fl.; No. 15. do. No. 3½ a. 9 fl.; No. 16. do. No. 3½ b. 9 fl.; No. 17. do. No. 3½ b. 9 fl.; No. 18. do. No. 4. 10 fl.; No. 19. do. No. 4½. 14 fl.; No. 20. do. No. 4½ a. 15 fl.; No 21. do. No. 4½ b. 14 fl.; No. 22. do. No. 5. 12 fl. 30 kr.; No. 23. do. No. 5 a. 16 fl.; No. 24. do. No. 5 b. 16 fl.; No. 25. do. No. 5½. 18 fl.; No. 26. do. No. 6. 18 fl.; No. 27. do. No. 6 a. 20 fl.; No. 28. do. No. 6½. 22 fl.; No. 29. do. No. 6½ a. 24 fl.; No. 30. do. No. 7. 25 fl.; No. 31. do. No. 7½. 25 fl.; No. 32. do. No. 8. 32 fl.; No. 33. do. No. 8 a. 20 fl.; No. 34. do. No. 9. 36 fl.; No. 35. do. No. 10. 38 fl.; No. 36. do. No. 11. 39 fl.; No. 37. do. No. 11½. 20 fl.; No. 38. do. No. 12. 25 fl.; No. 39. do. No. 13. 26 fl.; No. 40. do. No. 14. 30 fl.; No. 41. do. No. 15. 38 fl.; No. 42. do. No. 16. 40 fl.; No. 43. do. No. 17. 19 fl.; No. 44. do. No. A. 15 fl.; No. 45. do. No. B. 15 fl.; No. 46. do. No. D. 28 fl.; No. 47. do. No. E. 20 fl.; No. 48. do. No. E. 20 fl.

The usual lenght of 1 »Stab« is 7½ ft. Prus. The prices for ornamented cornices are 5 fl. more for 100 ft. than the above mentioned.

2847. Weber, G., & Co., manu. of polished wood-boxes, Esslingen. London 1862 honor. mention. Agt. Charles Mittler (Mittler & Eckhardt), 6 Grocer's Hall Court, Poultry.

Fancy wooden articles, one sample of each kind, prices p. piece: No. 1. work-box, gray, gold mounted 13/9½ in. 13 fl. 48 kr.; No. 2. do. cedar-wood, gold mounted 13 9½ in. 14 fl. 57 kr.; No. 3. do. ash-wood, gold mounted 10½/7½ in. 10 fl. 21 kr.; No. 4. do. gray, gold mounted 10½/7½ in. 11 fl. 30 kr.; No. 5. do. cedar-wood, gold mounted 8½/6 in. 8 fl. 3 kr.; No. 7. do. ash-wood, gold mounted 7/5 in. 5 fl. 10 kr.; No. 7. glove-box, gray do. 6 fl. 54 kr.; No 8. do. 6 fl. 54 kr.; No. 9. do. cedar-wood, do. 5 fl. 45 kr.; No. 10. cigar-box, gray to be opened do. 13 fl. 12 kr.; No. 11. do. 13 fl. 48 kr.; No. 12. do. cedar-wood, do. 10 fl. 21 kr.; No. 13. do. gray, do. 9 fl. 12 kr.; No. 14. tea-canister, ash-wood do. 2 divisions 8 fl. 36 kr.; No. 15. do. gray, do. 9 fl. 12 kr.; No. 16. do., without divisions 5 fl. 30 kr.; No. 17. tobacco-box, gray, do. 5 fl. 45 kr.; No. 18. watch-box, cedar-wood do. 3 fl. 10 kr.; No. 19. do., ash-wood do. 3 fl. 4 kr.; No. 20. cigarholder, pigeon-house do. 5 fl. 45 kr.; No. 21. ash-box, gray, gold mounted 46 kr.; No. 22. do. pail form, do. 52 kr.; No. 23. inlaid stocking-ball 52 kr.; No. 24. work box, gray, gold mounted 4 fl. 36 kr.; No. 25. do. 5 fl. 45 kr.; No. 26. writing-desk, do. 13 fl. 48 kr.; No. 27. glove-box, do. inlaid 2 fl.; No. 28. cigar-box do. 3 fl. 12 kr.; No. 29. pin-box do. 32 kr.; No. 30. needle-case do., 10 partitions 46 kr.; No. 31. play-plate (p. doz. 2 fl. 45 kr.) 14 kr.; No. 32. writing-desk 8 fl. 36 kr.; No. 33. work-box, 5 pieces 2 fl. 36 kr., 2 fl. 6 kr., 1 fl. 38 kr., 1 fl. 21 kr., 1 fl. 15 kr., together 8 fl. 56 kr.; No. 34. card-box, gray inlaid 3 fl. 27 kr.; No. 35. pen-box with ruler, inlaid 32 kr.; No. 36. sewing-cushion, gray inlaid 3 fl. 12 kr.; No. 37. watch-box do. 1 fl. 27 kr.; No. 38. tea-canister do. 1 fl. 55 kr.; No. 39. candle-stick do. 28 kr.; No. 40. needle-box, do. 5 partit. 35 kr.; No. 41. stamp-box, do. 4 partit. 19 kr.; No. 42. match-box, do. 35 kr.: No. 43. pen-box, do. 23 kr.; No. 44. stamp-box, do. 30 kr.; No. 45. do. 8 kr.: No. 46. needle-holder, inlaid (p. doz. 3 fl. 30 kr.) 18 kr.

2848. Wolbach, Wilhelm, manu. of gilt cornices, Ulm. London 1862 honor. mention. Agt. the Württemberg, Trading-Company at Stuttgart, represented by Sam. Sessing, 10 Huggin-Lane, Wood-street, Cheapside, E. C.

Gilt cornices of various sizes, also brown and black varnished, and white cornices prepared for real gilding, prices p. 100 ft. rhen. 4 sh. 4 d. to 4 £ 10 sh.

2849. The Württemberg trading-company, Stuttgart. London 1862 medal. Agt. s. No. 2848. (s. No. 2693 a, 2724 a, 2771 a, 2822.)

Folding table-mats of plum-wood and maple; one sample of each; prices p. dozen if not otherwise said: round No. 3. 3 sh. 9 d.; No. 4. 5 sh. 8 d.; No. 6. 12 sh.; No. 8. 1 £ 7 sh.; English No. 1. 5 sh.; No. 3. 7 sh.; No. 6. 11 sh. 6 d.; No. 8. 14 sh.; curved No. 0. 5 sh.; No. 1. 8 sh. 4 d.; No. 2. 14 sh. 3 d.; long No. 1. 7 sh.; No. 3. 15 sh. 3 d.; No. 4½. 1 £ 4 sh.; No. 5. 1 £ 12 sh.; oval No. 1. 5 sh. 3 d.; No. 3. 7 sh. 4 d.; No. 6. 10 sh. 6 d.; No. 8. 1 £ 2 sh.

Folding table mats, Guyana à filets: long No. 2½. 17 sh. 8 d.; round No. 3. 5 sh. 8 d.; palixander à filets long No. 2½. 1 £ 1 sh., do. ordinary No. 7. 18 sh. 1 d.

Chess boards: No. 1. 12 sh. 3 d.; No. 1½. 15 sh. 8 d.; No. 2. 19 sh.; No. 2½. 1 £ 1 sh.; No. 4. 1 £ 18 sh. 6.; No. 5. 1 £ 8 sh.

Draught board 19 sh.; 2 sets of boxes of four pieces each, p. set 12 sh. 3 d.; 2 tea caddies 1 £ 15 sh.

Rules: No. 1. 2 sh. 6 d.; No. 2. 2 sh. 9 d.; No. 3, 3 sh.

Needle boxes: No. 1. 12 sh. 2 d.; No. 2. 12 sh. 2 d.; No. 3. 10 sh.; No. 4. 10 sh.; No. 5. 10 sh.; No. 6. 7 sh. 8 d.; No. 7. 10 sh.; No. 8. 10 sh.; No. 9. 7 sh. 8 d.

Stamps-box No. 6263 b. 2 sh. 9 d.; glove do. 1 £ 4 sh. 4 d.; folding table-mat, p. piece 14 sh.; chess-board with Engl. figures No. 2. 3 £ 2 sh.; No. 2½. 3 £ 7 sh.; No. 3. 4 £ 12 sh.; chess-table with mosaic foot and plate p. piece 6 £; round table nutwood 13 in. Württ. measure 4 £ 9 sh.; do. 15 in. Württ. measure 5 £ 15 sh.; do. double 15 in. Württ. measure 7 £ 6 sh.; do. 17 in. Württ. measure 6 £ 16 sh.; chess table, curved, nutwood 6 £ 16 sh.; chess table, square 7 £ 16 sh.; do. oval 9 £ 18 sh.

CLASS XXXI.

IRON AND GENERAL HARDWARE.

2850. Bahnmayer, J. L., Esslingen. London 1862 honor. mention. Agt. Sevin, Chinery & Co., 155 Fenchurch-street.

Copper pastry-moulds; prices p. piece: mould for fish No. 1. straight 3 sh. 2 d.; bent No. 2. 2 sh. 11 d.; No. 3. 1 sh. 2 d.; No. 4. 10 d.; No. 5. and 6. mould for crabs 6 sh.; No. 7. and 8. do. grapes 5 sh. 4 d.; jelly moulds: No. 9. with pipe 5 sh. 10 d.; No. 10. do. 5 sh.; No. 11. without pipe 4 sh. 6 d.; No. 12. with do. 4 sh. 4 d.; No. 13. without do. 3 sh. 10 d.; No. 14. with do. 4 sh. 2 d.; No. 15. without do. 3 sh. 9 d.; No. 16. with do. 3 sh. 4 d.; No. 17. without do. 3 sh.; No. 18. with do. 3 sh. 4 d.; No. 19. without do. 3 sh.; garland-moulds, No. 20. with pipe 5 sh.; No. 21. do. 5 sh.; No. 22. do. 5 sh.; No. 23. large-star mould 8 d.; No. 24. small do. 2 d.; No. 25. shell-mould 5 d.; No. 26. melon-mould 7 d.; No. 27.

biscuit-mould 7 d.; No. 28. cock 5 £; No. 29. tart-mould 5 in. 2 sh.; tart-moulds with pipe: No. 30. 5½ in. 2 sh. 2 d.; No. 31. 6 in. 2 sh. 4 d.; No. 32. 6½ in. 2 sh. 8 d.; No. 33. 7 in. 3 sh.; No. 34. 7¾ in. 3 sh. 8 d.; No. 35. and 36. 8¼ in. 8 sh. 2 d.; No. 37. 8¾ in. 4 sh. 6 d.; No. 38. 9 in. 4 sh. 8 d.; No. 39. 9½ in. 5 sh. 1 d.; No. 40. 10 in. 3 sh. 18 d.; pastry-moulds without pipes No. 41. 6 in. 1 sh. 8 d.; No. 42. 7 in. 2 sh. 1 d.; No. 43. 8 in. 2 sh. 8 d.; melon-moulds No. 44. 6½ in. 1 sh. 10 d.; No. 45. 7 in. 2 sh.; No. 46. 7½ in. 2 sh. 3 d.; No. 47. 8 in. 2 sh. 6 d.; No. 48. 8½ in. 2 sh. 9 d.; No. 49. 9 in. 2 sh. 11 d.; biscuit-moulds No. 50. and 51. 3 in. 11 d.; No. 52. and 53. 3$\frac{2}{12}$ in. 1 sh. 1 d.

No. 54—63. Etuis containing 10 different calibers or sliding measures 4 £ 9 d.

Sheets of drawings, No. 64. of jelly-moulds; No. 65. copper-ware; No. 66. copper-moulds; No. 67. and 68. of calibers or sliding measures.

No. 69. show-board with brass-wares 3 sh. 10 d.

2851. Bührer & Kallenberg, manu., Ludwigsburg. Munich 1854 medal of honor; London 1862 honor. mention. Agt. the Württemberg Trading-Company at Stuttgart, represented by Sam. Sessing, 10 Huggin-Lane, Wood-street, Cheapside, E. C.

Collection of pastry-moulds.

2852. Deffner, C., manu., Esslingen. Berlin 1844 silver medal; London 1851 prize-medal; Munich 1854 great medal; London 1862 medal. Agt. J. W. Green (E. G. Zimmermann). (s. No. 2865.)

Japanned tin ware for use and ornament; japanned tin bird-cages; coffee-machines; lamps; brass and bronzed copper goods; silver plated ware.

2853. Manufactory of iron and iron-wire Erlau near Aalen. Munich 1854 medal of honor; London 1862 medal. Agt. s. No. 2851.

Prices p. cwt.; No. 1. 98 Pfd. vulcanized telegraph-wire 18 fl.; No. 2. 31 Pfd. ordinary wire 11 fl. 40 kr.; No. 3. 8¾ Pfd. fashion wire for spinning mills 20 fl.; No. 4. 8¾ Pfd. another fashion of the same 14 fl.; No. 5. 11 Pfd. flat wire No. 8/22. 15 fl. 30 kr.; No. 6. 10¾ Pfd. ordinary wire No. 6. 14 fl.; No. 7. 12¼ Pfd. galvanized wire No. 8. 14 fl. 40 kr.; No. 8. 10 Pfd. tinned do. No. 3. 45 fl.; No. 9. 8 Pfd. ordinary do. No. 12. 42 fl.; No. 10. 3½ Pfd. do. No. 18. 80 fl.; No. 11. 1½ Pfd. do. No. 20. 100 fl.; No. 12. 4 Pfd. do. No. 14. 57 fl.; No. 13. 2¼ Pfd. do. No. 17. 76 fl.; No. 14. 3¼ Pfd. do. No. 15. 64 fl.

Prices p. Pfd.; No. 15. 37 Pfd. sample of a chain; No. 16. 3 Pfd. hold-back twofold 4/0. 9½ kr.; No. 17. 2 Pfd. do. single with mail chain No. 4/0. 11 kr.; No. 18. 1 Pfd. 11 Lth. breast-chain with mail-chain No. 3/0. 12 kr.; No. 19. 1 Pfd. 13 Lth. breast-chain ordinary No. 4/0. 9½ kr.; No. 20. 3 Pfd. 26 Lth. chain for cows 3/0. 10 kr.; No. 21. 1 Pfd. 27 Lth. chain for calves No. 1. 14 kr.; No. 22. 1 Pfd. 26 Lth. halter-chain with 2 bradoon-chains No. 0. 15 kr.; No. 23. 1 Pfd. 14 Lth. double bridle No. 1. 18 kr.; No. 24. 20 Lth. hand bridle No. 2. 19 kr.; No. 25. 15 Lth. snaffle with bradoon No. 2. 19 kr.: No. 26. 9 Lth. snaffle with ring No. 2. 19 kr.; No. 27. 1 Pfd. 21 Lth. 2 trace-chains No. 3/0. 11 kr.; No. 28. 17 Lth. do. double No. 2/0. 12 kr.; No. 29. 1 Pfd. 13 Lth. bridle for cattle with nose-band and buckle No. 3. 24 kr.; No. 30. 1 Pfd. 20 Lth. Vienna-snaffle No. 4. 36 kr.; No. 31. 2 Pfd. 12 Lth. hold-back, tinned No. 3/0. 20 kr.; No. 32. 1 Pfd. 6 Lth. halter-chain with spring No. 1. 18 kr.

No. 33. 13 Lth. flat chain, tinned; No. 34. 1 Pfd. 30 Lth. hold-back for carriage, tinned 1 fl. p. piece; No. 35. 4 Pfd. 2 Lth. ice-chain No. 8/0. 9 kr. p. Pfd.

Prices p. cwt.: No. 36. 10 Pfd. 14 Lth. carpenter's tacks No. 24/78. 11 fl.; No. 37. 5 Pfd. 29 Lth. do. rough No. 23/66. 12 fl.; No. 38. 6 Pfd. 27 Lth. do. smooth No. 23/60. 11 fl.; No. 41. 2 Pfd. 14 Lth. locksmith's tacks No. 17/15. 13 fl. 15 kr.; No. 42. 1 Pfd. 30 Loth do. rough No. 15/12. 16 fl.; No. 43. 1 Pfd. 24 Lth. belt tacks, tinned No. 124. 27 fl.; No. 44. 1 Pfd. 8 Lth. locksmith's tacks rough No. 121. 16 fl. 30 kr.; No. 45. 26 Lth. Swiss nails No. 83. 24 fl. 15 kr.; No. 46. 1 Pfd. 3 Lth. Paris tacks No. 12. 26 fl. 30 kr.

No. 47. 1 Pfd. 1 Lth. brass-tacks p. Pfd. 11 kr.; No. 48. 1 Pfd. tacks for upholsterers p. cwt. 50 fl.; No. 49. 1 Pfd. 11 Lth. hinge-tacks 90 fl.; No. 50. 1 Pfd. 15 Lth. brass tacks p. m. 32 kr.; No. 51. 2 Pfd. 9 Lth. furniture-tacks, tinned p. Pfd. 30 kr.; No. 52. 1 Pfd. 29 Lth. Paris tacks No. 19. p. cwt. 17 fl. 30 kr.; No. 53. 1 Pfd. 13 Lth. yellow furniture-tacks 150 fl.; No. 54. 2 Pfd. 3 Lth. flat belt-tacks No. 124. 32 fl.; No. 56. 4 Pfd. 5 Lth. floor-tacks, small, rough 11 fl.; No. 58. 5 Pfd. 7 Lth. do. large, rough 11 fl. 30 kr.; No. 59. 7 Pfd. 16 Lth. smooth tacks for building purposes No. 22/56. 10 fl. 30 kr. Prices from No. 53—59. p. cwt.

2854. Erhard & Sons, Schwäbisch-Gmünd. Munich 1854 medal of honor; London 1862 medal. Agt. s. No. 2851. (s. No. 2867.)

Different kinds of articles of metal: 1. Toys: of brass, varnished, silvered and gilt, such as coffee-, tea- and furniture sets, portraits, glasses, candlesticks, lustres, watches, clocks, vases, jewellery for dolls etc. etc. 2. Fancy-articles: stamped and cast, as well as produced by the galvano-plastic process, viz: ash-boxes, letter-pressers, beer-glasses, feather-cleaners, smell-bottles, lamps, sewing-wax, seals, travelling-chessboards, jewel-boxes, napkin-holders, thermometers, watch-stands, tooth-pick and matches-stands. 3. Tinder-boxes: of brass, silvered and of genuine silver, with steel and flint; the same with phosphorized matches.

2855. Martin, mechanician and prop. of a grinding mill, Tübingen. Agt. s. No. 2851.

Smoothing-irons; prices p. dozen: 1 gas-smoothing-iron 5 £ 13 sh. 2 d.; 2 charcoal smoothing-irons with grid-irons 3 £ 18 sh. 2 d.: 2 charcoal smoothing-irons with chimney 2 £ 17 sh. 7 d.; 2 spirit smoothing-irons 3 £ 18 sh. 2 d.; 1 brass-smoothing-iron 6 £ 3 sh. 5 d.; 1 ordinary smoothing-iron, Württemb. fashion No. 1½. 1 £ 4 sh. 8 d.; 1 do. Bavaria fashion No. 2. 1 £ 7 sh. 5 d.; 1 do. Württemb. fashion No. 3. 1 £ 8 sh. 5 d.; 1 do. Württemb. fashion No. 0. 19 sh. 6 d.; 1 do. Rhenish fashion No. 4. 1 £ 10 sh. 2 d.; 1 do. Rhenish fashion No. 3. 1 £ 8 sh. 5 d.: 1 do. Württemb. fashion No. 2. 1 £ 6 sh. 4 d.: 1 stand, bronzed 2 £ 1 sh. 2 d.

2857. Rueff. Dr. A., professor of veterinary science at the Royal academy. Hohenheim near Stuttgart. Munich 1854 honor. mention; London 1862 honor. mention. Agt. s. No. 2851.

4 nose-rings for taming cattle, of the newest construction, being of such efficiency, that after their application every other instrument can be dispen-

sed with; head of a bull, cut out of wood, with nose-ring applied; 20 copies of printed instructions on the use of the nose-rings.

2858. STOTZ, A., manu., Stuttgart. London 1862 honor. mention.

Objects of malleable cast-iron.

2859. STRAUB & SCHWEIZER, manu. of silver-plated and bronced copper-ware, Geislingen. London 1862 medal.　(s. No. 2872.)

Bronzed copper-ware.

2860. VETTER, FRIEDR., manu. of bird-cages, fruitbaskets, tea-trays etc., Ludwigsburg. Munich 1854 medal of honor; Paris 1855 silver medal; London 1862 medal. Agt. the Württemberg-Trading-Company at Stuttgart, represented by Sam. Sessing, 10 Huggin-Lane, Wood-street, Cheapside, E. C.

1. Bird-cages; prices p. piece: 1. No. 37. 14 frs. 30 c.; 2. No. 48. 8 frs. 55 c.; 3. No. 43. brass-yarn 12 frs.; 4. No. 53. 5 frs. 80 c.; 5. No. 54. 6 frs. 60 c.; 6. No. 5. 7 frs. 50 c.; 7. No. 6. 10 frs. 50 c.; 8. No. 8. 14 frs.; 9. No. 56. large 11 frs. 80 c.; 10. No. 0. 2 frs. 85 c.; 11. No. 57. 3 frs. 45 c.; 12. No. 2. 3 frs. 57 c.; 13. No. 58. 8 frs. 15 c.; 14. No. 30. 6 frs. 85 c.; 15. No. 31. 6 frs. 85 c.; 16. No. 34. 7 frs.; 17. No. 39. 10 frs. 50 c.; 18. No. 40. 11 frs. 25 c.; 19. No. 59. small 3 frs. 15 c.

2. Tea-trays; prices p. dozen: No. 20. 16 in. black and white 15 frs. 45 c.; No. 21. 16 in. tortoise, with yellow arabesques 21 frs. 65 c.; No. 22. Sandwich fashion, 16 in., tortoise in gold 32 frs. 15 c.; No. 23. barroque fashion, 18 in. 41 frs. 35 c.; No. 24. do. 18 in., colored 54 frs. 20 c.; No. 25. oval fashion, 18 in., black, with birds 61 frs. 30 c.; No. 26. oval fashion 18 in., tortoise-bouquet 61 frs.; No. 27. do. 18 in., oval with roses 44 frs. 60 c.; No. 28. do. 18 in., oval, mahogany, ebony and gold 42 frs. 85 c.; No. 29. oval, Turkish, 24 in., red with medaillon 140 frs. 35 c.; No. 30. do. 24 in., green with roses 108 frs. 20 c.

3. Oil-gas-lamps; prices p. piece: Compositions No. 31. 4½ in. 6 frs.; No. 32. 5 in. 6 frs. 45 c.; No. 33. 5⅝ in. 7 frs. 5 c.; No. 34. 6 in. 7 frs. 50 c.; No. 35. 7 in. 9 frs.; No. 36. tin-foot, 6 in. 5 frs. 50 c.

4. No. 37. work-lamp of brass, 5 in. 8 frs.

2861. WAGNER, GOTTLOB, manu. of copperware, Esslingen. Paris 1855 bronze medal; London 1862 medal. Agt. s. No. 2860.

Pastry-moulds of copper; prices p. piece. Moulds for torts with pipe: No. 1. and 2. 3 fl. 36 kr.; No. 3. and 4. 3 fl. 18 kr.; No. 5. and 6. 3 fl. 6 kr.; No. 7. and 8. 2 fl. 48 kr.; No. 9. and 10. 2 fl. 42 kr.; No. 11. to 16. 2 fl. 33 kr.; No. 17. and 18. 2 fl. 24 kr.; No. 19. and 20. 2 fl. 3 kr.; No. 21. and 22. 1 fl. 54 kr.; No. 23. to 28. 1 fl. 48 kr.; No. 29. and 30. 1 fl. 36 kr.; No. 31. 1 fl. 30 kr.; No. 32. 1 fl. 24 kr.; No. 33. and 34. 1 fl. 18 kr.; No. 35 and 36. 1 fl. 12 kr.

Pastry-moulds without pipe: No. 37. 1 fl.; No. 38. 1 fl. 8 kr.; No. 39. 1 fl. 15 kr.; No. 40. 1 fl. 25 kr.; No. 41. 1 fl. 36 kr.; No. 42. 1 fl. 48 kr.

Melon-moulds: No. 43. 42 kr.; No. 44. 57 kr.; No. 45. 1 fl.; No. 46. to 48. 1 fl. 9 kr.; No. 49. and 50. 1 fl. 12 kr.: No. 51. and 52. 1 fl. 18 kr.; No. 53. and 54. 1 fl. 30 kr.; No. 55. and 56. 1 fl. 45 kr.; No. 57. and 58. 2 fl.

Biscuit-moulds, No. 59. to 70. p. doz 3 fl. 24 kr.; No. 71. to 82. do. 3 fl. 48 kr.

Jelly-mould without pipe, No. 83. 1 fl. 48 kr.; No. 84. do. with pipe 2 fl.; No. 85. do. 2 fl. 30 kr.; No. 86. do. 2 fl. 36 kr.; No. 87. do. 3 fl. Jelly-mould without pipe No. 88. 3 fl. 6 kr. Jelly-moulds, small, without pipe: No. 89. 22 kr.; No. 90. 20 kr.; No. 91. 14 kr.; No. 92. 20 kr.; No. 93. 20 kr.; No. 94. 21 kr.

Fish-moulds: No. 95. 1 fl. 54 kr.; No. 96. 1 fl. 45 kr.; No. 97. 48 kr.; No. 98. 36 kr.

Mould for grapes: No. 99. 3 fl. 12 kr.; do. for crabs No. 100. 3 fl. 36 kr.

CLASS XXXII.

STEEL AND CUTLERY.

2863. DITTMAR, BROTHERS, manu. of cutlery, Heilbronn. Berlin 1844 silver medal; London 1851 prize-medal; Munich 1854 great medal; Paris 1855 silver medal; London 1862 medal. Patents for razor-straps etc. Agt. s. No. 2860.
(s. No. 2727.)

Assortment of razors, straps, pocket and pen-knives, scissors, whetting-steels and different cutlery, scarificators, garden-implements etc., value 400 £.

2864. HAUEISEN & SON, manu., Stuttgart. London 1851 prize-medal; Munich 1854 medal of honor; Paris 1855 great medal of honor; London 1862 medal. Agt. C. Mittler (Mittler & Eckhardt), 20 St. Dunstans Hill.

I. Scythes. a. German: 2 briar-scythes 14 and 16 in.; 8 scythes, empire-form, 20, 22, 24, 26, 28, 32, 36 and 38 in.; 1 do. gray, heavy, 26 in.; 1 do. half-girded, 37 in.; 8 do. pointed Leipsic form, 25, 27, 30, 32, 35, 36, 40 and 42 in.; 9 do. round Leipsic form 25, 30, 32, 33, 36, 40, 42 in.; 1 do. Pomerania 42 in.

b. French: 12 scythes with short heels 20, 22, 24, 26, 29, 30, 32 in.; 17 do. with large heels 26, 28, 30, 32, 34, 36, 38 in.

c. Italian: 7 scythes straight 20, 22, 24, 26, 28, 30 in.; 1 do. large heel and with bill 26 in.; 2 do. large heels and with bills 29 and 34 in.

d. Swiss: 7 scythes with bills, 26, 28, 30, 32, 34, 35 in.; 6 do. half-girded 28, 30, 32, 35, 38 in.; 1 do. Morteau-form, 30 in.

e. Russian: 7 scythes, straight 26, 27, 28 and 44 in.; 4 do. half-large, 30 and 32 in.

f. Dutch: 3 scythes, large and heavy 30, 32, 34 in.; 2 do. curved, 25 and 27 in.

g. American: 5 scythes, large-sized, 34, 36, 37, 41 in.; 2 do. small-sized, 43 in.; 2 briar-scythes, 16 in.

II. Sickles. 75 sickles, round No. 000, 00, 0, ½, 1, 1½, 2, 2½, 3, 4, 5, 6, 8; 32 do. small, great and large, Swiss-form; 6 do. large Italian-form; 18 do. straight, small narrow No. 0000. 000. 00.

III. Straw-cutters. 2 blue straw-cutters 3½ lb.; 2 do. 3¾ lb.

CLASS XXXIII.

WORKS IN PRECIOUS METALS, AND THEIR IMITATIONS AND JEWELLERY.

2865. DEFFNER, CARL, manu., Esslingen. Berlin 1844 silver medal; London 1851 prize-medal; Munich 1854 great medal. Agt. J. W Green (E. G. Zimmermann).　(s. No. 2852.)

Silver-plated ware.

2867. ·Erhard & Sons, Schwäbisch-Gmünd. Munich 1854 medal of honor. Agt. the Württemberg Trading-Company at Stuttgart, represented by Sam. Sessing, 10 Huggin-Lane, Woodstreet, Cheapside, E. C. (s. No. 2854.)

Articles used in catholic worship viz: crosses, holywater-pots, holystands · etc.; church-articles, viz: communion-cups for common and private use, reliquaries, providing-crosses, monstranzes etc.: garnishments for books, albums, cartonnage and fine wooden articles, viz: corner and middle-pieces, clasps for prayer-books, albums, bibles and missals; bands, escutcheons, handles etc.; genuine silver galvanoplastic goods; riding-whip heads.

2868. Forster, Dom., manu. of silverware, Schwäbisch-Gmünd. · Agt. s. No. 2867.

Silver wares: No. 1. goblet 5 £ 9 sh. 5 d.; No. 2. do. 5 £ 2 d.; No. 3. do. 3 £ 17 sh. 11 d.; No. 4. wine-cup 1 £ 9 sh. 1 d.; No. 5. do. 2 £ 5 sh. 9 d.; No. 6. cigar-stand 2 £ 2 sh. 10 d.; No. 7. tooth pickstand 13 sh. 6 d.; No. 8. do. 14 sh.; No. 9. candle-stick 1 £ 19 sh. 3 d.; No. 10. bread-basket 4 £ 16 sh.; No. 11. candle-stick 2 £ 7 sh. 8 d.; No. 12. cup 15 sh. 3 d.; No. 13. cigarstand 1 £ 17 sh. 8 d.; No. 14. cup 1 £ 7 sh. 6 d.; No. 15. cigar-stand 1 £ 18 sh. 1 d.; No. 16. cup 2 £ 10 sh.; No. 17. cigar-stand 1 £ 10 sh. 10 d.; No. 18. cup 19 sh. 2 d.; No. 19. candle-stick 3 £ 1 sh.; No. 20. bread-basket 2 £ 5 sh. 1 d.; No 21. do. 3 £ 5 sh. 9 d.; No. 22. candle-stick 2 £ 11 sh. 9 d.; No. 23. goblet 1 £ 7 sh. 11 d.; No. 24. hand-candle-stick 1 £. 12 sh. 9 d.; No. 25. inkstand 1 £ 17 sh. 7 d.; No. 26. do. 2 £ 19 sh.; No. 27. hand-candle-stick 1 £ 16 sh. 1 d.; No. 28. salt-stand with glass-cup 18 sh. 2 d.; No. 29. do. 13 sh. 7 d.; No. 30. mustard-pot 1 £ 1 sh. 4 d.; No. 31. salt-stand with glass-cup 18 sh. 3 d.; No. 32. do. 12 sh. 4 d.; No. 33. cinnamoncaster 1 £ 6 d.; No. 34. salt-stand 1 £ 1 sh. 6 d.; No. 35. do. 15 sh. 5 d.

Napkin-rings: No. 36. 6 sh.; No. 37. 6 sh. 8 d.; No. 38. 7 sh.; No. 39. 14 sh. 2 d.; No. 40. 9 sh. 4 d.; No. 41. 16 sh. 4 d.; No. 42. 6 sh. 2 d.; No. 43. 7 sh. 4 d.; No. 44. 10 sh. 2 d.

Match-boxes: No. 45. 10 sh.; No. 46. 8 sh.; No. 47. 8 sh.; No. 48. 10 sh.

Sugar-basins: No. 49. 2 £ 6 sh. 10 d.; No. 50. 3 £ 4 sh. 5 d.; sugar-basin-stands No. 51. 1 £ 4 d.; No. 52. 19 sh. 4 d.; No. 53. 1 £ 4 sh. 7 d.

Cream-pots: No. 54. 1 £ 17 sh. 4 d.; No. 55. 1 £ 9 sh.; No. 56. 1 £ 5 sh. 6 d.; No. 57. sugar-box 2 £ 15 sh. 2 d.; No. 58. do. 4 £ 15 sh.; No. 59. coffee-pot 6 £ 18 sh.; No. 60. tea-pot 6 £ 2 sh. 7 d.; No. 61. sugar-basin 4 £ 15 sh. 2 d.: No. 62. cream-pot 2 £ 13 sh.; No. 63. waiter 2 £ 5 sh. 10 d.; No. 64. cream-pot 19 sh.; No. 65. cup 2 £ 8 d.; No. 66. fish-knife 19 sh. 3 d.; No. 67. cakeknife 1 £ 1 sh.; No. 68. fish-knife 18 sh. 5 d.; No. 69. cake-knife 17 sh. 8 d.; No. 70. fish-knife 15 sh. 1 d.

Cake-knives: No. 71. 17 sh. 10 d.; No. 72. 14 sh. 10 d.; No. 73. 14 sh. 3 d.; No. 74. 17 sh.; No. 75. 15 sh. 3 d.; No. 76. 15 sh.; No. 77. 15 sh. 2 d.; No. 78. 15 sh. 2 d.; No. 79. 15 sh. 3 d.; No. 80. 12 sh. 3 d.; No. 81. 15 sh. 6 d.

Card-cases: No. 82. 15 sh. 10 d.; No. 83. 19 sh. 4 d.

2870. Kott. D., manu. of silver-ware, Schwäbisch-Gmünd. London 1862 honor. mention. Agt s. No. 2867.

Works in silver: No. 1. chandelier; No. 2—3. candle sticks; No. 4. hand-candlestick, plain; No. 5—6. cake-baskets with handles; No. 7—16. coffee- and tea-services with plates; No. 17. plate; No. 19. cup with cover; No. 20—22. goblets; No. 23. egg-cup-stand; No. 24. salt-pepper- and mustard-stand; No. 25. sugar-basin; No. 26. inkstand; No. 27—30. cigar-stands; No. 31—32. toothpickstands; No. 33—35. table-bells; No. 36. dessert- and flower-stand; No. 37. flower-vase; No. 38. butter-box; No. 39. bottle-stand for vinegar, oil etc.; No. 40. salt-, pepper- and mustard-stand; No. 41. mustard-pot; No. 42. salt-box; No. 43—44. fish-knife and fork; No. 45. cake-trowel; No. 46. sugar-spoon; No. 47. fruit-spoon; No. 48—50. napkin-rings; No. 51—52. sugar-tongs; No. 53—55. tea-strainers; No. 56. confect-spoon; No. 57. saltspoon; No. 58—59. butter-knives; No. 60. saladfork and spoon; No. 61—64. children's rattles; No. 65—66. match-boxes.

2871. Ott & Co., manu., Schwäbisch-Gmünd. London 1862 medal. Agt. s. No. 2867. Jewellery.

2872. Straub & Schweizer, manu. of silver plated and bronzed copper-wares, Geislingen. London 1862 medal. Agt. s. No. 2867. (s. No. 2859.)

Silver-plated and bronzed copper-wares: No. 1. Victoria tray 30 in. richly chased 80 fl.; No. 2. a. 1 lb. 21 Loth. plaqué 40 in. No. 8., p. pound 3 fl. 50 kr.; b. 2 lb. 2 Loth. double 40 in. No. 9., p. pound 4 fl. 17 kr.; No. 3. punch-ladle with ivory handle 6 fl. 8 kr.; No. 4. tart-ladle with do. 5 fl.; No. 5. soup-ladle with do. 4 fl. 30 kr.; No. 6. fish-ladle with do. 5 fl. 20 kr.; No. 7. egg-cup stand with 6 cups and 6 spoons 18 fl. 16 kr.; No. 8. filtering jug, No. 2. plaqué 10 fl. 48 kr.; No. 9. inkstand, No. 4. with 2 glasses 5 fl. 36 kr.; No. 10. 2 chandeliers, Paris fash. 13 fl. 24 kr.; No. 11. tea-kettle, Germ. fash. No. 2. 32 fl.; No. 12. do. No. 3. 37 fl.; No. 13. communion-cup, small do. 8 fl. 6 kr.; No. 14. drinking-cup, large 5 fl. 45 kr.; No. 15. waxwinder-stand 2 fl. 40 kr.; No. 16. tea-pot, French, No. 3. 9 fl. 50 kr.; No. 17. cream-pot, fash. No. 3. 6 fl. 18 kr.; No. 18. slopbasin, No. 1. gilt 5 fl. 24 kr.; No. 19. Victoria sugar-box No. 2. 8 fl. 20 kr.; No. 20. drinkingcup No. 2. without foot, gilt 2 fl. 12 kr.; No. 21. do. with foot, gilt 2 fl. 45 kr.; No. 22. tea-strainer 2 fl.; No. 23. salver 18 fl. 48 kr.; No. 24. do. 6 fl. 54 kr.; No. 25. do. 5 fl. 28 kr.; No. 26. 2 chandeliers 4 fl. 40 kr.; No. 27. 2 do. 15 fl. 4 kr.; No. 28. 2 do. No. 1. 3 fl. 30 kr.; No. 29. coffee-pot No. 4. 14 fl.: No. 30. stand with 2 glasses 7 fl. 52 kr.; No. 31. altar-jug with gilt cross 35 fl.; No. 32. do. No. 4. 21 fl. 30 kr.; No. 33. bottle-cooler No. 4. 27 fl.; No. 34. sugar-vase, gilt in the interior 7 fl. 48 kr.; No. 35. bottle-cooler No. 2. of bronzed copper 8 fl.; No. 36. Berzeliuslamp No. 2. 14 fl. 24 kr.; No. 37. tea-kettle, No. 1. German 12 fl.; No. 38. coffee-pot, No. 4. plain 6 fl. 15 kr.; No. 39. cream-pot, No. 4. of bronzed copper 3 fl. 48 kr.; No. 40. dust-shovel, chased, plaqué 6 fl. 20 kr.; No. 41. duster with handle of bone 1 fl. 57 kr.; No. 42. Victoria breadbasket No. 2. 9 fl. Prices from No. 3. to. No. 42. p. piece.

2873. Wöhler, Ed.. & Co., manu. of jewellery, Schwäbisch-Gmünd. London 1862 honor. mention. Agt. s. No. 2867.

Jewellery.

Garnitures: 729. enamel and pearl 3 £ 10 sh.: 730. onyx 1 £. 15 sh.; 702. black enamel 1 £

12 sh. 3 d.; 677. flower and gems 17 sh. 6 d.; 698. coral-grape 1 £ 1 sh. 11 d.; 604. bird and fly 3 £ 15 sh. 10 d.; 605. flower 4 £ 4 sh. 2 d.; 719. do. 2 £; 737. coral-rose 1 £ 15 sh.; 727. hop-plant 2 £ 12 sh. 6 d.; 754. coral-flowers 1 £ 4 sh. 2 d.

Brooches: 575. gems 16 sh. 3 d.; 794. onyx, pearl and amethyst 1 £ 12 sh. 6 d.; 773. fruits and corals 18 sh. 4 d.; 745. glass-brooch with gems 1 £ 5 sh.; 818. brooch with ivy and gems 18 sh. 6 d.; 759. do. onyx 1 £ 2 d.; 633. glass-brooch with corals 1 £. 6 sh. 8 d.; 744. do. gems 1 £ 7 sh. 10 d.; brooches 618. flower and leaves 1 £. 3 sh. 4 d.; 697. leaf and beetle 2 £ 1 sh. 8 d.; 718. do. of corals 1 £ 6 sh. 5 d.; 708. glass-brooch with gems 1 £ 7 sh. 6 d.; 633. do. 1 £ 7 sh. 6 d.

12 pair of buttons: 768. coral 6 sh. 8 d.; 611½. grape 7 sh. 3 d.; 649. flower and gems 10 sh.; 737½. coral-rose 10 sh. 10 d.; 765. hop-plant and gems 10 sh.; 520. flower and gems 8 sh. 4 d.; 755. coral-flower 12 sh.; 686. flower 10 sh.; 520½. flower and coral 8 sh. 2 d.; 679. acorn-corals 8 sh.; 816. flower 6 sh. 9 d.; 764¼. gems 7 sh. 8 d.

3 garnitures of sleeve-buttons: 764. 11 sh. 4 d.; 764½. colored 11 sh. 4 d.; 763. garland and gems 10 sh. 6 d.; 573. 2 chemisette-buttons 3 sh. 4 d.

Crosses: 571. twig-shaped 6 sh. 1 d.; 742. coral-stem 4 sh. 5 d.; 610. gems and pearls 6 sh. 11 d.; 687. corals 4 sh. 2 d.; 648. do. 4 sh. 1 d.; 570. do. 5 sh. 7 d.; 610½. do. 5 sh. 4 d.; 687½. gems 6 sh. 11 d.; 571½. twig-shaped 2 sh. 6 d.; 571¼. do. 2 sh. 1 d.; 570½. do. 1 sh. 11 d.; 570¼. corals 3 sh. 11 d.; 671½. do. 1 sh. 7 d.

CLASS XXXV.
POTTERY.

2875. WEYSSER, CARL, manu., Liebenzell. Agt. Charles Mittler (Mittler & Eckhardt), 6 Grocer's Hall Court, Poultry.

Artificial whet-stones: No. 1/6. 6 samples ord., for chaff-knives, scythes and sickles; No. 7/12. 6 do. fine, only for scythes and sickles. Polishing powder for the use of brass-workers, steel-workers and for japanners etc. No. 13/14. 2 packets fine; No. 15. 1 packet ord.

TABLES OF REDUCTION

for

calculating the money, weights and measures occurring in the catalogue.

I. MONEY

The »Münzpfund« (pound used in the mint) of 500 Grammes of fine silver forms the common legal standard of coinage in the States of the Zollverein. From this pound 30 Thalers are coined. The Thaler is divided into 30 Silbergroschen or Neugroschen, the Silbergroschen into 12 and the Neugroschen into 10 Pfennige, the Florin into 60 Kreutzers. A Thaler is equal to a Florin and 45 Kreutzers, a Florin to $\frac{4}{7}$ Thalers; $3\frac{1}{2}$ Florins are consequently adequate to 2 Thalers.

Besides these kinds of coins there occur in the quotations of prices for woollen cloth » gute Groschen« which are now no longer coined and of which 24 à 15 Pfennige make a Thaler, and »Grotes« of which one in silver is equal to $\frac{1}{72}$ Thaler or 5 Pfennige.

In calculating the following tables the £ st. is taken at 6 Thalers and 20 Silbergroschen or 11 Florins and 40 Kreutzers, the Franc at $8\frac{1}{2}$ Sgr. or 28 Kreutzers and 1 Deut. The fractures of Pfennige, Deutes and Centimes, if less than $\frac{1}{2}$, are omitted, those of $\frac{1}{2}$ or above $\frac{1}{2}$ are reckoned as full.

a. Thaler-Value.

Thaler-Value.			South-german Value. (52½ fl. p. pound.)			English Value.			French Value.	
Rℓ	Sgr	Pf	G	Kr	Dt	L	sh	d	Fr	C
—	—	1	—	—	1	—	—	0,1	—	1
—	—	2	—	—	2	—	—	0,2	—	2
—	—	3	—	1	—	—	—	0,3	—	3
—	—	4	—	1	1	—	—	0,4	—	4
—	—	5	—	1	2	—	—	0,5	—	5
—	—	6	—	1	3	—	—	0,6	—	6
—	—	7	—	2	—	—	—	0,7	—	7
—	—	8	—	2	1	—	—	0,8	—	8
—	—	9	—	2	3	—	—	0,9	—	9
—	—	10	—	3	—	—	—	1,0	—	10
—	—	11	—	3	1	—	—	1,1	—	11
—	1	—	—	3	2	—	—	1,2	—	12
—	2	—	—	7	—	—	—	2,4	—	25
—	3	—	—	10	2	—	—	3,6	—	37
—	4	—	—	14	—	—	—	4,8	—	49
—	5	—	—	17	2	—	—	6,0	—	62
—	10	—	—	35	—	—	1	—	1	24
—	20	—	1	10	—	—	2	—	2	47
1	—	—	1	45	—	—	3	—	3	71
2	—	—	3	30	—	—	6	—	7	42
3	—	—	5	15	—	—	9	—	11	13
4	—	—	7	—	—	—	12	—	14	85
5	—	—	8	45	—	—	15	—	18	56
10	—	—	17	30	—	1	10	—	37	11
20	—	—	35	—	—	3	—	—	74	23
30	—	—	52	30	—	4	10	—	111	34
40	—	—	70	—	—	6	—	—	148	45
50	—	—	87	30	—	7	10	—	185	57
100	—	—	175	—	—	15	—	—	371	13
200	—	—	350	—	—	30	—	—	742	27
300	—	—	525	—	—	45	—	—	1113	40
400	—	—	700	—	—	60	—	—	1484	54
500	—	—	875	—	—	75	—	—	1855	67
1000	—	—	1750	—	—	150	—	—	3711	34

b. South-german Value. (Standard of 52½ Florins pr. pound.)

South-german Value. (52½ fl. p. pound.)			Thaler-Value.			English Value.			French Value.	
G	Kr	Dt	Rℓ	Sgr	Pf	L	sh	d	Fr	C
—	—	1	—	—	1	—	—	0,1	—	1
—	—	2	—	—	2	—	—	0,2	—	2
—	—	3	—	—	3	—	—	0,3	—	3
—	1	—	—	—	3	—	—	0,3	—	4
—	2	—	—	—	7	—	—	0,7	—	7
—	3	—	—	—	10	—	—	1	—	11
—	4	—	—	1	2	—	—	1,2	—	14
—	5	—	—	1	5	—	—	1,7	—	18
—	10	—	—	2	10	—.	—	3,4	—	35
—	15	—	—	4	3	—	—	5,1	—	53
—	20	—	—	5	9	—	—	6,9	—	71
—	25	—	—	7	2	—	—	8,6	—	88
—	30	—	—	8	7	—	—	10,3	1	6
1	—	—	—	17	2	—	1	8,6	2	12
2	—	—	1	4	3	—	3	5,1	4	24
3	—	—	1	21	5	—	5	1,7	6	36
4	—	—	2	8	7	—	6	10,3	8	49
5	—	—	2	25	9	—	8	6,9	10	61
10	—	—	5	21	5	—	17	1,7	21	22
20	—	—	11	12	10	1	14	3,4	42	43
30	—	—	17	4	3	2	11	5,1	63	65
40	—	—	22	25	9	3	8	6,9	84	86
50	—	—	28	17	2	4	5	8,6	106	8
100	—	—	57	4	3	8	11	5,1	212	15
200	—	—	114	8	7	17	2	10,3	424	31
300	—	—	171	12	10	25	14	3,4	636	46
400	—	—	228	17	2	34	5	8,6	848	62
500	—	—	285	21	5	42	17	1,7	1060	77
1000	—	—	571	12	10	85	14	3,4	2121	55

c. English Value.

English Value L.	sh.	d.	Thaler-Value ℛℓ	Sgr	₰	South-german Value (52½ fl. p. pound.) G.	Kr.	Dt.	French Value Fr.	C.
—	—	1	—	—	10	—	3	—	—	10
—	—	2	—	1	8	—	5	3	—	21
—	—	3	—	2	6	—	8	3	—	31
—	—	4	—	3	4	—	11	3	—	41
—	—	5	—	4	2	—	14	2	—	52
—	—	6	—	5	—	—	17	2	—	62
—	—	7	—	5	10	—	20	2	—	72
—	—	8	—	6	8	—	23	1	—	82
—	—	9	—	7	6	—	26	1	—	93
—	—	10	—	8	4	—	29	1	1	3
—	—	11	—	9	2	—	32	—	1	13
—	1	—	—	10	—	—	35	—	1	24
—	2	—	—	20	—	1	10	—	2	47
—	3	—	1	—	—	1	45	—	3	71
—	4	—	1	10	—	2	20	—	4	95
—	5	—	1	20	—	2	55	—	6	19
—	6	—	2	—	—	3	30	—	7	42
—	7	—	2	10	—	4	5	—	8	66
—	8	—	2	20	—	4	40	—	9	90
—	9	—	3	—	—	5	15	—	11	13
—	10	—	3	10	—	5	50	—	12	37
1	—	—	6	20	—	11	40	—	24	74
2	—	—	13	10	—	23	20	—	49	48
3	—	—	20	—	—	35	—	—	74	23
4	—	—	26	20	—	46	40	—	98	97
5	—	—	33	10	—	58	20	—	123	71
10	—	—	66	20	—	116	40	—	247	42
20	—	—	133	10	—	233	20	—	494	85
30	—	—	200	—	—	350	—	—	742	27
40	—	—	266	20	—	466	40	—	989	69
50	—	—	333	10	—	583	20	—	1237	11
100	—	—	666	20	—	1166	40	—	2474	23
200	—	—	1333	10	—	2333	20	—	4948	45
300	—	—	2000	—	—	3500	—	—	7422	68
400	—	—	2666	20	—	4666	40	—	9896	91
500	—	—	3333	10	—	5833	20	—	12371	13
1000	—	—	6666	20	—	11666	40	—	24742	27

d. French Value.

French Value Fr.	C.	Thaler-Value ℛℓ	Sgr	₰	South-german Value (52½ fl. p. pound.) G.	Kr.	Dt.	English Value L.	sh.	d.
—	1	—	—	1	—	—	1	—	—	0,1
—	2	—	—	2	—	—	2	—	—	0,2
—	3	—	—	3	—	—	3	—	—	0,3
—	4	—	—	4	—	1	1	—	—	0,4
—	5	—	—	5	—	1	2	—	—	0,5
—	6	—	—	6	—	1	3	—	—	0,6
—	7	—	—	7	—	2	—	—	—	0,7
—	8	—	—	8	—	2	1	—	—	0,8
—	9	—	—	9	—	2	2	—	—	0,9
—	10	—	—	10	—	2	3	—	—	1,0
—	20	—	1	7	—	5	3	—	—	1,9
—	30	—	2	5	—	8	2	—	—	2,9
—	40	—	3	3	—	11	1	—	—	3,9
—	50	—	4	1	—	14	1	—	—	4,9
1	—	—	8	1	—	28	1	—	—	9,7
2	—	—	16	2	—	56	2	—	1	7,4
3	—	—	24	3	1	24	3	—	2	5
4	—	1	2	4	1	53	1	—	3	2,8
5	—	1	10	5	2	21	2	—	4	0,5
6	—	1	18	6	2	49	3	—	4	10,2
7	—	1	26	7	3	18	—	—	5	8
8	—	2	4	8	3	46	1	—	6	5,6
9	—	2	12	9	4	14	2	—	7	3,1
10	—	2	20	10	4	42	3	—	8	1
20	—	5	11	8	9	25	3	—	16	2
30	—	8	2	6	14	8	2	1	4	3
40	—	10	23	4	18	51	1	1	12	4
50	—	13	14	2	23	34	—	2	—	5
100	—	26	28	4	47	8	1	4	—	10
200	—	53	26	8	94	16	1	8	1	8
300	—	80	25	—	141	24	2	12	2	6
400	—	107	23	4	188	32	2	16	3	4
500	—	134	21	8	235	40	3	20	4	2
1000	—	269	13	4	471	41	1	40	8	5

II. WEIGHT.

The »Zollpfund forms the unity of weight for the Zollverein and is equal to ½ Kilogramme or 500 Grammes. 100 Zollpfund make a »Zollcentner (hundredweight). The Zollpfund is divided into 30 Loths à 10 Quentchen, à 10 Cents, à 10 Corn. The proportion of this weight to the English avoir-du-pois weight — of which 1 Cwt. is equal to 4 Quarters, 112 Pounds or 1792 Ounces — results from the following table.

Zoll - Weight.		English avoir du poids W.		English avoir du poids W.		Zoll - Weight.	
Ctr.	℔	Cwt. = 112 lb.	Pounds.	Cwt. = 112 lb.	Pounds.	Ctr.	℔
—	1	—	1,102	—	1	—	0,907
—	2	—	2,205	—	2	—	1,814
—	3	—	3,307	—	3	—	2,722
—	4	—	4,409	—	4	—	3,629
—	5	—	5,512	—	5	—	4,536
—	6	—	6,614	—	6	—	5,443
—	7	—	7,716	—	7	—	6,350
—	8	—	8,819	—	8	—	7,257
—	9	—	9,921	—	9	—	8,165
—	10	—	11,023	—	10	—	9,072
—	11	—	12,126	—	11	—	9,979
—	12	—	13,228	—	12	—	10,886
—	13	—	14,330	—	13	—	11,793
—	14	—	15,432	—	14	—	12,701
—	15	—	16,535	—	15	—	13,608
—	16	—	17,637	—	16	—	14,515
—	17	—	18,739	—	17	—	15,422
—	18	—	19,842	—	18	—	16,329
—	19	—	20,944	—	19	—	17,236
—	20	—	22,046	—	20	—	18,144
—	30	—	33,070	—	30	—	27,215
—	40	—	44,093	—	40	—	36,287
—	50	—	55,116	—	50	—	45,359
—	60	—	66,139	—	60	—	54,431
—	70	—	77,162	—	70	—	63,503
—	80	—	88,186	—	80	—	72,574
—	90	—	99,209	—	90	—	81,646
—	—	—	—	—	100	—	90,718
1	—	—	110,232	1	—	1	1,604
2	—	1	108,464	2	—	2·	3,208
3	—	2	106,696	3	—	3	4,812
4	—	3	104,928	4	—	4	6,417
5	—	4	103,160	5	—	5	8,021

III. MEASURES.

a. Long-measures.

The long-measures, particularly of moment in the calculation of prices, are the foot and the Ell. The foot is divided into 12 Zoll (inches) à 12 Lines generally, in Baden, the Grand-Duchy of Hessia, Nassau and Württemberg, however into 10 Zoll à 10 Lines.

A foot is equal in			An ell is equal in		
Baden . . . . to 0,300 mètres	0,984 Engl. foot.		Baden . . . . to 0,600 mètres	0,656 yards.	
Bavaria . . . . » 0,292	»	0,957 »	Bavaria . . . . »· 0,833	»	0,911 »
do. Rhenish . . » 0,333	»	1,092 »	do. Rhenish . . . . . » 1,200	»	1,312 »
Brunswick . . . . . » 0,285	»	0,934 »	Brunswick . . . . : » 0,571	»	0,624 »
Frankfort on M. . . » 0,285	»	0,934 »	Frankfort on the M. . . » 0,547	»	0,598 »
Hanover . . . . . . » 0,292	»	0,957 »	do. Brab. ell » 0,699	»	0,764 »
Hessia (Grand-d.) . . » 0,250	»	0,820 »	Hanover . . . . » 0,584	»	0,638 »
do. (Elector.) . . » 0,288	»	0,943 »	Hessia (Grand-d.) . . . . » 0,600	»	0,656 »
Nassau . . . . » 0,300	»	0,984 »	do. (Elector.) 0,570		0,623 »
Oldenburg . . . . . » 0,296	»	0,270 »	Nassau . . . . » 0,600	»	0,656 »
Prussia . . . . » 0,314	»	1,029 »	Oldenburg . · . . . . » 0,581	»	0,633 »
Saxony . . . . » 0,282	»	0,922 «	Prussia (Berlin) . . . . . » 0,667	»	0,729 »
Württemberg . . . » 0,286	»	0,939 »	Saxony (Leipzig) . . . . » 0,565	»	0,617 »
			Württemberg . . . . . . » 0,614	»	0,671 »

A mètre is equal in		A metre is equal in	
Baden . . . . to 3,333 foot.		Baden . . . . to 1,666 ells.	
Bavaria . . . . » 3,424	»	Bavaria . . . . » 1,200	»
do. Rhenish . . . . » 3,003	»	do. Rhenish . . . . » 0,833	»
Brunswick . . . . » 3,508	»	Brunswick . . . . » 1,751	»
Frankfort on the M. . . . . » 3,508	»	Frankfort on the M. . . . . » 1,828	»
Hanover . . . . » 3,424	»	do. Brabant ell . . . . . . » 1,430	»
Hessia (Grand-d.) . . . . » 4,000	»	Hanover . . . . » 1,712	»
do. (Elector.) . . . . » 3,472	»	Hessia (Grand-d.) . . . . · » 1,666	»
Nassau . . . . » 3,333	»	do. (Elector.) . . . . » 1,754	»
Oldenburg . . . . » 3,377 »		Nassau . . . . » 1,666	»
Prussia . . . . » 3,184	»	Oldenburg . . . . » 1,721	»
Saxony . . . . » 3,546	»	Prussia (Berlin) . . . . » 1,499	»
Württemberg . . . . » 3,496	»	Saxony (Leipzig) . . . . » 1,766	»
		Württemberg . . . . » 1,628	»

A foot English is equal in		A yard is equal in	
Baden . . . . to 1,017 foot.		Baden . . . . to 1,524 ells.	
Bavaria . . . . » 1,041	»	Bavaria . . . . » 1,097	»
do. Rhenish . . . . » 0,913	»	do. Rhenish . . . . » 0,762	»
Brunswick . . . . » 1,067	»	Brunswick . . . . » 1,602	»
Frankfort on the M. . . . . » 1,067	»	Frankfort on the M. . . . . » 1,672	»
Hanover . . . . » 1,041	»	do. Brabant ell . . . . . . » 1,308	»
Hessia (Grand-d.) . . . . » 1,216	»	Hanover . . . . » 1,567	»
do. (Elector.) . . . . » 1,059	»	Hessia (Grand-d.) . . . . » 1,524	»
Nassau . . . . » 1,017	»	do. (Elector.) . . . . » 1,605	»
Oldenburg . . . . » 1,027	»	Nassau . . . . » 1,524	»
Prussia . . . . » 0,971	»	Oldenburg . . . . » 1,579	»
Saxony . . . . » 1,078	»	Prussia (Berlin) . . . . » 1,371	»
Württemberg . . . . » 1,063	»	Saxony (Leipzig) . . . . » 1,620	»
		Württemberg . . . . » 1,490	»

b. Dry-Measures.

1. For dry goods.

Prussian corn-measure.

1 Scheffel = 4 Viertel = 16 Metzen.
1 — = 4 —

English corn-measure.

Quarter.	Cooms.	Bushels.	Pecks.	Gallons.	Pottles.	Quarts.	Pints.	Gills.
1 =	2 =	8 =	32 =	64 =	128 =	256 =	512 =	2048
	1 =	4 =	16 =	32 =	64 =	128 =	256 =	1024
		1 =	4 =	8 =	16 =	32 =	64 =	256
			1 =	2 =	4 =	8 =	16 =	64
				1 =	2 =	4 =	8 =	32
					1 =	2 =	4 =	16
						1 =	2 =	8
							1 =	4

Proportion of the Prussian and English corn-measure.

1 Scheffel Prussian = 0,189 Quarters or 12,096 Gallons.	‖	1 Quarter = 5 Scheffel 4,64 Metzen.
1 Viertel — = 0,047 — or 3,024 —		1 Bushel = — 10,58 —
1 Metze — = 0,011 — or 3,024 Quarts.		1 Gallon = — 1,32 —
		1 Quart = — 0,33 —

Proportion of the Prusian and French corn-measure.

1 Scheffel Prussian = 54,962 Litres.	‖	1 Hectolitre = 100 Litres = 1,819 Scheffel.
1 Viertel — = 13,740 —		1 — = 0,018 —
1 Metze — = 3,435 —		

2. Measures for liquids (wine-measures).

BADEN:

1 Ohm = 100 Maass = 400 Schoppen = 150 Litres = 33,014 Gall. or 33 Gall. » Quarts 1,12 Pints.
1 — = 4 — = 1,5 — = 0,330 — or » — 1 — » — 2,56 Gills.
1 Hectolitre = 100 Litres = 66,666 Bad. Maass or 66 Maass 2,66 Schoppen.
1 — = 0,666 — or » — 2,66 —
1 Gallon = 4 Quarts = 8 Pints = 32 Gills = 3,028 Maass.
1 — = 2 — = 8 — = 0,757 — = 3,028 Schoppen.
1 — = 4 — = 0,378 — = 1,512 —
1 — = 0,094 — = 0,378 —

BAVARIA:

1 Eimer = 60 Maasskannen = 240 Quartel = 64,141 Litres = 14,117 Gall. or 14 Gall. » Quarts » Pints 3,744 Gills.
1 — = 4 — = 1,069 — = 0,235 — or » — » — 1 — 3,520 —
1 — = 0,267 — = 0,059 — or » — » — » — 1,880 —
1 Hectolitre = 100 Litres = 1,559 Eimer or 1 Eimer 33 Maasskannen 2 Quartel.
1 — = 0,015 — or » — » — 3,711 —
1 Gallon = 4 Quarts = 8 Pints = 0,07 Eimer or 4,248 Maasskannen.
1 — = 2 — = 0,017 — or 1,062 —
1 — = 0,008 — or 0,531 — or 2,124 Quartel.

GRAND-DUCHY OF HESSIA:

1 Ohm = 80 Maass = 320 Schoppen = 160 Litres = 35,215 Gall.
  1 — = 4 — = 2 — = 0,440 — or 1 Quart 1 Pint 2 Gills.
    1 — = 0,5 — = 0,110 — or » — » — 3,52 —

1 Hectolitre = 100 Litres = 50 Maass.
  1 — = 0,5 — = 2 Schoppen.

1 Gallon = 4 Quarts = 2,271 Maass or 2 Maass 1,08 Schoppen.
  1 — = 0,567 — or » — 2,268 —

NASSAU:

1 Ohm = 80 Maass = 320 Schoppen = 135,576 Litres = 29,839 Gall.
  1 — = 4 — = 1,694 — = 0,372 — or 1,488 Quart.
    1 — = 0,423 — = » — or » — 2,976 Gills.

1 Hectolitre = 100 Litres = 59,007 Maass.
  1 — = 0,590 — or » Maass 2,36 Schoppen.

1 Gallon = 4 Quarts = 2,679 — or 2 — 2,71 —
  1 — = 0,669 — or » — 2,67 —

PRUSSIA:

1 Oxhoft = 1½ Ohm = 3 Eimer = 6 Anker = 180 Quart = 206,106 Litres = 45,363 Gall.
  1 — = 2 — = 60 — = 68,702 — = 15,121 —
    1 — = 1,145 — = 0,252 — or 1,008 Quart.

1 Hectolitre = 100 Litres = 87,333 pr. Quart.
  1 — = 0,873 —

1 Gallon = 4 Quarts = 3,967 —
  1 — = 0,991 —

WÜRTTEMBERG:

1 Eimer (Helleiche) = 16 Imi = 160 Maass = 640 Quart = 293,927 Litrés = 64,692 Gall.
  1 — = 10 — = 40 — = 18,370 — = 4,043 —
    1 — = 4 — = 1,837 — = 0,404 —
      1 — = 0,459 — = 0,101 —

1 Hectolitre = 100 Litres = 54,435 Maass or 54 Maass 1,720 Quart württ.
  1 Litre = 0,544 — or » — 2,176 —

1 Gallon = 4 Quarts = 2,473 — or 2 — 1,720 —
  1 — = 0,618 — or » — 2,474 —

SECOND DIVISION.

(No. 153. of the offic. Catalogue.)

The Manufactory of Argol & Cognac-Oil

of

C. LICHTENBERGER

at **Hambach** near **Neustadt** on the Haardt (Pfalz).

1.	Cognac-Oil prime . . . . the ½ Kilo	5 £	17	sh.
2.	Cognac-Oil second . . . . " ½ "	3 "	15	"
3.	Cognac-Oil rough . . . . . " ½ "	3 "	7	"
4.	Oenanth acid rectified . . . " ¼ "	16 "	14	"
5.	Oenanth acid rough . . . . " ¼ "	12 "	10	"
6.	Wine-lees-brandy . . . . the 1000 Litre	37 "	10	"
7.	Cognac-Spirit . . . . . " 1000 "	66 "	14	"
8.	Argol from the lees of wine (90 % holding) the 50 Kilo . .	5 "	—	"
9.	Tartaric lime (96 % holding) the 50 Kilo	4 "	4	"

Obtainable only of **F. M. Pokorny,** 15 Fish Street Hill, City, LONDON, Sole Consignee, to whom samples may be applied for and orders transmitted for execution.

(No. 2059. of the offic. Catalogue.)

Our

METAL AND GLASS LETTERS

are intended

for public buildings,

𝔐𝔢𝔯𝔠𝔞𝔫𝔱𝔦𝔩𝔢 𝔉𝔦𝔯𝔪𝔰,

Hotel-Offices,

and

Door-Plates.

In the orders it must be specified, if the name is to be executed entirely in CAPITAL LETTERS, or in capital and SMALL LETTERS, and if the name shall be fixed on a wooden, Zinc, or stone ground.

KOCH & BEIN. BERLIN.
Metall- & Glas-Buchstaben-Fabrik.

Koch & Bein, manufacturiers de lettres de verre et de métal, Berlin.

INDUSTRIE AUSSTELLUNG

LONDON · PARIS

KOCH & BEIN Manufacturers of glass & metal letters. BERLIN.

J. G. BIELING General Agent 19 Gloucester St. Queen's Square, W. C. LONDON.

PETERSBURG · BERLIN

R LONDON MAY 1862. G

Koch & Bein, Metall- und Glas-Buchstaben-Fabrikanten, Berlin.

The drawing before us is a copy of our large Tableau at the great International Exhibition: it is executed in **letters of metal and of glass** of different colours.

The **E** in the middle stands for

EUROPE;

the letters at the four corners are the Initial letters of

THE FOUR CHIEF COUNTRIES OF EUROPE

with their national colours and the names of the Capitals of each. The name of our firm is on the frame of the Tableau in the languages of these four countries.

(No. 1317. of the offic. Catalogue.)

C. SCHLICKEYSEN,

BERLIN,

Engineer and manufacturer of tiles, pipes and bricks.

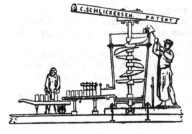

Patent Universal machine for: **bricks, tiles, pipes, turf,** mortar, for mixing in the sugar manufactory, pressing and forming chicory, and preparing graphite in the manufactury of pencils. Cost from £5 to £500.

Mr. SCHLICKEYSEN undertakes the building of whole brick-kilns for summer and winter work, with horse or steam power. This machine will be kept going during the whole of the exhibition, and works up materials sent, at the expense of the sender.

[4.]   (No. 2067. of the offic. Catalogue.)

The Ironworks of Count Einsiedel at

LAUCHHAMMER

in Prussian Saxony

is an establishment of very large extent and various manufacture. Favoured by excellent opportunities and arrangements, it produces, for moderate prices, every variety of cast-iron articles, both raw and worked, which are for the greatest part the origin of this factory.

The following are the principal branches of manufacture:

Enamelled cooking apparatus, distinguished by lightness and elegance of form in a great selection, the enamel being durable and free from admixture of lead; stoves ovens, fenders, fire-irons, warming apparatus for halls, rooms etc., in great variety, especial attention being given to chimneys and to chimney-pieces of a light and tasteful design; minute and delicate castings consisting of trinkets, standishes, chandeliers, letter-pressers, tinder-boxes, watch-stands, figures etc., in an abundant assortment and of careful execution; balustrades, monuments, constructions used in architecture, as winding stairs, verandahs, balconies, columns etc.; articles for gas-illuminations, particularly gas-delivering pipes, candelabra, lamps, gas-arms, gas-chimneys etc.; utensils of convenience, castings for machines, and other objects of a similar kind.

The **bronze-foundry at Lauchhammer** is celebrated far and wide, the greatest artistic works, as statues, busts, portraits etc., being cast and chased with great skill and care. „The colossal monument to be erected to Luther at Worms after the sketch of E. Rietschel" is being executed at this foundry.

Drawings and price-currents are exposed to view in London by Messrs. Bauerrichter & Co., 41 Charter-house Square, these gentlemen having kindly undertaken to give every necessary information upon the subject. For further details we beg to direct inquiries to be made to *"The Ironworks of Count Einsiedel at Lauch-hammer, near Mückenberg, in Prussian-Saxony."*

[5.]   (No. 982. of the offic. Catalogue.)

William Hartmann,

Commission-merchant,

Cologne
(RHENISH-PRUSSIA).

Rhenish Argols; Crystallized Argols (made from the lees of wine); *Acetate of lead* (brown, yellow and white); *Acetic Acid; Verdrigris* (of better quality than the French); *Creosote.*

[6.]   (No. 1923. of the offic. Catalogue.)

HEINR. AUG. SCHÖLLER SONS,

Paper manufacturers

AT DÜREN.

These establishments, which have been in existence for 84 years, are provided with all the latest inventions and improvements, among which may be mentioned two large English paper and pulp machines and twenty-seven Dutch machines, working day and night with steam and water power equal to more than 200 horses. Every species of paper, including the finest sorts, are turned out of these paper mills.

[7.]   (No. 26. of the offic. Catalogue.)

Xaver v. Kilian

of

WALDSHUT (Grand Duchy of Baden),

possessor of the

"BLEICHE WALDSHUT."

The stone, which is procured here, consists of pure quartz sand, very hard and close, and is exceedingly well-adapted for grinding corn, rice, &c., the husks of which it separates with great nicety leaving the flour quite free from foreign admixture. By the completeness of the process accordingly it is possible to obtain groats in equal proportion with the fine white flour.

In grinding varnish (porcelain or earthenware) and coloured surfaces, especially ultramarine, this stone offers a peculiar advantage in consequence of the purity of its grain, in as much as the grinding power is preserved undisturbed, while the detritus that may perhaps fall from the stone in working is melted to pure glass in the fire.

These stones can be delivered of any size that may be required, there being always a **thousand** or more kept ready in stock.

P. J. LANDFRIED,

OF RAUENBERG

(Wiesloch Railway Station, Grand Duchy of Baden).

Branch business at DIELHEIM and MÜHLHAUSEN.

Manufacture of Palatine Cigars.

800 persons are employed in this establishment. The annual production amounts nearly to 50 Millions of Cigars. The assortment exhibited is a selection of first rate qualities, made of the best tobacco of the Baden-Palatinate.

PRICES IN ENGLISH MONEY.

	Shillings	Pence	
Imperials	25	—	The prices here noted are per 1000 Cigars, including the cedar-wood cases, but excluding the packing cases, delivered franco to the Heidelberg station. Each sort is to be had in an inferior quality, in ordinary case, at moderate prices.
¼ Regalia	21	8	
Opera, thick made, in bundles	18	4	
Londres Regalia	17	6	
Trabucos, thick made	17	6	
Regalia, thin made	16	8	
Opera, in bundles	15	10	
Trabucos	15	10	
½ Regalia	15	—	
Entractos	15	—	
Londres, in bundles	14	2	
do. do.	13	9	
Manilla, round, in bundles	13	4	
do. ☐ pressed, do.	13	4	
Comunes (P. J. L.)	11	8	
do. do. ☐ pressed in bundles	12	6	
do. do. a smaller sort	11	8	
Ladies Segars	11	8	
Londres Regalia of another size . . .	17	6	
Comunes, stitched with silk	12	6	
Manilla, round, with tin-foil at the ends, in bundles	15	—	

J. PINTUS & Cᵒ.,
Ironfounders, Agricultural & General Machinists,
Brandenburg a. Havel near Berlin, Prussia.

(Agents in *London:* Mess. A. Heintzmann & Rochussen, 9, Friday Street, Cheapside.) — Depôt in *Berlin:* Bauschule.

☞ J. Pintus & Co. manufacture all kinds of Agricultural and other Machines both for home use and exportation on the most approved Systems and Models of all Nations. J. P. & Co. have obtained 23 first Prizes, Gold, Silver and other Medals etc., and are ready to supply Agriculturists as well as the Trade with the best modern Machinery and Implements of solid make and highly finished at moderate prices. Illustrated Catalogues sent free on application. The Catalogue prices include all parts necessary for working; no extras charged. — Delivery to all parts of the world.

☞ **An Agent always present in the Building who will give every information.** ☜

Pauline Bessert-Nettelbeck,
Berlin, Kronenstrasse 52.

Gold, silver, silk embroidery establishment.

London Agent: Mr. Louis Andrée,
care of Mr. Wm. Klein,
3 Fowkes Buildings, Great Tower Street E. C.

S. UHLMANN,
HOPMERCHANT,
FÜRTH.
ENGLISH OASTHOUSES AT FÜRTH AND HAMBURGH.

AGENT: MORITZ AUERBACH,
10 TOKENHOUSE YARD LONDON.

"GEORGHÜTTE near ASCHERSLEBEN"
(PRUSSIA).
Mineral oil and paraffine works.

Paraffine, refined, white, and inodorate, of various quality; melting-point from 48⁰ to 58⁰ Cels. (118⁰ to 136⁰ Fahr.).

Solar oil (paraffine oil), best quality, specific gravity 0,830 to 0,835; boiling-point from 250⁰ to 350⁰ C. (482⁰ to 662⁰ F.)

Photogene, best quality, specific gravity 0,800 to 0,810; boiling-point from 140⁰ to 250⁰ C. (284⁰ to 482⁰ F.)

[13.]

(No. 1308. of the offic. Catalogue.)

The Cast-Steel Works

of

FRIED. KRUPP

at

ESSEN,

Rhenish - Prussia.

The following articles are manufactured and supplied in the rough or finished state, from solid, cast and forged blocks of

cast-Steel:

Patent Railway tires without a weld; of any size and shape.
Railway carriage axles.
Straight and crank axles for locomotives.
Complete sets of wheels (axles, disc wheels and tires).
Crank pins, piston rods, connecting rods, slide bars.
Railway carriage and locomotive springs.
Paddle and propeller shafts, straight and cranked; of any size.
Ship screws and anchors.
Rolls of every description.
Gun barrels.
Pieces of Ordnance of every calibre.
Pump rods for mines up to 60 feet length with couplings.
Tool steel.

One of the most important manufactures of these works are the tires made from a solid forged piece without a weld patented in 1853, the patent being still available in England and Prussia. More than **40,000** tires are already supplied from the works, and are used on almost all the important lines of the Continent and in England, as well as on East Indian and American Railways. The original price was 120 sh. per cwt., but owing to the great improvements made in the process, the price was gradually reduced to 85 sh. per cwt.

Cast-Steel guns were first manufactured in these works in 1847, and since 1856 — until which period experiments were chiefly carried on — more than 1000 guns have been turned out. The largest gun ever made here was of 9 inch bore and weighed when entirely finished 8 tons. This is now exhibited in London.

Considerable numbers of large paddle and propeller shafts have been supplied more especially to the Austrian Lloyd at Trieste, the Bremen Lloyd, the Hamburgh American Steam-packet Company, the Danube Steamboat Company, the French Navy, some also being furnished to Messrs. Penn & Co. at Greenwich, and the Royal Mail Company. The largest intermediate (crank) shaft forged here under a hammer of fifty tons weighs when finished 15 tons; however any larger size may be executed.

The price of guns and axles varies according to description, size, and number required, and is usually quoted per piece.

However £ 15 per cwt. may be considered as a rough average for guns and Locomotive crank-axles when finished, and £ 12 per cwt. for large intermediate shafts with one crank only. Straight shafts for steamers, ready turned £ 5—£ 7½.

Further printed particulars on the articles of these works may be had at their exhibition place.

Krupp's cast-Steel-works are represented at
London by Mr. *Alfred Longsdon*, Civil-Engineer, 9 Chepstow-place, Camberwell New-road S.
Paris by M. *Fried. Krupp*, 12 rue de l'échiquier.
Berlin by M. *Carl Meyer*, Schellingstrasse No. 3.
Vienna by M. *A. Strecker*, Civil-Engineer, Fünfhaus No. 225/6.
Amsterdam by M. *Aug. Köster*, Rokin E. No. 89.
St. Petersburg by M. *Carl Barth*, Civil-Engineer, New Kalinkinbridge, house Zimmermann.
New-York by Mr. *Thos. Prosser*, Civil-Engineer, 28. Platt street.

[14.]

(No. 1311. of the offic. Catalogue.)

THE PLASTIC COAL FACTORY,

BERLIN, 15. ENGEL-UFER.

The Proprietors H. LORENZ and TH. VETTE

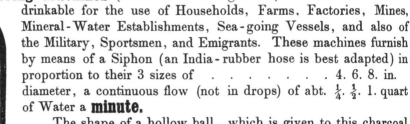

manufacture **Filtering-Machines** (Reservoir-Filtering-Balls) to render bad Water drinkable for the use of Households, Farms, Factories, Mines, Mineral-Water Establishments, Sea-going Vessels, and also of the Military, Sportsmen, and Emigrants. These machines furnish by means of a Siphon (an India-rubber hose is best adapted) in proportion to their 3 sizes of . . . . 4. 6. 8. in. diameter, a continuous flow (not in drops) of abt. $\frac{1}{4}$. $\frac{1}{2}$. 1. quart of Water a **minute.**

The shape of a hollow ball, which is given to this charcoal makes the Apparatus conveniently portable, presents proportionately the largest surface for filtering, and makes the application to any kind of vessel possible.

Patent.

The cleaning of the same is easily effected by any body.

Prospectuses in all languages can be had at our Agent, Mr. CHARLES EGAN, 9 Tower Hill, East.

List of Prices

delivered in Berlin, exclusive of Package, for cash.

4 in. filtering ball,	with 2 feet hose in Japanned box for Travelling, ... each £0. 5. 0.
4 „ do. do.	with 2 feet hose and tin-cock, (in cane netting 6 d. more) . . . . . . „ „ 0. 4. 7.
6 „ do. do.	with 4 feet hose and tin-cock, (in cane netting 1 s. more) . . . . . . „ „ 0. 8. 7.
8 „ do. do.	with 4 feet hose and tin-cock, (in cane netting 1 s. 6 d. more) . . . „ „ 0. 11. 7.

[15.]

(No. 2197. of the offic. Catalogue.)

H. L. PERLBACH,

DANTSIC,

AMBER MERCHANT AND MANUFACTURER OF AMBER-BEADS.

Agent: MEYER LEVIN, *London. 17 South Street Finsbury.*

A most rare and excellent piece of Amber of 12 ℔. weight, 3 inches thick and 14 inches in circumference. It was found in an Amber mine in the village of Gluckau, near Dantsic, at a depth of 70 feet, and its worth has been estimated at £600. On account of its size, smoothness, good quality, and fine colour, it surpasses all similar productions, which have hitherto been brought to light from the bowels of the earth, and may be considered as the only specimen now extant in such perfection and size. As a great natural curiosity it deserves the attention of all naturalists and is to be recommended as a valuable acquisition to National Museums or to private Cabinets of natural curiosities.

Further: A piece of Amber of 4 ℔. weight, one of 2 ℔. weight, and sundry samples of assortments of small Amber in which a considerable transmarine trade is carried on.

Samples of sundry assortments of Amber-Beads exhibited to shew the kinds used for exportation to the African Markets, as Senegal, Gambia etc. — For prices, information etc., apply to the above named agent, who is also charged with the sale of the large piece of Amber, described above.

KLETT & Co.

NUREMBERG.

Engineers. Bridge Constructors. Ironfounders. Manufacturers of Railway material — Waggons, Wheels, Turntables &c. Wire nail makers &c.

Model of the new Railway-Bridge over the Rhine near Mayence constructed on the system of M. Pauli,

which differs from others in being the result of a scientific development of the principle of beams of equal resistance, supposing the top and bottom girder to be acted upon by equal forces at every point of their length, thus requiring equal sections throughout.

These principles and conditions determine the shape of the beam in a most exact manner, and it may be proved, that no other combination gives a construction of beams requiring less material.

The model represents one Arch (or Span) of the Rhine-bridge, in a scale of 1:10, the whole Bridge having 4 Principal arches, over the stream, of 101,9^m (332,4 feet) clear space each, and 63° obliquity, followed by 6 of 33,5^m (109,9 feet), 13 of 15^m (49,2 feet), 2 of 25^m (82,0| feet), and 7 of 15^m (49,2 feet) clear span, giving together 32 Arches, with a total length of 1028,6^m = 3375 feet = 0,64 mile between the abutments.

The Bridge table is to be for a double line, or two pair of Rails, the supporting beams of the one being independent of those of the other, with Footpaths also, on either side for Public traffic. The traffic load bearing always upon the joint-points of the supporting beams, they are nowhere affected by bending forces; all inner strains have the same direction with the resisting pieces and cause only tension or compression.

The simple disposition of the filling between the top and the bottom girder (compression bar, and tension bar) being composed of vertical posts and diagonal ties, allows the calculation of the inner strains of the beams with great accuracy, so that the dimensions may be determined with complete safety.

By this plan there will in no part be dead masses to be supported, necessitating a heavier construction without essentially adding to strength.

The immediate supports of the Railway are arranged in such a manner, that the Rails rest directly over Irongirders; the wooden sleepers serving only as an elastic medium to lessen the influence of shocks as much as possible, so that even neglected repair of the woodwork will occasion no danger to the traffic.

M. Pauli's System allowing a strict execution according to the forces in action as above mentioned, the manner of joining the single pieces must be of equal security. From this motive warm rivetting at the principal joints of construction was rejected, and bolts and nuts employed instead; the shafts of which are turned conically $\frac{1}{100}$, and fitted exactly into the holes, drilled out and tapered accordingly.

This manner of joining always gives a greater security against shearing (amounting from 6—8 pCt.) compared with rivets put in hot.

As there is no force acting longitudinally upon the bolts and nuts, but that of shearing only, there is no action upon the nuts, and consequently no loosening of them, the threads of the screws not even being hammered.

Every beam rests at each end by planed plates on a transverse cylindrical face, curved according to a radius large enough to allow the beam to deflect without causing any onesided pressure of great intensity upon the supporting plates. To prevent a dislocation of those plates resting upon one another in any direction, they are fitted with projections on both sides, acting like teeth.

The lower ones of the aforesaid supporting plates are attached to chairs, of which one is fixed, while the other is moveable on cylinders, or parts of such, to allow an alteration in length by change of temperature.

In course of construction all flat iron of the beams which has to resist tension is stretched by a peculiar machine as much as to 1200 Kilogr. per square centimetre (7,62 tons per square inch) of section, being at the same time struck by hammer blows. By this operation the elasticity of the iron is augmented, so that, within the said limits of tension, no permanent lengthening will take place from this moment save an elastic one only.

To protect the iron from oxidation, all scale and rust is taken away as much as possible by macerating it in acids, scouring and washing, and putting it afterwards in this state into seething oil.

All moisture is extracted by the high temperature of the oil (300° Celsius), so that a coat of oil varnish adheres firmly to the clean iron. Upon this a ground colour of iron minium is given.

In adjustment, all ties are stretched by a peculiar contrivance as much as $\frac{1}{6000}$ of their length, in order to destroy all curvities possibly left, so that the ties will be in action at the beginning of any alteration of the form in consequence of the traffic.

In determining the maximum working strain, the effect of the shocks caused by the traffic load was thought necessary to be taken into consideration, and their amount to be estimated in proportion to the width of the span. The effect of the shocks was considered as a multiple of the traffic load.

In calculating the maximum working strain in the system, the dimensions were taken, such that a charge considered resting, and being composed partly of the permanent proper weight, partly of the multiple of the traffic load, which represents the shocks, would equalise the limit of elasticity.

The proportion of the bearing power, up to the limit of elasticity, less the proper weight to the resting traffic load, is undoubtedly a true scale for judging of the security of any construction.

In calculating the dimensions of bridge beams, the maximum traffic load being given, this proportion is invariably adapted for all spans.

If we call

p_0 the permanent weight per unit of length,
p_1 the traffic load   do.   do.
R the maximum of working strain caused by the traffic load resting,
Rg the limit strain of elasticity, and
N the relative bearing power as above,

we have

$$\frac{p_0 + Np_1}{p_0 + p_1} = \frac{Rg}{R};$$

and further N being determined

$$R = Rg \, \frac{p_0 + p_1}{p_0 + Np_1}.$$

The traffic load for Railway-Bridges is considered to consist of:

3 Locomotives with tenders ahead, each Locomotive 30 tons, Tender 15 tons weight, with $13,5^m$ (44,3 feet) length from buffer to buffer. Any number of Waggons of two Axles, with 8 tons weight on each axle, standing 3^m (9,8 feet) apart, the whole length to be 6^m (19,7 feet).

In calculating for every joint of the bearing beams this Train is to be shifted about, till the highest momentum is found, and by these the form of Pauli's curve, as well as the dimensions of sections are obtained.

In the following table the uniformly distributed weights per unit of length are contained, which will give the same momentums for the middle of the opening as the above mentioned train in its most unfavourable position will do. Applying these weights to a great number of existing bridges, and chiefly

those of English Engineers, the relative bearing power was determined, and as a mean the number 3 was found.

It is in applying this figure and the train above mentioned, that the Railway-Bridges, of which the execution is entrusted to the Exhibitors, are calculated, in case no extra load caused by the traffic is to be taken into account (as by Mr. Engerth's heavy Engines).

For the determination upon the relative bearing power, and for the purpose of removing all doubts about the amount of the maximum of working strain to be adopted in the construction of Railway-Bridges, it would be very desirable that Engineers should come to an agreement about this standard figure.

The beam on M. Pauli's system was complained of, as being too light and therefore not to have body enough to check the shocks of the traffic load. But it will easily be understood that, with the same amount of cost, greater masses may be put into a bridge on M. Pauli's system, as into a lattice work or such of sheet iron, a bed of gravel beneath the rails being all that is required.

This plan is evidently more rational than providing a greater capacity to resist shocks, by the accumulation of masses of iron in the girders, which at the same time disturbs the correctness of the construction.

The weights and cost of the iron work of the bridge over the Rhine near Mayence are to be for a single track, including elevation and scaffolding.
A.   4 principal arches   1,430 tons   £ 42,910
B.   6 arches next to these   235 »   » 6,780
C.   22 Smaller ones   300 »   » 8,670

The Exhibitors completed since 1857, besides the bridge over the Rhine on M. Pauli's system: 46 Railway-Bridge arches for single tracks — from $8-54^m$ (26—177 feet) clear span each, making a total of 858^m (2816 feet) and 6 openings for road bridges, having in total a clear span of $142,1^m$ (466 feet).

The following table gives the weights and cost of Railway-Bridges, in accordance with the above mentioned load and a threefold bearing power, up to the limit of elasticity, this being at 1600 Kil. per square centimetre, or 10—16 tons per square inch.

Scale of Prices &c.

for Railway-Bridges for one Track according to M. Pauli's System.

The prices are calculated according to the rates of the materials of Rhineish Ironworks.

Clear span between piers		Distance of supporting points	Permanent load		Traffic load		Working Strain		Cost per arch	
mètres	feet	mètres	per mètre	per foot	per mètre	per foot	Kilogr. per ☐centimtr.	tons per ☐mètre	£	ℛ
			tons	tons	tons	tons				
9,4	31	10	0,87	0,28	4,80	1,57	592	3,76	162	1,080
19,2	63	20	1,13	0,37	4,03	1,32	625	3,97	480	3,200
28,9	94	30	1,39	0,46	3,68	1,21	654	4,15	930	6,200
38,7	127	40	1,64	0,54	3,53	1,15	675	4,29	1,580	9,600
48,4	158	50	1,88	0,62	3,40	1,11	690	4,38	2,340	15,600
58,2	190	60	2,20	0,72	3,28	1,08	720	4,57	3,370	22,470
67,9	223	70	2,48	0,81	3,18	1,04	745	4,73	4,520	30,130
77,7	254	80	2,73	0,88	3,09	1,01	772	4,90	5,760	38,400
87,7	286	90	2,98	0,97	3,00	0,98	797	5,06	7,160	47,730
97,2	320	100	3,26	1,07	2,93	0,96	820	5,21	8,790	58,600
106,9	351	110	3,52	1,16	2,88	0,94	842	5,35	10,550	71,030
126,4	415	130	4,89	1,44	2,81	0,92	900	5,72	15,850	102,330
146,0	478	150	5,26	1,72	2,77	0,91	938	6,08	22,200	148,000

J. P. GOEBEL & SON,

GROSSALMERODE NEAR HESSE-CASSEL.

Manufactory of Plumbago-crucibles and Hessian-crucibles.

WAREHOUSE OF CLAY-PIPES.

Exhibited Objects.

1) 1 Plumbago-Crucible of 600 mark, which, according to the testimony of the Royal Prussian Mint at Berlin, has undergone 30 meltings and notwithstanding is still fit for use.

2) 1 Plumbago-Crucible of 600 mark, which, according to the testimony of the mint at Frankfort on Maine, has undergone 42 meltings, and is still fit for use.

3) A Plumbago-Crucible of 600 mark not used.

4) An assortment of plumbago-crucibles.

5) An assortment of Hessian crucibles.

6) An assortment of cheap clay-pipes.

Our Plumbago-crucibles bear the melting of silver and brass 40—50 times, of copper and packfong 20—30 times, and of cast-Steel 6—7 times; they are made in every form and size.

J. WEIMAR,

SCULPTOR,

BERLIN, ORANIENSTR. 118.

CAST MARBLE.

This, my new invention, has the properties both of plaster of Paris and of marble. Of the plaster of Paris: because it can be cast, and reproduces the finest lines of the original model; of the marble: because it has the same hardness, transparency, crystalization, and capability of receiving polish. It gives the most natural imitation of every kind of white as well as coloured marble, and by a particular treatment may be made to represent the clear transparency of alabaster. The resistance against the influences of the temperature, the striking likeness to marble and alabaster, the cheapness of the materials, and their easy manipulation, its close affinity to glass, porcelain, wood and metal gives it the preference over all works of sculpture, architecture, stucco, and the embellishment of rooms, vessels (fireplaces, glasses &c.), as a cheap and perfect substitute for marble and plaster of Paris, and by its cheapness, allows even the man of moderate means to surround himself with works of fine art. The cleaning of this marble is effected in the same way as with natural marble, with soap and a hard brush.

The above-mentioned inventor intends to sell the patents of several countries, and gives particulars at the London Exhibition in the Prussian department.

P. J. Landfried, Künzle, & Co.

Heidelberg
(GRAND-DUCHY OF BADEN).

Great assortment of Rhine, Palatinate, and Baden wines (white and red), of the most renowned vintages and at very moderate prices.

PRICE CURRENT.

IN CASK.

Baden Wines.		The Baden aum of 33⅓ Gallons or 200 Bottles or 150 Litres. L. Sterl. From — To
Markgräfler	1857. 1858.	8—12
Zeller (red Wine) . . .	1857. 1858.	9—12
Affenthaler (red Wine) .	1857. 1858.	10—18
Lützelsachser (red Wine)	1857. 1858.	9—12

Palatinate Wines.

Deidesheimer, Wachen-heimer	1857.	
Ungsteiner, Dürkheimer .	1858.	10—24
Ruppertsberger, Feuer-berger	1859.	
Forster Traminer and Riesling . . .	1857. 1858.	12—25
Forster Freundstück . .	1857.	25—35
Forster Ungeheuer . . .	1857.	30—45

Rhenish Wines.

Rüdesheimer, Rüdes-heimer Ruland and Rüdesheimer Berg	1857. 1858.	15—35
Geisenheimer, Rauen-thaler, Neroberger	1857. 1858.	25—40
Hochheimer	1857. 1858.	25—45
Marcobrunner, Stein-berger	1857. 1858.	30—50
Rüdesheimer Berg Cabinet and Rüdesheimer Hinter-haus	1857. 1858.	40—60
Oberingelheimer (red Wine)		12—20
Assmannshäuser (red Wine)		15—24

NB. Wines delivered in Casks not under ten Gallons.

IN BOTTLE.

Per Dozen in Baden Bottles

White Wines. 1857. 1858.

	L. Shilling
Markgräfler	1. —
Ruppertsberger Traminer . . .	1. 5.
Ungsteiner	1. 10.
Forster Riesling	1. 10.
Forster Riesling Auslese . . .	1. 15.
Forster Riesling Auslese	2. —
Feuerberger Traminer Auslese .	1. 15.
Forster Sorgenbrecher	2. 10.
Forster Monster	3. —
Rüdesheimer Ruland	1. 10.
Rüdesheimer	1. 15.
Rüdesheimer Berg	2. —
Geisenheimer.	2. —
Neroberger	2. 5.
Rauenthaler	2. 10.
Hochheimer	3. —
Marcobrunner	3. 5.
Steinberger Cabinet	3. 10.
Rüdesheimer Hinterhaus . . .	3. 10.
Johannisberger	5. 10.

Red Wines. 1857. 1858.

	L. Shilling
Zeller	1. 8.
Affenthaler	1. 12.
Oberingelheimer.	1. 12.
Assmannshäuser	1. 15.

Delivered in London free of charge: duty casks and packing included

Hobrecker, Witte, & Herbers,

at

HAMM, Westphalia,

PRUSSIA,

Manufacturers of Wire, Wire-Nails etc., enjoy particular advantages which enable them to distance all competition. **Hamm** is situated in the Centre of the richest Iron and Coal District at the junction of four principal Railways, with which Messrs. **Hobrecker & Co.'s** Establishment is connected by their own line of Rails. They are thus enabled to economize all intermediate expenses of Carriage and obtain the best German Pig-Iron delivered at their Factory, at the smallest possible cost of transport.

Their Establishments contain 21 Puddling Furnaces, 3 Welding Furnaces, 4 Steam-Hammers, 3 Rolling-Mills, and 3 endless wire-drawing Mills, which employ constantly 76 large and 300 small wire drums, independent of 120 Card wire drums, while the waste heat from the different Furnaces is turned to account in generating Steam for the motive-power of the Establishment.

They use the finest German Iron to manufacture, round, half-round, flat, square and half-square wire, and by means of 42 Nail-Machines, they supply Wire-Nails, or Paris points, from the smallest size to a dimension of 16 Inches long by $\frac{1}{2}$ Inch thick.

Messrs. **Hobrecker, Witte, & Herbers** devote themselves especially to the production of Telegraph-Wire, either for conducting, joining, or covering, bright, annealed, or japanned, and also of Cast Steel Telegraph-Wire for immersion and for the combination of different electric currents. Their resources for the development of this branch of their business are on a Scale of magnitude, sufficient to meet the most extensive demands for Telegraph-Wire which may be made upon them.

PRIZE - MEDALS.

BERLIN 1844.   BERLIN 1844.
LONDON 1851.   LONDON 1851.
MUNICH 1854.   MUNICH 1854.
PARIS 1855.   PARIS 1855.

H. M. ENGELER & SON,
furnishers to the royal Prussian court.

BRUSH MANUFACTORY.

Export — Wholesale and Retail —

BERLIN,
Behrenstrasse No. 36. Opera place.

We recommend our **brushes of all sorts**; toilet-brushes in wood and ivory; all kinds of brushes for the house and stable as well as for every technical use; brushes for room, decorative, and oil painting etc.

Quality the very best.
Lists of price are to be had.

As a new invention we recommend:

The Steel Hair Brush.

This brush made of wire, penetrates the thickest hair, and, even if the hair is thin, acts upon the skin of the head most softly and agreeably, and clears it of all scurf, thus preventing the hair from becoming gray or falling off.

The magnetic Steel Hair Brush

has a strong magnet inside. In addition to the qualities of the simple Steel hair brush, it has that of furthering the growth of the hair by its salutary effects on the roots, and has equally a great influence upon the reascension of the nutritive matter in the capillary tubes. It is very effectual against rheumatic and nervous headaches, and after being used leaves a pleasant freshness of sensation on the skin of the head.

The electro-galvanic Steel Hair Brush

has the same qualities and effects as the magnetic Steel hair brush, but in greater force. It is, however, necessary to arrange the battery every day, which is in its interior.

The magnetic and electro-galvanic Steel Hair Brush

preserves the skin in a healthy state, and especially prevents all rheumatic and nervous diseases of the body

G. W. SUSSNER OF NUREMBERG.
PATENT CRETA POLYCOLOR.

Notice.

In order to prevent mistakes, I earnestly request the Artistic public not to confound my Creta Polycolor with Creta Laevis, Pastel, or any other coloured pencils, the mixture of my material (Composition of oil) being perfectly new and of a very peculiar nature. These pencils, combining extraordinary softness with consistency, retain the finest point and have the peculiar quality, that drawings or paintings of them — like oil-paintings — continue permanent and imperishable on paper, which quality no other dry colour-material hitherto known is possessed of.

I refer to the subjoined testimonials to which the names of several of the most distinguished and celebrated Authorities in Art of Germany, France, and Belgium are annexed, to whom I submitted the Creta Polycolor in order to make a particular examination. I shall therefore abstain from further recommendations.

Now to put the public on their guard against receiving a spurious article, I beg to observe that my Creta Polycolor — like leadpencils — are cased in natural cedar-wood, bearing the stamp:

G. W. SUSSNER PATENT CRETA POLYCOLOR,

and are assorted and packed in cloth-covered cases, containing 48, 36, 24, 18 or 12 pointed pieces each, and are to be had at all Printsellers' and drawing-material warehouses.

Testimonials.

We the undersigned have tried the Creta Polycolor, invented by M. G. W. Sussner of Nuremberg, and discovered in this new invention a most valuable addition to the materials for the drawing world, which, for the benefit of all artists, we wish will meet with a great circulation.

Munich, January 19th 1859.

Professor W. v. Kaulbach, Director of the Royal Academy of fine Arts of Bavaria.

Moritz v. Schwind, Professor. Foltz, Professor. Dietz, Court Painter of Bade.

Lange. Schoen. Dr. v. Hefner-Alteneck. Julius Lange. Reiff. Kirchner. Ed. Schleich. J. Bernhardt. J. Heigel. Morgenstern. Hagn. G. Seeberger. W. Boshart. J. Koch. H. Marr. K. Altmann. K. E. Doepler. W. Asselborn. Joh. Kirner. B. Savery. E. Kundy. Frd. Gärtner. P. Sporer. Aerttinger.

II. To the undersigned the Creta Polycolor of the inventor, M. G. W. Sussner of Nuremberg, appears a very recommendable acquisition for topographic Sbetches, rough drawings of maps in various tones of colours, and for supplement works in maps on the ground — on account of their superior firmness on the paper, which surpasses by far that of the general Colour-Crayons, containing less harshness and krittleness than is generally found in these sorts of pencils.

Munich, January 25th 1859.

H. von Schintling,
Colonel and Director of the topographic bureau.

III. We the undersigned avail ourselves with pleasure of this opportunity, to declare that the invention of the Creta Polycolor, of M. G. W. Sussner of Nuremberg, promotes very much the way of drawing in a manner both pleasing and useful, adding but our wish, that they may be universally adopted.

Munich, January 31th 1859.

Jos. Resch. A. Kölbl. Höfer. A. Doll. M. Huber. M. F. Heil. Jos. Blanz. M. Stieler. A. Lier. Noere. W. Philippi. Rothbart. Kirstein. Lang. de Normy. J. F. Spengel. Marées. Metzener. F. Kreutzer. Kökert. Brausewetter. Reichard.

IV. I have made use of the pencils, called Creta Polycolor, which you have submitted to me.

In the kind of pastel I use, they are extremely useful in expressing delicacy of tone in the flesh tints and brillant points in the accessories.

With a little practice I believe the most beautiful results could be obtained by this sort of pencils, which are capable of an extremely fine point without losing their shine, and easily allow of their being erased, if one wishes to diminish the intensity of the tone or to restore it light and lively.

They can be advantageously substituted for black lead pencils in drawing portraits and sketches from nature.

I cannot praise too much the firm of M. G. W. Sussner on account of the happy results obtained by them, and which are offered to artists and amateurs.

Paris, March 19th 1859.

Vidal.

Perfectly agreeing in the judgment of M. Vidal, with pleasure I join my name to his.

Winterhalter.

V. At the request of M. G. W. Sussner of Nuremberg, Manufacturer, we the undersigned members of the Academic Counsel and Professors of the Royal Academy of Fine Arts certify with pleasure that the pencils called: Creta Polycolor (coloured chalk), invented by M. G. W. Sussner, and which have been submitted to a trial by us, perfectly answer the demands which are required in coloured pencils, the material of which is the best of the sort that has been known to the public; consequently they may be considered of great advantage in the Arts relating to drawing.

Dresden, April 15th 1859.

Ernst Rietschel. Julius Schnorr v. Carolsfeld. Julius Hübner. Ludwig Richter. Eduard Bendemann. Ernst Hänel. Gustav Heine.

VI. It is with pleasure we certify to M. G. W. Sussner, Manufacturer of Nuremberg, that the pencils invented by him, and called Creta Polycolor, have been submitted to a trial by us; they are of excellent quality and perfectly answer what is required in a coloured pencil. By this invention he will render an important service to artists and designers.

Anvers, Mai 30th 1859.

C. Guffens. Jan Sweerts. Van Lerius.

VII. I highly approve of the judgment of my colleagues on the subject of the Creta Polycolor, invented by M. G. W. SUSSNER of NUREMBERG, and I wish that this useful and beautiful invention may be fully appreciated by all competent persons as it deserves to be. Coburg, August 4th 1859.

Rothbart, Court Architect.

VIII. The Creta Polycolor, invention of M. G. W. SUSSNER of NUREMBERG, is a valuable substance and of great utility in drawing portraits. It is desirable that this discovery be generally known and used. Berlin, August 25th 1859.

Julius Schrader,
Professor and member of the senate of the Royal Academy of fine Arts.

IX. It is with great satisfaction that I have found in the Creta Polycolor a substance long desired, which, beside its other qualities, is distinguished by the precision of its touch, the fineness and freshness of its tone, and allows of its being equally well used for landscape sketches. Berlin, August 25th 1859.

Ferdinand Bellermann,
Professor.

X. It is with the greatest satisfaction that I am enabled to speak of the coloured pencils or Creta Polycolor, which you have invented, and have had the goodness to submit to my inspection in order to make a trial of, as a new acquisition of great importance for drawing.

Beside the different advantages which they offer to those who draw, I would mention above all as one of the principal, that your pencils can be advantageously used in drawing from nature and in portraying firmly upon paper the effects of light and atmosphere, which often change very suddenly.

For this reason as well as for the value of the invention in general, I join my fellow artists in congratulating you with all my heart. Berlin, August 26th 1859.

Eduard Hildebrand,
Royal Court Painter and Professor.

XI.

EXTRACT
from the
REVUE DES BEAUX-ARTS,
TRIBUNE DES ARTISTES,
fondée
sous les auspices de la Société libre des Beaux-Arts

21. supplying.
Paris, November 1st 1859.
Works of the Committee.
Opinion on Mr. Sussner's "Creta Polycolor" pencils.

The committee, appointed to inspect the value of the style, the quality and use of the Creta Polycolor pencils of M. SUSSNER, have acknowledged after several trials that they can be of excellent use, especially if one wishes to seize instantly any point according to nature, be it the effect of light and colour, be it a group of trees, or any thing else in landscape, the notice of which appeals but imperfectly to the remembrance of the artist.

If in an artistic view the pencils of M. SUSSNER cannot be compared with the pastel in brightness and vigour of colour in the trials which we have made, this arises undoubtedly from our being so little accustomed to their usage. However we can affirm that, if the pencils Creta Polycolor are not indispensable to what is properly called Art, they can be advantageously employed in different ways for portraits, landscapes, flowers etc. They may be used by the tourist artist who will thus avoid the embarrassment of a box of colours; they also ought to be adopted in schools to familiarize the pupils with the first elements of colours, and on account of the cleanliness of their use.

The Referee,
Lenoir.

XII. The undersigned have found in the pencils invented by M. G. W. SUSSNER of NUREMBERG, and called Creta Polycolor (Oil coloured chalk), an excellent material for drawing, and can fully recommend them to all Artists and Amateurs of the Fine Arts on account of their extreme usefulness and good quality. Dusseldorf, December 17th 1859.

E. Bendemann, Director of the Royal Academy.
A. Achenbach, Professor.   W. Camphausen, Professor.
C. Sohn, Professor.   C. Scheuren, Professor.
C. Bewer.   C. Hübner.   A. Weber.   C. Böttcher.
A. Schmitz.   F. Heunert.

XIII.
Extract from a German Newspaper, entitled:
Neueste Nachrichten.

To the interests of Artists as well as Amateurs of design, we cannot but particularly recommend the excellence of the pencils called Creta Polycolor, manufactured by M. G. W. SUSSNER of NUREMBERG, which admit of a very extended use upon the Canson, the Papier Pellée, in Photography, and the Retouching of impression of colours; we are firmly convinced that no other substance could be employed for drawing with so much diversity, and so surprising in its effects as the before-mentioned Creta Polycolor. They are particularly distinguished from all other coloured pencils in their oily and indelible consistency, in the employment of water-colours with or after the use of pencils, lithographic chalk etc. The clear strength of a watercolour and the depth of an oil painting can be attained by these pencils, which are not easily broken and do not crumble when used; they can be employed either for the simplest sketch, or for the most complete picture. These pencils are a great resource to the Artist; the Amateur could also avail himself of this composition which would spare him many difficulties he would otherwise encounter in painting. Munich, January 29th 1859.

J. G.

XIV. The undersigned have, according to the wish expressed by M. G. W. SUSSNER of NUREMBERG, inspected his new discovery, the Creta Polycolor, and assert that these pencils surpass all other pastel or chalk pencils before discovered, and are adapted in every respect for artistic use.

We can therefore recommend this discovery, as it fully deserves, to the attention of artists. Stockholm, June 16th 1860.

C. G. Qvarnström,
Director of the Royal Academy of fine Arts.
John Boklund, Professor.   N. Anderson, Professor.
J. M. Stöck, Landscape Painter and Professor.
C. Th. Staaff, Royal Court Painter.

(No. 232. du Catalogue offic.)

G. W. SUSSNER A NUREMBERG
PATENT CRETA POLYCOLOR.

Observation.

Pour éviter tout malentendu, je prie l'artiste ou l'amateur de ne point confondre mes Creta Polycolor avec le Creta Laevis, les Pastels ou tout autre crayon de couleur déjà connus, attendu que l'amalgame de ma composition est à l'huile, tout nouveau et entièrement particulier à moi. Ces crayons extraordinairement tendres et cependant d'une consistance solide, se conservent aigus et ils ont surtout la propriété de fixer un dessin ou toute peinture sur un papier aussi ineffaçablement qu'un véritable tableau à l'huile, propriété, qui jusqu'à présent a manqué à toutes les espèces de crayons de couleur sèche.

Me référant aux certificats ci-joints de plusieurs des Autorités artistiques les plus célèbres de l'Allemagne, de la France et de la Belgique, auxquelles j'ai soumis les Creta Polycolor pour en faire un examen spécial, je m'abstiendrai en conséquence de toute recommandation ultérieure.

Pour empêcher donc toute méprise de la part du public, j'ose observer ici, que mes Crayons Creta Polycolor sont, comme tous les autres crayons, coulés dans des tuyaux ronds en bois de cèdre naturel, munis de la marque:

G. W. SUSSNER PATENT CRETA POLYCOLOR

et renfermés taillés dans un étui de cambric au nombre de 48, 36, 24, 18 et 12 pièces. On peut se les procurer chez tous les marchands d'objets de bureau, de dessin et de peinture.

Certificats.

I. Les soussignés ont mis à l'épreuve les Creta Polycolor, inventés par Mr. G. W. Sussner à Nuremberg, et voient dans cette nouvelle invention un enrichissement de la matière, qui ne peut être assez appréciée pour les arts concernant le dessin, à laquelle dans l'intérêt des artistes ils souhaitent qu'elle ait bientôt une propagation très-étendue.

Munich, le 19 janvier 1859.

Professeur W. v. Kaulbach, Directeur de l'Académie Royale des Beaux-Arts de Bavière.
Moritz y. Schwind, Professeur.   Foltz, Professeur.
Dietz, Peintre de la Cour du Grand-Duc de Bade.
Lange. Schoen. Dr. v. Hefner-Alteneck. Julius Lange. Reiff. Kirchner. Ed. Schleich. J. Bernhardt. J. Heigel. Morgenstern. Hagn. G. Seeberger. W. Boshart. J. Koch. H. Marr. K. Altmann. K. E. Doepler. W. Asselborn. Joh. Kirner. B. Savery. E. Kundy. Frd. Gärtner. P. Sporer. Aerttinger.

II. Les Creta Polycolor, inventés par Mr. G. W. Sussner à Nuremberg, paraissent être au soussigné très-recommandables soit pour des croquis topographiques que pour des dessins de carte faits sur le terrain à la volée en divers tons de couleurs et pour des suppléments, à cause de leur consistance sur le papier, supérieure aux crayons de pastel ordinaires, et de leur peu de rudesse et de fragilité.

Munich, le 25 janvier 1859.

H. von Schintling,
colonel et Directeur du bureau topographique.

III. C'est avec plaisir que les soussignés saisissent l'occasion de témoigner qu'ils ont reconnu les crayons dits: Creta Polycolor, inventés par Mr. G. W. Sussner à Nuremberg, comme étant très-utiles et propres au dessin; ils ne peuvent qu'émettre le vœu qu'ils soient bientôt répandus partout.

Munich, le 31 janvier 1859.

Jos. Resch. A. Kölbl. Höfer. A. Doll. M. Huber. M. F. Heil. Jos. Blanz. M. Stieler. A. Lier. Noere. W. Philippi. Rothbart. Kirstein. Lang. de Normy. J. F. Spengel. Marées. Metzener. F. Kreutzer. Kökert. Brausewetter. Reichard.

IV. J'ai fait usage des crayons dits: Creta Polycolor, que vous m'avez soumis.

Dans le genre de pastel que je traite, ils m'ont été très-utiles pour exprimer des délicatesses de ton dans les chairs et les points brillants des accessoires.

Avec un peu de pratique je crois qu'on pourrait obtenir de très-beaux résultats par cette sorte de crayons: ils se prêtent à la plus extrême sérénité de pointe sans perdre de leur éclat et cèdent facilement à l'action du grattoir, lorsqu'on veut atténuer l'intensité du ton ou bien ramener des lumières vives.

On peut les substituer avantageusement à la froide mine de plomb dans les portraits, croquis d'après nature.

Je ne saurais donc, Monsieur, trop louer la maison G. W. Sussner en présence des résultats heureux qu'elle vient d'obtenir et qu'elle offre aux artistes et aux amateurs.

Paris, le 19 mars 1859.

Vidal.

Étant parfaitement d'accord avec le jugement de Mr. Vidal, je joins avec plaisir mon nom au sien.

Winterhalter.

V. A la sollicitation de Mr. G. W. Sussner, Fabricant à Nuremberg, les soussignés membres du Conseil Académique et Professeurs à l'Académie Royale des Beaux-Arts certifient avec plaisir, que les crayons dits: Creta Polycolor (craie de couleur) inventé par Mr. G. W. Sussner, et qui ont été soumis par nous à une épreuve, répondent parfaitement à ce que l'on exige des crayons de couleur, et que la matière est ce qu'il y a de mieux de ce que l'on connait jusqu'à présent en ce genre; l'on peut en conséquence les considérer comme un avantage précieux pour les arts concernant le dessin. Dresde, le 15 avril 1859.

Ernst Rietschel. Julius Schnorr v. Carolsfeld. Julius Hübner. Ludwig Richter. Eduard Bendemann. Ernst Hänel. Gustav Heine.

VI. C'est avec plaisir que nous certifions à Mr. G. W. Sussner, Fabricant à Nuremberg, que les crayons, dits Creta Polycolor, inventés par lui, et qui nous ont été soumis pour être essayés, sont de qualité excellente, répondent parfaitement à ce que l'on exige des crayons de couleur, et qu'ils rendront un service notable aux artistes, notamment aux dessinateurs.

Anvers, le 30 mai 1859.

C. Guffens.   Jan Sweerts.   Van Lerius.

VII. Je me joins aussi au jugement de MM. mes collègues au sujet du Creta Polycolor, inventé par Mr. G. W. Sussner à Nuremberg, et je souhaite que cette belle et utile invention soit appréciée par toutes les personnes compétentes comme elle le mérite.

Cobourg, le 4 août 1859.

Rothbart,
Constructeur et conseiller du Duc de Cobourg.

VIII. Les Creta Polycolor, invention de Mr. G. W Sussner à Nuremberg, sont une substance nouvelle de la plus grande utilité pour dessiner le portrait. Il est à désirer, qu'ils soient bientôt connus et répandus partout. Berlin, le 25 août 1859.

Julius Schrader,
Professeur et membre du Sénat de l'Académie Royale des Arts.

IX. C'est avec la plus vive satisfaction que j'ai trouvé dans les Creta Polycolor une substance désirée depuis longtemps, laquelle entre autres se distingue par la précision du trait, la finesse et la fraîcheur du ton, et dont on peut ainsi se servir avantageusement pour des croquis de paysages.

Berlin, le 25 août 1859.

Ferdinand Bellermann,
Professeur.

X. C'est avec une vive satisfaction que je puis vous dire, que je reconnais comme une acquisition très-importante pour le dessin, les crayons de couleur (Creta Polycolor), que vous avez inventés, et que vous eûtes l'obligeance de soumettre à mon inspection pour en faire l'essai.

Outre les différents avantages que ceux-ci offrent aux dessinateurs, je me plais surtout à mentionner le suivant comme un des principaux: Que vos crayons peuvent être employés surtout avantageusement pour dessiner d'après nature, retenant ferme sur le papier les effets de la lumière et de l'air qui changent souvent très-promptement.

A cet égard ainsi qu'à la valeur de cette invention en général, je ne puis que vous en féliciter ainsi que nous autres artistes de tout mon cœur.

Berlin, le 26 août 1859.

Eduard Hildebrand,
peintre de la Cour Royale et Professeur.

XI.
EXTRAIT
DE LA REVUE DES BEAUX-ARTS
TRIBUNE DES ARTISTES
fondée
sous les auspices de la Société libre des Beaux-Arts.

21e livraison.
Paris, le 1er novembre 1859.

Travaux du Comité.
Rapport sur les crayons „Creta Polycolor" de Mr. Sussner

La commission appelée à se prononcer sur la valeur des tons, de la qualité et de l'emploi des crayons Creta Polycolor de Mr. Sussner, a reconnu, après plusieurs essais, qu'on peut en tirer un excellent parti, surtout lorsqu'on veut saisir presque instantanément d'après nature, soit un effet de lumière et des couleurs, soit un groupe d'arbres ou tout autre motif de paysage que des notes ne rappelleraient qu'imparfaitement au souvenir de l'artiste.

Si, au point de vue de l'art, les crayons de Mr. Sussner ne peuvent être comparés au pastel et comme éclat et comme vigueur de ton dans les essais que nous avons faits, cela tient sans doute à notre peu d'habitude de leur usage. Cependant nous pouvons affirmer que si les crayons Creta Polycolor ne sont pas indispensables à l'art proprement dit, ils s'emploient avec avantage dans les divers genres, les portraits, les paysages, les fleurs etc. Ils doivent être acceptés par le touriste-artiste qui s'évite l'embarras d'une boîte à couleurs; ils doivent être également adoptés dans les pensions pour familiariser les élèves avec les premiers éléments de la couleur, et aussi pour la propreté de leur emploi.

Le rapporteur, Lenoir.

XII. Les soussignés ont reconnu dans les crayons Creta Polycolor (crayons de couleur à l'huile) inventés par Mr. G. W. Sussner à Nuremberg, une substance excellente pour les arts concernant le dessin, et vu leur utilité et leur bonté efficace ne peuvent en conséquence que les recommander tout particulièrement à tous les artistes ainsi qu'à tous les amateurs s'occupant des Beaux-Arts.

Dusseldorf, le 17 décembre 1859.

E. Bendemann, Directeur de l'Académie Royale.
A. Achenbach, Professeur.   W. Camphausen, Professeur.
C. Sohn, Professeur.   C. Scheuren, Professeur.
C. Bewer.   C. Hübner.   A. Weber.   C. Böttcher.
A. Schmitz.   F. Heunert.

XIII.
Extrait d'un Journal Allemand, intitulé:
„Neueste Nachrichten".

Dans l'intérêt des artistes ainsi que dans celui des amateurs de dessin, nous ne pouvons que recommander tout particulièrement l'excellence des crayons dits Creta Polycolor, fabriqués par Mr. G. W. Sussner à Nuremberg, vu que pouvant en faire un usage très-varié sur le Canson, le Papier Pellée, sur les Photographies, les Retouches de gravures coloriées, nous sommes fermement convaincus qu'aucune substance ne peut être employée pour le dessin d'une manière aussi diversifiée et aussi surprenante dans ses effets que les susdits Creta Polycolor. Ils se distinguent particulièrement sur tous les crayons de couleur en ce que, outre leur consistance moëlleuse et ineffaçable, en employant les couleurs d'acquarelle sur ou sous le crayon, la craie de lithographie etc., ils atteignent la force luisante d'une aquarelle et le fond d'un tableau à l'huile, qu'ils ne se cassent ni ne se brisent, et qu'ils peuvent être employés soit pour esquisser légèrement, soit aussi pour le tableau le plus complet. Autant ils sont à l'artiste de profession d'une grande ressource, autant l'amateur aimera aussi à se servir d'une substance, qui lui épargnera beaucoup de difficultés d'autres genres qui se rencontrent dans la peinture.

J. G.

Munich, le 29 janvier 1859.

XIV. Les soussignés ont sur demande essayé et mis à l'épreuve les Creta Polycolor, inventés par Mr. G. W. Sussner à Nuremberg, et trouvé qu'ils surpassent tous les autres crayons de ce genre connus jusqu'à présent; ces crayons sont particulièrement propres à l'emploi dans les beaux-arts, c'est pourquoi nous recommandons cette nouvelle invention aux artistes, comme elle le mérite de bon droit.

Stockholm, le 16 juin 1860.

C. G. Qvarnström,
Directeur de l'Académie des Beaux-Arts.
Jean Boklund, Professeur.   N. Anderson, Professeur.
J. M. Stöck, Peintre de paysage et Professeur.
C. Th. Staaff, Peintre de la Cour Royale.

(No. 2831. of the offic. Catalogue.)

Heidenheim a. B. (Wurttemberg, Germany).

With the successful introduction of HENRY VÖLTER'S "**Wood-pulp**" as a substitute for rags, the problem of employing wood or woody fibres for manufacturing paper and paper-like fabrics has at last found its practical solution. — In an economical point of view and with reference to industry in general, this invention has actually proved to be of the utmost importance, and Patents having been secured in most of the countries of Europe and the U. S. of America, the inventor is now fully prepared to furnish and sell either the ready Machinery or grant rights and licenses to Others as the cases may be.

The great variety of papers on exhibition by "HENRY VÖLTER'S SONS" sub No. 2831. for writing, printing, and wrapping purposes, the tissue papers, the articles for paper-hangings, bills etc. etc. are principally made of the "Wood-pulp" containing more or less of, and from 20 to 80% of this material; their manufacture is now carried on upon a prodigious scale at the extensive establishments of VÖLTER'S in Heidenheim and Gerschweiler, the articles turned out never failing to find a ready and excellent market, wherever put up.

In connection with such facts it may also be proper to state here, that no extra labour or pains have been bestowed upon the samples exhibited, but that they have been promiscuously taken from amongst the ordinary and daily issues of the Mills. — The self-cost of rag-pulp in the ordinary qualities taken into consideration (averaging now ca. 7 Thalers pr. 100 ℔ in Germany) — the manufacturing expenses of "VÖLTER'S WOOD-PULP" as a substitute, will not exceed even 3 Thalers.

For particulars see the German Catalogue of the "Zollverein", also the specifications at No. 2831. in the Exhibition rooms.

Sole agency for sale of the papers, the "wood-pulp" Machinery, Patents and Licenses:
Address: The traders society of Wurttemberg (*Wurttembergische Handelsgesellschaft*), represented by Mr. S. SESSING, 10 Huggin-Lane, Wood Street, London. Further and direct information relating **to the Patent rights and the different Machines** may also be obtained by addressing the inventor M. **HENRY VÖLTER, city of Heidenheim a. B., Wurttemberg,** and **as regards the papers and articles exhibited,** Messrs. **HENRY VÖLTER'S, SONS, of the same place.**

(No. 2828. of the offic. Catalogue.)

CHARLES ROMETSCH'S
PATENT METAL-SLATES MANUFACTORY.
STUTTGART.

First Inventor and manufacturer of Patent Metal-Slates for schools, Offices etc. PRIZE-MEDALS were awarded to M. CHARLES ROMETSCH, at the Exhibitions of LONDON, PARIS, MUNICH, DRESDEN, and STUTTGART.

At the last named exhibition a prize of 30 Ducats was moreover given for the superior chemical qualities of his Patent-Slates.

Agents: Messrs. FRED. HAAS & Co., 34 Jewin Street, London.

(No. 2794. of the offic. Catalogue.)

J. M. Eckart
of
Ulm a. D.

Manufacturer of varnished muslins, cap-peaks, cap-straps, japanned leather-sashes, of very superior and pliant oil-cloth, especially adapted for coverings and articles for the military, and very serviceable for the edging of stuffs and other purposes; of japanned sheep's leather, keeping perfectly smooth and even in spite of the roughest treatment.

(No. 2817. of the offic. Catalogue.

J. M. OTTENHEIMER & SONS.
Manufacturers of Woven Corsets.
STUTTGART.
Branche-Office 27 Rue de Cléry, PARIS.

Being the Owners of the patented and only Invention to produce our Article by Power-Looms, we are enabled, not only to establish a more regular and equal Article, but also to effect great economies, which we shall be very happy to share with our customers.

Notice.

Living in the midst of the most important Hop districts of Bavaria, on the South-North Railway, near Spalt, the undersigned are in a position to execute orders in every variety of Spalt, and other sorts of Bavarian Hops.

To purchasers, we beg leave to recommend our services, engaging to send, on request, in addition to the Hop intelligence contained in the *General Bavarian Hop Gazette* a communication by letter on the state of the Stock, Crop, Price and Quality of Spalt Hops.

Concerning the Quality of our Product we take the liberty of directing your attention to the unsmoked Hop exhibited by our producers in the London Industrial Exhibition of 1862.

Messrs. Albrecht & Comp.

Roth (in Bavaria).

Messrs. GROSSBERGER & KURZ

of

NUREMBERG

are represented in Great Britain and Ireland by their agent,

C. E. ELLIOTT,

5 Aldermanbury Postern, City, London E. C.

They manufacture all kinds of Black Lead and Red Chalk Pencils, Black Chalk, Colored Pencils *(Creta Polycolor, Creta Laevis)*, and Pastel Colored Pencils for drawing in Water Colors, and beg leave to say that persons of the highest distinction in Art have everywhere borne testimony to the excellence of their articles. Their recently invented Pastel Colored Pencils for drawing in Water Colors are the most valuable things of their kind: framed in Cedar, they may be used equally as crayons or water colors, and the colors are fast and permanent. If used as water colors, the colors must be put on dry and rendered fluid by a brush dipped in water: a procedure which may be said greatly to facilitate sketching from Nature. Or, each color may be dissolved in water and so used. This new species of Colored Pencils may, there-fore, be highly recommended more particularly for the use of Military Academies, Surveyors, Architects, Engineers &c. The prices of our various articles in all qua-lities are such as cannot fail to give satisfaction to purchasers.

The cheapest sorts of Lead Pencils in White Wood . . .	from 1	sh.	to 2	sh.	
Common Lead Pencils in Cedar Wood	»	$1\frac{1}{2}$	»	» $3\frac{1}{2}$	»
Middling	»	4	»	» 6	»
Fine	»	6	»	» 9	»
Polygrade Lead Pencils	»	$7\frac{1}{2}$	»	» 39	»
Lead Pencils for Stenographers	»	22	»		
Hiexagonal Pencils for Artists with points of German Silver	£ 2	8	»		
Dtto for Portfolios	£ 2	2	»		
Pocketbook Pencils	from 2		»	» 14	»
Pencils for Builders and Carpenters	»	3	»	» 9	»
Colored Pencils *(Creta Polycolor, Creta Laevis)* in Cedar wood, not polished	»	11	»		
Ditto, polished and of the same color as the chalk . . .	»	15	»		
Pastel Colored Pencils for drawing in Water Colors, framed in Cedar wood	£ 2	5	»		

All the above prices have reference to the gross à 12 doz.

The agent already mentioned supplies detailed lists of the prices of all our articles.

G. SIGL,

General Engineer, Boiler Maker, Iron Founder etc. etc.,

Berlin 29, Chaussée Street,

Vienna 39 & 42, Michelbairisch Ground, and

Vienna Neustadt,

desires to call the attention of Engineers & others to the various improved Machines, which he is making.

Prices and full particulars will be sent on application.

The first named establishment at **Berlin,** of which a specimen of its patented lithographic fly-presses is to be seen in the Machinery Department of this year's Exhibition, applies itself particularly to the fabrication of small printing presses, moved by hand, and printing fly-presses (800 of the latter having been sent, hitherto, to all parts of the world) lithographic fly-presses, smoothing and bundling presses, satin laminating rollers, paper machines (20 of the latter being issued-cylindrical paper-mill — cutting machines), as well as to the manufactory of all machinery and apparatuses that are necessary for the fabrication of paper. Besides these, the Berlin Establishment is in the habit of furnishing all kinds of tools, amongst which, a stock of lathes, planing, boring, punching and shaping machines, constructed according to the newest improvements, offers an abundant choice.

The **Vienna Establishment,** 39 & 42 Michelbairisch Ground, being in conjunction with a large Iron-foundry and special works for making boilers, builds all kinds of steam-engines, locomotives, and marine engines, arranges mill works, powder and sugar-mills, as well as India-rubber and other manufactories, and applies itself in the same manner as the Berlin establishment to the fabrication of tools of every kind, printing presses, small and fly-presses in the same selection.

The third establishment in **Vienna Neustadt** occupies itself, in conjunction with the Vienna works, exclusively with the fabrication of Locomotives, having sent out of its works 120 of them up to the present time.

More than 1800 workmen are employed in these three establishments which are supplied with the most excellent machines, and requisite steam-power.

The aforesaid being the leading articles, which the three works are constantly making, and to the improvement of which the enterpriser's attention has been specially directed, those requiring such machines will find the above unsurpassed, whether as regards principle, material, or workmanship.

(No. 154. of the offic. Catalogue.)
Fr. Mittler: Augsburg.
Agent: Messrs. ENGELBACH & WOLFGANG, 6 Little Pat Lane, London.

Mittler's Harmless Green,
free from all poisonous admixture.
(Patented for the Kingdom of Bavaria.)

This colour demands the particular attention of the public from the perfectly harmless nature of ts composition, its peculiar brilliancy, fast and durable qualities, and its easy application to all practical purposes. For many years the want has been experienced of a green colour combining all these various qualities, and this is now supplied in the present invention of Herr **Mittler.**

From its composition this Green may be Called a Chromoxide-Hyrdate, and is prepared in shades and qualities the most various, from the brightest yellow-green to dark-green; it is free from all admixture of Arsenic, Copper, Lead, and from every kind of poisonous metallic combinations. It remains entirely unaffected by the light, or by the direct action of the sun's rays, and can accordingly be employed with equal advantage as a **water, oil,** or **chalk** colour; it is particularly adapted for the manufacture of papering, blinds, tinted paper, flowers, as also for decorations and the painting of rooms; it is fit for all kinds of oil and varnish painting, and can be applied with the greatest success in the composition of al fresco works of art.

This colour has not the glaring bright look of the poisonous Schweinfurt and New greens, but, on the contrary, is of a beautiful, liquid, yet brilliant tone, eminently grateful to the eye, as may be seen at a glance on inspection of the samples here exhibited. Another quality recommending its general adoption is its power of covering other colours; and its price from 12 Thalers upwards the 50 Kilo is very moderate.

The correctness of the above description is corroborated by numerous high official opinions and testimonials.

[48.] (No. 2108 a. of the offic. Catalogue.)

WILD & WESSEL

OF

BERLIN.

Referring to the particulars given in the first part of the Catalogue of the lamps and parts of lamps exhibited by us, we beg to observe that our collection includes only those articles which have proved thoroughly practical and serviceable in general use, and are distinguished for their good quality and cheapness.

We beg to recommend our large choice of Moderateur lamps, and brass oil lamps with regulator of 5 different Sizes.

Paraffin lamps: of which we only exhibit the burners as the most essential part of this lamp to which we have given our best attention for some years. We have succeeded in constructing burners, which may be applied to any kind of paraffin oil in trade.

By the newest mechanical improvements in our manufactory we are enabled to charge the very lowest prices. The prices are to be understood as wholesale and for cash taken in Berlin.

MEDAL of the exhibition at Paris 1855.

TWO MEDALS of the exhibition for Pomerania at Stettin 1857

Martin Grashoff,

Horticulturist,

Cultivator of seeds, plants and flowers,

MEDAL of the exhibition at Halberstadt 1856.

Honorary diploma of the exhibition at Berlin 1860.

Wholesale Seeddealer and Farmer of a royal domain at

Quedlinbourg and Westerhausen, Department of Magdebourg, Kingdom of Prussia.

The prices are accommodated to the produce of the crop

---

1. Vegetable-seeds for nursery. grouns; for general consumption at table.
2. Vegetable-seeds for forcing.
3. Vegetables for the kitchen and for general consumption.
4. Root-plants for kitchen and table.
5. Lettuces for winter and summer, for consumption at table.
6. Various sorts of roots for general consumption at table.
7. Various sorts of Rapes for general nourishment.
8. Pod-plants with and without pods, in the green state, for the general table.
9. Vegetables for drying and preserving, as nutritious matter for winter and for a longer time, for general consumption at table.
10. Coffee substitutes and Beet roots, the richest sorts in sugar for the preparation of sugar.
11. Beetroots for agriculture, with many leaves and roots, a means for increasing the milk and for the fattening of horned cattle.
12. Root-plants, Turnip-cabbage, a milky food for horned cattle and sheep.
13. Carrots for economical consumption for horses, sheep, and horned cattle.
14. Various sorts of herbs and roots for sheep, with an exuberant growth of leaves; a sound food.
15. Economical seeds for the production of food, and various sorts of grass for meadows and parks.
16. Legumens, rich in meal.
17. Oil-seeds.
18. Seeds of colour-plants.
19. Seeds of officinal plants.
20. Seeds of venomous plants.
21. Different kinds of corn for winter and summer in grains and ears.
22. Seeds for pleasure gardens, and wood shrubs.
23. Potatoes for eating at table.
24. Potatoes from China, James's potatoes.
25. Flower-seeds for the embellishment of gardens and pleasure-grounds.
26. Finer flower-seeds for chamber and pot-bloom.
27. Seeds of ornamental plants in the open ground.
28. Seeds of climbing flowers for the embellishment of walls and bowers, and for the decoration of chambers.
29. Ornamental gourds.
30. Eatable gourds.

The annexed catalogues give special information.

(No. 219. of the offic. Catalogue.)

ALOIS DESSAUER.

Colored Paper and Glue Manufactory,

ASCHAFFENBURG, BAVARIA.

Prize Medals at Berlin, Münich and other exhibitions.

Represented in Great Britain by Mr. FREDERICK RUDOLPH,
188 Gresham House, Old Broad Street, London E. C.

All sorts of colored papers for bookbinders, fancy-box manufacturers, lithographers, and for photographic purposes; plain colored, embossed and fancy papers, printed with colors, Gold and Silver metal.

Marblepapers of all Kinds, Nonpareil, Spanish-Shell, and fancy sorts; embossed and plain Morocco papers; watered papers, white and tinted; white enamelled and flint-glazed surface papers; colored Cabel-papers Steel-blue, Bronze-blue, Bronze-brown etc.

Every Kind of superfine enamelled unchangeable Cardboards, white and tinted, for visiting and show cards; enamelled Boards for Buttons; a variety of paper imitations of wood; strawplait and artificial flower papers.

Best light yellow Glue, and white Parchments Glue.

[51.] (No. 2735. of the offic. Catalogue.)

EBLE'S HOROSCOPE,
or
Indicator of Time,

is mentioned by the most celebrated astronomers of Germany, such as Littrow, Zech, Reuchle, Grunert etc., as by far the most convenient, the most simple, and the cheapest of all instruments which serve to determine time, and is specially recommended by them for the use of schools; for it merely suffices to place a diopter against the sun in order to determine the time at any moment of the day.
To be had of

Rud. Engler, Ellwangen, Württemberg.

Agency at London: Mr. Samuel Seffing, 11 Huggin Lane, Wood Street, Cheapside London E. C. Commissioners: Mr. Robert Hoffmann, Leipsic; Paul Neff, Stuttgart.

[52.] (No. 215. of the offic. Catalogue.)

G. C. BEISSBARTH, SON,

NUREMBEG (BAVARIA).

MANUFACTURER
OF
ARTISTS' AND PAINT BRUSHES
OF EVERY DESCRIPTION.

The reputation of our Manufactory, as the very first and largest of Germany, has been confirmed by all eminent Artists and Academies of Art, as well as by the Prize-Medals, first class, received at the grand Exhibitions of Industry of Paris 1855, Munich 1854, and Nuremberg 1852.

A. W. FABER.

LEAD PENCIL MANUFACTORY

IN

STEIN near NUREMBERG.

HOUSE AT PARIS:
12 BOULEVARD DE STRASBOURG.

HOUSE AT NEW-YORK
133 WILLIAM STREET.

AGENCY AT LONDON:
9 FRIDAY STREET, CHEAPSIDE, E. C.

This establishment, being the largest of the kind in existence and employing from 500 to 600 workmen, has by its unremitting efforts for the last century succeeded in producing the best Pencils, made of prepared Black-Lead, and has in consequence obtained a well established reputation throughout the whole civilized world. At all the great industrial exhibitions in Germany, England, France, and the United States it has been rewarded with prize-medals of the highest distinction, and its producions are exclusively patronised by the greatest artists of the day. It furnishes Lead Pencils from 1 sh. to 25 sh. a Gross (or 12 dozen) and will, therefore, satisfy every want and defy all competition.

During several years the manufacturer has occupied himself in producing Pencils of Siberian Black-Lead. The difficulties he had to master in manufacturing these Pencils were not only to produce Pencils equal in quality to those of the Cumberland Lead, but also free from all those defects, which are still inherent in the very best quality of the latter. Aided by long experience, improvements and discoveries during a number of years, and by his incessant studies and experiments in this new material, the proprietor at length was enabled to produce Pencils from this Siberian Black Lead, that combine all the advantages, formerly never united in his finest Pencils of artificial composition, nor in those made of pure Cumberland Lead.

These new Siberian Lead Pencils having all the excellent qualities, which until now have been attained only with those of the Cumberland Lead, still surpass the latter in the following qualities:

1st. They are entirely free from those objectionable substances ever found in the Cumberland Lead;

2ly. The point is firmer and has a greater power of endurance;

3ly. Their degrees of hardness and softness are and remain invariably the same, a problem which the English manufacturers for centuries have in vain tried to solve;

4ly. The Lead in the Siberian Lead is always one piece throughout.

The manufacturer having succeeded in producing the best Pencils, now resolved that, although he had obtained superiority in quality, he would also add the great desideratum of cheapness.

The soft Pencils of Cumberland Lead have been paid for as high as £ 12 10 sh. per Gross; the hard numbers £ 4. Comparing these prices with Siberian Pencils we find that they are at least 100 % cheaper, than the former Cumberland Pencils.

These new Siberian Lead Pencils are at the Present Exhibition for the first time offered to the public.

Regarding the prices, the Manufactory sells:

			per Gross or 12 dozen	
Lead Pencils in white wood			1 sh. —	d
do. in Cedar wood			1 » 4	»
do. do. in 3 different degrees soft, middling, hard			2 » 8	»
do. do. middlefine in 4 different degrees No. 1—4 polished			5 » 10	»
do. do. fine, in 4 different degrees, No. 1—4 polished			7 » 6	»
A. W. FABER's Polygrade Lead Pencils in 4 different degrees No. 1—4			8 » 4	»
The Same black polished-Gold, or yellow polished-Silver in 7 different degrees			11 » 8	»
The Same hexagon, in 5 different degrees			15 » —	»
The Same superfine, black polished-Gold, or yellow polished-Silver in 4 different degrees .			20 » —	»
The Same Finest and Best, round, in 9 different degrees, red polished-Gold			25 » —	»
Colored Pencils, assorted in 48 and 100 colors			13 » 4	»
The Same Finest, polished-Gold			20 » —	»
Black Chalk Pencils, in white wood, soft and hard			5 » —	»
do. in Cedar do.			6 » 8	»
Red Chalk Pencils in different qualities, from . . . . 5 sh. to			16 » 8	»
Slate Pencils in white wood			2 » —	»
do in Cedar wood			4 » 4	»
New Artist's Pencils, with moveable Lead			48 » 4	»
New Polygrade Lead Pencils of Siberian Black Lead			30 » —	»

can, therefore, satisfy every want or requirement, and defy any competition.

ENGELMANN & C⁰., STUTTGART,
Manufacturers of Sparkling Wines and Liquors.

Prices are free of freight and cost of package.

Sparkling wine, 1. Quality, the dozen bottles 28 Sh. = Frs. 36. *do.*, 2. Quality, the dozen bottles 25 Sh. = Frs. 32. ½ bottles are 2 d. dearer. — *Maraschino*, *Absynthe*, *Vanille*, *Chartreuse*, *Eau d'or*, the dozen bottles 26 Sh. = Frs. 32. 50.; the gallon 9 Sh. 4 d. = Frs. 12. — *Anisette*, *Créme de Canelle*, *Cumin*, *Fleur d'Orange*, *Menthe*, *Mocca*, *Roses*, *Curaçao*, *Noyaux*, *Maagbitter*, *Parfait-amour*, the dozen bottles 22 Sh. = Frs. 28. 25.; the gallon 8 Sh. 6 d. = Frs. 11. — *Double-Cumin*, *Quince* and all the above detailed sorts 2. Quality, the dozen bottles 16 Sh. 6 d. = Frs. 21. 25.; the gallon 6 Sh. = Frs. 7. 75. — *Arac*, *Rum*, *Groc*, *Winepunch-Essence* with Ananas and Orange, the dozen bottles 24 Sh. = Frs. 31.; the gallon 11 Sh. 8 d. = Frs. 15. — *Kirschgeist*, *Heidelbeergeist*, *Himbeergeist*, the dozen bottles 25 Sh. = Frs. 32.; the gallon 9 Sh. = Frs. 11. 50.

NB. Samples and references to be obtained of Mr. SESSING, Agent of the exhibitors of Württemberg.

In the German catalogue the prices are inscribed in French and German measure and money.

J. F. BÜRKLE.

MANUFACTURER OF CHEMICALS,
GROSSHEPPACH (WÜRTTEMBERG).

Recommends his sulphur barrel-brand (Schwefelschnitten) intended to correct and purify beverages and foul barrels. The manufacturer has been supplying with it most of the German and other states of Europe as well as those of North-America for the last 23 years.

His article is perfectly free of arsenic, and can be had with or without aromatic ingredients. By burning it in the casks, wine, beer and other liquors are not only preserved from ropiness and acidity, but also highly improved in flavour. By this means any drinks that may have got out of order can be restored in 8 or 14 days.

Prices.

Barrel-Brand 1 Cwt. equal to 50 Kilogr. with aromatic ingredients 70 fl. or £5 16 sh. 8 d.
ditto without aromatic ingredients 43 fl. 20 kr. or £3 12 sh. 3 d.
Bürkle's Sealing-wax 1 Cwt. from 50 fl. to 400 fl.
Bürkle's Tooth-powder 20 fl., 40 fl., 60 fl. the 100 boxes.
Bürkle's Aromatic Lotion the 100 bottles at 25 and 50 fl.
Bürkle's Oil for killing bugs the 100 bottles at 40 and 80 fl. or £3 6 sh. 8 d. and £6 13 sh. 4 d.
Bürkle's Hair balsam the 100 bottles at 100 fl. or £8 6 sh. 8 d.
Bürkle's Tooth-tincture the 100 bottles at 50 fl. and 100 fl. or £4 3 sh. 4 d. and £8 6 sh. 8 d.

LANCASTER, KLEEMANN & CO.,
MACHINE-MAKERS
IN
OBERTÜRKHEIM
(WÜRTTEMBERG),

beg to recommend their Power Looms as: Common Calico-Looms, Looms with drop boxes from one to eight shuttles, and Petticoat Looms, all for 24—66 inches (French) reed space, and made after the system patented by us, recommending itself especially by the soft picking and by the easy motion of the treadles as well as by other improvements.

When particularly desired, we also build Looms according to the English system, and assure our Customers, that though the prices are as low as possible, nevertheless the Looms will be found solid, strong, and well made.

[62.] (No. 1680. of the offic. Catalogue.)

RITZ & VOGEL,

Cloth Manufacturers, Aix la Chapelle

(Rhenish Prussia).

ASSORTMENT.

1. Doeskins.

a) **Summer goods** (for European use and export to South-America &c.) 10 Qualities at from 3 sh. to 6 sh. 6 d.

b) **Medium-weight** and **heavy goods** (for European use and export to the United States, the River Plate, the West of South-America &c.) at from 3 sh. to 6 sh. 6 d.

The above quoted prices are for $\frac{3}{4}$; double the prices for $\frac{6}{4}$.

2. Plain and twilled Cloths

(for European use and exportation) 14 qualities of every weight and fineness at from 3 sh. 6 d. to 19 sh. for $\frac{6}{4}$ or $\frac{7}{4}$.

3. Tricots.

3 qualities at from 6 sh. to 9 sh. for $\frac{6}{4}$.

4. Repps.

3 qualities at from 4 sh. 6 d. to 6 sh. for $\frac{6}{4}$.

5. Moskowa's Doubles

and other articles with plain or beaver back at from 7 sh. 6 d. to 11 sh. 6 d. for $\frac{6}{4}$.

6. Particular arrangements

have been made for the production of **fine colours,** such as scarlet, white, amaranth, yellow &c., both for military and other use: **every** quality and shade is to be had. The qualities and colours wanted for military use in Europe are generally in stock. — Particular attention is paid to the qualities for exportation to the Orient, Central- and South-America, East-India, China and Japan, and **Spanish Stripes, Habit, Medium** and **Heavy Cloths** are continually manufactured in every desired width and quality, at from 3 sh. 6 d. to 19 sh. for $\frac{6}{4}$ or $\frac{7}{4}$.

7. Cloth for Piano-Manufacturers,

scarlet and white, in every width and finish at from 11 sh. to 19 sh. for $\frac{6}{4}$.

All the above quotations are per Yard. Samples and prices-lists are at disposal if required.

Price - Current

of

C. CONRADTY'S

BRONZE - COLOURS MANUFACTORY

at

NUREMBERG.

BRONZE - COLOURS.

№	4000			2000			1000			800			600			500			400			300			200			100		
	£.	sh.	d.	£.	sh.	d.	£.	sh.	d.	£.	sh.	d.	£.	sh.	d.	£.	sh.	d.	£.	sh.	d.	£.	sh.	d.	£.	sh.	d.	£.	sh.	d.
Real gold à ducat	—	15	—																											
Real silver à mark	2	15	—																											
English silver à ½ Kilo	3	3	4	2	16	8																								
Argentan	1	13	4	1	6	8	1	3	4																					
Rosa and salmon	2	16	8	2	13	4	2	10	—																					
Flora, green-gold, citron flora	1	10	—	1	8	4	1	6	8	1	6	5	1	3	4	1	1	8	1	—	—		16	8		16	8		11	8
English green-gold, citron and pale citron	1	8	4	1	5	—	1	3	4	1	1	8	1	—	—		18	4		16	8		15	—		13	4		10	—
Redbrown, scarlet, crimson, lilac, carmine, violet, flesh-coloured and pale gold A., Rich gold B., English white, light-green	1	6	8	1	3	4	1	1	8	1	—	—		18	4		16	8		15	—		13	4		11	8		8	4
Dark-green, grass-green and ducat	1	3	4	1	1	8	1	—	—		16	8		15	—		13	4		11	8		10	—		8	4		6	3
Pale yellow, high yellow, orange and silver composition	1	—	—		18	4		16	8		15	—		13	4		11	8		10	—		8	4		6	8		5	—

by ¼ and ½ ounce papers packed up.

FAINT COLOURS BRONZE - POWDER.

Green-gold and copper . . à ½ Kilo . . . . . 11 sh. 8 d.

Yellow and white . . . . ” . . . . . 6 ” 8 ”

BROCADE FOR PAPER HANGING.

	fine.	extraf.	superf.	Paper hanging.
No. 2½, 3, 4 and 7	5 sh. 5 d.	6 sh. 3 d.	6 sh. 18 d.	10 sh. — d.
Citron, high yellow, orange .	5 ” 10 ”	6 ” 8 ”	8 ” 4 ”	11 ” 8 ”
Green, new silver	6 ” 8 ”	6 ” 18 ”	9 ” 2 ”	13 ” 2 ”

½ Kilo.

Delivered here, on drafts of three months, as by Invoice, for cash 2 % discount.

ALBERT EPPNER & COMP.

WATCH MANUFACTURERS.

WATCH MAKERS TO HIS MAJESTY THE KING
AND
HIS ROYAL HIGNESS THE CROWN - PRINCE OF PRUSSIA.

OUR

MANUFACTORY

IS IN

LAEHN

IN

SILESIA

ORDERS

can also be
forwarded to our

AGENTS
IN BERLIN
BEHRENSTR. 31.

IN BRESLAU
JUNKERSTR. 32.

IN COLOGNE
MINORITENSTR. 119.

IN BREMEN
BAHNHOFSTR. 1.

EPPNER & Co. offer to the public notice their manufacture of watches in which the technical and mechanical workmanship are executed with equal exactitude at the lowest wholesale prices. Orders to the amount of 12 dozen only can be received for the cheap description of watches

A. Gold chronometer	75 — 300	Thalers.
Ditto anchor repeating - watch	70 — 250	»
Watches of the I. quality	65 — 150	»
II. »	36 — 90	»
III. »	25 — 35	»
IV. »	18 — 24	»
Cylinder watches	15 — 38	»
B. Silver anchor watches I. quality	26 — 45	»
II. »	14 — 25	»
III. »	8 — 12	»
Cylinder watches	5 — 10	»

Our address during the Exhibition in **London** is:
Mssrs. LION M. COHN and PHALAND & DIETRICH of Berlin,
General Commercial Agency of Prussia and other German States for the London Exhibition,
LONDON, 20. St. Dunstan's Hill, City.

(No. 2487. of the offic. Catalogue.)

PATENT LEATHER-PLAITINGS

of

Amalie Neubert,
Leipsic.

The undersigned begs leave to attract the attention of the public to these articles. The plaiting and the decoration consist entirely of leather. The materials are prepared after my own invention and the articles plaited and decorated by myself. The objects recommend themselves by their novelty, elegance, and durability. The objects exhibited are to be sold. Direct applications by letters post-paid to

AMALIE NEUBERT, LEIPZIG.

[72.] (No. 2491. of the offic. Catalogue.)

L. Hausding

at

CHEMNITZ,

manufactures, besides the usual middling sorts of umbrellas as here exhibited,

cotton umbrellas from 15 Sh. p. Doz.,
silk umbrellas from £ 3 6 Sh. p. Doz.,
cotton parasols from 8 Sh. p. Doz.,
silk en-tous-cas from £ 2 2 Sh.,
silk "marquises" from £ 1 16 Sh. p. Doz.

and all other articles belonging to this branch of industry at the most moderate prices.

[73.] (No. 2332. of the offic. Catalogue.)

HUGO SCHICKERT,

at

DRESDEN,

manufactures Balances after the Construction of his exhibited one, and after other loads up to 10 grams, at the same time warranting a sensibility of a millionth parth of the highest load.

[74.] (No. 2500. of the offic. Catalogue.)

Schmidt and Harzdorf

at

HARTMANNSDORF,
near Chemnitz,

have only exhibited cotton hose on account of the smallness of the space, but they manufacture all other woollen articles of hosiery, drawers, shirts, gloves &c. &c., particularly for exportation and more especially for the South-American market.

[75.] (No. 2397. of the offic. Catalogue.)

FACILIDES & WIEDE,
PLAUEN
(Saxony).

Our establishment was erected in the year 1847 and began the spinning of angola yarns in 1848. — The collection sent for the exhibition does not represent all the different kinds of angola yarns spun by us, but principally the qualities most consumed in England and Scotland. The "Ombré" and "Iris" mixtures on the card in the étui show what perfection is attainable as regards the combination of mixtures.

The prices quoted are such as are usually paid. But should the abnormal rates of cotton prices last still longer, we shall soon be compelled to raise our prices.

Besides the spinning manufacture we possess our own dyeing and scouring (for yarns) establishment under the direction of Mr. L. IRMSCHER.

MANUFACTURES
OF THE
SAXON CAST-STEEL WORKS
AT
DÖHLEN near DRESDEN.

REPRESENTED BY ADOLPH von KANIG,
11. HIGH ROW, KNIGHT'SBRIDGE 11.

Frogs for railroads of the same construction as those ordinarily employed, but with the difference that the parts most subjected to wearing out consist of cast-steel. The latter, fastened on the cast iron block by screw-bolts and capable of being easily removed and supplied by new ones, as soon as they are worn, give evident advantages to these frogs over the wholly cast iron ones which, after the least damage, must be rejected as quite useless. On the Bavarian railways a great number of such frogs are in application. It is true they involve greater expense at first, but will notwithstanding prove themselves profitable by the incalculable durability and great advantages they offer.

Cranks, finished, for locomotive engines, forged out of only one piece of cast-steel.

Connecting Rods, finished, for locomotive engines.

Axle-Trees for steam-engines, forged and turned for the journals.

The manufactory furnishes similar pieces up to the weight of 5,000 lbs. and more.

Tension-Springs with welded ears.

Supporting Springs with rolled ears and groovings to prevent the displacing of the leaves.

Volute Springs for buffers and for drawing apparatus.

Rolling Mill for goldsmiths etc., with hardened rollers of $3\frac{9}{16}$ inches length by $2\frac{1}{2}$ inches diameter.

Rollers, hardened and highly polished, of $9\frac{1}{2}$ inches length by $4\frac{1}{2}$ inches diameter for the manufacturing of plates out of precious metals and for plating.

Anvils with cast-steel head-pieces.

Knives for paper-cutting machines.

Circular Knives for paper-machines.

Knives, two-edged, for cylindrical paper-mills.

Knives for the ground-works of the same.

Knives for rag-cutting machines.

Knives for leather-splitting machines.

Fractures of Cast-Steel in the following qualities:

 Cast-Steel with an alloy of wolfram for working very hard metals;

 Ditto, still harder;

 Quality most employed for tools, like chisels, borers etc.;

 Quality for objects which require a great hardness and high polish, as coin and medal stamps;

 Softest and very tenacious quality for axles, shafts etc.;

 Quality for springs, soft and elastic;

 Harder quality for piston-rods etc.;

 Still harder quality for knives etc.;

 Welding Cast-Steel for steeling large surfaces; further for anvils, miners' tools etc.;

 Cast-Steel by nature hard enough to cut red-hot metals; for rollers etc.

EXPLANATIONS
OF THE MACHINES EXHIBITED BY
RICHARD HARTMANN
OF
CHEMNITZ (SAXONY).

1. The LOCOMOTIVE ENGINE has been executed on the model of those engines that are running on the Saxon railways in the mountainous districts with a remount from 1 to 40 and curves of a radius of 275' Engl.

The fore-axle is in these engines laid in a moveable frame, and a retouching of the tires is only necessary after having traversed 14000 English miles. The dimensions of these locomotives are:

area of the grate	$11\frac{1}{2}'$ □
fire surface of the 148 boiler tubes.	790' □
direct fire surface	70' □
diameter of the piston	— 15"
stroke of the piston	— 22"
diameter of the coupled wheels	$4\frac{1}{2}'$
pressure on the rails by each pair of coupled wheels	10^{tons} 8
pressure on the rails by the fore-wheels	6 » 4
weight of the machine with water and coals	28 8

2. The CONDENSING STEAM-ENGINE of 20 horse power with high pressure and variable expansion, $3\frac{1}{2}$ atmospheres steam-pressure, the cylinder with an iron mantle cast in one piece, shaft of wrought iron, crank and fly-wheel. Diameter of the cylinder 1' $4\frac{1}{2}$", stroke of the piston 3' $1\frac{1}{2}$" by 36 revolutions pr. minute.

3. Of THE TOOLS, those under a., b. and c. are patented in England, France and Germany, and serve principally for making screws and nuts which are made by these machines much more exactly, accurately, and cheaper than by any other method of working. The screws and nuts leave the machine quite finished and can be employed without any retouching. The tools of these machines are very simple and can easily be kept in condition. These machines offer the same advantages in working the squares of spinning-cylinders etc., and in the exhibition opportunity is offered to see their working and the pieces worked.

The **traversing drilling machine** d. is worth consideration for its double use in boring long and round holes, and it may be employed with the greater advantage as it is universal in establishments where a machine cannot be exclusively occupied by traversing drilling.

The **radial drilling machine** e., much simpler than ordinary radial drilling machines is in general not only sufficient but much more commodious and convenient for many special articles, principally for boring holes into the wheels for railway purposes which are mounted on the axles. The planed foundation plate on the one side of the machine serves for boring cylinder, and lid bolts for locomotives and steam-engines, for frames, and main-cylinders for carding-engines. Upon the upright moveable table pieces can be fixed which are commonly worked on ordinary drilling machines. On this table is cast an upright triangle for fixing pieces, the front sides of which are bored, and which can have any length, when in the front (side) of the triangle a channel is made. When this channel is lengthened to the opposite side of the foundation plate, pieces of great circumference and length may be bored, because the arm of the machine is turning in 270", contrary to ordinary radial drilling machines of 180°—190°.

The **slotting machine** f., and the **slide lathe** g., are of a most simple, solid and convenient construction for different purposes.

4. The CARDING MACHINES in their external appearance, and their solid, practical, and elegant construction give already a testimony of practical progress. They are entirely constructed of iron, that is to say, the frame as well as the main and doffing cylinder, strippers and clearers are so made; this construction all of iron is quite new in this branch, and has afforded the same satisfaction as in the cotton branch.

The commodious and safe position of the pillows and other parts, the arrangement for easily detaching the mechanism of the slubbing machine, and the arrangement of simultaneous grinding of the main and doffing cylinders with a grinding roller, the combination of the chupper with the doffer, both being removeable simultaneously and each for itself, the advantage of universal mutability of the position of the table and feeding rollers, in proportion to the height of the clothed main-cylinder, and the perfect arrangement of the particular chupper-motion as well as other improvements: all these advantages will give general satisfaction, in as much as they ensure a higher production and less attendance.

In general the new chupper motion is most practical. Every experienced man in this branch knows that the vibrating motion of the chupper of former construction was the obstacle to an increased quickness of revolution, and caused vibrations which became troublesome to the whole machine.

By avoiding heavy swinging parts and by employing shorter and lighter chupper shafts, by the diminution of eccentricity at the crank shafts, in general by the avoidance of inequalities of weight, thus much has been attained by the slubbing machine, that every chupper has its own chupper shaft and eccentric, finally by the employment of conical bolts and shafts and conical boxes for regulation, by which is avoided the wearing out of pillows, — by all these improvements a perfection is given to the chupper motion, which by an increase of revolutions warrants a good regular working and a great duration as well as an increase of the production of the machines.

JORDAN & TIMAEUS

OF

DRESDEN.

The Main Etablishment in Dresden, founded 1823,

[Branch in BODENBACH in Bohemia established 1854 (see Austria)]

possesses, in addition to a

MANUFACTORY OF COFFEE SUBSTITUTES,

a Manufactory of Chocolate and Confectionary

with 4 Steamengines of 40 horse power, 200 working men and 20 Clerks, applying a great variety of Machinery of the newest invention. — The chief places of consumption are found in the States of the Zollverein, and in Denmark, Switzerland, Russia, and Transatlantic States.

There is a permanent exhibition of the articles produced by the firm in Dresden in the Palaisplatz No. 6, they are universally renowned for their cheapness and excellence, and have received from the undermentioned Exhibitions at different times the following medals in acknowledgement:

1837 from the Exhibition in Leipsic	1842 from the Exhibition in Mayence
1844 „ „ „ „ Berlin	1845 „ „ „ „ Leipsic
1850 „ „ „ „ Leipsic	1854 „ „ „ „ Munich,

as well as one Gold medal from King Antony of Saxony in 1837.

In regard to the variety of their assortment, which is so as to suit all classes of purchasers and consumers, and which is constantly increasing, the firm refers to the Price Current annexed in the Exhibition Building, offering a selection from:

I. Chocolates.

a.	28 Kinds of **Vanille-Chocolate**	1 *sh.* 2 *d.* to 3 *sh.* 4 *d.* pr. ℔ English	
b.	30 „ „ **Spice-Chocolate**	— *sh.* 8 *d.* „ 1 *sh.* 4 *d.* „
c.	16 „ „ **Health-Chocolate** without spice	— *sh.* 10 *d.* „ 1 *sh.* 6 *d.* „
d.	15 „ „ **Homoeopathic** or **Medicated Chocolate,** carefully prepared from prescriptions of celebrated physicians	1 *sh.* 2 *d.* „ 2 *sh.* 11 *d.* „
e.	22 „ „ **Cocoa paste** without sugar and spice . . .	1 *sh.* 1 *d.* „ 2 *sh.* 11 *d.* „
f.	8 „ „ **Soluble Cocoa**	1 *sh.* — *d.* „ 1 *sh.* 10 *d.* „
g.	2 „ „ **Racahoûts,** an excellent drink for children, women in childbed, invalids	1 *sh.* — *d.* „ 2 *sh.* 6 *d.* „
h.	34 „ „ **Eating Chocolates**	1 *sh.* — *d.* „ 3 *sh.* 4 *d.* „
i.	27 „ „ **Chocolate Pastilles** and **Pralinés**	1 *sh.* 2 *d.* „ 3 *sh.* 9 *d.* „
k.	44 „ „ **Desert Chocolates** for the Table	1 *sh.* 8 *d.* „ 10 *sh.* — *d.* „
l.	1500 **Designs** of **Fancy Chocolate figures** weighing from ⅛ Ounce to 30 ℔ each	2 *sh.* 6 *d.* „ 5 *sh.* — *d.* „

II. Confectionary.

α.	90 Kinds of **Caramel-Bonbons**	from £ 2 11 *sh.* to 6£ 5 *sh.* pr. Cwt. English
β.	14 „ „ **Rocks**	„ £ 4 14 *sh.* „ 5 £ 10 *sh.* „
γ.	30 „ „ **Drops**	„ £ 3 7 *sh.* „ 4 £ 14 *sh.* „
δ.	50 „ – – **Dragées**	„ £ 3 13 *sh.* „ 9 £ 6 *sh.* „
ε.	75 „ „ **sundry Confectionary**	„ £ 3 3 *sh.* „ 9 £ 6 *sh.* „

The articles exhibited can only form a very small portion of our assortment. The affixed English Prices are those for Imperial Pounds and Hundredweights free of freight to Hamburg. Cash Purchasers 1½ pCt. Discount. We beg to draw attention to the fact, that having already an extensive connexion with transmarine ports, we are enabled to afford every facility to the purchaser, and accommodate ourselves in regard to package and embellishment to the wishes of the consumers.

(No. 1453. of the offic. Catalogue.)

Charles  Grimm

𝕸𝖚𝖘𝖎𝖈𝖆𝖑 𝕴𝖓𝖘𝖙𝖗𝖚𝖒𝖊𝖓𝖙 𝕸𝖆𝖐𝖊𝖗 𝖙𝖔   𝕳𝖎𝖘 𝕸𝖆𝖏𝖊𝖘𝖙𝖞 𝖙𝖍𝖊 𝕶𝖎𝖓𝖌 𝖔𝖋 𝕻𝖗𝖚𝖘𝖘𝖎𝖆

𝕭𝕰𝕽𝕷𝕴𝕹   Kurstrasse No. 15.

Manufacturer of all kinds of stringed instruments after the system of the greatest Italian masters, and according to his own experience founded upon years of study.

Copies of celebrated old Instruments are made in such perfection that they cannot be distinguished from the originals.

My Instruments are at present in the possession of the following Virtuosi: *Laub, Ries, Ganz, O. Bernhardt, Liebig, Steffens, Backmann, Stahlknecht, Wohlers,* and of a number of distinguished amateurs.

Repairs of all kinds, especially of good instruments, are undertaken with every guarantee.

(No. 2535. of the offic. Catalogue.)

JULIUS GÜNTHER

IN

WALDHEIM (SAXONY).

1 Cupboard of nutwood . . . . . price:	Thlr.	1200	or	£ 180	—	Sh.
1 Lady's library do.	»	200	»	» 30	—	»
1 Dressing-glass do.	»	30	»	»	4 10	»

These articles are furnished with very rich carvers' work, and have been manufactured at the Royal house of correction in Waldheim (Saxony).

In the same style the whole furniture is always at hand; the designs of them will be found lying on the exhibited goods.

The dressing glass was made by an imprisoned mechanician.

Agent: Mr. VICTOR BAUER, No. 1. Ironmonger Lane, London E. C.

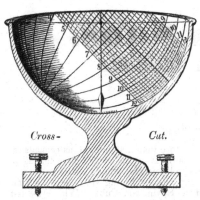

Cross-   *Cut.*

[81.]   (No. 1408. of the offic. Catalogue.)

H. Schmeisser's
patent hemispherical Sun-dial,

manufactured and exhibited by

A. Meissner
of
BERLIN,
Friedrich-Street No. 71.

The **hemispherical Sun-dial** consists of a hollow semi-globe, representing the firmament, on which the daily course of the Sun during the whole year is clearly shown. Across the top of the instrument two threads are placed crosswise, so that the point of intersection is exactly the centre of the globe. The shadow of this cross being thrown into the hollow globe, the **point** of **intersection** will there follow the same course, which the Sun pursues on the firmament, so as to describe on each day a certain given circle, which in there traced for it by the design, thus showing the time each day to the minute. — The Sun-dial is correctly set up in the different degrees of latitude by means of a small pendulum suspended from the cross point. — A small index is engraved on the instrument to regulate the difference between the actual time shown and the mean time.

The special **advantages** of the hemispherical Sun-dial are, that it is not needful for the correct setting up of the same to know the points of the compass, but only the exact **degree** of **latitude** in and **the date** on which the instrument is set up*, the direction of the meridian is then immediately shown; the observation is now correct to within single minutes from sunrise to sunset, as the sharply defined shadow is always thrown perpendicularly an the surface of the semi-globe, even when the Sun stands at its lowest point, thus moving always at the same even pace; each Sun-dial can be used within twenty degrees of latitude. — These advantages qualify the hemispherical Sun-dials to be used every where as **standard-dials**. They are as yet constructed in two ways, the one for use between 20 and 42, the other between 42 and 63 degrees of N. Latitude. — The casing of the instrument is made of porcelain and closed at the top with a glass plate, so that it can be used in the open air and in any climate. —

Price: nine Prussian Thalers.

* For full particulars see the description sold with the instrument.

[82.]   (No. 2502. of the offic. Catalogue.)

TH. VOECKLER & Co.
MEISSEN.

Exhibitor of **Wallosine.**

This is the name of a chemical prepared substitute for **Whale-bone for umbrella-ribs.**

A practical experience of several years and an attentive observation of the umbrellas made with ribs of "Wallosine" warrants the inventor in saying, that no other existing substitute for whalebone is equal to this in durability, and lasting elasticity.

The impregnation of the rods, that gives them their vital elasticity, is united with the substance so closely, that the duration is warranted for the longest time.

The rods of Wallosine are not affected by water; they become by rain somewhat harder and stiffer, bat not softer, as whalebone does. Wallosine is lighter than whalebone. The cost, of production is about $\frac{1}{4}$ of the price of whalebone.

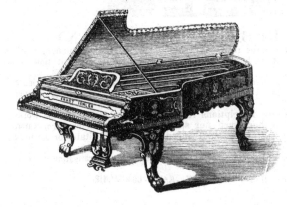

[83.]   (No. 2344. of the offic. Catalogue.)

ERNST IRMLER'S
Pianoforte Manufactory
in Leipzig.

(Munich Prize-Medal 1854.)

All kinds of Grand and square pianos as well as pianinos, with English and German action. Having constantly endeavoured for 16 years to place before the musical public the most perfect instruments in tone and action, their durability has been made so so great, that they withstand the influences of all climates, and are guaranteed for the term of five years.

(No. 1456. of the offic. Catalogue.)

Adolph Ibach Sons,

piano forte and organ builders

at

BARMEN,

RHENISH - PRUSSIA.

Owners of several prize medals.

The above concern has been in existence since the year 1794. More than 200 musical instruments are annually manufactured and sent away to all parts, even to North and South-America, everywhere finding an equal demand and sale on account of their solid make and the melodious and powerful sound in which they excel.

The construction of organs began with the year 1826, 106 organs having since been built for churches, chapels, concert-rooms etc., of which 18 were delivered in Spain and Cuba. Within the last six years alone 42 organs were manufactured.

The organs distinguish themselves both by an easy and elastic touch, and by the delicate and characteristic intonation of the various parts together with a peculiar fulness and power of sound of the complete works, in testimony of which a great number of competent judges have pronounced a most favourable opinion. — The few names we mention may serve to corroborate the above statements:

The royal director of music, M. BACH at Berlin,
the director, M. E. NEUMANN at Lisbon,
the organist, M. A. HAUPT at Berlin,
the organist and director of music at the cathedral of Merseburg, M. ENGEL,
the professor of music, BREIDENSTEIN at Bonn,
the professor L. BISCHOFF at Cologne,
the organist of the cathedral of Valencia, Dr. PASCUAL PEREZ at Valencia etc.

At the former exhibitions of London and Paris we did not introduce our manufactures; for the present exhibition we have forwarded under No. 1456. of the Catalogue

a harpsichord for concert-rooms and an oblique piano,

both of them instruments particularly adapted for damp and hot climates.

(No. 360. of the offic. Catalogue.)

THE HARBURG INDIA RUBBER COMB COMPANY

beg to invite the attention of the Public to their Combs manufactured of hardened India-Rubber, which material surpasses in quality all other substances that have been employed up to the present time for this purpose.

The extensive scale on which their works have been constructed, and the scientific skill and experience which they have brought to bear, enable them to supply a superior Article at a very moderate price cheaper than common Ivory and Buffalo Combs.

The Combs produced by them excel all other descriptions for the care bestowed upon the workmanship and the superiority of the material; they neither warp nor split, can be cleaned in warm water, are of uncommon durability, and being exceedingly soft and elastic, afford a perfect substitute for Tortoise shell Combs, to which they are in many respects preferable.

This article, on account of its cheapness, will be of interest for **export** to transatlantic parts, as it will never be subjected to the injurious influence of any climate.

Each Comb bears the stamp of the Fabric: HARBURGER GUMMI-KAMM-COMPAGNIE.

Harburg India Rubber Comb Company.

PER DOZEN.

DRESSING COMBS.

	6,	6½,	7,	7½,	8 Inches
Octagonal ... No.	601.		701.		801.
Roach ...... »	602.	652.	702.	752.	802.
» ...... »		653.		753.	803.
Curvated .... No.		654.		754.	804.

	5/6.	6/3.	7/.	8/6.	10/.
Roach ..... No.	612.	662.	712.	762.	812.

	5/.	5/6.	6/. .	7/.	8/.
. »	622.	672.	722.	772.	822.
Octagonal .... No.	621.		721.		821.
Curvated .... No.		674.		774.	824.

	4/.	4/6.	5/3.	6/.	7/.
Straight ..... No.	625.	675.	725.	775.	825.

	3/3.	3/6.	4/.	4/6.	5/.	
» ..... »	605.	655.	705.	755.	805.	905.

	4/3.	4/6.	5/.	5/6.	6/.	9/.
» fine .. »	606.	656.	706.	756.	806.	906.

	5/.	5/6.	6/.	7/.	7/6.	10/6.
» wide....... No.	716.	766.	816.	866.		

	7/6.	8/.	9/.	10/.
»	726.	776.	826.	876.

	5/6.	6/.	6/6.	7/.
» »	736.	7".		7/6.
» »	745.	7½".		5/.
» quill back .... »	717.		817.	867.
	9/.		11/.	12/.

Fluted .... No.	707.	757.	807.
	8/6.	9/.	11/.

Fancy shapes .... No.	777.	7½".	12/.
»	827.	8".	14/.
»	708.	7".	9/.
»	809.	8".	13/.

Toilet Combs .... No.	1.	475.	485.	490.
	12/.	6/.	7/.	9/6.
No.	477.	487.	497.	
	5/.	6/.	7/.	

POLL COMBS plain .. No. 121. 121a. 124. 126.
4/6. 3/6. 3/6. 2/9.
quill back No. 122. 123. 123a. 125. 128.
5/. 5/. 4/6. 3/9. 3/9.
Fancy back .... No. 127. 7/.

FINE TOOTH-COMBS.

No.	110.	111.	112.	113.	114.	115.	116.
	2/6.	2/9.	3/.	3/3.	3/4.	3/6.	3/9.
»	80.	81.	82.	83.	84.		
	2/9.	3/.	3/3.	3/6.	3/9.		
No.	212.	213.	214.	215.	216.		
	3/6.	3/9.	4/.	4/3.	4/6.		
»	14.		24.		34.		
	4/.		4/6.		5/.		
No.	25.		35.	45.			
	5/.		5/6.	6/.			
No. 12.	22.	32.					
	4/6.	5/3.	5/6.				
» 13.	23.	33.	43.				
	3/.	3/6.	4/.	4/6.			

POCKET COMBS

No.	1.	2.	3.A	4.A	5.	6.
	6/.	5/.	6/.	6/.	6/.	6/.
No.	3.	3½.	4.	4½.		
per Gross	27/.	30/.	33/.	36/.		
No.	305.	355.	405.	455.		
	24/.	27/.	30/.	33/.		
»	525.	575.	505.	555.		
per Doz.	3/.	3/2.	3/3.	3/9.		
No.	103.	104.	105.	106.	107.	
	3/6.	4/.	5/.	5/6.	6/.	

PUFF COMBS .... 3 — 3½". per Gross **24/.**
3 — 3½". No. 2. **18/.**

SIDE COMBS .... 3 — 4". **18/.**
3 — 4". No. 2. **12/.**

NECK COMBS .... 5 — 6". **24/.**
5 — 6". No. 2. **18/.**

TWIST COMBS plain .. No. 51. 52. 53. 54.
per Doz. 6/. 5/3. 4/6. 5/3.
Fancy shapes No. 60. 61. 62. 63. 64. 67. 69.
7/6. 8/. 10/6. 7/. 9/. 15/. 20/.

BONNET COMBS .... No. 71. **6/.**

BONNET PINS ..... No. 1. 2. 3. 4.
per Gross 24/. 24/. 30/. 18/.

HAIR PINS » » **9/.**

Agents in LONDON:

BENDA BROTHERS, 96 Newgate Street,
who will give any imformation as to Discounts etc.

H. THOMAS

ENGINEER AND IRON-FOUNDER

established 1845

Bendlerstrasse 16. BERLIN. *Grabenstrasse 31.*

Manufacturer of Machinery of all Descriptions for dressing Woollen-Goods.

Washing Fulling Raising Shearing Brushing Drying and Finishing
Machines etc. of the most improved Construction.

Hydraulic-Presses, Shoddy-Mills, Steam-Pumps, Cranes for all Purposes,
Lathes, Planing Machines, Tools etc.

Iron-Castings of all kinds

for Civil-Engineering, Building-Contrivances, Stable-Fittings and ornamental Works.

Obtained Prize-Medals for Shearing Machinery

LEIPZIG 1850. LONDON 1851. PARIS 1855.

H. SCHOMBURG & Co.,
20. Moabit, Berlin.

Double - clock 1862.

Isolator pruss. 1862.

Manufacturers of Solid China,

which is not liable to break or crack at boiling point, or even an increased degree of heat.

Coffee and Tea making Machines

of a new invention, entirely of China, can be used over any Gas-cock apparatus or Alcohol lamp.

Isolators, Isolating coupling-boxes
and
porous Cylinder of China-clay
for telegraphs.

Door and Window trimmings.

Bell-pull handles, Gas-burners
(Argand's system incision burner.)

Knife-sharpeners, Pen-cleaners etc. etc.

as also all kinds of China-ware for table and household use, white and decorated at the lowest wholesale prices.

Knife - sharpeners of China.

Oldest Manufacturers and Proprietors by inheritance of the Brilliant-gilding

on China and Crockery-ware, Glass and Earthen-ware (the production of a bright gilding direct from the fire, without polishing, for which the purchaser of the right for France took out a patent in 1850).

Price-currents, patterns, samples, and terms at
R. SCHOMBURG & Co., 90 Cannon Street, City, London.

FERDINAND SCHÜTT,
WINE MERCHANT
AT
AFFENTHAL, community of EISENTHAL, district of BÜHL, Grand Duchy of BADEN.

PRICES.

Affenthaler red wine, year 1859, the Baden Ohm (100 Baden quart) loco Affenthal 97 fl. = $55\frac{1}{2}$ Thlr.
Do. year 1861, loco Affenthal 70 fl. = 40 Thlr.
Wholesale and retail.
Cherry brandy, first qual., the Baden quart free of charge London 1 fl. 45 kr. = 1 Thlr.

CARL HILLER & SON,
Hosiery Manufacturers,
Chemnitz
(SAXONY),

are obliged by the limited space to exhibit only a small part of their numerous articles. The latter consisting of all kinds of Hosiery of every quality, thickness, size, colour, and pattern, are made on every description of machine and handloom.

Rhenish Prussia — *Rheinberg* — Lower Rhine.

Boonekamp of Maag-Bitter,

celebrated aromatic stomachic Bitters,

known under the device:

"Occidit qui non servat"

manufactured by

H. UNDERBERG-ALBRECHT.

Purveyor to

His Majesty King WILLIAM I. of Prussia,
His Majesty King MAXIMILIAN II. of Bavaria,
His Royal Highness the Prince FREDERICK of Prussia,
His Royal Highness the Prince CARL ANTON of Hohenzollern-Sigmaringen.
Imported into Russia by ukase of His Majesty the emperor ALEXANDER II.,
and into France by decree of His Majesty the Emperor NAPOLEON III.

These unrivalled stomachic Bitters (cordial drops) had no sooner been invented and manufactured by H. UNDERBERG-ALBRECHT than they grew celebrated far and wide for their fine flavour and superior wholesome qualities. This beverage is esteemed as an advantageous remedy against many diseases resulting from impurity of the blood, and a powerful preservative against **fever, dysentery, diarrhea, scurvy,** and **sea-sickness.** It excites appetite, restores it when lost, and also relieves constipation and other disorders in the bowels. This liquor has been approved of by the most eminent medical authorities, and found by experience to possess such sanative qualities, and so many people have used it with the greatest success, that it has now become an indispensable article in every house at home and abroad.

Mr. H. UNDERBERG-ALBRECHT therefore begs to call the attention of the export houses in this country to his manufacture. Its demand increases from year to year, and its excellent qualities have secured it a place amongst the most saleable articles for exportation. It does not suffer the least damage either by long warehousing, or by the longest voyage, nor is it subjected to the influence of climate. It is exported to all ports of the world and stores are continually provided in the following places, viz:

Alexandria, Adelaide, Amapola, Arequipa, Batavia, Bahia, Bombay, Buenos-Ayres, Capetown, Calcutta, Caracas, Cairo, Cincinnati, Colombo, Constantinople, Copenhagen, Demerara, Guayaquil, Hongkong, Havannah, Hobart Town, St. John's NB., Kanagawa, La Guayra, Lima, St. Louis, Leghorn, La Union, Madras, Mauritius, Manilla, Macassar, Malta, Malmoe, Melbourne, Montevideo, New York, New Orleans, Nagasaki, Naples, Port Natal, Port Elizabeth, Porto Alegre, Padang, Pernambuco, Puerto-Cabello, Rio de Janeiro, Rio Grande do Sul, Rosario, Samarang, Sydney, San Francisco, Singapore, Shanghai, Soerabaija, Smyrna, Syra, Tampico, St. Thomas, Valparaiso, Vera Cruz.

Mr. H. UNDERBERG-ALBRECHT begs to caution the consumers of his celebrated manufacture and the public in general against the numerous adulterations. Most of them are nothing but common Bitters as sold in every common public house and their composition is some times very unwholesome. The falsificators do not hesitate to imitate his label and seal and even his signature, and thus deceive the public in an impudent manner. Mr. H. UNDERBERG-ALBRECHT is the inventor and sole distiller of the **Boonekamp of Maag-Bitter** known under the device: *"Occidit qui non servat"* and all other Bitters sold under this denomination, without the facsimile of his signature on the label and his seal on the cork, are not genuine.

The **Boonekamp of Maag-Bitter** is to be had in quart and pint bottles, and also in small ones, called flagons.

For samples and further information please inquire of:

M. HERMAN RÜBECK, 12 little Tower Street, LONDON E. C.
General agent for the United Kingdom;

or address the undersigned in any language whatever.

Seal.

William Rieger

LONDON
26 Lambeth Hill, Doctors' Commons E. C.

and FRANKFORT O|M.

TOILET-SOAPS AND PERFUMERY

for

Home Consumption and Export.

Stand No. 309, Zollverein Department.

The objects exhibited form a great collection of every description of Toilet-Soaps and Perfumery in as great a variety in the finer and superior qualities, as in the lower and cheaper ones, both for home consumption and exportation to all countries and climates. The number and diversity of the articles are too great for a detailed specification; a few only are here named.— Among these, the first rank is held by the

TRANSPARENT CRYSTAL-SOAP.

it is shewn in large blocks, balls, tablets for washing, round shaving-sticks and cakes &c. This soap can be confidently recommended as the purest, mildest and best of all Toilet-Soaps, the one, which alone should be used by ladies with a delicate skin as also for tender infants. — The manufacturer calls the particular attention of the public to his transparent Crystal

Exhibition Tablets and Exhibition Sticks.

Eau de Cologne triple rectifiée,

in bottles and flasks of various sizes and shapes, cut crystal and wicker bottles &c.: It is recommended as the finest and most exquisite quality that can possibly be produced. — At the exhibitor's stand it s constantly gushing forth from two fountains for the use of the public, who are invited to moisten their handkerchiefs with it and compare it with other productions of the same name in order to ascertain the superiority of William Rieger's. —

In Essences for the handkerchief the manufacturer has prepared for the Exhibition a new article well worth attention, viz: his

Extraits d'odeurs

AUX FLEURS ANIMÉES.

The taste displayed in the getting up of the flasks, filled with the most delicate and strongest essences, and the exquisite artistical embellishments of the labels have attracted the attention and obtained the praise of connoisseurs. — Among the Pommades and other preparations for the growth and the embellishment of the hair exhibited to public inspection the

NUTRITIVE GOLDEN OIL

is a speciality of William Rieger's. It combines the virtues of a good Pommade with those of the best Hair-Oils and meets with general approbation. — Another article which deserves mention is

William Rieger's
Vinaigre de Toilette cosmétique et sanitaire
(Toilet Vinegar.)

This indispensable cosmetic for a Lady's toilet is distinguished for the richness of its aromatic ingredients and for its most beneficial and grateful effects. — It holds equal rank with the manufacturer's

Eau de vie de Lavande double ambree

whose virtues are highly appreciated especially in the East Indies. —

Extra Scented Violet Toilet Powder:	**Royal Rose Tooth Paste:**
Detergent Vegetable Hair Wash:	**Crême d'amandes ameres pour la barbe**

and many other articles which cannot be enumerated. Any desired particulars or information will be given by the manufacturer's agent at stand No. 309, Zollverein Department. —

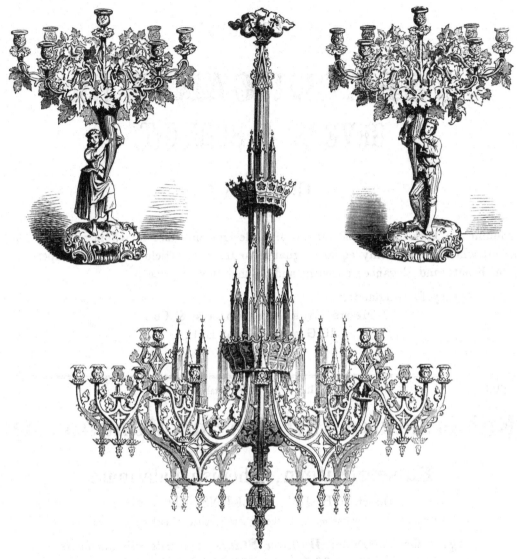

GOTHIC LUSTRE

BY

KISSING & MÖLLMANN,

ISERLOHN (Prussia).

Exhibition of lustres and all kinds of bronze articles, during the Jubilate- an
Michaelis-fairs, 43 Peter street, third floor, Leipsic.

Manufacturers and dealers

of stamped and cast bronze-ware, brass cast-ware, packfong and brass stamped-ware, umbrella-furniture
and thimbles, iron, brass and tombac-wire, latten-plate, wire-nails, iron-chains, and
all kinds of iron, steel, and small-wares.

MANUFACTORY

of

TURKISH CARPETS.

GEVERS & SCHMIDT,

of:

GOERLITZ

(Prussia),

manufacture Turkish (Smyrna) carpets of all designs and sizes in whole pieces without seam, which are not only equal in quality to the real article but remain superior to it in beauty and elegance, notwithstanding their lower price.

Agency for England:.

Messrs. EICHHOLZ, MORRIS, & Co.,

16 Gresham Street E. C.

LONDON.

[94.] (No. 152. of the offic. Catalogue.)

Kaiserslauter Ultramarine Manufactory

at

Kaiserslautern, Rhenish palatinate.

Director: HERM. WILKENS, Dr. phil.

Author of several works upon the theory and practice of making Ultramarine.

Agent for LONDON: *William Staats*, Dowgate Hill Chambers, 38 Dowgate Hill E. C.

For twenty years the above-mentioned director has been engaged in the preparation of Ultramarine according to his first invented method of muffle burning. In consequence of this long experience he has been successful in always attaining a uniform excellence in the material, and that to such an extent that the various shades may be produced according to pleasure.

This certainty of manufacture in a large business allows of the colours being mixed with a white alloy, which produces a cheap blue possessing brilliancy and a power of covering. These sorts of Ultramarine naturally have a great advantage over other cheap kinds of inferior colours which must necessarily be sold at a low price, and as a proof of this, reference may be made to the large and rapid sale experienced by the establishment for some years.

Besides the fine sorts available for printing upon carpets and other materials, and for the manufacture of paper hangings, the mixed sorts above-mentioned are accordingly recommended, and the conviction is entertained that every one who tries and investigates them, will recognise them as in every way serviceable and advantageous.

Prices and samples may be always procured at our agents in London.

E. SCHERING,

21. Chaussée Street,

BERLIN.

Manufactory of Chymical Productions.

Honourable mention at PARIS 1855.

WHOLESALE.

Represented at

LONDON: by BRUNO BRESLAUER, 6 Onslow Crescent, Brompton, and
R. SCHOMBURG & Co., 90 Cannon Street, City.
VIENNA: by BRUNO BRESLAUER, 978 City.
ST: PETERSBURGH: by A. WEYERT.
HAMBURGH: by THEOD. ULEX.

The principal article turned out from my chymical works on their first establishment in 1855 consisted of **potash of Iodide,** since which time I have been enabled to produce large quantities of the most satisfactory quality, and compounded after various receipts.

In addition to this article, I directed my attention to the branch of *pharmaceutic preparations*, including: **Acid. Acetic. Fort., Argent. Nitras,** as also **Baryte, Calc, Zinc, Bismuth, Ferr.** etc. etc., **salts,** and **Glycerine, Potass Bromide, Iodine re-sublimed** etc. etc.

With regard to the chymicals for *technical purposes*, I have to draw particular attention to the **Cyanide of potash,** which, in accordance with Liebig's directions, I have succeeded in producing of a pure brilliant white, and fine quality. I may further include in this class **White Glycerine,** which has been largely called for of late years by the public. The manufacture of this has been carried to a high degree of perfection, as well as that of **Baryte, Strontian,** and other kinds of **salts.**

It was at an early period that I gave full attention to *photographic preparations*. It not having been necessary to work up the mother-waters, already consumed in the preparation of Potash of Iodide and Bromide, I was enabled to furnish the **smaller Bromide** and **Iodine** in a state of extreme purity. I also succeeded in producing a **Pyrogallic Acid** of far superior excellence to any that my competitors are able to equal. All the other preparations connected with photography, such as, **Cotton-Collodion. Lithion, Fluor,** as well as **Gold** and **Argent salts** etc. etc. have been always so carefully and honestly worked up in order to obtain the purest and most superior quality, that my productions have gained credit not only throughout Germany, but in all European, and even in a large number of transatlantic countries. Indeed the demand is really so animated (and, I have no doubt, increases with every day), that I am continually compelled to consider the necessity of making an adequate extension in the range of my laboratories.

It was not till a few months ago that I began to turn my attention to the manufacture of *photographic paper* on a more extensive scale, and I have the satisfaction to say at the present time, that this article is more extensively used than ever, and is highly valued by foreign as well as home consumers.

After the testimony thus afforded by their general reception, I can sincerely recommend to both merchants and dealers in photographic and technical preparations, the productions of my manufactory, assuring those who will honour me with their orders, that I shall do all in my power to give full satisfaction, and merit the confidence conferred upon me.

D*

(No. 1437. of the offic. Catalogue.)

PRIZE MEDAL FOR INDUSTRIAL PURSUITS IN PRUSSIA,

GUSTAV BECKER,

Pendulum (Regulator) Manufactory,

FREYBURG

in Silesia.

GUSTAV BECKER recommends his regulators of different sizes and kinds, going for 8 days and also for a mouth, with second beats; pendulums with and without stone and stone casings, levers, clocks with and without striking apparatus. Cases of various designs.

(No. 2736. of the offic. Catalogue.

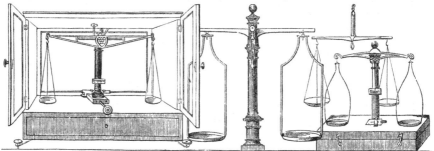

AUGUST SAUTER, EBINGEN, WURTTEMBERG.
MANUFACTURER OF BALANCES.
"Wurttemberg-Fortschritts-Medaille."

(No. 2466. of the offic. Catalogue.)

ALEXANDER SCHUMANN,
Oil-Cloth Manufactory,
at LEIPSIC.

This manufactory, which has been established upwards of 35 years, is represented by corresponding agents in all great cities on the continent, and whose productions are well known among the export trade, continues to supply the following articles:

Carpets in flower designs, and floor carpeting.

oiled-fustian, plain in every shading, imitations of every kind of wood, coloured and bross-coloured printing upon dark and light ground,

oiled-muslins
oiled-cambrios } in every shape and designed as above.

INDIA RUBBER WORKS
of

Albert Cohen, Vaillant & Co.,
HARBURG, Kingdom HANOVER.

Hanover 1859 great golden medal of honor.

Represented in exhibition and by Messrs. **A. Faber & Co.**, Marklane.

Doz.	Nomination of the Articles.	Dessin No	à doz. pair £	s.	d.
	A. India rubber overshoes. (Table I.)				
	a. Ordinary mens				
2	covered with tricot lining	1.	1	4	—
2	halfcovered idem	2.	1	4	—
2	slippers idem		1	4	—
2	sandals idem	3.	1	4	—
½	different shapes of extralight quality		1	4	—
	covered with tricot plush lining	1.	1	16	—
	halfcovered idem	2.	1	16	—
	slippers idem		1	16	—
	sandals idem	3.	1	16	—
	covered with woollen lining	1.	1	16	—
	fancy shapes with diff. linings.		1	16	—
	idem		1	16	—
	idem		1	16	—
	idem		1	16	—
	idem		1	16	—
	idem		1	16	—
	idem		1	16	—
	idem		1	16	—
	idem		1	16	—
	idem		1	16	—
	b. Ordinary for women				
2	covered with tricot lining	20.	—	18	—
2	halfcovered idem	21.	—	18	—
2	slippers idem	22.	—	18	—
2	sandals idem	23.	—	18	—
2	covered sandals idem	24.	—	18	—
2	sandals with three straps	25.	1	1	—

Doz.	Nomination of the Articles.	Dessin No	à doz. pair £	s.	d.
2	womens overshoes with heels		1	1	—
	covered, with tricot plush lining	20.	1	7	—
	halfcovered idem	21.	1	7	—
	slippers idem	22.	1	7	—
	covered sandals idem	24.	1	7	—
	sandals with several straps	25.	1	7	—
	sandals ordinary	23.	1	7	—
	overshoes with deep heels		1	7	—
	c. Ordinary for boys				
1	covered tricot lining	6.	1	1	—
1	½ covered idem	2.	1	1	—
1	sandals idem	3.	1	1	—
	covered tricot plush lining	6.	1	10	—
	½ covered idem	2.	1	10	—
	sandals idem	3.	1	10	—
	covered woollen lining	6.	1	10	—
	½ covered idem	2.	1	10	—
	sandals idem	3.	1	10	—
	covered fancy lining	6.	1	10	—
	½ covered idem	2.	1	10	—
	sandals idem	3.	1	10	—
	d. Ordinary for misses				
2	covered tricot lining	20.	—	15	—
2	½ covered idem	37.	—	15	—
2	sandals idem	28.	—	15	—
2	idem covered idem	24.	—	15	—
2	idem 3 straps idem	25.	—	18	—
	covered tricot plush lining	20.	1	2	6
	½ covered idem	37.	1	2	6

Albert Cohen, Vaillant & Co. Harburg.

doz.	Nomination of the Articles.	Dessin No.	doz. pair £	s.	d.
¼	sandals tricot plush lining	23.	1	2	6
¼	idem covered idem	24.	1	2	6
¼	idem 3 straps idem	25.	1	2	6
¼	covered woollen lining	20.	1	2	6
¼	½ covered idem	37.	1	2	6
¼	sandals idem	23.	1	2	6
¼	idem covered idem	24.	1	2	6
¼	idem 3 straps idem	25.	1	2	6
¼	covered fancy lining	20.	1	2	6
¼	½ covered idem	3„.	1	2	6
¼	sandals idem	23.	1	2	6
¼	idem covered idem	24.	1	2	6
¼	idem 3 straps idem	25.	1	2	6
	e. Ordinary for children				
2	covered	20.	—	12	—
2	½ covered	37.	—	12	—
	f. Footholders for men . .	40.	1	1	—
2					
	B. Selfacting overshoes, to be put on and out without need of hands. (Table I.)				
	a. For men	Dessin No.			
2	covered tricot lining	7.	1	12	—
2	½ covered idem	8.	1	12	—
2	spring slippers idem	9.	1	12	—
2	spurshoes idem	4.	1	12	—
¼	covered tricot plush lining	7.	2	5	—
¼	½ covered idem	8.	2	5	—
¼	slippers idem	9.	2	5	—
¼	spurshoes idem	4.	2	5	—
¼	covered woollen lining	7.	2	5	—
¼	½ covered idem	8.	2	5	—
¼	slippers idem	9.	2	5	—
¼	spurshoes idem	4.	2	5	—
¼	covered fancy lining	7.	2	5	—
¼	½ covered idem	8.	2	5	—
¼	slippers idem	9.	2	5	—
¼	spurshoes idem	4.	2	5	—
	b. for women				
1	covered tricot lining	26.	1	4	—
1	½ covered idem	27.	1	4	—
1	slippers idem	28.	1	4	—
1	with heels idem		1	4	—
¼	covered plush lining	26.	1	16	—
¼	½ covered idem	27.	1	16	—
¼	slippers idem	28.	1	16	—
¼	with heels idem		1	16	—
¼	covered woollen lining	26.	1	16	—
¼	½ covered idem	27.	1	16	—
¼	slippers idem	28.	1	16	—
¼	with heels idem		1	16	—
¼	covered fancy lining	26.	1	16	—
¼	½ covered idem	27.	1	16	—
¼	slippers idem	28.	1	16	—
¼	with heels idem		1	16	—
	Selfacting mens				
2	with brass springs assorted in shapes		2	8	—
	Russian boots				
2	with tricot lining	13.	2	14	—
2	woollen lining	13.	3	3	—

piece	Nomination of the Articles.	Dessin No.	per pair £	s.	d.
	India rubber boots	Dessin No.			
4	long boots with woollen lining .	14.	—	12	—
4	idem „ tricot lining . .	14.	—	10	6
4	long fishing boots woollen lining	15.	1	—	—
4	idem tricot lining .	15.	—	18	—
4	Napoleon boots woollen lining . .		—	13	3
4	idem tricot lining . . .		—	12	—
2	gaiters woollen lining		—	7	6
2	idem tricot lining		—	6	—
2	fancy, boots long, woollen lining		—	18	—
2	idem tricot „ .		—	15	6
2	idem fishing, woollen „ .		1	6	—
2	idem „ tricot „ .		1	4	-
2	idem Napoleon woollen „ .		—	19	6
2	idem „ tricot „ .		—	18	—
2	gaiters idem woollen „ .		—	10	6
2	idem tricot „ .		—	9	—
2	enameled india rubber leather boots woollen lining		—	—	—
2	enameled india rubber leather boots tricot lining		—	—	—
	at the same prices as ord. india rubber boots.				
	C. Shoes to wear alone. (Not overshoes.) (Table I.)				
doz.	**Enameled india rubber leather shoes**	Dessin No.	à doz. pair		
¼	mens and boys, ordinary shape .	10.	1	12	—
¼	idem slippers with rosettes		1	12	—
¼	women, ordinary shape	29.	1	4	—
¼	idem slippers with rosettes . .		1	4	—
¼	childrens and misses, ordinary shape	29.	1	1	—
¼	idem with rosettes .		1	1	—
	Varnished india rubber lace shoes				
½	for men		1	12	—
½	for women		1	4	—
½	lace gaiters for women	33.	1	12	—
½	idem „ men		1	16	—
	Clothshoes with rubber soles and edgings				
½	lasting gaiters for men with laces	35.	2	8	—
½	idem „ „ elastic	35.	3	—	—
½	idem for women „ laces	34.	1	16	—
½	idem „ „ elastic	34.	2	5	—
½	idem „ laces with heels	34.	1	19	—
½	idem „ elastic without „	34.	2	8	—
½	feltgaiters „ with laces	36.	1	16	—
½	idem for men „ heels	36.	2	8	—
½	laceboots of cotton velvet . . .	35.	1	18	6
½	idem „ satin . . .	35.	1	18	6
½	idem „ moleskin . .	35.	1	18	6
½	idem „ canvass . .	35.	1	18	6
½	feltshoes for men	5.	2	8	—
½	idem „ women	32.	1	16	—
½	idem „ children	32.	1	12	—
½	feltclothshoes for men	5.	2	2	—
½	idem „ women . . .	32.	1	12	—
⅓	cotton velvetshoes for women . .	32.	1	7	—
⅙	idem „ men . . .	5.	1	16	—
⅓	cotton satinshoes „ women . .	32.	1	7	—
⅙	idem „ men . . .	5.	1	16	—
⅓	cotton moleskinshoes for women	32.	1	7	—
⅙	idem „ men . .	5.	1	16	—
⅓	cotton flannelshoes „ women	32.	1	7	—
⅙	idem „ men .	5.	1	16	—

Albert Cohen, Vaillant & Co. Harburg.

Nomination of the Articles.	à doz. pair £ s. d.	Nomination of the Articles.	per mètre quarré £ s. d.
Hamp canvasshoes. Dessin No. 12.	1 7 —	**Piece** No. 3. Woollen	
NB. Of all these shoes ⅓ is made without rubber edgings ⅔ with edgings.		1 grey cloth with black rubber	— 5 2
		4 different qualities for horse covers and travelling rugs.	
All mens - and womensshoes made in three different widths: wide, middle and narrow.		No. 4. Silk	
		2 black silk with black rubber . . .	— 8 4
		1 brown idem . . .	— 9 —
Childrens - and missesshoes in only two widths: middle and wide.		1 grey idem . . .	— 9 —
		1 check silk with grey rubber . . .	— 9 —
		1 drap idem . . .	— 9 —
D. Waterproof cloth.		b. Double texture.	
a. Single texture.	per mètre quarré	No. 1. Cotton	
Piece No. 1. Cotton		1 brown fine percal	— 3 —
2 brown percal thin with black rubber	— 2 4	1 white idem · . . .	— 3 —
1 white do. » idem	— 2 4	1 brown and black percal . . .	— 3 —
1 black do. » idem	— 2 4	1 white stout percal	— 3 —
1 white do. » with white rubber	— 2 4	1 brown idem . . .	— 3 —
1 grey do. » with grey rubber	— 2 4	1 grey idem . . .	— 3 —
1 brown do. stout with black rubber	— 2 7	1 brown fine twilt . . .	— 3 3
1 black do. » idem	— 2 7	1 white idem . . .	— 3 3
1 white do. » idem	— 2 7	1 brown and black twilt . . .	— 3 3
1 » do. » with white rubber	— 2 7	1 white stout twilt . . .	— 3 6
1 » do. » with grey rubber	— 2 7	1 brown idem . . .	— 3 6
1 grey do. » idem	— 2 7	1 grey idem . . .	— 3 6
2 brown twilt thin with black rubber	— 2 7	1 white twilt and tops	— 3 6
1 white do. » idem	— 2 7	1 brown idem . . .	— 3 6
1 black do. » idem	— 2 7	1 brown twilt and black tops . . .	— 3 6
1 white do. » with white rubber	— 2 7	1 grey idem . . .	— 3 6
1 grey do. » with grey rubber	— 2 7	1 white canvass	— 3 6
1 brown do. stout with black rubber	— 3 —	No. 2. Orleans	
1 a c k do. » idem	— 3 —	1 extrafine	— 6 5
1 bli te do. » idem	— 3 —	1 fine	— 6 —
1 wh do. » with white rubber	— 3 —	No. 3. Cashmere	
1 » do. » with grey rubber	— 3 —	1 black	— 10 —
1 »y do. » idem	— 3 —	1 brown	— 10 —
2 grewn tops » with black rubber	— 3 —	1 grey	— 10 —
1 grete do. » idem	— 3 —	No. 4. Silk	
1 brok do. » idem	— 3 —		
1 blac k e do. » with white rubber	— 3 —	1 black on one side and check on the other	— 13 5
1 whit do. » with grey rubber	— 3 —	No. 5. Woollen	
1 grey canvass with black rubber . . .	— 3 2	1 grey and brown	— 8 —
1 brown do. idem	— 3 2	1 grey on both sides	— 8 —
1 grey moleskin with grey rubber . . .	— 3 8		
1 brown do. with black rubber . . .	— 3 8	**E. Waterproof clothing.**	
1 red do. idem	— 4 —	(Table 1.)	
1 brown indian cloth thin with black rubber	— 5 —	a. Overcoats.	
1 black do. » idem	— 5 —	No. 1. Fine	p. Piece sh.
1 grey do. » with grey rubber	— 5 —	1 silk . . . . dessin No. 2. from	24—36
1 white do. » with white rubber	— 5 —	1 twilt . . . . . » » »	12—21
2 black and white checks with black rubber	— 3 —	1 india cloth . . . » » »	12—21
2 black twilt with black rubber	— 3 8	1 orleans . . . » » »	12—21
1 grey » with grey rubber . . .	— 3 8	1 fine cashmere . . » » »	24—36
No. 2. Orleans		1 silk . . . . dessin No. 8.	24—36
2 black ordinary black rubber . . .	— 3 4	1 twilt . . . . . »	12—21
2 » fine idem . . .	— 4 —	1 india cloth . . . »	12—21
2 » extrafine idem . . .	— 5 —	1 orleans . . . »	12—21
1 grey ordinary idem . . .	— 3 4	1 fine cashmere . . . »	24—36
1 » fine idem . . .	— 4 —		
1 » extrafine idem . . .	— 5 —		
1 » fine white rubber . . .	— 4 —		

Albert Cohen, Vaillant & Co. Harburg.

Piece	Nomination of the Articles.		sh.	sh.
1	twilt	dessin No. 11.	12	21
1	silk	» »	24	36
1	india cloth	» »	12	21
1	orleans	» »	12	21
1	fine cashmere	» »	24	36
1	silk	dessin No. 12.	24	36
1	twilt	» »	12	21
1	india cloth	» »	12	21
1	orleans	» »	12	21
1	fine cashmere.	» »	24	36

No. 2. Ordinary

Piece			sh. d.	sh.
1	percal	dessin No. 2.	4.10	16
1	twilt	» »	8.—	18
1	stout	» »	4.10	16
1	orleans	» »	12.—	21
1	drap moleskin	» »	12.—	21
1	check	» »	9.—	21
1	percal	dessin No. 8.	4.10	16
1	twilt	» »	8.—	18
1	stout	» »	4.10	16
1	orleans	» »	12.—	21
1	drap moleskin	» »	12.—	21
1	check	» »	9.—	21
1	percal	dessin No. 11.	4.10	16
1	twilt	» »	8.—	18
1	stout	» »	4.10	16
1	orleans	» »	12.—	21
1	drap moleskin	» »	12.—	21
1	check	» »	9.—	21
1	percal	dessin No. 12.	4.10	16
1	twilt	» »	8.—	18
1	stout	» »	4.10	16
1	orleans	» »	12.—	21
1	drap moleskin	» »	12.—	21
1	check	» »	9.—	21
1	percal	dessin No. 18.	4.10	16
1	twilt	» »	8.—	18
1	stout	» »	4.10	16
1	orleans	» »	12.—	21
1	drap moleskin	» »	12.—	21
1	check	» »	9.—	21
1	percal	dessin No. 17.	4.10	16
1	twilt	» »	8.—	18
1	stout	» »	4.10	16
1	orleans	» »	12.—	21
1	drap moleskin	» »	12.—	21
1	check.	» »	9.—	21

No. 3. Extrastrong

Piece			sh. d.	sh.
1	twilt	dessin No. 2.	8.—	18
1	canvass	» »	8.—	18
1	stout	» »	4.10	16
1	moleskin	» »	12.—	21
1	cloth	» »	18.—	36
1	twilt	dessin No. 8.	8.—	18
1	canvass	» »	8.—	18
1	stout	» »	4.10	16
1	moleskin	» »	12.—	21
1	cloth	» »	18.—	36
1	twilt	dessin No. 20.	8.—	18
1	canvass	» »	8.—	18
1	stout	» »	4.10	16
1	moleskin	» »	12.—	21
1	cloth	» »	18.—	36
1	twilt	dessin No. 21.	8.—	18

Piece	Nomination of the Articles.		sh. d.	sh.
1	canvass	dessin No. 21.	8.—	18
1	stout	» »	4.10	16
1	moleskin	» »	12.—	21
1	cloth	» »	18.—	36
1	twilt	dessin No. 22.	8.—	18
1	canvass	» »	8.—	18
1	stout	» »	4.10	16
1	moleskin	» »	12.—	21
1	cloth.	» »	18.—	36

b. Hoods for the said overcoads.

Piece		
1	silk	
1	twilt	
1	india cloth	
1	orleans	
1	cachmere	
1	percale	
1	twilt	
1	stout	
1	orleans	
1	drap moleskin	
1	check	
1	twilt	
1	canvass	
1	stout	
1	moleskin	
1	cloth.	

c. Leggins, short and long.
(Dessin No. 10 N. and 12 N.)

Piece		
2	silk	
2	twilt	
2	india cloth	
2	orleans	
2	drap moleskin	
2	cashmere	
2	percal	
2	twilt	
2	stout	
2	orleans	
2	check	
2	twilt	
3	canvass	
3	stout	
3	moleskin	
3	cloth.	

d. Caps and hats
(Dessin No. 4., 6., 9 N.)

Piece		
24	of different models.	

F. Enameled india rubber leather.
(Table II. b.)

Piece		
2	stout, one ply	
2	twilt	
2	stout and twilt, two plies	
2	2 stout and 1 twilt, three plies	
2	3 » » 1 » four	
1	died	
1	idem.	
6	Cart and waggon covers	
3	Carriage laps with all the necessary rings and straps belonging to them	
3	Horsecovers	
6	Common carpetbags	
6	superior idem	

Albert Cohen, Vaillant & Co.   Harburg.

	Nomination of the Articles.	£ s. d.			Nomination of the Articles.	£ s. d.
piece			doz.			
3	guncovers		12	nipples with shield		
3	mailbags		3	thumb-stalls		
3	poiltebags		12	nipples quality 1. No. 1.		
3	pouches dessin No. 7.		12	idem " 1. " 2.		
3	portmanteaux and trunks dess. No. 1. 2. 3. 4.		12	idem " 1. " 3.		
3	mailbags for carriage		12	idem " 2. " 1.		
	of liquids (mexican expedition)		12	idem " 3. " 2.		
3	knapsacks for cavalry and infantry		12	idem " 3. " 3.		
3	kneecovers for horses		1	stopples dessin No. 25—29.		
			piec.3	kitchen-pails " 8.		
3	horsepads		3	fire-buckets " 6.		
3	cartridge boxes		doz.1	tabacco-bags " 5. u. 9.		
3	umbrella cases		piec.3	chamber-pot " 10.		
			3	travelling-wrinat " 10.		
	G. Airproof articles.		doz.1	ear-tubes (speaking-trumpets) dessin 31.		
	(Table II. a.)	per piece	1	belts dessin No. 32.		
2	Aircushions round No. 1. dessin No. 12.—	6 —	1	robe-holder		
2	" " " 2. " 14.—	7 5	1	chest-expander " 12.		
2	" ½ " " 1. " 12.—	6 —	1	door-springs of different shapes		
2	" " " 2. " 14.—	7 5		dessin 16. 15. 18.		
2	" square " 1. " 9.—	6 —	1	billard cushions of different shapes		
6	" with back					
	size No. 1.	—11 3		**I. India rubber mats and carpets.**		
	" " 2.	—13 5	piec.6	Door mats		
	" " 3.	—15 8	6	idem with inscriptions		
12	Neckcushions, 2 of every kind dess. No. 11.		6	stair-mats		
2	twilled No. 1.	6 —	6	idem ornamental		
2	" " 2.	7 5	6	carpets for offices, billard-rooms,		
2	first qnality rubber No. 1.	6 —		halls etc.		
2	idem " 2.	7 5				
2	common rubber No. 1.	6 —		**K. India rubber toys.**		
2	idem " 2.	6 —		(Table III.)		
4	pillows in 2 sizes and shapes dess. 5. 6.—	7 5	1	Assortment of small balls with and without		
6	railroadbeds cloth outside dess. No. 4.—	18 —		hole and solid balls		
		per pound	1	assortment of grey and colored (airfilled)		
3	mattresses " 4.	5 5		balloons, footballs		
2	idem with pillows " 18.	5 5				
		per piece				
2	swimmbelts shape 1. dessin No. 10. 13.—	9 —	1	ball whistles		
2	" " 2. " 10. 13.—	12 —		with rilbon and with toothring		
2	" " 3. " 10. 13.—	15 —	1	bettles		
2	heaters, of twilled No. 1.					
2	idem idem " 2.			Large and highly elegant assortment of cry-		
2	round washing and bathing tubs 1.			puppets		
2	idem idem 2.			(Table IV.		
2	idem idem 3.			fancy balloons		
2	sitbath No. 1. dessin No. 2.			fruits		
2	" " 2. " 2.			stars		
1	bathing-vessel No. 1. dessin No. 3.			globes etc.		
1	" " 2. " 3.			red balloons with gas-filling		
3	sickcushion of rubber dessin Mo. 15.			idem with trumpets		
1	pontoons with tubes			Doll-heads of all sizes		
1	idem " balloons					
1	fishing and lifeboats			Printed price lists of all these toys will		
1	" " lifecloakes			be found in the exhibition, and with		
2	life-coats common			Messrs. Aug. Faber & Co.		
2	idem superior					
3	life-jackets					
	H. Articles for domestic use.			**L. Articles for arts.**		
	(Table II. b.)			Models drawing		
12	elastic bands of all sizes			complete assortment of Parthenons-		
13	sleevepads			Basreliefs		

Albert Cohen, Vaillant & Co. Harburg.

Nomination of the Articles.	Nomination of the Articles.

M. Rubber for tecnical purposes.
(Table V. a. b.)
For railroads marine and industrial purposes.

1. Department.
India rubber sheel and rings for steam-packing.
(Table V. b.)

1. Quality:
washers for condensors and airpomps, cold and hot-water pomps, pomps for destillery, cremeries etc.; where the 1. quality of rubber is necessary to the regularity of motion and to the security of the machinery this rubber is required. Quality warranted for washers.
Para No. 1.
Specific weight 1,030.

2. Quality:
for the same purpose, but not warranted in service.
Java A.
Specific weight 1,300
withe Java B.
Specific weight 1,400.

3. Quality:
hot and cold water and steam-packing, valves for sugar raffineries and de-stilleries etc.
Mineral G. B.
AB.
Specific weight 1,500.

4. Quality:
for ordinary packing.
W. 11. Specific weight 1,600.

5. and 6. Quality:
E. & F. cheap quality
for ordinary joints shock and traction.
Specific weight 1,600.

2. Department.
Buffers.
Dessin 5. 8. 9. 11. 19. 24.

1. Quality:
Very elastic.
density 1,030
market *A.*

2. Quality:
Highly recommended to railroads for shock, drawhooks safety chains etc.
density 1,030
market *G.*

3. Quality:
For the same purposes. Quality for germany
density 1,480
market *BW.*

4. Quality:
For the same purposes and also em-ploied in germany
density 1,600
market *AB.*

5. and 6. Quality:
density 1,600
market *E* and *F*
cheap quality for the same purpose.
All these sheets and valves can find application in every kind of indu-stry and are made af any length and thickness wanted, and. of 7 feet or less width.

3. Department.
Different articles.

Locomotive-steam-hose of all shapes and sizes
India rubber hose
Pure rubber tubing   dessin No. 2.
Hose with spiral wire   3.
Varnished and spiralised gas pipes
Rope packing of pure rubber des. 10. 12.
  idem   with canvass
India rubber dissolution
Packing dops for sugar raffinery
Valves for centrifugal engines
Cones for vacuum joints
Machine beltings with hemp on cotton canvass of all dimensions dessin No. 1.
Waterproof persennings and carriage-covers
  Of all these articles special price cur-rents and sketches are delivered on demand.

N. Surgial articles.

Injections pomps
  idem   syringes
Surgical cushions
Ombilical cushions
Pissuariums
Nipples
Crutch-cushions
Linen conservers.

O. Articles for military and marine use.

Tents
Knapsacks
Pouches   dessin No. 7. 11.
Powder-bags
Persennings for baggage carts
Bottles
Camps beds
Camp blankets
Mulebags for water carriage
Pontoons
Buffers for guns
Sheet-Packing for gunboats

P. India rubber thread and elastic neavings made with it.

Large assortment of all these tecnical articles of every description and all sizes will be found in the exhibition and with Messrs. AUG. FABER & Co.

Printet Price lists of all these articles will be found in the exhibition and with Messrs. AUG. FABER & Co.

India rubber works of the manufactories of ALBERT COHEN, VAILLANT & Co. Harburg.

Table I.

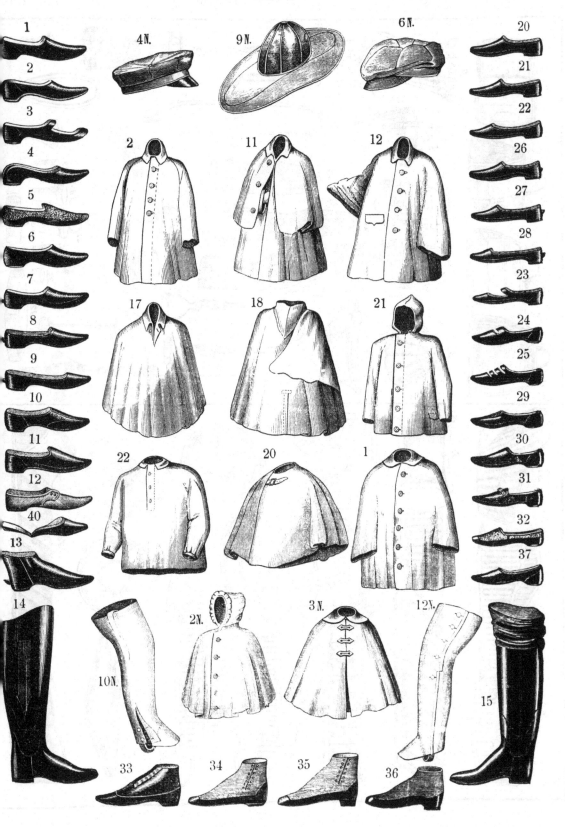

India rubber works of the manufactories of ALBERT COHEN, VAILLANT & Co. Harbu

a. Airproof articles.   Table II.   b. India Rubber leather and Articles for domestic

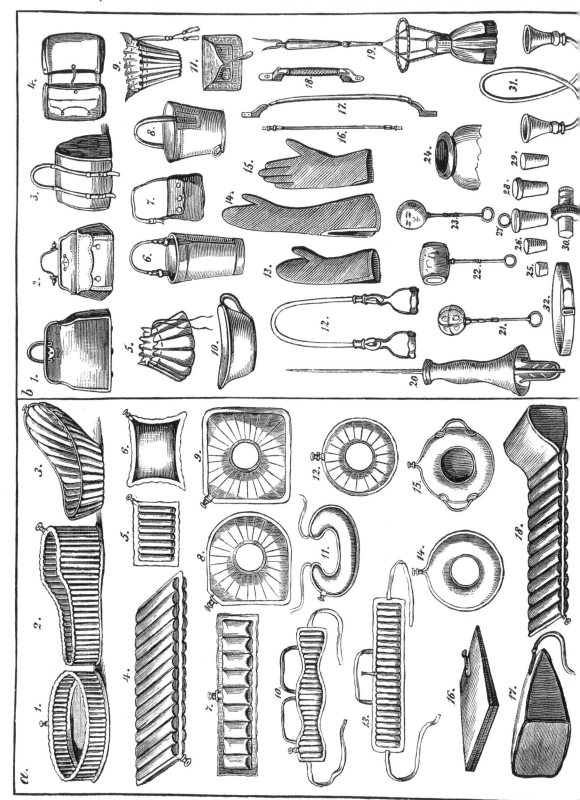

India rubber balls of the manufactories of ALBERT COHEN, VAILLANT & Co. Harburg.

Table III.

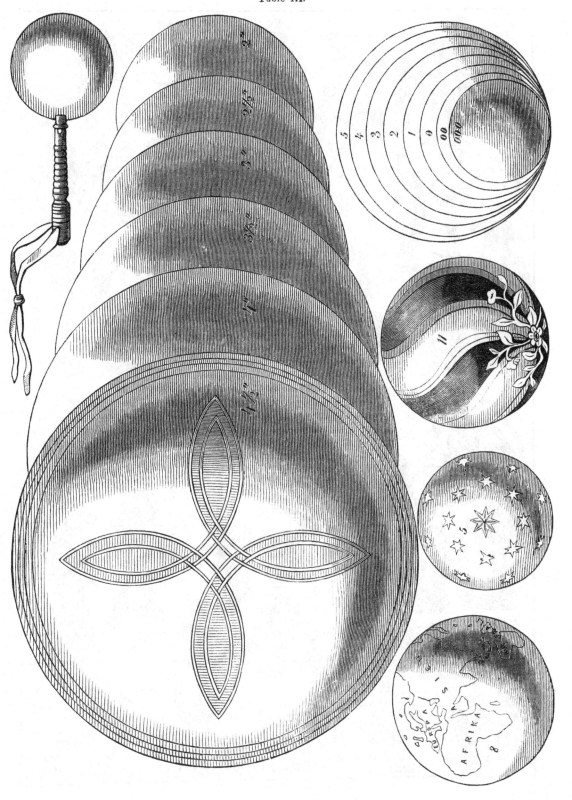

India rubber toys of the manufactories of ALBERT COHEN, VAILLANT & Co. Harburg.

Table IV.

ber for tecnical purposes of the manufactories of **ALBERT COHEN, VAILLANT & Co. Harburg.**

Table V.

Rubber for tecnical purposes of the manufactories of ALB. COHEN, VAILLANT & Co. Harbu

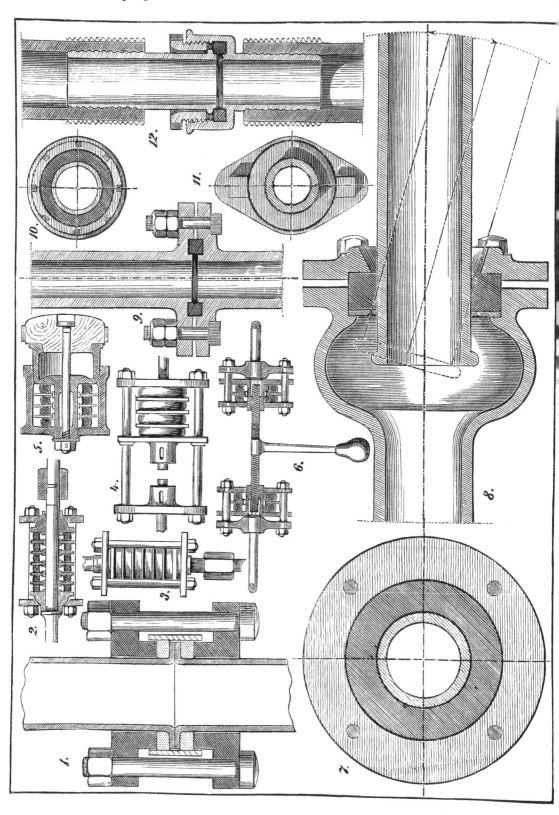

ROYAL

PORCELAIN-   MANUFACTORY

AT BERLIN.

The ROYAL PORCELAIN-MANUFACTORY AT BERLIN was founded by FREDERICK II. in the last century and was intended as a model-establishment for the cultivation of the fine arts in this branch of industry. The establishment is at the same time expected to yield a net produce in order to cover as much as possible the interest on the capital fund. It now works with 7 kilns and employs about 300 persons, whose total produce amounted during the later years on an average to half a million of articles and to the value of 150,000 prussian dollars. The superintendence and management is entrusted to Director **Kolbe**, under whom Dr. **Elsner**, as chemist, M. **Mantel** as master-modeller, and M. **Looschen**, as head-painter, are actively employed in their different departments.

Among the objects sent to the exhibition deserve to be particularly noticed: Two vases about 5 feet high, in the form of amphoras with figures of elves and fairies a similar vase with the figure of Borussia; two vases in the Medicean form, one with the triumphal procession of Bacchus, the other with Homer's Nausicaa followed by Ulysses, the paintings executed from the originals of professors **Steinbrueck, Daege, Bendemann** and **Schroetter**; an oval basin with views of the royal castle of Babelsberg and its environs, and several vases with landscapes; two wine-coolers with groups of children after **Rubens,** painted in the manner of the ancient majolica; several lustres and table-plates after models of the age of **Frederick the Great.** Of the three illustrations of the catalogue, the first represents some of the above mentioned vases and other ornamental vessels; the second exhibits on the upper half a newly modelled table-ornament in the **renaissance-**style, "Amor and Psyche", the middle division being $3\frac{1}{2}$ feet high, and the lower department containing several ornaments and vessels, for most part in the **classic** style; the third illustration shows a number of table-ornaments in the **rococo-**taste.

The exhibited articles are to be sold: for prices and particulars apply to the agent appointed for that purpose by the royal manufactory.

ROYAL PORCELAIN - MANUFACTORY AT BERLIN.

ROYAL PORCELAIN-MANUFACTORY AT BERLIN.

ROYAL PORCELAIN - MANUFACTORY AT BERLIN.

ALPHABETIC LIST OF THE EXHIBITORS.

The arabic number refers to the page of the first, the roman to that of the second division.

A.

Abelsdorff, Alb. & Jul. 112
Achilles, W. 4
Actien - Gesellschaft for mining and smelting, Stolberg and Nordhausen 48
Actien - Gesellschaft for mining and the manufacture of lead and zinc, Stolberg and in Westphalia 48
Adam, Gerh. (Wesel) 82. LXVIII
Adam, J. M. (Rennweg) 10
Ade, Eduard 170
Adenau XXII a. local-department of the Rhenish-Prussian agricultural society 65
Adler, E. (Neustadt) 84. 87
Adler, Brothers (Neustadt) 84. 87
Adler, Markus (Berlin) 74
Albert, Joseph 13
Albrecht, F. (Bromberg) 76
Albrecht, J. (Mentz) 33
Albrecht, Josef (Freiburg) 6
Albrecht, R. (Chemnitz) 135
Aleff 45
Alff, Franz 61
Allendörffer, J. C. 32
Altenbeken Hüttenwerk 48
Altenberg, Zwitterstocks-Faktory 130
Alves, Heinr. 101. 113. 128
Amalia 48
Amende, R. 54
Ammon, Johann Paul 16
Am Schwaben 48
Ananias 48
Andrae & Grüneberg 54
André, Carl August 21
Andreae, Christoph 87. 88
Anhalt, association for chemical manufactures 3
Appel, Heinrich, & Co. 58
Arendt, Eduard 88
Arndt, Brothers 18
Arnheim, S. J. 116
Arnoldi, H. 151

Arnoldt, Wilhelm 150
Aron, J. (Berlin) 89
Aron, Joseph (Berlin) 94
Aron & Jacobi 113
Aschrott, H. S. 46
Assmann, Julius 133
Atzrott & Co. 48
Augustin, A. 125
Ausfeld, Hermann 149
Ax, Heinrich 89
Axmann, R. 59

B.

von Babo, L. 6
Bach, J. (Fürth) 18
Bach, J. G. (Leipzig) 141
Bachem & Co. 48
Backhaus, Brothers 153
Bader, Adolph Friedrich 5
Badish association for the cultivation and commerce of tobacco 5
Baevenroth, C. F. 59
Bahnmayer, J. L. 174
Baltic association for the promotion of agriculture at E dena 65
Bardenheuer, C., & Co. 148
Barheine, Rudolph Robert (R. Barheine) 74
Barre, E. 59
Bartels, Brothers 87
Barth, Stephan & Comp. (Würzburg) 11
Barth, W. (Bernsdorf) 81. 124
Bartsch, Friedr., Sons 101
Baruth, count Solms' glass works administration 125
de Bary, Julius 36
Basch, W., & Co. 113. 116
Bassermann, Herrschel & Dieffenbacher 5
Bauch, C. A. (Rosswein) 135
Bauch, T. (Rosswein) 135
Baudouin, E., & Co. 115
Bauer, Adolph 16
Baumann, C. 161
Baumgaertel, C. F., & Son 134

vom Baur, J. H., Son 87
Baute & Co. 59
Bautenberger Einigkeit 48
Bechstein, C. (Berlin) 82
Bechstein, Friedrich Wilhelm Carl (Berlin) 82
Beck, Daniel 139
Becke, C. G. 100
Becker, A. (Munich) 15
Becker, Ch. (Pforzheim) 9
Becker, Gustav (Freyburg) 81. LII
Becker, K. (Pforzheim) 9
Becker & Auerbach 89
Beckeroder Hütte 24
Beckert, Richard 135
Beckhard & Sons 59
Beckmann, A. 123
Beer, C. 151
Beermann (Riesenbeck) 48. 52
Beermann, B. (Münster) 76
Beha, J. A. 7
Behrend, B. (Cöslin) 106
Behrend, G. (Hirschberg) 54
Behrend, Gustav (Hirschberg) 54
Behrens, C. (Alfeld) 29
Beisert, Adolph 59
Beissbarth, G. C., Son 15. XXVIII
Beissel, Steph. Widow & Son 122
Beck & Salzmann 164. XXXI
Belgisch - Rheinische Gesellschaft for mining upon Ruhr 48
Belle, Richard 78
Bellingrath, C. H., & Linkenbach 89. 99
Bembé, Anton 113
Benckiser, J. Adam (Pforzheim) 5
Benckiser & Co. (Pforzheim) 9
Benda, G., & Co. 150
Bender, Wilh., sen. 65. 85
Bendorf, central - silkworm and winding - establishment 87

Benjamin, M. A. 101
Benneke & Herold 54. 65
Benz, August 170
Benzing, Rapp & Co. 159
Berg, W. 116
Berger, M. (Peitz) 89
Berger, Rud. (Cöthen) 4
Berger & Co. (Witten) 76
Berger & Co. (Witten) 48
Bergischer Gruben- und Hütten-Verein 48
Bergmann & Co. 98
Beringer, A. 54
Berlin, joint-stock company for the manufacture of railway requisites 70
Berlin, manufactory of asphaltic paste-board for roofing 75
Berlin, plastic coal factory 72. VIII
Berlin, Geh. Ober-Hofbuchdruckerei 107
Berlin, Royal Prussian Ministry of trade and public works 74
Berlin, Royal iron-foundery 116
Berlin, Royal »Gesundheitsgeschirr« manufactory 126
Berlin, Royal porcelain-manufactory 125
Berlin, patent paper manufactory 106
Berlin, Renaissance - Commandit - Gesellschaft für Holzschnitzkunst 114
van Berlo, J. & A. 103
Bernstein & Lichtenstein 89
Bernstorff, C., & Eichwede 29
Berolzheimer & Illfelder 15
Bertelsmann & Niemann (Bielefeld) 87. 89
Bertelsmann & Son (Bielefeld) 85. 87
Bessert - Nettelbeck, Pauline 99. VI
von Beulwitz 61
Beyrich, Ferdinand 54. 80
Bialon, J. 71

Berlin, printed by R. Decker printer to the court of His Majesty the King of Prussia.

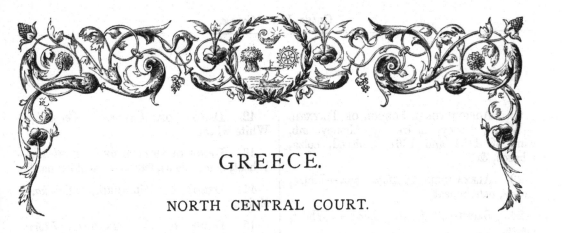

GREECE.

NORTH CENTRAL COURT.

CLASS I.

1. AGRICULTURAL SCHOOL OF TIRYNTH.—Limestone.

2. ALEXANDROPOULOS, C. *Tripolis.*—Black and variegated stone.

3. CAPARIAS, J. G. *Panormas.*—Black and white marble of Tinos.

4. CENTRAL COMMITTEE, ATHENS.—Pair of hand mills; hewn mill-stones; specimens of chromium ore, and of Naxos emery stone.

4A. CLEANTHES, *Athens.*—Specimens of verde antique marble from Tinos.

5. DELENDAS, P. G. *Theræ.* — Theraic porcelain.

6. DEMOS OF CERONELEON, *Mantoudion.*—Magnesite or white stone.

7. DEMOS OF CHALCIS, *Chalcis.*—Magnesite or white stone.

8. DEMOS OF DORION, *Soulima.* — Soft soap-clay.

9. DEMOS OF PANORMOS, *Panormos.*—Green and white marble paper presses.

10. HOME DEPARTMENT, ATHENS.—Seventeen specimens of marbles from Peloponnesus.

11. LANDERER, H., *Athens.*—Specimens of minerals, metals, and vegetable substances.

11A. MALAKATES, J. *Athens.*—Specimens of marbles.

12. MELAS, B. *Athens.*—Sulphur from Milo.

13. MICHALACHACOS, C. *Panitza.*—Red or porphyry marble of Laconia.

14. PETRINOS, N. *Tripolis.* — Marble paper-press.

14A. PHYTALÆ, L. & G. *Athens.*—Specimens of marbles.

15. SPIROPOULOS, C. *Gargaliani.*—Specimen of ceramic clay.

CLASS II.

16. CONGOS, G. *Patras.*—Liquorice.

17. DEMOS OF CYTHNOS, *Cythnos.* — Mineral waters.

18. DEMOS OF PHARIS, *Xerocampi.*—Siderites or wild tea.

19. DEMOS OF PROSCHION, *Platanos.*—Wild tea.

20. DEMOS OF PSOPHIS, *Livartzi.*—Wild tea.

21. DEMOS OF SPARTA, *Sparta.*—Orange flower water.

21A. LANDERER, H. *Athens.*—Collection of plants.

CLASS III.

22. AGRICULTURAL SCHOOL OF TIRYNTH.
—Wheat, barley, maize, rye, honeycomb,
wine of 1851 and 1861, aniseed, colza,
tobacco, &c.

23. ALEXANDRIS, P. *Mataranga.*—Maize,
barley, oats, wheat.

24. BACHLOS, J. J. *Cyme.* — Wheat,
maize.

25. BASILIADES, B. N. *Nauplia.*—Wine,
spirits of wine, almonds.

26. BOUTOUNAS, B. N. *Karichi.*—Haricot
beans.

26A. CAIRIS, L. M. *Andros.*—Soft wheat
and figs.

27. CARTEROULIS, C. K. *Kalamæ.*—
Wine.

28. CONDYLIS, G. *Andros.*—White wheat.

29. CONGOS, G. *Patras.*—Currants, &c.

30. CONSTANTINIDES, D. *Villia.*—Wheat,
chick-beans, garlick.

31. CONVENT OF MEGASPILION, *Megaspilion.*—Haricot beans.

32. DEMOS OF ÆGÆON, *Limne.*—Wheat,
chick-beans, &c.

33. DEMOS OF ÆGINA, *Ægina.* —
Almonds, barley.

34. DEMOS OF AMBRACHIA, *Carvassaras.*
—Wheat, corn meal, &c.

35. DEMOS OF APODOTIAS, *Great Lombo-
tina.*—Wheat, lentils.

36. DEMOS OF ASOPON, *Molae.*—Wheat,
barley, beans, figs, tobacco, &c.

37. DEMOS OF ATALANTE, *Atalante.*—
Wheat.

38. DEMOS OF CALAVRETA, *Calavreta.*—
Broad haricot beans, barley.

39. DEMOS OF CALTHEZON, *Vlackoker-
assia.*—Wheat, barley.

40. DEMOS OF CASTORION, *Castanea.*—
Dari.

41. DEMOS OF CERONELEON, *Mantoudion.*
—Wheat, maize, haricot beans, broad beans,
vetches, oats, dari, lentils, rye.

42. DEMOS OF CERPENI, *Cerpeni.* —
White wheat.

43. DEMOS OF CHALCHIDEON, *Chalchis.*—
Wheat, chick-beans, lentils, haricot beans, &c.

44. DEMOS OF CLETORIA, *Mazeika.*—
Maize.

45. DEMOS OF COLOCYNTHION, *Flomo-
chori.*—Lupins, vetches, and peas.

46. DEMOS OF CORYTHION, *Stenon.*—
Wheat, barley, colyander, peas, haricot and
French beans, maize, dari, vetches, linseed.

47. DEMOS OF DAPHNESEON, *Livanates.*
Wheat, barley, oats, vetches.

48. DEMOS OF DORION, *Soulima.*—Wheat,
lentils, barley.

49. DEMOS OF DRIMIAS, *Dadi.*—Wheat,
maize, barley.

50. DEMOS OF ELLATIA, *Vroulia.*—Barley.

51. DEMOS OF ELLATIA, *Baischine.* —
Wheat and almonds.

52. DEMOS OF ELOUS, *Apedea.*—Wheat,
barley, haricot beans, tobacco, and maize.

53. DEMOS OF ERANIS, *Philiatra.*—Currants, wheat, wine.

54. DEMOS OF GERONTHRON, *Geraki.*—
Tobacco.

55. DEMOS OF HERACLEOTON, *Moscocho-
rion.*—Wheat, barley, sesame, chick-beans,
maize.

56. DEMOS OF HYPATE, *Hypate.*—
Tobacco.

57. DEMOS OF IDOMENIS, *Xerachia.*—
Wheat.

58. DEMOS OF IPERCHIAS, *Aga.*—Tobacco.

59. DEMOS OF ISTIAEON, *Xerochorion.*—
Wheat, beans, lentils, tobacco.

60. DEMOS OF LAMIA, *Lamia.*—Wheat,
barley.

61. DEMOS OF LARYMNA, *Martinon.*—
Wheat.

62. DEMOS OF LELENTION, *Steny.*—Walnuts.

63. DEMOS OF LETRINON, *Pyrgos.*—Wheat, maize, wine, and currants.

64. DEMOS OF LIVADIA, *Livadia.*—Rice, tobacco, wheat, maize, aniseed, sesame, and barley.

65. DEMOS OF LYCOUSOURAS, *Isaris.*—White maize, wheat, dari, haricot beans.

66. DEMOS OF MACRACOMIS, *Varybombe.*—Wheat.

67. DEMOS OF MALEVRION, *Panitza.*—Haricot beans.

68. DEMOS OF MEDEA, *Merbacca.*—Tobacco.

69. DEMOS OF MEGALOUPOLIS, *Synanon.*—Indian corn, &c.

70. DEMOS OF MISSOLONGHI.—Tobacco.

71. DEMOS OF MONEMBASIA, *Monembasia.*—Wheat, barley, beans, honey, &c.

72. DEMOS OF MYCENÆ, *Couzopodi.*—Tobacco.

72A. DEMOS OF MYRTOUDION, *Lechena.*—Linseed.

73. DEMOS OF NASSON, *Nasson.*—Indian corn meal, and dari.

74. DEMOS OF NAUPLIA, *Nauplia.*—Agegartus raisins, and currants.

75. DEMOS OF ŒNOUNTOS, *Vrestena.*—Wheat.

76. DEMOS OF ORCHOMENION, *Scripou.*—Wheat, barley, beans, and maize.

77. DEMOS OF ORCHOMENION, *Levidion.*—Wheat, barley, maize, lentils, dari, beans.

78. DEMOS OF PHALARON, *Stylis.*—Maize.

79. DEMOS OF PELANIS, *Agorgena.*—Wheat from American seed.

80. DEMOS OF PLATAMODOUS, *Gargagliani.*—Lentils, haricot beans, vetches.

81. DEMOS OF PROSCHION, *Platanos.*—Wheat, maize.

82. DEMOS OF PTELEOS, *Pteleos.*—Tobacco.

83. DEMOS OF SALAMIS, *Salamis.*—Honey, oil.

84. DEMOS OF SICHYONOS, *Chiatoa.*—Haricot beans.

85. DEMOS OF SPARTA, *Sparta.*—Wheat.

86. DEMOS OF TEGEA, *Achouria.*—Beans, lentils, barley, maize, wheat, oats, vetches.

87. DEMOS OF THERMOPYLÆ, *Molos.*—Red wheat and cocoons.

88. DEMOS OF THIAMON, *Valmada.*—Wheat, maize, wine, and haricot beans.

89. DEMOS OF TRINASSON, *Scala.*—Wheat.

90. DEMOS OF TRIPYLES, *Sarachinada.*—Wheat.

91. DEMOS OF TYMPHRISTOU, *Mavrillon*—Haricot beans.

92. DEMOS OF VION, *Neapolis.*—Wheat, barley, figs, peas, vetches.

93. DEMOS OF ZARACOS, *Rigea.*—Wheat, barley, honey.

94. DIALETTES, G. *Missolonghi.*—Raisins, bottargo, maize, oats, barley, wheat.

95. DOUROUTES, A. *Athens.*—Flour.

96. ECONOMIDES, G. *Oropos.* — Wheat, barley, and tobacco.

97. EMMANUEL, A. *Ægina.*—Barley.

98. EPARCH OF DORIS, *Lidorikion.*—Haricot beans, maize, wheat.

99. ERIOTIS, N. G. *Ægina.*—Barley.

100. EVANGELIS, P. & NICHOLAOU, P. *Calamos.*—Honey.

101. GAMELIARIS, N. *Argos.*—Tobacco.

102. GEORGACOPULOS, G. A. *Cyparissia*—Sesame, honey-seed.

103. GEORGANDAS, G. *Athens.*—Wine.

104. GHICAS, D. N. *Ægina.*—Honeycombs.

105. GHICAS, G. *Livinatae.*—Wheat.

106. IOANNOU, S. *New Corinth.*—Currant wine, currants, and almonds.

107. IRIOTTIS, N. G. *Ægina.*—Barley.

108. KEZEAS, V. *Dolæ.*—Honey.

109. LAMBRINIDES, L. G. *Argos.*—Currants.

109A. LANDERER, H. *Athens.*—Wheat, &c.

110. LANGADAS, A. N. *Naxos.*—Wine.

111. Manousos, P. *Cyparissia.*—Barley.

112. Marcopulos, Th. *Calamae.*—Figs.

113. Mastakas, D. A. *New Pelli.*—Honey.

114. Milakis, A. *Patras.*—Wine.

115. Mostras, M. N. *Xerochorion.*—Muscat wine.

115A. Nicolopoulos, G. D. *Gargagliani.*—Wine.

116. Œtylon, Bishop of, *Areoupolis.*—Almonds.

117. Panagiatopoulos, C. N. *Gargagliani.*—Sultana raisins and currants.

118. Petrides, D. *Styles.*—Green olives.

119. Petropulos, G. *Tripolis.*—Sparkling and common wine.

120. Petrou, E. *Atalante.* — Tobacco, wheat.

121. Provincial Committee of Locris, *Atalante.*—Indian corn.

121A. Printesis, X. *Syra.*—Naked barley.

122. Simandiras, I. *Argos.*—Wine.

123. Sirochilos, A. *Livanatae.*—Wheat, oats.

124. Sotiropulos, P *Ætolicon.*—Wheat, barley, maize, oats, raisins.

124A. Spiropulos, C. *Garghagani.*—Broad beans.

125. Stephanopulos, C. G. *Tripolis.*—Wheat.

126. Sterianopulos, L. *Sarmousakli.*—Wheat.

127. Theodosios, A. *Megara.*—Honey

128. Thomaras, C. *Cyparissia.* — Almonds.

128A. Varthalites, G. T. *Syra.*—Barley.

129. Vouchèrer, G. G. *Piræus.*—Spirits of wine.

130. Voulpiotis, N. *Baischine.*—Tobacco, wheat.

131. Voutounas, B. N. *Caryke.*—Fruit.

132. Wine Company, Patras.—Champagne and other wines.

132A. Wine Company, *Thera, Santorin.*—Wines.

133. Zacharopulo, M. *Cyparissia.* — Currants.

134. Zimbourachi, P. A. *Stamna.*—Wheat, maize, oats, barley.

CLASS IV.

135. Agricultural School, Tirynth.—Acacia wood, cotton, wool, silk-worms, wax.

136. Alexandris, P. *Mataronga.* — Fustic wood and oil.

137. Boudounas, B. *Lycosura.*—Cotton wool, and fustic.

138. Bournakis, J. *Leonidion.*—Cochineal.

139. Calamaras, N. C. *Ægina.*—Sponges.

140. Caramoussas, P. *Steni.* — Fustic wood.

141. Central Committee, Athens.—Fustic, sumach, tragacanth, and specimens of the forest timber of Greece.

142. Committee of Troezinia, *Poros.*—Green olive oil, common oil.

143. Constantinides, D. *Villia.*—Oil, resin.

144. Demarch of Mycenae, *Coutzopodi.*—Valonea.

145. Demos of Ægition, *Lidorichion.*—Tallow.

146. Demos of Ægeon, *Limne.*—Resin, naphtha, fustic wood.

147. Demos of Ægina, *Ægina.*—Madder-roots, sponges, oil.

148. Demos of Apodotia, *Great Lobotina.*—Cocoons.

149. Demos of Aroania, *Sopotou*—Silk.

150. DEMOS OF CASTORION, *Castanea.*—Galls.

151. DEMOS OF CERONELEON, *Mantudion.*—Ash, maple, pine, elm, oak, platane woods, &c. Resin, naphtha, Colophany resin-tar.

152. DEMOS OF CLEITORIA, *Mazeika.*—Yellow berries.

153. DEMOS OF CROKEON, *Levetzova.*—Yellow sumach, fustic, &c.

154. DEMOS OF DEONDEON, *Eubea.*—Fustic.

155. DEMOS OF DRYMIAS, *Dadi.*—Cotton.

156. DEMOS OF ERANIS, *Philiatra.*—Oil, soap.

157. DEMOS OF GYTHEON, *Gytheon.*—Wax, valonea.

158. DEMOS OF IRACLEOTON, *Moschochorion.*—Cotton wool.

159. DEMOS OF LIVADIA, *Livadia.*—Cotton wool.

160. DEMOS OF MALEVREON, *Panitza.*—Wax and valonea.

161. DEMOS OF MELITENES, *San Nicholas.*—Wax, galls.

162. DEMOS OF MONEMBASIA, *Monembasia.*—Wax, oil.

163. DEMOS OF MYRTOUDION, *Lechena.*—Linseed.

164. DEMOS OF NAUPLIA, *Nauplia.*—Sponges.

165. DEMOS OF PHELLIAS, *Torani.*—Wax, galls.

166. DEMOS OF PLATOMODUS, *Gargagliani.*—Cocoons and linseed.

167. DEMOS OF PROSCHION, *Platanos.*—Cocoons.

168. DEMOS OF TRIPYLES, *Sarakinada.*—Valonea.

169. DEMOS OF VOEON, *Neapolis.*—Sponges.

170. DEMOS OF VOUPRASION, *Vouprasion.*—Galls.

171. DEMOS OF VRYSEON, *Anavryte.*—Yellow berries, sumac, fustic.

172. DEMOS OF ZARACOS, *Richea.*—Wax, fustic.

173. DOUROUTES, A. *Athens.*—Silk, oil.

174. DURAND, F. *Andros.*—Silk.

175. ECONOMOU, C. *Leonidion.*—Cocoons.

176. ELIADES, P. *Gytheon.*—Oil.

177. EMMANUEL, A. *Ægina.*—Madder-roots.

178. EVANGELIS, C. *Athens.*—White and yellow wax, and candles.

179. FELLS & Co., *Calamae.*—Silk.

180. GEORGANDAS, G. *Athens.*—Oil.

181. KEZEAS, V. *Doloe.*—Oil, silk.

182. LAPIÈRE, J. *Salessi.*—Valonea.

183. LONDOS, A. C. *Patras.*—Cotton wool.

184. MARCOPOULOS, TH. *Calamae.*—Cocoons.

185. MERLIN, M. *Athens.*—Cotton from American seed.

186. ŒTYLON, BISHOP OF, *Areoupolis.*—Oil.

187. PETRITES, A. N. *Ægina.*—Madder-roots, oil.

188. POTIROPOULOS, B. *Calamae.*—Oil.

189. REMBOUTZIKAS, A. *Lower Achaïa.*—Valonea.

190. RHALLIS, L. *Piræus.*—Silk.

191. SABBAS, G. *Mantoudion.*—Madder.

192. SOTIROPULOS, T. *Etolicon.*—Linseed.

193. THEOLOGOS, N. *Piræus.*—Soap.

194. TOMARAS, C. *Cyparissia.*—Cotton.

195. VRYSAKIS, C. & TH. *Athens.*—Vegetable red dye, kermes.

196. ZAFIRACOS, N. *Gytheon.*—Valonea.

197. ZOUCHLOS, *Leonidion.*—Fustic and oil.

———

209. PALEOLOGOS, D. *Steni.*—Box-wood.

CLASS VII.

198. DEMIDES, C. *Athens.*—Specimens of type casts.

CLASS VIII.

199. FETICHANES, B. *Athens.*—A calico printing machine.

200. GREEK STEAM NAVIGATION Co. *Syra.*—A steam-engine, 4¾ h.p., working shafts in the Western Annex, near column 32.

CLASS X.

204. CLEANTHES, *Athens.*—Two verde antique pillars, two round tables, and one chimney-piece of the same marble from Tinos.

205. COSSOS, I. *Athens.*—Marble busts, &c.

207. MALAKATES, F. *Athens.* — Marble busts and a marble paper press.

208. MALAKATES, I. *Athens.*—A funeral monument, a fountain, and two chimney-pieces, all of Pentelic marble.

210. PHYTALAE, L. & G. *Athens.* — Table of Pentelic marble, marble busts, &c.

CLASS XI.

214. CONSTANDOULAKIS, D. M. *Hydra.*—Military and naval officers' belts.

215. DEMOS OF NAUPLIA, *Nauplia.*—Gold lace and silver cords.

CLASS XIV.

216. CONSTANTIN, D. *Athens.*—Specimens of photography.

217. MARGARITIS, F. *Athens.* — Photographs.

CLASS XVIII.

218. CARAMERZANES, D. *Dadi.*—Cotton towels and a bed cover.

219. DIALETIS, J. *Missolonghi.* — Table-cloth, bed-sheet, towels, pillow-cases, and handkerchiefs.

220. GEORGIOU, P. *Chalcis.*—Cotton towels.

221. KOUTZOUKOS, A. D. *Chalcis.* — Cotton fabrics.

222. PHOTINOS, P. *Patras.*—Cotton twist.

223. PISPIRINGOS, G. *Patras.* — Cotton cords.

CLASS XX.

224. CONSTANDOULAKIS, D. M. *Hydra.*—Barège dresses and silk fabrics.

225. DEMOS OF ERANES, *Philiatra.*—Silk fabrics.

226. DEMOS OF SPARTA, *Sparta.*—Silk fabrics.

227. FRANGOULIS, N. *Avlonarion.*—Barège dresses, handkerchiefs, scarves, silk fabrics, &c.

228. MELANE MONACHE, *Calamae.* — Barège handkerchiefs, mosquito curtains, &c.

CLASS XXI.

229. CONSTANDOULAKIS, D. M. *Hydra.*—Scarf, and various mixed fabrics.

CLASS XXII.

230. ALEXANDRIS, P. *Mataranga.* — A carpet.

231. DEMOS OF CALAVRYTA, *Calavryta.*—A carpet.

232. DEMOS OF DISTOMON, *Distomon.*—A carpet.

233. DEMOS OF LAMIA, *Lamia.*—A carpet in two breadths.

234. DEMOS OF RACHOVA, *Rachova.*—A carpet.

235. DEMOS OF STRATON, *Adamas.*—Bed coverlet.

236. DEMOS OF TRIPOLIS, *Tripolis.*—A carpet.

237. DEMETRIOU, M. G. *New Peli.*—A carpet.

238. DEVA, ELLEN, *New Peli.*—Three carpets.

239. STYLOUDY, S. *New Peli.* — Three carpets.

240. ZERVAS, M. & E. *Krieza.*—A carpet.

CLASS XXIV.

241. DEMARCH OF HERMOUPOLIS.—Lace.

242. DEMOS OF SPARTA, *Sparta.*—Lace collars and sleeves.

243. PISPIRINGOS, G. *Patras.*—Lace, &c.

243A. ISANTYLAS, C. *Athens.* — An embroidered table-cover.

———

256. CINGLIS, *Athens.*—Gold tassels, legging ties, &c.

257. CHALMOUCUPOULOS, Z. *Lamia.*—A gold and silk knit cap.

259. DEMOS OF LAMIA, *Lamia.*—A goat's hair capot, a mantlet, a worsted tunic, and a pair of leggings.

260. DEMOS OF LELANTION, *Steni.* — A female peasant's dress.

275. ANDREOU, A. & ZENOU, T. *Athens.*—A richly embroidered gold table cover.

281. EDIFIDES, I. *Athens.*—Lace bag.

CLASS XXV.

244. Yalistras, S. *Athens.* — Furs of various kinds, and tobacco pouches.

———

247. Giatsis, Ch. *Missolonghi.*—Goatskins.

251. Plataniotis, N. *Chalcis.*—Coloured skins.

CLASS XXVI.

245. Anastasiou, G. *Chalcis.* — Bull hides.

246. Fachiris, E. G. *Hermoupolis.*—Bull hides.

249. Lagouras, G. *Hermoupolis.*—Bull hide.

252. Roumbakes, E. *Hermoupolis.*—Bull hide.

CLASS XXVII.

255. Alexandris, P. *Mataranga.*—A pair of Greek sandals.

258. Crianopoula, Q. *Syra.* — Printed head covering.

261. Demos of Nauplia, *Nauplia.*—Red silk sashes, and silk fabric.

262. Galinos, G. *Athens.*—Black dress coat.

263. Gialeli, A. *Syra.* — A handkerchief.

264. Ioannou, N. P. *Athens.*—Red fez caps.

265. Melane, Monache, *Calamae.* — Jackets, burnous, collars, sleeves, and sashes.

266. Nicolaidis, A. *Andros.*—A handkerchief.

267. Nicolaou, D. *Athens.* — Gentleman's and lady's velvet embroidered costumes.

268. Papacostas, C. N. *Athens.*—Richly embroidered Greek costumes for a gentleman, lady, and a boy.

269. Pizouras, G. *Chalcis.*—A silk belt.

270. Theodorou, G. *Athens.*—Silk embroidered gentleman's dress.

271. Tsantylas, C. *Athens.*—A lady's jacket, mantlets, and other embroidered goods.

272. Varia, G. *Hermoupolis.*—A head-covering.

———

248. Gregoriadis, H. *Athens.*—Boot lasts.

250. Levandis, G. *Athens.*—Boot lasts.

253. Tatos, Al. *Andros.*—A pair of man's boots.

254. Zivarakis, Bros. *Hermoupolis.*—Boots of various kinds.

CLASS XXVIII.

273. PENDEFRES, P. *Athens.*—Account books.

213. SKIADOPOULOS, P. *Athens.*—Specimens of wood carving, and of the electrotype process.

274. VLASTOS, HEIRS OF S. G. *Athens.* —Printed books.

CLASS XXIX.

274A. CONDES, H. *Athens.*—A manuscript book of Greek caligraphy.

282. MASTORES, M. *Gargagliani.*—Specimen of writing.

CLASS XXX.

202. ALBANOPOULOS, A. *Chalcis.*—Wooden vase.

276. ATHANASIOU, I. *Poros.*—A brass lamp.

203. BARALES, M. *Chalcis.* — Cubical barrel.

206. DEMOS OF THYAMON, *Valmada.*—A tobacco pipe.

277. SARGENTES, M. *Chalcis.*—A small wooden box.

278. TSANGLIS, EM. *Hermoupolis.* — A brass tap.

———

201. AGATHANGELOS, *Athens.* — "The Coming of our Lord," in carved wood.

211. PAPAGEORGIOU, C. *Athens.*—Wood carving.

212. PRAOUDAKES, M. *Athens.*—Wood carving.

CLASS XXXI.

279. COCOREMBAS, P. *Tripolis.*—Butcher's knife.

CLASS XXXIII.

280. ANTONIADES, TH. *Chalcis.*—A gold ring.

HAWAIIAN OR SANDWICH ISLANDS.

NORTH-EAST COURT.

Samples of cocoa-nut oil, from Fanning's Island.
Manufactured by H. English & Co.
Annual product, 120 to 150 tons.

HAYTI.

NORTH-EAST COURT.

THE HAYTIAN GOVERNMENT.
Cotton.
Cocoa.
Arrow-root.
Coffee.
Sugar.
Plantain meal.
Pitre.
Tea.
Tanning beans.
Holly gum.
Pulse.
Sesame, or " hoholo."
A large block of mahogany.

Two mahogany wardrobes.
One mahogany wash-stand.
Water-bottles and drinking vessels in lignum-vitæ.
Two mahogany pestles.
Eight earthenware syphon-pitchers.
Gut-whips.
Wooden locks.
A polished stone for a table top.
Saddles.
Plated bit.
Specimens of minerals.
Iron ore.
A *toque avec macoite*, or a *sac-paille*.

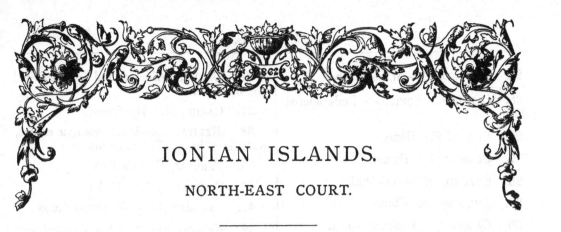

IONIAN ISLANDS.

NORTH-EAST COURT.

A.—CEPHALONIA.

1. CEPHALONIAN COMMITTEE.—A collection of natural productions and manufactured articles, viz. :—

Lime and limestone, chalk, sand, red earths, and other cements ; fossils, &c. ; mineral waters.

Various specimens of timber, and miscellaneous vegetable productions.

Cereals—Wheat, barley, Indian corn, oats, and rye.

Pulse — Peas, beans, French beans, vetches, and avrios.

Textile materials—Flax, wool, silk, and cotton.

Oil, honey, almonds, walnuts, stone pine seeds, carob pods, capers, onions, garlic.

Jaws of sheep, with gold tint from Pilaros.

Woollen cloth, basket of rye-straw, embroidered handkerchiefs, slippers, woollen socks, sashes.

Model of a net.

2. LUSI, G.—Quince and gourd preserves.

3. MANZAVINO, D.—Boots.

4. CANGILLARI, C.—Dress tail-coat.

5. MORAITI, G.—Chest of drawers, chair, and small boxes.

6. ANGELLATO, P.—A table.

7. GASI, D. M.—A carpet.

8. BERDCHÉ, M. D. M.—A carpet.

9. BENEDETTO, A.—Banisters for balustrades.

10. CORAFAN, G.—Gold head-pins and ring.

11. INGLESSI, T.—Gown of aloe lace, doyleys and collars.

12. BAGGALI, C.—Counterpane, burnous, polkas, and sleeves.

13. CURCUMELI, V.—An apron and gloves.

14. METAXA, E.—An apron.

15. CARITATO, S.—A tobacco bag.

16. INGLESSI, E.—Polkas.

17. VULISMA, A.—Two toilette tables.

18. MONTALDO, A.—Coffee-pot and tray.

19. DALAPORTA, G.—Wax candles.

B.—CERIGO.

20. CERIGO COMMITTEE.—Mineral waters, everlasting flowers, sponges, marbles, slate, stalactites, clays, petrified bones, toys.

21. STAI, DR. N.—Oil and olives.

22. VARIPATI, G. — Oil, honey, salted quails, four kinds of grain, beans, two sorts of skins.

23. ALEXANDRACHI, A.—Wine

24. CALIGERO, A.—Wine, honey.

25. CALIGERO, SOFIA. — Embroidered collar.

26. PISANO, Z.—Hams.

27. COSACHI, P.—Hams.

28. COSACHI, M.—Cloth stuff.

29. AVIERINO, S.—Hams.

30. CALUCI, G.—Hams, almonds.

31. MAVROMATI, G.—Capers.

32. MAVUDI, P —Cheese.

33. FAZEA, A.—Cheese (ricotta secca).

34. MORMORI, DR. E. — Barley meal, sweet herbs, lime.

35. MORMO, F.—Baskets.

36. CARIDI.—Almonds and figs, medicinal and other herbs.

37. CARIDI, M.—Dyed wool.

38. MELITA, E.—A silk sash, a carpet, waterproof bags, silk and cotton towels.

39. FARDULI, C.—Baskets.

40. CONDOLENI, M.—Baskets.

41. CASTRISSO, P.—Waterproof bags.

42. NEOPHOTISTO, G.—Dress sacket and kerchiefs.

43. PLUSACHI, S.—Embroidered collar.

44. AVEPERINA, P.—Embroidered collars and handkerchiefs.

45. DIACOPULO, S.—Embroidered hand-kerchiefs.

46. VESE, G.—Boxes of olive wood.

C.—CORFU.

47. ALEXIS. — Montenegrin and Greek caps and costumes.

48. TAYLOR, N., worked by COSTI, S.—Albanian bags, cases, cuffs, caps, and costume.

49. COSTI, S.—Embroidery.

50. PSORULA. — Albanian shoes, jacket and waistcoat, and other embroidery.

51. RUBAN, M.—Embroidered cushions, pouches, and costumes.

52. PAVIA. — Silversmith's work and jewellery, rings, &c.

53. MOSEO, PAPA.—Gold necklaces and other jewellery.

54. FLORIAS, A.—Filigree-work.

55. COSTA, PAPA.—Silver lamp, sacramental cup, &c.

56. PAUDIN, P. BROS. — Silversmith's work and jewellery.

57. BIASI, G. DE.—Knitted collars and cuffs.

58. SARACHINO, SA.—Gilt and silk sashes, &c.

59. MULATO, A.—Silk sashes.

60. PAPASTERI, E. — An embroidered dress.

61. MANDUCHIO SCHOOL.—Embroidered collars, cuffs, &c.

62. CURCULO, A. — Embroidered hand-kerchief.

63. CALOGEROPULO, SA.—Worked pocket-handkerchief.

64. GALDIES, G.—Specimens of small cabinet-work in jujube, acacia and olive wood, &c.

65. PETROVICH, S. — Inlaid works in native wood.

66. ALTAR, G.—Inlaid cabinet work in native wood.

67. DIMOCASTO, N.—Inlaid paper-cutter.

68. SCHURLURLING, ROSA. — Vinewood walking-sticks.

69. GALLO, G.—Whip handles.

70. MONTANARI.—Carved wooden frames.

71. ZERVO, A.—Flax.

72. ZERVO, N. — Straw hat and blue cloth.

73. SERVO, C.—Cotton, raw and carded; wool, raw and carded.

74. MUTHOLLAND.—Boots and shoes.

75. DUSMANI, CT. SIR A. L.—Marble, stalactites, &c.

76. SARACHINO, SA.—77. DIODACHI, G.—Blue cloths.

78. POLLITÀ. — Cotton cloths, yellow veils, and machine for extracting cotton.

79. PALURIA, A.—80. COSTANDI, C.—81. CALICHIA, N.—82. BALBI, BARON.—83. CORATA, S.—Tobacco.

84. CURCUMELLI, SIR D.—Tobacco, papyrus cloth, woods, grain, oil, wine, and vinegar.

85. SAULI.—Tobacco, olive oil, and rice.

86. VENTURA.—Pack-saddles, wine, and tobacco.

87. ZANONI.—Stuffed birds, &c.

88. MAMAS, C.—Model of an oil mill and a lamp.

89. WORSLEY.—Collection of insects.

90. VASSILACHI, E.—Wine casks, baskets, olive oil, wine, and preparations of pork.

91. DELVINIOTTI.—Dyes.

92. DESSILA.—Medicinal plants.

93. COLLAS, BROS.—Almond and castor oil, medicines, and articles of toilette.

94. ALEXACHI, P.—Wax.

95. ALEXACHI, DR. — Rose-water, &c., and opium.

96. ZANINI, SA.—Flowers.

97. GIRONCI.—A lock.

98. BUFFA.—Model of staircase.

99. BARBIROLI.—Chronometer.

100. MIGLIARESSI. — 101. CAMBISO. — 102. PROVATÀ, C.—103. CAPO D'ISTRIA, CT.—Olive oil.

104. CARMENI.—Liqueurs.

105. THE PACKET ESTABLISHMENT. — Flags of the Islands.

106. CATHOLIC SISTERS OF CHARITY.—Bead embroidery.

D.—ITHACA.

107. DENDRINO, T.—Rug.

108. DOVA, G.—White marble.

109. GIANNIOTTI, G.—Sponges.

110. XANTHOPULO, P.—Anchor of wood.

111. CUZZUVELI, P.—Silk and cotton stockings.

112. PROCOPI, G.—Silk and cotton stockings.

113. ALIMERIATI, P.—Silk and cotton stockings.

114. PROCOPI, G.—Silk stockings and raw silk.

115. ZANNETTI, P.—Wine and oil.

116. PROCOPI, G. — Red and white Rosolio, and Rachi of Mastica.

117. FERENDINO, S.—Currants.

118. CENTRAL COMMISSION.—Coral.

E.—PAXO.

119. THE PAXO COMMITTEE.—Pitch and mineral water.

120. VEGLIANITI, ALOISIO.—Olive oil.

121. VEGLIANITI, ANASTASIO.—Olive oil.

122. VEGLIANITI, ANDREA.—Olive oil.

123. VEGLIANITI, ATT.—Stone trough.

124. BOGDANO, A.—Olive oil.

125. CARUSO, F.—Olive oil.

156. MORICHI, C.—Olive oil.

127. ARGIRO, V.—Olive oil.

128. MIZZIALI, DR.—Olive oil and preserved olives.

129. LECCA, S.—Olive oil.

130. PETRO, A. P. M.—Olive oil.

131. MACRÌ, N.—Red and white wine.

132. VLACOPULO, M.—Willow fibre cloth.

133. VLACOPULO, G.—Lintel.

134. MACÌ, S.—A female dress.

F.—SANTA MAURA.

135. SANTA MAURA COMMITTEE.—Conglomerate, limestone, marble, sandstone, timber, cochineal, flax (dressed and undressed), seeds, salt, cotton, and a carpet from Caria.

136. ARCHELLE.—Silk, raw and manufactured.

137. PEZZALI, THE MISSES.—Embroidery.

138. VRIONI, P.—Dye stuffs.

139. VALAORITI, CAV. A.—Cotton from Madurì.

140. WOLFF, H. D.—A carpet.

G.—ZANTÈ.

141. ZANTÈ COMMITTEE.—Mineral waters. A collection of minerals, raw silk, corn, pulse, linseed, sesamum, and coriander seed.

142. STRAVAPODI, DR.—Currants and raisins.

143. ARVANATACHI, G.—144. DOMENEGHINI, DR. G.—145. PLANTERO, G.—146. SOLOMOS, COUNT SIR D.—147. COMUTO, G.—148. FLAMBURIARI, CT. SIR D.—149. BARFF, T.—150. MERCATI, CT. D.—151. CARAMALICHI, N.—Currants.

152. WODEHOUSE, THE HON. COL. B.—Red wine of 1851 and 1857. Sponge.

153. CHIEFALINO, D.—White wine of 1859.

154. CATEVATI, P.—Red wine of 1859.

155. CONOFAO, D.—White wine of 1860.

156. ROSSOLIMO, C. — White wine of 1857.

157. ROSSI, G.—Currant wine, 1857 and 1861.

158. RISURCA, A. — Porous jars and bottles.

159. PAPADATO, N.—Manufactured silk, hemp-cord, flax, and cotton.

160. ARGASSARI, D.—Manufactured articles and raw silk.

161. POLITI, N.—Silk handkerchiefs.

162. DRAGENA, I.—Manufactured articles, silk fringe and braid.

163. PAPANDRICOPULO.—Cotton dresses, table-cloths, bed-ticking, and towels.

164. GROSSU, M.—Towels.

165. ANDRAMIOTI, A.—Cotton socks, embroidered handkerchiefs.

166. PASSARIA.—Spirit made from currants.

167. CALAVATI, G.—168. RAFTANI, C.—169. ZAZICHI, D.—170. ANDULICO, C.—171. PILICAS, F.—Soap.

172. Costandacopulos, D.—Soap and bees' wax.

173. Mopurgo, R.—Soap and purified oil.

——

174. Veja, Dionisio.—A festa at Corfu, showing the Romaika or ancient Pyrrhic dance; a picture of Greek costume; "Madonna."

175. Aspioti.—A picture as used in Greek churches.

176. Manzavino.—Interior of a Greek church.

177. Pieri.—Picture of Greek costume.

ITALY.

SOUTH CENTRAL COURT, SOUTH CENTRAL GALLERY, AND WESTERN ANNEX.

The names of provinces, according to the new administrative division of the kingdom, are placed within parentheses, thus *(Florence)*.

CLASS I.

1. ADRAGNA, BARON G. *Trapani.*—Salt.

2. AGOGNA AND BROVELLO MINING CO. *Palanza (Novara).*—Argentiferous lead ore.

3. ALDISEO, G. & Co. *Turin (Iron-works at Bard, Val d' Aosta).*—Iron rod: gun barrels, wrought while cold.

4. ANGHIRELLI, G. *Montalcino (Sienna).*—Floating bricks, made of "mountain meal" (siliceous skeletons), from Monte Amiata.

5. ARRIGONI, A. *Varese (Milan).*—Peat.

6. BARBA TROYSE, G. *Spezia (Genoa).*—Ores of manganese. Decomposed metamorphic jasper, for hydraulic cement (pozzolana).

7. BELTRANI, G. *Trani (Terra di Bari).*—Limestone and tufa.

8. BELTRAMI, COUNT P. *Cagliari.*—Lead slags of ancient smelting works, and pig of lead obtained from them. Ores of lead, manganese, iron, and copper. Refractory clay, hydraulic lime, lignite.

9. BENTIVOGLIO, CAV. C. *Modena.*—Lime.

10. BIRAGHI, G. & Co., *Milan.*—Lignite.

11. BOLOGNA MINERAL CO. *Bologna.*—Copper pyrites and purple copper ore from the mine of Bisano, with gangue, &c. Serpentine, steatite, &c.

12. BOTTINO CO. *Leghorn.*—Argentiferous and other ores of lead, from the mine of Bottino. Specimens illustrating the smelting process. Lead, refined copper, and refined silver. Model of the apparatus employed for transporting the ores.

13. BOUGLEUX, F. *Leghorn.*—Millstones of quartzose rock.

14. BOUQUET & SERPIERI, *Cagliari (Smelting-works at Domusnovas and Flumini, Cagliari).*—Ancient slags. Pigs of lead from Domusnovas and Flumini Maggiore.

15. BRESCIA ATHENÆUM.—Specimens of the rocks and ornamental stones of the province.

16. BUCCI, G. *Campobasso (Molise).*—Campobasso marble.

17. BURGARELLA, A. *Trapani.*—Sulphur.

18. CAGLIARI SUB-COMMITTEE FOR THE EXHIBITION.—Specimens of building stone.

19. CALZA CRAMER, G. *Grugliasco (Turin).*—Peat pressed into moulds.

20. CHIAVARI ECONOMIC SOCIETY, *Chiavari (Genoa).*—Slates for various purposes, from Lavagna quarries, Chiavari.

21. CHIAVENNI SUB-COMMITTEE FOR THE EXHIBITION.—Gypsum, and limestone, from Madesimo, near Chiavenna.

22. CHIOSTRI, L. *Pomarance (Pisa).*—Specimens of rocks found in the neighbourhood of Pomarance and Libbiano: geological map, &c.

23. COCCHI CAV. BROTHERS, *Florence.*—Jasper from Giarreto, Val di Magra; limestone, hydraulic lime, gypsum, plaster of Paris.

24. COJARI AVV. V. *Fivizzano (Massa-Carrara).*—Copper ore from S. Giorgio mine, Ajola, Fivizzano; statuary, and other marble, from Equi, Fivizzano.

25. CORBI ZOCCHI, C. *Sienna.*—Sienna earth.

26. CORNELIANI, L. *Milan (Iron-works at Premadio, near Bormio, Sondrio).*—Iron ore, pig iron, wrought iron.

27. COSTANZO, C. *Catania.* — Sulphur, crude and manufactured, from the solfatara of Cugno, &c.

28. COSTANZO, L. *Catania.*—Sulphuriferous marl, from the solfatara of Estricello, &c.

29. CURIONI, PROF. G. *Milan.*—Casting sand for iron furnaces.

30. DAMIOLO, S. *Pisogna (Brescia).*—Iron ore, pig iron, &c.; specimens illustrating metallurgical processes.

31. DELPRINO, M. *Vesina, near Acqui (Alexandria).* — Limestone, and gypsum found around Acqui.

32. DE MORTILLET, G. *Milan.*—Magnesiferous and hydraulic lime, from different parts of Italy; sand, cements, gypsum; plans of works.

33. DINI, P. *Camaiore (Lucca).*—Manganese ore from Fontanaccio.

34. DODERLEIN, PROF. P. *Modena.*—Collection of minerals, of the Provinces of Modena and Reggio, with catalogue, &c.

35. DOL, BALDASSARRE, LESSEE OF THE COMMACCHIO ROYAL SALT WORKS, *Turin.*—Crystallized salt, table salt, curing salt.

36. D'URSO, F. P. *Salerno (Principato Citeriore).*—Building stones from Paterno, &c.

37. FERRATA & VITALE, *Brescia.*—Millstones, grindstones, and scythe stones.

38. FERRO, F. *Cagliari.* — Antimony, from Su Suergiu mine, Villasalto; lead ore, from Ringraxius mine.

39. FLORENCE ROYAL NATURAL HISTORY MUSEUM, GEOLOGICAL DEPARTMENT (Director, PROF. I. COCCHI).—Collection of ornamental stones, marbles, breccias, alabaster, serpentine, chalcedony, jasper, building materials, clays, &c.; iron ore from Val d'Aspra, lead ore from Castellaccia, &c.

40. FOIGO, G. *Chiavenna, Sondrio.*—Collection of Chiavenna minerals.

41. FORNOVE JUNTA FOR THE EXHIBITION, *Parma.* — Specimens of Petroleum from Neviano de' Rossi, &c.

42. FRANEL, E. & Co. *Turin.*—Copper, and copper pyrites from Ollomont mine, Aosta; specimens illustrating mode of dressing the ores, &c.

43. GABRIELE, A. *S. Bartolomeo in Galdo (Benevento).*—Spathose iron ore (carbonate of iron), from S. Bartolomeo; peat.

44. GANNA, S. *Turin.*—Flagstone, from the quarries of Guessi, &c. (Pinerolo.)

45. GARUCCIU, CAV. G. M. *Iglesias (Cagliari).*—Lead ore, from Is Cortis de Pubusinu mines, near Iglesias.

46. GAVIANO, A. *Lanusei (Cagliari).*—Lignite, from S. Sebastiano a Secis (Cagliari).

47. GENNAMARI & JUGURTUSU MINING Co. *Arbus.*—Argentiferous lead ores, &c., from Jugurtusu and Gennamari (*Cagliari*); quartz gangue, with copper pyrites, blende, &c.

48. GIOVANNINI, BROTHERS, *Carmignano (Florence).* — Tables in serpentine, from Monteferrato, near Prato.

49. GLISENTI & RAGAZZONI, *Brescia.*—Iron ore, cast iron, wrought iron, and steel of Val Trompia and Val Sabbia.

50. GOUIN, LÉON, & Co. *Iglesias (Cagliari).*—Argentiferous lead ore (galena), mine of S. Giorgio, near Iglesias; rock accompanying the vein.

51. GRASSI, BROTHERS, *Schilpario (Bergamo).*—Iron ore and pig iron, from Valle di Scalve (Bergamo).

52. GREGORINI, A. *Lovere (Bergamo).*—Iron ore, pig iron, steel; specimens illustrating the manufacture of steel by the Bergamese process.

53. GUERRA, BROTHERS, *Massa (Massa-Carrara).*—Specimens of marble, from the Val del Palazzuolo, &c. (*Massa*); various articles in marble.

54. GUIDOTTI, F. *Lucca.*—Umber.

55. GUPPY & PATISSON, *Naples.*—Cold-bent iron bars.

56. HAUPT, T. *Florence.*—Plans of the mines of Tuscany; synoptical tables.

57. HENFREY & FRANEL, *Iglesias (Cagliari).*—Argentiferous lead from Monte Cour, &c.

58. ITALIAN MARBLE Co., HÄHNER & Co. *Leghorn.*—Specimens of marble from the hills above Massa.

59. JACOBELLI, A. *S. Lupo (Benevento).* —Specimens of marble, from Vitulano and Pietraroia (Benevento).

60. JERVIS, W. P. *of the Royal Italian Central Committee.*—Specimens of magnetic iron ore from Forno (*Massa-Carrara*), discovered by exhibitor.

61. LIPARI JUNTA FOR THE EXHIBITION (*Messina*).—Minerals.

62. MACERATA SUB-COMMITTEE FOR THE EXHIBITION.—Lignite.

63. MAFFEI, CAV. N. *Volterra (Pisa).*— Collection of minerals; articles in breccia; chalcedony, from Monte Rufoli.

64. MAGGI, SANTI, & BECCHINI, *Montalcino (Sienna).*—Sienna earth, raw and burned; "mountain meal," with floating bricks, and bricks for polishing metals made of it.

65. MAGRI, D. *Bologna.*—Gypsum, plaster of Paris, scagliola.

66. MALMUSI, CAV. C. *Modena.*—Clay for pottery, marl from Biamana.

67. MANNA, E. *Iglesias (Cagliari).*—Lead ore, from the mine of S. Benedetto, near Iglesias.

68. MARCHESE, E. *Cagliari.*—Collection of minerals from the island of Sardinia.

69. MASSA, C. *Casale, Novara.*—Limestone and gypsum.

70. MASSERANO, G. *Biella, Novara.*— Syenite from La Balma in Quittanga, near Biella.

71. MASSOLENI, M. *Genoa.*—Millstones, made at Genoa of fragments of stone imported from France.

72. MASSONE & MUSANTI, *Genoa.*—Argentiferous lead ore; galena with gangue, spathose iron, copper pyrites, &c.; lead desilverised by Pattinson's process.

73. MASSONE, CAV. M. *Cagliari.*—Minerals from a lead mine, commune of Lula, near Nuoro.

74. MELIS, S. *Cagliari.* — Slags from smelting works existing last century near Villacidro; pig of lead made from them, &c.

75. MESSINA SUB-COMMITTEE FOR THE EXHIBITION. — Lignite, marbles, metalliferous and other minerals from the province of Messina.

76. MILESI, A. *Bergamo.*—Iron ore, cast iron; steel capable of being welded like iron, and of scratching glass.

77. MINERALOGICAL MUSEUM, *Naples* (Director, PROF. SENATOR SCACCHI).—Collection of minerals from the southern provinces of Italy.

78. MODENA AGRICULTURAL INSTITUTION.—Marls, earths, &c.

79. MONTE ALTISSIMO MARBLE CO. *Florence.*—Statuary, and other marble, from the quarries of Monte Altissimo. They were worked by Michel Angelo.

80. MONTEPONI MINING CO. *Genoa.*— Lead ore, and other minerals, from Monteponi mine (Sardinia).

81. NAPLES ROYAL FOUNDRY.—Ores of iron and lead, pig iron, coal.

82. NICOLAI, P. A. (President of Monteponi Mining Co. Sardinia), *Genoa.*—Carbonate of lead, sulphuret of lead, and galena, from S. Giovanni mine.

83. NOCERA, MUNICIPALITY OF (*Umbria*). —Travertine, and various kinds of limestone partaking more or less of the character of marble.

84. OLLOMONT MINING CO.—*Aosta (Turin).*—Copper ore, copper.

85. ORFINI, COUNT, *Fuligno (Umbria).*— Tables of veined Alberese (cretaceous limestone).

86. PEDEVILLA, F. *Tortona (Alexandria).*—Limestone, and lime.

87. PELLICCIA, L. *Naples.*—Felspar and refractory clay, from Parghelia (Calabria Ulteriore II.).

88. PÉTIN GAUDET & Co. *Cagliari.*— Magnetic iron ore, &c., from Perda Niedda mine, near Domusnovas, and Perda Sterria mine, Domus de Maria; gangue, &c.

89. PISTILLI, F. *Campobasso (Molise).*— Iron pyrites from Salcito (Molise).

90. PLATINIA, P. & Co. *Catania.*—Sulphur in cakes, &c.; sulphur melted with the heat naturally evolved from the solfatara.

91. PONTICELLI, G. *Grosseto.*—Stalactite from Poggetto, near Grosseto.

92. QUARTAPELLE, R. *Teramo (Abruzzo Citeriore I.*)—Lignite, clay, pozzolana, gypsum, limestone.

93. RAVENNA SUB-COMMITTEE FOR THE EXHIBITION.—Salt from Cervia, gypsum.

94. ROMAGNA SULPHUR MINING CO. *Bologna.*—Sulphur in rolls, &c.

95. ROMAN IRON MINING CO. *Rome.*— Ores of iron, and refractory bricks from Tolfa; specimens of cast and wrought iron from Terni; various kinds of iron wire from Tivoli, &c.

96. ROSSI, F. & N. *Lucca.*—Hydraulic lime.

97. SADDI, S. & Co. *Cagliari.*—Lead ore from Arcilloni mine, commune of Burcei; lead ore from Su Bacci di S. Arrideli mine, commune of S. Vito.

98. SADUN & ROSSELLI, *Sienna.*—Ore of mercury, metallic mercury, &c., from S. Fiora mine, Sienna; plan of mine.

99. SANTI, C. *Montalcino (Sienna).*— "Mountain meal," with floating bricks, and other articles, made of it.

100. SANTINI, G. *Seravezza (Lucca).*— Statuary, and other marble, &c., from Campanice, Vaglisotto.

101. SARDINIAN SALT WORKS, *Genoa.*— Bay salt from Cagliari salt works.

102. SCACCHI, PROF. SENATOR A. *Naples.* —Artificial crystals.

103. SCLOPIS, BROTHERS, *Turin.*—Pyrites, Giobertite, &c.; plans of the mine of Brosso (Turin).

104. SERPIERI, E. *Cagliari.* — Ancient slags from Domusnovas and Villamassargia, and Flumini Maggiore; pigs of lead obtained from them, &c.

105. SERRA, L. *Iglesias (Cagliari).* — Granite, from Arbus and Guspini; trachite, from Carloforte.

106. SISTO, BARON, *Catania.*—Cake of sulphur, from Muglia.

107. SPANO, L. *Oristano (Cagliari).*— Ores of iron.

108. STREIFF, G. & Co. *Bergamo.*—Ores of copper and lead from the Valsassina (Bergamo), 1,300 feet above the plain.

109. TALACCHINI, A. *Milan.* — Copper pyrites from the mine of Nibbio, near Palazza, Lago Maggiore.

110. TIMON, CAV. A. *Cagliari.*—Lignite from Terras de Collu, commune of Gonnesa, near Iglesias.

111. THOVAZZI, C. *Fornovo (Parma).*— Petroleum.

112. ROYAL TUSCAN IRON MINES AND FOUNDRIES, *Leghorn.*—Ores of iron from the mines of Rio, Capo Calamita, Terra Nera; cast and wrought iron from Follonica.

113. VELLANO, S. *Turin.* — Glass and emery paper, emery cloth, emery leather; machines for pounding and sifting glass.

114. VICTOR EMMANUEL MINING CO.— *Miggiandone (Novara).*—Copper ore.

115. VILLA, A. &. G. *Milan.*—Classified collection of minerals, from the cretaceous system of Brienza.

116. VINCENTINI, COUNT P. O. *Rieti (Umbria).*—Calcareous sandstone, coralline, and yellow breccia, from Monte Alviano, near Rieti.

117. VOLTERRA SALT WORKS, *Volterra.* —Specimens of salt.

118. ZICCARDI, N. *Campobasso (Molise).* —Gypsum, from Ripalomosi, near Campobasso.

———

119. ACERBI, G. *Torpiana (Genoa).*— Copper ore, serpentine, decomposed jasper for making hydraulic cement.

120. BARBAGALLI, S. *Catania.*—Sublimed sulphur.

121. BOURLON & Co. *Pisa.*—Iron ore, from Monte Valerio.

122. COJOLI, E. *Leghorn.*—Copper ore, from Caggio; lignite, from Monterufoli.

123. COSSU, P. *Domusnovas (Cagliari).*— Lead ore, from Acquabona.

124. CRIVELLI, C. *Tortona (Alexandria).* —Flagstone, from Sorli.

125. DE BOISSY, MARCHIONESS T. *Setimello (Florence).*—White marble, serpentine, granite, ochres from Elba.

126. DEL GRECO, F. *Arezzo.*—Marl for manure.

127. D'ERCHIA, A. *Monopoli (Terra di Bari).*—Limestone, tufaceous stone.

128. FEDERICI, Dr. M. *Arcola, Sarzana (Genoa).*—Manganese ore, from Arcola.

129. FOGGIA ROYAL ECONOMIC SOCIETY *(Capitanata).* — Collection of marble, from the Gargano; alabaster, travertine.

130. FORESI, L. *Portoferrajo, Elba (Leghorn).* — Copper and antimony ores, from Elba ; plans of the mineral deposits.

131. FORLÌ SUB-COMMITTEE FOR THE EXHIBITION.—Sulphur, gypsum, lignite.

132. GELICHI, T. *Florence.* — Calcareous serpentine (oficalce), and calcareous exphotide, from Colle Salvetto.

133. GIACOMELLI, P. *Lucca.*—Steel.

134. GIUDICE, G. *Molo (Girgenti).* — Sulphur, raw and refined ; gypsum, and selenite.

135. LA FONTANA MINING CO. *Domusnovas (Cagliari).*—Lead ore, from Monte Cervus.

136. LICCIARDELLO, S. *Catania.*—Powdered sulphur for vines.

137. LICATA, MUNICIPALITY OF (*Girgenti*). —Dolomite.

138. LUCCA SUB-COMMITTEE FOR THE EXHIBITION.—Collection of marbles of the province.

139. MAZZIOTTI, BARON, & CO. *Turin.*—Oxide of manganese, from Framura mine.

140. NAPLES SUB-COMMITTEE FOR THE EXHIBITION.—Sand from the Bay of Naples, rich in iron ore.

141. NOCITO, DR. G. *Girgenti.*—Sicilian agates ; bituminous tertiary schists, from the neighbourhood of Girgenti.

142. NURCHIS, R. *Domusnovas (Cagliari).* —Lead ore, from Buoncammino.

143. PAPARELLA, G. *Tocco, near Chieti (Abruzzo Citeriore).*—Petroleum.

144. PERELLI, G. *Laurino (Principato Citeriore).*—Marble from Laurino and Laurito.

145. PIRAZZI, MAFFIOLA, & CO. *Piedimulera (Novara).*—Native gold, from Val Toppa.

146. PIROLI, PROF. A. *Parma.*—Iron and copper ores, lignite, &c.

147. PODESTÀ, B. *Sarzana (Genoa).*—Copper ore, from Bracco mine.

148. PODESTÀ, D. *Sarzana (Genoa).*—Bardiglio and black marble, from the neighbourhood of Sarzana.

149. ORREGONI, A. *Varese (Como).*—Peat, from near the Lake of Varese.

———

2078. ALBIANI TOMEI, CAV. F. *Seravezza (Lucca).*—Slab of "bardiglio" marble, from La Cappella quarry, near Seravezza ; marble squares for pavements, from Solais quarry, near Seravezza.

2079. FALLICA, A. *Catania.*—Crystallised sulphur of Girgenti, and celestine of Catanisetta.

2080. PARMA UNIVERSITY, NATURAL HISTORY MUSEUM OF.—Minerals and fossils, from the provinces of Parma and Placenza.

2081. PATE, BROS. *Leghorn.*—Regulus of antimony, from S. Stefano (*Grosseto*).

2082. RACALMUTO, MAYOR OF (*Girgenti*). —Rock salt, from Racalmuto.

2083. RACCHI, DR. G. *Casalduni (Benevento).*—Lignite, from Casalduni and Pagliara.

2084. REGGIO (CALABRIA) SUB-COMMITTEE FOR THE EXHIBITION. — Magnetic iron ore, from Aspremonte ; antimonial nickel ; ores of argentiferous lead and manganese ; amianthus ; marble.

2085. REGGIO (EMILIA) AGRICULTURAL ASSOCIATION.—Marl, plaster of Paris, lime, gypsum ; earth rich in nitrogenous matter.

2086. RICCARDI DI NETRO, CAV. E. *Turin.* —Iron and copper ores, from Traversella ; electric apparatus, for separating magnetic oxide of iron from copper pyrites ; dressed ore, some of it concentrated by the electric apparatus.

2087. REMEDI, MARQUIS A. *Sarzana (Genoa).*—Copper ore, from Marciano.

2088. ROCCA, GUERRIERO, & CO. *Levanto (Genoa).*—Copper pyrites, from La Francesca mine, Levanto.

2089. ROCCHETTA, MUNICIPALITY OF (*Massa-Carrara*). — Umber, serpentinous breccia, jasper for making hydraulic cement, manganese ore.

2090. RUSCHI, AVV. P. *Sarzana (Genoa).* —Red marble, from Monte Caprione.

2091. SARAGONI & TURCHI, *Cesena (Forlì).* —Sulphur.

2092. SCACCHI, D. *Gravina (Terra di Bari).*—Building stone.

2093. SCOVAZZO - CAMMERATA, BARON Rocco, *Catania.*—Sulphur.

2094. SLOANE, HALL, BROS. & COPPI,

Florence.—Copper ores, from Monte Catini, near Volterra (*Pisa*) ; copper.

2095. SPEZIA, BROS. *Pestarena* (*Novara*).—Gold, from Pestarena, in Val d' Anza.

2096. TANCREDI, P. *Trebiano, Sarzana* (*Genoa*).—Manganese ore, from Graziola and Guarcedo.

2097. TURIN ENGINEERING SCHOOL (*Scuola d'applicazione per gl' Ingegneri*).—Galena, from Monteponi mine (*Cagliari*); copper pyrites, from St. Marcel mine (*Turin*) ; copper pyrites, from Ollomont, Aosta (*Turin*); nickel pyrites, from Varallo; nickeliferous matt; lignite, from Cadibona, with plans of the mine.

2098. COCCHI, PROF. I. *Florence.*—Geological collection, from the mountains of Spezia, with sections.

2099. CURRO, *Catania.*—Flowers of sulphur.

2100. FLORI, A. *Forlì.*—Sulphur.

2101. GALLIGANI, DR. G. *Seravezza* (*Lucca*).—Marble table-tops and vase.

2102. GINORI-LISCI, MARQUIS, *Florence.*—Alabaster, lignite and copper ores, from near Volterra (Pisa).

2103. HÄHNER, CAV. *Leghorn.*—Marble, from Massa Carrara ; copper and lead ores, from Val di Castello (*Lucca*) ; cinnabar, from Ripa (*Lucca*).

2104. PESARO AGRICULTURAL ACADEMY.—Iron ore, from Monte Nerone.

2105. RICHARD, G. *Milan.*—Turf, natural, and compressed without machinery, for distillation.

2106. ROYAL ENGINEER CORPS, *Turin.*—Topographical map of the former kingdom of Sardinia.

2107. ROYAL ENGINEER CORPS, *Naples.*—Topographical map of the Neapolitan provinces of the kingdom of Italy.

2108. RODRIGUEZ, *Lipari.*—Sulphur.

2109. SCALIA, L. *Palermo.*—Sulphur ore, from Rabbioni mine, Serradifalco (*Caltanisetta*) ; sulphur.

CLASS II.

SUB-CLASS A.

150. ALBERTI, F. *Naples.*—Collection of chemical products.

151. ARROSTO, G. *Messina.*—Citric acid, prepared by a new process; acid of the bergamotte citron ; citrate of lime, &c.

152. ASQUER, CAV. A. *Cagliari.*—Soda produced from ashes of plants growing near Cagliari.

153. BOTTONI, DR. C. *Ferrara.*—Crystallized cream of tartar.

154. CAGLIARI SUB-COMMITTEE FOR THE EXHIBITION.—Sulphate of magnesia, crude soda from vegetable ashes, cream of tartar, potassa in cakes and powder.

155. CONVENT OF THE PADRI SERVITI, *Sienna.*—Bicarbonate of potassa, prepared with carbonic acid spontaneously evolved from mineral waters.

156. CORSINI, L. *Florence.*—Blacking, varnish for leather.

157. CURLETTI, A. *Milan.*—Crude potassa from vegetable ashes, raw and partly purified carbonate of soda, caustic soda, saltpetre.

158. DE LARDAREL, HEIRS OF COUNT, *Leghorn.*—Boracic acid, from the Lagoons, near Volterra ; natural productions of the Lagoons.

159. FANNI, F. *Cagliari.*—Crude soda.

160. FERRONI, G. *Florence.*—Blacking.

161. GASPARE, M. *Teramo* (*Abruzzo Ulteriore, I.*).—Cream of tartar.

162. GHIBELLINI, D. & V. *Persiceto* (*Bologna*).—Japan for ironwork.

163. LODINI, BROTHERS, *Persiceto* (*Bologna*).—Japan for ironwork.

164. MAFFEI, G. *Reggio, Emilia.*—Heliolene, a colourless and inodorous spirit for lamps.

165. MAJORANA, BROTHERS, BARONS DI NICORRA, *Catania.*—Crude soda.

166. MARINI, G. *Arezzo.*—Potassa.

167. MARRA, E. *Salerno (Principato Citeriore).* — Crude carbonate of potassa, from incinerated grape skins; purified potassa.

168. MASSEI, C. *Giulia (Abruzzo Ulteriore I.).*—Cream of tartar.

169. MELIS, B. *Quartu, near Cagliari.*—Crude soda, obtained from vegetable ashes.

170. MIRALTA, BROTHERS, *Savona (Genoa).*—Tartaric acid, cream of tartar.

171. ORSINI, ORSINO, & NEPHEW, *Leghorn.*—Specimens of saltpetre.

172. PARODI, P. *Savona (Genoa).*—Cream of tartar.

173. PETRI, G. *Pisa.* — Reduced iron, and protiodide of iron, preserved from oxidation by a new process. Solidified cod-liver oil.

174. RIATTI, V. *Reggio, Emilia.*—Cyanide of aluminium and iron; aluminate of soda, and chloro-sulphuric acid.

174A. SCERNO, E. *Genoa.*—White lead.

175. SCLOPIS, BROTHERS, *Turin.*—Sulphuric and other acids; green vitriol, blue vitriol, double sulphate of iron and copper: Epsom salts.

176. SINISCALCO, M. *Salerno (Principato Citeriore).* — Crude cream of tartar, from wine; purified cream of tartar.

177. SUPPA & CASOLINO *Trani (Terra di Bari).*—Nitre; cream of tartar, crude, and purified.

178. TOVO, F. *Turin.*—Coral reduced to the state of mucilage, and capable of being coloured and moulded.

179. VERCIANI, A. *Lucca.*—Illustration of a process for staining ivory indelibly.

———

180. BELTRANI, G. *Trani (Terra di Bari).*—Crude cream of tartar.

181. CASASCO, G. *S. Antonino, Susa (Turin).*—Essence of peppermint.

182. CIUTI, N. & SON, *Florence.*—Photographic chemicals.

183. DE BELLIS, G. *Castellana (Terra di Bari).*—Cream of tartar.

184. DI CEVA, MARQUIS, COLONEL G. B. *Nocetto (Genoa).* — Saltpetre; refined sulphur.

185. DE VITA, N. *Giffone Velle Piane (Principato Citeriore).*—Crude and purified soda.

186. DURVAL, E. *Monterotondo (Grosseto).*—Common and purified boracic acid, borax, sulphate of ammonia.

187. FORLÌ SUB-COMMITTEE FOR THE EXHIBITION.—Sulphuric acid.

188. GULLI, G. *Reggio, Calabria (Ulteriore I.)*—Essence of bergamotte.

189. LEONI, A. *Leghorn.*—White lead.

———

2110. LOFARO, B. *Reggio (Calabria Ulteriore I.)*—Essence of bergamotte citrons, lemons, citrons, bitter oranges; concentrated citric acid.

2111. MELISSARI, F. S. *Reggio (Calabria Ulteriore I.).*—Essence of bergamotte citrons, bitter lemons, Portugal lemons, bitter oranges, and mandarin oranges.

2112. MINERVINOM, *Benevento.*—Crude Benevento potassa, prepared from the skins of white grapes.

2113. REGGIO (CALABRIA) SUB-COMMITTEE FOR THE EXHIBITION.—Crude and refined cream of tartar.

2114. SPANO, L. *Oristano (Cagliari).*—Hard and soft charcoal.

2115. TARTARONE, G. *Giffone Velle Piane (Principato Citeriore).*—Refined potassa.

2116. TORRISI, M. *Trecastagni (Catania).*—Cream of tartar.

2117. AMANTINI, *Urbino.*—Specimens of lacquered iron.

2118. CAMPISI, A. *Melitello (Catania).*—Citric acid.

2119. MESSINA SUB-COMMITTEE FOR THE EXHIBITION.—Potash alum; citric acid.

SUB-CLASS B.

190. ABBAMONDI, PROF. N. *Solopaca (Benevento)*. — Sulphurous water from Jalisa, acidulous waters from Villa, sulphurous aluminous water from S. Antonio: all rising from an extinct volcanic crater.

191. ALDROVANDI, M. *Bologna*.—Antiscorbutic water.

192. ARROSTO, G. *Messina*. — Mineral waters from the neighbourhood of Messina, with memoir.

193. BARRACCO, BARON, *Cotrone (Calabria Ulteriore II.)*.—Liquorice root, liquorice.

194. BERTOLOTTI, P. *Bologna*—Felsina water, a perfumed cosmetic.

195. BOLOGNA PROVINCIAL DEPUTATION. —Thermo-mineral waters of La Poretta (*Bologna*).

196. CAGLIARI SUB-COMMITTEE FOR THE EXHIBITION. — Saline, chalybeate, and alkaline mineral waters of the province of Cagliari.

197. CARINA, PROF. A. *Lucca*.—Mineral waters from the Baths of Lucca, with their natural deposit.

198. COJARI, AVV. V. *Soliera, near Fivizzano (Massa-Carrara)*.— Sulphurous water from Equi, Fivizzano.

199. COMI, R. *Giulia (Abruzzo Ulteriore I.)*.—Liquorice juice.

200. CUGUSI, E. *Cagliari*. — Mineral waters from Domusnovas.—Thermo-mineral water from Siliqua, Sardaru, Fordongianus, and Villacidro; chalybeate waters from Capoterro.

201. DE ROSA, R. *Atri (Abruzzo Ulteriore I.)*.—Liquorice juice.

202. DUFOUR, BROTHERS, *Genoa*.—Mannite, quinine, cinchonine, cinidine, cinchonidine, and chinoline.

203. FAVILLI, G. *Pontesserchio (Pisa)*.— Unoxidizable protiodide of iron, reduced iron.

204. FOTI, S. *Acireale (Catania)*.—Thermo-mineral water, called S. Venere del Pozzo, near Acireale.

205. GALLIANI & MAZZA, *Milan*.—Mannite, castor oil.

206. GARELLI, DR. G. *Turin*.—Samples of the thermo-sulphurous and alkaline waters of Valdieri (Cuneo); medicinal deposit and scum of those waters.

207. GENNARI, PROF. P. *Cagliari*.—Collection of Italian medicinal plants.

208. GIORGINI, PROF. G. *Parma*.—Saline mineral water containing iodine, from Sassuolo (Modena): pamphlet on its analysis, and medical properties.

209. GIORGINI, DR. G. *Radicofani (Sienna)*.—Mineral waters, from S. Casciano dei Bagni.

210. GRASSI, P. *Acireale (Catania)*. — Mineral waters from S. Tecla, near Acireale.

211. LIPARI JUNTA FOR THE EXHIBITION (*Messina*).—Thermo-mineral water from the Lipari Islands.

212. MACERATA SUB-COMMITTEE FOR THE EXHIBITION.—Collection of mineral waters.

213. MADESIMO MINERAL WATER CO. *Madesimo (Sondrio)*.—Mineral water, from Madesimo.

214. MONTINI, P. *Fabriano (Ancona)*.— Seltzer water, seidlitz water, Vichy water, magnesian water, chalybeate water, soda water.

215. MORELLI, G. *Rogliano (Calabria Citeriore)*.—Manna of the ash (*Fraxinus Ornus*), several kinds.

216. NAPLES SUB-COMMITTEE FOR THE EXHIBITION.—Liquorice.

217. NOCERA, MUNICIPALITY OF (*Umbria*).—Acidulous mineral water from Nocera; Samian or Nocera earth, used for polishing metals.

218. ORSI, A. *Montalcino (Sienna)*.—Sulphurous alkaline mineral water, from Collalli, near Montalcino.

219. PATUZZI, L. *Limone (Brescia)*.— Citron water.

220. PELLAS, C. F. *Genoa*.—Cod-liver oil; calcined, and fluid magnesia; acid to render carbonate of magnesia effervescent.

221. PERI, G. *Milan*.—Anti-rheumatic oil.

222. PIGHETTI, A. *Salò (Brescia)*.—Citron water.

223. POLI, G. B. *Brescia.*—Pills made from the deposit of the principal mineral waters of France; anti-mercurial sarsaparilla pills.

224. RICCI, G. *Turin.*—Digestive pastiles, tincture of mint.

225. RIOLO, MUNICIPALITY OF (*Ravenna*).—Mineral waters.

226. REGGIO (EMILIA) SUB-COMMITTEE FOR THE EXHIBITION.—Saline, chalybeate, and sulphurous mineral waters.

227. RUSPINI, G. *Bergamo.*—Sulphurous saline, and saline ferruginous waters containing iodine; other mineral waters; and crystallized mannite, obtained by a new process.

228. SCERNO, E. *Genoa.*—Sulphate and citrate of quinine, sarsaparilla.

229. SCOLA, B. *Turin.*—Gelatine capsules, containing medicines.

230. SPANO, L. *Oristano* (*Cagliari*).—Mineral waters, from the thermal, tepid, and cold springs of Fordongianus; deposit of these springs.

231. TURIN ROYAL ACADEMY OF MEDICINE.—Mineral waters of the ancient Sardinian provinces; medicinal deposit formed by several of them; catalogue.

232. VERATTI, C. *Bologna.*—Cashew-nut pastiles.

233. VERGA, DR. A. *Milan.* Saline mineral water containing iodide, lately discovered near Miradolo (Pavia); pamphlet regarding it.

234. AMICARELLI, V. *Montesantàngelo* (*Capitanata*).—Manna.

235. BELLIA, S. *Caltagirone* (*Catania*).—Sulphurous mineral waters.

236. BRASINI, BROS. *Forlì.*—Purgative mineral waters.

237. CASTAGNACCI, A. *Florence.*—Compound lichen pastiles, sulphur pastiles, vermifuge sweetmeats.

238. CONTESSINI, F. & Co. *Leghorn.*—Salts of quinine; caffeine: mannite; santonine; morphine.

239. CORRIDI, G. *Leghorn.*—Mannite; sulphate and citrate of quinine; santonine, castor oil.

240. CROPPI, C. *Forlì.*—Mineral waters.

241. GIORDANO, D. *Cetara* (*Principato Citeriore*).—Manna.

242. MAJORANA, G. & TORNABENE, F. *Catania.*—Mineral waters.

243. PIGNATELLI, V. *Cosenza* (*Calabria, Citeriore*).—Liquorice.

244. POLENGHI, C. S. *Fiorano* (*Milan*).—Mallow yarn for surgical use.

245. PONDI, G. *Palagonia* (*Catania*).—Chalybeate and acidulous mineral waters.

246. REGGIO (CALABRIA) SUB-COMMITTEE FOR THE EXHIBITION.—Liquid and solid nitrate of magnesia.

247. TORRI, DR. F. *Pisa.*—Mineral waters from the baths of Pisa and Asciano.

248. VITI, MARQUIS A. *Orvieto* (*Umbra*).—Acidulous and chalybeate waters.

249. VALERI & Co. *Lagnano* (*Brescia*).—Castor-oil and castor-oil seeds; residuum of castor-oil manufacture, for manure.

CLASS III.

SUB-CLASS A.

260. ADRAGNA, G., BARONE DI ALTAVILLA, *Trapani.*—Canary-seed, and linseed.

261. ALEXANDRIA, SUB-COMMITTEE FOR THE EXHIBITION.—Collection of various kinds of wheat cultivated in the province.

262. ALGOZINO, S. *Leonforte* (*Catania*).—Spelta wheat.

263. ANZALONE, F. *Catania.*—Kidney beans.

264. ARESU, S. *Selargius* (*Cagliari*).—Wheat in the ear.

265. BELLESINI, BROTHERS, *Imola, Bologna.*—Specimens of rice, to illustrate its cultivation; Chinese, Novara, and American rice.

266. BELTRAMI, COUNT P. *Cagliari.*—Wheat in the ear, rice, and beans.

267. BELTRANI, G. *Trani (Terra di Bari).*—Dried figs, raisins, sweet and bitter almonds, apricot kernels, and olives.

268. BERGAMI, P. *Ferrara.*—Wheat and Indian corn.

269. BOLOGNA AGRICULTURAL SOCIETY.—Collection of corn, and other seeds of the province of Bologna.

270. BURESTI, F. *Arezzo.*—Wheat, rye, and Indian corn.

271. CAGLIARI SUB-COMMITTEE FOR THE EXHIBITION.—Wheat, semolina, flour, bran, starch, barley, Indian corn, sorgho, beans, pease, chick-pease, mustard seed, canary seed, almonds, nuts of various kinds, fennel, raisins, and dried figs.

272. CAMPOBASSO SUB-COMMITTEE FOR THE EXHIBITION.—Various kinds of wheat and leguminous seeds, Indian corn, lentils, beans, and kidney beans.

273. CAO DI S. MARCO, COUNT, *Cagliari.*—Wheat, barley, chick-pease, beans, and lentils.

274. CASALI, A. *Calci, Pisa.*—Corn and flour.

275. CASAZZA, CAV. A. *Ferrara.*—Wheat, Indian corn, oats, barley, chick-pease, and grass seeds.

276. CASERTA PROVINCIAL AGRICULTURAL SOCIETY, *Terra di Lavoro.* —Various kinds of wheat, barley, oats, beans, and pease; ground nuts.

277. CASSOLA, AVV. C. *Vercelli (Novara).*—American rice, from Anitre, near Vercelli.

278. CATANIA, HERMITS OF S. ANNA.—Chick-pease.

279. CELI, PROF. E. *Modena.*—Collection of various grass seeds grown in the province.

280. CHELLI, F. *Leghorn.*—Starch, made by a new process.

281. CHERICI, N. *S. Sepolcro, Arezzo.*—Various kinds of wheat; Indian corn, flour, and semolina; Indian corn bread and polenta; chestnuts, chestnut flour, and polenta.

282. CHERICI, CLELIA, *S. Sepolcro, Arezzo.*—Dried edible fruit of the *carlina acaulis.*

283. CHIETI SUB-COMMITTEE FOR THE EXHIBITION, AND OTHER EXHIBITORS (*Abruzzo Citeriore*).—Various kinds of wheat (some exhibited by G. LAUCIANO and D. D'ONOFRIO); risciola, winter rye, barley, oats, Indian corn, beans, kidney beans (some exhibited by D. BUTOLO and DR. G. BONETTI) ; chick-pease (some exhibited by V. JULIANI VILLEMAGNA); lentils, lupins, pease, vetches, flax seed, hemp seed, and potatoes.

284. CONSIGLIO, M. *Lentini, Noto.*—Almonds.

285.—CONVENT OF S. FRANCESCO D'ASSISI, *Catania.*—Barley.

286. DE' GIUDICI, A. *Arezzo.*—Wheat, oats, Indian corn, beans, chick-pease, millet, and lupins.

287. DI NISSA, MARQUIS G. *Cagliari.*—Sweet and bitter almonds, carob and other beans, pine cones.

288. DRAMMIS, BARON S. *Scandale (Calabria, Ulteriore II.).*—Various kinds of wheat ; Peruvian barley.

289. DROUIN, G. *Naples.*—Flour, bran, and bread.

290. FERRARA ROYAL CHAMBER OF COMMERCE.—Indian corn and Indian corn flour: wheat of various kinds and wheat flour ; oats, rice, chick-pease, and beans.

291. FIORENTINI, G. *Castrocaro, Florence.*—Sainfoin.

292. GABRIELE, DR. A. *S. Bartolomeo in Galdo (Benevento).*—Cereals and leguminous seeds.

293. GIORDANO, E. *Salerno (Principato Citeriore).*—Red and white yams, introduced into Italy by the exhibitor.

294. GRASSI, A. *Giarre (Catania).*—Almonds cultivated on the flanks of Mount Etna.

295. GREFFI, A. *Monterchi, Arezzo.*—Wheat.

296. GUACCI, F. *Campobasso (Molise).*—Varieties of wheat.

297. GUIDA, G. & G. *Gargarengo, Vicolonga (Novara)* —Various kinds of wheat, barley, oats, rye, Indian corn, millet, sorgho, pease, chick-pease, beans, and lupins.

298. LAI, L. *Lanusei (Cagliari).*—Ground nuts, beans, dried figs, and dried prunes.

299. Lega, M. *Brisighella (Ravenna.)*—Aniseed.

300. Ligas, A. *Selargius (Cagliari.)*—Wheat in the ear.

301. Lipari Junta for the Exhibition, *Messina.*—Raisins and capers.

302. Lucca Sub-Committee for the Exhibition. — Collection of cereals and other agricultural produce of the province.

303. Macerata Sub-Committee for the Exhibition.—Various kinds of wheat and Indian corn.

304. Majorana Brothers, Barons di Nicorra, *Catania.*—Indian corn, oats, canary seed, beans, pease and chick-pease, lentils, sesame seed, mustard seed, hemp and flax seed, and acorns.

305. Mameli, F. *Selargius (Cagliari).*—Wheat in the ear, and beans.

306. Mancuso, M. *Catania.* — Wheat, canary seed, oats, beans, chick-pease, clover, flax, and mustard seed.

307. Marini Demuro Avv. T. *Cagliari.*—Wheat, barley, beans, pease, chick-pease, and figs.

308. Marozzi, E. *Pavia.*—Cleaned and uncleaned rice ; clover seed.

309. Massone, Cav. M. *Cagliari.*—Pease, and millet seed.

310. Mazzurana, F. *Trent.*—Carraway seed.

311. Melis, B. *Quartu (Cagliari).* — Wheat, barley, and beans.

312. Meloni, A. *Quartu (Cagliari).*—Dried figs, and raisins.

313. Minutoli Tegrimi, Count E. *Lucca.*—Rice obtained without rotation of crops ; Chinese and American rice, cleaned and uncleaned.

314. Modena Royal Botanical Gardens.—200 varieties of Indian corn, collected by the late Professor G. Brignoli.

315. Modena Sub-Committee for the Exhibition.—Cereals, and vegetable seeds.

316. Monari, C. & C. *Bologna.*—Several kinds of cleaned rice.

317. Monterisi, G. *Bisceglie (Terra di Bari).*—Pease and lentils.

318. Montori, G. *Colonnella (Abruzzo Ulteriore I.).*—Agricultural products.

319. Moscero, G. *Cosenza (Calabria Citeriore).*—Dried figs.

320. Murru-Murru, A. *Sanluri (Cagliari).*—Buck wheat.

321. Natoli, A. D. *Patti (Messina).*—Clover seed.

322. Niedda di S. Margherita, Count P. & Brother, *Cagliari.*—Stalks of sorgho for forage.

323. Padri Benedettini Monks, *Monte Cassino (Catania).*—Various kinds of wheat ; barley and beans.

324. Pagarelli, Dr. L. *Castrocaro (Florence).*—Aniseed and coriander seed.

325. Palumbo, O. *Trani (Terra di Bari).*—Lentil and beans.

326. Palumbo, P. *Cava (Principato Citeriore).*—Potato and other kinds of starch ; dextrine obtained from potato starch.

327. Panna, Dr. G. *Quartu, near Cagliari.*—Dried figs.

328. Pantano, F. P. *Asaro (Catania).*—Chick-pease.

330. Pasi, G. *Ferrara.*—Various kinds of beans.

331. Pavanelli, G. *Ferrara.* — Wheat, maize, and rape seed.

332. Piccaluga, G. *Cagliari.*—Pistachio nuts.

333. Pistilli, F. *Campobasso (Molise).*—Wheat and chick-pease.

334. Pittau, M. *Sanluri (Cagliari).*—Chick-pease.

335. Quercioli Brothers, *Modigliana (Florence).*—Aniseed.

336. Ramo, S. *Laconi (Cagliari).*—Wheat.

337. Ravenna Sub-Committee for the Exhibition.—Wheat, oats, rice, and beans.

338. Revedin, Count G. *Ferrara.*—Wheat and Indian corn.

339. Rundeddu, R. *Selargius (Cagliari).*—Beans, chick-pease, and pease.

340. Sanna, V. *Selargius (Cagliari).*—Beans, figs.

342. SAVONA JUNTA FOR THE EXHIBITION, *Genoa.*—Collection of leguminous seeds.

343. SCHLAEPFER, WENNER, & CO. *Salerno* (*Principato Citeriore*). — Potato starch: dextrine and gum from potato starch.

344. SHERRA, DR. L. *Iglesias* (*Cagliari*). —Wheat, barley, Indian corn, linseed, and raisins.

345. SPANO, L. *Oristano* (*Cagliari*).— Barley, wheat, chick-pease, beans, lucerne seeds, and pine cones.

346. TARDITI & TRAVERSA, *Brà, Turin.* —Various kinds of semolina, and flour.

347. TELLINI, V. *Calci* (*Pisa*).—Wheat flour.

348. TURIN AGRICULTURAL ACADEMY.— Collection of agricultural produce; beans, chick-pease and pease, lupins, and fenugreek. Artificial fruit.

349. VACCARO, L. *Cosenza* (*Calabria Citeriore*).—Seeds of the ground pistachio.

350. ASSOM VILLASTEL, *Turin.*—Hemp-seed.

———

351. AREZZO SUB-COMMITTEE FOR THE EXHIBITION.—Collection of the agricultural products of the province.

352. ARRANGA, G. *Serracapriola* (*Capitanata*).—Wheat.

353. ASCOLI SUB-COMMITTEE FOR THE EXHIBITION. — Collection of cereals, leguminous seeds, &c.

354. BARONE, BROS. *Foggia* (*Capitanata*). —Wheat.

355. BARACCO, BROS. *Catanzaro* (*Calabria Ulteriore I.*).—Agricultural products.

356. BOCCARDO, BROS. *Candela* (*Capitanata*).—Saragolla, Carlantino, and Carosella wheat.

357. BOFONDI, COUNT P. *Forlì.*—Cereals.

358. CAMMARATA SCOVAZZO, BARON ROCCO, *Palermo.*—Wheat.

359. CAPELLI, CAV. *Foggia* (*Capitanata*). —Wheat.

360. CARBONE, F. *Catania.*—Kidney beans.

361. CASERTA SUB-COMMITTEE FOR THE EXHIBITION.—Collection of cereals, leguminous seeds, potatoes, yams, beetroot, &c.

362. CASSANO, F. *Gioja* (*Terra di Bari*). —Cereals, mustard and fennel seeds.

363. CHIARINI, P. *Faenza* (*Ravenna*).— Cleaned rice of different kinds.

364. CICCHESE, P. *Campobasso* (*Molise*). —Wheat.

365. CICCHESE, R. *Campobasso* (*Molise*). —Flour.

366. COSTANTINO, G. *S. Marco dei Cavoti* (*Benevento*).—Leguminous seeds.

367. DANZA, D. *S. Agata* (*Capitanata*). —Wheat.

368. DANZA, G. *S. Agata* (*Capitanata*). —Wheat.

369. DE FIDIO, G. *Casaltrinità* (*Capitanata*).—Wheat.

370. DE LUCA, P. *Catania.*—Pistachio nuts.

371. DEL BUONO, E. *S. Agata* (*Capitanata*).—Wheat.

372. DE LEO, A. *Casaltrinatà* (*Capitanata*).—Wheat.

373. DELL'ÆRMA, V. *Castellano* (*Terra di Bari*).—Figs.

374. DEMURTAS, E. *Lanusei* (*Cagliari*). —Wheat, leguminous seeds.

375. FANTINI, *Bertinori* (*Forlì*).—Aniseed.

376. FIAMINGO, S. *Giuno* (*Catania*).— Indian corn, rye, French beans, almonds, &c.

377. FOGGIA SUB-COMMITTEE FOR THE EXHIBITION.—Oats, barley, beans.

378. FORLÌ SUB-COMMITTEE FOR THE EXHIBITION.—Almonds, aniseed.

379. GARAU CARTA, L, *Sanluri* (*Cagliari*). —Almonds.

380. GIULIANI, L. *S. Marco in Lamis* (*Capitanata*).—Carosella wheat.

381. GIUDICE, G. *Favara* (*Girgenti*).— Wheat, almonds, linseed.

382. GULINELLI, COUNT G. *Ferrara.*— Cereals.

383. LECCE SUB-COMMITTEE FOR THE EXHIBITION. — Cereals, leguminous seeds, almonds, raisins, walnuts, &c.

384. LOFARO, B. *Reggio (Calabria)* — French beans.

385. LUCERA, MAYOR OF *(Capitanata).*—Wheat.

386. MELE, N. G. *S. Agata (Capitanata).*—Wheat.

387. MERCATILE, COUNT M. *Ascoli.*—Cereals.

388. MILAN CHAMBER OF COMMERCE.—Collection of the agricultural produce of the province; semolina, bran, flour.

389. NERI, A. *Bologna.*—Cleaned rice.

390. ORTONA, MUNICIPALITY OF *(Abruzzo Citeriore).*—Figs.

391. PACCA, MARQUIS G. *Benevento.*—Wheat and leguminous seeds.

392. PATERNO CASTELLO, PRINCESS M. *Catania.*—Wheat, leguminous seeds, &c.

393. PATERNÒ CASTELLO, MARQUIS DI S. GIULIANO, *Catania.*—Peas, &c.

394. PARMA SUB-COMMITTEE FOR THE EXHIBITION.—Cereals, grapes.

395. PASCAZIO, V. *Mola (Terra di Bari).*—Carob-beans, figs.

396. PELLEGRINO, D. *Castaltrinità (Capitanata).*—Wheat.

397. PESARO AGRICULTURAL SOCIETY.—Collection of cereals, leguminous seeds, castor oil seeds, &c.

398. REGGIO (CALABRIA) SUB-COMMITTEE FOR THE EXHIBITION.—Prickly pears, figs.

399. REGGIO (EMILIA) AGRICULTURAL ASSOCIATION.—Collection of the agricultural produce of the province.

———

2120. REGGIO (EMILIA) AGRICULTURAL SOCIETY.—Wheat, millet, rice, beans, pease, chestnut and bean flour.

2121. ROMEO, L. *Acquaviva (Terra di Bari).*—Aniseed.

2122. RUBINO, M. *Foggia (Capitanata).*—Wheat.

2123.—SANTORO, G. *S. Agata (Capitanata.)*—Wheat.

2124. SARAGATU, AVV. P. *Sanluri (Cagliari).*—Wheat, starch.

2125. SARCINA, N. R. *Castaltrinità (Capitanata).*—Wheat.

2126. SCOCCHERA, *Canosa (Terra di Bari).*—Castor oil seeds.

2127. SINISCALCO, BROS. *Foggia (Capitanata).*—Wheat.

2128. TORRI, L. *Bondeno (Ferrara).*—Wheat.

2129. TREJAVILLA, A. *Cerignola (Capitanata).*—Wheat.

2130. TROIA, MAYOR OF, *Capitanata.*—Wheat.

SUB-CLASS B.

400. BELLENTANI, G. *Modena.*—Shoulder ham, sausages, lard, Italian paste, and tomato sauce.

401. BENEDETTI, P. BROTHERS, *Faenza (Ravenna).*—Italian paste.

402. BIANCHI, G. & C. *Lucca.*—Italian paste, &c.

403. BOSCARELLI, A. *Cosenza (Calabria Citeriore).*—Preserved larks.

404. BOTTAMINI, B. *Bormio (Sondrio):*—Honey.

405. CALDERAI, A. *Florence.*—Sausages, &c.

406. CESARI, L. *Torre Annunziata (Naples).*—Paste for soup.

407. DOZZIO, G. *Belgioioso (Pavia).*—Cheese.

408. DRAGHI, D. *Placenza.*—Salt meat, known as *coppe.*

409. FANNI, F. *Cagliari.*—Tunny eggs and mullet eggs *(Bottarghe).*

410. FERRARA ROYAL CHAMBER OF COMMERCE.—Cheese and sausages.

411. FORNI, A. *Bologna.* — Bologna sausages *(mortadella)*, &c.

412. FRANZINI, B. *Pavia,*—Cheese.

413. GABRIELE, DR. A. *S. Bartolomeo in Galdo (Benevento).*—Cheese and honey.

414. GATTI, A. *Cosenza (Calabria Citeriore).*—Mushrooms.

415. GIULIANI, V. *Turin.*—Chocolate.

416. GUELFI, G. *Navacchio (Pisa).*—Biscuits.

417. JACCHINI, G. A. *Alexandria.*—Sausages, &c.

418. LAMBERTINI, G. *Bologna.* — Sausages, &c.

419. LANCIA, BROTHERS, *Turin.*—Preserved lard; salt, and preserved food.

420. LAVAGGI, G. *Augusta (Noto).*—Hyblean honey.

421. LIUZZI, C. *Reggio, Emilia.*—Sheep's-milk cheese, from Loriano.

422. LUPINACCI, BARON L., BROTHERS, *Cosenza (Calabria Citeriore).* — Butter and cheese.

423. MAJORANA, BROTHERS, BARONS DI NICORRA, *Mostarda (Catania).*—Honeycomb, and honey; cheese, butter, preserved olives, artichokes.

424. MALMUSI, CAV. C. *Modena.*—Honey.

425. MARINI DEMURO AVV. T. *Cagliari.*—Honey, and saffron.

426. MILAZZO JUNTA FOR THE EXHIBITION.—Salt fish.

427.—MODENA SUB-COMMITTEE FOR THE EXHIBITION.—Sheep's-milk cheese.

428. NUNS OF S. LUCIA, *Cagliari.*—Bunch of flowers in sugar.

429. ORRÙ, S. *Burcei (Cagliari).*—Bitter honey.

430. ORSI, R. & Co. *Bologna.*—Bologna pork sausages.

431. PAOLETTI, F. *Pontedera (Pisa).*—Italian paste and biscuits.

432. PAOLETTI, G. *Pontedera (Pisa).*—Italian paste, biscuits, and flour.

433. PAOLETTI, O. *Florence.*—Biscuits.

434. PARMA SUB-COMMITTEE FOR THE EXHIBITION. — Cheese, Italian paste, and salame.

435. PETRUCCELLI, C. *Castelfranco (Benevento).*—Sheep's-milk and other cheese.

436. RAINOLDI, G. *Milan.* — Salt and smoked pork.

437. REVEDIN, COUNT G. *Ferrara.*—Cheese.

438. SAGLIOCCA, G. *Pietro Elcina (Benevento).*—Cheese.

439.—SALTARELLI, A. *Pisa.* — Candied fruit.

440. SAMOGGIO, G. *Bologna.* — Hog's lard.

441. SONA, C. *Alexandria.*—Mostarda, a kind of preserve.

442. SPANO, L. *Oristano (Cagliari).*—Pickled olives, smoked mullet, salted eels, honey, &c.

443. TORRICELLI, A. *Florence.*—Chocolate.

444. VALAZZA, G. *Turin.*—Sardines, and tunny.

445. VALERI, A. *Ferrara.*—Peach preserve.

446. ZANETTI, G. *Bologna.* — Bologna sausages, and *capocollo.*

———

447. AMICARELLI, D. V. *Montesantangelo (Capitanata).*—Honey.

448. ASTENGO, C. *Savona (Genoa).*—Vermicelli and other kinds of Italian paste for soup.

449. BARBETTI, S. *Fuligno (Umbria).*—Chocolate, and sweetmeats.

450. BARACCO, BARON A. *Naples.* — Cheese, olives, chestnuts.

451. BARRACCO, BROS. *Cotrone (Calabria Ulteriore II.).*—Calabrian cheese.

452. BELTRAMI, COUNT P. *Cagliari.*—Cow's-milk cheese.

453. BERGAMI, P. *Ferrara.*—Cheese.

454. BERNARDI, BROS. *Borgo a Buggiano, Val di Nievole (Lucca).*—Italian paste for soup.

455. BIBIANO (MUNICIPALITY OF), *Reggio, Emilia.*—Old cheese.

456. BIFFI, P. *Milan.*—Confectionery.

457. BODINO, L. *Genoa.*—Chocolate.

458. BOLLINI, G. *Alexandria.* — Dried salt pork, bondiola.

459. BOSIO, D. *Alexandria.*—Pork sausages.

460. BOTTAMINI, B. *Bormio (Sondrio).*—Honey.

461. BRASINI, BROS. *Forlì.*—Chocolate.

462. CAGLIARI SUB-COMMITTEE FOR THE EXHIBITION.—Maccaroni, vermicelli, and other kinds of paste for soup; tunny and mullet eggs; sweet and bitter honey.

463. CAPASSO, F. *Benevento.*—Benevento torrone, &c.

464. CASERTA SUB-COMMITTEE FOR THE EXHIBITION (*Terra di Lavoro*).—Maccaroni, vermicelli, &c.

465. CASSANO, F. *Gioja* (*Terra di Bari*). —Cheese.

466. CARPANETO & GHILINO, *Genoa.*— Preserved food and fruit.

467. CICCHESE, P. *Campobasso* (*Molise*). —Maccaroni, semolina made of Saragolla wheat.

468. CICCHESE, R. *Campobasso* (*Molise*). —Italian paste, semolina.

469. CIOPPI, L. & S. *Pontedera, Pisa.*— Italian paste for soup, salame.

470. COSTANTINO, G. *S. Marco de' Cavoti* (*Benevento*).—Cheese.

471. DE GAETANO, F. *Gallico* (*Calabria Ulteriore I.*).—Maccaroni and vermicelli.

472. DE GORI, COUNT A. *Sienna.* — Goat's milk cheese.

473. DEMURTAS, E. *Lanusei* (*Cagliari*).— White Italian paste for soup, cheese.

474. FARINA, BROS. *Baronissi* (*Principato Citeriore*). — Provoloni and Caciocavallo cheese.

475. FOGGIA SUB-COMMITTEE FOR THE EXHIBITION. — Hand and machine-made Italian paste for soup; pickled capers.

476. FORLÌ SUB-COMMITTEE FOR THE EXHIBITION.—Sausages.

477. FORNO MATTEI, A. *Prato* (*Florence*). —Biscuits.

478. GALASSO, G. *Benevento.*—Benevento torrone.

479. GIORDANO, D. *Ceara* (*Principato Citeriore*).—Salt anchovies.

480. LANZARINI, BROS. *Bologna.*—Pork sausages, &c.

481. LEMBO, P. *Minori* (*Principato Citeriore*).—Collection of Italian paste.

482. LOFARO, B. *Reggio* (*Calabria Ulteriore I.*).—Pickled olives.

483. MELISSARI, F. S. *Reggio* (*Calabria Ulteriore I.*).—Dried and pickled olives.

484. MOSCATO, BROS. *Salerno* (*Principato Citeriore*).—Caciocavallo cheese.

485. PACCA, MARQUIS G. *Benevento.*— Cheese, honey.

486. PALUMBO, O. *Trani* (*Terra di Bari*). —Olives, tomato sauce.

487. PASCAZIO, V. *Mola* (*Terra di Bari*). —Tomato sauce.

488. RAMIREZ, G. *Reggio* (*Calabria Ulteriore I.*).—Preserved tomatoes.

489. REGGIO (CALABRIA) SUB-COMMITTEE FOR THE EXHIBITION.—Sheep's-milk cheese, honey, eleozaccaro.

490. REGGIO (EMILIA) AGRICULTURAL ASSOCIATION.—Ham, shoulder ham, coppa, sausages.

491. SALERNO SUB-COMMITTEE FOR THE EXHIBITION (*Principato Citeriore*). — Dried figs, pears, chestnuts, &c.; Cilento sausages.

492. SPEZI, D. *Fuligno* (*Umbria*).—Confectionery and chocolate.

493. TRUCILLI, V. *Salerno* (*Principato Citeriore*).—Provole, a buffalo-milk cheese; butter preserved in an envelope of cheese.

494. VIVARELLI, C. *Pistoja* (*Florence*).— Sheep's-milk cheese.

———

495. CAMPOBASSO SUB-COMMITTEE FOR THE EXHIBITION.—Biscuits.

496. DAMIANI, C. *Portoferajo, Elba* (*Leghorn*).—Biscuits.

497. IANNICELLI, M. *Salerno* (*Principato Citeriore*).—Italian paste.

498. MARINELLI, E. *Parma.*—Collection of Italian paste.

499. MATTEI, A. *Prato* (*Florence*). — Rusks.

500. SGARIGLIA, *Dalmonte.* — Pickled olives.

Sub-Class C.

505. Agnini, T. *Modena.*—Rosolio, rinfresco.

506. Agnello, Baron, *Siculiana* (*Girgenti*).—Wine.

507. Agozzotti, Avv. *Modena.*—Lambrusco wine.

508. Alonzo, A. *Catania.*—White wine.

509. Anselmi, B. *Verona.*—Wine.

510. Albino, P. *Campobasso* (*Molise*).—Wine.

511. Alessi, G. *Messina.*—Tobacco.

512. Allemano, Brothers, *Asti* (*Alexandria*).—Wines.

513. Almerici, Marquis G. *Cesena* (*Ferrara*).—Wine.

514. Anghirelli, G. *Montalcino* (*Sienna*).—Wines.

515. Asquer, Viscount di, *Flumini* (*Cagliari*).—Red wine.

516. Ballor, G. *Turin.*—Liqueurs, &c.

517. Baracco, N. & Co. *Turin.*—Liqueurs.

518. Bartholini, C. *Cosenza* (*Calabria Citeriore*).—Wines.

519. Bellentani, G. *Modena.*—Vinegar.

520. Beltrani, G. *Trani* (*Terra di Bari*).—Wines.

521. Bernardi, F. *Sienna.*—Red wine.

522. Bologna Royal Tobacco Manufactory.—Tobacco, snuff, and cigars.

523. Bonnet, G. *Comacchio* (*Ferrara*).—Wine.

524. Bonolis, F. *Teramo* (*Abruzzo Ulteriore*).—Effervescing wines.

525. Boratto, D. *Alexandria.* — Vermouth.

526. Borlasca, C. *Govi* (*Alexandria*).—Red wine.

527. Botti, A. *Chiavari* (*Genoa*).—Wine.

528. Bozzo, M. *Benevento.*—Tobacco.

529. Braggio, Count F. (*Strevi Alexandria*).—Wines.

530. Brescia Sub-Committee for the Exhibition.—Wine.

531. Cadoni, A. *Quartu* (*Cagliari*). —Wine.

532. Cagliari Royal Excise Office.—Tobacco, snuff, and cigars.

533. Cagliari Sub-Committee for the Exhibition.—Wines, spirits of wine, vinegar, and tobacco.

534. Caimi, F. *Sondrio.*—Wine.

535. Cara, Cav. G. *Cagliari.*—Wines.

536. Caramora, P. *Asti* (*Alexandria*).—Liqueurs and wine.

537. Casazza, Cav. A. *Ferrara.*—Wines.

538. Castagnino, I. *Imola* (*Bologna*).—Wine.

539. Castiglione Bendinelli, *Novi* (*Alexandria*).—Wines.

540. Castiglione delle Stiviere Junta for the Exhibition (*Brescia*).—Tobacco and snuff.

541. Cerrone, G. *Teramo* (*Abruzzo Ulteriore I.*).—Effervescing wines.

542. Cherici, N. *S. Sepolcro* (*Arezzo*).—Wines.

543. Clarkson, S. V. *Mazzara* (*Trapani*).—Wine.

544.—Cobianchi & Ardizzoli, *Boca* (*Novara*).—Wines.

545. Cocchi, F. *Reggio, Emilia.*—Rosolio.

546. Cocozza, C. *Benevento.*—Wine.

547. Codigoro, Municipality of, *Ferrara.*—Red wines.

548. Cojari, Avv. V. *Fivizzano* (*Massa and Carrara*).—Wines.

549. Conti, B. *Pontedera, Pisa.*—Wines, &c.

550. Cucchi, T. *Parma.*—Effervescing wine.

551. D'Antonio, S. *Ornano* (*Abruzzo Ulteriore I.*).—Wines.

552. De Angelis, M. *Isola* (*Abruzzo Ulteriore I.*).—Wine.

553. DE GORI, COUNT A. *Sienna.*—Wines.

554. DEL PRINO, DR. M. *Vesime, near Acqui (Alexandria).*—Wines.

555. DEMICHELI, G. *Novi (Alexandria).*—Wine.

556. DEMURTAS, E. *Lanusei (Cagliari).*—Wines.

557. DENEGRI, G. *Novi (Alexandria).*—Wine, some of it made of dried grapes.

558. DE RUBERTIS. L. *Lucito (Molise).*—Wine, made of dried grapes.

559. DI BLASIO, F. *Bagnoli (Molise).*—Wine.

560. FANTINI, G. *Comacchio (Ferrara).*—Wine.

561. FAVARE VERDIRAME, V. *Mazzara (Trapani).*—Wine.

562. FERRARINI, DR. A.—*Reggio (Emilia).*—Wines.

563. FLORENCE ROYAL TOBACCO MANUFACTORY.—Tobacco, snuff, and cigars.

564. FLORIS COIANA, P. *Cagliari.*—Wines.

565. GARAU CARTA, L. *Sanluri (Cagliari).*—Wine.

566. GAVIANO, A. *Lanusei (Cagliari).*—Wines.

567. GENTA, AVV. P. *Caluso (Turin).*—Wines.

568. GINNASI, COUNT D. *Imola (Bologna).*—Wine.

569. GRISALDI DEL TAIA, DR. C. *Sienna.*—Wine.

570. GROSSO, E. *Turin.*—Liqueurs.

571. JACCHINI, B. *Modena.*—Wine made without grapes.

572. LAI, L. *Lanusei (Cagliari).*—Wines.

573. LIPARI JUNTA FOR THE EXHIBITION (*Messina*).—Wines.

574. LORU, CAV. A. *Cagliari.* — Wine vinegar.

575. LUCCA ROYAL TOBACCO MANUFACTORY.—Tobacco, snuff, and cigars.

576. LUPINACCI, BARON L. & BROS. *Cosena (Calabria Citeriore).*—Wine, &c.

577. MADONNA, G. *Isola Abruzzo (Ulteriore I.).*—Wine.

578. MALMUSI, CAV. C. *Modena.*—Aromatic vinegar, 200 years old.

579. MARCHI, L. ROYAL ESTATE OF S. LORENZO, *Volterra (Pisa).*—Gin, and spirit from the arbutus.

580. MARINI DEMURO AVV. T. *Cagliari.*—Wine, and wine vinegar.

581. MARINI, P. *Cagliari.*—Wines.

582. MARYRETTI, G. *Savona (Genoa).*—Liqueurs.

583. MASSA AVV. C. *Casale (Alexandria).*—Wines.

584. MASSONE, M. *Cagliari.*—Red wine, and wine vinegar.

585. MAZZAROSA, MARQUIS G. B. *Lucca.*—Wine.

586. MELIS, B. *Quartu (Cagliari).*—Wine.

587. MELLUSI, G. *Torrecusi (Benevento).*—Wine.

588. MERLO, G. B. *Castelnuovo, Bormida (Alexandria).*—Wines.

589. MILAN ROYAL TOBACCO MANUFACTORY.—Tobacco, snuffs, and cigars.

590. MILIANI, F. *Peccioli (Pisa).*—Wines.

591. MILAZZO JUNTA FOR THE EXHIBITION.—Wine.

592. MODENA ROYAL TOBACCO MANUFACTORY.—Tobacco, snuff, and cigars.

593. MODENA SUB-COMMITTEE FOR THE EXHIBITION.—Wines, liqueurs, and vinegar.

594. MONCALVO, D. *Bisio, near Novi (Alexandria).*—Wines.

595. MONTEMERLO, E. *Novi (Alexandria).*—Red wine.

596. MONTERISI, G. *Bisceglie (Terra di Bari).*—Wines.

597. MONTINI, P. *Fabriani (Ancona).*—Wine and liqueurs.

598. MORANDO, I. & SONS, *Sampierdarena (Genoa).*—Liqueurs.

599. MORIANI, CAV. N. *Florence.*—Wines.

600. MORTINI, L. *S. Bartolomeo in Galdo (Benevento).*—Wine.

601. MURGIA, G. *Sanluri (Cagliari).*—Wines.

602. MURRU MURRU, A. *Sanluri* (*Cagliari*). —Wine.

603. NAPLES SUB-COMMITTEE FOR THE EXHIBITION.—Wine.

604. OREGGIA, C. *Savona* (*Genoa*).—Wine.

605. ORLANDO, G. *Pescolamazza* (*Benevento*).—Wine.

606. OUDART, L. *Genoa.*—Wines.

607. OVADA, MUNICIPALITY OF (*Alexandria*).—Red wines.

608. PACIFICO, G. *Salerno* (*Principato, Citeriore*).—Wines.

609. PALUMBO, O. *Trani* (*Terra di Bari*). —Wines, brandy, and rum.

610. PARENTE, C. *Monterocchetto* (*Benevento*).—Red wine.

611. PARENTE, G. *Ceppaloni* (*Benevento*). —Red wine.

612. PARMA SUB-COMMITTEE FOR THE EXHIBITION.—Wine and liqueurs.

613. PASOLINI, G. *Imola* (*Bologna*). — Wine.

614. PATRICO, DR. V. *Trapani.*—Wine.

615. PAVANELLI, G. *Ferrara.*—Red wine.

616. PENNACCHI, F. *Orvieto* (*Umbria*).— Wine.

617. PERINI, P. *Desenzano* (*Brescia*).— Liqueurs.

618. PERRA, A. *Cagliari.*—Wines.

619. PERUSINO, V. *Sandamiano d'Asti* (*Alexandria*).—Wines and vinegar.

620. PICCHIO, COUNT P. *Alexandria.*— Wines.

621. PIZZI, L. *Petrella* (*Molise*).—Wines.

622. PLACENZA SUB-COMMITTEE FOR THE EXHIBITION.—Collection of wines of the province.

623. POTENZIANI, HEIRS OF, *Rieti* (*Umbria*).—Wine.

624. PRAMPOLINI, A. *Reggio, Emilia.*— Vinegar made with mother a century old.

625. PRUNAS, CAV. R. *Bosa* (*Cagliari*).— Wines and vinegar.

626. RAPPIS, P. *Andorno Cacciorna* (*Novara*).—Ratafia.

627. RAVIZZA, G. BROTHERS, *Orvieto* (*Umbria*).—Wine.

628. REGGIO (EMILIA) SUB-COMMITTEE FOR THE EXHIBITION.—Collection of wines of the province.

629. RICASOLI, BARON B. *Florence.*—Wines.

630. RICCARDI STROZZI, C. *Florence.*—Wines.

631. RICCI, G. B. *Asti* (*Alexandria*).—Wines.

632. RICCI, L. *Bruno, near Aqui* (*Alexandria.*—Red wine.

633. RIDOLFI, MARQUIS C. *Florence.*—Wine.

634. RONCHI, P. *Florence.*—Vinegar.

635. ROTA & CO. *Alexandria.*—Beer.

636. SAGLIOCCA, G. *Pietra Elcina* (*Benevento*).—Wine.

637. SALIS, F. *Lanusei* (*Cagliari*). — Wines.

638. SANNA, V. *Selargius* (*Cagliari*).—Wines.

639. SANTI, DR. C. *Montalcino* (*Sienna*). —Wines and liqueurs.

640. SANTOSPACO, N. *Castiglione alla Pescara* (*Abruzzo Ulteriore I.*).—Muscat wine.

641. SATTA FLORIS, R. *Cagliari.*—Wines.

642. SAVORINI, F. *Persiceto* (*Bologna*).— Liqueurs.

643. SCAZZOLA, G. D. *Cascine, near Alexandria.*—Wines.

644. SERRA, DR. L. *Iglesias* (*Cagliari*).— Wine.

645. SESIMA, V. *Alexandria.*—Liqueurs.

646. SIRIGU, G. *Cagliari.*—Vermouth.

647. SPANO, L. *Oristano* (*Cagliari*).— Wine, alcohol, and tobacco.

648. SPENSIERI, G. *Ferrazzano* (*Molise*). —Wine.

649. SUPPA & CASOLINO, *Trani* (*Terra di Bari*).—Spirit of aniseed, &c.

650. TARTAGLIOZZI, G. *Isola* (*Abruzzo Ulteriore I.*).—Wine.

651. TORO, B. F. & E. *Chieti* (*Abruzzo Citeriore.*)—Centerba.

652. TORRI, L. *Bondino* (*Ferrara*).—Red wine.

653. TOTORO, N. *Archi* (*Abruzzo Citeriore*).—Boiled wine.

654. TURIN ROYAL TOBACCO MANUFACTORY.—Tobacco, snuff, and cigars.

655. ULRICH, D. *Turin.*—Vermouth, &c.

656. VALLINO BROTHERS, *Brà* (*Cuneo*).—Wines.

657. VARVALLO, F. *Asti* (*Alexandria*).—Wines.

658. VENTURA, V. *Castiglione alla Pescara* (*Abruzzo Ulteriore I.*).—Muscat wine.

659. VICENTINI, P. O. *Rieti* (*Umbria*).—Imitation champagne.

660. VIETRI, D. A. *Salerno* (*Principato Citeriore*).—Wines.

661. VITTONE, F. *Milan.* — Wines and liqueurs.

662. ZICCARDI, V. *Foiano* (*Benevento*).—Wine.

———

663. ARRANGA, G. *Serracapriola* (*Capitanata*).—Wine.

664. BALSAMO, G. N. *Catania.*—Tobacco leaves.

665. BARBAGALLO, S. *Catania.*—Wine.

666. BARI SUB-COMMITTEE FOR THE EXHIBITION (*Terra di Bari*).—Liqueurs.

667. BAZZIGER, L. & C. *Sassuolo.*—Liqueurs.

668. BERGAMI, P. *Ferrara.* — Common red wine.

669. BERTI, F. & G. *Rubbiera* (*Reggio, Emilia*).—Aniccione and alkermes.

670. BIFFI, P. *Milan.*—Liqueurs.

671. BOCCARDO, BROS.—*Candela* (*Capitanata*).—Wine.

672. BONI, E. *Modena.* — Vinegar 150 years old.

673. BUELLI, E. *Bobbio* (*Pavia*).—Collection of wines.

674. CAMPOLONGHI, G. B. *Parma.*—Rosolio.

675. CANTON, G. *Turin.*—Liqueurs.

676. CASALTRINITÀ, MAYOR OF (*Capitanata*).—Wine.

677. CASERTA SUB-COMMITTEE FOR THE EXHIBITION (*Terra di Lavoro*).—Wine.

678. CASSANO, F. *Gioja* (*Terra di Bari*).—Wine.

679. CASSINESI BENEDICTINE MONKS, *Catania.*—Wine.

680. CESENA, C. *Bari* (*Terra di Bari*).—Red wine.

681. CIANI, G. *Bisceglia* (*Terra di Bari*).—Wine.

682. COLLENZA, E. *Valenzano* (*Terra di Bari*).—Malvasìa wine.

683. COPPOLI, MARQUIS R. *Perugia* (*Umbria*).—Wine.

684. CORRIDI, G. *Leghorn.*—Alcohol.

685. COSTARELLI, M. *Catania.*—Wine.

686. D'AMBROGIO, L. *Deliceto* (*Capitanata*).—Wine.

687. D'AMBROSIO, V. *Sansevero* (*Capitanata*).—Wine.

688. DELL' ERMA, N. *Castellana* (*Terra di Bari*).—Wine.

689. DELL' ERMA V. *Castellana* (*Terra di Bari*).—Wine.

690. DELLA BELLA, D. *Vico* (*Capitanata*).—Muscat wine.

691. DEL TOSCANO, MARQUIS, *Catania.*—Wine.

692. DE MARTINO, G. *Salerno* (*Principato Citeriore*).—Wine.

693. DI GROSSI, G. *Riposto* (*Catania*).—Aniseed liqueur.

694. EBOLI, N. *Bari* (*Terra di Bari*).—Liqueurs.

695. FERRARA ROYAL CHAMBER OF COMMERCE.—Wine.

696. FASCIA, *S. Marco la Catola* (*Capitanata*).—Wine.

697. FERRAROTTO, G. *Catania.*—Muscat wine.

698. FERRI VITO, N. *Canneto* (*Terra di Bari*).—Musaglica wine.

699. FIAMINGO, G. B. *Riposto* (*Catania*).—Wine, alcohol.

700. FLORIO, BROS. *Asti* (*Alexandria*).—Wine, vermouth.

701. FORLÌ SUB-COMMITTEE FOR THE EXHIBITION.—Wine.

702. FREJAVILLE, A. *Cerignola* (*Capitanata*).—Wine.

703. GASPARRI, A. *Biccari* (*Capitanata*).—Wine.

704. GERVASIO, G. *Canneto* (*Terra di Bari*).—Zagarese wine.

705. GIORDANO, G. *Salerno* (*Principato Citeriore*).—Alcohol from the Arbutus unedo and Helianthus tuberosus.

706. GIOVINE, G. B. *Canelli* (*Alexandria*).—Wine.

707. GIULIANI, L. *S. Marco in Lamis* (*Capitanata*).—Wine.

708. GIUSTI, G. *Modena.*—Balsamic vinegar, 150 years old.

709. GIVONI, V. *Catania.*—Wine, vinegar.

———

2140. GENOESE ZERBI, D. *Reggio* (*Calabria Ulteriore I.*).—Wine.

2141. GUARNASCHELLI, CAV. G. *Broni* (*Pavia*).—Wine, vinegar.

2142. GUIDI, C. *Volterra* (*Pisa*).—Wine.

2143. LADERCHI, A. *Faenza* (*Ravenna*).—White wine.

2144. LELLA, G. *Messina.*—Wine.

2145. LOFARO, B. *Reggio Calabria* (*Ulteriore I.*)—Wine.

2146. LOMBARDI, *Sansevero* (*Capitanata*).—Wine.

2147. LUCERA, MAYOR OF (*Capitanata*).—Wine.

2148. MAJORANA, BROS., BARONS OF NICORRA, *Catania.*—Wine, vinegar, rum, collection of tobacco.

2149. MANCUSO, M. *Catania.*—Wine.

2150. MANGINI, F. *Modena.*—Vinegar, a century old.

2151. MARCHI, P. *Florence.*—Alkermes.

2152. MARTINI, L. *S. Bartolomeo in Galdo* (*Benevento*).—Wine.

2153. MASSA-CARRARA ROYAL TOBACCO MANUFACTORY.—Snuff, tobacco, cigars.

2154. MASSELLI, A. *Sansevero* (*Capitanata*).—Wine, and vinegar.

2155. MASSETTI, COUNT P. *Florence.*—Wine and vinegar.

2156. MELISSARI, F. S. *Reggio* (*Calabria Ulteriore I.*)—Wine.

2157. MENGAZZI, F. *Cesena* (*Forlì*).—Absinthe.

2158. MERLONI, BROS. *Bertinoro* (*Forlì*).—Wine.

2159. MESSINA SUB-COMMITTEE FOR THE EXHIBITION.—Wine.

2160. MONCADA, A. *Catania.*—Wine.

2161. MUNELLI GALILEI, L. *Pontedera* (*Pisa*).—Wine, vermouth.

2162. MURATO-SOLI, P.—Balsamic vinegar, 132 years old.

2163. NAPLES AND CAVA ROYAL TOBACCO MANUFACTORY.—Tobacco, cigars.

2164. NESII, A. *Reggio Calabria* (*Ulteriore I.*).—Wine.

2165. NOVA, D. A. *S. Agata* (*Capitanata.*—Wine.

2166. ORTANO, MUNICIPALITY OF (*Abruzzo Citeriore*).—Rosolio.

2167. PACCA, MARQUIS G. *Benevento.*—White and red wine.

2168. PAGANO, M. A. *Pisciotta* (*Principato Citeriore*).—Wine.

2169. PAGLIANO, F. *Asti* (*Alexandria.*)—Wine, and vinegar.

2170. PALIZZI, BARON C. *Reggio* (*Calabria Ulteriore I.*).—Wine.

2171. PARLATORE, E.—*Florence.*—Spirits and fruit of arbutus (Arbutus unedo).

2172. PARMA ROYAL TOBACCO MANUFACTORY.—Snuff, tobacco, cigars.

2173. PASCAZIO, V. *Mola* (*Terra di Bari*).—Wine.

2174. PETROSEMILO, A. *Ortona.* — Liqueurs.

2175. PICCARDI, G. *S. Casciano.*—Wine.

2176. PRATI, G. *Alexandria.*—St. Bernard's elixir.

2177. QUADRAT, L. *Genoa.*—White and red wine.

2178. Ravenna Sub-Committee for the Exhibition.—Wines.

2179. Reggio (Calabria) Sub-Committee for the Exhibition.—Wine, tobacco.

2180. Salimbeni, L. *Modena.*—Wine; aromatic vinegar 100 years old; liqueurs.

2181. Salvagnoli Marchetti, Cav. A. *Corniola, Empoli (Florence).*—Wine.

2182. Sant' Agostino Monastery, *Catania.*—Wine.

2183. Sta. Anna, Hermits of, *Catania.* —White wine.

2184. S. Francesco Monastery, *Catania.* —Wine.

2185. S. Placido Monastery, *Catania.* —White and red wine.

2186. S. Scolastica Monastery, *Bari (Terra di Bari).*—Liqueurs.

2187. Santoro, G. *S. Agata (Capitanata).* —White wine.

2188. Saraceno, V. *Catania.*—Vinegar.

2189. Savorelli, Marquis A. *Forli.*—Wine.

2190. Scocchera, S. *Canosa (Terra di Bari).*—White and red wine.

2191. Scuderi, F. M. *Catania.*—Zambria liqueur, alcohol.

2192. Sestri Ponente Royal Tobacco Manufactory, *near Genoa.*—Tobacco, cigars.

2193. Siniscalco, M. *Salerno (Principato Citeriore).*—Rectified spirits of Arbutus unedo, Asphodelus ramosus, and Pancratium maritirum; rum.

2194. Sisto, Baron A. *Catania.*—Wine.

2195. Spano, Cav. P. *Oristano (Cagliari).* —Vernaccia wine.

2196. Sylos Labini, V. *Bitonto (Terra di Bari).*—Wine.

2197. Tarantini, N. *Corato (Terra di Bari).*—Wine.

2198. Tarello, M. *Viverone (Novara.)*— Wine made of dried grapes.

2199. Tesi, L. *Pistoja.*—Wine, vermouth.

2200. Trapani, G. *Gallico (Calabria Ulteriore I.).*—Wine.

2201. Vagliasindi, F. *Catania.*—Vinegar.

2202. Zerbini, P. *Modena.*—Lambrusco wine.

2203. Li-Gresti, *Catania.* — Zambu liqueur.

CLASS IV.

Sub-Class A.

710. Alfani, C. *Nocera.*—Olive oil.

711. Astengo, Brothers, *Savona (Genoa).*—Manufactured wax.

712. Baffoni, V. *Fermo (Ascoli).*—Grapestone oil.

713. Bancalari, L. *Chiavari (Genoa).*— Olive oil.

714. Bartolini, *Cosenza (Calabria Citeriore).*—Olive oil.

715. Baracco, Brothers, *Cotrone (Calabria Ulteriore II.).*—Olive oil.

716. Beltrani, G. *Trani (Terra di Bari).*—Olive oil.

717. Bellella, Cav. E. *Capaccio (Principato Citeriore).*—Olive oil.

718. Botti, A. *Chiavari (Genoa).*—Olive oil.

719. Botteghi, A. *Chiavari (Genoa).*— Olive oil.

720. Carobbi, G. *Florence.*—Wax cakes, candles, tapers, and torches: spermaceti cakes, and candles.

721. Cattaneo, G. B. *Chiavari (Genoa).* —Olive oil.

722. Conti, B. *Villa Saletta, Pontedera (Pisa).*—Olive oil.

723. CONTI, E. BROTHERS, *Leghorn.*—Soap.

724. DANIELLI & FILIPPI, *Buti (Pisa).*—Olives, olive oil, olive kernels, &c.

725. DANZETTA, BARON, & BROTHERS, *Perugia (Umbria).*—Olive oil.

726. DE CESARE, A. *Penne (Abruzzo Ulteriore I.).*—Olive oil.

727. DE GORI, A. *Sienna.*—Olive oil; wax.

728. DEMURTAS, E. *Lanusei (Cagliari).*—Olive oil and wax.

729. DE RUBERTIS, L. *Lucito (Molise).*—Olive oil.

730. DUNANT, G. M. *Milan.*—Soap and perfumery.

731. FRANCIOSI, P. *Terriciola (Pisa).*—Olive oil.

732. FURLANI, G. *Florence.*—Soap.

733. GABRIELE, DR. A. *S. Bartolomeo in Galdo (Benevento).*—Collection of oils from the neighbourhood.

734. GAVIANO, A. *Lanusei (Cagliari).*—Olive oil.

735. GHIGO, C. *Saluzzo (Cuneo).*—Wax candles.

736. GIORDANO, G. *Naples.*—Collection of oils.

737. GIUSTI, N. *Pisa.*—Olive oil, wax.

738. GRISALDA DEL TAIA, *Sienna.*—Olive oil.

739. LUPINACCI, BARON L. & BROTHERS, *Cosenza (Calabria Citeriore).*—Olive oil.

740. MAJORANA, BROTHERS, BARONS DI NICORRA, *Catania.*—Unbleached wax; olive oil; hard soap.

741. MANGANONI, L. *Milan.*—Stearine candles, large tapers, and soap.

742. MARCHI, L. *Volterra (Pisa).*—Pistachio nut oil.

743. MASTIANI SCIAMANNA, MARQUIS C. *Pisa.*—Olive oil.

744. MAZZULLO G. *Mandanici (Messina).*—Olive oil.

745. MAZZAROSA, MARQUIS G. B. *Lucca.*—Olive oil.

746. MESSINA SUB-COMMITTEE FOR THE EXHIBITION.—Collection of the oils of the province of Messina.

747. MILAZZO JUNTA FOR THE EXHIBITION.—Olive oil.

748. MILIANI, F. *Peccioli (Pisa).*—Olive oil.

749. MILAN CHAMBER OF COMMERCE.—Collection of the oils manufactured in the province of Milan.

750. MINUTOLI TEGRIMI, COUNT E. *Lucca.*—Olive oil.

751. MODENA SUB-COMMITTEE FOR THE EXHIBITION.—Grape-stone and other oils.

752. MOSCERO, G. *Cosenza (Calabria Citeriore).*—Oil.

753. NOBERASCO & AQUARONI, *Savona (Genoa).*—Soap.

754. OREGGIA, C. *Savona (Genoa).*—Olive oil.

755. ORLANDO, G. *Pescolamazza (Benevento).*—Olive oil.

756. OTTOLINI BALBANI, COUNTESS C. *Lucca.*—Olive oil.

757. PANCANI, BROTHERS, *Florence.*—Olive oil and other soaps.

758. PENSA, F. *Teramo (Abruzzo Ulteriore I.).*—Wax candles.

759. PIERI PECCI, G. *Sienna.*—Olive oil.

760. PISTIS, G. *Elini (Cagliari).*—Olive oil.

761. PRUNAS, CAV. R. *Bosa (Cagliari).*—Olive oil.

762. REGGIO (EMILIA) AGRICULTURAL SOCIETY.—Linseed and other oils.

763. RICCARDI STROZZI, C. *Florence.*—Olive oil.

764. SARDINI, G. *Lucca.*—Olive oil.

765. SCUDERY, A. *Messina.*—Olive oil.

766. SPANO, L. *Oristano (Cagliari).*—Olive and other oils; wax.

767. SQUARCI, E. *Leghorn.* — Stearine candles.

768. TACCHI, G. *Bergamo.*—Wax candles, torches, and cakes.

769. TALENTI, COUNT L. *Lucca.*—Olive oil.

770. TURCHI, L. & Co. *Ferrara.*—Olive and other oils ; toilet soap.

771. VACCARO, L. *Cosenza (Calabria Citeriore).*—Oil obtained from the ground nut.

———

772. AGAZZOTTI, AVV. F. *Modena.*—Walnut oil.

773. ALBIANI, CAV. F. *Pietrasanta (Lucca).*—Olive oil.

774. ARPINI, CAV. E. *Ascoli.*— Olive oil.

775. ARRANGA, G. *Serracapriola (Capitanata).*—Maccaroni, capers.

776. BARBAGALLO, S. *Catania.*—Hard and soft soap, tallow, linseed oil.

777. BARBATO, N. *S. Agata (Capitanata).*—Olive oil.

778. BASTONI, V. *Turin.*—Grape-stone oil.

779. BAZZIGER, L. & Co. *Sassiuolo (Modena).*—Liqueurs.

780. BERGAMO SUB-COMMITTEE FOR THE EXHIBITION.—Wax candles.

781. BERNARDI, F. *Sienna.*—Olive oil.

782. BOCCARDO, BROS. *Candela Foggia (Capitanata).*—Olive oil.

783. CAGLIARI SUB-COMMITTEE FOR THE EXHIBITION.—Olive, almond, and lentisc oil ; wax.

784. CANOSA, MUNICIPALITY OF *(Terra di Bari).*—Olive oil.

785. CANTALLAMESSA, I. *Ascoli.*—Olive oil.

786. CARDUCCI, A. *Taranto (Terra di Otranto).*—Olive oil.

787. CASALTRINITÀ, MAYOR OF *(Capitanata).*—Olive oil.

788. CASERTA SUB-COMMITTEE FOR THE EXHIBITION *(Terra di Lavoro).*—Olive and ground-nut oil ; oil of the Holcus cernuus and Sorghum saccharatum.

789. CASTORINA, *Catania.*—Soft soap.

790. COSENTINO, S. *Catania.*—Soft soap.

791. COSTANTINO, G. *S. Marco de' Cavoti (Benevento).*—Olive oil.

792. CORSINI, S. *Casciano (Pisa).*—Olive oil.

793. D'AMBROSIO, L. *Deliceto (Capitanata).*—Olive oil.

794. DE BIASE, G. *S. Marco la Catola (Capitanata).*—Olive oil.

795. DE CATALDIS, O. *Giffoni (Principato Citeriore).*—Olive oil.

796. D'ERCHIA, A. *Monopoli (Terra di Bari).*—Olive oil, soap.

797. DELLA BELLA, D. *Vico (Capitanata).*—Olive oil.

798. DELL 'ERMA, V. *Castellana (Terra di Bari).*—Common olive oil.

799. DELLI SANTI, F. *Manfredonia (Capitanata).*—Olive oil.

800. DI RIGNANO, MARQUIS, *Foggia (Capitanata).*—Olive oil.

801. FREJAVILLE, A. *Cerignola (Capitanata).*—Olive oil.

802. GASPARRI, A. *Biccari (Capitanata).*—Olive oil.

803. GIOIA, A. *Corato (Terra di Bari).*—Olive oil.

804. GIOVANDONATO, O. *Benevento.*—Olive oil.

805. GIRARDI, M. *Turin.*—Olive, hazelnut, walnut, colza, linseed, and castor oil.

806. IDONE, G. *Lecce (Terra di Otranto).*—Olive oil.

807. LAMENACO, L. & G. *Corato (Terra di Bari).*—Olive oil.

808. LANZA, BROS. *Turin.* — Stearine, and stearine candles.

809. LOFARO, F. *Catania.*—Soft soap.

810. MACERATA SUB-COMMITTEE FOR THE EXHIBITION.—Olive and olive-kernel oil.

811. MANNI, D. *Tocco (Abruzzo Citeriore).*—Fine olive oil.

812. MASETTI, COUNT P. *Florence.*—Olive oil.

813. MASSELLI, A. *S. Severo (Capitanata).*—Olive oil.

814. MELISSARI, F. S. *Reggio (Calabria Ulteriore I.).*—Olive oil.

815. MERCATILE, M. *Ascoli.*—Olive oil.

816. MILELLA, G. *Bari* (*Terra di Bari*).—Fine olive oil.

817. MUNNELLI, GALILEI, L. *Pontedera* (*Pisa*).—Olive oil.

818. MAZZUCHETTI, E. *Turin.*—Castor oil.

819. NALDINI, B. *Florence.*—Olive oil.

———

2205. NIEDDA DI STA. MARGHERITA, COUNT P. *Cagliari.*—Olive oil.

2206. NOVI, D. *S. Agata* (*Capitanata*).—Olive oil.

2207. ORTONA, MUNICIPALITY OF (*Abruzzo Citeriore*).—Olive oil.

2208. PACCA, MARQUIS G. *Benevento.*—Olive oil.

2209. PALIZZI, BARON C. *Reggio* (*Calabria Ulteriore I.*).—Olive oil.

2210. PALUMBO, O. *Trani* (*Terra di Bari*).—Gum of the olive tree; tamarisk.

2211. PAOLELLA, G. *Castelluccio, Val Maggiore* (*Capitanata*).—Olive oil.

2212. PASCAZIO, V. *Mola* (*Terra di Bari*).—Olive oil.

2213. PAULUCCI, MARQUIS G. B. *Forlì.*—Oil from the seeds of the *Kolreuteria paniculata.*

2214. PESARO AGRICULTURAL SOCIETY.—Olive, and olive-kernel oil.

2215. PESCI, G. *Fuligno* (*Umbria*).—Olive oil.

2216. PICCARDI, GIUSEPPE, *S. Casciano.*—Olive oil.

2217. PORTO MAURIZIO SUB-COMMITTEE FOR THE EXHIBITION.—Olive oil.

2218. REGGIO (CALABRIA) SUB-COMMITTEE FOR THE EXHIBITION.—Olive oil.

2219. RICASOLI, BARON B. *Brolio* (*Sienna*).—Olive oil.

2220. RIGNANO, MARQUIS OF, *Foggia* (*Capitanata*).—Olive oil.

2221. ROSPIGLIOSI, PRINCE, *Pistoja* (*Florence*).—Olive oil.

2222. SANSONE, P. *Cagnano* (*Capitanata*).—Olive oil.

2223. SANTORO, G. *S. Agata* (*Capitanata*).—Olive oil.

2224. SAULLI, L. *Pisciotta* (*Principato Citeriore*).—Olive oil.

2225. SAVORELLI, MARQUIS A. *Forlì.*—Stearine, and stearine manufactures.

2226. SCOCCHERA, S. *Canosa* (*Terra di Bari*).—Olive oil.

2227. SERRACAPRIOLA, MUNICIPALITY OF (*Capitanata*).—Olive oil.

2228. SERVENTI (HEIRS OF), *Parma.*—Wax, and wax candles.

2229. SGARIGLIA, M. *Ascoli.*—Olive oil.

2230. SYLOS LABINI, V. *Bitonto* (*Terra di Bari*).—Olive oil, and olive-kernel oil.

2231. TESI, L. *Pistoja* (*Florence*).—Olive, and other oils.

2232. VALERI & Co. *Legnago* (*Verona*).—Cold-drawn castor oil, &c.

2233. AICARDI, F. & Co. *Bari* (*Terra di Bari*).—Fine, filtered, and common olive oil.

2234. LAURI, COUNT J.—Olive oil.

2235. PISA SUB-COMMITTEE FOR THE EXHIBITION.—Collection of common, best, and washed olive oil of the province of Pisa.

SUB-CLASS B.

820 BARRACCO, BROTHERS, *Cotrone* (*Calabria Ulteriore II.*).—Sheep's fleeces.

821. BENTIVOGLIO, CAV. C. *Modena.*—Sheep's fleeces.

822. BERTONE, *Turin.*—Sheep's fleeces.

823. BINDA GRUGNOLA & Co. *Milan.*—Combs.

824. BUSSOLATI, BROTHERS, *Parma.*—Silk-worm cocoons.

825. CAGLIARI SUB-COMMITTEE FOR THE EXHIBITION.—Coral, fished on the coast of Sardinia; guano; burned dung.

826. CAMPI, COUNT G. *Dovadola* (*Florence*).—Silkworm cocoons.

827. CARRO, MARIANNA, *Cagliari.*—Bunch of flowers, made of shells.

828. CHISOLI, A. *Brignone* (*Bergamo*).—Cocoons and thrown silk.

829. CUCCHI, T. *Parma.*—Cocoons.

830. DONINI, S. *Bologna.* — Animal manure.

831. FINO, L. *Turin.*—Albumen, hematosine, &c.

832. GABRIELE, DR. A. *S. Bartolomeo in Galdo (Benevento).*—Wool.

833. GARAU, CARTA L. *Sanluri (Cagliari).*—Opened cocoons.

834. MACERATA SUB-COMMITTEE FOR THE EXHIBITION.—Sheep's fleeces.

835. MILAN.—ROYAL LOMBARD SCIENTIFIC INSTITUTION.—Collection illustrating the metamorphoses of the silkworm.

836. MONTALTI, E. *Bologna.*—Gelatine and glue.

837. ORLANDO, G. *Pescolamazza (Benevento).*—Sheep's fleeces.

838. PANICHI, *Perugia (Umbria).*—Carded wool.

839. PIZZETTI, F. *Parma.*— Silkworm cocoons, silkworm moths, and raw silk.

840. PONTICELLI, G. *Grosseto.*—Long merino wool.

841. PUPILLI, G. *Pontedera (Pisa).*—Glue.

842. SICCARDI, BROS. *Mondovi (Cuneo).*—Silkworm cocoons.

843. SOMMARIVA, B. *Palermo.*—Glue.

844. SPANO, L. *Oristano (Cagliari).*—Wool, silkworm cocoons, &c.

845. TONI, F. *Perugia (Umbria).*—Cocoons and raw silk.

846. VEGNI, L. *Città di Castelio (Umbria).*—Glue.

847. VETERE, G. *Gerchiara (Calabria Citeriore).*—Wool.

———

848. CASSANO, F. *Gioia (Terra di Bari).*—Wool.

849. COSTANTINO, G. *S. Marco de' Cavoti (Benevento).*—Wool.

850. GALANTI, PROF. A. *Perugia (Umbria).*—White Chinese cocoons.

851. GIOVANETTI, G. BROS. *Pisa.*—Bone buttons.

852. PACCA, MARQUIS G. *Benevento.*—Wool.

853. VACCARO, L. *Cosenza (Calabria Citeriore).*—Calabrian cantharides (*Cantharis vesicatoria*).

SUB-CLASS C.

N.B.—All the specimens of cotton are placed together in the collection formed by the Royal Italian Commission.

880. ANZI, DON M. *Bormio (Sondrio).*—Collection of lichens.

881. AUGIAS, S. *Tempio (Sassari).*—Three kinds of archil.

882. AVENTI, COUNT F. M. *Ferrara.*—Hemp.

883. AYMERICH, I. *Cagliari.*—Bark, and acorns of the cork oak.

884. BAFFONI, V. *Fermo (Ascoli).*—Rapeseed cakes, &c.

885. BARATELLI, BARON, *Ferrara.* — Green hemp.

886. BARTOLINI, C. *Cosenza (Calabria Citeriore).*—Indigenous cotton.

887. BARTOLINI, F. *Corigliano (Calabria Citeriore).*—Prepared flax.

888. BELLELLA, G. *Salerno (Principato Citeriore).*—Madder.

889. BELLELLA, CAV. E. *Capaccio (Principato Citeriore).*—Madder.

890. BELTRAMI, CAV. P. *Cagliari.*—Woods, charcoal, and cork.

891. BELTRANI, G. *Trani (Terra di Bari).*—Mustard seed, castor oil seed, and cotton.

892. BENZI, T. *Carpi (Modena).*—Straw plait.

893. BERNARDUSI, M. *Ferrara.*—Green hemp.

894. BIAVATI, P. *Crevalcore (Bologna).*—Green hemp.

895. BOLOGNA AGRICULTURAL SOCIETY.—Preparation of the *Botys silicealis*, an insect

that blights the hemp plant: by Prof. Bertoloni.

896. BONORA, *Ferrara*.—Raw hemp.

897. BOTTER, PROF. F. *Bologna*.—Collection illustrating the cultivation of hemp in the Emilia; hemp plants blighted and diseased; hemp seed oil, &c.

898. BURGARELLA, A. *Trapani*.—Sumac.

899. CAGLIARI SUB-COMMITTEE FOR THE EXHIBITION.—Flax, hemp, lentisc, madder, sumac, saffron, gum, cork, castor oil fruit, white and tawny yellow Siamese cotton (*Gossypium Siamense*); straw baskets.

900. CALANDRINI, PROF. F. *Florence*.—Collection of 185 species of wood, indigenous or acclimatised in Tuscany.

901. CAMPOBASSO SUB-COMMITTEE FOR THE EXHIBITION.—Hemp-seed.

902. CAVALIERI, P. *Ferrara*.—Hemp.

903. CERTANI, A. *Bologna*.—Hemp-seed and hemp.

904. CHERICI, N. S. *Sepolcro* (*Arezzo*).—Specimens of timber woods; woad (*Isatis tinctoria*), woad plant and seed.

905. CREMONA SUB-COMMITTEE FOR THE EXHIBITION. — Collection of building and other woods, grown in the province.

906. CRIPPA, IDA, *Florence*.—Pine cones, pine seeds, and pine seed oil.

907. D'ALESSIO, G. *Capaccio* (*Principato Citeriore*).—Madder.

908. DE LUCA, P. S. *Giovanni in Fiore* (*Calabria Citeriore*).—Pitch, pine-resin, turpentine, and spirits of turpentine.

909. FACCHINI, BROTHERS, *Bologna*.—Dressed hemp.

910. FAVARA, T. *Trapani*. — Sumac leaves; white and tawny Siamese cotton (*Gossypium Siamense*), and seeds of the same.

911. FERRARA ROYAL CHAMBER OF COMMERCE.—Hemp-seed and hemp.

912. FERRARA AGRICULTURAL SCHOOL.—Green hemp.

913. FIORELLI, G. *Salò* (*Brescia*).—Sumac leaves.

914. FRASSELLI, BROTHERS, *Castrocaro, Florence*.—Flax.

915. FRÖLICH & CO. *Castellamare* (*Naples*).—Garancine and madder.

916. GAROVAGLIO, S. Director of the Botanic Garden, *Pavia*.—Collection of dried lichens.

917. GATTI, A. *Cosenza* (*Calabria Citeriore*).—Thread and cloth, made of the fibres of the broom.

918. GENNARI, P. *Cagliari*.—Herbarium of 200 dried rare Sardinian plants, &c.

919. GRANOZIO, D. *Salerno* (*Principato Citeriore*).—Madder, and raw cotton.

920. GUIDA, BROTHERS, *Gargarengo* (*Novara*).—Colza, rape-seed, flax, and lupins.

921. GULINELLI, COUNT G. *Ferrara*. — Hemp stalks.

922. LORU, PROF. CAV. A. *Cagliari*.—Saffron, tamarisk leaves, and bulbs of the saffron-crocus.

923. LUPINACCI, BARON L. BROTHERS, *Cosenza* (*Calabria Citeriore*).—Flax stalks and partly dressed flax.

924. MACCAFERRI & CO. *Bologna*.—Specimens of hemp, softened by machinery invented by the exhibitor.

925. MACERATA SUB-COMMITTEE FOR THE EXHIBITION. — Sumac leaves; merino and *vissana* wool; collection of woods.

926. MAFFEI, CAV. N. *Volterra* (*Pisa*).—Juniper berries, myrtle and lentisc leaves, collection of woods.

927. MAGGIORANA, F. *Milan*.—Sumac.

928. MAJORANA, BROTHERS, BARONS DI NICORRA, *Catania*. — Sumac, mustard, and other seeds; collection of Sicilian grown cotton.

929. MARATTI, V. *Benevento*.—Woods.

930. MERCATILI, COUNT M. *Ascoli*.—Hemp.

931. MODENA SUB-COMMITTEE FOR THE EXHIBITION.—Collection of the various kinds of timber of the province, bark, &c.

932. MOZZANO, MUNICIPALITY OF (*Ascoli*).—Cotton seed.

933. MUSICÒ, D. *Messina*.—Fibre of the American aloe.

934. NAPLES SUB-COMMITTEE FOR THE EXHIBITION.—Madder; white Siamese and

common cotton; dried specimen of the cotton plant.

935. NAPLES ROYAL FOUNDRY.—Specimens of the woods of the south of Italy.

936. NIEDDA DI STA. MARGHERITA, COUNT P. & BROTHER, *Cagliari.*—Sumac.

937. PACIFICI, COUNT D. *Ascoli.*—Hemp.

938. PACIFICO, G. *Salerno (Principato Citeriore).*—Madder.

939. PAGANELLI, DR. E. *Castrocaro (Florence).*—Saffron, and saffron-crocus seed.

940. PALLOTTA, C. S. *Giuliano di Sepino (Molise).*—Hemp-seed and hemp.

941. PALLOTTA, S. *Orvieta (Umbria).*—Dressed hemp.

942. PALUMBO, O. *Trani (Terra di Bari).*—Asclepias, or vegetable silk.

943. PASI, G. *Ferrara.*—Hemp-seed and dressed hemp.

944. PASOLINI, G. *Imola (Bologna).*—Hemp, raw and dressed.

945. PAVANELLI, G. *Ferrara.*—Hemp.

946. PICCALUGA, G. *Cagliari.*—Collection of woods of exotic trees acclimatised and grown around Cagliari.

947. PIÙ, F. *Lanusei (Cagliari).*—Madder.

948. RAVENNA SUB-COMMITTEE FOR THE EXHIBITION.—Wood from the pine forests around Ravenna.

949. RAMO, S. *Laconi (Cagliari).*—Holly, and cork bark.

950. RIETI COMMITTEE FOR THE FLORENCE EXHIBITION (*Umbria*).—Castor-oil seeds, gall-nuts, mustard, and various gums.

951. REGGIO (EMILIA) AGRICULTURAL SOCIETY.—Plait of willow bark, and plait made of the glumes or husks of Indian corn.

952. RENUCCI, V. *Parma.*—Straw for plaiting.

953. REVEDIN, COUNT G. *Ferrara.*—Hemp.

954. RIZZOLI, R. *Bologna.*—Dressed hemp.

955. SAGLIOCCA, G. *Pietralcina (Benevento).*—Wild madder.

956. SALADINI, COUNT M. *Ascoli.*—Hemp.

957. SERRA, L. *Iglesias (Cagliari).*—Linseed.

958. SONDRIO, ADMINISTRATION OF THE FORESTS OF.—Indigenous woods of the Valtellina, &c.

959. SPANO, L. *Oristano (Cagliari).*—Collection of woods, gums, archil, bark, seeds, berries, manufactured articles, and their materials, &c.

960. SACCONI, COUNT E. *Ascoli.*—Hemp.

961. TEDESCHI, L. I. *Reggio, Emilia.*—Brooms and their materials.

962. TIMON, CAV. A. *Cagliari.*—Various kinds of building wood.

963. TORRI, L.B. *Ferrara.*—Hemp-stalks.

964. VARSI, G. *Oristano (Cagliari).*—Cork.

965. VINCI, M. *Carpi (Modena).*—Straw plait.

966. VINCENZI, P. *Carpi (Modena).*—Straw plait.

967. VONWILLER & CO. *Naples.*—Cork.

———

968. AMICARELLI, V. *Montesantangelo (Capitanata).*—Pine resin.

969. ARNAUDON, PROF. G. *Turin.*—Collection of woods; dyeing and tanning substances.

970. ASCOLI SUB-COMMITTEE FOR THE EXHIBITION.—Weld (*Reseda luteola*).

971. AVELLINO SUB-COMMITTEE FOR THE EXHIBITION.—Collection of woods.

973. BARACCO, BROS. *Cotrone (Calabria Ulteriore II.).*—Flax from Siconia.

974. BERTERO, A. & GALLA, G. B. *Carmignola (Turin).*—Hemp for ropes.

975. BERTONI, MARQUIS, *Turin.*—Flax.

976. BIANCAVILLA, MUNICIPALITY OF (*Catania*).—White Siamese cotton, cultivated at Biancavilla.

977. BRINDISI SUB-COMMITTEE FOR THE EXHIBITION (*Terra di Otranto*).—Best white and yellow Siamese cotton.

978. BISCARI, PRINCE OF, *Catania.*—White Siamese cotton, with the seeds, cultivated at Paterno.

979. CANOSA, MUNICIPALITY OF (*Terra di Bari*).—White cotton (*G. herb*).

980. CASERTA SUB-COMMITTEE FOR THE EXHIBITION (*Terra di Lavoro*).—Raw and dressed hemp, tow; safflower, madder, &c.

981. CATANZARO SUB-COMMITTEE FOR THE EXHIBITION (*Calabria Ulteriore II.*).— White and tawny yellow Siamese cotton (*Gossypium Siamense*), cultivated in the province, and seeds of the same.

982. CAMPOBASSO SUB-COMMITTEE FOR THE EXHIBITION.—Oak wood, &c.

983. FINZI, M. *Carpi* (*Modena*).—Willow bark for making chip bonnets.

984. FOGGIA SUB-COMMITTEE FOR THE EXHIBITION (*Capitanata*).—White and tawny Siamese cotton (*G. Siamense*); common white cotton (*G. herbaceum*).

985. FOGGIA ROYAL ECONOMIC SOCIETY (*Capitanata*).—Riga and Calabrian flax.

986. FORLI SUB-COMMITTEE FOR THE EXHIBITION.—Hemp.

987. GIRGENTI SUB-COMMITTEE FOR THE EXHIBITION.—White Siamese cotton; white cotton.

988. GIUDICE, GASPARE, *Girgenti.*—Powdered sumac.

989. GUIDI, C. *Volterra* (*Pisa*).—Leaves of the *Martino*, for tanning; bird-lime.

990. HENKEL, L. *Florence.*—Waterproof cloth, flannel and cambric; waterproof shawl.

991. LIBRA, F. *Catania.*—White cotton.

992. LICATA, MUNICIPALITY OF (*Girgenti*). —White cotton (*G. herbaceum*).

993. MAZZARA, MUNICIPALITY OF (*Trapani*).—Cotton seed.

994. MENFI, MUNICIPALITY OF (*Girgenti*). —White and tawny yellow cotton.

995. MONTALLEGRO, MUNICIPALITY OF (*Girgenti*).—White cotton (*G. herb.*).

996. MUGGIONI, A. *Placenza.*—Roots of the *Andropogon Ischœmum*, for making brooms, brushes, &c.

997. ORLANDO, G. *Pescolamazza* (*Benevento*).—Wool.

998. ORTONA, MUNICIPALITY OF (*Abruzzo Citeriore*).—Oak wood.

999. PACE, V. *Castrovillari* (*Calabria Citeriore*).—White and tawny yellow Siamese cotton (*G. Siam.*).

————

2240. PASQUI, G. *Forlì.*—Hops.

2241. REGGIO AGRICULTURAL ASSOCIATION (*Reggio, Emilia*).—Collection of woods; flax, rape, and colza seed.

2242. REGGIO (CALABRIA) SUB-COMMITTEE FOR THE EXHIBITION.—Sumac, broom tow, and thread; broom thread.

2243. ROYAL ITALIAN COMMISSION.—Collection of Italian grown cotton, made by the Royal Italian Commission, and described under the head of the different producers. See Nos. 294, 886, 891, 899, 910, 919, 928, 934, 967, 976, 977, 978, 979, 981, 984, 987, 991, 992, 993, 994, 995, 999, 2244, 2245, 2247, 2248, 2249, 2251, 2253, 2266, 2267, 2268.

2244. SCIACCA, MUNICIPALITY OF (*Girgenti*),—White cotton.

2245. SALERNO SUB-COMMITTEE FOR THE EXHIBITION (*Principato Citeriore*).—White Siamese and common white cotton.

2246. SAVONA JUNTA FOR THE EXHIBITION, *Savona* (*Genoa*).—Wooden hoops of various kinds.

2247. SCOCCHERA, S. *Canosa* (*Terra di Bari*).—White cotton (*G. herb.*) and cotton seed; madder.

2248. SICULIANA, MUNICIPALITY OF (*Girgenti*).—White cotton (*G. herb.*).

2249. SINASTRA, C. *Noto.* — White Siamese cotton, with the seeds; tawny yellow Siamese cotton, with the seeds.

2250. SINISCALCO, M. *Salerno* (*Principato Citeriore*).—Pancratium maritimum, preserved in spirits, fecula of the same; teazles, &c.

2251. TAORMINA, MUNICIPALITY OF (*Messina*).—White Siamese cotton.

2252. TRUFFELI DI TREVIGLIO.—Flax.

2253. UGO, G. MARQUIS DELLE FAVARE, *Catania.*—White Siamese and common white cotton ; common cotton, with the seeds.

2254. BALDI, G. *Florence.*—Lasts for boots and shoes.

2255. BUGGIANO, A. *Placenza.*—Roots of *Andropogon Ischæmum,* for making brooms.

2256. CATANIA SUB-COMMITTEE FOR THE EXHIBITION.—Brooms and rope.

2257. COSTANTINO, T. *Ascoli.*—Flax.

2258. FORCALLI, G. *Salò (Brescia).*—Sumac.

2259. GIORDANO, E. *Salerno (Principato Citeriore).*—Flax.

2260. ITALIAN CRYPTOGAMICAL SOCIETY, *Genoa.*—Herbarium of Italian Cryptogamous plants.

2261. MUNAFÒ, G. *Sicily.*—Sumac.

2262. PESARO AGRICULTURAL ACADEMY. —Woods, fungi for making tinder, &c.

2263. PICCHI, P. *Leghorn.* — Cork and corks.

2264. SEMMOLA, CAV. F. *Naples.* — Woods.

2265. SACCONI, COUNT E. *Ascoli.*—Hemp.

2266. PATERNÒ, MUNICIPALITY OF, *Catania.*—White Siamese and common cotton.

2267. CATANIA, MUNICIPALITY OF. — White Siamese cotton.

2268. DILGH, E. & Co. *Catania.* — White Siamese and common white cotton, and seeds of the same.

SUB-CLASS D.

2273. FONSIO P. *Palermo.* — Essential oil of lemons and oranges.

2274. GARDNER, ROSE, & Co. *Palermo.* —Essential oils of lemons and oranges.

2275. PRANZINI, L. *Florence.* — Perfumery.

CLASS V.

1000. FUSINI, V. *Pavia.*—Atmospheric railway, with valveless tube

1001. GRIMALDI, F. *London.*—Rotatory steam-boiler.

1002. PIETRARSA ROYAL WORKS, *Portici (Naples).*—Locomotive with 6 coupled wheels.

1003. TURIN ENGINEERING SCHOOL. — Model of a locomotive in section.

1004. VANOSSI, G. *Chiavenna (Sondrio).* —Steam engines, for railways and steamers, on a new principle.

1005. VELINI & Co. *Verzaro (Milan).*— Model of locomotive tender, applicable to steep inclines.

1006. VINCENZI, E. *Modena.*—Model of an electric signal to prevent the collision of railway trains.

1007. AGUDIO, T. *Turin.*—Model of machinery for railway trains on inclined planes, set in motion by two stationary engines.

1008. LUÉ, A. *Milan.*—Wooden model of a horse tram-road ; model of a new system of rails, capable of being employed successively on four sides.

1009. SIPRIOT, C. *Milan.*—Tarpaulin for railway wagons.

CLASS VI.

1014. BERTI, P. *Milan.*—State carriage, with silver mountings; harness, with silver and chased steel mountings for the same.

CLASS VII.

SUB-CLASS A.

1020. BONELLI, CAV. G. *Turin.*—Electric-loom, for weaving any kind of material.

1021. BOSSI, L. *Milan.*—Beam for warping silk, &c. with screw movement.

1022. DELPRINO, M. *Sesime (Alexandria).*—Model of new method of winding cocoons.

1023. DELAPIERRE, F. *Naples.* — Steel card for silk manufacture.

1024. FORNARA, G. *Turin.*—Combs for weaving.

1025. FRIGERIO, G. *Molteno (Como).*—New apparatus for preparing cocoons for winding.

1026. GUPPY & PATISSON, *Naples.* — Spinning machine.

1027. SANROME, M. BROTHERS, *Como.*—Cards for weaving.

1028. SILVATICI, G. *Vico Pisano (Pisa).*—Cards for wool and cotton.

1029. VINCENZI, E. *Modena.*—Loom.

1030. CAMPI, COUNT G. *Bellosguardo, Florence.*—New system of alternate motion for spinning machines.

1031. MANGANO, A. & SON, *Messina.*—Frame for winding and throwing silk.

1032. ROSSI, P. & Co. *Bibbiena (Arezzo).*—Spinning frames and spindles.

SUB-CLASS B.

1036. CIANFERONI, A. *Florence.*—Blocks for printing oil-cloth.

1037. DEI, F. *Florence.* — Brass and wooden blocks for printing cotton stuffs and handkerchiefs.

1038 PAVASI, G. *Pavia.*—Universal tap for manufacturing screws.

1039. PISA SUB-COMMITTEE FOR THE EXHIBITION.—Weaving machine and implements employed in the manufacture of cotton.

1040. RICCI, R. *Leghorn.*—Chest of carpenter's tools.

1041. VEROLE, P. *Turin.* — Universal fixed tool-holder, for planing machines.

1042. ROSSI, P. & Co. *Bibbiena (Arezzo).*—Cheap cooperage and other wood-work.

1043. TREVES, M. *Florence.*—Universal tap for manufacturing screws of any size.

1044. SOMMEILLER, GRANDIS, & GRATTONE, *Turin.* — Machinery employed in making the tunnel through Mont Cenis (drawing); topographical map and section of Mont Cenis, indicating the line of railway.

CLASS VIII.

1050. BERNARD, A. *Naples.* — Light-house reflectors.

1052. CORTI, D. *Milan.*—New portable pump.

1053. FUSINA, V. *Pavia.*—Machines for filling up cart-ruts, and clearing away snow from streets; gear applicable to rotatory motion under various circumstances; model of pontoon for railway bridges; jets for fire-engines, &c.

1054. LANCIA, G. *Turin.*—Meat chopping machine, machine for filling sausages.

1055. MACRY, H. & Co. *Naples.*—Steam-engine cylinder.

1056. MURATTI, COL. A. *Naples.*—Model of a crane.

1057. PEREZ, V. C. *Lanciano (Abruzzo Citeriore).*—Drawing of water mill.

1058. PIETRARSA ROYAL WORKS, *Portici (Naples).*—Toothed wheels; steam-case and admission valve for a large steam-engine; large cast-iron shaft for screw propeller, &c.

1060. ACQUADIO, B. *Biella (Novara).*—Firework apparatus.

1061. GAUTHIER, A. *Turin.*—New system of corking machine.

1062. LEVINSTEIN & Co. *Milan.*—Lustring machine for giving a gloss to dyed silk.—*See* pages 60 and 61.

1063. ROYAL POST-OFFICE.—Two mechanical letter-boxes, as used in Piedmont, in which the postman cannot change or see the letters.

1064. TEODORANI, S. *Forli.*—Lever of the first order.

1065. TOVO, *Vinadio (Turin).* — Apparatus for giving alarm in case of fires.

1066. TURCHINI, *Florence.* — Mechanical letter-box, in which the postman cannot change the letters.

CLASS IX.

1078. BATTAGLIA, G. *Cermignana (Como).*—Apparatus for winding cocoons with a single fire.

1079. BOLGÉ, T. *Brescia.*—Flax scutching machine; apparatus for compressing hay, and cleaning corn, &c.

1080. BACCIOLANI, C. L. *Modena.*—Harrows.

1081. BALDANTONI, G. & BROTHERS, *Ancona.*—Machine for thrashing Indian corn, wine press, straw cutter, and corking machine.

1082. BARGIONI, G. *Florence.*—Hempen, rush, and wicker work bags used in expressing olive oil, &c.

1083. BERTELLI, G. *Bologna.* — Reversible plough.

1084. BERTONE DI SAMBUY, GENERAL, MARQUIS E. *Turin.*—Ploughs.

1085. BOLOGNA AGRICULTURAL SOCIETY.—Model of Bolognese hemp farm, sowing machines, tanks for macerating hemp, &c.

1086. BORELLO, S. & BOANO, A. *Asti, Alexandria.*—Wine press, with screw workable in both directions.

1087. BOTTER, PROF. F. *Bologna.*—Collection illustrating instruments used in the culture of hemp.

1088. CAMBINI, E. *Florence.*—Apparatus for sulphurating vines.

1089. CASUCCINI, P. *Sienna.*—Agriculturists' levels.

1090. CERTANI, A. *Bologna.*—Plough for deepening the furrow made by an ordinary plough.

1091. CIAPETTI, B. *Castelfiorentino (Florence).*—Tuscan cast-iron plough, machine for thrashing Indian corn, cart, harrow, &c.

LEVINSTEIN & Co. *Milan.*—Lustring machine for giving a gloss to dyed silk.

The inventors of this Machine have found out and applied, by means of it, quite a new method of Lustring Silk. The steam being thrown directly upon it, a perfect lustre and a sensible extension of the thread are secured; while, by the old system, in which the steam was introduced within the cylinders, not only was the lustre imperfect, but the original length of the silk was not in the smallest degree increased.

This Machine, so valuable on account of its simplicity, is capable of lustring 200 kilogrammes of sewing silk, and 130 of organzine and trame daily.

Those interested in the matter are earnestly invited to visit the Italian Silk Department, where they will find silks in every colour and of every quality, and those admirable results of the application of this Machine which have obtained the medal of honour.

The public can see it at work every Wednesday and Saturday, from three till five o'clock in the afternoon; at which time, any information or explanations regarding it which may be desired will be given. It is patented in all European countries.

Gl' inventori con questa Macchina hanno ritrovato ed adattato un sistema tutto nuovo per lucidare la seta. Il vapore andando direttamente sulla medesima, non solo procura un lucido perfetto, ma benanche possono tendere si fili di modo sensibile; mentre che col vecchio sistema il vapore andava a cadere nei cilindri, e non solo non si otteneva mai un lucido perfetto, ma la seta non aumentava per niente la primitiva sua lunghezza.

Questa Macchina è apprezzabile per la sua semplicità, e sene possono lucidare giornalmente k^{mi} 200 di seta da cucire, ossia k^{mi} 130 organzini e trame.

Noi invitiamo pertanto coloro che ne hanno interesse a volere visitare il Dipartimento Italiano per le sete, ed ivi troveranno ogni qualità di seti in tutti i colori, ed i resultati ottenuti nell' applicazione di quella Macchina coi vantaggi sopra spiegati, per il che ci venne rilasciata la medaglia di merito.

Ogni Mercoledi e Sabbato, dalle 3 alle 5 ore pom^{e}, questa macchina verrà posta in esercizio a vista del pubblico, e saranno all' uopo dati gli schiarimenti e le istruzioni necessarie a coloro ne facessero rechiesta.

LEVINSTEIN & CO. *Milan—continued.*

Les inventeurs ont produit avec cette Machine une révolution radicale dans le lustre et l'étirage de la soie. Ils font frapper la vapeur directement sur les matteaux, et obetiennent ainsi les avantages bien supérieurs à ceux de l'ancien système qui consistait à chauffer l'intérieure des cylindres. Non seulement la machine à lustrer Levinstein & Cie. donne plus de brillant à la soie, mais à cause de l'état d'humidite qu'elle lui procure elle en rend l'étirage plus facile.

Quant au travail c'est là surtout que l'on trouve le bénéfice de l'invention. 200 kilos. de soie à coudre ou 130 kilos. d'organsin et de trame, sont étirés, lustrés, et prennent un brillant et une souplesse extraordinaires, indescriptibles, dans l'espace d'un seul jour, grâce à la simplicité du jeu de la machine.

On en trouve toujours de toutes prêtes dans nos ateliers de Milan et de Lyon, malgré les innombrables commandes qui nous arrivent, de toutes les parties du Monde.

Venez au Palais de l'Exposition tous les Mercredis, et tous les Samedis, et vous la verrez travailler. Le matteau de soie le plus obscur prendra sous vos yeux l'éclat le plus éblouissant et des chatoyements splendides. On vous donnera en même temps tous les renseignements et toutes les informations désirables sur ce merveilleux travail.

Nous vous invitons à visiter notre exposition de soie au département Italien. Tout le lustre qui la distingue, et qui nous a valu la médaille d'honneur, est le résultat du travail de cette admirable Machine.

———

Diese Lustrir Maschine ist patentirt in allen Ländern Europas.

Die Erfinder haben mit dieser Maschine ein ganz neues System Seide zu glänzen eingeführt indem durchden Dampf, der auf die Seide fällt, solche nicht allein einen besseren in Glanz annimmt, sondern auch das Strecken der Fäden erlaubt, während bei dem altem System, wo der Dampf in die Cylinder geht, die Seide ohne zerreissen sich kaum strecken läszt und obendrein nachher wieder zusammen geht. Ausserdem zeichnet sich diese Maschine durch ihre Einfachheit aus, und kann man damit 200 Kilo Näh-Seide oder 130 K. Organzine & Trame per Tag glänzen, während auf der Maschine alten Systemes man nur 30 K. per Tag glänzen kann.

Alle Mittwoch und Sonnabend von 3-5 Uhr Nachmittags wird auf dieser Maschine lustrirt und jegliche Auskunft ertheilt.

Wir laden die sich Interessirende ein, unsere auf dieser Maschine lustrirte Seiden im Italienischem Seiden Departement zu besichtigen, wofür uns die Preis Medaille zuerkannt wurde.

1092. CROSETTI, P. *Asti (Alexandria).*—Measures for wine.

1093. DE CAMBRAY DIGNY, COUNT L. G. *S. Piero a Sieve (Florence).*—Digny's ploughs, with helicoidal wing.

1094. DE FASSI, *Milan.*—Drawing of a machine for cleaning rice.

1095. DELLA BEFFA, G. *Genoa.*—Thrashing machine.

1096. DELPRINO, M. *Vesime (Alexandria).*—Silkworm nursery.

1096A. DUINA, A. *Brescia.*—Ploughshares, &c.

1097. FÀA DI BRUNO, A. *Alexandria.*—Farmers' walking-stick, serving as a level, plumb-line, square, &c.

1098. FACCHINI, BROTHERS, *Bologna.*—Model of a machine for softening hemp; combs used in dressing it.

1099. FISSORE, G. B. *Tortona (Alexandria).*—Dombasle's ploughs, with spare shares.

1100. GELLI & DELLE PIANE, *Pistoia (Florence).*—Bresciana.

1101. GIUNTINI, O. *Piccioli (Pisa).*—Short-handled cast-iron Tuscan, and other ploughs.

1102. GUPPY & PATISSON, *Naples.*—Hydraulic and screw presses, for making olive oil.

1103. JACUZZI, G. B. *Pistoia (Florence).*—Bresciana.

1104. KRAMER, E. *Milan.*—Models illustrating the Lombard system of farming and irrigation.

1105. LEOLI, N. *Brescia.*—Spade and shovel.

1106. MACCAFERRI, D. *Bologna.*—Model of apparatus for softening raw hemp.

1107. MAFFEI, CAV. N. *Volterra (Pisa).*—Models of bee-hives.

1108. MARCHI, L. *Volterra (Pisa).*—Machine for compressing faggots.

1109. MILAN ROYAL LOMBARD SCIENTIFIC, LITERARY, AND ARTISTICAL INSTITUTION.— Model of exit gate, for irrigatory canals.

1110. MORI, G. *Greve (Florence).*—Two-pronged fork.

1111. PAGNONI, A. *Ferrara.*—Model of an apparatus for breaking ·hemp-stalks by hand.

1112. PASQUI, G. *Forlì.*—Agricultural implements.

1113. PIZZARDI, MARQUIS, & BROTHERS, *Bologna.*—Machine for chopping up horns and hoofs for manure.

1114. ROSSI, A. *Bologna.*—Clod breaker.

1115. SAJNO, F. *Milan.*—Apparatus for hatching silkworms' eggs.

1116. SANTINI, L. *Fuccechio (Florence).*— Spades and hoe.

1117. SPANO, L. *Oristano (Cagliari).*— Agricultural cart, and plough.

1118. SPINA SANTALA, F. *Acireale (Catania).*—Ploughshares and scythes.

1119. STAFFUTI, O. *Pesaro.*—Corking machine.

1119A. SUPERCHI, P. *Parma.*—Parmisan plough on wheels.

1120. TORELLI, D. *Luco (Abruzzo Ulteriore II.)*—Spade, hoe, and shears for sheep-shearing.

1121. VAIRO, G. *Messina.*—Reaping machine.

1122. VIDO, F. *Codogno (Milan).*—Cart for soft ground.

———

1123. CONROTTO, C. *Turin.*—Machine for packing silk; apparatus for killing the silkworm chrysalis; silkworm cocoons.

1124. FUSINA, V. *Pavia.* — Thrashing machine for Indian corn.

1125. LUCHINI, G. *Florence.* — Copper churn.

1126. MUSSIARI, DR. G. *Parma.*—Parmesan sub-soil plough.

CLASS X.

SUB-CLASS A.

1140. ALTOVITI AVILA, CAV. F. *Florence.*—Bricks and brick-pavements.

1141. CALZA, A. *Spezia (Genoa).*—Metamorphic-manganesiferous jasper from Beverone, near Spezia, with hydraulic cement made of it.

1142. CARAFA DI NOIA, P. *S. Giovanni a Teduccio (Naples).*—Tubes, tiles, cornices, &c., of terra cotta; refractory bricks.

1143. COLONNESE, F. & G. *Naples.*—Enamelled bricks and tiles; pipes for water-closets.

1144. GALLIGANI, DR. G. *Seravezza, Lucca.*—Ofiocalce and marble vase.

1145. GUALA, G. *Turin.*—Architectural model in 24 pieces.

1146. GUERRA, COUNT P. *Massa Carrara.*—Slabs of polished marble.

1147. MOLINARI & DESCALZI, *Genoa.*—Drawing of apparatus for submarine constructions, and design for a port.

1148. PELAIS, G. *Pistoia (Florence).*—Argillaceous limestone, and hydraulic cement prepared from it.

1149. PETIT-BON, G. *Parma.*—Hollow bricks.

1150. PIANA, G. *Bologna.*—Model of an improved tiled roof.

1151. RONDANI, T. *Parma.*—Tiles, hollow bricks, &c.

1152. SEMMOLA, CAV. F. *Naples.*—Collection of the building materials of the Neapolitan provinces.

1153. SPANO, L. *Oristano (Cagliari).*—Tiles and bricks.

1154. Treves, M. *Florence.*—Artificial marbles for pavements, &c.

1155. Zecchini, I. *Milazzo (Messina).*—Artificial marbles.

———

1156. Armao, G. *S. Stefano di Camastra (Messina).*—Bricks.

1157. Gai, F. *S. Mato, Pistoia (Florence).*—Terra-cotta tiles of a new form; terra-cotta for paving.

1158. Lee, G. *Sarzana (Genoa).*—Bricks, tiles, &c. of different kinds; water pipes.

1159. Leoncini, Bros. *Rotta (Pisa).*—Collection of bricks and tiles.

1160. Paradossi, O. *Leghorn.*—Drawings of a drawbridge.

1161. Puliti, C. *Pelago (Florence).*—Double glazed gas pipes; water pipes.

1162. Savona Junta for the Exhibition, *Savona (Genoa).*—Refractory clay; lime, bricks.

1163. Taiani, G. *Vietri (Principato Citeriore).*—Bricks for ornamental flooring.

1164. Valerio, C. *Turin.*—Drawing of a new kind of graving dock for tideless seas, with manuscript report.

Sub-Class B.

1165. Atenolfi, Prince of, *Castelnuovo (Naples).*—Draining pipes.

Sub-Class C.

1166. Borella & Boiano, *Asti (Alexandria).*—Model of a shop front, to form a private entrance at night.

1167. Bacci, F. *Impruneta (Florence).*—Ornamented vase, flower pots, cornices, roses, capitals, &c.

1168. Brunetti, G. *Florence.*—Model of a mechanical self-supporting staircase.

1169. Brusa, G. B. *Milan.*—Model of a new kind of stove.

———

2282. Campana, Marquis G.—Collection of artificial stone and marble work-tables, vases, statues, pedestals, sphynx, Egyptian figure, &c.

2283. Caproni, G. *Perugia (Umbria).*—Architectural design.

2284. Della Valle, P. *Leghorn.*—Scagliola and inlaid work on terra-cotta.

2285. Galizioli, B. *Brescia.*—Two frescoes on linen removed from a wall.

2286. Lega, M. *Forlì.*—Artificial marbles.

2287. Mattarelli, G. *Lecco (Como).*—Model of Milan Cathedral, in inlaid wood.

2288. Piegaja, R. *Lucca.*—Terra-cotta model of Mediæval cornices, &c.

2289. Rabbini, Cav. A. *Turin.*—Collection of statistical documents.

CLASS XI.

Sub-Class A.

1170. Binda, A. *Milan.*—Silk, military, and other cravats; scarves, military buttons.

1171. Jamoli, G. *Turin.*—Plumes used by *Bersaglieri*, and naval officers.

Sub-Class B.

1180. Excoffier, G. *Asti (Alexandria).*—Model of a military observatory.

1181. Galli, G. *Milan.*—Oiled cloth.

1182. Sipriot, C. *Milan.*—Oiled cloth for artillery.

———

1183. Henkel, Luigi, *Florence.*—Waterproof military tent.

Sub-Class C.

1190. BERNARDI, P. *Rimini*.—Muskets.

1191. COLOMBO, C. M. *Milan*.—Fowling-piece, revolvers, and side arms.

1192. COMINAZZI, M. *Gardone (Brescia)*. —Pistol and gun barrels.

1193. DE STEFANO, *Campobasso (Molise)*. —Sword, sabre, and foil.

1194. FUSEO, F. *Vitulano (Benevento)*.— Double-barrelled pistol, both barrels in one piece, &c.

1196. IZZO, A. *Naples*.—Double-barrelled guns : and revolver rifle, the barrel made with iron wire.

1197. LANCIA, G. *Turin*.—Breech-loading cannon, &c.

1198. LABRUNA, G. *Naples*.—Sword.

1199. MARELLI, A. *Milan*.—Gun barrel, long gun, and pistol.

1200. MAZZA, S. *Naples*.—Rifled revolver, fowling-pieces, rifles, &c.

1201. MEROLLA, S. *Naples*.—Double barrelled guns.

1202. MINOTTINI, G. & LANCETTI, F. *Perugia (Umbria)*.—Engraved sword, and gun in inlaid case.

1203. MURATTI, COL. A. *Naples*.—Machine for compressing fulminating powder into percussion caps ; model of a carriage for a mortar.

1204. PARIS, M. *Brescia*.—Musket, and fowling-piece in several parts.

1205. PILLA, G. *Benevento*. — Six-barrelled pistol, each barrel going off separately by once pulling the trigger.

1206. PRIORA, G. *Milan*.— Revolvers.

1207. RISSONE, L. *Parma*. Shot of various kinds.

1208. SICHLING, A. *Turin*. — Sabres, swords, hunting knives, and steel for making damasked blades.

1209. SQUINZO, L. *Cagliari*.—Revolvers.

1210. TORRE ANNUNZIATA ROYAL MANUFACTORY OF ARMS.—Locks, damasked barrel, and sabre blade.

1211 TOSCHI, A. *Lugo (Ravenna)*.—Two fire-arms, one to fire six, and the other sixty times.

1212. TRAVAGLINI, C. *Pisa*.—Eight-gun naval battery, workable by six men.

1213. BEVILACQUA, P. *Campobasso (Molise)*.—Gun barrel.

1214. DI TURO, C. *Campobasso (Molise)*. —Walking sticks containing arms.

1215. FABBRICA SOCIALE, *Brescia*. — Rifles.

1216. LANDI, G. *Salerno (Principato Citeriore)*.—Model of a revolver cannon.

1217. MONGIANA ROYAL METALLURGICAL WORKS, *Mongiana (Calabria Ulteriore II.)*.— Arms.

1218. ROYAL ARSENAL, *Turin*.—Model of Cavalli's steel-protected battery, as employed at the siege of Gaeta ; Cavalli's breech-loading cannon ; field and mountain pieces, with Cavalli's carriage.

1219. ROYAL MANUFACTORY OF ARMS, *Turin*.—Arms.

CLASS XII.

Sub-Class B.

1225. TAGLIACOZZO, P. *Rome*.—Combination of river barges, capable of being put together so as to form a single sea-going steam-boat.

CLASS XIII.

1226. BIFEZZI, G. *Naples.*—Telegometer, for surveying and mensuration.

1227. BONELLI, CAV. G. *Turin.*—Typo-electric telegraph, capable of transmitting 500 messages hourly; four compositors' tables for the above.

1228. CASSANI, E. *Milan.*—Spectacles.

1229. CASUCCINI, P. *Sienna.*—Level.

1230. FÀA DI BRUNO, A. *Alexandria.*—Ellipsograph.

1231. GONNELLA, T. *Florence.* — Calculating machines, for whole numbers and fractions, with descriptive pamphlet.

1232. JEST, C. *Turin.*—Collection of philosophical instruments for schools.

1233. LUCIFERO, T. *Messina.*—Constant chloride of sodium battery.

1234. MARCHI, U. *Florence.*—Self-registering maximum and minimum thermometer.

1235. AMICI, PROF. G. B. *Florence.*—Achromatic refractor, diameter 17 7-10th inches; parabolic speculum for a large telescope; ocular micrometer, with double image, for measuring the diameter of planets, and similar small angular distances; reflecting sextant, and repeating circle; telescopes; surveying cross, without parallax; levels; several kinds of camera lucida; polarizing, and other microscopes; microscope camera lucida for drawing small objects, &c.

1236. BANDIERI, G. *Naples.*—Chemical balance, electro-dynamic apparatus, &c.

1237. MANUELLI, G. *Reggio, Emilia.*—Economic pile, with charcoal diaphragm, for obtaining a constant current.

1238. PAVIA UNIVERSITY, PHILOSOPHICAL MUSEUM OF.—Electrophorus, condenser, electrometer, Voltaic piles: interesting as having belonged to Volta. Prof. Belli's apparent hygrometer, double-action psicrometer, and air-pump; Cantoni's calorimeter and thermographs.

1239. ROBERTO, P. *Naples.*—Genometer, for making any kind of metrical scales.

1240. MARONI, M. *Milan.*—Modification of Morse's telegraph.

1241. MILESI, A. *Bergamo.*—Electrical apparatus for taking votes in large public meetings.

1242. MINOTTO, G. *Turin.*—New constant pile.

1243. MURE, BROS. *Turin.*—Standard for taking the height of recruits; steel standard meter.

1244. SELLA, COMMENDATORE Q. *Turin.*—Tripsometer, for measuring the co-efficient of friction.

CLASS XIV.

1245. DURONI, A. *Milan.*—Photographs.

1246. MAZA, E. *Milan.*—Photographs.

1247. MODENA SUB-COMMITTEE FOR THE EXHIBITION.—Photographs.

1248. RANCINI, C. *Pisa.*—Photographic miniature of a fresco in the Composanto, Pisa.

1249. RONCALLI, A. *Bergamo.*—Photographs of microscopic objects executed directly.

1250. VAN LINT, E. *Fine Art Studio, Pisa.*—Photographs.

1251. ALINARI, BROS. *Florence.*—Views of Florence; portfolio of photographs of paintings in the galleries of Florence, Venice, and Vienna.

1252. CHIAPELLA, F. M. *Turin.*—Photographs on silk.

1253. FRATACCI, C. *Naples.*—Views.

CLASS XV.

1255. BERNARD, A. *Naples.* Church clock.

1256. DECANINI, C. *Florence.*—Escapement for watches.

1257. MANUELLI, G. *Reggio, Emilia.*—Escapement for watches.

1258. OLETTI, P. *Turin.*—Small astronomical chronometer.

CLASS XVI.

1265. AIELLO, S. *Naples.*—Strings for musical instruments.

1266. BOCCACCINI, A. *Pistoia (Florence).*—Improved and common drums.

1267. DE MEGLIO, L. *Naples.*—Grand piano.

1268. FORNI, E. *Milan.* — Flutes and clarinets.

1269. FUMMO, A. *Naples.*—Piano-melodium with two rows of keys, vertical piano-melodium, new kind of flute.

1270. MARZOLO, G. *Padua.*—Organ with melographic apparatus, for repeating and printing any music played.

1271. PANNUNZIO, D. *Agnone (Molise).* — Model of a brass bridge for pianofortes.

1272. RUGGIERO, C. *Naples.*—Straight horn, in B flat and A flat.

1273. SIEVERS, F. *Naples.*—Pianos and piano-stools.

1274. VINATIERI, F. & SONS, *Turin.*—Flutes and clarinets.

———

1275. BOLGÉ, T. *Brescia.*—Drum.

1276. PELITTI, G. *Milan.*—Wind instruments.

1277. PIETRASANTA, L. & SONS, *Lucca.*—Wind instruments.

CLASS XVII.

1285. ARIANO, G. *Turin.*—Veterinary surgical instruments.

1286. BARBERIS, A. *Turin.*—Surgical, veterinary, and dentists' instruments.

1287. BELTRAMI, G. *Placenza.*—Surgical instruments.

1288. BERTINARI, G. *Turin.*—Surgical instruments.

1289. COMERIO, BROS. *Brescia.*—Artificial legs, orthopœdic apparatus, apparatus for fractured limbs, ruptures, &c.

1290. FERRERO, G. *Turin.*—Bandages for curing inguinal and crural hernia.

1291. GADDI, PROF. CAV. M. *Modena.*—Injections of the auditory organs of man, quadrupeds, and birds.

1292. GIORDANO, S. *Turin.*—Surgical instruments.

1293. LOLLINI, P. & P. *Bologna.*—Collection of surgical instruments.

1294. MONTI, ELVIRA, & Co. *Florence.*—Herniary bandages, and surgical apparatus of various kinds.

1295. OBIGLIO, L. *Turin.* — Artificial teeth, variously mounted.

1296. PAVIA UNIVERSITY.—Surgical instruments.

1297. PAVIA UNIVERSITY, ANATOMICAL MUSEUM OF. — Preparation of the facial nerves; preparation of an abnormal human trunk, remarkable for the transposition of both thoracic and abdominal viscera; various injections and preparations

1299. SERGI, P. *Messina.*—Surgical instruments.

1300. TUBI, G. *Milan.*—Orthopœdic shoes for horses.

—

1301. BRIZIANO, A. *Milan.* — Sticking and corn plasters.

1302. OLMETA, A. *Cagliari.*—Dentists' instruments.

1303. PARMA VETERINARY INSTITUTION (ANATOMICAL MUSEUM OF THE).—Preparation of the muscular system of the dog.

1304. PARMA VETERINARY INSTITUTION (PATHOLOGICAL MUSEUM OF THE).—Monstrous fœtus of a cow.

CLASS XVIII.

1307. ALEXANDRIA PENITENTIARY.—Cotton goods.

1308. CALAMINI, M. & Co. *Pisa.*—Coloured cotton goods.

1309. CREMONCINI, A. *S. Vivaldo (Florence).*—Cotton counterpanes.

1310. CANTONI, C. *Milan.*—Cotton yarn, cotton stuffs, fustians, calicoes, dimity, damask, &c.

1311. COBIANCHI, P. & SON, *Intra (Novara).*—Cotton yarn.

1313. LUALDI, E. *Brescia.*—Cotton yarn.

1314. MILAN CHAMBER OF COMMERCE.—Samples of the cotton manufactures of the province of Milan.

1315. MODENA SUB-COMMITTEE FOR THE EXHIBITION.—Cotton cloth with coloured threads.

1316. MORELLI, F. *Florence.*—Coloured cotton stuffs for trousers and dresses.

1317. OSCULANI PIROVANO & Co. *Monza (Milan).*—Cotton trousering, damask, and fustians, &c.

1318. PERSICHETTI, S. *Ancona.*—Cotton sail cloth.

1319. PIATTI & Co. *Placenza.*—Cotton stuffs.

1320. PISA SUB-COMMITTEE FOR THE EXHIBITION.—Cotton stuffs manufactured in the province of Pisa.

1321. SCHLAEPFER, WENNER, & Co. *Naples.*—Unbleached calico, and printed cotton goods.

1322. STEINAUER, J. A. *Chiavenna (Sondrio).*—White and coloured wadding.

1323. THOMAS, A. *Milan.*—Machine-made cotton stuffs.

1324. VONWILLER & Co. *Naples.*—Cotton yarn.

1325. ZEPPINI, F. *Pontedera (Pisa).*—Cotton counterpanes.

—

1326. CAMPANA, J. & F. *Gandino (Milan).*—Cotton counterpanes.

1327. HOZ & FONZOLI, *Terni (Umbria).*—Cotton goods.

1328. LAZZARI, R. *Lucca.*—White cotton mosquito gauze.

CLASS XIX.

1329. BERTERO, A. & GALLO, G. B. *Carmagnola (Turin).*—Dressed hemp for cordage.

1330. BORZONE, G. *Chiavari (Genoa).*—Fringed towelling.

1331. CAGLIARI SUB-COMMITTEE FOR THE EXHIBITION.—Linen fabrics made by peasants.

1332. CAMPOBASSO SUB-COMMITTEE FOR THE EXHIBITION (*Molise*).—Linen for household purposes.

1333. COSTA, GIULIA, *Chiavari (Genoa).*—Linen.

1334. De-Angelis, Brothers, *Naples*.—Sail-cloth.

1335. Devoto, L. *Chiavari (Genoa)*.—Linen.

1336. Ferrara Royal Chamber of Commerce.—Hand-made ship's cable, marlines, ropes, and sail-cloth.

1337. Lupinacci, Baron L. & Brothers. *Cosenza (Calabria Citeriore)*.—Hand-made linen.

1338. Mezzano, P. *Celle (Genoa)*.—Fishing nets.

1339. Milan Chamber of Commerce.—Collection illustrating the linen manufacture of the province.

1340. Morelli, G. *Cosenza (Calabria Citeriore)*.—Hand-made napkin, linen for shirting.

1341. Noberasco, L. *Savona (Genoa)*.—Sail-cloth.

1342. Osculati, Pirovano, & Co. *Monza (Milan)*.—Flaxen trousering.

1343. Padoa, P. *Cento (Ferrara)*.—Sail-cloth, sacking, and bed ticking.

1344. Parthenope Manufacturing Co.

Naples.—Flax and hemp, with linen and yarn made of them.

1345. Pellegrinetti, F. *Florence.*—Linen, damask, &c.

1346. Persichetti, S. *Ancona*.—Yarn, twine, ropes, cables, &c.

1347. Polenghi, C. *S. Fiorano (Milan)*.—Flax, hand-made linen, &c.

1348. Quadri, E. *Naples*.—Tarred rope made by machinery; hemp, scutched without steeping or preparation, &c.

1349. Reggio Agricultural Society, *Reggio, Emilia*.—Hempen and flaxen cloth, and yarn.

1350. Sanguinetti, F. *Chiavari (Genoa)*.—Linen.

1351. Ziliani, G. B. *Brescia*.—Nets.

1352. Ferrigni, G. *Leghorn.*—Ropes and cordage.

1353. Reggio (Emilia) Agricultural Association.—Bruised, scutched, and carded flax and hemp.

1354. De Angelis, Brothers, *Naples*.—Ships' cables.

CLASS XX.

1365. Abbate, P. *Parma*.—Raw silk, spun by a steam-engine.

1366. Acquaviva, Count C. *Giulia (Abruzzo Ulteriore I.)*.—Raw silk.

1367. Alexandria Penitentiary.—Brocade.

1368. Andreis, V. *Trincotto a Racconigi*.—Organzine.

1369. Arcangioli, A. *Pistoia (Florence)*.—Hanks of raw silk.

1370. Ascoli, A. *Terni (Umbria)*.—Hanks of good raw silk obtained from diseased cocoons.

1371. Assom, Brothers, *Villa Stellone. (Turin)*.—Cocoons, &c.

1372. Baldini, L. *Perugia (Umbria)*.—Raw silk.

1373. Bancalari, G. *Chiavari (Genoa)*—Raw silk.

1374. Barozzi, Antoinetta, *Milan*.—Waste of carded silk.

1375. Bavassano, G. B. *Alexandria*.—Raw silk.

1376. Belletti, G. *Bologna*.—Spun silk, and silk veils.

1377. Bellini, G. *Osimo (Ancona)*.—Raw silk.

1378. Bellino, Brothers, *Turin*.—Raw silk, with cocoons.

1379. Beretta, Cav. D. *Ancona*.—Raw silk and silk waste.

1380. Beretta, Brothers, *Parlenghe (Brescia)*.—Silk.

1381. BERIZZI, S. *Bergamo.*—Raw silk and organzine.

1382. BERTARELLI, C. *Cremona.*—Raw silk, brocade, galloon.

1383. BEVILACQUA, M. & SON, *Lucca.*—Raw silk, brocade, galloon.

1384. BINDA, A. *Milan.*—Waistcoating.

1385. BOLMIDA, BROTHERS, *Turin.*—Raw and carded silk, organzine.

1386. BOLOGNINI RIMEDIOTTI, *Pistoia (Florence).*—Raw silk.

1387. BOZZOTTI, C. *Milan.*—Raw silk, tram, and sewing silk.

1388. BRACCO, M. & SONS, *Turin.*—Raw silk and organzine.

1389. BRUNI, F. *Milan.*—Organzine and tram.

1390. CARRADORI, COUNT G. *Osimo, near Ancona.*—Raw silk.

1391. CASISSA, SONS, *Novi (Alexandria).*—Raw silk.

1392. CECCONI & SANTINI, *Lucca.*—Floss silk for embroidering.

1393. CERIANA, BROTHERS, & NOÉ, *Turin.*—Organzine, &c.

1394. CHABANON, A. *Portici (Naples).*—Galloon.

1395. CHICHIZOLA, G. & Co. *Turin.*—Silk velvets.

1396. COLLER, D. *Portici (Naples).*—Silk ribbons.

1397. COLLER L. *Portici (Naples).*—Silk ribbons.

1398. COLOMBO, F. *Ceva (Cuneo).*—Raw silk.

1399. COMBONI, BROTHERS, *Limone (Brescia).*—Raw silk.

1400. COMPAGNO, P. *Cosenza (Calabria Citeriore).*—Organzine.

1401. CONTI, A. & Co. *Fossombrone (Pesaro and Urbino).*—Raw silk.

1402. CONTI, F. *Milan.*—Raw silk, tram, and organzine.

1403. CORNA, G. *Pisogne (Brescia).*—Raw silk.

1404. CORTI, BROTHERS, *Milan.*—Raw silk and silk yarn.

1405. COZZA, COUNT G. *Orvieto (Umbria).*—Raw silk.

1406. CRESTINI, D. *Asinalunga (Sienna).*—Silk yarn.

1407. DE ANTONI, *Milan.*—Carded silk waste.

1408. DE FERRARI, BROTHERS, *Genoa.*—Silk velvet; silver and gold brocades.

1409. DE FILIPPI, MERZAGORA, & Co. *Arona (Novara).*—Silk waste for spinning.

1410. DE GORI, COUNT A. *Sienna.*—Raw silk.

1411. DELPRINO, M. *Vesime (Alexandria).*—Raw silk, wound by a particular method.

1412. DEMEO, F. *Messina.*—Silk ribbons and stuffs.

1413. DENEGRI, G. *Novi (Alexandria).*—Wansey silk and organzine waste.

1414. DEVINCENZI, G. *Notaresco (Abruzzo Ulteriore I.).*—Raw silk.

1415. DITTAIUTI, COUNT G. *Osimo (Ancona).*—Raw silk.

1416. FARAGLIA, M. *Terni (Umbria).*—Raw silk.

1417. FERRARI, F. *Codogno (Milan).*—Raw silk, and tram.

1418. FERRI, BROTHERS, *Grosseto.*—Raw silk.

1419. FONTANA, B. & Co. *Turin.*—Organzine.

1420. FOSSI & BRUSCOLI, *Florence.*—Raw silk.

1421. FRANCHI, BROTHERS, *Brescia.*—Raw silk and organzine.

1422. GADDUM, E. F. *Torre Pellice (Turin).*—Raw silk and organzine.

1423. GALATTI, G. *Messina.*—Raw silk and cocoons.

1424. GAVAZZI, P. *Milan.*—Raw silk, organzine, and tram.

1425. GIAMBARINI, A. *Bergamo.*—Various kinds of tram.

1426. GIARDINIERI, BROTHERS, *Osimo (Ancona).*—Raw silk.

1427. GRANOZIO, D. *Salerno (Principato Citeriore).*—Silk, spun by machinery.

1428. GRASSI, F. *Vicofaro (Florence).*—Raw silk.

1429. JAEGER, G. *Messina.*—Raw silk.

1430. KELLER, A. *Turin.*—Silk, raw and manufactured.

1431. LANZANI, L. & BROTHER, *Milan.*—Silk, from the inner coarse part of the cocoon, manufactured by hand and machine.

1432. LARDINELLI, B. *Osimo (Ancona).*—Raw silk.

1433. LAZZARI, R. *Lucca.*—Silk mosquito gauze.

1434. LEVINSTEIN & Co. *Milan.*—6000 tints of sewing silk; organzine and tram.

1435. LAZZARONI, P. *Milan.*—Raw silk and tram.

1436. MACERATA SUB-COMMITTEE FOR THE EXHIBITION.—Raw silk.

1437. MAFFIO, BROTHERS, *Sondrio.*—Raw silk.

1438. MAGNANI, E. *Florence.*—Raw silk.

1439. MASSINA, L. *Calvenzano (Brescia).*—Raw silk and cocoons.

1440. MAZZERI, P. *Milan.*—Sewing silk, organzine, and tram.

1441. MIRABELLI, F. *Cosenza (Calabria Citeriore).*—Raw silk.

1442. MODENA, A. *Reggio, Emilia.*—Silk yarn.

1443. MORESCO & MOLINARI, *Genoa.*—Velvets.

1444. NEFETTI, A. *Sta. Sofia (Florence).*—Raw silk.

1445. NIERI & LENCI, *Lucca.*—Raw silk.

1446. NOVELLIS, C. G. *Savigliano (Cuneo).*—Organzine.

1447. OTTAVIANI, BROTHERS, *Messina.*—Raw silk.

1448. PADOA, P. *Cento (Ferrara).*—Silk yarn.

1449. PADOVANI, BROS., *Codogno (Milan).*—Raw silk.

1450. PALAZZESCHI, G. *Città di Castello (Umbria).*—Raw silk and silk yarn.

1451. PASQUI, CAV. Z. *Florence.*—Raw silk.

1452. PASTACALDI, F. *Pistoia (Florence).*—Raw silk.

1453. PERIPETTI, C. *Placenza.*—Raw silk.

1454. PIATTI & Co. *Placenza.*—Raw silk.

1455. PIAZZONI, G. B. *Bergamo.*—Raw silk.

1456. PICCALUGA, E. F. *Gavi (Alexandria).*—Raw silk.

1457. PIRI PECCI, COUNT G. *Sienna.*—Hanks of raw silk.

1458.—PIZZNI, A. M. *Rossiglione.*—Raw silk and organzine.

1459. PORRO, P. *Milan.*—Raw silk, tram, and organzine.

1460. PREISWERK, G. & SON, *Milan.*—Organzine and tram.

1461. RAMPOLDI, D. *Como.*—Woven picture, in colours.

1462. RAVENNA SUB-COMMITTEE FOR THE EXHIBITION.—Raw and spun silk.

1463. RONCHETTI, BROTHERS, *Milan.*—Raw silk, organzine, and tram.

1464. ROSSI, M. *Sondrio.*—Raw silk and silk yarn.

1465. ROSSINI, G. *Terni (Umbria).*—Raw silk.

1466. ROTA, A. *Chiari (Brescia).*—Raw silk.

1467. RUBINACCI, S. *Naples.*—Raw silk and sewing silk.

1468. SALARI, D. *Fuligno (Umbria).*—Raw silk.

1469. SARI, B. *Lucca.*—Raw silk.

1470. SCOLA, G. *Villa d'Adda (Bergamo).*—Raw silk.

1471. SEGRE, S. *Vercelli (Novara).*—Raw silk.

1472. SENNOCHI, G. *Placenza.*—Raw silk.

1473. SINIGAGLIA, S. *Lugo (Ravenna).*—Raw silk.

1474. SOLARI, M. *Chiavari (Genoa).*—Raw silk.

1475. SORLINI, A. *Ospitaletto (Brescia).*—Raw silk.

1476. SPEDALIERE, P. *Portici (Naples).*—Galloons.

1477. STEINER & SONS, *Bergamo.*—Raw silk, tram, and organzine.

1478. SURTERA SOPRANSI, M. *Codogno (Milan).*—Raw silk.

1479. TALLACCHINI, BROTHERS, *Milan.*—Raw silk, organzine, tram, and grenadine.

1480. TESI, L. *Pistoia (Florence).*—Raw silk.

1481. TODI VECCHI, *Reggio, Emilia.*—Raw silk, sewing silk, and tram.

1482. TOMASSONI, G. *Jesi (Ancona).*—Raw silk.

1483. VALVO, P. *Portici (Naples).*—Silk ribbons and stuffs.

1484. VANNUCCI, G. *Pistoia (Florence).*—Raw silk.

1485. VIOLA, G. *Cairo, near Savona (Genoa).*—Raw silk.

1486. ZAMERA, HEIRS OF, *Brescia.*—Raw silk.

1487. ZUPI, BROTHERS, *Cerisano (Calabria Citeriore).*—Raw silk.

1488. ZUPPINGER, SIBER, & CO. *Bergamo.*—Raw silk, tram, and organzine.

———

1489. BACCHINI ROSSI L. *Perugia (Umbria).*—Silk shawls made by a hand frame without any loom.

1490. CAMPANI, I. & F. *Gaudino (Milan).*—Counterpanes made of waste silk.

1491. COPPOLO, A. *Reggio, Calabria.*—Raw silk.

1492. CRISTOFANI & SON, *Florence.*—Figured stuffs and armoisine.

1493. DE CIANI, D. *Trent, Tyrol.*—Raw and thrown silk.

1494. DE FERRARI, G. (late F.) *Genoa.*—Velvet.

1495. DIENA, MARQUIS G. *Modena.*—Raw silk, cocoons.

1496. GIOVANELLI, A. & D. *Pesaro (Pesaro and Urbino).*—Raw silk.

1497. HALLAM, T. *Villa S. Giovanni (Calabria Ulteriore I.).*—Raw silk.

1498. HUTH, P. *Como.*—Dyed black silk.

1499. IMPERATORE, G. (late B. & SONS), *Intra (Novara).*—Thrown silk.

1500. LOFARO, A. *Reggio, Calabria.*—Raw silk.

1501. LOFARO, G. *Reggio, Calabria.*—Raw silk.

1502. MOSCHETTI, G. A. *Boves (Cuneo).*—Raw silk, organzine, cocoons.

1503. NAPLES SUB-COMMITTEE FOR THE EXHIBITION.—Raw silk ; dyed silk yarn.

1504. RIZZI, BROS. *Pisogne (Brescia).*—Raw silk.

1505. SCIARRONI, M. *Reggio, Calabria.*—Raw silk.

1506. SOLEI, B. *Turin.*—Silk stuffs for furniture and decoration.

1507. VALAZZI, L. *Pesaro (Pesaro and Urbino).*—Raw silk.

1508. VERZA, BROS. (late C.), *Milan.*—Raw and thrown silk.

1509. VIALI & MASSETTI, *Fano (Pesaro and Urbino).*—Raw silk.

1510. VIGANOTTI, G. *Milan.*—Galloon.

1511. ZANARDINI, P. *Pisogne (Brescia).*—Raw silk.

1512. ZAMOLI, L. *Cesena (Forlì).*—Raw silk.

CLASS XXI.

1515. ANTONGINI BROTHERS, *Milan.*—Raw and coloured woollen yarn.

1516. CAGLIARI SUB-COMMITTEE FOR THE EXHIBITION.—Woollen cloth, &c., made by peasants of Aritzo.

1517. CASTELLI, C. *Milan.*—Woollen counterpane.

1518. HOZ & FONZOLI, *Terni (Umbria).*—Goods of mixed cotton and wool.

1519. LUPINACCI, BARON L. & BROTHERS, *Cosenza (Calabria Citeriore).*—Hand-made woollen cloth.

1520. MORELLI, F. *Florence.*—Textures of cotton and wool.

1521. MORELLI, G. *Rogliano (Calabria Citeriore).*—Knitted wool.

1522. ORLANDO. G. *Pescolamazza (Benevento).*—Woollen cloth.

1523. OSCULATI, PIROVANO, & Co. *Monza (Milan).*—Textures of cotton and wool, of cotton, wool, and silk, and of cotton and flax; cotton and wool damask.

1524. PIRAS, MARIA, *Samassi (Cagliari).*—Sardinian wallet.

1525. ROSSI, F. *Milan.*—Woollen stuffs.

1526. SELLA, BROTHERS, *Turin.*—Cloth, of various kinds and colours.

1527. SELLA, M. *Biella(Novara).*—Cloth, velvet, flannel, &c.

1528. SPANO, L. *Oristano (Cagliari).*—Woollen wallets; counterpanes of wool, cotton, and silk, and of linen and cotton.

1529. THOMAS, A. *Milan.*—Machine-made stuffs, of mixed cotton and wool; cotton and flax thread, of several colours.

1530. CALAMINI, M. & Co. *Pisa.*—Plaid woollen shawls.

1531. COSTANTINO, G. *S. Marco de' Cavoti (Benevento).*—Woollen cloth.

1532. CROCCO, C. & L. *Genoa.*—Woollen hosiery.

1533. FLORENCE WORKHOUSE.—Woollen counterpanes, flannel.

1534. GIANNATASIO, G. *S. Cipriano (Principato Citeriore).*—Blankets.

2291. ALEXANDRIA PENITENTIARY. — Mixed fabrics.

CLASS XXII.

1535. CAMPRA, C. *Graglia, near Biella (Novara).*—Woollen carpets.

1536. CASTELLI, C. *Milan.*—Carpets.

1537. MILAN BLIND ASYLUM. — Carpet made by the blind.

1538. PIRAS, V. *Samossi (Cagliari).*—Carpet.

1539. CIANFERONI, A. *Florence.* — Oil cloth.

1540. GALLI, G. *Milan.*—Oil cloth.

CLASS XXIII.

1545. BOSIO, F. & Co. *Castello di Lucento (Turin).*—Cotton yarn, dyed different colours.

1546. CECCONI & SANTINI, *Lucca.*—Wool dyed in 10 colours, for embroidery.

1547. FOLETTI, WEISS, & Co. *Milan.*—Turkey-red cotton yarn.

1548. SANTILLI, B. *Isernia (Molise).*—Specimens of cloth simultaneously dyed a different colour on either side.

1549. WISER, S. *Modena.*—Fleece, the wool coloured in various tints.

1550. HUBER & KELLER, *Pisa.*—Turkey-red cotton yarn.

CLASS XXIV.

1555. BAFICO, ANGELA, *Chiavari (Genoa).*—Several kinds of point lace.

1556. BASSETTI, ANTOINETTA, *Sienna.*—Embroidered sleeves.

1557. BINDA, A. *Milan.*—Trimmings of silk, and of mixed silk and cotton.

1558. CALANDRA, CAMILLA, *Savigliano (Cuneo).*—Embroidered white counterpane.

1559. CUCCHIETTI, C. *Busca (Cuneo).*—Embroidery in worsted and silk, representing Mary Queen of Scots at Langsyde.

1560. FIESCHI (CONSERVATORY), *Genoa.*—Embroidered cambric handkerchiefs, collar, &c.

1561. FUMMO, MARIA, *Naples.*—Embroidered pocket-handkerchief.

1562. GARBESI, ERSELIA, & ANGELA, *Vorno (Lucca).*—Silk shawl, embroidered to imitate ancient lace.

1563. MARTINI, L. *Milan.*—Embroidery, silk and brocade stuffs.

1564. MARTINI, ERSELIA, *Milan.*—Embroidery.

1565. NAPLES (ROYAL INSTITUTION OF CARMINELLO).—Embroidered handkerchiefs.

1566. PARLANTI, E. *Florence.*—Embroideries, representing a satyr and monkey, after Annibal Caracci, &c.

1567. TACCHINI, TERESA, *Modena.*—Chiaro-scuro embroidery, representing a halt at an inn; embroidery on cloth.

———

1568. BUONINI, M. *Lucca.*—Lace.

1569. BROCCI, D. & A. *Cantù (Como).*—Lace veil, mantle, &c.

1570. GENOA POOR ASYLUM (ALBERGO DEI POVERI).—Embroidered cambric handkerchief; shirts; towels with lace border.

1571. LANDUZZI, F. *Bologna.*—Embroidery.

1572. LEPORATTI, E. *Pistoia (Florence).*—Embroidered silk.

1573. MARINO, P. *Turin.*—Trimmings.

1574. PANIZZI, M. *Parma.*—Embroidered cambric handkerchief.

———

2293. CAMPODONICO, E. *Genoa.*—Lace.

2294. PETTI, E. *Campobasso (Molise).*—Embroidery on wool.

2295. PETRUCCI, A. *Lucca.*—Embroidered handkerchief.

2296. SERVI, E. *Florence.*—Black Thibet scarf, embroidered with silk.

2297. TECCHI, A. *Pisa.*—Invisible mending and darning on various kinds of stuffs.

2298. TRAFIERI, A. *Lucca.*—Embroidery on cambric.

2299. TESSADA, F. *Genoa.*—Lace shawl.

CLASS XXV.

SUB-CLASS A.

1574A. ORRU, S. & G. *Cagliari.*—Carpet made of stag, deer, and mufflone skins. Prepared skins.

1575. PRATTICO, F. *Naples.*—Bird's skin cap, and muff of Siberian fox fur.

1576. PILLONI, ARNETTA, *Cagliari.*—Prepared skins.

1577. SEVERI, A. *Reggio, Emilia.*—Furs.

1578. SPANO, L. *Oristano (Cagliari).*—Skins of Gangorra.

———

1579. WISER, S. *Modena.*—Sheepskin.

SUB-CLASS C.

1580. PICCINI, A. *Florence.*—Various kinds of brushes.

CLASS XXVI.

Sub-Class A.

1585. Berselli, Ciro, & Co. *Reggio, Emilia.*—Various kinds of leather.

1586. Bossi, E. *Naples.* — Coloured tawed skins for gloves.

1587. Carletti, L. *Chiavenna (Sondrio).*—Collection of tawed skins.

1588. Consigli, G. *Leghorn.*—Leather and buffalo hides.

1589. Deidda, A. *Cagliari.*—Sole leather, and shagreen kid skin.

1590. Del Sere, G. *Florence.*—Leather for various purposes.

1591. Donati & Co. *Sienna.*—Calf-skin and sole leather.

1592. Durio Brothers, *Turin.*—Leather prepared without lime.

1593. Jammy Bonnet, M. *Castellamare (Naples).*—Leather for various purposes.

1594. Mancini, A. *Arezzo (Umbria).*—Sheep-skin.

1595. Orrù, S. & G. *Cagliari.*—Lambskin for gloves; sole, and other leather.

1596. Parma Sub-Committee for the Exhibition.—Hides and leather.

1597. Pellerano, G. B. *Naples.*—Kidskins.

1598. Piella, G. *Pavia.*—Calf-skin and sole leather.

1599. Santoni, F. *Calci (Pisa).*—Sole leather, made with gelatine; calf and kidskin, &c.

1600. Tanning Co. *Modena.*—Calf-skin for shoes.

———

1601. Arnaudon, L. *Turin.*—Goat and sheep leather; morocco and varnished leather.

1602. Avellano Sub-Committee for the Exhibition. — Sheep skins, morocco leather, and parchment.

1603. Baldini, A. & Co.—Sole leather; white and black calf upper leather.

1604. Bolgé, T. *Brescia.* — Parchment for drums, &c.

1605. Capon, G. *Venice.*—Sole and calf shoe leather.

1606. Capretti, P. *Brescia.*—Hides.

1607. Ceresole, Bros. *Turin.* — Black and coloured leather; portmanteau and hog leather.

1608. Cioni, L. *Florence.* — Black and coloured varnished leather.

1609. De Fabritiis, Bros. *Teramo (Abruzzo Ulteriore I.).*—Leather.

1610. Derosa, P. *Benevento.* — Sole leather.

1611. Fiorini, G. *Darfo (Brescia).*—Calf hides.

1612. Fornari, Bros. *Fabriano (Ancona).*—Saddlery, waxed, and morocco leather.

1613. Gambazzi, P. *Brescia.*—Hides.

1614. Impacciatore, T. *Elice (Abruzzo Ulteriore I.).*—Leather.

1615. Lanza, Bros. *Turin.* — Saddlery and calf leather.

1616. Messina Sub-Committee for the Exhibition.—Collection of skins for gloves; gloves.

1617. Piacentini, Cecchi, & Co. *Pescia (Lucca).*—Leather.

1618. Ponci, S. *Sarteano (Florence).*—Parchment.

1619. Pracchi, A. *Lucca.*—Varnished leather.

1620. Romano, F. *Turin.*—Calf leggings; calf leather.

1621. Sorbi, L. *Leghorn.*—Sole leather; waxed and chamois leather; shagreen kid leather.

1622. Stickling, A. *Leghorn.* — Sole leather.

1623. Vignoli, *Forlì.*—Sheep skins, morocco leather.

SUB-CLASS B.

1625. BARBARO, L. *Naples.*—Bridle and harness.

1626. CORA, D. & SONS, *Turin.*—Harness, saddle and bridle, and saddle without bows.

1627. LICHTENBERGER, BROTHERS, *Turin.*—Saddle.

———

1628. ASTORRI, M. *Forlì.*—Harness.

1629. DEL PERO, G. B. *Brescia.*—Whip handles.

1630. MARINO, P. *Turin.*—Carriage trimmings.

1631. SANTI, TALAMUCCI, & SON, *Florence.*—Saddles.

SUB-CLASS C.

1635. MARZOCCHINI, C. *Pisa.*—Leather cigar-cases, match-boxes, and powder-flask.

———

1636. SANTI, TALMUCCI, & SON, *Florence.*—Cigar-cases, match-boxes, tobacco pouches, &c.

CLASS XXVII.

SUB-CLASS A.

1640. AZZI, BROTHERS, *Lucca.*—Stiff and pliable hare-skin hats.

1641. BELTRAMI, P. *Milan.*—Military folding hats; hats in process of manufacture.

1642. BORELLO, P. & BROTHERS.—*Biella (Novara).*—Felt hats made of mixed hair.

1643. CAMPOBASSO SUB-COMMITTEE FOR THE EXHIBITION.—Woollen hats for peasants.

1644. CAVIGLIONE, R. *Turin.*—Silk and gibus hats.

1645. FOSSATI, A. *Monza (Milan).*—Woollen, goat-skin, and hare-skin hats.

1646. GALISE, V. *Naples.*—Silk hats.

1647. PONZONE, A. *Milan.*—Military, flexible, and other hats.

1648. PUGLIESE, A. *Cagliari.*—Woollen caps.

1649. RAVENNA SUB-COMMITTEE FOR THE EXHIBITION.—Felt hats.

———

1650. LA FARINA, *Palermo.*—Collection of silk hats.

1651. MANTELLERO, S. *Sagliano d' Adorno (Novara).*—Otter, hare, and rabbit-skin hats; lambs'-wool hats.

1652. PEONE, BROS. *Leghorn.*—Felt hats.

1653. PIEROTTI, U. & A. *Florence.*—Felt hats.

SUB-CLASS B.

1655. CALZAROSSA, M. *Parma.*—Bonnets and head-dresses.

1656. CLEMENTE, B. *Teramo (Abruzzo Ulteriore I.).*—Straw bonnets.

———

1657. BRAZZINI, D. *Florence.*—Fiesole straw plait, horse-hair, and chenille bonnet-trimmings.

1658. CONTI, C. *S. Giacomo (Florence).*—Collection of straw plait and trimming; Tuscan straw hats and bonnets; straw plait cigar-cases.

1659. KUBLI, J. J. *Florence.*—Collection of Tuscan straw plait and trimmings; straw hats and bonnets; straw plait cigar-cases.

———

2314. MASINI, A. *Florence.*—Collection of straw plait; Tuscan straw hats, &c.

2315. NANNUCCI, A. *Florence.*—Tuscan straw hats, slippers, straw work.

2316. VYSE & SONS, *Prato (Florence).*—Tuscan plait, Tuscan straw hats, trimmings, &c.

SUB-CLASS C.

1660. ALEPPI, L. *Parma.*—Seamless cotton drawers, made with a common frame.

1661. BINDA, A. *Milan.*—Buttons of various materials.

1662. BOSSI, E. *Naples.*—Gloves.

1663. BRACHETTI, G. S. *Giovanni (Arezzo).*—Reversible trousers and convertible clothes.

1664. DE ANGELIS, A. *Messina.*—Artificial flowers.

1665. DE MARTINO, G. *Naples.*—Ornamented parasols and umbrellas.

1666. DESSI MAGNETTI AVV. V. *Cagliari.*—Byssus of the Pinna, with thread, gloves, &c., made of it.

1667. GILARDINI, G. *Turin.*—Umbrellas.

1668. MESSINA SUB-COMMITTEE FOR THE EXHIBITION.—Gloves.

1669. MONTECCHI, E. A. *Parma.*—Artificial leaves and fruit.

1670. PELLERANO, G. B. *Naples.*—Gloves.

1671. PRATTICO, F. *Naples.*—Gloves.

1672. RANDACCIU, M. *Cagliari.*—Shawl made with the byssus of the Pinna.

1673. SALA, F. *Milan.*—Gloves.

1674. TACCHINI, LERTORA & Co. *Milan.*—Scarves, gloves, and cravats of silk; scarves of silk and cotton, and of silk and wool. Collection of buttons of every description.

1675. TESSADA, F. *Genoa.* — Cambric handkerchiefs, burnous, mantilla, shawls, &c.

1676. BERTI, A. *Florence.*—Gloves cleaned by a new method.

1677. CERNUSCHI, BROTHERS, *Milan.* — Silk, woollen, cotton, and other braid cord, and tape; buttons.

1678. DE BENEDETTI, BROTHERS, *Asti (Alexandria).*—Cotton shirts with coloured fronts.

1672. FESTA, G. of *Turin* [41, *Somerset St. Portman Sq. London*].—Stays.

2317. GIOVANETTI, G. & SONS, *Pisa.*—Bone buttons.

SUB-CLASS D.

1680. BRUNO, G. *Turin.* — Boots and slippers.

1681. DELIA, P. *Leghorn.*—Boots and shoes.

1682. GALLI, N. *Pisa.*—Top boots and cavalry boots.

1683. PERRATA, S. *Savona (Genoa).*—Shoes.

1684. ROLANDO, A. *Turin.*—Ladies' silk boots; shoes.

1685. FLORENCE WORKHOUSE (*Pia casa di Lavoro*).—Boots and shoes.

1686. GNESI, G. *Florence.* — Boots and shoes.

1687. PASQUERO, D. *Castiglione (Turin).*—Boots and shoes.

1688. SALANI, A. *Leghorn.*—Boots.

CLASS XXVIII.

SUB-CLASS A.

1695. GHILIOTTI, B. *Pegli (Genoa).* — Hand-made paper.

1696. JACOB, L. & Co. *Milan.*—Specimens of paper, &c.

1697. MAGLIA, PIGNA, & Co. *Milan.*—Paper of various kinds.

1698. MARTELLI, D. *Florence.* — Fancy paper.

1699. MOLINO, P. A. *Milan.*—Pulp for paper manufacture, and paper.

1700. PICCARDO, A. *Genoa.*—Paper.

1701. PLONCHERI, G. *Chiavenna (Sondrio).*—Incombustible paper, made of amianthus.

1702. POLI, A. *Villa Basilica (Lucca).*—Straw paper and pasteboard.

———

1703. AVONDO, BROS. *Borgo Sesia (Turin).*—Drawing, writing, and office paper.

1704. MAGNANI, E. *Pescia (Lucca).*—Hand-made paper.

1705. MAGNANI, G. *Pescia (Lucca).*—Hand-made paper.

1706. MAFFIZOLI, A. *Toscolano (Brescia).*—Collection of drawing and writing paper.

1707. POLLERA, A. M. *Lucca.*—Hand-made paper.

1708. SORVILLO, N. *Naples.*—Drawing, writing, printing, lithographic, and other kinds of paper.

1709. VOLPINI, C. *Florence.*—Printing, drawing, letter, and other paper.

SUB-CLASS C.

1710. BENTIVOGLIO, CAV. C. *Modena.*—Nature printing, obtained by simple pressure.

1711. BERNARDONI, G. *Milan.*—Specimens of printing.

1712. BORZINO, U. *Milan.* — Chromo-lithography.

1713. CORNIENTI, G. *Milan.* — Lithographic portraits, drawn directly on stone.

1714. GAMBERINI, D. *Ravenna.*—Papyrus writing.

1715. GIOZZA, G. *Turin.*—New process for stereotyping, employing very thin moulds, which dry instantaneously.

1716. MECHITARISTI MONKS, *Venice.* — Prayer of S. Narsete, translated into 24 languages; Milton's "Paradise Lost," and other works, in Armenian.

1717. PARIS, A. *Florence.*—Chromo-lithography, &c.

1718. PROSPERINI, P. *Padua.* — Lithography and chromo-lithography; lithographic stones.

1719. RICCÒ, F. *Modena.*—Nature printing, image produced by simple pressure.

1720. SALARI, R. *Florence.*—Interesting fac-similes, written with the pen.

1721. VALLABREGA, G. *Bologna.* — Improved compositor's table.

———

1722. APPIANI & DUCCI, *Florence.*—Self-feeding cushion for stamps.

1723. BOLLINI, P. *Milan.*—Powder for making ink by simple addition of cold water.

1724. CANTI, G. *Milan.*—Music; catalogue of publications.

1725. GRAVINA, D. B. *Monreale (Palermo).* — Chromo-lithographic views of the cathedral of Monreale.

1726. LIVIZZANI, E. *Bologna.* — Silhouettes.

1727. TREVES, M. *Padua.*—Lithography and chromo-lithography.

1728. NOBILI, G. *Naples.*—A volume of chromo-lithographical views of the monuments of Pompeii.

SUB-CLASS D.

1730. BIANCONCINI, L. *Naples.*—Books bound in morocco.

1731. ELISEO, D. *Campobasso (Molise).*—Ornamental binding.

1732. FAGIUOLI, G. *Florence.*—Album for photographs, ornamented with Florentine mosaics; *Guerino il Meschino*, a code of the 15th century, bound in the style of the times.

CLASS XXIX.

Sub-Class A.

1739. Borsari.—Educational works for the deaf and dumb.

1740. Barbèra, G. *Florence.*—Books.

1741. Cellini, M. *Florence.*—Books.

1742. Della Beffa, G. *Genoa.*—Books.

1743. Ferraris, Dr. C. *Alexandria.*—Books.

1744. Gicca, A. *Turin.*—Books.

1745. Guidi, G. G. *Florence.*—Music.

1746. Jervis, W. P. *Royal Italian Central Committee.*—Collection of the Italian newspaper and periodical press.

1747. Le Monnier, Cav. F. *Florence.*—Forty-three volumes of the *Biblioteca Nazionale.*

1748. Lucca, F. *Milan.*—Musical publications.

1749. Marietti, G. *Turin.*—The Scriptures, stereotyped Latin translation.

1750. Naples Sub-Committee for the Exhibition.—Books on various subjects.

1751. Paganucci, Prof. L. *Florence.*—Two plates of the anatomy of the horse.

1752. Pendola, Prof. T. *Sienna.*—Books for the use of the deaf and dumb.

1753. Pinelli, L. *Florence.* — Nuovo Testamento, Diodati's translation; various Evangelical works.

1754. Puccinelli, M. *Lucca.*—Books.

1755. Rabbini, Cav. A. *Turin.* —Trigonometrical surveys, &c.

1756. Riccordi, T. *Milan.*—Music.

1757. Royal Commission of the Italian Exhibition of 1861.—Works descriptive of, and connected with the Italian Exhibition at Florence in 1861; photographs of the building, &c.

1758. Sanseverino, Count F. *Turin.*—Books.

1759. Timon, Cav. A. *Cagliari.*—Books, printed at Cagliari.

1760. Tron, G. *Turin.*—The Scriptures in Italian; collection of Evangelical works.

1761. Villa, A. & G. B. *Milan.*—Geological works.

———

1762. Abbate, G. *Messina.*—Specimen of caligraphy.

1763. Arcozzi Masimo, Avv. L. *Turin.* Treatise on the manner of rearing silkworms, with plates.

1764. Bologna Sub-Committee for the Exhibition.—Scientific periodicals.

1765. Borgo-Cavatti, G. *Cuneo.* — Treatise on ornamentation.

1766. Botarelli, P. *Valiano (Sienna).*—New system of caligraphy.

1767. Castiglioni, P. *Milan.*—Books.

1768. Cellini, M. *Florence.*—Books.

1769. Civelli Institution, *Milan.*—Geographical school atlas.

1770 Florence, Magliabecchian Library.—Collection of books.

1771. Lofaro Pietrasanta, D. Duke of Serradifalco, *Florence.*—The antiquities of Sicily illustrated; notices on the Cathedral of Monreale; album of picturesque views of the antiquities of Sicily.

1772. Mazzei, Cav. F. *Florence.*—Books.

1773. Migliaccio, R. *Salerno (Principato Citeriore).*—Books.

1774. Muzzi, L. *Florence.*—Phonic system of teaching to read and write Italian.

1775. Ramo, S. *Naples.*—Books.

1776. Rizzetti, Dr. G. *Cagliari.* —Medical and chemical works of the Exhibitor.

1777. Santerini, Bros. *Cesena (Forlì)* Caligraphy.

1778. Tenerelli, F. *Teramo (Abruzzo Ulteriore I.).* — A book — Easy method of learning to read Italian.

1779. Vigano, F. *Milan.*—The Exhibitor's works on political economy, &c.

———

2318. Lambruschini, Cav. Prof. R.— Various works on public instruction.

2319. Marzullo, *Palermo.* — Grammar for the deaf and dumb.

2320. Unione Typografica Torinese.— *Encyclopedia Populare Italiana ; Nuova Biblioteca Italiana ; Dizionario della Lingua Italiana, &c.*

Sub-Class B.

1780. Caimi, E. *Sondrio.*—Topographical model of the Pass of the Stelvio.

1781. Capurro, Rev. G. F. *Novi, Alexandria.*—Telegraphic alphabet, for teaching a large number of children without books.

1782. Fàa di Bruno, Cav. F. *Turin.*— Educational apparatus.

1783. Florence Royal Natural History Museum.—Botanical specimens. Collection of graminaceous plants. Specimens from the Royal Museum : Various kinds of rice, with semolina, and flour made of it. Indian corn, with semolina, flour, bread, and alcohol made from it. Various kinds of canary seed. Sorgho, with sugar, rum, alcohol, and bread made from it. Varieties of millet. Clothes, ropes, and matting, with the materials of which they were made. Brooms, baskets, plait, brushes, and wicker-work, with the materials of which they were made. Oats. Cane ; and bamboo cane, with various objects made of it ; bamboo preserve. Wheat, with flour, bran, starch, wafers, bread, biscuits, maccaroni, alcohol, &c., made from it, and articles manufactured with its straw. Barley, with flour, ale, and porter made from it, and articles made from its straw. Sugar-cane, sugar, sugar-candy, and rum. Other cereals.

1784. Gennari, Prof. P. *Cagliari.*— Minerals, and natural history specimens.

1785. Milazzo Junta for the Exhibition (*Messina*).—Collection of non-metalliferous minerals, shells, seaweeds, and fossils.

1786. Pavia University, Machinery Museum.—Drawings of machinery.

1787. Pisa, Natural History Museum, Royal University of.—Plaster and wax models of fossils.

1788. Turin Engineering School. — Models of fossils and crystals.

1789. Zappala, G. *Messina.*—Moses, a plaster cast.

1790. Calenzoli, C. & S. *Florence.*— Anatomical wax figures and preparations ; lymphatic system ; the eye, ear, brain, &c.

1791. Casella, Dr. G. *Laglio (Como).* —Fossils.

1792. Convent of the Signore delle Quiete, *Florence.*—Plan and elevation of the school attached to the convent.

1793. Florence Infant Asylums. — Photographic drawings, plans, regulations, &c.

1794. Florence, Laurentian Library. Views of the library, &c.

1795. Florence Municipal Schools.— Regulations and statistics, &c.

1796. Florence Riccardian Library. —Drawings.

1797. Florence Royal Academy of Fine Arts (Library of).—Collection of illustrated works.

1798. Florence, Royal Archæological Department for the Tuscan Provinces. —Photographic views of the archive office.

2321. Florence Artistic Society. — Collection of engravings.

2322. Florence, Library of the Hospital of Sta. Maria Novella.—Books.

2323. Florence, Royal Natural History Museum.—Views of the museum.

2324. Florence, Royal Gallery of Mosaics in Pietre Dure.—Views of the gallery ; historical notices.

2325. Florence Royal Lyceum and Gymnasium.—Plan and photographs of the building, &c.

2326. Florence Royal Marucellian Library.—Photographic views.

2327. Florence Royal Natural History Museum.—Wax preparation of the grape disease ; anatomical wax preparations of the rabbit.

2328. Florence Royal Normal Girls' Schools for the People.—Regulations and statistical notices ; photographic view.

2329. Florence Royal Normal Boys' School.—Plans, regulations, and statistics.

2330. FLORENCE ROYAL NORMAL GIRLS' SCHOOL.—Photographic views, regulations, and statistics of the school.

2331. FLORENCE S. MARCO DOMENICAN LIBRARY.—Photographic view of the library.

2332. FLORENCE WORKHOUSE. — Photographs; statistics; notices on the system of education adopted.

2333. MAZZEI, CAV. F. *Florence.*—Photographs of the restorations at the Palazzo del Podestà.

2334. MALATESTIAN LIBRARY, *Cesena* (*Forlì*).—Photographs of the library.

2335. MILAN SOCIETY FOR THE PROMOTION OF ARTS AND MANUFACTURES. — Diagrams of agricultural implements.

2336. PAVIA UNIVERSITY, ZOOLOGICAL MUSEUM OF.—Drawings illustrating the development and diseases of the silkworm, by Dr. Maestri.

2337. PIEROTTI, P. *Milan.* — Plaster casts, &c.

2338. PISA ROYAL UNIVERSITY.—Photographs and plans.

2339. PRATO ORPHAN ASYLUM, *Prato* (*Florence*).—Photographic views of the establishment.

2340. PRATO ROYAL LYCEUM, *Prato* (*Florence*).—Plans of the Lyceum.

2341. RANDACCIU, G. *Sassari.*—Anatomical wax figures.

2342. RAVENNA, CLASSENSE PUBLIC LIBRARY.—Plans of the library.

2343. RAVENNA ROYAL ACADEMY OF FINE ARTS.—Plans of the academy.

2344. RIPOLI, CONSERVATORIO DI, *Florence.*—Photographs.

2345. RONDANI, PROF. C. *Parma.*—Collection of Italian dipterous insects, with account of the same, &c.

2346 S. ANDREA WORKHOUSE, *Leghorn.*—Photographs of the establishment, statistics, &c.

2348. S. S. ANNUNZIATA ROYAL INSTITUTE FOR GIRLS, *Florence.*—Photographs of the establishment, &c.

2349. VIAREGGIO, PROPOSED HOSPITAL (*Lucca*).—A photograph, &c.

2350. VILLA, I. *Florence.*—New terrestrial planisphere, indicating the time for every longitude; new celestial planisphere, indicating the passage of stars for every terrestrial meridian; cosmographical diagrams, &c.; collection of photographs of the Exhibitor's artistic works.

SUB-CLASS D.

1799. BOLGI, T. *Brescia.*—Toys.

CLASS XXX.

SUB-CLASS A.

1800. BERTOLOTTI, G. *Savona* (*Genoa*).—Collection of marquetry tables.

1801. BETTI, F. *Florence.* — Table in Florentine mosaics.

1802. BIANCHINI, PROF. G. *Florence.*—Table in Florentine mosaics.

1803. BINAZZI, G. *Florence.*—Real and imitation mosaic tables.

1804. BOCCHIA, E. *Parma.*—Inlaid-wood wash-stand.

1805. BOSI, E. *Florence.* — Tables and ebony case, inlaid with mosaic and bronze work, &c.

1806. CANTIERI, G. *Lucca.*—Work-table inlaid with tortoiseshell, wood, and ivory.

1807. CANEPA, G. B. *Chiavari* (*Genoa*).—Chiavari chairs.

1808. CENA, G. *Turin.*—Inlaid table, &c.

1809. CIACCHI, J. *Florence.*—Inlaid pavement in Byzanto-Gothic style.

1810. COEN, M. *Leghorn.*—Walnut-wood sideboard.

1811. COSTA, A. *Lavagna (Genoa).*—Slate tables inlaid with various marbles.

1812. DELLA VALLE, P. *Leghorn.*—Inlaid scagliola tables.

1813. DELLEPIANE, L. *Savona (Genoa).*—Chairs.

1814. DE MARTINO, G. *Naples.*—Mahogany toilet table.

1815. DESCALZI, E. *Chiavari (Genoa).*—Chiavari chairs.

1816. DESCALZI, G. *Chiavari (Genoa).*—Chiavari chairs; inlaid table.

1817. ESCOUBAS, M. A. & SCOTTI, I. *Genoa.*—Oval inlaid tables.

1818. FLORENCE WORKHOUSE, *Florence.*—Veneered sideboard.

1819. FLORENCE ROYAL MANUFACTORY OF MOSAICS IN PIETRE DURE.—Florentine mosaic table, mosaics, vases of Egyptian porphyry, carved and inlaid chest; collection of 121 siliceous stones employed in the Royal Manufacture of Pietre-Dure mosaics.

1820. FRANCESCHI, E. *Florence.*—Looking-glass.

1821. FRULLINI, L. *Florence.*—Walnut-wood escritoire, style of the 14th century.

1822. GARGIULO, L. *Sorrento (Naples).*—Marquetry furniture, made with Italian woods.

1823. GIUSTI, PROF. P. *Sienna.*—Sculptured chests.

1824. GUALA, G. *Turin.*—Walnut-wood fire-screen, and bedstead.

1825. HOLMAN, R. *Florence.*—Wardrobe, ancient style.

1826. IANNICELLI, M. *Salerno (Principato Citeriore).*—Mosaic toilet table.

1827. LANCETTI, F. *Perugia (Umbria).*—Ebony table and casket, inlaid with woods, ivory, &c.

1828. LEVERA, BROTHERS, & Co. *Turin.*—Various articles of furniture, inlaid, &c.

1829. LURASCHI, A. *Milan.* — Billiard tables.

1830. LUCCHESI, BROTHERS, *Lucca.* — Small writing-table inlaid with metal, &c.

1831. MAINARDI, B. *Milan.*—Inlaid table, with mosaic painting.

1832. MARTINOTTI, G. & SONS, *Turin.*—Cornices, escritoires, tables, and sideboards, inlaid, gilt, &c.: folding travelling furniture.

1833. MERLINI, C. *Florence.*—Ebony chest, with compartments in Florentine mosaic.

1834. MONTELATICI, A. BROTHERS, *Florence.*—Round table in Florentine mosaic.

1835. MONTENERI, A. *Perugia (Umbria).*—Marquetry views of Rome, Venice, Florence, Naples, &c.

1836. MOROZZI, F. *Florence.*—Machine-cut veneer.

1837. MUSICO, D. *Messina.*—Chairs.

1838. ODIFREDI, G. *Leghorn.*—Toilet and writing-table, inlaid.

1839. PASQUINI, G. *Florence.*—Roll of nut-wood veneer, cut by circular saw.

1840. RIGHINI, C. *Milan.*—Furniture.

1841. ROVELLI, C. *Milan.* — Window blinds.

1842. SCALETTI, A. *Florence.* — Ebony coffer, inlaid, &c.; inkstand.

1843. SCOTTI, I. *Genoa.*—Inlaid tables.

1844. SGUEZZO, V. *Savona (Genoa).*—Chairs.

1845. TORRINI, G. & VIECCHI, C. *Florence.*—Tables in Florentine mosaics.

1846. VITI, CAV. A. *Volterra (Pisa).*—Tables in coloured and indurated alabaster, in imitation of Florentine mosaics.

1847 ZORA, G. *Turin.* — Inlaid wood-flooring.

———

1848. BARBETTI, A. & SONS, *Florence.*—Carved walnut-wood bookcase, carved bench, &c.

1849. BIGAGLIA, CAV. P. *Venice.*—Inlaid tables.

1850. BUCCI, R. *Ravenna.*—Inlaid table.

1851. CHALON & ESTIENNE, *Florence.*—Inlaid wood flooring, models of inlaid flooring.

1852. CORRIDI, P. *Leghorn.* — Inlaid table-top.

1853. FONTANA, D. *Milan.*—Monumental inlaid escritoire.

1854. GATTI, G. *Rome.*—Inlaid cabinet and looking-glass frame.

1855. GHIRARDI, G. *Brescia.*—Furniture.

1856. GRANDVILLE, M. *Sorrento (Naples).*—Marqueterie work.

1857. INGEGNERI, P. *Scilla (Calabria Ulteriore I.).* — Terra-cotta figures in the Abruzzan and Sicilian costume.

1858. MORESCHI, G. A. *Brescia.*—Table.

1859. NOVI, C. *Brescia.*—Inlaid table.

1860. SALVIATI, DR. A. *Venice.*—Roman and Venetian mosaic tables.

1861. TANGASSI, CAV. BROS.—*Volterra (Pisa).*—Inlaid alabaster table.

1862. ZAMPINI, L. *Florence.*—Imitation Chinese lacquered folding screen.

SUB-CLASS B.

1870. BACCI, F. *Florence.* — Cornices, brackets, Corinthian capital, &c.

1871. BARBENSI, G. *Florence.*—Napoleon III. on horseback, executed in Florentine mosaics; table in Florentine mosaics.

1872. BENSI, C. *Volterra (Pisa).*—Candelabra in veined alabaster; alabaster vases.

1873. CHERICI, G. & BROTHERS, *Volterra (Pisa).*—Alabaster vases, and candelabrum.

1874. FRANCESCHI, E. *Florence.*—Ornamental cornices, brackets, and frames.

1875. FRULLINI, L. *Florence.*—Figures in nut-wood; bracket, with grotesque animals' heads; ebony jewel-case, with relievo figures.

1876. GIUSTI, PROF. P. *Sienna.*—Sculptured and carved frames.

1877. LOMBARDI, A. *Sienna.*—Carved walnut-wood frame.

1878. PARENTI, G. *Volterra (Pisa).*—Paper weights, and alabaster vase.

1879. PAPI, L. *Florence.*—Carved frame, purchased by H.M. the King of Italy.

1880. PICCHI, A. *Florence.*—Cornices in ebony, &c.

1881. RENZONI, A. *Pisa.*—Model of the church of S. Maria della Spina Pisa, and of the leaning tower of Pisa, in alabaster.

1882. TANGASSI, CAV. C. & BROTHERS, *Volterra (Pisa).*—Articles in alabaster.

1883. VITI, CAV. A. *Volterra (Pisa).*—Alabaster statuettes.

1884. VAN LINT, E. *Pisa.*—Alabaster ornament.

1885. ZAMBELLI, G. B. *Milan.*—Wood carving.

1886. AMBROGIO, G. S. *Alessandro (Brescia).* — View of Solferino, carved in cork.

1887. BILLOTTI, DR. P. *Turin.*—Copies of paintings executed on marble.

1888. BRILLA, A. *Savona (Genoa).*—Crucifix.

1889. COLLETTI, M. *Florence.*—Carved frames.

1890. FIESCHI (CONSERVATORY), *Genoa.*—Bouquet, artificial and wild flowers.

1891. GARASSINO, V. *Savona (Genoa).*—Ivory statuette; painting in inlaid woods.

1892. GARNIER VALLETTI, F. *Turin.*—Artificial fruit.

1893. LIPPI, A. *Pietrasanta (Lucca).*—Statuary marble platter.

1894. MICALI, G. & SONS, *Leghorn.*—Alabaster work.

1895. NEGRONI, G. *Bologna.*—Epergne, with chased work.

1896. NORCHI, E. *Volterra (Pisa).*—Alabaster vases and statuettes.

1897. PACINOTTI, F. *Florence.*—Engraved marbles.

1898. SARTORI, G. *Venice.* — Carved picture frames, brackets, &c.

1899. TRABALLESI, P. *Florence.*—Bacchus sleeping in a barrel; plaster of Paris model, in one piece.

CLASS XXXI.

Sub-Class A.

1900. ALFANO, A. & G. *Naples.*—Bedstead.

1901. ANGIOLILLO, G. A. *Campobasso (Molise).*—Lock.

1902. AZZERBONI, C. *Pontassieve (Florence).*—Locks.

1903. BALDANTONI, G. B. *Ancona.*—Iron bedstead.

1904. BARGIANI, F. *Pisa.*—Shoes for race-horses.

1905. BECCALOSSI, F. *Brescia.* — Hand and machine-made nails, screws, hammers, &c.

1906. BOLZANI, S. *Milan.*—Wire gauze.

1907. CALEGARI, V. *Leghorn.*—Cast-iron work, &c.

1908. CAMPOBASSO SUB-COMMITTEE FOR THE EXHIBITION.—Door locks, bells for cattle.

1909. CECCHETTA, P. *Pisa.* — Horse-shoes.

1910. CESARI, G. *Cremona.*—Strong-box in hammered iron-work, with cast-iron and bronze ornaments.

1911. CIANI, G. *Florence.*—Lock, with countless combinations; other locks.

1912. CIMA, G. B. *Lecco (Como).* — Hardware.

1913. COBIANCHI, V. *Omegna (Novara).*—Iron wire.

1914. FLORENCE WORKHOUSE. — Bedsteads, table, chairs, and other articles of iron.

1915. FORNARA, G. *Turin.* — Various kinds of wire gauze and fire guards.

1916. FRANCI, P. *Sienna.*—Iron gate, made with the hammer.

1917. GHIBELLINI, BROTHERS, *Persicet (Bologna).*—Japanned iron table and sofa.

1918. GUPPY & PATISSON, *Naples.* — Collection of iron and brass nails.

1919. LONDINI, BROTHERS, *Bologna.*—Iron bedstead.

1920. MACRY, HENRY, & Co. *Naples.*—Cast-iron candelabrum; ornamental castings.

1921. MOMBELLI, G. *Milan.*—Pins and nails.

1922. MOSSONE, G. B. *Andorno, Cacciorna (Novara).*—Locks for strong-boxes and shops.

1924. PIETRARSA ROYAL WORKS, *Portici (Naples).*—Ornamental iron castings.

1925. RUSCONI, A. *Breno (Brescia).*—Ploughshares, frying-pans, &c.

1926. SIMION, G. *Pescia (Lucca).* — Frame for paper manufacture.

1927. SPANO, L. *Oristano (Cagliari).*—Gate-lock used in Sardinia.

————

1928. BOLGÉ, T. *Brescia.*—Iron wire-work.

1929. BEVILACQUA, P. *Campobasso (Molise).*—New system of locks.

1930. DE LA MORTE, F. *Naples.*—Ornamental iron castings.

1931. IGNESTI, F. *Florence.* — Helmet hammered from a single piece of iron.

1932. THEODORANI, S. *Forlì.*—Steelyards.

1933. ROVELLI, C. *Milan.*—Iron sofa, stool, and chairs.

Sub-Class B.

1935. BECCALOSSI, F. *Brescia.*—Candlesticks and other articles in brass.

1936. CAMMILLETTI, A. *Perugia (Umbria).*—Candelabrum and other articles in bronze.

1937. MANUELLI, G. *Prato (Florence).*—Domestic utensils in copper.

1938. MARINELLI, T. *Agnone (Molise).*—Bronze bell, for a belfry, producing an entire semitonic octave. ————

1939. PENZA, F. *Naples.*—Ornamental castings.

1940. GUPPY & PATISSON, *Naples.* — Brass work.

SUB-CLASS B.

1941. GIANI, V. *Como*.—Bronze statue of Balilla, cast at the Royal Arsenal, Turin.

SUB-CLASS C.

1945. COLOMBO, N. *Milan*.—Leaden and tinned articles.

1946. KRAMER & Co. *Milan*. — Lead pipes.

1947. SIMMOLA, CAV. F. *Naples*. — Metal-work. ——

1948. DECQPPET, L. *Turin*.—Patent lead pipes.

1949. LAU, A. *Naples*. — Metal tea service, &c.

CLASS XXXII.

SUB-CLASS A.

1960. CERIE, C. *Lucca*.—Steel bit and spurs.

SUB-CLASS B.

1965. BARBERI, A. *Turin*.—Various objects in cutlery.

1966. BECCALOSSI, F. *Brescia*.—Knives, forks, and files.

1967. DUINA, A. *Brescia*.—Cutlery.

1968. GRAVINA, M. *Campobasso* (*Molise*).—Scissors, knives, razors, &c.

1969. OLMETTA, A. *Cagliari*. — Large knife.

1970. SANTANGELO, S. *Campobasso* (*Molise*).—Clasp-knives, scissors, razors, &c.

1971. SELLA, L. *Massareno* (*Novara*).—Knives, penknives, razors, &c.

1972. SPETRINI, L. *Campobasso* (*Molise*).—Chased dagger.

1973. VENDITTI & TERZANO, *Campobasso* (*Molise*).—Knives, scissors, penknives, &c.

1974. VILLANI, R. *Campobasso* (*Molise*).—Damasked and chased table and dessert knives.

1975. VINEIS, G. B. *Biella* (*Novara*).—Hunting knife, scythes.

1976. VINEIS-BARON & BROTHERS, *Mongrando* (*Novara*).—Scythes.

1977. VINEIS, C. & BROTHERS, *Mongrando* (*Novara*).—Scythes.

1978. VINEIS, M. G. *Mongrando* (*Novara*).—Scythes.

1979. VINEIS, S. & BROTHERS, *Mongrando* (*Novara*).—Scythes.

1980. VINEIS, T. & BROTHERS, *Bologna*,—Scythes.

1981. VINEIS, T. & NEPHEWS, *Mongrando* (*Novara*).—Scythes.

——

1982. BUFFI, G. *Scarperia*. — Collection of cutlery.

1983. DE STEFANO, BROTHERS, *Campobasso* (*Molise*).—Cutlery.

1984. SPINA, SANTALA, F. *Acireale* (*Catania*).—Scythes.

1985. TORO, P. A. *Campobasso* (*Molise*).—Cutlery.

CLASS XXXIII.

1990. Ambrosini, G. *Naples.* — Coral necklace, ear-rings, brooches, &c.

1991. Calvi, G. *Ripateatina (Abruzzo Citeriore).*—Jewel case with 5 stones from Vesuvius.

1992. Della Valle, P. *Leghorn.*— Brooches in scagliola, in imitation of the Florentine mosaics.

1993. Forte, E. *Genoa.*—Filigree work.

1994. Fusco, G. *Naples.*—Red coral, and lava work.

1995. Ghezzi, A. & Sons, *Milan.*—Engraved silver chalice, candelabra, &c.

1996. Grisetti, E. *Milan.*—Gold brooch set with stones.

1997. Guida, C. *Trapani.*—Red coral work.

1998. Labriola, F. *Naples.* — Various articles in tortoiseshell, inlaid with gold.

1999. Masini, G. *Naples.*—Lava ornaments, electrotype articles, bronze and silver articles.

2000. Minottini, G. *Perugia (Umbria).* —Reliquary with engraved work.

2001. Parazzoli, L. *Milan.* — Ring serving as perpetual calendar.

2002. Peluffo, V. *Cagliari.*—Peasants' jewellery.

2003. Pieroni, A. *Lucca.* — Figure in silver.

2004. Pierotti, P. *Milan.* — Galvanoplastic imitations of ancient shields, helmets, &c.

2005. Rocca, R. & A. *Modena.*—Table ornament.

2006. Scaletti, A. *Florence.* — Church ornament.

2007. Avolio & Sons, *Naples.* — Collection of coral and lava work.

2008. Bologna Royal Mint. — Medals and coins.

2009. Borani, Cav. *Turin.*—Swords presented by the Tuscans to General Alfonso della Marmora; another sword presented by the Legations and Marches; another sword

presented as a remembrance of the Piedmontese troops in the Crimea. Laurel crown, with precious stones, presented to General Cialdini, after the taking of Gaeta.

2010. Capurro, N. *Pisa.*—Books.

2011. Castellani, *of Rome and London.* —Large collection of archæological jewels from existing originals.

2012. Ercolani, E. *Florence.*—Copy of St. John by Donatello, beaten on a plate of metal.

2013. Florence Royal Mint.—Medals.

2014. Germani, Dr. G. *Cremona.*—Engraved quartz and hyacinth.

2360. Nannei, G. *Florence.* — Turned silver cup.

2361. Naples Royal Mint.—Coins and medals.

2362. Pane, M. *Naples.*—Wrought silver work.

2363. Rogai, L. *Florence.*—Steel dies.

2364. Royal Italian Committee for the Exhibition.—Swords, presented to H.M. the King of Italy, by the citizens of Rome; sword presented by several cities of Central Italy; sword presented by the armourers of Mongiana.

2365. Salviati, Avv. A. *Venice.*—Silver filigree model of the church of St. Mark's, &c.

2366. Santarelli, Prof. E. *Florence.*— Collection of medals.

2367. Sichling, A. *Turin.*—Sword, the property of H.M. the King of Italy.

2368. Torroni, G. & Vecchi, C. *Florence.* —Florentine mosaic paper weights, brooches, &c.

2369. Turin Royal Mint.—Collection of medals.

2370. Penna, S. & C. *Leghorn.*—Cameos.

2371. Bassi, B. *Macerata.*—Cameos.

2372. Finizio, G.—Cameos, shell and lava.

2373. Laodicini, G.—Shell cameos.

2375. Cortellazzo, A. *Vicenza.*—Sword belonging to the King of Italy; repoussé iron-plate, inlaid with silver gilt.

CLASS XXXIV.

SUB-CLASS A.

2015. BIGAGLIA, CAV. P. *Venice.*—Collection of artificial aventurine, and other glass work.

2016. FRANCINI, G. *Florence.*—Painted glass window, style of the 14th century.

2017. PACINOTTI, F. *Florence.*—Engraving on coloured glass.

2018. SALVIATI, Dr. A. *Venice.*—Collection of artificial chalcedony, mosaics, enamelled, and other glass work.

2019. TRARI, M. *Bologna.*—The three Graces of Canova, engraved on gilt glass.

———

2020. FRANCISCI, FATTORINI, & MORETTI, *Todi (Umbria).*—Painted glass windows.

2021. BERTINI, *Milan.*—Painted glass window.

SUB-CLASS B.

2030. BRUNO, G. *Naples.*—Glass shades.

2031. CARAFA DI NOIA, P. *S. Giovanni a Teduccio, Naples.*—Bon-bons for acids.

2032. NARDI, R. & SON, *Montelupo (Florence).*—Wine and oil flasks, protected by straw.

2033. VENICE UNITED MANUFACTORIES.—Large collection of Venice beads.

2034. MORGANTINI & BERNARDINI, *Ravenna.*—Blown glass.

2035. MENCACCI, M. & Co. *Lucca.*—Flasks protected by straw; coloured glass.

2036. SEVOULLE, B. & Co. *Vietri (Principato Citeriore).*—Window glass; glass shades.

———

CLASS XXXV.

2040. ARMAO, G. *S. Stefano di Camastra (Messina).*—Imitation Pompeian and Egyptian vases.

2041. BELTRAMI, COUNT P. *Cagliari.*—Terra-cotta stove, &c.

2042. CALVETTI, AVV. G. *Pianezza (Turin).*—Terra-cotta flower vase.

2043. COLONNESE, F. & G. *Naples.*—Terra-cotta flower vases, imitations of Etruscan and Greco-Siculean vases.

2044. FERNIANI, COUNT A. *Faenza (Ravenna).*—Two majolica vases, painted in the ancient style.

2045. GALEAZZO, G. A. *Castellamonte (Turin).*—Economical earthenware fire-grates.

2046. GINORI LISCI, MARQUIS L. *Florence.*—Collection of porcelain; collection of majolica, in imitation of that of Urbino and Pesaro, of the 14th and 15th century; imitation Lucca della Robbia ware; majolica for common use; earthenware; stoves.

2047. MARRAS, F. *Assemini (Cagliari).*—Collection of pottery.

2048. OLIVIERI & FERRO, *Savona (Genoa).*—Tobacco pipes.

2049. RICHARD & Co. *Milan.*—Porcelain services: white and coloured earthenware; crucibles, &c.; garden vases; fire bricks.

2050. RONDANI, T. *Parma.*—Earthenware diaphragms and cells for galvanic batteries; earthenware furnaces for jewellers.

2051. SAVONA JUNTA FOR THE EXHIBITION (*Genoa*).—Collection of common stoves.

2052. SPANO, L. *Oristano (Cagliari).*—Collection of terra-cotta stoves.

———

2053. BERTE & STROBEL, *Parma.*—Fire pan; clay and quartz employed in the manufacture of the pans.

2054. Carocci, Fabbri, & Co. *Gubbio (Umbria).*—Vases, plates, &c. with historical figures.

2055. Furlani, G. & Co. *Florence.*—Terra-cotta stove.

2056. Mossa, Bros. *Pianezza (Turin).*—Water pipes, flower pots, &c.

CLASS XXXVI.

Sub-Class B.

2070. Cora, D. & Sons, *Turin.*—Portmanteau.

2071. Ghezzi, *Milan.* — Portmanteaux, trunks, &c.

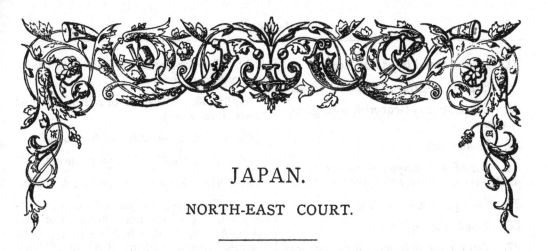

JAPAN.

NORTH-EAST COURT.

1.   ALCOCK, R. Esq., H. M. Envoy Extraordinary and Minister Plenipotentiary at the Court of the Tycoon.

A.—*Specimens of Lacquer Ware, Lacquering on Wood, Inlaid Wood and Lacquer mixed, Lacquer on other materials, as Ivory, Shells, Tortoiseshell, &c.*

INLAID WOODS.

1.   Very curious old lacquered cabinet in colours, with enamelled figures in basso relievo; richly chased bronze and metal handles.

2.   Old lacquered cabinet, with gold leaves and figures.

3.   Old cheffonier cabinet, with intaglios and carved figures in basso relievo inlaid; a very unique specimen.

4.   Small toilet cabinet of red and carved lacquer, with carved ivory handles; the same sort of lacquer known in China as "*Soochow ware*."

5.   Large cabinet on hard wood, elaborately carved and lacquered, with cameos of finely-chiselled ivory, and embossed metal handles; very old, and perfectly unique of its kind.

6.   Small carved cabinet of plain unlacquered cedar wood; brought from Osaca.

7.   Small toilet cabinet and cheffonier; very fine lacquer.

8.   Small cabinet drawers for toilet.

9.   Ditto, ditto.

10.   Lacquered cabinet stand.

11.   Ditto, with globe-top imitation cracked china, fitted up elaborately with all the requisites for a luncheon party, plates, dishes, cups, and saké bottle.   Cabinets of lacquer and basket work mixed.

12.   Cabinet model of a Japanese tea-house and garden, used for luncheon baskets.

13.   Ditto, ditto, another specimen.

14.   Ditto, ditto.

15.   Another cabinet model of a Japanese tea-house, &c.

16.   Ditto, ditto.

17.   Small cabinet of inlaid straw work.

18.   Basket and lacquer cabinet.   Cabinets of inlaid wood.

19.   Small cabinet of inlaid wood.

20.   Cabinet drawers of inlaid wood, from Hakoni, which open out, and form a portable writing-table.

21.   Cabinet of inlaid wood.

21A.   Ditto of carved wood, with folding top.

22.   Small lacquer etagère, old pattern.

23.   Cabinet drawers of inlaid wood, forming a portable writing-table.

24.   Trays and stands.

25.   Large circular lacquered tray.

26.   Another specimen of large circular lacquered tray.

27.   Small square lacquered tray.

28.   Two lacquered round dishes.

29.   Richly lacquered circular tray.

30. Nest of round trays, old lacquer.

31. Small oblong tray of common lacquer.

32. Large tea tray lacquered, with graceful pattern.

33. Smaller ditto.

34. Set of old lacquered square trays, with Pompeian pattern and carved borders.

35. Four pairs of ditto, ditto.

36. Stand of same pattern.

37. Ditto, more elegant in form, but less classic.

38. Small lacquered tray, old, and fine in quality.

39. Set of two square trays, old lacquer.

40. Set of three, ditto.

41. Small round lacquered salver.

42. Oblong stand for tea or saké.

43. Specimen of old and fine lacquer; small box.

44. Card tray or bowl, with richly lacquered and gold figures.

45. Set of three lacquered trays, with graceful pattern.

46. Set of six square lacquered trays.

47. Old lacquered small tray.

48. Small tray, of modern lacquer.

49. A lacquered stand.

50. Ditto, ditto.

51. Gold-lacquered waiter, of the finest kind, and with crest of Daimio (Ando-Tsnsima-no-Kami, Minister for Foreign Affairs).

52. Six small lacquer stands.

53. Old lacquered box, with circular cover.

54. Pair of lacquer spill stands.

55. Pair of lacquer bon-bon boxes.

56. Gold-lacquered nest of boxes.

57. Small circular red lacquer box, with vine leaf.

58. Two small lacquered jars.

59. Large lacquered pot pourri bowl, with cover.

60. Two lacquered red saucers and small lacquered cups for tea.

61, 62, 63. Three sets of richly lacquered bowls (three each).

64, 65. Two sets ditto, of five each.

66. Three small red lacquer saucers.

67. Three lacquer cups and a saké ladle.

68. Three large lacquer saké cups, a present from one of the late Governors of Foreign Affairs, Hori-oribeno-no-Kami, with the usual accompaniment of a slip of dried fish, tied up with a red and white string made of paper, in token of humility, as coming from a race of poor fishermen.

69. Fine lacquered cup and cover, bamboo pattern.

70. Lacquer cup and cover.

71. Two silver-lined lacquer goblets.

72. Round lacquered bowl, with cover.

73. Ditto, ditto.

74. Ditto, smaller, with cover.

75. Ditto, ditto.

76. Ditto, ditto.

77. Daimio's rice bowl, of the richest gold lacquer.

78. Daimio's saké pot, with crest of Makino-bizen-no-Kami.

79. Set of three rich lacquered cups, of rare pattern and colour.

80, 80A. Two large lacquered bowls.

81. Set of three lacquered bowls, on stands of the richest kind.

82. Lacquered cup, saucer, and stand; fine specimens made for Kino-shita-chi-kugono-Kami, a Daimio.

83. Set of eight lacquered cups and saucers, without stands.

84. Ingenious luncheon service, in form of a melon.

85. Ditto, in form of a gourd.

86. Lacquer fish, in which an entire fish is served at great feasts.

87. Large lacquer set of trays, for same use.

88. Large lacquer rice bowl.

89. Lacquer cup, with cover.

90. Lacquered bowl, with cover.

91. Lacquered basin, with cover for soup.

92. Set of three lacquered basins, with encrusted exterior.

93. Richly lacquered bon-bon box.

94. Gold lacquered pencil box.

95. Ditto, ditto.

96. Card tray and counter boxes.

97. Lady's toilet box, with smaller cases inside for cosmetics, &c.

98. Ditto, ditto.

99. Note and envelope case, richly ornamented.

100. Butterfly lacquer box.

101. Set of ring drawers for toilet table.

102. Lacquer box, with cover.

103. A card box, with tray, of fine old lacquer.

104. Two round common lacquer trays.

105. Lady's small toilet box, of gold lacquer.

106. Lady's lacquered comb.

107. Gold oblong toilet box, with trays.

108. Pentagonal lacquer box.

109. Ring box.

110. Fan-shaped ring box, gold lacquer.

111. Small square ditto.

112, 113, 114. Glove boxes (used by Japanese for letters and despatches).

115, 116. Ditto, ditto, red lacquer.

117. Ditto, ditto, gold and silver lacquer.

118. Glove box, Tycoon's pattern.

119. Small lacquer square box.

120. Round lacquered box.

121. Pair of lacquer work baskets, with handles.

122. Toilet case, of rich lacquer.

123. Small card box, of lacquer.

124. Carved writing and pencil case, with silver medicine box.

125. Set of three pencil trays.

126. Officer's lacquered hat.

127. Toilet bottle, with lacquered case.

128. Pair of lacquered boxes.

129. One ditto, ditto.

130. Lacquered card box, with tray.

131. Ditto, of fine lacquer.

132. Square box, old lacquer, of Pompeian style.

133. Daimio's luncheon box.

134. Set of three models of Daimio's wardrobes or coffers.

135. Tobacco apparatus for fire, &c.

136. Ditto, ditto, of inlaid wood.

137. Fan-shaped work-box, with trays.

138. Lacquered box.

139. Gentleman's dressing-case.

140. Lacquered table, Dutch model (Nagasaki ware, made for European market formerly).

MODELS OF JAPANESE LADIES' TOILET APPARATUS.

141. Lacquered clothes horse.

142. Lacquered stand for looking-glass.

143, 144. Lacquered drawer for paints, on which a toilet glass also fits.

145. Bowls for water, and washing of mouth.

146. Can for water, and foot pan.

147, 148. Ditto, sizes usually employed.

149. A stand for Daimios' swords.

150. A stand for Daimios when dressing.

151. Rich lacquered trousseau box, for a Daimio's wife, crest of Juabu-tangono-Kami.

152. Dressing-case box for ditto.

153. Ditto, richly lacquered and ornamented.

154. Another pattern for ditto, made for the Regent, Ikomono-no-Kami, murdered in 1860.

155. Ditto, with drawers, bearing the crest of the Prince of Siconsin.

156. Small toilet cabinet and cheffonier, of superior lacquer.

157. Daimio's despatch box, with crest.

158. Ditto, dressing box to match (Iko-mono-Kami; the late Regent).

159. Despatch box, made for Kiogoko-manga-tono-Kami.

160. Daimio's pillow, with box, made of very fine lacquer, and of great value.

161. Carved bamboo spill stand, very old and finely chiselled.

162. Common pillow, wanting the cushion of same size, usually secured on the top.

163. Japanese travelling pillow, imitation leather, with Japanese lock.

164. Lacquered square box.

165. Oblong letter box, fine old lacquer.

166. Red lacquered writing box.

167. Lacquered writing box.

168, 169, 170. Lacquer despatch boxes, of small size.

171. Carved small box.

172. Square despatch box, Tycoon's pattern.

173. Heart-shaped lacquer box.

174. Red lacquered note or envelope box, stork pattern.

175. Envelope or cigar box, a fine specimen of the oldest and best lacquer, with richly embossed and enamelled tray.

176. Square lacquered flat box.

177. Bon-bon box, of gold lacquer, with embossed top.

178. Lacquer box.

179. Small antique nest of drawers, Pompeian pattern.

180. Small nest of workbox trays, gold lacquer.

181. Small lacquer box, containing all the table service for a meal.

182. Small toilet box, with drawers.

183. Small lacquer box for a popular game, with squares.

184. Small lacquered box, with cover.

185. Ditto, with cards.

186. Lacquer box.

187. A pair of red embossed lacquer boxes.

188. Lacquered square box, with tray.

189. Square lacquer box.

190. Round lacquer box.

191. Despatch box, lacquered in gold relief on polished wood, more highly valued than any other kind of lacquer ware.

191A. Lacquer on wood, a tobacco pouch in form of a purse.

LACQUER AND ENAMEL ON IVORY; TORTOISE-SHELL, MOTHER-OF-PEARL, &C.

192. Pair of shells, with enamelled fish.

193. Tortoiseshell lacquered nest of boxes.

194. Ditto, ditto.

195. Ditto, ditto.

196. Tortoiseshell small ring case.

197. Tortoiseshell saucer, gold figures.

198. Four tortoiseshell boxes, with enamelled landscapes, and embossed ornaments in gold.

199. Two tortoiseshell lacquered saucers.

200. Tortoiseshell lacquered box.

201. Tortoise.

202. Tortoiseshell pedestal, and two tortoises in ivory also.

203. Two in the natural shell, a work of great ingenuity and finish.

204. Pair of lacquered tortoiseshell spill stands, with finely chiselled bronze figures embossed on the sides.

205. Imitation mother-of-pearl lacquer saké bottle.

206. Ditto, ditto.

207. A Japanese medicine case, forming a series of boxes, a fine specimen of enamel on ivory, with the figures in raised relief; of great delicacy and minuteness of finish.

208. Another specimen of the same kind, with different ornamentation.

209. Ditto, of gold lacquer, with em-

bossed ivory and enamelled figures, of great finish and artistic excellence.

(Nearly every officer carries one of these at his girdle; and, like the clasps and brooches to their tobacco-pouches and portfolios (see Nos. 472 to 474), all that Japanese art can accomplish is lavished upon such objects without stint.)

210. Set of inlaid and enamelled ivory buttons.

211. Set of silver ditto.

211A. Enamelled case, with balanced cup, for compass (or incense).

212. Very valuable gold-lacquered luncheon box (contributed by Lord John Hay, H.M.S. *Odin*).

INLAID AND CARVED WOODWORK.

In the mountain ranges, especially of Hakoni, all the inhabitants, during the winter months, as in Switzerland, employ their spare time in turning, carving, and inlaying woodwork.

The following specimens were collected on the spot:—

213. Two Hakoni turned boxes; one containing small spinning tops, and the other nests of eggs.

214. Small tray, with six ring boxes.

215. A Japanese conjuring box.

216. Carved wood box.

217. Two inlaid boxes; one for despatches, and the other for ink, slab, pencils, &c.

218. Note-paper box.

219. Carved despatch box, antique.

220. Inlaid box, containing small specimens.

221. Ditto, smaller, with bronze mounting.

222. Maple wood despatch box.

223. Curiously inlaid writing-case.

224. Inlaid box.

225. Inlaid wood ring box.

226. Carved square box, with dragon inlaid, very antique.

227. Inlaid wood despatch box.

227A. Ingenious Chow-chow box, with tray, &c., to slide out; specimen of perfect fitting and workmanship even in the commonest article.

228. Eight specimens of Hakoni ware chop-stick cases, &c.

229. Box, with cover.

230. Inlaid wood cabinet, with book-stand on top.

231. Set of four decanter stands, lacquered.

232. Ditto, rustic, with rough bark of tree on.

233. Despatch box of inlaid wood.

234. Despatch box and tray of curiously grained wood, lacquered inside.

235. Telescope fishing-rod of bamboo.

236. Two bread trays.

237. Specimens of men's and women's clogs, used habitually in the streets in bad weather.

238. Specimens of Japanese wood, furnished by the Japanese Government.

B.—*Specimens of Straw and Basket Work.*

The Japanese expend great labour and ingenuity in straw work, combined with lacquer, in cabinets and boxes, often imitating inlaid woods with marvellous closeness, and interweaving very elaborate designs of birds, men, landscapes, &c. In all kinds of basket work they excel also, producing many objects of elegant design, and for a vast variety of purposes; by the aid, occasionally, of lacquer inside, they make flower vases and vessels to carry water. Some mixed straw and lacquer ornamented cabinets have already been catalogued under the preceding class of lacquer ware. (*See* Cabinets.)

239. Large despatch box, imitation inlaid wood, with figures and landscape in bright-coloured straw; a very fine specimen of the work.

240. Set of three, fitting into each other, and beautifully ornamented with storks.

BASKET AND BAMBOO ON RATAN WARE.

241. Basket-covered case.

242. Straw toilet box.

243. Set of three oblong bamboo baskets, with handles and covers.

244. Set of two ditto, fine ratan.

245. Set of two circular bamboo baskets.

246. Set of two ditto, with handles.

247. Flat circular ditto.

248. Round ratan, with filigree lid.

249. Octangular ditto.

250. Tiffin basket, with lid.

251. Square ditto.

252. Small basket-covered nest of three drawers.

253. Two basket circular trays, with bamboo edges.

254. Two ditto, ditto, without edging.

255. Flower basket, with inside lacquered.

256. Tiffin or luncheon set of boxes enclosed in basket-covered case.

257. Another tiffin set of boxes.

258. Saké bottle, basket-covered.

259. Tray, with handle, for washing cups.

260. Writing-case, with drawers.

261. Basket tobacco pouches.

262. Set of two figured basket inlaid boxes.

263. Separate boxes, different sizes.

264. Set of ditto, ornamented with storks.

265. Straw-worked glove boxes.

266. Small toilet set of drawers, for rings, &c.

267. Small writing box.

268. Envelope case.

269. Firefly cages (five).

270. Basket vase, to hang on the wall, for flowers.

271. Inlaid basket nest of drawers and trays.

272. Basket-covered despatch box, lacquered top (with three straw pencil boxes inside).

273. Basket-work note-paper case.

274. Set of three ratan and lacquer bread-baskets.

275. Pair of basket vases, for suspending flowers.

276. Two luncheon baskets.

277. Pair of lady's sandals.

278. Pair of gentleman's ditto.

279. Tobacco pouch, with grotesque button.

280. Portemonnaies (three specimens).

281. Two baskets for flowers.

282. Four sets of small baskets.

283. Pair of pencil boxes.

284. Japanese straw hat for fine weather.

285 to 297. Thirteen different kinds of cigar cases, made chiefly of straw, ratan, &c. One or two of silk, made for Europeans only.

C.—*Specimens of China, Porcelain, and Pottery (from Yedo and Yokohama).*

VASES AND JARS.

298. A pair of large vases, richly coloured, with medallions and stands.

299. A pair of smaller ditto, fair porcelain.

300. Ditto, ditto.

301. Ditto, ditto.

302. Two flower baskets.

303. Two small scent bottles.

304. Pair of gold vine-leaf jars.

305. Pair of gourd-shaped and enamelled porcelain jars.

306. Large blue porcelain jar, with lacquered pattern.

307. Pair of white porcelain flower jars, with lacquered edges.

308. Ditto, ditto.

309. Ditto, with blue pattern.

310. Ditto, blue and gold, used for saké by the Japanese.

311. Jar, with cover and Daimio's crest.

312. Gourd-shaped blue porcelain jar (18 inches).

313. Pair of fine porcelain jars (2 feet).

314. Blue preserve jars.

315. Enamelled porcelain jar, with cover.

EGG-SHELL CHINA.

316. Set of three fine blue and gold bowls.

317. Set of six saucers, with tortoises that float when water is poured in.

318. Seven small cups.

319. Ditto, ditto, blue and gold.

320. Set of egg-shell cups and saucers, covered with fine bamboo (saké cups).

321. Two ditto.

322. Porcelain bowl, with portraits or designs of European and Japanese ladies.

323, 324. Two china saucers, with boat and moveable window.

325. Fine egg-shell china bowls.

326. Ditto.

327. Set of six egg-shell porcelain saucers.

Brought from Osaca.

328. Three tazza specimens of blue enamelled egg-shell porcelain.

329. Ditto, ditto.

330. Ditto, with designs of figures.

LACQUERED CHINA—EUROPEAN MODELS.

331. Pair of lacquered cups and saucers.

332. Set of six ditto; coffee cups, with handles.

333. Ditto, ditto (from European models).

334. Ditto, ditto.

335. Ditto, ditto.

336. Ditto, ditto.

337. Lacquered china cup, saucer, and cover; Japanese pattern.

338. Two richly enamelled porcelain bowls.

339. Blue porcelain sponge basin.

340. Pair of porcelain bowls, with lacquered outsides.

341. Circular tray and bowl, richly ornamented.

342. A twelve-inch porcelain bowl.

343. A two-feet porcelain salver.

344. Purple porcelain bowl and cover.

345. Tooth-powder pot, with cover.

346. Ditto, ditto.

347. Small blue jar, with cover—Bird of Paradise.

348. Blue porcelain preserve jar.

349. Pair of miniature teapots.

350. Set of four porcelain bowls.

351. Set of three porcelain dishes.

352. Two porcelain bowls.

353. Ditto, for live fish.

354. Two figures of Japanese, in coloured china—a merchant and his wife.

355. A grotesque—a baboon with a moveable head, dressed up as a bonze, or monk.

356. Two small grotesques, with moveable heads and projecting tongues.

357. Cock and hen, with chickens; a good specimen of workmanship.

358. Japanese friar and his companion.

Pottery brought from Osaca.

359. Blue ware cup, in the form of a lotus leaf.

360. Crackled china dish, with a lacquered fish.

361. Oblong ditto, plain inside, gold fish outside.

362. Larger ditto, with imitation bark of tree outside, and Japanese characters sunk in gold.

363. Circular ditto, on stand, with gold leaves in centre.

364. Square basket of crackled china.

365. Palissy dish, with raised and coloured flowers.

366. Quaint saké pot, with cup and stand, of yellow ornamented pottery.

367. Fantastic-shaped teapot, of brown ware.

368. Small saké jar, with figure of executioner carrying a criminal's head on a pole.

369. Japanese Palissy bowl, with flowers in relief (found at Yokohama).

370. Flower stand, of very rare kind; the inside of crackled china, and the outside made to imitate the bark of a tree. (No other specimen could be found in Osaca, Yedo, or Nagasaki. This came from Nagasaki, and is probably very old.)

371. Three small pottery flower-pots (three).

Specimens of Porcelain and Pottery brought from *Saki*, a town between Osaca and Yedo, very celebrated for its china ware made in the neighbourhood; hence porcelain is called frequently Monosaki; literally, Saki products, or things of Ocasaki.

372. Porcelain inkstand.

373. Two blue porcelain preserve jars, with covers.

374. Two saké jars.

375. Blue porcelain citron-shaped jars.

376. Purple porcelain teapot.

377. Five cups to match.

378. Another set of five, of different pattern. (The Japanese mak^ all sets in fives or tens.)

379. Set of ten white egg-shell ditto.

380. Ditto, with coloured pattern.

381. Two fine white porcelain bowls.

382. Set of three ditto.

383. Ditto, ditto.

D.—*Bronzes and all Work in Metals.*

384. Pair of large bronze jars, with alto-relievo designs and dragon handles.

385. Pair of Etruscan-shaped vases, on elaborately wrought dragon stand.

386. Pair of ditto, with basso-relievos.

387. A cylindrical vase, with alto-relievo dragon, and representation of waves.

388. An old bronze vase, with handles of antique and Etruscan form.

389 to 393. Five smaller specimens of modern vases.

394 to 396. Specimens of silver inlaid vases.

397. Imitation rustic rock-work, and shell for stand to a vase or cup.

398. Quaint device of rock-work and water, with gold fish supporting a large funnel-shaped vase for flowers.

399. Tripod stand for flat vases or bowl, with basso-relievo figures.

400. Smaller ditto, of antique form.

401. A fine specimen, with storks in basso-relievo.

402. Bronze basket, wreathed with leaves and flowers, a fine specimen, of very old date.

403. Bronze cage.

404. Paper-weight.

405. Vine-leaf hanging bronze flower jar-

406. Hanging vase for flowers.

407. Ditto, in the shape of a gourd, incrusted with leaves and a scarabæus, used to hang against the pillars of a room with flowers.

408. Ditto, ditto.

409. Ditto, of more modern art, with graceful grouping of leaves.

410. Hanging vase of fine bronze, with moveable handles for packing.

411. Antique and elaborately-chased bronze vase for flowers, with dragon ornaments, said to contain a large alloy of silver.

412, 413. Flower-stands.

414. Vase for flowers, in imitation of a basket, antique.

415. Ditto, ditto.

416. Bronze dragon paper-weight.

417. Box in the form of a brinjall, or egg fruit, a common vegetable in the East.

418. Bronze paper-weight.

419. Bronze embossed tazza and stand.

420. Paper and spill stand.

421. Bronze mirror.

422. Metal ditto.

423. Bronze lobster.

424. Bronze incense stand.

425 to 428. Bronze models of tortoises, and a crab; good specimens of imitation art.

429. A fine specimen of modern bronze manufacture, brought from Osaca. A brazier (or charcoal chafing pan), with alto-relievo figures on panels on each side.

430, 431. Small bronze ditto, for the hands, to be placed on the table in cold weather when writing.

432 to 436. Specimens of bronze candlesticks (candles made of vegetable wax and paper wicks added).

(The Japanese are great admirers of storks, and these generally supply the model.)

437. Antique tripod candlestick with double cup.

438. Two folding and portable candlesticks, ingeniously constructed.

439. A fine and elaborately-carved bronze hanging night-lamp, found at Osaca.

440 to 445. Bronze cups of quaint and antique forms.

446 to 449. Bronze receptacles for pipe ashes. 448 is a candlestick also.

450. A fine specimen of ancient date, representing a mussel-shell and rock.

451. Bronze grotesque, representing a man looking down a well; for tobacco ashes; to be placed on a tray.

452. An ancient bronze, representing a saké tub, and the attendant, with a ladle, standing on the brim.

453 to 459. Various specimens of in-genious devices for pen and pencil rests (453, 454, 456), very curious specimens of mixed metals and graceful in design; 457–459 represent miniature Japanese gardens, with their dwarf trees, bridges, &c.

460 to 465. Six specimens of similarly ingenious devices in bronze, to supply the water required to rub down the Indian ink in writing. Several of these are as graceful in design as they are quaint in conception. The gourd suggests the idea of 462-3, entwined with leaves, and a scarabæus very perfectly rendered; 464 is taken from the brinjall, a common vegetable. Toads, tortoises, dragons, fruits, all are laid under contribution. The educated Japanese seem never to weary in the invention of fantastic designs for all the accessories to the brain work of their life—*writing.*

METAL BUTTONS.

Specimens of fine artistic work, graven and embossed in mixed metals.

466. A set of twelve.

467. Ditto.

468. Set of eleven.

469. Set of six.

470. One.

471. Set of gold inlaid and enamelled fittings for a Japanese lady's toilet case.

472, 473, 474. Three nests of trays, containing a collection of two hundred and fifty specimens of cast and chiselled metal brooches, used as fastenings to portfolios, tobacco pouches, &c., on which the Japanese seem to lavish the highest artistic skill they can command, whether in design or execution; and they display great excellence in both. The intaglio graving on the metal surface, the mixture of metals in the most intricate designs of figures in relief, and the expression of these, with all their minuteness, have rarely been excelled or rivalled by the most celebrated workers in metal, whether of the Middle Ages or our own.

475. Two inlaid jointed bracelet clasps.

476. Metal lacquered box for tobacco ashes.

477. A selection of twenty-five charms, small scent bottles, boxes, &c., of inlaid and mixed metals. Ivory or basket ware beautifully finished.

478. Miniature metal case, for writing materials.

479. Two chased tobacco pipes.

480. Equestrian statue (very rare)— Warrior examining his arrow.

481. Antique incense bowl, elaborately carved with embossed leaves.

482. Antique tazza, with dragon in relief.

482A. A set of lady's silver head ornaments, and two pairs of chopsticks.

483. Japanese bell, a very antique specimen of their bell metal, when it was said to be at its best.

484. Double dragon bronze, resting on a mythic animal.

485. Inlaid bronze can, with cover for water.

486. Pair of inlaid bronze jars.

487. Bronze model of bat, with extended wings, for hanging in a room, with incense.

488. Finely-modelled dragon, for pen-rest, antique.

CUTLERY AND WORKMEN'S TOOLS.

489. Set of carpenter's tools.

ARMS AND ARMOUR.

490. Japanese shirt, and cap of mail.

491. Japanese long two-handed sword.

492. Short sword of one of the party of assassins which attacked the British Legation on the 5th of July, 1861 (with leather purse containing cash, and seal attached, belonging to the same).

493. Fire brigade head-dress, helmet and leather surcoat worn by officers.

494. Models of Japanese fencing apparatus, mask, armour, of bamboo, and quarter-staff. Second tray contains the insignia of rank carried before a Daimio.

495. Models of insignia on a larger scale, as these are set up in a Daimio's residence.

496. Model of a Japanese matchlock (with two lacquered pipe sticks); contributed by J. MACDONALD, Esq., of Her Majesty's Legation.

497, 498, 499. Specimens of lacquered Japanese bows, in use among Japanese officers of rank.

500. Bundle of arrows.

501, 502, 503. Specimens of lacquered bows in use among boys, the sons of officers.

504. Bundle of arrows for the same.

505. Japanese officer's quiver, with ring for spare bowstrings.

506. Officer's smaller sized quiver.

507, 508. Specimens of officer's archery gloves.

509. Set of Daimio's ladies' bows and arrows.

MINERALS.

510. Specimens of Japanese copper mines, the copper in bar

511. Ditto of lead.

512. Ditto lead ore from the mines of Hakodate.

513. Specimens of lava, brought from Fusiyama.

514. Ditto of coal, brought from the mines of Fezin.

515. Specimen of stone resembling Turkey stone, from Hakodate, and used for sharpening tools.

516. Imitation glass cylinder for blinds, made of rice.

517. A collection of Japanese coins, gold, silver, and copper.

E.—*Paper—Its raw materials—Paper for rooms—for writing—for handkerchiefs— for packing—for imitation leather—Papier mache, &c.*

518. Specimens of Japanese paper of various colours, brought from a manufactory at Atami, made from the bark of trees (several different shrubs are employed); specimens of one kind attached.

519. Handkerchief, of paper, made in various patterns, for ladies.

520. Japanese rain coat, of oiled paper.

521. Reticule, made of paper imitation leather, with metal clasp.

522. Imitation leather despatch box.

523. Ditto, ditto.

524. Two smaller ditto.

525. Tobacco pouch ditto.

526. Small cabinet paper screen.

Large ditto, ditto.

527, 528. Umbrella and parasol fans, convertible at pleasure (from Osaca).

529. Common fans with Indian ink designs.

530. Two fans.

531 Numerous specimens of paper, made to imitate leather, with prices marked at which they are sold in Japan (to foreigners).

532. Book of patterns for papering rooms, with specimen of the mineral substance used to imitate silver, which is both cheap, and not liable to change colour, or tarnish; contributed by J. MACDONALD, Esq., of Her Majesty's Legation.

533. Specimens of Japanese lanthorns, made of bamboo and paper, and compressible. The name of the owner, or his arms and crest, are always printed on these.

534. A Japanese umbrella, of oiled paper.

535. Paper money of Japan—a packet of bank-notes from 30 cash to 500 each—a farthing to sixpence. These are only in circulation in part of Kinsin.

536. Specimens of paper, together with portion of the tree from which it is principally made; contributed by the Japanese Government.

F.—*Textile Fabrics—Silk Crapes—Silks and Satins—Tapestry—Printed Cottons —Articles of Clothing, made from the bark of a tree.*

537. Box of Japanese crape scarfs, of peculiar fabric.

538. Piece of red crape, beautiful in colour.

539. Three rolls of Japanese silk.

540. Four pieces of washing silk, for dresses.

541. Narrow piece of Japanese silk.

542. Silk net watch pocket.

543. Silk covered box.

544. Seven pieces of tapestry for screens (from Yedo).

545. Four pieces of tapestry embroidered.

546. Two printed on silk.

547. One ditto, on crape; brought from Osaca.

548. Japanese lady's work-box, with reels of silk.

549. Gentleman's portfolios for carrying writing paper, handkerchiefs, &c.

550. Two pieces of soft cotton lining, of the texture of lint, and a possible substitute.

551. Six specimens of Japanese summer fabrics from Arimatz, a place celebrated for their manufacture.

552. Two specimens of a fabric made from the bark of a creeper, called in Japanese Kowo-o (purchased at Cakigawa).

553. Sample of the bark as prepared for the manufacturer.

554. Rain cloak and cowl, made of same material.

555. Ditto, made of reeds.

556. Cable of human hair, considered stronger and more indestructible than rope made of any other material, and it is expensive in proportion.

G.—*Works of Art—Carvings in Ivory, Wood, and Bamboo—Paintings—Illustrated Works—Lithochrome—Prints—Models.*

557. Mussel-shell carved in wood.

558. Fruit carving in bamboo wood. Twenty-five specimens of the best ivory carvings, showing great mastery of the chisel and power of expression.

559. Eight books, specimens of maps, illustrated works, &c.

559A. Twenty-four volumes of ditto.

560. Japanese play bills.

560A. Leaves from a Japanese scholar's writing exercise.

561. Two boxes of lithochrome printing, on a peculiar fabric of crape paper.

562. Book of fire-brigades in Yedo, with the crests and insignia, detail of city wards, &c.

563. Specimens of figures by a native artist.

564. Map of Yedo.

565. Itinerary of the Tokaido, or grand route to all the Imperial towns.

566. Map of Japan in 66 Provinces (2 vols.).

567. Specimens of Japanese lithochrome.

568. Printing of old date; representing a pilgrimage to Fusiyama, the new foreign settlement at Yokohama.

569, 570. Two maps of Fusiyama—the volcanic mountain, with the various stations in the ascent.

571. Further specimens of lithochrome printing, consisting of two hundred illustrations of the manners, costume, and architecture of the Japanese.

572. · Specimens of story-books—popular literature, written in the Hirakana character, for women and children and the less educated classes, as easier to read than the Giosho, or other styles of writing.

572A. Specimen of Japanese official writing—a letter from the Ministers of Foreign Affairs announcing the despatch of a diplomatic mission to England.

H.—*Educational Works and Appliances—Books of Science—Copies of European Models and Instruments—Toys, &c.*

573. "Sole Treasury of Eternal Usefulness," a Japanese Encyclopædia.

574. Explanation of the products of the earth, mountains, and sea in Japan, with numerous plates and illustrations, in four volumes of five parts each.

575. Secrets of the cultivation of silk, with plates (3 volumes).

575A. A treatise on chemistry, with plates, in 21 volumes.

576, 577. Two volumes of Japanese fishes, with descriptions. A series of coloured plates of ditto, and another set of birds.

578. Work on chemistry, with plates, in 21 numbers.

579. Red-book of Japan, giving a list of all the public officers, their names, rank, &c.

580. Quadrant and sun-dial.

581. Compass in metal case.

582. Ditto.

583. Pedometer, of Japanese workmanship.

584. Japanese clock, ditto.

585. Thermometer, ditto.

586. Telescope, ditto.

Toys, &c.

587, 588, 589. Three specimens of spinning tops. The first is spun along a string

by the aid of a small ring attached to it. The second, on being lifted up, becomes a lantern. The third is filled with a spinning family of young tops.

590. Five spinning tops.

The Japanese are great top spinners and kite fliers, and have an infinite variety of both; but in kites they are inferior to the Chinese, while in tops they are unrivalled.

591. Conjuring-box, and divers Japanese puzzles.

    a. Any one or more of the pieces to be reversed and the lid put on, when an adept will tell with certainty the number turned.

    b. Two of the three figures to be taken out and the lid replaced, the one remaining to be certainly told.

    c. A small coin to be placed on one of the squares, and covered with a cup or the hand. The conjuror to make the coin disappear without removing the cover.

592. Children's crackers—the figures are first moistened, and then thrown into the fire to explode.

593. Stereoscopic view of tea-house in Japan.

594. Delusion glasses—making an object to be touched appear far removed from the place where it actually is.

595. Four boxes of artificial flowers which expand in water (made of pith).

596. Two toy tortoises.

597. Model of Japanese lantern.

Bucket of a well, running on a pulley.

598. Strings of beads, made of vegetable wax.

599. Toy-box with a Japanese courtesan in full costume, embossed in silk.

600. A series of clay figures, coloured to represent Japanese costume and life.

601. Set of draped clay figures, scene in an interior.

602. Ditto, a group of Japanese female musicians, as they perform at the tea-houses.

603. Buffalo, carved out of the bone of a whale.

I.—*Miscellaneous.*

604. Box of very precious tea, in powder.

605. Specimens of tooth-powder (26), in the manufacture of which the Japanese may be supposed to excel; for if the matrons have the blackest teeth in the world, the maidens and the men generally have the whitest.

606. A collection of Japanese shells.

607. Three guard-chains made of bamboo.

608. Three inlaid and enamelled ladies' pipes, bamboo and silver-mounted.

609. Three solitaires of elaborately-carved steel.

610. One rose-pattern bracelet, of mixed metal.

611. A set of Japanese coins.

612. Handsome cabinet.

613. A very rare cabinet, with porcelain inlaid tablets.

614. Two bronzes (Yaconins, in full costume, mounted on buffaloes).

615. VYSE, CAPT. F. H., *H.M. Consul at Yokohama.*—Screens and vases, ivory curios and charms, a porcelain bottle, and lacquer china cups, saucers, salvers, and bowls.

616. MYBURGH, DR. F. G.—198 Japanese medicines, and a collection of surgical instruments.

617. CRAWFORD, MRS.—A Japanese table.

618. COPLAND, C.—A cabinet.

619. NEAVE, MR.—Lacquered boxes.

620. REMI SCHMIDT, & CO.—Raw silks and cocoons.

Inlaid cabinet work and lacquer ware.

Books, arms, bronzes, porcelain, &c.

621. BARTON, DR. A.—Old and choice lacquer and bronzes, swords, &c., from Japan.

622. BARING, MESSRS. — Two polished spheres of rock crystal.

623. HAY, COMMODORE LORD JOHN, C.B. —A collection of raw silk.

LIBERIA.

NORTHERN COURTS, UNDER THE STAIRS, NEAR ENTRANCE TO HORTICULTURAL GARDENS.

1. LIBERIAN COMMISSION.—Coffee; cocoa; sugar; molasses; spices; preserves; meal; rice; starch; bacon; textile materials and fabrics; native weapons, implements, &c., in wood and iron; skin pouches; basket-work; earthenware; oil; timber; mineral ink, &c.

2. BARNARD, J. L. *Cornhill.*—Samples of medicinal oil.

MADAGASCAR.

NORTH-EAST COURT.

1. MAURITIUS GOVERNMENT.—Packets of gums, specimens of unwrought iron, iron ore, blocks of wood.

2. MIDDLETON, LT.-COL.—An iron chair, presented by H.M. King Radama II. A spear, a collection of spoons, and two musical instruments.

3. MELLISH, E. — Specimens of spades, knives, and tools; horn spoons, shoes, stockings, and silk lambas.

4. E. N. — Specimens of spades, axes, choppers, and chisels; silk lambas, and one of cotton.

5. MARINDIN, F. A.—A spade.

6. BEDINGFELD, F. — A spear, and a cotton lamba.

8. CALDWELL, J.—A collection of tools, a mat, rafia cloth and cord, and a cotton lamba.

9. ELLIS, THE REV. W.—Lambas for different purposes; matting, boxes, cloth, &c., made from the rafia palm; an iron lamp.

10. MORRIS, J.—A native straw box, containing five others.

11. LONDON MISSIONARY SOCIETY.—Brass scales for weighing money, iron and silver weights (the latter made from dollars); silver ore, native ornaments, &c., of silver and beads; baskets made from a reed, grass for basket work; a pair of shoes, two wooden dishes, a sword and scabbard, knives, a walking-stick, an amulet of beads.

12. ———. A branch of the rafia palm, with a cone.

THE NETHERLANDS.

NORTH-WEST COURT.

CLASS I.

1. BOUVY, J. J. *Amsterdam.* — Refined salt.

2. BRANDHOFF ISSELMAN, J. J. *Leyden.*— Dutch table salt.

CLASS II.

3. BOSSON, K. G. W. DE, *Dordrecht.*—Preparations of iron, and other chemicals.

4. DIEDERICHS, P. A. *Amsterdam.*—Box of water colours.

5. ELST AND MATTHES, VAN DER, *Amsterdam.*—Sublimed sal-ammoniac, and sulphate of ammonia.

6. NETHERLAND CARBONISATION FACTORY, *Hillegersberg, near Rotterdam.*—Carbonised peat.

8. FOCK VAN COPPENAAL, G., M.D. *Amsterdam.*—Chemicals for technical and pharmaceutical use. Pure acids, ethers, weights made of aluminium.

10. GARANCINE AND MADDER MANUFACTORY, *Tiel.*—Garancine, and other products of madder.

11. GROOTES, BROS. D. & M. *Westzaan.*—Powder-blue, cocoa, and chocolate.

12. GROOTE & ROMENY, *Amsterdam.*—Chemicals for photography.

13. GRINTEN, L. VAN DER, *Venlo.*—Tartrate of potash and iron, in crystals.

14. JACOBS, AZ. S. & A. *Zwolle.* — Samples of lac, varnish, and standoil.

15. JONG, H. DE, *Almelo.*—Varnish, with specimens of application.

16. LENSING COLLARD, H. *Leeuwarden.*—A map of 1739, and engravings, restored. Ink and varnish.

17. MENDEL BOUR & Co. *Amsterdam.*—Samples of garancine.

18. MEIJER, J. W. *Workum, Friesland.*—Butter and cheese preserver. Eau de Cologne. Cheese-dye; antifebrile herbs.

19. MULLER, M. & Co. *Utrecht.*—Chrysammic acid (polychromic acid).

20. NOORTVEEN & Co. *Leyden.*—Chromic and Dutch green, yellow chromate. Paris blue, lac, varnish, and standoil.

21. OCHTMAN, VAN DER VLIET, & Co. *Zierikzee.*—Madder, and its products.

22. RENTERGHEM, C. A. VAN, & Co. *Goes.*—Madder, garancine, fleur de Garance, alizarine, and colorine.

23. SPRUYT & Co. *Rotterdam.*—Purified cod-liver oil. Salts of ammonia. Samples of ink.

24. TACONIS, P. *Joure, Friesland.*—Dyestuff (Friesland green), and oil.

26. VERHAGEN & Co. *Goes.*—Madder and garancine products.

28. VRIESENDORP, C. A. & SONS, *Dordrecht.*—Lac and varnish, &c.

29. FRANKEN & Co. *Amsterdam.*—Enamel coating for iron.

30. SOETENS, C. *the Hague.*—Charcoal manufactured by a new process from Dutch peat.

CLASS III.

31. AGRICULTURAL SOCIETY.—*Culemborg.* —Preserved provisions.

32. ANDEL, T. VAN, *Gorinchem.*—Wheat, raw and in the various stages of manufacture.

33. APKEN & SON, *Purmerend.* — Preserved sweetmeats.

34. BOGAARD, J. VAN DEN, & Co. *Gennep, Limburg.*—Wheat flour.

35. BOLS, ERVEN LUCAS LOOTSJE, *Amsterdam.*—Liqueurs.

36. BONT & LEYTEN, DE, *Amsterdam.*—Sugar, chocolate, and liqueurs.

37. BOOTZ, H. *Amsterdam.* — Liqueurs and elixirs.

38. CATZ & SON, *Pekela.* — Stomachic elixir.

39. CHARRO, F. DE, *Amersfoort.*—Dutch tobacco.

40. DYEL, L. VAN DER, & SON, *Weesp.*—Chocolate powder.

41. DIEVELAAR, H. & DE BREUK, '*s Hertogenbosch.*—Chicory and pea-coffee.

42. DOYER & VAN DEVENTER, *Zwolle.*—Liqueurs.

44. DUYVIS, J. *Koog-on-Zaan.*—Urling's starch and glue from the refuse.

46. DUTCH SUGAR-REFINING SOCIETY, *Amsterdam.*—Sugar in loaves and crushed.

47. EBERSON, H. P. *Arnhem.*—Liqueurs.

48. EGBERTS, B. H. *Dalfsen.*—Chicory.

49. ELLEKOM & VISSER, VAN, *Amsterdam.* — Preserved provisions. Fruits in vinegar.

50. FOCKINK, W. *Amsterdam.*—Liqueurs.

51. GORTER, H. S. *Dokkum.*—Friesland clover seed.

52. GRIENDT & LUYTEN, VAN DER, *Rotterdam.*—Preserved provisions, in tin boxes, with crystal bottoms of the exhibitor's invention.

53. HAAGEN, BROS. VAN, *Utrecht.*—Cigars.

54. HENKES, J. H. *Delftshaven.*—Dutch gin.

55. HEYLIGERS, T. & SON, *Schiedam.*—Dutch gin.

56. HOPPE, P. *Amsterdam.*—Liqueurs, rectified and unrectified spirits.

58. HUNCK, H. P. *Amsterdam.*—Chocolate powder, flavoured with spices.

61. KAKEBEEKE, Gz. J. H. C. *Goes.*—Samples of Zealand wheat, wheat flour, bran, wheaten groats, &c.

64. KOPPEN, H. *Leerdam.* — Cigars, manufactured from Java tobacco. Dutch tobacco.

65. KORFF, F. *Amsterdam.* — Chocolate powder, cocao-butter or vegetable-fat.

66. LANS, H. & SON, *Haarlem.*—Samples of beer.

67. LEVERT & Co. *Amsterdam.*—Liqueurs.

69. MACKENSTEIN, A. F. & SON, *Amsterdam.*—Cavendish tobacco.

70. NICOLA KOECHLIN, & Co. *The Hague.* —Wheat-meal, wheat, and rye-flour.

71. OBENHUYSEN, D. P. A. *Amsterdam.* —Samples of vinegar.

74. OOLGAARD, D. & SON, *Harlingen.*—Liqueurs.

75. OOLGAARDT, J. *Amsterdam.* — Preserved fruits and sweetmeats.

76. PATERS, P. L. *Leyden.*—Dutch preparations of buckwheat.

78. RALAND, G. A. *Deventer.*—Preserved fruits and vegetables, pickles, &c.

79. REEKERS, L. & Co. *Haarlem.*—Raisin-vinegar.

80. REYNVAAN, A. J. *Amsterdam.*—Tobacco, snuff, and cigars.

82. RÖNTGEN, J. E. *Deventer.*—Liqueurs.

83. ROYAL LIQUEUR DISTILLERY, STIBBE, BROS. *Kampen.*—Liqueurs and bitters.

84. SCHONEVELD & WESTERBAAN, *Gouda.*—Potato-flour; white, yellow, and brown syrups, and sago.

87. ULRICH, J. S. & C. *Rotterdam.*—Ship-bread, table and dessert biscuits.

89. ZUIJLEKOM LEVERT, VAN, & Co. *Amsterdam.*— Purified and rectified alcohol, amyl-alcohol, gin, brandy, liqueurs.

90. AA, T. J. VAN DER, *Rotterdam.*—Liqueurs.

91. GEVERS DEIJNOOT, D. R. *near The Hague.*—Cheese.

92. NIEUWENHUIS & Co. *Rotterdam.*—Preserved provisions.

CLASS IV.

93. ALBERDINGK, F. & SONS, *Amsterdam.*—Oil.

94. BENOIST, J. L. & MOOI, J. *Heerenveen.*— Sealing or bottle-wax of various colours.

96. BOUSQUET, J. & Co. *Delft.*—Samples of soap.

97. CLERCQ, H. DE, *Haarlem.*—Guttapercha: collection, illustrative of its applications.

98. CROMMELIN, R. *Renkum.*—Rape-seed cakes and oil.

99. DIEMONT, J. J. & SON, *Amersfoort.*—Raw and ground oak bark.

100. DOBBELMAN, BROS. *Nijmegen.*—Samples of soap.

101. DORSSEN, GZ. G. VAN, *IJsselstein.*—White and grey hoops.

102. DUTCH STEARINE MANUFACTORY, *Amsterdam.*—Materials for the manufacture of stearine; stearine; stearine candles; oleine; spermaceti candles, &c.

106. JANSSENS, E. B. F. *Weert (Limburg).*—Hard and soft soap.

107. KAMER, W. P. VAN DE, *Middelburg.*—Eau de Zélande.

108. KROL, G. J. & Co. *Zwolle.*—Bone-black.

109. LOBRY & PORTON, *Utrecht.*—Cod-liver oil.

111. ROBETTE & DRAIJER, *Gouda.*— Grease, glue, gelatine for weaving purposes, phosphate of lime, animal manure.

112. ROYAL WAX CANDLE MANUFACTORY, *Amsterdam.*—Products of stearine, stearine candles, oleine.

113. RÖNTGEN, C. A. *Deventer.*—Oils, seeds and kernels of indigenous plants.

114. SMITS, WIDOW P. & SON, *Utrecht.*—Bone-black; sulphuric and nitric acid; sulphate of iron.

115. SOCIETY FOR THE MANUFACTURE OF VARNISH, COLOURS, &c. MOLIJN & Co. *Rotterdam.*—Lac, lacquered panels. Tools used for the imitation of wood.

116. STEARINE CANDLE MANUFACTORY, *Gouda.*—Materials, fat-acids, stearic acid, candles, &c.

118. VERKADE, E. G. *Zaandam.*—Rape and linseed oil, raw and purified. Train oil, raw and purified.

119. VIRULY, J. P. & Co. *Gouda.*—Soft green soap.

120. VISSER, E. E. *Amersfoort.*—Yellow wax.

122. WOLFF, M. *Amersfoort.* — Yellow wax.

123. HILST, L. V. D. *The Hague.* — Corks.

CLASS VI.

124. BEYNES, J. J. *Haarlem*, coach-builder and railway carriage maker.

[Obtained a silver medal at the International Exhibition of Industry, Haarlem, 1861.]
A berline, with round corners, from which the rumble behind can be easily removed.—Landau carriage.

125. DEVENTER, J. S. VAN, *Zwolle.*—Carriage (Victoria), with a new kind of springs.

126. HERMANS, M. L. & Co. *The Hague.*—Carriages.

127. PREYER, J. HZ. B. *Amsterdam.*—A Victoria carriage.

128. ROECKX, H. W. *Maastricht.*—Chiselled springs for a gala carriage.

CLASS VII.

131. HUNCK, WIDOW J. T. & SON, *Amsterdam.*—Tools for diamond-cutters.

132. WATSON, G. & SON, *Rotterdam.*—Rasps and files.

CLASS VIII.

135. PASTEUR, W. C. & Co. *Rotterdam.*—Fire-engine, model of others, and gasometer.

136. PECK, BROS. *Middelburg.*—Miniature fire-engine, with model of new metallic pistons, &c.

137. ARENDS, ALUMENUM E. *Amsterdam.*—Fire-engine.

138. HAM, T.—Fire escape.

139. YSERMAN, J. M. *The Hague.*—A centralisator for fire-engines.

CLASS IX.

140. BRAKELL VAN DEN ENG, BARONET. F. L. W. VAN, *Lienden.*—Model of sheep-house.

142. GELUK, J. AZ. A. *Tholen.*—Clod-crusher, Cheddam wheat, red chaffed wheat, madder.

143. JENKEN, W. *Utrecht.*—Foot plough, with dredges, coulter, and disk-coulter. Cylindric mangle.

144. RIPHAGEN & Co. *Hattem.*—Cultivator, with levers; turning harrow.

145. SARPHATI, DR. S. *Amsterdam.*—Models of agricultural implements; peat ash.

146. SIX, SIR P. H. *'sGraveland.*—Meadow-sledge (model).

147. STARRE, G. VAN DER, *Sloten, North-Holland.*—Model of a Dutch hay or corn rick.

148. STARING, DR. W. C. H. *Haarlem.*—Three-wheeled tipping cart and whipple-trees.

149. STOUT, G. *Tiel.*—Paddy thrashing machine, with sieves.

150. VOGELVANGER, E. *Hulst.*—Self-acting harrow, for one or two horses.

CLASS X.

151. Andriessens, J. *Roermond.*—Samples of pipes manufactured from pewter, &c.; sheet of pressed lead.

152. Esta, F. F. van, *Harlingen.*—Blue enamelled tiles; hard, red waterproof bricks.

153. Haart, C. H. de, *Utrecht.*—Agate marble vases; alabaster; black marble chimney-piece.

154. Heukelom, N. van, *Erlecom, near Nijmegen.*—Artificial paving-stones; samples of bricks; specimens of Dutch masonry.

155. Hulst, J. van, *Harlingen.*—White tiles.

158. Mirandolle, C. *Fijenoord, near Rotterdam.*—Painted wood, creosoted, and not creosoted.

159. Muller & Co. *Valkenburg.*—Samples of bricks.

160. Rhee, S. J. *Haarlem.*—Model of a joined beam; staircase.

161. Truffino, W. F. K. A. *Leyden.*—Bricks; Rhine stones.

162. Verweyde, C. C. *Amsterdam.*—Model of a moveable vertical window-blind.

164. Cuijpers & Stolzenberg, *Roermond.*—Sculpture in wood.

165. Venema, L. *Hertogenbosch.*—Freda or baldequin, in Portland stone.

CLASS XI.

166. Department of War, *The Hague.*—Breech of a 6-pounder, filled with brass, by Maritz, to be bored as a 4-pounder.

167. Stevens, P. *Maastricht.* — Fire arms.

CLASS XII.

170. Hoogen, J. van den, *Dordrecht.*—Hemp and iron rigging.

171. Maas, A. E. *Scheveningen.*—Model and description of a life-boat.

172. Vervloet, W. A. *Rotterdam.*—Apparatus for hauling ships upon a slip.

CLASS XIII.

176. Becker & Buddingh, *Arnhem.*—Brass balances and beams.

177. Eder, S. J. *Rotterdam.*—Scales for letters.

178. Eijk, Dr. J. A. van, *Amsterdam.* Thermometer adapted for public lectures.

179. Emden, A. van, *Amsterdam.*—Compass, the rose floating in spirit.

180. Geissler, W. *Amsterdam.*—Hygrometer, Prof. von. Baumhauer's principle, &c.

181. Holleman, F. A. *Oisterwijk.*—Hygrometer for domestic use.

185. Olland, H. *Utrecht.*—Balance of aluminium for chemical use; sea barometer.

186. Spanje, T. van, *Tiel.*—Electro-magnetic clock, with commutator.

187. SOCIETY " DE ATLAS," *Amsterdam.* —Water-meters for high and ordinary pressure, Prof. von. Baumhauer's principle.

188. WELLINGHUYSEN, WIDOW, R. J. & SON, *Amsterdam.*—Gas lamps for chemical use, Prof. V. Baumhauer's principle.

CLASS XIV.

190. EIJK, DR. J. A. VAN, *Amsterdam.*— Photographic copies of etchings by Rembrandt, &c.

191. SANDERS, VAN LOO, *Amsterdam.*— Photographs on dry collodion.

CLASS XV.

191A. HOWHU, A. *Amsterdam.*—Chronometer with secondary compensation, of the exhibitor's invention; chronometer with ordinary compensation.

CLASS XVI.

192. BERGEN, A. H. VAN, *Midwolde, Groningen.*—Musical bells; Turkish cymbals.

193. BEVERSLUIS, P. *Dordrecht.*—Semi-melodiums.

194. CUYPERS, J. F. *The Hague.*—Inlaid pianino for hot climates.

198. OSCH, E. P. VAN, *Maastricht.*—Brass wind-instruments.

CLASS XVII.

203. BERGHUIS, J. *Groningen.*—Skeletons of various animals.

204. KOENAART, A. J. C. *The Hague.*— Moveable set of teeth.

205. LINDEN, J. & SON, *Rotterdam.*—

Surgical, obstetrical, orthopœdical, and anatomical instruments.

206. SCHMEINK, BROS. A. & B. *Arnhem.* —Collection of surgical, anatomical, orthopœdical instruments and knives.

CLASS XVIII.

209. CATE, HZ. H. TEN & CO. *Almelo.*— Raw and bleached cotton manufactures, calicoes, shirtings, cambrics, drills, table-cloth, striped dimities.

210. SCHAAP, L. & A. *Amersfoort.* — Plain and striped dimities.

211. VELTMAN, BROS. *Amsterdam.* — Cotton blankets, quilts.

212. VISSER & CO. *Amersfoort.*—Linen and cotton Amersfoort, Marseilles.

CLASS XIX.

216. CATZ, J. B. VAN & SON, *Gouda.*— Horse reins and yarns.

217. DUINTJER, J. J. *Veendam.*—Rope rigging, &c., made by machinery. Iron joining-screws.

218. ELIAS, J. *Stryp, near Eindhoven.*— White linens.

219. GALEN, B. VAN, *Gouda.*—Specimens of yarns for netting, packing, sail-cloth, &c. Strings and reins.

220. GORTER, H. S. *Dokkum.*—Bundle of flax.

221. KAMERLING, Z. & SON, *Almelo.* — Damask collation table-cloth. Napkins of flax-yarn, woven in one piece.

222. KLEYN, A. VAN DER & SON, *Gouda.* —Hand-spun reins.

223. KORTENOEVER, J. *Gouda.* — Yarns for sailmakers, shoemakers, &c. Hemp and flax.

224. LEEUW, M. DE & SON, *Boxtel.* — Table-cloth of damask with heraldic designs.

225. LEYDEN, D. VAN & SON, *Krommenie.* —Sail-cloth of Dutch hemp and handspun yarn.

226. LEYDEN & DERKER, VAN, *Krommenie.* —Sail-cloth.

229. VERSLUYS, L. J. *Amsterdam.* — Twilled fire-engine hose, of tanned and untanned hemp.

230. VISSER, S. E. & SON, *Amersfoort.*— Mixed yarn and cotton twillings and Marseilles.

CLASS XX.

232. SARTINGEN, S. *Horst, near Venlo.*—Silk taffetas, satin, satin de Chine, velvet.

CLASS XXI.

235. KERSTENS, J. A. A. *Tilburg.*—Wool-dyed pilot-cloth, Amadou. Beaver frieze. Baize.

236. KRANTZ, J. J. & SON, *Leyden.*— Woollen cloth, &c.

237. LEDEBOER, L. V. & SONS, *Tilburg.*— Wool-dyed and mixed pilot-cloth. Baize and castor, piece dyed.

238. MEER, P. VAN DER & SONS, *Leyden.* —Leyden worsteds.

239. POSTHUMUS & CO. *Maastricht.*—Felts for paper-making.

240. SCHELTEMA, JZ. J. *Leyden.*—Woollen blankets. Wrappers. Horse blankets.

241. WILLINK, J. *Winterswyk.* — Mixed woollen and cotton cloth.

242. WYK, BROS. VAN & CO. *Leyden.*— Woollen blankets.

243. IJSSELSTEYN, J. F. J. *Leyden.* — Woollen blankets. Serge and pilot-cloth.

244. ZAALBERG, J. C. & SON, *Leyden.* — Woollen blankets.

245. ZUURDEEG, J. & SON, *Leyden.*— Woollen blankets.

CLASS XXII.

248. HEUKENSFELDT, J. *Delft.*—Imitation Smyrna carpet, knitted and woven from wool.

249. KRONENBERG, W. F. *Deventer.* — Deventer Smyrna carpet, woven in one piece.

250. PRINS, WIDOW L. J. *Arnhem.* — Scotch carpet. Carpets of cow-hair and wool.

CLASS XXIII.

252. JANSSENS, BROS. & Co. *Roermond.*— Turkey-red dyed yarns.

254. ROOIJEN, H. VAN, *Utrecht.*—Silk dyed with chrysammic acid. Silk dyed with an extract of soot.

CLASS XXIV.

258. BLOCK, A. DE, *Amsterdam.*—Gold and silver embroidery, military ornaments, galloon, &c. Raw materials.

259. OLIE, C. H. *Amsterdam.*—Embroidered fire-screen and table.

260. OVEN, BROS. VAN, *The Hague.* — Mantle richly embroidered with gold.

261. RENIER VAN WILLES, MAD. A. S. *Sluis (Zeeland).*—Lace-work.

262. TEULINGS & Co. *'sHertogenbosch.*— Linen lace (Bosscher band), &c.

263. KRAFT & Co. *Amsterdam.*—Coach livery lace, &c.

CLASS XXV.

264. CATZ, P. S. & Co. *Amsterdam.* — Rough, drawn, and curled horse-hair.

265. DEVENTER, J. S. VAN, *Zwolle.*—Fur carpet made of indigenous skins.

266. DIRKS, H. J. *Dordrecht.*—Brooms, brushes, &c.

267. JONKER, BROS. *Amsterdam.*—Brushes.

268. DR. VAN LITH DE JEUDE, *Utrecht.* —Fur carpet.

269. NYMAN, H. *'sHertogenbosch.* — Brushes and prepared swine-hair.

270. REUS, PZ. N. *Dordrecht.*—Swine-hair and brushes.

271. SIEBERG, J. *Amsterdam.*—Artificial hair-work.

272. THIJSSEN, W. *Tiel.* — Foot, bath, and horse-brushes, &c.

273. GREEVE & SON, *Amsterdam.* — Fur carpet.

CLASS XXVI.

275. HÄGER, J. G. *The Hague.*—Saddle with pockets and boxes for military surgeons.

276. HAVEKOST, T. H. *Amsterdam.* — Head-stalls.

CLASS XXVII.

279. BOASSON, A. *Middelburg.*—Hygienic corsets.

280. CIERENBERG, A. *Zwolle.*—Stockings, knitted in blue and black worsted.

281. DONCKERWOLCKE, J. B. F. *Amsterdam.*—Shirts, flannel waistcoats, and trousers.

282. DONKER, H. *Ijsselmonde.*—Wooden shoes.

286. HOLSBOER, P. & SON, *Arnhem.*— Waterproof Russian-leather shooting-boots and bottines, &c.

287. LEENARTS, J. *Utrecht.*—Silk hats, indigenous genet skins.

288. VELDE, H. VAN DER, *Zwolle.*—Boots and bottines for deformed feet.

289. VIGELIUS, H. & Co. *Rotterdam.*— Kid gloves and skins.

CLASS XXVIII.

290. ABRAHAMS, BROS. *Middelburg.* — Ledger, journal, diaries.

292. DEPARTMENT OF WAR (topographical section), *The Hague.*—Copies of topographical, geological, and other maps.

293. EMRIK & BINGER, *Haarlem.*—Lithography in crayon tints and chromo-typography.

294. ES, GÉRARD D. VAN, *Amsterdam.*— Nature printing. Specimens of wood-cut printing.

295. ENSCHEDÉ J. & SONS, *Haarlem.*— Types, specimens of printing, stereotype plates, &c.

297. LAMPERT, P. *Middelburg.*—Map of the Isle of Walcheren, lithographed for the Department of War.

298. LHOEST, LAMMERS, & Co. *Maastricht.* —Various samples of paper.

299. METZLER & BASTING, *Amsterdam.*— Printings in colours, with gold and silver.

301. PANNEKOEK, NEUY, *Heelsum.*—Specimens of paper.

302. PANNEKOEK, T. *Heelsum.*—Specimens of paper.

303. RINCK, F. W. *The Hague.*—Album for photographs, &c.

305. SMULDERS, J. & Co. *The Hague.*— School maps of the Netherlands. Map of Palestine, in chromo-lithography. Stamped works.

306. SYTHOFF, A. W. *Leyden.* — Books printed in the Japanese, Chinese, and other languages.

307. TETTERODE, N. *Amsterdam.*—Printings in various Oriental types.

309. WYT, M. & SONS, *Rotterdam.* — Specimens of printing.

311. SPANIER, E. *The Hague.* — Lithographs.

312. MULLER, F. *Amsterdam.*—Dictionaries in Oriental languages; anastatic printing, &c.

CLASS XXIX.

313. BACKER, J. *Arnhem.*—Slabs for entomologists.

314. BRUINSMA, J. J. *Leeuwarden.*—Specimens illustrating the anatomy and development of the silk-worm.

315. BURG, J. L. VAN DEN, *Amsterdam.*— Plaster busts of Paul Potter and Van der Helst.

I

CLASS XXX.

317. ANSLYN, H. J. *Haarlem.*—Ladies' work-table, teaboard, &c.

318. BORZO, WIDOW, J. & SONS, *'sHerto-genbosch.* — Looking-glasses and models of frames.

319. DRILLING, A. *Amsterdam.* — Inlaid etagère-tables.

321. HEYMANS, W. G. F. *Amsterdam.*—Tea-table painted in imitation of fine wood.

322. HORRIX, BROS. *The Hague.*—Saloon furniture.

323. KERREBYN, J. *Haarlem.*—Mosaic-painted table.

324. KEMMAN, J. H. W., *Amsterdam.*—Bookcase convertible into a desk; lady's bureau.

325. KLOESMEYER, W. F. *Amsterdam.*—Lacquered table with a carved foot.

326. MICHIELS, J. & WEUSTENRAED, J. *Maastricht.*—Oil cloth.

327. MOEN, P. *Baambrugge.* — Painted mosaic inlaid table.

328. PENNOCK, D. J. *Middelburg.* — Horsehair mattress and beddings.

329. PIETERSE, P. & Co. *Gouda.*—Envelopes of straw and rush for packing bottles.

330. RUTTEN, J. H. *Maastricht.*—Paper-hangings.

331. SALA, D. & SONS, *Leyden.*—Frames manufactured by machinery.

333. VIERLING, M. *The Hague.*—Mahogany buffet—etagère.

334. WILHELM, A. J. *Deventer.*—Fire-proof writing-table with secret lock.

335. WIJDOOGEN, JR. J. *Amsterdam.*—Paper-hangings.

336. ZIRKZEE, J. E. *Leyden.*—Transparent painted window-blinds.

337. BOSCH, A.—Inlaid tea-table and trays.

CLASS XXXI.

339. BEKKERS, WIDOW J. & SON, *Dordrecht.*—Japanned goods.

340. ENTHOVEN, M. J. & SON, *Zalt-Boemel.*—Nails and culinary utensils of tinned iron.

342. HESSELS, WIDOW J. P. *Dordrecht.*—Tinplate ewer and basin, painted in imitation of porcelain.

343. HEUS, H. DE & SON, *Utrecht.*—Specimens of rolled and hammered copper.

344. HUBERS, F. W. *Deventer.*—Cocoanut fibre matting and cloth; woven wire.

346. LEEFERS, J. L. *The Hague.* — Circular fire grate (model).

347. NOOYEN, L. J. *Rotterdam.*—Japanned goods.

349. PRESBURG, M. J. & Co. *Nijmegen.*—German silver and copper tobacco-boxes, &c.

351. RUYVEN, A. H. VAN, *Amersfoort.*—Nails of 45 sorts.

356. VOLKERS, JR. J. *Zwolle.*—Domestic and culinary utensils.

357. WALL BAKE, VAN DEN, *Utrecht.*—The arms of the Netherlands in cast iron.

359. ZWART, D. H. *Oudewater.*—Collection of horse-shoes for preventing and healing diseases.

CLASS XXXIII.

361. BONEBAKKER, A. & SON, *Amsterdam.*—Silver and gold articles.

362. COSTER, M. E. *Amsterdam.*—Rough, cut, and polished diamonds.

363. GREVINK, G. *Schoonhoven.*—Silver-smiths' work.

365. LITTEL, W. & Co. *Schoonhoven.*—Silver and goldsmiths' work.

366. LOON, H. W. VAN, *Rotterdam.*—Netherlanders' gold and silver head-dresses; silver articles.

CLASS XXXIV.

370. BOUVY, J. H. B. J. *Dordrecht.*—Bent glass.

371. CASTRO, D. H. DE, *Amsterdam.*—Glass cups engraved with diamond.

CLASS XXXV.

374. BOSCH, N. A. *Maastricht.*—Earthenware.

375. DRAAISMA, D. *Deventer.*—Water-closets, kilnstones, &c.

376. LAMBERT, G. & Co. *Maastricht.*—Earthenware, stoneware, and terra-cotta.

378. MAAS, B. VAN DER, *Gouda.*—Tobacco and cigar-pipes.

379. PRINCE, J. & Co. *Gouda.*—Tobacco and cigar-pipes.

380. REGOUT, F. *Maastricht.*—Articles of glass, crystal, and fine earthenware.

381. SPARNAAY & SONS, *Gouda.*—Tobacco and cigar-pipes.

383. WANT, AZ. P. J. VAN DER, *Gouda.*—Tobacco-pipes.

CLASS XXXVI.

385. PILGER, L. *Amsterdam.*—Trunks.

DUTCH COLONIES.

387. HOWARD, *Java.*—Specimens of Cinchona barks, grown in Java, and products obtained from them.

388. NETHERLAND COMMISSION.—Collection of produce from the Netherland Colonies, for the European market.

389. MINING DEPARTMENT AT JAVA.—Minerals and metals from India.

390. CARIMON TIN MINE COMPANY.—Tin ore and tin from Carimon.

391. PREANGER GOVERNMENT.—Iron ore, coals, and flint from the Preanger, West Java.

392. NETHERLANDS COMMISSION.—Gypsum from Cheribon, Java.

393. PREANGER GOVERNMENT.—Sulphur from West Java.

394. ———Naphtha from West Java.

395. PADANG GOVERNMENT.—Gold dust from Corintje.

396. TEYSMAN, T. E. *Buitenzorg.*—Pakoe-kidang.

397. PADANG GOVERNMENT.—Camphor baros, camphor wood, dragon's blood.

398. TEYSMAN, T. E. *Buitenzorg.*—Substances used for food, prepared at Java.

399. AMBON GOVERNMENT. — Sago in several forms.

400. PADANG GOVERNMENT. — Cassia lignea, from the west coast of Sumatra.

401. TEYSMAN, T. E. *Buitenzorg.*—Oils.

402. MAKASSAR GOVERNMENT.—Macassar oil.

403. PADANG GOVERNMENT. — Dammar oil.

404. PREANGER GOVERNMENT.—Oils.

405. MOLUCCO GOVERNMENT.—Sereh and clover oil.

406. NETHERLAND GOVERNMENT IN INDIA.—Leaves and flowers from plants cultivated at Java.

407. PADANG GOVERNMENT.—Ivory, horn of the rhinoceros.

408. PREANGER GOVERNMENT.—Malemsala, wax and gummi.

409. TEYSMAN, T. E. *Buitenzorg.*—Wax and gums.

410. MAKASSAR GOVERNMENT. — Getah soesoh.

411. PADANG GOVERNMENT.—Samples of getah pertja from the interior of Sumatra.

412. WEBER, L. *Buitenzorg.* — Gambir (Nauslia Gambier), from Java.

413. PADANG GOVERNMENT. — Gambir from the West Coast of Sumatra, and amboina moeda (dye stuff).

414. PREANGER GOVERNMENT. — Specimen of wood.

415. MOLUCCO GOVERNMENT.—Wood.

416. WEBER, L.—Prepared material for paper.

417. TEYSMAN, A. E.—Flax from Koffo and yuta, cultivated at Java; gemoeti.

418. ROYAL NETHERLAND COMMISSION.—Flax and thread of Koffo.

419. WEBER, L.—Flax.

420. TEYSMAN, T. E.—Fruits and leaves, dried, &c.

421. PREANGER GOVERNMENT.—Benzoin and gum malabar from Java.

422. PADANG GOVERNMENT. — Benzoin from Sumatra W. Coast.

423. TEYSMAN, T. E. — Stangee from Makassar and Amboina.

424. DEPARTMENT OF PUBLIC WORKS AT JAVA.—Model of the irrigating sluices in the Kedirie river, near Soerabaya.

425. MAKASSAR GOVERNMENT.—Stangee and doepa from Makassar.

426. ROCHUSSEN, J. J.—A rifle from Borneo.

427. PREANGER GOVERNMENT.—A knife or sword, used by the Javanese.

428. PADANG GOVERNMENT. — Swords and krisses from the Malays of Sumatra, manufactured by Datoe Panghoelve, Radja from Soengipoeor and Agam.

429. TEYSMAN, A. E.—Cleaned cotton from Buitenzorg.

430. PREANGER GOVERNMENT. — Cotton cloth and yarn from the west part of Java.

431. SAMARANG GOVERNMENT. — Shirtings in the various stages of manufacture for dyeing, and tools used.

432. SAMARANG GOVERNMENT.—Sarong and kainpanjangs.

433. CHERIBON GOVERNMENT.—Kainpanjangs, slindangs, and cloth, batikked and woven.

434. ROCHUSSEN, J. J.—Three sarongs from Makassar, and cloth, batikked at Buitenzorg.

435. TEYSMAN, A. E.—Three sarongs from Buitenzorg, cloth of bamboe and bark.

436. PADANG GOVERNMENT.—A sarong of silk from Sumatra.

437. SAMARANG GOVERNMENT.—Sarongs and kainpanjangs of silk.

438. ROCHUSSEN, J. J.—Silk cloth batikked, and a crape shawl embroidered at Batavia.

439. SAMARANG GOVERNMENT.—Carriage harness and shoes.

440. ——— Samples of embroidery.

441. PREANGER GOVERNMENT. — Two bedayas, two capala Java, and fruits carved in stone by Javanese.

442. SAMARANG GOVERNMENT.—Two bedayas, carved in stone by Javanese.

443. ROCHUSSEN, J. J.—Two Hindoo gods, carved in ivory at Bali.

444. PREANGER GOVERNMENT. — Baskets of rattan from the west part of Java.

445. ROCHUSSEN, J. J. — Baskets and spoons for rice ; box for pickles.

446. PALEMBANG GOVERNMENT. — Lacquered wood.

447. CHERIBON GOVERNMENT.—Precious stones, used by the Javanese.

448. SAMARANG GOVERNMENT.—A pair of gold bracelets, with jewels.

449. BATAVIA COMMISSION. — A silver box for sirh (tampat sirh).

450. PADANG GOVERNMENT. — Filigree and other gold and silver works.

451. ROCHUSSEN, J. J.—A cigar box from Soerakarta.

452. POOLMAN, W.—Filigree and other gold and silver works.

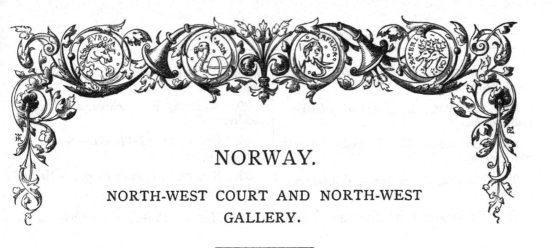

NORWAY.

NORTH-WEST COURT AND NORTH-WEST GALLERY.

CLASS I.

1. AAS, A. *Throndhjem.*—Copper ores.

2. BRÜNECH, J. *Bergen.*—Copper ores.

3. DAHLL, J. *Kragero.*—Magnetic pyrites, about one-fourth nickel; and apasite, chiefly phosphate of lime.

4. DAHLL, T. *Kragero.*—A collection of Norwegian minerals.

5. EGERSUND MINING CO. *Egersund.*—Ilmenite.

6. ESMARK, M. *Tönsberg.*—Norwegian minerals, discovered by Esmark.

7. GOVERNMENT SILVER MINE OF KONGSBERG.—Silver ores, pugworks, and smelting products.

8. GRÖNNING, B. *Bergen.*—Limestone, burnt lime, and bricks.

9. HOLE, H. *Langöén.*—Model of a lime-kiln.

10. IRON WIRE MANUFACTORY OF KJELSAAS, *near Christiania.*—Wire of different metals.

11. JOHANNESEN, *Bergen.*—Ores and other minerals.

12. KJEER, N. *Næs Iron Works at Tvedestrand.*—Feldspath and quartz.

13. PETERSEN, F. *Stavanger.*—Copper ore and magnetic pyrites, containing nickel.

14. REINERTSEN, H. *Christiansand.*—Limestone and lime from Södal.

15. ROSCHER, *the nickel-mine of Ringerige.*—Magnetic pyrites, one-fiftieth nickel; and nickel-metal, one-half nickel.

16. THE CEMENT MANUFACTORY OF LANGÖ, *Christiania.*—Cement and lime.

17. THE COPPER MINE OF SELBO, *Throndhjem.*—Copper ores and smeltings.

18. THE IRON WORKS OF ÖIENSJÖFOS, *Christiania.*—Limonite, and the iron produced from it.

19. THE LEAD AND SILVER MINE OF RANEN, *in Helgeland, Throndhjem.*—Ores of lead and silver.

20. THE ROLLING WORKS OF LEEREN, *Throndhjem.*—Rolled copper from the mines of Röraas.

21. WEDEL-JARLSBERG, BARON H. *The Iron Works in Bærum.*—Iron ores and smeltings.

22. WEDEL-JARLSBERG, BARON H. *The Iron Works at Moss.*—Iron smeltings.

AALL & SON, *Næs Iron Works, near Tvedestrand.*—Magnetic iron ore.

CLASS II.

23. BUCHNER, *Kaupanger.*—Oil of turpentine.

24. CHROME MANUFACTORY OF LEEREN, *Throndhjem.*—Salts of chromium.

25. FINCKENHAGEN, *Hammerfeet.* — Medical cod-liver oil.

26. IBENFELDT, S. *Aalesund.*—Medical cod-liver oil.

27. KNUTZON, A. *Christiansund.*—Medical cod-liver oil.

28. KRAMER, J. *Bergen.*—Medical cod-liver oil.

29. MANUFACTORY OF LYSAKER, *Christiania.* — Pyrites, and superphosphate of lime.

30. MÖLLER, P. *Christiania.* — Medical cod-liver oil.

31. STEEN, D. *Christiania.*—Soaps, and cod-liver oil.

32. SVENDSEN, L. & S. *Bergen.*—Medical cod-liver oil.

——

HAUSEN, F. *Aalesund.*—Cod-liver oil.

CLASS III.

33. BÖ, P. *Gausdal, near Lillehammer.*—Cheese from the sweet milk of goats.

34. BORNHOLDT, G. I. *Christiania.*—Prepared mustard.

35. LIND, MISS, *Christiania.*—Anchovies.

36. MÖLLER, P. *Christiania.*—Norwegian wild medical plants.

37. NORDBYE, J. *Christiania.* — Cheese made of the sweet milk of goats.

38. NORMAN, *Tromsöe.* — Specimen of grass grown in Finmark.

39. SCHÜBELER, DR. F. C. *Christiania.*—Cereals and other vegetable products of Norway.

40. SMITH, MRS. G. *Christiania.*—Anchovies.

41. THE AGRICULTURAL SOCIETY OF TROMSÖ.—Cereals from Finmarken (latitude 70 deg. north) :—

ANCUNDSEN, T. *Tromsöe.*—Seeds of wild vetches.

CHRISTENSEN, C. P. *Trondenæs.*—Rye.

KROGSENG, *Maalselven.*—Barley.

LUDVIGSEN, J. H. *Gisund.*—Barley.

MAURSUND, *Tromsöe.*—Barley.

OXAAS, *Lyngen.*—Barley.

POULSEN, P. *Dyró.*—Potatoes.

SCHJÓLBERG, *Sengen.*—Rye, barley, oats, and pease.

STENSOHN, D. C. *Tromsöe.*—Barley.

STENSOHN, D. C. *Tromsöe.*—Rye.

STENSOHN, D. C. *Tromsöe.*—Turnip-seeds.

STRÓM, LARS, *Dyró.*—Rye.

STRÓM, LARS, *Dyró.*—Oats.

42. THORNE, CHR. AUG. *Drammen.*—Hermetically preserved articles of food; anchovies.

CLASS IV.

43. FROST, *Tranöe.*—Eider-down.

44. IHLEN, J. S. *Christiania.*—Deals and boards of fir and Norway spruce-fir.

45. KLINGENBERG, H. *Throndhjem.* — Furs of Norwegian wild animals.

46. NISSEN, A. *Kirkesnæs.*—Wild reindeer horns.

47. THE AGRICULTURAL SOCIETY OF TROMSÖ.—Samples of forest products, from Finmarken (latitude 70 deg. north).

48. WEDEL-JARLSBERG, BARON H. *Christiania.*—Deals and boards of fir and Norway spruce-fir.

CLASS VI.

49. GRÖNNEBERG, I. H. *Drammen.*—A cariole and sledge.

50. HEFFERMEHL, *Drammen.*—A sledge and apron.

51. NORMAN, *Tromsöe.*—A Fin sledge.

52. STEEN, A. *Drammen.*—Sledges, and a cariole.

CLASS VII.

53. JACOBSON, O. *Christiania.*—Model of a mangle.

54. NILSEN, I. *Christiania.*—Sliding-rest, with double cutting apparatus for metallic cogwheels.

55. THE MECHANICAL WORKS OF AKER, *Christiania.*—Lund's ball-press.

56. THE MECHANICAL WORKS OF THE ROYAL NAVY, *Horten.*—A set of screw taps.

CLASS VIII.

57. THE MECHANICAL WORKS OF AKER, *Christiania.* — A high pressure expansive steam-engine, 6 h. p.

58. THE MECHANICAL WORKS OF NYLAND, *Christiania.* — A steam-engine; cocks and valves for water and steam; a turning-lathe.

59. THE MECHANICAL WORKS OF THE ROYAL NAVY, *Horten.* — A donkey-engine steam-pump.

60. THE WORKS OF LAXEVAAG, *Bergen.*—Machinery for ropemaking.

CLASS IX.

61. BISETH, *Christiania.*—Models of agricultural implements; a cultivator for root-crops; Biseth's double mould-board plough, &c.

62. JACOBSON, O. *Christiania.*—A chaff-cutter and a Norwegian revolving harrow.

63. ROLFSEN, W. *Bergen.* — An iron plough.

64. ROSING, C. W. *Frederikstad.* — An iron plough, with a steel mould-board.

65. SCHÜBELER, DR. F. C. *Christiania.*—A hand-cultivator, invented by the exhibitor, and other horticultural implements.

66. SKJETNE, G. *Tiller by Throndhjem.*—A double plough, constructed by exhibitor.

67. WINGAARD & BOUILLY, *Bergen.* — A chaff-cutter.

CLASS X.

68. THE ROYAL OFFICE FOR THE PUBLIC ROADS.—Models of a wheelbarrow and ballast waggon.

69. ——— System of the Norwegian highways.

CLASS XI.

70. BUE, E. *Lillehammer.*—A gun.

71. HJELMELAND, O. *Bergen.*—A revolver rifle, a gun, and a pistol with four barrels.

72. LARSEN, H. *Drammen.*—A breech-loading rifle and a revolver rifle.

73. TENDEN, R. *Bergen.*—A rifle.

74. THE BOARD OF ORDNANCE, *Christiania.*—Collection of arms for the Norwegian army, a hunting-rifle and a field-carriage.

CLASS XII.

75. BALCHEN, H. *Bergen.*—Specimens of ropemakers' work.

76. BENDIXEN, C. *Stavanger.* — Ship's augers, constructed by exhibitor.

77. BRUNCHORST, C. H. *Bergen.*—Model of a ship.

78. CHRISTOFFERSEN, *Bergen.*—A grapnel.

79. DEKKE, A. *Bergen.*—Models of ships.

80. ELLERTSEN, *Bergen.* — Model of a sloop.

81. GRAN, J. *Bergen.*—Models of ships.

82. HOEL, *Christiania.*—Model of a ship, with several new inventions.

83. HOLMBOE, O. *Wefsen.*—Model of a Nordland boat.

84. JORDAN, H. *Bergen.*— Specimens of ropemaker's work.

85. THE MECHANICAL WORKS OF THE ROYAL NAVY, *Horten.*—Anchors and chains: models of Norwegian boats.

CLASS XIII.

86. ENGER, CHR. *Christiania.* — Instruments for surveying.

87. LUNDGREN, J. *Bergen.*—Optical instruments.

88. QVAMME, L. *Bergen.*—Instruments for surveying, invented by exhibitor.

CLASS XIV.

89. SELMER, M. *Bergen.*—Photographs of Norwegian national dresses, and of Norwegian scenery.

CLASS XV.

90. CHRISTOPHERSEN, *Christiania.* — An astronomical clock.

91. IVERSEN, J. *Bergen.*—A chronometer.

92. PAULSEN, M. *Christiania.*—An astronomical clock.

CLASS XVI.

93. BRANTZEG, P. *Christiania.*—A mono-chord, and an upright piano.

94. HALS, BROTHERS, *Christiania.* — A grand and an upright piano.

95. HELDAL, A. *Bergen.*—A Hardanger fiddle, used by the Norwegian peasantry.

CLASS XVII.

96. GALLUS, *Christiania.*—Surgical instruments.

97. STEGER, A. *Christiania.*—A medical syringe apparatus.

98. SUNDBY, H. *Christiania.*—An artificial leg.

CLASS XVIII.

99. HALVOR SCHOU, *Christiania.*—Striped cotton stuffs and checks.

100. HOLST, MRS. *Bergen.*—Quiltings.

101. JEBSEN, P. *Bergen.*—Dyed shirtings and twills, bleached and grey calico.

102. LILLEHAMMER COTTON SPINNING MILL, *Lillehammer.*—Grey cotton yarn.

103. RICHTER, MISS I. *Inderöen.*—Bed-quilts.

104. RICHTER, MISS G. *Inderöen.*—Quilting.

105. THE SPINNING MILL of NYDALEN, *Christiania.*—Cotton yarn.

106. THE SPINNING MILL OF FOSS, *Christiania.*—Cotton yarn.

CLASS XIX.

107. CHRISTIANIA SAILCLOTH FACTORY, *Christiania.*—Linen-yarn and sailcloth.

108. ECHE & SON, *Bergen.* — Window-blinds.

109. KNOPH & BLICHFELDT, *Christiania.*—Hosiery.

110. RICHTER, MISS I. *Inderöen.*—Damask.

111. RICHTER, MISS G. *Inderöen.*—Damask.

112. THE HOUSE OF CORRECTION AT BERGEN.—Fishing-nets for herrings.

CLASS XXI.

113. DONS, N. *Sengen.* — Home-made stuffs, from Finmark; woollen gloves, and specimen of wool.

114. HANSEN, H. & Co., *Bergen.* — Woollen stuffs.

115. SCHOU, HALVOR, *Christiania.*—Half-woollen stuffs.

CLASS XXII.

116. Brun, Miss A. B. *Fosnœss.*—Quiltings.

117. Kjelven, Miss G. *Sogn.*—Quiltings.

118. Knagenhjelm, *Kaupanger.*—Quiltings.

119. Westrem, *Kaupanger.*—Quiltings.

CLASS XXIV.

120. Neumann, Miss, *Bergen.*—Embroidery.

121. Wiencke, L. *Bergen.*—Trimmings.

CLASS XXV.

122. Eckhardt, I. M. *Christiania.*—hair-work.

123. Pettersen, *Bergen.*—Hair-work.

CLASS XXVI.

124. Amundsen, T. *Frederikshald.*—Skins and leather.

125. Bérgmann, G. *Drammen.*—Harness.

126. Berner, I. *Christiania.*—Skins and leather.

127. Brandt, C. F. *Bergen.*—Furs.

128. Christoffersen, O. *Vœrdalsoren.*—Skins.

129. Ellerhusen, J. *Bergen.*—Leather.

130. Endstrup, C. *Christiania.*—Harness.

131. Erichsen, E. *Bergen.*—Saddler's work.

132. Grönneberg, I. H. *Drammen.*—Harness.

133. Hallén, C. *Christiania.*—Leather.

134. Heyerdahl, C. *Christiania.*—Skins.

135. Rohde, E. *Bergen.*—Saddler's work.

136. Valeur, P. *Nordland.*—Harness for a reindeer.

CLASS XXVII.

137. Christiania Industrial Union for Indigent Women, *Christiania.*—Needlework and embroidery.

138. Eliassen, A. *Bergen.*—Boots.

139. Fougner, J. *Bergen.*—Hats.

140. Frey, M. *Christiania.*—Hats, and a seamless dress made of felt.

141. Hansen, C. *Christiania.*—Gentlemen's boots and shoes.

142. Hansen, H. F. *Christiania.*—Ladies' boots and shoes.

143. Lotz, P. *Bergen.*—Gloves.

144. Salvesen, J. *Bergen.*—Hats.

145. SOLBERG, N. *Christiania.*—Gentlemen's boots and shoes.

146. SVENSEN, F. *Bergen.*—Boots.

147. THISTED, A. *Bergen.*—Boots.

148. TRESSING, E. *Christiania.*—Boots.

149. TYSLAND, G. *Bergen.*—Boots.

CLASS XXVIII.

150. BERG, G. *Christiania.* — Printed books.

151. BEYER, F. *Bergen.*—Printed books, and bookbinding.

152. BROGGER & CHRISTIE, *Christiania.*—Printed books.

153. DAHL, J. *Christiania.* — Printed books.

154. GJERTSEN, E. *Bergen.*—Bookbinding.

155. HELDAL, J. *Bergen.*—Pocket-books.

156. HENRICHSEN, C. *Christiania.*—Playing cards.

157. HERMANN, L. *Bergen.*—Paste-work.

158. HJORTH, G. *Christiania.* — Paste-work.

159. JENSEN, H. *Christiania* —Printed books.

160. THE WORKS OF GAUSA, *Lillehammer.* —Pasteboard.

CLASS XXIX.

161. THE GOVERNMENT DEPARTMENT FOR THE CHURCH, *Christiania.*—Drawings and models of buildings and furniture for schools.

162. ———— Books and instruments for teaching generally.

163. ———— Appliances for physical education.

164. ———— Specimens of school work.

165. ———— Museums.

166. ———— Drawings and objects for the Philanthropic Congress.

CLASS XXX.

167. HENNEMOE, M. *Throndhjem.*—Furniture made of veined birch-wood.

168. LOSTING, I. *Bergen.*—Looking-glass with console.

CLASS XXXI.

169. BERGENDAHL, H. *Bergen.* — Lacquered work.

170. BONGE, R. *Bergen.*—Articles in iron and zinc.

171. FLOOD, G. *Bergen.*—Smith's-work.

172. JOHANNESEN, C. *Christiania.*—Lacquered work.

173. Nilsen, J. *Christiania.* — Door handles.

174. Opsahl, P. *Christiania.*—Locks.

175. Rokne, K. *Sogn.*—Augers.

176. Schmidt, E. *Bergen.*—Horse-shoes.

177. The Factory of Laxevaag, *Bergen.*—Various castings.

178. Wingaard & Bouilly, *Bergen.*—Various castings.

179. Wallendahl, B. *Bergen.*—A stove.

CLASS XXXII.

180. Eimstad, Aslak, *Bergen.*—Cutlery.

——

Aall & Son, *Næs Iron Works, near Tvdestrand.*—Polished blistered steel.

CLASS XXXIII.

181. Halvorsen N. *Bergen.*—Norwegian peasant's bridal ornaments, of silver.

182. Hammer, L. *Bergen.* — Norwegian peasants' ornaments, of silver.

183. Tostrup, I. *Christiania.*—Flower vase with flowers of silver, an epergne, a beer-can, a coffee-pot, and other articles.

CLASS XXXV.

184. Schwarzenhorn & Beyer, *Christiania.*—Paintings on china.

CLASS XXXVI.

185. Blytt, H. *Bergen.*—Artificial flies, for fishing.

186. Broström, G. W. *Bergen.*—Carving in wood.

187. Bucher, H. *Bergen.*—Lithographs.

188. Buck, *Oexfjord.*—Finmark dresses and a cradle.

189. Christie, W. *Bergen.* — Objects illustrating the life and industry of the Norwegian peasants.

190. Christiania Penitentiary, *Christiania.*—Carvings in wood.

191. Ege & Co. *Bergen.*—Matches.

192. Fandrem, *Karasjok.*—Fin-shoes.

193. Fladmoe, T. *Christiania.*—Carvings in wood and stone.

194. Hanno, F. *Christiania.* — A baptismal font of sandstone.

195. Hansen, O. *Christiania.* — Persiennes.

196. Haugse, O. *Bergen.*—Turners' work.

197. Iversen, I. *Sandefjord.*—A knife with sheath; carvings.

198. Jansen, N. *Bergen.*—Coopers' work.

199. Jórgensen, H. *Skjærvóe.* — Model of a mussel-dredge from Finmark.

200. KLUTE, I. *Christiana.* — Basket-makers' work.

201. KNUDSEN, A. *Bergen.*—Articles made of huldrabromms, a sickly development of the birch-tree.

202. KÜHNE, V. *Christiania.* — Basket-makers' work.

203. LARSEN, G. *Lillehammer.*—Carvings in meerschaum.

204. LARSEN, MISS L. *Christiania.*—Dried and pressed flower-work.

205. LÖBERG, O. *Bergen.*—Fishing-lines.

206. LOSTING, *Bergen.*— Illustrations of skin diseases (leprosy).

207. MÖLLER, E. *Bergen.* — Basket-makers' work.

208. NORMAN, *Tromsóe.* — Collection of national costumes from Finmark.

209. OLSEN, J. *Bergen.*—Novel window-frame.

210. PRAHL, G. *Bergen.*—School atlas.

211. SCHWENZEN.—Maps of coast survey.

212. SÖRUM, O. *Hadeland.*—A knife, with carved-wood sheath, and belt.

213. STEENSTRUP, CHR. *Horton.*—A moveable jet-piece for fire-engines; apparatus for fishing, sounding, and preventing rust in boilers; an economical adjustable oil can; an apparatus for communicating with the diver, by speaking, &c.

214. STOCHFLEDT, FR. *Bergen.*—Drawings of architectural ornaments, applicable to furniture.

215. THE COMMITTEE AT BERGEN, *Bergen.* —A collection of fishing implements, and models of fishing-boats.

216. THE HOUSE OF CORRECTION OF AGERSHUUS, *Christiania.*—Articles in polished granite and porphyry: and objects illustrating the life and customs of the Laplanders.

217. THE TOPOGRAPHICAL SURVEY OFFICE. —Maps of some Norwegian provinces.

218. VALEUR, P. *Nordland.*—Finmark dresses, and a cradle.

219. VEDELER, E. *Bergen.*—National costumes, after a picture of A. Tidemand.

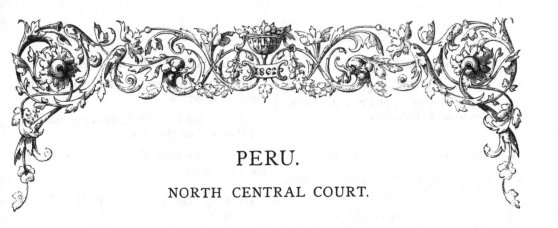

PERU.

NORTH CENTRAL COURT.

KENDALL, H. Consul for Peru, 11, *New Broad St. E.C.*—

1. Silver ores, from the mines at Cerro de Pasco.

2. Quicksilver ore, from the mines of Huancavelica.

3. Nitrate of soda, or cubic nitre, from Iquique.

4. Borax from the same district.

5. Guano from the Chincha Islands.

6. Remarkable specimen of guano, containing a pure white crystallised ammoniacal salt.

7. Cascarilla, or Peruvian bark, from Huanuco.

8. Coca-leaf from Huanuco; it supports Indian miners and couriers under great fatigue with very little food.

9. Matico-leaves, a very powerful styptic and astringent.

10. Sarsaparilla from the department of Piura.

11. Large maize from Cuzco, in grain and in the ear.

12. Rice from Lambayeque and San Pedro.

13. Coffee from the mountains near Huanuco.

14. Samples of sheep's wool from Junin and Puno.

15. Alpaca wool from Arequipa and Puno.

16. White alpaca skin.

17. Llama and Vicuña, or Vigonia wools.

NAYLOR, J. E. Consul for Peru, *Liverpool.*—

18. Silver vessel, from the ruins of the ancient Indian city of Grand Chimu.

19. Silver hammer, seals, buttons, and medals, from the same ruins.

20. Drinking cup mounted in silver.

21. Poncho of Peruvian cotton, made by Indians.

22. Small piece of poncho cloth, taken from an Indian grave some centuries old.

23. Jaquima, or halter of plaited hide, mounted in silver.

24. Wooden stirrups mounted in silver.

25. Ancient stirrups of carved wood mounted in silver.

26. Crupper of stamped leather.

27. Four huaqueros, or earthen jars.

28. A face carved in wood, from the ruins of the Temple of the Sun at Pachacamac.

ELIAS, DON D. *Estate at Nasca.*—

29. Cotton from Nasca, producing 1,200,000 lbs. yearly: it might be 3,000,000 lbs.

30. Cochineal from Nasca; annual produce, 50,000 lbs.

31. Wine from Hoyos, near Pisco, similar to Madeira wine; annual produce, 100,000 gallons.

32. Wine from Urrutia estate, near Pisco; Annual produce, 200,000 gallons.

33. Aguardiente Italia, produced at Yca, near Pisco; a liqueur made from Muscatel grape.

HOYLE, DON J. *Truxillo.*—

34. Blankets or coverlets made from cotton, of great antiquity, and recently found in the ruins of an Indian city.

KENDALL, MRS. *The Limes, Mortlake.*—

35. A silver ornament, known in Lima as "Briscado" work, made by the nuns.

36. Specimens of silver filigree work, made in Huamanga.

37. A piece of plata-piña, or crude porous silver, moulded in the shape of a dog.

EASTTED, W. & Co. 27, *Regent Street.*—

38. Wines from Moquegua resembling sherry and Muscat.

———

EGUSQUIZA, DA. PAULA, *of Lima.*—

39. Two straw hats, plaited by the Indians of Moyobamba.

40. Six cigar cases, made in the Department of Yca.

41. Silk and cotton poncho, as made by the Indians of Cuzco before the Conquest.

42. Cotton coverlet, made by the Indians of Eten.

43. Silk braces and garters, made by the Indians of Huamanga.

44. Two filigree turkeys, made at Ayacucho.

45. Four bottles of Moscatel wine, from the Falconi estates at Yca.

46. Four bottles of wine, sherry class, from the same.

47. Two samples of chocolate of Cuzco cocoa.

48. Coffee from Carabaya, Department of Arequipa.

49. Coffee from Huanuco, Department of Junin.

50. Quinua, a grain indigenous to all the Sierra of Peru.

51. Pallares, a large kind of bean, from Yca.

DE LA JARA, DA. M. A., of *Lima.*

52. Two filigree baskets, with dishes, made at Ayacucho.

53. Cotton and silk table-cover, made in the Province of Huaylas.

54. Cotton from the valley of Viso, Province of Jauja.

GALVEZ, HIS EXCELLENCY DN. PEDRO, *Paris.*—

55. Silver ores from different mines.

56. Felt of Vicuna wool.

MAIZ, DN. THOMAS MORENO.—

57. Coca leaf (vide No. 8).

COTES, SRA. ALTHAUS DE.—

58. Cloth Poncho.

59. Counterpane from Lambayeque.

60. Towelling from Loja, Caxamarca, Arequipa, and Yca.

61. Embroidery work from Moquegua.

62. Specimens of lace from Camana, Arequipa, and Carameli.

63. Purse from Cuzco, in figure of an Indian.

64. Relics of the nuns at Cuzco.

65. Garters made at Cuzco.

66. Collar used by the Indian women of San Jose.

67. Fine lace cap.

68. Vicuna gloves from Cuzco.

69. Cloth made in Arequipa.

HAYNE, MRS. *Gloucester Square.*—

70. Painting, by a native artist at Cuzco, representing portraits of the Incas, from Manco Capac to Atahualpa, with historical notes.

71. Jewellery, forming a nun's rosario, made in Lima.

72. Collection of metallic ores.

WENT, MRS. *Cleveland Square.*—

73. Silver filigree baskets and flowers.

74. Specimens of lace and work from Arequipa.

PORTUGAL.

SOUTH CENTRAL COURT, AND SOUTH CENTRAL GALLERY.

The Portuguese Commissioners having forwarded their Catalogue alphabetically arranged according to Christian instead of Surnames, it has not been considered expedient to alter the arrangement, owing to the necessity of preserving the consecutive order of the numbers.

CLASS I.

1. A. J. Botelho, *Evora.* — White marble.

2. A. A. da Silveira Pinto, *Porto.*—Ores of lead, and other mining products.

3. A. A. Ribeiro, *Bragança, Macedo de Cavalleiros.*—Amianthus.

4. A. F. Larcher, *Portalegre.*—Ochres, and moulding sand.

5. A. J. Ferreira, *Santarem, Rio Maior.*—Salt.

6. A. J. M. Relvas, *Portalegre.*—Hydraulic lime.

7. A. P. da Costa, *Bragança, Miranda do Douro.*—Copper pyrites, oxide of tin, and metallurgic products.

8. Commission (District) of Aveiro.—Clays used for earthenwares, salt and lime, anthracite coal.

9. Commission (District) of Coimbra.—Limestone, marbles, clays, millstones, and slates.

10. Commission (District) of Evora.—Copper ore.

11. Commission (District) of Faro.—Sulphuret of antimony, copper pyrites.

12. Commission (Filial) of Alandroal.—*Alemtejo, Evora.*—Copper ore.

13. Commission (Filial) of Borba, *Alemtejo, Evora.*—Marbles.

14. Commission (District) of Leiria.—Marble and ochres.

15. Commission (District) of Portalegre.—Marbles and granites.

16. Company (Perseverança Mining), *Porto.*—Antimony and tin ores.

17. Company of Lezirias, *Lisboa.*—Peat.

18. C. de Almeida e Sousa, *Coimbra, Penacova.*—Models of mill-stones.

19. Count de Farrobo, *Lisboa.*—Anthracite coal.

20. ———————————— Lignites.

21. Dejeante, L. B. *Lisboa.*—Marbles.

22. Direction of the Public Works of Angra, *Açores.*—Volcanic grit, &c., used in making hydraulic cements.

23. Direction of the Public Works of Coimbra.—Hydraulic lime.

24. Direction of the Public Works of Leiria.—Capital of a column, &c., in soft limestone.

25. E. Deligny, *Paris.* — Cupriferous iron pyrites.

26. E. I. Parreira, *Açores, Angra.*—Ochres.

27. Feuerheerd, D. M. *Porto.*—Ores of lead, zinc, and copper, metallurgic products, &c.

28.—Fidié, A. M. A. G. *Faro.* — Sulphuret of antimony, limestone, &c.

29. F. P. M. Furtado, *Beja, Moura.*—Copper pyrites with calcareous spath.

30. GARCIA, G. A. C. *Bragança.*—Bars of tin.

31. G. CROFT, *Lisboa.*— Lignites, and magnetic iron.

32. G. J. DE SALLES, *Lisboa.*—Limestone, marbles, &c.

33. INSPECTOR OF PUBLIC WORKS AT AÇORES, *Lisboa.*—Pozzolana from the Azores.

34. INSPECTOR OF THE SECOND MINING DISTRICT, *Lisboa.*—Magnetic iron ore, micaceous oxide of iron, marbles, lignites, &c.

35. J. PRING, JUN. L. F. DEFFERARI E A. C. DE ALMEIDA, *Lisboa.*—Lead ore and blende.

36. J. F. BRAGA, *Lisboa.*—Marbles.

37. J. G. ROLDAN, *Lisboa.*—Tin ore.

38. J. J. RAMOS, *Evora, Redondo.*—Marl.

39. J. M. S. GOUVEIA, *Coimbra, Figueira da Foz.*—Salt.

40. J. PEREIRA, *Leiria.*—Common limestone and marble.

41. J. A. DOS SANTOS, *Lisboa.*—Marble.

42. J. U. PERES, *Coimbra, Penella.*—Marble.

43. J. F. P. BASTO, *Lisboa.* — Ores of copper, lead, and zinc.

44. J. J. DE LEMOS SOUSA E C. E A. L. BATALHA, *Lisboa.*—Cupriferous iron pyrites.

45. J. J. DA SILVA PEREIRA CALDAS, *Braga.*—Granite, magnetic iron ore.

46. J. M. DE FIGUEIREDO ANTAS, *Bragança, Vimioso.*—Marble.

47. J. R. TOCHA, *Evora, Extremoz.*—Copper ores, &c.

48. LACERDA, R. V. DE SOUSA, *Leiria, Alcobaça.*—Gypsum, coal.

49. L. DE ABREU MAGALHAES FIGUEIREDO, *Guarda, Ceia.*—Slate.

50. L. F. DEFFERARI, *Lisboa.*—Lead ore.

51. M. J. V. NOVAES, *Lisboa, Setubal.*—Salt.

52. M. MONTEIRO, *Santarem, Rio Maior.*—Flints.

53. MARQUIS DA BEMPOSTA, L. G. C. DE SALLE, *Lisboa.* — Asphaltic limestone, bituminous grit, and pitch.

54. O. MACHADO & IRMAOS, *Lisboa.*—Pozzolana.

55. S. A. DA CUNHA, *Santarem.*—Lime.

56. V. B. PIRES, *Faro.*—Salt.

57. V. B. PIRES, JUN. *Faro.*—Ores of copper.

58. VISCOUNT DE BRUGES, *Açores.*—Freestone, pumicestone, filtering stone, &c.

59. VISCOUNT DE VILLA MIAOR, *Lisboa.* Sulphate of barites.

———

1131. A. P. DA FONSECA VAZ, *Santarem.*—Lime, and phosphate of lime.

1132.—J. N. REBELLO VALENTE, *Aveiro.*—Slate from Serra da Gualva; quartz from which glass is made; clay.

1133. M. M. DA SILVA, *Porto Vallongo.*—Slate.

CLASS II.

60. A. A. ANDRADE, *Porto.*—Pharmaceutical products.

61. A. SIMOES, *Coimbra.*—Ricinus-seed.

62. A. J. CARDOSO, *Portalegre.*—Medicinal seeds, roots, herbs, &c.

63. A. X. DA SILVA, *Lisboa, Villa Franca de Xira.*—Black mustard.

64. B. J. P. DA MOTA, *Coimbra.*—Calcined bones.

65. B. J. DE SOUSA, *Aveiro.*—Lesser centaury.

66. C. J. PINTO, *Lisboa.*—Pharmaceutical products.

67. C. A. C. GARCIA, *Bragança.*—Charcoal.

68. COMMISSION (CENTRAL), *Aveiro, Feira.*—Charcoal, common ergot, and squalus-liver oil.

69. COMPANY OF TEJO AND SADO LESI-RIAS, *Lisboa, Villa Franca de Xira.*—Mustard.

70. F. J. DA COSTA, *Vizeu, Lamego.*—Eau de Cologne.

71. GENERAL PORTUGUESE WOOD ADMINISTRATION, *Leiria.*—Turpentine, acids, essences, resin, charcoal, &c.

72. GENERAL SOCIETY OF CHEMICAL PRODUCTS, *Lisboa.*—Chemical products.

72A. G. T. DE MAGALHAES COLLAÇO, *Coimbra, Soure.*—Marsh-mallow root.

73. I. DE CIRIA, *Coimbra, Figueira da Foz.*—Fish oil.

74. J. A. D. GRANDE, *Portalegre.*—Syrup of currants; charcoal.

75. J. F. NORBERTO, *Lisboa.*—Pharmaceutical products.

76. J. P. DUARTE, *Beja.*—Oil of bitter almonds.

77. J. C. PUCCI, *Lisboa.*—Orange flower water.

78. J. C. DA COSTA, *Coimbra, Louza.*—Cantharides, poppy and mustard seed.

79. J. F. DA SILVA, *Lisboa.*—Watch-makers' oil, nitrate of potash, jelly, &c.

80. J. G. DA CRUZ VIVA, *Faro.*—Essential oil of orange peel.

81. J. M. DOS SANTOS, *Evora, Montemór o Novo.*—Syrups of currants and wallflowers.

82. M. M. DA TERRA BRUM, *Horta.*—Fish oils.

83. M. DA CUNHA DE ABREU, *Braga.*—Tallow candles.

84. M. J. DE SOUSA FERREIRA, *Porto.*—Pharmaceutical products.

85. VISCOUNT DE BRUGES, *Angra do Heroismo.*—Tortoise oil, camomile, mustard seed, &c.

———

1134. COMMISSION (CENTRAL PORTUGUESE), *Aveiro, Feira.*—Charcoal.

1135. COUNT OF SAMODAES, *Porta, Lamego.*—Charcoal.

1136. DOMINGOS SANT' AGATHA, *Lisboa.*—Artificial guano.

1137. J. FERREIRA, *Coimbra, Louza.*—Charcoal.

1138. J. R. CORREIA BELEM & BRO. *Lisboa.*—Pharmaceutical products.

1139. T. E. AYLORES, JUN. *Lisboa.*—Verdigris.

1140. Y. V. D'ALMEIDA, *Santarem.*—Mustard.

1141. M. F. DA SILVA, *Porto.*—Fish oil.

1142. VISCOUNT OF TAVEIRO, *Coimbra.*—Mustard.

CLASS III.

86. A. F. GARCIA DE ANDRADE, *Horta.*—Brandy.

87. A. M. PEREIRA, *Portalegre, Alter do Chão.*—Wheat.

88. A. B. FERREIRA, *Aveiro, Mealhada.*—Wine geropiga.

89. AGRICULTURAL INSTITUTE OF LISBOA.—Specimens of the various kinds of maize, wines, and wheat produced in Portugal.

90. A. J. DE MORAES, *Bragança, Mogadouro.*—Wine, olive oil.

91. A. SIMOES, *Coimbra.*—Maize.

92. A. J. PEREIRA, *Faro.*—Wine of 1845 and 1846.

93. A. P. SOARES, *Vizeu, Carregal.*—Cheese, brandy.

94. A. ALLEN, *Porto.*—Port wine.

95. A. DA FONSECA COUTINHO, *Portalegre.*—Acorn coffee.

96. A. DE BRITO MONTOSO, *Portalegre, Monforte.*—Wheat and rye.

97. A. DA FONSECA CORSINO, *Castello Branco, Covilha.*—Wine brandy.

98. A. G. PONCES, *Beja.*—Wheat.

99. A. R. Coutinho, *Santarem, Almeirim.*
—Rice.

100. A. de Almeida Pereira, *Beja, Aljustrel.*—Wheat.

101. A. A. de Moraes Campilho, *Bragança, Vinhaes.*—Wheat, potatoes.

102. A. A. da Costa Simoes, *Aveiro, Mealhada.*—Wine.

103. A. A. Ribeiro, *Bragança, Macedo de Cavalleiros.*—Rye, wheat, and honey.

104. A. B. de Moraes, *Bragança, Villa Flor.*—Wine, brandy.

105. A. C. de Figueiredo, *Lisboa, Alcacer do Sal.*—Wheat.

106. A. de Carvalho, *Portalegre, Aviz.*—Red wine, French beans, and maize in double ears.

107. A. Ceia, *Portalegre.* — Cherry brandy.

108. A. C. Perdigao, *Beja, Vidigueira.*—Cheese.

109. A. da Costa Lima, *Santarem, Almeirim.*—Maize, French beans.

110. A. D. Miranda, *Santarem.*—Wheat.

111. A. E. B. Freire, *Beja.*—White wine of 1846 and 1860.

112. A. do Espirito Santo Ferreira, *Coimbra.*—Rye, rye flour, brandy.

113. A. F. Larcher, *Portalegre.*—Maize, rice, oil, honey.

114. A. F. da Motta, *Santarem, Thomar.*—Wine.

115. A. da Fonseca Esguelha, *Lisboa, Villa Franca de Xira.*—Chick-beans, beans.

116. A. Fortio, *Portalegre, Fronteira.*—Chick-beans.

117. A. F. R. M. Tavares, *Bragança, Freixo de Espada á Cinta.*—Chestnuts.

118. A. Franco, *Beja, Aljustrel.* — Wheat.

119. A. F. Carvao, *Santarem, Chamusca.*—White wine of 1854, olive oil.

120. A. G. C. Cotrim, *Lisboa, Azambuja.*—Wheat, maize, French beans.

121. A. G. P. Salazar, *Coimbra, Penella.*—Maize, plums, walnuts, plum brandy.

122. A. H. dos Santos, *Santarem.*—Wheat and maize.

123. A. I. Pereira, *Evora, Redondo.*—Brandy of sugar cane.

124. A. J. da Costa, *Portalegre, Gaviao.*—Olive oil, honey.

125. A. Joao, *Beja, Odemira.*—Rice.

126. A. J. de Castro, *Evora, Montemór o Novo.*—Rice.

127. A. J. da Fonte, Jun. *Santarem, Torres Novas.*—Rye, French beans.

128. A. J. G. de Barahona, *Beja, Aljustrel.*—Brandy.

129. A. J. Lopes, *Beja, Moura.*—Goats'-milk cheese, wheat.

130. A. J. Moreira, *Beja, Ferreira.*—Chick-pease.

131. A. J. P. de Matos, *Faro.*—Wine, dried figs.

132. A. J. P. de Campos, *Evora.*—Wheat, sausages, cheese.

133. A. J. de Sousa, *Beja, Ferreira.*—Red wine.

134. A. J. G. Branco, *Lisboa, Alcacer do Sal.*—Rice.

135. A. J. M. Relvas, *Portalegre, Crato.*—Honey.

136. A. J. Nogueira, *Faro.*—Wheat, rye, maize.

137. A. J. dos Santos, *Coimbra, Oliveira do Hospital.*—Arbutus-berry brandy.

138. A. de Lemos Teixeira de Aguilar, *Porto, Pesqueira.*—Port wine.

139. A. L. Ferreira, *Coimbra, Tábua.*—Raisins.

140. A. L. de Gusmao, *Portalegre, Alter do Chão.*—Olive oil.

141. A. L. Duarte, *Beja.*—Wheat.

142. A. L. da Silveira, *Santarem.*—Wheat, maize, chick-pease, French beans.

143. A. M. P. Rodrigues, *Beja, Serpa.*—Wheat, olive oil, barley, sheep's-milk cheese.

144. A. M. da Cunha e Sá, *Portalegre.*—Red wine.

145. A. M. S. C. Fajardo, *Castello Branco, Belmonte.*—Cereals, olives, &c.

146. A. M. C. Bello, *Portalegre, Castello de Vide.*—Vinegar, honey, wine, cheese; olive oil.

147. A. M. Murteira, *Portalegre, Campo Maior.*—Barley.

148. A. M. Paiva, *Lisboa, Villa Franca de Xira.*—Wheat.

149. A. de Matos, *Portalegre, Campo Maior.*—White wine.

150. A. M. P. Cabral, *Bragança, Macedo dos Cavalleiros.*—Potatoes.

151. A. M. Almeida, *Lisboa, Alcacer do Sal.*—Wheat flour.

152. A. M. Godinho, Jun. *Beja, Alvito.*—Potatoes.

153. A. N. dos Reis, *Lisboa.*—Wheat, wine, cyder.

154. A. P. Salgado, *Lisboa.*—Wheat.

155. A. P. C. M. Certa, *Lisboa, Alcacer do Sal.*—Wines and liqueurs.

156. A. P. N. de Vellez, Jun. *Portalegre.*—Potatoes.

157. A. Pereira, *Lisboa, Barreiro.*—Wine.

158. A. P. de Carvalho, *Faro.*—Wine.

159. A. P. da Silva, *Lisboa, Villa Franca de Xira.*—Potatoes.

160. A. Pinto, *Coimbra, Montemór o Velho.*—Wheat.

161. A. Pires, *Portalegre, Marvão.*—Hazel-nuts, wheat.

162. A. R. Pereira, *Lisboa, Azambuja.*—Barley, chick-beans, and chick-pease.

163. A. de Sampaio Coelho e Sousa, *Guarda.*—Olive oil, white wine.

164. A. S. Franco, *Portalegre, Fronteira.*—Cheese.

165. A. de Sousa Zuzarte Maldonado, *Portalegre, Fronteira.*—Corn, wine, cheese, &c.

166. A. T. de Sousa, *Vizeu, Lamego.*—Wines.

167. A. T. de Sousa Franco, *Evora, Portel.*—Olive oil.

168. A. V. de Almeida Fernandes, *Santarem, Benavente.*—Chick-beans, wine.

169. A. X. da Silva, *Lisboa, Villa Franca de Xira.*—Wheat.

170. A. C. Branco, *Lisboa, Alcacer do Sal.*—Wine.

171. A. C. da Costa Barbosa, *Santarem, Azambuja.*—Chick-pease.

172. A. F. Bairrao Ruivo, *Santarem, Abrantes.*—Wheat, French beans, maize.

173. A. M. de Almeida Garcia Fidié, *Faro.*—French beans, maize, barley, carob.

174. A. Pereira, *Santarem, Torres Novas.*—Maize, French beans, barley, chick-beans, &c.

175. A. P. Bretes, *Santarem, Torres Novas.*—White wine, brandy, wheat, almond.

176. A. V. Falcao, *Fundão.* — Red wine.

177. B. J. C. das Neves e Costa, *Coimbra, Pampilhosa.*—Hams.

178. B. J. de Assis e Brito, *Beja.*—Wheat.

179. B. J. de Araujo, *Beja, Vidigueira.*—Honey.

180. B. M. da Cunha Osorio, *Portalegre, Elvas.*—Preserved olives, olive oil.

181. Baron de Prime, *Vizeu.*—Walnuts.

182. Baron de Seixo, *Porto.* — Port wine.

183. Baron de Viamonte da Boa Vista. —*Villa Real, Sabrosa.*—Port wine.

184. B. F. Jorge, *Aveiro, Mealhada.*—Wine.

185. B. I. D. de Almeida, *Coimbra, Tábua.*—Maize, chick-pease, French beans, &c.

186. B. J. de Mello Pinto, *Braga.*—Rye, maize, maize flour, French beans, &c.

187. B. Coelho, *Beja, Ourique.*—Honey.

188. B. J. C. das Neves, *Coimbra.*—French beans, chestnuts, honey, oil, wine, &c.

189. B. P. F. T. Coelho, *Braga, Celorico de Basto.*—Wheat, maize, French beans, walnuts.

190. B. P. R. Parente, *Portalegre, Fronteira.*—Barley, wheat.

191. B. do Canto Medeiros, *Ponta Delgada, Villa Franca.*—Wine.

192. B. J. de Sousa, *Aveiro, Coimbra.*—French beans, potato starch, honey, &c.

193. B. P. de Aragao, *Vizeu, Lamego.*—Maize, arbutus-berry brandy.

194. B. da P. Figueira, *Beja, Cuba.*—Wheat, raisins.

195. B. da Silva Consolado, *Santarem, Abrantes.*—French beans, walnuts, dried figs.

196. C. J. Ferreira, *Bragança, Vinhaes.*—Walnuts.

197. C. Vellez, *Portalegre, Aviz.*—French beans.

198. C. Pecegueiro, *Coimbra, Montemór o Velho.*—Lupines, beans, French beans, chick-beans.

199. C. da Silveira Barbosa, *Santarem, Almeirim.*—French beans.

200. C. Ferreira, *Coimbra.*—Pease.

201. C. M. da Silveira Almendro, *Santarem, Almeirim.*—Rice.

202. C. Pereira, *Santarem, Thomar.*—Dried plums.

203. C. A. Carneiro, *Bragança, Freixo de Espada á Cinta.*—Almonds, walnuts, honey, &c.

204. C. A. C. Garcia, *Bragança.*—Corn, flour, potatoes, &c.

205. C. J. Gomes, *Castello Branco.*—French beans.

206. Collares & Irmao, *Lisboa.*—Preserved food for long voyages.

207. Commission (Central), *Aveiro, Ilhavo, Ovar, Feira.*—Cereals, fruits, oils, fish, honey, and other edibles.

208. Commission of Alandroal, *Evora, Alandroal.*—Corn, oil, and cheese.

209. Commission of the District of Angra do Heroismo.—Corn, fruit, potatoes, brandies, &c.

210. Commission (Filial) of Borba, *Evora.*—Cheese, olive oil, wine.

211. Commission of the District of Coimbra.—Maize, maize flour, vegetables, and fruit.

212. Commission of the District of Evora.—Bacon.

213. Commission of the District of Faro.—French beans, and various fruits.

214. Commission of Ponta Delgada.—Maize, millet, sweet potatoes, French beans, nuts.

215. Company of Alto-Douro, *Porto.*—Port wines.

216. Company of Tejo and Sado Lesirias, *Santarem, Benavente, Lisboa.*—Wheat, rye, French beans, lentils, chick-pease, canary-grass, rice, maize.

217. C. de Almeida, A. e Sousa, *Coimbra.*—French beans, maize, maize flour.

218. C. F. de Menezes, *Beja.*—Vinegar, filtered olive oil.

219. C. G. de Campos, *Portalegre, Aviz.*—Rye, barley, maize, chick-beans.

220. Count d'Atalaia, *Santarem, Almeirim.*—Rye.

221. Count d'Azambuja, *Lisboa.*—Port wine.

222. Count da Graciosa, *Aveiro, Anadia.*—Orange brandy, orange wine, geropiga, white wine.

223. Count de Samodaes, *Porto, Lamego, Varzea, Armamar.*—Cereals, fruits, wine, &c.

224. Countess de Samodaes, *Porto, Marco de Canavezes.*—Wine and preserved fruits.

225. Count do Sobral, *Santarem, Almeirim.*—Honey, wine, wheat, chick-pease.

226. Dabney, *Horta.*—Wine.

227. D. A. M. de Almeida, *Portalegre, Alter de Chao.*—Maize.

228. D. J. da Costa Leao, *Bragança, Vinhaes.*—Red wine.

229. Delgado & Pereira, *Evora, Montemór o Novo.*—Wine.

230. D. A. Tallé Ramalho, *Evora.*—Cereals, seeds, cheese, wine, dried fruit, &c.

231. D. A. Fiuza, *Evora.*—Wheat, rye, barley.

232. D. A. de Freitas, *Coimbra.*—Wheat, flour, biscuits, vermicelli, maccaroni, &c.

233. D. J. Fialho, *Beja, Alvito.*—Wheat.

234. D. J. GONÇALVES, *Portalegre.*—Pork sausages.

235. D. P. DA SILVA, *Beja, Serpa.*—Wheat.

236. E. A. MORA, *Santarem, Sardoal.*—Wine, olives, fruits, wheat, maize, &c.

237. E. A. DE SOUSA, *Bragança, Vinhaes.*—Brandy, wine, honey, chick-pease, wheat.

238. E. LARCHER, *Portalegre.*—Olive oil, vinegar.

239. E. DO CANTO, *Ponta Delgada.*—Wine.

240. D. E. M. M. SMITH, *Lisboa, Olivaes.*—White wine.

241. E. C. A. C. DE VASCONCELLOS, *Portalegre, Elvas.*—Wine, olives, wheat, &c.

242. F. A. P. DE ARAUJO, *Beja, Serpa.*—Sheep's milk cheese.

243. F. J. F. S. DE CARVALHO, *Beja.*—Wheat flour.

243A. F. M. DE AQUINO FIALHO, *Beja, Vidigueira.*—Wheat.

244. F. T. DA SILVA, *Santarem, Thomar.*—Raisins.

245. F. M. S. PINÇAO, *Beja, Aljustrel.*—Olive oil.

246. F. DE SOUSA FERREIRA, *Coimbra.*—Araucaria Brasilienses kernels.

247. F. M. B. PINTO, *Vizeu Armamar.*—Wine.

248. F. A. GOMES, *Beja, Alvito.*—Honey.

249. F. A. SOBRINHO, *Beja, Alvito.*—French beans, maize, panicum, almonds.

250. F. DA SILVA ROBALLO, *Castello Branco, Idanha a Nova.*—Wheat.

251. F. T. FERRAO, *Vizeu, Carregal.*—Geropiga, wine.

252. F. M. L. GONÇALVES, *Beja, Alvito.*—Olive oil, brandy.

253. F. DE C. THEMEZ, *Vizeu, Lamego.*—Honey, wine, olive oil.

254. F. RITA (D.) *Beja.*—Rye.

255. F. SANCHES GUERRA (D.) *Bragança, Freixo de Espada á Cinta.*—Sheep's milk cheese.

256. F. DE ABREU CALADO, *Portalegre, Aviz.*—Wheat, maize, goat and sheep's milk cheese.

257. F. A. DA S. PRADO, *Beja, Odemira.*—Wheat.

258. F. A. BERNARDES, *Bragança, Mogadouro.*—Wheat, barley.

259. F. A. CARNEIRO DE MAGALHAES, *Bragança, Moncorvo.*—French beans.

260. F. A. DINIZ, *Coimbra.*—Almonds.

261. F. A. DA GAMA, *Santarem, Chamusca.*—French beans.

262. F. A. MALATO, *Portalegre.*—Side of bacon.

263. F. A. MARQUES, *Coimbra.*—Walnuts.

264. F. A. MENDES, *Coimbra, Montemór o Velho.*—Wheat.

265. F. A. DA SILVA GRENHO, *Evora, Montemór o Novo.*—Olive oil.

266. F. A. MAGRO, *Castello Branco, Idanha a Nova.*—Honey.

267. F. DE ASSIS CALEJO, *Bragança, Mogadouro.*—Chick-pease.

268. F. DE ASSIS LEDESMA E CASTRO, *Bragança.*—Olives, pease, potatoes, French beans.

269. F. DE ASSIS SOBRINHO, *Beja.*—Rice flour.

270. F. BARRETO CASTELLO BRANCO, *Portalegre, Alter do Chao.*—Wheat.

271. F. BERNARDES DE SARAIVA, *Coimbra, Condeixa.*—Wine, pear brandy.

272. F. DE BRITO, *Portalegre.*—Pork sausages.

273. F. CABRAL PACHECO METELLO DE NAPOLES, *Guarda, Celorico.*—Wheat, wine.

274. F. C. GIRALDES, *Castello Branco, Idanha a Nova.*—Rye.

275. F. C. DE MORAES CARVALHO E MACHADO, *Bragança, Mogadouro.*—Wine.

276. F. C. N. DE CARVALHO, *Portalegre, Fronteira.*—Wheat.

277. F. CORREIA DA COSTA, *Coimbra.*—Acorns.

278. F. DA COSTA FIALHO, *Beja, Alvito.*—Wheat.

279. F. E. DA SILVA BARROS, *Bragança.*—Wine.

280. F. F. Carneiro, *Beja, Vidigueira.*—Wine.

281. F. F. de Oliveira, *Beja, Serpa.*—Wine.

282. F. de Freitas Macedo, *Santarem.*—Fruit, pickles, wine, oil, and condiments.

283. F. H. Ripado, *Portalegre, Elvas.*—Olives.

284. F. J. de Sousa, *Portalegre, Fronteira.*—Red wine, olive oil.

285. F. J. da Costa, *Coimbra.*—French beans, and various liqueurs.

286. F. Leite Basto, *Braga, Cabeceiras de Basto.*—Brandy.

287. F. de Lemos Ramalho A. Coutinho, *Coimbra.*—Sorgho.

288. F. Magalhaes Mascarenhas, *Coimbra, Louza.*—Millet.

289. F. M. de Campos Carvalho Pacheco, *Lisboa, Alcacer do Sal.*—Wheat.

290. F. M. da Costa, *Braga.*—French beans, potatoes.

291. F. M. Loureiro, *Beja, Serpa.*—Wild honey.

292. F. M. Felgueiras Leite, *Bragança, Mogadouro.*—Wheat, barley, potatoes.

293. F. Marques de Figueiredo, *Coimbra.*—Olives, liqueurs, dried fruits, wine, &c.

294. F. M. Ferreira Serrano, *Santarem.*—Honey, maize.

295. F. M. Palma, *Beja.*—Wheat, maize.

296. F. Neto Pratas, *Santarem, Chamusca.*—Sheep's milk cheese.

297. F. de Paula da Fonseca Esguelha, *Lisboa.*—Barley, lentils, canary grass.

298. F. de Paula Parreira, *Beja, Serpa.*—Wheat, aniseed.

299. F. de Paula Risques, *Portalegre, Alter do Chao.*—Wheat.

300. F. P. da Veiga, *Vizeu, Lamego.*—Wheat, vinegar, olive oil.

301. F. Pereira da Costa, *Vizeu, Tondella.*—Maize.

302. F. Pessanha de Mendonça, *Beja.*—Wheat, pease.

303. F. R. de Carvalho, *Coimbra.*—Rice.

304. F. R. da Silva, *Coimbra, Penella.*—Sheep's milk cheese.

305. F. da Silva, *Portalegre, Aviz.*—Chick-beans, oats, maize.

306. F. da Silva Lobao, *Portalegre, Arronches.*—Wheat, chick-pease.

307. F. da Silva Paes, *Portalegre, Aviz.*—Wheat, maize.

308. F. de Sousa, *Vizeu, Carregal.*—Honey.

309. F. Tavares de Almeida Proença, *Castello Branco.*—Wheat, rye, French beans, olives, olive oil.

310. F. T. Sequeira de Sá, *Beja, Vidigueira.*—Potatoes.

311. F. V. da Costa Cardoso, *Portalegre, Aviz.*—Cheese.

312. F. X. de Moraes Pinto, *Bragança, Mirandella.*—Olive oil.

313.—F. X. de Moraes Soares, *Villa Real, Chaves.*—Cereals, fruit, wine, vegetables, &c.

314. F. X. da Motta Portocarrero, *Santarem, Thomar.*—Olive oil.

315. F. X. Neves, *Bragança, Mogadouro.*—Wheat.

316. F. da Silva Oliveira, *Aveiro, Mealhada.*—Wine.

317. G. Roxo Larcher, *Portalegre.*—Lard.

318. G. T. de Magalhaes Collaço, *Coimbra, Soure.* — Cereals, dried fruits, pickles, &c.

319. G. C. Garcia, *Beja, Almodovar.*—Barley and cheese.

320. G. A. Teixeira, *Bragança, Vinhaes.*—Wheat, rye, maize, &c.

321. G. F. Pereira Nunes, *Coimbra.*—Cereals, maize, seeds, &c.

322. G. C. Herlitz & Co. *Lisboa.*—Food preserved in olive oil.

323. H. C. de Macedo, *Braga, Guimaraes.*—Maize, French beans, wine.

324. H. J. F. de Lima, *Bragança.*—Wheat, rye, wine, &c.

325. H. L. de Aguiar, *Beja, Cuba.*—Wheat.

326. I. A. DA GAMA, *Santarem, Chamusca.*—Maize.

327. I. F. G. RAMALHO, *Evora.*—Wheat.

328. I. X. DE ORIOL PENA, *Santarem, Torres Novas.*—Olive oil, walnuts.

329. I. L. B. GODIM, *Beja.*—Sweet wine.

330. I. COLLACO (D.) *Beja.* — Chick-pease.

331. I. M. R. VALENTE, & T. ARCHER, *Porto.*—Port wine.

332.—J. R. M. COELHO, *Coimbra, Figueira da Foz.*—Wine.

333. J. BARATA, *Evora, Mora.*—Wine.

334. J. C. DE TORRES, *Evora.*—Wheat.

335. J. G. CURADO E SILVA, *Lisboa, Villa Franca de Xira.*—Chick-pease.

336. J. J. DA MOTTA, *Beja.*—Olive oil, raisins.

337. J. ALVES DE SÁ BRANCO, *Lisboa, Alcacer do Sal.*—Wheat, olive oil.

338. J. A. D. GRANDE, *Portalegre.*—Wheat and other cereals; wine, fruit, brandy, vinegar, olive oil, coffee, honey, &c.

339. J. A. DE CAMPOS, *Bragança, Moncorvo.*—Wheat, maize, almonds, olives, wine.

340. J. A. CARDOSO, *Evora, Extremoz.*—Wheat.

341. J. A. MARQUES ROSADO, *Evora, Redondo.*—Wine.

342. J. A. DE OLIVEIRA E SILVA, *Evora, Montemór o Novo.* — Chick-pease, French beans.

343. J. A. ROCO, *Bragança, Mogadouro.*—Wheat.

344. J. A. RODRIGUES, *Coimbra.*—French beans.

345. J. B. DE CARVALHO, *Bragança.*—Wine.

346. J. B. CASIMIRO, *Bragança, Mirandella.*—Cereals, honey, olive oil, &c.

347. J. B. DOUTEL, *Bragança, Vinhaes.*—Wine.

348. J. BARRETO DA COSTA REBELLO, *Portalegre, Aviz.*—Olive oil.

349. J. BATALHA, *Portalegre, Aviz.* —Rye, cheese.

350. J. B. PACHECO TEIXEIRA, *Braga, Celorico de Basto.*—Rye, chick-beans, French beans, millet.

351. J. DE BRITO PIMENTA DE ALMEIDA, *Beja, Moura.*—Wheat, barley, olive oil.

352. J. CARDOSO DE SOUSA, *Portalegre, Alter do Chao.*—Maize, French beans, cheese. Indian-fig brandy.

353. J. CARLOS, *Portalegre, Niza.*—Rye.

354. J. C. DE ARAUJO BASTO, *Braga, Cabeceiras de Basto.*—Wine.

355. J. DO CARMO RAPOSO, *Beja, Moura.*—Olives, olive oil.

356. J. DE CASTRO SAMPAIO, *Braga, Guimaraes.*—French beans.

357. J. DO CARAÇAO DE JESUS FIGUEIREDO, *Vizeu, Carregal.*—French beans, chick-pease.

358. J. DA COSTA CALLADO, *Portalegre, Alter do Chao.*—Wheat.

359. J. DE DEUS, *Portalegre, Aviz.*—Chick-pease, chick-beans.

360. J. FERREIRA DE CARVALHO, *Coimbra.*—Maize flour and maize.

361. J. FERREIRA DE LIMA, *Coimbra.*—Maize and maize flour.

362. J. GONÇALVES VIEIRA, *Braga, Cabeceiras de Basto.*—Wine.

363. J. H. PINHEIRO, *Castello Branco, Idanha a Nova.*—Wheat.

364. J. I. PEREIRA, *Bragança, Vinhaes.*—Chestnuts.

365. J. J. RAMOS, *Evora, Redondo.*—Ears of wheat, French beans, dried plums.

366. J. J. LE COCQ, *Portalegre, Castello de Vide.*—Sparkling and other wines.

367. J. J. DE SOUSA, *Beja, Serpa.*—Maize.

368. J. M. DA COSTA, *Santarem, Almeirim.*—French beans.

369. J. M. DA COSTA BARBOSA, *Santarem, Cartaxo.*—Chick-beans, French beans.

370. J. M. HENRIQUES, *Coimbra.*—Maize flour.

371. J. M. P. VASQUES, *Beja, Cuba.*—Wine.

372. J. M. DA SILVA, *Lisboa.*—Olive oil and dried plums.

373. J. DE MATOS DE FARIA BARBOSA, *Braga, Barcellos.*—Millet, seeds, confections, wine, &c.

374. J. N. P. GIRAO, *Faro.*—Wheat, dried fruit, wine, olive oil, &c.

375. J. N. R. VALENTE, *Aveiro, Oliveira de Azemeis.*—Wheat.

376. J. N. R. VALLADA, *Aveiro, Oliveira de Azemeis.*—Maize.

377. J. N. DA CONCEIÇAO, *Portalegre, Elvas.*—Preserved fruits.

378. R. DE OLIVEIRA CABEÇAS, *Braga, Guimaraes.*—Flour.

379. J. P. DE FARIA LACERDA, *Beja, Vidigueira.*—Wheat, olive oil, and olives.

380. J. P. CORDEIRO, *Portalegre, Elvas.* —Brandy, orange wine.

381. J. P. MARTINS, *Lisboa, Setubal.*— Wine.

382. J. P. ROXO, *Portalegre, Portalegre.* —Potatoes and pickles.

383. J. P. DA SILVA, *Portalegre, Aviz.*— Wheat.

384. J. P. GUIMARAES, *Vizeu, Carregal.* —Common wine.

385. J. P. FRAUSTO, *Portalegre, Marvao.* —Rye, walnuts.

386. J. P. T. CLERC, *Evora.*—Vinegar.

387. J. R. DE PAIVA LOBATO, *Portalegre, Fronteira.* — Wheat, olive oil, and cheese.

388. J. R. DE AZEVEDO, *Santarem.*— Wheat.

389. J. DE SACADEIRA ROBE-CORTE REAL, *Vizeu, Nellas.*—Wine.

390. J. S. X. LEITAO, *Portalegre, Aviz.* —Ears of maize.

391. J. DE SOUSA FALÇAO, *Santarem, Almeirim.*—Olives.

392. J. TAVARES DE AZEVEDO LEMOS, *Braga, Cabeceiras de Basto.*—Dried chestnuts.

393. J. T. P. DA MAIA, *Evora.*—Rye, wine, brandy.

394. J. V. DE ALMEIDA, *Santarem, Benavente.*—Wheat, maize, honey, and wine.

395. J. DE ALMEIDA CAMPOS, *Vizeu.*— Hazel-nuts, dried fruits.

396. J. A. DE SÁ BRANCO, *Lisboa, Alcacer do Sal.*—Olive oil.

397. J. A. MONTEIRO, *Portalegre, Monforte.*—Wheat.

398. J. A. MURTEIRA, *Portalegre, Campo Maior.*—Wheat.

399. J. A. PEREIRA DE MATOS, *Faro.*— Dried figs and fig brandy.

400. J. A. RIBEIRO, *Lisboa, Alcacer do Sal.*—Red wine and rice.

401. J. A. SIMOES, *Coimbra, Figueira da Foz.*—Olive oil, wine, brandy.

402. J. A. DA SILVA CORDEIRO, *Santarem.*—Red wine.

403. J. DA COSTA, *Beja, Cuba.*—Maize.

404. J. DA CRUZ FREIRE, *Coimbra, Cantanhede.*—Red wine.

405. J. F. DA CUNHA OSORIO, *Portalegre, Fronteira, Arronches.*—Wheat, olive oil.

406. J. F. FERNANDES, *Beja.* — Wheat and other cereals, cheese, wine, and brandy.

407. J. F. LEVITA, *Portalegre.*—Potatoes.

408. J. GAVINO DE VASCONCELLOS, *Santarem, Gollega.*—Wheat, maize, lupines, and beans.

409. J. I. CABRITO, *Beja, Cuba.*—Red wine.

410. J. I. DE SALDANHA MACHADO, *Santarem, Benavente.*—Wheat, barley, maize, and French beans.

411. J. JOSÉ, *Portalegre, Elvas.*—Wheat.

412. J. J. DE ARAUJO, *Lisboa, Barreiro.* —Wine.

413. J. J. DE CASTRO, *Portalegre, Aviz.* —Wheat and chick-beans.

414. J. J. DA COSTA, *Portalegre, Elvas.* —Wheat.

415. J. J. MARIA, *Santarem.*—Wheat.

416. J. J. CLEMENTE, *Portalegre, Monforte.*—Wheat.

417. J. M. PENTEADO, *Portalegre, Aviz.* —Cheese.

418. J. M. TELLES, *Portalegre, Aviz.*— Mixed wheat, olive oil.

419. J. M. Camacho, *Beja, Alvito.*—Honey, olive oil.

420. J. M. Ferreira Pestana, *Coimbra, Figueira da Foz.*—Wheat, maize, white wine, brandy.

421. J. M. Rodrigues, *Coimbra.*—Pineapple kernels.

422. J. M. R. de Brito, *Coimbra.*—Pease, French beans, and maize.

423. J. M. da Silva, *Evora, Extremoz.*—Olive oil.

424. J. M. L. de Carvalho, *Lisboa, Alemquer.*—Wine.

425. J. P. Duarte, *Beja.*—Chick-pease.

426. J. Pereira Claro, *Portalegre, Arronches.*—Cheese.

427. J. Pereira da Costa, *Portalegre, Fronteira.*—Wheat.

428. J. Ribeiro do Amaral, *Coimbra, Oliveira do Hospital.*—Dried pears.

429. J. R. do Nascimento, *Coimbra.*—Dried chestnuts.

430. J. Sotero Soares Couceiro, *Coimbra, Montemór o Velho.*—Orange vinegar.

431. J. de Sousa Guimaraes, *Porto.*—Port wine.

432. J. U. Peres, *Coimbra, Penella.*—Olive oil, arbutus-berry brandy.

433. J. Ferrao, *Castello Branco.*—Bucellas wine.

434. J. Maxima, *Coimbra, Soure.*—Dried cherries.

435. J. de Aguilar, *Porto, Pesqueira*—Wine.

436. J. A. de Lemos Trigueiros, *Vizeu, S. Joao das Areias.*—Olive oil.

437. J. de Andrade, *Portalegre, Arronches.*—Wheat and chick-pease.

438. J. A. da Cruz Camoes, *Evora.*—French beans, cheese, pork sausages.

439. J. A. de Castro, *Portalegre, Fronteira.*—Wheat and chick-pease.

440. J. A. Garcia Blanco, *Faro, Silves.*—Wheat and other cereals.

441. J. A. Junqueiro, Jun., *Bragança.*—Wine.

442. J. A. Lopes Maia, *Braga.*—Brandy.

443. J. A. da Silva, *Santarem, Abrantes.*—Wine.

444. J. A. de Oliveira, *Coimbra.*—Brandy.

445. J. A. Mendes Pereira, *Guarda, Pinhel.*—Wine.

446. J. A. S. de Aboim, *Beja.*—Wheat.

447. J. de Beires, *Vizeu, Lamego.*—Wheat and maize.

448. J. do Canto, *Ponta Delgada.*—Arrowroot starch.

449. J. C. Pucci, *Lisboa.*—Wine, liqueurs, potatoes, confections, olive oil, &c.

450. J. Carrilho Garcia, *Beja, Almodovar.*—Wheat, lupines, oats, and honey.

451. J. Carvajal Vasconcellos Gama, *Portalegre, Campo Maior.*—Cheese.

452. J. da Conceicao Guerra, *Portalegre, Elvas.*—Preserved fruit.

453. J. da Costa, *Portalegre, Fronteira.*—Wheat.

454. J. da Costa Malhau, *Santarem, Almeirim.*—White wine and chick-beans.

455. J. Correia Monteiro Gorjao, *Santarem, Torres Novas.*—Wine.

456. J. Cupertino da Fonseca e Brito, *Coimbra, Arganil.*—Dried pears and plums.

457. J. D. de Carvalho Monteior, *Coimbra.*—Potato flour.

458. J. Diogo, *Portalegre, Alter do Chao.*—Wheat.

459. J. D. Fazenda, *Beja, Vidigueira.*—Wine and brandy.

460. J. Eschrich, *Lisboa.*—Wine.

461. J. Farinha Relvas de Campos, *Santarem, Torres Novas.*—White wine.

462. J. Ferreira da Silva, *Lisboa.*—Wine, liqueurs, and potato-starch.

463. J. Fialho Coelho, *Beja, Moura.*—Olive oil and oats.

464. J. F. Levita, *Portalegre.*—Red wine.

465. J. F. da Cruz, *Coimbra.*—Wheat, flour, and biscuits.

466. J. F. da Gama Freixo, *Evora.*—Red wine.

467. J. F. DA SILVA, *Beja.* — Quince brandy, lentils, and beans.

468. J. DA GAMA CALDEIRA, JUN. *Portalegre, Crato.*—Chick-pease.

469. J. GONÇALVES, *Coimbra.* — Maize ears.

470. J. GONÇALVES DA CRUZ VIVA, *Faro.* —Maize, chick-beans, and brandies.

471. J. GONÇALVES DE S. THIAGO, *Santarem, Almeirim.*—Rice.

472. J. GUEDES COUTINHO GARRIDO, *Coimbra, Penella.*—Maize flour.

473. J. GUERREIRO DA LANÇA SOBRINHO, *Beja, Ferreira.*—Wheat and red wine.

474. J. I. PINTO GUERRA, *Bragança, Miranda do Douro.*—Rye, chick-pease, and almonds.

475. J. JERONYMO DE FARIA, *Portalegre, Aviz.*—Wheat.

476. J. J. CARDOSO, *Portalegre.*—Side of bacon and sausages.

477. J. J. DE CARVALHO, *Evora, Extremoz.* —Red wine.

478. J. J. CORREIA, *Ponta Delgada, Ribiera Grande.*—Wine.

479. J. J. DA COSTA, *Lisboa, Alemquer.*— Olive oil.

480. J. J. DA COSTA, *Braga, Guimaraes.*— Vinegar and pease.

481. J. J. FIUSA GUIAO, *Evora, Montemór o Novo.*—French beans.

482. J. J. G. A. DE VILHENA, *Beja, Ferreira.*—Wheat.

483. J. J. LAMPREIA, *Beja, Vidigueira.*— Brandy.

484. J. J. PINTO GUERRA, *Bragança, Miranda do Douro.*—Wheat.

485. J. J. RAMOS, *Evora.*—Wine.

486. J. J. RODRIGUES, *Bragança, Vinhaes.*—Potatoes.

487. J. J. DA SILVEIRA, *Santarem.*— Wheat, barley, and French beans.

488. J. J. DO VALLE, *Vizeu, Santa Comba Dao.*—Wheat, rye, and barley.

489. J. J. MARÇAL, *Coimbra, Montemór o Velho.*—Rice.

490. J. LEAL DE GOUVEIA PINTO, *Coimbra, Miranda do Corvo.* — Maize, maize flour, and olive oil.

491. J. LEITE GONÇALVES BASTO, *Braga, Cabeceiras de Basto.*—Wine.

492. J. DE LIMA GUIMARAES, *Santarem, Gollega.*—Wheat, maize, and French beans.

493. J. L. COELHO, *Portalegre, Aviz.*— Rye.

494. J. L. GUIMARAES, *Coimbra.* — Brandy and wine.

495. J. LOPES SERRA, *Coimbra, Montemór o Velho.*—Cereals, potatoes, and wine.

496. J. L. DA COSTA E SILVA, *Beja, Vidigueira.*—White wine and brandy.

497. J. M. FERREIRA, *Bragança, Vinhaes.* —Rye.

498. D. J. M. DE MENEZES DE ALARCAO, *Santarem, Coruche.* — Wheat, rice, acorns, marmalade, and confections.

499. J. M. DO MONTE, *Evora, Redondo.* —Wheat and other cereals; honey and wine.

500. J. M. A. CARDOSO, *Portalegre.*— Beans, fruits, honey, and olive oil.

501. J. M. AYRES DE SEIXAS, *Portalegre, Gaviao.*—Red wine.

502. J. M. DE BRITO, *Beja, Odemira.*— Maize and French beans.

503. J. M. DO COUTO GANÇOSO, *Evora.*— Honey.

504. J. M. DE FIGUEIREDO ANTAS, *Bragança, Vimioso.*—Rye.

505. J. M. DE FIGUEIREDO, *Guarda, Gouveia.*—Cheese.

506. J. M. DA FONSECA, *Lisboa.*—Wine.

507. J. M. HENRIQUES, *Coimbra, Poiares.* —Maize, honey, and olive oil.

508. J. M. L. FALCAO, *Beja, Odemira.*— Maize.

509. J. M. DE MATOS, *Portalegre, Campo Maior.*—Chick-pease.

510. J. M. DA MOTTA CERVEIRA, *Santarem, Almeirim.*—French beans.

511. J. M. RAPOSO, *Ponta Delgada.*— Wine.

512. J. M. ROXO LARCHER, *Portalegre.* —Gin and pickled olives.

513. J. M. DE SÁ PEREIRA E MOURA, *Santarem, Benavente.*—Wheat.

514. J. M. DA SILVA REGUINGA, *Santarem, Almeirim.*—Beans.

515. J. M. CORREIA BELLO, *Faro.*—Common wine.

516. J. M. GODINHO, *Beja, Alvito.*—Aniseed, fruit, honey, olive oil, wine, and brandy.

517. J. M. LEITAO, SEN. *Beja, Vidigueira.*—Wheat and wine.

518. J. M. LEITAO SOBRINHO, *Beja, Vidigueira.*—Wheat and wine.

519. J. M. LOPES, *Faro, Olhao.*—Dried figs.

520. J. DE MATOS MACHADO, *Portalegre, Ponte de Sor.*—Maize and rice.

521. J. M. DA SILVA AZEVEDO, *Belmonte, Castello Branco.*—French beans, olive oil.

522. J. DE MELLO PITTA, *Vizeu, Tarouca.*—Cereals, raisins, and wine.

523. J. MILITAO DE C. E SOUSA, *Beja.*—Canary-grass.

524. J. DE MORAES PINTO DE ALMEIDA, *Coimbra.*—Carolina rice.

525. J. N. DA MOTTA, *Coimbra, Penella.*—Cheese.

526. J. DO NASCIMENTO NORONHA, *Vizeu, Lamego.*—Wine.

527. J. P. CORTEZ DE LOBAO, *Beja, Serpa.*—Wheat, olive oil, and vinegar.

528. J. P. LEAL, *Beja.*—Wheat.

529. J. P. DE MIRA, *Evora, Redondo.*—Wheat, wheat-ears, and oats.

530. J. P. BURGUETE DE MAGALHAES E OLIVEIRA, *Santarem, Abrantes.*—Vinegar and wine.

531. J. P. DA SILVA, *Portalegre, Aviz.*—Wheat, oats, and chick-beans.

532. J. PERFEITO PEREIRA PINTO, *Vizeu, Lamego.*—Rye, French beans, and wine.

533. J. PIRES DE CARVALHO, *Portalegre, Aviz.*—Rice.

534. J. QUIRINO THADEO DE ALMEIDA, *Beja, Ourique.*—Wheat and olive oil.

535. J. R. CORTEZ DE LOBAO, *Beja, Serpa.*—Wheat.

536. J. R. L. DE CARVALHO, *Santarem, Torres Novas.*—Raisins.

537. J. RILVAS, *Portalegre, Fronteira.*—Wheat.

538. J. RIBEIRO MACHADO GUIMARAES, *Coimbra.*—Maize and pearl barley.

539. J. ROBALLO, *Castello Branco, Fundao.*—Olive oil.

540. J. R. P. BASTO, *Aveiro, Oliveira de Azemeiz.*—Wheat, wheat-ears, maize, and pine-apples.

541. J. R. DE OLIVEIRA, *Santarem, Torres Novas.*—Dried figs.

542. J. S. DE BRITO, *Coimbra, Tábua.*—Dried plums and figs.

543. J. SERRAO DO VALLE, *Beja, Odemira.*—French beans.

544. J. DA SILVA, *Lisboa, Alcacer do Sal.*—French beans and rice.

545. J. SOARES MASCARENHAS, *Faro.*—Wine, dried plums, geropiga, and vinegar.

546. J. SOARES TEIXEIRA DE SOUSA, *Angra do Heroismo, Vélas.*—Maize and wine.

547. J. DE SOUSA FALCAO, *Santarem, Constancia.*—Olive oil and wine.

548. J. DE VASCONCELLOS NORONHA, *Vizeu, Lamego.*—Wheat.

549. J. V. DE ALMEIDA, *Portalegre, Arronches.*—Cheese.

550. J. G. DE ALMENDRO, *Bragança, Vinhaes.*—Red wine.

551. J. A. BORGES DA SILVA, *Bragança, Vinhaes.*—Wheat, French beans, and ham.

552. J. COELHO PALHINHA, *Evora, Montemór o Novo.*—Honey.

553. J. M. BAIAO MATOSO, *Beja, Vidigueira.*—Honey and cheese.

554. KEMPES & CO. *Lisboa.*—Olive oil.

555. L. RODRIGUES, *Bragança, Vinhaes.*—Red wine.

556. L. V. PEREIRA PINTO GUEDES, *Bragança, Vinhaes.*—French beans and walnuts.

557. L. X. DE FIGUEIREDO, *Coimbra, Cantanhede.*—Brandy and wine.

558. L. PEREIRA DE CASTRO, *Braga, Cabeceiras de Basto.*—Wine.

559. L. DE ABREU MAGALHAES E FIGUEI-REDO, *Guarda, Çeia.*—Dried fruits.

560. L. A. DIAS, *Coimbra, Miranda do Corvo.*—Barley.

561. L. A. DE MAGALHAES, *Castello Branco, Fundao.*—Wine.

562. L. C. FERREIRA, *Porto, Figueira de Castello Rodrigo.*—Preserved olives, olive oil, almonds, and wine.

563. L. CARDOSO DE ALARACO, *Coimbra, Penella.*—Maize and maize flour.

564. L. J. RODRIGUES, *Santarem, Benavente.*—Wine.

565. L. J. DA ROSA LIMPO, *Portalegre.*—Salep-root flower.

566. L. M. DO SIDRAL, *Portalegre, Alter do Chão.*—Wheat.

567. L. DE PINA CARVALHO FREIRE FALCAO, *Castello Branco.*—Cheese.

568. L. PINTO TAVARES, *Castello Branco, Fundao.*—Olive oil.

569. L. X. DE BARROS CASTELLO BRANCO, *Portalegre.*—Olive oil.

570. M. J. TAVARES MENDES VAZ, *Coimbra, Montemór o Velho.*—Wine, geropiga.

571. M. M. DA TERRA BRUM, *Horta.*—Wine.

572. M. S. MACEDO, *Angra do Heroismo, Vélas.*—Butter.

573. M. A. PEREIRA, *Faro.*—Wheat.

574. M. A. DO RIO, *Lisboa, Belem.*—Olive oil.

575. M. A. GIRALDES, *Bragança Vimisio.*—Chick-pease and wheat.

576. M. A. L. NAVARRO, *Bragança, Freixo de Espada á Cinta.*—Sausages and raisins.

577. M. A. DE ALMEIDA VALLEJO, *Santarem, Abrantes.*—Pine-apple kernels.

578. M. B. PESTANA GONTAO, *Portalegre, Niza.*—Wheat and other cereals, wine, and cheese.

579. M. BRANDAO, *Coimbra, Cantanhede.*—Wheat and red wine.

580. M. C. CABRAL VELHO DE LEMOS CALHEIROS, *Santarem, Benavente.*—Wheat and lentils.

581. M. DE CARVALHO, *Coimbra.*—Maize and maize flour.

582. M. C. G. COELHO, *Lisboa.*—Acorn coffee.

583. CORDEIRO, *Coimbra, Montemór o Velho.*—Husked barley.

584. M. DA CRUZ AMANTE, *Coimbra.*—Wheat and wheat flour.

585. M. D. BAPTISTA, *Portalegre, Ponte de Sor.*—Maize and husked rice.

586. M. DAS DORES NUNES, *Portalegre, Elvas.*—Wine.

587. M. FERREIRA DE AZEVEDO, *Aviero, Mealhada.*—Wine.

588. M. FERREIRA BRETES, *Santarem, Torres Novas.*—Wine.

589. M. FREIRE DE ANDRADE, *Coimbra.*—Vinegar.

590. M. GAIFAO BELLO, *Santarem, Macao.*—Wheat, maize, honey, olive oil, and wine.

591. M. GOMES, *Santarem, Thomar.*—Wine.

592. M. G. GAMEIRO CARDOSO, *Santarem, Torres Novas.*—Dried figs and raisins.

593. M. GUERRA, *Bragança.*—Dried plums and brandy.

594. M. DE GUIMARAES DE ARAUJO PIMENTEL, *Braga, Celorico de Basto.*—Wine.

595. M. H. DE M. FEIO, *Beja.*—Aniseed.

596. M. J. MOCINHA, *Portalegre, Campo Maior.*—Rye.

597. M. J. FIALHO TOJA, *Evora, Portel.*—Cheese.

598. M. J. GUERRA, *Bragança Freixo de Espada á Cinta.*—Almonds.

599. M. J. DE OLIVEIRA, *Bragança, Miranda do Douro.*—Alcohol, sausages.

600. M. J. DE OLIVEIRA ESTACAL, *Lisboa, Barreiro.*—Wine.

601. M. J. DE BIVAR GOMES DA COSTA, *Faro.*—Wheat and other cereals, fruits, wine, olive oil, &c.

602. M. J. COUTINHO, *Santarem, Almeirim.*—Wheat.

603. M. J. FERREIRA DA SILVA GUIMARAES, *Braga, Guimaraes.*—Lupines, French beans, and maize.

604. M. J. DE PINHO SOARES DE ALBERGARIA, *Leiria.*—Honey, olive oil.

605. M. J. DE VIEIRA DE NOVAES, *Lisboa, Setubal.*—French beans, lupines, and pease.

606. M. L. VARELLA, *Evora Arraiollos.*—Cheese.

607. M. L. AÇO, *Beja.*—Barley.

608. M. LUIZ, *Coimbra.* — Chick-pease and chick-beans.

609. M. DE MAGALHAES COUTINHO, *Coimbra, Cantanhede.*—Wine.

610. M. M. CALISTO, *Coimbra.*—French beans.

611. M. MARQUES, *Santarem, Thomar.*—Wine.

612. M. MARQUIS DE FIGUEIREDO, *Coimbra.* — Beans, maize, panicle, seed, and vinegar.

613. M. MARQUES & IRMA, *Bragança, Freixo de Espada á Cinta.*—Almonds in confection.

614. M. M. DA SILVA, *Lisboa, Villa Franca de Xira.*—Wheat.

615. M. NUNES MOUZACO, *Castello Branco, Belmonte.*—Wheat and French beans.

616. M. NUNES SERRAO, *Beja, Alvito.*—Rye, maize, and French beans.

617. M. P. DE OLIVEIRA, *Bragança.*—Wine.

618. M. DE OLIVEIRA, *Santarem, Chamusca.*—Rice.

619. M. PINTO DE ALBUQUERQUE, *Coimbra, Arganil.*—Olive oil.

620. M. DOS PRAZERES E SILVA, *Braga, Guimaraes.*—French beans and chestnuts.

621. M. A. CAIEIRO, *Beja, Serpa.*—Red wine.

622. M. ANTUNES, *Coimbra, Miranda do Corvo.*—Honey.

623. M. C. DE ASSIS ANDRADE (D.), *Beja.*—Walnuts.

624. M. CANDIDA DA FONSECA (D.), *Vizeu.*—Sweetmeats.

625. M. DA CONCEIÇAO, *Braga.*—Confections.

626. M. G. FRANCO (D.), *Portalegre, Fronteira.*—Wheat.

627. M. J. FERRAO CASTELLO BRANCO (D.), *Coimbra, Louza.*—Walnuts and dried plums.

628. M. DA NAZARETH, *Coimbra.*—Cereals and brandy.

629. M. R. FERREIRA DE CASTRO, *Braga, Guimaraes.*—Wine.

630. M. J. DE SOUSA FEIO, *Beja.*—Wheat, chick-beans, chick-pease, maize, cheese, and olive oil.

631. M. J. DE MEDEIROS, *Ponta Delgada, Villa Franca.*—Dragon-wort starch.

632. MARQUIS OF ALVITO, *Beja, Alvito.*—Olives.

633. M. J. RAPOSO, *Beja, Moura.*—Wheat, oats, and cheese.

634. M. MARQUIS AYRES DE SEIXAS, *Portalegre, Gaviao.*—Maize and rice.

635. M. A. MALHEIROS, *Porto.*—Wine.

636. M. A. MARRECO, *Coimbra, Miranda do Corvo.*—Maize and maize flour.

637. M. C. MARTINS, *Portalegre, Aviz.*—Brandy.

638. M. J. DA FONSECA ESGUELHA, *Lisboa, Villa Franca de Xira.*—Maize.

639. M. L. DA SILVA ATHAIDE, *Leiria.*—Maize and wine.

640. M. OSORIO CABRAL DE CASTRO, *Coimbra.*—Cereals, fruits, wine, and olive oil.

641. M. R. DE CARVALHO, *Portalegre.*—Wine.

642. N. J. PEDROSO, *Santarem, Chamusca.*—Maize, French beans, and wheat.

643. N. OF BELLAS, *Coimbra.*—Preserved fruits.

644. N. OF CELLAS, *Coimbra.*—Preserved apricots.

645. N. OF FERREIRA, *Vizeu.*—Preserved plums.

646. N. OF S. DOMINGOS, *Braga, Guimaraes.*—Preserved fruits.

647. N. DE SANT' ANNA, *Coimbra.*—Preserved fruits.

648. N. DE SANTA ROSA DE GUIMARAES, *Braga, Guimaraes.*—Preserved fruits.

649. N. DE SEMIDE, *Coimbra, Miranda do Corvo.*—Confection of turnip.

650. N. C. DE MATOS FERRAO CASTELLO BRANCO, *Coimbra, Louza.*—Maize, maize flour, and olive oil.

651. O. S. LEITE, *Lisboa.*—Dried fruit.

652. P. FERREIRA, *Bragança.*—Confection.

653. P. J. DE MESQUITA, *Coimbra, Tábua.*—Olive oil.

654. P. L. DE OLIVEIRA VELHO, *Santarem, Abrantes.*—French beans.

655. P. M. COELHO MACHADO, *Portalegre.*—Cheese.

656. P. VIEIRI GORJAO, *Santarem.*—Olive oil.

657. P. A. DA SILVA REBELLO COELHO VASCONCELLOS MAIA, *Braga, Povoa de Lanhoso.*—Wheat, barley, chestnuts, honey, olive oil, and wine.

658. R. LARCHER, *Portalegre.*—French beans.

659. R. B. SOBRINHO, *Beja, Alvito.*—Wheat and chick-beans.

660. R. GONÇALVES MONIZ, *Beja, Alvito.*—Wheat, honey, and brandy.

661. R. J. SOARES MENDES, *Santarem, Abrantes.*—Olive oil and wine.

662. R. WIGHAM & Co. *Porto.*—Wine.

663. R. PEREIRA MENDES, *Santarem, Thomar.*—French beans and maize.

664. R. SOARES CASTELLO BRANCO, *Faro, Lagoa.*—Dried figs.

665. R. DE VITERBO (D.), *Vizeu, Lamego.*—Pears.

666. S. PIRES DE OLIVEIRA, *Lisboa.*—Red wine.

667. S. GIL TOJO, *Beja, Vidigueira.*—White wine.

668. S. GIL TOJO BORJA DE MACEDO, *Evora, Portel.*—Wine.

669. S. J. DA GUERRA, *Bragança, Freixo de Espada á Cinta.*—Wheat and olives.

670. S. DE MELLO FALCAO TRIGOSO, *Lisboa, Torres Vedras.*—Wheat, olive oil, maize, and wine.

671. S. PINHEIRO, *Portalegre, Campo Maior.*—Red wine.

672. S. REI, *Coimbra.*—Rice.

673. S. A. DA CUNHA, *Santarem.*—French beans.

674. S. FARIA GARCIA, *Lisboa, Alcacer do Sal.*—Honey.

675. T. A. DE CARVALHO E ALMEIDA, *Braga, Cabeceiras de Basto.*—Wine.

676. T. DA COSTA, *Coimbra, Tábua.*—Dried pears.

677. T. GUERREIRO, *Beja.*—Wheat.

678. T. J. DUARTE, *Coimbra, Figueira da Foz.*—Butter.

·679. T. J. DA SILVA & Co. *Angra do Heroismo.*—Maize.

679A. TOBACCO Co. *Lisboa.*—Snuff and cigars.

680. V. B. PIRES, *Faro.*—Cereals, fruits, honey, wine, &c.

681. V. B. PIRES, JUN. *Faro.*—Walnuts and fig brandy.

682. V. J. DE ALCANTARA, *Coimbra, Oliveira do Hospital.*—Cheese.

683. V. J. DIAS, *Coimbra, Miranda do Corvo.*—Honeycomb.

684. V. A. MONTEIRO, *Santarem, Chamusca.*—Chick-pease.

685. VISCOUNT DE BRUGES, *Angra do Heroismo.*—Cereals, butter, confection, brandy, &c.

686. VISCOUNT DA ESPERANÇA, *Beja, Cuba.*—Cereals, fruits, cheese, honey, wine, brandy, &c.

687. VISCOUNT DA FOZ, *Evora, Extremoz.*—Red wine.

688. VISCOUNT DE GUIAES, *Vizeu, Lamego.*—Honey, olive oil, and wine.

689. VISCOUNT DE OLEIROS, *Castello Branco.*—Cheese.

690. VISCOUNT DE SÁ, *Santarem, Almeirim.*—Wheat, rye, maize, French beans, and barley.

691. VISCOUNT DE TAVEIRO, *Coimbra.*—Cereals, fruits, seeds, &c.

692. VISCOUNT DO TORRAO, *Beja, Alvito.*—Wheat and rice.

693. VISCOUNTESS DE ALPENDURADA, *Porto.*—Red wine.

694. VISCOUNTESS DE FONTE BOA, *Santarem.*—Wheat and olive oil.

695. VISCOUNTESS DA VARZEA, *Vizeu, Lamego.*—Dried plums and figs.

696. V. J. PEREIRA FRANCO, *Guarda.*—Wheat, rye, barley, beans, cherries, vinegar, and olive oil.

697. W. M. DE CARVALHO, *Coimbra.*—Walnuts.

———

1143. A. M. DOS SANTOS, *Porto.* — Brandy.

1144. A. L. REBELLO DA GAMA, *Porto.*—Wine, honey.

1145. A. ALLEN, *Porto.*—French beans, pease, maize, wheat.

1146. ANGELINA ROSA CARNEIRO, *Porto, Santo Thyrso.*—Wine.

1147. A. B. DE ALMEIDA SOARES LENCASTRE, *Porto.*—French beans, maize, honey.

1148. A. B. FERREIRA, *Porto.*—Port wine.

1149. A. E. R. DE SOUSA PINTO, *Porto.*—Rye, maize, hazel-nuts, vinegar, &c.

1150. A. F. BAPTISTA, *Porto.*—Oil of olives.

1151. A. F. MENEZES, *Porto.*—Port wine.

1152. A. G. DA COSTA, *Porto.* — Dried fruits, honey.

1153. A. J. CARREIRA, *Lisboa.*—Dried fruits, pease and beans, olives.

1154. A. J. AYRES DE MENDOÇA, *Faro, Olháo.*—Wine.

1155. A. P. SOARES DA COSTA, *Porto.*—Wine.

1156. A. DE SOUSA CARNEIRO, *Porto.*—Agricultural products, flour, wine.

1157. A. T. DE QUEIROZ, *Porto.*—Dried fruits.

1158. A. V. DE TOVAR MAGALHAES E ALBUQUERQUE, *Guarda.*—Dried pears, cheese.

1159. BARON OF VAZZEA, *Porto.*—Acorns.

1160. B. DO C. DE M. PEREIRA, *Porto.*—Maize, walnuts, hazel-nuts.

1161. B. J. JACOME, *Vianna do Castello, Espozende.*—Wheat.

1162. C. DE A. GUIMARAES, *Lisboa.*—Collares wine.

1163. COMMISSION (CENTRAL PORTUGUESE), *Aveiro, Anadia.*—Wine from Bairrada, vinegar.

1164. COUNT OF ARROCHELLA, *Braga, Guimaraes.*—French beans, wheat, barley, walnuts.

1165. COUNT OF SAMODAES, *Porto, Lamego.*—Wheat, flour, chick-beans, dried figs and raisins, honey, potatoes, French beans.

1166. COUNT OF VILLA REAL, *Lisboa.*—Wine.

1167. E. A. DE SOUSA, *Bragança, Vinhaes.*—French beans.

1168. E. DE SEQUEIRA, *Lisboa.*—Collares wine.

1169. F. A. DA ROCHA, *Lisboa, Setubal.*—Alimentary preserves.

1170. F. M. B. P. DE CARVALHO, *Lamego, Peso da Regua.*—Wine.

1171. F. DE O. CALHEIROS, *Lisboa.*—Oil of olives.

1172. F. R. BATALHA, *Lisboa.*—Coffee from Madeira.

1173. F. X. DE MORAES SOARES, *Villa Real, Chaves.*—Cheese.

1174. FREIRAS BENEDICTINAS, *Porto.*—Preserved fruits.

1175. G. P. M. AGUIAR, *Porto.*—Wine and vinegar.

1176. G. G. DE CARVALHO, *Porto.*—Oil of olives.

1177. H. THOMÁS, *Lisboa.*—Collares wine.

1178. I. F. DE CARVALHO, *Lisboa.*—Collares wine.

1179. J. MARTINS & SON, *Lisboa.*—Muscadine wine.

1180. J. E. DE BRITO E CUNHA, *Porto.*—French beans.

1181. J. N. REBELLO VALENTE, DR. *Aveiro, Oliveira de Azemeis.*—Walnuts.

1182. J. P. ARAUJO DE SEQUEIRA, *Lisboa.*—Collares wine.

1183. J. DE V. CARNEIRO MENEZES, *Porto.*—French beans, walnuts, oil of olives, and wine.

1184. J. A. Ferreira, *Porto.*—Wheat, French beans.

1185. J. A. Rodrigues Coimbra, *Porto.*—Wheat, maize, French beans, wine.

1186. J. F. Pereira, *Porto.*—Walnuts, pine kernels.

1187. J. L. Martins, *Porto.*—Wheat, rye, and maize.

1188. J. A. da Cunha Macedo, *Porto.*—Wine.

1189. J. A. da Silva, *Porto.*—French beans.

1190. J. B. Mascarenhas, *Faro, Villa Nova de Portimao.*—Dried figs.

1191. J. B. Vaz, *Porto.*—Hazel-nuts.

1192. J. F. da Silva, *Porto.*—Dried whiting, and dried roach.

1193. J. F. Pinto, *Porto.*—French beans.

1194. J. G. Sobrinho, *Porto.*—Rye.

1195. J. J. T. da Costa Guimaraes, *Porto.*—Agricultural products, honey, oil of olives, and wine.

1196. J. L. Fernandes, *Porto.*—Maize.

1197. J. M. de Sousa Rodrigues, *Porto, Santo Thyrso.*—Rye, French beans, chestnuts, and acorns.

1198. J. M. da V. Cabral Sampaio, *Villa Real.*—Port wine.

1199. J. da Ribeiro, *Porto.*—Walnuts and French beans.

1200. J. R. de Azevedo, *Santarem, Benavente.*—Maize.

1201. J. R. C. Belem & Bro. *Lisboa.*—Preserves and liqueurs.

1202. J. de S. Magalhaes Cabral, *Porto.*—Oil of olives.

1203. L. P. de Castro, *Braga, Guimaraes.*—Rye, millet, oil of olives, &c.

1204. L. A. M. da Silva Araujo, *Porto, Santo Thyrso.*—French beans and wheat.

1205. M. A. Pereira, *Faro.*—Mendobi.

1206. M. A. P. de Sampaio & Bro. *Villa Real.*—Port wine of different years.

1207. M. A. Fernandes, *Porto.*—Wheat, barley, maize, and French beans.

1208. M. D. P. de Aragão (D.), *Faro.*—Almonds.

1209. M. J. F. da Silva Guimarães, *Braga Guimaraes.*—Acorns.

1210. M. F. F. da Costa Araujo, *Porto, Santo Thyrso.*—Vinegar.

1211. M. G. Bello, *Santarem, Maçao.*—Rye.

1212. M. Marques da Silva, *Porto, Vallongo.*—Agricultural products.

1213. M. M. Alves, *Porto.*—Oil of olives and honey.

1214. M. P. da Silva, *Porto.*—Maize.

1215. M. P. P. de S. Villas Boas, *Porto.*—Brandy.

1216. M. dos P. e Silva, *Braga, Guimaraes.*—Chestnuts.

1217. M. da R. G. Camoẽs, *Porto.*—Agricultural products, dried fruits, wine and vinegar, oil of olives, and honey.

1218. M. V. Guedes de Ataide (D.), *Porto.*—French beans, walnuts, barley, dried figs, and maize.

1219. M. J. de Sousa Feio, *Beja.*—Sorgho.

1220. R. Wanzeller, *Porto.*—Wheat and French beans.

1221. S. de M. Falcão Trigoso, *Lisboa, Torres Vedras.*—Virgin oil.

1222. S. P. de Mesquita, *Porto.*—Maize, wheat, wine, oil of olives, chestnuts, and French beans.

1223. S. R. Ferreira, *Porto.*—French beans.

1224. V. B. Pires Junior, *Faro.*—Wine.

1225. V. N. Corado, *Lisboa.*—Collares wine.

CLASS IV.

698. ABBOT DE CRESPOS, *Braga.*—Flax.

698A. A. CAETANO, *D'Oliveira, Bragança.*—Soap.

699. AGRICULTURAL INSTITUTE OF LISBOA.—Portuguese wool and silk.

700. A. FERNANDES, *Bragança, Vinhaes.*—Flax.

700A. A. MOREIRA DOS SANTOS, *Porto.*—Soap.

701. A.P. PEREIRA (D.), *Coimbra, Louza.*—Potato starch.

702. A. A. RIBEIRO, *Bragança, Macedo de Cavalleiros.*—Cocoons.

703. A. C. BANHA, *Beja.*—Cocoons.

704. A. I. PEREIRA, *Evora, Redondo.*—Virgin and white wax.

705. A. J. DURAES CASTANHEIRA, *Vizeu, Carregal.*—Black wool.

706. A. J. GONÇALVES BRANCO, *Alcacer do Sal, Lisboa.*—Cork.

707. A. J. DA SILVA FRANCO, *Leiria, Peniche.*—Archil.

708. A. M. PACHECO RODRIGUES, *Beja, Serpa.*—Wool.

709. B. PERES, *Beja, Ferreira.*—Virgin wax.

710. B. J. PINTO DA MOTTA, *Coimbra.*—Pitch.

711. B. J. DE SOUSA, *Aveiro, Coimbra.*—Wax, goat's hair, raisins, and berries.

712. C. PECEGUEIRO, *Coimbra.*—Linseed.

713. C. J. DE MATOS VEIGA, *Beja.*—Mace-reed flower.

714. CARRILHO, J. G. *Beja, Almodovar.*—Wax.

715. C. J. MORA E VAZ, *Bragança, Mogadouro.*—Wood.

716. C. A. C. GARCIA, *Bragança.*—Sumach and cocoons.

717. COMMISSION OF THE ALANDROAL, *Evora, Alandroal.*—Mace-reed.

718. COMMISSION OF THE DISTRICT OF ANGRA DO HEROISMO.—Mat-weeds.

719. COMMISSION (CENTRAL) PORTUGUESE, *Aveiro, Coimbra Vagos.*– Beehive, linseed, Armenian bole, wool, and seeds; mace-reed, thistle, and clay.

720. COMMISSION OF THE DISTRICT OF COIMBRA.—Wood, lentisk, and linseed.

721. COMMISSION OF THE DISTRICT OF EVORA.—Wood.

722. COMMISSION OF THE DISTRICT OF FARO.—Saffron and cork.

723. COMMISSION OF THE DISTRICT OF GUARDA, *Subugal.*—Flax.

724. COMMISSION OF PONTA DELGADA.—Wood and bark.

725. COUNT DE ARROCHELLA, *Braga, Guimaraes.*—Wood and cork.

726. COUNT DE SAMODAES, *Porto, Lamego.*—Silk and cocoons, vine ashes, elderberries.

727. COUNT DO SOBRAL, *Santarem, Almeirim.*—Wax and olive husks.

728. D. J. DA CUNHA GUIMARAES, *Coimbra.*—Ox-horn.

729. D. J. DA COSTA LEAO, *Bragança, Vinhaes.*—Cocoons.

730. DELGADO & PEREIRA, *Evora, Montemór o Novo.*—Woollen fleece.

731. D. A. TALLÉ RAMALHO, *Evora, Redondo.*—Woollen fleeces, and thistle to curdle milk.

732. D. GARCIA, *Faro, Silves.*—Cork.

733. D. M. DE SÁ MORAES, *Bragança Vinhaes.*—Wool.

734. E. A. DE SOUSA, *Bragança, Vinhaes.*—Hops and wax.

735. E. LARCHER, *Portalegre.*—Sumach.

736. F. M. S. PINÇAO, *Beja, Aljustrel.* —Linseed.

737. F. DE SOUSA ROMEIRAS, *Evora, Montemór o Novo.*—Cork-tree bark.

738. F. A. GOMES, *Beja Alvito.*—Virgin wax.

739. F. CABRAL PAES PINTO, *Vizeu, Sernancelhe.*—Raw silk and cocoons.

740. F. DE FREITAS MACEDO, *Santarem.*
—Linseed, lavender, and elder-berry.

741. F. J. WENCESLAU, *Portalegre, Gaviao.*—Juniper berries.

742. F. MAIA, *Coimbra, Montemór o Velho.*—Linseed.

743. F. DE MAGALHAES, *Coimbra.*—Cork.

744. F. M. MAGALHAES MASCARENHAS, *Coimbra, Louza.*—Cork and flax.

745. F. M. FELGUEIRAS LEITE, *Bragança, Mogadouro.*—White wool, wood, and cork.

746. F. MARQUIS DE FIGUEIREDO, *Coimbra.*—Flax.

747. F. T. DE ALMEIDA PROENÇA, *Castello Branco.*—Black and white wool.

748. GENERAL PORTUGUESE WOOD ADMINISTRATION, *Leiria.* — Carbonized and resinous billet, pitch, tar, colophany, and wood.

749. ROBINSON, G. *Portalegre.*—Cork.

750. G. T. DE MAGALHAES COLLAÇO, *Coimbra, Soure, Louza.*—Juniper, aloe, mastic-tree leaves, and hops.

751. G. F. PEREIRA NUNES, *Coimbra, Oliveira do Hospital.*—Ray-grass.

752. J. R. M. COOK, *Coimbra.*—Cork.

753. J. J. DA MOTTA, *Beja.* —Cocoons.

754. J. A. DIAS GRANDE, *Portalegre.*—Cork-tree bark, virgin wax, wood, linseed, and juniper berries.

755. J. A. DE CAMPOS, *Bragança, Moncorvo.*—Flax, hemp, and wood.

756. J. A. GARCIA, *Coimbra.*—Madder.

757. J. B. CASIMIRO, *Bragança, Mirandella.*—Wood.

758. J. B. DE MIRA, *Beja.*—Cocoons.

759. J. M. AFFONSO, *Bragança, Mogadouro.*—Linseed.

760. J. DE MATOS DE FARIA BARBOSA, *Braga, Barcellos.*—Woollen fleeces.

761. J. N. PESTANA GIRAO, *Faro.*—Guinea aloe, cotton seeds.

762. J. N. REBELLO VALENTE, *Aveiro, Oliveira de Azemeis.*—Wood.

763. J. A. PEREIRA DE MATOS, *Faro.*—Virgin and white wax.

764. J. FERREIRA, *Coimbra, Louza.*—Willow bark.

765. J. GONÇALVES FINO, *Coimbra.*—Wood.

766. J. P. DA SILVA PINÇAO, *Beja, Aljustrel.*—Flax.

767. J. ALVES, *Portalegre, Gaviao.*—Cork.

768. J. A. C. DE CARVALHO, *Beja, Aljustrel.*—Flax.

769. J. A. SOUSA, *Portalegre.*—Pastil.

770. J. A. DA CRUZ CAMOES, *Evora.*—Woollen fleece.

771. J. A. DIAS ROMANO, *Bragança, Vimioso.*—Flax.

772. J. B. TEIXEIRA DE ALMEIDA, *Bragança, Mogadouro.*—Wood.

773. J. DO CANTO, *Ponta Delgada.*-Archil.

774. J. CARVAJAL VASCONCELLOS GAMA, *Portalegre, Campo Maior.*—Woollen fleece.

775. J. D. DE CARVALHO MONTENEGRO, *Coimbra, Louza.*—Wood.

776. J. J. DA COSTA, *Braga, Guimaraes.*—Linseed.

777. J. LEAL GOUVEIA PINTO, *Coimbra, Miranda do Corvo.*—Cork, lees of olive oil.

778. J. M. LOPES FALCAO, *Beja, Odemira.*—Cork.

779. J. M. DE ALARCAO E MENEZES (D.), *Santarem, Coruche.*—Bark, wood, and cork.

780. J. M. FERREIRA, *Bragança, Vinhaes.*—Cocoons.

781. J. M. HENRIQUES, *Coimbra, Poiares.*—Wax.

782. J. M. CORREIA BELLES, *Faro.*—Nankin cotton.

783. J. PARREIRA CORTEZ DE LOBAO, *Beja, Serpa.*—Woollen fleece.

784. J. R. P. BASTO, *Aveiro, Oliveira de Azemeis.*—Woollen fleece.

785. J. SERRAO DO VALLE, *Beja, Odemira.*—Cork.

786. J. DA SILVA, *Lisboa, Alcacer do Sal.*—Cork.

787. J. SOARES TEIXEIRA DE SOUSA, *Angra do Heroismo, Vélas.*—Wool.

788. J. T. Rabito, *Portalegre, Niza.*—Woollen fleece.

789. J. Coelho Palhinha, *Evora, Montemór o Novo.*—Wax.

790. L. A. de Carvalho, *Coimbra, Louza.*—Linseed and flax.

791. L. da Cunha Martins, *Guarda, Manteigas.*—Fleeces.

792. M. A. Geraldes, *Bragança, Vimioso.*—Wool.

793. M. Guerra, *Bragança.*—Cocoons.

794. M. J. Ferreira da Silva Guimaraes, *Braga, Guimaraes.*—Flax.

794A. M. G. D'Oliveira, *Lisboa.*—Soap.

795. M. M. Holbeche Correia, *Santarem.*—Seed, olive-husks, aloe-threads, goat's hair, thistle-flower and seed.

796. M. N. Furtado, *Bragança.*—Potato-starch.

797. M. J. Ferrao Castello Branco (D.), *Coimbra, Louza.*—Cocoons.

798. M. da Nazareth, *Coimbra.*—Flax.

799. M. Osorio Cabral de Castro, *Coimbra.*—Flax, tow, and linseed.

800. N. C. Matos Ferrao Castello Branco, *Coimbra, Louza.*—Cherry-tree wood and oak-bark.

801. P. A. da Silva Rebello Coelho Vasconcellos Maia, *Braga, Povoa de Lanhoso.*—Cocoons and flax.

802. P. L. Guimaraes, *Braga, Guimaraes.*—Silk.

803. R. B. Sobrinho, *Beja, Alvito.*—Raisins.

804. R. L. de Mesquita Pimentel, *Angra do Heroismo.*—Raw silk, blue silk, and bored cocoons.

804A. J. Du Guerra, *Bragança.*—Soap.

805. S. de Mello Falcao Trigoso, *Lisboa, Torres Vedras.*—Sheep's wool.

806. T. de Mello Serrao, *Beja, Ourique.*—Fleeces.

807. V. B. Pires, *Faro.*—Linseed.

808. V. B. Pires, Jun. *Faro.*—Madder and oak-bark.

809. Viscount de Bruges, *Angra do Heroismo.*—Hemp, moss, herbs, seeds, roots, &c.

810. Viscount da Esperança, *Beja, Cuba.*—Wool, cork, and wood.

811. Viscount de Guiaes, *Vizeu, Lamego.*—Silk.

812. Viscount de Taveiro, *Coimbra.*—Cocoons, flax-seed, and hops.

813. Viscount de Torrao, *Beja, Alvito.*—Elder-berries.

814. V. J. Pereira Franco, *Guarda.*—Linseed, and elder-berries.

815. W. M. de Carvalho, *Coimbra.*—Bark and root of walnut tree.

———

1226. Abbot of Crespos, *Braga.*—Cocoons.

1227. A. C. de Sousa e Sá, *Porto.*—Linseed, tow, and flax.

1228. A. P. Monteiro, *Porto.*—Wood.

1229. A. Grant, *Porto.*—Wood.

1230. A. Allen, *Porto.*—Wool.

1231. A. B. de A. S. Lencastre, *Porto.*—Wood, and virgin wax.

1232. A. E. R. de S. Pinto, *Porto.*—Wood.

1233. A. M. Arriscado, *Braga, Espozende.*—White and black wool.

1234. A. P. Soares da Costa, *Porto.*—Wood.

1235. A. de S. Carneiro, *Porto.*—Wood.

1236. Arsenal of Marine, *Lisboa.*—Wood.

1237. B. do C. de Maria Pereiro, *Porto.*—Flax.

1238. Biester Falcáo & Co. *Lisboa.*—Corks.

1239. Castro Silva & Sons, *Porto.*—Soap.

1240. Commission (Central Portuguese), *Aveiro.*—Vegetable ashes, and flax.

1241. Commission of Evora, *Evora.*—Wax.

1242. Count of Arrochella, *Braga, Guimaraes.*—Cork.

1243. D. A. F. Ramalho, *Evora, Redondo.*—Wool.

1244. E. I. Parreira, *Angra.*—Wool.

1245. F. M. da Silva Pinçao, *Beja.*—Flax.

1246. F. B. Castello Branco, *Portalegre.*—White wool.

1247. F. R. Batalha, *Lisboa.*—Archil.

1248. J. P. Valverde, *Porto, Miranda do Corvo.*—Cocoons.

1249. J. N. R. Valente, Dr. *Aveiro, Oliveira de Azemeis.*—Flax.

1250. J. de V. C. de Menezes, *Porto.*—Wood.

1251. J. V. de Almeida, *Santarem, Benavente.*—Mustard.

1252. J. A. Rodrigues Coimbra, *Porto.*—Wood and cork.

1253. J. C. F. Correia Falcao, *Castello Branco.*—White wool.

1254. J. J. Teixeira da Costa Guimaraes, *Porto.*—Flax and wax, wood.

1255. J. M. de Sousa Rodrigues, *Porto, Santo Thyrso.*—Linseed and flax.

1256. J. Marianni, *Porto.*—Cocoons.

1257. J. de S. Magalhaes Cabral, *Porto.*—Linseed and flax.

1258. Kempes & Co. *Lisboa.*—Soap.

1259. L. Huet Bacellar, *Porto.*—Cocoons.

1260. M. Francisco, *Castello Branco.*—Wool.

1261. M. J. da Silva, *Porto.*—Flax.

1262. M. L. da C. Vasconcellos, *Braga.*—Wool.

1263. M. M. da Silva, *Porto, Vallongo.*—Flax, wax.

1264. M. P. de Carvalho, *Porto.*—Black wool.

1265. M. P. P. de Sousa Villas Boas, *Porto.*—Wood.

1266. M. da C. do Amaral, *Braga, Guimaraes.*—Wood.

1267. M. da R. Gonçalves de Camoes, *Porto.*—Wood, wool, raw silk, linseed and flax, cocoons and silk, virgin wax, husks of maize, &c.

1268. P. M. de Macedo, *Porto.*—Wax.

1269. R. Wanzeller, *Porto.*—Flax.

1270. Viscount of Taveiro, *Coimbra.*—Wood.

1271. Widow of J. B. Burnay, *Lisboa.*—Oils.

CLASS VII.

816. A. da Costa Pereira, *Braga.*—Wooden hoops for sieves.

817. J. M. F. Thomas, *Coimbra.*—Pulleys and blocks.

818. National Printing Office, *Lisboa.*—Printing apparatus.

819. V. J. de Castro, *Lisboa.*—Typographic plate.

1272. D. J. de A. Bubone, *Lisboa.*—Block and tackle.

CLASS VIII.

820. B. Potier, *Lisboa.*—Thrashing machine.

821. Perseverance Co. *Lisboa.*—Steam-engine, for the manufacture of olive-oil.

1273. L. F. de S. Cruz, *Porto, Cedofeita.*—Wheel for rudder and binnacle, on a new principle; rolling machine, hydraulic machine.

CLASS IX.

822. A. POLYCARPO BAPTISTA, *Lisboa.*—Garden-tools.

823. A. SERRA, *Portalegre.*—Agricultural implements.

824. COMMISSION (CENTRAL), *Aveiro, Oliveira de Azemeis, Ovar.*—Shovel to turn grain, rake, sieve, and hatchet.

825. COUNT DE SAMODAES, *Porto, Lamego.*—Hatchet.

826. F. DE FREITAS MACEDO, *Santarem.*—Model of agricultural implement.

827. J. FELICIANO, *Coimbra.*—Sieve.

828. J. G. FERREIRA, *Coimbra.*—Casks.

829. J. R. P. BASTO, *Aveiro, Oliveira de Azemeis.*—Shovel and rake for maize.

830. O. DA COSTA, *Santarem, Anciao.*—Hammer and hoe.

831. PERSEVERANCE Co. *Lisboa.*—Distilling apparatus.

———

1274. J. M. DA SILVA, *Guimaraes.*—Garden tools.

CLASS X.

832. COMMISSION (CENTRAL), *Aveiro, Oliveira de Azemeis.*—Model of syphon.

833. L. LOUGE, *Lisboa, Setubal.*—Refractory bricks.

CLASS XIII.

834. INDUSTRIAL INSTITUTE OF LISBOA, *Alcantara.*—Philosophical instruments.

CLASS XIV.

835. M. N. CODINHO, *Lisboa.*—Photograph of a picture drawn with a pen by Exhibitor.

1358. J. N. SILVEIRA, *Lisboa.*—Photography.

———

CLASS XVI.

836. F. A. TEIXEIRA DE CARVALHO, *Braga.*—Viol.

837. P. J. TEIXEIRA, *Braga.*—Violin.

CLASS XVII.

838. A. POLYCARPO BAPTISTA, *Lisboa.*—Surgical instruments.

CLASS XVIII.

839. A. G. Porto, *Portalegre, Niza.*—Crochet work.

840. A. J. Vieira Rodrigues Fartura, *Angra do Heroismo.*—Cotton quilt.

841. A. P. de Oliveira, *Braga.*—Cotton drills.

842. B. Daupias & Co. *Lisboa.*—Cotton under-waistcoats.

843. Co. of Cotton Fabrics, *Lisboa.*—Cotton fabrics.

844. J. J. Antunes, *Braga.* — Cotton drills.

845. J. P. Dabney, *Horta.* — Cotton stockings.

846. J. D. da Costa, *Braga.*—Cotton drills.

847. Lisbon Spinning and Weaving Co. *Lisboa, Belem.*—Cotton fabrics, ropes, and strings.

848. L. Beraud, *Lisboa.* — Stockings, shirts, and caps.

849. M. A. Rodrigues, *Braga, Villa Verde.*—Striped stuff and aprons.

850. M. da Graça Alfaia, *Portalegre, Niza.*—Crochet work.

851. Spinning Co. of Crestuma, *Aveiro, Villa da Feira.*—Woof and warp.

———

1275. A. M. de Aguiar Alvaro, *Porto Bomjardim.*—Yellow nankeen.

1276. Viscount of Bruges, *Angra do Heroismo.*—Cotton thread prepared for fishing-nets.

CLASS XIX.

852. Commission (Central), *Aveiro.*—Linen.

853. Commission of the District of Guarda, *Guarda, Sabugal.*—Skein of thread.

854. Commission of Ponta Delgada.—Cords made of New Zealand flax.

855. F. J. de Oliveira, *Braga, Guimaraes.*—Sewing thread.

856. G. F. Pereira Nunes, *Coimbra, Oliveira do Hospital.*—Cloth and linen.

857. J. do Pilar, *Braga.*—Linen.

858. Lisbon Spinning & Weaving Co. *Lisboa, Belem.*—Thread.

859. M. dos Desamparados Soares, *Braga.*—Knitted flax stockings.

860. R. Vieira, *Braga.*—Linen.

861. Torres Novas National Spinning & Weaving Co. *Lisboa.*—Flax and hemp fabrics.

862. Viscount de Bruges, *Angra do Heroismo.*—Flax fabrics, and hemp cord.

———

1277. A. C. de Sousa e Sá, *Porto.*—Sacking and linen.

1278. B. do C. de Maria Pereira, *Porto.*—Linen.

1279. J. N. Rebello Valente, Dr. *Aveiro.*—Linen.

CLASS XX.

863. A. J. Barbosa Araujo, *Braga.*—Shaded satin and damasks.

864. A. M. G. da Silva Ramos, *Braga.*—Damasks, satins, and silk ribbons.

865. B. J. C. das Neves, *Coimbra, Pampilhosa.*—Silk.

866. C. A. C. Garcia, *Bragança.*—Silk.

867. Count do Farrobo, *Lisboa, Villa Franca de Xira.*—Silk, woof, &c.

868. Count de Samodaes, *Porto, Lamego.*—Silk.

869. Cordeiro & Irmao, *Lisboa.*—Silks and velvets.

870. E. M. Ramires, *Lisboa.*—Silks and satins.

871. J. L. de Almeida Araujo, *Vizeu.*—Silk thread.

872. J. J. Ferreira de Mello e Andrade, *Braga, Povoa de Lanhoso.*—Skein of silk.

873. M. J. Guerra, *Bragança, Freixo de Espada á Cinta.*—Silk for sieves.

874. J. da Silva Pereira de Vasconcellos, *Braga.*—Velvets.

875. M. J. Rodrigues, *Bragança, Mirandella.*—Silk.

876. R. L. de Mesquita Pimentel, *Angra do Heroismo.*—Sewing-silk and silk caps.

———

1280. A. de A. Peres, *Porto.*—Spun silk.

1281. J. P. Valverde, *Miranda do Corvo, Porto.*—Spun silk.

1282. J. M. Brandao, *Porto.*—Spun and twisted silk.

1283. J. Marianni, *Porto.*—Organzine, woof, spun and twisted silk.

1284. L. Huet Bacellar, *Porto.*—Spun silk.

CLASS XXI.

877. A. J. de Lima, *Braga.* — White wool.

878. A. Pinto de Oliveira, *Braga.*—Kerseymere.

879. B. Daupias & Co. *Lisboa, Belem.*—Woollen and worsted fabrics, &c.

880. Central Commission, *Aveiro.*—Fids.

881. Commission (District) of Faro.—Cloth.

882. Co. of Woollen Manufactures of Campo Grande, *Lisboa, Olivaes.*—Cloth.

883. Corsino, Irmao, & Co. *Guarda.*—Blankets.

884. E. I. Parreira, *Angra do Heroismo.*—Quilt, and wool.

885. F. J. de Almeida, *Vizeu.* — Blankets.

886. F. M. Mascarenhas, *Coimbra, Louza.*—Cloth.

887. Igreja, Roldan, & Co. *Lisboa, Seixal.*—Cloth.

888. J. D. da Costa, *Braga.*—Coatings.

889. L. Cometudo, *Braga.* — Flannel shirt.

890. Larcher & Cunhados, *Portalegre.*—Cloth and kerseymeres.

891. Larcher & Sobrinhos, *Portalegre.*—Cloths and kerseymeres.

892. M. Antonia, *Guarda, Pinhel.*—Stockings.

893. Viscount de Bruges, *Angra do Heroismo.*—Woollen apron.

CLASS XXIII.

894.—A. C. Miranda & Co. *Lisboa.*—Shawls, handkerchiefs, and printed calicoes.

895. F. J. da Luz, *Lisboa, Cintra.*—Shawls.

896. Pinto & Co., *Lisboa, Belem.*—Bilbao shawls.

897. P. J. L. dos Anjos, *Lisboa, Belem.*—Handkerchiefs and printed calicoes.

CLASS XXIV.

898. A. E. DE SALLES (D.), *Angra do Heroismo.*—Satin towel, embroidered in embossed gold.

899. BARONESS DE PRIME, *Vizeu.*—Embroidered pincushion.

900. COMMISSION (CENTRAL), *Aveiro, Ovar.*—Lace.

901. F. P. DOS SANTOS, *Lisboa.*—Embroideries in gold.

902. G. CONDERT, *Lisboa, Setubal.*—Lace.

903. G. R. VIEIRA MACHADO (D.), *Braga.*—Embroidered towel.

904. J. P. DABNEY, *Horta.*—Embroidered handkerchief.

905. M. C. PRATA, *Lisboa, Setubal.*—Lace.

906. M. J. TEIXEIRA DE CARVALHO E SAMPAIO (D.), *Vizeu.*—Silk embroidery, representing Conway Castle, N. Wales.

907. M. L. DO AMARAL (D.), *Lisboa.*—Embroidery, representing the monument of Joseph I.

908. F. WILKINSON & Co. *Madeira, Funchal.*—Embroideries.

———

1285. FERREIRAS MADRUGAS (SENHORAS), *Horta.*—A representation of the English arms, made with the pulp of the fig tree.

1286. M. L. DA SILVA MAFRA, *Lisboa.*—Palace of Mafra, in embroidery.

CLASS XXV.

909. F. X. DE MORAES SOARES, *Villa Real, Chaves.*—Kid skin.

910. J. G. PEREIRA CALLADO, *Lisboa, Alfama.*—Morocco.

911. VISCOUNT DE BRUGES, *Angra do Heroismo.*—Skins and furs.

CLASS XXVI.

912. ABBOT DE CRESPOS, *Braga.*—Heifer skins.

913. A. J. DE PASSOS, *Braga.*—Goat and sheep skins.

914. A. DA FONSECA CARVAO PAIM, *Angra do Heroismo.*—Skins and hides.

915. COMMISSION (CENTRAL), *Aveiro, Oliveira de Azemeis.*—Saddle and tanned leather.

916. FONSECA & FERREIRA, *Porto.*—Tanned hide.

917. M. B. MONTEIRO, JUN. *Guarda, Pinhel.*—Tanned skins.

918. M. GUEIFAO BELLO, *Santarem, Maçao.*—Tanned skins.

919. TANNING FABRIC ASSOCIATION OF EXTREMOZ, *Evora.*—Leather.

CLASS XXVII.

920. A. ROXO, *Lisboa.*—Hats.

921. A. MOREIRA E SILVA, *Aveira, Oliveira de Azemeis.*—Hats.

922. COMMISSION OF THE DISTRICT OF COIMBRA, *Figueira.*—Shoes.

923. F. DA COSTA, *Vizeu.*—Shoes.

924. F. J. MAIA, *Braga.*—Caps.

925. F. LAURENCE, *Coimbra.*—Gloves.

926. J. P. DABNEY, *Horta.*—Shawls and ornament made from the threads of the Guinea aloe.

927. J. J. ROBALLO DA FONSECA, *Lisboa.* —Coat.

928. J. DA CUNHA ALVES DE SOUSA, *Braga.*—Boots.

929. J. CURRY DA CAMARA CABRAL, *Horta.*—Silk mantle with lace made from the threads of the Guinea aloe.

930. M. J. DE SOUSA, *Angra do Heroismo.* —Hat.

931. M. J. DE CARVALHO, *Aveiro, Oliveira de Azemeis.*—Hats.

932. M. L. DA SILVA, *Aveiro, Oliveira de Azemeis.*—Hats.

933. M. DIAS DE AFFONSECA, *Braga.*— Woollen hats.

934. R. L. PESSOA, *Coimbra, Figueira da Foz.*—Wooden shoes.

———

1287. J. JORGE, *Porto.*—Hessian boots of satin, silk, &c.; boots.

1288. K. KEIL, *Lisboa.* — Articles of clothing.

1359. A. M. DA SILVA, *Lisboa.*—Boots.

CLASS XXVIII.

935. COMMISSION (CENTRAL), *Aveiro, Feira.* —Paper and pasteboard.

936. E. M. M. SMITH (D.), *Lisboa, Olivaes.* —Paper.

937. FERIN, *Lisboa.*—Bookbinding.

938. GYMNASIUM GODINHO COLLEGE, *Lisboa.*—Caligraphy.

939. J. D. DE CARVALHO, *Lisboa.*—Caligraphy.

940. J. A. CABRAL DE MELLO, *Angra do Heroismo.*—Ode dedicated to H.R.H. Prince Alfred.

941. M. DIAS CESARIO, JUN. *Lisboa.*— Caligraphic frame.

942. M. N. GODINHO, *Lisboa.* — Caligraphy.

943. NATIONAL PRINTING OFFICE, *Lisboa.* —Printing and lithography.

944. RUAES PAPER MANUFACTORY, *Braga.* —Paper.

945. VISCOUNTESS DE VILLA NOVA DA RAINHA, *Santarem, Thomar.*—Paper.

———

1289. J. DE SÁ COUTO, *Aveiro, Feira.*— Paper.

1360. NOGUEIRA DA SILVA.·--Clichets in wood.

CLASS XXIX.

946. WEIGHTS AND MEASURES BOARD, *Lisboa.*—Models of educational instruments; coins, &c.

1361. ROYAL GEODESICAL COMMISSION, *Lisboa.*—Geodesical maps.

CLASS XXX.

947. District Commission of Angra do Heroismo.—Bay-tree wood flower-work.

948. I. Caetano, *Lisboa.*—Escutcheon of the Portuguese arms.

948a. Decombes & Co. Sociâty, *Lisboa.*—Articles made with the endless saw, and boring machine.

CLASS XXXI.

949. A. Gomes, *Braga.*—Horse bits and stirrups.

950. H. Schalck, *Lisboa.*—Buttons and nails.

951. J. Correia, *Braga.*—Nails.

CLASS XXXII.

952. A. P. Baptista, *Lisboa.*—Scissors, knives, and razors.

1290. J. M. da S. Guimaraẽs, *Porto.*—Scissors and a clasp-knife.

CLASS XXXIII.

1362. Mouzado & Bro. *Porto.*—Filigrane, and other works in silver.

CLASS XXXIV.

953. A. Michon C. Pierre, *Porto, Villa Nova de Gaia.*—Glass covers and plates.

CLASS XXXV.

954. A. A. da Lapa, *Portalegre, Elvas.*—Pots.

955. Commission (Central), *Aveiro.*—Earthenware.

956. Commission of the District of Coimbra.—Earthenware.

957. General Society of Chemical Products, *Porto, Villa Nova de Gaia.*—Earthenware (grit stone) for acids.

958. G. J. Howarth, *Lisboa.*—Stone china service.

959. J. Correia da Costa, *Coimbra.*—Earthenware.

960. J. A. Braamcamp, *Lisboa.*—Drainage tubes.

961. J. Francisco, *Vizeu, Tondella.*—Earthenware.

962. J. J. CESAR, *Coimbra.*—Vases and jars.

963. J. LUIZ, *Portalegre, Niza.*—Earthenware.

964. M. LOBO, *Portalegre, Crato.* — Earthenware.

965. VISCOUNT DE VILLA MAIOR, *Porto Villa Nova de Gaia.*— Earthenware.

———

1363. FERREIRA PINTO, *Lisboa.* Articles in hard porcelain.

CLASS XXXVI.

966. A. CORREIA DE LEMOS, DR. *Vizeu.* —Broom made of maize straw. Ivory carving.

967. A. P. CARDOSO CRUZ, *Braga.*—Cards.

968. A. DE BETTENCOURT, *Lisboa.*—Map of Portugal.

969. BIESTER, FALCAO, & CO. *Lisboa.*—Cork.

970. C. DA PURIFICAÇAO, REIS GUEDES (D.), *Lisboa.*—Flowers made of threads of the Guinea aloe.

971. COMMISSION (CENTRAL), *Aveiro, Feira, Ovar, Estarreja.*—Tobacco, halters, shackles, bed, and mattress.

972. COMMISSION OF THE DISTRICT OF COIMBRA, *Soure.*—Razor strop, made from the Guinea aloe.

973. COMMISSION OF THE DISTRICT OF EVORA.—Wax and cork.

974. C. DE ALMEIDA A. E SOUSA, *Coimbra, Penacova.*—Toothpicks.

975. COUNT OF SOBRAL, *Santarem, Almeirim.*—Horse-hair halter.

976. D. G. BLANCO, *Faro, Silves.*—Cork.

977. F. A. PEREIRA, *Braga.*—Objects in horn, &c.

978. F. A. DE VASCONCELOS, *Lisboa.*—Toothpicks.

979. F. DOMINGOS, *Santarem.*—Sofa and chair of mace-reed.

980. F. G. GOMES, *Portalegre.*—Girth-cloth, crupper, and halter.

981. F. LUDGERO MARQUES, *Lisboa.*—Spectacles.

982. F. M. B. PINTO DE CARVALHO, *Lamego.*—Bellows.

983. F. DA SILVA, MARQUES & CO. *Porto.*—Cork.

984. F. MARQUIS DE FIGUEIREDO, *Coimbra.*—Mat.

985. I. M. PROFIRIA DE CASTRO, *Lisboa.*—Wax flowers and fruits.

986. T. DABNEY, *Angra.*—Baskets.

987. J. R. M. COOKE, *Coimbra, Figueira da Foz.*—Cork.

988. J. A. DE OLIVEIRA, *Braga.*—Inkstands.

989. J. M. DA COSTA BARBOSA, *Santarem, Cartaxo.*—Mace-reed cord, and horsehair halter.

990. J. MARQUES DIAS, *Braga.*—Horsehair cloth for sieves.

991. J. J. DOS REIS, *Lisboa.*—Umbrellas, parasols, and sticks.

992. J. A. DE SOLEDADE, *Leiria, Peniche.*—Artificial flowers, made of shells and sea productions.

993. J. F. DA C. TERRA BERQUÓ, *Horta.*—Straw flowers.

994. J. HENRIQUES & FILHOS, *Coimbra, Poiares.*—Manufactured wax.

995. J. J. RIBEIRO, *Lisboa.*—Spectacles.

996. J. M. DE AZEVEDO GIRAO, *Santarem, Alpiaça.*—Straw cloak.

997. J. R. DA SILVA, *Braga.*—Strings for viols and violins.

998. M. DA COSTA, *Braga.*—Straw cloaks.

999. M. DA S. SOUSA, *Coimbra.*— Models of casks.

1000. M. M. HOLBECHE CORREIA, *Santarem.*—Secret lock; gourd-bark fruit-stand and sugar-pot; beehive.

1001. M. J. DA SILVA DOMINGUES (D.), *Lisboa.*—Glazed frame, containing a nosegay, and ornamented with threads of the Guinea aloe.

1002. MARIA DO O. *Coimbra, Penacova.* —Toothpicks.

1003. M. DA PIEDADE, *Santarem.*—Goat-skin dressed to carry wine.

1004. M. OSORIO CABRAL DE CASTRO, *Coimbra.*—Mats and straw cloaks.

1005. PERSEVERANCE Co. *Lisboa.* — Lead pipes and copper chocolate pots.

1006. P. A. BRANDAO, *Coimbra.*—Statue of Pedro V.

1007. R. J. ALMEIDA, *Lisboa.*—Mat.

1008. T. J. FERREIRA, *Lisboa.*—Mats.

1009. T. C. DE OLIVEIRA & FILHOS, *Lisboa.*—Billiard balls, &c.

1010. VISCOUNT DE BRUGES, *Angra do Heroismo.*—Cords and ropes made with different kinds of vegetable fibre.

1011. VISCOUNT OF TAREIRO, *Lisboa.*—Goat-skin bottles.

1012. Z. J. PINTO, *Braga.* — Box-tree pedestal.

1013. B. DA SILVA, *Lisboa.*—Mats.

1014. J. J. DE MACEDO, *Madeira, Funchal.* —Thirty-one samples of plaited straw.

———

1291. J. DE SÁ COUTO, *Aveiro, Feira.*— Corks.

1292. J. DA COSTA, *Portalegre.*—Basket.

1293. J. F. DA PIEDADE, *Porto.*—Umbrellas and parasols.

1294. J. M. DA C. BARBOSA, *Santarem, Cartaxo.*—Agricultural implements.

1295. J. M. DE C. N. L. E VASCONCELLOS, *Lisboa.*—Cocks for gas.

1296. J. P. CARDOSO & SON, *Porto.*— Shot, tin-foil, silver, and gold-leaf.

1297. L. B. DA SILVA, *Porto.*—Silver purses.

1298. L. CARNEIRO, *Arganil, Coimbra.* —Winnowing sieves.

1299. M. M. DA S. RAMOS, *Castello Branco, Covilha.*—Medals; image of the Virgin in boxwood.

1300. PRISONERS OF THE GAOL OF PORTALEGRE, *Portalegre.*—A mat.

1301. VALENTE, DR. J. N. R., *Aveiro, Oliveira de Azemeis.*—Ornamented vase of boxwood.

1302. VISCOUNT OF ESPERANÇA, *Beja.*— Model of a triangular harrow.

COLONIES.

CLASS I.

1015. A. H. DA COSTA MATOS, *Ilhas de S. Thomé e Principe.*—Red ochres and tabatinga.

1016. DEJEANTE, L. B., *Lisboa.*—Marbles from the Cape de Verde Islands.

1017. COMMISSION OF CABO VERDE, *Ilhas de Cabo Verde.*—Lime, sand, and salt.

1018. F. R. BATALHA, *Lisboa.*—Salt from Angola and Timor.

1019. J. N. DE SALLES, *Ilhas de Cabo Verde.*—Volcanic products.

1020. MARTINS & LIMA, *Ilhas de Cabo Verde.*—Salt.

1021. P. M. TITO & Co., *Ilhas de Cabo Verde.*—Volcanic products.

1022. B. J. BROCHADO, *Angola Mossamedes.*—Ochre in powder; iron and copper rings, iron and copper poniard.

1023. F. A. PONCE DE LEAO, *Angola, Mossamedes.* — Magnetic iron, copper ore, plastic stone.

1024. F. DA COSTA LEAL, *Angola Huilla.* —Copper bracelet.

1025. F. WELWITCH, *Lisboa.*—Micaceous iron, from Cacula; iron pyrites; pemba stone, from Angola, Cazengo.

1026. GOVERNOR OF DOMBE GRANDE, *Angola, Dombe Grande.*—Sulphur, copper ore, ochre.

1027. J. D. D'ALMEIDA, *Angola Mossamedes.*—Salt.

1028. J. J. D'ALMEIDA, *Angola, Golungo Alto.*—Salt.

1029. J. J. DE PAIVA, *Angoga, Mossamedes.*—Gypsum stone, copper ore, limestone.

1030. J. TEIXEIRA, XAVIER, *Angola, Benguella.*—Gypsum, rough and calcined, in masses and in powder; limestone.

1031. T. M. BESSONE, *Timor.*—Asphalt.

1032. ULTRAMARINE BOARD, *Lisboa.*—Copper ore from Angola, Cuio.

CLASS II.

1033. COMMISSION OF CABO VERDE ISLANDS, *Ilhas de Cabo Verde.*—Ashes.

1034. F. R. BATALHA, *Lisboa.*—Bindweed from Angola, and purgative cassia from Timor.

1035. G. DA CRUZ LIMA, *Ilhas de Cabo Verde.*—Palma-Christi seeds.

1036. J. F. ANTONIO SPENCER, *Ilhas de Cabo Verde.*—Senna plant.

1037. J. J. BOMTEMPO, *Ilhas de Cabo Verde.*—Seed.

1038. J. M. DE SOUSA E ALMEIDA, *Ilhas de S. Thomé e Principe.*—Palm oil and cocoa oil.

1039. J. DO PINO, *Ilhas de S. Thomé e Principe.*—Palm oil.

1040. M. J. DA COSTA PEDREIRA & J. V. DE CARVALHO, *Ilhas de S. Thomé e Principe.* —Cassia and cocoa oil.

1041. R. M. X. DE RAMOS, *Ilhas de S. Thomé, e Principe.*—Palm oil and cocoa oil.

1042. F. DA C. LEAL, *Angola Huilla.*—Flax seed.

1043. F. A. P. BAYAO, *Angola, Duque de Bragança.*—Sawdust fecula, roots, hemp.

1044. F. WELWITSCH, *Angola.*—Chemical substances and products.

1045. GOVERNOR OF ENCOGE, *Angola Encoge.*—Ginger, bastard saffron, and N-cassa bark.

1046. GOVERNOR OF MASSANGANO, *Angola, Massangano.*—Palm oil.

1047. GOVERNOR OF ZEUZA OF COLUNGO, *Angola, Zeuza do Colungo.*—Butua root, and palm oil.

1048. J. J. D'ALMEIDA, *Angola, Golungo Alto.*—Ginguba oil, dong a luto (a root).

1049. J. T. XAVIER, *Angola, Benguella.* —Alcohol, bunze (a plant).

1050. M. P. S. VENDUNEM, *Angola, Barra do Bengo.*—Palm oil, gimbunze, &c.

CLASS III.

1051. F. R. BATALHA, *Lisboa.*—Canary-almond, cinnamon, coffee, sago, &c. from Timor; cocoa from Prince's Island, cinnamon from Goa.

1052. J. J. DE CARVALHO, *Ilhas de S. Thomé e Principe.*—Cocoa.

1053. J. M. DE SOUSA E ALMEIDA, *Ilhas*

de S. Thomé e Principe.—Coffee, cocoa, manihot, balsam of S. Thomé.

1054. J. F. A. SPENCER, *Ilhas de Cabo Verde.*—Tamarinds.

1055. J. J. DE MELLO, *Ilhas de S. Thomé e Principe.*—Cocoa.

1056. J. M. de Freitas, *Ilhas de S. Thomé e Principe.*—Coffee, arrow-root, cocoa, and brandy.

1057. J. do Pino, *Ilhas de S. Thomé e Principe.*—Tapioca, coffee, and cocoa.

1058. J. Ribeiro da Cunha Azurar, *Ilhas de S. Thomé e Principe.*—Safú.

1059. M. J. da Costa Pedreira & J. Velloso de Carvalho, *Ilhas de S. Thomé e Principe.*—Tapioca, manihot flour, pulp of tamarinds, fruits, coffee, &c.

1060. M. dos Reis Borges, *Ilhas de Cabo Verde.*—Rice, coffee, Yuca manihot flour.

1061. P. R. Tavares, *Ilhas de Cabo Verde.*—Coffee.

1062. Ultramarine Board, *Lisboa.*—Cinnamon, rice, and coffee, from Timor.

1063. A. A. Sequeira Thedim, *Cabo Verde.*—Rum, and coffee.

1064. F. A. P. Bayao, *Angola, Duque de Bragança.*—Giéfu seeds, cola nuts.

1065. F. Welwitsch, *Angola.*—Maize, different seeds, maboca (a fruit).

1066. Governor of Benguella, *Angola, Benguella.*—Different seeds, and manihot starch.

1067. Governor of Caconda, *Angola, Caconda.*—Uindo (bark).

1068. Governor of Encoge, *Angola, Encoge.*—Fruits, seeds, and coffee.

1069. Governor of Zeula of Golungo, *Angola, Zeuza of Golungo.*—Rice, French beans, tapioca, and honey.

1070. J. A. G. Pereira, *Angola, Cazengo.*—Coffee.

1071. J. D. d'Almeida, *Angola, Mossamedes.*—Honey.

1072. J. J. d'Almeida, *Angola, Golungo Alto.*—Rum, and brandy extracted from maize.

1073. J. L. d'Albuquerque, *Angola, Bumbo.*—Sugar.

1074. J. T. Xavier, *Angola, Benguella.*—Brandy.

1075. M. P. dos Santos Venduneni, *Angola, Barra do Bengo.*—French beans.

1076. Ultramarine Board, *Angola, Encoge.*—Coffee.

CLASS IV.

1077. E. A. de Sousa, *Ilhas de Cabo Verde.*—Corals, lichen for dyeing, archil, and seed.

1078. F. de Alva Brandao, *Ilhas de S. Thomé e Principe.*—Bastard saffron root, and cotton.

1079. F. R. Batalha, *Lisboa.*—Archil, lichen, cotton, &c. from various Portuguese colonies; filaments of caroco, dye-wood, &c. from Timor and Mozambique.

1080. G. da Cruz Lima, *Ilhas de Cabo Verde.*—Seeds.

1081. Hortet Raymundo, *Ilhas de Cabo Verde.*—Archil.

1082. J. M. de Sousa e Almeida, *Ilhas de S. Thomé e Principe.*—Wood, archil, and white cotton.

1083. J. B. de Oliveira, *Ilhas de S. Thomé e Principe.*—Palm-tree wool, and cotton.

1084. J. R. da Cunha Azurar, *Ilhas de S. Thomé e Principe.*—Wood.

1085. L. J. Moniz, *Ilhas de Cabo Verde.*—Cotton.

1086. M. J. da Costa Pedreira & J. Velloso de Carvalho, *Ilhas de S. Thomé e Principe.*—Cotton, wood, and gum.

1087. M. dos Reis Borges, *Ilhas de Cabo Verde.*—Cotton.

1088. P. M. Tito & Co. *Ilhas de Cabo Verde.*—Archil.

1089. P. A. de Oliveira, *Ilhas de Cabo Verde.*—Cotton and indigo.

1090. R. de Sá Nogueira, *Ilhas de Cabo Verde.*—Cotton.

1091. T. da Silva Bastos Varella, *Ilhas de S. Thomé e Principe.*—Wood.

1092. Ultramarine Board, *Lisboa.*—Tobacco, archil, and wood, from Timor.

1093. Z. PEREIRA MAFRA, *Ilhas de S. Thomé e Principe.*—Wood.

1094. A. C. DE SOUSA E CUNHA, *Angola, Encoge.*—Resin.

1095. A. J. DE SEIXAS, *Angola.*—Wax from Loanda and Benguella.

1096. B. J. BROCHADO, *Angola Mossamedes.*—Wax.

1097. F. DA COSTA LEAL, *Angola, Bumbo.*—Samples of woods.

1098. F. A. PINHEIRO BAIJAO, *Angola, Duque de Bragança.*—Tobacco.

1099. F. WELWITSCH, *Angola.* — Gum tragacanth from Loanda; copal from Benguella; dragon's blood from Huilla; muance gum and cabella from Golungo; copal from Zenza do Golungo; mumbango gum from Ambaca; resin from Cazengo; fifty-two samples of wood.

1100. GOVERNOR OF BENGUELLA, *Angola, Benguella.* — Copal, archil, and dragons' blood.

1101. J. D. D'ALMEIDA, *Angola, Huilla.*—Tobacco and archil.

1102. T. M. BESSONE, *Moçambique.* Gum arabic and caoutchouc.

CLASS IX.

1103. M. P. DOS SANTOS VENDUNENI, *Angola, Barra do Bengo.*—Press for extracting palm oil.

CLASS XVIII.

1104. F. R. BATALHA, *Lisboa.* — Holes worked with cotton thread, spun by the savages of Agra.

1105. J. R. DE CARVALHO (D.), *Illas de Cabo Verde.*—Cotton cloth.

1106. ULTRAMARINE BOARD, *Lisboa.*—Cotton from Timor, raw and manufactured.

1107. B. F. DE FIGUEIREDO E CASTRO, *Angola Mossamedes.*—Cotton in the pod, and ginned.

1108. J. D. D'ALMEIDA, *Angola Mossamedes.*—Cotton in the pod.

1109. M. J. CORREA, *Angola, Moçambique.*—Yellow cotton in the pod.

CLASS XXIII.

1110. M. T. MOUTEL (D.), *Ilhas de Cabo Verde.*—Fabrics of cotton and silk.

1111. ULTRAMARINE BOARD, *Lisboa.* — Cloth of wool and silk, from Timor.

CLASS XXV.

1112. B. J. BROCHADO, *Angola, Mossamedes.*—Hart-hide.

1113. F. A. P. BAYAO, *Angola, Duque de Bragança.*—Fox, monkey, and deer-skins.

1114. G. DOS R. C. E BARROS, *Angola, Muxima.*—Deer-skins.

1115. GOVERNOR OF ZEUZA OF GOLUNGO, *Angola, Zeuza of Golungo.*—Stag-hide.

CLASS XXVI.

1116. F. DA C. LEAL, *Angola, Huilla.*—Tanned leather, hides.

CLASS XXXVI.

1117. F. R. BATALHA, *Lisboa.*—Cauris from Timor, carved walking-stick of sandal-wood.

1118. J. M. DE SOUSA E ALMEIDA, *Ilhas de S. Thomé e Principe.*—Tobacco and cigars.

1119. P. A. FERREIRA, *Ilhas de S. Thomé e Principe.*—Tobacco.

1120. ULTRAMARINE BOARD, *Lisboa.* —Ramé and cords made of it, caroco, and wild banana-tree, from Timor.

1121. F. A. PINHEIRO BAIAO, *Angola, Duque de Bragança.*—Pipes.

1122. F. DA COSTA LEAL, *Angola, Huilla.*—Bricks and tiles.

1123. F. WELWITSCH, *Angola.* —Filaments, and articles made of them ; feathers ; elephant's mane, and articles made of it ; subi sieve.

1124. G. DOS R. CLARO E BARROS, *Angola Muxima.*—Teeth of the hippopotamus.

1125. GOVERNOR OF CATUMBELLA, *Angola Catumbella.*—Palm wood.

1126. GOVERNOR OF ENCOGE, *Angola, Encoge.*—Gimbusu straw, and articles made of it.

1127. GOVERNOR OF ZENZA DO GOLUNGO, *Angola, Zenza do Golungo.* — Empacassa horns.

1128. J. J. D'ALMEIDA, *Angola, Golungo Alto.*—A bag made of the filaments of embondeira ; porcupine bristles.

1129. M. P. S. VENDUNEM, *Angola, Barra do Bengo.*—Empalanca and empacassa horns.

1130. T. M. BESSONE, *Moçambique.* — Filaments of Guinea aloe, and hemp.

ROME.

SOUTH-CENTRAL COURT.

---

1. Breviary; a present from H.H. the Pope to H.E. Card. Wiseman.

2. Ebony case for the above, which also forms a reading desk.

CLASS I.

3. BALDINI, HIS EXCELLENCY THE BARON COMMENDATORE, P. D. Minister of Commerce and Public Works.—Calcareous stones of Monticelli, St. Angelo in Capoccia, and Tivoli; pozzolane tuf; argillaceous earths, and plasters; with vases, artificial marbles, &c. made of them.

4. ―――― Travertini, peperini, lave macchi, with other building stones; marbles for decorations; millstones, refractory materials, asphalt, &c.

5. BONDI, G. & Co.—Argillaceous earth, and bricks made of it.

6. BONIZI, A.—Roman cement from Tolfa.

7. BONIZI, G. & Co.—Minerals.

8. SOCIETÀ ROMANA.—Ores of iron and metallurgical products; bricks.

CLASS II.

9. DE PAOLIS, A. B.—Potash.

10. GOVERNMENT ESTABLISHMENT OF THE ALUMS OF TOLFA.—Specimens of alum.

11. GOVERNMENT SALT WORKS (THE COM. BALDASSARE, Director).—Specimens of marine salt.

12. MOSAIC MANUFACTORY OF THE VATICAN.—Smalts.

12A. BARBERI, M.—Mosaics.

12B. MOGLIA, L.—Copy of the Madonna della Seggiola.

12C. BARBERI, L.—Mosaic table.

12D. TADDEI, L.—Mosaic table.

12E. BARZOTTI, B.—Mosaic representing St. Peter's at Rome.

12F. SIBBIO.—Panorama of Rome, in mosaic.

12G. ROSSI, A.—Mosaic table.

12H. DESTRADA, D.—Specimens of Etruscan, Roman, Greek, and Byzantine goldsmith's work.

12I. DIES, G.—Specimens of Etr uscan &c., goldsmith's work.

12K. ODELLI, A.—Cameos.

12L. SAULINI, T.—Cameos.

12M. LISTRUCCI, E.—Cameos.

12N. LISTRUCCI, B.—Cameos.

12O. PICKLER.—Engraved sardonyx.

12P. LUPI, F.—Engraved sardonyx.

12Q.　BIANCHI, G.—Bronze medal.

12R.　CELLI, V.—Bronze medal, &c.

12S.　VESPIGNANI, R.—Ebony frame.

12T.　MARCHETTI, L. & BAVADIN.—Door of the Vatican.

12U.　ERCOLI, P.—Bas-relief in ivory.

CLASS III.

12V.　JACOBINI, BROS.—Wine, vinegar, and oil.

CLASS IV.

13.　ANTONELLI, CONTE F.—Indian corn from the Pontine marshes.

14.　ARVOTTI, G.—Raw and spun silk.

15.　ERBA, B.—Asphalt, crude and prepared, for various purposes.

16.　CASTRATI, G. B.—Wax candles.

17.　MUTI PAPAZZURRI, MARQUIS S.—Stearine candles.

18.　ORTO AGRARIO OF THE ROMAN UNIVERSITY.—Cereals, and textile plants.

CLASS VII.

19.　SOCIETÀ ROMANA.—Crucibles, and furnace for laboratory.

CLASS VIII.

20.　GRAIZIOS, N.—Machine for showing new movement.

21.　ROSSI, P.—Oil mill.

CLASS IX.

22.　PFEIFFER, F.—Arched saws for gardeners, pruning tools.

23.　ROSSI, M. S. DE.—Stenographic and orthographic machine.

CLASS X.

24.　ROSSI, P.—Anti-concussion apparatus, applicable to various purposes.

CLASS XI.

25. BRAND, R.—Double-barrel gun, with appendages.

26. TONI, T.—Revolver gun.

CLASS XIII.

27. TESSIERI, PROF. P.—Medal-holder, for examining medals and gems.

CLASS XIV.

28. ANDERSON, G.—Photographic views of Rome, and of ancient and modern sculpture.

29. CUCCIONI, T.—Photographs of paintings by A. Carracci, and of the Roman Forum, Colliseum, Piazza of St. Peter's, &c.

30. DOVIZIELLI, P. — Photographs of paintings in the Farnesina, and of the Colliseum, Roman Forum, &c.

31. MACPHERSON, R.—Photographs.

32. ROCCHI, D.—Photographs.

CLASS XX.

33. ARVOTTI, G.—Silk fabrics, corded, coloured, embroidered with gold, &c.

PASQUALE, S.—Articles in silk; and silk fabrics, plain, coloured, enriched with gold, &c.

STEFANI, P.—Silk.
BIANCHI, A.—Silk.

CLASS XXIV.

35. ADMINISTRATION OF PRISONS. — Articles of lace of various kinds, made by the prisoners.

36. HOSPITAL OF S. MICHAEL.—Tapestry copied from an ancient mosaic, &c.

CLASS XXVIII.

37. ANGELINI, CAV. A.—Treatise on perspective.

38. BERTINELLI, G. — Richly bound Psalter.

39. OLIVIERI, L. — Monuments of the Lateran Museum, richly bound.

CLASS XXX.

40. Manzi, L. M.—Table made of a rare stone, found in the ruins of Rome; tables made of breccia found in Adrian's villa.

41. Martinori, P.—Articles in Oriental alabaster.

42. Muti Papazzurri, Marquis S.— Marble table, inlaid work; table of petrified lumachello.

43. Società Anonima dei Marmi Artificiali.—Tables, &c. in imitation malachite, lapis lazuli, porphyry, &c.; inlaid tables of imitation Oriental alabaster.

CLASS XXXII.

44. Pfeiffer, F.—Scissors and razors.

CLASS XXXIII.

45. Arvotti, G.—Roman pearls, and articles made of them.

45a. Pazzi, V.—Roman pearls.

CLASS XXXV.

46. Società Anonima dei Marmi Artificiali. — Pavements in imitation marbles, porphyry, granite, &c.

CLASS XXXVI.

47. Dies, G.—Tazzo of Giallo, and rosso antiquo.

48. Chialli, B.—Bronzed lamp, Pompeian style; other lamps.

49. Rainaldi, G.—Déjeunés and tazze of Egyptian alabaster: Pompeian lamp of rosso antiquo.

50. Beaitier, jewelled.

51. Box, with miniature of H. H. Pius VII.

52. Reliquary.

53. Two statuettes, copies of statues in front of St. Peter's.

54. Rainaldi, G. — Tazza in Oriental alabaster.

55. Lucatelli, G. — Trajan's column, and obelisks in rosso antico.

56. Monachesi, A.—Gothic table, and alabaster vases.

57. Prince Aldobrandini. — Etruscan and Chinese vases.

58. Ricciani, Sisters.—Flowers embroidered on cloth.

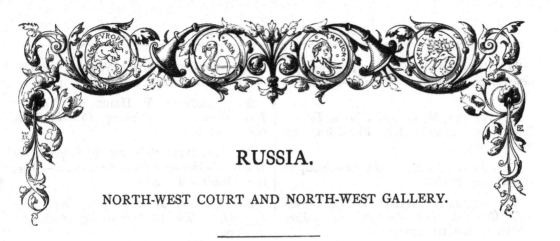

RUSSIA.

NORTH-WEST COURT AND NORTH-WEST GALLERY.

CLASS I.

1. ADMINISTRATION OF KOZAKS SETTLEMENT, *Orenburg.*—Collection of minerals.

2. ALIBERT, N. P.—Specimens and ornaments of black lead, blocks of nephrite, collection of other Siberian minerals.

3. BELOSSELSKI-BELOZERSKI, PRINCE K. *Katava Iron Works, Oofin Circ. Orenburg Gov.* —Steel, wrought and sheet iron.

4. BELOSSELSKI - BELOZERSKI, PRINCE, HEIRS OF, *Katava-Ivanofski Iron Works, Orenburg Gov.*—Collection of minerals, iron and steel.

5. BOGOSLOVSKI CROWN COPPER WORKS, *Perm Gov.*—Specimens of copper ore, and refined copper, &c. &c.

6. CABINET OF HIS IMPERIAL MAJESTY, *St. Petersburg.*—Collection of samples of polished hard stones, from different Russian quarries.

7. CAUCASIAN AGRICULTURAL SOCIETY, *Tiflis.*—Collection of ores, coal, and other Transcaucasian minerals.

8. DEMIDOF, P. P. *Nijne-Tagil Works, Perm Gov.*—Cast steel, sheet iron, and copper.

9. GOLITZYN, PRINCE S. *Nytvinsk Foundry, Okhansk Circ. Perm Gov.*—Iron ore, and cut iron bars.

10. GOOBIN, HEIRS OF, *Nijne-Sergin, and Michaïlof Mining Works, Perm Gov.*—Bar and sheet iron.

11. GRECHISHCHEF, *Michael, near Riazan.* —Lime and plaster.

12. KIRGHIZ DISTRICT ADMINISTRATION, *Orenburg.*—Chalk and common salt, from Lake Inder.

13. KNAUFS MINING CO. *Ossinsk and Perm Circuits, Perm Gov.*—Bar iron.

14. LATKIN, M. *Ustsissolsk, Vologda Gov.* —A grindstone, from the quarry of Mount Broossianaïa.

15. ADMINISTRATION OF CROWN DOMAINS, *Tchernigof Gov.*—China clay from Poloski village, near Glookhof.

16. MINING DEPARTMENT OF POLAND, *Warsaw.*—Coal, iron ore, calamine, fire-clay, cast iron, zinc sheets, &c. &c.

17. OORAL KOZAKS, *Orenburg.*—Samples of copper ore, chalk, and gypsum.

18. PASHKOF, A. *Bogoyavlensky Copper Mines, Orenburg Gov.*—Copper ores: lingots, sheets, and wire.

19. PASHKOF, M. V. *Preobrajensk Mining Works, Orenburg Gov.*—Geological specimens, plate and foil copper.

20. PASHKOF, HEIRS OF, *Voskressensk Iron Works, Orenburg Gov.*—Collection of ores, and other minerals.

21. PERMIAN MINING DISTRICT ADMINISTRATION. — Collection of ores and other minerals, and of metallurgical products.

22. POPOF, A. & N. BROS. *Kirghiz District, Siberia.*—Native copper, different ores, coal &c.

23. RACHETTE, V. *Nijnetagilsk, Perm Gov* —Models of universal high furnaces and cupola ovens.

24. ROCHEFORT, COUNTESS, *Olghinski Copper Works, Perm Gov.*—Copper ore and metallic copper.

B

25. SAMSONOF, S. & MAMONTOF, *Sernopol Semipalatinsk District, Siberia.*—Specimens of black lead.

26. SIDOROF, M. *Eastern Siberia, Tooroohansk Circ. Yenisseisk Gov.*—Black lead and other minerals.

27. TCHERKASSOF, N. *Moscow.*—Samples of peat and fire-clay.

28. VÖLKERSAM, BARON G. VON, *Papenhof, Coorland Gov.*—Samples of yellow amber, containing insects.

29. VOTKINSK CROWN WORKS, *Perm Gov.* —Assortment of iron and steel.

30. YAKOVLEF, P. HEIRS OF, *Navialof Iron Works, Yekaterinburg Circuit, Perm Gov.*—Bar iron.

31. YAKOVLEF, S. HEIRS OF, *Alapaef Iron Works, Verkhotoorsk Circ. Perm Gov.*—Sheet iron, bluish and bright.

32. ZEITLER, M. *Michalof Works, near Slawkof, Radom Gov.*—Fire-clay, coal, and iron ore.

CLASS II.

Sub-Class A.

33. CHETVERTAKOF, N. M. *Moscow.*—Red paint and white lead.

34. COLLEGE OF FORESTERS, *Lissino, Tsarskoe-Selo Circ. St. Petersburg Gov.*— Birch bark, oil, and tar.

35. EGGERS, *Reval.*—Acetate of lead, and vinegar.

36. EPSTEIN, A. & LEVY, *Warsaw.*—Copperas, Roman vitriol, white lead, Glauber's salt, and saltpetre.

37. HIRSCHENFELD, R. *Warsaw.*—Matches without phosphorus.

38. IRTEL, I. VON, *Tiflis.*—Raw and refined soda from Erivan.

39. KRAUSE, J. *Warsaw.*—Oil and spirit varnishes, oil colours, floor rubbing wax, and marking ink.

40. KRUSE, G. *Reval.*—Varnish for furniture.

41. LEPESHKIN, BROS. *Moscow.*—Garancin.

42. OOSSACHEF, B. *St. Petersburg.*— Colours.

43. REICHEL, A. *Somin Chemical Works, Borovitch Circ. Novgorod Gov.*—Birch bark, oil, and turpentine.

44. SANIN, V. J. *near Borovsk, Kalooga Gov.*—Chemical products.

45. SCHMIDT, K. E. & Co. *Svatoi Island, Caspian Sea.*—Paraffin.

46. SHIPOF, A. *Kineshma Circ. Kostroma Gov.*—Chemicals.

47. SPIES, LEWIS, *Tarkhomina, near Warsaw.*—Bone dust for manure, and artificial guano, &c.

48. TORNAU, BARON, & Co. *near Bakoo.*— Raw naphtha and naphthadehil, and products of their distillation.

49. VOLOSKOF, J. *Rjef, Tver Gov.*—Carmine.

Sub-Class B.

50. ANOKHIN, A. J. *St. Petersburg.*— Scents, pomatum, hair powder, and other cosmetics.

51. GLAZER, F. *Tiflis.*—Blossom of pyretrum, carneum, and Persian powder.

52. IRTEL, I. VON, *Tiflis.*—Blossom and seed of pyretrum, carneum, and Persian powder.

53. JDANOF, BROS. *St. Petersburg.*—Aromatic waters, perfumed and deodorising fluids.

54. NATANSON, J. & SHEEMAN, *Warsaw.*— Perfumery, toilet powders, cold cream, &c.

55. OORAL KOZAKS, *Orenburg.*—Roots of rhubarb, glycyrrhiza glabra, inula helenium, and dried salvia leaves.

CLASS III.

Sub-Class A.

56. AHMET, BEKIR-OGLOO, *Derekoi Vit. Alooshta District, Crimea.*—Walnuts.

57. ADMINISTRATION OF CROWN DOMAINS, *Vilno.*—Cereals.

58. ADMINISTRATION OF THE KOZAK SETTLEMENTS, *Orenburg.*—Collection of cereals, peas, poppy, lin and hemp seed.

59. AGRICULTURAL DEPARTMENT, *St. Petersburg.*—Collection of cereals, flour, groats, pulse, oil and grass seeds, nuts, chicory, and malt.

60. ALEXANOF, M. *Alexandropol Circ. Erivan Gov.*—Lentils.

61. ARUTINOF, A. *Alexandropol, Erivan Gov.*—Carmeline and lucerne seeds.

62. ARUTINOF, A. *Elisabethopol Circ. Tiflis Gov.*—Hemp seed.

63. ARUTINOF, A. *Guzander Vil. Alexandropol Circ. Erivan Gov.*—French beans and lentils.

64. ARUTINOF, VANO, *Boluis Vil. near Tiflis.*—Peas.

65. BOBIN, B. *Slavianka Vil. Telav Circ. Tiflis Gov.*—Millet.

66. BOBYSHEF, I. *Elisabethopol, Tiflis Gov.*—Linseed.

67. BOGOLUBSKI, S. *Protopresbyter, Nerchinsk Circ. Irkootsk Gov.*—Cereals, oil seeds, flour, and pine nuts.

68. BORISSOF, J. *Rogestvensk Village, Korotoiak Circ. Voronesh Gov.*—Canary seed.

69. BRUKHOVETSKY, N. *Markovka Village, Bogoochar Circ. Voronesh Gov.*—Buckwheat.

70. CAUCASIAN AGRICULTURAL SOCIETY, *Tiflis.*—Mountain rice (chaltik), and saffron.

71. CLAYHILL & SONS, *Reval.*—Samples of Estonian cereals.

72. COLONISTS OF ANNENFELD SETTLEMENT, *Tiflis Gov.*—Barley, common and spelt wheat, millet, peas, lentils, linseed.

73. COLONISTS OF EKATERINFELD SETTLEMENT, *near Tiflis.*—Barley, oats, samples of wheat, and French beans.

74. CORNIES, J. *Tashchinak Freehold, Melitopol Circ. Tauride Gov.*—Wheat, rye, millet, and lucerne seed.

75. DENGINK, A. *Kishenef, Bessarabia.*—Cereals, oil seeds, madder seed, &c.

76. DOROSHENKOF, P. *Beerooch, Voronesh Gov.*—Sunflower seeds.

77. EKATERINSTADT COLONISTS, *Saratof Gov.*—Common Russian and Turkish wheat.

78. ELIOZOF, H. *Kvarlee Vil. Telav. Circ. Tiflis Gov.*—Millet.

79. ERISTOF, E. *Goree, Tiflis Gov.*—Millet, common and French beans, peas, and linseed.

80. GENT, G. E. & Co. *Pskof.*—Linseed.

81. HARTMANN, *Riga.* — Assortment of cereals.

82. HASSAN-OGLOO, M. *Mashadee Vil. Bakoo Circ.*—Rice.

83. HENNER, T. *Helendorf Settlement, Alexandropol Circ. Erivan Gov.*—Beans.

84. IVANOF, G. *Mikhaïlovka Vil. Elisabethopol Circ. Tiflis Gov.*—Hemp-seeds.

85. KELBLER, J. *Tonkoroonofka Settlement, Saratof Gov.*—Wheat and rye.

86. KERBELAÏ-SADECH-MEKHTI, O. *Kschil-Agatch Vil. Lenkoran Circ. Bakoo Gov.*—Barley.

87. KHANAGOF, J. *Djelal-Ogloo Vil. Alexandropol Circ. Erivan Gov.*—Common and French beans.

88. KOOSHELEF-BEZBORODKO, COUNT NICHOLAS, *Illinsko, Anchekrak Estate, near Odessa.*—Samples of wheat.

89. KOVESHNIKOF, J. *Markovka Village, Bagoochar Circ. Voronesh Gov.*—Spring wheat.

90. LANDAU, G. *Warsaw.*—Rye meal.

91. LEVSHIN, T. A. *Pasheelino Village, Yefremof Circ. Toola Gov.*—Soft white peas.

92. LUH, J. *Paninsk Settlement, Saratof Gov.*—Sunflower seed.

93. LYSACK, B. P. *Kistero Village, Starodoob Circ. Tchernigof Gov.*—Hemp seed.

94. MANHOLD, C. *St. Petersburg.*—Flour of beans and peas, prepared without grinding.

95. MANOOKIANTZ, *Martyros, Elisabethopol Circ. Tiflis Gov.*—Spelt wheat.

96. MARIINSKAÏA MODEL FARM, *near Saratof.*—Millet.

97. MOROZOF, P. *Pantzerevka Village, Gorodischi Circ. Penza Gov.*—Green rye and buckwheat groats.

98 MUSTIALA AGRICULTURAL INSTITUTION, *Tavasthus Gov. Finland.*—Samples of wheat, oats, barley, buckwheat, and fir seeds.

99. NASHROOLEE-OGLOO, *Sopkooli, Lenkoran Circ. Bakoo Gov.*—Wheat.

100. NESTEROF, J. *Yasenkof Village, Nijnedevitzk Circ. Voronesh Gov.*—Hemp seed.

101. VOLOGDA ROWN DOMAINS, *Velikii-Oostug Circ. Vologda Gov.*—Rye and barley.

102. OOMANSKI, *Imperial Appanage Estates in the Gov. of Tver.*—Pennsylvania rye.

103. OORAL KOZAKS, *Orenburg Gov.*—Wheat, rye, oats, millet, and sunflower seed.

104. PETROOSSOF, O. *Alexandropol, Erivan Gov.*—Rape seed.

105. PETSCHKE, *Coorland.*—Assortment of cereals.

106. PLEININGER, G. *Marienfeld Settlement, near Tiflis.*

107. PLOTNIKOF, S. *Pessok Village, Novokhopersk Circ. Voronesh Gov.*—Millet and winter rye.

108. PNIOWER, J. & I. BROS. *Piotrkow, Warsaw Gov.*—Wheat flour and groats.

109. POLEJAEF, BROS. *Belozersk Circ. Novgorod Gov.*—Wheat flour, 1st, 2nd, and 3rd qualities.

110. POOSANOF, M. A. *Schigrof Circ. Koorsk Gov.*—Oats, buckwheat, and millet.

111. RIGA COMMITTEE FOR EXHIBITION.—Linseed, pease, &c.

112. ROTHHAR, A. *Tonkoroonojka Settlement, Saratof Gov.*—Barley.

113. SAMARIN, S. *Mikhaïlovka Vil. Elisabethopol Circ.*—Hemp seed.

114. SARKISSOF, A. *Zeiva Vil. Etchmiadzin Circ. Erivan Gov.*—Wheat.

115. SCHALE, A. *Marienfeld Settlement, near Tiflis.*—Oats.

116. SEESSOEF, P. *Novo Saratof Settlement, Elisabethopol Circ.*—Hemp seed.

117. SHVEELEE, N. M. *Kvarelec Vil. Telav Circ. Tiflis Gov.*—French beans.

118. SHVEELEE, S. *Ooriatuban Vil. Telav Circ. Tiflis Gov.*—Wheat.

119. SHESTOF, J. *Mshaga Village, Nijni Novgorod Circ.*—Linseed.

120. SOOLKHANOF, E. *Goree, Tiflis Gov.*—Oats and wheat.

121. SOUTH-EASTERN MODEL FARM, *Kazan Gov.*—Seeds of spring rye, buckwheat, millet, &c. &c.

122. STEIGERWALD, H. *Krasny-Yar Settlement, Saratof Gov.*—Turkish wheat.

123. STOROSHEF, D. *Trostianka Village, Ostrogoshsk Circ. Voronesch Gov.*—Winter wheat.

124. TOMILIN, J. *Pelagiada Vil. near Stavropol.*—Oats.

125. TSHOOKMALDIN, N. *Tiumen Circ. Tobolsk Gov.*—Cereals.

126. TURINE, A. *Konstantinovka Vil. Novobaïazet Circ. Erivan Gov.*—Oats.

127. VÖLKERSAM, BARON G. VON, *Papenhof, Coorland Gov.*—Samples of sunflower seeds and heads, and winter linseed.

128. YAKOVLEF, G. *Proossinich Village, Mohileff Gov.*—Linseed.

129. YAZDOONOF, G. *Dogkuz Vil. near Erivan.*—Ricinus and sesamum seeds.

130. SADYRIN, PH. *Kotelnich Circ. Viatka Gov.*—Cereals, peas, and grass seeds.

Sub-Class B.

131. BOCHAREF, M. I. *Yamskaïa Sloboda, near Koorsk.*—Buckwheat, flour, and groats.

132. BOGOLUBSKI, S. *Protopresbyter, Nerchinsk Circ. Irkootzk Gov.*—Groats.

133. BORISSOVSKY, M. *Moscow.*—Refined beet-root sugar.

134. CAUCASIAN AGRICULTURAL SOCIETY, *Tiflis.*—Dried apricots, plums, figs, raisins, and Tchoorkhel (dainties of the natives).

135. EKATERINHOF SUGAR-REFINING CO. *St. Petersburg.*—Refined sugar.

136. EPSTEIN, H. *Hermanovo and Lyshkowicy, Lowicz Circ. Warsaw Gov.*—Refined beet-root sugar.

137. FOONDOOKLEY, J. J. *Ossota Factory, Chighirin Circ. Kief Gov.*—Specimens of raw beet-root sugar.

139. HAUF, BARON, *St. Petersburg.*—Refined sugar.

140. JACKOWSKI & Co. *Przasnysz Circ. Plotsk Gov.*—Sugar.

141. KIRGHIZ DISTRICT ADMINISTRATION, *Orenburg.*—Dried mutton and smoked beef.

142. KLIKOVSKY, PROF. *Kazan.*—Crystallized honey.

143. KOOSHELEF-BEZBORODKO, COUNT N. *Novochigly Refinery, Bobrof Circ. Voronesh Gov.*—Refined beet-root sugar.

144. KRICH, K. *Reval.*—Estonian anchovy.

145. LANDAU, G. *Warsaw.*—Groats and biscuits.

146. MONAKHOF, A. *Klin, Moscow Gov.*—Potato syrup.

147. NATANSON, BROS. I. & J. *Goozof, Lovicz Circ. Warsaw Gov.*—Refined beetroot sugar.

148. NATANSON, S. & I. *Sanniki, Gastyn Circ. Warsaw Gov.*—Raw and refined beet-root sugar.

149. GERKE, BROS. L. *Gliadkovo Vil. Elatom Circ. Tambof Gov.*—Raw beetroot sugar.

150. KLAASSEN, F. *Ladekop Settlement, Berdiansk Circ. Tauride Gov.*—Dried Mirabelle plums.

151. OORAL KOZAKS, *Orenburg Gov.*—Caviar, isinglass, viasiga, balyk (dried fish).

152. PNIOWER, BROS. J. & I. *Piotrkow, Warsaw Gov.*—Groats.

153. RAVICZ, A. & Co. *Elsbetowo, Sedletz Circ. Lublin Gov. Poland.*—Refined beet-root sugar.

154. ROCHEFORT, COUNTESS OLGA, *Ossa Circ. Perm Gov.*—Lime-tree blossom honey, obtained by cold pressure.

155. ROTHERMUND, A. *Bobrik Refinery, Soomy Circ. Kharkof Gov.*—Raw beet-root sugar.

156. SOOMAKOF, T. *St. Petersburg.*—Tablets of portable veal soup.

157. SPIES, L. *Tarkhomino, near Warsaw.*—Pulverised bones for food.

158. TCHERKASSOF, N. *Moscow.*—Refined beet-root sugar.

159. TERENTIEF, M. *Novotorjsk Circ. Tver Gov.*—Wheat and rye starch.

160. WISNOWSKY, R. *Warsaw.*—Biscuits.

Sub-Class C.

161. ABHAZOF, PRINCE D. *Kakhetia.*—Red and white wine.

162. AGRICULTURAL DEPARTMENT, *St. Petersburg.*—Assortment of tobacco leaves from American, Turkish, and Russian seeds.

163. BEKIR, I. OGLOO, *Derekoi Vil. Alooshta District, Crimea.*—Samson tobacco leaves.

164. CHARENTON, B. I. *Akkerman and Chabog Vineyards, Bessarabia.*—Red and white wines.

165. AUTORHUFFEN, T. *St. Petersburg.*—Cigarettes.

166. BOSTANJOGLO, M. & SONS, *Moscow.*—Tobacco, cigars, and cigarettes.

167. GABAÏ & MIGRI, *St. Petersburg.*—Samples of tobacco from Turkish seeds.

168. GROOTE, VON, *Distillery, Livonia.*—Cumin liqueur.

169. HEINRICHS, F. *St. Petersburg.*—Tobacco, cigars, and cigarettes.

170. HELLER, T. H. *St. Petersburg.*—Cigars made by machinery.

171. INGLESY, A. *Tatareshty Vil. Orgey Vil. Bessarabia.*—Tobacco leaves.

172. JAPBA, B. *Moscow.*—Spirituous liqueurs.

173. ADMINISTRATION OF TCHERNIGOF CROWN DOMAINS, *Tchernigof.*—Bakoon tobacco and seeds.

174. MULLER, A. TH. *St. Petersburg.*—Cigars, cigarettes, and tobacco.

175. MUSTAPHA, HALIL-OGLOO, *Goorsoof Vil. Alooshta District, Crimea.*—Samson tobacco leaves.

176. ONANOF, *near Kootaïs.*—Turkish tobacco leaves.

177. RIGA COMMITTEE FOR EXHIBITION.—Assortment of tobacco.

178. SCHWABE, H. *Riga.*—Rectified spirits of wine and liqueurs.

179. SCHWEINFURTH & SEECK, *Riga.*—Wine.

180. STADLER, J. *Podstepnoi Settlement, Saratof Gov.*—Tobacco leaves from American seeds.

181. STRIEDTER, *Distillery, St. Petersburg.*—Raw, purified, and rectified spirits ; rum, cognac, and liqueurs.

182. TÖPFER, A. *St. Petersburg.*—Cigars and cigarettes.

183. UNGERN-STERNBERG, BARON VON, *Distillery, Estonia.*—Cumin liqueur.

184. WIGAND, C. *Krasny-yar-Settlement, Saratof Gov.*—Tobacco leaves.

185. WINKELSTERN, A. *Ernestinendrof Settlement, Saratof Gov.*—Tobacco leaves.

186. WICKEL, *Riga.*—Currant wines.

187. WISNOWSKY, R. *Warsaw.*—Liqueurs.

188. WORONZOF, PRINCE, *Crimea.*—Wines of his estate.

CLASS IV.

Sub-Class A.

189. BORODOOLIN, N. wax chandler, *St. Petersburg.*—Wax candles.

190. COMPANY OF THE ST. PETERSBURG STEARINE CANDLES, SOAP, AND OLEIN FACTORY.—Assortment of stearine candles.

191. COMPANY FOR PREPARING SZAR SOAP AND RUSSIAN COSMETICS.—Egg-yolk oil, soap cosmetics, &c.

192. EPSTEIN, A. & LEVI, M. *Warsaw.*—Block of stearine, and stearine candles.

193. KRESTOVNIKOF, BROS. *Kazan.*—Stearine candles.

194. METEOR OIL MILL, *Gorodische Circ. Penza Gov.*—Hemp-seed oil-cake.

195. NATANSON, J. & SIMON, *Warsaw.*—Common and scented soap.

196. NENNINGER, A. *Sevsk Circ. Orel Gov.*—Hemp and linseed oil, raw and refined.

197. ALFTHAN & Co. *Finland.*—Stearine candles.

198. STEAM OIL MILL Co. *Riga.*—Hemp and linseed oil refined, and linseed cake.

199. KUEMMEL & Co. *Odessa.*—Rape and linseed oil, and oil cake.

200. OIL STEAM MILL Co. *St. Petersburg.*—Linseed cake.

201. OORAL KOZAKS, *Orenburg.*—Photogen and solar oil, train oil, tallow, and soap.

202. PROKHOROF, A. *Belef-Toola Gov.*—Tallow.

203. SAPELKIN, V. A. *Vladimerovka Vil. near Moscow.*—Wax candles.

204. STEINER, *St. Petersburg.*—Samples of common soap.

205. TIMOFEEF, BROS. *Stary Oskol, Koorsk Gov.*—Common soap.

Sub-Class B.

206. ADMINISTRATION OF THE KOZAKS SETTLEMENTS.—Samples of sheep's wool.

207. AGRICULTURAL DEPARTMENT, *St. Petersburg.*—Collection of fleeces, raw and washed, wether, ewe, and goat, and goat's hair.

208. BABARYKIN, J. *Kholm, Pskof-Gov.*—Samples of bristles.

209. BOGOLUBSKI, S. *Protopresbyter, Nerchinsk Circ. Irkootsk Gov.*—Wool, goat's and camel's hair.

210. SHER, N. *St. Petersburg.*—Medallion carved in ivory.

211. DOLGANOF, J. *St. Petersburg.*—Ivory and tortoiseshell combs, &c.

212. DÖRING, *Livonia.*—Fleeces.

213. DORONIN, J. *Archangel.*—Basso-relievo, folding knives, and other articles carved in ivory.

214. ERISTOF, M. *Ossetia District, Caucasus.*—Wether, ewe, and lamb wool.

215. FOONDOOCLEY, J. J. *Reshbairaki Estates, Robrinetz Circ. Kherson Gov.*—Fleeces in raw state.

216. GRAND DUCHESS HELEN PAVLOVNA. —Merino fleeces and wool, from Her Imperial Highness's estate, Karlovka, Gov. of Poltava.

217. GROODININ, P. *Velikii-Looki, Pskof Gov.*—Assortment of bristles.

218. KABYZEF, M. K. *St. Petersburg.*— Bone black and bone dust.

219. KAZAKOF, L. *Velikii-Oostug, Vologda Gov.*—Assortment of bristles.

220. KELBI-OGLOO, I. *Nakhitchevan.* — Samples of wool.

221. MAMONTOF, A. *Moscow.*—Samples of bristles: okatka, 1st and 2nd quality.

222. VOLOGDA CROWN DOMAINS, *Velikii-Oostug Circ. Vologda Gov.*—Bristles, various.

223. PHILIBERT, A. *Atamanay Farm, near Ghenitchesk, Melitopol Circ. Tauride Gov.*— —Merino fleeces.

224. OORAL KOZAKS, *Orenburg.*—Wool, goat's and camel's hair, and glue.

225. RUSSIAN AMERICAN CO. *St. Petersburg.*—Walrus teeth.

226. SHVEELEE, G. B. *Kvarelee Vil. Telav Circ.*—Sheep's wool.

227. SHVEELEE, S. M. *Oorsatossany Vil. Telav Circ. Tiflis Gov.*—Wool.

228. SOOLTANOF, K. K. *Reshish-Rend Vil. Nakhichevan Circ. Erivan Gov.* — Combed wool.

229. STEPANOF, N. *Eysk.*—Washed wool.

230. VOKOOYEF, TH. *Mezen Circ. Archangel Gov.*—Walrus tooth.

231. VONSOWSKY, *Zeiva Vil. Echmiadzin Circ. Erivan Gov.* — Wool of a Kurtinsk wether.

Sub-Class C.

232. ABHAZOF, PRINCE D. *Kardansky Vil. Signah Circ.*—Safflower.

233. ABKHAZOF, PRINCE D. *Kakhetia.*— Safflower-seeds and blossom.

234. ADMINISTRATION OF THE KOZAK SETTLEMENTS, *Orenburg.*—Samples of flax, hemp, and wild madder root.

235. AGRICULTURAL DEPARTMENT, *St. Petersburg.*—Samples of flax from Poodosh and Vladimir, madder and statice roots.

236. BABARYKIN, J. *Kholm, Pskof Gov.*— Samples of flax.

237. BEK-HADJINOF, J. *Shemakha.*—Madder-root.

238. CAUCASIAN AGRICULTURAL SOCIETY, *Tiflis.*—Dendrological collection of eighty-two Transcaucasian trees and shrubs.

239. CLAYHILL & SONS, *Reval.*—Samples of Estonian flax.

240. LEESSIN, S. *Illinskoe Vil. Lookaïanov Circ. Nijni-Novgorod Gov.*—Mats of lime-tree bast.

241. FEDOROF, G. *Koshelevo, Rogachef Circ. Mohilef Gov.*—Lime-tree bast mats.

242. GENT, G. E. & Co. *Pskof.*—Samples of flax.

243. GRYZMALA, V. *Obrowec, Grubeszow Circ. Lublin Gov. Poland.*—Log of oak.

244. HADJI-DJAVAT-BEEK-ALI-OGLOO, *Koola, Gov. of Bakoo.*—Madder-root.

245. IVANOF, D. *Boody, Rogachef Circ. Mohilef Gov.*—Lime-tree bast mats.

246. PERCY, JACOBS, *Riga.*—Corks.

247. KARAPET-MIKIRTOOMOF, *Tiflis.* — Cotton, cleaned and hackled by machinery.

248. KARAPET-SHAGIANOF, *Nigri Vil. Ordoobat Circ.*—Samples of cotton, raw, clean, and hackled.

249. KARDAKOF, M. *Kotelnich Circ. Viatka Gov.*—Flax.

250. KAZAN-KHAN-MUSTAFA-OGLOO, *Kooba, Bakoo Gov.*—Madder-root.

251. KIRGHIZ DISTRICT ADMINISTRATION, *Orenburg.*—Wild madder root.

252. KOOZMIN, J. *Boody Village, Rogachef Circ. Mohilef Gov.*—Lime-tree bast mats.

253. KOPILOF, M. *Potchep Borough, Mglin Circ. Tchernigof Gov.*—Clean hemp.

254. KORNILOF, J. *Doorashkovo Village, Lookaïanof Circ. Nijni-Novgorod Gov.*—An oaken barrel.

255. KRIEGSMANN, A. *Riga.*—Samples of corks.

256. MASHADÉE-HADJI-ALI-OGLOO, *Kooba, Bakoo Gov.*—Madder-root.

257. MEDVEDEF, T. *Medvedef Village, Mohilef Gov.*—Flax.

258. MOROZOF, P. *Pantzerevka Village, Gorodishche Circ. Penza Gov.*—A bag in lime-tree bast.

259. MUSTIALA AGRICULTURAL INSTITUTION, *Tavasthus, Finland.*—Tanning bark used in Finland.

260. NEMILOF, A. M. *Orel.*—Samples of half-clean and clean hemp.

261. MALOKROSHECHNOY, I. *Pudosh Olonetz Gov.*—Flax.

262. VOLOGDA CROWN DOMAINS, *Sol Vychegodsk Circ. Vologda Gov.*—Vychegodsk flax.

263. OBRASTZOF, B. *Rshef, Tver Gov.*—Clean flax.

264. ONANOF, *near Kootaïs.* — Clean cotton.

265. OORAL KOZAKS, *Orenburg.* — Wild madder root.

266. PETCHORA TIMBER TRADE CO. *St. Petersburg.*—Larch logs, from Petchora River.

267. PHILEMONOF, T. *Semakof Settlement, Mohilef Gov.*—Flax.

268. POOZANOF, M. A. *Schigrof Circ. Koorsk Gov.*—Specimens of fine hemp.

269. PROKHOROF, A. *Belef-Toola Gov.*—Clean hemp.

270. ROTCHEF, I. *Mezen Circ. Archangel Gov.*—Larch-tree sponges.

271. SAVITCH, I.—Walnut wood, exported from Caucasus by the Russian Steam Navigation and Trading Company.

272. SCHEGLOF, N. *Scherschof Village, Nijni-Novgorod Gov.*—Fishing net.

273. SEIDLITZ, N. *Nookha, Tiflis Gov.*—Safflower.

274. SHESTOF, J. *Mshaga Village, Nijni-Novgorod Circ.*—Samples of flax.

275. SOROKIN, P. *Belef, Toola Gov.*—Samples of hemp.

276. SOROKIN, R. & S. BROS. *Belef, Toola Gov.*—Clean hemp.

277. SOROKIN, B. *Belef, Toola Gov.*—Hemp.

278. TCHOOKMALDIN, N. *Tiumen Circ. Tobolsk Gov.*—Hemp.

279. RUSSIAN AMERICAN INDIA RUBBER Co. *St. Petersburg.*—Elastic bands, tubes, interlayers, &c.

280. VASSILTCHIKOF, PRINCE A. *Vybit Village, Starorooss Circ. Novgorod Gov.*—Flax.

281. VOLKERSAM, BARON G. VON, *Papenhof, Coorland Gov.*—Samples of flax.

282. SADYRIN, PH. *Kotelnich Circ. Viatka Gov.*—Flax.

CLASS V.

283. DEMIDOF, P. P. *Nijni-Tagilsk Iron Works, Perm Gov.*—Rails, with bolts and fittings.

CLASS VI.

285. FROEBELIUS, T. *Carriage Factory, St. Petersburg.*—A town carriage.

286. JAKOVLEF, P. *St. Petersburg.* —Drosky and sledge for racing.

287. KOOPIDONOF, T. & A. BROS. *Moscow.*—Carriage-springs.

288. LIEDTKE, A. *Warsaw.*—Town carriage for four persons.

289. LUBLINSKI, L. *Warsaw.*—A pony-chaise.

290. MOKHOF, J. *Pagost Vil. Pereïaslof Circ. Vladimir Gov.*—Set of coach springs.

291. NELLIS, C. SEN. *Carriage Factory, St. Petersburg.*—A coach.

292. NELLIS, CH. JUN. *St. Petersburg.*—Drosky (egoitska).

293. POLIAKOF, J. *Mooravikha Village, Nijni-Novgorod Gov.*—Oaken wheels for a town carriage.

294. RENTEL, J. *Warsaw.*—Two-seated town carriage.

295. SCHWARTZE, H. *St. Petersburg.*—Two-seated town calash.

296. WAGNER, T. *St. Petersburg.*—Four-seated calash-landau.

CLASS VII.

297. BOSTANJOGLO, B. *Moscow.*—Crucibles.

298. TECHNICAL SCHOOL, ORDNANCE DEPARTMENT, *St. Petersburg.*—Models of different machinery.

CLASS VIII.

299. HECKER, H. *Engine and Agricultural Implement Manufactory, Riga.*—Linseed sorter, colour mills, decimal balances.

300. LIKHATCHEF, COL. *Yaroslaff.*—Machinery for making staves.

CLASS IX.

301. ADMINISTRATION OF KOZAK SETTLEMENT, *Orenburg.*—Share of a sokha (Russian plough).

302. CIEHOWSKY, R. *Linow, Sandomir Circ. Radom Gov.*—Ploughs, grubber, and triangular iron harrow.

303. KLIKOVSKY, PROF. *Kazan.*—Models of beehives.

304. KOSZARSKI, C. *Warsaw.*—Plough, invented by the exhibitor.

305. MUSTIALA AGRICULTURAL INSTITUTION, *Tavasthus Gov. Finland.*—Model of a common drying kiln for corn.

306. IANITZKY, *Boleslas, Kaligorka Estate, Cherson Gov.*—Implement for beet-root harvesting.

CLASS X.

307. BOSTANJOGLO, M. *Moscow.*—Bricks and artificial stones, made of Gshelsk clay.

308. CABINET OF HIS IMPERIAL MAJESTY.—Candelabra, vase, and column in jasper.

309. CIECHANOWSKI, J. *Cement Works, Gorodetz & Slawkow, Olkush Circ. Radom Gov. Poland.*—Cement, cement castings, sandstone.

310. EKATERINBURG STONE-POLISHING FACTORY.—Ornamental cups, inkstands, and letter weights, in jasper, porphyry, &c.

211. MUSTIALA AGRICULTURAL INSTITU-

TION, *Tavasthus Gov. Finland.*—Specimen of shingle-roofing, and shingles.

312. PETERHOF STONE-POLISHING FACTORY, *St. Petersburg Gov.*—Paper weights and vase in nephrite.

313. SCHMIDT, *St. Petersburg.*—Specimens of asphalt.

314. STURM, H. *Dorpat.*—Dutch tile mantel-piece.

315. TSEPENNIKOF, J. *St. Petersburg.*—Models of pneumatic and bath ovens.

316. ZIMARA, R. *St. Petersburg.*—Model of a pneumatic heating oven, patented 1860.

CLASS XI.

Sub-Class A.

317. JIGOONOF, N. *St. Petersburg.*—Epaulets and other military accoutrements.

318. SOULKHANHOFF, *Goree, Tiflis Gov.*—Cartridge box.

Sub-Class B.

319. NISSEN, W. *St. Petersburg.*—Camp bedstead, mattress, and cushion.

Sub-Class C.

320. AGADJANOF, A. *Alexandropol, Erivan Gov.*—Child's dagger.

321. BOORUNSOOZOF, M. *Akhalzih, Kootaïs Gov.*—Asiatic sabre and dagger.

222. THE COMMANDER OF BASHKIR MILITIA, *Orenburg.*—A wooden cross-bow, with sheath, quiver, and 25 arrows.

323. CROWN STEAM-SHIP FACTORY OF THE PORT OF CRONSTADT.—Gun-carriage, designed by Colonel Pestitch.

324. HADJI-SEID-AGHI-SEID-OGLOO, *Nookha, Bakoo Gov.*—Dagger.

825. KOBATEE-IBRAKHIM-MAHMED-OGLOO, *Daghestan District, Caucasus.*—Dagger.

326. OBOOKHOF, COL. *Orenburg Gov.*—Twelve pounder cast-steel gun, and a steel ring sawn from the gun before polishing.

327. POPOF, J. *Tiflis.*—Fowling-piece, pistol, powder flask, sabre, dagger, and girdle.

328. TCHIFTALAROF, M. *Akhaltzyk.*—Rifle, fowling-piece, and pistol.

329. VISHNEVSKY, F. *St. Petersburg.*—A revolver, invented by the exhibitor.

330. YOOST-SAMAN-OGLOO, *Nookha, Bakoo Gov.*—Gun barrel.

331. ZLATOUST CROWN ARMOUR FACTORY, *Southern Ooral, Orenburg Gov.* — Sword-blades, swords and scythes, polished cast-steel breastplate.

CLASS XII.

Sub-Class A.

332. CROWN STEAM-SHIP FACTORY OF THE PORT OF CRONSTADT.—Ship fittings.

333. HAAKER, A. *St. Petersburg.*—Model of the man-of-war "Victory."

334. IJORA ADMIRALTY IRON WORKS, *near St. Petersburg.*—Specimens of chain cable and cat-block.

335. MANUFACTORY & MODEL ROOM OF NAVAL ARCHITECTURE, *St. Petersburg.*—Models of ships for the Imperial Russian navy.

336. YEGOROF, LIEUT. J. *St. Petersburg.*—Model of the 111-gun ship "Emperor Nicholas I."

CLASS XIII.

337. KADINSKY, PROF. K. *St. Petersburg.*—Logarithmical or calculating sliding rule.

338. NAUTICAL INSTRUMENT MANUFACTORY, *St. Petersburg.*—Sea barometer, compass with illuminator, and patent log.

339. PIK, J. *Warsaw.*—Model of hydraulic press, electro-galvanic apparatus, magnifying glass, spectacles in filagree setting, crystal thermometer, and apparatus for assaying.

340. REISSNER, PROF. *Dorpat.*—Microscopic objects.

341. STEINBERG, T. *Osseenovoy-Koost, near Saratof.*—Controlling apparatus for distilleries.

CLASS XIV.

342. DENIER, *St. Petersburg.*—Portraits.

343. FAJANS, M. *Warsaw.*—Photographs.

344. LEVITZKY, S. 22 *Rue de Choiseul, Paris.*—Portraits.

345. LORENS, A. *St. Petersburg.*—Photographic portraits and stereoscopic prints.

346. MIECZKOWSKI, J. *Warsaw.*—Photographic prints and visiting cards, on albumenised paper.

347. PETROFSKI, *St. Petersburg.*—Photographic copies: Bruni's picture "The Brazen Serpent," and Ch. Brulof's "Last Day of Pompeii."

348. ROSENBERG, *Riga.*—Coloured photographs, without after-touch (elaiögraphy).

349. RUMINE, G. 5, *Lower Gore, Kensington, London.*—Life-size portraits, photographed with carbon, on canvas, oil painted, &c.

350. SHPAKOFSKI, A. *St. Petersburg.*—Portraits.

CLASS XV.

351. SON, H. *Moghilef.*—A clock, with the inventor's new mechanism.

CLASS XVI.

352. BECK, P. *St. Petersburg.*—A pianoforte.

353. RUDERT, H. *Warsaw.*—Musical instruments.

CLASS XVII.

354. CROWN FACTORY FOR SURGICAL INSTRUMENTS, *St. Petersburg.*—Various sets of surgical instruments.

355. VARIPAEF, TH. *Pavlovo, Gorbatof Circ. Nijni-Novgorod Gov.*—Surgical instruments.

CLASS XVIII.

356. THE ADMINISTRATION OF THE ORENBURG KOZAK SETTLEMENTS.—A knitted cover, from English cotton yarn.

357. ARECKOF, G. *Novobayazet, Erivan Gov.*—Cotton stuffs.

358. BOORNAYEF, A. *Kazan.*—Long cloth.

359. BORISSOVSKY, M. *Pereslavl-Zalessky, Vladimir Gov.*—Cotton yarn.

360. CAUCASIAN AGRICULTURAL SOCIETY, *Tiflis.*—Cotton stuff, called "Noshoree."

361. DEINESS, J. *Norki Settlement, Saratof Gov.*—Cotton cloth (sarpinka).

362. FINLAYSON & Co. *Tammerfors, Finland.*—Cotton cloth and yarn.

363. KINDSVATER, O. *Splavnookha Settlement, Saratof Gov.*—Cotton cloth (sarpinka).

364. MZIREOOLOF, SOPHIA, *Tarsky District, Caucasus.*—Cotton yarn.

365. FORSSA, *Cotton Mill Company.*—Finland shirting.

366. SPADI, W. *Norki Settlement, Saratof Gov.*—Cotton cloth (sarpinka).

367. PYCHLAU, *Riga.*—Cotton twist.

368. REINEKE, A. *Popovka Settlement, Saratof Gov.*—Cotton cloth (sarpinka), and a checked head kerchief.

369. SCHEFER, J. *Goly Karaslish Settlement, Saratof Gov.*—Red checked cotton cloth; shirting.

370. SMIDT BROS. *Oost-Zolikha Settlement, Saratof Gov.*—Six pieces of cotton cloth (sarpinka).

CLASS XIX.

371. THE ADMINISTRATION OF THE ORENBURG KOZAK SETTLEMENTS. — Hand-spun thread; linen, plain and twilled.

373. ALEXANDROF, EUDOXIA, *Sopelki Village, Yaroslaf Circ.*—A piece of linen.

374. CAZALET, A. & SONS, *St. Petersburg.*—Rope yarn, bolt rope, cordage, white rope log lines, &c.

375. DOMBROWICZ, C. *Dobrovolia, Mariampol Circ. Augustowo Gov.*—Table-cloth, napkins, and towels.

376. HIELLE, C. & DITTRICH, CH. *Girardovo, Lovicz Circ. Warsaw Gov.*—Linen, table-cloths, napkins, and towels.

378. NEMILOF, A. *Orel.*—Hemp yarn.

379. LUCKS, C. *Halberstadt Settlement, Berdiansk Circ. Tauride Gov.*—Ropes.

380. RIGA LOCAL COMMITTEE. — Hemp yarn.

381. NOVIKOF, A. & J. *Briansk, Orel Gov.*—Hemp rope-yarn, white and tarred, and clean. hemp (mas plant).

382. STIEGLITZ, BARON, *near Narva, St. Petersburg Gov.*—Assortment of hemp and flax, sail cloth.

383. TSHOOKMALDIN, N. *Tiumen Circ. Tobolsk Gov.*—A towel and linen.

CLASS XX.

384. ABDOOL-BEK-HADJI-MEERAM-BECK-OGLOO, *Shemakha, Bakoo Gov.*—Red and yellow móv (silk stuff).

385. AGA-KISHI-BEK-MAHMED-HASSEIN-BECK-OGLOO, *Lenbaran Vil. Shoosha Circ. Bahoo Gov.*—Silk stuff (Djeedjim).

386. ANDRONIKOF, PRINCE S. *Bakoortzyk Vil. Signah Circ. Tiflis Gov.*—Raw silk and cocoons.

387. ATAKEESHEE-OGLOO, *Khachmaz Vil. Nookha Circ. Bakoo Gov.*—Raw silk.

388. BEGLIAROF, *Akoolissy Vil. Ordoobat Circ. Erivan Gov.*—Samples of raw silk.

389. CAUCASIAN AGRICULTURAL SOCIETY, *Tiflis.*—Silk stuff (móv) and blankets.

390. CLEMENTZ, LÖH, & Co. *Quellenstein, Livonia.*—Samples of silk.

391. DAVYDOFF, *Varagirt Vil. Ordoobat Circ. Erivan Gov.*—Raw silk and cocoons.

392. DEUTSCHMAN, A. *Tiflis.*—Raw silk.

393. FET-ALI-OGLOO, *Rostadir Vil. Shemacha Circ. Bakoo Gov.*—Raw silk.

394. KOMAROVSKY, COUNTESS MARIA, *Prilooki Circ. Poltava Gov.*—Raw silk.

395. KRIPNER, P. G.—Raw silk, cocoons.

396. MAHMUD-ALI-YUSSOOF-OGLOO, *Buck Degniz Vil. Nookha Circ. Bakoo Gov.*—Silk stuff, called "Djajeem."

397. MARTHI-BOODOOGHIA-SHVEELEE, *Kvareli Vil. Telav. Circ. Tiflis Gov.*—Raw silk.

398. MASHADI-ALI-ZARAB-OGLOO, *Shoosha, Bakoo Gov.*—Silk stuff, called "Kassabec."

399. MESHADI-HUSSEIN-OGLOO, *Shoosha, Bakoo Gov.*—Silk stuff, called "Aleeshee."

400. MORTIEROSSOF, M. *Nookha, Bakoo Gov.*—Raw silk and cocoons.

401. MOSCOW COMMERCIAL SCHOOL.—Cocoons and raw silk from common silk-

worms hatched, fed, and reared in the school-garden at Moscow, on white mulberry leaves.

402. NASSIR-ABDOOLARTZIZ-OGLOO, *Bakoo Gov.*—Red and black móv (silk stuff).

403. NISSEN, A. *near St. Petersburg.*—Silk stuffs, raw silk, and organzine spun from cocoons produced in the south governments of Russia.

404. PETAÏL-OGLOO-MULLAH, *Zuzzid Vil. Nookha Circ. Bakoo Gov.*—Raw silk.

405. STROOKOF, P. *Ekaterinslov Gov.*—Raw silk.

406. REDA-OGLOO, *Zengishali Vil. Koobinsk, Circ. Bakoo Gov.*—Raw silk.

407. SAPOJNIKOF BROS. *Astrakhan.*—Raw silk and cocoons.

408. SEMENOF, *Vartaly Vil. Nookha Circ. Bakoo Gov.*—Raw silk and cocoons.

409. SHOROEF, *Avak-Shemakha, Bakoo Gov.*—Blue móv (silk stuff).

410. VORONIN, BROS. & ALEXEYEF, B. *Nookha, Bakoo Gov.*—Raw silk and cocoons.

411. YOORÏEVA, MISS L. *Wladimirovka Village, Stavropol Gov.*—Raw silk.

CLASS XXI.

412. THE ADMINISTRATION OF THE OREN-BURG KOZAK SETTLEMENT.—Woollen yarn, cloths from camel's hair.

413. ALI-PANAH-OGLOO, *Dash-Sala-Ogloo Vil. Elisabethopol Circ. Tiflis Gov.*—Mafrash and palas.

414. ARMAND, E. *Moscow.*—Plain and figured Orleans, Paramata satin, cotton warp, and Cashmere, pure worsted.

415. BABKIN, BROS. HEIRESSES OF, *Koopavna, Bogorodsk Circ. Moscow Gov.*—Broadcloth, called "Mezeritski."

416. BAKHROOSHINA, N. S. & SONS, *Moscow.*—Broadcloth, black and blue.

417. CAUCASIAN AGRICULTURAL SOCIETY, *Tiflis.*—Ropes of woollen yarn.

418. CLEMENTZ, LÖH, & CO. *Quellenstein, Livonia.*—Woollen yarns and cloth.

419. COMMANDER OF BASHKIR MILITIA, *Orenburg Government.*—Bashkir cloth.

420. FIELDER, G. A. *Opatovka, Kalish Circ. Warsaw Gov.*—Broadcloths.

421. GOOCHKOF, E. & SONS, *Moscow.*—Plain worsted fabric.

422. HADJI-YOOSOF-SHABAN-OGLOO, *Zakatala Circ.*—Common Lesghin cloths, and shawl.

423. BAERG, H. *Halbstadt Settlement, Beridansk Circ. Tauride Gov.*—Cloth and flannel.

424. IOKISH, B. *Michalkof Manufactory, Moscow Circ.*—Cloth.

425. KOOBAREF, M. *Klintzy Borough, Soorash Circ. Tchernigof Gov.*—Broadcloth, black and grey.

426. KOPALIANTZ, N. *Alexandropol, Erivan Gov.*—Samples of woollen yarn (dyed).

427. MAHOM-AKBAR-OGLOO, *Zakatala Circ.*—Common Lesghin cloth, and shawl.

428. NAVRUS, G. *Gorskee Circ.*—Lesghin cloth shawl.

429. OORALIAN KOZAKS, *Orenburg Gov.*—Common cloth.

430. SCHEFER, J. *Goly Karaslish Settlement, Saratof Gov.*—Half-silk kerchief.

431. SCHEPELER, T. *Riga.*—Mixed fabrics.

432. SELIVERSTOF, COLONEL N. *Roomianstof Factory, Korsun Circ. Simbirsk Gov.*—Samples of cloth.

433. SOOLHANOF, *Goree, Tiflis Gov.*—Common cloths and tiftik; pouch and palas, with felted lining.

434. THILO, A. *Riga.*—Cloth.

435. TURPEN, E. *Nikolaevsky Worsted Manufactory, St. Petersburg.* — Wool and worsted.

436. UNGERN, BARON STENBERG, *Isle of Essel.*—Cloth.

437. VADYM-AHVERDI-OGLOO, *Arab Redine Vil. Shemakha Circ. Bakoo Gov.*—Palas.

438. VONSOVSKY, *Zciva Vil. Etchmiadzin Circ. Erivan Gov.*—Cloth of Koortinsk wool.

439. WERGAU, J. *Lodz, Lenczic Circ. Warsaw Gov.*—Woollen table covers.

440. WOEHRMANN & SON, *Zintenkhoff Factory, near Pernau, Livonia.*—Cloths.

CLASS XXII.

441. FLANDIN & Co. *Klin Circ. Moscow Gov.*—Carpets.

442. HASSAN-ALI-DJEBRAEL-OGLOO, *Sharbachee Vil. Shemakha Circ. Bakoo Gov.*—Carpet.

443. MUSTAPHA-HADJI-OGLOO, *Oodoola Vil. Shemakha Circ. Bakoo Gov.*—Carpet.

444. MUSTAPHA - KARA - MIRZA - OGLOO, *Ymam-Koolee Vil. Kooba Circ. Bakoo Gov.*—Carpet.

445. NOOR-ALI-FET-ALI-OGLOO, *Sharbachee Vil. Shemakha Circ. Bakoo Gov.*—Carpet.

446. TSHOOKMALDIN, N. *Tiumen Circ. Tobolsk Gov.*—A carpet, specimen of peasant women's work.

CLASS XXIII.

447. ABKHAZOF, PRINCE D. *Kakhetia.*—Silk, dyed with safflower.

448. ADAM, CH. *Th. Bittepage Manufactory, Shlisselburg Circ. St. Petersburg Gov.*—Chintzes.

449. STROGANOF SCHOOL OF TECHNICAL DRAWING, *Moscow.*—Napkins, horsecloth, and a carpet.

450. BARANOVA, ALEXANDRA, *Troitzko-Alexandrov Manufactory, Alexandrovsk Circ. Vladimir Gov.*—Dyed and printed cotton goods.

451. GOOTCHKOF, E. SONS OF, *Moscow.*—Woollen, cotton, and mixed stuffs.

452. GOOTCHKOF, J. *Moscow.*—Dyed and printed woollen, silk, and cotton mixed stuffs.

453. HUBNER, A. *Moscow.*—Chintzes.

454. PROKHOROF, BROS. *Moscow.*—Cotton print.

455. REZANOF, T. *Moscow.*—Printed cotton kerchiefs.

456. TZINDEL, E. *Moscow.*—Chintz.

457. ZOOBKOF, P. *Vosnecensk Borough, Shooya Circ. Vladimir Gov.*—Chintzes.

CLASS XXIV.

458. THE ADMINISTRATION OF THE ORENBURG KOZAK SETTLEMENTS. — Embroidered linen towels, towel trimmings, and laces.

459. ALMAZOF, S. *Torjok, Tver Gov.*—Articles of embroidered Morocco leather.

460. BAHCHINOF, G. *Alexandropol, Erivan Gov.*—Embroidered slippers.

461. CAUCASIAN AGRICULTURAL SOCIETY, *Tiflis.*—Embroidered table-cloth and arm-chair coverings.

462. DIATCHKOVA, P. *Torjok, Tver Gov.*—Girdles, sashes, caps, and aprons.

463. DUTACQ, MRS. M. L. *St. Petersburg.*—Crotchet-work carpet, both sides alike.

464. KARELIENA, *Torjok, Tver Gov.*—Embroidered boot and shoe fronts ; caps and cushions in morocco leather; velvet, satin, and leather patchwork.

465. KISSELEVSKY, CATHERINE. — Embroidered cushion.

466. KRZYWICKA, MARY, *Warsaw.*—Bed coverings and small table-covers in worsted work.

467. PROHASKA & STENZEL, *St. Petersburg.*—Table-cover of cloth patchwork.

468. SAPOJNIKOF, HEIRS OF, *Moscow.*—Gold and silver brocades.

469. TER-POGOSSOF, *Nookha, Bakoo Gov.*—Embroidered caparison.

470. WITKOWSKA, NATALIE, *Warsaw.*—A carpet in worsted work.

471. WORONCOW-WELIAMINOV, *Nepomucena, Warsaw.*—A carpet in worsted work.

CLASS XXV.

Sub-Class A.

472. AGRICULTURAL DEPARTMENT, *St. Petersburg.*—Collection of sheep, lamb, and goat skins, raw and tanned.

473. ALIBERT, N. P. *Irkootsk Gov. Siberia.*—Stuffed sables.

474. ARCHANGEL CHAMBER OF CROWN DOMAINS.—Dressed skins of wolf, fox, and swan.

475. BOOKIN, J. *Bolshoe - Moorashkino, Kniaghinin Circ. Nijni-Novgorod Gov.*—Black lamb and sheep skins.

476. DITZEL, N. *St. Petersburg.*—Carpet made from seal skins.

477. GOLOF, T. *Mezen Circ. Archangel Gov.*—Dressed bear skin.

478. OORALIAN KOZAKS, *Orenburg Gov.*—Furs and skins.

479. PHILIPOF, H. *Mezen Circ. Archangel Gov.*—Undressed blue fox skins.

480. POPOF, M. *Mezen Circ. Archangel Gov.*—Dressed skin of reindeer.

481. PRESNIAKOF, P. *Bolshoe-Moorashkino, Kniaghinin Circ. Nijni-Novgorod Gov.*—White and black lamb skins.

482. ROOSHNIKOF, J. *Mezen Circ. Archangel Gov.*—Blue fox skins, undressed.

483. ROTCHEF, I. *Mezen Circ. Archangel Gov.*—Young reindeer skins, dressed.

484. ROTCHEF, J. *Mezen Circ. Archangel Gov.*—Dressed skins of reindeer, and black and blue foxes.

485. RUSSIAN-AMERICAN Co. *St.Petersburg.*—Skins of sea-otter, fox, and seal.

486. SHRAPLAU, H. *Moscow.*—A carpet, and a morning gown made of fur.

487. TYRKASOF, N. *Mezen Circ. Archangel Gov.*—Dressed otter skins.

488. YERMILOF, J. *Ostasheva Village, Romano-Borisoglebsk Circ.Jaroslaf Gov.*—Tanned sheep-skin coats, called "Romanoffsky."

Sub-Class B.

489. ARTEYEF, S. *Mezen Circ. Archangelsk Gov.*—Eider down.

490. MIRONOF, G. *Troobino Village, Maloiaroslavez Circ. Kalooga Gov.*—Goose down, feathers, and wings.

Sub-Class C.

491. BEZROOKAVNIKOF-SOKOLOF, A. S. *St. Petersburg.* — Samples of horse-hair, raw, cleaned, and curled.

492. BRÄUTIGAM, MARY, *Hair-cloth and Curled-hair Manufactory, St. Petersburg.*—Samples of horse-hair.

493. KONDRATENKO, P. *St. Petersburg.*—Stage wigs.

CLASS XXVI.

Sub-Class A.

494. ARCHANGELSK CHAMBER OF CROWN DOMAINS.—White chamois dressed reindeer hides.

495. ARTEYEF, S. *Merzen Circ. Archangel Gov.*—Yellow chamois-dressed reindeer hides.

496. BAHROOSHINA, N. & SONS, *Moscow.*—Boot, enamelled, and morocco leather.

497. BAUERFEIND T. F. *Warsaw.*—Leather.

498. BROOSNITZIN, N. *St. Petersburg.*—Assortment of leather.

499. GORIACHKIN, G. *Klintzy Borough, Sooraj Circ. Tchjernigof Gov.*—Calf and morocco leather.

500. HEINRICH, *Riga.*—Leather.

501. HUBNER, N. *St. Petersburg.*—Leather and tanned skins.

502. KELBI - KHAN - OGLOO, *Nakhichevan, Erivan Gov.*—Deer-skin.

503. KHECHATOOROF, S. *Signah, Tiflis Gov.*—Sheep leather.

504. LIEDTKE, J. H. *Warsaw.*—Samples of leather.

505. MILLER, ERDMAN, *St. Petersburg.*—Boot-legs and vamps.

506. MESNIKOF, *Stavropol.* — Oxen and buffalo leather.

507. MILLER, C. *St. Petersburg.*—Boot-legs and vamps.

508. MILLER, J. T. *St. Petersburg.*—Boot-legs and vamps.

509. POPOF, M. *Mezen Circ. Archangelsk Gov.*—Chamois-dressed reindeer hides.

510. SHEVNIN, A. *St. Petersburg.*—Samples of leather.

511. SHOOVALOF, P. & T. BROS. Manufacturers, *Moscow.*—Boot leather.

512. TEMLER, C. & A. & SZWEDE, L. *Warsaw.*—Sole, saddle, and enamelled leather, calf and morocco leather, strap leather.

513. VASSALIEF, A. *Moscow.*—Samples of parchment.

514. SKOOBEEF TANNING Co. *Alexandria Circ. Kherson Gov.*—Leather.

515. YEREMEYEF, T. *Perm.* — Russian, leather (yooft), calf and morocco leather.

516. ZAREENOF, M. *Akhalzik, Kootais Gov.*—Dyed kid leather.

Sub-Class B.

517. CAUCASIAN AGRICULTURAL SOCIETY, *Tiflis.*—Kabardian saddle, bridle, and whip.

518. TSHOOKMALDIN, N. *Tiumen Circ. Tobolsk Gov.*—A common bridle.

519. KIRGHIZ DISTRICT ADMINISTRATION, *Orenberg Gov.*—Kirghiz saddlery.

CLASS XXVII.

Sub-Class A.

520. ALEXANDRO, A. *Kootais.*—Imeritian caps (papaneeki).

521. CAUCASIAN AGRICULTURAL SOCIETY, *Tiflis.*—National fur cap (papakha.)

522. DADASH-MOLLA-HOOSSEIN-OGLOO, *Shoosha.*—Persian cap.

523. KHECHATOOROF, A. *Signah, Tiflis Gov.*—Tooshinsk felted caps.

524. KIRGHIZ DISTRICT ADMINISTRATION, *Orenburg Gov.*—Kirghiz caps.

525. RODIONOF, T. *Novostarinska Village, Kniaghinin Circ. Nijni-Novgorod Gov.*—Lamb-skin caps.

526. ROTCHEF, I. *Mezen Circ. Archangelsk Gov.*—Samoyed woman's cap of dressed skins.

527. SOODAKOF, S. *Moscow.*—Hats.

528. SYROF, N. *Zaproodna Village, Kniaghinin Circ. Nijni-Novgorod Gov.*—Bockharian white lamb-skin cap.

529. ZIMMERMAN, F. *St. Petersburg.*—Hats, caskets, and caps.

Sub-Class B.

530. FLEROVSKY, D. *Tomsk, Siberia.*—A woman's head-dress (kokoshnik), made of birch-bark.

Sub-Class C.

513. ADMINISTRATION OF THE KOZAK SETTLEMENTS, *Orenburg* —Goat's-hair gloves, and worsted mittens.

532. AGRICULTURAL DEPARTMENT, *St. Petersburg.*—Fur clothes (malitzas) of the Zyrian and Samoyed.

533. ARTEYEF, J. *Mezen Circ. Archangel Gov.*—Samoyed coat of dressed reindeer skin.

534. AVASTERKOF, J. P. *Lookoyanof Circ. Nijni-Novgorod Gov.*—Woollen mittens.

535. CAUCASIAN AGRICULTURAL SOCIETY. —Caucasian dress and felt cloak.

536. CHANCERY OF THE GOVERNMENT GENERAL OF ORENBURG & SAMARA.—Hand-spun and knit goat's-hair scarfs.

537. CHILINCHAROF, M. *Akhalzik, Kootais Gov.*—Black wrapper, with gold and silver embroidery.

538. THE COMMANDER OF THE BASHKIR MILITIA, *Orenberg Gov.*—A belt with pouch.

539. KASSUM-YVOZBASH-ALI-OGLOO, *Leubaran Vil. Shoosha Circ. Bakoo Gov.*—Black shawl.

540. KIRGHIZ DISTRICT ADMINISTRATION, *Orenberg.*—Kirghiz belts.

541. MAHMET-KEEZI, WIDOW, *Sala-Ogly Vil. Elisabethopol Circ. Tiflis Gov.*—A pouch.

542. MALAVAL, O. *St. Petersburg.*—Specimens of gloves.

544. OORALIAN KOZAKS, *Orenberg Gov.*—Sashes.

545. PAJER, M. *Marsaw.*—Plaited corset, invented by the exhibitor.

546. ROTCHEF, I. *Mezen Circ. Archangel Gov.*—Samoyed coats of dressed skins.

547. STOLZMAN, A. *Warsaw.* — Ladies' sashes.

548. TSHOOKMALDIN, N. *Tiumen Circ. Tobolsk Gov.*—Mittens of sheep-skin.

549. YOOSBASHEF, A. *Khachmaz Vil. Nookha Circ. Bakoo Gov.*—Silk trousers.

Sub-Class D.

550. HUBNER, N. *St. Petersburg.*—Boots, shoes, and over-shoes ; boots made of plaited leather.

551. LERCH, W. *St. Petersburg.*—Boots and goloshes.

552. LOOJIN, A. *Korchef Circ. Tver Gov.*—Boots.

553. MASHADI-KALI-OGLOO, *Shoosha, Bakoo Gov.*—Shoes (koshi).

554. MOKRIAKOF, A. *Korchef Circ. Tver Gov.*—Boots.

556. ROTCHEF, I. *Mezen Circ. Archangel Gov.*—Samoyed boots and half-boots.

557. RUSSIAN - AMERICAN INDIA - RUBBER Co. *St. Petersburg.*—Caoutchouc goloshes and half-boots.

558. LAUBE, W. *St. Petersburg.*—Pair of boots.

559. SELIVERSTOF, P. *Pochinok, Nijni-Novgorod Gov.*—Fell boots.

560. SHIRMER, E. *Moscow.*—Boots.

561. SITNOF, P. *Ankino Village, Korchef Circ. Tver Gov.*—Boots and goloshes.

562. SOOLKHANOF, *Goree, Tiflis Gov.*—Worsted socks.

563. STOLAREF, A. *Korchef Circ. Tver Gov.*—Boots.

564. STOLAREF, S. *Korchef Circ. Tver Gov.*—Boots.

565. SVEDONTSEF, A. *Korchef Circ. Tver Gov.*—Boots.

566. TER-POGOSSOF, *Nookha, Bakoo Gov.*—Pachichi (boot fronts).

567. TONENKOF, J. *Baykovo Village, Look-oyanof Circ. Nijni-Novgorod Gov.*—Plaited bark shoes.

568. TSELIBEYEF, T. *St. Petersburg.*—Boots and shoes.

569. TSHOOKMALDIN, N. *Tiumen Circ. Tobolsk Gov.*—Peasants' shoes of sheep-skin.

570. VANEZOF, M. *Tiflis.*—Common and travelling boots.

571. VOKOOYEF, T. *Mezen Circ. Archangel Gov.*—Boots of reindeer skin.

572. YOOSBASHIANETZ, G. *Damboolak Vil. Nookha Circ. Bakoo Gov.*—Worsted socks.

CLASS XXVIII.

Sub-Class A.

573. EPSTEIN, J. *Sorewka, Gostyn Circ. Warsaw Gov.*—Samples of paper.

574. TROITZKO-KONDROVSKI PAPER FACTORY CO. *Medyn Circ. Kalooga Gov.*—Assortment of paper.

575. VARGOONIN, A. & P. BROS. *Nevski Paper Factory, St. Petersburg.*—Samples of paper.

Sub-Class B.

576. GERKE, A. *St. Petersburg.* — Red marking and printing ink.

577. KRAUSE, J. *Warsaw.*—Samples of sealing-wax.

Sub-Class C.

578. FAJANS, M. *Warsaw.*—Lithographs and chromo-lithographs.

579. KANTOR, A. *Warsaw.*—Samples of bookbinding.

580. LEMAN, T. *St. Petersburg.*—Samples of types and punches.

Sub-Class D.

581. HAAG, C. *St. Petersburg.*—Album.

CLASS XXIX.

Sub-Class A.

582. NAMANSKI, A. *St. Petersburg.*—Table for facilitating the demonstration of elementary arithmetical rules.

583. NOWOLECKY, A. *Warsaw.* — Illustrated historical alphabet, printed on glazed calico.

584. ROGOJSKI, J. *Kelce, Radom Gov.*—Two synoptic tables, facilitating the study of chemistry.

585. ZOLOTOF, B. *St. Petersburg.*—Set of his publications for elementary instruction.

Sub-Class B.

586. GOTLUND, *Helsingflors.*—Collection of mushrooms, dried by the exhibitor's peculiar process.

587. HAN, C. *Dorpat.*—Figures in national dresses.

588. HEISER, H. *St. Petersburg.* — Collection of ethnological and zoological models.

589. IMPERIAL MINING DEPARTMENT, *St. Petersburg.*—A standard set of Russian monies, weights, and measures.

590. OOSPENSKY, P. *St. Petersburgh.*—Collection of stuffed birds.

591. UNIVERSITY OF DORPAT. — Wax models of fruits, and parts of the human body.

CLASS XXX.

Sub-Class A.

592. FREIBERG, A. *Billiard Factory, St. Petersburg.* — Billiard, improved structure, patented.

593. PETERHOF STONE-POLISHING FACTORY, *near St. Petersburg.*—Inlaid tables, cupboard, chair, jardinière, &c.

Sub-Class B.

594. CAMUSET & Co. *St. Petersburg.*—Paper-hangings.

595. COMMANDER OF THE BASHKIR MILITIA, *Orenburg.*—Scoop and cup made from maple root.

596. FLEROVSKY, D. *Irkootsk, Siberia.*—Dish carved in birchwood (root and bark).

597. SALZMAN, J. *Warsaw.*—A wood carving, after a French bronze by MENE.

598. TOLSTINSKI, J. *St. Petersburg.*—Paper hangings.

599. VETTER, A. & Co. *Warsaw.*—Paper hangings.

CLASS XXXI.

Sub-Class A.

600. BOORAKOF, A. *Tver.*—Nails, hand-made.

601. KOROLEF, G. *Tarki Village, Gorbatof Circ. Nijni-Novgorod Gov.*—Padlocks.

602. NASONOF, N. *Oostug, Vologda Gov.*—A chest with secret locks.

603. RAÏVOLOVO WORKS, *near St. Petersburg, Finland.*—Stonecutter's and locksmith's tools, and locks.

604. SHEBAROF, J. *Tarki Village, Gorbatof Circ. Nijni-Novgorod Gov.*—Padlocks.

605. SEID - IBRAHIM - KOORTAY - OGLOO, *Shoomy Vil. Alooshta District, Crimea.*—Spades.

606. TER-KREEKOROF, D. *Shoosha.*—Iron bridle bit.

607. VARYPAEF, T. *Pavlovo, Gorbatof Circ. Nijni-Novgorod Gov.*—Locks.

608. VORONOF, A. *Ooliagii Vil. near Olonetz.*—Spades and drainage implements.

609. WOJNICKI, *Warsaw.*—Padlocks.

Sub-Class B.

610. CHOPIN, F. *St. Petersburg.*—Monument representing the Empress Catherine II.

611. HAYDOOKOF, A. *St. Petersburg.*—Specimens of tin and lead foil.

612. JONOVA, *Toola.*—Tea-urns.

613. KUMBERG, J. *St. Petersburg.*—Lamps and lanterns used in the Imperial navy.

614. LOMOF, E. J. *Toola.*—Tea-urns in tombac (yellow metal).

615. MORAND & Co. *St. Petersburg.*—A group in bronze—The overthrow of idols in Russia, in the tenth century; and the model of a monument to Admiral Lazaref; both after designs and moulds of Professor Pimenof.

616. ROCHEFORT, COUNTESS OLGA, *Perm Gov.*—Medallion of Peter the Great, bronze cast, chased, and gilt.

617. ROODAKOF, S. *Toola.*—Tea-urns and coffee-pot.

618. SAMGHIN, D. *Moscow.*—Church bells.

619. TCHERNIKOF, N. *Toola.*—Tea-urns and church bells.

CLASS XXXII.

Sub-Class B.

620. BAKANOF, M. *Reebino Vil. Gorbatof Circ. Nijni-Novgorod Gov.*—Penknives.

621. DOORACHKIN, M. *Vorsma Village, Gorbatof Circ. Nijni-Novgorod Gov.*—Edge tools.

622. IVANOF, J. *Tiumen Circ. Tobolsk Gov.*—A mortice chisel.

623. JARKOF, P. *Reebino Vil. Gorbatof Circ. Nijni-Novgorod Gov.*—Penknives.

624. KIREELOF, B. *Reebino Vil. Gorbatof Circ. Nijni-Novgorod Gov.*—Pocket and pen knives.

625. KIREELOF, M. *Reebino Vil. Gorbatof Circ. Nijni-Novgorod Gov.*—Penknives.

626. KIRGHIZ DISTRICT ADMINISTRATION, *Gov. of Orenburg.*—Kirghiz knives.

627. LEPESHKING, J. *Martova Vil. Gorbatof Circ. Nijni-Novgorod Gov.*—Penknives.

628. LEVIN, T. *Dolotkovo Village, Gorbatof Circ. Nijni-Novgorod Gov.*—Cutlery.

629. OSMAN, OMER-OGLOO, *Goorsoof Vil. Alooshta District, Crimea.*—Axes.

630. MYSHIN, T. *Pavlovo Village, Gorbatof Circ. Nijni-Novgorod Gov.*—Scissors.

631. REESEF, J. *Boolatnikova Vil. Gorbatof Circ. Nijni-Novgorod Gov.*—Penknives and files.

632. VARIPAEF, T. *Pavlovo Gorbatof Circ. Nijni-Novgorod Gov.*—Cutlery and edge tools.

633. VOROTILOF, A. *Pavlovo, Nijni-Novgorod Gov.*—Table and pen knives.

634. ZAVIALOF, BROS. *Vorsma Village, Gorbatof Circ. Nijni-Novgorod Gov.*—Cutlery, files, &c.

CLASS XXXIII.

635. ALEXEEF, V. *Moscow.*—Gold and silver thread, wire, and spangles.

636. ARAPET-AGADJANOF, *Shoosha.*—Silver buckle for a belt.

637. BEK-TARKHANOF, J. *Shoosha.*—Silver buttons worn by Tartar women.

638. BELIBEKOF, M. *Tiflis.*—Enamelled silver cup, tray, and koolas.

639. BOIANOFSKY, C. *St. Petersburg.*—Silver vases, basket, and ostensorium.

640. GALAMKAROF, K. *Akhaltzyk.*—Box in filigree.

641. GOOBKIN, S. *Moscow.*—Works in silver for ecclesiastical and for household use.

642. MORAND & Co. *St. Petersburg Foundry and Galvano-Plastic Establishment.*—A

SAZIKOFF, J. P. *Moscow and St. Petersburg.*   Agents in London, Messrs. A. SKWARCOW
& Co. 24, *Leadenhall Street, E.C.*

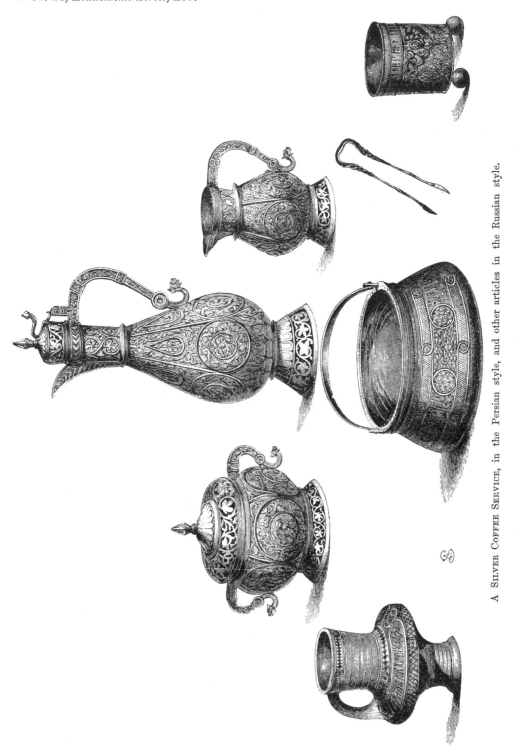

A SILVER COFFEE SERVICE, in the Persian style, and other articles in the Russian style.

In the above engraving are reproduced several pieces of plate for domestic use ; the coffee service is in the Persian style, the other articles are essentially Russian, and remind one by their shape of the **primitive simplicity** of the ancients.

SAZIKOFF, J. P. *continued.*

Articles manufactured for the Church of his IMPERIAL HIGHNESS THE GRAND DUKE MICHAEL OF RUSSIA.

Manufactured and exhibited by J. P. SAZIKOFF, of Moscow and St. Petersburg.

Agents in London : MESSRS. A. SKWARCOW & Co. 24, Leadenhall Street, E.C.

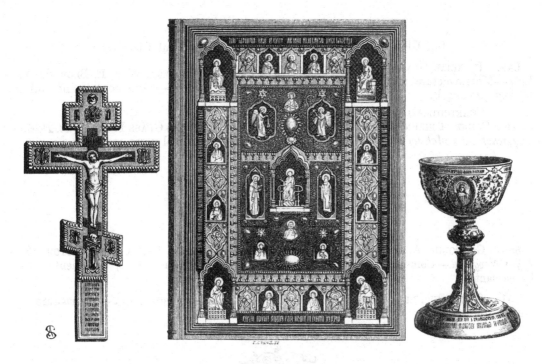

CROSS, CHALICE, and BIBLE used in the celebration of the Greek Ritual, made by order of His Imperial Highness the Grand Duke Michael for use in his private chapel.

These articles, represented in the annexed engraving, have been justly admired. Executed in the Byzantine style, and richly ornamented with precious stones, they are conspicuous by the admirable finish of the workmanship, which challenges criticism.

A clock offered as a testimonial to His Excellency Count Mouravieff Amoursky, is a beautifully designed piece of art and of excellent workmanship. On a pedestal which serves as a case for the clockwork, reposes an allegorical figure of the river Amoor. On the four sides of the pedestal are fixed medallions representing four phases of the expedition which was crowned by the acquisition to Russia of an extensive and fertile province that of the Amoor. At the foot of the pedestal are two statues, one representing a Russian soldier and the other a Chinese, exchanging gifts. The execution of this group is remarkable by the truth of the costumes and the artistic execution of the details. On the dial is the date of the treaty concluded by the Count with the Chinese, and the twelve letters employed in reproducing the date, form the twelve hours of the day.

M. Sazikoff likewise exhibits a quantity of statuettes reproducing types of Russian character ; all are essentially artistic and elegant in their design.

The best proof of the merit of M. Sazikoff's productions lies in the fact that purchasers eagerly contested for the possession of them immediately they were exhibited, the several articles being very suitable for presents and souvenirs of the Exhibition. They are very moderate in price.

group, representing St. George, by Professor Pimenof.

643. THE REV. E. POPOV, 32, *Welbeck-st.* —The New Testament in chaste silver-gilt binding ; a crucifix in chaste silver-gilt, made by Sazikoff, Moscow.

644. SAZIKOF, V. *St. Petersburg & Moscow.* —Ornamental articles in silver, cups, goblets, &c. (*See pages* 20 *and* 21.)

645. VERKHOVZEF, T.—Silver plates for a cabinet, with portraits of Tzars.

CLASS XXXIV.

Sub-Class A.

646. IMPERIAL GLASS WORKS, *St. Petersburg.*—Mosaic pictures representing St. Nicholas and two angels.

647. TCHETCHER GLASS WORKS (belonging to the COUNT TCHERNISHOF—KROOGLIKOF), *Rogatchef Circ. Mohilef Gov.*—Window glass.

Sub-Class B.

648. HORDLICZKA, W. & E. BROS. *Czechy, Gov. of Lublin.* — Ornamental and other glass.

649. IMPERIAL GLASS FACTORY, *St. Petersburg.*—Ornamental vases and glass.

CLASS XXXV.

650. GOLDOBIN, A. S. *Bov Village, near Nijni-Novgorod.*—Common earthenware used by peasants.

651. GRECHISHCHEF, M. *near Riazan.*—Glazed tiles.

652. IMPERIAL CHINA MANUFACTORY, *St. Petersburg.*—China figures, ornaments, and tea services.

653. KOOZNETSOF, *Riga.*—China-ware.

CLASS XXXVI.

Sub-Class B.

654. KADALOF, T. *Agrapino Village, Lookaianof Circ. Nijni-Novgorod Gov.*—Plaited bark pouch.

655. KEPHAR - MASHADI - DJAFAR - OGLOO,

Oodoola Vil. Shemakha Gov.—A travelling pouch.

656. NISSEN, W. *St. Petersburg.*—Portmanteaus, travelling bags, and pouches.

657. STOLZMAN, A. *Warsaw.*—Leather cases and trunks ; basket, portfolio.

ADDENDA.

CLASS I.

658. DMITRIEF, N. *near Odessa.*—South Russian prairie soil.

CLASS II.

Sub-Class A.

659. HESEN, A. *Moscow.*—Specimens of phosphoric matches.

660. PITANCIER, G. & Co. *Odessa.*—Sulphuric and hydrochloric acids, soda, copperas, and other chemical products.

CLASS III.

Sub-Class A.

661. ARGOOTINSKY-DOLGOROOKOF, PRINCE N. *Tiraspol Circ. Cherson Gov.*—Winter wheat and maize.

661A. ABBEY, W. *near St. Petersburg.*—A truss of Timothy grass.

662. BENEDSKI, A. *Katerinovka Vil. near Odessa.*—Winter wheat.

663. BONELLIS, W. *Schoensee Settlement, Berdiansk Circ. Tauride Gov.*—Samples of wheat flour.

664. IVASHCHENKO, N. *Odessa.*—Wheat, rye, and French beans.

665. KOORT, SEID-HUSSEIN-OGLOO, *Derekoi Vil. Alooshta District, Crimea.*—Nuts.

666.—LOBODA, P. *Pavlovsk Vil. Alexandrovsk Circ. Ekaterinoslaf Gov.* — Wheat (Arnaootka).

667. ROMANDIN, *Kishenef, Bessarabia.*—French beans, maize.

668. SEID-OMER-ABDOOL-OGLOO, *Derekoi Vil. Alooshta District, Crimea.*—Nuts.

669. SEMENIUTA, ARTEMIUS, *Goolaïpole, Alexandrovsk Circ. Ekaterinoslaf Gov.*—Linseed.

670. SHABELSKI, C. *Rostof Circ. Ekaterinoslaf Gov.*—Wheat and wild rape seed.

671. STROONNIKOF, T. & V. BROS. *Gov. of Tver and St. Petersburg.*—Wheat-flour and semoulia.

672. TRITHEN, O. *Odessa.*—Samples of wheat, maize, rye, and wild rape seed.

673. VINEIEF, *Nikolaevka Vil. Bobrynetz Circ. Cherson Gov.*—Winter wheat, wheat-flour, and seed of Triticum repens.

674. VOLKOF, T. *Jerebetz Vil. Alexandrovsk Circ. Ekaterinoslaf Gov.*—Ghirka wheat.

675. YOORIN, A. *Konstantinofka Vil. Novobaiazet Circ. Erivan Gov.*—Oats.

Sub-Class B.

676. DENGINK, A. MANAGER OF THE KISHENEF HORTICULTURAL SCHOOL, *Bessarabia.*—Dried plums, pears, and apples.

677. SHABELSKI, C. *Rastof Circ. Ekaterinoslaf Gov.*—Isinglass.

678. WIEBE, P. *Youshanly Freehold, Berdiansk Circ. Tauride Gov.*—Ham, cheese, dried apples, pears, and plums.

Sub-Class C.

679. CORAY, TN. & TB. *Ackerman Circ. Bessarabia.*—Red and white wine.

680. DECHESKOOL, V. *Galegit Vil. Orgey Circ. Bessarabia.*—Samples of tobacco.

681. DENGINK, A. *Manager of the Kishenef Horticultural School, Bessarabia.* — Tobacco.

682. FOONDOOCKLEY, J. *Goorsoof Estate, Southern Crimea.*—Wine.

683. MAGARATCH MODEL VINEYARDS, *near Ialta, Southern Crimea.*—Wines.

684. MALMBERG, C. *Moscow.*—Cigars and cigarettes.

685. MATVEÏEF, M. *Moscow.*—Snuff.

686. PHILIBERT, A. *Limena Vil. near Aloopka, Southern Crimea.*—Wines.

687. ROMANDIN, *Kishenef, Bessarabia.*— Red and white wine, and tobacco leaves.

688. ROTHE, R. *near Odessa.*—Wine.

689. SEID - OMER - ZMAIL - OGLOO, *Aootky Vil. Alooskta District, Crimea.*—Samson tobacco leaves.

690. TRITHEN, O. *Odessa.*—Red and white wine.

691. WOLFSCHMIDT, A. *Riga.*—Liqueurs and fine brandies.

CLASS IV.

Sub-Class A.

692. PITANCIER, G. & Co. *Odessa.*—Stearic and oleic acids, stearine candles and soap.

Sub-Class B.

693. SHABELSKI, C. *Rastof Circ. Ekaterinoslaf Gov.*—Goat's hair.

694. TRITHEN, O. *Odessa.*—Samples of washed wool from merinos, half-bred and common Crimea sheep.

695. WIEBE, P. *Youshanby Freehold, Berdiansk Circ. Tauride Gov.*—Washed merino fleeces.

696. YEAMES, W. & Co. *Rostof-on-the-Don,* and *Eysk, near Azof, Wool-washing Establishment.*—Donskoy, short and combing wools.

Sub-Class C.

697. ALABIN, K. *Peasant, Kotchkoorof Vil. Lookyanof Circ. Gov. of Nijni-Novgorod.*— Mats of lime-tree bast.

698. CORNIES, D. *Orlof Settlement, Berdiansk Circ. Tauride Gov.*—Baskets.

699. RIAZAN CROWN DOMAIN ADMINISTRATION.—Hemp and hempseed.

700. RIGA LOCAL COMMITTEE FOR THE INTERNATIONAL EXHIBITION.—Collection of woods; samples of flax and hemp.

701. SHABELSKY, *Port Caton Vil. Rastov Circ. Cherson Gov.*—Flax.

CLASS X.

702. RAÏVOLOVO WORKS, *near St. Petersburg, Finland.*—Military engineering tools.

CLASS XI.

Sub-Class A.

703. SYTOF, J. *St. Petersburg.*—Epaulets, shoulder-knots, &c.

704. TOOMASSOF, K. *Stavropol.*—Silver dagger and girdle.

CLASS XII.

Sub-Class C.

705. RaÏvalovo Works, *near St. Petersburg, Finland.*—Naval implements and tools.

CLASS XVII.

706. Shimanofsky, *Professor of the University of Kief.*—Surgical instruments of the exhibitor's invention.

CLASS XX.

707. Bephani, P. *Yeremovka Vil. near Novogeorgiefsk, Cherson Gov.*—Raw silk.

708. Dengink, A. *Manager of the Kishenef Horticultural School, Bessarabia.*—Raw silk and cocoons.

709. Hamm, M. *Orlof Settlement, Berdiansk Circ. Tauris Gov.*—Raw silk.

710. Tchorba, *Cherson Gov.*—Raw silk.

CLASS XXI.

711. Peltzer, G. *Moscow.*—Woollen stuffs for paletots, &c.

712. Sokolof, N. & M. Bros. *Moscow.*—Cloths.

713. Stepoonin, A. *Klintsee Borough, Tchernigof Gov.*—Cloths.

714. Vinogradof, A. *Nijni-Novgorod.*—Paletot and kerchiefs, made of goat's hair.

CLASS XXIV.

715. Sytof, J. *Moscow.*—Gold and silver cloth, thread, &c.

CLASS XXV.

Sub-Class A.

716. Sidorof, M. *Eastern Siberia, Tooroohansk Circ. Yenisseisk Gov.*—Furs and skins.

CLASS XXVI.

Sub-Class A.

717. Koossof, J. Sons of, *St. Petersburg.*—Leather and morocco.

718. Sidorof, M. *Eastern Siberia, Tooroohansk Circ. Ycnisscisk Gov.*—Leather made of deer-skin, plain and embroidered.

Sub-Class B.

719. Kerbelay Zoolal Ogloo, *Shoosha, Caucasus.*—Harness for Georgian saddle and bridle.

720. Lucks, C. *Halberstadt Settlement, Berdiansk Circ. Tauride Gov.*—Horse trappings, whips, &c.

CLASS XXVII.

Sub-Class A.

721. CHOORKIN, *Moscow.*—Hats.

722. KALLO, U. *Odessa.*—Cap adapted to different seasons.

723. TOOMASSOF, *Karapet, Stavropol.*—Fur cap (papakha).

CLASS XXVIII.

Sub-Class C.

724. PAULY, VON, *St. Petersburg.*—Ethnological essays, "Les peuples de la Russie."

CLASS XXX.

Sub-Class A.

725. MERCKLING, *Odessa.*—Walnut cabinet, modern style.

Sub-Class B.

726. GOETSCHY, T. *St. Petersburg.*—Paper hangings.

727. ABROSIMOF, *Great Okhta, St. Petersburg.*—A shrine carved in cedar wood.

CLASS XXXI.

Sub-Class B.

728. CLODT, BARON, *St. Petersburg.*—Bronze groups.

729. LIBERIH, *St. Petersburg.*—Hunting groups.

SIAM.

NORTH-EAST COURTS.

1. SCHOMBURGK, SIR R. H. *H.M. Consul at Bangkok.*—Silk petticoats, worn by the Lao and Burmese females, do. of a Karen, ornamented with Job's tears (seeds of Coix); cotton coverlets of Lao and Karens, travelling bags, cotton thread and spindle, fauteuils of carved teak-wood, collection of walking-sticks, samples of paper, native-made Lao cutlasses, and a collection of trade products ; medicines.

2. MARKWALD, A. & Co. *Bangkok.*—A collection of Siam products, consisting of silk, cotton, and other fibres, rice and pulse, resins, dye-stuffs, and tanning substances, sugars, coffee, woods, and animal products.

3. SIMMONDS, P. L. 8, *Winchester-st. London.*—Specimens of horns, feathers, edible birds'-nests, shark-fins, beche-de-mer, stick-lac.

4. HAMMOND, W. P.—A collection of ornamental articles.

5. BOWRING, SIR J.—Various articles, illustrating Siamese art.

SPAIN.

SOUTH-WEST COURT AND SOUTH-WEST GALLERY.

[The Numbers are those given in the Spanish Catalogue.]

CLASS I.

1. ABAD, V. *Nijar, Province of Almeria.*—Peroxide of manganese.

2. ABUEL, B. *Province of Camarines—Norte, Paracale, Philippine I.*—Loadstone.

3. ADMINISTRATION OF FINANCE, *Jaen.*—Salt, in the rough and in cakes.

4. ADMINISTRATION OF FINANCE OF SARAGOSSA, CHIEF OF, *Remolinos*—Rock-salt.

5. ALBIÑANA, J. F. *Tarragona.*—Marbles.

6. ALCALDE (MAYOR) OF CABRA.—Marbles.

7. ALCALDE OF OVIEDO.—Collection of marbles; sulphate of lime, solid and crystallised.

8. ALCALDE OF VALDES.—Sheet-iron.

9. ALCALDE OF YABUCOA, *Puerto-Rico.*—Granite.

10. ALVÁREZ, S. *Huelva.*—Sulphate of iron.

11. ARGÜELLES, V. *Piloña.*—White marble.

12. ARROYO AZNAR, M. *Valence.*—Calamine ore, from Benahada; sulphate and carbonate of lead, from Sierra de Gados; silver ore, from Sierra Almagrera; sulphate of antimony, from the district of Oria.

13. AZTIZ, M. R. *Aldar Navarre.*—Calamine.

14. ASTORGA, MARQUIS OF, *Madrid.*—Huelva marine salt.

15. ASTURIANA ROYAL COMPANY, *Arnao, Asturias.*—Specimens of coal, zinc in sheets, white zinc.

16. ASTURIANA UNITED SOCIETY, *Mieres.*—Calamine, quicksilver, cinnabar.

17. AURRE, G. *Siero, Asturias.*—Hydroxide of iron.

18. AYUNTAMIENTO (TOWN CORPORATION) OF BERNARDOS, *Segovia.*—Slates.

19. AYUNTAMIENTO OF CARTHAGENA.—Stones, gypsum.

20. AYUNTAMIENTO OF COLUNGAS, *Asturias.*—Breccia marbles.

21. AYUNTAMIENTO OF SIERO, *Asturias.*—Hydraulic cement, refractory clay.

22. AYUNTAMIENTO OF MORON.—Marbles, jasper, limestones.

23. AYUNTAMIENTO OF ORANLUNGA.—Oxide of iron.

24. AYUNTAMIENTO OF VELILLA, *Soria.*—Red and yellow ochre.

25. BALBAS CASTRO, J. *Magallanes, Philippine I.*—Copper pyrites with grey sulphurous copper on quartz and kaolin, sulphurous copper and black oxide, native and arsenical copper with antimonial grey copper, various minerals.

26. BALLESTEROS, S. *Oviedo.*—Bituminous slate.

27. BARRON, T. *Almeria.*—Sulphate and carbonate of copper.

28. BATIER, L. *Santander.*—Calamine, sulphate of iron, oxide of iron.

29. BERINERE & CO. *Santander.*—Calamine.

30. BLANCO, J. *Seville.*—Marbles.

31. BLANQUEZ, J. *No. 1, Calle de Luchana, Madrid.*—Magnesite.

32. BOIVIN JENTY & CO. *Maestu, Alava.*—Asphaltic rock, mastic and bituminous cakes, composed, the former of rock, pure, reduced to powder and boiled without mixture of any other asphalt. These mastics are employed for making superior black varnish, and are a substitute for Jews' asphaltum.

33. BOULAY, L. *Cabrales, Asturias.*—Copper ores.

34. BRAVO, J. M. *St. Eulalia de Oscos, Asturias.*—Sheet-iron.

35. BRIGUILER, C. *Lena, Asturias.*—Specimens of oxide and sulphate of antimony.

36. BROCK, F. *Linares, Jaen.*—Argentine galenas, lead produced in an English reverberatory furnace.

37. BURGOS, J. *Cordova.*—Copper ore from the mine Duquesa, coal from various mines.

38. BURGOS, J. *Almeria.*—Barytes.

39. CARRIAS, BLANCO, & CO. *Almeria.*—Carbonate and silicate of zinc.

40. CASALES, J. *Balaguer, Lerida.*—Iron.

41. CASTILLO, M. *Huelva.*—Marbles.

42. CASTRO, M. *Magallanes.*—Coal.

43. CIFUEGOS & CO. B. *Belmonte.*—Malleable iron in bars.

44. CILLERUELO, M. *Santander.*—Iron.

45. COLLANTES, A. *Madrid.*—Carbonate of copper and ferruginous copper pyrites.

46. COLLANTES, A. *Aller.*—Coal and coke.

47. COMPANY OF SPAIN GENERAL MINING.—Sulphate of lead.

48. COMPANY OF GUADALQUIVIR, *Seville.*—Huelva iron.

49. COMPANY OF GUADALQUIVIR, *Seville.*—Coal.

50. COMPANY ST. MICHAEL ARCHANGEL, *Madrid.*—Iron and copper pyrites, calcined ore, cementation copper, black ditto, granulation and fine ditto, from the province of Huelva.

51. CONCHA, J. M. DE LA, *Manilla.*—Specimen of mineral from Mount Camanchile, where it is found lying on the surface, containing 35 per cent. pure iron. Samples of cast and wrought iron obtained from said mineral.

52. CORDERO, E. M. *Constantina.*—Galena.

53. DAGUERRE DOSPITAL, *Seville.*—Copper pyrites.

54. DEPUTATION OF ALAVA, PROVINCIAL.—Common salt, alcohol.

55. DEPUTATION OF ALAVA, GENERAL.—Salt, lead ore, copper and blende, lignite, bituminous coal.

56. DUNCAN SHAW, *Cordova.*—Lead.

57. DIRECTION OF AGRICULTURE, INDUSTRY, AND TRADE, *Madrid.*—Topographical and geological map of the Province of Oviedo, by Dn. G. Schultz.
Geological maps of four provinces of Spain, by Dn. Casiano de Prado.

58. DUCLERC, E. *Madrid.*—Specimens of coal.

59. DURO & CO. *Langreo, Asturias.*—Coke, ferruginous sand, cast iron, malleable iron bars, round and square.

60. ESTABLISHMENT, NATIONAL, OF MINES OF ALMADEN.—Large samples of cinnabar, varying in richness.

61. ESTABLISHMENT OF RIO-TINTO, NATIONAL, *Huelva.*—Sixty specimens of cupreous iron pyrites, native copper, fine ditto, intermediate products in the conversion of copper ore into copper.

62. FACTORY (IRON WORKS) *The Angel, Marbella, Málaga.*—Magnetic iron ore.

63. FACTORY (SALT-WORKS) OF TORREVIEJA.—Common salt.

64. FACTORY (WORKS) OF ST. JUAN DE ALCARAZ, *Albacete.*—Metallurgical products.

65. FACTORY (WORKS) OF TRUBIA, NATIONAL.—Coals, coke, iron ore.

66. FALCONI, J. *Almeria.*—Carbonate of lead.

67. FERNÁNDEZ, V. *Mieres, Asturias.*—Coal and coke.

68. FIGUERAS, E. *Madrid.*—Coal and coke.

69. FIGUEROA, Y. *Jaen.*—Lead in bars, from St. Miguel mine.

70. FORCADA, A. *Lérida.*—Iron.

71. FORCADA, U. *Almatres, Lérida.*—Coal.

72. FOSSEY & Co. *St. Sebastian.*—Castings.

73. FRANCO, M. *Maceda, Soria.*—Asphalt, tar.

74. FRASSINELLI, R. *Cangas de Onis, Asturias.*—Red ochre, coal, peat, amber.

75. GALLEGO, M. *Palencia.*—Coal from Jóven-Ildephonso mine.

76. GIL & Co. *Langreo, Asturias.*—Iron.

77. GOVERNOR OF THE PROVINCE OF SEVILLE.—Peroxide of manganese.

78. GOVERNOR OF THE PROVINCE OF GUIPUZCOA.—Marl and cement from works of La Fé, raw and calcined marl and cement from those of Yraeta, cement from La Garrucha, marbles and minerals.

79. GOVERNOR OF THE PROVINCE OF MALAGA.—Minerals from the district of Ojen, sulphuret of nickel from the district of Carratraca, ferruginous oxide, Marbella magnetic iron, galena from Virgen del Pilar mine, district of Cutar and Sierra Narja ; bar of lead from Don Guillermo Strachan's Works, carbonate of lead from the plains of La Plata, carbonate of copper and serpentine of Coin.

80. GÓMEZ SALAZAR, Y. *Almeria.*—Minerals and products from the mines and quarries of Almeria, Sierras de Gados, Almagrera, Filabres, and Alhamilla, Cabo de Gata, Pechina, St. Lucar, Roquettes, Yeron, Somontin, Nacimiento, Nijar, Berja, Mojacar, Fijala, Lucainena, Oria, Macael, Rioja, Huercal-Overa and Benahada ; cast iron from La Garrucha Works.

81. GOROZABE, A. *Vitoria.*—Blende, copper.

82. GUILLEN CALOMARDE, G. *Huelva.*—Native sulphur.

83. GURRUCHAGA & Co. *Zumaya.*— Cement, finished and unground ; cement pure, and with mixture of sand ; raw and calcined stone ; plaster.

84. GUTIERREZ QUEVEDO, *Santander.*—Peat.

85. HARTLEY, ZAFRA, & Co. *Huelva.*—Cementation copper, natural and artificial ; sulphur in flowers and rolls, amianthus, pyrolusites, cupreous iron pyrites, galena, red jaspar.

86. HEIM, G. *Quiros.*—Coals of various sorts, iron ore, cast steel.

87. HENARES, CORREDOR, & Co. *Montoro, Cordova.*—Galena, lead.

88. HEREDIA'S SONS, M. *Malaga.*—Soft iron in sheets, iron in square blocks, fine iron for ornamental and building purposes, lead, white lead, litharge, silver in the rough.

89. HEREDIA'S IRON WORKS, *Malaga.*—Forged iron, copper nails and tacks, wire.

90. HERNANDEZ, L. *Seville.* — Cupreous iron pyrites.

91. HERRERO, J. *Vitoria.*—Asphalt from the mine Dolores, cement, plaster, raw and calcined stones.

92. HERRERO, T. *Espinar, Segovia.*—Crystallisations of quartz.

93. HUELIN, M. *Garrucha, Almeria.*—Sulphate of antimony.

94. ICETA, M. *San Sebastian.*—Cement.

95. IBARRA, M. *Valence.*—A show-case with minerals.

96. IBARRA & Co. *Bilbao.*—Refined iron of various sizes, made from metallic sponge ; malleable iron of different shapes and dimensions.

97. IBARRA & Co. *Biscay.*—Samples of minerals, metallic sponge, forged and in its natural state ; cast-iron ingots, refined iron bars made from metallic sponge, puddled iron ditto.

98. INSPECTOR OF MINES OF THE ALMERIA DISTRICT.—Earths, phosphates, porphyries, marbles.

99. INSPECTOR OF MINES OF THE BADAJOZ.—Copper, sulphate of lead.

100. INSPECTOR OF MINES OF THE BARCELONA DISTRICT.—Hydraulic and argillaceous limestones, pit-coal and coke from St. Juan de las Abadesas mines, pit-coal from Erill-Castell, iron ore, antimonial galena, blende, barytes, nickel, pyrites, lignites, rocksalt.

101. INSPECTOR OF MINES OF THE BURGOS DISTRICT.—Coal, coke, galena ; grey and other copper ; manganese, iron, sulphate of soda, tin.

102. INSPECTOR OF MINES OF THE CACERES DISTRICT.—Limestone.

103. INSPECTOR OF MINES OF THE CORDOVA DISTRICT.—Pit-coal, coke, iron, lead, lead in bars, galena, calamine, grey copper, marbles, mercury, cinnabar.

104. INSPECTOR OF MINES OF GALICIA.—Refractory clay, semi-pegmatite, quartz steatite, serpentines, marbles, lignites, cupreous iron pyrites, hydrated oxide of iron, oxide of manganese, oxide of tin, metallic tin.

105. Inspector of Mines of the Granada District.—Iron ore, copper, zinc, lead, marbles, nitre, salt.

106. Inspector of Mines of the Guadalajara District.—Silver ore, hydraulic cement, coke, iron ore, wrought iron, alabaster, salt.

107. Inspector of Mines of the Madrid District.—Kaolin, grey hæmatite, marbles, limestones, magnesite, cupreous pyrites, blende.

108. Inspector of Mines of the Murcia District.—Iron and lead ores, sulphur, alum, building-stone, marbles.

109. Inspector of Mines of the Province of Oviedo.—Pit coal, sulphuret of tin, copper, lead, iron, and manganese.

110. Inspector of Mines of the Province of Santander.—Salt, bituminous schist.

111. Inspector of Mines of the Valencia District.—Marbles, alabaster, lead, copper, salt, gypsum, sulphur.

112. Inspector of Mines of the Valladolid District.—Hydrated oxide of iron, coal, fire bricks, metallurgic products, iron, slates, marbles, grey copper, coke.

113. Inspector of Mines of the Viscay District.—Lignites, magnetic iron ore, siderose, calamine, blende, galena, cupreous pyrites, sulphate of soda, asphalt, iron, lead, marbles.

114. Inspector of Mines of the Saragossa District.—Grey copper, sulphate of soda, pit-coal, asphalt, marbles, manganese, colouring earths.

115. Iris Amarillo (*Mining Company*), *Cava-Alta, Madrid.*—Ochre, sulphuret of iron.

116. Junta (Board) of Agriculture, Industry, and Trade of Granada.—Coals.

117. Junta of Agriculture and Industry of Jaen.—Bone-white marbles, black, slight black, and honey-colour.

118. Junta of Agriculture, Industry, and Trade of Oviedo.—Hydraulic limestone, calcareous rocks, slates and marbles.

119. Lagigas, F. de las, *Province of Bulacan, Cupan and Macabuetum Mines, Philippine Islands.*—Oxidised magnetic iron, and the same with red oxide; iron; barytes and grey smelting.

120. Llana, J. *Castro Urdiales.*—Refractory earths and cements.

121. Llanos, R. *Vitoria.*—Coals.

122. Leuchtenberg, Grand Duchess of, *Russia—Siero and Langreo Mines.*—Coals, washed and unwashed, coke.

123. Manterola, Cortazar, & Co. *St. Sebastian.*—Hydraulic lime.

124. Massia, E. *Tortosa.*—Jasper.

125. Marron, V. M. *Valladolid.*—Tin ore.

126. Martinez, J. *Loria.*—Sulphur.

127. Mercia & Co. *Huelva.*—Fine and common copper in ingots, sulphur, argillaceous slate with a per-centage in native copper of 8½, samples of native copper, cupreous iron pyrites, fine manufactured copper.

128. Miguel, L. *Soria.*—Espejon marbles.

129. Mora, F. *Alicante.*—Marbles.

130. Morentin, F. M. *Allo, Navarre.*—Sulphate of soda.

131. Municipality of Luquillo, *Puerto-Rico.*—Gold-dust.

132. Noriega, F. *Onis, Asturias.*—Grey copper, carbonate of copper, malachite.

133. Oria, R. *Cangas de Tineo, Asturias.*—White marbles.

134. Orozco, G. *Almeria.*—Silicated carbonate of zinc, fluor-spar.

135. Orozco, R. *Almeria.*—Argentiferous lead and iron ores.

136. Orta, J. J. *Huelva.*—Red and yellow ochre.

137. Ortigosa, M. *Valverde del Camino.*—Cupreous iron pyrites.

138. Pedroso, Mining Company of, *Seville.*—Iron ore.

139. Pelayo, M. *Buiceres, Asturias.*—Coals.

140. Peña, M. *Espejo, Soria.*—Iron ore.

141. Pérez, B. *Soria.*—Argentiferous lead.

142. Pérez, E. *Almeria.*—Argentiferous lead ore.

143. Pérez, Cardenal A. *Zamora.*—Foundry earths.

144. Picazo, A. *Saragossa.*—Salt obtained by evaporation.

145. Porta, F. *Alicante.*—Mercury, copper.

146. Prieto & Acha, R. *Rivadesella, Asturias.*—Spar, crystallized.

147. PROVINCE OF NAVARRE.—Sulphate of soda and calamine.

148. PUESTA, F. *Seville.*—Pyrolusites.

149. REDONDO, J. *Pereruela-Zamora.*—Refractory earths.

150. RESTOY, A. *Fondon, Almeria.*—Lead ore, sulphurous.

151. REYES, J. *Huelva.*—Pyrolusites.

152. RICKEN, J. *Huelva.*—Pyrolusites.

153. RODRÍGUEZ, V. *Valencia.*—A show-case of polished marbles.

154. ROJAS BROTHERS, *Isle of Cebu, Philippine I.*—Coal, coke.

155. RUBIN, F. *Llanera, Asturias.*—Refractory *pudinga.*

156. RUIZ REYES, M. *Almeria.*—Nickel, cobalt, and cinnabar ores.

157. SANS, D. *Riara, Segovia.*—Earths, paving slates, grindstones.

158. SERGANT, E. *Huelva.*—Manganese.

158a. SERRAPIÑERA, A. *Havannah.*—Coal, iron, copper.

159. SERRANNO, C. *Corunna.*—Chromium.

160. SITCHÁ, J. *Valladolid.*—Gypsum.

161. SOCIETY OF SABUÑOS, BELGIAN, *Langreo, Asturias.*—Coals.

162. SOCIETY " CÁNDIDO CONDE & CO." *Saragossa.*—Rock-salt.

163. SOCIETY "THE LITTLE BLACKSMITH," *Seville.*—Cupreous iron pyrites, fine copper.

164. SOCIETY "EL PORVENIR DE ASTURIAS," *Madrid.*—Cinnabar, mercury, pit-coal.

165. SOCIETY OF BELMEZ AND ESPIEL, *Carboniferous and Metalliferous Fusion, Cordova.*—Coals, coke.

166. SOCIETY "ST. ANN," COAL MINING, *St. Martin del Rey Aurelio, Asturias.*—Coals.

167. SOCIETY "LA CONCEPCION," *Seville.*—Cupreous iron pyrites, fine Huelva copper.

168. SOCIETY "LA JUSTA," *Langreo, Asturias.*—Coals and coke.

169. SOCIETY "LA PODEROSA," *Seville.*—Cupreous iron pyrites, fine Huelva copper.

170. SOCIETY "LOS SANTOS," *Belmez Cordova.*—Pit-coal, coke, iron.

171. SOCIETY "LA VALIENTE," *Madrid.*—Sulphate of lead.

172. SOCIETY "THE UNITED MINES OF LAS GUARDAS CASTLE," *Seville.*—Cupreous iron pyrites, sub-sulphate of iron, copper, sulphur.

173. SOCIETY "N. Sra DE LOS REYES," *Seville.*—Cupreous iron pyrites, calcined ore, natural and artificial cementation copper.

174. SOCIETY "N. Sra DE LA SALUD," *Seville.*—Cupreous iron pyrites, and fine Huelva copper.

175. SOCIETY "PALACIOS DE GOLONDRINAS," *Caceres.*—Silver, zinc, lead, and copper ores.

176. SOCIETY, PROTECTOR, *Cienpozuelos, Madrid.*—Sulphate of soda, anhydrous, calcined, and as a hydrate; crystallised gypsum.

177. SOCIETY, SEVILLIAN, *Caceres.*—Silver, zinc, lead, and copper ores, from the Giralda mine.

178. SOCIETY "ST. TELMO," *Seville.*—Cupreous iron pyrites, cementation copper and fine Huelva ditto.

179. SOCIETY "UNION ASTURIANA," *Mieres, Asturias.*—Mercury.

180. SOCIETY "UNION DEL COMERCIO," *Seville.*—Specimens of ferruginous copper pyrites.

181. SOCIETY OF ASTURIAS, COAL AND METALLURGIC MINING, *Mieres.*—Coal, coke, iron, fibrous hæmatite, oxide of iron and red iron, cast and malleable iron, rolled hoops and inoxidable iron.

182. SOCIETY, MINING, "AMISTAD," *Alicante.*—Mercury.

183. SOCIETY OF ESTREMADURA, ARGENTIFEROUS MINING, *Caceres.*—Silver, zinc, lead, and copper ores.

184. SOCIETY, MINING, "CHISPA," *Abulense, Avila.*—Copper ore.

185. SOCIETY, MINING, "EL CONSUELO," *Chinchon, Madrid.*—Sulphate of soda, barilla, soda, fuller's earth.

186. SOCIETY, MINING, "FELIZ HALLAZGO," *Alicante.*—Iron.

187. SOCIETY, MINING, "FRATERNIDAD," *Saragossa.*—Common salt.

188. SOCIETY, MINING, "LA CAMPURREANA," *Santander.*—Copper ore.

189. SOCIETY, MINING, "LA LEALTAD," *Santander.*—Calamine.

190. SOCIETY, MINING, "PROVIDENCIA," *Santander.*—Calamine.

191. SOTO, M. *Orihuela.*—Pure copper for fusion, marbles, alabaster, gypsum.

192. TADEO, N. M. "*Province of Camarines—Norte, Pueblo de Mambalao, Philippine I.*—Samples of iron ore.

193. TEUREIRO, N. *Madrid.*—Gypsum.

194. TIVALLER DE VELART, J. M. *Barcelona.*—Iron.

195. TORRES MUÑOZ LUNA, R. *Madrid.*—Apatite, from Jumilla.

196. UNZUETA, T. *Villarrubia de Santiago, Toledo.*—Sulphate of soda, natural and calcined, gypsum.

197. VALLE, A. DEL, *Malaga.*—White and blue marble slabs.

198. VILLAFRANCA, M. DE, *Mazarron, Murcia.*—Alum.

198a. VILLAYTRE, MARQUIS OF, *Santiago de Cuba.*—Copper ores.

199. VIÑAS, T. *Lerida.*—Lead.

200. VRIE, R. *Cangas de Tineo, Asturias.*—White marble.

CLASS II.

201. AYUNTAMIENTO (TOWN CORPORATION) OF SARREANS, *Orense.*—Leeches.

202. AYUNTAMIENTO OF COCA, *Segovia.*—Spirits of turpentine, varnish, resin.

203. BACH, F. L. *Barcelona.*—Six jars with a vegetable panacea.

204. BALLESTEROS, S. *Oviedo.*—Oil extracted from bituminous slate.

205. BERRENGS, H. *Gracia, Barcelona.*—Mercurial products, red-wood lac, verdegris, morphia, minium.

206. BONNET, J. *Barcelona.*—Chemicals.

207. BOTT, E. *Oviedo.*—Pharmaceutical products.

208. CALLEJA, S. *Villaviciosa, Asturias.*—Pharmaceutical products.

209. CANALES, HEREDEROS DE, *Malaga.*—Essence of lemon, citric acid.

210. CASTELLÓ, N. *Areyns, Barcelona.*—Verdegris.

211. CARRASCOSA, E. *Ariza, Saragossa.*—Opium.

212. CARREÑO, J. *Seville.*—Stearine.

213. CISNEROS LANUZA, J. *Seville.*—Essential oil of orange peel.

214. CROS, J. T. *Barcelona.*—Acetate of alumina, ferruginous and calcareous pyrolusites, sulphate of alumina, acetate of lime, nitric and other acids, salts with basis of soda, iron, lead, &c.

214a. DE OCAN CATALAN, J.—Saffron.

215. FRIAS, P. J. *Balearic Islands.*—Lichen.

216. GARCIA DE VINUESA, J. *Seville.*—Artificial guano.

217. GOVERNOR OF ORENSE.—Bone black, potato fecula.

218. GÓMEZ SALAZAR, Y. *Velez-Rubio, Almeria.*—Essence of aniseed.

219. GONZALEZ ESQUIVEL, J. G. *Province of Manilla.*—Gogo, a vegetable substance, efficacious as a medicine, and a substitute for common washing soap.

220. GRAU, J. *Seville.*—Black and coloured ink, boot blacking, and preparation for removing stains.

221. INURRIA, J. *Seville.*—Liquorice.

222. ZOBEL, J. *Philippine I.*—Drugs.

223. MANJARRES, R. *Seville.*—Products extracted from the Logrosan phosphorite, ditto from algæ (sea-weeds) collected on the beach at St. Lucar de Barrameda.

224. MANZANO, L. B. *Salamanca.*—Chemicals.

225. MARQUES MATA, R. *Barcelona.*—Spanish opium, and a variety of pharmaceutical products.

226. MAYER & BARTRINA, 1 *Prado, Madrid.*—Lithographic and typographic inks, varnish for lithographs.

227. MENJÍBAR, J. *Bilbao.* — Artificial saltpetre.

228. NOEL VASSEROT & Co. *Seville.*— Liquorice.

229. PADRÓ, T. *Barcelona.*—Hair-dye of his own invention.

230. RIVAS, FRAY MANUEL, & MATEO RUSELL, *Procince of Cavite, Philippine I.*— Medicinal palissander.

231. ROYO, M. *Valencia.*—White and red lead.

232. ROBILLARD, F. *Valencia.*—Essences.

233. SALVÁ CARDELL, M. *Llurmayor, Baleares.*—Verdigris.

234. SIMON, T. *Madrid,* 3, *Caballero de Garcia.*—Zeiodetite, a newly discovered substitute for lead.

235. TAMARIT, MARQUIS OF, *Tarragona.* —Soda and barilla.

236. TOLOSA, R. *Madrid.*—Artificial barilla.

237. VILLANUEVA, G. *Province of Batangas, Philippine I.*—Itiban bark, suitable for making balsam.

238. VINUESA COUSINS, *Seville.* — Artificial guano.

CLASS III.

239. ADALID, J. *Seville.* — Barley, oil, wheat.

240. ADANERO, COUNT OF, *Caceres.* — Cheese.

241. AGREDA, J. A. *Jerez.*—Wines.

242. AGUADO MUÑOZ, F. *Madridanos. Zamora.*—Chick peas, soft; also hard, for seed.

243. AGUDO, F. *Cehegin, Murcia.*—Brandy.

244. AGUIRRE, T. *Corunna.*—Chocolate.

245. AGUIRRE, S. *Soria.*—Honey.

246. AGUIRRE & Co. *Manilla.*—Refined sugar.

247. AICART, V. *Valencia.*—Wine, vinegar, large pea-nuts from Peru, acclimatised in 1861 ; smooth-podded tares, brandy.

248. AIGE, A. *Lerida.*—Figs.

249. ALBARGONZALEZ, R. *Gijon.* — Preserves, cider.

250. ALBERIDE, J. *Tarragona.*—Common wine.

251. ALBERREACH, J. *Reus, Tarragona* —Oil.

252. ALBERT, J. *Valencia.*—Maize and wine.

253. ALCALÁ, B. DE, *Huesca.*—Wine.

254. ALCALÁ, B. *Cordoba.* — Oil, pomegranates.

255. ALCALDE (MAYOR) OF GRADO, *Oviedo.* —Chestnuts, nuts, walnuts, round beans, long beans.

256. ALCALDE OF BEMBIMBRE, *Leon.* — Chestnuts, nuts.

257. ALCALDE OF MORCIN, *Oviedo.*—Round beans.

258. ALCALDE OF VILLAGRANDE, *Leon.*— Acorns, beech-tree seed.

259. ALCALDE OF VILLASABARIEGO, *Leon.* —Short wheat.

260. ALEMANY, A. *Tortosa.*—Oil.

261. ALFARO SANDOVAL, J. *Albacete.* — Wheat.

262. ALFONSO, R. *Jativa.*—Starch.

263. ALFONSO, J. *Valencia.*—Oil.

264. ALGUE, P. *Tarragona.* — Brandy, ratafia.

265. ALMECH, E. *Saragossa.*—Oil.

266. ALOMAS, J. *Tarragona.*—Wine.

267. ALONSO, G. *Moraleja del Vinoso, Zamora.*—Common wines.

268. ALONSO, F. *Scville.*—Maize.

269. ALONSO DE PRADO, M. *Leon.*—Flour.

270. ALONSO DE PRIDA & Co. *Leon.*— Chocolate.

271. ALONSO, J. *Mojados, Valladolid.*— Madder, raw and in powder.

272. ALOS, J. A. BALAGUER, *Lerida.*—Wheat, hemp-seed.

273. ALVAREZ, J. & BROTHERS, *Havannah.*—Cigars.

274. ALVAREZ, CALLEJA S. *Villaviciosa.*—Apples.

275. ALVAREZ, G. *Ricote, Murcia.*—Common wine.

276. ALVAREZ, N. *Binetres, Tarragona.*—Soft and "esperanza" almonds.

277. ALZUETA, J. *Peralta, Navarre.*—Wine.

278. ALZUGARAY, WIDOW OF, & Co.—Flour.

279. AMORES, M. *Seville.*—Agricultural produce.

280. AMOROS, J. *Tarragona.*—Wine, nuts.

281. AMOROS, R. *Tarragona.*—"Rancio" wine.

282. ANDRES, B. *Rio Seco de Soria.*—Garlic.

283. ANEZCAR & Co. *Pamplona.*—Flour.

284. ANGLADA, T. *Tabucoa, Puerto-Rica.*—Sugar.

285. ANGUERA, C. *Reus, Tarragona.*—Common wines, dried fruits.

286. ANGUERA, A. *Reus Tarragona.*—Haricot beans, and sundry varieties of maize.

287. ANGUERA, R. *Tarragona.*—Superior wines.

288. ANTOLI, Y. *Valencia.*—Old wine.

289. ARANA, L. *Villanueva de Puente, Navarre.*—Wheat.

290. ARANDA, J. *Havannah.*—Cigars.

291. ARBONES, J. *Almatret, Lérida.*—Aniseed.

292. ARDERIUS, M. *Liñola, Lérida.*—Camomile.

293. ARÉVALO, J. *Matapozuelos, Valladolid.*—White wine.

294. ARIAS, M. *Saragossa.*—Flour.

295. ARIAS, S. *Zamora.*—Common wine.

296. ARMEL, JULIÁ, Y FERRER, *Lares, Puerto-Rico.*—Coffee.

297. ARMERO, A. *Seville.*—Beans, oil.

298. ARROU, L. *Llubi, Balearic Islands.*—Capers.

299. ARROYO, P. *Navalmoral, Toledo.*—Oil.

300. AUÑON, J. *Seville.*—Beans.

301. AURORA, LA, FLOUR MILLS, *Rio Seco.*—Flour.

302. AVEDILLO, Y. *Moralejo del Vino, Zamora.*—Common wine.

303. AVILA, C. *Lepe, Huelva.*—Figs.

304. AYALA, L. DE, *Guadalcanal, Seville.*—Wheat, barley, one-grained wheat, beans.

305. AYUNTAMIENTO (TOWN COUNCIL) OF GUADAIRA.—Agricultural produce.

306. AYUNTAMIENTO OF CARRION, *Palencia.*—Haricot beans, chick peas, linseed.

307. AYUNTAMIENTO OF COCA, *Segovia.*—Pine-nut kernels, with and without shells.

308. AYUNTAMIENTO OF GARROBILLAS, *Cáceres.*—Sausages.

309. AYAMANS, COUNT OF, *Llunmayor, Majorca.*—Brown wheat.

310. AYUNTAMIENTO OF MEDINA DEL CAMPO, *Valladolid.*—Wine, wheat, vetches.

311. AYUNTAMIENTO OF MONTANCHEZ, *Cáceres.*—Hams.

312. AYUNTAMIENTO OF MORON DE LA FRONTERA.—Oil, chick peas, olives.

313. AYUNTAMIENTO OF MULA, *Murcia.*—Wheat.

314. AYUNTAMIENTO OF RIAÑO, *Leon.*—Wild lentils.

315. AYUNTAMIENTO OF RIO SECO DE *Soria.*—Fine-grained wheat.

316. AYUNTAMIENTO OF VILLANUEVA DE SALMANCIO, *Valladolid.*—Wheat, barley.

317. AYUNTAMIENTO OF SANTA EULALIA, *Baleares.*—Almonds.

318. AYUNTAMIENTO OF SAN SERVERA, *Balearic Islands.*—Beans.

319. AYUNTAMIENTO OF SOTO DE SAN ESTEBAN, *Soria.*—Common barley, white wheat.

320. AYUNTAMIENTO OF TORDEHERMOSOS, *Valladolid.*—White wheat.

321. BAETA, M. *Saragossa.*—Lucerne seed, beans.

322. BAIGES, P. *Tarragona.*—Oils.

323. BAILLES, J. *Cordoba.*—Shumac.

324. BALLESTER, L. *Barcelona.*—Wines.

325. BALLESTER, D. J. *Lérida.*—Barley.

326. BALLESTEROS, P. *Saragossa.*—Oil.

327. BANCO, M. *Tafalla.*—Brandy.

328. BARCIÑO, T. *Tarragona.*—Oils.

329. BARENGS, T. *Tarragona.*—Common wine.

330. BARRERA, J. *Tudela.*—Oil.

331. BARRERA, L. *Almonte, Huelva.*—Wine.

332. BARROETA, L. H. *Biscay.*—Preserves.

333. BARTOMEU, J. *Reus.*—Almonds, oils.

334. BATELLAS, F. *Tarragona.*—Oils.

335. BATTLE, A. *Tarragona.*—Oils.

336. BEIGADA, J. *Verin, Orense.*—Garlic, chesnuts.

337. BELDA, A. *Valencia.*—Wines, oils, black barley grown on dry land; common barley, dwarf wheat and "Jeja" wheat, grown on dry land; sweet "Sorghum," from China; common almonds, bitter almonds, and smooth-podded tares.

338. BELDA, RAÑO, & Co. *Bocayvente, Valencia.*—Brandy.

339. BELLOT & ESTEVE, *Onil.*—Almonds.

340. BELMONTE, A. *Corrales, Zamora.*—Chick peas.

341. BELVIT, R. *Reus.* — "Esperanza" almonds, wines.

342. BENITO, C. *Avila.*—Barley, wheat, chick peas.

343. BENJUMEA, P. *Seville.*—Wheat.

344. BERASTEDI, L. *Lacunza, Navarre.*—Haricot beans, maize.

345. BERENGUER, J. B. *Valencia.*—Early rice, muscat rice, red "arisnegro" Tangan-rock wheat, haricot beans, various kinds of maize.

346. BERNALDEZ, WIDOW OF, *Villanueva, Seville.*—Wines.

347. BERNALDEZ, L. *Seville.*—Wines.

348. BERNALDEZ, F. *Seville.*—Wines.

349. BERNER, J. *Elche.*—Figs, almonds.

350. BERRO, E. A. *Jaen.*—Oil.

351. BESSO, F. *Tarragona.*—Brandy.

352. BETEGON, A. *Loadilla de Rio Seco, Palencia.*—Red wheat, chick peas, chickling vetches.

353. BEYNON, STOCKEN, & SONS, *London.*—Havannah cigars.

354. BIOSCA, T. *Fuenta la Higuera.*—Honey.

355. BLANCH, M. *Castellbell.*—White and rich wine.

356. BLANQUER, M. *Saragossa.*—Red wine.

357. BLAT, F. *Almusafes, Valencia.*—Rice.

358. BOIX, M. *Lérida.*—Beans, large beans, maize.

359. BOFIL, R. *Orihuela, Alicant.*—Oil, ground pepper.

360. BOLT & TOLOSA, J. *Caravaca, Murcia.*—Red wheat grown on dry land.

361. BONAPLATA, N. *Seville.* — Wines, semola, barley, oils.

362. BONET, S. *Tarragona.*—Wines.

363. BORJA, T. *Tarragona.*—Wines.

364. BORRAS, J. *Reus, Tarragona.*—Haricot beans, barley, rancio wine, brandy, oil.

365. BORRAS, M. *Tarragona.*—Common wine.

366. BOST, J. *Caravaca, Murcia.*—Wheat.

367. BOVER, P. *Tarragona.*—Common wine.

368. BROSCA, T. *Valencia.*—Oil, wheat, honey.

369. BRUCART, J. *Manresa.*—Wine, haricot beans.

370. BRUCE, HAMILTON, & Co. *Teneriffe.*—Wines.

371. BRUMENGO, J. *Rota, Cadiz.*—Sweet wine, "tintilla," "muscatel," "pajarete."

372. BUCHACA, A. *Valencia.* — Common wine.

373. BULL & WHEAL, *London.*—Sherry wines.

374. BURGALESA, LA, FLOUR MILLS, *Burgos.*—Flour.

375. CAAMIÑA ARIAS, J. *Puente Domingo Flores, Leon.*—Haricot beans.

376. CABALLERO, T. *Barajas de Melo, Cuenca.*—White wheat.

377. CABALLERO, J. *Chinchon, Madrid.*—Wine, brandy.

378. CABALLERO, J. *Madrid.*—Wine.

379. CABALLERO, MARQUIS OF, *Valladolid.*—Wines.

380. CABAU, J. M. *Aguadilla, Puerto-Rico.*—Tobacco in leaf and thread.

381. CABRE, WIDOW OF, *Riudons, Tarragona.*—Haricot beans.

382. CAILÁ, T. *Tarragona.*—Full-bodied wine.

383. CALARIANA, V. *Reus.*—Nuts, lentils, wines.

384. CALATRAVA, F. GARCIA, *Alcovendas, Madrid.*—Australian barley.

385. CANELLAS, BROTHERS, & Co. *Tarragona.*—Brandy.

386. CALCANO, J. *Seville.*—Pastes.

387. CALVO, P. *Fregeneda, Salamanca.*—Oil; almonds.

388. CALVO, R. *Casaseca de las Chanas, Zamora.*—Haricot beans, chick peas.

389. CALVO, T. *Moraleja del Vino, Zamora.*—Common wine.

390. CALZADA, G. *Alcalá de Henares.*—Wines.

391. CALZADA, J. *Astorga.*—Chocolate.

392. CALZADA, T. *Seville.*—Oils.

393. CALZADILLA, M. *Seville.*—Olives " de la Reyna."

394. CAMACHO, A. *La Palma, Huelva.*—Wines.

395. CAMARA, M. *Seville.*—Beans, wheat, barley, maize, oil, olives.

396. CAMERO DE ALDEA, T. *Sanzoles, Zamora.*—Common wine.

397. CAMPOFRANCO, MARQUIS OF, *Alba, Balearic Islands.*—Wine.

398. CAMPS, J. *Tortosa.*—Oils.

399. CANDALIZA, A. *Andujar.*—Wheat, barley, peas, chick peas, beans.

400. CANELLAS, BROTHERS, & Co. *Tarragona.*—Wines, brandies.

401. CANOVAS, J. *Caravaca, Murcia.*—Potatoes, oil, honey.

402. CANTOS, F. M. *Albacete.*—Hard red wheat, grown on dry land; "jeja" wheat.

403. CANTONS, P. *Reus.* — "Esperanza" almonds, black smooth-podded tares, oils, wines.

404. CAPELLA, WIDOW OF, *Barcelona.*—Chocolate, wines.

405. CARBONELL, P. *Tarragona.*—Oils.

406. CARDENAS, A. *Valencia.* — White wine.

407. CARDIER, —, *Oviedo.*—Coffee roasted, in grain, and ground.

408. CAREY, BROTHERS, *Saragossa.*—Wines.

409. CARMONA, J. *Avila.*—White wheat.

410. CARO, BROTHERS, & Co. *Tarragona.*—Wines.

411. CARO, R. *Constantina, Seville.*—Wheat.

412. CARRABIAS, J. *Salamanca.*—Wheat.

413. CARROSCOSA, Y. *Buñol, Valencia.*—Oils.

414. CARREÑO, S. *San Cristobal de la Vega, Segovia.*—Wheat.

415. CARRERAS, A. *Mahon, Minorca.*—Wheat, cheese.

416. CARRERAS, J. *Balearic Islands.* — Red wine.

417. CARRETERO, P. *Cordoba.*—Wines.

418. CARRETERO, R. *Mercadal, Minorca.*—"Jeja" wheat.

419. CARVAJAL, H. *Zamora.*—Almonds.

420. CASADO, S. *Olmedo, Valladolid.* — Wheat.

421. CASANOVA, J. *Havannah.*—Farinaceous tubercle, called "namé."

422. CASAJUANA, J. *Castellgali, Barcelona.*—Oil.

423. CASAS, BROTHERS, *Reus.* — White beans, black smooth-podded tares, wines, oils.

424. CASARAMOS, MARQUIS OF, *Alcalá de Guadaira.*—Oils.

425. CASARES, R. *Toro.*—Sweet and red wine.

426 CASO, M. *Infiesto, Asturias.*—Bacon.

427. CASTEL, A. *Pous, Barcelona.*—Wines.

428. CASTELL DE PONS, *Esparraguera, Barcelona.*—Wine, oil, olives, vinegar.

429. CASTELLET, B. *Tarrasa.*—Wine.

430. CASTELLO, P. *Madrid.*—Wine.

431. CASUYES, J. *Tarragona.*—Oil.

432. CATALA, A. *Jabea, Alicant.*—Raisins, almonds.

433. CATALAN DE OCON, *Monral del Campo-Teruel.*—Saffron.

434. CEBALLOS, B. *Nava del Rey, Valladolid.*—Wines.

435. CEMELI, F. *Saragossa.*—Wine, oil.

436. CENTENO, G. *Acebo, Caceres.*—Oil.

437. CERVERA, C. *Valencia.*—Raisins.

438. CINOS, A. *Villamayor, Malaga.*—Wines.

439. CISNEROS, L. *Constantina, Seville.*—Wines, brandy, vinegar, beans.

440. CLAPES, J. *Santa Eulalia, Iviza.*—Almonds.

441. CLEMENTS, J. *Malaga.*—Almonds, raisins.

442. COLAS, L. *Velamazan, Soria.*—Honey.

443. COLLANTES, A. *Madrid.*—Smooth-podded tares, peas, beans, maize, common barley, Australian barley, wheat, oats.

444. COLLDFORNS, J. *Rellinas, Barcelona.*—Wine.

445. COLOM, J. *San Lucar de Barrameda.*—Muscatel wine, "lachryma" wine, oils.

446. COMESAÑA, P. *Seville.*—Wheat, oils.

447. COMMITTEE OF EXHIBITORS OF SEVILLE.— Cumin seed, mustard, coriander seed, marjoram, carraway seed, lentils, chesnuts, acorns.

448. COMMISSION FROM MANILLA, *Philippine Islands.*—Cigars, cigarettes made of China paper; cigarettes made of Alcoy paper.

449. CONTE, J. A. *Cordoba.*—Plums.

450. CORBACHO, A. *Montellano, Seville.*—Wheat, barley.

451. CORDOBA, M. *Tortosa.*—Hazel-nuts, "negreta," "morella," "grifolla."

452. CORRO DE BRESCA, *Malaga.*—Muscatel wine, raisins.

453. COSTA, J. *Havannah.*—Preserves.

454. COTONER, F. *Baleares.*—"Albaflor" wine.

455. COWEN & Co. *London.*—Havannah cigars.

456. CREIBACH, B. *Valdeuro, Castellon.*—Oil.

457. CRESPO, F. *Estepa.*—Oil.

458. CRESPO, M. *Nava del Rey, Valladolid.*—Barley.

459. CRESPO, J. *Seville.*—Barley.

460. CRISTIA & VARNES, *Cassá de la Selva.*—Haricot beans.

461. CUESTA, M. *Villada, Palencia.*—Wheat.

462. CUETARA, E. DE LA, *Palencia.*—Flour of first and second quality.

463. DALMAN, J. *Reus.*—Maize, brandy.

464. DAMETO, A. *Palma.*—Millet.

465. DAVIDSON & Co. *Santa Cruz de Teneriffe.*—Wine.

466. DELGADO, V. *Zalamca Real, Huelva.*—Honey.

467. DEOCON, F. *Segorve.*—Oil.

468. DIAGO, T. *Havannah.*—Sugar.

469. DIAZ, B. *Tudela.*—Wine, brandy.

470. DIAZ, E. *Huelva.*—Chick peas, barley, hard and soft-shell almonds, round almonds, oats.

471. DIAZ, T. *La Palma, Huelva.*—Chick peas.

472. DIAZ, P. *Tudela.*—Wines.

473. DIAZ OBREGON, M. *Seville.*—Brandies, liquors.

474. DIAZ & Co. R. *Havannah.*—Cigars.

475. DIEZ, V. *Alicant.*—Wines.

476. DEPUTATION, PROVINCIAL, FROM ALAVA.—A collection of cereals, and other alimentary substances, composed of the following varieties: " raspudo " wheat, white wheat, Valencia wheat, flour of first quality, from the mills of Molinuevo & Co. of Vittoria; rye, black oats, Australian barley, vetches, white round haricot beans, ditto large, ditto red, ditto from shoots; small beans, large beans; lentils; yellow maize, red maize, purple maize, long-eared maize; capers; white and red wines, oils, cyder, syrups, cheese, butter.

477. DOMENECH, M. *Esparraguera, Barcelona.*—Wines, oils.

478. DOMINGO, C. *Játiva, Valencia.*—Almonds, edible cyperus.

479. DOMINGO, D. *Reus.*—Common wine, vinegar, brandy, oil, smooth-podded tares.

480. DOMINGO, C. *Tarragona.*—Common wine.

481. DOMINGUEZ, C. *Lillo, Leon.*—Butter.

482. DOMPER, G. *Atimuran, Tayabas, Philippine Islands.*—Wines.

483. DORADO & Co. *St. John, Puerto-Rico.*—Chocolate.

484. DUFF, GORDON, & Co. *Jerez de la Frontera.*—Amontillado wine, sherry wine, Pedro Ximenez wine.

485. DULCE MARIA, LA, FLOUR MILLS, *Arando de Duero.*—Flour.

486. ELVIRA, *Moraleja del Vino, Zamora.*—Common wine.

487. ENRIQUE, J. A. *Toro.* — Brandy, spirits, dry wines.

488. ESCOFET, J. *Saragossa.*—Brandy.

489. ESCUDERO, M. *Orihuela, Alicant.*—White wine.

490. ESCUELA, SCHOOL, OF AGRICULTURE OF ALAVA.—Agricultural produce.

491. ESCUER, ORDAZ, & Co. *Huesca.*—Wheat, flour.

492. ESLABA, J. *Tudela.*—Wine, oil.

493. ESPARCIA, P. *Albacete.*—Long-eared white wheat.

494. ESTARICO, R. *Valencia.*—Wines.

495. ESTEBAN, P. *Valencia.* — Common wine.

496. ESTELLEZ, G. *Valencia.*—Red wine, oil, wheat.

497. ESTELLER, G. *Benicarlo, Castellon.*—Wine.

498. ESTEPA, J. *Urrea de Jalon, Saragossa.*—Fine-grained wheat.

499. ESTEVEZ, H. *Zamora.*—Wine, wheat, almonds, hard and soft.

500. EUGENIO, R. *Saragossa.*—Oil, wheat, maize.

501. ESTRANY, M. *Tortosa.*—Rice, oil.

502. FLOUR MILLS OF " NUESTRA SEÑORA DE LOS REMEDIOS," *Malaga.*—Flour made from white and hard wheat.

503. FAJARDO, J. L. *Constantina, Seville.*—Wheat, barley, oil.

504. FAJO & Co. *Havannah.*—Cigars.

505. FEBRER, J. *Benicarlo, Castellon.*—Wines.

506. FERNANDEZ, A. *Bullas, Murcia.*—Wine, brandy, saffron.

507. FERNANDEZ, J. *Masroig, Tarragona.*—" Esperanza " and soft almonds, " choyote," a vegetable from America, one plant alone of which has produced in three years 1,408 pods.

508. FERNANDEZ, M. *Tafalla, Navarre.*—Wheat, barley.

509. FERNANDEZ DE CORDOBA, M. *Constantina, Seville.*—Wheat, barley, oil.

510. FERNANDEZ, ELVIRA L. *Sanzoles, Zamora.*—Wines.

511. FERNANDEZ, V. *Cáceres.*—Oil.

512. FERNANDEZ & VENTOSA, *Corunna.*—Wheat, flour.

513. FERRANDIZ, J. *Valencia.* — Wine, vinegar, brandy.

514. FERRER, B. *Seville.*—Oil, chick peas, beans, vetches, one-grained wheat.

515. FIERRO, T. *Santa Cruz de la Palma.*—Preserves, Malmsey wine, arrow-root.

516. FIBALLER, J. M. *Tarragona.*—Oil, smooth-podded tares, haricot beans, hazel nuts, almonds.

517. FLORES, A. *Moguer, Huelva.*—Alcohol.

518. FOMOLLAR, COUNT OF, *Barcelona.*—Wines.

519. FONOLLERAS, F. *Gerona.* — Mixed wheat, "jega" wheat, large beans, beans, wines, oils.

520. FOLDR, L. *Tarragona.*—Oils.

521. FONT, B. *Tarragona.*—Common wine.

522. FONT, P. R. *Tarragona.*—Oils.

523. FONT, J. *Murcia.*—Olives.

524. FONTELLAS, M. DE, *Navarre.*—Oil.

525. FORCES, M. *Saragossa.*—Red and muscatel wines.

526. FORNES, M. *Saragossa.*—Wines.

527. FOSALBA, —, *Pierola, Barcelona.* — Wines, oils.

528. FREIXAS, T. *Castellgali, Barcelona.*—Red wine.

529. FUENMAYOR, P. *Seville.*—Wines.

530. FUENMAYOR, V. *Caltoja, Soria.*—Honey.

531. FUENTES DEL SAUCE, COUNT OF, *Cordoba.*—Oil, figs.

532. FUENTES DEL SAUCE, COUNT OF, *Constantina, Seville.*—Oil, wheat, barley.

533. FUENTES, A. *Cordoba.* — Montilla wine.

534. FUENTES, HORCAS A. *Tarragona.*—Common wine.

535. GAGO, ROPERUELOS M. *Zamora.*—Barley, cheese.

536. GAITAN, F. *Tudela.*—Oils.

537. GALES, V. *Esparraguera, Barcelona.*—Wine, vinegar.

538. GALI, A. *Tarraso.* — Full-bodied wines, dry ditto, semi-sweet ditto, sweet ditto.

539. GALI, J. *Fals, Barcelona.*—Oil, wine.

540. GALNIDO, A. *Seville.*—Agricultural produce.

541. GAMERO, M. *Seville.*—Canary seed.

542. GARCES, R. *Huesca.*—Barley, maize.

543. GARCIA, —, *Santa Olaya de Olivares.*—Vetches.

544. GARCIA, ACEÑA P. *Seville.*—Starch.

545. GARCIA, S. *Lora del Rio, Cordoba.*—Wheat, vetches.

546. GARCIA, GOMEZ DE LA SERNA, *Belalcazar, Cordoba.*—Wheat, chick peas, barley, acorns.

547. GARCIA, D. *Guadalajara.*—White wheat, wine.

548. GARCIA, F. *Avila.*—White wheat.

549. GARCIA, J. *Barco de Avila.*—White haricot beans.

550. GARCIA, J. R. *Valencia.*—Various kitchen-garden seeds.

551. GARCIA, R. *Onil, Alicant.*—Wines, spirits.

552. GARCIA, R. *Madrid.*—Chocolate.

553. GARCIA, ALFONSO A. *Moral de Orbigo, Leon.*—White wheat.

554. GARCIA, CALATRAVA F. *Alcovendas, Madrid.*—Wheat, chick peas, barley, wine.

555. GARCIA MORENO, J. *Valencia.*—Wine, oil, figs.

556. GARCIA, M. *Pedroso, Seville.*—Wheat, barley.

557. GARRIGA, J. *Manresa.*—Wines.

558. GASPAR, J. *Reus.*—Oils, smooth-podded tares.

559. GASSET, J. *Tarragona.*—Brandy.

560. GAYEN, J. *Malaga.* — Preserves, olives.

561. GENER, P. *Havannah.* — Concentrated sugar.

562. GILL, F. *Tarragona.*—Wines.

563. GIL Y BORRAS, F. *Reus.*—Wines, oil, almonds.

564. GINER, J. *Valencia.*—White haricot beans, millet corn, wheat, hemp seed.

565. GIOL, T. *Reus.*—Wines, oils.

566. GISBERT, T. A. *Onil, Alicant.*—Oils.

567. GISBERT, B. *Linon, Majorca.*—Wheat.

568. GOVERNOR OF ALICANT.—Dry almonds, oil, wheat, brandy, preserves, raisins, white maize, figs, starch; wines, Malmsey, Fondellot, common white, muscatel, old, common, full-flavoured.

569. GOVERNOR OF BARCELONA.—Wines.

570. GOVERNOR OF CUENCA.—Honey.

571. GOVERNOR OF GUADALAJARA. — Honey from la "Alcarria."

572. GOVERNOR OF NAVARRE. — Oils, brandy, maize, haricot beans, flour, nuts.

573. GOVERNOR OF ORENSE.—White, red, claret, and common white wines; fine vinegar, white and coloured haricot beans, maize.

574. GOVERNOR OF SEVILLE.—Brandy; wheat, maize, chick peas, haricot beans, beans, peas, shumac, starch, semolá.

575. GODIA, F. *Alpical, Lérida.*—Lentils.

576. GOMBAN, T. *Reus.*—Wines, almonds.

577. GOMEZ DE SALAZAR, J. *Almeria.*—Pedro Ximenez wine, figs, palmettoes.

578. GOMEZ, J. *Jerica, Castellon.*—Wines.

579. GOMEZ, J. *Orense.*—Beans.

580. GOMEZ, M. *Jerica, Castellon.*—Brandy.

581. GOMEZ, V. *Saragossa.*—Wines.

582. GOMEZ, ALONSO, J. *Serradilla, Cáceres.*—Wines.

583. GOMEZ DE ALIA, G. *Navalmoral, Toledo.*—Oil.

584. GOMEZ DE BARREDA, F. *Seville.*—Oils.

585. GOMIS, T. *Manresa.*—Wines.

586. GONZALEZ, J. *Moron.*—Agricultural produce.

587. GONZALEZ ARCAIMA, J. *Carthagena.*—Wines.

588. GONZALEZ, DUBOSCQ, & Co. *Jerez.*—Amontillado, old and soft, pale, "oloroso" wines; a model of their factory.

589. GONZALEZ DE MESA, A. *Laguna, Canarias.*—Black Mexican beans.

590. GONZALEZ DEL VALLE, A. *Havannah.*—Cigars of all sizes, from the factory of "*Hija de Cabañas y Carvajal,*" of the Havannah.

591. GONZALVO, G. *Huesca.* — Common wine.

592. GORMAN & Co. *Port St. Mary* — Sherry and Montilla wines; a model of their factory.

593. GOYENETA, M. *Seville.*—Agricultural produce.

594. GRANJA PROVINCIAL (MUNICIPALITY) OF LEON.—Haricot beans, slate-coloured haricot beans, "Alolbas," Polish oats.

595. GRAU, J. *Reus.*—Wines, oils, Modan almonds.

596. GREMIO (FRATERNITY) OF SAUSAGE-MAKERS OF CANDELARIO.—Small and large sausages.

597. GÜELL, J. *Almacella, Lérida.*—Wheat.

598. GUELVES, A. *Tarragona.* — Full-flavoured wines.

599. GUENDULAIN, COUNT OF, *Tafalla, Navarre.*—Wines.

600. GUIMATS, J. *Gratallops, Reus.*—Soft almonds, black figs, red "rancio" wine, lachryma wine.

601. GUILLE, BROTHERS, *Barcelona.* — Wines, brandies.

602. GUILLEM, J. *Saragossa.*—Honey.

603. GUILLEM, L. *San Esteban de Litera, Huesca.*—Wines.

604. GUILLEM, M. *Huesca.*—Vinegar.

605. GUILLOT, G. *Havannah.*—Cigarettes, from the manufactory "Para Vm."

606. GUITO, M. *Gerona.*—Wine.

607. GUISI, BROTHERS, *Tarragona.*—Common wine, vinegar, brandy, oil.

608. GUTIERREZ, C. *Corrales, Zamora.*—Imitation port wine, peas.

609. HAMILTON & GRIEVE, *London.*—Sherry wines.

610. HARTLEY, ZAFRA, & Co. *Huelva.*—"Mazagan" beans.

611. HELVANT, S. *Jerez.*—Very old Amontillado sherry wine, Manzanilla wine.

612. HEREDIA, M. *Malaga.*—Sugar, oil.

613. HERNANDEZ, A. *Murcia.*—Oils.

614. HERNANDEZ, Y. *San Frontis, Zamora.*—Common wine.

615. HERNANDEZ, F. *Avila.* — White wheat, chick peas.

616. HERNANDEZ, V. *Miranda del Castañar, Salamanca.*—Oils.

617. HERNANDEZ, R. *Orihuela, Alicant.*—Starch.

618. HERNAN-SANZ, E. *Iscar, Valladolid.* —Pine-nut kernels, with and without shells.

619. HERRERA, S. A. *Seville.*—Wheat.

620. HERP, J. *Manresa.*—Wines, oil.

621. HIDALGO, E. *Jerez.*—Pedro Ximenez, muscatel, manzanilla wines.

622. HIDALGO, ISLA R. *Cordoba.*—Artificial egg-plant.

623. HIDALGO, J. *Rola.*—Old tintilla wine, new ditto.

624. HIGUERA, BARBAJERO, R. DE LA, *Toro.* —Red wine, sweet and dry; common red wine.

625. HIGUERA, A. DE LA, *Seville.*—Oils.

626. HOMS & DE PRAT, M. *Manresa.*—Red wine.

627. HOTARRA, E. *Tarragona.*—Rose wine, white ditto.

628. IBARRA, M. *Valencia.*—Wine, vinegar, oil.

629. IBIRICA, H. *Tudela.*—Oil.

630. ILLAN, J. *Zamora.*—Common wine.

631. INGENIERO DE MONTES (SUPERINTENDENT OF WOODS AND FORESTS), *Leon.*—Acorns from the *quercus sessiliflora*, ditto from the *quercus pubescens*, beech seed.

632. INGUANZO, F., *Rivadesella, Asturias.* —Chesnuts.

633. IÑIGO, Z. *Saragossa.*—Beans, wines.

634. ISCART, T. *Tarragona.*—Common wine.

635. JIMENEZ, A. *Seville.*—Crackers.

636. JIMENEZ, A. *Cascante, Navarre.*—Wines.

637. JIMENEZ, G. *Avila.*—Chick peas.

638. JIMENEZ, D. P. *Madrid.*—Arganda wine.

639. JIMENEZ DE TEJADA, BROTHERS, *Moguer, Huelva.*—Wines, spirits.

640. JIMENO, G. *Saragossa.*—Oil.

641. JIMENO, G. C. *Valladolid.*—Evergreen peas.

642. JIMENO & AZPEITIA, *Ateta, Saragossa.*—Wines, brandy, liquors.

643. JORDANA, J *Albalate, Teruel.*—Wines, Indian corn meal.

644. JUDES, M. *Constantina.*—Wheat.

645. JUNCO, P. *Rivadesella, Asturias.*—Apples.

646. JUNTA (BOARD OF AGRICULTURE) OF BURGOS.—Rye from Pineda de la Sierra, short wheat from Villagonzalo de Pedernales, "Mahamuz" wheat, bare barley, barley for horses, barley for asses, common ditto, oats, "arberjas," chick peas, beans, lentils, red fine-grained wheat from Burgos, lentils from Villalta, vetches from Delorado, kidney beans, maize from Miranda de Ebro.

647. JUNTA OF CASTELLON.—Wine, brandy, figs, raisins.

648. JUNTA OF CORUNNA.—Wheat, rye, oats, barley, chick peas, peas, beans, maize, millet, French beans, nuts and chesnuts, pressed and salted herrings, salt beef, alimentary preserves.

649. JUNTA OF GRANADA.—Oil, wine, vinegar, wheat, barley.

650. JUNTA OF OVIEDO.—Hams, beans, vetches, nuts, apples, chesnuts.

650a. JURADO, M. D. *Seville.*—Essence of aniseed.

651. JUVES, F. *Manresa.*—Wheat, garlic.

652. LACAMBRA, J. *Saragossa.*—Pastes.

653. LACASA, J. A. *Jaen.*—White and yellow maize.

654. LACAVE, J. P. *Seville.*—Olives "de la Reyna."

655. LACORTE, J. A. *Cabra.*—Wines, vinegar, honey, figs, plums.

656. LAGORIO, F. DE P. *Carthagena.*—Wines.

657. LAGUARTA, V. *Savayes, Huesca.*—Spirits.

658. LA HIGUERA, A. *Seville.*—Agricultural produce.

659. LAHOZ, V. *Saragossa.*—Flour.

660. LA INDUSTRIAL HARINERA (FLOUR MILLS), *Barcelona.*—Flour.

661. LAMAS, M. *Cebreros, Avila.*—White and red wine.

662. LA NAVARRA, *Tudela.*—Vinegar, brandy.

663. LARA, J. PEDROSO, *Seville.*—Canary seed.

664. LASALA, V. *Valencia.*—Wine, oil, syrup.

665. LASERNA, J. J. *San Juan del Puerto, Huelva.*—Strong red wheat.

666. LASERNA, J. P. *Moguer, Huelva.*—Orange wine.

667. LASERNA, M. *Seville.*—Wheat.

668. LASIERRA, J. *Quinzano, Huesca.*—Barley, wheat.

669. LASTRA, J. *Seville.*—Oil.

670. LEON BENDICHO, J. *Almeria.*—Maize.

671. LEMBEYE, J. *Corunna.*—Maize.

672. LEZCANO, M. *Saragossa.*—Oil, wine.

673. LINARES, F. J. *Seville.*—Wheat, haricot beans, vetches.

674. LOPEZ, J. A. *Alagon, Saragossa.*—Oil.

675. LOPEZ, A. *Corcos, Valladolid.*—Aniseed.

676. LOPEZ, S. *Malaga.*—Flour.

677. LOPEZ, M. *Caravaca, Murcia.*—Millet.

678. LOPEZ, M. *Madrid.*—Chocolate.

679. LOPEZ, S. *Alama, Murcia.*—Figs.

680. LOPEZ ARRUEGO, M. *Velilla de Ebro, Saragossa.*—Dried fruit.

681. LOPEZ CABALLERO, S. *Murcia.*—Wheat, nuts, figs, almonds, rice, honey.

682. LORENZANA, R. *Leon.*—Linseed, beans, fine haricot beans.

683. LORENZO, T. *Reus.*—Wines, brandies, oils.

684. LORETO, MARQUIS OF, *Seville.*—Oils.

685. LOSADA DE RAMIREZ, D. *Villesca, Toledo.*—Wines, oils, wheat, chick peas.

686. LOZANO, A. *Fermosella, Zamora.*—Oils.

687. LOL, J. F. *Liñola, Lérida.*—Camomile.

688. LUELMO, A. *Moralejo del Vino, Zamora.*—Wines.

689. LUXAN, L. *Arenas, Avila.*—Oils, chesnuts.

690. LUNA, COUNT OF, *Elche, Alicant.*—Muscatel wine, oils, wheat, barley, figs, almonds.

691. LUZON, J. *Albacete.*—Saffron.

692. LLADO, M. *Campos, Balearic Islands.*—"Jeja" wheat.

693. LLAURADO, F. *Reus.*—Smooth-podded tares.

694. LLAURADO, J. *Tarragona.*—Common wine.

695. LLAVERIA, J. A. *Tarragona.*—Wine.

696. LLAVERIA, E. *Tarragona.*—Brandies.

697. LLEVAT, A. *Tarragona.*—Common wine.

698. LLORENS, A. *Tarragona.*—"Probadillo" wine.

699. LLUCH, A. *Gerona.*—Wines, oils.

700. MAESTRE, G. *Petrel, Alicant.*—Wines.

701. MALDONADO, M. *Ciudad Real.*—White Gijon wheat.

702. MALEGUÉ, J. *Reus.*—Wines, brandy, nuts, hazel nuts.

703. MARCA, E. *Tarragona.*—Oils.

704. MARCH, M. *Castellbell, Barcelona.*—Wines.

705. MARCO, M. *Espluch, Huesca.*—Wheat.

706. MAROTO, E. *Carmona, Toledo.*—Wine, white wheat, chick peas.

707. MARQUET, B. *Saragossa.*—Beans, vetches.

708. MARRACO, P. *Saragossa.*—Flour.

709. MARRACO, J. *Saragossa.*—Flour.

710. MARRON, B. *Seville.*—Oil.

711. MARSELLA, B. *Bullas, Murcia.*—Wine, brandy.

712. MARRACO, M. *Madrid.*—Chocolate.

713. MARTI, J. *Reus.*—Wines, raisins, plums.

714. MARTI, R. *Reus.*—Oils, brandies.

715. MARTIN, A. *Corca.*—Oil, chick peas.

716. MARTIN DE LA SIERRA, J. *Daimiel.*—Millet corn.

717. MARTIN, M. *Olmedo, Valladolid.* — White wines.

718. MARTIN, P. *Povoleda, Tarragona.*— Wines, vinegar.

719. MARTINEZ, A. *Jaen.*—White wines.

720. MARTINEZ, F. *Tarragona.*—Beans.

721. MARTINEZ, F. *Tarragona.*—Honey.

722. MARTINEZ, WIDOW OF R. *Murchante, Navarre.*—Wines.

723. MARTINEZ DE EUGENIO, J. *Navalmoral, Toledo.*—Oils.

724. MARTINEZ GUTIERREZ, J. *San Lucar de Barrameda.*—Manzanilla, heavy, amontillado, white, Pedro Ximenez, pajarete, and muscatel wines.

725. MARTINEZ YBOR, V. *Havannah.*— Cigars.

726. MARTINEZ MADRID, J. *Carthagena.*— Wines.

727. MARTIN SUAREZ, —, *Seville.*—Agricultural produce.

728. MARTORELL, P. *Ciudadela, Balearic Islands.*—Wines.

729. MASANET, A. *Muro, Balearic Islands.*—Lentils, peas, and twenty-one varieties of haricot beans.

730. MAS Y MACIO, J. *Tarragona.*—Red wine.

731. MASPONS, T. *Vinaroz, Castellon.*—Alcohol.

732. MASSIA, E. *Tortosa.*—Oil.

733. MATEOS, C. *Cedreros, Avila.*—White and muscatel wines.

734. MATEOS, V. *Cedreros, Avila.* — Brandies.

735. MATEU, T. *Bacaritas, Barcelona.*—Wines.

736. MATTIEPHEN, FURTON, & Co. *London and Xerez.*—Wines.

737. MAYOL, M. *Balearic Islands.* — Raisins, figs, smooth-podded tares.

738. MAZARRON, M. *Madrid.*—Valdepeñas wine.

739. MEAUX, F. *Las Piedras, Puerto-Rico.*—Leaf tobacco.

740. MEDRANO, T. *Almonacid de Zorita, Guadalajara.*—Oil.

741. MELA, G. *Madrigal, Avila.*—Wines.

742. MENDEZ DE VIGO, F. *Oviedo.*—Bottled cider.

743. MENZA, J. *Lérida.*—Oils.

744. MERCADAL, M. *Mahon.* — Wines, beans.

745. MERCADER, F. *Tarragona.*—Common wines.

746. MERIC & Co. Steam Chocolate Mills, *Madrid.*—Chocolate, alimentary substances.

747. MERINO, M. *Berlanga, Soria.* — Potatoes.

748. MEZQUIRA, T. *Valencia.*—Common wines.

749. MESTRES, F. *Tarragona.*—Red wine.

750. MICHANS, A. *Lecumberri, Navarre.*—March beans.

751. MIGUEL, J. *Tarraga, Lérida.* — Vetches.

752. MIRAFLORES, COUNT OF, *Seville.*—Agricultural produce.

753. MIRALLES, ARAGON, & Co. *Barcelona.*—Wines.

754. MIRET & TERSA, *Barcelona.*—Beer.

755. MIURA, A. *Seville.*—Chick peas.

756. MOLPECERES, V. *Olmedo, Valladolid.*—Wheat.

757. MONTFORT, A. *Barcelona.*—Oils.

758. MONFORT, F. *Torrente de Cima, Huesca.* — Oil, wheat, haricot beans, figs, almonds, beans, dry sweetmeats.

759. MONJAS (NUNS) OF SAN PELAYO, *Oviedo.*—Preserves.

760. MOMPEAU, J. C. *Callosa de Segura, Alicant.*—Wines.

761. MOMPO, J. *Valencia.*—Wines.

762. MAUPOEY, P. *Valencia.*—Oils, muscatel raisins, edible cyperus, pea-nuts, palmetto.

763. MONSERRAT, F. A. *Maella.*—Figs.

764. MONSERRAT, R. *Puigpelat, Tarragona.*—Fine wheat, soft almonds, hazel nuts, maize, wines.

765. MONTANER, J. *Reus.*—Wines, vinegar, brandy, oils, red smooth-podded tares, black ditto, black hazel nuts, dried peaches.

766. MONTAGUT, C. *Tarragona.*—Common wine.

767. MONTANES, J. *Tarragona.* — Dry white wine, muscatel.

768. MONTERO, S. *Seville.*—Agricultural produce.

769. MONTOLIN, P. M. *Tarragona.*—Wine, oil, olives, barley, beans, smooth-podded tares, sixty-six varieties of haricot beans.

770. MONTIEL, M. *Trigueros, Huelva.*—Wines.

771. MONTILL, L. *Gerona.*—Wines.

772. MONTORO Y HERMANOS, *Havannah.*—Cigars.

773. MORALES DE LOS RIOS, A. *Lora del Rio.*—Oils.

774. MORELL, P. *Balearic Islands.*—Oils.

775. MORENO, A. *Guadalcanal, Seville.*—Wheat.

776. MORENO & GUERRERO, *Guadalcanal.*—Oils.

777. MORRAS, M. *Tafaya, Navarre.*—Wines, wheat.

778. MOTILLA, M. DE LA, *Seville.*—Agricultural produce.

779. MOYANO SANCHEZ, P. *Nava del Rey, Valladolid.*—Wheat.

780. MULET Y MAS, A. *Balearic Islands.*—Muscatel wine, Malmsey, orange wine.

781. MUNOZ, P. *Valladolid.*—Sausages.

782. MURILLO, M. *Lérida.*—Wax.

783. MURUVÉ, M. *Los Palacios, Seville.*—Oils, chick peas.

784. NADAL, J. *Ludarell, Lérida.*—Rye.

785. NARANJO, L. *Caralla, Seville.* — Brandies.

786. NASARRÉ, A. *Lupinen, Huesca.* — Oils, wheat, beans.

787. NAVARRO, T. *Cordoba.*—Wine.

788. NEBOT, S. *Son Servera, Balearic Islands.*—Figs, thirty-two varieties of haricot beans.

789. NEUMAN & SANDERS, *Puerto Plata.*—Leaf tobacco.

790. NOCEDO, F. *Orense.*—Maize.

791. NOËL, VASSEROT, & Co. *Seville.*—Liquorice.

792. NUÑEZ, J. *Corunna.*—Flour.

793. ODENA, BROTHERS, *Reus.*—Soft almonds, hazel nuts.

794. ODENA, P. *Tarragona.*—Oils.

795. ODENA, J. *Tarragona.* — Common wine.

796. OJEITO, J. *Fregeneda, Salamanca.*—Wines.

797. OLANIEL, A. *Jativa Valencia.*—Rice.

798. OLEZA, J. *Balearic Islands.*—Unclarified vinegar.

799. OLIVA, D. *Reus.*—Wines.

800. OLIVA, T. *Salamanca.*—Wheat.

801. OLIVER, B. *Ciudadela, Balearic Islands.*—Honey.

802. OLMEDO, L. *Borlanga, Soria.*—Beans.

803. OLLER, F. *Cassá de la Selva, Gerona.*—Wheat, beans.

804. ORDUÑA, C. *Cascante, Navarre.*—Liquors.

805. ORTEGA, L. *Reus.*—Wines, oils, vinegar, fruits.

806. ORTIZ, V. *Brozas, Cáceres.*—Wines.

807. ORNS, M. *Manresa.*—Wines.

808. ORNS & Co. *Huesca.*—Spirits of wine.

809. OSACAR, G. *Santesteban, Navarre.*—Maize, French beans.

810. OSORNO, WIDOW OF, *Villanueva, Seville.*—Wines.

811. OTERIM, J. *Guizo de Luinia, Orense.*—Turnips.

812. OTERO, J. *Verin, Orense.*—Onions.

813. PAGES, A. *Tarragona.*—Oils, common wine.

814. PALACIO, J. M. *Espeluy, Jaen.*—Wines, oils, wheat, honey, olives.

815. PALAZUELOS, WIDOW OF, *Toledo.*—Oils, wheat, oats, olives.

816. PAMIES, J. M. *Tarragona.*—Oils.

817. PANES, D. *Manzanilla, Huelva.*—Wine.

818. PARADELL, J. *Sanpedor, Barcelona.*—Wine.

819. PARDO, J. *Valencia.*—Indian corn and lentils.

820. PARES, J. *Collbato, Barcelona.*—Wine.

821. PARTAGAS, J. *Havannah.*—Cigars.

822. PASCUAL, B. *Reus.*—Wine, oil, brandy, smooth-podded tares, dried fruits.

823. PASCUAL, J. *Tarragona.*—Common wine.

824. PASCUAL, P. *Bruch, Barcelona.*—Wine.

825. PASCUAL, T. *Cosuenda, Saragossa.*—Wine.

826. PASSETTI, J. *Malaga.* — Preserves, different sorts.

827. PATERNA, MARQUIS OF, *Caballos, Seville.*—Oil, barley.

828. PAULES, J. *Saragossa.*—Oil.

829. PAYO, M. *Vermilio de Sayago, Zamora.*—Rye.

830. PAZ & Co. *Medina del Pomar.*—Flour.

831. PEDROSA, J. *Esparraguera, Barcelona.*—Wine, oil.

832. PELAEZ, M. *Valencia.*—Starch (coarse grain and powdered).

833. PELLICER, J. *Tarragona.* — Wine, brandy, vinegar.

834. PEÑA, F. *Saragossa.*—Wine.

835. PENILLAS, J. *Bollullos del Condado, Huelva.*—White wine.

836. PEÑAFLOR, COUNT OF, *Seville.*—Oil, barley, wheat.

837. PEREIRA, J. *Seville.*—Wheat, barley.

838. PÉREZ, J. *Tudela.*—Oil.

839. PÉREZ, L. *Cedreros, Avila.*—Wine.

840. PÉREZ, F. *Orense.*—Hazel nuts.

841. PÉREZ, D. R. F. *Ibi, Alicant.*—Oil, almonds.

842. PÉREZ, J. *Orense.*—Indian corn.

843. PÉREZ BAERLA, M. *Magallon, Saragossa.*—Wine, oil, strong wheat, and fine-grained ditto.

844. PÉREZ DE LOS COBOS, C. *Jumilla, Murcia.*—Oil.

845. PÉREZ DE LOS COBOS, P. *Jumilla, Murcia.*—Oil.

846. PÉREZ MARCO, F. *Rellen, Alicant.*—Oil, raisins, figs, almonds.

847. PÉREZ PAULENO, D. *Fregeneda, Salamanca.*—Oil of almonds.

848. PÉREZ BILLORIA, R. *Fregeneda, Salamanca.*—Oil, wine, olives, almonds.

849. PÉREZ ZAMORA, A. *Puerto de la Orotaba, Canary Islands.*—Arrow-root.

850. PÉREZ ZAMORA, A. *Puerto de la Orotabo, Canary Islands.*—Havannah cigars.

851. PERIBANES, L. *Saragossa.*—Oil.

852. PI, R. *Tarragona.*—Red wine.

853. PIMENTEL & Co. *Valladolid.*—Wines.

854. PINOS, M. *Saragossa.*—Honey.

855. PLA, J. *Tafalla.*—Spirits of wine.

856. PLA, F. *Reus.*—Wines and spirits.

857. POEY, J. *Havannah.* — Muscovado sugar.

858. PONS, L. *Mahon.*—Honey.

859. PONS, P. *Tarragona.*—Oil.

860. PORTA, M. *Saragossa.*—Liqueurs.

861. PORTILLA, D. *Seville.*—Semola.

862. PORTO, J. DE, *Havannah.* — Paper cigars of the brand " *Mi fama por el Orbe vuela.*"

863. POZUELO, T. *Daimiel.*—Millet corn.

864. PRATS, J. *Fals, Barcelona.*—Chick peas, wine.

865. PRESAS, S. *Corunna.*—Sardines.

866. PRIEGO, J. *Doña Mencia, Cordoba.*—Wine, brandy, wheat.

867. PRIEGO MARMOL, J. *Tarragona.*—Wines.

867a. PROHENS, D. *Balearic Islands.*—Essence of aniseed.

868. PUGA, WIDOW OF, & SONS, *Zamora.*—Wines, liqueurs.

869. PUIG, D. *Seville.*—Oil, beans.

870. PUIGGENER, J. *Oleza, Barcelona.*—Wines.

871. PUJADAS, P. *Tudela.*—Wines, wheat.

872. PUJOL, T. *Tarragona.* — Wines, brandy, oil.

873. QUER, M. *Aytoria, Lérida.*—Wines.

874. QUINZAFORTEZA, J. *Majorca.*—Thirty-two various kinds of almonds.

875. QUINTANA, M. *Balearic Islands.*—Olives.

876. RABASSO, J. *Vendrell, Tarragona.*—Wines, brandy.

877. RAMIREZ, F. *Saragossa.*—Oils.

878. RAMIREZ DE ARRELLANO, J. *Buñuel, Navarre.*—Oils.

879. RAMIREZ, BROTHERS, *Saragossa.*—Liquorice in paste.

880. RAMIREZ DE LOSADA, D. *Yllescas, Toledo.*—" Chamorro " wheat.

881. RAMOS, CALONGES, *Seville.*—Beans.

882. REPIELD, E. *St. Sebastian.*—Cider.

883. REQUESO, M. *Zamora.*—Barley.

884. RESO, G. *Balearic Islands.*—Red and Malmsey wine.

885. REYERO, B. *Leon.*—Honey.

886. RIBOT, J. *Petra, Balearic Islands.*—Wheat.

887. RICAÑO & MILIAN, M. *Havannah.*—Rappee snuff.

888. RIEGO & MÁRMOL, J. *Cordova.*—Shumac.

889. RICO, D. *Madrid.*—Muscatel wine.

890. RIERA, J. *Huelva.* — Peas, wheat, barley.

891. RIERA, J. *Paya, Barcelona.*—Wines.

892. RIESCO, J. P. *Alicant.*—Wines, oils.

893. RINCON, A. *Tembleque, Toledo.* — Wheat, cheese.

894. RINCON, M.C. *Seville.*—Barley, wheat, beans.

895. RIPALDA, M. *Pampeluna.*—Wines.

896. RIPOLL, M. *Saragossa.*—Wines, barley.

897. RIPOLL, M. *Balearic Islands.* — Twenty-six various kinds of almonds.

898. RIBAS, M. *Tarragona.*—Red wines.

899. RIBAS, M. & MATTHEW RUSELL, *Negro Island, Philippine Islands.*—Cocoa nuts.

900. RIBAS, J. *Tarragona.*—Common wine.

901. RIVES, A. *Manresa.*—Beans.

902. ROCA, B. *Palma.*—Alimentary products, oils.

903. ROCA DE TOGORES, BARON OF, *Orihuela, Alicant.*—Oils, three months wheat, dwarf maize.

904. ROCA, BROTHERS, *Murcia.*—Ground pimento.

905. ROCAMORA, J. *Reus.*—Wines, oils, black haricot beans.

906. RODRIGUEZ, J. *Seville.*—Maize.

907. RODRIGUEZ, B. *Manzanilla.*—Amontillado and manzanilla wine.

908. RODRIGUEZ, C. *Orense.*—Beans.

909. RODRIGUEZ LORENZO, J. *Toro.*—Honey.

910. RODRIGUEZ MODENES, T. *Cordova.*—Montilla wine.

911. RODRIGUEZ TEJEDOR, A. *Toro.*—Wines.

912. ROJALS, T. *Tarragona.*—Brandy.

913. ROMAN, WIDOW AND SONS OF, *Leon.*—Chocolate.

914. ROMERO, M. *Seville.*—Brandy, wheat.

915. ROMERO DE LAVANDERA, M. *Malaga.*—White wine.

916. ROMO, T. M. *Hinojosa, Salamanca.*—Almonds.

917. ROTOVA, C. D. *Valencia.*—Dates.

918. ROYO, M. *Valencia.*—Wines, oils.

919. RUBERT, J. *Balearic Islands.*—Oils.

920. RUBIO, J. *Tarragona.*—Wines.

921. RUBIO VELAZQUEZ, M. *Malaga.* —Wines.

922. RUIZ LEGANES, J. *Madrid.*—Wines, brandy.

923. RUIZ, M. *Burgo de Osma, Soria.*—White haricot beans.

923a. RUIZ, ZORRILLA, & Co. *Soria.*—Flour.

924. SABATER, R. *Montroig, Tarragona.*—Oils, haricot beans "de abundancia."

925. SACALL, F. *Tarragona.*—Brandy.

926. SADABA, F. *Palencia.*—Liqueurs.

927. SAGASTI, J. *Tudela.*—Oil.

928. SALA, E. *Saragossa.*—Flour.

929. SALVA, M. *Palma.*—Oils.

930. SALVADOR, C. *Villaralgo Zamora.*—Wine, wheat.

931. SALVADOR, G. *Tafalla.*—Brandy.

932. SALVADOR, J. *Tortosa.* — Oils, soft almonds of various sizes.

933. SALVADOR, M. *Tarragona.*—Wine.

934. SALLES, V. *Rellinas, Barcelona.*—Wines.

935. SAMPOL, P. J. *Balearic Islands.*—Oils.

936. SAN ADRIAN, M. D. *Monteagudo, Navarre.*—Oils.

937. SAN ANDRES, M. WIDOW OF, *Puerto de la Cruz, Canary Islands.*—Arrow-root.

938. SANCHEZ, A. *Murcia.*—Oil.

939. SANCHEZ CHICARRO, A. *Leon.*—"Clarete" wine, honey.

940. SANCHEZ, P. J. *Elche, Alicant.*—Wine, brandy.

941. SANCHEZ VIDA, T. *Guadalcanal, Seville.*—Oil, beans.

942. SANCHO, A. *Gandia.*—Sugar-cane.

943. SANS, E. *Tarragona.*—Vinegar.

944. SAN ROMAN, A. *Fermoselle, Zamora.*—Oil, wine, brandy, spirits of wine.

945. SANTAMARIA, F. *San Juan del Puerto, Huelva.*—Wines.

946. SANTANA, C. *Salamanca.* — Wheat, barley, rye, smooth-podded tares, peas, vetches.

947. SANTIAGO, J. *Zamora.* — Wines, brandies, spirits.

948. SARDEN, J. *Valencia.*—Common wine.

949. SATORRAS, A. *Tarragona.*—Common black barley, almonds of various kinds, figs, smooth-podded tares, full-bodied wines, oils.

950. SANTONJA, B. *Valencia.*—Wines.

951. SANTILLAN, P. *Riaza, Segovia.*—Wild-goat flesh, dried.

952. SANTOS, A. *Leon.*—Flour of first quality.

953. SANZ, E. *Olmedo, Valladolid.*—Wheat.

954. SANZ, M. *Zamora.*—Common wine.

955. SARDER, J. *Valencia.*—Wine.

956. SAVALL, B. *Tarragona.*—Oils.

957. "SECCION DE FOMENTO" (BOARD OF TRADE), *Valladolid.*—Chick peas.

958. "SECCION DE FOMENTO," *Castellon.*—Beans, smooth-podded tares.

959. SEGOVI, J. *Tarragona.*—Oils.

960. SEMITIER, COUNT OF, *Tortosa.*—Oils, smooth-podded tares.

961. SERNA, J. *Renda.*—"Alonso" wheat.

962. SERPA, J. *Seville.*—Wines.

963. SERRA, J. *Lérida.*—Plums.

964. SIMÓ, P. *Porrera, Tarragona.*—Wines, brandy.

965. SIMO, B. *Tarragona.*—Full-bodied wines.

966. SIMO, J. *Tarragona.* — Full-bodied wines.

967. SIMO, A. *Tarragona.*—Superior wines.

968. SOBERANO DE MENGO, D. *Tarragona.*—"Rancio" wine.

969. SOBRADIEL, COUNT OF, *Saragossa.*—Oils.

970. SOCIEDAD ECONÓMICA (INDUSTRIAL SOCIETY) OF MURCIA.—Pepper, white maize.

971. SOLDEVILA, R. *Corbus, Lérida.*—Oat

972. SOLAR DE ESPINOSA, BARON, *Jumilla, Murcia.*—Oils; red "jeja" wheat, grown on dry land, white wheat, raspinegro wheat, grown on moist land.

973. SOLER, J. *Mahon.*—"Jeja" wheat.

974. SOLER, D. *Lérida.*—Raisins.

975. SOLER, R. *Manresa.*—Wines.

976. SOLER, BROTHERS, *Tarragona.* — Wine, brandy.

977. SOLER Y SELLES, B. *Reginas.*—Muscatel wine.

978. SORÁ, J. *Iviza.*—Wheat.

979. SORIANO, — *Madrid.*—Sweet acorns from Murcia.

980. SOTERO, MARTINEZ, J.—Chick peas, three-month barley, Castilian barley, Gijon and white wheat.

981. SOTO, M. *Orihuela, Alicant.*—White maize.

982. SOULERE, J. B. *Tarragona.*—"Rancio" wine ; vinegar, oil.

982a. STEBOT, A. *Reus.*—Wines.

983. SUAREZ CENTI, J. *Valladolid.*—Flour.

984. SUBDELEGACION AGRICOLA DE SAN ISIDRO, *Reus.* — Wines, brandies, vinegar, oils, cereals, vegetables, dried fruits.

985. SUBIRAT, J. *Pierola, Barcelona.*—Wines.

986. SUELVES, J. DE, *Tortosa.*—Oil, maize, vetches.

987. SULLA, J. *Tremp, Lérida.* — Wine, brandy.

988. SUSINE & Co.*Havannah.*—Cigarettes, cut and pressed tobacco for cigarettes, from the factory "La Honradez."

989. TAMARIT, MARQUIS OF, *Tortosa.*—Wines, oils, rice, black figs.

990. TARAZA, P.*Piera, Barcelona.*—Wines, spirits.

991. TEJADA, L. R. D. *Viso.*—Wheat, one-grained wheat, beans.

991a. TEJADA, BROTHERS, *Huelva.*—Alcohol.

992. TEJERINA, B. *Vegas del Condado, Leon.*—Beans.

993. TENA, C. *Guadalcanal, Seville.*—Oils, sumach.

994. TERO, G. *Alicant.*—Figs.

995. TERNERO Y PEÑA, *Guadalajara.*—Flour of superior quality.

996. THORICES, F. R. *Moguer, Huelva.*—Wines, vinegar.

997. TERUERO, J. *Marchenas, Seville.*—Oils.

998. TIO, M. *Valencia.*—Wines.

999. TIPPING, W. R. *London.* — Sherry wine.

1000. TOME, T. *Corrales, Zamora.*—Spirit of wine.

1001. TOMAR, Y. *Torre de Segre, Lérida.*—Wines.

1002. TORELLA, J. *Castellgali, Barcelona.*—Wine.

1003. TORRES, C. *Arahal, Seville.*—Chick peas, haricot beans.

1004. TORRES, J. *Corbin, Lérida.*—Canary seed.

1005. TORRES, L. *Guadalcanal, Seville.*—Wheat, barley, "alberjones."

1006. TORRES, MARQUIS OF LAS, *Seville.*—Agricultural products.

1007. TORRAJA, J. *Tarragona.*—Common table wine.

1008. TOUS, M. *Palma.*—Almonds.

1009. TOUS, M. D. *Seville.*—Agricultural products.

1010. TRAPERO, F. *Rota, Cadiz.*—Tintilla, muscat, and paxarette wines.

1011. TRAVEZ, G. *Toro.*—Almonds.

1012. TRELL, P. *Adra, Almeria.*—Pedro Ximenez wine.

1013. TRIAS, P. J. *Estorlas, Baleares.*—Soft wheat, "macandon" ditto, pulmonary lichen, ash-colour ditto.

1014. TRUSAT, R. *Tarragona.*—Wines.

1015. TUMAYA, R. *Tarragona.*—Common wine.

1016. UPMANN & Co. H. *Havannah.*—Cigars.

1017. UCEDA, M. *Berlanga, Soria.* — Onions.

1018. URRUTIA, M. *Tudela.*—Oil.

1019. URSAIZ, J. *Lucena del Puerto, Huelva.*—Muscat, Pedro Ximenez, and orange wines.

1020. VADO, M. *Guadalajara.*—Oil, white wheat.

1021. VALERO, J. *Elche, Alicant.*—Muscat wine.

1022. VALLE, G. *Recana.*—White wheat.

1023. VALLE, V. *Rivadesella, Asturias.*—Cider.

1024. Valles, A. *Castillsabas, Huesca.*—Wine.

1025. Valls, J. A. *Seville.*—Oils; one-eared wheat and dwarf maize, grown on dry lands; beans, honey.

1026. Various Landed Proprietors of Villena, *Alicant.*—Wines and spirits.

1027. Vargas, F. *Cassalla, Seville.*—Oil, chick peas.

1028. Vargas, R. *Cordoba.* — Wine, brandy, honey.

1029. Vargas, R. *Tarragona.*—Wines.

1030. Various Landed Proprietors of Huelva.—Brandies, chick peas, beans, honey, cheese.

1031. Vazquez, J. *Seville.*—Wheat, barley.

1032. Vazquez, L. R. *Moraleja del Vino, Zamora.*—Common wine.

1033. Vela, J. *Rota, Cadiz.*—Tintilla wines.

1034. Ventayo, D. *Barcelona.*—Superior wines.

1035. Vera, G. *Balearic Islands.*—Red wines, figs.

1036. Vera, M. *Havannah.*—Coffee from Las Lomas.

1037. Verdeja, J. A. *Seville.*—Chick peas.

1038. Verdugo, J. *Toledo.*—White wheat.

1039. Veruet, J. *Bardellos, Tarragona.*—Brandy, almonds.

1040. Vicente, P. *Saragossa.* — Wines; white haricot beans, fine-eared maize, early maize.

1041. Victor, J. & Co. *Xerez.*—Wines.

1042. Vidal, A. *Carthagena.*—Sorghum.

1043. Vidal, G. *Carthagena.* — Compressed figs.

1044. Vidal, R. *Saragossa.* — Wheat, flour.

1045. Vidal Brothers, *Tarragona.* — Wines.

1046. Vides, F. *Seville.*—Crackers.

1047. Vila, F. *Almatret, Lérida.*—Almonds.

1048. Vila, J. *Valencia.*—Wines.

1049. Vila, Widow of F. de, *Tarragona.*—Oils, brandies.

1050. Vilanova, P. *Alcalá, Castellon.*—Wine, oil.

1051. Vilanosa, D. *Tarragona.*—Vinegar.

1052. Vilaplana, P. *Onil, Alicant.*—Preserves.

1053. Vilches, F. *Almeria.*—Wines.

1054. Villalcazar (Marquis of), *Salamanca.*—Flour.

1055. Villalta, M. *Jaen.* — Aniseed, "canihueco" wheat.

1056. Villalonga, M. *Palma, Balearic I.*—Oil, honey.

1057. Villalonga, M. *Llunmayor, Balearic I.*—White wheat.

1058. Villalonga, M. *Porreras, Balearic I.*—Brandy.

1059. Villanueva, M. *Tudela.*—Wine, oil.

1060. Villanueva, M. *Tafaya.*—Wine.

1061. Villapanes, Count of, *Seville.*—Wheat, barley.

1062. Villapineda, Count of, *Seville.*—Agricultural produce.

1063. Villeri, J. M. *Zamora.*—Wine.

1064. Villores, M. D. *Alcalá, Castellon.*—Wine, oil, raisins.

1065. Vinagre, J. *Toro.*—Almonds.

1066. Vinez, J. *Reus.*—Wines.

1067. Viñas & Pamies, J. *Tarragona.*—Red and Malmsey wine, brandy, vinegar.

1068. Viñals, B. *Flassa, Gerona.*—Rye, Indian corn, beans.

1069. Viosca, F. *Fuente la Higuera.*—Oil.

1070. Yegros, S.—Oil, wine.

1071. Yuste, A. *Valladolid.*—Sausages.

1072. Zaforteza, J. *Palma.*—Almonds.

1073. Zayas, J. *Seville.*—Wheat.

1074. Zerpa, J. *Villanueva, Seville.*—Wines.

1075. Zorrilla & Co. *Burgo de Osma, Soria.*—Flour.

1076. ZOZAYA, M. *Errazu, Navarre.*—White beans.

1077. ZUBIRI, *Tafalla, Navarre.*—Wheat.

1078. ZULVAGA, F. *Corunna.*—Preserves.

1079. ZULUETA, J. *Havannah.*—Sugar.

CLASS IV.

1080. ABAD, L. *Province of Jecos-Sur, Philippines.*—Sapan wood.

1081. ALONSO, Y. *Mojados, Valladolid.*—Madder.

1082. ALFONSO, R. *Játiva.*—Starch.

1083. ALVAREZ PINILLA, *Corunna.*—Soap.

1084. ARAZO, BROTHERS, *Valencia.*—Cork.

1085. ARBONES, J. *Almatret, Lérida.*—Aniseed.

1086. ARGUINDEY, WIDOW OF, *Puerto de Bejar, Salamanca.*—Glue.

1087. ARNAO, J. *Belloch, Lérida.*—Barilla.

1088. AYNES & Co. *Valladolid.*—Madder, garancine.

1089. AYUNTAMIENTO (TOWN CORPORATION) OF CARTHAGENA.—Sugar-cane, palm-leaf brooms, hemp grass.

1090. BALBAS CASTRO, J. *Province of Laguna, Philippines.*—Barks suitable for making paper.

1091. BALLER, J. *Cordova.*—Sumach.

1092. BARRION, F. *Albay, Bacacay Town, Philippines.*—Hemp.

1093. BASTANT DE CONSTANTIN, L. G. *Barcelona.*—Starch from the yarrow plant.

1094. BELDA, A. *Valencia.*—Bastard saffron.

1095. BERNER, F. *Elche, Alicante.*—Almond gum, green almond bark.

1096. BORDERIAS, V. *Montmesa, Huesca.*—Camomile.

1097. BOYERO, J. *Cáceres.*—Corks.

1098. BRIZUELA, J. M. *Salamanca.*—Oil of aniseed, crystallized.

1099. CABRERA & Co. *St. Juan, Puerto Rico.*—Tallow candles.

1099a. CALDERA, M. *Philippine Islands.*—Fibres.

1100. CAMBRELEN, V. *Province of Laguna, Philippines.*—Barks, suitable for making paper, arboreous fibres, sundry samples of gutta percha from district of La Infanta.

1101. CANS, P. DE, *Gerona.*—Cork.

1102. CARREÑO, J. *Seville.*—Stearine candles, soap.

1103. CASADO, M. *Malaga.*—Sugar-cane.

1104. CASAS DE GONZALEZ, DOÑA CALISTA, *Province of Lebú, Philippines.*—Aromatic lozenges for perfuming.

1105. CATALAN, J. *Monreal del Campo, Teruel.*—Brown saffron.

1106. CIFRA, P. *St. Cruz de Teneriffe.*—Aloes.

1107. CARRERO, F. *Gerona.*—Royal arms of England in cork.

1107a. CONQUISTA, MARQUIS DE LA, *Cáceres.*—Fleece.

1107b. CASTELL & SERRA, *Barcelona.*—Sculpture in wood.

1108. CORTES GOVÁNTES, L. *Province of Zambales, Philippines.*—Tortoise-shell, ambabay bark, woods, hemp, reeds, &c.

1109. COMPANY, GURRI's, *Barcelona.*—Oak and yew tree.

1110. COMMISSION FROM CENTRAL HAVANNAH.—Woods.

1111. COMMISSION FOR THE EXHIBITION, *Almeria.*—Hemp-grass.

1112. CONRADI, T. *Seville.*—Soap.

1113. CUERPO (CORPS) OF ENGINEERS OF WOODS AND FORESTS, *Madrid.*—Collection of woods, comprising 312 different species, prepared after Rossmessler's method.

1114. CUNNINGHAM,—,*Seville.*—Liquorice.

1115. CURA (PAROCHIAL PRIEST), *Province of La Pamponga, Philippines.*—"Palasan," measuring 100 yards in length.

1116. COURT & SON, *St. Juan de Aznalfarache, Seville.*—Perfumery.

1117. CRÉDITO MOBILIARIO BARCELONÉS, *Barcelona.*—Woods from Muniello, Asturias.

1118. DELGADO, V. *Zalamea-la-Real, Huelva.*—Wax.

1119. DOMPER, G. *Tayabas, Philippines.*—Resin extracted from the pili-tree of Mount Atimonan, hemp ropes from Mauban and Cabo Negro, vase from Maminga.

1120. DURAN, A. *Almeria.*—Coloquintida.

1121. ESTER, J. *Seville.*—Soap.

1122. FERNANDEZ, M. *Almeria.*—Hemp.

1123. FIERRO, J. *St. Cruz de la Palma, Canary Islands.*—Cochineal, woods.

1124. FLUJÁ, F. *Elche, Alicant.*—Palmwood.

1125. FONOLLERAS, E. *Gerona.*—Cork in the rough, boiled ditto, wooden tobacco-pipes.

1126. GAGO, E. *St. Clara de Avedilla, Zamora.*—Sumach.

1127. GAGO ROPERUELOS, M. *Zamora.*—Camomile.

1128. GALIANO, M. *Seville.*—Casks.

1129. GALLARDO, L. *Barcelona.*—Starch from the yarrow plant.

1130. GARCIA, J. R. *Valencia.*—Cochineal.

1131. GARCIA, M. L. *Cortejana, Huelva.*—Corks.

1132. GARCIA ACEÑA, P. *Seville.*—Starch.

1133. GASTON, V. *Seville.*—Corks.

1134. GARRET, SAENZ, & Co. *Malaga.*—Stearine cakes, candles, and tapers.

1135. GEVENZÚ & SON, *Cascante, Navarre.*—Stearine candles.

1136. GIMÉNEZ, BROTHERS, *Mora, Toledo.*—Hard soap.

1137. GIRONIERA, P. *Philippines.*—Bark of indigenous trees.

1138. GOVERNOR OF THE PROVINCE OF ALICANT.—Almond gum, samples of palmwood and straw, starch, green almond bark, soda ashes.

1139. GOVERNOR OF THE PROVINCE OF CUENCA.—Bastard saffron.

1140. GOVERNOR OF THE PROVINCE OF SEVILLE.—Starch.

1141. GOVERNOR OF THE DISTRICT OF LEYTE, *Dulac Town, Philippines.*—Hemp.

1142. GÓMEZ SALAZAR, Y. *Almeria.*—Hemp grass in its natural state and in paste, various articles made from the same.

1143. GONZALEZ ESQUIVEL, J. G. *Marianas, Philippines.*—Saffron.

1144. GOTZENS, DELOUSTAL, & Co. *Barcelona.*—Soaps made from both animal and vegetable substances, soap adapted to salt water.

1145. GRACIAN & Co. *Malaga.*—Soap.

1146. GUERRA, A. *Valladolid.*—Madder.

1147. GUERRERO & SONS, WIDOW, *Mora, Toledo.*—Hard soap.

1148. GUEVARA, E. *Province of Bantangas, Philippines.*—Buri-leaf for hats.

1149. GUIMENO, J. *Saragossa.*—Soap.

1150. GUZMAN, V. DE ——.—Barks.

1151. HENNA, J. *Ponce, Puerto Rico.*—Oils of sesame, cocoa nut, pea nut, and others; "higuereta" seed.

1152. HERAS, M. *Zamora.*—Aniseed oil.

1153. HITA, R. M. *Havannah.*—Oils of cocoa nut and pea nuts.

1154. INSPECTORS OF WOODS AND FORESTS OF GERONA.—Articles in cork.

1155. INSPECTORS OF WOODS AND FORESTS OF THE PROVINCE OF SARAGOSSA.—Sample of woods, roots, liquorice.

1156. INURRIA, J. DE, *Seville.*—Liquorice in paste.

1157. INSPECTORS OF WOODS AND FORESTS OF LEON.—Woods.

1158. JUNTA (BOARD) OF AGRICULTURE, INDUSTRY, AND TRADE OF GRANADA.—Sugar cane.

1159. JUNTA (BOARD) OF AGRICULTURE OF PALENCIA.—Indigenous woods.

1160. JUNTA (BOARD) OF AGRICULTURE OF CORUNNA.—Soap.

1161. JUNTA (BOARD) OF AGRICULTURE OF CORUNNA.—Soap.

1162. LACAVE, J. P. *Seville.*—Samples of corks and bung corks.

1163. LETE, N. *Province of La Union, Philippines.*—A log of sapan wood.

1164. LIZARBE, P. *Berlanga, Soria.*—Stearine candles.

1165. LLANOS, FRAY ANTONIO, *Province of Bulacan, Philippines.*—Reeds.

1166. LÓPEZ, A. *Corcos, Valladolid.*—Aniseed.

1167. LÓPEZ DE AYALA, L. *Guadalcanal.*—Sumach.

1168. LUEBAN DE ST. MIGUEL, A. *Camarines-Norte, Philippines.*—Resin from the ladiangao tree.

1169. LUZ BURGALESA, MANUFACTORY OF, *Burgos.*—Stearine candles, soap.

1170. MARTICORENA, R. *St. Sebastian.*—Soap and candles.

1171. MIRANDA, A. *Oviedo.*—Woods from the forests of Quiros.

1171a. MONTERO, S. *Alcala del Rio.*—Fleece.

1172. MONTFORT, F. *Torrente de Cinca, Huesca.*—Aniseed.

1173. MORA, B. *Porrera, Baleares.*—Saffron.

1174. MORÁ, F. *Villanueva de los Castillejos, Huelva.*—Wax.

1174a. MIÑON, F. *Leon.*—Flax.

1174b. NATAL, J. *Leon.*—Flax.

1175. OLIVER, F. *Seville.*—Corks.

1176. OROZ, F. *Pampeluna.*—Linseed oil.

1177. PADILLA, F. *Almeria.* — Spanish hemp-grass, in paste.

1177a. PALAZUELO, VISCOUNT OF, *Toledo.*—Fleece.

1177b. PEREZ MARCO, D. *Belen.*—Fleece.

1178. PELAEZ, M. *Salamanca.*—Starch.

1179. PEREZ CARDENAL, *Zamora.*—Dye-wood, lavender water.

1180. PERLA, F. *Madrid.*—Candles and soap.

1181. PRIEGO MÁRMOL, J. *Cordova.*—Sumach.

1182. PROVINCE OF NAVARRE, *Pampeluna.*—Candles, wax and imitation matches.

1183. QUEMADA, J. *Cuellar, Segovia.*—Madder (pulverized).

1184. RAMIREZ, J. F. *Saragossa.*—Liquorice.

1185. ROYAL CROWN OF SPAIN.—Woods and cork.

1185a. RIPALDA, COUNT OF, *Valencia.*—Hemp.

1186. RIVAS, F. *Saragossa.*—Sabine wood.

1187. RIVAS, FRAY MANUEL, & MATEO RUSELL, *Cavite, Philippines.*—Goat skin for making strings.

1188. RIBOT, J. *Gerona.*—Cork carving, representing the Cathedral of Brussels.

1189. RICO, E. *Madrid.*—Soap.

1190. RISSECH, J. *Gerona.*—Samples of cork.

1191. ROBILLARD, F. *Valencia.*—Essences of flowers and aromatic herbs.

1192. ROSA, J. *Gerona.*—Corks.

1193. ROYERO, J. *Salorica, Cáceres.* —Corks.

1194. RUSELL, M. & FRAY MANUEL RIVAS, *Cavite, Yndan Town, Philippines.* — Fibres from Cabo Negro, buri-leaf.

1195. ST. ANDRES, COUNT OF, *Puerto de la Cruz, Canaries.*—Cochineal.

1195a. SACRISTAN, R. *Segovia.*—Fleece.

1196. SANCHO, A. *Gandia, Valencia.*—Sugar cane.

1197. ST. ROMAN, A. *Farmosella, Zamora.*—Water hemlock, hemlock seed, digitalis leaves.

1198. SERRANO, L. *Zalamea-la-Real, Huelva.*—Wax.

1199. SERRAPIÑANA, A. *St. Domingo.*—Wax, grindstones, vegetable gums.

1200. SOL, J. *Liñola, Lérida.*—Camomile.

1201. SOPEÑA & CO. *Burgos.*—Candles, wax, soap.

1202. SPENCER & RODA, *Almeria.*—Barilla.

1203. SOTELO, J. *Malaga.*—Soap.

1204. TAMARIT, MARQUIS OF, *Tortosa.*—Soda and barilla.

1205. TENA, C. *Guadalcanal.*—Sumach.

1206. TECSON, J. *Batangas, Philippines.*—Mecate from Cabo Negro, buri-leaf.

1207. TORRES, F. *Bulacan, Philippines.*—Ropes.

1208. TRABER, G. *Toro.*—Gum from fruit trees.

1209. TRIAS, P. *Esporlas, Baleares.*—Fibres of ash-tree.

1210. VARIOUS OF HUELVA.—Wax and candles.

1211. VALDES, T. *Valladolid.*—Natural and pulverized madder.

1212. VALERO, J. *Elche, Alicant.*—Soda and barilla.

1212a. VERA, R. *Soria.*—Fleece.

1213. VIDAL, A. *Carthagena.*—Barilla.

1214. VIDES, F. *Seville.*—Corks.

1215. VILCHES, F. *Almeria.*—Palmetto.

1216. VILLARIN, B. J. *Laguna, Philippines.*—Creeper without knots, measuring eight yards in length.

1217. WIEDEN, F. *Seville.* — Cork in sheets.

1218. YURRITA, J. *Tolosa, Guipuzcoa.*—Seville wax matches with imitation tops.

1219. ZAMORA, P. *Ylocos-Sur, Philippines.*—Gums, aromatic ditto, sapan wood.

1220. ZOZAYA, M. *Errazo, Navarre.*—Wax.

CLASS V.

1221. ASTRUA, D. *Cordova.*—A waggon wheel, constructed so as to move without friction at the axle.

1222. GALLARDO, L. *Barcelona.*—Model of a locomotive moved by hydrogen gas.

1223. SOIGNY, A. DE, *Aviles, Asturias.*—Model of rails.

CLASS VI.

1224. GALLEGOS.—Carriages.

1224a. SANIDAD MILITAR (SANITARY DEPARTMENT).—An ambulance waggon.

CLASS VIII.

1225. BERGUE, M. DE, *Barcelona.*—Punching machine of his own invention.

1226. BERNER, F. *Elche, Alicant.* — A press for various purposes.

1227. BRIDGEMAN, E. *Tarragona.*—A gas-meter.

1228. CIERVO & CO. *Barcelona.*—A gas-meter for 60 burners, gas apparatus.

1229. GOVERNMENT OF PROVINCE OF ALICANT.—An apparatus for pressing and extracting fluids.

1230. MALABOUCHE, F. *Valencia.* — A letter-weigher, lever and scale.

1231. TOSSER & CO. *St. Sebastian.*—Machinery.

CLASS IX.

1232. ASPE, J. *Seville.*—Agricultural implements.

1233. ESCUELA DE VETERINARIA (VETERI-NARY SCHOOL) OF CORDOVA.—Hippometer for measuring horses.

1234. LAGIGAS, F. DE LAS, *Manilla.*—Ploughs from the foundry of Acle.

CLASS X.

1235. FIVALERT DE VELLART, *Barcelona.*—Design for a village school.

1236. GOVERNOR OF THE PROVINCE OF SEVILLE.—Design for a monument to Murillo.

1237. HERRERO, J. *St. Sebastian.*—Architectural ornaments.

1238. LECUMBERRI, C. *Madrid.*—Model for a lunatic asylum.

1239. MANTEROLA CORTAZAR & Co. *St. Sebastian.*—Building materials.

1240. MUÑOZ, A. *Madrid.*—Models for masonry.

1241. PICKMAN & Co. *Seville.*—Bricks and tiles.

1242. ROGENT, E. *Barcelona.*—Design for the University of Barcelona.

1243. SEDÓ, J. *Valencia.*—Model of a wooden bridge.

1244. N. N. *Tarragona.*—Coloured bricks.

CLASS XI.

1245. CARBONELL, A. *Alcoy, Alicant.*—Military uniform.

1246. "CUERPO DE ARTILLERIA," *Toledo.*—Rifled artillery, projectiles, fire-arms, daggers, and side arms.

1247. "DIRECCION GENERAL DEL ARMA DE INFANTERIA," *Toledo.*—Uniform for a fusilier of the Barbastro Battalion of Chasseurs, No. 4; ditto of the Provincial Battalion of Madrid, No. 43.

1248. "DIRECCION DE LA GUARDIA CIVIL," *Toledo.*—Uniforms for a foot and a mounted guard.

1249. "DIRECCION GENERAL DE ESTADO MAYOR," *Madrid.*—Topographical plans.

1249a. FACTORY OF TRUBIA, ROYAL.—Large rifled cannon and projectiles.

1250. FACTORY OF ARMS OF OVIEDO, ROYAL.—Rifled cannon, fire arms, and projectiles.

1251. FACTORY OF ARMS OF OVIEDO.—Minié's rifled guns and carbines.

1252. FURNES, J. *Bich.*—Miniature uniform for an infantry officer.

1253. LORENZALE, M. *Madrid.*—Military articles.

CLASS XII.

1254. ARSENAL OF CARTHAGENA.—Sailcloth and rigging.

1255. BEECH, J. *Manilla.*—Hemp rigging.

1256. BRAÑA, B. *Corunna.*—Sail-cloth.

1257. EUSTER LABHARD & Co. *Manilla.*—Hemp rigging.

1258. FABRA, C. *Barcelona.*—Fishing nets.

1259. FONTANILLA, M. *Philippines.*—Sail-cloth.

1260. GOVERNOR OF THE PROVINCE OF ALICANT.—Sail-cloth and cordage.

1261. JUNTA (BOARD) OF AGRICULTURE, INDUSTRY, AND TRADE OF CASTELLON.—Cordage.

1262. LOZANO VERA, E. *Orihuela, Alicant.*—Sail-cloth.

1263. MASDEN, M. *Malaga.* — Cast-iron diving apparatus.

1264. MATEO, J. DE, & M. FONTANILLA, *Ylocos-Norte, Lavag Town, Philippine Islands.*—Sail-cloth.

1265. MESA, SANTA, FACTORY, *Manilla.*—Cordage.

1266. ORTIZ, F. *Ylocos-Norte, Vigan Town, Philippines.*—Sail-cloth.

1267. PALAZUELOS, VISCOUNT OF, *Toledo.*—Cordage.

1268. PERICA, WIDOW OF, & SONS, *Palma, Baleares.*—Cordage, canvas, &c.

2169. ROJAS, BROTHERS, *Manilla.*—Cordage.

1270. RODRÍGUEZ, J. *Elche, Alicant.*—Cordage.

1271. SOCIEDAD ECONÓMICA DE MURCIA.—Sail-cloth.

CLASS XV.

1272. JUSTE VILLANUEVA, T. — Pocket chronometer.

1273. LOSADA, J. R. *Madrid and London.*—Chronometer clocks, astronomical pendulum, pocket chronometers, and other watches

CLASS XVI.

1274. ALBERT, J. *Valencia.*—Guitar and other strings.

1275. CABALLET, L. *Seville.* — Vertical piano.

1276. GUARRO, M. *Barcelona.*—Two vertical pianos.

1277. MONTANO, V. *Madrid, 3, Calle de San Bernardino.*—A grand pianoforte and a piccolo.

1278. PIAZZA, C. *Seville.*—Vertical piano.

CLASS XVII.

1279. BOUSQUET, A. *Valencia.* — Set of teeth.

1280. DIRECCION DE SANIDAD MILITAR.—Sanitarium.

1281. GALLEGOS, J. *Madrid.*—Artificial arm.

1282. PONS SOUBIRAT, Y. *Barcelona.*—Show-case with lints.

1283. RILAS, R. *Valencia.*—Set of teeth.

1284. RONAULT, BROTHERS, *Madrid.*—Orthopædic apparatus.

1285. TORREZ MUÑOZ LUNA, R. *Madrid.*—Medico-surgical case.

1285a. VILAR DE PSAYLA, *Madrid.*—Artificial teeth.

CLASS XVIII.

1286. BELTRAN, C. *St. Miguel, Ylocos-Norte, Philippines.*—Table-linen.

1287. BUSTAMANTE, S. & R. *Lavag, Ylocos-Norte, Philippines.*—Petticoat.

1288. CABEL, P. *St. Nicholas, Ylocos-Norte, Philippines.*—" Coyote."

1289. CÁRDENAS CHAVEZ, M. *Havannah.*—Cotton.

1290. CASTILLO, M. DEL, *Seville.*—Cotton.

1291. CIFRA, P. *St. Cruz de Teneriffe, Canary Islands.*—Cotton.

1292. COMMISSION, CENTRAL, *Havannah.*—Cotton.

1293. COSTA, J. *Havannah.*—Cotton.

1294. FERRER, J. & Co. *Barcelona.*—Madapolam, white and printed cambric, brillantine, organdie, jaconet, printed shawls.

1295. GONUS, T. *Manresa.*—Cotton.

1296. GÜEL & Co. *Barcelona.*—Cotton velvet, various colours and designs.

1297. JUNTA (BOARD) OF AGRICULTURE, INDUSTRY, AND TRADE OF GRANADA.—Cotton.

1298. LARA, M. *Valladolid.* — Coloured meltons.

1299. LARA VILLARDELL & SONS, *Valladolid.*—Cotton goods.

1300. LLANOS, T. *Cienfuegos, Cuba.* — Cotton.

1301. LEAÑO, B. *Bintan, Ylocos-Norte, Philippines.*—Striped goods.

1302. LUNA, COUNT OF, *Elche, Alicant.*—Cotton in the pod.

1303. MISA, J. *Lavag, Ylocos-Norte, Philippines.*—Petticoat.

1304. RUIZ DE LA PARRA, G. *Lacavada, Santander.*—Unbleached cotton goods.

1305. RUSELL, M. & FRAY MANUEL RIVAS, *Yndan, Prov. de Cavite, Philippines.*—Cotton towels.

1306. SAVAS, D. *Batag, Ylocos-Norte, Philippines.*—Gingham.

1307. SERRAPIÑANA, A. *St. Domingo.*—Cotton.

1308. SOLER, V. *Rellina.*—Raw cotton and cotton seeds.

1309. TAMAYO, U. *Lavag, Ylocos-Norte, Philippines.*—Striped goods.

1010. VILLANUEVA, F. *Province of Batangas, Philippines.*—Raw cotton.

CLASS XIX.

1311. ALCALDE (MAYOR) OF AGREDA, *Soria.*—Raw hemp and flax.

1312. ALFARO, F. *Valladolid.*—Damasked linen.

1313. ALONSO, T. *Orense.*—Flax.

1314. ALOS, T. A. *Balaguer, Lérida.*—Hemp.

1315. ANTON GUTIERREZ, B. *Zamora.*—Flax.

1316. BARRION, F. *Bacacay, Province of Albay, Philippines.*—Hemp.

1317. BECTH, J. *Buria, Philippines.*—Hemp.

1318. CALDERA, M. *Ligas, Albay, Philippines.*—Hemp.

1319. CARRETERO, F. *Seville.* — Hemp, table and other linen goods.

1320. COMMISSION FOR THE EXHIBITION, *Almeria.*—Articles made from Spanish hemp.

1321. CHICO, A. *Cehegin, Murcia.*—Hemp.

1322. CORTES GOVÁNTES, L. *Batangas, Philippines.*—Hemp, and other agricultural products.

1323. ESCUDERO, M. *Orihuela, Alicant.*—Hemp.

1324. ESTEPA, T. *Urrea de Talon, Saragossa.*—Flax.

1325. ESTÉVEZ, J. C. *Orense.* — Table linen.

1326. FERREIRO, J. *Orense.*—Flax.

1327. FRUTOS, A. *Seville.*—Thread goods.

1328. GAGO, ROPERUELOS, *Zamora.*—Flax.

1329. GARCIA, ALONSO, A. *Moral de Orbigo.*—Flax.

1330. GOVERNOR OF THE PROVINCE OF ALICANT.—Dressed flax.

1331. GOVERNOR OF THE PROVINCE OF GRANADA.—Hemp.

1332. GOVERNOR OF THE LEYTE DISTRICT, *Dulac Town, Philippines.*—Hemp.

1333. GOVERNOR OF THE PROVINCE OF ORENSE.—Raw flax, table-linen, towels, napkins, and Gallician linen goods.

1334. GUIER, J. *Valencia.*—Hemp.

1335. JIMENO, J. *La Milla del Rio, Leon.*—Flax.

1336. JUNTA (BOARD) OF AGRICULTURE, INDUSTRY, AND TRADE OF CASTELLON.—Hemp goods.

1337. JUNTA (BOARD) OF AGRICULTURE, INDUSTRY, AND TRADE OF GRANADA.—Hemp and flax.

1338. LAFUENTE, R. M. *Valladolid.*—Flax.

1339. LOZANO VERA, E. *Orihuela, Alicant.*—Hemp.

1340. MACÉRES, J. *Orihuela, Alicant.*—Hemp.

1341. MALDONADO, M. *Orense.*—Gallician linen.

1342. MAÑA & YSERT, *Barcelona.*—Drills.

1343. MARTÍNEZ, F. *Tarragona.*—Hemp, flax.

1344. MATA & CO. *Barcelona.* — Blue striped damasked linen ticking, a hammock.

1345, MORA, J. *Callosa de Segura, Alicant.*—Hemp.

1346. MUÑOZ, F. *Leon.*—Flax.

1347. OLAN, A. *Játiva.* — Samples of hemp.

1348. PALAZUELOS, VISCOUNT OF, *Toledo.*—Hemp.

1349. PARRA, J. R. *Santibanez de Toro, Zamora.*—Flax.

1350. PUIG, F. *Barcelona.*—Sewing thread.

1351. RIPALDA, COUNT OF, *Valencia.*—Raw hemp.

1352. RIVAS, FRAY MANUEL, & MATEO RUSELL, *Yndan, Province of Cavite, Philippines.*—Hemp.

1353. ROCA DE TOGORES, B. *Orihuela, Alicant.*—Flax.

1354. RUBIO & RODRIGUEZ, *Seville.* — Thread goods.

1355. SADÓ, J. *Barcelona.* — Damasked napkins, table-linen, and towels.

1356. SANZ, B. *Burgo de Osma, Soria.*—Raw flax.

1357. TAMARIT, MARQUIS OF, *Tarragona.*—Hemp.

1358. VACA, M. *Hospital de Orbigo, Leon.*—Flax.

1359. VALERO, J. J. *Valencia.*—Hemp.

1360. VEGAS, A. *Hospital de Orbigo, Leon.*—Flax.

1361. VIDAL, SEMPRUM, & CO. *Vallodolid.*—Unbleached linen.

CLASS XX.

1362. ABAJA, P. V. *Vigan, Ylocos-Sur, Philippines.*—Silk tapestry.

1363. ACOSTA, M. *Lavag, Ylocos-Norte, Philippines.*—Silk handkerchiefs.

1364. ALMANZA, D. *Murcia.*—Sewing silk.

1365. BLANCA, J. *Macabebe, Pampanga, Philippines.*—Silk tapestry.

1366. BELTRAN, C. *St. Miguel, Ylocos-Norte, Philippines.*—Silk handkerchiefs.

1367. BONELL, T. R. *Valencia.*—Brocade of silk and gold.

1368. BUNNANG, A. *St. Miguel, Ylocos-Norte, Philippines.*—Silk handkerchiefs.

1369. CANTÓ, WIDOW OF, & SONS, *Alcoy, Alicant.*—Samples of silk.

1370. CARRERE, E. *Grans, Huesca.*—Silk and silk gauze.

1371. CASTILLO, M. *Seville.*—Silk handkerchiefs, brocaded silk (various colours), shawls, damask goods, &c.

1372. CASTRO, Q. *Bulacan, Philippines.*—Silk tapestry.

1373. DULDULAO, F. *St. Miguel, Ylocos-Norte, Philippines.*—Silk handkerchiefs.

1374. ESCUDER, WIDOW OF, & SONS, *Barcelona.*—Figured silk goods, glacé ditto, &c.

1375. FIERRO, T. *St. Cruz de la Palma, Canary Isles.*—Spun, woven, and raw silk.

1376. FONTANILLA, M. *Lavag, Ylocos-Norte, Philippines.*—Silk handkerchiefs.

1377. GALINDEZ, T. *Pavay, Ylocos-Norte, Philippines.*—A mantelet.

1378. GARIN, M. *Valencia.* — Tapestry, goods in silk and gold.

1379. GARRIGA, L. *Manresa.*—Raw and spun silk, &c.

1380. GOVERNMENT OF THE PROVINCE OF ALICANT.—Samples of thrown silk.

1381. GOVERNMENT OF THE PROVINCE OF SEVILLE.—Dyed silks.

1382. GÓMEZ SALAZAR, Y. *Almeria.*—Spun silk and cocoons.

1383. HOMS, T. *Barcelona.*—Glacé and other silks.

1384. HORTAL, J. *Salamanca.*—Cocoons.

1385. JUNTA (BOARD) OF AGRICULTURE, INDUSTRY, AND TRADE OF CASTELLON.—Raw silk.

1386. LALDUA, A. *Lavag, Ylocos-Norte, Philippines.*—Handkerchiefs.

1387. LÓPEZ UBEDA, T. *Valencia.*—Spun silk.

1388. LOZANO VERA, E. *Orihuela, Alicant.*—Spun silk.

1389. MARIN, B. *Murcia.*—Cocoons.

1390. MARTÍNEZ & Co. M. *Seville.*—Silks.

1391. MOLINA-NUÑEZ, F. *Orihuela, Alicant.*—Spun silk.

1392. MORENGO-MIGUEL, V. *Madrid.*—Valencia silk.

1393. PEÑAFIEL, E. *Murcia.*—Silk, spun and thrown by a single process; cocoons.

1394. PÉREZ-VILLORIA, R. *Salamanca.*—Spun silk.

1395. PORTALES & Co. *Talavera de la Reina.*—Spun silk.

1396. PUJAL (WIDOW OF) & Co. *Valencia.*—Spun and thrown silk.

1397. RALLITA, R. *Pavay, Ylocos-Norte, Philippines.*—A mantelet.

1398. REAL, J. *Seville.*—Galloons.

1399. REIG, J. *Barcelona.*—Piedmontese crape shawls, plain and figured, of various qualities.

1400. ROSELLÓ, J. *Reus, Tarrogona.*—Silk and velvet goods, damasks, &c.

1401. RUBIO, BROTHERS, *Valencia.*—Damask, velvet, and silk goods.

1402. SANTONJA, F. *Barcelona.*—Silk and velvet ribbons, satin ditto, glacé and other kinds ditto, suitable for decorations.

1403. SATORRAS, A. *Tarragona.* — Raw silk.

1404. SABER & TENA, M. *Valencia.*—Silk tapestry.

1405. SOCIEDAD ECONÓMICA DE MURCIA.—Cocoons, thrown and dyed silk.

1406. SORNY (WIDOW OF) & SONS, *Valencia.*—Moiré silk ribbons.

1407. TORNES ORTIZ, M. *Valencia.*—Spun and thrown silk.

1408. TRELL, J. *Berja, Almeria.*—Cocoons.

1409. VILLANUEVA, G. *Batangas, Philippines.*—Silk tapestry.

1410. VILLANUEVA, E. *Lavag, Ylocos-Norte, Philippines.*—Handkerchiefs.

1411. VILUMARA, BROTHERS, & Co. *Barcelona.*—Glacé and other silk goods.

1411*a.* VICTOR, F. *Valencia.*—Silks.

CLASS XXI.

1412. AMAT, TRIES, & VIETA, *Tarrasa and Barcelona.*—Black and coloured cloth, black and blue satin cloth, fancy woollen goods, beaver.

1413. ANDRES, J. B. *Alcoy, Alicant.*—Woollen goods.

1414. ARROYO, G. *Palencia.*—Shawls.

1415. BERNARDO, M. *Bernardos, Segovia.*—Coarse cloth, worsted goods.

1416. BONAPLATA, N. *Seville.* — Worsted of various colours.

1417. BUXAREU & MASOLIVER, *Barcelona.*—Hosiery, under-garments, woollen fabrics for making gloves, &c. ; merino wool.

1418. BUXEDA, D. & CO. *Barcelona.*—Cloth, beaver, ratteens, plain and figured zephyrs, woollen velvet, satin cloth, &c.

1419. CANTÓ, WIDOW OF, & SONS, *Alcoy, Alicant.*—Woollen cloth.

1420. CAMMANY, J. *Barcelona.* — Cloth, satin cloth, fancy woollen goods.

1421. CASANOVAS & SONS, *Sabadell.*—Pilot cloth, fancy woollen goods.

1422. CASANOVAS & BOSCH, A. *Sabadell.*—Plain and fancy woollen goods.

1423. CASTELL, BROTHERS, *Seville.*—Spun wool of various colours.

1424. COMA, T. *Barcelona.*—Spun worsted.

1425. COMA & CO. *Barcelona.*—Merino shawls and Cashmere ditto.

1426. DURAN & CO. *Sabadell and Barcelona.*—Cloth ; satin cloth and beaver ditto, various colours ; trouserings, and other fancy goods.

1427. CONQUISTA, MARQUIS OF LA, *Trujillo, Cáceres.*—Saxon and merino wool.

1428. GALI & CO. *Tarrasa.*—Broadcloth, light ditto, zephyrs, kerseymeres, pilot cloth, trouserings, and other fancy woollen goods.

1429. GARCIA, A. *Cordova.* — Woollen cloth.

1430. GOVERNOR OF THE PROVINCE OF ALICANT.—Raw wool, fancy woollen goods.

1431. GOVERNOR OF THE PROVINCE OF SEVILLE.—Dyed worsted.

1432. GÓMEZ RODULFO, G. *Bejar, Salamanca.*—Broadcloth.

1433. GÓMEZ SERNA, A. *Cordova.*—Various kinds of wool, worsted.

1434. GONZALEZ, M. *Valdeavellano de Tera, Soria.*—Articles in fine and mixed Saxon wool.

1435. GOMEZ SALAZAR, T. *Almeria.*—Raw Dahlia wool.

1436. GORINA, Y. *Sabadell.*—Broadcloth, beaver, &c.

1437. HERRERO, D. *Almeria, Zamora.*—Worsted.

1438. JORDA, F. *Alcoy, Alicant.*—Woollen goods.

1439. LOSADA, D. R. *Ilescas, Toledo.*—Churra wool.

1440. MAÑÁ & YSERT, *Barcelona.*—Ratteens, kerseymeres, and other fancy woollen goods.

1441. MAIGUEZ, T. *Valencia.*—Raw wool.

1442. MARÓ, M. *Villalpando, Zamora.*—Wool of native sheep.

1443. MONTERO, S. *Alcalá del Rio, Castel Blanca.*—Fine merino wool.

1444. PALAZUELOS, VISCOUNT OF, *Toledo.*—Worsted.

1445. PEREZ MARCO, *Alicant.*—Wool.

1446. PINTO DA COSTA, R. *Lumbrales, Salamanca.*—Shawls, sackcloth.

1447. SACRISTAN, R. *Segovia.*—Wool.

1448. SALLARES, J. *Sabadell, Barcelona.*—Cloth, beaver, trouserings, and other woollen goods.

1449. SAMUEL, E. *Seville.*—Worsted.

1450. SANTOS, N. & CO. *Tolosa, Guipuzcoa.*—Beaver cloth, black and coloured cloth, fancy woollen goods, and satin cloth.

1451. SERRET & PALAU, *Sabadell.*—Shawls of various designs and colours.

1452. SOLÁ & CO. *Barcelona.*—Shawls, silk and woollen fabrics for ladies' use, articles for waistcoats, fancy woollen goods.

1453. SPENCER & RODA, *Almeria.*—Raw wool.

1454. TASTEL, BROTHERS, *Seville.*—Spun wool, of different colours.

1455. TELLO, J. V. *Valencia.*—Woollen cloth.

1456. TEROL & GISBERT, *Alcoy, Alicant.*—Fine woollen cloth.

1457. TURULL, P. *Sabadell.*—Fancy woollen goods and beaver cloth.

1458. VERA, R. *Soria.*—Fine and merino wool, worsted.

1459. VIETA & CO. *Tarrasa.*—Broadcloth, satin cloth, and fine woollen fabrics.

CLASS XXII.

1460. BOCHS, J. *Province of Pangasinan, Philippine Islands.*—Mats.

1461. CORTES GÓVANTES, L. *Zambales, Philippine Islands.*—Mats.

1462. QUIBLIER & CO. *Barcelona.*—Carpets, moreen, articles for hangings.

1463. SOCIETY, ECONÓMICA, OF MURCIA.—Blankets.

1464. TELLO TICULART, J. V. *Valencia.*—Blankets, counterpanes, &c.

1465. TORRES, A. *Calumpit, Bulacan Province, Philippine Islands.*—Buri-leaf mats.

1466. VIDAL, J. *Baleares.*—Carpets.

CLASS XXIII.

1467. ACHON, J. *Barcelona.* — Cotton prints.

1468. RICART & SON, J. *Barcelona.*—Cotton prints.

CLASS XXIV.

1469. ANDINO, C. *St. Juan, Puerto-Rico.*—Open-work linen sheet.

1470. BOSCH, N. *Bayamon, Puerto-Rico.*—Open-work towel.

1471. CABAÑERAS, J. A. *Barcelona.*—Silk lace shawls, lace mantillas and flounces.

1472. CAMMANY & VOLART, *Barcelona.*—Chantilly lace shawl, veils, silk lace, &c.

1473. FITER, J. *Barcelona.*—Black veils, black lace for parasols, silk lace for mantillas; lace shawls, mantillas, &c.

1474. GARCIA-ATIENZA, *Valencia.* — Embroidery.

1475. HIDALGO, R. *Palma de los Rios, Cordova.*—Embroidery.

1476. JUNTA (BOARD) OF AGRICULTURE OF CORUNNA.—Silk embroidery on paper.

1477. MARGARIT LLEONART, J. *Barcelona.*—Silk lace mantle, black silk lace belts, point-lace collars, and sundry articles in guipure, lace, and blonde.

1478. MONTEIRO, N. *London.*—Linen-tape guipure work.

1479. PADRON DE MACAYA, N. *Aguadilla, Puerto Rico.*—Gold thread embroidery.

1480. PÉREZ, —, *Jaen.*—Embroidery.

1481. ROIG MARCH, Y. *Barcelona.* — A veil-mantilla, embroidered in chiaro-oscuro; imitation Brussels veils, &c.

1482. SERRANO, B. *Linares, Jaen.*—Embroidery.

1483. SUÁREZ, T. *Corunna.*—Embroidery.

CLASS XXV.

1484. ANQUIDAD, F. *Corunna.*—Kid skins.

1485. BOSCH RIVAS, J. *Vich.*—Calf-skin for gloves, chamois.

1486. BOSCH, J. *Vich.*—Kid skins.

1487. CORTES GOVÁNTES, L. *Leyte, Philippine I.*—Dressed skins.

1488. DIAZ, J. *Seville.*—Gazelle furs.

1489. GELI, F. & P. *Seville.*—Kid skins.

1490. GOVERNOR OF THE PROVINCE OF SEVILLE.—Sundry dyed skins.

1491. GRAU, J. *Seville.*—Brushes.

1492. ORTELES, C. *Barcelona.*—Strings, bracelets, brooch, and other articles in hair.

1493. PERRIER, Y. *Seville.*—Kid skins.

1494. ROMERO, J. *Seville.*—Brushes.

1495. RUSELL, M. & M. RIVAS, *Cavite, Philippine I.*—Skins for making strings.

CLASS XXVI.

1496. ARGUINDEY, WIDOW OF, & SONS, *Bejar, Salamanca.*—Sole and other leather.

1497. CADENA, J. *Malaga.*—Horse harness.

1498. CARABIAS, J. *Salamanca.*—Skins.

1499. FERNÁNDEZ, WIDOW OF, *Seville.*—Skins.

1500. GOVERNOR OF THE PROVINCE OF ALICANT.—Leather for sieves.

1501. GOVERNOR OF THE PROVINCE OF SEVILLE.—Spanish leather.

1502. GARCIA DORADO, G. *Valladolid.*—Saddle and harness.

1503. MANZANO, A. *Seville.*— Spanish leather.

1504. MARTICORENA, R. *St. Sebastian.*—Calf skins.

1505. MUÑOZ, M. *Placencia, Cáceres.*—Sole leather.

1506. RAMOS, J.` *Navalmoral, Cáceres.*—Sole leather.

1507. RODRÍGUEZ, A. *Corunna.*—Skins.

1508. ROMERO, P. *Fuente Pelayo, Segovia.*—Skins.

1509. TEJA, J. *Corunna.*—Sole leather.

1510. YGLESIAS, M. BROTHER, *Seville.*—Morocco leather and dressed skins.

CLASS XXVII.

1511. ABAD, P. *Valencia.*—Water-colour landscapes for fans.

1512. ABAYA, V. *Ylocos-Sur, Vigan, Philippine I.*—Chinese wood, cane, and bamboo ditto, worked over by the natives.

1513. ACEBEDO, A. *Valencia.*—Hats.

1514. ANQUIDAD, F. *Corunna.*—Gloves.

1515. BALADÓ, F. *Barcelona.*—Artificial cambric bouquets; garland of flowers; embroidered cambric basket, ornamented with flounces and Valenciennes lace, &c.

1516. BALBAS CASTRO, T. *Laguna Philippine I.*—Lanete cane.

1517. BOCHS, J. *Pangasinan, Philippine I.*—Buri-leaf hats.

1518. BOSCH, J. *Vich.*—Gloves.

1519. CASTILLO, M. *Malaga.*—Gloves.

1520. COLOMINA, J. *Valencia.*—Fans in sandal, ebony, box-wood, &c.

1521. CREIXACH, V. *Valencia.*—Sandals.

1522. CUBEDO, J. *Valencia.*—Lamb and kid-skin gloves.

1522a. DIAZ, J. *Burgos.*—Hats.

1523. DOMPER, G. *Tayabas Province, Philippine I.* — Malatapay cane, palm-leaf and other hats.

1524. ENSEÑAT, M. *Barcelona.*—Artificial flowers.

1525. ESPESO, D. *Valladolid.* — Brass buttons.

1526. ESQUERDO, T. *Villajoyosa, Alicant.*—Palm-leaf hats.

1527. FORTUN, J. *Saragossa.*—Hats, and illustrative designs of their manufacture.

1528. GAUDE, E. *Valladolid.*—Lamb and kid-skin gloves.

1529. GELI, F. & P. *Seville.*—Gloves.

1529a. GIL, A. *Burgos.*—Felt hats.

1530. GOMEZ, J. *Madrid.*—Various kinds of hats.

1531. GUERRERO, J. B. *Eccija.*—Hats.

1532. HORNA, M. *Zamora.*—Waterproof hats.

1533. JUNTA (BOARD) OF AGRICULTURE OF CORUNNA.—Boots and shoes.

1534. MARTICORENA, N. *St. Sebastian.*—Fine and common walking-sticks.

1535. MARTIN, F. *Valencia.*—Fans.

1536. MARTIN, M. J. *Salamanca.*—Shoes for peasants, boots.

1537. MASFARNER, J. *Valencia.*—Gloves.

1538. MENDIETA & CO. *Riara, Segovia.*—Pins.

1539. MOLINÁ, S. *Jaen.*—Boots and shoes.

1540. MITJANA, F. *Malaga.*—Fans.

1541. PERRIER, J. *Seville.*—Gloves.

1542. RAY, J. *Corunna.*—Hats.

1543. REINALDO, J. *Madrid.*—Boots and shoes.

1544. ROJAN, WIDOW OF, & SONS, *Seville.*—Brass buttons.

1545. TEJEDOR, R. *Bermillo de Sayago, Zamora.*—A sayaguese mantelet.

1546. VILCHES, D. *Seville.* — A pair of boots for Andalusian majo.

1547. VILLARIN, B. T. *Laguna, Philippine I.*—Walking-stick of "ojos del sol" wood.

1548. YGLESIAS & BROTHERS, M. *Seville.*—Boots and shoes.

CLASS XXVIII.

1549. ARZA, EIZMENDI, & CO. *Alegria and Tolosa, Guipuzcoa.*—Writing paper of various sorts, printing ditto, fine and common.

1550. CARBÓ, L. *Riba, Saragossa.*—Mill-boards.

1551. FABRICANTES DE ALCOY, *Alicant.*—Smoking paper.

1552. FORT, E. *Alcoy, Alicant.*—Cigarette paper.

1553. JUNTA (BOARD) OF AGRICULTURE, INDUSTRY, AND TRADE, *Granada.* — Continuous paper.

1554. LIZARBE & CO. *Cascante, Navarre.*—Smoking paper.

1555. LÓPEZ, H. 16 *Calle de St. Ann, Madrid.*—Smoking paper, prepared by a new process.

1556. MASUSTEGUI, M. *Castellano.* — Paper.

1557. PEÑA, J. M. *Candelario, Salamanca.*—Continuous paper.

1558. PEÑA, J. B. *Candelario, Salamanca.*—Fine paper.

1559. PROVINCE OF NAVARRE, *Pampeluna.*—Paper for cigarettes.

1560. RIPALDA, E. *Alegria and Tolosa, Guipuzcoa.*—Paper for cigarettes.

1561. ROMANI & MIRO, *Barcelona.* — Drawing papers, cards.

1562. ROMANI & OLIVELLA, *Barcelona.*—Paper of all kinds.

1563. ROMANI & TARRES, *Barcelona.*—Drawing and other paper, paper for cigarettes.

CLASS XXIX.

1564. Arens, P. *Barcelona.*—Apparatus for teaching the blind to write, and a pamphlet illustrative of the method to be adopted.

CLASS XXX.

1565. Altamira Codina, A. *Barcelona.*—Mosaic in wood, composed of 4,826 pieces.

1566. Administration of the Alhambra, *Granada.*—Arabesques.

1567. Aleman, V. *Alicant.*—Bedstead, changeable into a column.

1568. Arraez, A. *Madrid.*—Arabesques.

1569. Ballesteros, S. 34 *Carrera de St. Geronimo, Madrid.*—Paper-hangings.

1570. Botana, J. *St. Eulalia de Dena, Pontevedra.*—Eagle on the point of darting on a snake, in mother-of-pearl.

1571. Castells y Serra, *Barcelona.*—Carved wood furniture.

1572. Dubuison & Co. *Seville.*—Gilt iron bedstead.

1573. Pérez, F. *Valencia.*—Mosaic cabinet work representing portrait of Cervantes, composed of 300,000 pieces.

1574. Zuloaga, E. *Madrid.* — Artistic furniture in steel, chased and embossed.

CLASS XXXI.

1575. Canales, D. *Barcelona.*—Tin articles.

1576. Cotarelo, J. *Oviedo.*—Inoxidizable nails.

1577. Factory (National Works) of Trubia.—Iron, steel, files.

1578. Malabouche, F. *Valencia.*—Scales and steelyards.

1579. Mata, L. *Valencia.*—Nails.

1580. Pérez, J. B. *Jaen.*—A section of iron railing.

1581. Ysaura, Brothers, *Barcelona.*—Silver-plated Gothic lustre.

CLASS XXXIII.

1582. Elena, M. *Salamanca.*—Jewellery and filigree-work.

1583. Gómez, L. *Salamanca.*—Filigree-work.

1584. Soler Perich, *Barcelona.*—Diamond enamelled gold bracelets, ear-rings, brooches, &c.

1585. Tello, F. *Salamanca.*—Silver filigree-work.

1586. Ysaura, Brothers, *Barcelona.*—Collection of medals.

F

CLASS XXXIV.

1587. CIFUENTE, POLA, & CO. *Gijon.*—Glass, hollow and flat.

1588. COLLANTES, A. *Santander.*—Glass.

1589. FRADERA GOL, J. *Barcelona.*—Artificial glass eyes.

CLASS XXXV.

1590. ALCALDE (MAYOR) OF MUELAS, *Zamora.*—Refractory clay.

1591. ALFONSO, M. *St. Miguel, Teneriffe.*—Delfts for flooring.

1592. BACAS, P. *Andujar.*—Articles in white clay.

1593. BAIGNOL, BROTHERS, & CO. *St. Sebastian.*—Porcelain.

1594. CUBERO, J. *Malaga.*—Malaga clay figures.

1595. GARCIA, B. *Tamames, Salamanca.*—Common earthenware.

1596. GOVERNOR OF THE PROVINCE OF JAEN.—Articles in white clay.

1597. GONZALEZ, A. *Malaga.*—Malaga clay figures.

1598. GONZALEZ VALLS, R. *Valencia.*—Ornamental glazed tiles and delfts for flooring.

1599. GUTIERREZ DE LEON, A. *Malaga.*—Malaga clay figures.

1600. JUAN SEVA, M. *Madrid.*—A large earthen jar.

1601. MIGUEZ, J. *Seville.*—Ornamental glazed tiles.

1602. NOLLA & SAGRERA, *Valencia.*—Ornamental glazed tiles.

1603. OJEDA GONZALEZ, ANZA, & SOTO, *Seville.*—Ornamental glazed tiles, delfts for flooring, mouldings, &c.

1604. PICKMAN & CO. *Seville.*—Fine earthenware, comprising Parian and Bisque china.

1605. OLEA, J. *Zamora.*—Refractory clay.

1606. RODRIGUEZ, R. *St. Juan de Aznalfarache, Seville.*—Fine earthenware.

1607. SÁNCHEZ CABALLERO, L *Malaga.*—Specimens of pottery.

1608. YGARROTES OR SAUVAGES, *District of Abra, Philippine I.*—Pipes and other articles in clay.

CLASS XXXVI.

1609. BOSCH, J. *Philippine I.*—Palm-leaf cigar cases.

1610. MISSIONARY TO PANPANGA, *Philippine I.*—Cigar cases made by native negroes.

1611. TECSON, —, *Batangas, Philippine I.*—Palm-leaf cigar cases.

1612. VILLARIN, B. J. *Lagnua, Philippine I.*—Cigar cases and other articles in wood.

APPENDIX.

ARENAL, J. DEL.—Aniseed liquor.

ARIETA, J.—Dry pale light wine.

AYUNTAMIENTO DE SAN MANCIO.—Red and white barley.

BARRENENGOA, DÁMASO.—Chocolate.

BELLASO HERMANOS.—Dry pale light wine.

CAMP, PELAYO DE.—Corks.

DE COSTA Y Cª.—Honey.

DEPARTAMENTO DE LA GUERRA, *Madrid.*—Collection of national uniforms, and military maps of the war in Africa.

DIEZ DE RIVERA, A. *Granada.*—Olive oil.

ESTRADA Y PANES, *Huelva.* — Aniseed liquor.

FÁBRICA DE PÓLVORA DE MURCIA, *Soria.*—Sulphur and saltpetre.

GARCIA YÑIGUEZ, *Moquer.*—Alcohol.

GARCHOTENEA, S. Y Cª, *Zumaya.*—Raw or natural cement.

LA ROSALIA.—Olive-oil soap.

LORENZANA, R. *Leon.*—Flax.

MARCO, M. *Esplus, Huesca.*—Camomile.

MARCO Y CORONAS, *Saragossa.*—Flour.

PIÑAL, R. Y Cª, *Seville.*—Quicksilver.

PROVINCIA DE OVIEDO.—Drugs and chemicals.

SOCIEDAD DE AGRICULTURA DE REUS, *Tarragona.*—Wine.

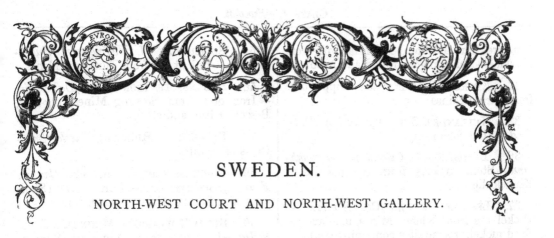

SWEDEN.

NORTH-WEST COURT AND NORTH-WEST GALLERY.

The Swedish Alphabet has three letters more than the English Alphabet, viz., Å, Ä, and Ö. These letters were originally the diphthongs Ao, Ae, and Oe, and follow in order after the letter Z, at the end of the Alphabet.

CLASS I.

1. ADELSWÄRD, S. E. BARON, *Åtvidaberg, Ostgothland.*—Specimens illustrative of the processes used at Åtvidaberg Copper Works.

2. ADLERS, A. *Yxhult, Nericia.*—Manger of lime-stone.

4. ARBORELIUS, E. G. *Gagnef, Elfdal, Dalecarlia.*—Vases, urns, chess boards, bracelets, knife handles, and other articles of porphyry.

5. ASCHAN, N. N. *Wrigstad Ohs, Smaland.*—Alloy of nickel and iron, from Klefva Nickel Works.

6. BALDERSNÄS IRON WORKS, *Dalsland.*—Bars of steel-iron, generally used in making cemented steel for files, saw-blades, &c.

7. BESKOW, J. W. *Wrethammar, Nericia.*—Iron ores from Stora Blanka Mine.

8. BISPBERG MINING Co. *Dalecarlia.*—Specimen of iron ore, weighing 16 cwt.

9. DANNEMORA MINING Co. *Upland.*—Iron ore, and the strata in which it is found.

10. ROCKHAMMAR IRON WORKS, *Lindesberg, Nericia.*—Iron ores.

11. EKMAN, C. *Finspong, Ostgothland.*—Iron ore, pig-iron, and bar-iron.

12. EKMAN, G. *Lesjöforss, Wermland.*—Pig-iron, with analysis attached.

13. ELFSTORP IRON WORKS, *Nericia.*—Iron ore, pig-iron; bar-iron, bent and twisted cold; scoria.

15. FREDRIKSBERG COPPER Co. *Smaland.*—Copper ore and copper.

16. GAMMALKROPPA AND SÄLBODA IRON WORKS, *Wermland.*—Series of iron ores from Persberg Mines, pig-iron, and bar-iron drawn cold into double knots.

17. GARPENBERG IRON WORKS, *Dalecarlia.*—Iron ore from Långvik and Bispberg Mines, pig-iron, and bar-iron.

18. GULDSMEDSHYTTE SILVER WORKS, *Nericia.*—Ores of lead (sulphurets), lead.

19. GUSTAF AND CARLBERG COPPER WORKS, *Jemtland.*—Copper ores, refined copper, copper-sheets and wire.

20. GYSINGE IRON WORKS. *Gefle.*—Pig-iron, steel-iron in bars.

21. HAMILTON, H. BARON, *Boo, Nericia.*—Bar-iron, steel, and peat.

22. HARMENS, AF, H. O. HEIRS OF, *Berga, Calmar.*—Bog-iron ore, pig-iron, specimens of porphyry.

23. HEIJKENSKÖLD, S. *Granhult, Nericia*—Pig-iron.

24. HELLSTRÖM, V. *Dalkarlshyttan, Nericia.*—Iron ores from Stripa Mine.

25. HERMANNSSON, C. COUNT, *Ferna, Westmanland.*—Specimens of bar-iron and steel.

26. HORNDAL IRON WORKS, *Dalecarlia*—Iron ores from Bispberg,, Holm, and Westersjö mines, pig-iron, and bar-iron.

28. JOHANNESSEN, W. *Sörbytorp, Scania.*—Marl.

29. KILLANDER, F. *Hook, Smaland.*—Iron ore from Taberg Mine, pig-iron, scoria, bar-iron, iron-wire.

31. KLOSTER IRON WORKS, *Dalecarlia.*—

Iron ore from Rällingsberg Mine, pig-iron, Bessemer steel in ingots, rolled, and in sheets.

32. KOCKUM, F. H. *Malmö.*—Copper ores from Virum Mines.

33. KULLANDER, C. G. *Kjellsviken, Dalsland.*—Slate for roofing.

36. LANDSTRÖM, C. *Osterplana, Westgothland.*—Stone articles from the quarries at Kinnekulle.

37. LESSEBO NICKEL CO. *Smaland.*—Nickel ore from Klefva Mines, mat and refined nickel, iron pyrites containing gold.

38. LIDEN, J. *Töksmark, Wermland.*—Cooking utensils of Talc stone.

39. LILLIECREUTZ, J. BARON, *Aminne, Smaland.*—Bog-iron ore and pig-iron.

40. LJUSNARSBERG COPPER WORKS, *Nericia.*—Copper ores, mat, black, and refined copper.

41. LUNDBOM, P. J. *Löfnäs, Nericia.*—Iron ores from Nya Kopparberg, pig-iron, bar-iron.

42. LUNDHQVIST, G. A. *Nyköping.*—Sheet-iron and cold-bent iron bars, iron pyrites; roll sulphur.

44. MARE, DE, G. *Tofverum, Westerwik.*—Bog-iron ore, pig-iron.

45. MOSSGRUFVE MINING CO. *Westmanland.*—Iron ore from Norberg Mines.

46. MÖLNBACKA IRON WORKS, *Wermland.*—Steel-iron.

47. NORDENFELDT, O. *Björneborg, Wermland.*—Pig-iron, bar-iron, steel.

48. BREFVEN IRON WORKS, *Nericia.*—Pig-iron, bar-iron, steel-iron.

49. PETRÉ, T. HEIRS OF, *Hofors Iron Works, Gefle.*—Iron ores, iron, and steel.

50. RETTIG, C. A. *Kihlaforss, Gefle.*—Bar-iron for gun-barrels, steel made from pig-iron and bar-iron, razors.

51. ROCZYCKI, C. BARON, *Kolmarden, Ostgothland.*—Manufactures of marble.

52. RAMEN IRON WORKS, *Wermland.*—Series of ores from mines in Wermland, pig-iron, bar-iron.

53. RÖNNÖFORSS IRON WORKS, *Jemtland.*—Bog-iron ore, pig-iron, bar-iron.

55. SEDERHOLM, J. *Näfveqvarn, Sudermannia.*—Cobalt ore from Tunaberg Mines.

56. SILJANSFORSS IRON WORKS, *Dalecarlia.*—Iron ores from Sörskog Mines, pig-iron, Bessemer iron and steel.

58. STABERG & BJÖRLING, *Stockholm.*—Cross of granite.

59. STOCKENSTRÖM, VON A. *Åker Cannon Foundry, Sudermannia.*—Iron ores from Skotvang and Förola Mines; cannon iron.

60. STORA KOPPARBERG MINING CO. *Dalecarlia.*—Iron ores from Windtjern, Skinnaräng, Tuna Hästberg; copper ores from Fahlu Mine; iron and copper.

61. SUBER & SJÖGREEN, *Bruzaholm, Smaland.*—Series of bog-iron ore and its products.

62. SVANA IRON WORKS, *Westmanland.*—Pig-iron and cold-bent bar-iron.

63. SÖDERFORSS IRON WORKS, *Upland.*—Pig-iron, bar-iron.

64. UDDEHOLM IRON WORKS, *Wermland.*—Pig-iron, bar-iron, steel, and iron-wire.

65. ULFF, C. R. *Wikmanshyttan, Dalecarlia.*—Pig-iron, bar-iron; steel produced by the method of Uchatius.

66. VIEILLE MONTAGNE ZINC MINING CO. *Ammeberg, Nericia.*—Ores of zinc (sulphurets), and minerals from Åmmeberg Mines.

68. WERMLAND, LOCAL COMMITTEE, *Carlstad.*—Iron ores from Persberg and other Mines; scoria.

69. ZETHELIUS, W. *Surahammar, Westmanland.* — Specimens of iron and steel, iron plates and sheets, boiler plate weighing nearly half a ton.

70. ÖREBRO LOCAL COMMITTEE, *Nericia.*—Collection of iron ores from Nora.

71. ÖSTBERG, C. *Elfkarleö. Upland.*—Pig-iron, bar-iron, blistered and hammered steel.

72. ÖSTERBY IRON WORKS, *Upland.*—Pig-iron, bar-iron, blistered and hammered steel.

———

73. CARLSDAL IRON WORKS, *Nericia.*—Pig-iron, steel.

74. FREDRIKSBERG IRON WORKS, *Dalecarlia.*—Bar-iron, pig-iron, blooms, steel.

75. GORANSSON, F. *Hogbo, Gefle.*—Iron ores, pig-iron, Bessemer iron, steel, and manufactures.

77. HELLEFORSS IRON WORKS, *Nericia.*—Pig-iron, bar-iron.

78. SCHRAM, J. R. *Singo, Grisslehamn.*—Marble table.

79. SMEDJEBACKEN ROLLING MILL, *Dalecarlia.*—Bar-iron.

80. SAGMYRE NICKEL WORKS, *Dalecarlia.*—Granulated nickel copper.

80*a.* OSTERHOLM, O. *Stockholm.*—Monument of granite.

80*b.* CARLEN, MISS, *Stockholm.*—Letterpress of marble, from the Isle of Gottland; petrifactions.

80*c.* SALA BERGSLAG, *Sala.*—Silver ore.

80*d.* STRIBERG MINING CO. *Nericia.*—Iron ore from the Asboberg Mine.

80*e.* ASPELIN, T. *Fagersto Iron Works, Westmanland.*—Pig-iron.

80*f.* CRONEBORG, W. *Elfsbacka Iron Works, Wermland.* — Pig-iron, blooms, bar-iron; manufactures, as shovels, saw-blades, chains, &c.

80*g.* GAMMELSTILLA IRON WORKS, *Province of Gefle.*—Steel and wire iron; gun.

80*h.* HÄFLA IRON WORKS, *Ostgothland.*—Rolled bar-iron, and steel.

80*i.* KEILLER, J. & Co. *Gothenburg.*—Copper ores, copper, pig-iron, steel, lead and iron ores, pyrites.

80*k.* MARE, A. DE, *Aukarsrun, Westerwick.*—Iron for scythes.

80*l.* ROCKESHOLM IRON WORKS, *Nericia.*—Iron ores from Blanka and Stoll Mines; pig-iron.

80*m.* WITSBERG, H. H. *Borghaum, Ostgothland.*—Frieze of Silurian limestone.

80*n.* MOTALA IRON WORKS, *Ostgothland.*—Bow of a steamer, showing the quality of the iron plate.

80*o.* BERG, —, *Stockholm.*—Granite urn.

CLASS II.

Sub-Class A

81. ASSOCIATED COPPER-SMELTERS, FAHLUN, *Dalecarlia.*—Granulated copper, oxide of copper, copperas, sulphur.

82. DJURO TECHNICAL MANUFACTORY, *Norrköping.*—Sulphate of copper.

83. DUFVA, C. F. *Stockholm.*—Blacking

84. FRIESTEDT, A. W. *Stockholm.*—Steamed bone-dust for manure; animal charcoal; blacking for horses' hoofs.

86. HAMILTON, COUNT H. D. *Bonsater, Westgothland.*—Alum, sulphate of iron, red ochre.

87. HASSELGREN, O. S. *Rådaneforss, Dalsland.*—Peat, carbonized peat; steamed bone-dust manure.

88. HEDENBERG, v. L. A. *Umeå.*—Salt petre made by peasants.

89. HJERTA, L. J. & MICHAELSSON, J. *Stockholm.*—Sulphuric acid.

90. JÖNKÖPING LUCIFER-MATCH CO. *Jönköping.*—Safety matches, &c.

The exhibitors manufacture all kinds of lucifer-matches for export to England and all parts of the world, warranted to stand any climate, and not to fail, known under the mark, *Jonkopings Tändstickor.* They are the largest importers to England of the celebrated match-boxes called "Swedish slides:" and the inventors of the renowned *safety matches,* which ignite only when rubbed on a surface specially prepared for the purpose. They obtained the *Silver Medal at the Paris Exhibition,* 1855.

91. KYLBERG, H. *Ryssbylund, Calmar.*—Poudrette.

92. KYLBERG, I. *Såtenäs, Westgothland.*—Poudrette.

93. LEWENHAUPT, COUNT C. M. *Claestorp, Sudermannia.*—Turpentine (raw and rectified), pinol, resin, lamp-black, &c.

94. LOFVERS ALUM WORKS, *Isle of Oland.*—Alum slate, alum.

95. MAJERAN, J. M. *Jonkoping.*—Fibrous matter prepared from pine needles; extracts therefrom.

96. PIPER, COUNT C. E. *Andrarun, Scania.*—Alum, sulphate of iron, red ochre, alum wash.

97. STORA KOPPARBERG MINING Co. *Dalecarlia*—Iron pyrites, red ochre.

98. WAERN, C. F. & Co. *Gothenburg.*—Superphosphate of lime.

99. WALLBERG, F. *Stockholm.*—Specimens of steamed bone-dust.

100. WESTERWIK LUCIFER-MATCH Co. *Westerwik.*—Various kinds of matches.

101. WIKLAND, P. *Swartsång, Wermland.*—Carbonized wood.

---

102. DYLTA SULPHUR WORKS, *Nericia.*—Sulphur, green vitriol, red ochre, jeweller's rouge.

103. HAZELIUS, A. K. *Stockholm.*—Sulphate of ammonia.

104. HJERSSE, J. W. *Ozebzo.*—Safety matches.

105. NYBLDUS, C. G. *Stockholm.*—Soda water, and blacking.

Sub-Class B.

106. CAVALLI, J. G. *Gothenburg.*—Pharmaceutical preparations, compressed medicinal vegetables.

107. DUFVA, C. F. *Stockholm.*—Eau de Cologne from Swedish spirits.

108. HYLIN & Co. *Stockholm.*—Perfumery.

111. PAULI, F. *Jonkoping.*—Perfumery.

112. WIKSTRÖM, Z. *Stockholm.*—Various kinds of pulverized drugs.

---

113. ALEXIS, T. *Stockholm.*—Eau de Cologne.

114. WILLMAN, L. *Ystad.*—Tooth powder.

CLASS III.

Sub-Class A.

116. AGRICULTURAL SOCIETY, OSTGOTHLAND, *Linköping.*—Grain produced in the province.

117. AGRICULTURAL SOCIETY, NORRA MORE, *Calmar.*—Grain, seeds of forage and other plants.

118. AMEEN, H. *Hogsta, Stockholm.*—Rye.

120. BOOS, J. *Gräfsnäs, Westgothland.*—Oats.

122. DUGGE, H. *Latorp, Nericia.*—Autumn wheat.

124. ERIKSSON, E. J. *Umea.*—Wheat.

125. EVERS, O. *Strom, Gothenburg.*—Wheat, oats, clover-seed.

126. FORSELL, C. A. *Mellösa, Nericia*—Pease.

127. FRIESTEDT, A. W. *Stockholm.*—Steamed bone-dust, used as food for cattle.

129. GYLLENCREUTZ, Y. *Gyljen, Lulea.*—Wheat, grown ten miles south of the Arctic circle.

132. HOFSTEN, S. E. v. *Sund, Nericia.*—Wheat and wheat-flour, oats.

133. KEY, E. *Sundsholm, Calmar.*—Wheat, rye, oats, pease.

134. KLOCKHOFF, O. D. *Ytterstforss, Umea.*—Autumn wheat, grown at 65° lat.

138. LIGNELL, C. A. *Carlslund, Jemtland.*—Wheat, rye, barley.

140. MALMSTEN, J. A. *Jonkopiag.*—Potato-flour.

142. NENSEN, J. A. *Umea.*—Ears of wheat.

143. NYBERG, J. *Brefven, Nericia.*—Autumn and spring wheat, rye.

144. ROSSANDER, F. J. *Isle of Oland.*—Wheat, rye, barley, oats.

145. SAHLMARK, C. P. *Ljungsnäs, Westgothland.*—Barley.

146. SPRENGTPORTEN, BARON, J. W. *Sparreholm, Sudermannia.*—Wheat, oats, grass-seeds.

147. SÖDERSTROM, A. G. *Hernosand.*—Barley, rye.

148. UNANDER, F. *Yttertafle, Umea.*—Timothy seeds.

149. WEDBERG, A. *Yxe, Nericia.*—Wheat, black tares.

151. AKERBLOM, C. M. *Norrkoping.*—Starch prepared from wheat.

152. ÅNGSTROM, J. *Berglunda, Umea.*—Four-rowed barley.

---

153. KIHLMAN, O. S. *Tofta, Gothenburg.*—Yellow beans.

154. ACADEMY, EXPERIMENTAL FIELD OF THE ROYAL AGRICULTURAL, *Stockholm.*—Spring wheat, vetches.

155. AGRICULTURAL SOCIETY OF DALECARLIA, *Fahlun.*—Collection of grains and seeds.

155a. LUNDSTROM, C. D. *Gothenburg.*—Oats.

155b. HAGENDAHL, C. A. *Ozebzo.*—Seeds of clover and Timothy.

155c. WESTERBOTTEN AGRICULTURAL SOCIETY, *Umea.*—Spring wheat.

Sub-Class B.

156. DUFVA, C. F. *Stockholm.*—Chocolate, mustard.

157. FROMELL, C. J. *Gothenburg.* — Confectionery.

160. JUHLIN-DANNFELT, C *Stockholm.*—Cranberries and blackberries, preserved in sugar.

161. KYLBERG, H. *Ryssbylund, Calmar*—Cheese.

163. LANDSKRONA SUGAR MANUFACTURING CO. *Landskrona.*—Beet-root sugar.

164. LINDBLOM, A. E. HEIRS OF, *Ronneby, Carlskrona.*—Articles prepared from potatoes.

165. LUNDGREN, P. W. *Stockholm.*—Starch, treacle, gum, &c. prepared from potatoes.

166. OLOFSSON, A. *Ase, Jemtland.*—Goat's-milk cheese.

168. WAHLBOM, G. B. *Köhlby, Calmar.*—Cheese.

169. WEDBERG, MRS. L. *Nora.*—Cranberries, cloudberries, and raspberries, preserved in sugar.

Sub-Class C.

171. BODACH, C. *Stockholm.*—Cigars.

172. BOMAN, J. A. & Co. *Gothenburg.*—Manufactured tobacco.

173. BRINCK, HAFSTROM, & Co. *Stockholm.*—Cigars, snuff, roll tobacco.

175. KOCKUM, F. H. *Malmo.*—Manufactured tobacco.

176. KYLBERG, H. *Ryssbylund, Calmar.*—Tobacco grown by the exhibitor.

178. LITTSTRÖM, E. G. *Fahlun.*—Refined corn brandy.

179. LUNDGREN, P. W. *Stockholm.*—Wines prepared from fruits.

180. PRYTZ & WIENCKEN, *Gothenburg.*—Manufactured tobacco.

181. ROSEN, E. A. & STROMBERG, *Stockholm.*—Cigars, tobacco, and snuff.

182. STERNHAGEN & Co. *Gothenburg.*—Snuff.

---

186. HELLGREN W. & Co. *Stockholm.*—Cigars, snuff, tobacco.

187. HOGSTEDT & Co. *Stockholm.*—Punch of various kinds.

190a. CEDERLUND, BROS. *Stockholm.*—Punch.

190b. ENGLANDER, J. N. *Wexio*—Wine prepared from berries.

190c. LINDGREN, J. M. T. *Christinehamn.*—Refined corn-brandy, punch.

CLASS IV.

Sub-Class A.

192. HIERTA, L. J. & MICHAELSSON, J. *Stockholm.*—Stearine candles and soap.

193. HYLIN & Co. *Stockholm.* — Soap; flower vase made of soap.

195. MONTEN, L. *Stockholm.* — Stearine candles.

196. PAULI F. *Jonkoping.*—Various kinds of soap, &c.

Sub-Class B.

201. CELSING, L. G. v. *Helleforss, Sudermannia.*—Fleeces of marino sheep.

202. ROSSING, J. A. *Gothenburg.* — Silk, produced by worms fed on *Scorzonera humilis* and *hispanica.*

203. SAHLSTROM, C. G. *Jonkoping.*—Albumen extracted from fish roe.

204. SWEDISH SILKWORM BREEDING SOCIETY, *Stockholm.*—Swedish silk and cocoons.

206. UNANDER, F. *Yttertafle, Umea.* — Reindeer horns.

207. RAMSTEDT, MISS F. *Stockholm.*—Wax flower.

208. ÅNGSTROM, J. *Lycksele, Umeå.*—Reins and shoes made from gut and skin of reindeer; rope of pine-roots.

209. LANGEMEIER, W. *Stockholm.*—Mats of Esparto, Mexican fibre, and horse-hair.

210. NILSON, P. U. *Jemtland.*—Horns of elk.

210a. WAHLBORN, G. B. *Calmar.*—Wool.

Sub-Class C.

211. BALDERSNÄS IRON WORKS, *Dalsland.* —Red deals.

213. DICANDER, A. J. *Uppbo, Westgothland.* —Speckled birch.

214. HAMILTON, COUNT H. *Mariedal, Westgothland.*—Red and white deals; sections of a pine tree.

215. HARMENS, AF, H. O. HEIRS OF, *Berga, Calmar.*—Pine, fir, oak, beech, and juniper.

216. HASSELGREN, O. S. *Radaneforss, Dalsland.*—Pine and fir seeds.

217. KOSTA GLASS WORKS, *Smaland.*—Pine and fir, pine seeds.

219. LEMAN, H. M. *Gothenburg* —Push mats.

220. LINDQVIST, A. P. *Bergsgården, Dalsland.*—Pine and fir seeds.

221. MAGNUSSON, N. P. *Grankulla, Isle of Oland.*—Ropes made of pine-bast.

222. MOLNBACKA IRON WORKS, *Wermland.* —Pine.

226. SPARRE, COUNT E. T. *Torpa, Westgothland.*—Pine.

227. SPRENGTPORTEN, BARON J. W. *Sparreholm, Sudermannia.*—Fir, pine, birch, and aspen.

231. WESTERWIK LOCAL COMMITTEE, *Westerwik.*—Oak, ash, birch, pine, and fir.

233. GYLLENKROK, BARON T. *Oby, Wexio.* —Basket-work.

234. LANDGVIST, R. *Wenersborg.*—Mirror-frame, carved and gilt.

235. LUNDMARK, A. *Stockholm.*—Brackets of carved walnut.

236. RYDBECK, J. O. *Stockholm.*—Pipe carved in wood with Runic characters.

237. TAUTZ, H. *Stockholm.*—Gilt framing.

CLASS V.

241. LUNDHQVIST, G. A. *Nykoping.* — Wrought-iron railway wheels, axle and tires of puddled steel, axles and tires.

242. MORGARDSHAMMAR IRON WORKS, *Dalecarlia.*—Rail-spikes.

243. BREFVEN IRON WORKS, *Nericia.*—Rail-spikes.

CLASS VI.

251. NORMAN, J. E. *Stockholm.*—Carriage.

252. OLSEN, N. *Gothenburg.*—Carriage.

253. SJOSTEEN, B. F. & TILLBERG, L. *Stockholm.*—Sledge with fur apron.

254. WEGELIN, J. F. & OSTBERG, J. W. *Stockholm* —Carriages.

CLASS VII.

Sub-Class B.

258. BOLINDER, J. & C. G. *Stockholm.*—Iron planing machine, copying press.

259. FROMELL, C. J. *Gothenburg.*—Chocolate mill.

260. GUSTAFFSON, C. *Hellstorp, Jönköping.*—Machine for making rolled nails.

263. SLÖÖR, J. *Stockholm.*—Potatoe-peeling machine.

CLASS VIII.

268. HALLSTRÖM, O. G. *Köping.*—Rotatory steam engine.

269. HOLMGREN, J. *Sala.*—Model of a blowing machine.

271. LINDAHL & RUNER, *Gefle.*—High-pressure steam engine.

272. MUNKTELL, T. *Eskilstuna.*—Groat-grinding mill, forcing pump.

273. SCHEUTZ, E. *Stockholm.*—Rotatory steam engine.

———

274. FRESTADIUS, A. W. *Bergsund, Stockholm.*—Marine steam engine; the steam is allowed to expand to nearly five times its volume, in a separate compartment, before going to the condenser.

CLASS IX.

276. CENTRAL COMMITTEE FOR SWEDEN, *Stockholm.*—Plans and perspectives of farms and farm buildings, agricultural maps, photographs.

277. CELSING, v. L. G. *Helleforss, Sudermannia.*—Plough, clover thrashing machine.

278. DAHLGREN, V. *Stockholm.*—Maps, indicating the division of the land belonging to a village.

280. HARMENS, AF, HEIRS OF, H. O. *Berga, Calmar.*—Iron halter for bulls.

281. HOLMGREN, J. *Sala.*—Churn.

283. LYCKEBY IRON FOUNDRY, *Bleking.*—Rotatory harrow, plough.

284. ROSSING, J. A. *Gothenburg.*—Forestry tools and implements.

286. SEDERHOLM, J. *Näjvequarn, Sudermannia.*—Plough.

287. ULTUNA AGRICULTURAL SCHOOL, *Upland.*—Iron plough.

288. ÖFVERUM IRON WORKS, *Calmar.*—Dressing machine, chaff cutter, ploughs, carrot drills, clover sowing machine.

———

290. EKMAN, P. J. *Stockholm.*—Models of a kiln and of a new method of stacking hay.

291. GUSSANDER, P. U. *Gammelstilla, Gefle.*—Set of dairy utensils.

292. HAGLUND, P. *Näfveqvarn, Nyköping.*—Grain and corn-crushing mill.

292a. LAGERBERG, COUNT F. *Stockholm.*—Milk can, with warming apparatus.

292b. LINDQVIST, C. M. *Stockholm.*—Churn, sowing and planting machines.

292c. ODELBERG, A. *Enskede, Stockholm.*—Model of a kiln.

292d. PALMAER, C. W. *Forsvik, Westgothland.*—Drain-tile machine—Schlosser's principle.

292f. AGRICULTURAL SOCIETY OF DALECARLIA, *Fahlun.*—Ploughs.

292g. ODELBERG, J. A. *Flemmingsburg, Stockholm.*—Model of open barn, for corn and hay.

CLASS X.

293. BARK & WARBURG, *Gothenburg.*—Doors, window frame with wainscoting, inlaid floor, joinery work.

294. CELSING, v. L. G. *Helleforss, Sudermannia.*—Gas welding furnaces, for gas from coal or peat.

295. GEFLE PROVINCIAL LOCAL COMMITTEE, *Gefle.*—Model of a roof formed of shingles.

296. KLEMMING, C. *Stockholm.*—Water-closet.

297. KYLBERG, J. *Sätenäs, Westgothland*—Draining tiles.

301. MARINO, J. *Stockholm.*—Water-closet; model of a binn for sweepings, with conveniences.

302. NABSTEDT, H. *Gothenburg.*—Water-closets.

303. RINGNER, A. *Gothenburg.*— Glazed porcelain stoves.

306. ÅKERLIND, O. H. *Stockholm.*—Glazed porcelain stoves.

310. LINDSKOG, S. & FRYCKHOLM, P. B. *Gothenburg.*—Water-closet.

310a. CRONEBORG, W. *Elfsbacka Iron Works, Wermland.*—Shingles, roof of shingles.

CLASS XI.

Sub-Class C.

311. ANKARCRONA, J. *Husqvarna, Smaland.*—Rifled gun, adopted in 1860 for the Swedish army, adapted for ranges between 600 and 4,000 feet.

312. CELSING, L. G. v. *Hellefors, Sudermannia.* — Five-pounder rifled cannon, invented by C. Engström, R.S.N. ; grenade.

313. EKMAN, C. *Finspong, Ostgothland.*—Breech-loading cannon, invented by C. Engström, R.S.N. ; six-pounder field cannon.

315. MARÉ, A. DE, *Aukarsrum, Calmar.*—Bombs.

316. SASSE, E. *Carlskrona.*—Models for cannon foundries of breech-loading and other cannon, projectiles for rifled bore.

317. STOCKENSTRÖM, A. v. *Åker, Sudermannia.*—Rifled breech-loading field cannon, with suitable conical ball.

318. SVENGREN, J. *Eskilstuna.*—Swordblades.

321. CARL GUSTAFFS STAD GUN FACTORY, *Eskilstuna.*—Infantry rifles.

CLASS XII.

Sub-Class A.

323. GRÖNDAHL, C. E. M. *Stockholm.*—Drawing and model of a ship.

Sub-Class C.

326. FURUDAL IRON WORKS, *Dalecarlia.*—Chains and chain-cables of iron.

327. SÖDERFORSS IRON WORKS, *Upland.*—Anchor.

CLASS XIII.

331. BYSTRÖM, O. F. *Stockholm.*—Hydropyrometer, for measuring high temperatures in furnaces.

332. HULTGREN, F. A. *Stockholm.*—Hydrometer.

333. KINDBOM, J. P. *Calmar.*—Photometer.

334. KLINTIN, J. F. *Stockholm.* — Logs, compass, vacuometer, manometer, spring balance.

336. MEURTHIN, J. *Gefle.*—Miner's compass.

337. WIBERG, M. *Malmö.*—Calculating machines.

338. EGGERTZ, V. *Fahlun.*—Apparatus for the analysis of iron ores.

339. LYTH, G. W. *Stockholm.*—Goniometer.

343. REYMYRE GLASS WORKS, *Ostgothland.*—Specimens of glass for chemical purposes.

340a. JONSSON, A. *Stockholm.*—Spherical compasses.

340c. ARWIDSSON, T. *Stockholm.*—Deviation cards, correcting the deviations of ships' compasses.

340d. BERG, F. J. *Stockholm.*—Goniometer.

340e. CENTRAL COMMITTEE FOR SWEDEN.—Complete sets of Swedish coins, weights, and measures.

340f. WREDE, GENERAL BARON F. *Stockholm.*—Philosophical instruments, illustrating the systems of waves of light; planimeter.

340g. ADERMAN, C. O. *Stockholm.*—Thermo-alcholometer, alcholometers, maximum and minimum thermometers.

CLASS XIV.

342. UNNA & HÖFFERT, *Gothenburg.*—Photography.

343. CARLEMAN, C. G. V. *Stockholm.*—Photographs.

344. MANDEL, *Stockholm.* — Photo-lithographic impressions.

CLASS XV.

346. CEDERGREN, J. T. *Stockholm.*—Transparent clock, enamelled works.

347. LINDEROTH, MRS. B. *Stockholm.* — Watch wheels and movements.

348. LINDEROTH, G. W. *Stockholm.*—Marine chronometers, astronomical pendulum watch.

350. MOLLBERG, L. R. *Stockholm.*—Balances and spiral springs for marine and pocket chronometers.

CLASS XVI.

355. MALMSJÖ, J. G. *Gothenburg.*—Pianoforte.

356. SÄTHERBERG, A. F. *Norrköping.* — Pianino.

CLASS XVII.

362. GEORGII, A. *London.*—Hygienic and medical gymnastic apparatus.

364. STILLE, A. *Stockholm.*—Surgical instruments.

CLASS XVIII.

368. ANDERSSON, A. *Fritsla Hagen, West-gothland.*—Cotton goods woven in hand-looms.

369. ANDERSSON, MRS. C. *Gefle.*—Woven counterpane.

370. ANDERSSON, S. *Kinna Sanden, West-gothland.*—Cotton goods woven in hand-looms.

371. BERG COTTON MANUFACTURING CO. *Norrköping.*—Domestic fabrics, bleached and unbleached; mole-skins.

372. BERG, J. T. *Nääs, Westgothland.*—Cotton goods.

373. EVERS, A. H. *Ahleforss, Westgothland.*—Cotton yarn.

374. HOLMEN CO. *Norrköping.*—Cotton yarns; domestic fabrics.

375. HAKANSSON, H. *Fritsla Högen, West-*

gothland.—Cotton goods woven in hand-looms.

376. JOHANSSON, M. *Källäng, Westgothland.*—Jacquard-woven cotton coverlet.

377. KAMPENHOF CO. *Uddewalla.*—Cotton sail-cloth, domestic fabrics, cotton yarns.

378. LARSSON, S. *Stämmemad, Westgothland.*—Cotton goods woven in hand-looms.

380. MALMÖ MANUFACTURING CO. *Malmö.*—Cotton yarn.

381. NORRKÖPING COTTON-WEAVING CO. *Norrköping.*—Cotton stuffs.

382. ROSENLUND COTTON-SPINNING CO. *Gothenburg.*—Cotton-yarn and stuffs.

383. GIBSON, W. & SONS, *Jonsered, Gothenburg.*—Cotton yarn.

CLASS XIX.

386. ALMEDAHL CO. *Gothenburg.*—Samples of flax, linen thread, flax and tow yarns, damask, diaper, and other goods.

389. BRODIN, B. C. *Gefle.*—Linen, damask, and diaper table-cloths, napkins, and towels.

388. BRODIN, E. C. *Gefle.*—Linen, damask, and diaper table-cloths and napkins.

390. DAHLQVIST, J. *Själevad, Hernosand.*—Table-cloths and napkins woven in hand-looms.

391. FRISK, E. *Hudikswall.*—Flax canvas woven in hand-looms.

392. GIBSON, W. & SONS, *Jonsered, Gothenburg.*—Yarns, sail-cloth, cordage, &c.

393. HERNÖSAND LOCAL COMMITTEE, *Hernösand.*—Flax, linen, thread, and yarn.

394. STENBERG, G. WIDOW OF, *Jönköping.*—Table-cloths and napkins.

395. SUNDSTRÖM, F. *Norrköping.*—Cordage and rope-yarns.

396. TROLLE, W. *Gidån, Hernösand.*—Linen goods.

CLASS XX.

401. ALMGREN, K. A. *Stockholm.*—Plain and figured silk, dyed in Sweden.

402. CASPARSON & SCHMIDT, *Stockholm.*—Silk, plain and figured.

403. MEYERSON, L. *Stockholm.*—Silk.

CLASS XXI.

407. APPELBERG, C. L. *Norrtelje.*—Waterproof woollen cloth.

408. BEHRLING, P. *Stockholm.*—Woollen shawls and blankets.

409. BOETHIUS & KLING, *Norrköping.*— Woollen goods of various kinds.

410. CARLSVIK MANUFACTURING Co. *Stockholm.*—Woollen, mixed woollen, and silk stuffs, yarns, &c. partly dyed.

412. DRAG WOOLLEN MANUFACTURING Co. *Norrköping.*—Woollen cloth of different kinds.

413. ELLIOT, S. & Co. *Stockholm.*—Shawls, woollen and mixed.

414. FÜRSTENBERG, L. & Co. *Gothenburg.* — Shawls and neckerchiefs, woollen and mixed.

415. JOHANSSON, M. *Källäng, Westgothland.*—Table-covers of mixed cotton-wool and silk; shawls of wool, and mixed wool and cotton.

417. LUNDSTRÖM, C. F. *Jönköping.* — Woollen and mixed stuffs.

The exhibitor's power-loom weaving, dying, and finishing works, established in 1857, manufacture orleans, paramatta, alpaca, lustre, satin, serge, madonna, lasting, serge de berry, coatings, grograms, cords, casinette, silk mixtures, alpaca and mohair stuffs, and other varieties of worsted, woollen, and silk mixtures, plain, checked, or figured, in all colours. *He is the inventor of a new dyeing process.*

418. MEYERSON, L. *Stockholm.*—Stuffs of mixed silk and cotton.

419. NORRKÖPING CORDEROY-WEAVING Co. *Norrköping.* — Woollen goods of various kinds.

420. QVIST, A. *Norrköping.* — Woollen goods.

422. SMEDJEHOLMEN WOOLLEN MANUFACTURING Co. *Norrköping.*—Woollen cloth of different kinds.

424. WAHLQVIST CLOTH MANUFACTURING Co. *Wexiö.*—Woollen cloth used by the Swedish army and navy.

425. WAHREN, H. *Norrköping.*—Woollen goods of various kinds.

426. WELLENIUS, J. F. *Norrköping.* — Woollen cloth.

———

427. ANDERSON, T. W. *Ernsta, Upsal.*— Woollen and linen yarn, homespun.

CLASS XXIII.

431. LUNDSTRÖM, C.F. *Jönköping.*—Stuffs, dyed by a method invented by the exhibitor.

CLASS XXIV.

436. BENSOW & Co. *Stockholm.*—Lace of various kinds.

437. HOLMGREN, W. *Gothenburg.*—Table-cover, netted with silk and embroidered.

———

438. BÄCK, MISS C. *Gothenburg.*—Embroidered baby linen.

440. HARTVICK, MRS. A. *Wadstena.* — Bone laces.

CLASS XXV.

Sub-Class A.

442. BASK, C. J. *Stockholm.*—Hearth-rug of fur, representing the arms of Sweden.

444. FORSSELL, D. & Co. *Stockholm.*— Ladies' jackets of martin, otter, and minever; furs, muffs, and collars.

445. NILSSON, P. U. *Jemtland.*—Skins of the bear, elk, and rein-deer.

Sub-Class C.

452. LUNDGREN, F. F. O. *Stockholm.*— Various brushes.

453. LÖFSTRÖM, P. & Co. *Gothenburg.*—Wigs.

455. ZINN, C. M. *Stockholm.*—Wigs of different kinds.

456. CARLSSON, C. A. *Stockholm.*—Brushes of various kinds.

457. LARSDOTTER, F. A. *Stockholm.*—Bracelets, collars, rings, &c. made of hair.

CLASS XXVI.

Sub-Class A.

462. COLLIANDER, J. A. *Gothenburg.*—Leather.

Sub-Class B.

467. KROOK, T. F. *Malmö.*—Harness, saddle, embossed leather straps.

468. SKOGLUND, C. *Stockholm.*—Lady's saddle, with bridle.

CLASS XXVII.

Sub-Class A.

471. ANGRESIUS, F. R. *Gothenburg.*—Hats.

472. ERICSSON, A. & Co. *Stockholm.*—Hats of various kinds.

Sub-Class C.

476. BERENDT, S. JUN. & Co. *Stockholm.*—Shirts.

477. BROCK, G. F. *Gothenburg.*—Gloves and skins.

479. MÖLLER, P. D. *Malmö.*—Gloves.

480. SALOMAN, J. *Gothenburg.*—Ready-made linen.

481. SÖDERBERG, J. *Stockholm.*—Hosiery goods.

482. WIECHEL, G. *Norrköping.*—Hosiery goods.

483. LINDBERG, Z. *Gothenburg.*—Shirts.

484. MÖLLER, J. P. *Lund.*—Gloves.

485. ANDERSSON, M. C. *Stockholm*—Laplanders' costumes.

485*a.* KAEDING, F. *Stockholm.*—Uniforms of several regiments of the Swedish army.

485*b.* WIDGREN, H. *Stockholm.*—Buttons.

Sub-Class D.

486. FJASTAD, P. C. *Stockholm.*—Boots and shoes.

487. HELLMAN, J. *Norrköping.*—Boots, overshoes, fur boots for ladies.

488. SUNDQVIST, L. M. *Gothenburg.*—Boots and shoes.

489. TRANÉ, F. *Stockholm.*—Boots and shoes.

490. TÖRNQVIST, C. W. *Åby, Scania.*—Shooting boots, ankle shoes.

491. LUNDSTEDT, O. *Stockholm.*—Boots and shoes.

CLASS XXVIII.

Sub-Class A.

493. BOCK, C. A. *Klippan, Scania.*—Various kinds of paper.

495. BRANDT, J. C. *Stockholm.*—Playing cards.

496. DEUTGEN, E. *Nyqvarn, Stockholm.*—Paper.

497. HUERLIN, G. *Stockholm.* — Playing cards.

498. LITHOGRAPHIC SOCIETY IN NORRKÖPING, *Norrköping.*—Playing cards, stationery, lithography.

500. MUNKTELL, J. H. HEIRS OF, *Gryksbo, Dalecarlia.*—Filtering paper for chemical purposes.

501. ROSENDAHL MANUFACTURING Co *Gothenburg.*—Various kinds of paper; firwood used in making paper, and pulp formed of it.

502. STENSHOLM MANUFACTURING Co. *Smaland.*—Specimens of paper, coloured and not coloured.

Sub-Class B.

506. NYBLÆUS, C. G. *Stockholm.*—Samples of ink.

507. SCHWARTZ, C. H. F. *Stockholm.*—Ornamental pasteboard box.

Sub-Class C.

510. MANDELGREN, N. M. *Paris.*—Scandinavian monuments, &c. executed in chromolithography.

511. MEYER & KÖSTER, *Gothenburg.*—Specimens of lithography.

512. BERLING, F. *Lund.*—Specimen of type.

Sub-Class D.

513. BECK, F. *Stockholm.*—Specimens of bookbinding.

CLASS XXIX.

Sub-Class A.

516. BONNIER, A. *Stockholm.* — Various maps of Sweden.

517. ERDMANN, A. *Stockholm.*—Geological specimens and maps.

518. LJUNGGREN, G. *Stockholm.*—Economical and statistical maps of different districts of Sweden.

519. HAHR, A. *Stockholm.*—Topographical and statistical maps.

520. WARBERG, E. *Stockholm.* — Charts published by the Hydrographic Office.

Sub-Class B.

521. CHALMER'S POLYTECHNICAL SCHOOL, *Gothenburg.* — Electro-magnet and dynamometer, made by the pupils of the school.

522. DIRECTION OF THE ROYAL INDUSTRIAL SCHOOL, *Stockholm.*—Drawings, &c. done by the pupils of the school.

523. FAHLGREN, C. J. *Askersund.*—Typograph, by which blind persons can correspond with one another.

525. LYCKEBY IRON FOUNDRY, *Bleking.*—School furniture.

526. MANILLA INSTITUTE FOR BLIND, DEAF, AND DUMB, *Stockholm.*—Various apparatus for the blind; articles made by them.

527. SILJESTRÖM, P. *Stockholm.*—Various furniture and apparatus for schools; gymnastic apparatus.

527*a.* MENTZER, LIEUT. *Stockholm.* — Globes, maps, models, and other school apparatus.

528. BAGGE, J. S. *Fahlun.*—Model of J. Cæsar's bridge over the Rhine.

529. ENGDAHL, C. L. *Stockholm.*—Model and description of new method of teaching writing.

CLASS XXX.

Sub-Class A.

532. PETTERSSON, C. *Gothenburg* —Writing-table of Jacaranda.

533. CLIFFORD, J. *Stockholm.*—Writing-table.

534. GUNDBERG, J. W. *Stockholm.*—Fire-screen.

535. LANGEMEIJER, W. *Stockholm.*—Sofa and chairs.

Sub-Class B.

537. LINDGREN, P. A. *Söderköping.*—Various articles carved in wood.

539. MALMSTEN, J. A. *Jönköping.* — Painted blinds.

541. NORLING, L. W. *Jönköping.*—Paper-hangings of various kinds.

542. SÖDERSTRÖM, A. J. *Ulricehamn.*—Paper-hangings.

543. ULANDER, M. *Gothenburg.*—Lady's work-table, the top inlaid with cork, birch-bark, and sponge.

544. ISBERG, S. (PEASANT WOMAN), *Motala.*—Goblet, carved in wood.

545. TAUTZ, H. *Stockholm.*—Candelabrum of papier-maché.

545a. DALIN, P. *Stockholm.*—Picture-frame of oak.

545b. MINEUR, C. G. *Stockholm.* — Hangings of pressed linen, and satin painted.

CLASS XXXI.

Sub-Class A.

546. BACKMAN, J. F. *Stockholm.*—Iron safe.

548. BOLINDER, J. & C. G. *Stockholm.*—Kitchen ranges, flat-iron warming apparatus.

549. CHRISTIANSSON, G. & SON, *Askersund.*—Hand-hammered nails and tacks.

550. FORSGREN, A. *Lindesnœs, Dalecarlia.*—Universal turn-screw.

551. GUNNEBO IRON CO. *Gunnebo, Westerwik.*—Wood screws, and nails of iron and brass, iron wire, iron springs for sofas.

552. GUSTAFSON, C. *Hellstorp, Jönköping.*—Specimens of rolled nails.

The exhibitor manufactures "rolled nails:" and is *Inventor and Patentee* for England, France, Belgium, Sweden, &c. of *nail-rolling machinery*, producing, direct from bar-iron or other metals in bars, all sizes of nails, which combine a new and advantageous shape, excellent quality, and low cost of production, in a way which supersedes all sorts of nails hitherto manufactured. Patent rights granted. Patent agents: for England, Messrs. Carpmael & Co. 24 Southampton Buildings, Chancery Lane, London; for France and Belgium, Mr. Ch. Perpigna, 13 Rue du Cherche-Midi, Paris.

553. HEDLUND, J. *Eskilstuna.*—Locks of different kinds.

555. JOHANSSON, J. M. *Målön, Smaland.*—Iron vices.

556. JÄDER IRON MANUFACTURING CO. *Westmanland.*—Iron manufactures.

559. LYCKEBY IRON FOUNDRY, *Bleking.*—Garden sofa of iron, flat-iron warming apparatus.

560. MOBERG, A. *Stockholm.* — Kitchen-range, with movable top plate, and copper tank for water.

561. ROBSON, A. *Aspa Iron Works, Nericia.*—Nails.

562. STENMAN, F. *Eskilstuna.*—Locks.

563. BERGSTRÖM, J. W. *Stockholm.* — Screw-stock on a new principle, invented by O. J. Aqvilon.

564. PETTERSSON, M. *Stockholm.*—Buttons.

565. BERG, A. *Eskilstuna.*—Hinges made by machinery.

565a. HEDENGRAU, A. *Eskilstuna.*—Yarn windle.

565b. SCHUBERT, J. G. *Stockholm.* — Embossed iron plate vase, executed by J. Stenberg.

565c. WALÉN, J. *Eskilstuna.*—Locks.

Sub-Class B.

567. HEDENSTRÖM, SONS, *Westeräs.*—Candlesticks, coffee pots.

569. WESTERBERG, E. *Gusum, Ostgothland.*—Brass-wire gauze, brass-wire, pins, insect pins.

Sub-Class C.

573. KOCKUM, F. H. *Ronneby, Bleking.*—Pans and kettles of tin-plate, iron and copper nails.

CLASS XXXII.

Sub-Class B.

576. ANDERSSON, H. *Malmö.*—Knives and scissors.

577. DAHLBERG, H. *Eskilstuna.* — Steel wares, etched and gilt.

578. GUNNEBO IRON MANUFACTURING CO. *Gunnebo, Westerwik.*—Chaff-knives, scythes.

579. HEDLUND, L. *Eskilstuna.*—Scissors, &c.

580. HELJESTRAND, C. *Eskilstuna.* — Razors.

582. STAHLBERG, L. F. *Eskilstuna.* — Knives and forks.

358. SVENGREN, J. *Eskilstuna.*—Cutlery.

584. OBERG, C. O. *Eskilstuna.*—Files.

584a. SVALLING, E. *Eskilstuna.*—Knives.

586. ÖSTERBERG, C. V. *Eskilstuna.* — Knives.

CLASS XXXIII.

590. FERON, L. C. *Stockholm.*—Various chased silver articles.

591. FYRWALD, C. J. M. & Co. *Stockholm.*—Gold-wire articles.

593. LARSON, L. & Co. *Gothenburg.* — Punch-bowl, modelled by Qvarnström, property of the Par Bricol Society.

597. DUFVA, A. G. *Stockholm.*—Drinking-horn, with electro-plate ornaments; electro-plate tea-urn.

CLASS XXXIV.

Sub-Class B.

599. BROMÖ GLASS WORKS, *Westgothland*—Specimens of their manufactures.

601. BRUSEWITZ, F. *Limmared, Westgoth-land.*—Specimens of manufactured glass, and of the materials employed.

602. KOSTA GLASS WORKS, *Smaland.*—Chalice, paten, and other articles of coloured glass and crystal.

CLASS XXXV.

606. GEIJER, B. R. HEIRS OF, *Rörstrand, Stockholm.*—Dinner services of porcelain, &c.

607. GUSTAFSBERG PORCELAIN MANUFAC-TURING Co. *Stockholm.*—Porcelain tables, dinner service, &c.

608. HÖGANÄS COLLIERY, *Scania.*—Terra-cotta goods.

SWITZERLAND.

NORTH-WEST COURT AND NORTH-WEST GALLERY.

CLASS I.

1. BRUNNER, PROF. C. *Berne.*—Metallic manganese, obtained by a new process.

2. GYPSUM SOCIETY, *Klosters, Grisons.*—Stucco, and casts made with it.

3. LEONI, C. *Coire.* — Marble from the Splugen ; a marble monument from the Grisons.

4. LINDENMANN & Co., *Bünzen, Argovie.*—Pressed peat in pipe form.

5. THEOBALD, PROF. G. *Coire.*—Minerals found in Grisons.

6. TUGGINER, WEY, & Co. *Soleure.*—Refractory fire-clay.

CLASS II.

10. CHAUTEN, *Geneva.*—Vegetable elixir for sea-sickness, &c.

11. COURT, AMI, *Geneva.*—Purple colour for painting on enamel or porcelain.

12. GIMPER, G. *Züric.*—Ethers, oils, and other pharmaceutical products.

13. LACROIX, J. M. *Geneva.*—A specific for the toothache.

14. LAUTERBURG, F. *Berne.*—Waterproof mineral varnish.

15. MATTHEY, A. O. *Locle.* — Diamond powder for polishing steel.

16. MULLER, J. J. & Co. *Basle.*—Extracts of dyewood ; aniline and its dyeing derivatives.

17. ST. MORITZ, SOCIETY OF, *Coire.* — Mineral water of St. Moritz.

18. TARASP, SOCIETY OF, *Coire.*—Mineral water of Tarasp.

---

19. DIEDEY, M. *Lausanne.*—" Eau des Alpes," a hair preserve.

CLASS III.

20. BAUER, C. A. *St. Gall.*—Various objects in gum tragacanth.

21. BECK-LEU, F. *Bekenhof, near Sursee, Lucerne.*—Honey, Alpenschotten sugar, &c.

22. BÉLENOT, F. F. *Mouruz, near Neufchâtel.*—Red Neufchâtel wine of 1857.

23. BÉRAUD, MARC, & Co. *Vevey*—White wine of Yvorne of 1800 to 1846; wine from a Neufchâtel vine.

24. BERTHOLET & Co. *Vevey.* — Swiss cigars.

25. BILLE, BOREL, & Co. *Thielle, Neufchâtel.*—Havana cigars.

26. BOUVIER, BROTHERS, *Neufchâtel.* — Neufchâtel champagne.

27. BOYMOND, *Geneva.*—Pectoral lozenges, Geneva drops, cordials.

28. BUNDNERISCHE WEINBAUGESELLSCHAFT, *Coire.*—Wine from the Grisons.

28a. CHERVAZ, G. *Vetroz, near Sion.*—Wine from Valais, white and red.

29. CHESSEX, R. *Montreux, Vaud.*—White wine from a Rhenish vine grown at Montreux.

30. CLÉMENT & Co. *Vevey.*—Swiss cigars.

31. CORBOZ-DUBOUX, BROTHERS, *Epesses, Vaud.*—White wine, Lavaux of 1859; white wine of 1857 and 1858.

32. DÉGLON, *Lausanne.*—Helvetian cigars.

33. DELAJOUX, JOHN, *Vevey.*—White wine of 1854, grown at Vevey.

34. BOUVIN, C. JUN. *Sion.*—Wines of the Valais.

35. DESPLANDS & RICHARDET, *Vevey.*—White wine, Lacôte of 1859, and Vevey of 1859.

36. DUBOIS, C. *Vevey.*—White wine from Vevey, 1854 and 1859.

37. DUCHOSAL, *Director at Vernets, Geneva.*—Samples of cereals.

38. DUPLAN & MONNERAT, *Vevey.*—White wine, Yvorne, Clos du Rocher of 1848 and 1858.

39. EICHENBERGER, J. J. *Beinwyl, Argovia.*—Cigars.

40. ELLES, H. *Vevey.*—White Markobrunner, grown at Vevey in 1856 and 1858.

41. FANKHAUSER, J. C. *Lausanne.*—Chocolate.

42. FASSBIND, G. *Arth, Schwytz.*—Cherry-waters.

43. FELDMUHLE, *Rorschach, St. Gall.*—Alimentary pastes.

44. FERT & MOSSU, *Geneva.*—A stomachic made of wine and aromatics.

45. FINAZ, *Geneva.*—Pectoral paste, liquor-ice juice, cashoo from Bologna, wine of Yvorne, &c.

46. FRÖHLICH, G. *Steffisbourg, Berne.*—Cherry-water.

47. FROSSARD-MULLER & SON, *Payerne, Vaud.*—Gruyere and Emmenthal cheese.

48. GERBER, C. SON, *Steffisbourg, Berne.*—Cherry-water.

49. GILLIARD, ELISE, & Co. *Fleurier, Neufchâtel.*—Green extract of absinthe.

50. KLAUS, C. *Locle.*—Pectoral paste, confectionery, pharmaceutical pastiles, &c.

51. KOEBEL, A. *Sion, Valais.*—Wine from Sion, and cherry-water.

52. KOHLER, AMÉDÉ, & SON, *Lausanne.*—Chocolate.

53. LEGLER, G. *Couvet.*—Extract of absinthe, &c..

54. MASSON, F. *Grandson.*—Cigars.

55. ORMOND & Co. *Vevey.*—Swiss cigars.

56. PASCHOUD & FREYMAN, *Vevey.* — Sparkling wine of Vaud, wine of Yvorne of 1854, Château du Châtelard of 1854, Lavaux of 1854, Lacôte of 1834.

57. LEUBA, A. *Columbier.* — Extract of absinthe.

58. REYMOND & WARNÉRY, *Payerne.*—Cigars and tobacco.

59. DE RIVAZ, A. *Ardon, Valais.*—Wines of Amique.

60. ROSSELET-DUBIED, *Couvet.*—Extract of absinthe, and cherry-water, &c.

61. RUFENACHT, D. *Vevey.*—White wine of Yvorne of 1854.

62. SCHERER, BROTHERS, *Meggen, Lucerne.*—Cherry-water of 1857.

63. SPINTZ, DOCTOR N. *Berzona, Tessin.*—Iodined chocolate.

64. STADELMANN, J. *Escholzmatt, Lucerne.*—Crystallized sugar of milk.

65. SUCHARD, PH. *Neufchâtel.*—Chocolate.

66. VIDOUDEZ & Co. *Lausanne.*—Cigars.

67. VAUTIER, BROTHERS, *Grandson.*—Cigars.

68. VEILLARD, COLONEL A. *Aigle, Vaud.*—White wine of Yvorne of 1859.

69. WASSALI, F. *Coire.*—Alpine honey, from near the sources of the Rhine.

70. WEBER, F. *Geneva.*—Syrup of lichens.

71. WEBER, S. & SON, *Menzikon.*—Cigars.

72. WICKY-VOGEL, *Schüpfheim.*—Sugar of milk.

73. WIEDMER, M. *Basle.*—Cigars.

74. ZIEGLER-PELLIS, *Winterthur.*—Wine of the canton of Züric, 1811, 1834, and 1859.

75. MANGE, P. *St. Cergues, Vaud.*—Gentian-water.

76. DELOES, COLONEL A. *Aigle.*—White wine of Aigle, 1859.

CLASS IV.

80. KLARIR, J. *Appeuzele.*—A box in carved wood.

81. GRUNINGEN, VON, J. G. *Saanen, Berne.*—Wood for sounding-boards, ingrained maple, ash, and fir-wood.

82. JAEGER & Co. *Brienz.*—Looking-glass frame, clock-case, &c. in carved wood.

83. MENGOLD, G. W. *Coire.*—Fruits from the Grisons in wax.

84. MICHEL, KASPAR, & Co. *Brienz.*—A bouquet-holder, pen-holders, &c. in carved wood.

85. MONNIER & PIGUET, R. *Nyon.*—Toilet soap, pomatum, perfumes.

86. SIEBER, J. H., *Züric.*—Cups, chessmen, brooches, paper knives, walking sticks, &c. in ivory.

87. TAVEL, GROSS, & Co. *Gunten, Berne.*—Wood for sounding-boards.

88. WALD, A. H. J. (Bazar Suisse), *Thoune.*—A cabinet and other articles in carved wood.

89. WIRTH, E. BROS. *Brienz.*—Carved wood fancy articles.

CLASS VII.

95. BEUGGER, J. *Winterthur.*—Can-roving frame for cotton.

96. EGLI, J. *Lucerne.*—Boot-trees and lasts.

97. FREULER, M. *Glarus.*—Instruments for the engravers of apparatus for printing cotton cloths.

98. HAAS TYPE-FOUNDRY, *Basle.*—Type-founding machine, cutting machine, German types, and book of proofs.

99. LEDOUX, A. *Geneva.*—Lithographic press; apparatus for endless impression.

100. WAHL & SOCIN, *Basle.*—Silk-ribbon looms.

101. WEINGART, J. *Amerzwyl, Berne.*—Barrels of oak-wood.

102. LANDOLT, EMIL, & MILSTER, J. *St. Georgen, St. Gall.*—Hydraulic press, for alimentary pastes.

103. MASCHINEN-BAUANSTALT,*Frauenfeld.*—A stitching and folding machine, and a newspaper-folding machine.

CLASS VIII.

104. ESCHER, WYSS, & Co. *Züric.*—Two compound cylinder marine engines for shallow river navigation.

105. BELL, T. & F. *Lucerne.* —Paper machines.

106. COLLADON, PROF. D. *Geneva.*—Floating hydraulic wheel, for the moving principle in a manufactory, or for raising water.

107. DESCHAMPS, ZACHARY, *Geneva.*—An assaying balance.

108. GRABHORN, B. *Geneva.*—An assaying balance.

111. SCHINZ, C. *Züric.*—Model of a glass furnace.

112. SULZER, BROS. *Winterthur.*—Washing-machine for bleachers, dyers, and stuff-printers.

IMPROVED MACHINE FOR WASHING TEXTILE FABRICS FOR BLEACHERS AND MANUFACTURERS.

In this machine the goods pass through without being subjected to any injurious strain, and undergo an energetic beating process while suspended from the squeezing bowls, an accumulation and opening up of the goods takes place in the water-trough each time after leaving the squeezing bowls and beaters, so that this machine can be used for the lightest as well as the heaviest class of goods; it completely supersedes the dash-wheel, doing the work of seven or eight of these latter machines, and at the same time effecting a great economy in manual labour, space and power. The cleansing process is much more thorough, complete and expeditious, than in any machine yet tried, and this, without in the least injuring the most delicate stuffs.

113. THIEMEYER, R. *St. Gall.*—Balances of different qualities.

CLASS IX.

118. BEAUMONT, H. *Bouthillier-de, Geneva.* —Ploughs on a new principle.

119. MARTIN-DUNOYER, *Trélex, Vaud.*— Light plough, for every description of soil.

120. MENZEL, PROF & GRABERG, *Züric.* —New contrivances to guide the bees in constructing their combs.

CLASS X.

125. BARGEZI, U. *Soleure.*—Marble slab and table.

126. COLLADON, D. *Geneva.*—Machine for boring stones or the ground.

127. DAPPLES, E. *Civil Engineer, Lausanne.*—Sheet-iron screws for wooden and cast-iron screw-piles.

These screws are intended to replace the cast-iron screws now in use. They can be made of any diameter; the pieces are cut out of flat sheet-iron, bent to shape on a mandril, and then riveted together. The point can be made with 3, 4, 5, 6, 7, 8, 10, or more sides. The more sides it has, the less resistance it will afford to the turning of the screw. On wooden piles, the screw is fixed by two vertical ears and bolts. On cast-iron piles or columns it is fixed in the same manner, or bolted on to a flange. The best mode is to have only the blade of wrought-iron, bolted or riveted on to a suitable flange of the column. The column itself is terminated by a conical point. These screws have many advantages over cast-iron screws. At equal strength they are thinner, and of equal section on their whole width. The blade affords less resistance to the turning: 1st, because it displaces less ground; 2d, because it is a thinner wedge; 3d, because it meets fewer stones; 4th, because, on meeting a stone, either the stone or the screw will be less displaced. These screws are lighter, more transportable, stronger, more economical to make, and easier to screw in. For large diameters, the central part of the blade is made of two or three plates riveted together.

128. LEHMANN, J. A. *Sargans, St. Gall.* —A kitchen range, with accessories.

129. PFISTER & SCHIRMER, *St. Gall.*— Cooking stoves, one with, the other without, flues.

130. STAIB, L. F. & Co. *Geneva.*—Caloriferes of cast-iron.

CLASS XI.

133. DEMARTINES, CH. *Lausanne.*—Swiss military head dress.

134. ERLACH, D', R. & Co. *Thoune.*—Rifle and carbine, with barrels of cast steel.

135. KOHLER, L. *Boudry, Neufchâtel.*—The carbine used by the Swiss soldiers.

136. PETER, J. *Geneva.*—Carbines.

137. RIGGENBACH, N. *Olten.*—Carriage of wrought-iron for a 4-pound rifle cannon.

138. SAUERBREY, V. *Basle.*—Double-barrelled guns, one after Le Faucheur; pistols; military carbines.

139. SWISS-BELGIAN INDUSTRIAL SOCIETY, *Schaffhausen.*—The carbine and rifle used by the Swiss soldiers.

141. VERET, COL. J. *Nyon.*—New rifle for infantry.

140. WOLF-BERNHEIM & Co. *Geneva.*—Uniforms of Swiss officers and privates.

141. GOLDSCHMIED, J. J. *Züric.*—Diastimeter for military use.

CLASS XII.

146. SÉNÉCHAUD, L. *Montreux.*—Canoe of sheet iron.

CLASS XIII.

150. AMSLER - LAFFON, *Schaffhausen.*—Reversing levels, planimeters with radial movement.

151. ATELIER DES TÉLÉGRAPHES SUISSES, *Berne.*—Several systems of telegraphic apparatus; electric writing chronograph and thermometer.

152. DAGUET, T. *Fribourg.*—Large disk of flint glass, and of crown glass, for optical purposes.

154. GUTKNECHT, J.J. *Landquart.*—Apparatus to measure spirits and water under any pressure; dry gas-meter.

155. GYSI, F. *Aarau.*—Collection of drawing instruments in German silver; mathematical cases.

156. HIPP, M. *Neufchâtel.*—Telegraphic apparatus, electric clocks, regulators.

157. HOMMEL-ESSER, *Aarau.*—Case of mathematical instruments; pair of compasses, with 29 divisions; pocket case.

158. KERN, J. *Aarau.*—German silver drawing compasses; mathematical cases; engineers' and surveyors' instruments.

159. MONNIER, J. F. *Bevaix, Neufchâtel.*—Barometer and thermometer.

CLASS XIV.

164. GEORG, *Basle and Geneva.*—Photographs of pictures in the Basle Museum.

166. PONCY, F. *Geneva.*—Photographs.

167. VUAGNAT, *Geneva.* — Photographic visiting cards, &c.

CLASS XV.

171. ASSOCIATION OUVRIÈRE, *Locle.*— Astronomical clock, and watches of different kinds and qualities.

171a. AUBERT, BROS. *Sentier, Vaud.*— Movements of watches, a chronometer, two repeaters.

172. AUDEMARS, L. *Le Brassus.*—Watches and gold cases.

173. BAUME & LEZARD, *Geneva, and* 21 *Hatton Garden, London.*—Watches.

174. BAUMEL & SON, *Geneva.*—Files and other tools for watchmakers and jewellers.

175. COURVOISIER, F. *Chauxdefonds.*— Gold and silver watches, pocket chronometers.

176. BOREL & COURVOISIER, *Neufchâtel.* —Gold and silver watches.

177. BOURGEAUX & DELAMURE, *Geneva.* —Screw plates and screws for watches, &c.

178. BREITLING-LAEDERICH, *Chauxdefonds.* —Various watches.

179. CAVIN, F. *Couvet.*—Machine for cutting and rounding the teeth of watch and clock wheels.

180. CORNIOLEY & SON, *Geneva.*—Springs for watches.

181. COURVOISIER, A. *Chauxdefonds.* —Horological articles.

182. COURVOISIER, BROS. *Chauxdefonds.* —Gold and silver watches.

183. DARIER, HUGHES, & CO. *Geneva.*— Horological appliances, watch keys, &c.

184. DEVAIN, J. *Locle.*—Jewels for watchmakers.

185. HENRY, J. *Chauxdefonds.* — Dialplates for watches.

186. EHNHUUS, H. 53 *Frith Street, Soho, London.*—Machines and tools for watchmakers, watch-work in various stages of manufacture.

187. GOLAY-LERESCHE, *Geneva.*—Mechanical case with chronometer, thermometer, barometer, watches, &c.

188. GRANDJEAN, H. & CO. *Locle.*—Marine and pocket chronometers, watches.

189. GRANGER, J. M. *Geneva.*—Enamelled dial-plates..

190. GROSCLAUDE, C. H. & CO. *Fleurier.* —Marine chronometer, gold and silver watches.

191. GUYE, U. *Fleurier.*—Files and machine to round the teeth of wheels.

192. HENRI, J. *Chauxdefonds.*—Dial-plate four lines in diameter, with microscopic inscriptions; different dial-plates.

193. HIRSCHI, E. *Chauxdefonds.*—Timepiece springs, &c.

194. HUGUENIN-THIÉBAUD, *Locle.* — Anchors and wheels for the escapement of watches.

195. JACCOTTET, P. E. *Travers.*—Separate parts of watches.

196. JACKY, BROS. *Chauxdefonds.*—Horological articles.

197. JACOT & SANDOZ, *Locle.* — Gold watches.

198. JEANJAQUET, C *Locle.* — Watch springs.

199. JEANRENAUD, AMI, SEN. *Travers.*— A watch, free escapement, anchor, duplex, dead seconds, lever d'entretien.

200. JEANRENAUD, G. H. *Fleurier.*—Rubies adapted for watch-making, &c.

201. INGOLD, P. F. *Chauxdefonds.*—Files to rectify the teeth of watch wheels.

202. JOHANN, A. *Chauxdefonds.*—Chronometer with diamond crown, &c.

203. JUNOD, EUG. & CO. *Chauxdefonds.* —Chronometer and watches.

204. JUNOD, L. E. *Lucens.*—Jewels for watch-making, sapphires for wire-drawing, &c.

205. KEIGEL & BOREL, *Couvet.*—Machines and tools for watch repairers.

206. LANG & PADOUX, *Geneva.*—Chronometers and half chronometers with certificates from the Observatory of Geneva.

207. LECOULTRE, U. *Sentier.*—Pinions for watches, chronometers, &c.

208. LEHMANN, C. *Bienne.*—Gold and silver watches.

209. LERESCHE-GOLAY, J. *Vallorbes.*—Tools and materials for watch-making.

210. LESQUEREUX, L. *Locle.*—Watch springs.

211. MAIRET, SYLVAIN, *Locle.*—Keyless clock, minute repeater, pocket chronometer, gold watches.

212. MARCHAND, P. A., *Chauxdefonds.*—A gold repeater and skeleton watch.

213. MATTHEY, A. O. *Locle.*—Polishing apparatus, fancy clock, boxes, electric clock, &c.

214. MATTHEY-DORET, P. *Locle.*—Chronometers and a minute repeater, &c.

215. MONTANDON, C. A. *Locle.*—Chronometers and repeaters.

216. MONTANDON, BROS. *Locle.*—Ladies' and boys' watches.

217. MOULINIÉ & LEGRANDROY, *Geneva.*—Box with artificial singing birds; watches, &c.

218. MULLER, A. *Locle.*—Electric chronograph, printing seconds and hundredth parts of seconds.

219. MÜLLER, BROS. *Geneva and London.*—Geneva watches, keys, tools, files, and watch materials.

220. MULLERTZ, J. *Locle.*—Pocket chronometers.

221. NARDIN, U. *Locle.*—Chronometers, remontoirs, repeater, dead seconds, stop, &c.

222. PERREGAUX, H. *Locle.*—Marine chronometer, and chronometer indicating mean temperature.

223. PERRET, A. *Locle.*—Repeaters and other watches, with new movements.

224. PERRET, J. *Chauxdefonds.*—Chronometer and watches.

225. PETITPIERRE, D. L. *Couvet.*—Horological tools; turn-benches to make, mend, round and replace pivots.

226. PIAGET, H. *Verrières.*—Gold, silver, and other watches, cylinder and anchor.

227. PIGUET, BROS. *Sentier, Vaud.*—Horological articles, striking clocks, minute repeaters, chronometers, &c.

228. RAUSS, A. *Geneva.*—Enamelled dial-plates for complicated watches.

229. RAYMOND & ROUMIEUX, *Geneva.*—Springs for chronometers, &c.

230. REYMANN, F. SON, *Geneva.*—Springs for watches with horizontal movements, English watches, &c.

231. LECOULTRE, A. *Sentier, Vaud.*—Millionimeter.

232. RINGGIER & Co. *Zofingen.*—Watches.

233. ROBERT-NICOUD, C. A. *Chauxdefonds.*—Chronometers, auxiliary fusees, &c.

234. ROSSEL & SON, *Geneva.*—Horological articles and jewellery.

235. ROULET, G. *Locle.*—Silver watches, anchor and cylinder.

236. SANDOZ, BROS. *Ponts-martel, Neufchâtel.*—Various watches.

237. SORDET, H. & SON, *Geneva.*—A watch four lines in diameter, a chronometer, half chronometer with thermometer, &c.

238. VAUTIER, SAMUEL, & SON, *Carouge.*—Files and gravers for watchmakers and jewellers; tools for engravers and chasers.

239. VEILLON-EMERY & Co., *Lausanne.*—Chronometers, plain and complex watches.

240. YONNER, C. A. *Verrières.*—Stones for superior watches, screw-plates, &c.

241. PROST, C. *Vercy.*—Pocket chronometer, nickel movement, &c.

CLASS XVI.

247. BRUGGER, F. L. *Motier-Travers, Neufchâtel.*—Accordion.

248. GREINER, T. & BREMOND, B. *Geneva.*—Musical box.

249. HERTIG, J. *Berne.* — Brass instruments.

250. HUNI & HUBERT, *Züric.*—Pianos.

251. KARRER, S. *Teuffenthal, Argovie.*—Musical box.

252. KÖLLIKER & TROST, *Züric.*—Cottage piano.

253. MULLER, BROS. *Geneva and London.*—Musical boxes.

254. NEEF, G. *Feuerthalen, Schaffhausen.*—A church harmonium.

255. SPRECHER & CO., *Züric.*—A grand and a cottage piano.

CLASS XVII.

259. APPIA, DR. L. *Geneva.*—Apparatus for the transport of the wounded.

260. GOLLIEZ, DR. *Lutry, Vaud.*—Tribulcon, an instrument suited to newly-invented missiles.

261. WINKLER, JOSEF, *Berne.* — Microscopic specimens of teeth and bones, fossil remains from lacustrine formations.

CLASS XVIII.

266. ANDEREGG, T. *Wattwyl.* — Cotton cloth dyed in grain, ginghams, checks, handkerchiefs.

267. BEUGGER, J. *Winterthur.*—Glass case containing bobbins of cotton.

268. BREITENSTEIN, J. & CO. *Zofingen.*—Cotton fabrics, machine and hand work.

269. BUNTWEBEREI, *Wallenstadt.*—Checks, ginghams, handkerchiefs, and scarfs.

270. EGLI, H. *At the Neuhof, Fischenthal, Züric.*—A piece of cotton cloth.

271. HEINIGER, J. *Burgdorf*—Canvas for embroidery.

272. HÜSSY, J. R. *Safenwyl.* — Cotton goods.

273. KELLER-STEFFAN, J. G. *Wattwyl.*—Handkerchiefs, gingham, &c.

274. MECHANISCHE-WEBEREY, *Altstaetten, St. Gall.*—Handkerchiefs and cotton goods.

275. MÜLLER, J. B. & CO. *Wyl.*—Checks.

276. NEF, J. J. *Herisau.*—Worked muslins.

377. RAMSAUER-AEBLY, J. U. *Herisau.*—White book muslins, white tarlatans.

378. RASCHLE, ABM. *Wattwyl.*—Coloured cotton goods and handkerchiefs.

279. SCHEFER, J. U. *Speicher.*—White, spotted, and figured book muslin.

269a. SCHMID, HENGGELER, & CO. *Neu-and Unter-Aegeri.*—Cotton yarns, raw stuff, &c.

280. SPINNEREI, *On the Lorze, Baar, Zug.*—Spun cotton from Surat, &c. with the raw material.

281. STEIGER, SCHOCH, & EBERHARD, *Herisau.*—Swiss muslins, curtains, dresses, piece goods, &c.

282. SCHMID, H. *Gattikon, Züric.*—Fine yarns.

283. WIEDMER, J. J. *Seon, Argovie.*—Cotton tissues and spring goods made by machine.

CLASS XIX.

287. SCHMIED, BROS. *Burgdorf.*—Plain and figured linen ticking, bleached and raw.

CLASS XX.

290. BAER & SPINNER, *Ryfferswyl, Züric.* —Black marceline, black gros du Rhin, &c.

291. BAUMANN, SEN. & Co. *Züric.*—Black and coloured taffeta, velveteen, marceline.

292. BAUMANN & STREULI, *Horgen.*—Gros de Naples and other silk goods.

293. BISCHOFF & SONS, *Basle.*—Plain and figured silk ribbons.

294. BISCHOFF, C. & J. *Basle.*—Black silk, satins, serges, and taffetas.

295. BLEULER & KELLER, *Kussnacht, Züric.*—Poult de soie of different quality.

296. BODMER, H. *Züric.*—Silk gauze.

297. BOELGER, M. *Basle.*—Spun silk for various purposes.

299. BRUNNER, H. *Züric.*—Gros de Naples, satin woven with cotton, &c.

300. BRUPBACHER-MÜLLER, *Züric.*—Silk and satin goods.

302. BARY, DE J. & SON, *Basle.*—Plain and fancy ribbons, &c.

303. DOLDER, A. *Meilen, Züric.*—Gros du Rhin, &c.

304. DREYFUS, I. SONS OF, *Basle.*—Silk and velvet ribbons.

305. DUFOUR & Co. *Thal, St. Gall.*—Silk bolting cloths for mills.

306. EGLI, J. C. *Richterswyl, Züric.*—Gros du Rhin and other silks.

307. FICHTER & SONS, *Basle.*—Silk ribbons.

309. HERZGO & Co. *Aarau.*—Silk ribbons.

310. HÖHN & STÄUBLI, *Horgen, Züric.*—Gros de Naples and other silks.

311. HOTZ & Co. *Oberrieden, Züric.*—Gros du Rhin, &c.

312. HÜNI-STETTLER, *Horgen, Züric.*—Gros de Naples, reps, gros du Rhin, &c.

313. HÜRLIMANN-TRÜMPLER & Co. *Waedenswyl, Züric.*—Poult de soie, checked.

314. MÜLLER-HAUSHEER, *Adlisweil, Züric.*—Black gros du Rhin.

315. NAEF, J. R. & SON, *Cappel.*—Lustring, foulards, gros de Chine, &c.

316. NAEGELI & Co. *Horgen, Züric.*—Gros de Naples, black gros du Rhin.

317. NAEGELI & WILD, *Züric.*—Poult de soie, gros du Rhin, checked.

318. NOTZ & DIGGELMANN, *Züric.*—Checked jaspés, chinés, armures.

319. PREISWERK, DIETRICH, & Co. *Basle.*—Satin ribbons and taffetas, Japan and China silks, &c.

320. REIFF-HUBER, *Wiedikon, Züric.*—Bolting cloth.

322. RUTSCHI & Co. *Züric.*—Satin de Chine, satin de Luxor, gros du Rhin.

323. RYFFEL & Co. *Stäfa, Züric.*—Silks, satins, taffeta, &c.

324. SARASIN & Co. *Basle.*—Silk ribbons.

325. SCHMIED, BROTHERS, *Thalweil, Züric.*—Foulards for handkerchiefs and dresses.

326. SCHNEEBELI, C. SONS OF, *Züric.*—Satins for various purposes.

327. SCHUBIGER & Co. *Utznach, St. Gall.*—Silk goods, taffeta for umbrellas.

329. SCHWARZENBACH, J. J. *Kilchberg, Züric.*—Gros du Rhin, &c.

330. SCHWARZENBACH-LANDIS, J. *Thalweil, Züric.*—Poult de soie and taffetas.

331. SIEBER, J. F. *Neumunster, Züric.*—Black gros du Rhin, satin-faced and satin throughout.

332. STAPFER, HÜNI, & Co. *Horgen, Züric.*—Florence, marceline, gros de Naples, gros du Rhin.

333. STAPFER, JOHN, SONS, *Horgen.*—Gros de Naples, poult de soie, checks.

335. STEHLIN & ISELIN, *Neiderschönthal, Basle.*—Long spun silk-yarn.

336. STEHLI-HAUSHEER, R. *Lunnern, Zug.*—Marceline, poult de soie, serge de Malaga, &c.

337. STOCKER, J. C. *Züric.*—Marcelines and other silk goods.

338. STÜNZI & SONS, *Horgen.* — Gros d'Alger glacé, black poult de soie, &c.

339. SUREMANN & CO. *Meilen, Züric.*—Satin de Chine, gros du Rhin.

340. TRÜDINGER & CO. *Basle.* — Fancy ribbons.

341. WEBER, FRIEDRICH, *Hausen-on-the-Albis, Züric.*—Gros du Rigi, &c.

342. WIEDMER-HÜNI, J. J. *Horgen.* — Poult de soie, gros du Rhin.

344. WIRZ & CO. *Seefeld, near Züric.*—Black gros du Rhin.

345. ZINGGELER, BROTHERS, *Wädensweil.*—Poult de soie, gros du Rhin.

CLASS XXI.

353. BALLY & SCHMITTER, *Aarau.*—Elastic textures for boots.

354. BREITENSTEIN, J. & CO. *Zofingen.*—Plain and figured tissues in colours, wool and cotton.

355. HUBLER & SCHAFROTH, *Burgdorf.*—Samples of unravelled wool, and woollen thread spun from that material.

356. HÜSSY, J. R. *Safenwyl, Argovie.*—Half-woollen stuffs.

357. WILDI & HAUSER, *Langenthal, Berne.*—Half-woollen stuffs.

CLASS XXII.

361. GRÄNICHER, SAMUEL, *Zofingen.*—Oil-cloths for floors, tables, &c.

CLASS XXIII.

364. HUNERWADEL, A. *Rapperswyl.*—Fabrics in Turkey-red.

366. KAUFMANN, G. *Triengen, Lucerne.*—Skeins of cotton yarn.

367. LUCHSINGER, ELMER, & OERTLI, *Glaris.* — Indiennes, handkerchiefs, and sarongs.

368. RIKLI, A. F. *Wangen, Berne.*—Turkey-red, brown, and pink coloured yarns.

369. SUTER, J. R. *Zofingen, Argovie.*—Turkey-red coloured cotton yarn.

CLASS XXIV.

374. ALTHERR, J. C. *Speicher, Appenzell.*—Edgings and insertions, machine embroidered muslin curtains and dresses.

375. BAENZIGER, J. *Thal, St. Gall.*—Embroidered muslin, jaconet, batistes, and tulle.

376. EHRENZELLER, F. *St. Gall.*—Embroidered curtains.

377. INDERMUHLE-BUHLER, MRS. *Thoune.*—Embroidered net-work curtains and table covers.

378. MONTANDON, MISS J. *Motiers-Travers, Neufchâtel.*—Table-covers, collars, cuffs, and pincushion-covers of crochet-work.

379. GORINI, J. B. *St. Gall.*—Embroidered handkerchiefs.

380. NEF, J. J. *Herisau.*—Embroidered curtains, balzorine, shawls, &c.

381. RITTMEYER, B. & Co. *St. Gall.*—Insertions, trimmings, and curtains, embroidered by machine.

382. RAUCH, R. *London (St. Gall).*—Embroidered curtains.

383. STÄHELI, WILD C. *St. Gall.*—Fine embroideries.

384. STEIGER-SCHOCH & EBERHARD, *Herisau.*—Embroideries on tulle and muslin.

385. TANNER, B. & H. & KOLLER, *Herisau.*—Embroidered white curtains, silk antimacassars.

CLASS XXV.

390. RECHEWERTH, E. *Züric.* — A rug, made with the skins of the heads, &c. of animals.

391. ROOS, G. *Lausanne.*—Specimens of the various Swiss furs.

392. ROTH, J. *Wangen, Berne.*—Curled horsehair.

CLASS XXVI.

396. MERCIER, J. J. & BROS. *Lausanne.*—Leathers and skins, wools, and horse-hair.

397. MERZ, J. *Herisau.*—Dressed skins.

398. PÜNTER, J. J. *Uerikon, Züric.*—Sole-leather, &c.

399. RAICHLEN, L. *Geneva.*—Leather and skins.

401. TESSE, F. *Lausanne.*—White calf-skins.

402. ZÜND, J. U. *Staefa, Züric.*—Hemp hose for fire-engines.

CLASS XXVII.

405. ANDEREGG, T. *Wattwyl.*—Shirts, and shirt-fronts.

406. BROSSY-PLUMETTAZ, JULIE, *Payerne.*—Flowers and feathers in straw.

407. ECOFFEY, MARIE, *Vevey.*—A crinoline skirt, and stays.

408. FISCHER, BROS. *Meisterschwanden, Argovie.*—Braids of hair, and of manilla and hair; trimmings.

409. FISCHER, J. L. SONS OF, *Dottikon, Argovie.*—Collection of straw and manilla plaits, cigar-cases, &c.

410. HESS-BRUGGER, *Amrisweil, Thurgovie.*—Fancy sleeves, mitts, &c. of English-spun diamond-cotton.

411. ISLER, ALOIS, & Co. *Wildegg, Argovie.*—Plaits, trimmings, and ornaments in straw.

412. ISLER, JACQUES, & Co. *Wohlen, Argovie.*—Straw and horse-hair trimmings, and ornaments.

413. LANDERER-ZWILCHENBART, E. *Wohlenschwyl, Argovie.*—Edgings; plaits in straw, hemp, hair, and silk; trimmings, hats, baskets.

414. RUMPF, C. C. *Basle.*—Silk shirts, under waistcoats, drawers, &c.

415. SCHAERLY - BAERISWYL, J. B. *Fribourg.*—Bonnets of Fribourg straw, and plait used in their manufacture.

416. STOLL, J. C. *Schaffhausen.*—Straw hats.

417. SUDAN, MRS. H. & Co. *Fribourg.*—Straw bonnets.

418. THEDY-GREMION, *Enney, near Bulle, Fribourg.*—Straw plaits from Gruyère.

419. CHIESA, J. *Loco and Morges.*—Straw plaits.

CLASS XXVIII.

422. DRESCHER, Y. *Lithographical Institute, Züric.*—Lithographs; pictures in galvano-plastic, relief.

423. GONIN, F. *Nyon.*—Galvanic engraved plates for printing earthenware.

424. RIETER-BIEDERMANN, *Winterthur*—Printed music, &c.

425. SIEGFRIED, *Wipkingen, Züric.*—Engravings in aqua-tinta.

CLASS XXIX.

429. BECK, E. *Berne.*—Maps in relief.

430. BERTSCH, H. *St. Gall.*—Apparatus for natural philosophy.

431. HUTTER, PROF. *Berne.*—Elementary course of drawing.

433. VORUZ, A. *Lausanne.*—Mathematical works.

434. WURSTER, J. & Co. *Topographical Office, Winterthur.*—Geological maps.

CLASS XXX.

438. CARRAZ, J. B. *Porrentruy, Berne.*—Buffets of oak, sofa, drawing-room table.

439. GAY, E. & F. *Aigle.*—Brushes, inlaid works of wood, table-mats.

440. LAVANCHY-FARAUDO, *Lausanne.*—Bureau.

441. MONNIER & Co. *Aigle.*—Inlaid and other floorings.

442. WETLI, M. *Berne.*—Bureau with springs.

CLASS XXXI.

446. BOSSI, G. *Locarno.*—A lock.

447. HARDMEYER, J. *Züric.*—Hammers for dressing mill-stones; articles in cast-steel.

448. KNECHT, J. *Glaris.*—Brass model for calico printing.

449. MADLIGER, J. *Langenthal, Berne.*—A needle perforated lengthwise.

450. MATTHEY, A. O. *Locle.*—Gold and silver electro-plating.

451. NEUHAUS & BLÖSCH, *Bienne.*—Iron wire for wool-cards, and springs.

452. STEINER, A. *Lausanne.*—Iron folding bedstead and spring mattress.

453. KUGLER-DELEIDERRIEZ, *Lausanne.*—Candlesticks, flambeaux, and lamps, of brass and German silver.

CLASS XXXII.

456. MATTHEY, A. & SONS, *Locle.*—Steel for watch springs, &c.

457. SCHNEIDER, C. F. *Geneva.*—Complicated pocket-knives.

CLASS XXXIII.

459. BERTHOUD, L. *Fleurier.*—Rubies, for wire-drawing, for astronomical purposes.

460. BESSON, J. *Neufchâtel.*—Chasings in steel, statuette, &c.

462. DUBOIS, A. *Chauxdefonds.*—Engravings in gold; castings in silver, enamellings, jewellery, pictures, &c.

463. FAVRE-BULLE, J. C. *Chauxdefonds.*—An oval engraved gold plate.

464. FAVRE-BULLE, L. E. *Locle.*—Decorations for watch-cases.

465. GRANDJEAN-PERRENOUD, H. *Chauxdefonds.*—Engraved decorations, letters, enamel, and diamonds, for watchmaking.

467. HUBER, J. J. *Geneva.*—Movable jewellery.

468. JUNOD, Z. *Locle.*—Jewels, pallets for chronometers, sapphire-gravers and files.

469. KUNDERT, F. *Chauxdefonds.*—Castings and plates in gold and silver, engravings, and precious stones for watchmaking and jewellery.

471. LANG & PADOUX, *Geneva.*—Brooches, bracelets, with landscapes or portraits, from photographs, on enamel.

472. MASSET, L. & SON, *La Motte.*—Frame with perforated rubies for watch-makers; wire-drawing sapphire and diamonds for gold-wire drawers.

CLASS XXXV.

476. GONIN, F. & BURNAND, A. *Nyon.*—Earthenware; imitations of porcelain; and articles for cooking.

477. LERBER, M. DE, *Romainmotier, Vaud.*—Pipes for wells and drains; bricks.

478. ZIEGLER-PELLIS, *Winterthur.*—Articles in terra-cotta.

479. HEIMBERG, BERNE, SOCIETY OF THE POTTERS.—Common pottery.

CLASS XXXVI.

481. CHAMUSSY, J. B. *Geneva.*—Trunks.

482. ISENRING, J. G. *Geneva.*—Articles for travelling.

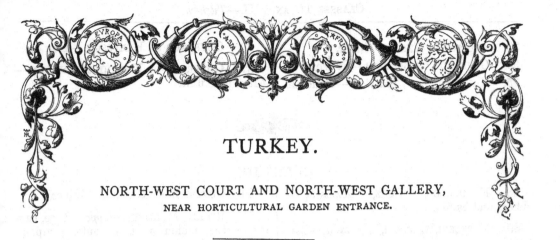

TURKEY.

NORTH-WEST COURT AND NORTH-WEST GALLERY,
NEAR HORTICULTURAL GARDEN ENTRANCE.

CLASS I.

1. MEHMED, HADJI, *Constantinople.* — Gold and silver leaf.

2. OHANES, *Constantinople.* — Gold and silver thread, and laces.

3. SIMEON, *Constantinople.*—Diamond for cutting glass.

4. GOVERNOR OF ZANOUN, *Amassia.* — Flint stone.

5. DIMITRI, *Amassia.*—Silver, silver ore, and ingot from the Jolnoon mines.

6. GOVERNOR OF BALIKESSER. — Plaster, stone, mill-stone, and other stones.

7. GOVERNOR OF MONASTIR. — Iron ore, and iron.

8. RESHID, MEHMED, *Aleppo.* — Yellow, black, and red stones, and specimen of their use.

9. GOVERNOR OF PHILIPPOLI.—Lime, lime-stone, flint, and kufgie stones ; saltpetre.

10. GOVERNOR OF MOUNT KHANIA.— Stones, and iron ore.

11. GOVERNOR OF TARASHJA. — Mineral earth.

12. GOVERNOR OF GUMULJINA.—Minerals and alum.

13. GOVERNOR OF CYPRUS. — Copper ore, minerals.

14. GOVERNOR OF LAPISKY.—Clay, emery, potter's earth.

15. GOVERNOR OF AVRATHISSAR.—Silver ore.

16. GOVERNOR OF CARADAGH.—Silver ore, gold dust.

17. GOVERNOR OF NEVREKOL.—Iron ore.

18. GOVERNOR OF LESCOFCHA.—Iron ore and bar iron.

19. GOVERNOR OF MOSSUL.—Mineral.

20. GOVERNOR OF KUTAHIEH.—Écume de mer (mineral).

21. GOVERNOR OF JANIK.—Iron-stone, and potter's earth.

22. NOOVALLAH, *Conia.* — Saltpetre (rough).

23. GOVERNOR OF TIRNOVA.—Copper.

24. MUSTAPHA, DERVISH, *Bosnia.*—Copper.

25. GOVERNOR OF BOSNIA.—Lead ore and minerals.

26. GOVERNOR OF DAMASCUS. — Copper and iron ore, and stone.

27. GOVERNOR OF ADANA. — Lead and copper ore.

28. GOVERNOR OF LEBANON. — Pit coal, iron ore, and other minerals.

29. GOVERNOR OF TREBISOND. — Copper and silver ore.

30. GOVERNOR OF CASTAMOUNI.—Copper ores.

31. GOVERNOR OF ANGORA.—Copper ore and mineral salt.

32. GOVERNOR OF CRETE.—Minerals.

33. APOSTOL OF CRETE.—Silver.

34. GOVERNOR OF BROUSSA.—Marble.

35. KOURSHID, AGA, *Lazistan.*—Minerals.

36. GOVERNOR OF KERESSON.—Iron ore.

37. GOVERNOR OF KARAHISSARSHARKY — Alum and flint.

CLASS III.

38. KURDOGLOO, HAFIZ, *Amassia.* — Wheat and bamia.

39. MEERLEVI, CHEIKHI, *Amassia.*—Barley.

40. HASSAN, HAJI, *Amassia.*—Plums.

41. AKHASSAN, HAFIZ, *Amassia.*—Lentils.

42. GULLEKREOGLOU, *Amassia.*—Peas.

43. BAKAL, KIUMIL, *Amassia.*—Kidney beans.

44. BAKAL KEHREMAN, *Amassia.*—Haricot beans.

45. HAJI, EFENDI, *Amassia.*—Aubergines.

46. GOVERNOR OF AMASSIA.—Beans and millet.

47. OSMAN, EFENDI, *Amassia.*— Haricot beans and maize.

48. KIORTASH-OGLY, *Amassia.*—Mehleb.

49. BOOROON-OGLY, KIRKOR, *Amassia.*— Salep.

50. BOORNOGLY, KERORK, *Keupri.* — Wines.

51. NICHOLAS, *Amassia.*—Litharge and lead.

52. GOVERNOR OF CA ASSY.—Wheat, barley, rye, sesame, &c.

53. GOVERNOR OF KOSAK.—Pistachio.

54. GOVERNOR OF BALIKESSER.—Different sorts of corn.

55. GOVERNOR OF LOMA.—Raisins.

56. GOVERNOR OF BALIKESSER.—Olive and sesame oils.

57. MEKHOORY, *Saïda.*—Pumpkin and aubergine seeds, and various other seeds.

58. GOVERNOR OF LATAKIA. — Various seeds, olives, sesame oil, honey, olive oil, and salt.

59. HASSAN, AGA, *Monastir.*—Plums.

60. GOVERNOR OF KESREE.—Wines.

61. RESHID, MEHMED, *Aleppo.*—Rice, barley, lentils, melon seed, walnuts, junipers, fennel, and liquorice &c.

62. RESHID, MEHMED, *Aleppo.*—Wines, spices, and dried fruits.

63. GOVERNOR OF PHILIPPOLI. — Rice, wheat, barley, rye, millet, beans, peas, &c.

64. GOVERNOR OF PRASDIN.—Raky.

65. RUSTEM, USTA, *Cozandjik.* — Chestnuts, dried cherries, and wheat.

66. GOVERNOR OF DJERBAN.—Wheat and barley.

67. GOVERNOR OF SHAEER.—Butter, olive, and sesame oils.

68. GOVERNOR OF TRIPOLI.—Olives, walnuts, and almonds.

69. GOVERNOR OF AKKIA.—Raky, wine, olives, and olive oil. Various grains and seeds.

70. GOVERNOR OF BELADBECHAN.—Raky, wine, olives, and olive oil. Various grains and seeds.

71. GOVERNOR OF DRAMA.—Wheat, barley, maize, and rice.

72. GOVERNOR OF GUMULGINA.—Wheat, barley, maize, and rice.

73. GOVERNOR OF CAVALA.—Wheat, barley, rye, and maize.

74. GOVERNOR OF SARISHABAN.—Wheat, barley, rye, and maize.

75. GOVERNOR OF CURASSOU. — Wheat, barley, rye, and maize.

76. GOVERNOR OF CYPRUS.—Wines, and various grains and seeds.

77. GOVERNOR OF DARDANELLES.—Wheat, sesame, peas.

78. GOVERNOR OF AIVALIK.—Olive oil.

79. GOVERNOR OF LAPISKY.—Grape molasses.

80. GOVERNOR OF SALONICA.—Various grains and seeds.

81. DARBILA, *Salonica.*—Samples of flour.

82. SOTIN-OGLOU, MICAL, *Salonica.* — Wines.

83. GOVERNOR OF SIROS.—Various grains and seeds.

84. GOVERNOR OF NISH.—Various grains and seeds.

85. GOVERNOR OF MOSSUL.—Wheat, millet, and corn.

86. GOVERNOR OF KUTAHIEH. — Corn, seeds, oils, fruits.

87. GOVERNOR OF ANGORA.—Wheat.

88. GOVERNOR OF ANDRINOPLE.—Bamia, sweetmeats, dried fruit, and oils.

89. GOVERNOR OF JANIK.—Corn, seeds, and fruit.

90. ABDULLAH, OOSTA, *Conia.* — Kernel and seed oils.

91. APOSTOL, *Conia.*—Wine and raky.

92. GOVERNOR OF NAPLOUSE.—Corn and seeds, oils.

93. GOVERNOR OF TOOLCHA.—Wheat, barley, millet.

94. GOVERNOR OF BOSNIA.—Dried plums.

95. GOVERNOR OF VIDIN. — Strawberry wine.

96. GOVERNOR OF DAMASCUS.—Samples of corn and seeds.

97. BEYAZID, EFENDI, *Damascus.*—Fruits, honey, pickles, molasses, sugars, &c.

98. GOVERNOR OF ADANA.—Wheat, barley, and sesame.

99. GOVERNOR OF ISLIMIA.—Wines and syrups.

100. GOVERNOR OF LEBANON. — Corn, honey, oils, olives, molasses, fruits, and seeds.

101. SELAMOUNI.—Spirits.

102. BEDROS, *Lebanon.*—Wines.

103. HOWSY, H. *Lebanon.*—Wines.

104. RESHID, H. *Lebanon.*—Wines.

105. GOVERNOR OF SCUTARI OF ALBANIA.—Corn and seeds ; wines and oils.

106. GOVERNOR OF MARDIN. — Dried fruits.

107. GOVERNOR OF CONIAH.—Corn and seeds, dried fruits, honey, and molasses.

108. GOVERNOR OF CASTAMOUNI.—Rice, saleb, and dried plums.

109. GOVERNOR OF ANGORA.—Wheat and barley.

110. GOVERNOR OF CRETE. — Wines, syrups, and spirits, dried fruits, salep, oils, honey, pickles, and seeds.

111. AGHACO, *Crete.* — Wheat, barley, beans, peas, and almonds.

112. SAMI BEY, *Crete.*—Butter, herrings, honey, and olives.

113. MONOULAKI, *Crete.*—Olives.

114. MICHALAKI, *Crete.*—Wines.

115. TANASH, *Broussa.*—Red and white wines.

116. GOVERNOR OF BROUSSA.—Olive oil.

117. KOURSHID, AGA, *Lazistan.*—Millet.

CLASS IV.

118. RIZA, ALI, *Constantinople.* — Wax candles.

119. PAPAZSGLOO, YANKO, *Gueva.*—Silk.

120. IBRAHIM, AGA, *Constantinople.* — Soap.

121. TOROSOGLOU, BOGOS, *Constantinople.*—Brutia silk.

122. GOVERNOR OF BROUSSA.—Silk, silkworm cods, silk tufts.

123. BROTTE, *Constantinople.*—Brutia silk.

124. ZADIKOGLOU, CURUBIS, *Uskulub.*—Silk.

125. HUSSEIN, *Hadjikioy.*—Hemp.

126. ISMAIL, HAJI, *Medzoza.*—Cotton.

127. GULBEKRE - OGLOO, *Amassia.* — Poppy.

128. BOYAJI, MEYER, *Amassia.*—Madder-roots.

129. KISHLAJICKLY - OGLY, *Amassia.* — Pérétek.

130. VIRMISHLI, AHMED, *Amassia.* — Earth used for the manufacture of grape molasses.

131. BOOROON-OGLY, KIRKOR, *Amassia.*—Opium.

132. ALI, AGA, *Uskulub.*—Yellow berries.

133. GOVERNOR OF USKUP.—Salt.

134. HASSAN, *Keupri.*—Tobacco.

135. PERGAMA, *Kirkagaj.*—Cotton.

136. MEHMED-OOSTA, *Artemid.*—Soap.

137. GOVERNOR OF BALIKESSER.—Poppy seed, cotton, nuts, madder roots, madder seed, tobacco, and opium.

138. KHORAJAN, MADAM, *Beyrouth.* — White and yellow silk.

139. KURI, N. *Beyrouth.*—White silk.

140. ANASSIR, HOJA, *Beyrouth.*—Yellow silk.

141. SHOUAN, BENI, *Beyrouth.*—Soap.

142. MEKHOORY, *Saïda.*—Cotton in shell.

143. GOVERNOR OF LATAKIA. — Tobacco, soap, sponges, and cotton.

144. MUSTAPHA, HAJI, *Monastir.*—Fine flax.

145. GOVERNOR OF MONASTIR.—Madder roots.

146. RESHID, MEHMED, *Aleppo.* — Gall-nuts, wool, wax, soap, yellow berries, tobacco, cotton, and tallow.

147. HUSSEIN, HAJI, *Philippoli.*—Silk.

148. GOVERNOR OF DJERBAN.—Opium.

149. GOVERNOR OF TRIPOLI.—Silk.

150. MANASTIL, *Shaeer.*—Soap.

151. GOVERNOR OF SHAEER.—Sponges.

152. GOVERNOR OF TRIPOLI.—Wax.

153. GOVERNOR OF MOUNT KHANIA.—Various kinds of grain and seeds.

154. GOVERNOR OF TRIPOLI. — Various seeds.

155. GOVERNOR OF BELAD BECHAN.—Tobacco.

156. GOVERNOR OF DRAMA. — Cotton wool, cotton in shell, tobacco.

157. GOVERNOR OF GUMULJINA.—Silk.

158. GOVERNOR OF CAVALA.—Tobacco and cotton.

159. GOVERNOR OF CHARISHABAN. — Tobacco and cotton.

160. GOVERNOR OF CURASSOU.—Tobacco.

161. GOVERNOR OF CYRUS.—Raw silk and cotton.

162. GOVERNOR OF DARDANELLES.—Gall-nuts.

163. GOVERNOR OF AÏVALIK.—Acorns.

164. YACOFINI, *Salonica.*—Raw silk.

165. GOVERNOR OF SALONICA.—Silk cods.

166. GOVERNOR OF AVRATHISSAR.— Silk cods.

167. GOVERNOR OF SIROS.—Cotton, tobacco, flax.

168. HALINI, AGA, *Smyrna.*—Silk.

169. GOVERNOR OF NISH.—Yellow dye and dye wood; silk.

170. GOVERNOR OF CAISERIA.— Yellow berries.

171. GOVERNOR OF KUTAHISH.—Tobacco, madder roots, mastic, silk, silk cods, gall-nuts, valonea.

172. GOVERNOR OF ANDRINOPLE.—Silk, silkworm cods, tobacco, wool, hemp, cotton, madder roots.

173. MELEDIK, YORGI, *Janik.*—Raw silk.

174. KHOORSHID, *Janik.*—Wool.

175. AHMED, *Janik.*—Tobacco.

176. GOVERNOR OF JANIK.—Hemp.

177. GOVERNOR OF NAPLOUSE.— Cotton, nuts, tobacco.

178. GOVERNOR OF TOOLCHA.—Wool.

179. MEHMED & MUSTAPHA, *Kurdistan.*—Raw silk and cotton.

180. GOVERNOR OF BOSNIA. — Flax and tobacco.

181. GOVERNOR OF VIDIN.—Wheat and maize.

182. BOOYANA, *Vidin.*—Raw silk.

183. GOURGU, *Vidin.*—Raw silk.

184. ABOUSALEH, *Damascus.*—Hemp.

185. HALIM, EL DEVAL, *Damascus.*—Wax.

186. GOVERNOR OF DAMASCUS.—Tobacco, shumac, madder roots.

187. DJERDJIS, *Damascus.* — Raw silk and cotton.

188. GOVERNOR OF MAGNESIA.—Raw silk, tobacco, cotton nuts, madder roots ; earth used in the manufacture of grape molasses.

189. GOVERNOR OF ADANA. — Cotton, tobacco.

190. GOVERNOR OF SCUTARI OF ALBANIA.—Raw silk and silk cods.

191. EDDÉ, M. *Lebanon.*—Raw silk.

192. GOVERNOR OF LEBANON. — Sugarcane.

193. MEDAWAR, N. *Lebanon.*—Raw silks.

194. SHAHDAN, Y. *Lebanon.*—Raw silks.

195. HALIL, E. *Lebanon.*—Raw silks.

196. GOVERNOR OF LEBANON.—Sponges, tobacco, and cocoons.

197. GOVERNOR OF SCUTARI OF ALBANIA. —Tobacco.

198. GOVERNOR OF CONIA.—Yellow berries, hemp seed, mastic, madder roots, shumac, opium, wax, valonia, gall-nuts, glue.

199. GOVERNOR OF CASTAMOUNI.—Saffron, hemp, wax, goats' wool and linseed.

200. GOVERNOR OF CRETE. — Saffron, opium, red and yellow dyeing earth, wax, tobacco, raw silk, cocoons, and dye roots.

201. AGHACO, *Crete.*—Raw silk.

202. SAMI BEY, *Crete.*—Various kind of soaps.

203. MANULAKI, *Crete.*—Raw silks.

204. GOVERNOR OF BROUSSA. — Walnut wood.

205. CONSTANTIN, *Broussa.*—Raw silk.

206. KULEYAN, OVAGHIM, *Broussa.*—Raw silk.

207. KIRMIZIAN, GARABET, *Broussa.*—Raw silk.

208. BEGOGLOO, APOSTOL, *Broussa.*—Cocoons.

209. GOVERNOR OF BROUSSA. — Valonia and gall-nuts.

210. MOOSSA, *Keresson.*—Raw hemp.

211. GOVERNOR OF KARAHISSAR SAHIB.—Opium, poppy seed, and linseed.

CLASS VII.

212. BEYAZID, EFENDI, *Damascus.* —Weaving machine.

CLASS VIII.

213. BALYAN, S. *Constantinople.*—Locomotive engine.

CLASS IX.

214. GOVERNOR OF DRAMA.—Rice cleaning machine.

215. GOVERNOR OF CARADAGH.—Machine for separating gold dust from sand.

216. HALIM, EL DEVAL, *Damascus.*—Agricultural implements.

CLASS XI.

217. IMPERIAL MANUFACTORY. — Pistols and rifles, for the army and navy.

218. MEHMED BEY, *Tirnova.*—Rifle.

219. ABDULHADI, *Saïda.*—Sword belt.

220. GOVERNOR OF HASKIOY.—Ramrods.

221. SHIMOUN, *Mossoul.*—Musket.

222. HASSAN-OOSTA, *Conia.*—Gun.

223. DERVISH, *Bosnia.*—Cartouch boxes and slippers.

224. SALIH, RAMO, & ABRO, *Bosnia.*—Musket battery.

225. DEVARA, MEHMED, *Damascus.*—Model of gun and carabine.

226. GOVERNOR OF SCUTARI OF ALBANIA.—Cartridge box and musket batteries.

227. GOVERNOR OF TREBISOND.—Gunpowder box.

228. HALIL-OOSTA, *Trebisond.*—Musket battery.

229. AIDIN, AGA, *Uskub.*—Pistols, gun, and rifles.

230. GOVERNOR OF CRETE,—Pistols, with gold ornaments.

231. BEDIGOGLOU, *Karahissar Sahib.*—Musket battery, &c.

217*a.* HETOOM.—Swords, inlaid with gold.

CLASS XVI.

232. TOPLEE, *Constantinople.* — Turkish musical instruments.

233. AVEDIS, *Constantinople.* — Kettledrums.

CLASS XVIII.

234. SALIH, *Constantinople.* — Cotton stuff.

235. BATO, *Constantinople.* — Printed calico.

236. ZEITCHA & MARIDRITCHA, *Constantinople.*—Printed calicoes.

237. IMPERIAL MANUFACTORY, *Constantinople.*—Printed calicoes, and calicoes.

238. YORDUM-OOSTA, *Amassia.*—Printed calico.

239. IBRAHIM-OOSTA, *Mersicon, Amassia.*—Cotton stuff for furniture.

240. OSMAN-OOSTA, *Mersicon, Amassia.*—Cotton stuff.

241. DALTABAN-OGLOU, *Mersicon, Amassia.*—Bath towels.

242. HASSAN, HADJ, *Balikesser.*—Printed calico for turban.

243. BEKRI-OOSTA, *Balikesser.*—Printed calico.

244. SONDASH, YONIS, *Saïda.*—Calico and bath towel.

245. GOVERNOR OF SAÏDA.—Cotton edging and button.

246. HICO, *Philippolis.*—Bath and other towels.

247. GOVERNOR OF PHILIPPOLIS.—Printed calicoes.

248. GOVERNOR OF SHAEER. — White cotton cap.

249. GOVERNOR OF TARASHJA.—Printed calico.

250. GOVERNOR OF BELADBREHAN. — Calico.

251. GOVERNOR OF DRAMA.—Cotton yarn and cloths.

252. HOFIF, HAJI, *Denisli.*—Cotton stuffs.

253. SELIM, HAJI, *Denisli.*—Cotton bath towel.

254. MUSTAPHA, HAJI, *Denisli.*—Calico.

255. ALIKZADA, MEHMED, *Denisli.* — Calico.

256. GOVERNOR OF CYPRUS.—Printed calico.

257. ELENCO, *Salonica.*—Bath sheets and towels.

258. GOVERNOR OF SIROS.—Cotton fabrics.

259. GOVERNOR OF SMYRNA.—Cotton yarn.

260. GOVERNOR OF NISH.—Calicoes.

261. HOOVA, YOUSSUF, *Mossul.*—Cotton stuff.

262. JERGES, SHEMAS, *Mossul.*—Printed calicoes.

263. GOVERNOR OF KUTAHIA.—Cotton goods.

264. GOVERNOR OF ANGORA.—Cotton goods.

265. GOVERNOR OF ANDRINOPLE.—Cotton goods.

266. GOVERNOR OF TIRNOVA.—Cotton goods (alaja).

267. COFO, *Bosnia.*—Calico.

268. THEODORA, *Vidin.*—Sheeting.

269. GEBRIL, OSMAN, *Damascus.*—Cotton stuffs.

270. GASSOBY, ABDULLAH, *Damascus.*—Cotton yarn.

271. HAMAS, MUSTAPHA, *Damascus.*—Cotton cloth.

272. ABDELKADER, SEID, *Damascus.*—Cotton towellings.

273. AHMED, SEID, *Damascus.*—Cotton stuff.

274. KLEUHDI, MEHMED, *Damascus.*—Cotton stuff.

275. GOVERNOR OF MAGNEZIA.—Cotton stuff.

276. EDDÉ, M. *Lebanon.*—Cotton stuffs.

277. RESHID HABIB, *Lebanon.*—Cotton stuffs.

278. BOUTROS, ELIAS, *Lebanon.*—Cotton stuffs.

279. HABBAS, A. *Lebanon.* — Cotton sheeting.

280. GOVERNOR OF CRETE.—Cotton socks.

281. KADRI, *Crete.*—Cotton sheeting.

282. HANUSSAKI, MURAD, & RIGEL, *Crete.*—Cotton bagging.

283. KOURSHID, AGA, *Lazistan.*—Cotton yarn.

284. PARSEKT, VARTAN, *Karahissar, Sharky.*—Calico and cotton yarn.

285. ANONIA, *Karahissar, Sharky.* — Cotton towels.

CLASS XIX.

286. HIDAYET, EFENDI, *Constantinople.*—Embroidered ablution towels.

287. IMPERIAL MANUFACTORY, *Héréké.*—Damasked linen.

288. IMPERIAL MANUFACTORY, *Prevesa.*—Linen cloth.

289. BOYAJI, MEGERDICH, *Hajikeny, Amassia.*—Linen cloth.

290. KESHISH, *Hajikeny, Amassia.*—Towels.

291. GOVERNOR OF LATAKIA.—Various linen fabrics.

292. MUSTAPHA, HAJI, *Monastir.*—Flax thread.

293. GOVERNOR OF TARASHJA.—Linen cloth.

294. CHAOUSH, OOSTA, *Denizli.*—Linen for shirting.

295. GOVERNOR OF CYPRUS.—Curtain cloths.

296. GOVERNOR OF SIROS.—Linen cloth.

297. GOVERNOR OF NISH.—Hemp cloth, ropes, and pack-thread.

298. AYSHA, *Conia.*—Linen yarn and cloth.

299. ABOUSALEH, *Damascus.*—Ropes and pack-thread.

300. GOVERNOR OF TREBISOND.—Linen cloths.

301. CHERVISH, JANIK, *Broussa.*—Handkerchiefs and towels.

302. ESKISHERLI, OVANES, *Broussa.*—Bath towels.

303. EMIN, AGA, *Keresson.*—Ropes.

CLASS XX.

304. SALIH, *Constantinople.*—Silk stuff.

305. GOMIDAS, *Constantinople.* — Embroidered gauze for gown.

306. HIDAYET, EFENDI, *Constantinople.*—Embroidered silk covering for coffee board; silk cloth for shirts.

307. SALIH, EFENDI, *Constantinople.*—Silk bath towel.

308. PAPAZOGLOO, YANKO, *Gueva.* — Velvet coffee board.

309. IMPERIAL MANUFACTORY, *Héreké.*—Damasked silks, velvets, silk stuffs for curtains, ribbons, silk cord, satin and silk gauze.

310. BOYHOSS, *Constantinople.* — Silk stuffs.

311. KURDBEKRE, NOURI, *Amassia.*—Silk shirtings.

312. GHANIZADE, MEHMED, *Amassia.*—Brocade.

313. ABDULDJELIL, *Beyrouth.* — Silk shawls, and other stuffs.

314. JERDJIS, CONDJI, *Beyrouth.*—Silk stuffs.

315. DJERDJÉSAN, *Beyrouth.*—Silk fabrics.

316. ALY, COOTLY, *Beyrouth.*—Red silk belt.

317. DJERDJER, SARÉJI, *Beyrouth.*—Silk stuff (cootni).

318. YAWAB, OUSTA, *Saiola.*—Edging, girdle, and fringe.

319. DJALIN, ILIAS, *Saiola.* — Twisted white silk.

320. HOOKDAR, *Saiola.*—Silk tufts.

321. BILSANAK, ARSLAN, *Saiola.* — Silk ribbon.

322. DANOTO, ANDON, *Aleppo.* — Silk fabrics.

323. KHATIOGLI, ANDON, *Aleppo.* — Various silk fabrics.

324. GOVERNOR OF PHILIPPOLI. — Silk shirting.

325. MOHAMMED, SEID, *Tripoli.* — Silk sheets and trimmings.

326. MOHMED-ALI, OUSTA, *Roustchouk.*—Silk bobbins.

327. GOVERNOR OF SMYRNA. — Silk shirting.

328. HOOVA, HODJA, *Mossul.*—Silk stuffs.

329. GOVERNOR OF ANDRINOPLE.—Silk goods.

330. CHERMEKLY, *Kurdistan.*—Silk stuffs.

331. MUSTAPHA, *Kurdistan.*—Silk stuffs.

332. ADJEM, AHMED, *Damascus.*—Silk shawl.

333. SEMADI, SALIH, *Damascus.*—Silk fabric.

334. BAGDADI, ABDULLAH, *Damascus.*—Silk belts and silk fabrics.

335. SHEIKUL, HERAKA, *Damascus.*—Silk fabrics.

336. RESHID, SHEIK, *Damascus.*—Silk fabrics.

337. KURISH, HUSSEIN, *Damascus.*—Silk stuffs.

338. GOVERNOR OF MAGNEZIA.—Silk girdles.

339. GOVERNOR OF SCUTARI OF ALBANIA. —Silk sheeting.

340. EDDÉ, M. *Lebanon.*—Silk stuffs.

341. MESHAKA, J. *Lebanon.*—Silk stuffs.

342. LATIF, M. *Lebanon.*—Silk stuffs.

343. RESHID, ABOUSHEKIR, *Lebanon.*— Silk fabrics.

344. KERKEBESH, *Lebanon.*—Silk millinery.

345. EMIN, EFENDI, *Trebisond.* — Silk mercery.

346. ALEMDAR, ALI, *Castamouni.*—Silk mercery.

347. GOVERNOR OF CRETE.—Silk cloths and mercery.

348. VASILIKY, *Crete.*—Silk mercery.

349. SOFALAKY, *Crete.*—Silk shirting, &c.

350. ISFAKIANAKY, *Crete.*—Silk sheetings.

351. HASSAN, AGA, *Crete.*—Silk girdle.

352. CHERVISH, JANIK, *Broussa.* — Silk stuffs and mercery.

353. ESKISHEHRLI, OVANES, *Broussa.*— Silk fabrics.

354. HAYRABIT, OVANES, *Broussa.*—Silk fabrics.

355. FOTAGI, AHMED, *Broussa.* — Silk fabrics, and silk and silver fabrics.

356. BELEDIJI, SALEH, *Broussa.* — Silk stuffs.

357. CHERVISH BEDROS, *Broussa.*—Silk stuffs.

358. SALEH, HAJI, *Broussa.*—Silk and silver cloth.

359. FOOCHIJI, MEHMED, *Broussa,*—Silk and silver cloth.

360. KATINCO, *Broussa.*—Silk purse.

CLASS XXI.

361. SALIH, *Constantinople.*—Cotton and silk stuff.

362. RIZA, ALI, *Constantinople.*—Cotton and silk stuff.

363. HIDAYET, EFENDI, *Constantinople.*— Cotton and silk fabrics.

364. AHMED, HAJI, & MEHMED, *Constantinople.*—Silk and cotton fabrics for furniture; woollen stuff called ehram.

365. IMPERIAL MANUFACTORY, *Héréké.*— Damasked woollen.

366. IMPERIAL MANUFACTORY, *Constantinople.*—Socks.

367. GOVERNMENT MANUFACTORY, *Con-* *stantinople.*—Woollen cloth for the army; blankets, red caps (fez), and white do.

368. PARSEGH, *Constantinople.* — Stuffs made with goats' wool.

369. HASSAN, EFENDI, *Amassia.*—Cotton and silk shirtings.

370. FATMA, *Amassia.*—Cotton and silk shirtings.

371. KESHISHOGLOU, STEPAN, *Amassia.*— Woollen hosiery.

372. GOVERNOR OF CARASSY.—Woollen aba.

373. GOVERNOR OF BALIKESSER.—Hosiery.

374. JERDGIS, CONDJI, *Beyrouth.*— Silk and cotton stuff.

375. DJERJES, *Beyrouth.*—Silk and cotton stuff.

376. GOVERNOR OF SOPHIA.—Woollen socks and cloth.

377. GOVERNOR OF MONASTIR.—Woollen socks.

378. DANOTO, ANTOON, *Aleppo.*—Cotton and silk fabrics.

379. KHATI, ANTOON, *Aleppo.*—Cotton and silk fabrics.

380. SALIM, NAOUM, *Aleppo.*—Cotton and silk fabrics.

381. YOUSSOUF, *Aleppo.*—Cotton and silk fabrics.

382. RISCULAH, *Aleppo.*—Cotton and silk fabrics.

383. GOVERNOR OF PHILIPPOLI.—Woollen stuffs.

384. GOVERNOR OF PRASDIN.—Woollen shiaks.

385. GOVERNOR OF PHILIPPOLI.—Cotton and silk stuffs.

386. GOVERNOR OF HUSKIVY.—Woollen fabrics.

387. IBRAHIM-OGLY, AHMED, *Djerban.*— Woollen ehrams.

388. AHMED-OGLY, MURAD, *Djerban.*— Woollen ehrams.

389. KEHIASGLY, HUSSEIN, *Djerban.*— Woollen ehrams.

390. GOVERNOR OF AKKIA.—Felts.

391. GOVERNOR OF DRAMA. —Woollen hosiery, and other fabrics.

392. GOVERNOR OF GUMULGINA.—Woollen fabrics.

393. SOLIMAN, EFENDI, *Denizli.*—Cotton and silk stuff.

394. KEHRIMAN, USTA, *Denizli.*—Cotton and silk stuff.

395. GOVERNOR OF CYPRUS.—Cotton and silk stuff.

396. LUTHFULLAH, *Salonica.*—Silk and cotton bath towels.

397. MARIA, *Salonica.* — Woollen and cotton.

398. CATINA, *Salonica.*—Silk and cotton cloth.

399. AGHOSSOS, *Salonica.*—Woollen aba.

400. GOVERNOR OF AORATHISSAR.—Woollen aba.

401. GOVERNOR OF SIROS. — Woollen socks.

402. GOVERNOR OF NISH.—Woollen aba and socks.

403. ABDULLAH, *Mossul.*—Cotton and silk stuff.

404. ABDULLAH & YUSSOUF, *Mossul.*— Woollen stuffs.

405. GOVERNOR OF KUTAHIEH.—Woollen stuffs and hosiery.

406. GOVERNOR OF ANGORA.—Woollen goods.

407. GOVERNOR OF ANDRINOPLE.—Woollen goods.

408. SHERIFA, *Conia.*—Hosiery.

409. GOVERNOR OF TIRNOVA.—Silk and woollen stuff.

410. KHOSKOBJI, OGLOU, *Kurdistan.*— Cotton and silk stuffs.

411. TUJAR, OGLOO, *Kurdistan.*—Cotton and silk stuffs.

412. BEDO, OGLOO, *Kurdistan.*—Cotton and silk stuffs.

413. JEVHER, AGHA, *Bosnia.*—Cotton and silk goods.

414. SULEYMAN, *Bosnia.*—Woollen aba.

415. THEODORA, *Vidin.*—Cotton and silk sheeting, and woollen socks.

416. GOVERNOR OF VIDIN.—Hosiery and felt.

417. ADJEIM, AHMED, *Damascus.*—Shawls.

418. KEMALI, SAÏD, *Damascus.*—Cotton and silk goods.

419. HUSSEIN, SHEÏK, *Damascus.*—Cotton and silk goods.

420. MUHYEDDIN, *Damascus.*—Silk and cotton stuffs.

421. GASSOBY, ABDULLAH, *Damascus.*—

422. CURABY, MEHMED, *Damascus.*—Silk and cotton sheeting.

423. KADREEL, HAM, *Damascus.*—Cotton and silk towelling.

424. KASSIM, HAJI, *Damascus.*—Woollen socks.

425. DEVARER, MEHMED, *Damascus.*—Wool yarn.

426. ABDOULLAH, *Damascus.*—Cotton and silk stuff.

427. HAMAS, MUSTAPHA, *Damascus.*—Wool yarn.

428. ABDULKADER, SEID, *Damascus.*—Silk and cotton towelling.

429. GOVERNOR OF ISLIMIA.—Felts.

430. EDDÉ, M. *Lebanon.*—Cotton and silk stuffs.

431. HABIKA, J. *Lebanon.*—Cotton and silk stuffs.

432. NACASH, H. *Lebanon.*—Shawl pattern cloth.

433. HALEBI, J. *Lebanon.*—Printed stuff.

434. KERAM, H. *Lebanon.*—Mixed cloth.

435. RESHID, MICHEL, *Lebanon.*—Silk and cotton shirtings.

436. GOVERNOR OF ANGORA.—Woollen hosiery.

437. VALENDIZO, P. *Crete.*—Woollen hosiery.

438. GOVERNOR OF CRETE.—Cotton and silk shirting.

439. SOFALAKI, *Crete.*—Silk and cotton shirting.

440. GOVERNOR OF CRETE.—Woollen felt.

441. CHERVISH, JANIK, *Broussa.*—Cotton and silk fabrics.

442. ESKISHEHRLI, OVANES, *Broussa.*—Silk and cotton, and wool and silk fabrics.

443. HAYRABET, OVANES, *Broussa.*—Silk and cotton goods.

CLASS XXII.

444. GOVERNOR OF BROUSSA.—Carpets.

445. GOVERNOR OF SOPHIA.—Carpets.

446. DANOTO, ANDON, *Aleppo.*—Prayer carpet.

447. GOVERNOR OF PHILIPPOLI.—Carpets.

448. GOVERNOR OF HASKIOY.—-Carpets.

449. GOVERNOR OF AKKIA.—Carpets.

450. TURCOMEN, *Akkia.*—Carpets.

451. GOVERNOR OF BELADBECHAN. — Carpets.

452. ISACHAGHA, *Salonica.*—Carpets.

453. GAVRIL, *Salonica.*—Carpets.

454. GOVERNOR OF SMYRNA.—Carpets.

455. GOVERNOR OF NISH.—Carpets.

456. GOVERNOR OF ANGORA.—Carpet.

457. MUSTAPHA & MEHMED, *Conia.*—Carpets.

458. GOVERNOR OF VIDIN.—Carpet.

459. GEBRIL, OSMAN, *Damascus.*—Cotton carpet.

460. HAFIF, HAJI, *Saroukhan.*—Carpet.

461. MUSTAPHA, *Saroukhan.*—Carpet.

462. IBRAHIM, EFENDI, *Saroukhan.*—Carpet.

463. SANDOUKJI, ZADÉ, *Saroukhan.*—Carpet.

464. YAVASH, ZADÉ, *Saroukhan.*—Carpet.

465. SHEVKI, EFENDI, *Saroukhan.*—Carpet.

466. KADIR, OGLOU, *Saroukhan.*—Carpet.

467. KARASHAHIN, *Saroukhan.*—Carpet.

468. KIATIB, ZADÉ, *Saroukhan.*—Carpet.

469. OSTAOGLOU, *Saroukhan.*—Carpet.

470. BAKGEVAN, HASSAN, *Saroukhan.*—Carpet.

471. BEKIR, ZADÉ, *Saroukhan.*—Carpet.

472. HUSSEIN, HAJI, *Saroukhan.*—Carpet.

473. ABDOULLAH, ZADÉ, *Saroukhan.*—Carpet.

474. HEKIM, ZADÉ, *Saroukhan.*—Carpet.

475. SARADJ, MUSTAPHA, *Saroukhan.*—Carpet.

476. MUSTAPHA, NEVAHI, *Saroukhan.*—Carpets.

477. BOURDOURLOU, ZADÉ, *Saroukhan.*—Carpets.

478. PAMOUK, ALI ZADÉ, *Saroukhan.*—Carpets.

479. GOVERNOR OF COOLA.—Carpets.

480. GOVERNOR OF ADANA.—Carpets.

481. GOVERNOR OF ANGORA.—Carpets.

482. CATHERINA, *Crete.*—Carpets.

483. MARIA, *Crete.*—Woollen batallia.

484. MURAD, HAJI, *Crete.*—Carpet.

485. HALIL, CHAVOUSH, *Crete.*—Carpet.

486. REGEL, *Crete.*—Carpet.

487. CHERVISH, JANIK, *Broussa.*—Embroidered carpets.

488. NOURI, SULEIMAN, *Broussa.*—Carpet.

489. MEHMED, ALI, *Keresson.*—Carpet.

490. SULEIMAN, OOSTA, *Karahissar Sahib.*—Carpets.

491. GOVERNOR OF KARAHISSAR SAHIB.—Carpets.

CLASS XXIV.

492. MICHAEL, *Constantinople.*—Embroidered gauze and silk stuffs.

493. HIDAYET, EFENDI, *Constantinople.*—Embroidered handkerchiefs, girdles, &c.

494. YAKUDJIOGLOU, KIRKOR, *Constantinople.*—Embroidered caparison.

495. MERIEM, *Constantinople.*—Embroidered gauze, and cambric for gowns.

496. TAKOUHI, *Constantinople.*—Embroidered handkerchief, silk lace, and purse.

497. YANKO, *Constantinople.*—Lace for hair-dress.

498. MUSTAPHA, EFENDI, *Constantinople.*—Embroidered coffee board covers.

499. MERYEM, *Constantinople.*—Embroidered cambric and scarf.

500. OSMAN, HADJI, *Constantinople.*—Silver lace for head-dress.

501. YELDISYAN, RAPAEL, *Constantinople.*—Silk lace.

502. MOOMKESSER, KESSY, *Beyrouth.*—Silk lace, and silk and silver embroidery.

503. BELSOUK-OGLI, *Saïda.*—Cotton lace.

504. BERZELAY, HAÏM, *Saïda.*—Edgings.

505. GOVERNOR OF LATAKIA.—Silk edging.

506. GOVERNOR OF PHILIPPOLI. — Lace and embroidery.

507. MEHMED, SEID, *Shaeer.*—Fringes.

508. GOVERNOR OF SIROS.—Lace and edging.

509. AVRAM, *Bosnia.* — Macrama, with gold.

510. KALVA, MAHMOUD, *Damascus.*—Cotton and silver embroidery.

511. GOVERNOR OF MAGNESIA.—Edging and lace.

512. GOVERNOR OF SCUTARIA OF ALBANIA.—Silver lace, silk lace, edgings, &c.

513. NASRANI, H. *Lebanon.*—Embroidered articles.

514. MARIAM, *Trebisond.*—Veil for ladies.

515. GOVERNOR OF CRETE.—Silk and gold lace.

516. EMINÉ, *Crete.*—Handkerchiefs.

517. NASLI, *Crete.*—Handkerchiefs.

518. ESKISHÉHIRLÍ, OVANE, *Broussa.*—Embroidered upholstery and tapestry.

519. OSMAN, BEY, *Broussa.*—Carriage cushions.

520. FATMA, *Broussa.* — Embroidered handkerchiefs.

521. CHERVISH, JANIK, *Broussa.*—Silver door curtain.

CLASS XXV.

522. ARTIN, *Merzikan, Amassia.*—Fitchet skin.

523. KRABELIK, ARTIN, *Merzikan, Amassia.*—Fox, wolf, jackal, cat, hare, and goat-skins.

524. AJABI, MOHIR, *Saïda.*—Hair bags.

525. GOVERNOR OF PHILIPPOLI.—Hair-cloth and hair bags.

526. MEHMED, *Djerban.*—Hair sack.

527. GOVERNOR OF DRAMA.—Fur skins.

528. GOVERNOR OF GUMULJINA. — Fur skins.

529. GOVERNOR OF CAVALA.—Fur skins.

530. GOVERNOR OF SHARISHABAN.—Fur skins.

531. GOVERNOR OF PRENESHTA.—Hair bags.

532. GOVERNOR OF SMYRNA.—Hair sacks.

533. GOVERNOR OF NISH.—Fur skins and goat's hair.

534. GOVERNOR OF ANGORA.—Goat skins and goat's hair.

535. GOVERNOR OF JANIK.—Fur skins.

536. EGHIA, OOSTA, *Conia.*—Furs and skins.

537. GURGUI, *Bosnia.*—Fur skins.

538. GOVERNOR OF COOLA. — Morocco leather.

539. GOVERNOR OF MAGNEZIA.—Morocco leather and fur skins.

540. GOVERNOR OF ADANA.—Fur skins.

541. GOVERNOR OF LEBANON.—Hair bags and fur skins.

542. GOVERNOR OF CASTAMOUNI.—Fur skins.

543. BEKIR, *Keresson.*—Fur skins.

CLASS XXVI.

544. BOUCHAIN, *Constantinople.*—Harness.

545. IMPERIAL TANNERY, *Constantinople.*—Kid skins, sole leathers, moroccos and other leathers.

546. MUSTAPHA, *Constantinople.*—Saddle and harness.

547. YOOSOOF, BABA, *Hajikioy, Amassia.*—Stirrups.

548. HASSAN, HAJI, *Hajikioy, Amassia.*—Morocco leathers.

549. AHMED, HAJI, *Merzikan, Amassia.*—Sole leather.

550. MEHMED, AGA, *Carassy.*—Harness.

551. BEKTASHI-OGLI, SALIH, *Balikesser.*—Morocco leather.

552. HAFIZ, HADJ, *Balikesser.*—Morocco and other leather.

553. MUMKESSER, KESSY, *Beyrouth.* — Morocco and other leather; bridle.

554. GOVERNOR OF BEYROUTH.—Harness.

555. HALIL, SEID, *Saïda.*—Silk bridle.

556. GOVERNOR OF LATAKIA.—Red morocco leather.

557. GOVERNOR OF SOPHIA.—Halter and whips.

558. DANSTO, ANDON, *Aleppo.*—Harness.

559. RESHID, MEHMED, *Aleppo.*—Morocco leather.

560. AHMED, HAJI, *Philippolis.*—Morocco leather.

561. CALFA, SOLIMAN, *Philippolis.*—Morocco leather.

562. CALFA, HASSAN. *Philippolis.*—Morocco leather.

563. GOVERNOR OF PHILIPPOLIS.—Leathers of various kinds.

564. CARA, HUSSEIN, *Djerban.*—Morocco leather.

565. RAHIM, HAJI, *Djerban*.—Morocco leather.

566. MEHMED, *Djerban*.—Morocco leather.

567. GOVERNOR OF TRIPOLI.—Harness.

568. GOVERNOR OF SHAEER.—Leathers.

569. AHMED, HAJI, *Shaeer*.—Leathers.

570. GOVERNOR OF DRAMA.—Morocco and other leathers.

571. GOVERNOR OF GUMULJINA.—Morocco and other leathers.

572. GOVERNOR OF PRENESHTA. — Sole leather.

573. GOVERNOR OF CARASSOU.—Morocco and other leathers.

574. ALI & SADIK, *Denizli*.—Morocco and sole leathers.

575. HUSSEIN, USTA, *Salonica*.—Morocco leather.

576. GOVERNOR OF SIROS.—Morocco and other leathers.

577. GOVERNOR OF NISH.—Morocco and other leathers.

578. GOVERNOR OF CAISERIA.—Morocco leathers.

579. GOVERNOR OF KUTAHIEH.—Morocco and other leathers.

580. GOVERNOR OF ANDRINOPLE.—Morocco and other leathers.

581. AHMED, HAJI, *Conia*.—Leathers.

582. EUMER, OOSTA, *Conia*.—Saddle.

583. GOVERNOR OF TIRNOVA. — Horse cloths.

584. GOVERNOR OF NAPLOUSE.—Morocco leather.

585. ABDOULLAH & IBRAHIM, *Bosnia*.—Horse cloth.

586. BAGDADI, ABDOULLAH, *Damascus*.—Silk harness, horse cloth, and saddle.

587. MELLAH, MUSTAPHA, *Damascus*.—Morocco and other leathers.

588. OMAR, HAJI, *Damascus*. — Horse cloth.

589. DJERDJES, *Damascus*.—Morocco leathers.

590. BADRANI, H. *Lebanon*. — Morocco leathers.

591. MUZIA, EFENDI, *Trebisond*.—Pistol holsters.

592. GOVERNOR OF TREBISOND.—Wallet.

593. HASSAN, ABDI, *Castamouni*.—Morocco leather.

594. ALEMDAR, MEHMED, *Castamouni*.—Morocco leather.

595. SELIM, *Crete*.—Pack saddle.

596. MUSTAPHA, *Crete*.—Harness.

597. AZAMAKI, *Crete*.—Bit, stirrups, &c.

598. AGHACO, *Crete*.—Morocco and other leathers.

599. HALIL, AGA, *Broussa*. — Morocco leather.

600. GOVERNOR OF BROUSSA.—Morocco leather.

601. GOVERNOR OF KARAHISSAR SAHIB.—Horse cloth.

CLASS XXVII.

602. MARDIROS, HADJI, *Constantinople*.—Embroidered velvet; ladies' dress.

603. JANIKOGLOU, ARTIN, *Constantinople*.—Velvet, cap, head-dress, tassels, and belt.

604. ZOITCHA & MARDRITCHA, *Constantinople*.—Embroidered slippers.

605. GOVERNOR OF JANINA.—Silver embroidered velvet; Albanian cloths.

606. GOVERNOR OF HAMA.—Silk mashlah, embroidered with gold.

607. YANKO, *Constantinople*. — Groom's livery, and a suit for pumpers.

608. STEPAN, OOSTA, *Balikesser*.—Boots and shoes.

609. KERKES, ILIAS, *Beyrouth*.—Silk shirt.

610. PALOUSE, ALI, *Beyrouth.*—Shoes and slippers.

611. ZEHAB, SELIM, *Beyrouth.*—Shoes.

612. GOVERNOR OF LAZIKIA.—Boots and shoes.

613. TAHARI, *Monastir.*—Woollen cloaks.

614. GOVERNOR OF MONASTIR.—Cotton, and wool and cotton shirts.

615. DANOTO, ANDON, *Aleppo.*—Neck-cloths and gown, with gold thread.

616. DIMITRAKI, *Philippoli*—Caps.

617. DEJUCIJ, MEHMED, *Tripoli.*—Slippers.

618. GOVERNOR OF TRIPOLI.—Boots and shoes.

619. GOVERNOR OF SHARISHABAN.—Woollen cloak.

620. GOVERNOR OF CYPRUS.—Boots.

621. COSTANDI, *Salonica.*—Embroidered bath shirts.

622. GOVERNOR OF NISH.—Red shoes, fur caps, and red cap.

623. AHMED, *Mossul.*—Shoes.

624. GOVERNOR OF ANDRINOPLE.—Boots, shoes, and slippers.

625. IBRAHIM, OOSTA, *Conia.*—Woollen cloak and cap.

626. MEHMED, HAJI, *Conia.*—Boots and shoes.

627. AHMED, HAJI, *Conia.*—Boots and shoes.

628. GOVERNOR OF TIRNOVA.—Silk head stall.

629. GOVERNOR OF NAPLOUSE.—Cloaks (mashlak).

630. SALIH & ABDULLAH, *Bosnia.*—Shoes.

631. GIBRIL, OSMAN, *Damascus.*—Cloaks (silk and silver).

632. BEDIR, MEHMED, *Damascus.*—Cloak.

633. DJEBRI, ABDELGHANI, *Damascus.*—Wooden shoes, inlaid with mother-of-pearl.

634. MEHMED ARNAOUD, *Damascus.*—Wooden shoes, inlaid with mother-of-pearl.

635. BOURDJIK, *Damascus.*—Boots, shoes, and slippers.

636. GHANOUM, *Damascus.*—Wool and cotton cloak.

637. DJERDJES, *Damascus.*—Silk cloak.

638. GOVERNOR OF SCUTARI OF ALBANIA.—Embroidered cloak.

639. SULEYMAN, *Lebanon.* — Woollen cloaks.

640. ARAB, H. J. *Lebanon.*—Shoes.

641. MUSSA, ALI, *Lebanon.*—Shoes.

642. ISKENDER, *Lebanon.*—Shoes.

643. SHAKIR, H. *Lebanon.*—Embroidered cloaks, &c.

644. NASRANI, H. *Lebanon.* — Slippers (silk), and embroidered caps and bags.

645. HATEM, E. *Lebanon.*—Red slippers and shoes.

646. DAKI, E. *Lebanon.*—Slippers.

647. BEROKES, Y. *Lebanon.*—Jackets.

648. GOVERNOR OF MARDIN.—Woollen cloak.

649. MUZIA, EFENDI, *Trebisond.*—Lady's head-dress.

650. SELIM, *Crete.*—Boots.

651. MANOL, *Crete.*—Boots.

652. LAZOGLOU, AHMED, *Broussa.*—Slippers.

653. MEHMED, ALI, *Lazistan.*—Shoes.

654. AHMED, OOSTA, *Karahissar Sahib.*—Sandals, inlaid with silver.

CLASS XXVIII.

655. KOORSHID, HADJI, *Constantinople.*—Paper and pens.

656. KIAMIL, EFENDI, *Constantinople.*—Inkstand, sand-box, &c.

I

657. BIDAT, EFENDI, *Constantinople.* — Satchel.

658. MUHENDISYAN, *Constantinople.* — Specimen of printing and engraving.

659. BROGHOLIOS, *Constantinople.*—Ala-baster inkstand and sand-box, cistern piece, and fan handle.

660. ELMADJI, MEHMED, *Roustchouk.*—Inkstand.

661. MUZIA, EFENDI, *Trebisond.*—Ink-stand.

CLASS XXX.

662. HUSSEIN, HADJI, *Constantinople.*—Combs and spoons made of tortoise-shell, ebony, coral, rhinoceros horn, ivory, and mother-of-pearl.

663. RIZA, EFENDI, *Constantinople.*—Per-fumery, pastils, and ornaments made with perfumed substances.

664. SHABAN, EFENDI, *Constantinople.*—Cocoa inkstand and coffee-cup.

665. FALINO, *Constantinople.*—Table-cover.

666. MELKIZET, ELEAZAR, *Constantinople.* —Table-cover.

667. RIZA, ALI, *Constantinople.*—Embroi-dered velvet bolster and amulet case.

668. STAVRI, *Constantinople.*—Drawer and stool of mother-of-pearl.

669. OSTA, ABDULLAH, *Constantinople.*—Chest of mother-of-pearls.

670. SAHAK, *Constantinople.* — Chest of mother-of-pearls.

671. OSMAN, HADJI, *Constantinople.*—Em-broidered prayer carpet.

672. GUMUSHIAN, BEDROS, *Constantinople.* —Backgammon; ivory chess board.

673. FRANER, OOSTA, *Carassy.* — Olive wood chest.

674. ISMAIL, HADJ, *Balikersır.* — Arm bolster.

675. DONATO, ANDON, *Aleppo.*—Table-covers with gold thread, and coffee board cover.

676. HICO, *Philippolis.*—Table-cover.

677. GOVERNOR OF HASKIOY.—Spoons.

678. GOVERNOR OF COZANDJIK. — Per-fumed oils and waters.

679. GOVERNOR OF TRIPOLI.—Silk and silver tassels, fringes, and others.

680. MELIHA, MICHAEL, *Tripoli.*—Hazel chest.

681. GOVERNOR OF ANDRINOPLE.—Cabi-net ware, perfumed soaps and oils.

682. ABDULLAH, OOSTA, *Besnia.*—Spoons with inlaid ornaments.

683. AHMED, *Widin.*—Porte cigar.

684. GEBRI, OSMAN, *Damascus.*—Camp stools.

685. MEHMED, NEDGIM, *Damascus.*—Stool inlaid with mother-of-pearls.

686. ABOU, AHMED, *Damascus.*—Stool inlaid with mother-of-pearls.

687. ABDOULLAH, OOSTA, *Damascus.*—Chest.

688. GOVERNOR OF DAMASCUS.—Trunks inlaid with mother-of-pearls.

689. RESHID, ABOUSHEKIR, *Lebanon.*—Ottoman and cushions, &c.

690. SOLIMAN, HADJI, *Trebisond.*—Trunk.

691. GOVERNOR OF CRETE.—Perfumery.

692. NOURI SULEIMAN, *Broussa.*—Stuff for sofa and chairs.

693. ANDON, HAJI, *Broussa.*—Specimen of sofa furniture.

CLASS XXXI.

694. SOLIMAN, HAFIS, *Constantinople.*—Yellow tin lantern and lamps.

695. HUSSEIN, AGA, *Constantinople.*—Brass mangal and boore.

696. KEVORK, *Constantinople.*—Lanterns.

697. RAMAZAN, OOSTA, *Balath.*—Coffee-mill.

698. GOVERNOR OF LAZIKIA.—Hardware.

699. GOVERNOR OF KUTAHIEH.—Horse-shoe and nails.

700. GOVERNOR OF JANIK. — Horseshoe and nails.

701. ALI, OOSTA, *Conia.*—Coffee-mill.

702. IBRAHIM, OOSTA, *Bosnia.* — Copper wares.

703. MUSTAPHA, DERVISH, *Bosnia.*—Copper picknick plate.

704. SULEYMAN & NAZIL, *Bosnia.*—Coffee-mills.

705. BEYAZID, EFENDI, *Damascus.* — Horseshoe.

706. SMEK, AGHA, *Trebisond.* — Copper wares.

707. MINASOGLOU, BOGHOS, *Trebisond.*—Copper wares.

708. USENGHIGI, H. *Castamouni.*—Copper wares.

709. APOSTOLOS, *Crete.*—Brass lamp.

710. YARDAN, *Broussa.*—Copper plate and cover.

CLASS XXXII.

712. KHOORSHID, HADJI, *Constantinople.*—Pen-knife and scissors.

713. SADDEDDIN, *Beyrouth.*—Knives.

714. GOVERNOR OF SOPHIA. — Scissors, with inlaid gold ornaments.

715. SEID, USTA, *Haskioy.*—Knives.

716. GOVERNOR OF SMYRNA.—Knives.

717. GOVERNOR OF KUTAHIEH.—Knives.

718. GOVERNOR OF TIRNOVA.—Cutlery.

719. HUSSEIN, HAJI, *Bosnia.*—Daggers and knives.

720. GOVERNOR OF BOSNIA.—Knife.

721. SHERIFF, *Vidin.*—Scissors, knife, &c.

722. OSMAN, *Trebisond.*—Daggers.

723. KILLADDEM, *Trebisond.* — Scissors, knives, forks, &c.

724. MUSTAPHA, *Crete.*—Cutlery.

725. BEKIR, OOSTA, *Broussa.* — Scissors, inlaid with gold

CLASS XXXIII.

726. MELKIZET, ELIAZIR, *Constantinople.*—Silver ewer and basin, rose-water pot, coffee board, cups for sweetmeats, coffee pot and saucers.

727. LUDFY, *Constantinople.*—Silver case.

728. DJIBOOR, TOOBRE, *Beyrouth.*—Gold and silver jewellery.

729. SHOUBIZI, NOOM, *Beyrouth.*—Silver saucer.

730. ARTUR, YORGI, *Vidin.* — Silver articles.

731. FILAP, ARSENIO, *Vidin.* — Silver cigar-case.

732. Governor of Tarashja. — Silver saucer and cigar-case.

733. Arakel, *Conia.*—Jewellery.

734. Governor of Tirnova. —Watch-case.

735. Goro, *Vidin.*—Silver cups.

736. Gourgi, *Vidin.*—Silver cigar box and glass.

737. Damar, Halil, *Damascus.*—Silver saucers.

738. Abou, Noukoul, *Lebanon.*—Silver wares.

739. Soulmoun, H. *Lebanon.*—Jewellery.

740. Hilmi, Efendi, *Trebisond.*—Silver jewellery.

741. Nicola, *Trebisond.*—Silver plate and jewellery.

742. Mehmed, *Crete.*—Silver watch and chain.

CLASS XXXV.

743. Moorad, Aga, *Constantinople.* — Pipe bowls.

744. Mohis, Ibrahim, *Beyrouth.*—Pipe and narguilé bowls.

745. Mekhouri, *Beyrouth.* —Cup, and earthenware pot.

746. Governor of Beyrouth.—Potter's earth.

747. Mekhoory, *Saïda.*—Earthenware.

748. Ajabi, Mohir, *Saïda.* — Wooden thread.

749. Governor of Latakia.—Earthenware.

750. Mehmed, Seid, *Shaeer.*—Pipe bowls.

751. Elmadji, Mehmed, *Roustchouk.*—Pipe bowls, &c.

752. Mehmed Ali, Usti, *Roustchouk.*—Earthenware pottery.

753. Mustapha, Usta, *Roustchouk.* —Water filter.

754. Gumby, Tadey, *Dardanelles.*—Soup plates.

755. Ahmed, Mustapha, *Dardanelles.*—Earthenware.

756. Governor of Kutahieh.—Earthenware.

757. Governor of Andrinople.—Water bowls, and other china.

758. Mehmed, *Widin.*—Pipe bowls.

759. Ahmed, *Widin.*—Pipe bowls.

760. Hassan, Ghazi, *Widin.*—Narguila bowls.

761. Ahmed, Oosta, *Widin.*—Bowls.

762. Ali, *Widin.*—Water cup.

763. Ibrahim, *Widin.*—Cup and cover.

764. Beyazid, Efendi, *Damascus.* —Bricks.

CLASS XXXVI.

765. Bedross, *Constantinople.* — Amber mouthpiece and pipes.

766. Shaban, Efendi, *Constantinople.* —Ebony pipe.

767. Janik-Oglou, Artin, *Constantinople.*—Tobacco purse.

768. Mazloom - Ogloo, Khatchadour, *Constantinople.*—Silk and shawl purses.

769. Riza, Ali, *Constantinople.*—Amber mouthpiece.

770. Ahmed, Oosta, *Constantinople.* —Narguila, with inlaid gold ornaments.

771. Zoitcha & Maridritcha, *Constantinople.*—Amber mouthpiece.

772. Izzet, Haji, *Constantinople.*—Narguilé pipes.

773. DEDA, BESMI, *Constantinople.* — Ebony staves.

774. RUSTEM, USTA, *Haskioy.*—Pipes.

775. GOVERNOR OF ANDRINOPLE.—Pipes and narguilas.

776. GOVERNOR OF BOSNIA. — Narguilé bowls.

777. REMLA, MEHMED, *Damascus.*—Cigar mouthpiece, and pipes.

778. ABOU, ALI, *Damascus.*—Pipes.

779. MEHMED, HAJI, *Damascus.* — Narguilas.

780. ABDELKADER, SHAKIR, *Damascus.*—Narguilas.

781. HAJI, AHMET, *Islimia.* — Ornamental pipe.

782. EMIN, BABA, *Islimia.*—Ornamental pipe.

783. ALI, HADJI, *Islimia.*—Ornamental pipe, and cigaret holder.

784. OSMAN, *Trebisond.* — Ornamental pipes, pipe bowl, and cigar holders.

785. RAZVAN, AGA, *Crete.*—Sticks.

786. SULUK, IBRAHIM, *Broussa.*—Jessamine pipes.

787. AHMET, OOSTA, *Karahissar Sahib.*—Crutchet and club, ornamented with silver.

UNITED STATES.

SOUTH-EAST COURT.

CLASS I.

1. FEUCHTWANGER, J. W.—42 *Cedar-st. New York.*—1,000 specimens of American minerals.

2. MEADS, T.—Cabinet of minerals from Lake Superior.

3. NEW JERSEY ZINC CO. *Newark, New Jersey.*—Specimens of zinc ores, with their products; pig and bar iron; steel.

3A. MOSHEIMER, J. *Neveda Territory.*—Specimens of gold, silver, quicksilver, copper ores, native sulphur and borax.

3B. PRECHT, DR. C. *San Francisco.*—Specimens of crystallised gold, and California marble.

CLASS II.

4. BAGLEY, M. H. *New York City.*—Crystal carbon oil for lamps.

5. PEASE, S. F. *Buffalo, New York.*—Samples of carbon and oils for lamps and lubrication.

6. HALE, A. *Lyons, New York.*—Essence of peppermint.

7. PARISH, E. (for the College of Pharmacy of Philadelphia), *Pennsylvania.*—Native roots and drugs.

———

77. RHODES, B. M. *Baltimore, M.*—Barrels super-phosphate lime.

CLASS III.

8. HOWLAND, O. *Utica, New York.*—Samples of cereals, clover, and timothy seed.

9. HICKER, BROS. *New York City.*—Samples of flour.

10. STEBBINS & CO. *Rochester, New York.*—Flour.

11. ONONDAGA SALT CO. *Syracuse, New York.*—Samples of table and curing salt.

12. GLENCOVE STARCH CO. *New York City.*—Samples of mazena or corn starch.

13. WADDELL, J. *Springfield, Ohio.*—Indian corn in the ear.

———

78. OSWEGO STARCH CO. *Oswego, New York.*—Samples of prepared corn.

CLASS IV.

14. GLENCOVE STARCH CO. *New York City.*—Samples of starch.

79. DUTCHER & ELLERY, *New York.*—American hops.

80. OSWEGO STARCH CO. *Oswego, New York.*—Samples of starch.

81. WILKINS & CO. *New York.*—Bristles and hair.

CLASS V.

15. LAWRENCE & WHITE, *Melrose, New York.*—Lock-nut, and ratchet-washer, giving to nuts and bolts the firmness and safety of rivets.

16. HOLMES, J. E. *New York City.*—Model of improved pneumatic despatch.

17. ROGERS' LOCOMOTIVE WORKS, *Patterson, New York.*—Lithographs and photographs of locomotives.

82. HOADLEY, J. C. *Lawrence, Mass.*—Model of trucks for locomotives.

83. TRAIN, G. F. *Boston, Mass.*—Model street tramway carriage.

84. REMINGTON, E. & SONS, *Ilion, New York.*—Revolving stereoscope machine.

85. ROWARD, A. H. *Alleghany City, Pa.* Represented by SCHENLEY, E. W. H. 14 *Princes'-gate.*—Car bumper.

86. WARD, W. H. *Auburn, New York.*—Self-centering railway turn-table.

CLASS VI.

18. BREWSTER & CO. *New York City.*—A phaeton and a road waggon.

19. BLANCHAND & BROWN, *Dayton, Ohio.*—Buggy and waggon spokes.

CLASS VII.

19A. WHEELER & WILSON, *New York City.*—American sewing machine. (*See page* 121.)

20. SINGER, J. M. *New York City.*—Sewing machine.

21. WILCOX & GIBBS, *New York City.*—Sewing machine.

21A. HOWE SEWING MACHINE CO. *New York.*—Sewing machine.

22. GOODWIN, C. R. *Boston, Mass.*—Machine for sewing leather, soles of boots and shoes, &c.

23. WRIGAT, H. & CO. 55 *Friday-st. E.C.*—Tape braiding, and tape sewing machine.

24. SMITH, A. *West Farms, New York.*—Power-loom, for weaving tufted piled fabrics; an entire row of tufts (108 or more) are placed by one operation.

24a. RICHARDS, W. D. *Boston, Mass.*—Machinery for sole-cutting and heel-trimming of boots and shoes.

87. CROSBY, C. O. *Boston, Mass.* and 55 *Friday-st. London.*—Machines for preparing tape and joint trimmings, and crimped rufflings.

WHEELER & WILSON, *New York City.*—American sewing machine.

LOCK-STITCH SEWING MACHINE.

THE WHEELER & WILSON CELEBRATED LOCK-STITCH SEWING MACHINES, with crystal cloth presser, new style hemmer, binder, corder, and all other recent improvements.

The lock-stitch sewing machine will gather, hem, fell, bind, or stitch with great rapidity and perfect regularity, beauty, and durability; the work is stronger than when done by hand, and will not ravel when cut. The machines are simple in design, combining find mechanism with elegance of model and finish; do not get out of repair, and are suitable alike for household purposes and the manufacturer's work room. The speed is from 1,000 to 2,000 stitches per minute, and they are so simple and easy of management, that a child may work them.

The following are two out of many testimonials to the efficiency of these machines :—

"LADY MALDEN begs to inform Messrs. Wheeler & Wilson that the sewing machine has given her the greatest satisfaction."

"For manufacturing purposes they are admirably adapted, being suitable for all kinds of work; and are easily managed. Having tried all the machines now in use, we consider yours far superior to any of them. To prove their durability, we are still using the same machines delivered by you eight years since.

(*Signed*) "W. BLENKIRON & SON.
"123, *Wood Street, Cheapside.*"

Illustrated prospectus gratis and post free, and instruction gratis to every purchaser.

Offices and sale rooms, 139 Regent Street, London, W.

CLASS VIII.

25. SANFORD & MALLORY, *New York City.*—Flax and fibre dressing machines.

26. WEMPLE, P. H. *Albany, New York.*—Spacing and boring machine.

27. NEAR, C. *New York City*—Self-registering dynamometer.

28. WORTHINGTON. R. H. & LEE, W. *New York City.*—Duplex pump.

29. PORTER, C. T. *New York City.*—Stationary engine, and governors.

30. ANDREWS, W. D.—Centrifugal pump, and oscillating engines.

31. LEE & LARNED, *New York City.*—Steam fire-engine.

32. DENNISON, C. H. *Rhode Island.*—Wilcox's hot-air engine.

33. BLAKE, BROS. *Newhaven, Conn.*—Stone-breaking machine.

34. SANBORN, G. H. *Boston, Mass.*—Rope and cord machine.

34a. PORTER, C. T. *New York City.*—Horizontal non-condensing engine indicator for high velocities ; governors.

35. SANBORN, G. H. *Boston, Mass.*—Iron refrigerator.

36. GORE, J. C. *Jamaica Plains, Mass.*—Belt shifter.

36a. REDSTONES & Co. *Indianapolis, Indiana.*—Model of portable engine.

37. SANBORN, G. H. *Boston, Mass.*—Spindle banding machine ; rope and cord machine.

38. ECKELL, J. J. *New York City.*—Combination press and compress.

39. STEELE, H. *Jersey City, New Jersey.*—Pumping engine.

40. HANSBROW, *California.*—Pumps.

41. SWEET, S. *New York.*—Newspaper addressing machine.

42. DEGNER, F. O.—Card and job printing press.

43. SANBORN, G. H. *Boston, Mass.*—Gas regulator.

44. HOLLOWAY & SONS. — American clocks.

————

44a. WARKER & EPPENSTEIN, *New York.*—Apparatus for preparing mineral and soda-water.

44b. SANBORN, G. H. *Boston, Mass.*—Gas-pipe tongs ; machines used in book-binding.

44c. PACKER, H. H. *Boston, Mass*—Improved ratchet drill.

44d. GIBSON, A. G. *Worcester, Mass.*—Improved carriage coupling.

44e. DICKINSON, *New York.* — Model of diamond mill dress.

44f. WALCOTT.—Samples of button hole cutters. (Agents—WARNER & Co. 108, *Minories, London.*)

————

88. EDGAR, T. W. *Espy, Pennsylvania.*—Washing machine.

89. PARKER, D. *Canterbury, New Hampshire.*—Washing machine.

90. FOOTE, A. M. *New York* (WHEELER & WILSON, 139 *Regent-st.*) — Lock umbrella stand.

91. GODDARD, C. S. *New York.*—Mestizo burring picker.

ERICSSON'S Caloric engine, PESANT. BROS. London Agents.

SICKLES, E. *New York.*—Steam steering apparatus. (*See pages 123 to 125.*)

ROSS, J. *Rochester, New York.*—Conical burr stone mills.

CONROY, E. *Boston, Mass.*—Cork-cutting machine.

Sickles, E. *New York.* —Steam steering apparatus.

POWER STEERING MACHINE.

The First Power Steering Machine.

A machine, taken off the steamer *Augusta*, while she was lying in the port of New York, is shown at the International Exhibition, though it is roughly made, and the experience gained by its use would now justify the construction of more highly finished specimens. In steering by power the steersman moves the wheel nearest the machine, thus causing it at the same time to move the hand wheel and the rudder correspondingly, the machine consisting principally of two cylinders working at right angles upon a shaft geared into a large wheel that is fastened by a friction plate lined with wood, and set by a screw to any desired pressure on the steering apparatus. The wheel turned by the steersman is connected with the valve motion of the working cylinders, so that either air, water, steam, or other fluid under pressure, will move the rudder as the steersman moves the valve motion in turning the steering wheel. A screw clutch is provided that disconnects the machine from hand steering apparatus placed in front of it, so that hand steering can at all times be resorted to.

"*New York, June* 30, 1862.

"*To the* Jury *of* Class No. 8,
London Exhibition.

"At the request of Mr. F. E. Sickels, inventor of the Steam Steering Apparatus, I submit a statement of my knowledge of the invention, and application of the machine made by him.

"My attention was first called to it by Mr. Sickels, who had the machine on exhibition at the Crystal Palace, in New York, in the year 1853, where it was in operation by steam-power, and gave evidence of its being a practical machine.

"In 1860, I was on board the *Great Eastern* on her first trial trip, and observing the difficulty experienced in steering the ship, I spoke to Captain Harrison about Sickle's Steam Steering Apparatus, and I told him I would ascertain on my return to America, if it had been applied to any vessel. On my arrival at New York, I found that Mr. Sickles had applied the same apparatus which had been on exhibition, to the steamer *Augusta*, then in the port of New York, having recently returned

SICKLES, E. *continued.*

from the South, where she had been running for several months, and using the Steam Steering Apparatus with perfect success.

I made a trip in the *Augusta* and found the machine working satisfactorily in all respects. To throw the rudder hard over, while the vessel was under headway, it required the exercise of only so much manual labour as was necessary to operate the small valves, and by its use the steamer made a complete turn in three and a half minutes. Thereupon, I wrote to Captain Harrison, stating my experience, and advised the application of Sickels' Steam Steering Apparatus to the *Great Eastern*, as it would enable one man to control the rudder better than any number of men could with the present steering apparatus.

"I am informed by the brother of Mr. Sickels, that he took the machine off the *Augusta*, and sent it to London, where the inventor, Mr. F. Sickels, has it now on exhibition.

"JOSEPH J. COMSTOCK,
Steam Ship 'Baltic.' "

"*Fernandina, Flo. April* 10, 1860.

"I HEREWITH report the operation of 'Sickles' Steam Steering Apparatus,' now in use upon the steamer *Augusta*, running between Savannah, Geo. and Fernandina, Florida.

"This apparatus consists of a small steam engine with balance slide valves, applied to the ordinary steering wheel. The rudder has been increased to nearly twice the size of rudders in common use upon vessels of the class of the *Augusta*, and the steering chains are strengthened correspondingly.

"It was applied to the vessel at New York, and in February last, the *Augusta* left that port for Savannah, to commence service upon her present route. When off the Capes of the Chesapeake, we encountered a heavy gale, which lasted for twelve hours, and were so much disabled, that we put into Norfolk for repairs. The value of this improvement was clearly shown during the gale: it required but one man to steer, and this he did without the application of more power by him than could be exerted with one hand, it being only so much as was necessary to work the balance valves.

"The large rudder, with ample steam-power to operate it, gave to the pilot far greater control of the movements of the vessel, in every emergency of our situation, than could have been obtained without the use of this steering apparatus; and to this superior control of the vessel during the gale I attribute the fact that we saved the boat, and were able to reach Norfolk. When the rudder was struck by a heavy sea, the friction wheel enabled the rudder to yield to the pressure without breakage of the steering chains, or any possible injury to the pilot.

"Our route between Fernandina and Savannah is partly through a swamp, the channel being so very crooked, that a single-engine boat could not run it, with the ordinary steering apparatus, without the use of warping lines in many places; but, by the aid of 'Sickels' Steam Steering Apparatus' we have the movements of the vessel under such perfect control that we are enabled to run the route without difficulty. We can 'round to' in little more than the length of the vessel, and can rapidly put the rudder 'hard over' without slowing the engine. Our pilots are negroes, and, though wholly ignorant of any kind of machinery, they find no difficulty in operating the steering apparatus: being required to exert but little power in steering under any circumstances, they are delighted with this improvement. The friction wheel preserves the gearing and ropes from any possible injury, either by a heavy sea striking the rudder or from any carelessness of the pilot. Although provision is made for the instant disconnection of the small engine, thereby reducing it to the ordinary steering apparatus, we have never had occasion for its use.

"From my experience of the operation of this Steam Steering Apparatus both at sea and when running in crooked channels, I regard it to be an exceedingly valuable improvement, and believe that its adoption upon steam vessels would, by the greatly increased control which it affords to the pilot of the movements of the vessel, be the means of saving life and property.

"I. O. PHILLIPS,
Captain of Steamer 'Augusta.' "

The following is to show that the Steering Apparatus remained in good order until the vessel was laid up for repairs:—

"*New York, July* 1, 1862.

" To the JURY of CLASS No. 8,
LONDON EXHIBITION.

"THIS is to certify that in the year 1860 I went on board the steamer *Augusta*, which had just returned from the South, then in the port of New York, and she steamed round to my ship-yard for the purpose of having repairs made to her hull.

"On that occasion the Steam Steering Apparatus was used by the pilot, and my attention having been particularly directed to it, I observed the operation of the machine with great interest; it was evidently a successfully working machine, and the captain of the *Augusta* informed me that it had been in constant use for several months, while the vessel was running between Savannah and Fernandina, and that its great advantages were not to be questioned.

"HENRY STEERS,
Shipbuilder, New York."

STATEMENTS WITH REFERENCE TO STEERING VESSELS

By Power as applied through the ordinary hand-steering gear.	*By Hand.*
No labour is required of the steersman to keep the vessel upon her course, and he may therefore be expected to do so.	Several men are frequently required to work the rudder, and as they cannot work perfectly together, the vessel is liable to vary from her course; and it requires such an exertion of attention and labour when one man only is employed, that it is frequently not obtained.
With only one man at the helm, the power is at all times at hand to rapidly move the rudder, and thus contribute to avoid accidents in emergencies.	Accidents sometimes occur, because, with only one man at the helm, the rudder is not moved with sufficient quickness, and assistance arrives too late for the emergency.
As in heavy weather, the steersman and the steering gear is safe from any injury by the force of the sea striking the rudder, he is at liberty to confine his attention to keeping the vessel upon her course. The action of the mechanism that limits the power exerted to move or hold the rudder, is such, that while it permits the rudder to yield to the blows of heavy seas, so as not to endanger the steering-gear or the steersman, it also prevents the steersman from applying a breaking strain, and yet he can apply as much power to move or hold the rudder without labour to himself as can be safely employed.	As in heavy weather, all hands at the wheel are liable to be injured if the rudder is struck by a heavy sea, they are naturally more on the look-out for their own safety than they are to keep the vessel on her course. The men are weaker than the steering-gear, and are therefore liable to be injured first; but if any mechanical means is used to assist them to hold the wheel, the steering-gear may be injured instead of the men, the breaking point being the weakest one when a breaking force is applied.

SICKLES, E. *continued.*

EFFICIENCY OF THE APPARATUS FOR STEERING VESSELS
WHEN MADE TO

Steer by Power.

A consideration of the relation of speed and tonnage to power in large and small vessels, suggests an increase in the power applied to the rudders of large vessels to secure a proper proportion between the directing and propelling power, so as to ensure in-

Steer by Hand.

The power being limited, and the labour of men expensive, the whole apparatus of rudder and steering-gear must be light, and therefore as small as possible, and proportions that would otherwise be chosen to ensure safety and efficiency. must be dispensed

creased facility in steering them, as the increased expense of working a large rudder securely fastened to the vessel, with ample strength in the steering-gear throughout, would be small, as compared with the expense of working a small rudder with light fastenings and steering-gear.

with for those that bring the whole steering apparatus within the limits of a hand-machine, and the larger the vessel, the greater must be the surrender of efficient proportions to the difficulty of a want of power to operate them.

Present address : F. E. SICKLES, care of NEWTON & SON, 66 Chancery Lane, E.C.

CLASS IX.

45. WOOD, W. A. *Hoosir Falls, New York.* —Self-raking reaper, combined reaper and mower, and grass mowing machine.

46. RUSSEL & TREMAIN, *Fayetteville, New York.*—Reaping machine—a new mechanical device.

47. KIRBY & OSBORNE, *Auburn, New York.*—Reaper and mower.

48. REDSTONES, BROS. & CO. *Indianapolis, Indiana.*—Mowing and reaping machine.

49. COE, O. *Ozaukee, Wisconsin.*—Cultivating harrow.

50. DANE, J. F. & Co. *Springfield, Ohio.* —Steel ploughs.

51. BATCHELLOR & SONS, *Wallingford, Vermont.*—Hoes, forks, and rakes.

52. DOUGLAS AXE CO. *Mass.*—American cutlery, axes, &c.

53. BLANCHARD & BROWN, *Dayton, Ohio.* —Cotton planter.

54. WENTWORTH & JARVIS, *Burlington, Iowa.*—Automatic farm gate, and windmill water-elevator.

55. LEVI, A. *Beardsly, New York* —Hay and earth elevator.

56. PRICE, R. *Albany, New York.*—Churn, and mop.

56a. PRINDLE, D. R. *East Bethany, New York.*—Model of corn and bean planter.

92. McCORMICK, C. H. *Chicago, Illinois.*— Reaping, mowing, and self-raking machine. (*See pages 126 and 127.*)

93. PRINDLE, D. R. *East Bethany, New York.*—Agricultural caldron.

94. CENTRAL RAILWAY, *New Jersey.*— Horse-power machine.

CLASS X.

57. DERROM, A.—Model house, specimens of steam carpentry.

58. SCHOLL, J. G. *Port Washington, Wisconsin.*—Model life-boat.

CLASS XI.

58a. DERROM, A.—Pontoons, bateaux, &c.

COLT'S PATENT FIRE-ARMS' MANU-

FACTURING Co. *New Haven, Conn.*—Samples of guns, pistols, powder flasks, shot pouches, &c.

McCORMICK, C. H. *Chicago, Illinois.*—Reaping, mowing, and self-raking machine.

REAPING AND MOWING MACHINE.

McCORMICK'S COMBINED REAPING AND MOWING MA-
CHINE, with self-acting delivery in sheaves, ready for
binding, out of the track of the horses.

The fame of this machine is such throughout the
world, that little need here be said of it.

Its success in the United States has resulted in the
manufacture and sale to the present, in a single **manufactory, of 45,000** of them, without the *self-raking
attachment,* which is now its *peculiar and important*
characteristic, and only introduced within the **last two**
years.

McCormick, C. H. *continued.*

It may be repeated of the McCormick machine, that at the great "world's fair" of 1851, it took the world by surprise, in demonstrating, by its operations *in the field,* the practical success of the *reaping machine,* when the great council medal of the Exhibition was awarded for it by the Council of Juries, and it was acknowledged to be the most valuable invention in any contribution to the Exhibition.

Again, at the "Universal Exposition of all Nations" at Paris, in 1855, it received a like distinction, upon a similar test of its working power, in the award, by a committee of scientific as well as practical jurors, of the "*Grand Gold Medal of Honor*" for the best reaping and mowing machine, and as being the original type after which all others were made.

And again at the great national trial of reaping and mowing machines, in the United States, in 1857, by the United States Agricultural Society, the only one ever held by that society, the "*Grand Gold Medal of Honour*" was awarded for this as the best reaping machine on trial.

The *prize medal* of the present International Exhibition having been awarded for the splendid specimen of this machine as now on exhibition, with its automaton appendage, which has attracted so much attention, whilst numerous reports in the newspapers have appeared of its successful operation in different parts of England, the first trial of its powers, in competition with others, came off at Preston on the 2d (Sept.) instant, by the Lanca-shire Agricultural Society, at its great annual trial of these implements. Of nineteen *entrés,* only eight or nine of them entered the corn-field for the test, of which were Samuelson's revolving self-raker, and Wood's combined machine, that are also in the International Exhibition.

The test was a hard and a decisive one, first on a heavy and partly laid crop of oats, on deeply furrowed land; and next on a crop of heavy wheat, also on furrowed land. The result was the triumph of the McCormick machine, which received the award of the first premium, of £15, and the medal of the Society, as decidedly the best machine, the second premium having been awarded for a Scotch hand-delivery machine.

So decided was the success of McCormick's machine, that, at the great banquet of the occasion, at which Lord Derby presided, in a toast given to the successful competitors, while Mr. Patterson's name was coupled with the *stock* department, that of Mr. McCormick only was selected to represent the whole implement department.

During the present harvest this machine has operated triumphantly in England, Scotland, Italy, France, Hungary, Belgium, and Russia, in which latter country it has had the patronage of the Imperial Household.

Patents have been secured for this machine, in most of the corn-growing countries of Europe, where Mr. McCormick hopes to effect its early and general introduction.

Present address : Care, 166 Fleet Street, London.

CLASS XII.

58*b.* **Ward, W. H.** *Auburn, New York.*—Day and night signal telegraph, telegraph steering and signal lanterns, &c.

CLASS XIII.

58*c.* **Darling & Schwartz,** *Bangor, Maine.*—Specimens of steel scales and rules.

CLASS XVI.

Steinway & Sons, *New York City.*—Grand and square pianos. (*See page 128.*)

60. **Dunham, J.** *New York City.*—Boudoir piano.

61. **Hulskamp, G. H.** *New York City.*—Grand piano, with improved sounding board.

62. **Decker, M.** *New York.*—Piano and triolodeons.

———

95. **Hulscamp, G. H.** *New York City.*—Violins of new improved construction.

CLASS XVII.

63. **Yard, A. A.** *New York City.*—Samples of waterproof adhesive plaster and court plaster.

64. **Matthews, Prof. C.** *New York City.*—Fumigator.

STEINWAY & SONS, *New York City.*—Grand and square pianos.

STEINWAY & SONS'
GRAND, SQUARE, AND UPRIGHT PIANOFORTES.

The pianos exhibited by this firm obtained the prize medal (International Exhibition, 1862), for "powerful, clear, and brilliant tone, with excellent workmanship." Messrs. STEINWAY & SONS have also obtained medals wherever and whenever their instruments have been placed in competition with those of other manufacturers. They are the largest makers in America, probably in the world, finishing upwards of thirty-five pianofortes every week.

The peculiarities of the various instruments exhibited in the American Court by this firm, are that they combine the durability of first-class workmanship, and thoroughly seasoned material, with rare strength, and sympathetic quality of tone. The first are accounted for by the fact that the climate of America renders the best timber and the best labour absolutely necessary; clumsy work or unseasoned material would quickly fall to pieces in a temperature that ranges in a single year from twenty degree below zero to ninety degrees in the shade. For the rest, the manufacturers have brought many valuable improvements and novelties to the construction of their pianos, the fruits of philosophical research into the exhaustless theory of sound.

It is well known that pianos with wooden frames, and pianos with iron frames have their separate characteristics. The great industrial question has been how to combine the sweetness of the one with the power and brilliancy of the other. Those makers who have most nearly solved this problem, are precisely those who are held in best esteem by musicians. The plan adopted by Messrs. Steinway and Sons may be briefly described as follows: The metal frame of the instrument is cast in a single piece, by which means the intersections rendered necessary by the old plan of support-bars are no longer needed. This frame is so constructed, that it braces against the tuning block, keeping the latter steady, protecting it from transverse vibrations, and sustaining the instrument well in tune.

In the arrangement of the strings several highly important improvements have been made by Messrs. Steinway & Sons. Their pianos are overstrung (that is to say, the wires of the lower notes are laid over the wires of the upper notes) in such a way as to secure the greatest amount of sounding-board surface to each note, and thereby the maximum of power. By a new distribution of the bridges, too, they are enabled to give to the scale of the square or upright piano as much evenness and almost as much sonorousness as to that of the grand piano itself.

Messrs. Steinway & Sons are the inventors of the improved repetition action used in their instruments. The principal advantages of this mechanism are to be found in the independence of the "jack" and the free and unrestrained movement of the "nut." Experience has proved that all appendages to the "jack," the "nut," or the hammer, ultimately and inevitably result in a rattling kind of noise and an injury to the tone, whereas in Messrs. Steinway & Sons' action the mechanism ensures elasticity, promptness, and force of touch.

CLASS XVIII.

96. GARDNER, BREWER, & CO. *Boston, Mass.*—Bleached and brown shirtings.

97. MANCHESTER PRINT WORKS, *Manchester, New Hampshire.*—Cotton prints.

CLASS XXI.

98. MANCHESTER PRINT WORKS, *Manchester, New Hampshire.*—De laines and woollen hose.

CLASS XXIII.

99. MANCHESTER PRINT WORKS, *Manchester, N. H.*—De laines and prints.

CLASS XXVI.

100. KOHNSTAMM, J. *New York.*—Specimens of leather, and leather imitations.

(Represented by KOHNSTAMM, H. 33 *Dowgate-hill.*)

CLASS XXVIII.

101. DEXTER & Co. *New York.*—Books in the Indian language.

HARVEY, G. portfolios :—

101a. Grand golfo for keeping prints, &c

101b. Library portograph on wheels.

101c. Sutherland portfolio stand.

101d. Vitrifolio for drawings.

101f. STEVENS, H. 4 *Trafalgar-sq.*—Specimens of American books, photographs, &c.

101g. Low, S. SON & Co. *Ludgate-hill.*—Specimens of American books.

102. GUN & Co. *Strand, London.*—Specimens of American newspapers.

103. JEWETT, M. P. *Poughkeepsie, New York.* — Catalogues of female seminaries, United States.

CLASS XXIX.

64a. CHACE, J. H. & Co. *Portland, Maine.*—Map of the State of Maine.

104. TAINTA, S. & Co. *Philadelphia*

(HOLMES, HARRISON, & Co. *London*).—Washington map of the United States.

105. BATES, R. *Philadelphia, Pa.*—Appliances for the cure of stammering.

CLASS XXX.

106. PICKHARDT, J. F. C. *New York.*—Extension sofa bedstead.

107. BERTRAM, F. M. B. *New York City, and 4, Gower-street, London.*—Two reception easy rocking, or invalid reclining chairs.

CLASS XXXII.

108. BLACKWELL, W. 9, *Cranbourne-street, London.*—American (tailors') shears.

CLASS XXXIII.

109. WATTS, A. J. *New York.*—Crystal gold, crystal gold foil, and non-adhesive foil.

110. GREEN, W. H. *Meriden, onn.*—Revolving castor.

111. BROWN, J. A. & Co. *New York* (Represented by BROWN, B. F. 11 *Cullum-street, London*).—Samples of plated lockets.

CLASS XXXIV.

112. HARTELL & LETCHWORTH, *Philadelphia.*—Hermetically sealed glass jars for fruits, and specimens of preserved fruits.

113. AMERICAN BANK NOTE COMPANY, *New York.*—Specimens of bank note engraving.

CLASS XXXVI.

64*b.* McDANIEL. — Specimens of natural flowers.

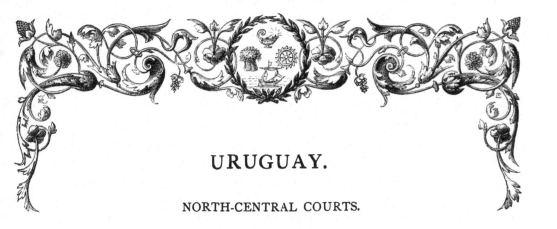

URUGUAY.

NORTH-CENTRAL COURTS.

1. MALLMAN & CO. *Soriano.*—Wool.

2. PRANGE, FELZ, & CO. *La Colonia.*—Wool.

3. DRABLE, BROS. *La Colonia.*—Wool.

4. WHITE, W. *La Colonia.*—Wool.

5. HORDEÑANA, SENOR.—Roots.

6. CABAL & WILLIAMS, *Salto.*—Hides; tiger and other skins.

7. HORDEÑANA, SENOR.—Medicinal roots.

8. CUNHA, S.—Earthen figure.

9. CAMARD, M.—Liqueurs.

10. MAUÁ, BARON DE, *Soriano.*—Wool.

11. BASCÓS & CO.—Flooring tiles, &c.

12. DEPARTMENT OF SAN JOSÉ.—Timber, wheat.

13. DEPARTMENT OF SORIANO.—Flooring tiles, coals, &c.

14. DEPARTMENT OF MALDONALDO.—Coloured marbles, wheat.

15. DEPARTMENT OF SALTO. — Timber, cotton, and harness bridle.

16. DEPARTMENT OF PAYSANDÚ.—Timber, wheat.

17. DEPARTMENT OF MINAS.—Lead, iron, copper, coals.

18. DEPARTMENT OF TACUAREMBO.—Tobacco, yerba mátte, gold and the stone which produces it.

19. MIGÑONE.—Spirit of Olniz, and wine feathers.

20. IVANICO, *Monte Video.*—Wheat.

21. SINISTRE.—Prepared beef.

22. DIAZ & LIMA, *Soriano.*—Wool.

23. OLIDEN, *Monte Video.*—Prepared beef.

24. NIN, P. *Monte Video.*—Cured beef.

25. PROUDFOOT, J. *Glasgow.*—Native saddle and trappings; timber.

26. VARIOUS DEPARTMENTS.—Wheat.

27. THE URUGUAYAN MILL.—Flour.

28. DEPARTMENT OF CANELONES.—Barley.

29. DEPARTMENT OF TACUAREMBO. — Flour, wine.

30. MARTIN, A. *Monte Video.*—Bridles, head-stalls.

31. BURZAGO & PIÑEYRUA, *Monte Video.*—Salted cow's-hide.

32. BURZACO & PIÑEYRUA, *Monte Video.*—Hung beef, liquified grease, and fat.

33. GRAÑELY, S. *Monte Video.*—Flour, wheat.

34. CRAWFORD, H.—Saddle and trappings.

35. RIVOLTA, *Monte Video.*—Saddle and fittings.

VENEZUELA.

NORTH-WEST COURT.

[The following have mostly been collected and forwarded by a Committee appointed by the Venezuelan Government, consisting of Senor Lino J. Revenga, Senor Charles Hahn, and Senor Carlos J. Marxen.

1. THE GOVERNMENT.—The arms of Venezuela, in feathers of the natural colours.

2. SCHEBBYE, *Venezuela.* — Bouquet of flowers in feathers.

3. SCHEBBYE, *Venezuela.* — Roses, in feathers.

4. GUADALUPE, NOVELL, *Venezuela.*—A coffee tree, in wax.

5. GUADALUPE, NOVELL, *Venezuela.*—Venezuelan fruits, in wax.

5a. GUADALUPE, NOVELL.—Branch of cotton tree, in wax.

6. DAVIS, F. L. 13 *Blandford-street.*—Stuffed birds.

7. DAVIS, F. L. 13 *Blandford-street.*—A case of butterflies.

8. DAVIS, F. L. 13 *Blandford-street.*—A totuma carved with a knife.

9. CALCANO, E. *Venezuela.*—Totuma carved with masonic emblems.

10. HEMMING, Mrs. F. H. 104, *Gloucester-place.*—Three Totumas, one painted.

11. MEYER, F. *Hamburgh.*—A hammock of Margarita cotton, made in that island.

12. ULSTRUP, N. P. B. *Venezuela.*—A hammock made by Indians.

13. ULSTRUP, N. P. B. *Venezuela.*—A hammock.

14. THE COMMITTEE. — A Venezuelan handkerchief, worked at Maracaibo.

15. THE COMMITTEE.—A linen handkerchief.

16. CALCANO, E. *Venezuela.*—A Caraccas shirt.

17. THE COMMITTEE.—A table of various woods.

18. AVILA, *Venezuela.*—A box of various woods.

19. ULSTRUP, N. B. P. *Venezuela.*—A box of Macanilla palm wood.

20. ULSTRUP, N. B. P. *Venezuela.*—A box.

21. ULSTRUP, N. B. P. *Venezuela.*—A picture frame.

22. MACHDOA, *Venezuela.* — Unbleached cotton cloth.

23. THE COMMITTEE.—Cotton wicks.

24. THE COMMITTEE.—Cables of fibres from Rio Negro.

25. THE COMMITTEE.—Rope of cocuiza plant, from Maracaibo.

26. MEYER, F. *Hamburgh.*—Cloth from the cocuiza plant.

27. THE COMMITTEE.—Coffee.

27a. NADAL, *Venezuela.*—Coffee.

27b. MEYER, F. *Hamburgh.*—Coffee.

28. THE COMMITTEE.—Cocoa.

28a. MEYER, F. *Hamburgh.*—Cocoa.

29. THE COMMITTEE.—Beans of various kinds.

30. THE COMMITTEE.—Indian corn.

31. THE COMMITTEE.—Dividivi, from Maracaibo.

32. THE COMMITTEE. — Cochineal, from the Caraccas.

33. STURUP, *Venezuela.*—Cebadilla.

34. STURUP, *Venezuela.*—Vanilla.

35. THE COMMITTEE.—Starch, from the root of the yuca plant.

36. THE COMMITTEE.—Tonquin beans.

37. STURUP, *Venezuela.*—Simarruba.

38. THE COMMITTEE.—Sereipa, from Guayana.

39. THE COMMITTEE.— Secua escandinava: an antidote against poisons, and preservative for iron against rust.

40. CONDE, F. *Venezuela.*—Curara: a cure for hæmorrhage, wounds, and ulcers.

41. CONDE, F. *Venezuela.*—Espino: a cure for hæmorrhage, wounds, and ulcers.

42. GIL, N. A. *Venezuela.*—Indian balsam.

43. STURUP, G. *Venezuela.*—Root of sarsaparilla.

44. STURUP, G. *Venezuela.*—Extract of sarsaparilla.

45. ESPINAL, M. *Venezuela.*—Pectoral oil, from sesame.

46. ESPINAL, M. *Venezuela.* — Sesame seed.

47. GATALAN & Co. *Venezuela.*—Bitters, from Maracaibo.

48. SYERS, BRAACH, & Co. *Venezuela.*—Bitters, from Angostura.

49. WARBURGH & Co. 16, *Devonshire-sq.*—Bitters, from Angostura.

50. THE COMMITTEE.—Preserved oranges.

51. THE COMMITTEE.—Preserved quinces.

52. THE COMMITTEE.—Preserved guayaba.

53. THE COMMITTEE.—Preserved guanàbana.

54. THE COMMITTEE.—Preserved quinces.

55. THE COMMITTEE.—Preserved peaches.

56. THE COMMITTEE.—Imitations of English soap.

57. THE COMMITTEE. — Imitations of Spanish soap.

58. THE COMMITTEE.—Candles of stearine, from Anauco.

59. BOLET, DR. *Venezuela.*—Wax.

60. HAHN, C. *Venezuela.*—White wax, from Carapita.

61. HAHN, C. *Venezuela.*—Yellow wax.

62. THE COMMITTEE.—Vegetable wax, with fruit and leaf of the tree which produces it.

63. THE COMMITTEE.—Sugar, from Guatire.

64. BARNOLA, J. *Venezuela.*—Chocolate, almonds.

65. BARNOLA, J. *Venezuela.*—Vanilla.

66. BARNOLA, J. *Venezuela.*—Cinnamon.

67. BARNOLA, J. *Venezuela.*—Cocoa.

68. MAYO, C. *Venezuela.*—Chocolates.

69. THE COMMITTEE.—Leaf tobacco, from Guanape.

70. THE COMMITTEE.—Leaf tobacco, from Cumanacoa.

71. THE COMMITTEE.—Cigars, from Cumanacoa.

72. THE COMMITTEE.—Cigars, from Caraccas.

73. THE COMMITTEE.—Cigars, from Carapita.

74. THE COMMITTEE.—Cigars, from Turmero.

75. GARRIDO, F. *Venezuela.*—Snuff.

76. GARRIDO, F. *Venezuela.*—Snuff.

77. GARRIDO, F. *Venezuela.*—Snuff.

78. THE COMMITTEE.—Tacamahaca.

79. THE COMMITTEE.—Resin of Algarrobo.

80. THE COMMITTEE.—Wool, from Coro.

81. HAHN, C. *Venezuela.*—Goat skins, from Coro.

82. HAHN, C. *Venezuela.* — Deer skins, from Caraccas.

83. RUETE, RÖHL, & Co. *Venezuela.*—Plantain leaf, from which paper is made.

84. RUETE, RÖHL, & Co. *Venezuela.*—Cottons, from the valleys of Aragua.

85. RUETE, RÖHL, & Co. *Venezuela.*—Cotton.

86. STOLTERFOHT & Co. *Liverpool.*—Sea Island cotton, from Maracaibo.

87. BAZLEY, T. M.P. *Manchester.*—Sea Island cotton, from Maracaibo.

88. BAZLEY, T. M.P. *Manchester.*—Sea Island cotton, from Maracaibo (manufactured).

89. MEYER, F. *Hamburgh.*—Cotton, from Barquisimeto.

90. MEYER, F. *Hamburgh.*—Cotton.

91. MEYER, F. *Hamburgh.*—Cotton, from Puerto Caballo.

92. HAHN, C. *Venezuela.*—Wild cotton, from Upata Guayana.

92a. VARGAS, DR. *Venezuela.*—Eighty-three different kinds of wood.

92b. RUETE, RÖHL, & Co. *La Guayra.*—Thirty-one different kinds of wood.

92c. CONDE, F. *Caraccas.*—Fourteen different kinds of wood.

92d. TARTARET & Co. *Venezuela.*—Block of red gateado.

93. MARCANO & Co. and DR. BETANCOURT—Copper ores from Teques.

94. HEMMING, F. H. 104 *Gloucester-place.*—Copper assayed by Johnson and Matthey.

95. HAHN, C.—Iron ore.

96. ROMERO, J. P.—Silver and lead ore from Carupano.

97. RODRIGUEZ, DR.—Silver ore, mine of "Gran Pobre," Carupano.

98. HAHN, C.—Gold from Caratal, Guayana.

99. MEYER, F. *Hamburgh.*—Gold quartz, from Guayana.

100. HAHN, C.—Gold quartz, from Caratal, Canton of Upata, province of Guayana.

101. REVENGA, L. J.—Green marble from Caracas.

102. REVENGA, L. J.—Red marble.

103. REVENGA, L. J.—Rock crystal.

104. REVENGA, L. J.—Petrified vera wood.

ADVERTISEMENTS.

THE FRIEDRICH WILHELMSHUTTE, *near Siegburg, Rhenish Prussia.*—E. Langen's patent smoke-consuming fire-grate. *Agents in England.* Messrs. ROOSEN & CORNELSEN, 37 *Cross Street, Manchester.*

By Her Majesty's Royal Letters Patent.

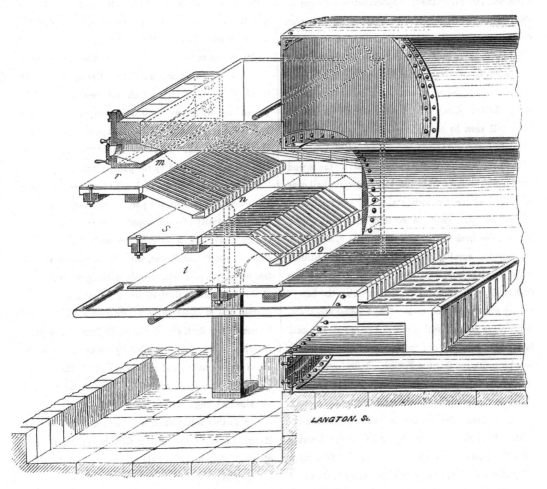

SMOKE-CONSUMING FIRE-GRATE.

LANGEN'S PATENT FIRE-GRATE has met with a more liberal reception than any other grate recently constructed for the purpose of attaining a perfect and smokeless combustion. During the last two years, no less than 1,500 of these grates have been applied to as many steam boilers in Germany, Austria, and France, giving everywhere entire satisfaction. The principle of Langen's Patent Grate is "*to supply the fresh or green coals under*

The Friedrich Wilhelmshutte, *continued.*

the burning ones," and to admit no other air into the fire space but that which enters at the feed-space through the fresh coals. The Patent Fire-Grate, as per annexed sketch, consists of two inclined and one horizontal grate, arranged one over the other, and of a perforated plate, on which the cokes are converted finally into cinders or slacks. Coals are supplied to the furnace through the openings m, n, o, each $3\frac{1}{2}$ in. high, which are left between the grates. To explain the working of the furnace, the grates are supposed to be covered with burning coals, which will also fill up the open spaces between the grates. The fireman now begins his work by throwing coals upon the three dead plates r, s, t; he then takes an iron crutch or rake, and forces the fuel deposited on the bottom-plate t, through the opening o, thereby pushing the burning coals forward on the horizontal grate, till they fall upon the perforated plate. The fire-bars of the horizontal grate will then be covered with fresh coals. By now also pushing the coals on the second plate s, forward through the opening n, the burning coals on the second grate will be removed from the bars and spread all over the fresh coals on the horizontal grate; and the bars of the second grate will be covered with fresh coals. Finally, the fresh coals on the top-plate r, are pushed through the opening m, thus removing the burning coals from the top bars, spreading them all over the fresh coals on the second grate, and covering the bars of the top grate with fresh coals. The fire-bars are, therefore, always covered with fresh and comparatively cold fuel, which greatly tends to their preservation; and it may be mentioned that the Friedrich Wilhelmshutte has not yet had occasion to renew any of their fire-bars, although their grates have now been constantly at work for three years. The earthy and all not combustible matter of the coal keeps on the surface of the burning coals, and slides down with the latter, till it falls upon the perforated plate, from where the slacks are removed into the flues of the boiler by moving the perforated plate backward and forward. When the flues become too full of slacks, they can be cleaned by removing the perforated plate.

The gases, slowly and regularly evolving from the coals, are mixed with air, before they pass into the fire-space through the burning layer of coals; and this fact accounts for a perfect and smokeless combustion without any artificial supply of air. It has been proved by experiments that Langen's Patent Grate does not admit more air into the fire-space than is actually necessary for the chemical composition of the products of a perfect combustion, while other grates, for the same purpose, require double that quantity; and yet it is a well-known fact, that the supply of air to the fire-space stipulates the quantity of heat (averaging a temperature of 300° C), which escapes through the chimney, and which, of course, is lost to the boiler. This circumstance accounts for a saving, with Langen's Patent Grate, of from 15 to 25 per cent. of fuel, which has been satisfactorily proved by numerous evaporating experiments. The Patent Grate is adapted for every shape and construction of boiler to an equal advantage, and old grates may in all cases be easily replaced by the patent ones.

Langen's Patent Grate has lately been provided with a self-acting feed motion, the purpose of which is to regulate the quantity of coals supplied to the furnace at a certain time, and to ease of the work of the fireman. One of these Patent Grates, with patent self-acting coal-feed motion, is now satisfactorily working at the works of the Friedrich Wilhelmshutte for some months; and another one is about to be constructed at the works of Messrs. Platt, Brothers, & Co. in Oldham. The practicability of introducing the fuel by mechanical means is out of the question, and as soon as some further experiments will have been completed, this new arrangement will also be brought before the public.

Any further information will be readily supplied by Messrs. ROOSEN & CORNELSON, Civil Engineers, 37 Cross Street, Manchester, the Patentees' sole Agents for Great Britain and Ireland.

Agents for Germany: The FRIEDRICH WILHELMS-HUTTE, near Siegburg.

Agents for France: Messrs. KARCHER & WESTERMANN, Ars-sur-Moselle.

John Maria Farina, *continued.*

The favourable reputation which my Eau de Cologne has acquired, since its invention by my ancestor, in the year 1709, has induced many people to imitate it, and in order to be able to sell their spurious article more easily, and under pretext that it was genuine, they procured themselves a firm of *Farina*, by entering into partnership with persons of my name, which is a very common one in Italy.

Persons, who wish to purchase *the genuine and original Eau de Cologne*, ought to be particular to see that the labels on the bottles have not only my name, *Johann Maria Farina*, but also the additional words *gegenüber dem Jülichs Platz*, that is, opposite the Jülichs Place, which is the distinguishing mark of my genuine and original article.

Travellers visiting Cologne, and desirous of buying my manufacture, are cautioned against being led astray by cabmen, waiters, guides commissionaires, and other parties who offer their services to them.

It happens too frequently that these persons conduct the stranger to the shop of one of the fictitious firms, where, notwithstanding advertisements to the contrary, they are remunerated with more than the third part of the price paid by the purchaser, who, of course, pays indirectly this remuneration by buying a bad article, and paying a high price for it.

Considering these impositions, and in order to caution the public against all imitations of my *Eau de Cologne*, I avail myself of the opportunity offered by this Exhibition, to display my only genuine manufacture, in the show case and fountain, and represented on the previou page, where my label and trade-mark, also the quality of my manufacture, may be examined.

Every information may be obtained in the Exhibition Building, and at my Agency, 7 *Old Jewry, E.C. London.*

Printed in the United States
By Bookmasters